Sixth Edition

Biology | Science for Life

WITH PHYSIOLOGY

Colleen Belk

UNIVERSITY OF MINNESOTA DULUTH

Virginia Borden Maier

ST. JOHN FISHER COLLEGE

330 Hudson Street, New York, NY 10013

Courseware Portfolio Management, Director: Beth Wilbur
Sponsoring Editor: Cady Owens
Courseware Director, Content Development: Ginnie Simione Jutson
Senior Development Editors: Debbie Hardin and Evelyn Dahlgren
Managing Producer: Michael Early
Content Producer: Tiffany Mok
Courseware Editorial Assistant: Alison Candlin
Rich Media Content Producers: Nicole Constantine and Libby Reiser
Full-Service Vendor: Cenveo® Publisher Services
Copyeditor: Jon Preimesberger
Compositor: Cenveo® Publisher Services
Design Manager: Maria Guglielmo Walsh
Interior Designer and Cover Designer: Elise Lansdon
Illustrators: Imagineeringart.com, Inc.
Rights & Permissions Project Manager: Cenveo® Publisher Services
Rights & Permissions Management: Ben Ferrini
Photo Researcher: Kristin Piljay
Manufacturing Buyer: Stacey Weinberger
Director of Product Marketing: Allison Rona
Product Marketing Manager: Christa Pesek Pelaez
Field Marketing Manager: Kelly Galli
Cover Photo Credits. Background image: Dimitrios/Shutterstock. Feature photos (clockwise, beginning with iceberg): Ragnar Th Sigurdsson/Arctic Images/Alamy; Pandora Studio/Shutterstock; Science Picture Co/Alamy Stock Photo; S-F/Shutterstock; Johner Images/Alamy Stock Photo; 10kPhotography/RooM the Agency/Alamy

Library of Congress Cataloging-in-Publication Data

Names: Belk, Colleen M, author. | Maier, Virginia Borden, author.
Title: Biology : science for life, with physiology / Colleen Belk, University of Minnesota-Duluth, Virginia Borden Maier, St. John Fisher College.
Description: Sixth edition. | New York, NY : Pearson, 2018. | Includes index.
Identifiers: LCCN 2017043101 | ISBN 9780134555430
Subjects: LCSH: Biology. | Physiology.
Classification: LCC QH307.2 .B43 2018b | DDC 570.1--dc23 LC record available at https://lccn.loc.gov/2017043101

(Student edition) ISBN 10: 0-134-55543-0
(Student edition) ISBN 13: 978-0-134-55543-0

www.pearson.com

2 2022

About the Authors

Colleen Belk and **Virginia Borden Maier** collaborated on teaching biology to nonmajors for more than a decade at the University of Minnesota Duluth. This collaboration has continued for an additional decade through Virginia's move to St. John Fisher College in Rochester, New York, and has been enhanced by their differing but complementary areas of expertise. In addition to the nonmajors course, Colleen teaches general biology for majors, genetics, cell biology, and molecular biology courses. Virginia teaches general biology for majors, evolutionary biology, zoology, plant biology, ecology, and conservation biology courses.

After several somewhat painful attempts at teaching the breadth of biology to nonmajors in a single semester, the two authors came to the conclusion that they needed to find a better way. They realized that their students were more engaged when they understood how biology directly affected their lives. Colleen and Virginia began to structure their lectures around stories they knew would interest students. When they began letting the story drive the science, they immediately noticed a difference in student engagement and willingness to work harder at learning biology. Not only has this approach increased student understanding, but it has also increased the authors' enjoyment in teaching the course—presenting students with fascinating stories infused with biological concepts is simply a lot more fun.

Preface

To the Student

Is it acceptable to clone humans? When does human life begin? What should be done about our warming planet? Who owns living organisms? What are our responsibilities toward endangered species? Having taught this course for nearly 45 combined years, we understand that no amount of knowledge alone will provide satisfactory answers to these questions. Addressing them requires the development of a scientific literacy that surpasses the rote memorization of facts. To make decisions that are individually, socially, and ecologically responsible, you must not only understand some fundamental principles of biology but also be able to use this knowledge as a tool to help you analyze ethical and moral issues involving biology. This is the aim of this textbook.

To help you understand biology and apply your knowledge to an ever-expanding suite of issues, we have structured each chapter of *Biology: Science for Life with Physiology* around a compelling story in which biology plays an integral role. Through the story you not only will learn the relevant biological principles but also will see how science can be used to help answer complex questions. As you learn to apply the strategies modeled by the text, you will also be strengthening your critical thinking skills.

Even though you may not be planning to be a practicing biologist, well-developed critical thinking skills will enable you to make better decisions about issues that affect your own life and form well-reasoned, fact-based opinions about personal, social, and ecological issues.

To the Instructor

You are probably all too aware that teaching nonmajors students is very different from teaching biology majors. You know that most of these students will never take another formal science course; therefore, your course may be the last chance for these students to appreciate how biology is woven throughout the fabric of their lives and to develop a deep understanding of the process of science. You recognize the importance of engaging nonmajors because you know that these students will one day be voting on issues of scientific importance, holding positions of power in the community, serving on juries, and making health care decisions for themselves and their families. This text is designed to help you reach your goals.

By now, most nonmajors biology instructors are aware that this book differs from other books in that we use a compelling storyline woven throughout the entire text of each chapter to garner student interest. Once we draw students in, we keep them engaged by returning to the storyline again and again until the end of the chapter, when students should be able to form their own data-driven opinions about each topic. Storylines are carefully crafted to allow the same depth and breadth of coverage as any other nonmajors biology text.

Our experience has taught us that students will not remember as many facts as we hope they will, but they can and do remember how to apply the scientific method to novel questions involving biology, and they can retain a strong appreciation for how science differs from other methods of understanding the world. To ensure our students leave our course with the ability to critically evaluate information they may come across, this text focuses heavily on the process of science, providing opportunities for students to practice applying the scientific method and analyze data at every opportunity.

New to the Sixth Edition

The positive feedback obtained in previous editions assured us that presenting science alongside a story works for students and instructors alike. In the sixth edition, we have added several new features, a new chapter, and several reorganized chapters. We also updated storylines and continued to improve popular features from previous editions as well as our supplements.

New Features: Got It?, Show You Know, Go Find Out, Make the Connection, and The Big Question

In this edition, we have added many active learning features to help engage student readers. Each text section includes a series of fill-in-the-blank **Got It?** questions to help students actively assess their content comprehension. The Chapter Review Summary now contains **Show You Know** questions to make reviewing the summary a more active process for students. **Go Find Out** includes activities students can perform on their own or in class in groups that challenge them to find information to answer contemporary questions. The Chapter Review ends with a **Make the Connection** exercise where students draw lines between statements about the storyline and the science in the chapter to help enhance their understanding. Lastly, each chapter ends with **The Big Question,** a feature that presents a topic, followed by some smaller questions—some answerable by science and some not. Once students determine which of the smaller questions science can answer, data is presented related to one of these questions. Students analyze the data in light of both the smaller question addressed and the big question that headlines the feature.

Revised Unit One Coverage and New Chapters

Because we have found that our students are interested in their own fertility, we have reorganized the mitosis and meiosis chapter into two separate chapters. **Chapter 6** still deals with mitosis and cancer, but a new **Chapter 7** now addresses human fertility and reproduction along with meiosis. **Chapter 8** discusses Mendelian genetics in a new storyline, addressing the development and use of newborn screening tests. A newly reorganized **Chapter 9** uses the storyline of wrongful convictions to help students learn about the inheritance of complex traits such as those used in identification of suspects by witnesses. In addition, the heritability section helps counter the notion that criminals are born not made, and the DNA profiling section explains how positive identification has been used to exonerate many wrongfully convicted individuals.

Updated Storylines

In addition to the new storylines listed above associated with content revisions, we've revised the storylines of some chapters without strongly modifying content. **Chapter 5** continues to address photosynthesis within a storyline about global climate change, but is updated to reflect humanity's response via the Paris Agreement. Our chapter on speciation (**Chapter 13**) still addresses the issue of supposed human races, but now through the lens of swimmer Simone Manuel's historic gold medal in the 2016 Olympics. The chapter covering climate and biomes (**Chapter 17**) now addresses the concept of the human "footprint." Our summary of the respiratory and cardiovascular systems (**Chapter 20**) addresses the known and unknown health issues of electronic cigarettes and the practice of "vaping." And the spread of Zika virus is now the storyline for **Chapter 23,** which describes the virus's effects on reproduction and embryonic development.

Supplements and Media

The supplements package continues to be updated and expanded by Judi Roux, Ed.D., a talented college instructor with years of classroom experience in nonmajors biology and colleague of Colleen Belk at the University of Minnesota Duluth. We think you will find that the supplements she has developed are brimming with ideas for how to reach this particular population of students. In addition to the Instructor's Manual (for use in traditional lectures as well as flipped classrooms) and a test bank, we also provide slides, animation, and videos to enrich instruction efforts. Available online, the *Biology: Science for Life with Physiology* resources are easy to navigate and support a variety of learning and teaching styles. Judi authored not only the Instructor Guide, but also many Mastering Biology Quiz and Test Items and the PowerPoint lectures as well.

New features in Mastering Biology include **figure walk-throughs** on tough topics, which provide students with the dynamic guidance of the authors to help them solidify their understanding of the concepts within challenging illustrations. **Ready-to-Go Teaching Modules** for select chapters provide instructors with assignments to use before and after class, as well as in-class activities that use Clickers or Learning Catalytics for assessment. Each Ready-to-Go Teaching Module also includes an **Instructor How-To** video, in which Colleen and Virginia provide additional background and helpful hints for presenting the content in the context of particular storylines.

We believe you will find that the design and format of this text and its supplements will help you meet the challenge of helping students both succeed in your course and develop science skills—for life.

We look forward to learning about your experience with *Biology: Science for Life with Physiology*, Sixth Edition.

Acknowledgments

Reviewers

Each chapter of this book was thoroughly reviewed several times as it moved through the development process. Reviewers were chosen on the basis of their demonstrated talent and dedication in the classroom. Many of these reviewers are already trying various approaches to actively engage students in lectures and to raise the scientific literacy and critical thinking skills among their students. Their passion for teaching and commitment to their students were evident throughout this process. These devoted individuals scrupulously checked each chapter for scientific accuracy, readability, and coverage level.

All of these reviewers provided thoughtful, insightful feedback, which improved the text significantly. Their efforts reflect their deep commitment to teaching nonmajors and improving the scientific literacy of all students. We are very thankful for their contributions.

Reviewers of the Sixth Edition

Oliver Beckers, *Murray State University*
Swapna Bhat, *University of North Georgia*
Stephanie Burdett, *Brigham Young University*
Michelle Cawthorn, *Georgia Southern University*
Kari Clifton, *University of West Florida*
Richard Cowart, *University of Dubuque*
Bryan Dewsbury, *University of Rhode Island*
Jeanette Gore, *University of Tampa*
Eileen Gregory, *Rollins College*
Jay Hodgson, *Armstrong State University*
Brenda Hunzinger, *Lake Land College*
Sarah Krajewski, *Grand Rapids Community College*
Kathy Kresge, *Northampton Community College*
Danielle McGrath, *San Jacinto College*
Jeanelle Morgan, *University of North Georgia*
Tyler Olivier, *San Jacinto College*
Brent Palmer, *University of Kentucky*
Monica Parker, *Florida State College at Jacksonville*
Jill Penn, *Georgia Gwinnett College*
Stephen Piccolo, *Brigham Young University*
Benjamin Predmore, *University of South Florida*
Isaiah Schauer, *Brazosport College*
Christine Simmons, *Southern Illinois University–Edwardsville*
Marialana Speidel, *Jefferson College*
Bishnu Twanabasu, *Weatherford College*
Susan Whittemore, *Gaston College*
Heather Woodson, *Gaston College*

Reviewers of Previous Editions

Daryl Adams, *Minnesota State University, Mankato*
Karen Aguirre, *Clarkson University*
Joseph Ahlander, *Northeastern State University*
Marcia Anglin, *Miami-Dade College*
Josephine Arogyasami, *Southern Virginia University*
Susan Aronica, *Canisius College*
Mary Ashley, *University of Chicago*
James S. Backer, *Daytona Beach Community College*
Ellen Baker, *Santa Monica College*
Gail F. Baker, *LaGuardia Community College*
Neil R. Baker, *The Ohio State University*
Andrew Baldwin, *Mesa Community College*
Thomas Balgooyen, *San Jose State University*
Tamatha R. Barbeau, *Francis Marion University*
Sarah Barlow, *Middle Tennessee State University*
Veronica Barr, *Heartland Community College*
Kelly Barry, *Southern Illinois State University*
Andrew M. Barton, *University of Maine, Farmington*
Katrinka Bartush, *University of North Texas*
Vernon Bauer, *Francis Marion University*
Paul Beardsley, *Idaho State University*
Donna Becker, *Northern Michigan University*
Tania Beliz, *College of San Mateo*
David Belt, *Penn Valley Community College*
Drew Benson, *Georgia Gwinnett College*
Steve Berg, *Winona State University*
Carl T. Bergstrom, *University of Washington*
Janet Bester-Meredith, *Seattle Pacific University*
Barry Beutler, *College of Eastern Utah*
Wendy Birky, *California State University, Northridge*
Donna H. Bivans, *Pitt Community College*
Lesley Blair, *Oregon State University*
John Blamire, *City University of New York, Brooklyn College*
Barbara Blonder, *Flagler College*
Susan Bornstein-Forst, *Marian College*
Bruno Borsari, *Winona State University*
James Botsford, *New Mexico State University*
Anne Bower, *Philadelphia University*
Robert S. Boyd, *Auburn University*
Bryan Brendley, *Gannon University*
Eric Brenner, *New York University*
Peggy Brickman, *University of Georgia*
Carol Britson, *University of Mississippi*
Carole Browne, *Wake Forest University*
Neil Buckley, *State University of New York, Plattsburgh*
Jamie Burchill, *California State University, Northridge*
Stephanie Burdett, *Brigham Young University*
Warren Burggren, *University of North Texas*

Rebecca Burton, *Alverno College*
Nancy Butler, *Kutztown University*
Suzanne Butler, *Miami-Dade Community College*
Wilbert Butler, *Tallahassee Community College*
David Byres, *Florida State College, Jacksonville*
Tom Campbell, *Pierce College, Los Angeles*
Cassandra Cantrell, *Western Kentucky University*
Merri Casem, *California State University, Fullerton*
Anne Casper, *Eastern Michigan University*
Deborah Cato, *Wheaton College*
Michelle Cawthorn, *Georgia Southern University*
Peter Chabora, *Queens College*
Bruce Chase, *University of Nebraska, Omaha*
Thomas F. Chubb, *Villanova University*
Gregory Clark, *University of Texas, Austin*
Kimberly Cline-Brown, *University of Northern Iowa*
Reggie Cobb, *Nash Community College*
Mary Colavito, *Santa Monica College*
William H. Coleman, *University of Hartford*
William F. Collins III, *Stony Brook University*
Walter Conley, *State University of New York, Potsdam*
Jerry L. Cook, *Sam Houston State University*
Melanie Cook, *Tyler Junior College*
Scott Cooper, *University of Wisconsin, La Crosse*
Erica Corbett, *Southeastern Oklahoma State University*
George Cornwall, *University of Colorado*
Angela Costanzo, *Hawaii Pacific University*
Charles Cottingham, *Frederick Community College*
James B. Courtright, *Marquette University*
Richard Cowart, *Coastal Bend Community College*
Angela Cunningham, *Baylor University*
Judy Dacus, *Cedar Valley College*
Judith D'Aleo, *Plymouth State University*
Deborah Dardis, *Southeastern Louisiana University*
Juville Dario-Becker, *Central Virginia Community College*
Garry Davies, *University of Alaska, Anchorage*
Melissa Deadmond, *Truckee Meadows Community College*
Edward A. DeGrauw, *Portland Community College*
Heather DeHart, *Western Kentucky University*
Miriam del Campo, *Miami-Dade Community College*
Veronique Delesalle, *Gettysburg College*
Lisa Delissio, *Salem State College*
Beth De Stasio, *Lawrence University*
Elizabeth Desy, *Southwest Minnesota State University*
Donald Deters, *Bowling Green State University*
Gregg Dieringer, *Northwest Missouri State*
Diane Dixon, *Southeastern Oklahoma State University*
Christopher Dobson, *Grand Valley State University*
Cecile Dolan, *New Hampshire Community Technical College, Manchester*
Matthew Douglas, *Grand Rapids Community College*
Lee C. Drickamer, *Northern Arizona University*
Dani DuCharme, *Waubonsee Community College*
Tcherina Duncombe, *Palm Beach Community College*
Susan Dunford, *University of Cincinnati*
Stephen Ebbs, *Southern Illinois University*
Douglas Eder, *Southern Illinois University, Edwardsville*
Steve Eisenberg, *Elizabethtown Community and Technical College*
Patrick J. Enderle, *East Carolina University*
William Epperly, *Robert Morris College*
Ana Escandon, *Los Angeles Harbor College*
Dan Eshel, *City University of New York, Brooklyn College*
Marirose Ethington, *Genesee Community College*
Donna Ewing, *McLellan Community College*
Deborah Fahey, *Wheaton College*
Chris Farrell, *Trevecca Nazarene University*
Michele Finn, *Monroe Community College*
Richard Firenze, *Broome Community College*
Lynn Firestone, *Brigham Young University*
Susan Fisher, *Ohio State University*
Brandon L. Foster, *Wake Technical Community College*
Richard A. Fralick, *Plymouth State University*
Barbara Frank, *Idaho State University*
Stewart Frankel, *University of Hartford*
Lori Frear, *Wake Technical Community College*
Jennifer Fritz, *The University of Texas at Austin*
David Froelich, *Austin Community College*
Suzanne Frucht, *Northwest Missouri State University*
Edward Gabriel, *Lycoming College*
Anne Galbraith, *University of Wisconsin, La Crosse*
Patrick Galliart, *North Iowa Area Community College*
Wendy Garrison, *University of Mississippi*
Janet Gaston, *Troy University*
Anthony Gaudin, *Ivy Tech Community College of Indiana–Columbus/Franklin*
Alexandros Georgakilas, *East Carolina University*
Robert George, *University of North Carolina, Wilmington*
Richard Gill, *Brigham Young University*
Tammy Gillespie, *Eastern Arizona College*
Sharon Gilman, *Coastal Carolina University*
Mac F. Given, *Neumann College*
Bruce Goldman, *University of Connecticut, Storrs*
Andrew Goliszek, *North Carolina Agricultural and Technical State University*
Beatriz Gonzalez, *Santa Fe Community College*
Eugene Goodman, *University of Wisconsin, Parkside*
Lara Gossage, *Hutchinson Community College*
Rebekka Gougis, *Illinois State University*
Tamar Goulet, *University of Mississippi*
Becky Graham, *University of West Alabama*
Mary Rose Grant, *University of Missouri, St. Louis*
John Green, *Nicholls State University*
Robert S. Greene, *Niagara University*
Tony J. Greenfield, *Southwest Minnesota State University*
Eileen Gregory, *Rollins College*
Bruce Griffis, *Kentucky State University*
Mark Grobner, *California State University, Stanislaus*
Michael Groesbeck, *Brigham Young University, Idaho*
Stanley Guffey, *University of Tennessee*

Mark Hammer, *Wayne State University*
Blanche Haning, *North Carolina State University*
Robert Harms, *St. Louis Community College*
Craig M. Hart, *Louisiana State University*
Jay Hatch, *University of Minnesota*
Patricia Hauslein, *St. Cloud State University*
Stephen Hedman, *University of Minnesota Duluth*
Bethany Henderson-Dean, *University of Findlay*
Julie Hens, *University of Maryland University College*
Peter Heywood, *Brown University*
Julia Hinton, *McNeese State University*
Phyllis C. Hirsh, *East Los Angeles College*
Elizabeth Hodgson, *York College of Pennsylvania*
Jay Hodgson, *Armstrong Atlantic State University*
Leland Holland, *Pasco-Hernando Community College*
Jane Horlings, *Saddleback Community College*
Margaret Horton, *University of North Carolina, Greensboro*
Laurie Host, *Harford Community College*
David Howard, *University of Wisconsin, La Crosse*
Michael Hudecki, *State University of New York, Buffalo*
Michael E. S. Hudspeth, *Northern Illinois University*
Laura Huenneke, *New Mexico State University*
Pamela D. Huggins, *Fairmont State University*
Sue Hum-Musser, *Western Illinois University*
Carol Hurney, *James Madison University*
James Hutcheon, *Georgia Southern University*
Anthony Ippolito, *DePaul University*
Richard Jacobson, *Laredo Community College*
Malcolm Jenness, *New Mexico Institute of Technology*
Carl Johansson, *Fresno City College*
Staci Johnson, *Southern Wesleyan University*
Ron Johnston, *Blinn College*
Thomas Jordan, *Pima Community College*
Jann Joseph, *Grand Valley State University*
Mary K. Kananen, *Penn State University, Altoona*
Arnold Karpoff, *University of Louisville*
Judy Kaufman, *Monroe Community College*
Michael Keas, *Oklahoma Baptist University*
Judith Kelly, *Henry Ford Community College*
Karen Kendall-Fite, *Columbia State Community College*
Andrew Keth, *Clarion University*
Trey Kidd, *University of Missouri, St. Louis*
David Kirby, *American University*
Stacey Kiser, *Lane Community College*
Dennis Kitz, *Southern Illinois University, Edwardsville*
Carl Kloock, *California State, Bakersfield*
Jennifer Knapp, *Nashville State Technical Community College*
Loren Knapp, *University of South Carolina*
Michael A. Kotarski, *Niagara University*
Sarah Krajewski, *Grand Rapids Community College*
Michelle LaBonte, *Framingham State College*
Phyllis Laine, *Xavier University*
Dale Lambert, *Tarrant County College*
Tom Langen, *Clarkson University*
Michael L'Annunziata, *Pima Community College*
Lynn Larsen, *Portland Community College*
Mark Lavery, *Oregon State University*
Brenda Leady, *University of Toledo*
Mary Lehman, *Longwood University*
Lorraine Leiser, *Southeast Community College*
Doug Levey, *University of Florida*
Lee Likins, *University of Missouri, Kansas City*
Abigail Littlefield, *Landmark College*
Andrew D. Lloyd, *Delaware State University*
Jayson Lloyd, *College of Southern Idaho*
Suzanne Long, *Monroe Community College*
Judy Lonsdale, *Boise State University*
Kate Lormand, *Arapahoe Community College*
Paul Lurquin, *Washington State University*
Kimberly Lyle-Ippolito, *Anderson University*
Douglas Lyng, *Indiana University/Purdue University*
Michelle Mabry, *Davis and Elkins College*
Stephen E. MacAvoy, *American University*
Molly MacLean, *University of Maine*
Charles Mallery, *University of Miami*
Cindy Malone, *California State University, Northridge*
Mark Manteuffel, *St. Louis Community College, Flo Valley*
Ken Marr Green, *River Community College*
Kathleen Marrs, *Indiana University/Purdue University*
Roger Martin, *Brigham Young University, Salt Lake Center*
Matthew J. Maurer, *University of Virginia's College at Wise*
Geri Mayer, *Florida Atlantic University*
T. D. Maze, *Lander University*
Steve McCommas, *Southern Illinois University, Edwardsville*
Colleen McNamara, *Albuquerque Technical Vocational Institute*
Mary McNamara, *Albuquerque Technical Vocational Institute*
John McWilliams, *Oklahoma Baptist University*
Susan T. Meiers, *Western Illinois University*
Diane Melroy, *University of North Carolina, Wilmington*
Joseph Mendelson, *Utah State University*
Paige A. Mettler-Cherry, *Lindenwood University*
Debra Meuler, *Cardinal Stritch University*
James E. Mickle, *North Carolina State University*
Craig Milgrim, *Grossmont College*
Hugh Miller, *East Tennessee State University*
Jennifer Miskowski, *University of Wisconsin, La Crosse*
Ali Mohamed, *Virginia State University*
Stephen Molnar, *Washington University*
James Mone, *Millersville University*
Daniela Monk, *Washington State University*
Linda Moore, *Georgia Military College*
David Mork, *Yakima Valley Community College*
Bertram Murray, *Rutgers University*
Ken Nadler, *Michigan State University*
John J. Natalini, *Quincy University*
Alissa A. Neill, *University of Rhode Island*
Dawn Nelson, *Community College of Southern Nevada*
Joseph Newhouse, *California University of Pennsylvania*
Jeffrey Newman, *Lycoming College*
Lori Nicholas, *New York University*

David L.G. Noakes, *University of Guelph*
Shawn Nordell, *St. Louis University*
Tonye E. Numbere, *University of Missouri, Rolla*
Jorge Obeso, *Miami-Dade College, North Campus*
Erin O'Brien, *Dixie College*
Igor Oksov, *Union County College*
Alex Olvido, *University of North Georgia*
Jennifer O'Malley, *Saint Charles Community College*
Kevin Padian, *University of California, Berkeley*
Arnas Palaima, *University of Mississippi*
Brent Palmer, *University of Kentucky*
Anthony Palombella, *Longwood University*
Murali Panen, *Luzerne County Community College*
Monica Parker, *Florida State College*
Marilee Benore Parsons, *University of Michigan, Dearborn*
Steven L. Peck, *Brigham Young University*
Javier Penalosa, *Buffalo State College*
Murray Paton Pendarvis, *Southeastern Louisiana University*
Shelly Penrod, *Lonestar College*
Krista Peppers, *University of Central Arkansas*
Rhoda Perozzi, *Virginia Commonwealth University*
John Peters, *College of Charleston*
Patricia Phelps, *Austin Community College*
Polly Phillips, *Florida International University*
Indiren Pillay, *Culver-Stockton College*
Francis J. Pitocchelli, *Saint Anselm College*
Nancy Platt, *Pima Community College*
Roberta L. Pohlman, *Wright State University*
Calvin Porter, *Xavier University*
Linda Potts, *University of North Carolina, Wilmington*
Robert Pozos, *San Diego State University*
Marion Preest, *The Claremont Colleges*
Anne-Marie Prouty, *Sam Houston State University*
Gregory Pryor, *Francis Marion University*
Rongsun Pu, *Kean University*
Narayanan Rajendran, *Kentucky State University*
Anne E. Reilly, *Florida Atlantic University*
Michael H. Renfroe, *James Madison University*
Laura Rhoads, *State University of New York, Potsdam*
Ashley Rhodes, *Kansas State University*
Gwynne S. Rife, *University of Findlay*
Todd Rimkus, *Marymount University*
Laurel Roberts, *University of Pittsburgh*
Wilma Robertson, *Boise State University*
Bill Rogers, *Ball State University*
William E. Rogers, *Texas A&M University*
Troy Rohn, *Boise State University*
Deborah Ross, *Indiana University/Purdue University*
Christel Rowe, *Hibbing Community College*
Yelena Rudayeva, *Palm Beach Community College*
Joanne Russell, *Manchester Community College*
Michael Rutledge, *Middle Tennessee State University*
Wendy Ryan, *Kutztown University*
Christopher Sacchi, *Kutztown University*
Kim Sadler, *Middle Tennessee State University*
Brian Sailer, *Albuquerque Technical Vocational Institute*
Jasmine Saros, *University of Wisconsin, La Crosse*
Ken Saville, *Albion College*
Michael Sawey, *Texas Christian University*
Louis Scala, *Kutztown University*
Debbie Scheidemantel, *Pima Community College*
Daniel C. Scheirer, *Northeastern University*
Beverly Schieltz, *Wright State University*
Nancy Schmidt, *Pima Community College*
Robert Schoch, *Boston University*
Julie Schroer, *Bismarck State College*
Fayla Schwartz, *Everett Community College*
Steven Scott, *Merritt College*
Gray Scrimgeour, *University of Toronto*
Roger Seeber, *West Liberty State College*
Mary Severinghaus, *Parkland College*
Allison Shearer, *Grossmont College*
Robert Shetlar, *Georgia Southern University*
Cara Shillington, *Eastern Michigan University*
Jack Shurley, *Idaho State University*
Bill Simcik, *Lonestar College*
Indrani Sindhuvalli, *Florida State College, Jacksonville*
Beatrice Sirakaya, *Pennsylvania State University*
Cynthia Sirna, *Gadsden State Community College*
Lynnda Skidmore, *Wayne County Community College*
Thomas Sluss, *Fort Lewis College*
Douglas Smith, *Clarion University of Pennsylvania*
Mark Smith, *Chaffey College*
Brian Smith Black, *Hills State University*
Gregory Smutzer, *Temple University*
Sally Sommers, *Smith Boston University*
Anna Bess Sorin, *University of Memphis*
Marialana Spiedel, *Jefferson College*
Bryan Spohn Florida, *Community College at Jacksonville, Kent Campus*
Carol St. Angelo, *Hofstra University*
Brooke Stabler, *University of Central Oklahoma*
Amanda Starnes, *Emory University*
Susan L. Steen, *Idaho State University*
Timothy Stewart, *Longwood College*
Jennifer Stovall, *Southcentral Kentucky Community & Technical College*
Shawn Stover, *Davis and Elkins College*
Bradley J. Swanson, *Central Michigan University*
Joyce Tamashiro, *University of Puget Sound*
Jeffrey Taylor, *Slippery Rock University*
Martha Taylor, *Cornell University*
Glena Temple, *Viterbo University*
Alice Templet, *Nicholls State University*
Tania Thalkar, *Clarion University of Pennsylvania*
Jeff Thomas, *California State University, Northridge*
Jeffrey Thomas, *University of California, Los Angeles*
Janis Thompson, *Lorain County Community College*
Nina Thumser, *California University of Pennsylvania*
Alana Tibbets, *Southern Illinois University, Edwardsville*
Martin Tracey, *Florida International University*
Sue Trammell, *John A. Logan College*

Jeffrey Travis, *State University of New York, Albany*
Michael Troyan, *Pennsylvania State University*
Robert Turgeon, *Cornell University*
Kimberly Turk, *Caldwell Community College*
Michael Tveten, *Pima Community College, Northwest Campus*
James Urban, *Kansas State University*
Brandi Van Roo, *Framingham State College*
John Vaughan, *St. Petersburg Junior College*
Martin Vaughan, *Indiana State University*
Mark Venable, *Appalachian State University*
Paul Verrell, *Washington State University*
Tanya Vickers, *University of Utah*
Janet Vigna, *Grand Valley State University*
Sean Walker, *California State University, Fullerton*
Don Waller, *University of Wisconsin, Madison*
Sandra Walsh, *The Citadel*
Mark Walvoord, *University of Oklahoma*
Tracy Ware, *Salem State College*
Jennifer Warner, *University of North Carolina, Charlotte*
Carol Weaver, *Union University*
Frances Weaver, *Widener University*
Derek Weber, *Raritan Valley Community College*
Elizabeth Welnhofer, *Canisius College*
Marcia Wendeln, *Wright State University*
Michael Wenzel, *Folsom Lake College*
Shauna Weyrauch, *Ohio State University, Newark*
Wayne Whaley, *Utah Valley State College*
Howard Whiteman, *Murray State University*
Jennifer Wiatrowski, *Pasco-Hernando Community College*
Vernon Wiersema, *Houston Community College*
Gerald Wilcox, *Potomac State College*
Peter J. Wilkin, *Purdue University North Central*
Heather Wilson-Ashworth, *Utah Valley University*
Robert R. Wise, *University of Wisconsin, Oshkosh*
Michelle Withers, *Louisiana State University*
Art Woods, *University of Texas, Austin*
Elton Woodward, *Daytona Beach Community College*
Kenneth Wunch, *Sam Houston State University*
Donna Young, *University of Winnipeg*
Michelle L. Zjhra, *Georgia Southern University*
John Zook, *Ohio University*
Michelle Zurawski, *Moraine Valley Community College*

The Book Team

The sixth edition has been energized by the work and ideas of our new editor Cady Owens. She has brought a fresh and insightful perspective that is much appreciated by both authors. We remain indebted to our editor for the previous three editions, Star MacKenzie, who was instrumental in helping us develop the revision plan for this edition. Our development editor for much of the sixth edition, Debbie Hardin, played a key role in shaping new and heavily revised chapters. Her talented successor, Evelyn Dahlgren, has drawn on her extensive experience to further improve the final product. We are also grateful for the steady hand of the Director of Courseware Portfolio Management Beth Wilbur, who is always thoughtful, responsive, and supportive of us and this project. We continue to feel fortunate to work with such a talented and devoted team.

This book is dedicated to our families, friends, and colleagues who have supported us over the years. Having loving families, great friends, and a supportive work environment has enabled us to make this heartfelt contribution to nonmajors biology education.

Colleen Belk
Virginia Borden Maier

"Because science, told as a story, can intrigue and inform the non-scientific minds among us, it has the potential to bridge the two cultures into which civilization is split—the sciences and the humanities. For educators, stories are an exciting way to draw young minds into the scientific culture."

—E. O. Wilson

Contents

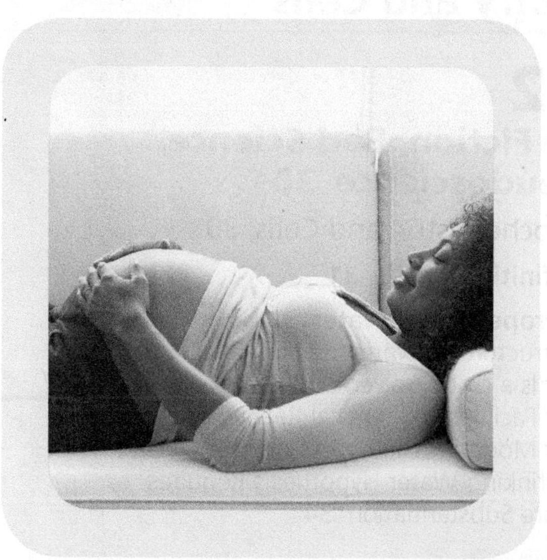

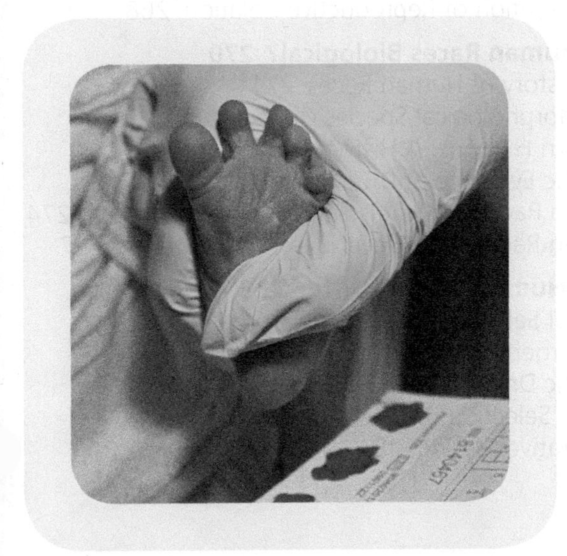

UNIT THREE

Evolution

CHAPTER 11

Where Did We Come From? 206

CHAPTER 12

An Evolving Enemy 238

CHAPTER 13

Understanding Race 262

UNIT FOUR
Ecology

Engage Students in Science with Stories That Relate to Their Lives

Biology: Science for Life with Physiology **weaves a compelling storyline throughout each chapter to grab student attention, exploring high-interest topics such as genetic testing, global warming, and the Zika virus. The authors return to the storyline again and again, using it as the basis on which they introduce the biological concepts behind each story.**

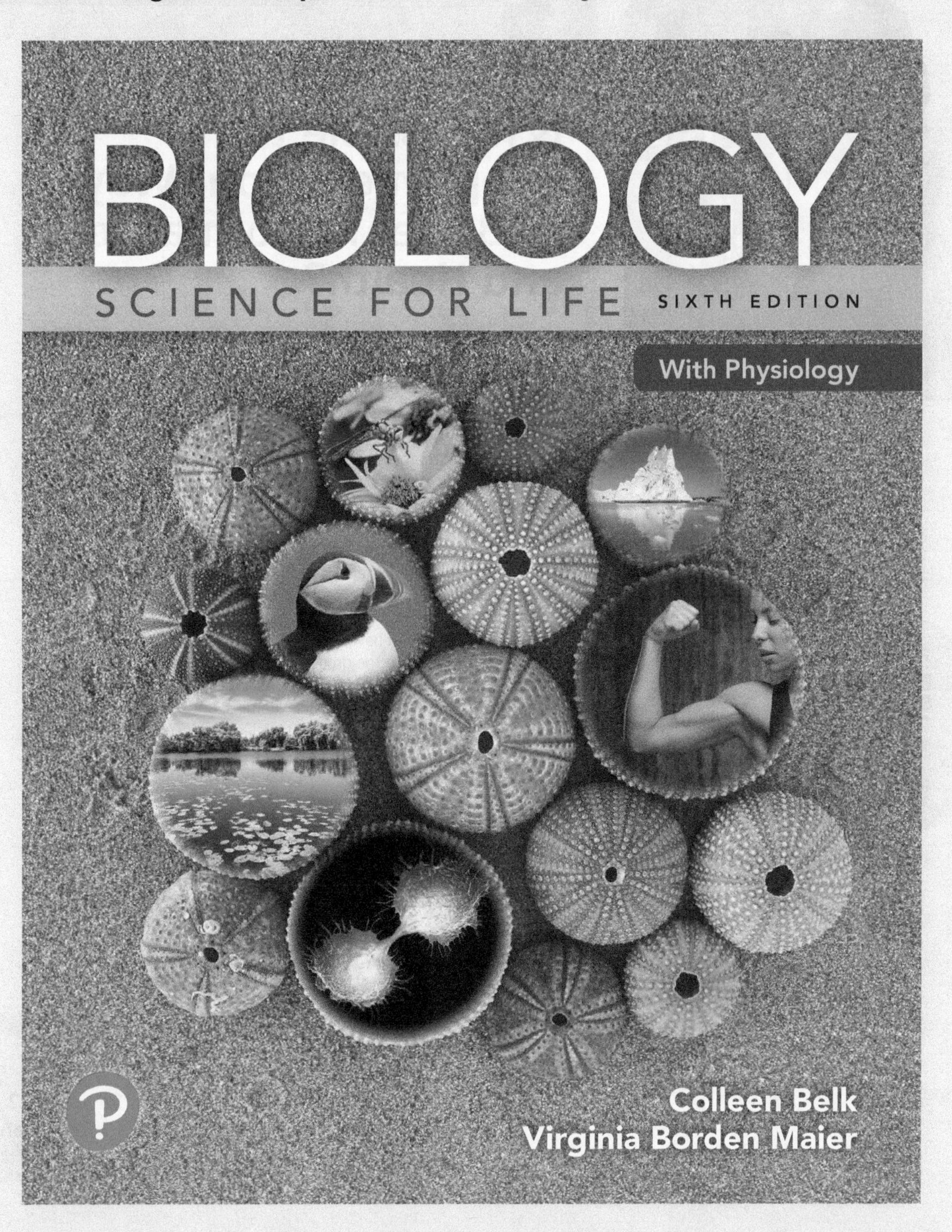

Capture Student Attention with

Each chapter weaves in a story based on a current issue or hot topic through which biological concepts, examples, and applications are presented and explained.

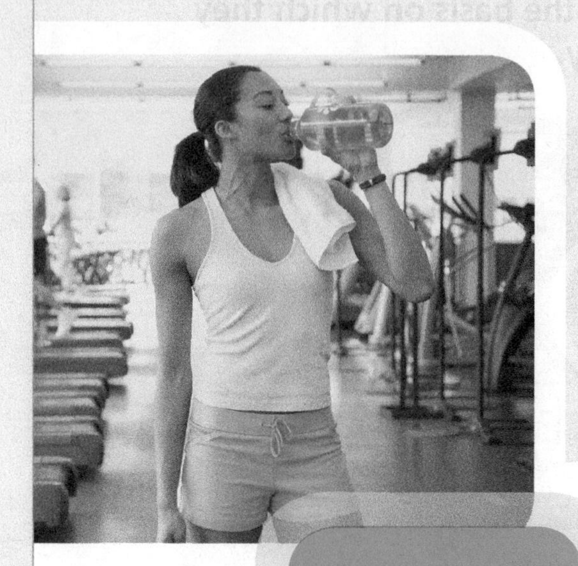

3

Is It Possible to Supplement Your Way to Better Performance and Health?

Do sports drinks enhance athletic performance?

Do nutritional supplements enhance academic performance or health?

Or is it more healthful to eat whole foods?

Nutrients and Membrane Transport

Gingko to improve your memory, kava to reduce stress, ginseng to boost energy, and melatonin to help you sleep. Sounds like a recipe for success for a busy student. For good measure, chase those supplements down with some coconut water to slow aging and prevent cancer. You may have heard claims about the health benefits of nutritional supplements like vitamins, minerals, herbs, yeast, and even enzymes. If these are truly good for you, why not replace some of the food you eat with products that have a longer shelf life than most foods? Instead of going to the grocery store every weekend, you could stock your pantry with energy drinks, vitamin-enriched waters, protein powders, nutrition bars, vitamins, and minerals. These can be bought in bulk and don't rot like fruits and vegetables. But are they as good for you as food?

Is it possible to supplement your way to enhanced academic performance or better health? It seems that most Americans think so—we spend around $6 billion a year on these items and more than two-thirds of us are taking at least one such supplement. Let's investigate whether these products are doing what we hope they are.

NEW! Storylines in the 6th edition:

- Chapter 7: Fertility
- Chapter 8: Does Testing Save Lives?
- Chapter 9: Biology of Wrongful Convictions
- Chapter 17: The Human Footprint

Relevant, Engaging Storylines

NEW! Make the Connection Activities tie the storyline of the chapter to the key scientific concepts behind it to ensure that students truly understand the relationship between the story and the science. These are also assignable in Mastering Biology.

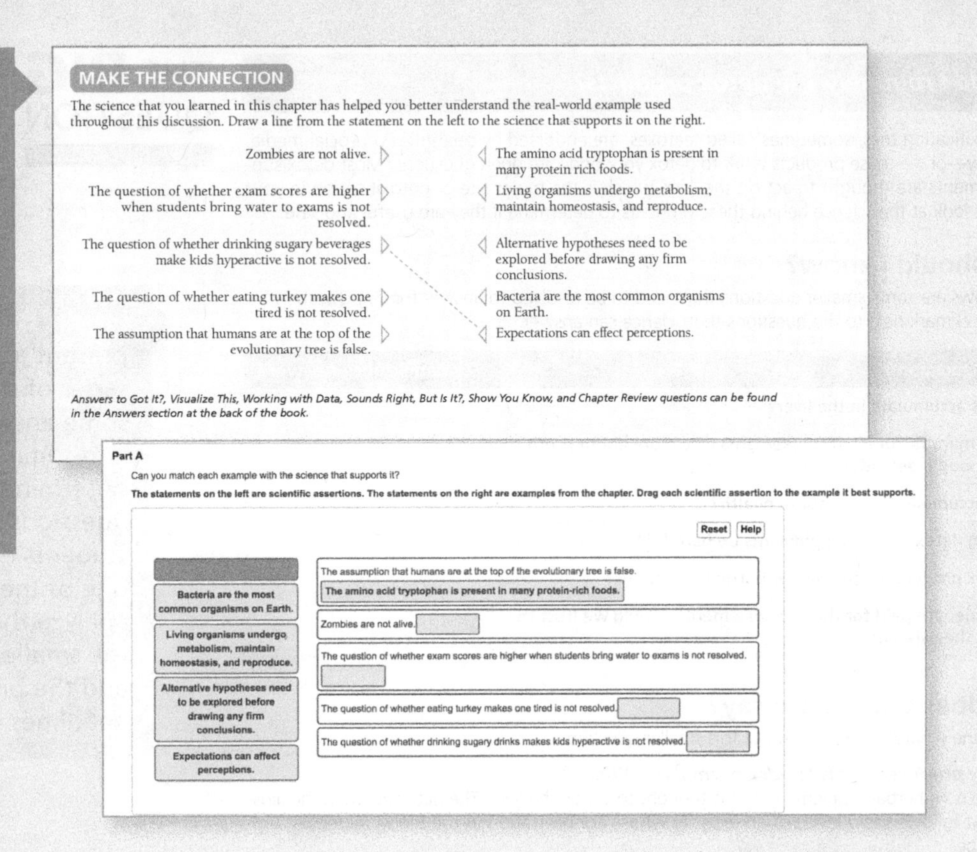

MAKE THE CONNECTION
The science that you learned in this chapter has helped you better understand the real-world example used throughout this discussion. Draw a line from the statement on the left to the science that supports it on the right.
Zombies are not alive.
The question of whether exam scores are higher when students bring water to exams is not resolved.
The question of whether drinking sugary beverages make kids hyperactive is not resolved.
The question of whether eating turkey makes one tired is not resolved.
The assumption that humans are at the top of the evolutionary tree is false.
The amino acid tryptophan is present in many protein rich foods.
Living organisms undergo metabolism, maintain homeostasis, and reproduce.
Alternative hypotheses need to be explored before drawing any firm conclusions.
Bacteria are the most common organisms on Earth.
Expectations can effect perceptions.
Answers to Got It?, Visualize This, Working with Data, Sounds Right, But Is It?, Show You Know, and Chapter Review questions can be found in the Answers section at the back of the book.
Part A
Can you match each example with the science that supports it?
The statements on the left are scientific assertions. The statements on the right are examples from the chapter. Drag each scientific assertion to the example it best supports.
Reset
Help
Bacteria are the most common organisms on Earth.
Living organisms undergo metabolism, maintain homeostasis, and reproduce.
Alternative hypotheses need to be explored before drawing any firm conclusions.
Expectations can affect perceptions.
The assumption that humans are at the top of the evolutionary tree is false.
The amino acid tryptophan is present in many protein-rich foods.
Zombies are not alive.
The question of whether exam scores are higher when students bring water to exams is not resolved.
The question of whether eating turkey makes one tired is not resolved.
The question of whether drinking sugary drinks makes kids hyperactive is not resolved.

Sounds right, but is it?

In the Harry Potter books and movies, many of the characters who knew Harry's parents tell him that he resembles his mother or note his similarity to his father in his willingness to bend the rules. To many fans, these comments make sense, because a child receives half of his genetic information from his mother and half from his father. Thus, it seems fair to say that:

Harry Potter has his mother's eyes.

Sounds right, but it isn't.

Answer the following questions to understand why.

1. Do you think it is more likely that the color and shape of a person's eyes are determined by one gene or many genes?
2. Did Harry receive copies of genes that determine eye color and shape from his mother?
3. Did he receive copies of genes that determine eye color and shape from his father?
4. Think back to the Punnett squares you've viewed and drawn. Do genes for only one or both parents likely influence eye color and shape?
5. Reflect on your answers to questions 1–4. Explain why the statement bolded above sounds right, but isn't.

Sounds Right, But Is It? activities at the end of each chapter address common student misconceptions about biology concepts.

Help Students Interpret and Apply Data

Should I routinely use detox products?

THE **BIG** QUESTION

Detoxification teas, sometimes called teatoxes, are endorsed by celebrities on social media. How—or if—these products work to detox your body are open questions. Most detox supplements are thought to act on the liver, which is the major site of detoxification in your body. Let's look at the science behind these products to determine if they are useful and safe.

What should I know?

What follows are some smaller questions that need to be resolved to answer the Big Question. Place a checkmark next to the questions that science can answer.

Smaller Questions	Can Science Answer?
Do toxins accumulate in the liver?	
Do most manufacturers of supplements care more about profit than helping people detoxify?	
Is toxin accumulation harmful to health?	
Can using detox teas or supplements be harmful?	
Are detox products helpful under normal conditions?	
If celebrities are paid for their endorsements, should we trust the products they are endorsing?	

What does the science say?

Let's examine what the data say about this smaller question:

Are detox products helpful under normal conditions?

Milk thistle is an herbal supplement that is thought to act on the liver. The data shown in the illustration that follows show levels of an enzyme whose concentration in the blood increases with liver damage.

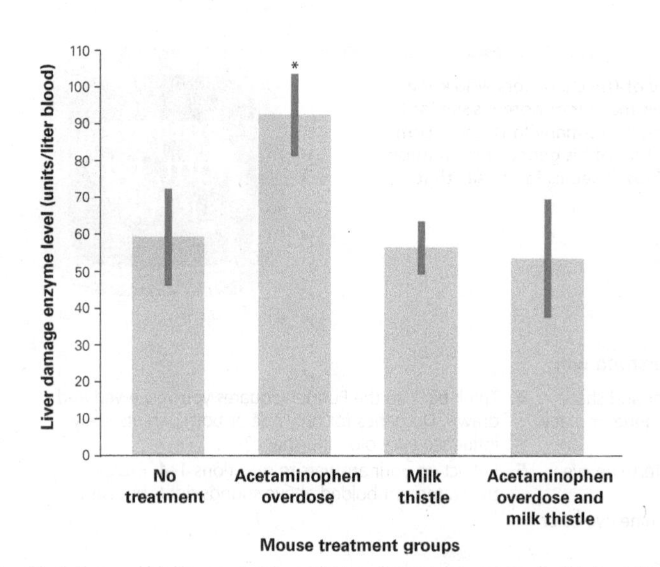

1. Describe the results. Does it appear that milk thistle helps prevent liver damage under normal conditions?
2. Given these data, do you think the smaller question is answered? If not, propose another study that would help answer this question.
3. Does this information help you answer the Big Question? What else do you need to consider?

Data source: N. Bektur, E. Sahin, C. Baycu, and G. Unver, "Protective Effects of Silymarin against Acetaminophen-Induced Hepatotoxicity and Nephrotoxicity in Mice," *Toxicology and Industrial Health* 32, no. 4 (2016): 589–600.

NEW! Big Question features present a topic, followed by a series of smaller questions—some answerable by science and some not. Once students determine which of the smaller questions science can answer, students explore data related to one of these questions. Students analyze the data in light of both the smaller question addressed and the big question that headlines the feature.

NEW! 10 GraphIt! Coaching activities help students read, interpret, and create graphs that explore real environmental issues using real data. They are presented in an entirely new mobile experience with accessible design.

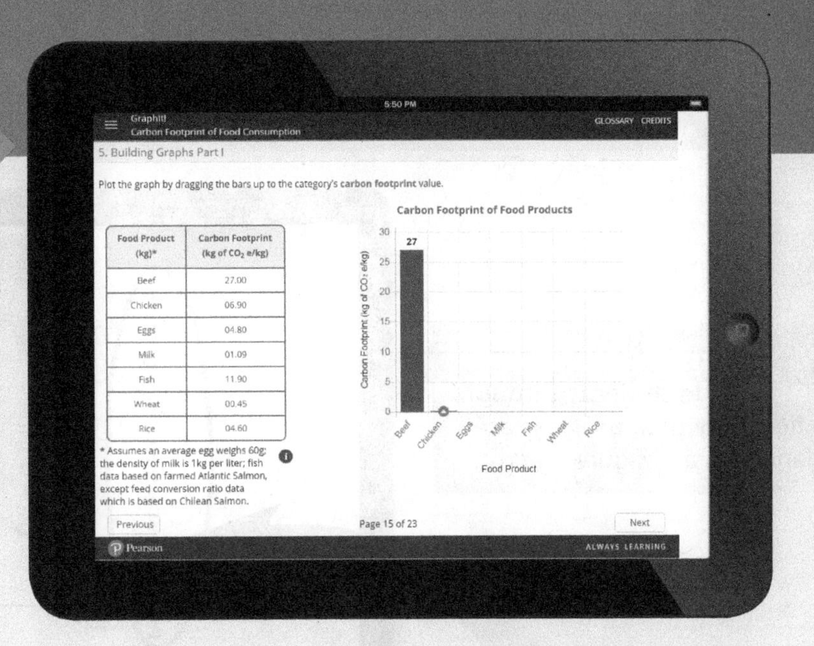

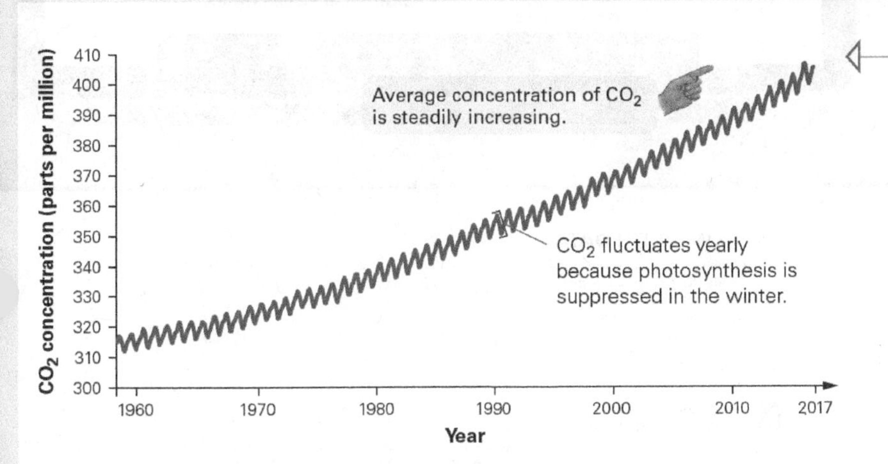

FIGURE 5.6 Increases in atmospheric carbon dioxide. Carbon dioxide levels from 1960 to present as measured by instruments at Mauna Loa observatory in Hawaii.

WORKING WITH **DATA**

What evidence in the graph demonstrates the increased rate of carbon dioxide accumulation from 2000 to 2017 compared with the 1960s?

Working with Data questions challenge students to analyze and apply their knowledge of biology to a graph or set of data.

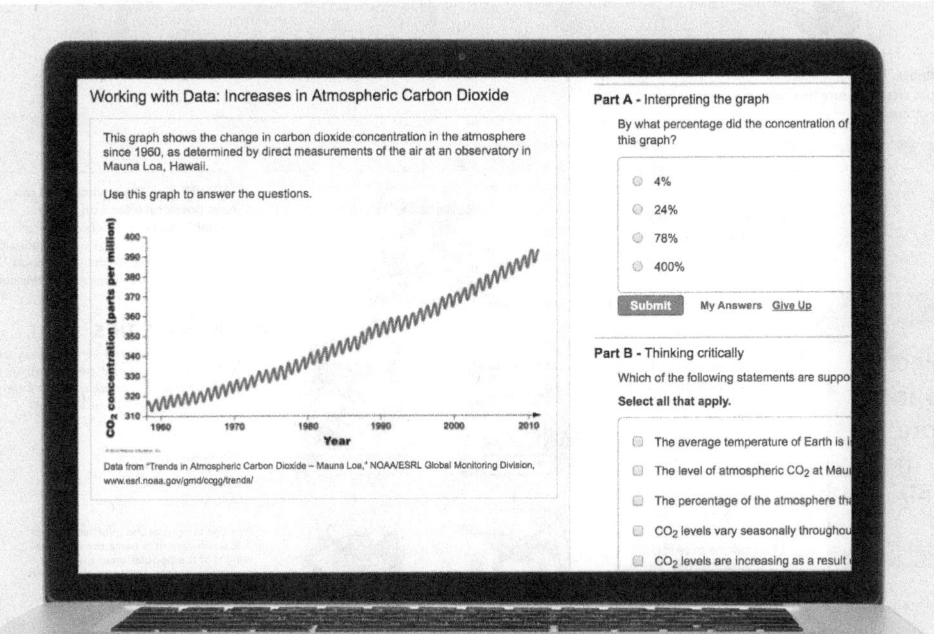

Select **Working with Data** questions are also assignable as activities in Mastering Biology.

Bring the Story to Life with

NEW! Figure Walkthrough videos guide students through key figures with author-narrated explanations and figure markups.

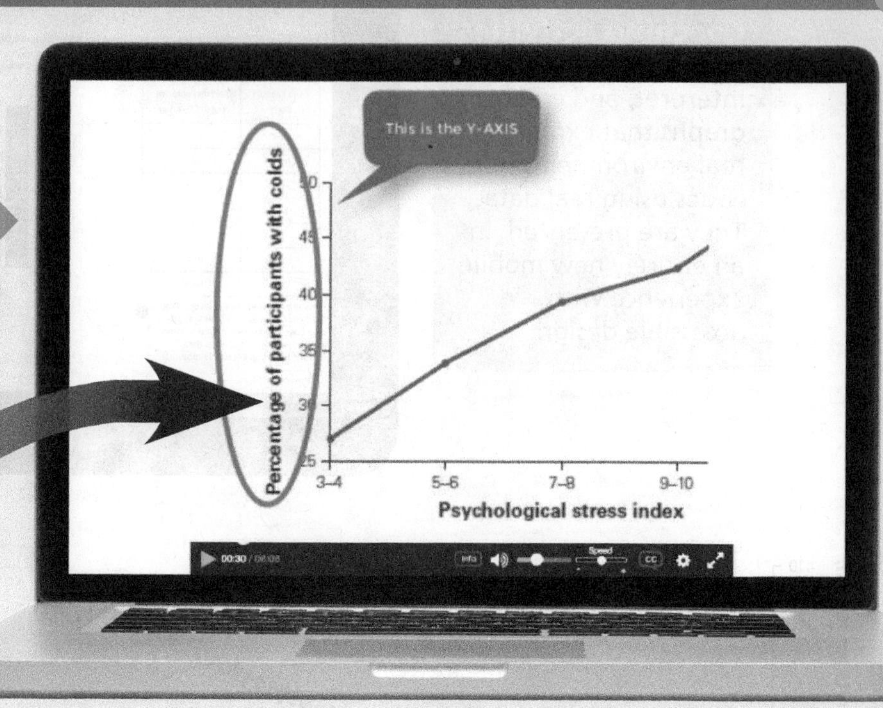

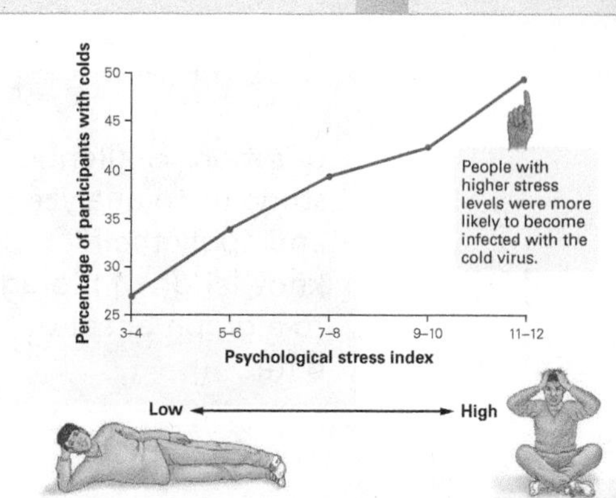

FIGURE 1.10 Correlation between stress level and illness. The graph indicates that people reporting higher levels of stress became infected after exposure to a cold virus more often than did people who reported low levels of stress.

WORKING WITH DATA

This graph groups people with similar but not identical, stress index measures. Why might this have been necessary? If people with stress indices 3 and 4 have the same susceptibility to colds, does this call into question the correlation? Why?

Watch **Correlation** in Mastering **Biology**

Visualize This questions within select figure legends encourage students to look more closely at figures to more fully understand their content.

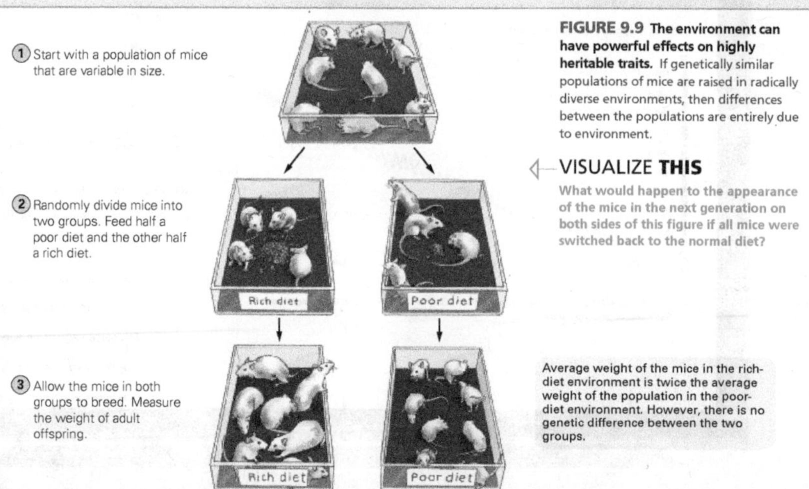

1 Start with a population of mice that are variable in size.
2 Randomly divide mice into two groups. Feed half a poor diet and the other half a rich diet.
Rich diet
Poor diet
3 Allow the mice in both groups to breed. Measure the weight of adult offspring.
Rich diet
Poor diet
Average weight of the mice in the rich-diet environment is twice the average weight of the population in the poor-diet environment. However, there is no genetic difference between the two groups.

FIGURE 9.9 The environment can have powerful effects on highly heritable traits. If genetically similar populations of mice are raised in radically diverse environments, then differences between the populations are entirely due to environment.

VISUALIZE THIS

What would happen to the appearance of the mice in the next generation on both sides of this figure if all mice were switched back to the normal diet?

Best-in-Class Artwork and Animations

several ingredients, in protein synthesis we use tRNAs that are dedicated to one specific ingredient.) The measuring spoons and cups bring the ingredients to the kitchen counter. Like the ingredients in a cake that can be used in many

FIGURE 10.6 Protein synthesis and cake baking. Making a protein in a cell is analogous to making a cake in your kitchen.

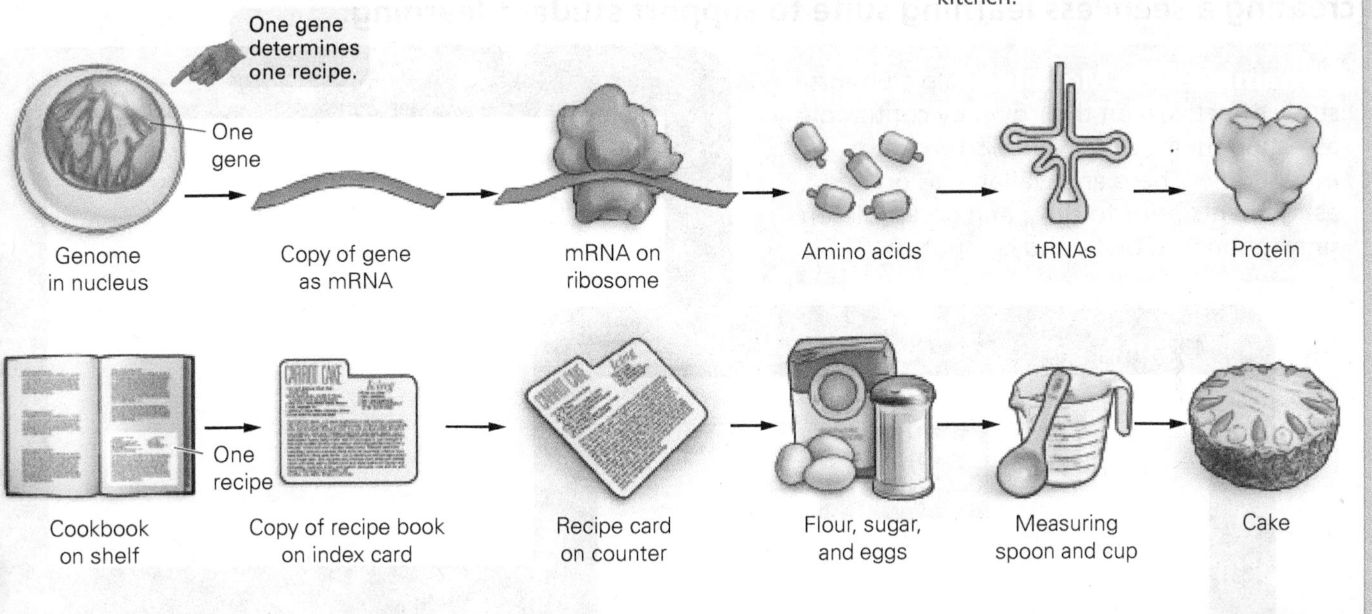

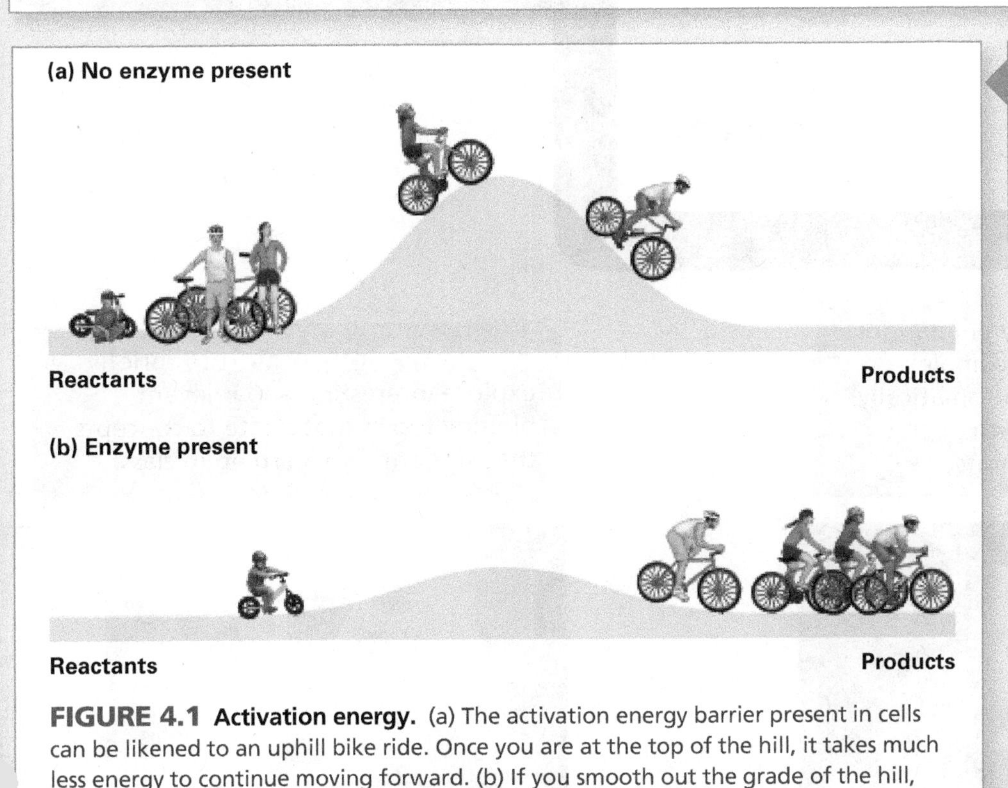

FIGURE 4.1 Activation energy. (a) The activation energy barrier present in cells can be likened to an uphill bike ride. Once you are at the top of the hill, it takes much less energy to continue moving forward. (b) If you smooth out the grade of the hill, more people will make it. In cells, there is an energy barrier that prevents chemical reactions from occurring. Adding an enzyme helps lower this barrier.

Visual Analogies simplify complex topics so students conceptualize and recall important concepts when they need them.

Personalize Learning with

Mastering™ Biology is an online homework, tutorial, and assessment platform that improves results by helping students quickly master concepts and skills. Features in the textbook and Mastering Biology work together, creating a seamless learning suite to support student learning.

Dynamic Study Modules help students study effectively on their own by continuously assessing their activity and performance in real time. These are available as graded assignments prior to class, and accessible on smartphones, tablets, and computers.

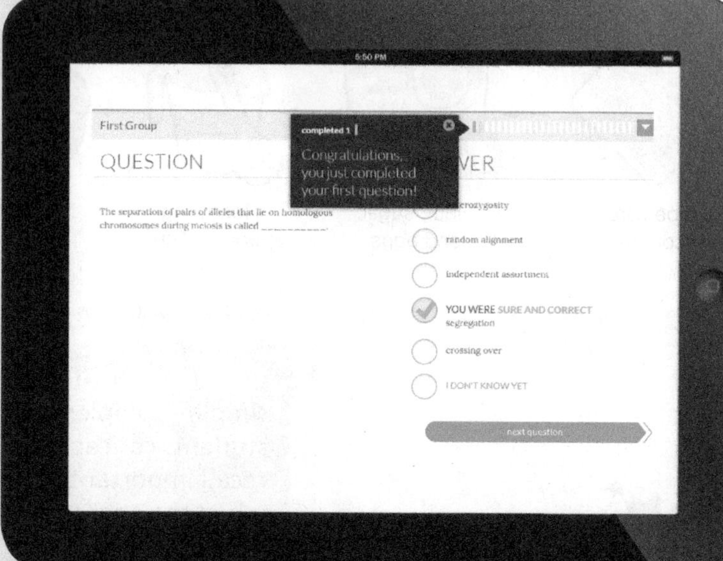

BioFlix™ 3D movie-quality animations help your students visualize complex biology topics and include automatically graded coaching activities with personalized feedback and hints.

Everyday Biology Videos briefly explore interesting and relevant biology topics that relate to concepts that students are learning in class.

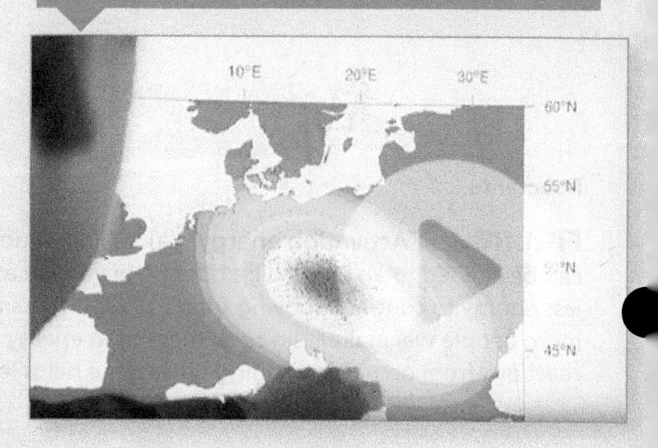

Mastering Biology

Evaluating Science in the Media: Soda Consumption and Aging

There has been much research on the effects of excessive sugar consumption over the last few decades. Not only are scientists interested in how the consumption of sugars affects long-term health and susceptibility to disease, but they are also concerned with how excessive sugar consumption may impact senescence, or aging, of cells.

If you wanted to learn more about the effects of sugar consumption on aging, where would you look for reliable information?

Suppose you did an Internet search and came upon this web site. These questions can help you evaluate the reliability of the information it provides.

Part A - First impression

Open the site in your browser and skim the article. Think about whether you believe the information presented or whether you have doubts about some of it.

On a scale of 0 to 6, where 6 is the most trustworthy, how would you rate this site? (Note that all responses will be marked as "correct" at this point.)

- 0-1 (not trustworthy at all)
- 2-4 (somewhat trustworthy; want to check some things)
- 5-6 (very trustworthy)

Submit My Answers Give Up

Correct

Your answer represents your first impression of the trustworthiness of this source. Now you will answer some specific questions and reevaluate this score at the end.

Part B - Authority

How can you know if the person or organization providing the information has the credentials and knowledge to speak on this topic? One clue is the type of web site it is--the domain name ".com" tells you that this site is owned by a commercial business.

Now scan the article to find the name and credentials of the person who wrote it.

Evaluating Science in the Media activities ask students to examine selected media (web sites, articles, videos) with a critical look at the sources and methods used to convey information.

Roots to Remember references appear in context within chapter discussions to help students learn the language of biology using word roots and include assignable activities in Mastering Biology.

Roots to Remember: Chapter 3

Knowing the meaning of prefixes and suffixes can help you understand biology terms.

Part A - Understanding roots

Can you match these prefixes and suffixes with their definitions?

Drag the roots on the left to the appropriate blanks on the right to complete the sentences.

Reset Help

1. The root **osmo-** means fluid.
2. The root **endo-** means inside.
3. The root **exo-** means outside or external.
4. The root **-plasm** refers to the movement of water.
5. The root **cyto- (or -cyte)** means cell or kind of cell.

Submit My Answers Give Up

Incorrect; Try Again

You filled in 2 of 5 blanks incorrectly. Osmotic shock can occur when there is a rapid change in water or solute concentration. What does the prefix *osmo-* mean?

ROOTS TO REMEMBER

The following roots of words come mainly from Latin and Greek and will help you to decipher terms:

cyto- means cell or a kind of cell. Chapter terms: *cytoplasm, cytoskeleton*

endo- means inside. Chapter terms: *endocytosis, endoplasmic reticulum*

exo- means outside. Chapter term: *exocytosis*

osmo- means water. Chapter term: *osmosis*

plasm means fluid. Chapter term: *cytoplasm, plasma membrane*

Bring Science to Life with

NEW! Ready-to-Go Teaching Modules make use of teaching tools for before, during, and after class, including new ideas for in-class activities. The modules incorporate the best that the text, Mastering Biology, and Learning Catalytics have to offer and can be accessed through the Instructor Resources area of Mastering Biology.

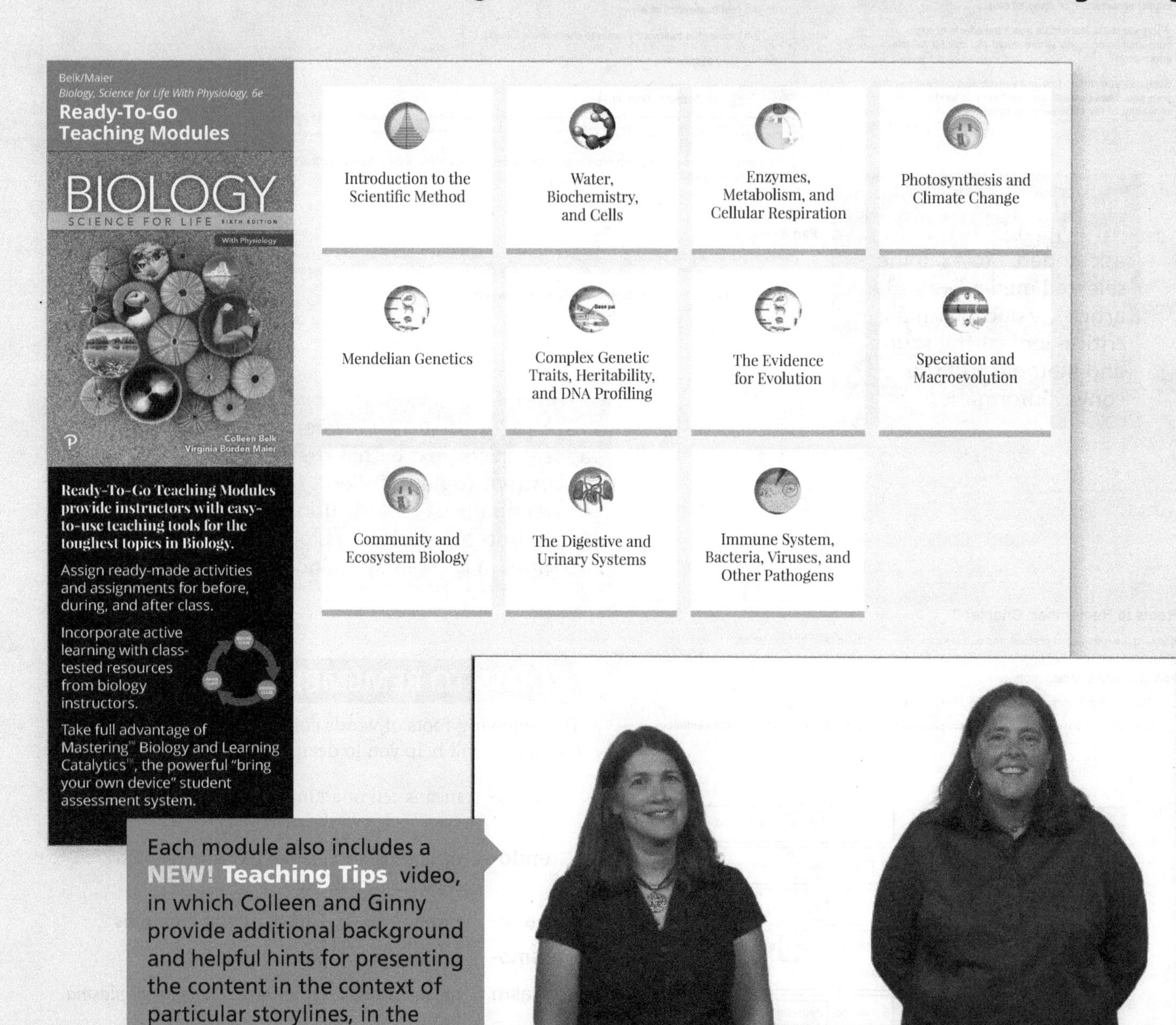

Belk/Maier
Biology, Science for Life With Physiology, 6e
Ready-To-Go Teaching Modules
BIOLOGY
SCIENCE FOR LIFE SIXTH EDITION
With Physiology
Colleen Belk
Virginia Borden Maier
Ready-To-Go Teaching Modules provide instructors with easy-to-use teaching tools for the toughest topics in Biology.
Assign ready-made activities and assignments for before, during, and after class.
Incorporate active learning with class-tested resources from biology instructors.
Take full advantage of Mastering™ Biology and Learning Catalytics™, the powerful "bring your own device" student assessment system.
Introduction to the Scientific Method
Water, Biochemistry, and Cells
Enzymes, Metabolism, and Cellular Respiration
Photosynthesis and Climate Change
Mendelian Genetics
Complex Genetic Traits, Heritability, and DNA Profiling
The Evidence for Evolution
Speciation and Macroevolution
Community and Ecosystem Biology
The Digestive and Urinary Systems
Immune System, Bacteria, Viruses, and Other Pathogens
00:30 / 06:06
info
Speed
CC

Each module also includes a **NEW! Teaching Tips** video, in which Colleen and Ginny provide additional background and helpful hints for presenting the content in the context of particular storylines, in the classroom or online.

Active Learning Resources

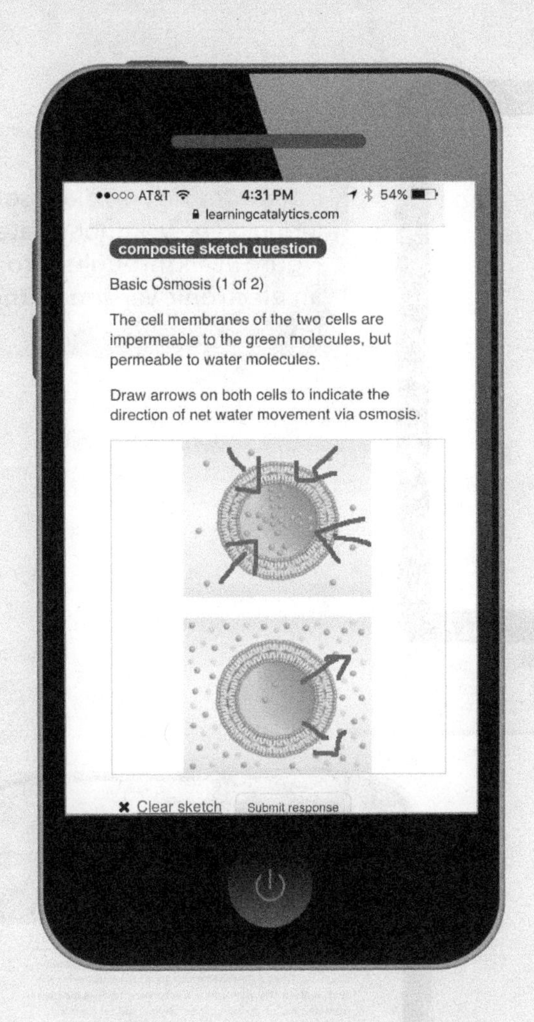

Learning Catalytics™ helps generate class discussion, customize lectures, and promote peer-to-peer learning with real-time analytics. Learning Catalytics acts as a student response tool that uses students' smartphones, tablets, or laptops to engage them in more interactive tasks and thinking.

- Help your students develop critical thinking skills.
- Monitor responses to find out where your students are struggling.
- Rely on real-time data to adjust your teaching strategy.

GO FIND OUT

1. Select one supplement you have wondered about and spend a few minutes doing some web-based research on whether the claims made on its label are backed up by scientific evidence.
2. Some cities have banned restaurants from using trans fats when cooking. Has such a ban been enacted in your city? Do you think the government should be involved in regulating the use of trans fats? Why or why not?

Bring ideas for active learning into your class with **NEW! Go Find Out** activities, located at the conclusion of each chapter.

Additional Resources:

- **"Flipped Classroom" Instructor's Manual** includes many activities that have been tested in the authors' own classes. Each text chapter is supplemented with a selection of in-class activities, suggestions for student "pre-work" outside of class, media references, and more. The new edition also includes implementation suggestions for the in-text "Go Find Out" activities.
- **PowerPoint presentations** centered around the storylines accompany each chapter to help instructors highlight the relevance of biology to everyday life.

Access the Text Anytime, Anywhere with Pearson eText

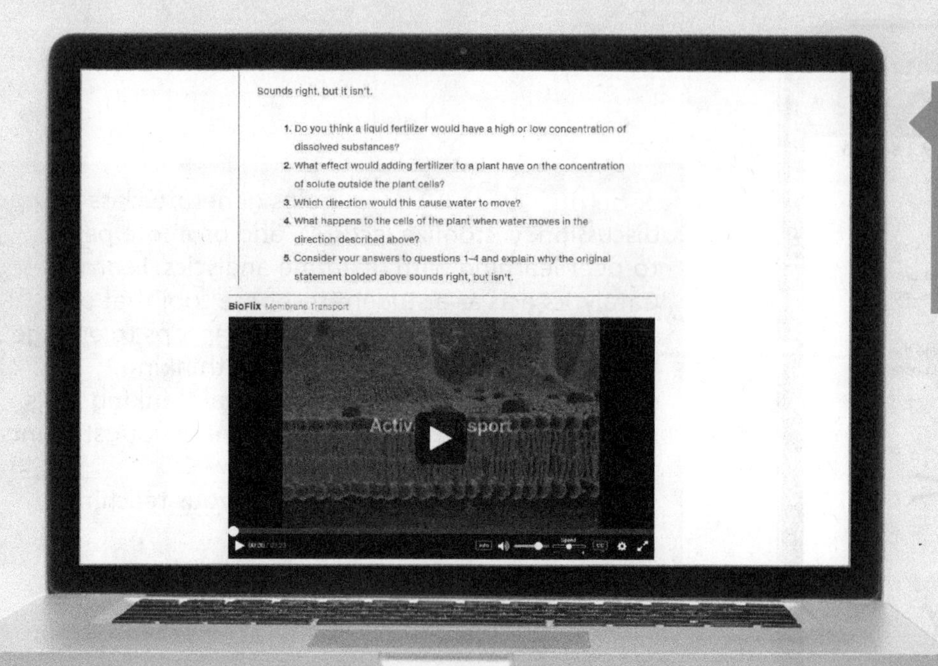

NEW! Pearson eText integrates rich media assets, including new author-created Figure Walkthroughs into an electronic version of the text.

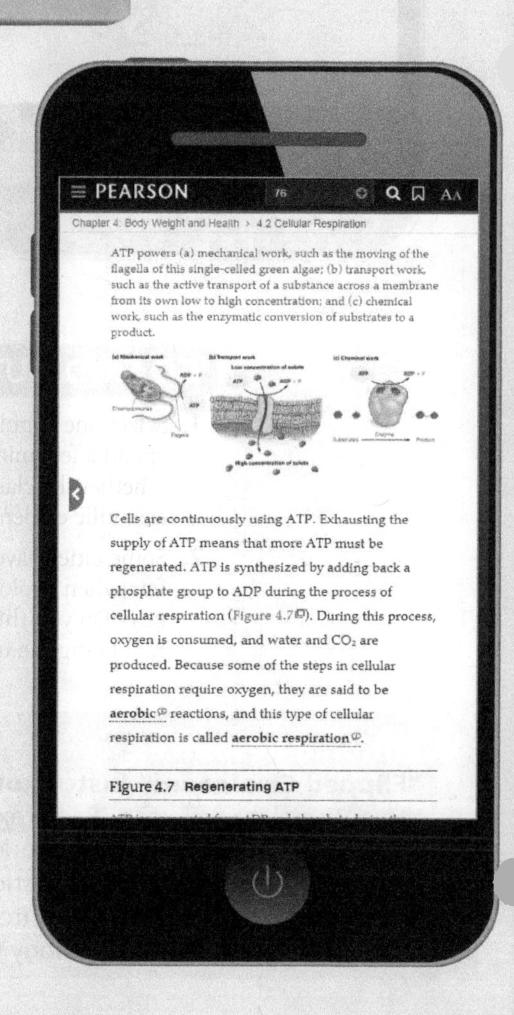

Pearson eText Mobile App offers offline access and can be downloaded for most iOS and Android phones/tablets from the Apple App Store or Google Play, providing:

- Seamlessly integrated videos and other rich media assets
- ADA accessibility (screen-reader ready)
- Configurable reading settings, including resizable type and night reading mode
- Instructor and student note-taking, highlighting, bookmarking, and search capabilities.

BIOLOGY

SCIENCE FOR LIFE

SIXTH EDITION

With Physiology

Colleen Belk
Virginia Borden Maier

1 Can Science Cure the Common Cold?

Another cold! What can I do?

Take massive doses of vitamin C?

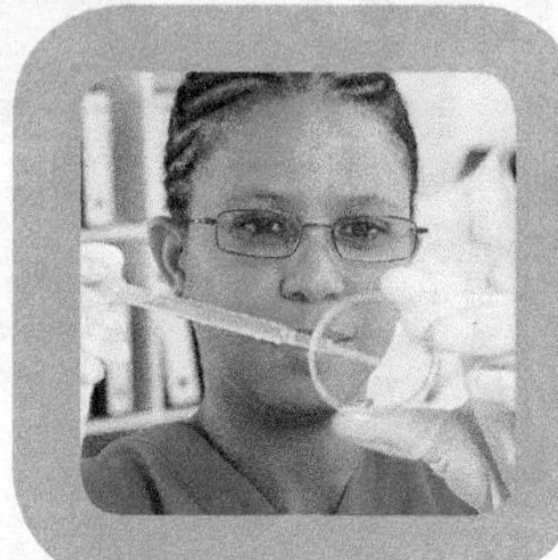

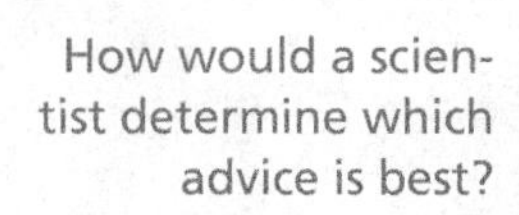

How would a scientist determine which advice is best?

Introduction to the Scientific Method

We have all been there—you just recover from one bad head cold and on a morning soon after you notice that scratchy feeling in your throat that signals a new one is about to begin. It is always at the worst time, too, when you have an important exam coming up, a term paper due, and a packed social calendar. Why are you sick yet again? What can you do about it?

If you ask your friends and relatives, you will hear the usual advice on how to prevent and treat colds: Take massive doses of vitamin C. Suck on zinc lozenges. Drink plenty of echinacea tea. Meditate. Spend more time with others. Get more rest. Exercise vigorously. Put that hat on when you go outside! You are left with an overwhelming list of options, often contradictory and some contrary to common sense. If you keep up with health news, you may be even more confused. One website reports that a popular over-the-counter cold treatment is effective, whereas a local TV news story details the risks of using this remedy and highlights its ineffectiveness. How do you decide what to do?

Faced with this bewildering situation, most people follow the advice that makes the most sense to them, and if they find they still feel terrible, they try another remedy. Testing ideas and discarding ones that don't work is a kind of "everyday science." We use this trial-and-error technique extensively, but it has its limitations—for example, even if you feel better after trying a new cold treatment, you can't know if your recovery occurred because the treatment was effective or because the cold was ending anyway.

Professional scientists conduct a more refined version of this process—using strategies that help eliminate other possible explanations for a result. And although some fields of science may use unfamiliar words or complicated and expensive equipment, the basic process for testing ideas is simple and universal to all areas of science. An understanding of this process can help you evaluate information about many issues that may concern and intrigue you—from health issues, to global warming, to the origin of life and the universe—with more confidence. In this chapter, we introduce you to the powerful process scientists use by asking the question we've considered here: Is there a cure for the common cold?

1.1 The Process of Science

The term *science* can refer to a body of knowledge—for example, the science of **biology** is the study of living organisms. You may believe that science requires near-perfect recall of specific sets of facts about the world. In reality, this goal is impossible and unnecessary—we do have reference books, after all. The real action in science is not memorizing what is already known but using the process of science to discover something previously unknown.

bio- means life.

-ology means the study of or branch of knowledge about.

This process—making observations of the world, proposing ideas about how something works, testing those ideas, and discarding (or modifying) our ideas in response to the test results—is the essence of the **scientific method.** The scientific method allows all of us to solve problems and answer questions efficiently and effectively. Can we use the scientific method to solve the complicated problem of preventing and treating colds?

The Nature of Hypotheses

When your mom says "wear a hat," that generates a question: Does wearing a hat in the winter actually prevent colds? That your mom believes the answer to this question is "yes" means that she has developed an understanding of how a body resists colds. This understanding is known as a **hypothesis**—that is, an idea about how things work (**FIGURE 1.1**). Science is the process of putting these ideas to the test.

hypo- means under, below, or basis.

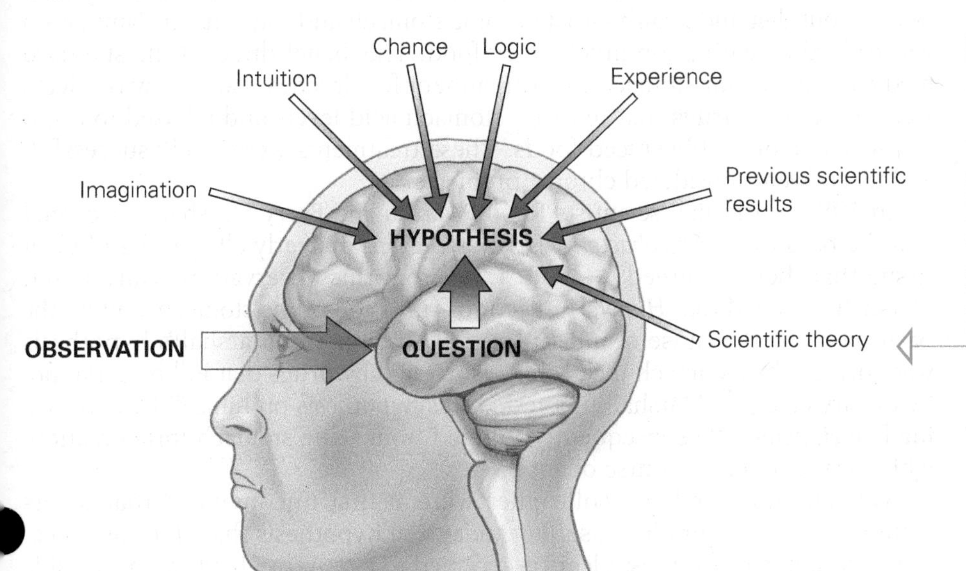

FIGURE 1.1 Hypothesis generation. All of us generate hypotheses. Many different factors, both logical and creative, influence the development of a hypothesis. Scientific hypotheses are both testable and falsifiable.

VISUALIZE THIS

Most colleges require students who are science majors to take courses in the humanities and social sciences, just as they require students in these majors to take science courses. What aspects of hypothesis generation listed in this figure are improved by study in the humanities and social sciences?

Hypotheses in biology come from knowledge about how the body and other biological systems work, experiences in similar situations, our understanding of other scientific research, and logical reasoning; they are also shaped by our creative mind. When your mom tells you to dress warmly to avoid colds, she is basing her advice on the following hypothesis: Becoming chilled makes you more susceptible to illness.

The hallmark of science is that hypotheses are subject to rigorous testing. Therefore, scientific hypotheses must be **testable**—it must be possible to evaluate a hypothesis through observations of the measurable universe. Not all hypotheses are testable. For instance, the statement that "colds are generated by disturbances in psychic energy" is not a scientific hypothesis because psychic energy has not been demonstrated to exist and thus cannot be measured by any known instrument.

A scientific hypothesis must also be **falsifiable;** that is, an observation or set of observations could potentially prove the hypothesis is false. The hypothesis that exposure to cold temperatures increases your susceptibility to colds is falsifiable; we can imagine an observation that would cause us to reject this hypothesis (for instance, the observation that people exposed to cold temperatures do not catch more colds than people protected from chills). Of course, not all hypotheses are proved false, but it is essential in science that incorrect ideas be discarded, which can occur only if it is *possible* to prove those ideas false. Lack of falsifiability is why hypotheses that require the intervention of a supernatural force cannot be tested scientifically. If something is **supernatural,** it is not constrained by any laws of nature, and therefore its behavior cannot be predicted using our current understanding of the natural world. Because a supernatural force can cause any possible result, hypotheses that rely on supernatural forces can never be falsified.

Finally, statements that are value judgments, such as, "It is wrong to cheat on an exam," are not scientific because different people have different ideas about right and wrong. It is impossible to falsify these types of statements. To find answers to questions of morality, ethics, or justice, we turn to other methods of gaining understanding—such as philosophy and religion.

(a)

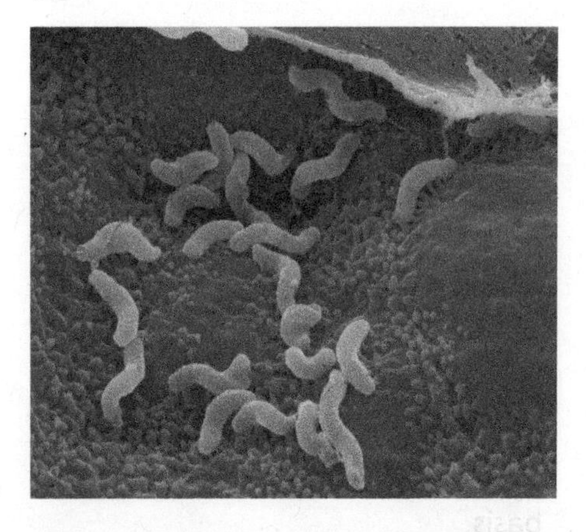

(b)

FIGURE 1.2 A scientific breakthrough. (a) *Helicobacter pylori* on stomach lining (image from electron microscope). (b) Robin Warren and Barry Marshall won the 2005 Nobel Prize in Medicine for their discovery of the link between *H. pylori* and ulcers.

Scientific Theories

Most hypotheses fit into a larger picture of scientific understanding. We can see this relationship when examining how research upended a commonly held belief about diet and health—that chronic stomach and intestinal inflammation was caused by eating too much spicy food. This belief directed the standard medical practice for stomach ulcer treatment for decades. Patients with ulcers were prescribed drugs that reduced stomach acid levels and advised to avoid eating acidic or highly spiced foods. These treatments were rarely successful, and ulcers were considered chronic problems.

In 1982, Australian scientists Robin Warren and Barry Marshall discovered that the bacterium *Helicobacter pylori* was present in nearly all samples of ulcer tissue that they examined (**FIGURE 1.2**). From this observation, Warren and Marshall reasoned that *H. pylori* infection—invasion of the stomach wall by the bacteria—was the cause of most ulcers. If Warren and Marshall's hypothesis was correct, then stomach ulcers are best treated by drugs that kill bacteria, not by dietary changes. Marshall first tested this hypothesis on himself by consuming live *H. pylori*. He subsequently suffered from acute stomach inflammation, which was cured by a course of antibiotics.

Warren and Marshall's colleagues were at first unconvinced that ulcers could have such a simple cause. But today, the hypothesis that *H. pylori* infection is responsible for most ulcers is accepted as fact. Why? First, no reasonable alternative hypotheses about the causes of ulcers (for instance, consumption of spicy foods) has been consistently supported by hypothesis tests; and second, Warren and Marshall's hypothesis has not been rejected—that is, there have been no carefully designed experiments that show that antibiotic treatment of *H. pylori* fails to cure most ulcers.

Third, the relationship between *H. pylori* and ulcers is considered fact because this understanding conforms to a well-accepted scientific principle—namely, the germ theory of disease. A **scientific theory** is an explanation for a set of related observations that is based on well-supported hypotheses from several different, independent lines of research. The basic premise of germ theory is that microorganisms (that is, organisms too small to be seen with the naked eye) are the cause of some or all human diseases.

The biologist Louis Pasteur first observed that bacteria cause milk to become sour. From this observation, he reasoned that these same types of organisms could injure humans. Later, Robert Koch demonstrated a link between anthrax bacteria and a specific set of fatal symptoms in mice, providing additional evidence for the theory. Germ theory is further supported by the observation that antibiotic treatment that targets particular microorganisms can cure certain illnesses—as is the case with bacteria-caused ulcers.

In everyday speech, the word *theory* is synonymous with untested ideas based on little information. In contrast, scientists use the term when referring to ideas that form the basis of their understanding of the world. The supporting foundation of all scientific theories is multiple hypothesis tests.

The Logic of Hypothesis Tests

One common hypothesis about cold prevention is that taking vitamin C supplements keeps you healthy. This hypothesis is very appealing, especially given the following generally accepted facts:

1. Fruits and vegetables contain a lot of vitamin C.
2. People with diets rich in fruits and vegetables are generally healthier than people who skimp on these food items.
3. Vitamin C is known to be an anti-inflammatory agent, reducing throat and nose irritation.

With these facts in mind, we can state the following testable and falsifiable hypothesis: *Consuming vitamin C decreases the risk of catching a cold.* This hypothesis makes sense given the statements just listed and the experiences of the many people who insist that vitamin C keeps them healthy.

The process we used to construct the hypothesis above is called **inductive reasoning**—combining a series of specific observations (here, statements 1–3) to discern a general principle. Inductive reasoning is an essential tool for understanding the world. However, a word of caution is in order: Just because the inductive reasoning that led to a hypothesis seems to make sense does not mean that the hypothesis is necessarily true.

induc- means to rely on reason to derive principles (also, to cause to happen).

Consider the ancient hypothesis that the sun revolves around Earth. This hypothesis was induced based on the observations that the sun rose in the east every morning, traveled across the sky, and set in the west every night. For almost all of history, this hypothesis was considered to be a "fact" by nearly all of Western society. It wasn't until the early seventeenth century that this hypothesis was overturned—as the result of Galileo Galilei's observations of Venus. His observations proved false the hypothesis that the sun revolved around Earth. Galileo's work helped to confirm the more modern hypothesis, proposed by Nicolaus Copernicus, that Earth revolves around the sun, and rotates as it does so.

deduc- means to reason out, working from facts.

So, even though the hypothesis about vitamin C is sensible, it needs to be tested to see if it can be proved false. Hypothesis testing is based on **deductive reasoning** or deduction. Deduction involves using a general principle to predict an expected observation. This **prediction** concerns the outcome of an action, test, or investigation. In other words, the prediction is the result we expect from a hypothesis test.

Deductive reasoning takes the form of "if/then" statements. That is, if our idea is correct, then we predict a specific outcome from a hypothesis test. A prediction based on the vitamin C hypothesis could be: *If* vitamin C decreases the risk of catching a cold, *then* people who take vitamin C supplements with their regular diets will experience fewer colds than will people who do not take supplements.

Deductive reasoning, with its resulting predictions, is a powerful method for testing hypotheses. However, the structure of such a statement means that hypotheses can be clearly rejected if untrue but impossible to prove if true. This shortcoming can be illustrated using the if/then statement concerning vitamin C and colds (**FIGURE 1.3**).

Consider the possible outcomes of a comparison between people who supplement with vitamin C and those who do not. People who take vitamin C supplements may suffer through more colds than people who do not; they may have the same number of colds as the people who do not supplement; or supplementers may in fact experience fewer colds. What does each of these results tell us about the hypothesis?

If, in a well-designed test, people who take vitamin C have more colds or the same number of colds as those who do not supplement, then the hypothesis that vitamin C provides protection against colds can be rejected. But what if people who supplement with vitamin C *do* experience fewer colds? If this is the case, then we can only say that the hypothesis has been supported and not disproven.

Why is it impossible to say from this experimental result that the hypothesis that vitamin C prevents colds is true? Because there are **alternative hypotheses** that explain why people with different vitamin-taking habits vary in their cold susceptibility. In other words, demonstrating the truth of the *then* portion of a deductive statement does not prove that the *if* portion is true.

Consider the alternative hypothesis that frequent exercise reduces susceptibility to catching a cold. And suppose that people who take vitamin C supplements are more likely to engage in regular exercise. If both of these hypotheses are true, then the prediction that vitamin C supplementers experience fewer colds than people who do not supplement would be true but not because the original hypothesis (vitamin C reduces the risk of colds) is true. Instead, people who take vitamin C supplements experience fewer colds because they are also more likely to exercise, and it is exercise that reduces cold susceptibility.

A hypothesis that seems to be true because it has not been rejected by an initial test may be rejected later because of a different test. This is what happened to the hypothesis that vitamin C consumption reduces susceptibility to colds. The argument for the power of vitamin C was popularized in 1970 by

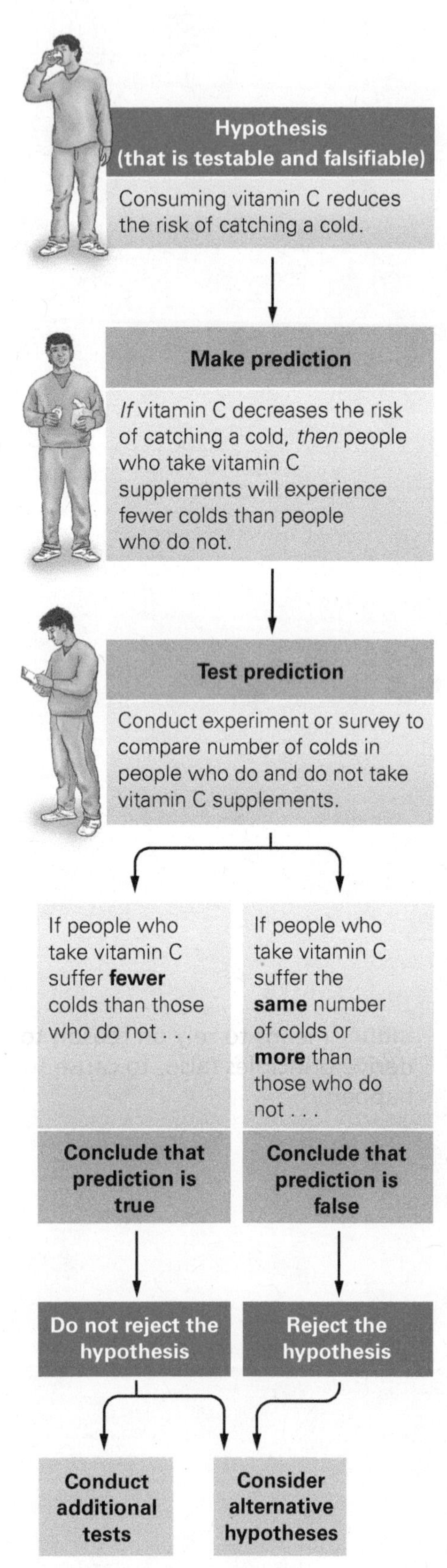

Hypothesis
(that is testable and falsifiable)
Consuming vitamin C reduces the risk of catching a cold.
Make prediction
If vitamin C decreases the risk of catching a cold, then people who take vitamin C supplements will experience fewer colds than people who do not.
Test prediction
Conduct experiment or survey to compare number of colds in people who do and do not take vitamin C supplements.
If people who take vitamin C suffer fewer colds than those who do not . . .
If people who take vitamin C suffer the same number of colds or more than those who do not . . .
Conclude that prediction is true
Conclude that prediction is false
Do not reject the hypothesis
Reject the hypothesis
Conduct additional tests
Consider alternative hypotheses

FIGURE 1.3 The scientific method. Tests of hypotheses follow a logical path. This flowchart illustrates the process of deduction as practiced by scientists.

VISUALIZE THIS

According to this flowchart, scientists should consider alternative hypotheses even if their hypothesis is supported by their research. Explain why this is the case.

Nobel Prize–winning chemist Linus Pauling. Pauling based his assertion—that large doses of vitamin C reduce the incidence of colds by as much as 45%—on the results of a few studies that had been published between the 1930s and 1970s. However, repeated, careful tests of this hypothesis have since failed to support it. In many of the studies Pauling cited, it appears that alternative hypotheses explain the difference in cold incidence between vitamin C supplementers and nonsupplementers. Today, most health scientists agree that the hypothesis that vitamin C prevents colds has been convincingly falsified.

The example of the vitamin C hypothesis also highlights a challenge of communicating scientific information. You can see why the belief that vitamin C prevents colds is so widespread. If you don't know that scientific knowledge relies on rejecting incorrect ideas, a book by a Nobel Prize–winning scientist may seem like the last word on the benefits of vitamin C. It took many years of careful research to show that this "last word" was, in fact, wrong.

Got It?

1. A(n) __________ is a proposed explanation for how things work.
2. A statement that is "falsifiable" must be able to be __________ __________.
3. A statement that is "testable" must be able to be evaluated through __________ of the known universe.
4. Deductive reasoning relies on testing the __________ of a hypothesis test.
5. If a hypothesis test returns the predicted results, the hypothesis is supported but not definitively __________.

1.2 Hypothesis Testing

The previous discussion may seem discouraging: How can scientists determine the truth of any hypothesis when there is a chance that the hypothesis could be falsified by a later test? Even if one of the hypotheses about cold prevention is supported, does the difficulty of eliminating alternative hypotheses mean that we will never know which approach is truly best? The answer is yes—and no.

Hypotheses cannot be proven absolutely true; it is always possible that the true cause of a phenomenon may be found in a hypothesis that has not yet been tested. However, in a practical sense, a hypothesis can be proven beyond a reasonable doubt. That is, when one hypothesis has not been disproven through repeated testing and all reasonable alternative hypotheses have been eliminated, scientists accept that the well-supported hypothesis is, in a practical sense, true. The hypothesis that *H. pylori* infection—and not spicy food—causes the majority of stomach ulcers is accepted as true. *Truth* in science can therefore be defined as *what we know and understand based on all currently available information*. But scientists remain open to the possibility that what seems true now may someday be proven false.

An effective way to test many hypotheses is through rigorous scientific experiments. Experimentation has enabled scientists to prove beyond a reasonable doubt that the common cold is caused by a virus. A virus is a microscopic entity with a simple structure—it typically consists of a short strand of genetic material and a few proteins encased in a relatively tough protein shell, sometimes surrounded by a membrane. A virus must infect a host cell to reproduce. Of the more than 200 types of viruses that are known to cause the common cold, most infect the cells in our noses and throats. The

VISUALIZE THIS

Find two points in this process where intervention by drugs or other treatment could disrupt either the virus or the immune response and therefore lead to fewer cold symptoms.

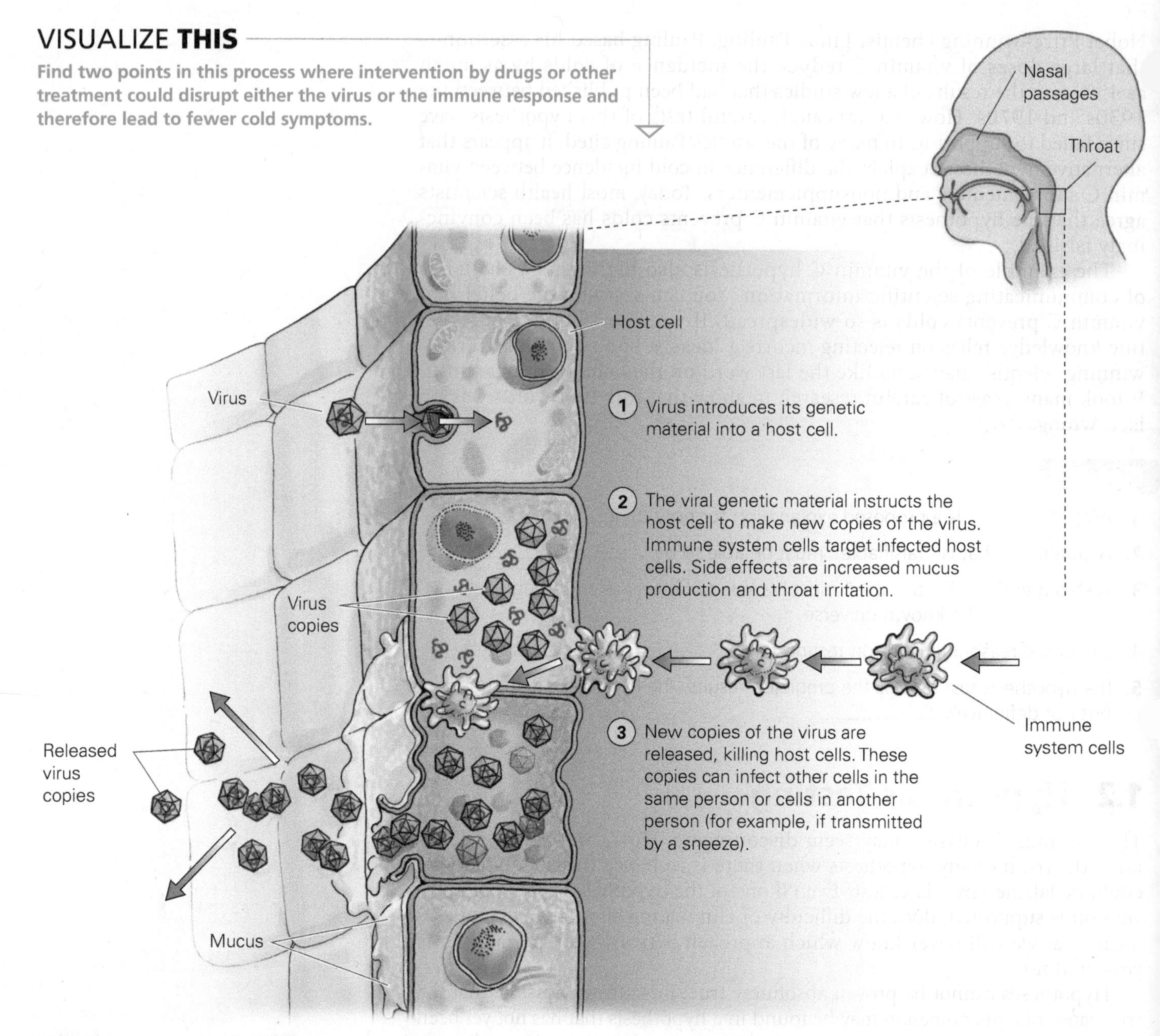

Nasal passages
Throat
Host cell
Virus
1 Virus introduces its genetic material into a host cell.
2 The viral genetic material instructs the host cell to make new copies of the virus. Immune system cells target infected host cells. Side effects are increased mucus production and throat irritation.
Virus copies
Immune system cells
Released virus copies
3 New copies of the virus are released, killing host cells. These copies can infect other cells in the same person or cells in another person (for example, if transmitted by a sneeze).
Mucus

FIGURE 1.4 A cold-causing virus. A rhinovirus causes illness by invading nose and throat cells and using them as "factories" to make virus copies. Cold symptoms result from immune system attempts to eliminate the virus.

sneezing, coughing, congestion, and sore throat of a cold appear to result from the body's protective response to a viral invasion, established by our immune system (**FIGURE 1.4**).

As you may know, if we survive certain viral infections, we are unlikely to experience a recurrence of the disease the virus causes. For example, it is extremely rare to suffer from chicken pox twice because one exposure to the chicken pox virus (through either infection or vaccination) usually provides lifelong immunity to future infection. However, for common viruses, like the one that causes flu, the large number of infections that occur each year means that there are many varieties of the virus. We require yearly flu vaccinations because the virus type that is most common changes slightly over time. The huge variety of cold viruses makes immunity to the common cold—and the development of a vaccine to prevent it—improbable. Scientists thus focus their experimental research about common colds on methods of prevention and treatment.

The Experimental Method

Experiments are sets of actions or observations designed to test specific hypotheses. Generally, an experiment allows a scientist to control the conditions that may affect the subject of study. Manipulating the environment allows a scientist to eliminate some alternative hypotheses that may explain the result.

Experimentation in science is analogous to what a mechanic does when diagnosing a car problem. There are many reasons why a car engine might not start. If a mechanic begins by tinkering with numerous parts to apply all possible fixes before restarting the car, she will not know what exactly caused the problem (and will have an unhappy customer who is charged for unnecessary parts and labor). Instead, a mechanic begins by testing the battery for power; if the battery is charged, then she checks the starter motor; if the car still doesn't start, she looks over the fuel pump; and she continues in this manner until identifying the problem. Likewise, a scientist systematically attempts to eliminate hypotheses that do not explain a particular phenomenon.

FIGURE 1.5 Testing hypotheses through observation. Not all hypotheses can be tested experimentally. Questions about the evolutionary history of life are tested by examining the data provided by the fossil record.

Not all scientific hypotheses can be tested through experimentation. For instance, hypotheses about how life on Earth originated or the cause of dinosaur extinction are usually not testable in this way. These hypotheses are instead tested using careful observation of the natural world. For instance, the examination of fossils and other geological evidence allows scientists to test hypotheses regarding the extinction of the dinosaurs (**FIGURE 1.5**).

The information collected by scientists during hypothesis testing is known as **data.** The data are collected on the **variables** of the test, that is, any factor that can change in value under different conditions. In an experimental test, scientists manipulate an **independent variable** (one whose value can be freely changed) to measure its effect on a **dependent variable.** The dependent variable may or may not be influenced by changes in the independent variable, but it cannot be systematically changed by the researchers. For example, to measure the effect of vitamin C on cold prevention, scientists can vary individuals' vitamin C intake (the independent variable) and measure their susceptibility to illness upon exposure to a cold virus (the dependent variable).

Data obtained from well-designed experiments should allow researchers to convincingly reject or support a hypothesis. This is more likely to occur if the experiment is controlled.

Controlled Experiments

Control has a specific meaning in science. A **control** for an experiment is a subject similar to an experimental subject except that the control is not exposed to experimental treatment. Controlled experiments are thus designed to eliminate as many alternative hypotheses as possible.

Once subjects are enlisted in an experiment, they are assigned to a control or an experimental group. If members of the control and experimental groups differ at the end of a well-designed test, then the difference is likely to be due to the experimental treatment.

Our question about effective cold treatments lends itself to a variety of controlled experiments on possible drug therapies. For example, an extract of *Echinacea purpurea* (a common North American prairie plant) in the form of echinacea tea has been promoted as a treatment to reduce the likelihood as well as the severity and duration of colds (**FIGURE 1.6**). A scientific

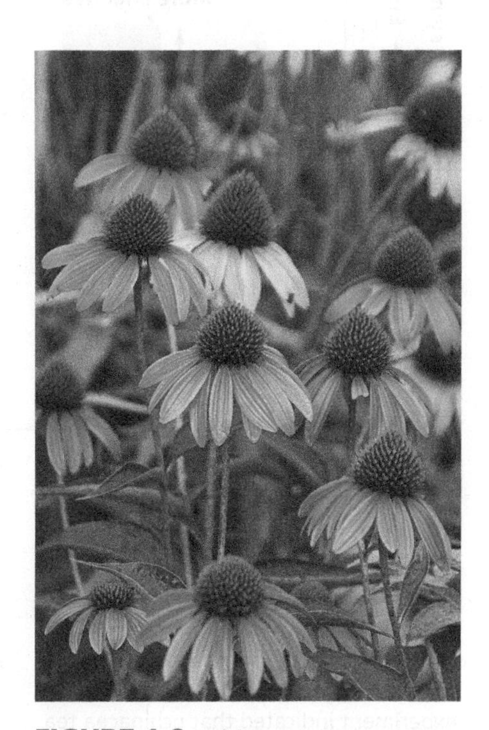

FIGURE 1.6 *Echinacea purpurea*, an American coneflower. Extracts from the leaves and roots of this plant are among the most popular herbal remedies sold in the United States.

WORKING WITH DATA

Imagine that the graph in part (b) was drawn so that the range of the vertical axis was from 2 to 5 rather than from 0 to 5. Does this affect how you might interpret the results?

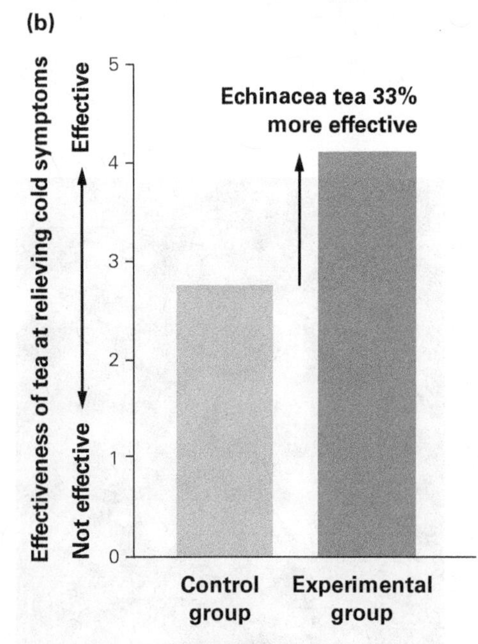

FIGURE 1.7 A controlled experiment. (a) In a controlled experiment testing echinacea tea as a treatment for colds, all 95 research participants were treated identically except for the type of tea they were given. (b) The results of the experiment indicated that echinacea tea was 33% more effective than the placebo in treating cold symptoms.

experiment on the efficacy of *Echinacea* involved asking individuals suffering from colds to rate the effectiveness of a tea in relieving their symptoms. In this study, people who used echinacea tea felt that it was 33% more effective. The "33% more effective" is in comparison with the rated effectiveness of a tea that did not contain *Echinacea* extract—that is, the results from the control group.

Control groups and experimental groups must be as similar as possible to each other to eliminate alternative hypotheses that could explain the results. In the case of cold treatments, the groups should not systematically differ in age, diet, stress level, or other factors that might affect cold susceptibility. One effective way to minimize differences between groups is the **random assignment** of individuals into groups. For example, a researcher might put all of the volunteers' names in a single pool, randomly draw out half, and designate the people drawn as the experimental group and the remaining people as the control group. As a result, each group should be a rough cross-section of the population in the study. In the echinacea tea experiment just described, members of both the experimental and control groups were female employees of a nursing home who sought relief from their colds at their employer's clinic. The volunteers were randomly assigned to either the experimental or control group as they came into the clinic.

The second step in designing a well-controlled experiment is to treat all subjects—or as human subjects are referred to, research participants—identically during the course of the experiment. In this study, all participants, whether in the control or experimental group, received the same information about the supposed benefits of echinacea tea, and during the course of the experiment, all participants were given tea to drink five to six times daily until their symptoms subsided. However, individuals in the control group received "sham tea" that did not contain *Echinacea* extract. Treating all participants the same ensures that no factor related to the interaction between participant and researcher influences the results.

The sham tea in this experiment would be equivalent to the sugar pills that are given to the control group during drug trials. Like other intentionally ineffective medical treatments, sham tea is a **placebo.** Using a placebo generates only one consistent difference between individuals in the two groups—in this case, the type of tea they consumed.

In the echinacea tea study, the data indicated that cold severity was lower in the experimental group compared with those who received a placebo. Because their study used controls, the researchers can be more confident that the groups differed only because of the effect of *Echinacea*. By reducing the likelihood that alternative hypotheses could explain their results, the researchers could strongly infer that they were measuring a real, positive effect of echinacea tea on colds (**FIGURE 1.7**). However, this one experiment cannot eliminate all alternative hypotheses for the results—including the possibility that the experimental and control groups differ just by chance.

The study described here supports the hypothesis that echinacea tea reduces the severity of colds. However, it is extremely rare that a single experiment will eliminate all possible alternative explanations and cause the scientific community to accept a hypothesis beyond a reasonable doubt. Dozens of studies, each using different experimental designs and many using extracts from different parts of the plant, have investigated the effect of *Echinacea* on common colds and other illnesses. Some of these studies have shown a positive effect, but many others have shown none. The most recent comprehensive review of the published studies on this topic concluded that it was unlikely that *Echinacea* reduces the duration of a cold.

Minimizing Bias in Experimental Design

Scientists and human research participants may have strong opinions about the truth of a particular hypothesis even before it is tested. These opinions may cause participants to influence the results of an experiment causing a **bias** that favors a particular outcome.

One potential source of bias is participant expectation. Individual experimental participants may consciously or unconsciously model the behavior they feel the researcher expects from them. For example, an individual who knew she was receiving echinacea tea may have felt confident that she would recover more quickly. This might cause her to underreport her cold symptoms. This potential problem is avoided by designing a **blind experiment** in which individual participants are not aware of exactly what they are predicted to experience. In experiments on drug treatments, this means not telling participants whether they are receiving the drug or a placebo.

Another source of bias arises when a researcher makes consistent errors in the measurement and evaluation of results. This phenomenon is called *observer bias*. In the echinacea tea experiment, observer bias could take various forms. Expecting a particular outcome might lead a scientist to give slightly different instructions about which symptoms constituted a cold to participants who received echinacea tea. Or, if the researcher expected people who drank echinacea tea to experience fewer colds, she might make small errors in the measurement of cold severity that influenced the final result.

To avoid the problem of experimenter bias, the data collectors themselves should be "blind." Ideally, the scientist, doctor, or technician applying the treatment does not know which group (experimental or control) any given research participant is part of until after all data have been collected and analyzed (**FIGURE 1.8**). We call experiments **double-blind** when both the research participants and the technicians performing the measurements are unaware of either the hypothesis or whether a participant is in the control or experimental group.

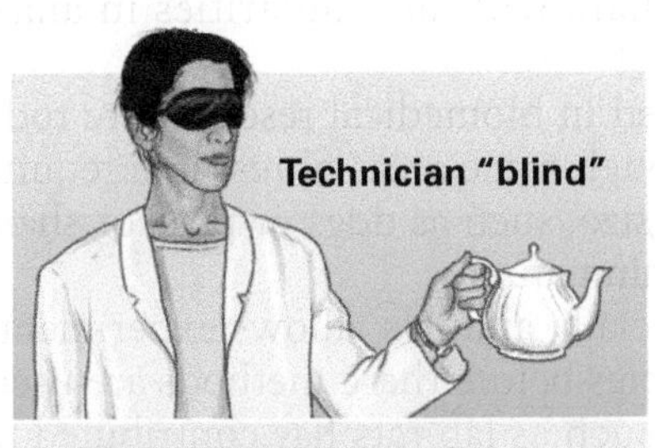

	Technician "blind"	Participant "blind"
What they know	• **Limited knowledge** of experimental hypothesis • **No knowledge** of which group participants belong to	• **Limited knowledge** of experimental hypothesis • **No knowledge** of which group he or she belongs to
How they behave	• **No difference** in instructions to participants • **No difference** in treatment of participants • **No difference** in data collection	• **Unbiased** reporting of symptoms or effects of treatment

FIGURE 1.8 Double-blind experiments. Double-blind experiments result in data that are more objective.

Blinding the data collector ensures that the data are **objective;** in other words, without bias. Double-blind experiments nearly eliminate the effects of human bias on results. When both researcher and participant have few expectations about the outcome, the results obtained from an experiment are more credible.

Using Correlation to Test Hypotheses

Double-blind, placebo-controlled, randomized experiments represent the gold standard for medical research. However, well-controlled experiments can be difficult to perform when humans are the research participants. The requirement that both experimental and control groups be treated nearly identically means that some people unknowingly receive no treatment. In the case of healthy volunteers with head colds, the placebo treatment of sham tea did not hurt those who received it.

However, placebo treatments are impractical or unethical in many cases. For instance, imagine testing the effectiveness of a birth control drug using a controlled experiment. This would require asking women to take a pill that may or may not prevent pregnancy while not using any other form of birth control!

Experiments on Model Systems. Scientists can use **model systems** when testing hypotheses that would raise ethical or practical problems when tested on people. For basic research that helps us understand how cells and genes function, the model systems are easily grown and manipulated organisms, such as certain species of bacteria, nematodes, and fruit flies, or even isolated cells from larger organisms that reproduce in dishes in the laboratory. In the case of research on human health and disease, model systems are typically other mammals and human cells (**FIGURE 1.9**). Nonhuman mammals are especially useful as model organisms in medical research because they are closely related to us. Like us, they have hair and produce milk for their young, and thus they also share with us similarities in anatomy and physiology.

(a)

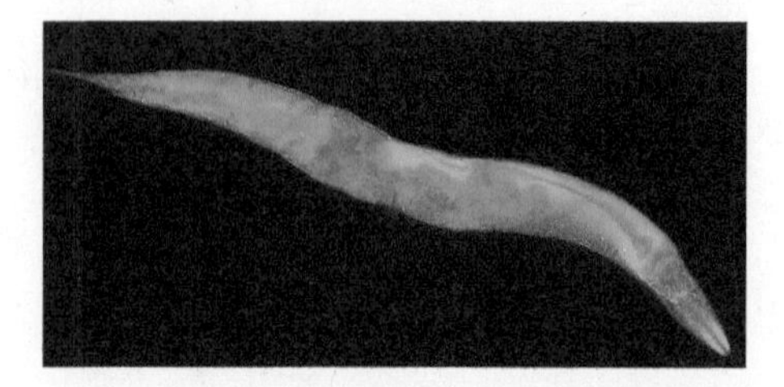

(b)

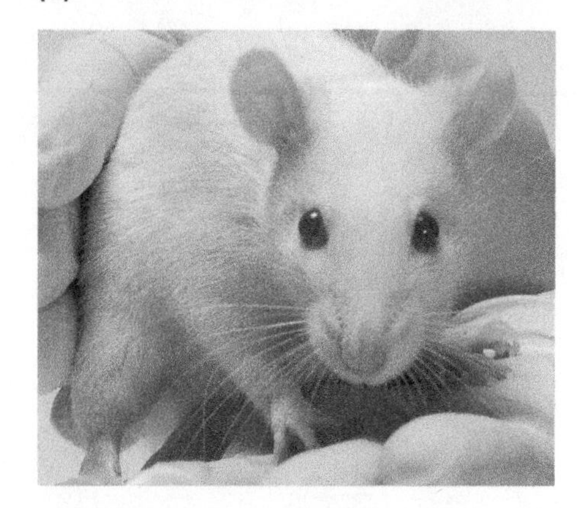

(c)

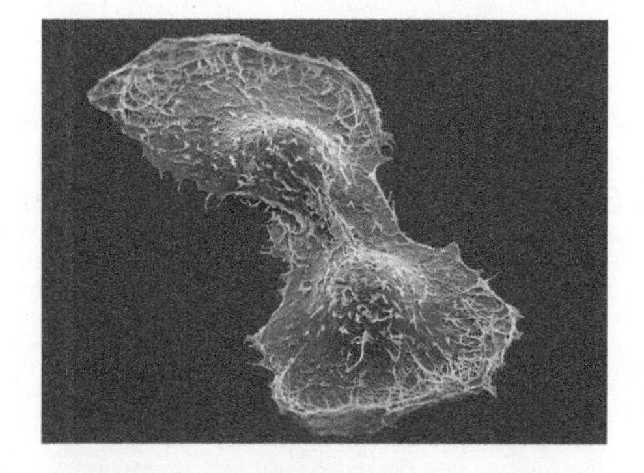

FIGURE 1.9 Model systems in science. (a) Research on the nematode *Caenorhabditis elegans* (also known as *C. elegans*) has led to major advances in our understanding of animal development and genetics. (b) The classic "lab rat" is easy to raise and care for and as a mammal is an important model system for testing drugs and treatments designed for humans. (c) HeLa cells are derived from a cancerous growth biopsied from Henrietta Lacks in 1951. Cancer cells have the unique property of being able to divide indefinitely, making them valuable for research on a wide variety of topics from HIV to human growth hormone.

The vast majority of animals used in biomedical research are rodents such as rats, mice, and guinea pigs, although some areas of research require animals that are more similar to humans in size, such as dogs or pigs, or share a closer evolutionary relationship, such as chimpanzees.

The use of model systems, especially animals, allows experimental testing on potential drugs and other therapies before these methods are used on people. Research on model organisms such as lab rats has contributed to a better understanding of nearly every serious human health threat, including cancer, heart disease, Alzheimer's disease, and AIDS (acquired immunodeficiency syndrome). However, ethical concerns about the use of animals in research persist and can complicate such studies. In addition, the results of animal studies are not always directly applicable to humans; despite a shared evolutionary history, animals can have important differences from humans. Testing hypotheses about human health in human beings still provides the clearest answer to these questions.

Looking for Relationships between Factors. Scientists can also test hypotheses using correlations when controlled experiments on humans are difficult or impossible to perform. A **correlation** is a relationship between two variables.

Suggestions about using meditation to reduce susceptibility to colds are based on a correlation between psychological stress and susceptibility to cold

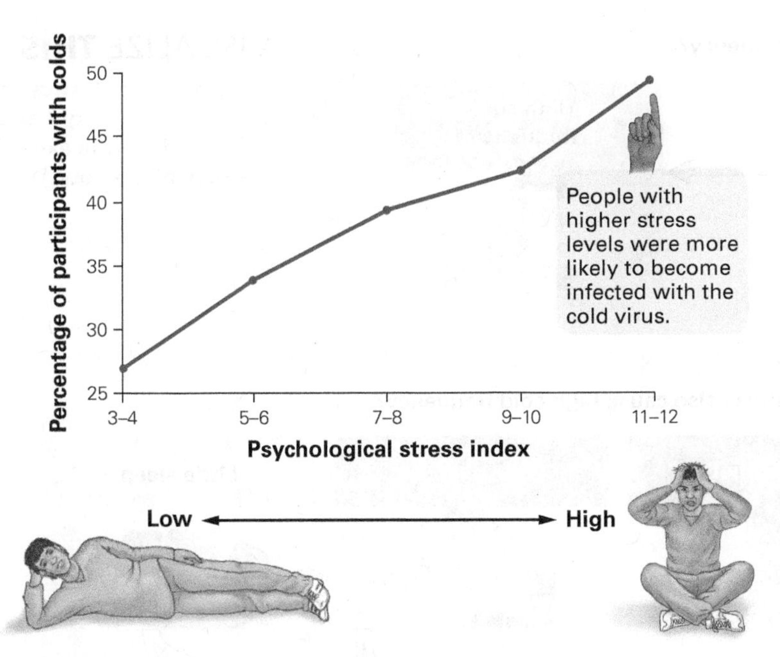

WORKING WITH DATA

This graph groups people with similar but not identical, stress index measures. Why might this have been necessary? If people with stress indices 3 and 4 have the same susceptibility to colds, does this call into question the correlation? Why?

FIGURE 1.10 Correlation between stress level and illness. The graph indicates that people reporting higher levels of stress became infected after exposure to a cold virus more often than did people who reported low levels of stress.

virus infections (**FIGURE 1.10**). This correlation was generated by researchers who collected data on research participants' psychological stress levels before giving them nasal drops that contained a cold virus. Doctors later reported on the incidence and severity of colds among participants in the study. Note that although the cold virus was applied to each participant in the study, the researchers had no influence on the stress level of the study participants—in other words, this was not a controlled experiment because people were not randomly assigned to different "treatments" of low or high stress.

Let's examine the results presented in Figure 1.10. The horizontal axis of the graph, or ***x* axis,** illustrates the independent variable. In this case, the variable is stress, and the graph ranks research participants along a scale of stress level—from low stress on the left edge of the scale to high stress on the right. The vertical axis of the graph, the ***y* axis,** is the dependent variable—the percentage of study participants who developed colds as reported by their doctors. Each point on the graph represents a group of individuals and tells us what percentage of people in each stress category had clinical colds.

The line connecting the five points on the graph illustrates a correlation—the relationship between stress level and susceptibility to cold virus infections. Because the line rises to the right, it illustrates a positive correlation. These data tell us that people who had higher stress levels were more likely to come down with colds when exposed to the virus. But does this relationship mean that high stress causes increased cold susceptibility?

To conclude that stress causes illness, we need the same assurances that are given by a controlled experiment. In other words, we must assume that the individuals measured for the correlation are similar in every way except for their stress levels. Is this a good assumption? Not necessarily. Most correlations cannot control for all alternative hypotheses. People who feel more stressed may have poorer diets because they feel time-limited and rely on fast food more often. In addition, people who feel highly stressed may be in situations where they are exposed to more cold viruses. These differences among people who differ in stress level may also influence their cold

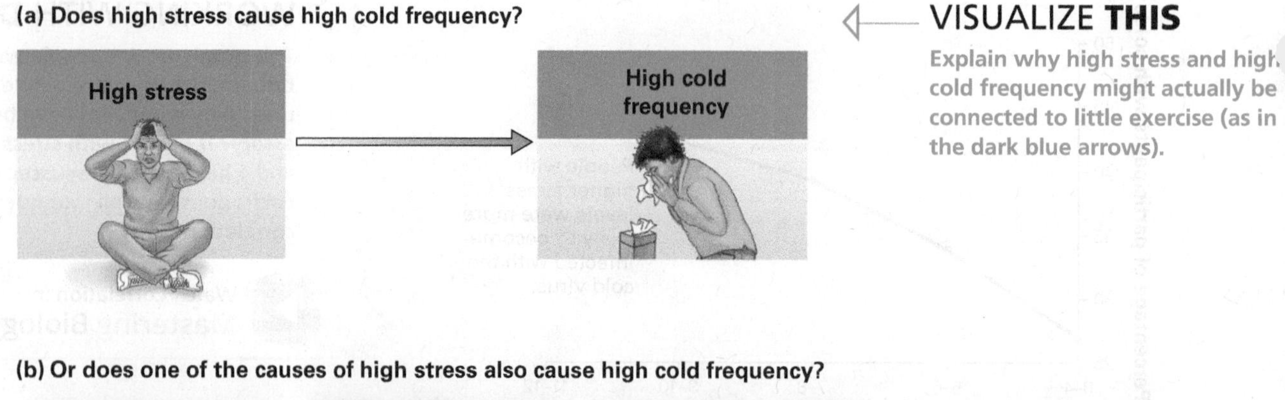

VISUALIZE THIS

Explain why high stress and high cold frequency might actually be connected to little exercise (as in the dark blue arrows).

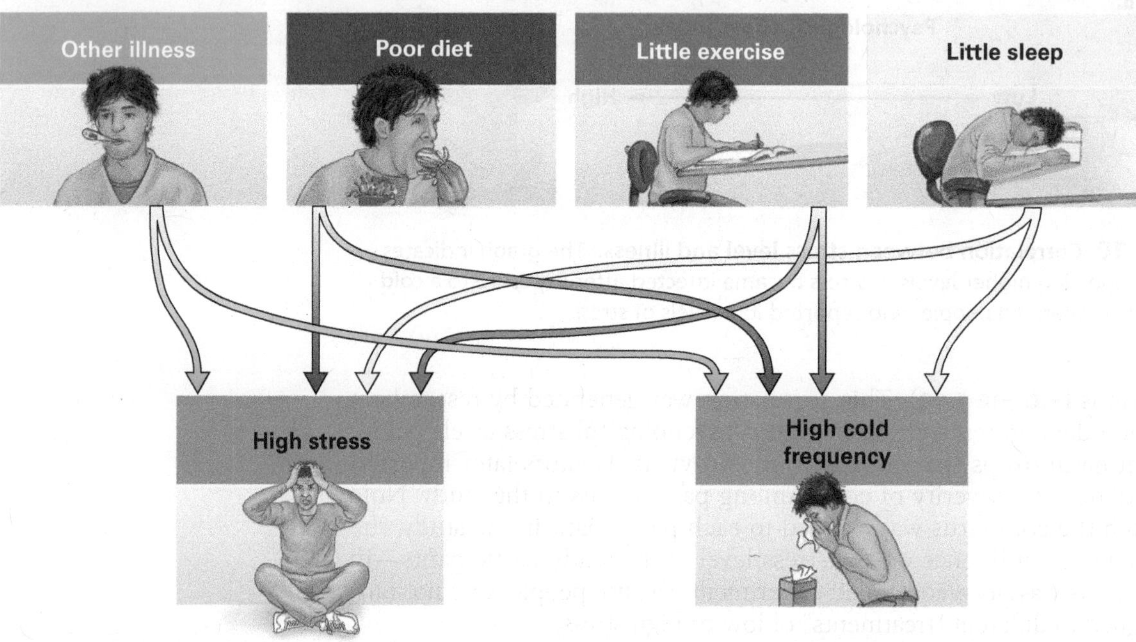

FIGURE 1.11 Correlation does not signify causation. A correlation typically cannot eliminate all alternative hypotheses.

susceptibility (**FIGURE 1.11**). Therefore, even with a strong correlational relationship between the two factors, we cannot strongly infer that stress causes decreased resistance to colds.

Researchers who use correlational studies can eliminate a number of alternative hypotheses by closely examining their research participants. For example, this study on stress and cold susceptibility collected data from participants on age, weight, sex, education, and their exposure to infected individuals. None of these factors differed consistently among low-stress and high-stress groups. Although this analysis increases our confidence in the hypothesis that high stress levels increase susceptibility to colds, people with high-stress lifestyles still may have important differences from those with low-stress lifestyles. It is possible that one of those differences is the real cause of disparities in cold frequency.

As you can see, it is difficult to demonstrate a cause-and-effect relationship between two factors simply by showing a correlation between them. In other words, correlation does not equal causation. For example, a commonly understood correlation exists between exposure to cold air and epidemics of the common cold. It is true that as outdoor temperatures drop, the incidence of colds increases. But numerous controlled experiments have indicated that chilling an individual does not increase his susceptibility to colds. Instead, cold outdoor temperatures mean increased close contact with other people (and their viruses), and this contact increases the number of colds we get during cold weather. Despite the correlation, cold air does not cause colds—exposure to viruses does.

Correlational studies are the main tool of **epidemiology,** the study of the distribution and causes of diseases. One commonly used epidemiological technique is a cross-sectional survey. In this type of survey, many individuals are both tested for the presence of a particular condition and asked about their exposure to various factors. The limitations of cross-sectional surveys include the effect of participant bias and poor recall by survey participants, in addition to all of the problems associated with interpreting correlations. **TABLE 1.1** provides an overview of the variety of correlational strategies used to study the links between our environment and our health.

Got It?

1. A control in an experiment is a subject that is treated identically to the experimental subject except that the ________ ________ is not applied.
2. The intent in a controlled experiment is to eliminate as many alternative ________ as possible.
3. Experimental ________ occurs when a researcher's expectation affects data collection.
4. Bacteria, nematodes, and mice are model systems that allow us to perform ________ experiments on hypotheses that are difficult to test on humans.
5. A correlation between two factors does not necessarily mean that one factor ________ a change to the other factor.

TABLE 1.1 Types of correlational studies.

Name	Description	Pros	Cons
Ecological studies	Examine specific human populations for unusually high levels of various diseases (e.g., documenting a "cancer cluster" around an industrial plant)	Inexpensive and relatively easy to do	• Unsure whether exposure to environmental factor is actually related to onset of the disease
Cross-sectional surveys	Question individuals in a population to determine amount of exposure to an environmental factor and whether disease is present	More specific than ecological study	• Expensive • Research participants may not know exposure levels • Cannot control for other factors that may differ among participants • Cannot be used for rare diseases
Case-control studies	Compare exposures to specific environmental factors between individuals who have a disease and individuals matched in age and other factors who do not have the disease	• Relatively fast and inexpensive • Best method for rare diseases	• Does not measure absolute risk of disease as a result of exposure • Difficult to select appropriate controls • Examines just one disease possibly associated with an environmental factor
Cohort studies	Follow a group of individuals, measuring exposure to environmental factors and disease prevalence	Can determine risk of various diseases associated with exposure to a particular environmental factor	• Expensive and time-consuming • Difficult to control for alternative hypotheses • Not feasible for rare diseases
Correlational experiment	Expose individuals who experience varying levels of an independent variable to an experimental treatment	Can control the amount of exposure to at least one environmental factor of interest	• Cannot eliminate alternative hypotheses • Only feasible for hypotheses for which an experimental treatment can be applied

1.3 Understanding Statistics

During a review of scientific literature on cold prevention and treatment, you may come across statements about the "significance" of the effects of different cold-reducing measures. For instance, one report may state that factor A appeared to reduce cold severity but that the results of the study were "not significant." Another study may state that factor B caused a "significant reduction" in illness. We might then assume that this statement means factor B will help us feel better, whereas factor A will have little effect. This is an incorrect assumption because in scientific studies, *significance* is defined a bit differently than it is in everyday language. To evaluate the scientific use of this term, we need a basic understanding of statistics.

VISUALIZE THIS

The snapshot from the 1960s is of a group of women who are considerably less racially and ethnically diverse than the more recent snapshot. How might this affect the comparison? Similarly, how might a lack of racial and ethnic diversity in a sample of a cold treatment affect our interpretation of results?

What Statistical Tests Can Tell Us

We often use the term *statistics* to refer to a summary of accumulated information. For instance, a baseball player's success in hitting is summarized by a statistic: his batting average, the total number of hits he made divided by the number of opportunities he had to bat. The science of **statistics** is a bit different; it is a specialized branch of mathematics used to evaluate and compare data.

An experimental test uses a small subgroup, or **sample,** of a population. Statistical methods can summarize data from the sample—for instance, we can describe the average length of colds experienced by experimental and control groups. In statistics, this average is known as the **mean. Statistical tests** can then be used to extend the results from a sample to the entire population.

When scientists conduct an experiment, they hypothesize that there is a true, underlying effect of their experimental treatment on the entire population. An experiment on a sample of a population can only estimate this true effect because a sample is always an imperfect "snapshot" of an entire population.

Consider a hypothesis that the average hair length of women in a college class in 1962 was shorter than the average hair length at the same college today. To test this hypothesis, we could compare a sample of snapshots from college yearbooks. If hairstyles were very similar among the women in a snapshot, you could reasonably assume that the average hair length in the college class is close to the average length in the snapshot. However, what if you see that women in a snapshot have a variety of hairstyles, from short bobs to long braids? In this case, it is difficult to determine whether the average hair length in the snapshot is at all close to the average for the class. With so much variation, the snapshot could, by chance, contain a surprisingly high number of women with very long hair, causing the average length in the sample to be much longer than the average length for the entire class (**FIGURE 1.12**).

A statistical test calculates the likelihood, given the number of individuals sampled and the variation within these samples, that the difference between two samples reflects a real, underlying difference between the populations from which these samples were drawn. A **statistically significant** result is one that is unlikely to be due to chance differences between the experimental and control samples, so it likely represents a true difference between the groups.

In the experiment with the echinacea tea, statistical tests indicated that the 33% reduction in cold severity observed by the researchers was statistically significant. In other words, there is a low probability that the difference in cold susceptibility between the two samples in the experiment is due to chance. If the experiment is properly designed, a statistically significant result allows researchers to infer that the treatment had an effect.

(a) Average hair length in this snapshot is shorter ...

- Little variability
- High probability of reflecting average of all women in the class

(b) ... than average hair length in this snapshot ...

- High variability
- Low probability of reflecting average of all women in the class

... so, is hair longer today than in 1962?

FIGURE 1.12 The role of statistics. Statistical tests calculate the variability within groups to determine the probability that two groups differ only by chance.

Factors that Influence Statistical Significance

We can explore the role that statistical tests play in more detail by evaluating another study on cold treatments. This study examined the efficacy of zinc lozenges on reducing cold severity.

Some forms of zinc can block common cold viruses from invading cells in the nasal cavity. This observation led to the hypothesis that consuming zinc at the start of a cold could decrease cold severity. Researchers at the Cleveland Clinic tested this hypothesis using a sample of 100 of their employees who volunteered for a double-blind study within 24 hours of developing cold symptoms. The researchers randomly assigned subjects to the experimental group, who received zinc lozenges, or the control group, who received placebo lozenges. Members of both groups received the same instructions for using the lozenges and were asked to rate their symptoms until they had recovered.

When the data from the experiment were summarized, the statistics indicated that the mean recovery time was over 3 days shorter in the zinc group than in the placebo group (**FIGURE 1.13**). On the surface, this result appears to support the hypothesis. However, recall the example of the snapshot of women's hair length. A statistical test is necessary because of the effect of chance.

The Problem of Sampling Error. The effect of chance on experimental results is known as **sampling error**—more specifically, sampling error is the difference between a sample and the population from which it was drawn. In any experiment, individuals in the experimental group will differ from individuals in the control group in random ways. Even if there is *no* true effect of an experimental treatment, the data from the experimental group will never be identical to the data from the control group.

For example, we know that people differ in their ability to recover from a cold infection. If we give zinc lozenges to one volunteer and placebo lozenges to another, it is likely that the two volunteers will have colds of different lengths. Even if the zinc-taker had a shorter cold than the placebo-taker, you would probably say that the test did not tell us much about our hypothesis—the zinc-taker might just have had a less severe cold for other reasons.

Now imagine that we had five volunteers in each group and saw a difference, or that the difference was only 1 day instead of 3 days. How would we determine if the lozenges had an effect? Statistical tests allow researchers to look at their data and determine how likely it is—that is, the **probability**—that the result is due to sampling error. In this case, the statistical test distinguished between two possibilities for why the experimental group had shorter colds: Either the difference was due to the effectiveness of zinc as a cold treatment or it was due to a chance difference from the control group. The results indicated that there was a low probability, less than 1 in 10,000 (0.01%), that the experimental and control groups were so different simply by chance. In other words, the result is statistically significant.

A statistical measure of the amount of variability in a sample is a number called the **standard error** of the mean; greater variability generates a larger standard

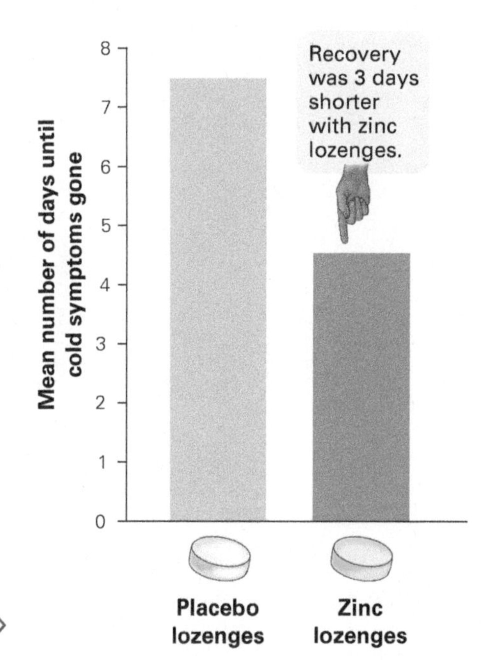

FIGURE 1.13 Zinc lozenges reduce the duration of colds. Fifty individuals taking zinc lozenges had colds lasting about 4½ days as opposed to approximately 7½ days for 50 individuals taking placebo.

WORKING WITH **DATA**

Sometimes graphs have lines and sometimes they have bars. Why do you think these two data points (average of 7.5 and average of 4.5) are not connected with a line?

VISUALIZE **THIS**

Increasing the size of the sample measured usually decreases the standard error. Why do you think this is the case?

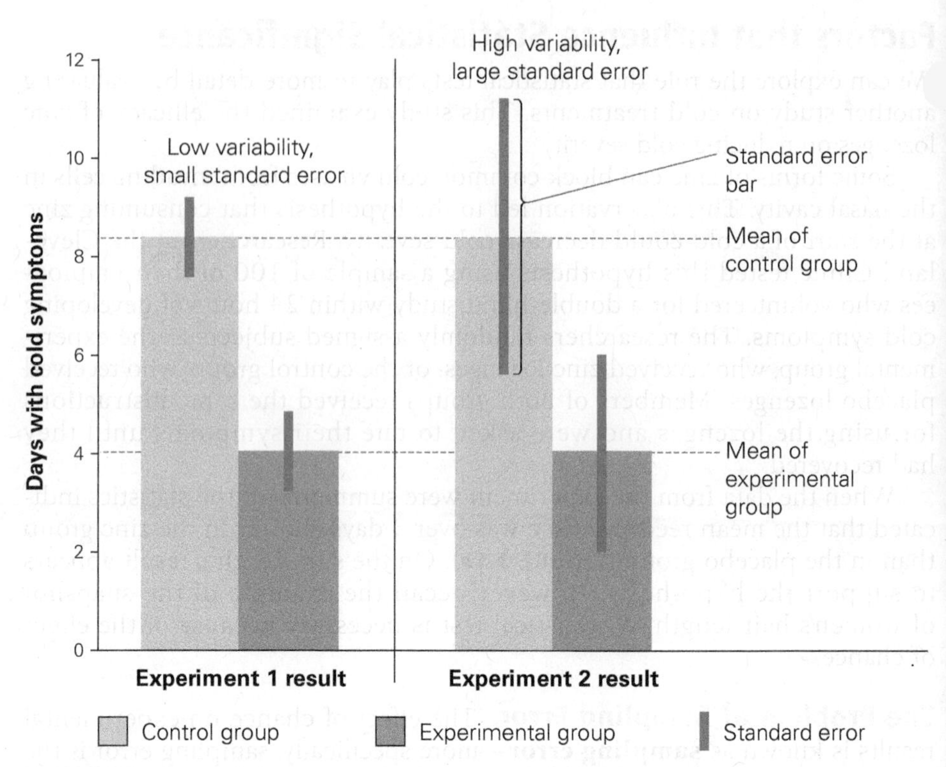

FIGURE 1.14 Standard error. The red lines on these bar graphs represent the standard error for each sample mean. A more variable sample has a larger standard error than a less variable sample. Even though the means are the same for both experimental and control groups in these two sets of bars, only the data summarized at left illustrate a significant difference because of the greater variability in the samples illustrated on the right side.

error (**FIGURE 1.14**). Put simply, the true population mean is very likely to lie somewhere in the range between the mean of the sample minus the standard error and the mean of the sample plus the standard error. You have likely heard the term *standard error* used before—when public opinion polls are released, this range is often referred to as the *margin of error*. Results with a small standard error are more likely to be statistically significant if the hypothesis is true.

Factors That Influence Statistical Significance. One characteristic of experiments that influences sampling error is **sample size**—the number of individuals in the experimental and control groups. If a treatment has no effect, a small sample size could return results that appear different simply because of a large sampling error. This was the case with the vitamin C hypothesis described at the beginning of the chapter. Subsequent tests with larger sample sizes allowed scientists to reject the hypothesis that vitamin C prevents colds.

Conversely, if the effect of a treatment is real but the sample size of the experiment is small, a single experiment may not allow researchers to determine convincingly that their hypothesis has support. For example, one experiment performed at a Wisconsin clinic with 48 participants indicated that echinacea tea drinkers were 30% less likely to experience any cold symptoms after virus exposure as compared with individuals who received a placebo tea. However, the small sample size of the study meant that this result was not statistically significant—based on this experiment, we still cannot say whether echinacea tea has a protective effect.

The more participants there are in a study, the more likely it is that researchers will see a true effect of an experimental treatment, even if the effect is very small. For example, a study of more than 21,000 men 6 years or older in Finland demonstrated that those who took vitamin E supplements had 5% fewer colds than those who did not take these supplements. In this case, the large

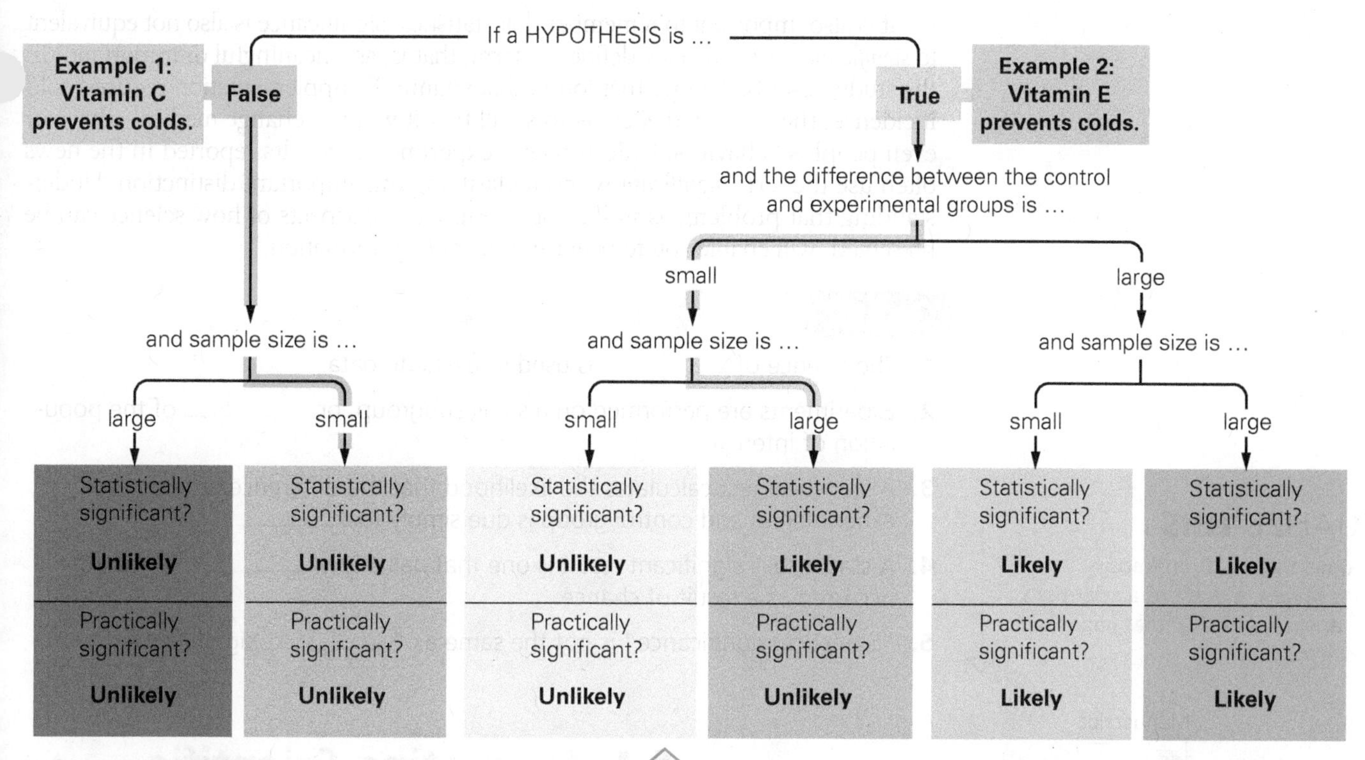

VISUALIZE **THIS**

The Nurse's Health Study, which has followed nearly 80,000 women for more than 20 years, has found that a diet low in refined carbohydrates such as white flour and sugar but relatively high in vegetable fat and protein cuts heart disease risk by 30% relative to women with a high refined carbohydrate diet. How would the results of this study map out on the flowchart illustrated here?

FIGURE 1.15 Factors that influence statistical significance. This flowchart summarizes the relationship between the true effect of a treatment and the sample size of an experiment on the likelihood of obtaining statistical significance.

sample size allowed researchers to see that vitamin E has a real, but relatively tiny, effect on cold incidence. In other words, this statistically significant result has little real-world *practical* significance. The relationship among hypotheses, experimental tests, sample size, and statistical and practical significance is summarized in **FIGURE 1.15**.

There is one final caveat to this discussion. A statistically significant result is typically defined as one that has a *probability* of 5% or less of being due to chance alone. If all scientific research uses this same standard, as many as 1 in every 20 statistically significant results (that is, 5% of the total) is actually reporting an effect that is *not real*. In other words, some statistically significant results are "false positives," representing a surprisingly large sampling error. This potential explains why one supportive experiment is not enough to convince all scientists that a particular hypothesis is accurate.

Even with a statistical test indicating that the result had a probability of less than 0.01% of occurring by chance, we should begin to feel assured that taking zinc lozenges will reduce the duration of colds only after additional hypothesis tests give similar results. In fact, scientists continue to test this hypothesis, and there is still no consensus about the effectiveness of zinc as a cold treatment.

What Statistical Tests Cannot Tell Us

All statistical tests operate under the assumption that the experiment was designed and carried out correctly. In other words, a statistical test evaluates the chance of sampling error, not observer error, and a statistically significant result is not the last word on an experimentally tested hypothesis. An examination of the experiment itself is required.

It is also important to remember that statistical significance is also not equivalent to *significance* as we usually define the term, that is, as "meaningful or important." In the study described above that found that vitamin E supplementation reduced cold incidence, the measured effect is so small that it will not change medical advice or even people's behaviors. Unfortunately, experimental results reported in the news often use the term *significant* without clarifying this important distinction. Understanding that problem, as well as other misleading aspects of how science can be presented, will enable you to better use scientific information.

Got It?

1. The science of ________ is used to evaluate data.
2. Experiments are performed on a small subgroup, or ________ of the population of interest.
3. A statistical test calculates the likelihood that the difference between an experimental and control group is due simply to ________.
4. A statistically significant result is one that has a low ________ of having occurred as a result of chance.
5. "Statistical significance" is not the same as "________ significance."

VISUALIZE **THIS**

How does having anonymous experts review a draft of a scientist's paper ensure that the final paper will be more reliable?

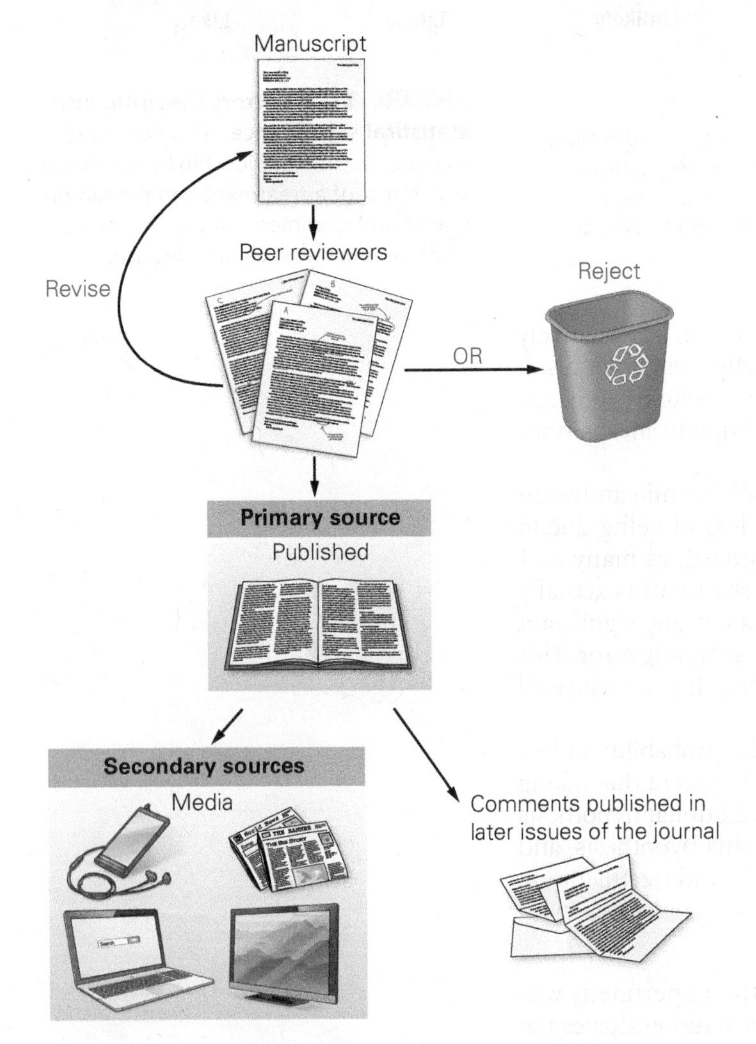

FIGURE 1.16 Primary sources: Publishing scientific results. Most scientific journals require papers to go through stringent review before publication.

1.4 Evaluating Scientific Information

The previous sections should help you see why definitive scientific answers to our questions are slow in coming. However, a well-designed experiment can certainly allow us to approach the truth.

Primary Sources

Looking critically at reports of experiments can help us make well-informed decisions about actions to take. Most of the research on cold prevention and treatment is first published as **primary sources,** written by the researchers themselves and reviewed within the scientific community (**FIGURE 1.16**). The process of **peer review,** in which other scientists critique the results and conclusions of an experiment before it is published, helps increase confidence in scientific information. Peer-reviewed research articles in journals such as *Science, Nature*, the *Journal of the American Medical Association*, and hundreds of others represent the first and most reliable sources of current scientific knowledge.

However, evaluating the hundreds of scientific papers that are published weekly is a task not one of us can perform alone. Even if we focused only on a particular field of interest, the technical jargon used in many scientific papers may be a significant barrier to our understanding.

Instead of reading the primary literature, most of us receive our scientific information from **secondary sources** such as books, news reports, and even advertisements. How can we evaluate information in this context? The following sections provide strategies for doing so.

Information from Anecdotes

Information about dietary supplements such as echinacea tea and zinc lozenges is often in the form of **anecdotal evidence**—meaning that the advice is based on one individual's personal experience. A friend's enthusiastic plug for vitamin C, because she felt it helped her, is an example of a testimonial—a common form of anecdote. Advertisements that use a celebrity to pitch a product "because it worked for them" are classic forms of testimonials.

You should be cautious about basing decisions on anecdotal evidence, which is not equivalent to well-designed scientific research. For example, many of us have heard anecdotes along the lines of the grandpa who was a pack-a-day smoker and lived to the age of 94. However, hundreds of studies have demonstrated the clear link between cigarette smoking and premature death. Although anecdotes may indicate that a product or treatment has merit, only well-designed tests of the hypothesis can determine its safety and efficacy.

Science in the News

Popular news sources provide a steady stream of science information. However, stories about research results often do not contain information about the adequacy of experimental design or controls, the number of subjects, or the source of the scientist's funding. How can anyone evaluate the quality of research that supports dramatic headlines such as those describing risks to human health and the environment?

First, you must consider the source of media reports. Certainly, news organizations will be more reliable reporters of fact than will entertainment tabloids, and news organizations with science writers should be better reporters of the substance of a study than those without (**FIGURE 1.17**). Television talk shows, which need to fill airtime, regularly have guests who promote a particular health claim. Too often these guests may be presenting information that is based on anecdotes or an incomplete summary of the primary literature.

Paid advertisements are a legitimate means of disseminating information. However, claims in advertising should be carefully evaluated. Advertisements of over-the-counter and prescription drugs must conform to rigorous government standards regarding the truth of their claims. Lower standards apply to advertisements for herbal supplements, many health-food products, and diet plans. Be sure to examine the fine print because advertisers often are required to clarify the statements made in their ads.

Another commonly used source for health information is the Internet. As you know, anyone can post information on this great resource. Typing in "common cold prevention" on a standard web search engine will return thousands of web pages—from highly respected academic and government sources to small companies trying to sell their products to individuals who have strong, sometimes completely unsupported, ideas about cures. Often it can be difficult to determine the reliability of a well-designed website, and sites such as Wikipedia may contain erroneous or misleading information. Here are some things to consider when using the web as a resource for health information:

1. Choose sites maintained by reputable medical establishments, such as the National Institutes of Health (NIH) or the Mayo Clinic.
2. It costs money to maintain a website. Consider whether the site seems to be promoting a product or agenda. Advertisements for a specific product should alert you to possible bias. Sites with a nonprofit domain name (such as .gov or .edu) are much less likely to be promoting a product.

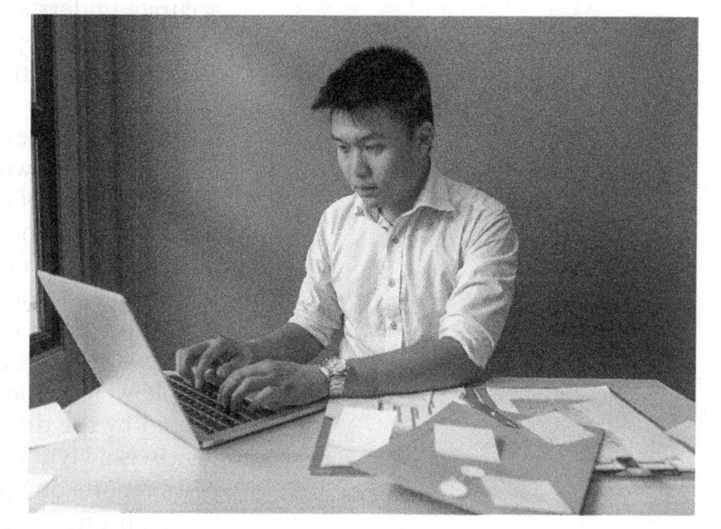

FIGURE 1.17 How we learn about science. Most of the information we hear about science does not come directly from professional scientists—it is translated for the nonspecialist audience by journalists and other professional communicators. Thoughtful and critical readers of these stories need to understand the scientific process to best evaluate their accuracy and reliability.

3. Check the date when the website was last updated and see whether the page has been updated since its original posting. Science and medicine are disciplines that must frequently evaluate new data. A reliable website will be updated often.
4. Determine whether unsubstantiated claims are being made. Look for references and be suspicious of any studies that are not from peer-reviewed journals.

Understanding Science from Secondary Sources

Once you are satisfied that a media source is relatively reliable, you can examine the scientific claim that it presents. Use your understanding of the process of science and of experimental design to evaluate the story and the science. Does the story about the claim present the results of a scientific study, or is it built around an untested hypothesis? Were the results obtained using the scientific method, with tests designed to reject false hypotheses? Is the story confusing correlation with causation? Does it seem that the information is applicable to nonlaboratory situations, or is it based on results from preliminary or animal studies?

Look for clues about how well the reporters did their homework. Scientists usually discuss the limitations of their research in their papers. Are these cautions noted in an article or television piece? If not, the reporter may be overemphasizing the applicability of the results.

Then, note if the scientific discovery itself is controversial. That is, does it reject a hypothesis that has long been supported? Does it concern a subject that is contentious? Might it lead to a change in social policy? In these cases, be extremely cautious. New and unexpected research results must be evaluated in light of other scientific evidence and understanding. Reports that lack comments from other experts may miss problems with a study or fail to place it i the context of other research. The "Big Question" feature in this chapter provides a guideline to evaluating reports of science in the news.

Finally, the news media generally highlight only those science stories that editors find newsworthy. As we have seen, scientific understanding accumulates relatively slowly, with many tests of the same hypothesis finally leading to an accurate understanding of a natural phenomenon. News organizations are also more likely to report a study that supports a hypothesis rather than one that gives less supportive results, even if both types of studies exist, because it makes for a better story.

In addition, even the most respected media sources may not be as thorough as readers would like. For example, a recent review published in the *New England Journal of Medicine* evaluated the news media's coverage of new medications. Of 207 randomly selected news stories, only 40% that cited experts with financial ties to a drug told readers about this relationship. This potential conflict of interest calls into question the expert's objectivity. Another 40% of the news stories did not provide basic statistics about the drug's benefits. Most of the news reports also failed to distinguish between absolute benefits (how many people were helped by the drug) and relative benefits (how many people were helped by the drug relative to other therapies for the condition). The journal's review is a vivid reminder that we need to be cautious when reading or viewing news reports on scientific topics.

Even after following all of these guidelines, you will still find well-researched news reports on several scientific studies that seem to give conflicting and confusing results. As you now know, such confusion is the nature of the scientific process—early in our search for understanding, many hypotheses are proposed and discussed; some are tested and rejected immediately, and some are supported by one experiment but later rejected by more thorough experiments. It is only by clearly understanding the process and pitfalls of scientific research that you can distinguish "what we know" from "what we don't know."

Is There a Cure for the Common Cold?

So will we ever find the best way to prevent a cold or reduce its effects? In the United States alone, more than 1 billion cases of the common cold are reported per year, costing many more billions of dollars in medical visits, treatment, and lost workdays. Consequently, there is significant cause to search for effective protection from the different viruses that cause colds.

A search of medical publication databases indicates that every year nearly 100 scholarly articles regarding the biology, treatment, and consequences of common cold infection are published. This research has led to several important discoveries about the structure and biochemistry of common cold viruses, how they enter cells, and how the body reacts to these infections.

Despite all of the research and the emergence of some promising possibilities, the best prevention method known for common colds is still the old standby—keep your hands clean. Numerous studies have indicated that rates of common cold infection are 20 to 30% lower in those who use effective hand-washing procedures. Cold viruses can survive on surfaces for many hours; if you pick up viruses on your hands from a surface and transfer them to your mouth, eyes, or nose, you may inoculate yourself with a 7-day sniffle (**FIGURE 1.18**).

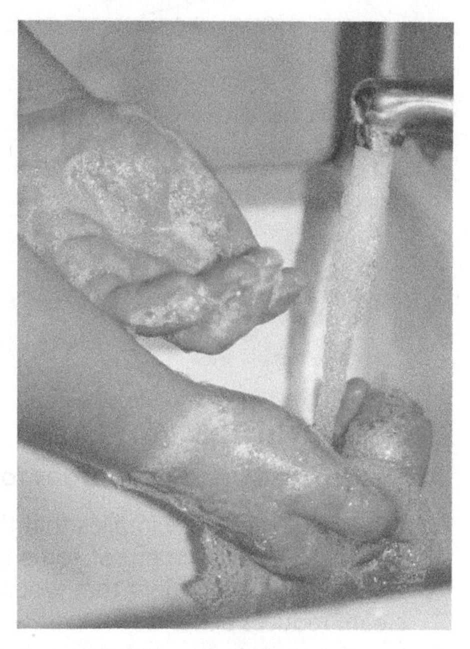

FIGURE 1.18 Preventing colds. According to the Centers for Disease Control and Prevention, the best defense against infection is effective hand washing. Scrub vigorously with a liquid soap for 20 seconds, then rinse under running water for 10 seconds. Alcohol-based hand gels (without water) containing at least 60% ethanol are less effective but can be used if soap and water are unavailable. Drying your hands with a paper towel or clean cloth while your hands are still wet with sanitizer appears to be the most effective technique for removing microorganisms.

Of course, not everyone gets sick when exposed to a cold virus. The reason one person has more colds than another might not be due to a difference in personal hygiene. The correlation that showed a relationship between stress and cold susceptibility appears to have some merit. Research indicates that among people exposed to viruses, the likelihood of ending up with an infection increases with high levels of psychological stress—something that many college students clearly experience.

Research also indicates that vitamin C intake, diet quality, exposure to cold temperatures, and exercise frequency appear to have no effect on cold susceptibility, although along with echinacea tea and zinc lozenges, there is some evidence that vitamin C may reduce cold symptoms a bit. Surprisingly, even though medical research has led to the elimination of killer viruses such as smallpox and polio, scientists are still a long way from "curing" the common cold.

Got It?

1. Research reports describing the outcome of a hypothesis test published in scientific journals are referred to as __________ sources.
2. Scientific papers are not published in most journals until after they have undergone the process of __________ review by other scientists in the field.
3. __________ evidence for hypotheses consist of individual's claims about "what worked for them" and is not equivalent to research published in journals.
4. Instead of reading primary literature, most of us receive our scientific information from __________ __________.
5. News stories are less likely to be written about research that provides __________ supportive results of a hypothesis test.

Sounds right, but is it?

Advertisements for skin care products often contain a phrase such as: "This product is clinically proven to. . . ." That statement may convince you to purchase a topical cream that claims to reduce acne, eliminate rough patches of skin, or lighten dark spots. You would be making this decision based on your belief in the following statement:

If a product is clinically proven to do what it advertises, that means it will work for you.

Sounds right, but it isn't.

Answer the following questions to understand why.

1. "Clinically" generally implies that the product was subject to an experiment of some type. Describe the ideal type of experiment that would test whether a cream reduces acne.
2. Imagine that the results of this experiment showed that that there was a statistically significant difference in the extent of acne between those patients using the "real" cream and those using the placebo. Is this difference in acne necessarily large?
3. Much of the time, a "clinically proven" claim relies on consumer evaluation of the product—in other words, surveys that ask whether acne was reduced as a result of using the product. What is the problem with this type of experimental approach?
4. Even if the experiment is well controlled and double-blinded, does the result of a single experiment prove a hypothesis?
5. Does this statistically significant result mean that everyone in the experiment who used the real cream had less acne than those who used the placebo?
6. Consider your answers to questions 1–5 and explain why the statement bolded above sounds right, but isn't.

THE BIG QUESTION

How do I know what to believe?

Every chapter in this book will examine whether science can answer a given "big question." Most examples of this feature will ask you to evaluate the kinds of questions science can and cannot answer related to a topic within the chapter. But for this introductory chapter, this feature is a little different. Here we present a tool to help you answer the Big Question "How do I know what to believe?" when you are evaluating science presented in the news.

The checklist in the table that follows helps you look for red flags that indicate that there may be problems with the reporting of the story or with the science itself. Although any issue that raises a red flag should cause you to be cautious about the conclusions of the story, it is rare that one issue should cause you to reject the story (and its associated research) out of hand. However, the fewer red flags raised by a story, the more reliable the report is likely to be. Use this checklist to evaluate this extract from a blog published by *Discover* magazine:[1]

> Researchers used a mouse-adapted version of the virus and tested it on cells taken from mouse airways, such as the nose and lungs. Researchers incubated the virus and airway cells at 91 degrees in one batch and at 98.6 degrees in another batch. To detect the cells' immune response researchers measured gene activation and chemical signaling within the cells.
>
> Researchers found that the cells stored at 98.6 degrees launched a more robust immune attack than the ones stored at 91 degrees. The findings, published in *Proceedings of the National Academy of Sciences*, indicate that exposure to cold air might lower our bodies' ability to fight the common cold. And although most of us can power through a cold, people with respiratory issues or impaired immune systems may have a harder time fighting off the virus.

[1] http://blogs.discovermagazine.com/d-brief/2015/01/06/catch-cold-being-cold/#.V2uzel-cG3A.

Question	Preferred Answer	Raises a Red Flag
1. What is the basis for the story?	Hypothesis test ☐	Untested assertion. No data to support claims in the article. ☐
2. What is the affiliation of the scientist?	Independent (university or government agency) ☐	Employed by an industry or advocacy group. Data and conclusions could be biased. ☐
3. What is the funding source for the study?	Government or nonpartisan foundation (without bias) ☐	Industry group or other partisan source (with bias). Data and conclusions could be biased. ☐
4. a. **If the hypothesis test is a correlation:** Did the researchers attempt to eliminate reasonable alternative hypotheses?	Yes ☐	No. Correlation does not equal causation. One hypothesis test provides poor support if alternatives are not examined. ☐
b. **If the hypothesis test is an experiment:** Is the experimental treatment the only difference between the control group and the experimental group?	Yes ☐	No. An experiment provides poor support if alternatives are not examined. ☐
5. Was the sample of individuals in the experiment a good cross-section of the population?	Yes ☐	No. Results may not be applicable to the entire population. ☐
6. Was the data collected from a relatively large number of people?	Yes ☐	No. Study is prone to sampling error. ☐
7. Were participants blind to the group they belonged to or to the "expected outcome" of the study?	Yes ☐	No. Research participant expectation can influence results. ☐
8. Were data collectors or analysts blinded to the group membership of participants in the study?	Yes ☐	No. Observer bias can influence results. ☐
9. Did the news reporter put the study in the context of other research on the same subject?	Yes ☐	No. Cannot determine if these results are unusual or fit into a broader pattern of results. ☐
10. Did the news story contain commentary from other independent scientists?	Yes ☐	No. Cannot determine if results are unusual or if the study is considered questionable by others in the field. ☐
11. Did the reporter list the limitations of the study or studies on which he or she is reporting?	Yes ☐	No. Reporter may not be reading study critically and could be overstating the applicability of the results. ☐

Chapter Review

Mastering Biology

Go to Mastering Biology to access eText 2.0, Dynamic Study Modules, and the Study Area, where you'll find practice quizzes, BioFlix™ animations, MP3 tutor sessions, current events, and more.

SUMMARY

Section 1.1

Describe the characteristics of a scientific hypothesis.

- Science is a process of testing hypotheses—statements about how the natural world works. Scientific hypotheses must be testable and falsifiable (pp. 3–4).

Compare and contrast the terms *scientific hypothesis* and *scientific theory*.

- A scientific theory is an explanation of a set of related observations based on well-supported hypotheses from several different, independent lines of research (pp. 4–5).

Distinguish between inductive and deductive reasoning.

- Hypotheses are often developed via inductive reasoning, which consists of making multiple observations and then inferring a general principle to explain them (p. 5).
- Hypotheses are tested via the process of deductive reasoning, which is when researchers make specific predictions about expected outcomes (pp. 5–6).

Explain why the truth of a hypothesis cannot be proven conclusively via deductive reasoning.

- Absolutely proving hypotheses is impossible because there may be other reasons besides the one hypothesized that could lead to the predicted result (pp. 6–7).

SHOW YOU KNOW 1.1 Consider this if/then statement: If your coworker Homer ate all the donuts, then there won't be any donuts left in the box that was placed in the break room 30 minutes ago. Imagine you find that the donuts are all gone—did you just prove that Homer ate them? Why or why not?

Section 1.2

Describe the features of a controlled experiment, and explain how these experiments eliminate alternative hypotheses for the results.

- Controlled experiments test hypotheses about the effect of experimental treatments by comparing a randomly assigned experimental group with a control group (p. 9).
- Controls are individuals who are treated identically to the experimental group except for application of the treatment (pp. 9–10).

List strategies for minimizing bias when designing experiments.

- Bias in scientific results can be minimized with double-blinded experiments that keep research participants and data collectors unaware of which individuals belong in the control or experimental group (p. 11).

Define correlation, and explain the benefits and limitations of using this technique to test hypotheses.

- Because performing controlled experiments on humans is considered unethical in some cases, scientists sometimes use correlations. The correlation of structure and function between humans and model organisms, such as other mammals, allows experimental tests of these hypotheses (p. 12).
- A correlation study can describe only a relationship between two factors, but it does not strongly imply that one factor causes the other (pp. 12–15).

SHOW YOU KNOW 1.2 In the echinacea tea experiment, a nonrandom assignment scheme might have put the first 25 visitors to the clinic in the control group and the next 25 in the experimental group. Imagine that in this version of the experiment, the experimental group did recover in fewer days than the control group. Describe an alternative hypothesis related to the nonrandom assignment of research participants that was not eliminated by this experimental design and that could explain these results.

Section 1.3

Describe the information that statistical tests provide.

- Statistics help scientists evaluate the results of their experiments by determining if results appear to reflect the true effect of an experimental treatment on a sample of a population (p. 16).
- A statistically significant result is one that is very unlikely to be due to chance differences between the experimental and control groups (p. 16).
- Even when an experimental result is highly significant, hypotheses are tested multiple times before scientists come to a consensus on the true effect of a treatment (pp. 18–19).

SHOW YOU KNOW 1.3 Imagine that average hair length was 60 centimeters (cm) (23.6 inches) in the class of 1965 and 65 cm (25.6 inches) in the class of 2018. Given the variation in styles in 2018, however, this information is not statistically significant. Can we say that the average hairstyle in 2018 is longer than the average in 1962?

Section 1.4

Compare and contrast primary and secondary sources.

- Primary sources of information are experimental results published in professional journals and peer-reviewed by other scientists before publication (p. 20).
- Secondary sources are typically news, books, or websites that summarize the scientific research presented in primary sources. Secondary sources are not peer-reviewed (p. 21).

Summarize the techniques you can use to evaluate scientific information from secondary sources.

- Anecdotal evidence is an unreliable means of evaluating information, and media sources are of variable quality; distinguishing between news stories and advertisements is important when evaluating the reliability of information. The Internet is a rich source of information, but users should look for clues to a particular website's credibility (pp. 21–22).
- Stories about science should be carefully evaluated for information on the actual study performed, the universality of the claims made by the researchers, and other studies on the same subject. Sometimes contradictory stories about scientific information are a reflection of controversy within the scientific field itself (p. 22).

SHOW YOU KNOW 1.4 Where would you look to find the latest research on common cold treatment and prevention? Explain why this is a good strategy for obtaining the best information possible.

ROOTS TO REMEMBER

The following roots of words come mainly from Latin and Greek and will help you to decipher terms:

bio- means life. Chapter term: *biology*

deduc- means to reason out, working from facts. Chapter term: *deductive reasoning*

hypo- means under, below, or basis. Chapter term: *hypothesis*

induc- means to rely on reason to derive principles (also, to cause to happen). Chapter term: *inductive reasoning*

-ology means the study of or branch of knowledge about. Chapter term: *biology*

LEARNING THE BASICS

1. Add labels to the figure that follows, which illustrates the characteristics of research participants in control versus experimental groups.

Control group	Experimental group
Experiencing early cold symptoms	________ ________
________ ________	Sought treatment from clinic
Received ________ tea	Received **echinacea** tea

2. Which of the following is an example of inductive reasoning?

A. All cows eat grass; **B.** My cow eats grass and my neighbor's cow eats grass; therefore, all cows probably eat grass; **C.** If all cows eat grass, when I examine a random sample of all the cows in Minnesota, I will find that all of them eat grass; **D.** Cows may or may not eat grass, depending on the type of farm where they live

3. A scientific hypothesis is ________.

A. an opinion; **B.** a proposed explanation for an observation; **C.** a fact; **D.** easily proved true; **E.** an idea proposed by a scientist

4. How is a scientific theory different from a scientific hypothesis?

A. It is based on weaker evidence; **B.** It has not been proved true; **C.** It is not falsifiable; **D.** It can explain a large number of observations; **E.** It must be proposed by a professional scientist

5. One hypothesis states that eating chicken noodle soup is an effective treatment for colds. Which of the following results does this hypothesis predict?

A. People who eat chicken noodle soup have shorter colds than do people who do not eat chicken noodle soup; **B.** People who do not eat chicken noodle soup experience unusually long and severe colds; **C.** Cold viruses cannot live in chicken noodle soup; **D.** People who eat chicken noodle soup feel healthier than do people who do not eat chicken noodle soup; **E.** Consuming chicken noodle soup causes people to sneeze

6. If I perform a hypothesis test in which I demonstrate that the prediction I made in question 5 is true, I have ________.

A. proved the hypothesis; **B.** supported the hypothesis; **C.** not falsified the hypothesis; **D.** B and C are correct; **E.** A, B, and C are correct

7. Control subjects in an experiment ________.

A. should be similar in most ways to the experimental subjects; **B.** should not know whether they are in the control or experimental group; **C.** should have essentially the same interactions with the researchers as the experimental subjects; **D.** help eliminate alternative hypotheses that could explain experimental results; **E.** all of the above

8. An experiment in which neither the participants in the experiment nor the technicians collecting the data know which individuals are in the experimental group and which ones are in the control group is known as ________.

A. controlled; **B.** biased; **C.** double-blind; **D.** falsifiable; **E.** unpredictable

9. A relationship between two factors, for instance, between outside temperature and the number of people with active colds in a population, is known as a(n) ________.

A. significant result; **B.** correlation; **C.** hypothesis; **D.** alternative hypothesis; **E.** experimental test

10. A primary source of scientific results is ________.

A. the news media; **B.** anecdotes from others; **C.** articles in peer-reviewed journals; **D.** the Internet; **E.** all of the above

11. A story on your local news station reports that eating a 1-ounce square of milk chocolate each day reduces the risk of heart disease in rats and that this result is statistically significant. This means that ________.

A. people who eat milk chocolate are healthier than those who do not; **B.** the difference between chocolate-eating and chocolate-abstaining rats in heart disease rates was greater than expected by chance; **C.** rats like milk chocolate; **D.** milk chocolate reduces the risk of heart disease; **E.** two ounces of milk chocolate per day is likely to be even better for heart health than 1 ounce

12. What features of the story on milk chocolate and heart health described in question 11 should cause you to consider the results less convincing?

A. The study was sponsored by a large milk chocolate manufacturer; **B.** A total of 10 rats were used in the study; **C.** The only difference between the rats was that human participants of the experimental group received chocolate along with their regular diets, and the human participants of the control group received no additional food; **D.** The reporter notes that other studies indicate milk chocolate does not have a beneficial effect on heart health; **E.** all of the above

ANALYZING AND APPLYING THE BASICS

1. There is a strong correlation between obesity and the occurrence of a disease known as type 2 diabetes—that is, obese individuals have a higher instance of diabetes than nonobese individuals do. Does this mean that obesity causes diabetes? Explain.

2. In an experiment examining vitamin C as a cold treatment, students with cold symptoms who visited the campus medical center either received vitamin C or were treated with over-the-counter drugs. Students then reported on the length and severity of their colds. Both the students and the clinic health providers knew which treatment students were receiving. This study indicated that vitamin C significantly reduced the length and severity of colds. Which factors make this result somewhat unreliable?

3. Brain-derived neurotrophic factor (BDNF) is a substance produced in the brain that helps nerve cells (neurons) to grow and survive. BDNF also increases the connectivity of neurons and improves learning and mental function. A 2002 study that examined the effects of intense wheel-running on rats and mice found a positive correlation between BDNF levels and running distance. What could you conclude from this result?

GO FIND OUT

1. What scientific research is taking place at your institution or those nearby? Visit the web pages of the science departments on your campus or other local colleges and research institutions to find out.

2. Much of the research on common cold prevention and treatment is performed by scientists employed or funded by drug companies. Often these companies do not allow scientists to publish the results of their research for fear that competitors at other drug companies will use this research to develop a new drug before they do. Should our society allow scientific research to be owned and controlled by private companies?

MAKE THE CONNECTION

The science that you learned in this chapter has helped you better understand the real-world example used throughout this discussion. Draw a line from the statement on the left to the science that supports it on the right.

Are there any effective strategies for preventing the common cold?	Scientific hypotheses must be able to be tested by observations of the natural world and able to be theoretically proved false.
A statement such as "my guardian angel protects me from colds" is not scientific.	If the outcome of a hypothesis test is what was predicted if the hypothesis was true, it still may be the case that the hypothesis is false.
The principle that certain diseases are caused by infectious agents is a scientific theory.	Statistical tests examine the data collected in an experiment and assign a probability that these data are different than what would be expected as a result of chance.
Someone who takes vitamin C to prevent colds has fewer colds than someone who doesn't. This doesn't prove that vitamin C prevents colds.	Statistical significance does not mean "practical significance."
If a large, randomly assigned group of individuals who take vitamin C do not have fewer colds, on average, than another group that does not take vitamin C, the hypothesis that vitamin C prevents colds is not true.	Science is a process of testing hypotheses, which are ideas about "how things work."
People with higher stress levels are more susceptible to cold virus infection, but this does not mean that stress causes increased susceptibility.	The process of peer review is an assurance that the information published in scientific journals meets a standard of rigor and accuracy.
A difference in recovery time from a cold for people who drink echinacea tea and those who drink a different herbal tea may appear large but may not be considered significant.	Controlled experiments that eliminate alternative hypotheses for results are the most accurate tests of those hypotheses.
Men who take vitamin E supplements have 5% fewer colds than men who do not take vitamin E. This result was statistically significant. However, vitamin E is not commonly sold as a cold preventative.	A correlation between two factors does not mean that one factor causes the other, because correlations cannot eliminate all alternative hypotheses.
Scientific studies that indicate that antibiotics are not an effective treatment for colds are more reliable than a blog post that asserts that antibiotics worked for the author of the post.	Certain ideas about how the world works are so well supported by evidence that they are considered central ideas in science.

Answers to **Got It?, Visualize This, Working with Data, Sounds Right, But Is It?, Show You Know,** and **Chapter Review** questions can be found in the **Answers** section at the back of the book.

2 Science Fiction, Bad Science, and Pseudoscience

Water, Biochemistry, and Cells

Can dead humans come back as zombies?

Do sugary drinks cause hyperactivity in children?

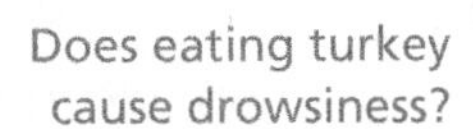

Does eating turkey cause drowsiness?

We are bombarded by information that is supposedly based in science. Although some of the so-called science is harmless entertainment loosely based in science, like the science fiction zombies seen on television or in movies, and although some bad science is believable only to the very young or uninformed, like planes and ships disappearing into the Bermuda Triangle, some dubious claims are plausible enough that we have a hard time knowing whether to believe them.

Here are some examples we will revisit later in the chapter: If your roommate tells you that you will do better on exams if you bring a bottle of water along, you might believe her. Many parents believe that giving children candy or sugary drinks makes them hyperactive. Is the relative who says he has to nap after eating turkey for Thanksgiving dinner just lazy, or does something in turkey make people tired?

These are pretty harmless examples. It won't hurt you to take water to an exam or to limit your intake of sweets or nap after Thanksgiving dinner. But what about spending money on crystals and magnets to treat disease or trying to cure cancer with alternative medicines instead of evidence-based treatments? Believing faulty science can lead to harm.

Sometimes people will claim that there is science backing a product or an idea when this is not actually the case. Such misstatements can be unintentional. For example, drawing conclusions from one study instead of waiting until a scientific consensus emerges can cause one to come to erroneous conclusions, as can relying on bad science, which results from experiments that are poorly designed or not properly controlled. In other cases, unscrupulous people will attempt to use science to lend legitimacy to a belief or ideology they want to promote. Dubious claims that are strengthened through the use of false or weakly supported assertions of scientific evidence are called pseudoscience.

In an effort to help develop the skill of assessing whether the science supporting an idea or product is sound, we will analyze examples of science fiction, weakly supported or bad science, and pseudoscience so that the qualities of each are more easily recognizable.

2.1 A Definition of Life

Zombies are imagined as human corpses that supposedly have been brought back to life. Often called the undead or walking dead, zombies are thought to survive by eating the flesh of living humans. If we suspend disbelief and assume that zombies exist, we know that zombies are not actually alive—that they are undead or walking dead is inherent in the definition of a zombie!—but they do share many characteristics seen in other living organisms.

Living organisms contain a common set of biological molecules and are composed of cells. Other attributes found in most living organisms include growth, some kind of movement, reproduction, response to external environmental stimuli, and **metabolism.** Metabolism includes all of the chemical processes that occur in cells, including the breakdown of substances to produce energy, the synthesis of substances necessary for life, and the excretion of wastes generated by these processes.

At first glance, zombies seem to exhibit some elements of this definition of life. They certainly can move, even though their movements are sometimes hindered by various injuries they have sustained. It might even appear that zombies can reproduce, because more zombies are created when existing zombies attack living humans. Likewise, zombies even respond to a limited number of stimuli, lurching toward humans for their next meal.

However, closer inspection leads us to question just how much zombies resemble living organisms. They do not grow; in most cases a child zombie does not become an adult zombie over time. And although the biting of humans by the zombies we see on TV or read about does create more zombies, this is not the same as the kind of reproduction living organisms are capable of. Living organisms pass genetic information to their offspring when they reproduce. It is also unlikely that zombies actually metabolize the flesh they eat. It seems that they are driven to eat human flesh more to spread zombiism than to provide nourishment to themselves.

An additional quality of living organisms is the ability to achieve **homeostasis,** the maintenance of a constant internal environment despite a constantly changing external environment (**FIGURE 2.1**). Popular movies seem to suggest that zombies are surprisingly good at maintaining their body temperature and blood pressure even with multiple flesh wounds and severed limbs. Unfortunately for the zombies, their limited homeostatic abilities do not prevent them from being in a progressive state of decline, with each assault bringing new injuries, piled on top of previously obtained unhealed injuries.

One last feature of populations of living organisms is that they can evolve—that is, change in average characteristics over time. Because zombies don't mate and cannot pass on their genes, evolution in zombies is not possible. With this

VISUALIZE **THIS**

The chickadee shown has a core body temperature of 107°F on a warm sunny day. Based on your understanding of homeostasis, what do you think its body temperature would be on a much colder day?

FIGURE 2.1 Homeostasis. The ability to maintain homeostasis requires complex feedback mechanisms between multiple sensory and physiological systems.

homeo- means like or similar.

in mind, we can conclude that zombies are not alive—and of course, zombies are the creation of science fiction!

Part of the reason we find zombies entertaining is because they do show some evidence of scientific plausibility. They exist in a state somewhere between the living and the dead; and the transmission of zombiism through a bite from an infected zombie is a mode of transmission we know to be at work in true infectious diseases. We will see that this confusion of scientifically plausible notions with our willingness to accept things that seem like they could be true can lead us to suspend our disbelief in cases where our entertainment is not the only thing at stake.

Got It?

1. Living organisms contain a common set of biological ________.
2. The basic structural unit of all living things is the ________.
3. Living organisms must be able to grow, reproduce, and respond to ________.
4. ________ includes all the chemical processes that occur in cells.
5. The ability to regulate metabolism and respond to changing conditions allows living organisms to maintain ________.

2.2 The Properties of Water

Let's return to another example of a questionable claim: Does drinking water during an exam help students perform better? When this hypothesis was tested by researchers at the University of East London, they found that this might just be the case (**FIGURE 2.2**). Before we analyze this study, let's first learn about some of the properties of water and other molecules so we can determine whether any of these properties can help explain the study's results.

Water is made up of two elements: hydrogen and oxygen. **Elements** are the fundamental forms of matter and are composed of atoms that cannot be broken down by normal physical means such as boiling. **Atoms** are the smallest units that have the properties of any given element. Ninety-two natural elemental atoms have been described by chemists, and several more have been created in laboratories. Hydrogen, oxygen, and calcium are examples of elements commonly found in biological systems. Each element has a one- or two-letter symbol: H for hydrogen, O for oxygen, and Ca for calcium, for example.

Atoms are composed of subatomic particles called **protons, neutrons,** and **electrons.** Protons have a positive electric charge; these particles and the uncharged neutrons make up the **nucleus** of an atom. The negatively charged electrons are outside the nucleus in an "electron cloud." Electrons are attracted to the positively charged nucleus. A *neutral atom* has equal numbers of protons and electrons (**FIGURE 2.3a–c**). Electrically charged **ions** do not have an equal number of protons and electrons (**FIGURE 2.3d**). In this case, the atom is not neutral but is instead charged.

WORKING WITH **DATA**

a. Which variable (bringing water to exam or exam score) is the independent variable (one whose value can be experimentally manipulated) and on which axis (*x* or *y*) do we find it?
b. If there is a statistically significant difference between the two treatment groups, how should that be shown on the graph?

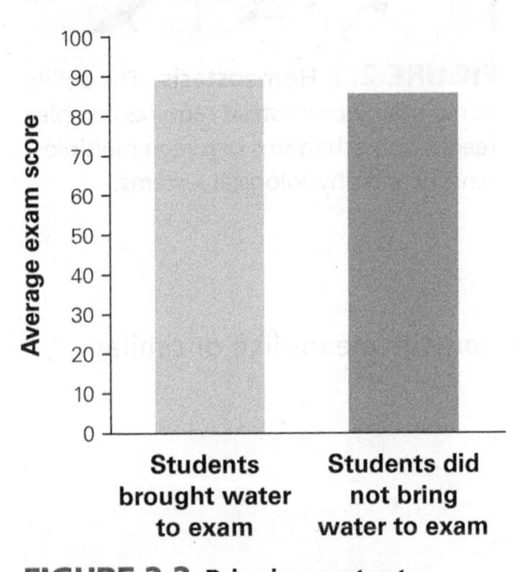

FIGURE 2.2 Bringing water to exams. This graph shows data from a study that compared exam scores of students who did and did not bring water with them to an exam.

The Structure of Water

The chemical formula for water (H_2O) indicates that it contains two hydrogen (H) atoms for every one oxygen (O). Water, like other molecules, consists of two or more atoms joined by chemical bonds. A molecule can be composed of the same or different atoms. For example, a molecule of oxygen consists of two oxygen atoms joined to each other, whereas a molecule of carbon dioxide consists of one carbon and two oxygen atoms.

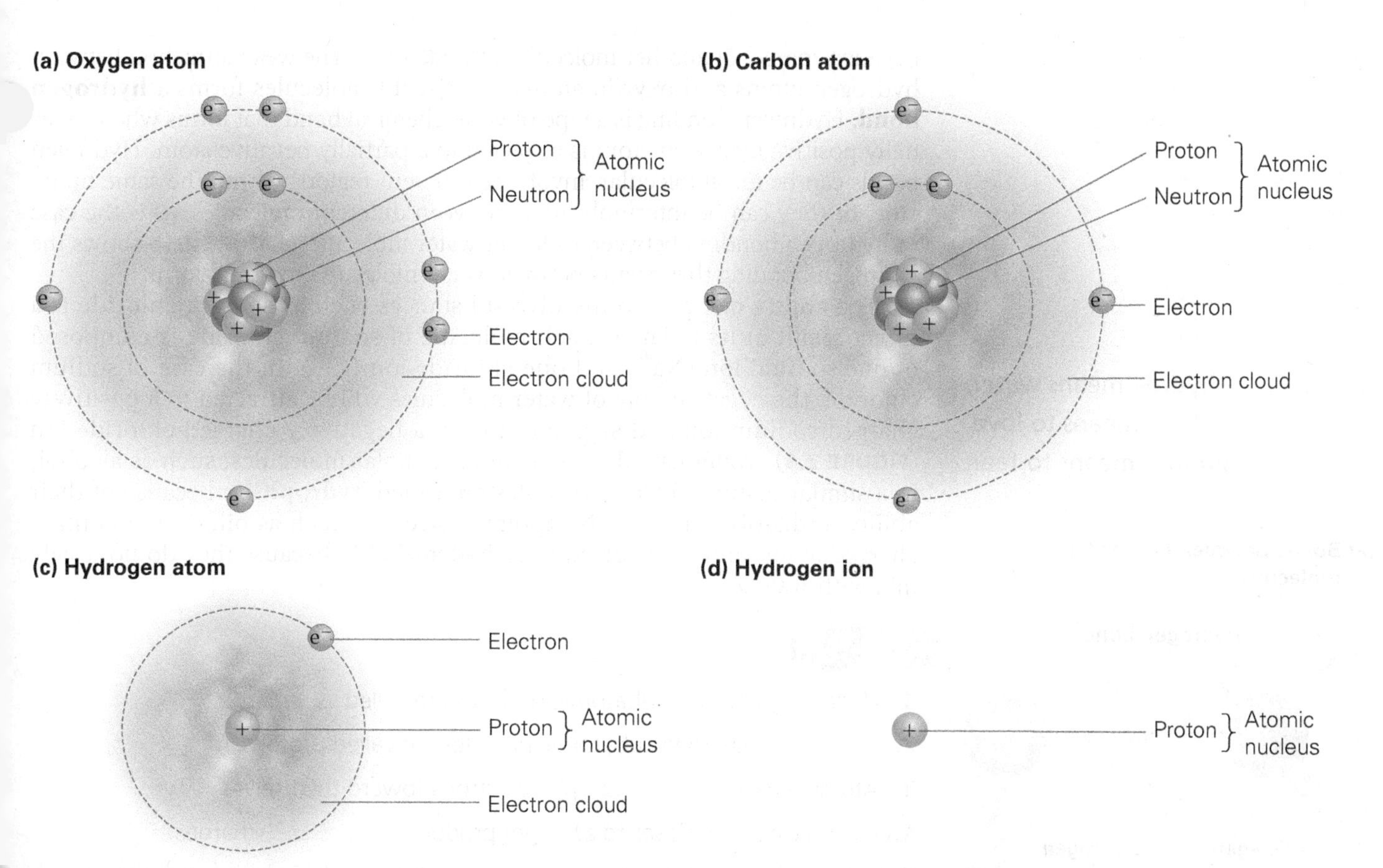

FIGURE 2.3 Atomic structure. An oxygen atom (a) contains a nucleus made up of eight protons, eight neutrons, and is orbited by eight electrons. Carbon (b), also a neutral atom, is composed of six protons, six neutrons, and six electrons. Hydrogen can exist as a neutral atom (c) with one proton and one electron or it can exist as a positively charged ion (d) composed of one proton only.

Water Is a Good Solvent

Water has the ability to dissolve a wide variety of substances. A substance that dissolves when mixed with another substance is called a **solute.** When a solute is dissolved in a liquid, such as water, the liquid is called a **solvent.** Once dissolved, components of a particular solute can pass freely throughout the water, making a chemical mixture or solution.

Water is a good solvent because it is **polar,** meaning that different regions, or poles, of the molecule have different charges. The polarity arises because oxygen is more attractive to electrons—that is, it is more **electronegative**—than most other atoms, including hydrogen. As a result of oxygen's electronegativity, electrons in a water molecule spend more time near the nucleus of the oxygen atom than near the nuclei of the hydrogen atoms. With more negatively charged electrons near it, the oxygen in water carries a partial negative charge, symbolized by the Greek lowercase letter delta and a negative sign: δ^-. The hydrogen atoms thus have a partial positive charge, symbolized by δ^+ (**FIGURE 2.4**). When atoms of a molecule carry no partial charge, they are said to be **nonpolar.** The carbon-hydrogen bonds, for example, share electrons equally and are nonpolar.

Water molecules tend to orient themselves so that the hydrogen atom (with its partial positive charge) of one molecule is near the oxygen atom (with its partial

VISUALIZE THIS

Toward which atom are the electrons of a water molecule pulled?

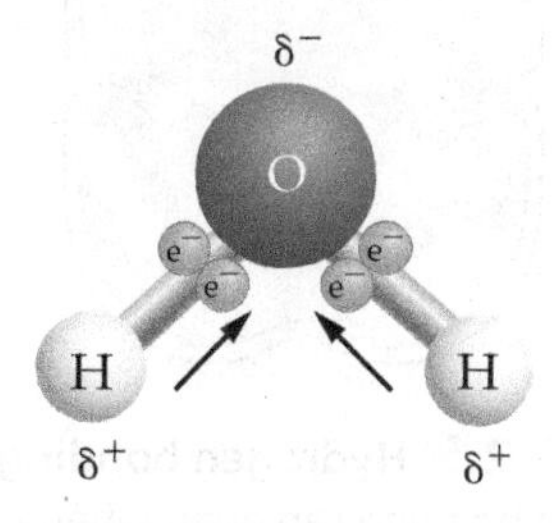

FIGURE 2.4 Polarity in water. Water is a polar molecule. Its atoms do not share electrons equally.

negative charge) of another molecule (**FIGURE 2.5a**). The weak attraction between hydrogen atoms and oxygen atoms in adjacent molecules forms a **hydrogen bond.** Hydrogen bonding is a type of weak chemical bond that forms when a partially positive hydrogen atom is attracted to a partially negative atom. Hydrogen bonds can be intramolecular, involving different regions within the same molecule, or they can be intermolecular, between different molecules, as is the case in hydrogen bonding between different water molecules. **FIGURE 2.5b** shows the hydrogen bonding that occurs between water molecules in liquid form.

The ability of water to dissolve substances such as sodium chloride is a direct result of its polarity. Each molecule of sodium chloride is composed of one sodium ion (Na^+) and one chloride ion (Cl^-). In the case of sodium chloride, the negative pole of water molecules will be attracted to a positively charged sodium ion and separate it from a negatively charged chloride ion (**FIGURE 2.6**). Water can also dissolve other polar molecules, such as alcohol, in a similar manner. Polar molecules are called **hydrophilic** because of their ability to dissolve in water. Nonpolar molecules, such as oil, do not contain charged atoms and are referred to as **hydrophobic** because they do not easily mix with water.

hydro- means water.
-philic means to love.
-phobic means to fear.

(a) Bonds between two water molecules

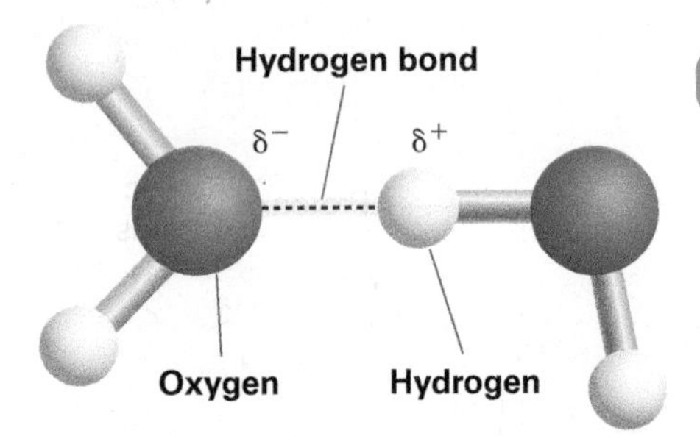

(b) Bonds between many water molecules

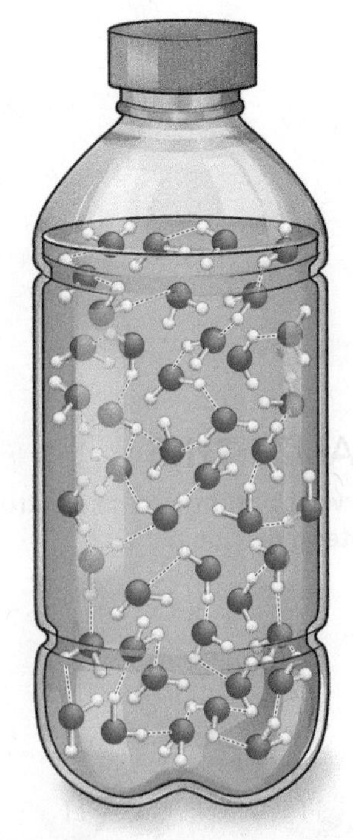

FIGURE 2.5 Hydrogen bonding. Hydrogen bonding can occur when there is a weak attraction between the hydrogen and oxygen atoms between (a) two or (b) many different water molecules.

Got It?

1. Negatively charged subatomic particles are called ________.
2. ________ charged subatomic particles are called protons.
3. Atoms that are ________ pull electrons toward themselves.
4. Pulling electrons closer to an atom produces ________ charges.
5. Polar molecules dissolve in water and are hydrophilic, whereas nonpolar molecules tend to be ________.

Water Facilitates Chemical Reactions

Because it is such a powerful solvent, water can facilitate **chemical reactions,** which are changes in the chemical composition of substances. Solutes in a mixture, called **reactants,** can come in contact with each other, permitting the modification of chemical bonds that occur during a reaction. The molecules formed as a result of a chemical reaction are known as **products.**

Water Moderates Temperature

When heat energy is added to water, the initial effect of heat is to disrupt the hydrogen bonding among water molecules. Therefore, this heat energy can be absorbed without changing the temperature of water. Only after the hydrogen bonds have been broken can added heat increase the temperature. In other words, the initial input of energy is absorbed.

The Drinking-Water Hypothesis Requires More Substantiation

Now that we have a better understanding of the properties of water, we can come back to our original question about whether drinking water can help students perform better on an exam. It is not too hard to imagine how some of the properties of water outlined earlier might lend it the ability to help regulate blood pressure, deliver dissolved nutrients to the body (including the brain), and maintai body temperature. It is also not hard to imagine that keeping blood pressure and temperature stable during a stressful exam and delivering nutrients to a hard-working brain might mean increased performance on an exam. Tempting as it is to stop there, a good scientist must not simply accept results because they are believable. A good scientist must also evaluate alterative hypotheses.

It might be the case that the more prepared students bring water to exams. Being more prepared for the exam in general could certainly increase exam scores whether or not someone drank water. The fact that alternate hypotheses may explain the data does not mean that the original hypothesis is incorrect. It only indicates that the original hypothesis requires more testing. In fact, the researchers that presented the study on water drinking and exam performance pointed out that the students bringing water to the exam tended to be upperclassmen, a variable that would need to be controlled for in subsequent studies. It is possible that upperclassmen, having gone through the experience of taking more exams, are more relaxed and perform better than other students, regardless of how much knowledge of the subject they may have.

Rigorous testing of alternate hypotheses is one major difference between bad science, unsubstantiated **pseudoscience,** and good science. Good scientists don't perform only experiments that will help confirm their hypothesis. They actively seek out ways to falsify their hypotheses, by developing and testing alternate hypotheses. In the case of drinking water improving performance on exams, alternate hypotheses will need to be tested to determine if this study's findings are supported.

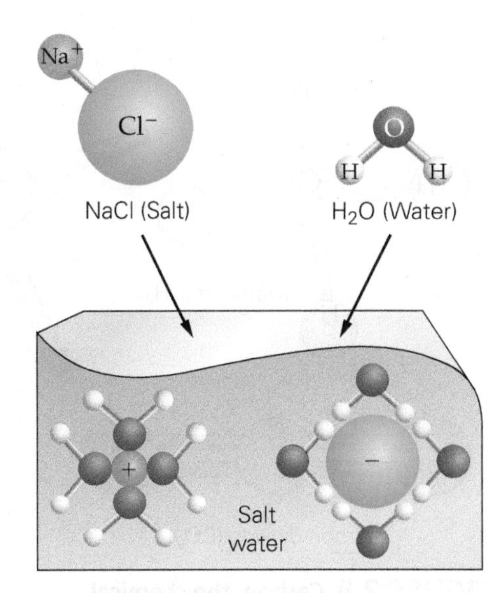

FIGURE 2.6 Water as a solvent. When salt is placed in water, the negatively charged regions of the water molecules surround the positively charged sodium ion, and the positively charged regions of the water molecules surround the negatively charged chlorine ion, breaking the bond holding sodium and chloride together and dissolving the salt.

pseudo- means false or fraudulent.

2.3 Chemistry for Biology Students

Many people find chemistry intimidating and shy away from trying to understand chemical principles. But even the simplest understanding of chemistry can help one sort out fact from fiction—and it's not that hard. Let's first discuss the basics of chemistry, and then use this understanding to consider whether the claims made about the rate of disappearance of ships and planes over the so-called Bermuda Triangle are based in science or pseudoscience.

The chemistry of biological systems is based on the element carbon. The branch of chemistry that is concerned with carbon-containing molecules is called **organic chemistry.** Carbon interacts with other elements to produce the more complex molecules that make up all living things. These interactions, or chemical bonds, follow a few simple patterns.

Chemical Bonds

Chemical bonds between atoms and molecules involve attractions that help stabilize structures. In general, this involves the sharing or transfer of electrons.

When electrons are transferred between positively and negatively charged ions they form **ionic bonds.** For example, an ionic bond forms when a sodium atom transfers an electron to a chloride atom to produce sodium chloride (NaCl). When the neutral sodium ion transfers its electron it becomes a charged sodium ion (Na^+), and the neutral chloride ion picking up the electron becomes negatively charged (Cl^-). These opposite charges are attracted to each other, forming the ionic bond.

Conversely, bonds involving the equal *sharing* of electrons form **covalent bonds.** A covalent bond forms when two atoms, each containing unpaired electrons, join together to form pairs of electrons. Covalent bonds are shown by a short line indicating a shared pair of electrons (**FIGURE 2.7a**). Double bonds involve two pairs of shared electrons. A double bond is symbolized by two parallel lines (**FIGURE 2.7b**).

Carbon is often involved in chemical bonding because of its ability to make bonds with up to four other elements. Like a Tinkertoy® connector, carbon has multiple sites for connections that allow carbon-containing molecules to take an almost infinite

(a) Methane

H
|
H—C—H
|
H

(b) Ethylene

H H
C=C
H H

FIGURE 2.7 Single and double bonds. (a) Covalent bonds are symbolized by a short line indicating a shared pair of electrons. (b) Double covalent bonds involve two pairs of shared electrons, symbolized by two parallel lines.

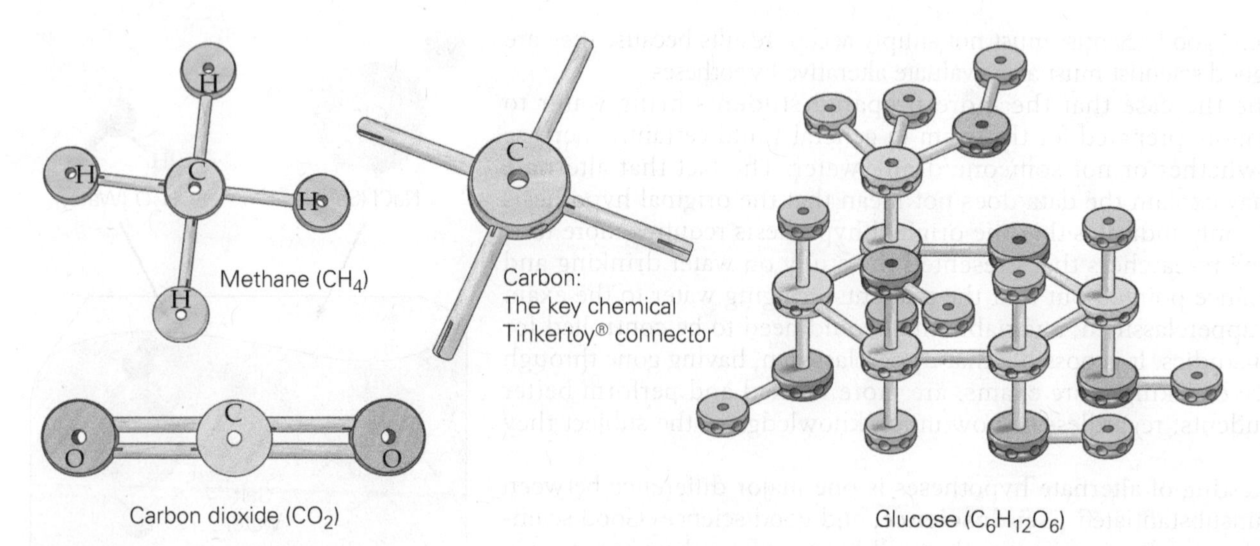

FIGURE 2.8 Carbon, the chemical Tinkertoy® connector. Because carbon forms four covalent bonds at a time, carbon-containing compounds can have diverse shapes.

variety of shapes (**FIGURE 2.8**). Because carbon has four unpaired electrons, it can form four single bonds, two double bonds, one double bond and two single bonds, and so on, depending on the number of electrons needed by the atom that is its partner.

Got It?

1. The element __________ is found in all living organisms.
2. Carbon can make bonds with many other elements to produce more complex __________.
3. The kind of chemical bond that occurs when atoms share electrons is the __________ bond.
4. The kind of chemical bond that involves a transfer of electrons is the __________ bond.
5. Carbon has __________ unpaired electrons.

The Bermuda Triangle Revisited

The Bermuda Triangle is a roughly triangular area off the coast of Florida (**FIGURE 2.9**). The fact that there is actually no evidence that ships and planes go missing from this area at a higher rate than other areas has done little to dampen fears of some prospective travelers.

During the early days of sea exploration, speculation about the cause of "disappearances" of ships focused on sea monsters, and then the idea of sea

VISUALIZE THIS

Although methane deposits do exist under the Bermuda Triangle, they also exist in the same or higher concentration at many other undersea locations. For the methane hypothesis to be supported, what must be true of other oceanic locations with similarly sized methane deposits?

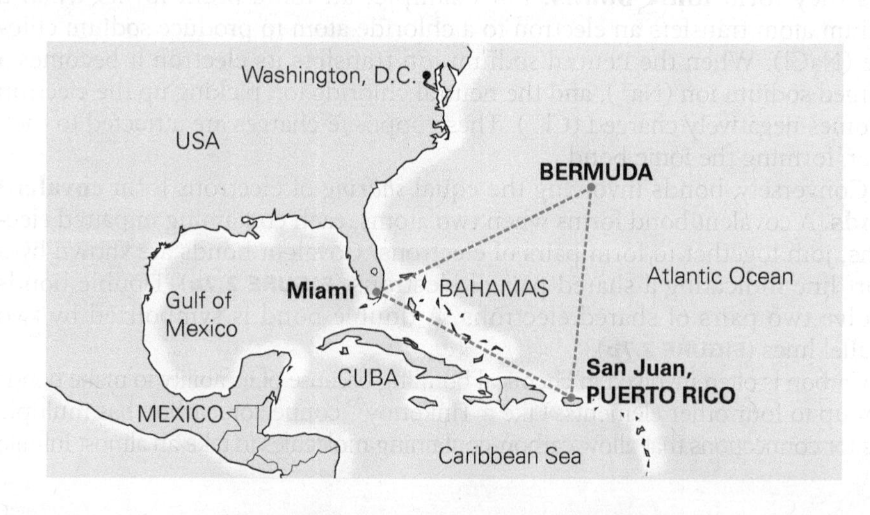

FIGURE 2.9 Underwater methane deposits. Current pseudoscience explanations for the "disappearance" of ships and planes over the Bermuda Triangle include an untested hypothesis about methane gas bubbles rising from the ocean floor to knock ships out of the water or igniting via lightning to cause planes to crash.

monsters was eventually replaced by notions about disruption in gravitational or magnetic fields. As understanding of gravity and magnetic fields increased, the allure of these explanations decreased—because neither gravity nor magnetic fields would explain wrecked ships or planes. One explanation of the alleged phenomenon of the Bermuda Triangle circulating now involves methane. This idea posits that methane deposits under the ocean in this region produce giant gas bubbles that rise up from the ocean floor and sink ships. If these same gas bubbles escape the ocean, rise into the atmosphere, and are struck by lightning, they could cause fires that bring down airplanes. Although there is no scientific evidence to support this hypothesis, its plausibility is enhanced, like that of all pseudoscience, by the inclusion of scientific principles—in this case chemistry.

2.4 Biological Macromolecules

Macromolecules are large organic molecules made of subunits. Those macromolecules present in living organisms include carbohydrates, proteins, lipids, and nucleic acids. Some macromolecules are synthesized by the body and some are obtained in the diet.

macro- means large.

Carbohydrates

Sugars, or **carbohydrates,** are the major source of energy for cellular processes and play important structural roles in cells. The simplest carbohydrates are composed of carbon, hydrogen, and oxygen in the ratio CH_2O. For example, glucose is symbolized as $6(CH_2O)$ or $C_6H_{12}O_6$. Glucose is a simple sugar, or **monosaccharide,** that consists of a single ring-shaped structure. Disaccharides are two rings joined together (**FIGURE 2.10a**). Sucrose is a disaccharide composed of glucose and fructose. Many plants contain sucrose, which is refined to produce table sugar. Fructose is a sugar found in fruits.

mono- means one.

Joining many individual **monomers** together produces **polymers.** Polymers of sugar monomers are called **polysaccharides** (**FIGURE 2.10b**). Plants use tough polysaccharides in their cell walls as a sort of structural skeleton. Likewise, the hard external skeletons of insects and spiders that make a crunching noise when

-mer means subunit.

poly- means many.

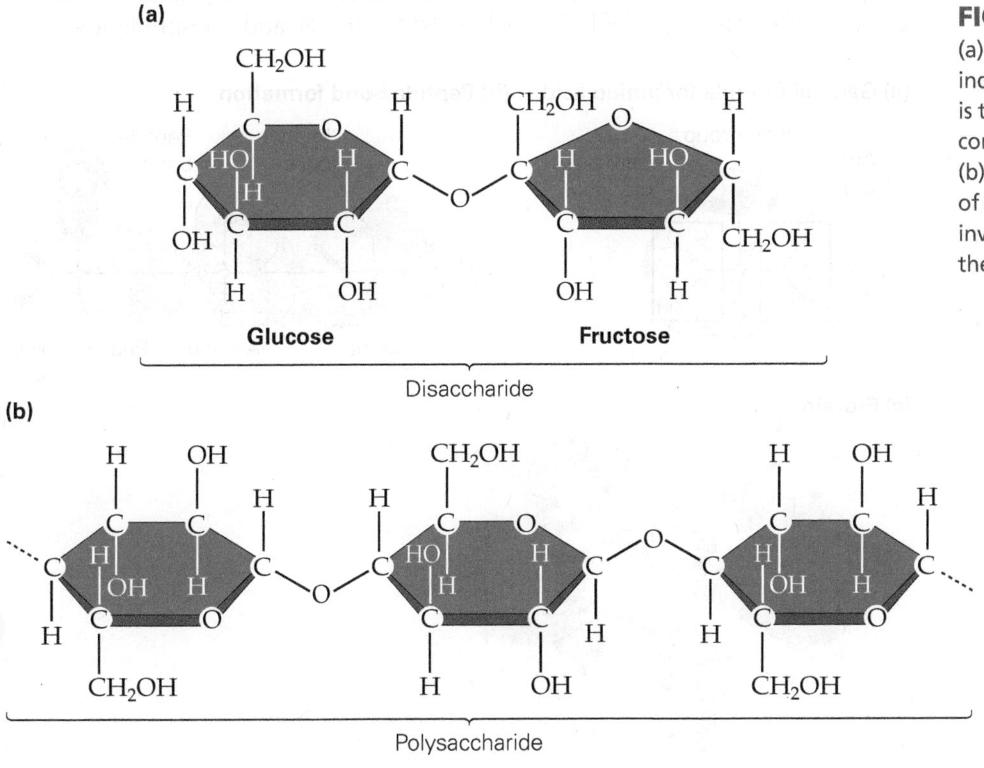

(a)

(b)

FIGURE 2.10 Carbohydrates. (a) Disaccharides are composed of two individual sugar monomers. Shown here is table sugar (sucrose), a disaccharide composed of glucose and fructose. (b) Polysaccharides are longer chains of sugars monomers. They are often involved in helping organisms maintain their structure.

damaged supply support to the insects and spiders. The cell wall that surrounds bacterial cells helps these tiny cells retain their structural integrity in a rapidly changing environment—and the cell wall, too, is rich in structural polysaccharides.

Proteins

Living organisms require **proteins** for a wide variety of processes. Proteins are important structural components of cells; they are integral to the makeup of cell membranes and they make up half the dry weight of most cells. Some cells, such as animal muscle cells, are primarily composed of proteins. Proteins called **enzymes** accelerate and help regulate all the chemical reactions that build up and break down molecules inside cells. The catalytic power of enzymes (their ability to help start chemical reactions) allows metabolism to occur under normal cellular conditions. Proteins can also serve as channels through which substances are brought into cells, and serve other important roles in cells.

Proteins are large molecules made of monomer subunits called **amino acids.** Like carbohydrates, amino acids are made of carbons, hydrogens, and oxygens; these form the amino acid's carboxyl group ($—COO^-$) at one end. On the other end of an amino acid is the nitrogen containing amino (NH_2^+) group. Various side groups are attached to the carbon between the amino and carboxyl group. Because the amino and carboxyl groups are the same in all amino acids, it is these side groups that give amino acids different chemical properties (**FIGURE 2.11a**).

Polymers of amino acids can be joined together in various sequences called polypeptides. **FIGURE 2.11b** shows three amino acids—valine, alanine, and phenylalanine—joined by covalent peptide bonds. Precisely folded polypeptides produce specific proteins in much the same manner that children can use differently shaped beads to produce a wide variety of structures (**FIGURE 2.11c**). Each amino acid side group has unique chemical properties, including being polar or nonpolar. Because each protein is composed of a particular sequence of amino acids, each protein has a unique shape and therefore specialized chemical properties.

Lipids

There are three different types of lipids, which are grouped together because they are partially or entirely hydrophobic organic molecules made primarily of hydrocarbons. The three types of lipids include fats, steroids, and phospholipids.

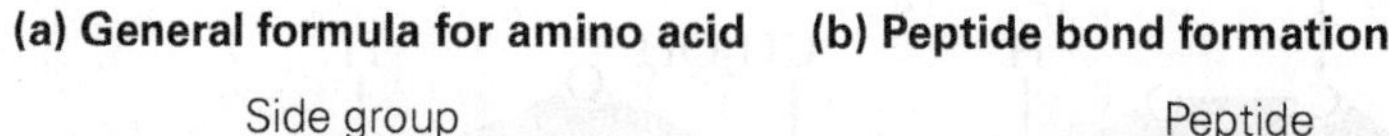

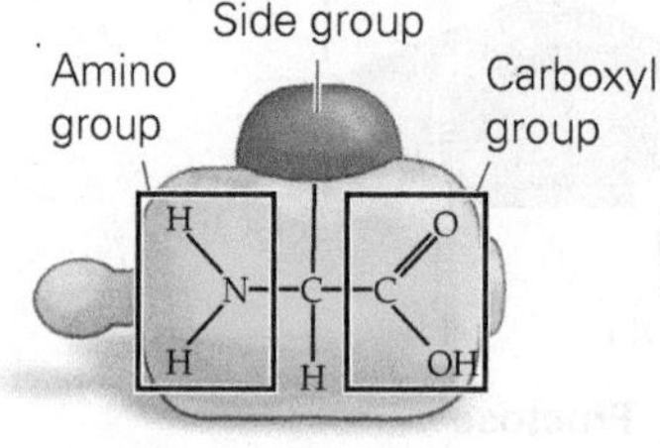

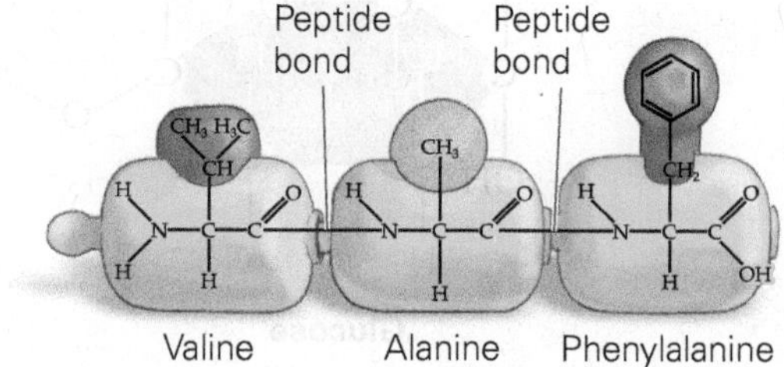

(c) Protein

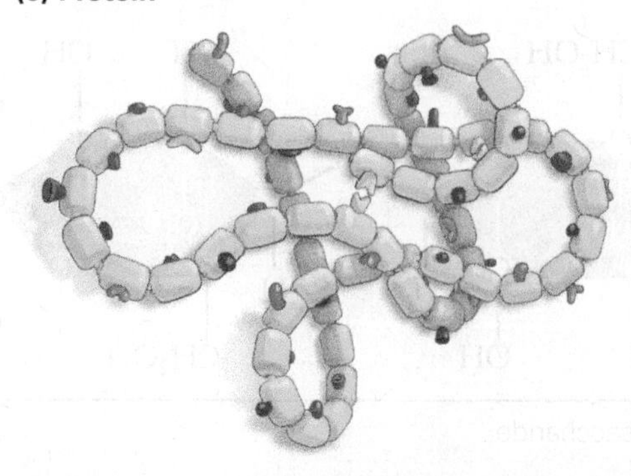

FIGURE 2.11 Amino acids, peptide bonds, and proteins. (a) All amino acids have the same backbone but different side groups. (b) Amino acids are joined together by covalent peptide bonds. Longer chains of these are called polypeptides. (c) Polypeptide chains fold upon themselves to produce proteins.

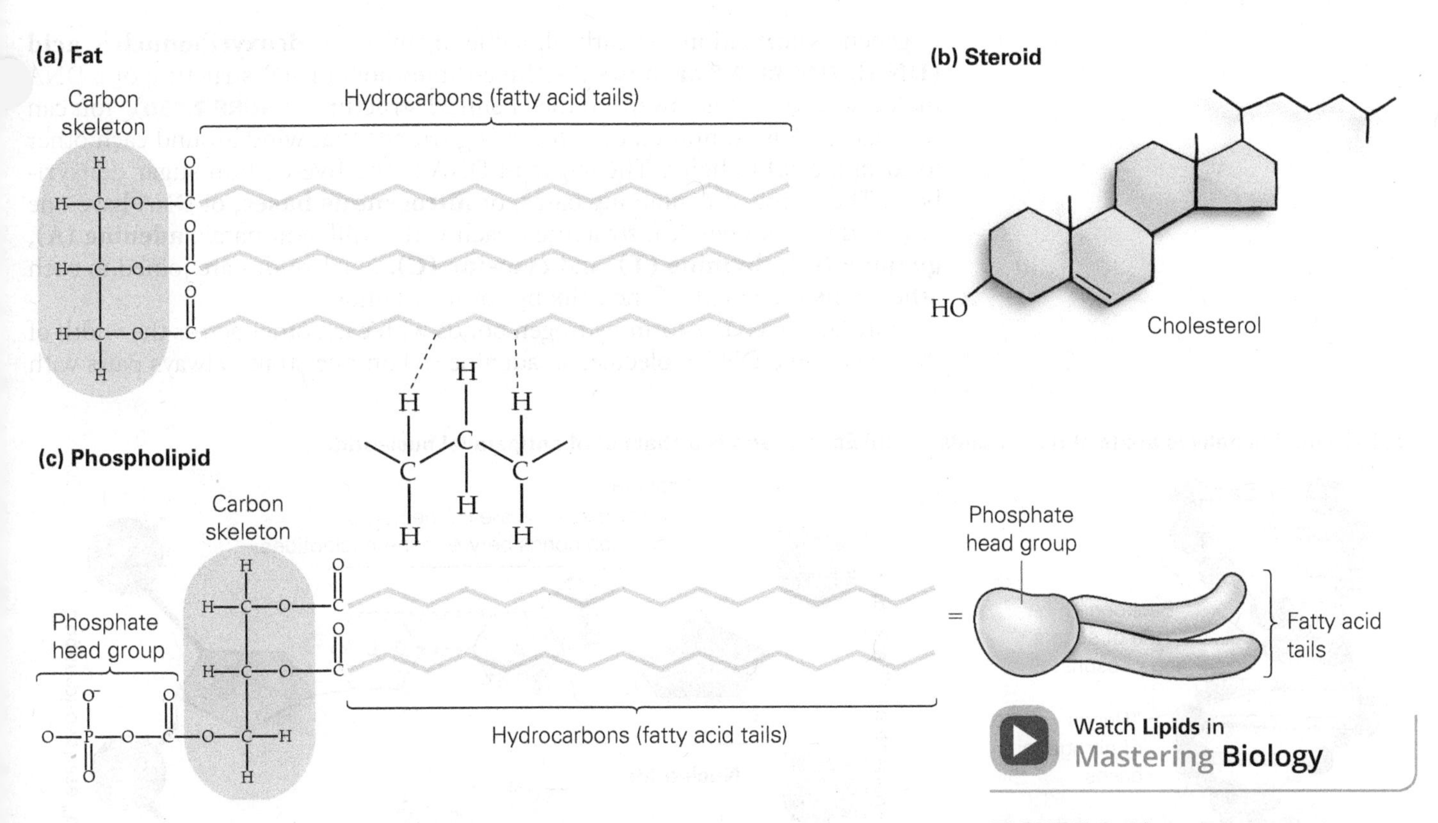

FIGURE 2.12 Three types of lipids. (a) Fats are composed of a three-carbon skeleton with three hydrocarbon-rich fatty acid tails attached. (b) Cholesterol is a steroid common in animal cell membranes. (c) Phospholipids are composed of a three-carbon skeleton with two fatty acids attached and one phosphate head group. The cartoon drawing to the right shows how phospholipids are often depicted.

Fat. The structure of a **fat** is that of a three-carbon skeleton molecule with three long hydrocarbon chains attached to each carbon in the skeleton (**FIGURE 2.12a**). **Hydrocarbons** are carbon- and hydrogen-rich. Like the hydrocarbons present in gasoline, these can be burned to produce energy. The long hydrocarbon chains are called **fatty acid** tails. Fats are hydrophobic and function in storing energy within living organisms.

Steroids. **Steroids** are composed of four fused carbon-containing rings. Cholesterol (**FIGURE 2.12b**) is one steroid that is probably familiar to you; its primary function in animal cells (plant cells typically do not contain cholesterol) is to help maintain the fluidity of membranes. Other steroids include the sex hormones testosterone, estrogen, and progesterone, which are produced by the sex organs and have effects throughout the body.

Phospholipids. **Phospholipids** are similar to fats in having a three-carbon skeleton molecule attached to hydrocarbons. However only two of the carbons are attached to fatty acid tails (not three, as you would find in a fat). The third bond in a phospholipid is to a phosphate head group. The phosphate head group consists of a phosphorous atom attached to four oxygen atoms and is hydrophilic. Thus, a phospholipid has a hydrophilic head and two hydrophobic tails (**FIGURE 2.12c**). Phospholipids often have an additional head group, attached to the phosphate, which also confers unique chemical properties on the individual phospholipid. Phospholipids are important constituents of the membranes that surround cells and that designate compartments within cells.

Nucleic Acids

Nucleic acids are composed of long strings of monomers called **nucleotides.** A nucleotide is made up of a sugar, a phosphate, and a nitrogen-containing base. There are two classes of nucleic acids in living organisms. **Ribonucleic acid (RNA)** plays a key role in helping cells synthesize proteins (and is discussed in detail in later chapters). The nucleic acid that serves as the primary storage of genetic information in nearly all living organisms is **deoxyribonucleic acid (DNA).** FIGURE 2.13a shows the three-dimensional (3D) structure of a DNA molecule and zooms inward to the chemical structure (FIGURE 2.13b). You can see that DNA is composed of two curving strands that wind around each other to form a double helix. The sugar in DNA is the five-carbon sugar deoxyribose. The nitrogen-containing bases, or **nitrogenous bases,** of DNA have one of four different chemical structures, each with a different name: **adenine (A), guanine (G), thymine (T),** and **cytosine (C).** Nucleotides are joined to each other along the length of the helix by covalent bonds.

Nitrogenous bases form hydrogen bonds with each other across the width of the helix. On a DNA molecule, an adenine (A) on one strand always pairs with

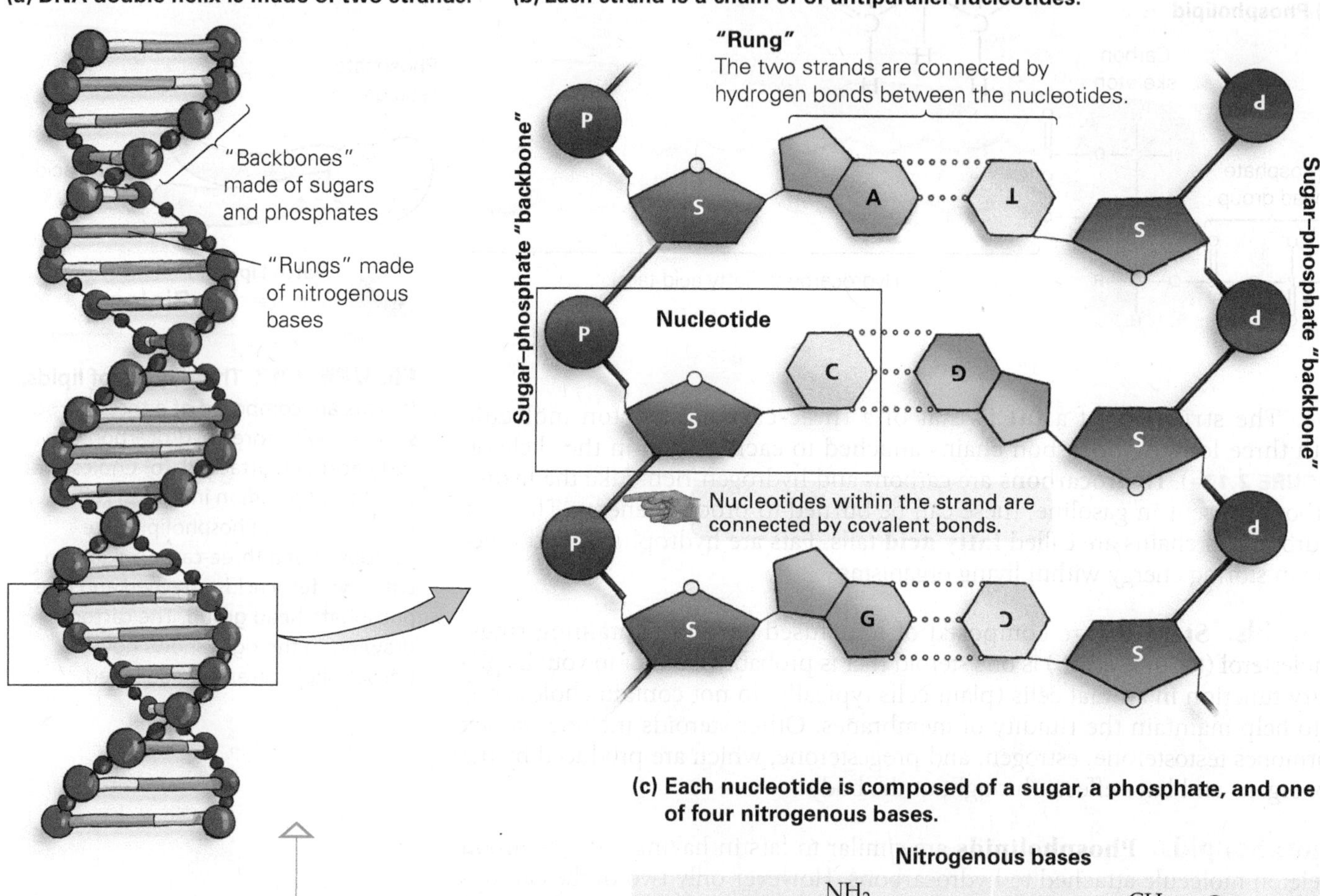
(a) DNA double helix is made of two strands.
"Backbones" made of sugars and phosphates
"Rungs" made of nitrogenous bases
(b) Each strand is a chain of of antiparallel nucleotides.
"Rung"
The two strands are connected by hydrogen bonds between the nucleotides.
Sugar–phosphate "backbone"
Nucleotide
Nucleotides within the strand are connected by covalent bonds.
Sugar–phosphate "backbone"
P
S
A
T
C
G
(c) Each nucleotide is composed of a sugar, a phosphate, and one of four nitrogenous bases.

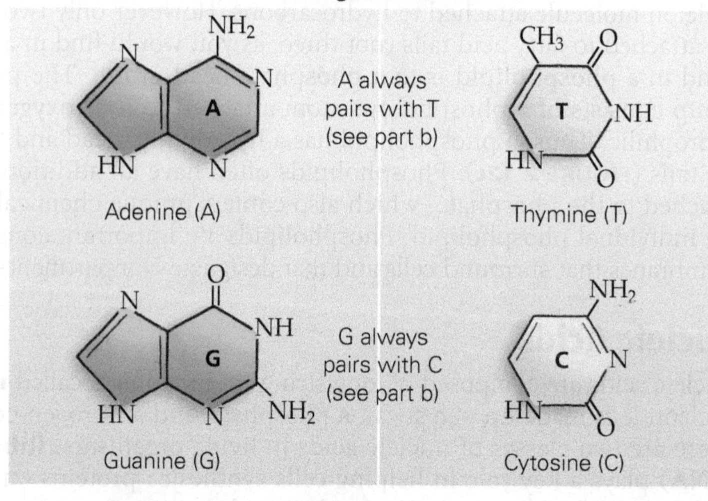
Nitrogenous bases
A always pairs with T (see part b)
Adenine (A)
Thymine (T)
G always pairs with C (see part b)
Guanine (G)
Cytosine (C)

VISUALIZE **THIS**

Look at part (a) of this figure. What do the large and small spheres along the DNA backbone represent?

FIGURE 2.13 DNA structure. (a) DNA is a double-helical structure composed of nucleotides. (b) Each strand of the helix is composed of repeating units of sugars and phosphates, making the sugar-phosphate backbone, and of nitrogenous bases across the width of the helix. (c) Each nucleotide is composed of a phosphate, a sugar, and a nitrogenous base.

a thymine (T) on the opposite strand (**FIGURE 2.13c**). Likewise, guanine (G) always pairs with cytosine (C). The term **complementary** is used to describe these pairings. For example, A is complementary to T, and C is complementary to G. Therefore, the order of nucleotides on one strand of the DNA helix predicts the order of nucleotides on the other strand. Thus, if one strand of the DNA molecule is composed of nucleotides AACGATCCG, then we know that the order of nucleotides on the other strand is TTGCTAGGC.

Each strand of the helix thus consists of a series of sugars and phosphates alternating along the length of the helix, the **sugar-phosphate backbone.** The strands of the helix align so that the nucleotides face "up" on one side of the helix and "down" on the other side of the helix. For this reason, the two strands of the helix are said to be antiparallel.

The overall structure of a DNA molecule can be likened to a rope ladder that is twisted, with the sides of the ladder composed of sugars and phosphates (the sugar-phosphate backbone) and the rungs of the ladder composed of the nitrogenous-base sequences A, C, G, and T.

Got It?

1. Proteins are composed of subunits called __________ __________.
2. Carbohydrates are composed of carbon, __________, and oxygen.
3. Fats are lipids that have three hydrocarbon rich __________ __________ tails.
4. Phospholipids are lipids that have __________ fatty acid tails and a phosphate group attached to the carbon skeleton.
5. __________ are lipids that are composed of four fused rings.

Dietary Macromolecules and Behavior

Most people accept as fact that hyperactivity in children can be induced by ingestion of sugary drinks or snacks, when there is not much scientific evidence to support this notion. There is, in fact, more scientific evidence to support the notion that parents expect their children to become hyperactive from ingesting sugar.

Similarly, have you ever heard an uncle claim that there is something in turkey that puts him to sleep, as he makes his way to the couch after Thanksgiving dinner? It is commonly believed that the presence of the amino acid tryptophan in the protein-rich meat of turkey causes drowsiness. This myth may have gotten some traction when scientists showed that sleep and mood can be affected by tryptophan. However, one look at the data (**FIGURE 2.14**) showing levels of tryptophan in other protein-rich foods should put this myth to bed! An alternative hypothesis that seems to be better supported is that eating a large amount of food, which certainly happens on Thanksgiving, leads to drowsiness.

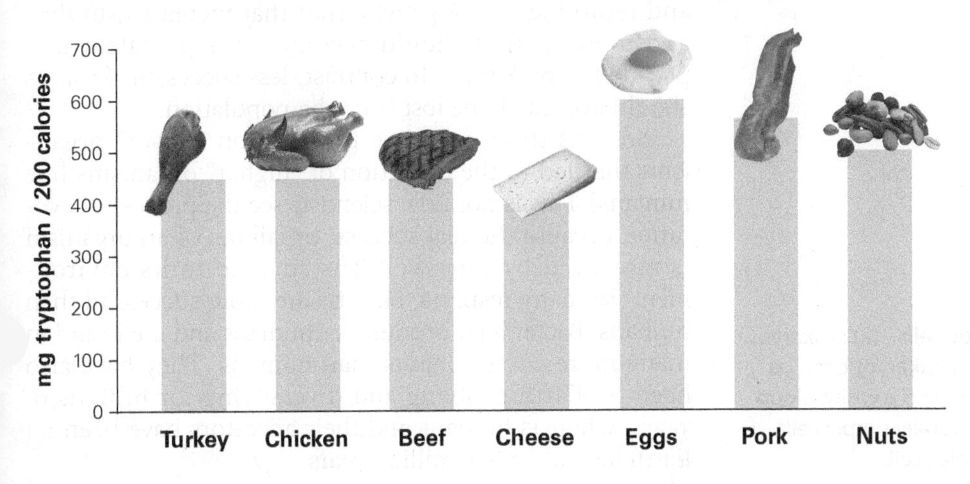

FIGURE 2.14 Tryptophan concentration in food. This graph shows that turkey is no more rich in the amino acid tryptophan than many other common foods.

2.5 An Introduction to Evolutionary Theory

Are humans evolving into higher beings with powers similar to those found in the X-men? The **theory of evolution** explains how a single common ancestor present on Earth 4 billion years ago could give rise to the more than 10 million different kinds of organisms found on Earth today. (Evolutionary theory will be covered in much greater detail in later chapters.) For now, we will try to understand whether the organisms on Earth today represent a series of steps in the ladder toward humanness. Does this ladder have a few more rungs that humans can climb, maybe gaining an extra gene or two that will provide superpowers similar to those found in the X-men?

To investigate this idea, let's see what happened to some of the earliest organisms to inhabit Earth—the tiny, single-celled bacteria (**FIGURE 2.15a**). Bacteria are classified as **prokaryotic** because they do not have a nucleus, a separate membrane-bound compartment to hold the genetic material. They also do not contain any membrane-bound internal compartments. They do, however, have a **cell wall** that helps them maintain their shape (**FIGURE 2.15b**) and protein-synthesizing structures called **ribosomes.**

The ancestor of bacteria also gave rise to a structurally more complex group of organisms called eukaryotes. **Eukaryotic** cells have a nucleus surrounding the genetic material (**FIGURE 2.15c**). In addition, eukaryotic cells contain membrane-bounded subcellular structures called organelles. Organelles will be covered in more detail later. For now, it is enough to understand that organelles work together to perform specific jobs inside cells that allow the cell to function properly. Eukaryotic organisms include single-celled amoebas and yeast as well as multicellular plants, fungi, and animals.

Although there are differences between prokaryotic and eukaryotic cells, there are also many similarities. In fact, all living organisms are made of cells that contain the same kinds of macromolecules and other structural features, most likely because they once shared a common ancestor.

The divergence and differences among groups of organisms arose as a result of **natural selection,** a process of gradual changes in the characteristics of populations over time. Natural selection was a major process in the diversification of life from a single ancestor because individual organisms vary from one another and because some of these variations increase the chances of survival and reproduction. A genetic trait that increases survival and reproduction should become more prevalent in a population over time. In contrast, less successful variants should eventually be lost from the population.

So, was there selective pressure on "lower" organisms that led to the evolution of "higher" organisms like humans? This is not how scientists see the process of evolution because the real success, or failure, of an organism is measured by how well it is adapted to its environment. In many respects, bacteria are more successful than humans. Bacteria far outnumber humans and are found in many more environments than humans. They have also been on Earth evolving and diversifying for billions of years, whereas humans and their ancestors have been on Earth for only a few million years.

(a) Different sizes: prokaryotic (red) vs. eukaryotic (white) cells

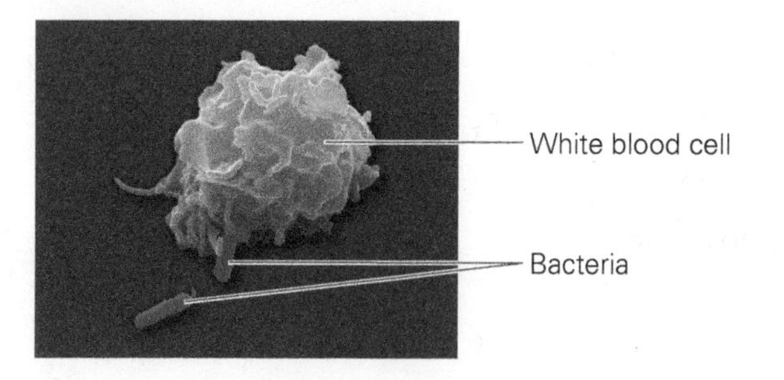

(b) Prokaryotic cell features

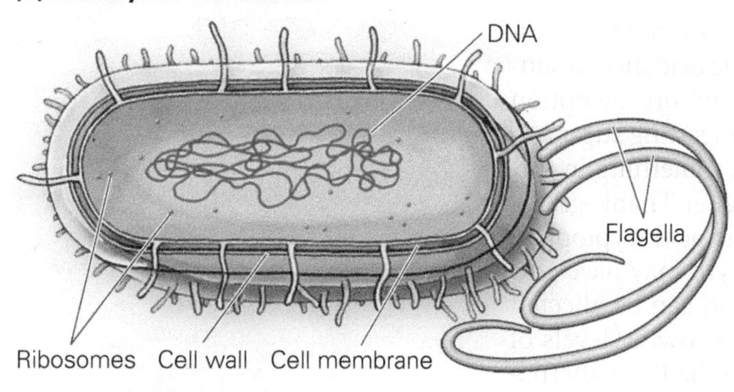

(c) Eukaryotic cell features

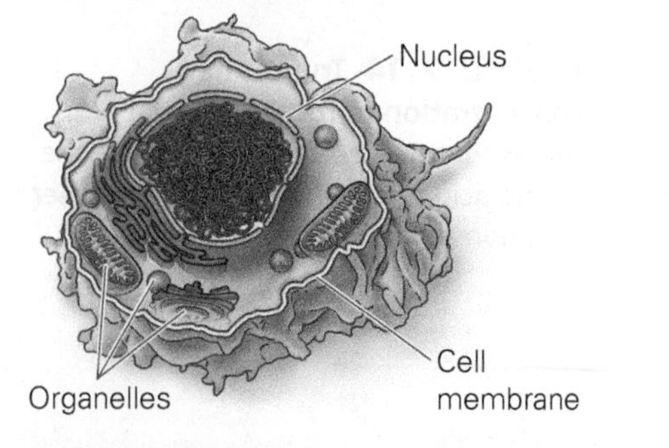

FIGURE 2.15 Prokaryotic and eukaryotic cells. (a) Prokaryotic cells are typically about 1/10 the diameter of a eukaryotic cell, as evidenced by the size of the two bacterial cells and a white blood cell shown here. (b) Prokaryotic cells are structurally simple cells. (c) Eukaryotic cells are more structurally complex cells.

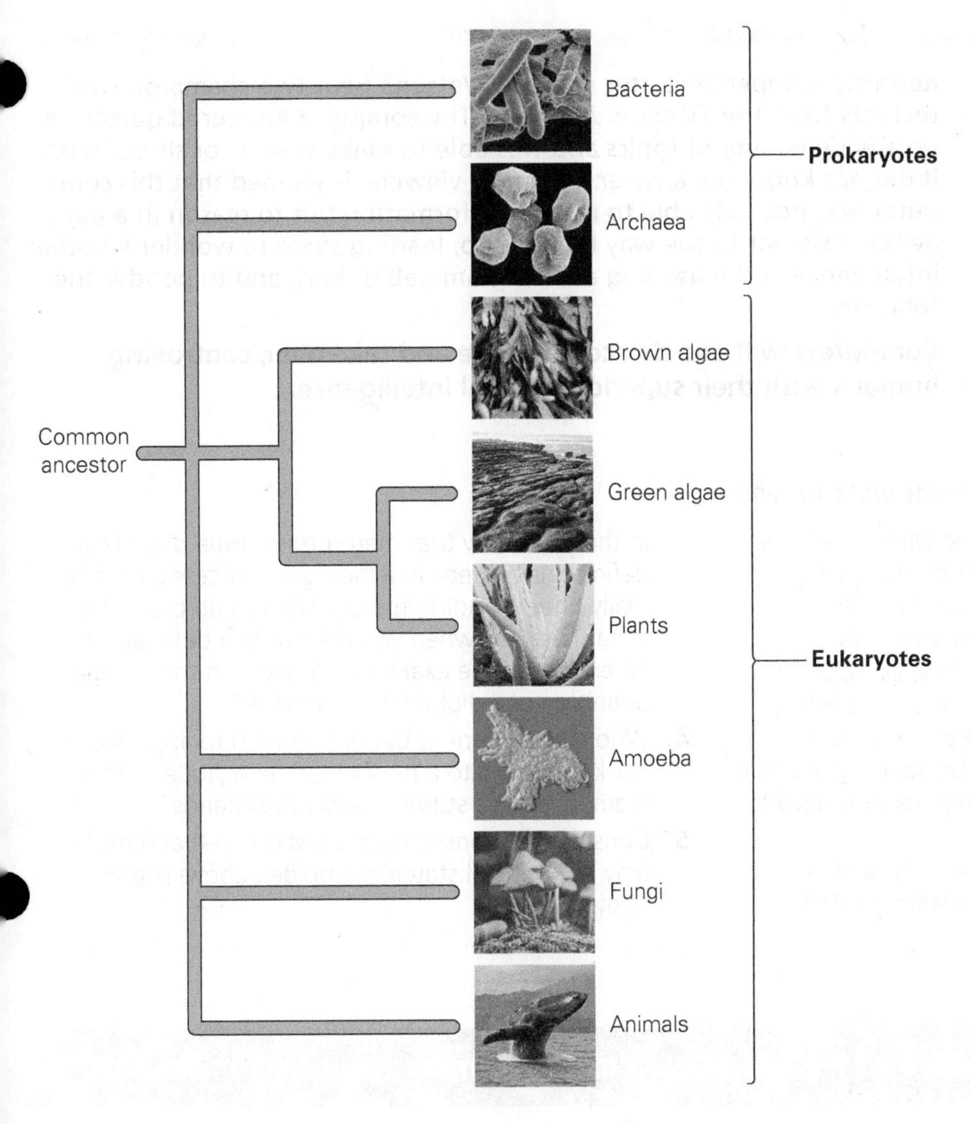

FIGURE 2.16 A common ancestor on the tree of life. All living organisms share basic characteristics and can be arranged into a branching tree of life based on more specific similarities. In this illustration, many groups are omitted for simplicity.

We have seen that evolution is not progressive—in other words, evolution is not a series of discrete steps toward increasing complexity. Because the process is not linear—representations of the tree of life are usually shown as horizontal and branched (**FIGURE 2.16**) instead of as a vertical ladder climbing upward. In this sense, bacteria that are well adapted to their environment are as highly evolved as humans that are equally well adapted to their environment. In fact, if a zombie apocalypse occurs and humans are wiped out and bacteria remain, bacteria would prove to be better adapted to their environment than humans.

Got It?

1. The earliest cells on Earth were single celled, structurally simple ________.
2. The more complex ________ cells arose from simpler cells.
3. All organisms on Earth arose from a single common ________.
4. One way evolution can occur is by ________ selection.
5. Evolutionary pressures select for organisms that are the best fit in their particular ________.

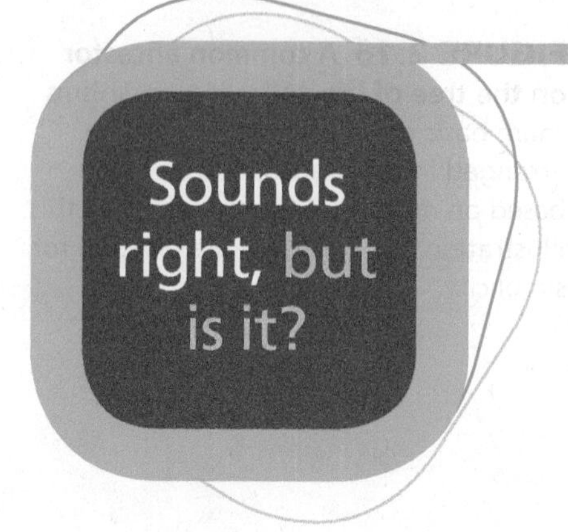
Sounds right, but is it?

Recently, a supercomputer named "Watson" beat two champion contestants from the TV show *Jeopardy*. The computer answered questions on a wide variety of topics and was able to make wagers or sit out when it did not know the answer. For many viewers, it seemed that this computer was not only able to retrieve information but to reason in a manner very similar to the way humans do, leading some to wonder whether intelligence and reasoning are programmable skills, and to ponder the following:

Computers will one day come to life and take over, controlling humans with their superior artificial intelligence.

Sounds right, but it isn't.

Answer the following questions to understand why.

1. Computers can retrieve data and calculate probabilities based on preprogrammed information. Using the characteristics of life provided in Section 2.1, list features of living organisms that computers lack.
2. Although chess-playing computers can respond to moves made by a human partner, to be autonomous, computers will need to be able to respond to many more stimuli. Do you think it is possible for programmers to predict all the different stimuli computers will need to be able to respond to in the future?
3. A key feature of human intelligence is the ability to learn. Retrieving stored data and learning differ in the same way that memorizing some biological definitions or steps in a biological process differs from really understanding the definitions and processes. What happens when you memorize a definition for an exam and the exam asks a question that uses a definition in a slightly different way?
4. Who do you think is better prepared to apply his or her knowledge to a new situation, a student who memorizes or a student who understands?
5. Consider your answers to questions 1–4 and explain why the original statement bolded above sounds right, but isn't.

THE BIG QUESTION

Does balanced reporting help us draw more accurate conclusions?

The idea that both sides of a contentious issue should be given equal coverage in news stories seems, at first blush, like a good one. This is called balanced coverage. Does balanced coverage lead us to the truth?

What should I know?

What follows are some smaller questions that need to be resolved to answer the Big Question. Please a checkmark next to the questions that science can answer.

Smaller Questions	Can Science Answer?
When there is a preponderance of evidence supporting one side of the issue and not much evidence on the other side, do people exposed to balanced coverage think that there is equal merit to both positions?	
Does the public benefit from exposure to all opinions, even those that are not evidence-based?	
Does presenting all sides of an issue help people make the best choices?	
Is giving more time to one side of an argument always biased coverage?	
Does repetition of an opinion lead listeners and readers to accept the opinion?	

What does the science say?

Let's examine what the data say about this smaller question:

Does repetition of an opinion lead listeners and readers to accept the opinion?

Researchers presented randomly selected college students with a made-up scenario. These students were told that a nearby state was deciding whether to set aside land to preserve open space for members of the public to enjoy. The students were also told that a focus group had met to discuss this issue, and then the students were given hypothetical statements made by participants of the focus group.

Three different groups of students read these statements in support of setting aside open space as follows:

Student group A read one supportive statement made by one participant.

Student group B read three different supportive statements made by one participant.

Student group C read three different supportive statements made by three different participants.

The students were then asked to rate from 1 (strongly disagree) to 7 (strongly agree) their level of support for setting aside open space for public use. The data from this experiment is shown below. There were significant differences between each group.

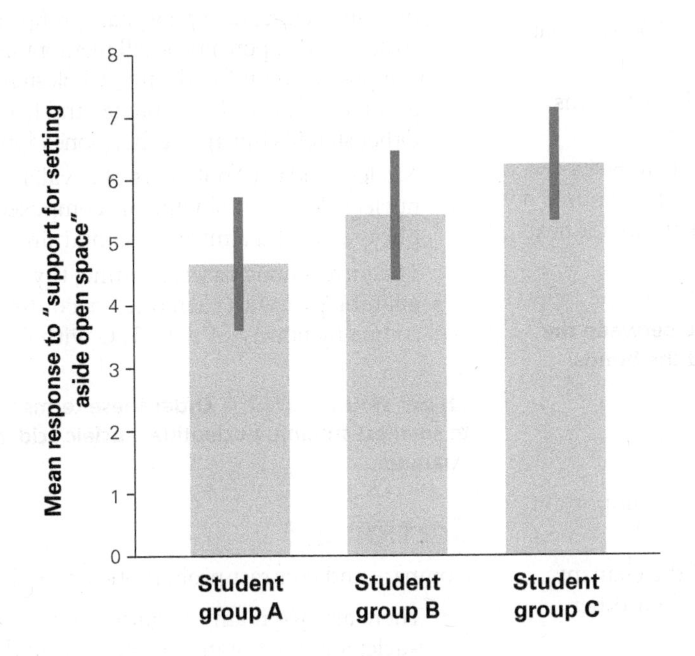

1. Describe the results. Which students were the most likely to agree with statements made by focus group participants? Which students were the least likely to agree?
2. Given these data, do you think the smaller question is answered? If not, propose another study that would help answer this question.
3. Does this information help you answer this Big Question? What else do you need to consider?

Data source: K. Weaver, S. Garcia, N. Schwarz, and D. Miller, "Inferring the Popularity of an Opinion from Its Familiarity: A Repetitive Voice Can Sound Like a Chorus," *Journal of Personality and Social Psychology* 92, no. 5 (2007): 821–833.

Chapter Review

Mastering Biology

Go to Mastering Biology to access **eText 2.0, Dynamic Study Modules,** and the **Study Area,** where you'll find practice quizzes, BioFlix™ animations, MP3 tutor sessions, current events, and more.

SUMMARY

Section 2.1

Describe the properties associated with living organisms.

- Living organisms contain a common set of biological molecules and are composed of cells. They are able to grow, metabolize substances, reproduce, and respond to external stimuli. In addition, living organisms can maintain homeostasis and evolve (pp. 31–32).

SHOW YOU KNOW 2.1 Humans are able to regulate body temperature, acid-base balance, fluid volume, and glucose and calcium concentrations within limited ranges. What property of living organisms are these examples of?

Section 2.2

List the components of water and some of the properties that make it important in living organisms.

- Water consists of one oxygen and two hydrogen atoms (pp. 32–33).
- Water is a good solvent in part because of its polarity and ability to form hydrogen bonds. Hydrogen bonding in water facilitates chemical reactions and allows for heat absorption (pp. 33–34).

SHOW YOU KNOW 2.2 Describe the difference between the chemical bonds within one water molecule and the bonds between adjacent water molecules.

Section 2.3

Summarize the chemical reasons that carbon is an important component of living organisms.

- Life on Earth is based on the chemistry of the element carbon, which can make bonds with up to four other elements (pp. 35–36).

Compare and contrast hydrogen, ionic, and covalent bonds.

- Hydrogen bonds are weak attractions between hydrogen atoms and oxygen atoms in adjacent molecules (p. 34).
- Ionic bonds form between positively and negatively charged ions (p. 35).
- Covalent bonds form when atoms share electrons. These tend to be strong bonds (p. 35).

SHOW YOU KNOW 2.3 What does not make chemical sense about the hypothetical molecule H_2CH_3?

Section 2.4

Describe the structure of carbohydrates, proteins, lipids, and nucleic acids and the roles these macromolecules play in cells.

- Carbohydrates function in energy storage and play structural roles. They can be single-unit monosaccharides or multiple-unit polysaccharides with sugar monomers joined by covalent bonds (pp. 37–38).
- Proteins play structural, enzymatic, and transport roles in cells. They are composed of amino acid monomers joined in different orders (p. 38).
- Lipids are partially or entirely hydrophobic and come in three different forms. Fats have a three-carbon skeleton, covalently bonded to three fatty acids. Fats store energy. Phospholipids are composed of a three-carbon skeleton, two fatty acids, and a phosphate group. They are important structural components of cell membranes. Steroids are composed of four fused rings. Cholesterol is a steroid found in some animal cell membranes and helps maintain fluidity. Other steroids function as hormones (pp. 38–39).
- Nucleic acids are polymers of covalently bonded nucleotides, each of which is composed of a sugar, a phosphate, and a nitrogen-containing base (pp. 39–40).
- The nitrogenous bases, adenine, thymine, cytosine, and guanine pair with each other according to the rules of complementarity (A with T; G with C) (pp. 40–41).

SHOW YOU KNOW 2.4 Order these terms from entire molecule to smallest subunit: nucleotide, nucleic acid, nitrogenous base, guanine.

Section 2.5

Compare and contrast prokaryotic and eukaryotic cells.

- There are two main categories of cells: Those with nuclei and membrane-bound subcellular organelles are eukaryotes; those lacking a nucleus and membrane-bound organelles are prokaryotes (pp. 42–43).

Provide a general summary of the theory of evolution.

- All living organisms are composed of cells that share the same organic chemistry and basic cellular features. These similarities provide support for the theory of evolution, which holds that all life on Earth derives from a single common ancestor (pp. 42–43).

SHOW YOU KNOW 2.5 Ribosomes are one of the few organelles found in prokaryotic cells. What structural component found in many eukaryotic organelles do ribosomes lack?

ROOTS TO REMEMBER

The following roots of words come mainly from Latin and Greek and will help you to decipher terms:

homeo- means like or similar. Chapter term: *homeostasis*

hydro- means water. Chapter terms: *hydrophilic, hydrophobic*

macro- means large. Chapter term: *macromolecule*

-mer means subunit. Chapter terms: *polymer, monomer*

mono- means one. Chapter terms: *monosaccharide, monomer*

-philic means to love. Chapter term: *hydrophilic*

-phobic means to fear. Chapter term: *hydrophobic*

poly- means many. Chapter term: *polymer*

pseudo- means false or fraudulent. Chapter term: *pseudoscience*

LEARNING THE BASICS

1. List the four biological molecules commonly found in living organisms.

2. List the structural features in a prokaryotic cell.

3. Add labels to the figure that follows, which illustrates the subatomic particles associated with a carbon atom.

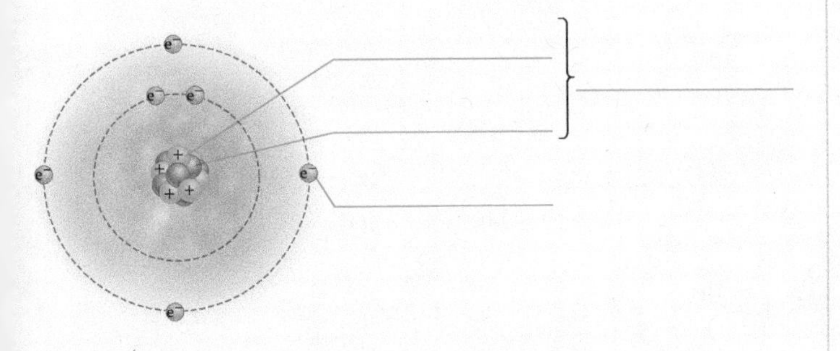

4. Water __________.

A. is a good solute; **B.** facilitates chemical reactions; **C.** serves as an enzyme; **D.** makes strong covalent bonds with other molecules; **E.** consists of two oxygen and one hydrogen atoms

5. Electrons __________.

A. are negatively charged; **B.** along with neutrons make up the nucleus; **C.** are attracted to the negatively charged nucleus; **D.** are not involved in ionic bonds; **E.** all of the above are true

6. Which of the following terms is least like the others?

A. monosaccharide; **B.** phospholipid; **C.** fat; **D.** steroid; **E.** lipid

7. Different proteins are composed of different sequences of __________.

A. sugars; **B.** lipids; **C.** fats; **D.** amino acids; **E.** carbohydrates

8. Proteins may function as __________.

A. genetic material; **B.** cholesterol molecules; **C.** fat reserves; **D.** enzymes; **E.** all of the above

9. A fat molecule consists of __________.

A. carbohydrates and proteins; **B.** complex carbohydrates only; **C.** saturated oxygen atoms; **D.** a carbon skeleton and fatty acids

10. Eukaryotic cells differ from prokaryotic cells in that only eukaryotic cells __________.

A. contain DNA; **B.** have a plasma membrane; **C.** are considered to be alive; **D.** have a nucleus; **E.** are able to evolve

11. Which of the following lists the chemical bonds from weakest to strongest?

A. hydrogen, covalent, ionic; **B.** covalent, ionic, hydrogen; **C.** ionic, covalent, hydrogen; **D.** covalent, hydrogen, ionic; **E.** hydrogen, ionic, covalent

12. Which of the following is not consistent with evolutionary theory?

A. All living organisms share a common ancestor; **B.** The environment affects which organism survives to reproduce; **C.** Natural selection always favors the same traits, regardless of environment; **D.** Humans are not necessarily the best adapted organisms.

ANALYZING AND APPLYING THE BASICS

1. Consider a virus composed of a protein coat surrounding a small segment of genetic material (either DNA or RNA). Viruses cannot reproduce without taking over the genetic "machinery" of their host cell. Based on this description and biologists' definition of life, should a virus be considered a living organism?

2. It was a commonly held notion that humans should consume eight glasses of water a day until recent scientific evidence called this practice into question. What chemical properties of water do you think contributed to the acceptance of this hypothesis?

3. Carbon, oxygen, hydrogen, and nitrogen are common elements found in living organisms. Which two of the four types of macromolecules contain all four of these elements?

GO FIND OUT

1. List some alternate explanations that should be explored before accepting that drinking sugary drinks causes hyperactivity in children.

2. Do some web-based research using scientifically sound resources on sugary drinks and hyperactivity in children. Does it appear that there is a scientific consensus on whether sugary drinks cause hyperactivity in children?

MAKE THE CONNECTION

The science that you learned in this chapter has helped you better understand the real-world example used throughout this discussion. Draw a line from the statement on the left to the science that supports it on the right.

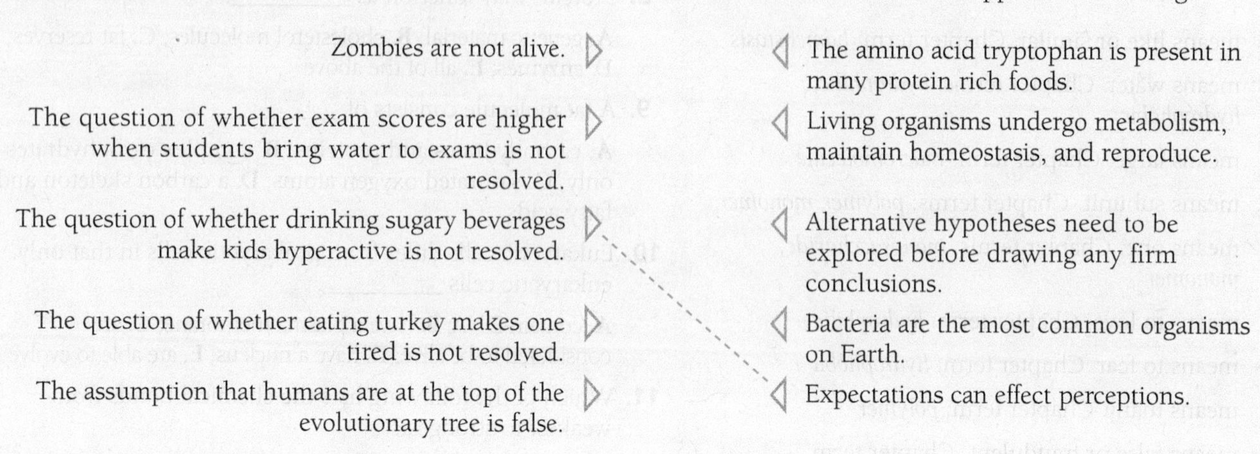

Zombies are not alive.
The question of whether exam scores are higher when students bring water to exams is not resolved.
The question of whether drinking sugary beverages make kids hyperactive is not resolved.
The question of whether eating turkey makes one tired is not resolved.
The assumption that humans are at the top of the evolutionary tree is false.
The amino acid tryptophan is present in many protein rich foods.
Living organisms undergo metabolism, maintain homeostasis, and reproduce.
Alternative hypotheses need to be explored before drawing any firm conclusions.
Bacteria are the most common organisms on Earth.
Expectations can effect perceptions.

Answers to **Got It?, Visualize This, Working with Data, Sounds Right, But Is It?, Show You Know,** and **Chapter Review** questions can be found in the **Answers** section at the back of the book.

3 Is It Possible to Supplement Your Way to Better Performance and Health?

Do sports drinks enhance athletic performance?

Do nutritional supplements enhance academic performance or health?

Or is it more healthful to eat whole foods?

Nutrients and Membrane Transport

Gingko to improve your memory, kava to reduce stress, ginseng to boost energy, and melatonin to help you sleep. Sounds like a recipe for success for a busy student. For good measure, chase those supplements down with some coconut water to slow aging and prevent cancer. You may have heard claims about the health benefits of nutritional supplements like vitamins, minerals, herbs, yeast, and even enzymes. If these are truly good for you, why not replace some of the food you eat with products that have a longer shelf life than most foods? Instead of going to the grocery store every weekend, you could stock your pantry with energy drinks, vitamin-enriched waters, protein powders, nutrition bars, vitamins, and minerals. These can be bought in bulk and don't rot like fruits and vegetables. But are they as good for you as food?

Is it possible to supplement your way to enhanced academic performance or better health? It seems that most Americans think so—we spend around $6 billion a year on these items and more than two-thirds of us are taking at least one such supplement. Let's investigate whether these products are doing what we hope they are.

3.1 Nutrients

The food and drink that we ingest provide building-block molecules that can be broken down and used as raw materials for growth, maintenance, repair, and as a source of energy. Another name for the substances in food that provide structural materials or energy is **nutrients.**

Macronutrients

Nutrients that are required in large amounts are called **macronutrients.** These include water, carbohydrates, proteins, and fats.

Water and Nutrition. Most animals can survive for several weeks with no nutrition other than water. However, survival without water is limited to just a few days. Besides helping the body disperse other nutrients, water helps dissolve and eliminate the waste products of digestion.

A decrease below the body's required water level, called **dehydration,** can lead to muscle cramps, fatigue, headaches, dizziness, nausea, confusion, and increased heart rate. Severe dehydration can result in hallucinations, heat stroke, and death.

Sweating is evaporation of water from the skin. This helps maintain body temperature. When water is low and sweating decreases, the body temperature can rise to a harmful level.

Every day, humans lose about 3 L of water as sweat, in urine, and in feces. A typical adult obtains about 1.5 L of water per day from food consumption, leaving a deficit of about 1.5 L that must be replaced.

Although it works quite well to replace this water with tap water, many people drink bottled water, in part because of concerns about the quality of tap water. However, the U.S. Food and Drug Administration (FDA), the government agency that sets the standards for bottled water and is responsible for ensuring food safety in general, uses the same standards applied by the Environmental Protection Agency (EPA) to ensure healthful tap water. In other words, water from both sources should be equally clean. In fact, nearly 40% of bottled waters actually contain water from municipal tap water.

Many people also choose to hydrate with sports drinks. But studies have shown that hydrating with water is better for athletes than hydrating with costly sports drinks that provide unneeded calories and additives of unproven benefit.

Bottled waters and sports drinks are also chosen for convenience and portability, but this comes at a cost to the environment. The billions of bottles used each year require 1.5 million barrels of oil to produce. Although the bottles are recyclable, 86% go to landfills each year. In addition, the energy required to transport water far exceeds the energy required to purify an equivalent amount of tap water.

Besides obtaining a healthy dose of water every day, people must consume foods that contain carbohydrates, proteins, and fats. (In Chapter 2, we explored the structure of these macromolecules, and now we focus on how they function in the body.)

Carbohydrates as Nutrients. Foods such as bread, cereal, rice, and pasta, as well as fruits and vegetables, are rich in macronutrients called carbohydrates. Carbohydrates are the major source of energy for cells and can be composed of single-unit monosaccharides or long chains of polysaccharides.

The single-unit simple sugars are digested and enter the bloodstream quickly after ingestion. Sugars found in milk, juice, honey, and most refined foods are simple sugars. Sugars that are composed of many subunits and arranged in branching chains are called **complex carbohydrates.** Complex carbohydrates are found in fruits, vegetables, breads, legumes, and pasta.

The body digests complex carbohydrates more slowly than it does simpler sugars because complex carbohydrates have more chemical bonds that must be broken during digestion. Endurance athletes will load up on complex carbohydrates for several days before a race to increase the amount of stored energy that will be available during competition. Ideally, these carbohydrates will be from food, rather than from dietary supplements (**FIGURE 3.1**).

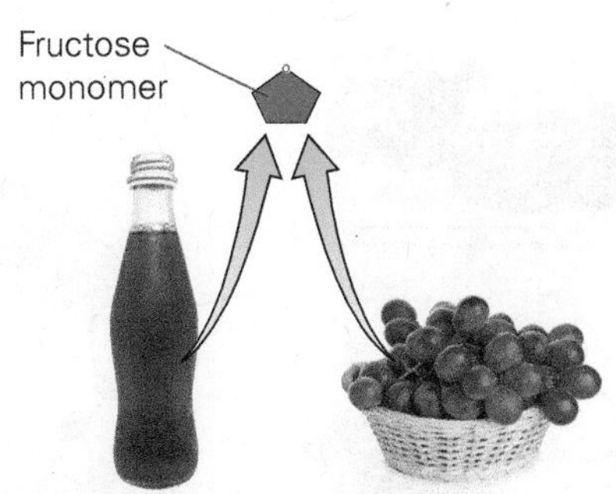

FIGURE 3.1 Carbohydrates in processed and whole foods. Sugary sports drinks can be rich in rapidly digested high fructose corn syrup. Fructose is also found in fruit, along with complex carbohydrates and fiber, in addition to other nutrients.

Nutritionists agree that most of the carbohydrates in a healthful diet should be in the form of complex carbohydrates, and that it is best to consume only minimal amounts of processed sugars. A **processed food** is one that has undergone extensive refinement and, in doing so, has been stripped of much of its nutritive value. For example, unrefined raw brown sugar is made from the juice of the sugarcane plant and contains the minerals and nutrients found in the plant. Processing brown sugar to produce refined brown sugar—like the kind most of us buy at the grocery store—results in the loss of these vitamins and minerals.

Foods that have not been stripped of their nutrition by processing are called **whole foods.** Whole grains, beans, and many fruits and vegetables are whole foods. In addition to being nutritious, whole foods also tend to be good sources of dietary fiber.

Fiber is an important part of a healthy diet. Also called *roughage,* fiber is composed mainly of those complex carbohydrates that humans cannot digest. For this reason, dietary fiber is passed into the large intestine, where some of it is digested by bacteria and the remainder gives bulk to feces. Although fiber is not a nutrient because it is not absorbed by the body, it is still an important part of a healthful diet. Fiber helps maintain healthy cholesterol levels and may decrease your risk of certain cancers.

Fiber bars are a common sight on the shelves of grocery and convenience stores. Although fiber bars can be rich in fiber, they are also often rich in processed sugars and nonnutritive additives, too. Because of this, it is better to get your fiber from whole food sources.

Proteins as Nutrients. Protein-rich foods include beef, poultry, fish, beans, eggs, nuts, and dairy products such as milk, yogurt, and cheese.

Your body is able to synthesize many of the commonly occurring amino acids that proteins are made of. Those your body cannot synthesize are called **essential amino acids** and must be supplied by the foods you eat. **Complete proteins** contain all the essential amino acids your body needs. Proteins obtained by eating meat are more likely to be complete than are those obtained by eating plants.

Individual plant proteins can often be missing one or more essential amino acids. This may be why different plant-based food pairings, like rice and beans, are so common all around the world. Vegetarians who eat a wide variety of plant-based foods will have little trouble obtaining all the essential amino acids.

People looking to gain muscle mass will sometimes supplement their diet with protein powders that can be mixed with milk or water to help rebuild protein-rich muscle after a strenuous workout. However, extra protein does not always lead to extra muscle. Like any nutrient, excess protein is stored as fat. So,

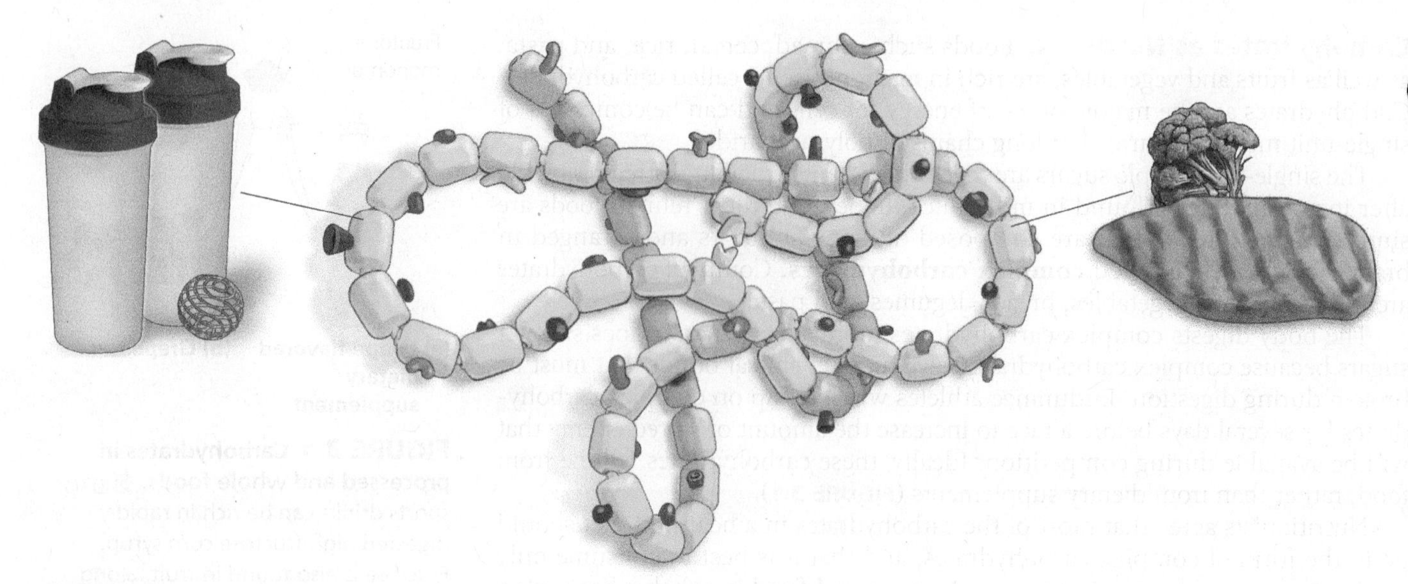

FIGURE 3.2 Protein in processed and whole foods. Nutritionists recommend that most people consume around 50 grams of protein a day. This can be obtained by drinking two 8-ounce protein shakes or eating one small chicken breast and one cup of broccoli.

if you eat too much protein, you will end up increasing body fat. In addition, diets that are too rich in protein can lead to health problems such as bone loss and kidney damage. If you are looking to build muscle, the most sensible thing to do is to combine your workout regimen with a healthful diet. Even the most active people, for example endurance athletes, can easily obtain enough protein by eating a healthy diet (**FIGURE 3.2**).

Fats as Nutrients. Fat is a source of stored energy. In fact, gram for gram, fat contains around twice as much energy as carbohydrates or protein. Foods that are rich in fat include meat, milk, cheese, vegetable oils, and nuts. Most mammals, including humans, store fat just below the skin to help cushion and protect vital organs, to insulate the body from cold weather, and to store energy in case of famine.

Recall that fats have long hydrocarbon-rich fatty acid tails. Your body can synthesize most of the fatty acids it requires. Those that cannot be synthesized are called **essential fatty acids.** Like essential amino acids, essential fatty acids must be obtained from the diet. Omega-3 and omega-6 fatty acids are essential fatty acids that can be obtained by eating fish. These fatty acids are thought to help protect against heart disease. Nutritionists recommend that people eat about 12 oz. of fish every week. Some people who don't eat much fish supplement their diets with fish oil capsules. However, fish contain other vitamins and minerals that fish oil supplements do not.

The fatty acid tails of a fat molecule can differ in the number and placement of double bonds (**FIGURE 3.3a**). When the carbons of a fatty acid are bound to as many hydrogen atoms as possible, the fat is said to be a **saturated fat** (saturated in hydrogen). When there are carbon-to-carbon double bonds, the fat is not saturated in hydrogen, and it is therefore an **unsaturated fat** (**FIGURE 3.3b**). The more double bonds there are, the higher the degree of unsaturation. When fat contains many unsaturated carbons, the fat is referred to as **polyunsaturated.** The double bonds in unsaturated fats make the structures kink instead of lying flat. This form prevents the adjacent fat molecules from packing tightly together, so unsaturated fat tends to be liquid at room temperature. Cooking oil is an example of an unsaturated fat. Unsaturated fats are more likely to come from plant sources, and the fats found in animals are typically saturated. Saturated fats, with their absence of carbon-to-carbon double bonds, pack tightly together to make a solid structure. This is why saturated fats, such as butter, are solid at room temperature.

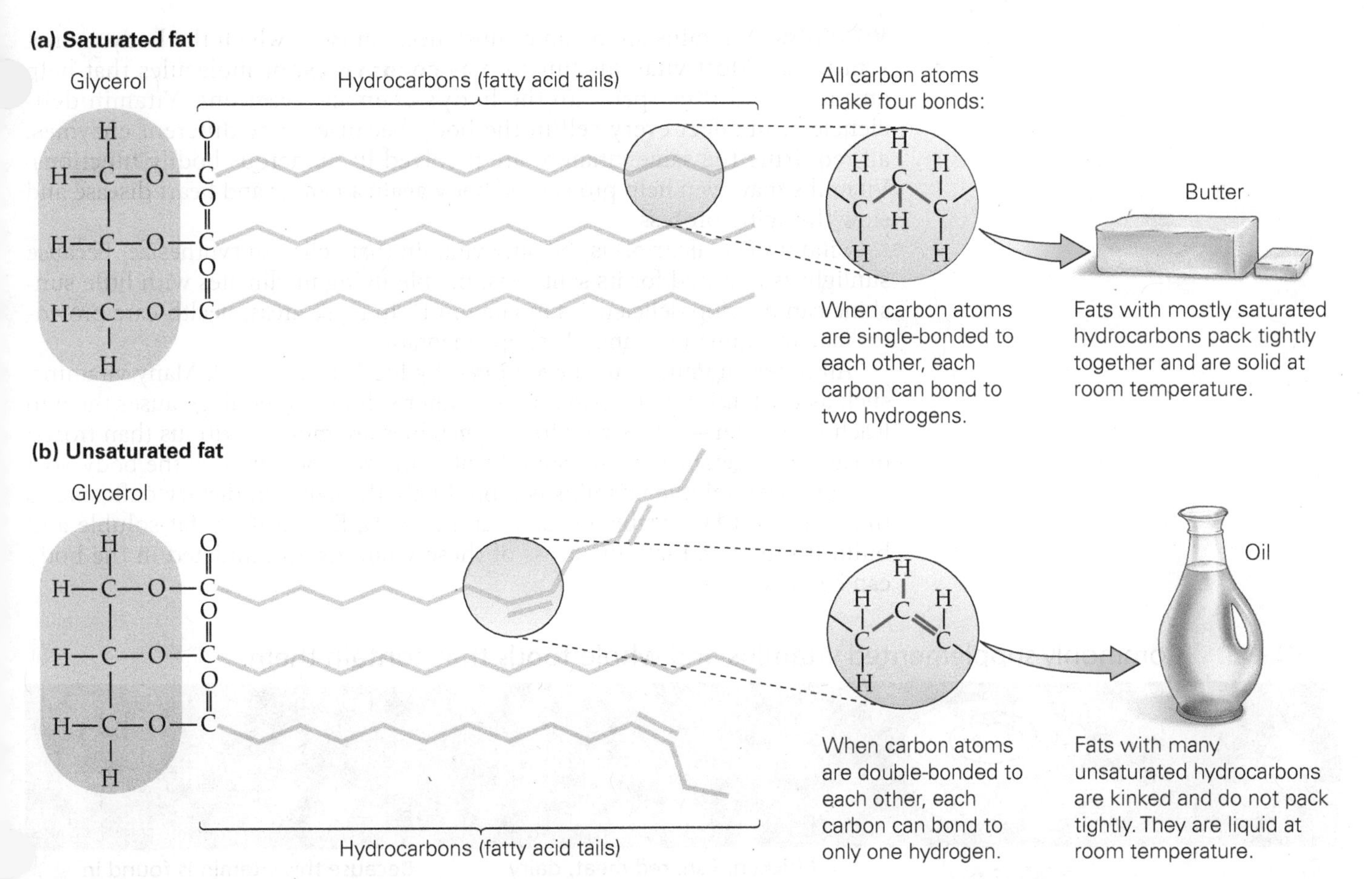

(a) Saturated fat
Glycerol
Hydrocarbons (fatty acid tails)
All carbon atoms make four bonds:
Butter
When carbon atoms are single-bonded to each other, each carbon can bond to two hydrogens.
Fats with mostly saturated hydrocarbons pack tightly together and are solid at room temperature.
(b) Unsaturated fat
Glycerol
Oil
When carbon atoms are double-bonded to each other, each carbon can bond to only one hydrogen.
Fats with many unsaturated hydrocarbons are kinked and do not pack tightly. They are liquid at room temperature.
Hydrocarbons (fatty acid tails)

FIGURE 3.3 Saturated and unsaturated fats. The types of bonds formed determine whether a fat will be (a) solid or (b) liquid at room temperature.

VISUALIZE **THIS**

Which of the fatty acid tails in this figure is polyunsaturated?

Commercial food manufacturers sometimes add hydrogen atoms to unsaturated fats by combining hydrogen gas with vegetable oils under pressure. This process, called **hydrogenation,** increases the level of saturation of a fat. This process solidifies liquid oils, thereby making food seem less greasy and extending shelf life. Margarine is vegetable oil that has undergone hydrogenation.

Trans fats are produced by incomplete hydrogenation, which also changes the structure of the fatty acid tails in the fat so that, even though there are carbon-carbon double bonds, the fatty acids are flat and not kinked. In contrast to most fats, trans fats are not required or beneficial—in fact, research has indicated they are unhealthy. If you choose to eat energy bars, protein bars, fiber bars, and other nutrition bars, it is best to look for those that do not contain trans fats.

The health risks of consuming foods rich in trans-fatty acids include increased risk of clogged arteries, heart disease, and diabetes. Because fat contains more stored energy per gram than carbohydrate and protein do, and because excess fat intake is associated with several diseases, nutritionists recommend that you limit the amount of all fats in your diet.

Micronutrients

Nutrients that are essential in minute amounts, such as vitamins and minerals, are called **micronutrients.** They are neither broken down by the body nor burned for energy.

Vitamins. Vitamins are organic substances, most of which the body cannot synthesize. Most vitamins function as **coenzymes,** or molecules that help enzymes, and thus speed up the body's chemical reactions. Vitamin deficiencies can affect every cell in the body because many different enzymes, all requiring the same vitamin, are involved in numerous bodily functions. Vitamins may even help protect the body against cancer and heart disease and slow the aging process.

Vitamin D (calcitriol) is the only vitamin that cells can synthesize. Because sunlight is required for its synthesis, people living in climates with little sunshine can develop deficiencies in vitamin D. In these areas, health-care providers may recommend vitamin D supplementation.

All other vitamins must be supplied by foods (TABLE 3.1). Many vitamins, such as B vitamins and vitamin C, are water soluble, so boiling causes them to leach into water—this is why fresh vegetables are more nutritious than frozen or canned vegetables. Water-soluble vitamins are not stored by the body, so a lack of water-soluble vitamins is more likely the cause of dietary deficiencies than a lack of fat-soluble ones. Vitamins A, D, E, and K are fat-soluble and build up in stored fat; an excess of these vitamins accumulated in the body can be toxic.

TABLE 3.1 Commonly supplemented vitamins and whole foods that contain them.

Vitamin	Food Sources	Notes
B_{12}	Chicken, fish, red meat, dairy	Because this vitamin is found in meat and dairy, some vegetarians and vegans are advised to supplement.
C	Most fruits, vegetables, and meats	Some people supplement vitamin C during cold season, although not much evidence supports this practice.
D	Milk, egg yolk, and soy	Because sunlight is required for the synthesis of this vitamin, people living in northern climates may be advised to supplement.
E	Almonds, many cooking oils, mangoes, broccoli, and nuts	Supplementing with high doses of vitamin E may increase risk of prostate cancer or stroke.
Folic Acid	Dark green vegetables, nuts, legumes (dried beans, peas, and lentils), and whole grains	Because deficiencies during pregnancy can lead to impaired spinal cord development in the fetus, women who are pregnant or planning on becoming pregnant are often advised to supplement.

The practice of taking multivitamin supplements has been called into question recently, especially when it comes to taking mega-doses of fat-soluble vitamins. Recent research shows that this practice may do more harm than good, potentially increasing risk of certain cancers, heart disease, and death.

Minerals. Minerals are substances that do not contain carbon but are essential for many cell functions. Because they lack carbon, minerals are said to be inorganic. They are important for proper fluid balance, in muscle contraction and conduction of nerve impulses, and for building bones and teeth. Calcium, chloride, magnesium, phosphorus, potassium, sodium, and sulfur are all minerals. Like some vitamins, minerals are water soluble and can leach out into the water during boiling. Also like most vitamins, minerals are not synthesized in the body and must be supplied through diet (TABLE 3.2).

Calcium is one of the more commonly supplemented minerals. Bodies need calcium to help blood clot, muscles contract, nerves fire, and bones to grow and stay healthy. When dietary calcium is low, calcium is lost from the bones, weakening them. If you are not getting enough calcium in your diet, around 1000 mg/day, many health-care providers will recommend supplements. Eight ounces of yogurt, 1.5 ounces of cheese, and one glass of milk together contain around 1000 mg of calcium.

TABLE 3.2 Commonly supplemented minerals and whole foods that contain them.

Mineral	Food Sources	Notes
Calcium	Milk, cheese, dark green vegetables, and legumes	Calcium is required for bone health. Those at risk for bone degeneration (osteoporosis) may be advised to supplement.
Magnesium	Spinach, fish, nuts, seeds, beans, and brown rice	Magnesium is required for skeletal and dental health. Supplementation is not typically recommended.
Potassium	Potatoes, tomatoes, bananas, avocados, and other fresh and dried fruits, dairy, whole grains, and meat	Potassium is required for muscle movement, nerve action, and proper kidney function. Potassium can help lower cholesterol, but a healthy diet does a better job of lowering cholesterol than supplements.

Antioxidants

In addition to vitamins and minerals, many whole foods contain molecules called **antioxidants** that prevent cells from damage caused by molecules that are generated by normal cell processes. These highly reactive molecules, called **free radicals,** have an incomplete electron shell, which makes them more chemically reactive than molecules with complete electron shells. Free radicals can damage cell membranes, the lining of arteries, and DNA. Antioxidants can inhibit the chemical reactions that involve free radicals and decrease the damage they do in cells. Antioxidants are abundant in fruits, vegetables, nuts, grains, and some meats.

After initial excitement about antioxidant supplements preventing heart disease, cancer, or slowing the aging process, we now know that the benefits of antioxidants seem to come only from those consumed in whole foods (TABLE 3.3). This may be because there needs to be a balance of antioxidants and free radicals for optimal health. When the balance gets shifted—let's say, when a person takes a high-dose antioxidant supplement—there are not enough free radicals to perform some beneficial functions we rely on them for, like killing new cancer cells, bacteria, or other invaders.

Got It?

1. Nutrients are substances in food that provide structural materials or __________.
2. The macromolecules required as nutrients include __________, carbohydrates, and proteins.
3. Whole foods are more healthful than __________ foods because they do not undergo nutrient stripping alterations.
4. __________ carbohydrates take longer to break apart than simple sugars.
5. __________ is an important dietary constituent that helps regulate bowel function but is not a nutrient because it is not metabolized.

TABLE 3.3 Commonly supplemented antioxidants and whole foods that contain them.

Antioxidants	Food Sources	Notes
Beta-carotene	Orange fruits and vegetables, including carrots, cantaloupe, squash, mangoes, pumpkin, apricots, collard greens, kale, and spinach	According to the National Center for Complementary and Alternative Medicine, supplementing with high doses of beta-carotene may increase the risk of lung cancer in smokers.
Flavanol	Cocoa and dark chocolate	Studies have not shown equivocally that supplementing provides any health benefit.
Lycopene	Red fruits and vegetables, including tomatoes and watermelon	To date, supplementing with lycopene has not been found to have any clear health benefit.

VISUALIZE **THIS**

List three things plant cells contain that animal cells do not.

(a) Animal cell

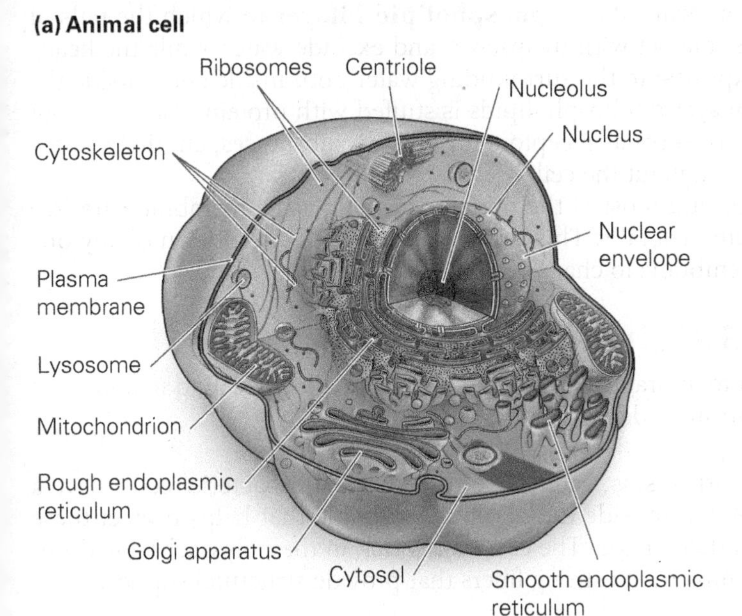

(b) Plant cell

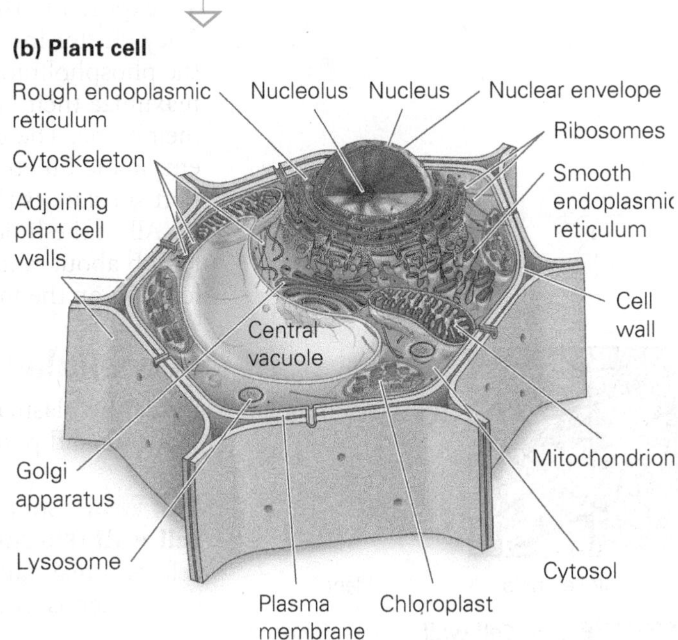

FIGURE 3.4 Plant and animal cells. These drawings of a generalized (a) animal cell and (b) plant cell show the locations and sizes of organelles and other structures.

3.2 Cell Structure

Some nutritional supplements are composed of extracts produced by grinding and breaking open plant cells. Others are made to affect particular subcellular structures in human cells. Because animal and plant cells evolved from a common ancestor, they share many of the same cellular structures and organelles. **Organelles** are to cells as organs are to the body. Each performs a specific job required by the cell and works in conjunction with other organelles to keep the cell functioning properly. Also inside cells is the **cytosol,** a watery matrix containing salts and many of the enzymes required for cellular reactions. The cytosol houses the organelles. The **cytoplasm** includes both the cytosol and organelles.

cyto- means cell.
-plasm means fluid.

Let's work our way in from the outside to the inside of an animal cell and a plant cell (**FIGURE 3.4**). and then examine the structure and function of various internal cell components before investigating some supplements made from plant cells and some that inhibit organelle function.

Plasma Membrane

All cells are enclosed by a structure called a **plasma membrane** (**FIGURE 3.5a**). The plasma membrane defines the outer boundary of each cell, isolates the cell's contents from the environment, and serves as a barrier that allows some nutrients into and out of the cell. Membranes that enclose structures inside the cell are usually referred to as cell membranes, and the outer boundary is the plasma membrane.

Membranes are **semipermeable,** meaning that they allow some substances to cross while preventing others from doing so. This characteristic allows cells to maintain a different internal composition from the surrounding solution.

Internal and external membranes are composed primarily of phospholipids (**FIGURE 3.5b**). The chemical properties of these lipids make membranes

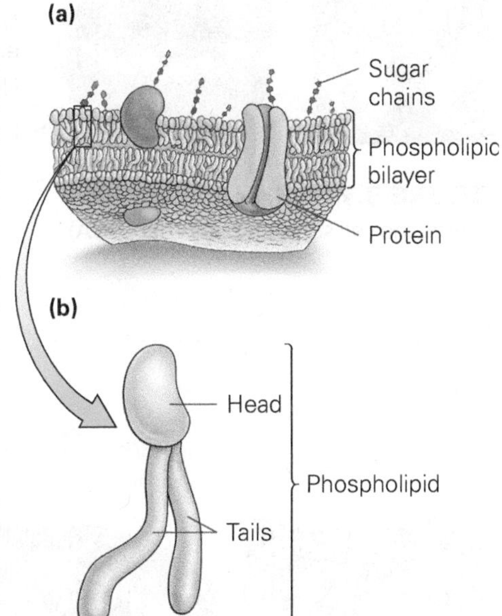

FIGURE 3.5 The plasma membrane. (a) All cells are surrounded by a plasma membrane. The phospholipids (b) that form a bilayer are each composed of a hydrophilic head and two hydrophobic tails.

flexible and self-sealing. When phospholipid molecules are placed in a watery solution, such as in a cell, they orient themselves so that their hydrophilic heads are exposed to the water and their hydrophobic tails are away from the water. They cluster into a form called a **phospholipid bilayer** in which the tails of the phospholipids interact with themselves and exclude water, while the heads maximize their exposure to the surrounding water both inside and outside the membrane. The bilayer of phospholipids is stuffed with proteins that carry out enzymatic functions, serve as receptors for outside substances, and help transport substances throughout the cell.

All of the lipids and most of the proteins in the plasma membrane are free to bob about, sliding laterally. This fluidity allows the composition of any one location on the membrane to change.

Subcellular Structures

Inside the plasma membrane are the structures that allow a cell to maintain its structure and perform its designated functions.

FIGURE 3.6 Cell wall. Plants, fungi, and bacteria have a cell wall outside the plasma membrane.

Cell Wall. Some organisms, such as plants, fungi, and bacterial cells, have a **cell wall** (FIGURE 3.6) outside the plasma membrane that helps protect these cells and maintain their shape. The cell wall is rich in the polysaccharide cellulose, which is assembled into strong fibers that provide structural support.

Nucleus. All eukaryotic cells contain a **nucleus** (FIGURE 3.7)—a spherical structure surrounded by two membranes, which together are called the **nuclear envelope.** The nuclear envelope is studded with nuclear pores that regulate traffic into and out of the nucleus. Inside the nucleus is chromatin, composed of DNA and proteins. The **nucleolus** inside the nucleus is where ribosomes are synthesized.

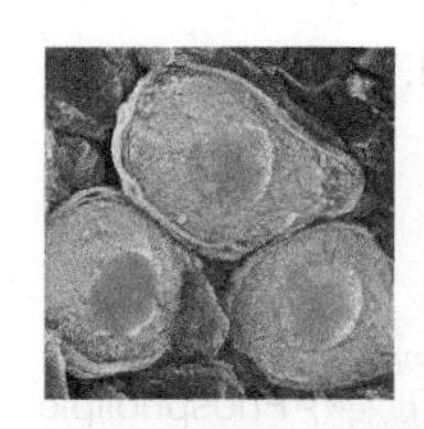

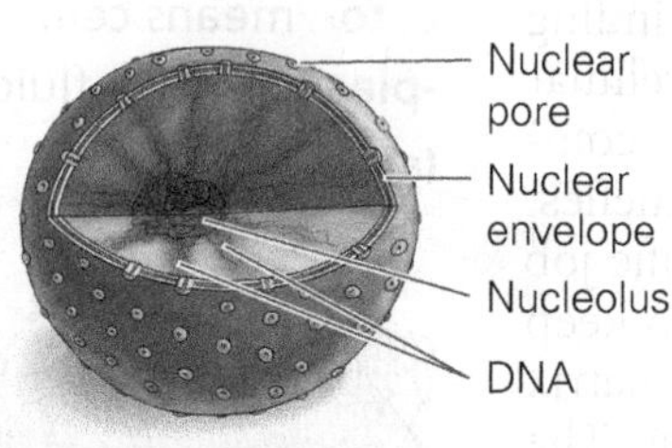

FIGURE 3.7 Nucleus. The nucleus houses the cell's genetic material (DNA).

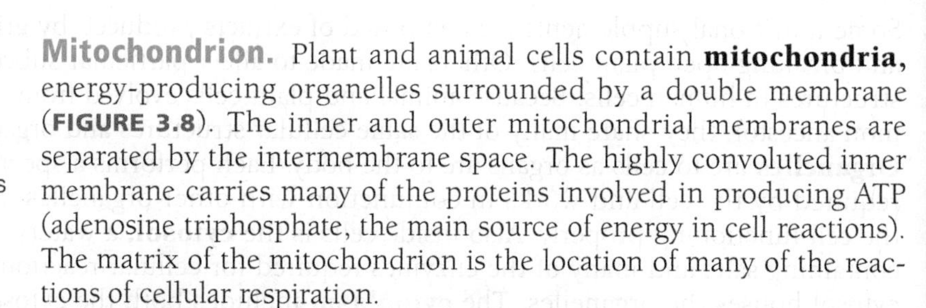

Mitochondrion. Plant and animal cells contain **mitochondria,** energy-producing organelles surrounded by a double membrane (FIGURE 3.8). The inner and outer mitochondrial membranes are separated by the intermembrane space. The highly convoluted inner membrane carries many of the proteins involved in producing ATP (adenosine triphosphate, the main source of energy in cell reactions). The matrix of the mitochondrion is the location of many of the reactions of cellular respiration.

Chloroplast. An important organelle present in plants and other photosynthetic eukaryotes, the **chloroplast** (FIGURE 3.9) uses the sun's energy to convert carbon dioxide and water into sugars. Each chloroplast has an outer membrane, an inner membrane, a liquid interior called the stroma, and a network of membranous sacs called

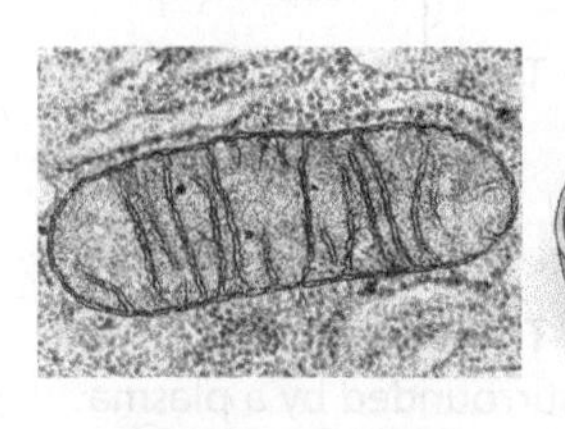

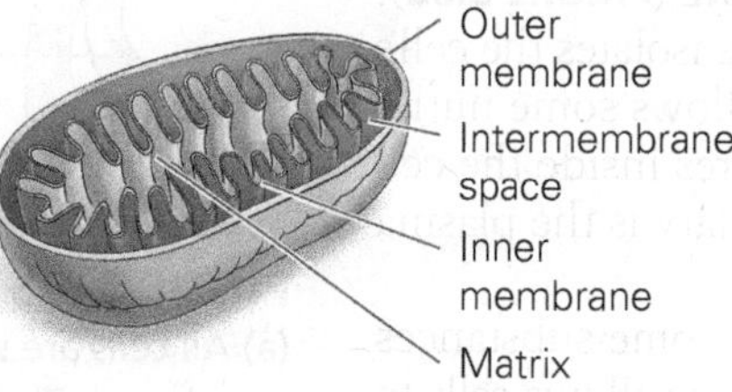

FIGURE 3.8 Mitochondrion. Plant and animal cells both contain mitochondria.

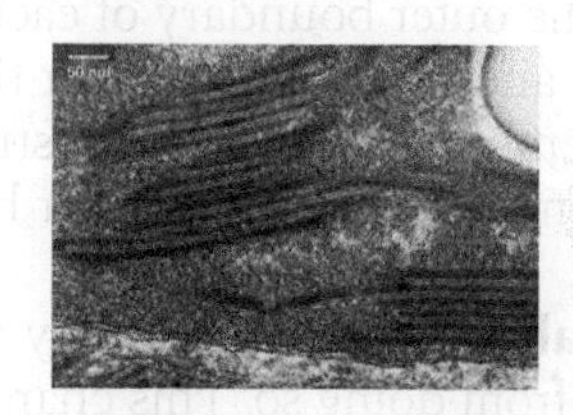

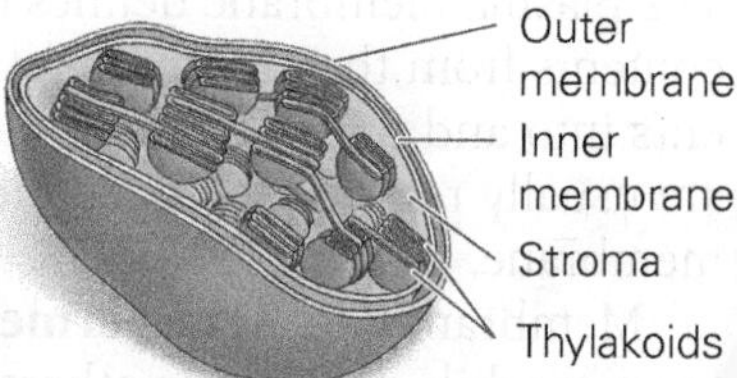

FIGURE 3.9 Chloroplasts. Cells and organisms that perform photosynthesis have chloroplasts.

thylakoids. Chloroplasts also contain pigment molecules. The pigment **chlorophyll,** like the pigments in your clothes, reflects some wavelengths of light and absorbs others. Chlorophyll reflects green light.

Lysosome. A **lysosome** (FIGURE 3.10) is a membrane-enclosed sac of digestive enzymes that degrade proteins, carbohydrates, and fats. The low pH of the lysosome sequesters these enzymes from the rest of the cell. Lysosomes roam around the cell and engulf targeted molecules and organelles for recycling.

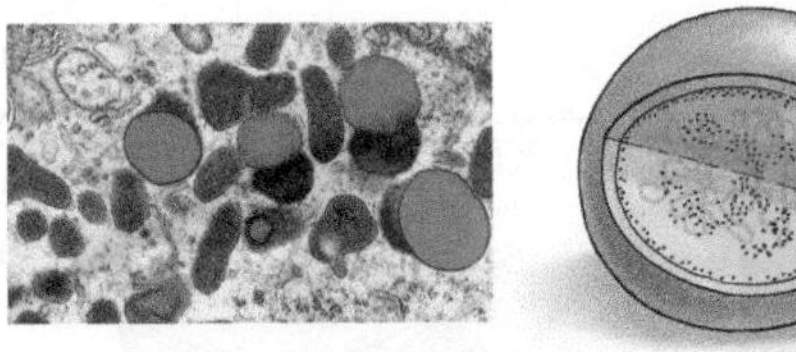

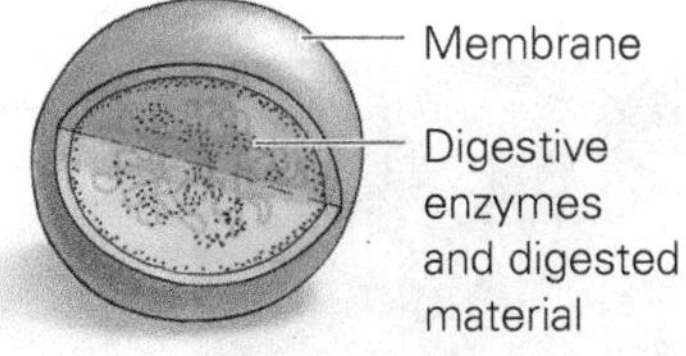

FIGURE 3.10 Lysosomes. Lysosomes are digestive organelles.

Ribosomes. Ribosomes (FIGURE 3.11) are workbenches where proteins are assembled. Composed of two subunits, they can be found floating in the cytoplasm or tethered to the **endoplasmic reticulum.**

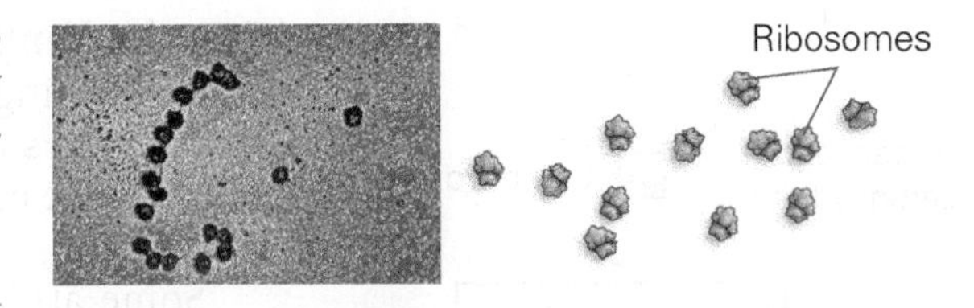

FIGURE 3.11 Ribosomes. Ribosomes, the sites of protein synthesis, are composed of two subunits.

Endoplasmic Reticulum. The endoplasmic reticulum (FIGURE 3.12), abbreviated ER, is a large network of membranes that flows outward from the nuclear envelope and extends into the cytoplasm of a eukaryotic cell. ER with ribosomes attached is called rough ER. Proteins synthesized on rough ER will be secreted from the cell or will become part of the plasma membrane. ER without ribosomes attached is called smooth ER. The function of the smooth ER depends on cell type but includes tasks such as detoxifying harmful substances and synthesizing lipids. Vesicles are pinched-off pieces of membrane from the ER that transport substances to the Golgi apparatus or plasma membrane.

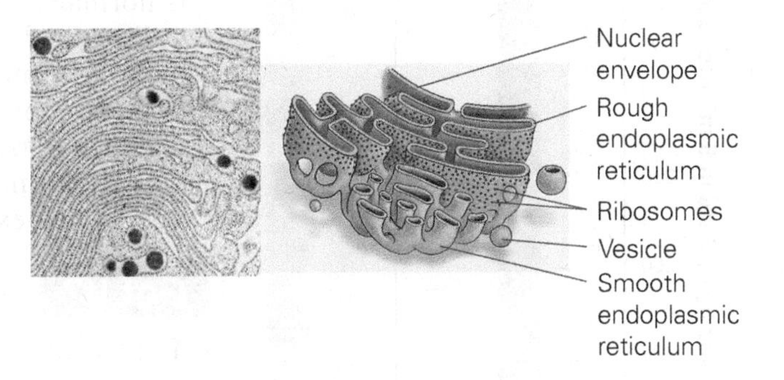

FIGURE 3.12 Endoplasmic reticulum. The ER is composed of membranous sacs and tubules.

Golgi Apparatus. The **Golgi apparatus** is a stack of membranous sacs (FIGURE 3.13). Vesicles from the ER fuse with the Golgi apparatus and empty their protein contents. The proteins are then modified, sorted, and sent to the correct destination in new transport vesicles that bud off from the Golgi apparatus sacs.

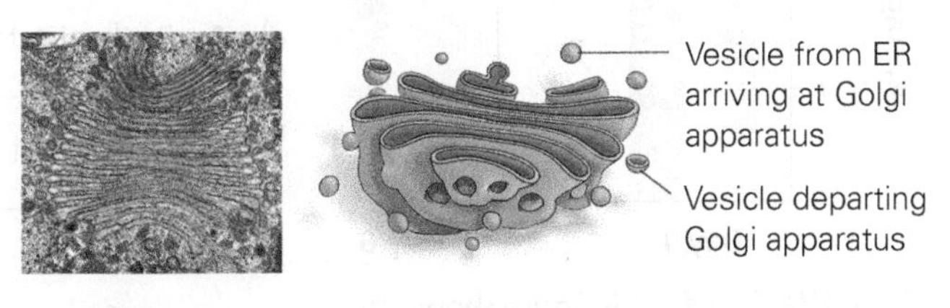

FIGURE 3.13 Golgi apparatus. The Golgi is composed of membranous sacs.

Centrioles. Centrioles are barrel-shaped rings composed of microtubules that anchor structures and help move chromosomes around when an animal cell divides (FIGURE 3.14). Centrioles are also involved in the formation of structures involved in motility—the cilia and flagella. Plant cells do not have centrioles.

Cytoskeletal Elements. Cytoskeletal elements (FIGURE 3.15) are protein fibers that make up the structure of the **cytoskeleton,** a framework that gives shape and structural support to cells.

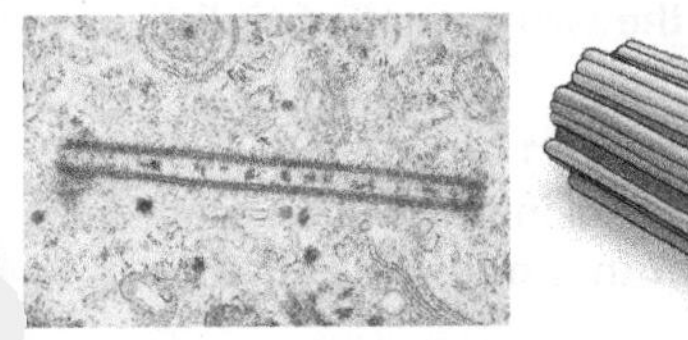

FIGURE 3.14 Centrioles. Centrioles help animal cells perform cell division.

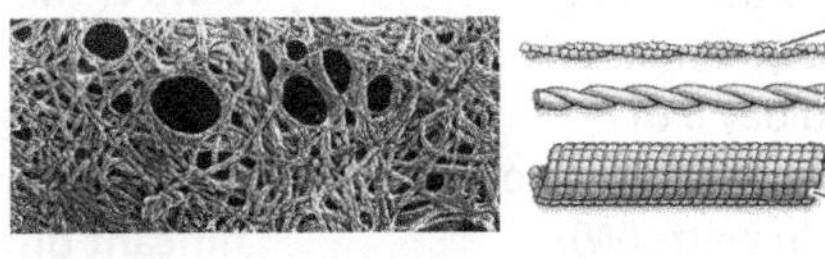

FIGURE 3.15 Cytoskeleton. This framework, composed of three different types of tubular supports, serve as cellular scaffolding.

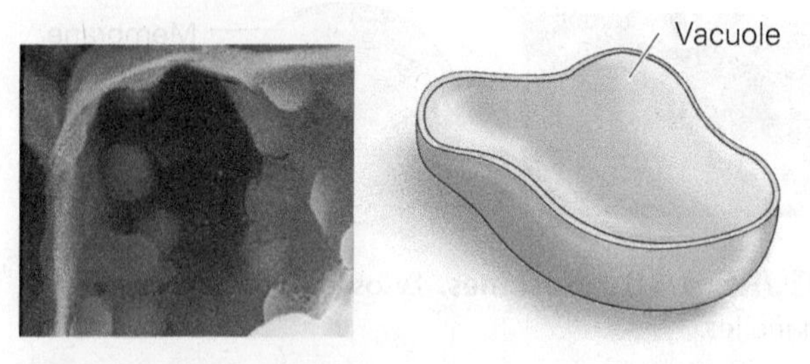

FIGURE 3.16 Central vacuole. This membranous organelle stores water and ions.

Subcellular structures like the nucleus or mitochondria are anchored by these elements, and some structures, like lysosomes, use the cytoskeleton like railroad tracks to travel from location to location inside the cell.

Central Vacuole. Plant cells have a large fluid-filled **central vacuole** (**FIGURE 3.16**) that contains a variety of dissolved molecules, including sugars and pigments that give color to flowers and leaves. Vacuoles also function to maintain pressure inside individual cells, which helps support the upright plant.

Supplements and Cell Structures. Chlorophyll supplements are marketed as being able to energize, detoxify, and help heal wounds. There is no evidence to suggest that any of these claims are true. In fact, because human cells do not contain chlorophyll, taking chlorophyll is not "supplementing" anything.

Some athletes use creatine to increase muscle mass or allow for more powerful bursts of energy. Creatine is used during protein synthesis on ribosomes and by mitochondria in the making of ATP. Although creatine is normally found in the body, the effects of supplementing have not been well studied. With side effects like kidney failure, it is safest not to supplement creatine above that which you normally obtain from your diet. It is also common for athletes to drink protein supplements while exercising to enhance performance and decrease recovery time. **FIGURE 3.17** shows the data from one controlled study testing the effects of protein supplementation during exercise in elite cyclists.

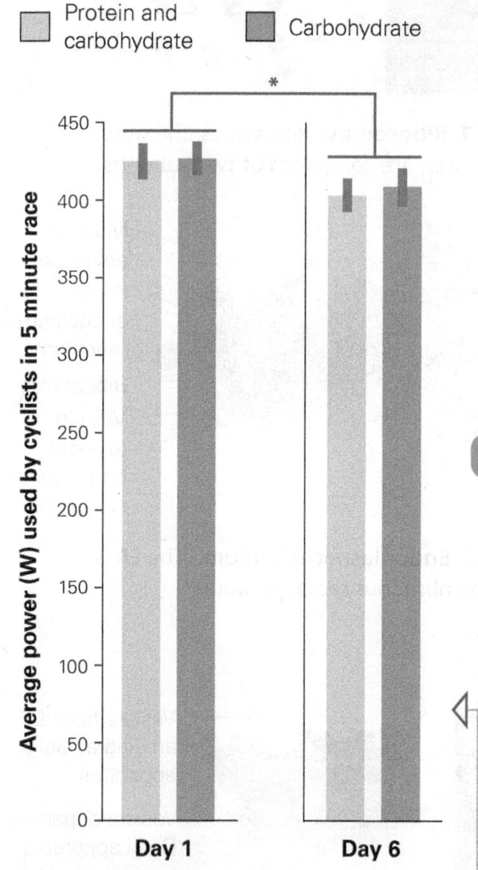

FIGURE 3.17 This figure shows the effects of ingesting liquid supplements during a 1-week training camp for a large number of elite cyclists. One group of cyclists were given only carbohydrate supplements and the other group was given carbohydrate and protein supplements. The data show the average power output during a test of 5-minute all-out cycling on day 1 and day 6 of training. Power is a measure of energy use over time and is measured in watts (W).

Data source: M. Hansen, J. Bangsbo, and J. Jensen, "Protein Intake During Training Sessions," *Journal of the International Society of Sports Nutrition* 13, no. 9 (2016).

Got It?

1. Proteins are made on workbench-like structures called ________
2. Ribosomes can be found floating free in the cytosol or attached to the ________ ________.
3. In plant cells, structural support is provided by the ________ ________, which is exterior to the plasma membrane.
4. Both plant and animal cells contain energy-producing organelles called ________.
5. Only plant cells are able to harvest energy from the sun using organelles called ________.

WORKING WITH DATA

a. **There appears to be a slightly lower power usage by cyclists who received protein in their supplements compared with those who did not. Scientists use an asterisk to show that a difference is statistically significant. Is there a significant difference between the two groups of cyclists on day 1?**

b. **Note the horizontal line connecting both bars at day 1 and both bars at day 6. The asterisk connecting these horizontal bars indicates that a significant difference was seen when comparing both groups on both days. What does this mean?**

c. **Did supplementing with protein and carbohydrate, versus just carbohydrate, help the cyclists reduce the amount of power required to complete the 5-minute cycling race?**

3.3 Transport Across Membranes

Whether ingested in food or supplemented in pills, once substances reach cells to be used they must traverse the plasma membrane that surrounds the cell. Molecules must cross the plasma membrane to gain access to the inside of the cell, where they can be used to synthesize cell components or be metabolized to provide energy for the cell. The chemistry of the membrane facilitates the transport of some substances and prevents the transport of others.

As we learned earlier, the plasma membrane that surrounds cells is composed of a phospholipid bilayer. The interior of the bilayer is hydrophobic. Hydrophobic substances can dissolve in the membrane and pass through it more easily than hydrophilic ones. In this sense, the membrane of the cell is differentially permeable to the transport of molecules, allowing some to pass through and blocking others.

Substances that can cross the membrane will do so until the concentration is equal on both sides of the membrane, a condition called equilibrium. Carbon dioxide, water, and oxygen move freely across the membrane. Larger molecules, charged molecules, and ions cannot cross the lipid bilayer on their own (**FIGURE 3.18**). If these substances need to be moved across the membrane, they must move through proteins embedded in the membrane. Proteins in the membrane can serve as channels to allow molecules to cross until their concentration is equal on both sides of the membrane or proteins can serves as pumps, driving molecules away from equilibrium.

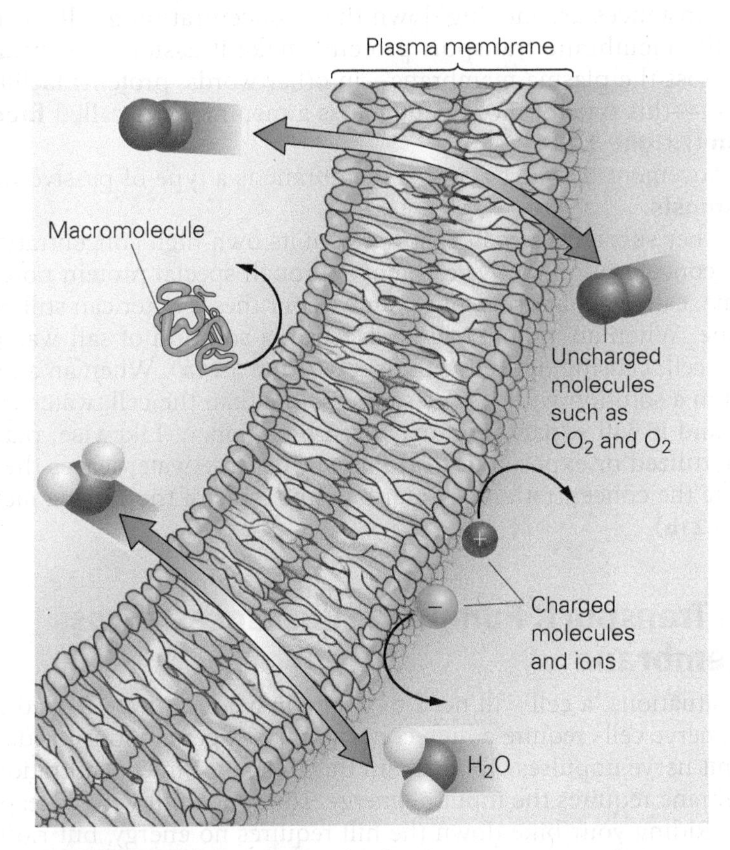

FIGURE 3.18 Transport of substances across membranes. The ability of a substance to cross a membrane is, in part, a function of its size and charge.

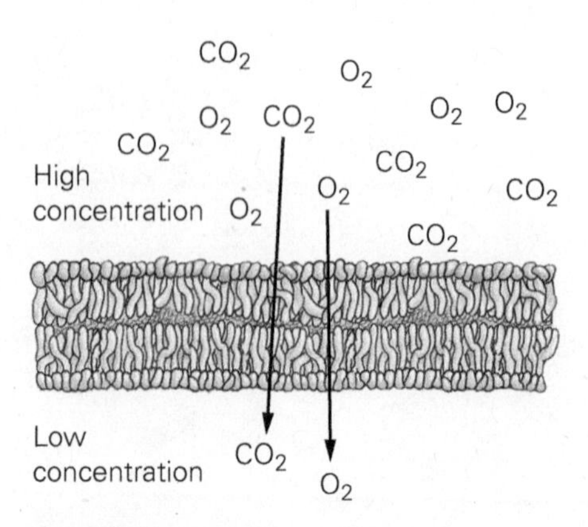

FIGURE 3.19 Simple diffusion. Simple diffusion of molecules across the plasma membrane occurs with the concentration gradient and does not require energy. Small hydrophobic molecules, carbon dioxide, and oxygen can diffuse across the membrane.

Passive Transport: Diffusion, Facilitated Diffusion, and Osmosis

All molecules contain energy that makes them vibrate and bounce against each other, scattering around like billiard balls during a game of pool. In fact, molecules will bounce against each other until they are spread out over all the available area. In other words, molecules will move from their own high concentration to their own low concentration. This movement of molecules from where they are in high concentration to where they are in low concentration is called **diffusion.** During diffusion, the net movement of molecules is from their own high to their own low concentration or *down* a concentration gradient. This movement does not require an input of outside energy; it is spontaneous. Diffusion will continue until there is equilibrium, at which time no concentration gradient exists, and there is no net movement of molecules.

Diffusion also occurs in living organisms. When substances diffuse across the plasma membrane, we call the movement **passive transport.** Passive transport is so named because it does not require an input of energy. The structure of the phospholipid bilayer that makes up the plasma membrane prevents many substances from diffusing across it. Only very small, hydrophobic molecules are able to cross the membrane by diffusion. In effect, these molecules dissolve in the membrane to slip from one side of the membrane to the other (**FIGURE 3.19**).

Hydrophilic molecules and charged molecules such as ions are unable to simply diffuse across the hydrophobic core of the membrane. Instead, these molecules are transported across membranes by proteins embedded in the lipid bilayer. This type of passive transport does not require an input of energy because substances are moving down their concentration gradient. Because the specific membrane transport proteins make it easier for substances to diffuse across the plasma membrane—in other words, proteins facilitate this movement—this type of movement across a membrane is called **facilitated diffusion** (**FIGURE 3.20**).

The movement of water across a membrane is a type of passive transport called **osmosis.**

osmo- means water.

Like other substances, water moves from its own high concentration to its own low concentration. Water can move through special protein pores in the membrane, called aquaporins, but even without these, water can still cross the membrane. When an animal cell is placed in a solution of salt water, water leaves the cell, causing the cell to shrivel (**FIGURE 3.21a**). When an animal cell is placed in a solution with less dissolved solute than the cell, water will enter the cell, and it will expand and may even break open. Likewise, plants that are overfertilized or exposed to road salt wilt because water leaves the cells to equilibrate the concentration of water on either side of the plasma membrane (**FIGURE 3.21b**).

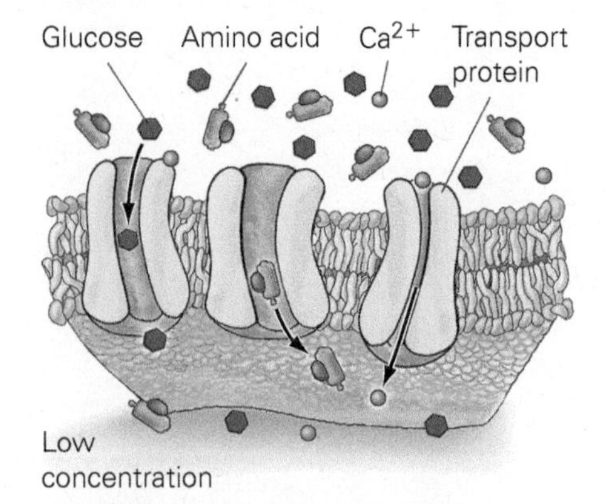

FIGURE 3.20 Facilitated diffusion. Facilitated diffusion is the diffusion of molecules assisted by substrate-specific proteins. Molecules move with their concentration gradient, which does not require energy.

Active Transport: Pumping Substances across the Membrane

In some situations, a cell will need to maintain a concentration gradient. For example, nerve cells require a high concentration of certain ions inside the cell to transmit nerve impulses. To maintain this difference in concentration across the membrane requires the input of energy. Think of a hill with a steep incline or grade. Riding your bike down the hill requires no energy, but riding your bike up the grade requires energy. At the cellular level, that energy is in the form of ATP. **Active transport** is transport that uses proteins, powered by ATP, to move substances up a concentration gradient (**FIGURE 3.22**).

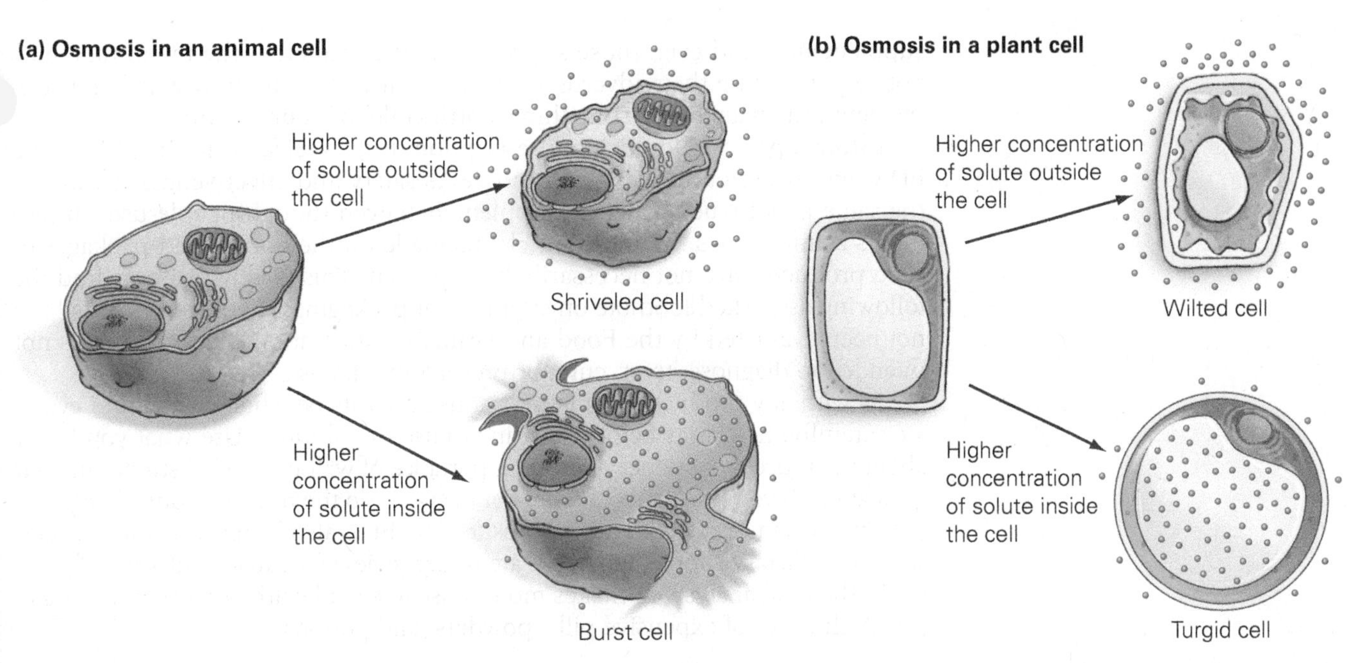

FIGURE 3.21 Osmosis. Osmosis is a special type of diffusion that involves the movement of water in response to a concentration gradient. Water moves toward a region that has more dissolved solute. (a) When water leaves an animal cell, it shrinks. (b) When water leaves a plant cell, the plant wilts instead of shrinks because of the support provided by the cell wall.

Exocytosis and Endocytosis: Movement of Large Molecules across the Membrane

Larger molecules are often too big to diffuse across the membrane or to be transported through a protein, regardless of whether they are hydrophobic or hydrophilic. Instead, they must be moved around inside membrane-bound vesicles (small sacs) that can fuse with membranes and then release their contents. **Exocytosis** (**FIGURE 3.23a**) occurs when a membrane-bound vesicle, carrying some substance, fuses with the plasma membrane and releases its contents into the exterior of the cell. **Endocytosis** (**FIGURE 3.23b**) occurs when a substance is brought into the cell by a vesicle pinching the plasma membrane inward.

Not all nutritional supplements can pass through the cell membrane. Supplements may be too large, or there may not be a transporter specific to the

Active transport
Low concentration
K^+
High concentration
ATP used
K^+ K^+ K^+ K^+

FIGURE 3.22 Active transport. Active transport moves substances against their concentration gradient and requires ATP energy to do so.

exo- means outside.
endo- means inside.

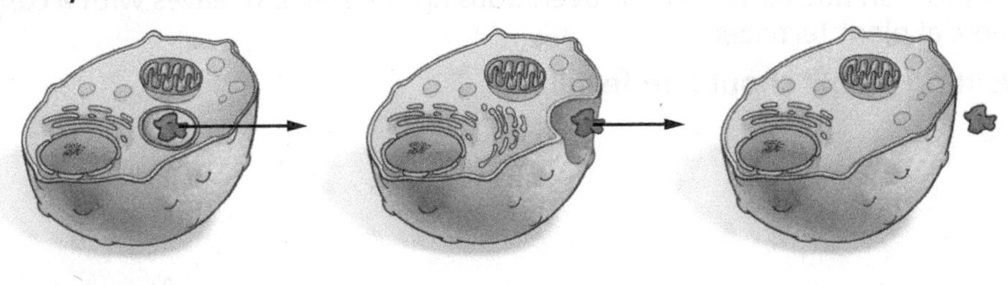

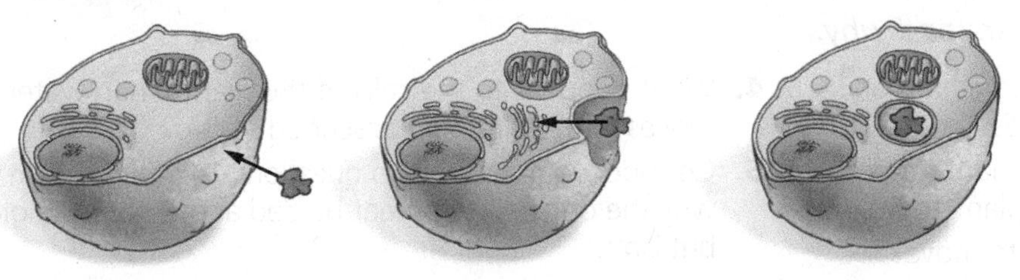

FIGURE 3.23 Movement of large substances. (a) Exocytosis is the movement of substances out of the cell. (b) Endocytosis is the movement of substances into the cell.

supplements. And even those supplements that can cross the membrane may not be performing their advertised function. How do you know whether a supplement is actually doing something worthwhile for your health?

Before a prescription or nonprescription drug is released to the public, the FDA requires scientific testing to prove its safety and effectiveness. However, the same is not true of dietary supplements, even though people use supplements as medicines. In addition, claims made on the bottles and packages of such products have not necessarily been proven. This is why you may find the following asterisked footnote on supplement packaging: "These statements have not been evaluated by the Food and Drug Administration. This product is not intended to diagnose, treat, cure, or prevent any disease."

As with any product that makes claims about its benefits, you should evaluate carefully the claims before deciding to use the product. Use what you know about the process of science and the importance of well-controlled studies in your evaluation. If you don't have the time or desire to do the research yourself, consult your physician or a trustworthy website. The FDA also keeps a record on their website of dietary supplements that are under review for causing adverse effects.

In the final analysis, it makes more sense to eat a healthy diet than it does to pop high doses of expensive pills, powders, and potions.

Got It?

1. Substances will move across a membrane until the concentration is equal on both sides of the membrane, a condition called ________.
2. It is difficult for many substances to cross cell membranes because the lipid bilayer has a ________ center.
3. Large molecules and ________ molecules cannot cross cell membranes unaided and require the help of proteins embedded in the membrane.
4. When substances move down their concentration gradient, no ________ is required.
5. To move a substance against its concentration gradient, protein ________ can be used.

Your parents gave you an African violet plant as a "dorm-warming" gift. The plant is not doing well. It is wilting and dropping leaves, even though you are watering it regularly. Your roommate believes that the plant must be short on nutrients and has been dousing the soil and leaves with a commercial plant fertilizer.

Wilting plants should be fertilized.

Sounds right, but it isn't.

Answer the following questions to understand why.

1. Do you think a liquid fertilizer would have a high or low concentration of dissolved substances?
2. What effect would adding fertilizer to a plant have on the concentration of solute outside the plant cells?
3. Which direction would this cause water to move?
4. What happens to the cells of the plant when water moves in the direction described above?
5. Consider your answers to questions 1–4 and explain why the original statement bolded above sounds right, but isn't.

THE BIG QUESTION

Should I routinely use detox products?

Detoxification teas, sometimes called teatoxes, are endorsed by celebrities on social media. How—or if—these products work to detox your body are open questions. Most detox supplements are thought to act on the liver, which is the major site of detoxification in your body. Let's look at the science behind these products to determine if they are useful and safe.

What should I know?

What follows are some smaller questions that need to be resolved to answer the Big Question. Place a checkmark next to the questions that science can answer.

Smaller Questions	Can Science Answer?
Do toxins accumulate in the liver?	
Do most manufacturers of supplements care more about profit than helping people detoxify?	
Is toxin accumulation harmful to health?	
Can using detox teas or supplements be harmful?	
Are detox products helpful under normal conditions?	
If celebrities are paid for their endorsements, should we trust the products they are endorsing?	

What does the science say?

Let's examine what the data say about this smaller question:

Are detox products helpful under normal conditions?

Milk thistle is an herbal supplement that is thought to act on the liver. The data shown in the illustration that follows show levels of an enzyme whose concentration in the blood increases with liver damage.

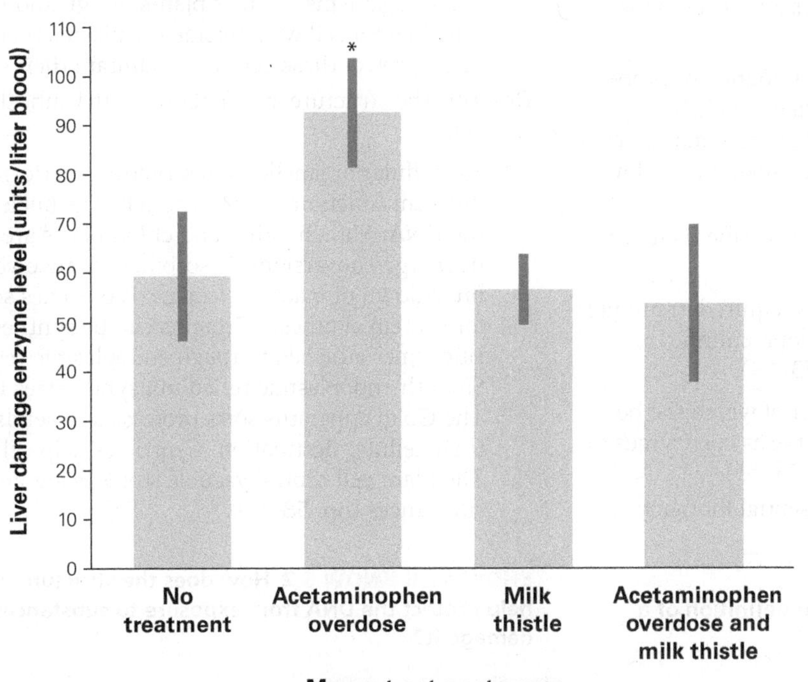

1. Describe the results. Does it appear that milk thistle helps prevent liver damage under normal conditions?
2. Given these data, do you think the smaller question is answered? If not, propose another study that would help answer this question.
3. Does this information help you answer the Big Question? What else do you need to consider?

Data source: N. Bektur, E. Sahin, C. Baycu, and G. Unver, "Protective Effects of Silymarin against Acetaminophen-Induced Hepatotoxicity and Nephrotoxicity in Mice," *Toxicology and Industrial Health* 32, no. 4 (2016): 589–600.

Chapter Review

Mastering Biology

Go to Mastering Biology to access **eText 2.0, Dynamic Study Modules,** and the **Study Area,** where you'll find practice quizzes, BioFlix™ animations, MP3 tutor sessions, current events, and more.

SUMMARY

Section 3.1

Describe the role of nutrients in the body.

- Nutrients provide structural units and energy for cells (p. 50).

Describe the function of water in the body.

- Water is an important dietary constituent that helps dissolve and eliminate wastes and maintain blood pressure and body temperature (p. 50).

Describe the major dietary macronutrients and discuss the functions of each.

- Macronutrients are required in large amounts for proper growth and development. Macronutrients include carbohydrates, proteins, and fats. All of these molecules are composed of subunits that can be broken down for use by the cell (pp. 50–53).

List the major dietary micronutrients and describe their functions.

- Micronutrients are dietary substances required in minute amounts for proper growth and development; they include vitamins and minerals (pp. 53–54).
- Vitamins are organic substances, most of which the body cannot synthesize. Many vitamins serve as coenzymes to help enzymes function properly (pp. 54–55).
- Minerals are inorganic substances essential for many cell functions (pp. 54–55).

SHOW YOU KNOW 3.1 Does water fit our definition of a nutrient? Why or why not?

Section 3.2

Describe the structure and function of the plasma membrane.

- To gain access to cells, nutrients move across the plasma membrane, which functions as a semipermeable barrier that allows some substances to pass and prevents others (pp. 57–58).
- The plasma membrane is composed of two layers of phospholipids, in which are embedded proteins (pp. 57–58).
- Some organisms, such as plants, fungi, and bacterial cells, have a cell wall outside the plasma membrane that helps protect these cells and maintain their shape (p. 58).

Describe the structure and function of the subcellular organelles.

- Subcellular organelles and structures perform many different functions within the cell. The nucleus houses the DNA. Mitochondria and chloroplasts are involved in energy conversions. Lysosomes are involved in the breakdown of macromolecules. Ribosomes serve as sites for protein synthesis. Proteins can be synthesized on ribosomes attached to rough endoplasmic reticulum. Smooth endoplasmic reticulum synthesizes lipids. The Golgi apparatus sorts proteins and sends them to their cellular destination. Centrioles help cells divide. The plant cell central vacuole stores water and other substances (pp. 58–60).

SHOW YOU KNOW 3.2 How does the structure of the nucleus help protect the DNA from exposure to substances that might damage it?

Section 3.3

Distinguish between passive transport and active transport.

- Passive transport mechanisms include unaided simple diffusion and diffusion facilitated by proteins. Passive transport always moves substances with their concentration gradient and does not require energy (pp. 61–63).
- Osmosis, the diffusion of water across a membrane, can involve the movement of water through protein pores in the membrane (pp. 62–63).
- Active transport is an energy-requiring process that uses proteins in cell membranes to move substances against their concentration gradients (pp. 62–63).

Describe the processes of endocytosis and exocytosis.

- Larger molecules move into (endocytosis) and out (exocytosis) of cells enclosed in membrane-bound vesicles (p. 63).

SHOW YOU KNOW 3.3 If a molecule is too big to diffuse across a cell membrane, will it always be prevented from entering the cell?

ROOTS TO REMEMBER

The following roots of words come mainly from Latin and Greek and will help you to decipher terms:

cyto- means cell or a kind of cell. Chapter terms: *cytoplasm, cytoskeleton*

endo- means inside. Chapter terms: *endocytosis, endoplasmic reticulum*

exo- means outside. Chapter term: *exocytosis*

osmo- means water. Chapter term: *osmosis*

plasm means fluid. Chapter term: *cytoplasm, plasma membrane*

LEARNING THE BASICS

1. What are the two main functions of nutrients?
2. Add labels to the figure that follows, which illustrates some molecules that can and cannot pass through cell membranes unaided.

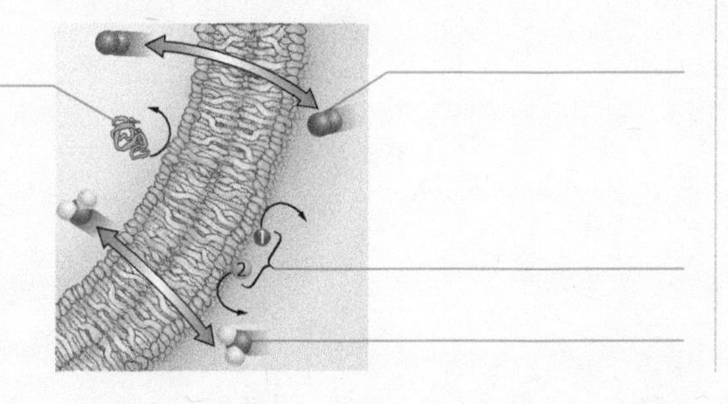

3. Macronutrients ______.

   A. include carbohydrates and vitamins; B. should make up a small percentage of a healthful diet; C. are essential in minute amounts to help enzymes function; D. include carbohydrates, fats, and proteins; E. are synthesized by cells and not necessary to obtain from the diet

4. Which of the following is not a function of water?

   A. dispersing nutrients throughout the body; B. helping prevent cancer; C. helping to regulate body temperature; D. helping to regulate blood pressure

5. Micronutrients ______.

   A. include vitamins and carbohydrates; B. are not metabolized to produce energy; C. contain more energy than fatty acids; D. can be synthesized by most cells

6. The main constituents of the plasma membrane are ______.

   A. carbohydrates and lipids; B. proteins and phospholipids; C. fats and carbohydrates; D. fatty acids and nucleic acids

7. A substance moving across a membrane against a concentration gradient is moving by ______.

   A. passive transport; B. osmosis; C. facilitated diffusion; D. active transport; E. diffusion

8. A cell that is placed in salty seawater will ______.

   A. take sodium and chloride ions in by diffusion; B. move water out of the cell by active transport; C. use facilitated diffusion to break apart the sodium and chloride ions; D. lose water to the outside of the cell via osmosis

9. Which of the following forms of membrane transport require specific membrane proteins?

   A. diffusion; B. exocytosis; C. facilitated diffusion; D. active transport; E. facilitated diffusion and active transport

10. Which of the following cannot pass through the membrane without the help of a membrane protein?

    A. carbon dioxide; B. water; C. oxygen; D. charged molecules

ANALYZING AND APPLYING THE BASICS

1. A friend of yours does not want to eat meat, so instead she consumes protein shakes that she buys at a nutrition store. Can you think of a dietary strategy that would allow her to be a vegetarian while not consuming protein shakes?

2. Studies have shown that as trans fat consumption increases, so does heart disease. If you were to use a line graph to show this relationship, which variable would be on the x axis and what would the line showing this positive correlation look like?

3. A vitamin-enriched water manufacturer claims that its product has all the vitamin C of an orange. What substances would you be missing from the vitamin water that you would get by eating an orange?

GO FIND OUT

1. Select one supplement you have wondered about and spend a few minutes doing some web-based research on whether the claims made on its label are backed up by scientific evidence.

2. Some cities have banned restaurants from using trans fats when cooking. Has such a ban been enacted in your city? Do you think the government should be involved in regulating the use of trans fats? Why or why not?

MAKE THE CONNECTION

The science that you learned in this chapter has helped you better understand the real-world example used throughout this discussion. Draw a line from the statement on the left to the science that supports it on the right.

Drinking tap water is a better way to stay hydrated, and to protect the environment, than drinking sports drinks. ▷	◁ Processed foods are high in trans fats.
The best way to meet your body's need for dietary fiber is to eat whole foods. ▷ - - - - - -	◁ Processed products are often high in calories, can be expensive, and contain unnecessary sugars and other additives.
The best way to meet your body's need for protein is to eat whole foods. ▷	◁ Excesses of this macromolecule in your diet can cause kidney damage.
The best way to meet your body's need for lipids is to eat whole foods. ▷	◁ Large doses of this micronutrient can cause many health problems.
The best way to meet your body's need for vitamins is to eat whole foods. ▷	◁ Processed fluids are high in sugar and calories and most of the bottles they come in are never recycled.
Antioxidants are best supplied by eating whole foods. ▷	◁ Taking supplements may shift the balance against the formation of free radicals that are required for some beneficial functions.

Answers to **Got It?, Visualize This, Working with Data, Sounds Right, But Is It?, Show You Know, and Chapter Review** questions can be found in the **Answers** section at the back of the book.

4 Body Weight and Health

Enzymes, Metabolism, and Cellular Respiration

Are overweight people less healthy than thin people?

Overweight people—are they unhealthy and lacking in self-control? If they could learn to control their appetites and exercise more, would their weight and health problems be cured? Why is it so hard for so many overweight and obese people to lose weight and keep it off with diet and exercise, even for those who demonstrate self-discipline and drive in many other areas of their lives?

The popular television series *America's Biggest Loser* chronicles the struggles of obese Americans who diet and exercise excessively to lose weight rapidly. One contestant, Sean Algaier, started the competition at 444 lb. He lost an impressive 155 lb during the contest, and others who have competed on the show have also shed astonishing amounts. But are contestants able to keep the weight off?

A study by scientists at the National Institutes of Health followed 14 contestants from Algaier's cohort who agreed to participate in the study. Almost all of those who participated in the follow-up study regained all or most of their weight postcompetition.

Extreme dieting can result in impressive weight loss.

Such weight gain is also typical of most people who lose weight on their own. Most diets fail and even dieters who do lose weight end up feeling like actual losers when they can't keep the weight off.

Yet, we continue to encourage not only contestants on this show but also ourselves and those we care about to lose weight because we so strongly believe that being thin will make us not only happier but also healthier. Are we right about this?

But is being overweight really that unhealthy?

No good can come of denying the negative health consequences of obesity, including increased risk of heart disease, stroke, and type 2 diabetes. But how overweight does one need to be to be at risk of developing these diseases? In addition, many people assume that it is better to be underweight than overweight. Is this true, or is it based on our subjective ideas of beauty? To better understand what science tells us about weight and health, we will first look at the biological factors affecting how much body fat one stores.

4.1 Enzymes and Metabolism

The amount of fat that a given individual will store depends partly on how quickly or slowly he or she breaks down food molecules into their component parts. **Metabolism** is a general term used to describe all of the chemical reactions occurring in the body.

Enzymes

All metabolic reactions are regulated by proteins called **enzymes** that speed up, or **catalyze,** chemical reactions in cells. Enzymes can help break down or build up substances, or build more complex substances from simpler ones. The enzymes that help your body break down the foods you ingest liberate the energy stored in the food's chemical bonds. Enzymes are usually named for the reaction they catalyze and their names end in the suffix **-ase**. For example, sucr**ase** is the enzyme that breaks down the table sugar sucrose.

-ase is a common suffix in names of enzymes.

To break chemical bonds, molecules must absorb energy from their surroundings, often by absorbing heat. This is why heating chemical reactants will speed up a reaction. However, heating cells to an excessively high temperature can damage or kill them, in part because proteins begin to break down at high temperatures. Enzymes do not require heat to catalyze the body's chemical reactions; they break chemical bonds without damaging or killing cells.

Activation Energy. The energy required to start the metabolic reaction serves as a barrier to catalysis and is called the **activation energy** (**FIGURE 4.1**). If not for the activation energy barrier, all of the chemical reactions in your cells would occur all the time, whether the products of the reactions were needed or not. Because most metabolic reactions need to surpass the activation energy barrier before proceeding, they can be regulated by enzymes. In other words, a given chemical reaction will occur when the correct enzyme is present. How do enzymes decrease the activation energy barrier?

FIGURE 4.1 Activation energy. (a) The activation energy barrier present in cells can be likened to an uphill bike ride. Once you are at the top of the hill, it takes much less energy to continue moving forward. (b) If you smooth out the grade of the hill, more people will make it. In cells, there is an energy barrier that prevents chemical reactions from occurring. Adding an enzyme helps lower this barrier.

Induced Fit. The chemicals that are metabolized by an enzyme-catalyzed reaction are called the enzyme's **substrate.** Enzymes decrease activation energy by binding to their substrate and placing stress on its chemical bonds, decreasing the amount of initial energy required to break the bonds. The region of the enzyme where the substrate binds is called the enzyme's **active site.** Each active site has its own shape and chemical climate. When the substrate binds to the active site, the enzyme changes shape slightly to envelop the substrate. This shape change by the enzyme in response to substrate binding results in stress being placed on the bonds of the substrate, which then alters the substrate. This process is called the **induced fit** model of enzyme catalysis. When the enzyme changes shape, it binds to the substrate more tightly, making it easier to break the substrate's chemical bonds. In this manner, the enzyme helps convert the substrate to a reaction product and then resumes its original shape so that it can perform the reaction again (**FIGURE 4.2**).

Each enzyme catalyzes a particular reaction—a property called **specificity.** The specificity of an enzyme is the result of its shape and the shape of its active site. Different enzymes have unique shapes because they are composed of different amino acids in varying sequences. The 20 amino acids, each with its own unique side group, are arranged in distinct numbers and orders for each enzyme, producing enzymes of all shapes and sizes, each with an active site that can bind with its particular substrate.

Enzymes mediate all of the metabolic reactions occurring in an organism's cells. Because enzymes, like all proteins, are coded for by genes, the amount of body fat a person stores is affected by many factors. Some factors can be controlled—for example, how much food you eat and how much you exercise. Other factors cannot be controlled—for example, which genes you inherit and how fast or slow your metabolic enzymes work.

Metabolism

The speed and efficiency of many different enzymes will lead to an overall increase or decrease in the rate at which a person can break down food. Thus, when you say that your metabolism is slow or fast, you are actually referring to the speed at which enzymes catalyze chemical reactions in your body.

A person's **metabolic rate** is a measure of his or her energy use. This rate changes according to the person's activity level. For example, we require less energy when asleep than we do when awake. The **basal metabolic rate**

VISUALIZE THIS

Is the enzyme itself permanently altered by the process of catalysis? Explain your answer.

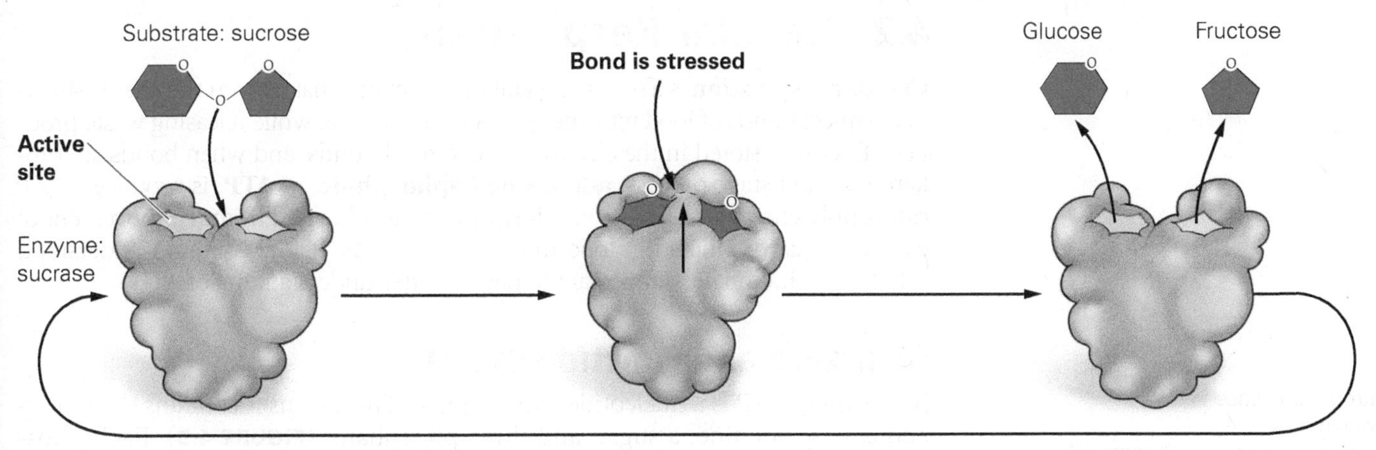
Substrate: sucrose
Active site
Enzyme: sucrase
Bond is stressed
Glucose
Fructose
1 The shape of the substrate matches the shape of the enzyme's active site.
2 Initial substrate binding to the active site changes the shape of the active site, inducing the substrate to fit even more snugly in the active site, stressing the bonds of the substrate.
3 The shape change splits the substrate and releases the two subunits. The enzyme is able to perform the reaction again.

FIGURE 4.2 Enzymes. The enzyme sucrase is cleaving (splitting) the disaccharide sucrose into its monosaccharide subunits.

represents the energy use of a resting wakeful person. The average basal metabolic rate is 70 calories per hour, or 1,680 calories per day. However, this rate varies widely among individuals because many factors influence each person's basal metabolic rate, including exercise habits, biological sex, and genetics.

During exercise, metabolic rate increases. In addition to burning calories during exercise, metabolic rate remains elevated for a period of time after exercise. The length of time the metabolic rate is elevated is a function of the intensity of the exercise.

Your biological sex also influences your metabolic rate. Males require more calories per day than females do because testosterone, a hormone produced in larger quantities in males, increases the rate at which fat breaks down. Men also have a higher percentage of muscle than women, which requires more energy to maintain than fat does.

Other genetic factors play a substantial role in determining body weight. Two people of the same size and sex, who consume the same number of calories and exercise the same amount, will not necessarily store the same amount of fat. Some people are simply born with lower basal metabolic rates. Genes that influence a person's rate of fat storage and utilization, like all genes, are passed from parents to children.

To be completely metabolized, food must be broken down by the digestive system and then transported to individual cells via the bloodstream. Once inside cells, food energy can be converted into chemical energy by the process of cellular respiration.

Got It?

1. Enzymes are proteins that ________, or speed up, chemical reactions.
2. Enzymes only catalyze one reaction, a property called ________.
3. To catalyze a reaction, the enzyme binds its ________ and bends it slightly, placing stress on chemical bonds.
4. Enzyme-catalyzed reactions are prevented from occurring at a rapid rate at all times by the presence of a(n) ________ energy barrier.
5. The ________ metabolic rate is the energy use of a resting wakeful person.

4.2 Cellular Respiration

Cellular respiration is a series of metabolic reactions that convert the energy stored in chemical bonds of food into energy that cells can use, while releasing waste products. Energy is stored in the electrons of chemical bonds, and when bonds are broken in a multistage process, **adenosine triphosphate,** or **ATP,** is produced. ATP can supply energy to cells because it stores energy obtained from the movement of electrons that originated in food into its own bonds. Before trying to understand cellular respiration, it is important to have a better understanding of ATP.

Structure and Function of ATP

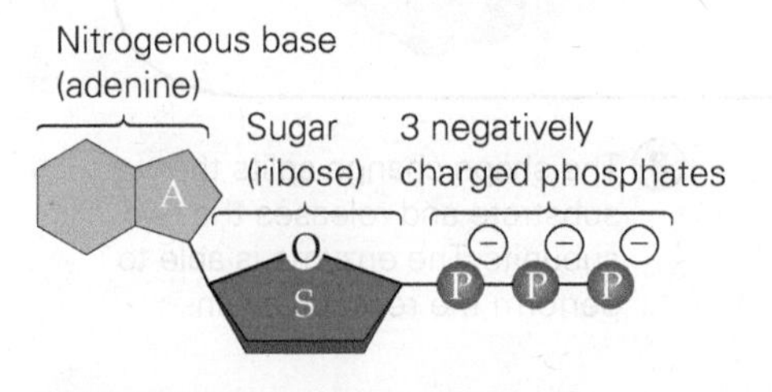

FIGURE 4.3 The structure of ATP. ATP is a nucleotide triphosphate.

Structurally, ATP is a nucleotide triphosphate. This means it contains the nitrogenous base adenine, a sugar, and three phosphates (**FIGURE 4.3**). Each phosphate in the series of three is negatively charged. This series of negative charges repel each other and results in an energy that is similar to the energy you can feel when you try to hold the negative poles of two magnets together.

Removal of the terminal phosphate group of ATP releases energy that can be used to perform cellular work. In this manner, ATP behaves much like a coiled spring. To think about this, imagine loading a dart gun. Pushing the dart into the gun requires energy from your arm muscles, and the energy you exert will be stored in the coiled spring inside the dart gun (**FIGURE 4.4**). When you shoot the dart gun, the energy is released from the gun and used to perform some work—in this case, sending a dart through the air.

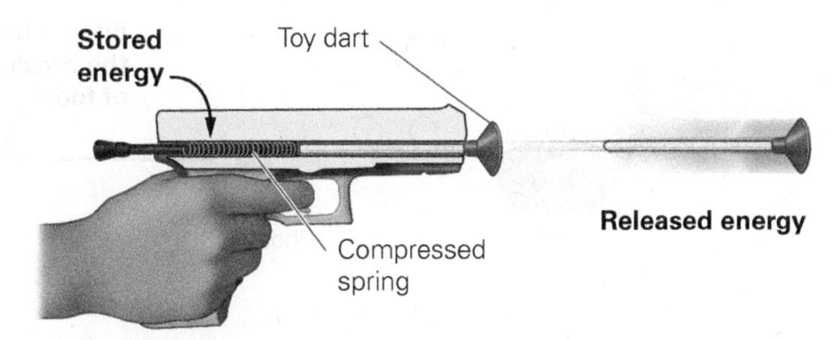

FIGURE 4.4 Stored energy. A dart gun uses energy stored in the coiled spring and supplied by the arm muscle to perform the work of propelling a dart.

The phosphate group that is removed from ATP can be transferred to another molecule. This process, called **phosphorylation,** energizes the molecule that receives the phosphate. When a molecule, say an enzyme, needs energy, the phosphate group is transferred from ATP to the enzyme, and the enzyme undergoes a change in shape that allows the enzyme to perform its job. After removal of a phosphate group, ATP becomes **adenosine diphosphate (ADP)** (**FIGURE 4.5**). The energy released by the removal of the outermost phosphate of ATP can be used to help cells perform many different kinds of work. ATP helps power mechanical work such as the movement of cells, transport work such as the movement of substances across membranes during active transport, and chemical work such as the making of complex molecules from simpler ones (**FIGURE 4.6**).

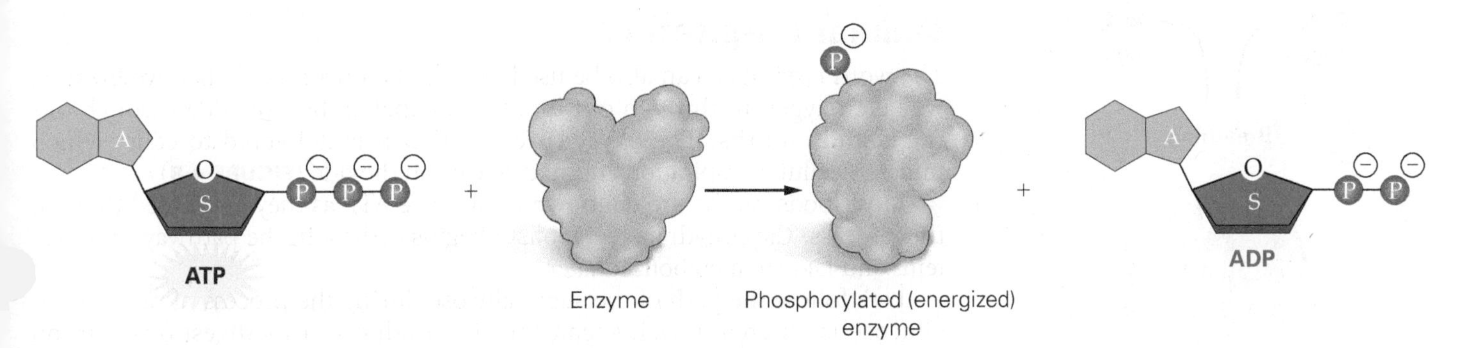

FIGURE 4.5 Phosphorylation. The terminal phosphate group of an ATP molecule can be transferred to another molecule, in this case an enzyme, to energize it. When ATP loses a phosphate, it becomes ADP.

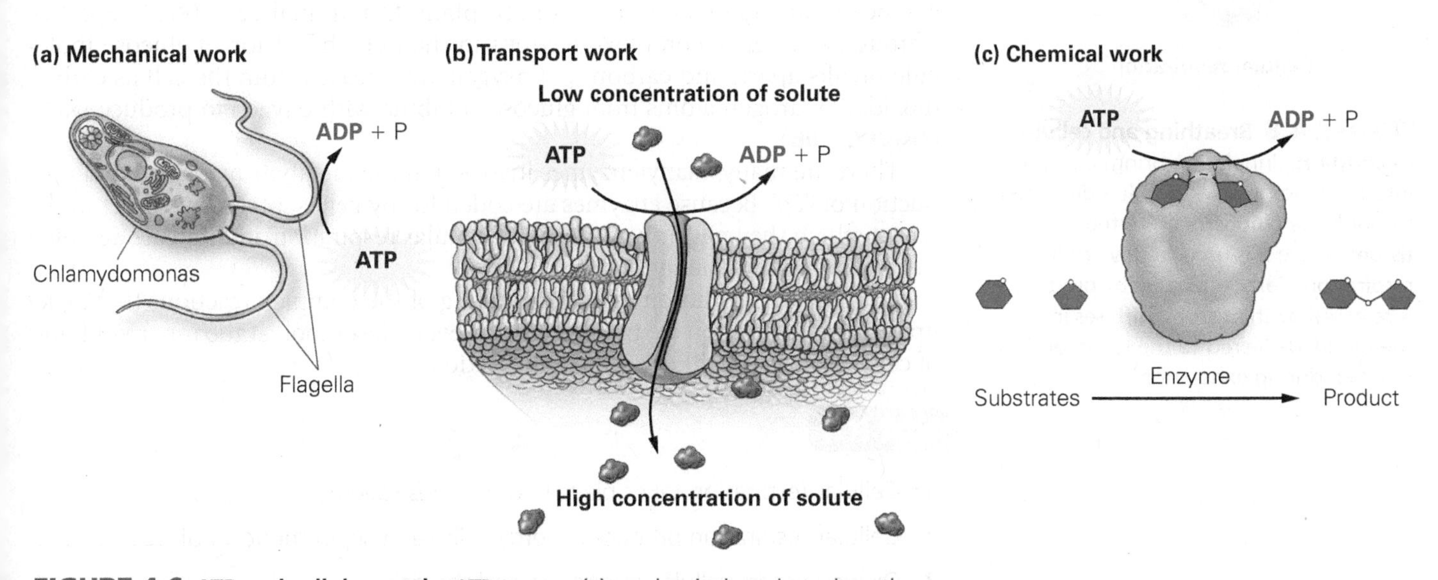

FIGURE 4.6 ATP and cellular work. ATP powers (a) mechanical work, such as the moving of the whiplike flagella of this single-celled green algae; (b) transport work, such as the active transport of a substance across a membrane from its own low to high concentration; and (c) chemical work, such as the enzymatic conversion of substrates to a product.

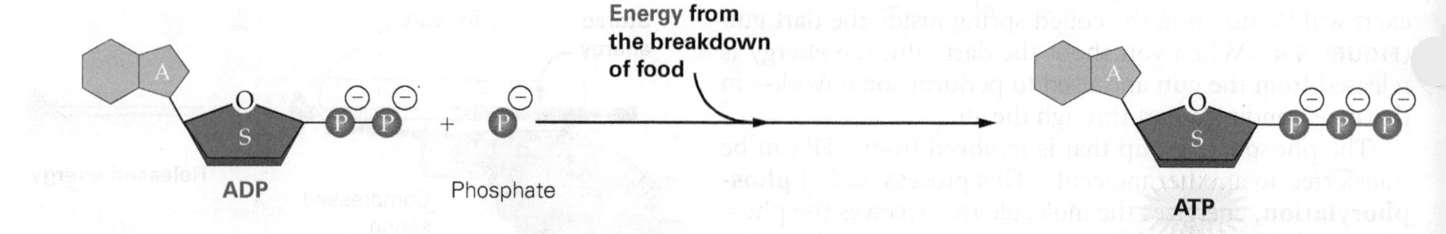

FIGURE 4.7 Regenerating ATP. ATP is regenerated from ADP and phosphate during the process of cellular respiration.

Cells continuously use ATP. Exhausting the supply of ATP means that more ATP must be regenerated. ATP is synthesized by adding back a phosphate group to ADP during the process of cellular respiration (**FIGURE 4.7**). During this process, oxygen is consumed, and water and CO_2 are produced. Because some of the steps in cellular respiration require oxygen, they are said to be **aerobic** reactions, and this type of cellular respiration is called **aerobic respiration.**

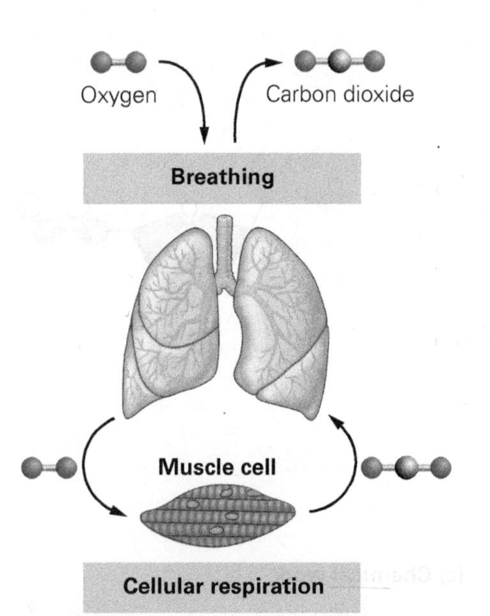

FIGURE 4.8 Breathing and cellular respiration. Inhalation brings oxygen into the lungs, from which it is delivered through the bloodstream to the tissues where it is used to drive cellular respiration. Carbon dioxide by-product is released by the tissues, diffuses into the blood, delivered to the lungs, and released during exhalation.

Cellular Respiration

The word *respiration* can also be used to describe breathing. When we breathe, we take oxygen in through our lungs and expel carbon dioxide through our noses and mouths. The oxygen we breathe in is delivered to cells, which undergo cellular respiration and release carbon dioxide (**FIGURE 4.8**).

Most foods can be broken down to produce ATP as they are routed through this process. Carbohydrate metabolism begins earliest in the pathway, and proteins and fats are metabolized later.

Let's follow the path of the sugar glucose during the process of respiration. Glucose is an energy-rich sugar, but the products of its digestion—carbon dioxide and water—are energy poor. So where does the energy go? The energy released during the conversion of glucose to carbon dioxide and water is used to synthesize ATP.

Many of the chemical reactions in this process occur in the mitochondria, a subcellular organelle found in both plant and animal cells (**FIGURE 4.9a**). Through a series of complex reactions in the mitochondrion, a glucose molecule breaks apart, and carbon and oxygen are released from the cell as carbon dioxide. Hydrogen atoms from glucose combine with oxygen to produce water (**FIGURE 4.9b**).

There are many, many enzymes involved in the metabolism of food and production of ATP. Because enzymes are coded for by genes, organisms differ in the rate at which they perform each step of cellular respiration and the rate at which they store unused energy as fat.

To gain a more thorough understanding of the complex reactions that make up cellular respiration we need to take an even closer look at the inner workings of cells, which we will do in the next section.

Got It?

1. Cellular respiration takes place in organelles called ________.
2. Cellular respiration produces energy cells can use, in the form of ________.
3. Oxygen enters cellular respiration and the gas ________ is produced.
4. Cellular respiration requires oxygen, making it a(n) ________ process.
5. Enzymes are proteins that are coded for by ________.

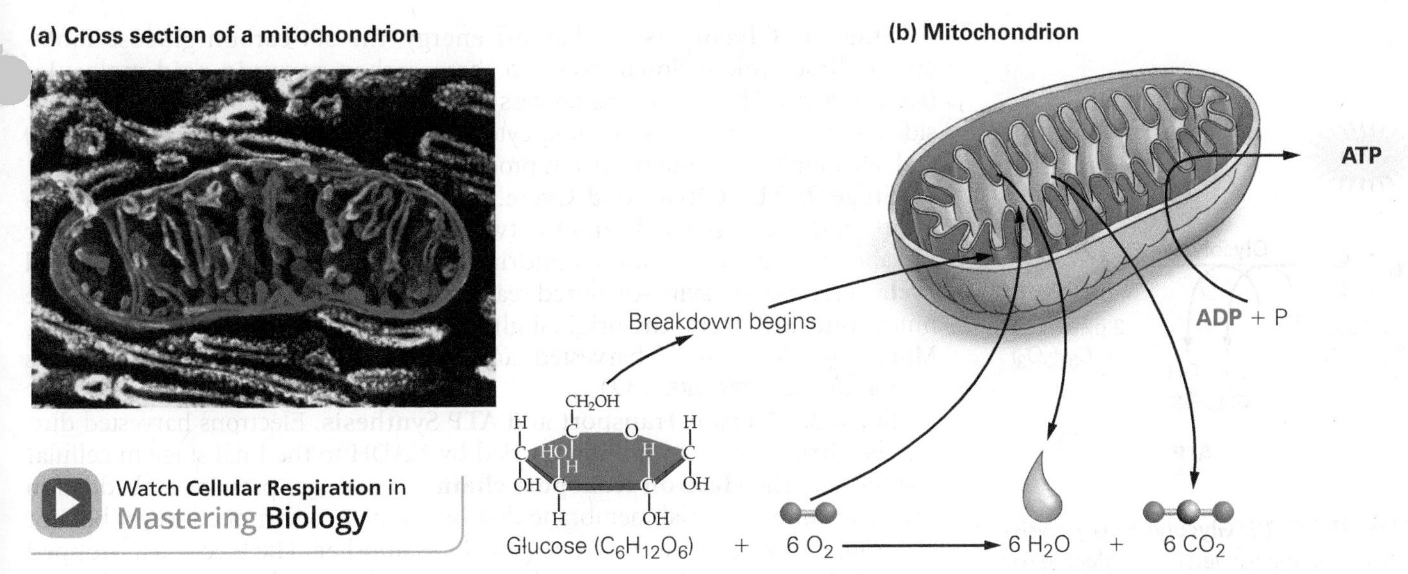

Watch **Cellular Respiration** in Mastering **Biology**

FIGURE 4.9 Overview of cellular respiration. (a) Much of the process of cellular respiration takes place within the mitochondrion. (b) The breakdown of glucose by cellular respiration requires oxygen and ADP plus phosphate. The energy stored in the bonds of glucose is harvested to produce ATP (from ADP and P), releasing carbon dioxide and water.

Stages of Cellular Respiration

Cellular respiration occurs in three stages. Electrons are removed from glucose during the early stages and are used to make ATP in the final stage. These electrons do not simply float around in the meantime because this would damage cell structures. Instead, they are carried by molecules called *electron carriers*. One of the electron carriers used by cellular respiration is a chemical called **nicotinamide adenine dinucleotide (NAD$^+$).** You can think of this molecule as a sort of taxicab for electrons. The empty taxicab (NAD$^+$) picks up electrons. A hydrogen atom contains one H^+ ion and one electron. When NAD$^+$ picks up a hydrogen atom it becomes NADH. The full taxicab (NADH) carries electrons to their destination, where they are dropped off, and the empty taxicab (once again NAD$^+$) returns for more electrons (**FIGURE 4.10**). NADH will deposit its electrons for use in the final step of cellular respiration.

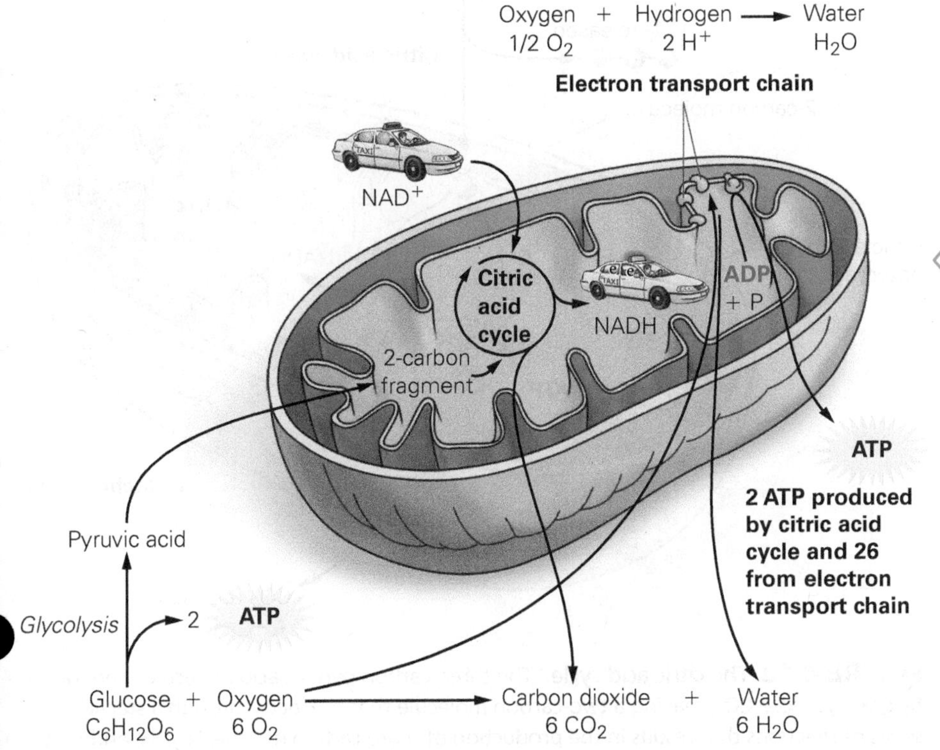

FIGURE 4.10 Electron carriers. Electron carriers in the cell are like taxicabs, shuttling electrons from the original glucose molecule to the final stage of respiration.

VISUALIZE THIS

NAD$^+$ (the empty cab) picks up one positively charged hydrogen atom and how many electrons to give the neutral NADH molecule?

Stage 1: Glycolysis. To harvest energy, the six-carbon glucose molecule is first broken down into two three-carbon **pyruvic acid** molecules (**FIGURE 4.11**). This part of the process of cellular respiration actually occurs outside the mitochondria, in the fluid cytosol. **Glycolysis** does not require oxygen and, after an initial input of energy, produces a net total of two molecules of ATP.

glyco- means sugar.

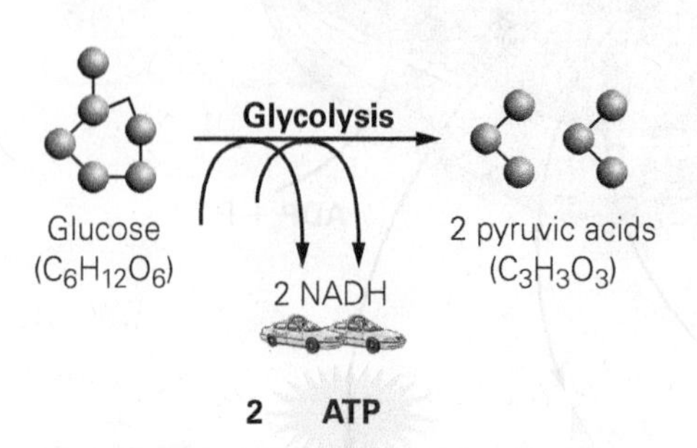

FIGURE 4.11 Glycolysis. Glycolysis, the enzymatic conversion of glucose to two pyruvic acids, also produces a small amount of NADH and ATP.

Stage 2: The Citric Acid Cycle. After glycolysis, the pyruvic acid loses a carbon dioxide molecule and the two-carbon fragment that is left is further metabolized inside the mitochondria. This fragment enters the **citric acid cycle,** a series of enzyme-catalyzed reactions that take place in the matrix of the mitochondrion. Here, the original glucose molecule is further broken down. More of its electrons are harvested, and the remaining carbons are released as carbon dioxide (**FIGURE 4.12**).

Stage 3: Electron Transport and ATP Synthesis. Electrons harvested during the citric acid cycle are now carried by NADH to the final stage in cellular respiration. The **electron transport chain** is a series of proteins embedded in the inner mitochondrial membrane that functions as a sort of conveyer belt for electrons, moving them from one protein to another. The electrons, dropped off by NADH molecules generated during glycolysis and the citric acid cycle, move toward the bottom of the electron transport chain toward the matrix of the mitochondrion, where they combine with oxygen to produce water.

Each time an electron is picked up by a protein or handed off to another protein, the protein moving the electron changes shape. This shape change allows the movement of hydrogen ions (H^+) from the matrix of the mitochondrion to the intermembrane space of the mitochondrion. Therefore, although the proteins in the electron transport chain are moving electrons down the electron transport chain toward oxygen, they are also moving H^+ ions across the inner mitochondrial membrane and into the intermembrane space. This movement decreases the concentration of H^+ ions in the matrix and increases their concentration within the intermembrane space. Whenever a concentration

VISUALIZE THIS

How many turns of the citric acid cycle would it take to release four of the original six carbons of glucose as carbon dioxide?

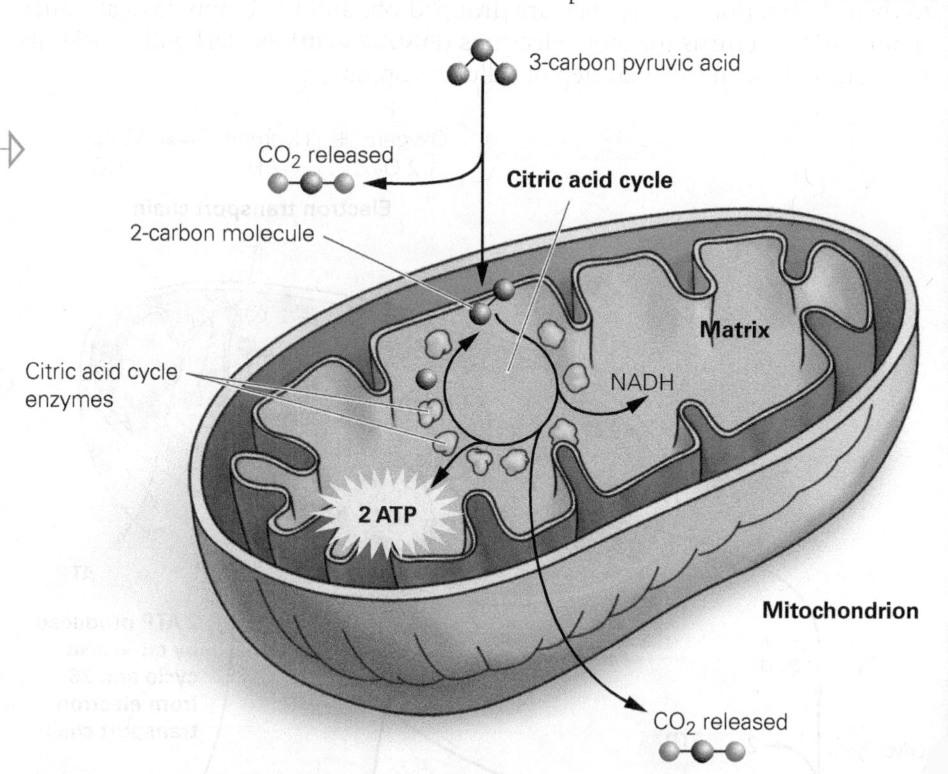

FIGURE 4.12 The citric acid cycle. The three-carbon pyruvic acid molecules generated by glycolysis lose CO_2, leaving a two-carbon molecule that proceeds through a stepwise series of reactions that results in the production of more carbon dioxide, NADH, and ATP.

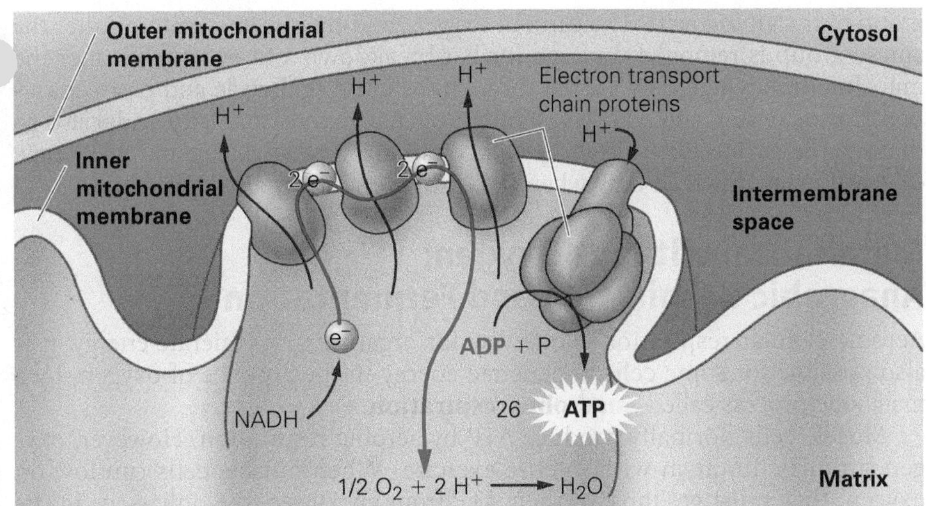

FIGURE 4.13 The electron transport chain of the inner mitochondrial membrane. NADH brings electrons to the electron transport chain. As electrons move through the proteins of the electron transport chain, hydrogen ions are pumped into the intermembrane space. Hydrogen ions flow back through an ATP synthase protein, which converts ADP to ATP. In this manner, energy from electrons added to the electron transport chain is used to produce ATP.

VISUALIZE **THIS**

What is meant by 1/2 O_2 in this figure?

gradient exists, molecules will diffuse from an area of their own high concentration to an area of their own low concentration. Because charged ions cannot diffuse across the hydrophobic core of the membrane, they escape through a protein channel in the membrane called **ATP synthase,** which allows the conversion of the energy in the proton gradient into energy the cell can use.

Although you may not understand exactly how it happens, you are probably aware that hydroelectric dams convert the energy present in water as it rushes through a mechanical turbine into electricity. You can think of the ATP synthase as converting forms of energy in a similar manner. This enzyme uses the energy generated by the rushing H^+ ions to synthesize 26 ATP from ADP and phosphate (**FIGURE 4.13**).

Metabolism of Other Nutrients

Most cells can break down not only carbohydrates but also proteins and fats. **FIGURE 4.14** shows the points of entry during cellular respiration for proteins and fats. Protein is broken down into component amino acids, which are then used to synthesize new proteins. Most organisms can also break down proteins to supply energy. However, this process takes place only when fats or carbohydrates are unavailable. In humans and other animals, the first step in producing energy from the amino acids of a protein is to remove the nitrogen-containing amino

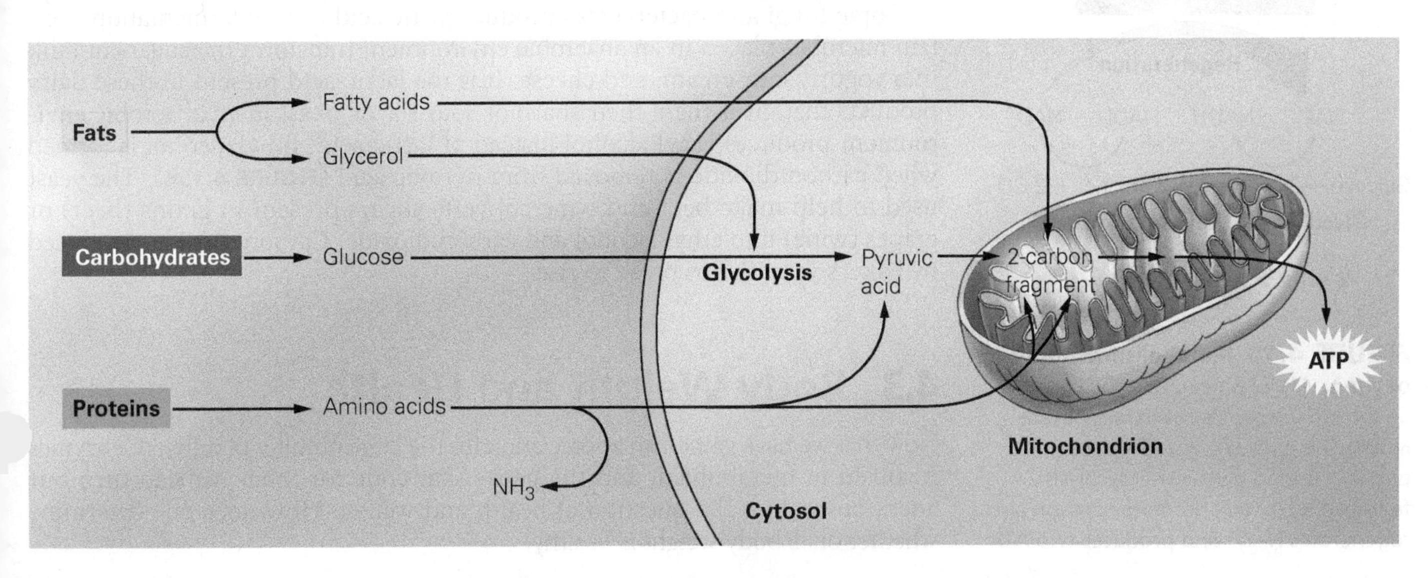

FIGURE 4.14 Metabolism of other macromolecules. Carbohydrates, proteins, and fats can all undergo cellular respiration; they just feed into different parts of the metabolic pathway.

group of the amino acid. The carbon, oxygen, and hydrogen remaining after the amino group is removed undergo further breakdown and eventually enter the mitochondria, where they are fed through the citric acid cycle and produce carbon dioxide, water, and ATP. The subunits of fats (glycerol and fatty acids) also go through the citric acid cycle and produce carbon dioxide, water, and ATP. Some cells will break down fat only when carbohydrate supplies are depleted.

Metabolism without Oxygen: Anaerobic Respiration and Fermentation

an- means absence of.

Aerobic cellular respiration is one way for organisms to generate energy. It is also possible for some cells to generate energy in the absence of oxygen, by a metabolic process called **anaerobic respiration.**

Muscle cells normally produce ATP by aerobic respiration. However, oxygen supplies diminish with intense exercise. When muscle cells run low on oxygen, they must get most of their ATP from glycolysis, the only stage in the cellular respiration process that does not require oxygen. When glycolysis happens without aerobic respiration, cells can run low on the electron carrier NAD^+ because it is converted to NADH during glycolysis. When this happens, cells use a process called **fermentation** to regenerate NAD^+.

Fermentation cannot, however, be used for very long because one of the by-products of this reaction leads to the buildup of a compound called lactic acid. Lactic acid is produced by the actions of the electron acceptor NADH, which has no place to dump its electrons during fermentation because there is no electron transport chain and no oxygen to accept the electrons. Instead, NADH deposits its electrons by giving them to the pyruvic acid produced by glycolysis (**FIGURE 4.15a**). Lactic acid is transported to the liver, where liver cells use oxygen to convert it back to pyruvic acid.

This requirement for oxygen—to convert lactic acid to pyruvic acid—explains why you continue to breathe heavily even after you have stopped working out. Your body needs to supply oxygen to your liver for this conversion, sometimes referred to as paying back your oxygen debt. The accumulation of lactic acid also explains why fatigued muscles will "burn" and the phenomenon called "hitting the wall." Anyone who has ever felt as though her legs were turning to wood while running or biking knows this feeling. When your muscles are producing lactic acid by fermentation for a long time, the oxygen debt becomes too large, and muscles shut down until the rate of oxygen supply outpaces the rate of oxygen use, restoring proper feeling to your legs.

Some fungi and bacteria also produce lactic acid during fermentation. Certain microbes placed in an anaerobic environment transform the sugars in milk into yogurt, sour cream, and cheese. It is the lactic acid present in these dairy products that gives them their sharp or sour flavor. Yeast in an anaerobic environment produces ethyl alcohol instead of lactic acid. Ethyl alcohol is formed when carbon dioxide is removed from pyruvic acid (**FIGURE 4.15b**). The yeast used to help make beer and wine converts sugars present in grains (beer) or grapes (wine) into ethyl alcohol and carbon dioxide. Carbon dioxide, produced by baker's yeast, helps bread to rise.

(a) Human muscle

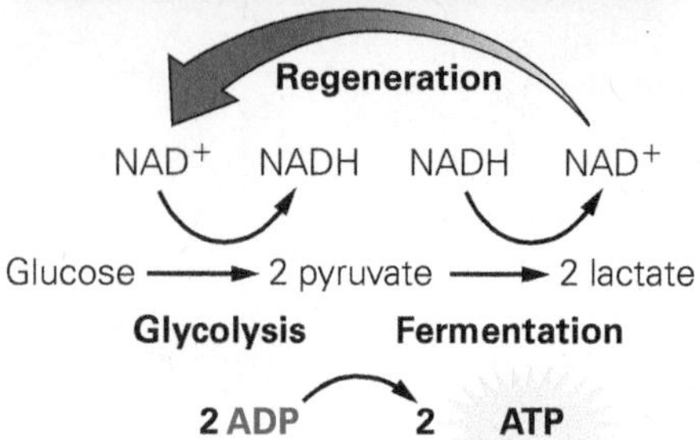

(b) Yeast

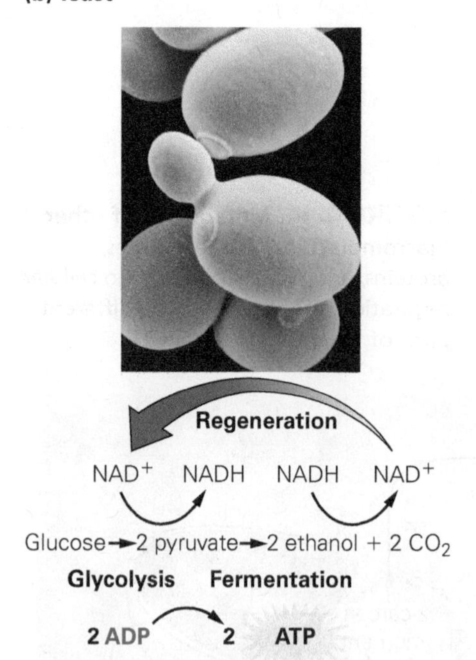

FIGURE 4.15 Metabolism without oxygen. Glycolysis can be followed by (a) lactate fermentation to regenerate NAD^+. This pathway also produces two ATP during glycolysis. Glycolysis followed by (b) alcohol fermentation also regenerates NAD^+ and produces two ATP.

4.3 Body Weight and Health

Now that we have gained an appreciation for the large number of different enzymes involved in metabolism, and the genes that code for them, we can turn our attention back to the question of health and weight. How does one determine whether one's body weight is healthy?

Body Mass Index

One method is to use a tool called the Body Mass Index, or BMI (TABLE 4.1). Your BMI is a value calculated by using your height and weight as an estimate of weight-related risk of illness and death. The BMI classifies individuals into five separate categories: underweight, normal weight, overweight, moderate obesity, or severe obesity.

However, the BMI categories are not as accurate as we would like them to be. Studies show that as many as one in four people may be misclassified by BMI tables because this measurement provides no means to distinguish between lean muscle mass and body fat. For example, an athlete with a lot of muscle will weigh more than a similar-sized person with a lot of fat because muscle is more dense than fat. Therefore, a person can be very fit but be classified as overweight using this table.

Underweight Is Unhealthy

If your BMI is below 18.5, you are at risk for **anorexia,** or self-starvation, a disease rampant on college campuses. Estimates are that one in five college women and one in 20 college men restrict their food intake so severely that they are

WORKING WITH DATA

What is the BMI of a person who is 6 feet tall and 200 lb?

TABLE 4.1 Body mass index (BMI). BMI values are calculated based on height (*y*-axis) and weight (in lb inside the table).

4'10"	91	96	100	105	110	115	119	124	129	134	138	143	167	191
4'11"	94	99	104	109	114	119	124	128	133	138	143	148	173	198
5'0"	97	102	107	112	118	123	128	133	138	143	148	153	179	204
5'1"	100	106	111	116	122	127	132	137	143	148	153	158	185	211
5'2"	104	109	115	120	126	131	136	142	147	153	158	164	191	218
5'3"	107	113	118	124	130	135	141	146	152	158	163	169	197	225
5'4"	110	116	122	128	134	140	145	151	157	163	169	174	204	232
5'5"	114	120	126	132	138	144	150	156	162	168	174	180	210	240
5'6"	118	124	130	136	142	148	155	161	167	173	179	186	216	247
5'7"	121	127	134	140	146	153	159	166	172	178	185	191	223	255
5'8"	125	131	138	144	151	158	164	171	177	184	190	197	230	262
5'9"	128	135	142	149	155	162	169	176	182	189	196	203	236	270
5'10	132	139	146	153	160	167	174	181	188	195	202	209	243	278
5'11"	136	143	150	157	165	172	179	186	193	200	208	215	250	286
6'0"	140	147	154	162	169	177	184	191	199	206	213	221	258	294
6'1"	144	151	159	166	174	182	189	197	204	212	219	227	265	302
6'2"	148	155	163	171	179	186	194	202	210	218	225	233	272	311
6'3"	152	160	168	176	184	192	200	208	216	224	232	240	279	319
6'4"	156	164	172	180	189	197	205	213	221	230	238	246	287	328
BMI	**19**	**20**	**21**	**22**	**23**	**24**	**25**	**26**	**27**	**28**	**29**	**30**	**35**	**40**

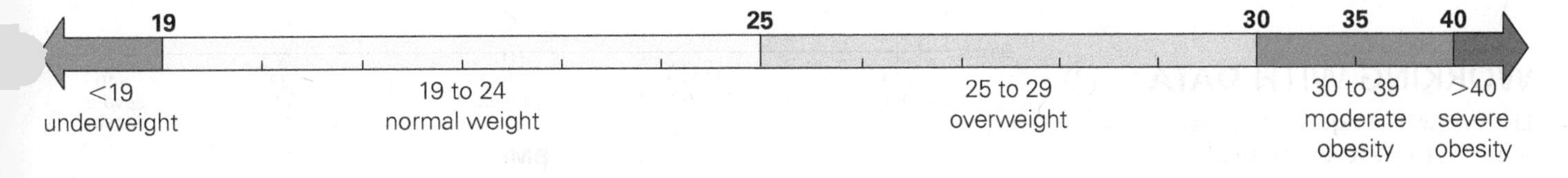

essentially starving themselves to death. Self-starvation can also occur when people allow themselves to eat—sometimes very large amounts of food (called binge eating)—but prevent the nutrients from being turned into fat by purging themselves, by vomiting or using laxatives. Binge eating followed by purging is called **bulimia.**

Anorexia has long-term health consequences. The disease can starve heart muscles to the point that altered rhythms develop. The lack of body fat in anorexia can also lead to the cessation of menstruation, a condition known as amenorrhea, which occurs when a protein called leptin, secreted by fat cells, signals the brain that there is not enough body fat to support pregnancy. Hormones (such as estrogen) that regulate menstruation are therefore blocked, and menstruation ceases. Amenorrhea can be permanent and causes sterility in a substantial percentage of people with anorexia. The damage done by the lack of estrogen is not limited to the reproductive system; bones are affected as well. Estrogen secreted by the ovaries during the menstrual cycle acts on bone cells to help them maintain their strength and size. Anorexics have reduced development of bone and put themselves at higher risk of broken bones as a result of a condition called **osteoporosis.**

If your BMI falls within the normal weight range you have no reason to worry about health risks from body weight. In fact, studies show that even those in the overweight and moderate obesity categories have less to worry about than we once thought. Many studies, including a recent meta-analysis (study of hundreds of studies), show that being overweight is actually associated with less mortality than being normal weight. **FIGURE 4.16** shows the relative risk of death for weight classes. Relative risk of death is the probability of death occurring in a population with the condition versus those without the condition.

Although moderate obesity is associated with slightly increased mortality, it turns out that subdividing the moderate obesity category into two subcategories gives a different result. The BMI classification for moderate obesity spans the 30 to 40 range. The trend today is to subdivide that category into Grade 1 obesity, spanning the 30 to 34.9 range and Grade 2 obesity and a BMI in the 35 to 39.9 range. This subdivision makes sense in light of the fact that Grade 1 obesity is not associated with any increased health risks, whereas Grades 2 and 3 are associated with the health issues we typically think of as occurring more frequently in the obese. Under this scheme, a BMI of 40 or higher is Grade 3 obesity.

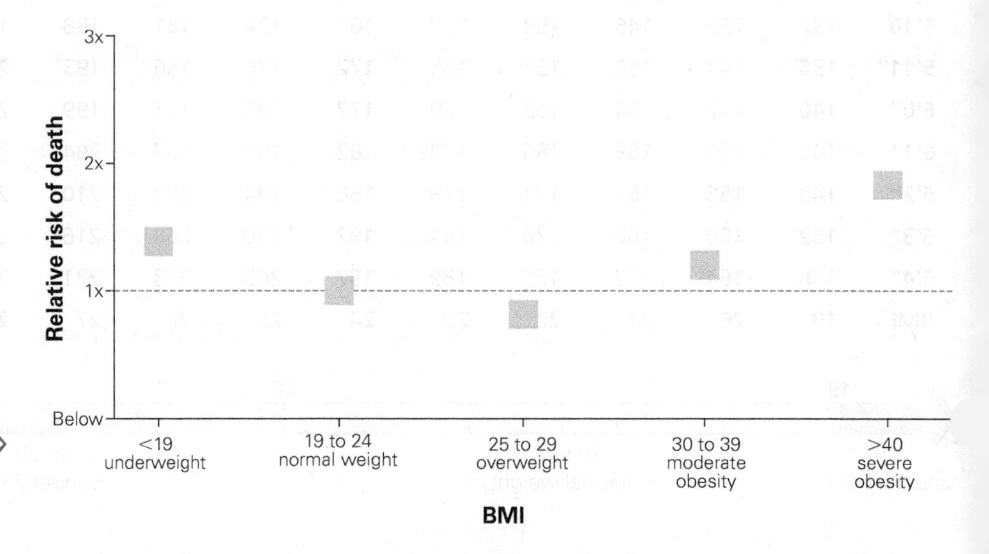

FIGURE 4.16 Risk of death and BMI. This graph shows the risk of death by any cause among adults in different BMI categories compared with the risk of death among normal-weight adults. Normal-weight adults are the control and therefore have a relative risk of death of 1×. The blue boxes indicate the average relative risk of death for each BMI classification.

WORKING WITH DATA

List the BMI categories in order of greatest to least risk of death.

So what do we do with this information? Body weight is very difficult for many individuals to control and the health consequences of overweight, while very real, don't seem to start until a person becomes quite obese. One way to prevent obesity in the first place may be to not participate in the kind of extreme calorie restriction and exercise regimens that contestants in *America's Biggest Loser* use to shed pounds. Study after study has shown that extreme dieting actually causes a body to become more efficient at using calories, and that maintaining weight loss is the exception rather than the norm (**FIGURE 4.17**).

For example, a study published in the peer-reviewed journal *Obesity* showed that the leptin levels of *America's Biggest Loser* contestants may have been permanently altered by their extreme dieting. Leptin is a hormone that controls hunger. Contestants had normal leptin levels before they started dieting, but by the season's finale, their leptin had diminished markedly, causing constant hunger. As the former contestants gained weight, their leptin increased a bit, but 6 years later was still not back to the level of leptin present at the beginning of the diet.

It also appears that the body alters the metabolic rate in response to extreme dieting, causing metabolism to slow. Most of the contestants studied now burn calories at a slower rate than expected. Six years after the show, Sean Algaier was burning around 450 fewer calories a day than would be expected for a man his size, and he weighed more than he did at the start of the competition. Adaptations in leptin levels and metabolic rates help explain why those who participate in extreme dieting have such difficulty maintaining their weight loss.

WORKING WITH DATA

(a) What percentage of the 13 contestants weigh the same or less 6 years after the season finale? (b) What percentage of the 13 contestants in the study weigh more than they did after the season finale?

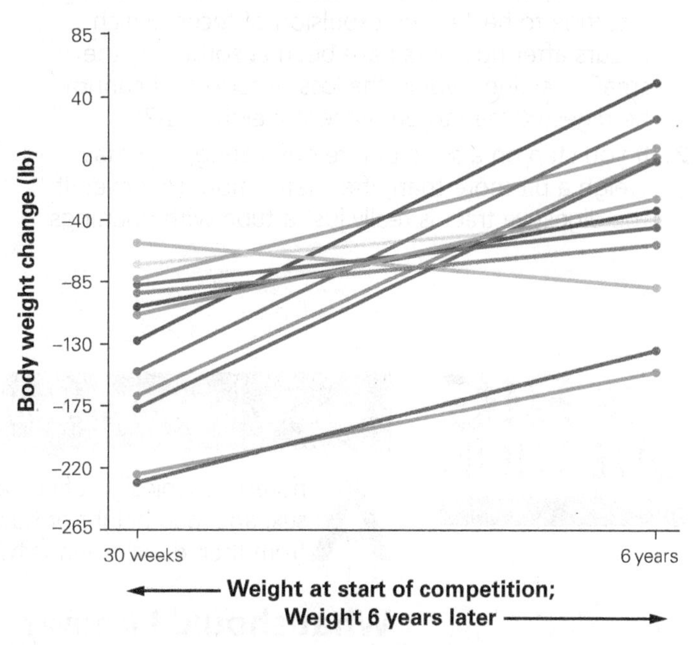

FIGURE 4.17 Biggest losers 6 years later. Body weights of 13 contestants from one season of *America's Biggest Loser* are shown at the season finale and again 6 years later.

Got It?

1. Self-starvation, or ________, can lead to permanent sterility.
2. Binge eating followed by purging is called ________.
3. The protein hormone ________ regulates hunger and alerts the brain if there is not enough body fat to support a pregnancy.
4. It appears that extreme ________ can alter leptin levels, leading to increased difficulty with maintaining weight loss.
5. People who have lost weight by extreme dieting may require ________ calories to maintain their weight than other similar sized people.

One ill-conceived strategy sometimes used for rapid weight loss is consumption of laxatives, which are designed to soften fecal matter, making it easier to defecate. Because defecation can make one feel lighter or less full, some people think that using laxatives can help them to lose weight. This abuse of laxatives can permanently damage the large intestine, leading to dependence on laxatives for bowel movements, dehydration, and increased risk of colon cancer.

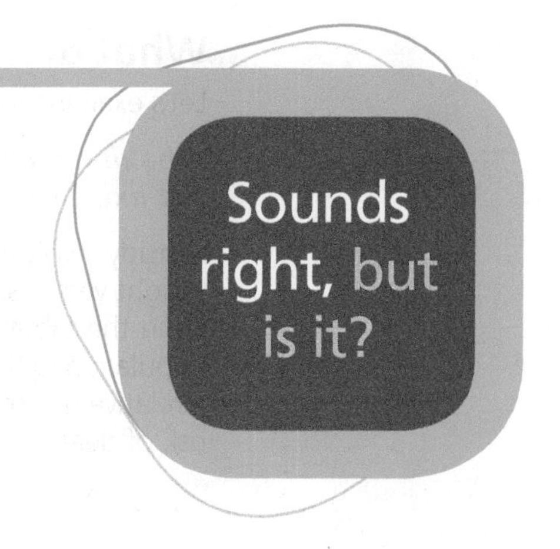

Laxatives cause permanent weight loss.

—Continued next page

Sounds right, but it isn't.

Answer the following questions to understand why.

1. Fecal matter is composed of fiber and other indigestible food remnants. It is present in the large intestine, until defecation. Laxatives act on the large intestine to hasten the expulsion of feces, which occurs after nutrients have been absorbed by the small intestine. Would the loss of food that cannot be digested lead to permanent weight loss?
2. If you step on a scale before defecating, you may weigh a bit more than after defecation. However, if the digestive tract is really just a tube with openings at the mouth and anus, is food that does not get absorbed by the small intestine ever really part of your body?
3. Losing weight involves the loss of fat. What gas do you breathe out when fat is metabolized?
4. In light of your answer to the previous question, how might exercise help with weight loss?
5. Consider your answers to questions 1–4 and explain why the original statement bolded above sounds right, but isn't.

How unhealthy is anorexia?

Anorexia is typically defined as having a body weight that is 15% below average for your sex, age, and height or having a BMI of 18.5 or less. Around 4% of anorectics will die from their disease annually.

What should I know?

What follows are some smaller questions that need to be resolved to answer the Big Question. Place a checkmark next to the questions that science can answer.

Smaller Questions	Can Science Answer?
Is having anorexia worse for your health than smoking cigarettes, drinking alcohol, or not exercising?	
Can a person with anorexia cause permanent damage to his or her health?	
Can a culture that places a high premium on thinness cause anorexia?	
Are underweight bodies unattractive?	
Is anorexia a mental health disorder?	

What does the science say?

Let's examine what the data say about this smaller question:

Is having anorexia worse for your health than smoking cigarettes, drinking alcohol, or not exercising?

A study involving more than 11,000 adults age 25 or older considered the effects of being underweight versus smoking status, alcohol consumption levels, and exercise levels. The data is presented in the table below. Recall that relative risk of death is the probability of death occurring in a population with the condition versus those without the condition. For example, nonsmokers have a relative risk of 1.0. Those who smoke cigarettes have a higher relative risk of death. The relative risk of death for a BMI of 18.5 or under is 1.73.

Behavior	Relative Risk of Death (95% confidence interval)
Smoking Cigarettes	
Never smoked	1.0
Occasional smoker	1.61
Daily smoker	2.27
Daily Alcohol Consumption	
One alcoholic beverage	0.96
One to two alcoholic beverages	1.0
More than three alcoholic beverages	1.31
Physical Activity	
Active	1.0
Moderately active	1.1
Inactive	1.43

1. Describe the results. Does it appear that anorexia is worse for your health than smoking cigarettes, drinking alcohol, or not exercising?
2. Given these data, do you think the smaller question is answered? If not, propose another study that would help answer this question.
3. Does this information help you answer the Big Question? What else do you need to consider?

Data source: H. Orpana, J. Berthelot, M. Kaplan, D. Feeny, B. McFarland, and N. Ross, "BMI and Mortality," *Obesity* 18, no. 1 (2010): 214–18.

Chapter Review

Mastering Biology

Go to Mastering Biology to access eText 2.0, Dynamic Study Modules, and the **Study Area,** where you'll find practice quizzes, BioFlix™ animations, MP3 tutor sessions, current events, and more.

SUMMARY

Section 4.1

Describe the structure and function of enzymes.

- Enzymes are proteins that catalyze specific cellular reactions. The substrate binds to the enzyme's active site and is converted to product (p. 70).

Explain how enzymes decrease a reaction's activation energy barrier.

- The binding of a substrate to the enzyme's active site causes the enzyme to change shape (induced fit), placing more stress on the bonds of the substrate and thereby lowering the amount of energy required to start the reaction (pp. 70–71).

Define metabolism.

- Metabolism includes all the chemical reactions that occur in cells to build up or break down macromolecules (pp. 71–72).

SHOW YOU KNOW 4.1 What would happen if you ingested a prescribed drug that blocked the active site of an enzyme?

Section 4.2

Describe the structure and function of ATP.

- ATP is a nucleotide triphosphate. The nucleotide found in ATP contains a sugar and the nitrogenous base adenine (p. 72).
- Cells use ATP to power energy-requiring processes (p. 73).
- Breaking the terminal phosphate bond of ATP releases energy (p. 74).

Describe the process of cellular respiration from the breakdown of glucose through the production of ATP.

- Cellular respiration begins in the cytosol, where a six-carbon sugar is broken down into two three-carbon pyruvic acid molecules during the anaerobic process of glycolysis (pp. 74–76).
- The pyruvic acid molecules then move across the two mitochondrial membranes, where they lose a CO_2. The remaining two-carbon fragment then moves into the matrix of the mitochondrion, where the citric acid cycle strips them of CO_2 and electrons (pp. 74–76).
- Electrons removed from chemicals that are part of glycolysis and the citric acid cycle are carried by electron carriers, such as NADH, to the inner mitochondrial membrane; there they are added to a series of proteins called the electron transport chain. At the bottom of the electron transport chain, oxygen pulls the electrons toward itself. As the electrons move down the electron transport chain, the energy that they release is used to drive protons (H^+) into the intermembrane space. Once there, the protons rush through the enzyme ATP synthase and produce ATP from ADP and phosphate (pp. 74–77).
- When electrons reach the oxygen at the bottom of the electron transport chain, they combine with the oxygen and hydrogen ions to produce water (p. 77).

Explain how proteins and fats are broken down during cellular respiration.

- Proteins and fats are also broken down by cellular respiration, but they enter the pathway later than does glucose (pp. 77–78).

Describe how anaerobic respiration differs from aerobic respiration.

- Anaerobic respiration is cellular respiration that does not use oxygen as the final electron acceptor (p. 78).

SHOW YOU KNOW 4.2 How does aerobic exercise like running allow the heart to pump more oxygen-containing blood per beat? How does aerobic conditioning help prevent the buildup of lactic acid during exercise?

Section 4.3

Discuss the relevance of body weight as a predictor of healthfulness.

- Obesity and severe underweight are associated with many negative health risks. Moderate overweight is not (pp. 78–81).

SHOW YOU KNOW 4.3 Is the relationship between body weight and calories burned during exercise a negative or positive correlation?

ROOTS TO REMEMBER

The following roots of words come mainly from Latin and Greek and will help you to decipher terms:

an-	means absence of. Chapter term: *anaerobic*
-ase	is a common suffix in names of enzymes. Chapter term: *sucrase*
glyco-	means sugar. Chapter term: *glycolysis*

LEARNING THE BASICS

1. What is meant by the term *induced fit*?
2. Add labels to the figure that follows, which illustrates the breakdown of a disaccharide inside a cell.

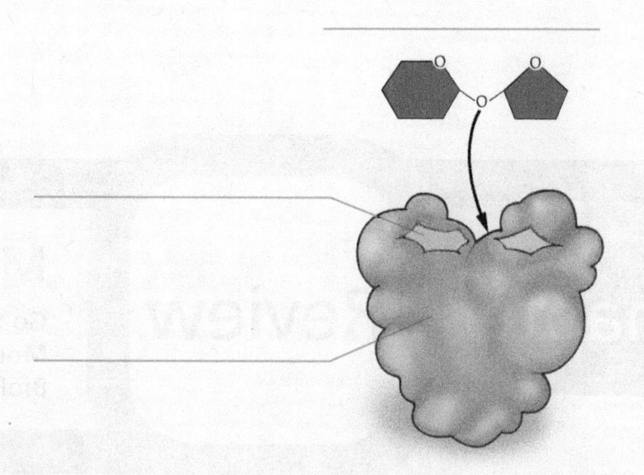

3. What are the reactants and products of cellular respiration?
4. Which of the following is a *false* statement regarding enzymes?

   A. Enzymes are proteins that speed up metabolic reactions; B. Enzymes have specific substrates; C. Enzymes supply ATP to their substrates; D. An enzyme may be used many times.

5. Enzymes speed up chemical reactions by ________.

A. heating cells; B. binding to substrates and placing stress on their bonds; C. changing the shape of the cell; D. supplying energy to the substrate

6. What would happen if activation energy barriers didn't exist?

A. Substrates would not bind properly to enzymes; B. Chemical reactions in the body would never occur; C. Enzyme function would not be affected; D. Metabolic reactions would proceed even if their products were not needed.

7. Cellular respiration involves ________.

A. the aerobic metabolism of sugars in the mitochondria by a process called glycolysis; B. an electron transport chain that releases carbon dioxide; C. the synthesis of ATP, which is driven by the rushing of protons through an ATP synthase; D. electron carriers that bring electrons to the citric acid cycle; E. the production of water during the citric acid cycle

8. The electron transport chain ________.

A. is located in the matrix of the mitochondrion; B. has the electronegative carbon dioxide at its base; C. is a series of nucleotides located in the inner mitochondrial membrane; D. is a series of enzymes located in the intermembrane space; E. moves electrons from protein to protein and moves protons from the matrix into the intermembrane space

9. Most of the energy in an ATP molecule is released ________.

A. during cellular respiration; B. when the terminal phosphate group is hydrolyzed; C. in the form of new nucleotides; D. when it is transferred to NADH

10. Anaerobic respiration ________.

A. generates proteins for muscles to use; B. occurs in yeast cells only; C. does not use oxygen as the final electron acceptor; D. uses glycolysis, the citric acid cycle, and the electron transport chain

ANALYZING AND APPLYING THE BASICS

1. A friend decides he will eat only carbohydrates because carbohydrates are burned to make energy during cellular respiration. He reasons that he will generate more energy by eating carbohydrates than by eating fats and proteins. Is this true?

2. If you could follow one carbon atom present in a carbohydrate you ingested and burned for energy, where would that carbon atom end up?

GO FIND OUT

1. A friend of yours is embarking on a diet. This friend qualifies as overweight on a BMI chart but exercises regularly and eats a healthy, well-balanced diet. Would you encourage this friend to go on a diet? Why or why not?

MAKE THE CONNECTION

The science that you learned in this chapter has helped you better understand the real-world example used throughout this discussion. Draw a line from the statement on the left to the science that supports it on the right.

Diabetes, heart disease, and stroke are consequences of being very overweight.	Genes encode proteins.
In two different individuals, the same enzyme can operate at a different speed.	Leptin levels may be permanently lowered by extreme dieting.
Females require more body fat than males.	Grade 2 obesity is correlated with health risks.
The BMI chart mischaracterizes many people.	Muscle is more dense than fat.
Dieting may cause permanent changes to hormones that control hunger, making it harder to keep weight off after dieting.	Basal metabolic rates differ between the sexes.

Answers to **Got It?, Visualize This, Working with Data, Sounds Right, But Is It?, Show You Know,** and **Chapter Review** questions can be found in the **Answers** section at the back of the book.

5 Life in the Greenhouse

Photosynthesis and Climate Change

The 2015 Paris accord represented unprecedented agreement among nearly 200 world leaders.

In late 2015, high-level representatives of the governments of nearly all of Earth's nations met in Paris for a historic dialogue. The end result was an agreement signed on December 12 to combat global warming. The 195 signatory countries agreed to eliminate human-derived greenhouse gas emissions by 2050. As the agreement is being implemented, developing countries are to receive financial assistance to help cut emissions; and aid will be given to those nations that lack the resources to mitigate the negative effects of climate change.

In dozens of cities throughout the world, hundreds of thousands of activists and ordinary citizens participated in peaceful and dramatic marches in the days leading up to the conference. The size of these rallies may have given world leaders the confidence to act—indeed, many of these leaders were exhilarated by the hard-won final product. "This is truly a historic moment," said Ban Ki-moon, the United Nations secretary general. "For the first time, we have a truly universal agreement on climate change, one of the most crucial problems on earth." Many environmental activists were more circumspect, noting that the accord did not specify any particular actions to combat the growing threat of climate change.

Global warming has already caused significant damage, including devastating storms and sea level rise.

Secretary General Ban Ki-moon's description of climate change as one of Earth's "most crucial problems" is not an exaggeration. With what could sound like a small increase—the 1°C temperature rise recorded since the beginning of the Industrial Revolution—dramatic climate shifts have already occurred. Mountain glaciers and polar ice caps are disappearing while sea levels rise; storms have become more severe and destructive; and insects carrying disease have changed their distribution patterns, threatening human populations. Scientists say that anything above the target set by the Paris accord, a 2°C (3.6°F) increase over the historical average, will doom us to a future of even more severe consequences. These consequences include even more severe storms and resulting widespread food and water shortages.

Can plants and forests help protect Earth from passing the 2°C threshold that scientists say would bring catastrophic change?

Given what scientists say about the severity of the crisis and the consequences of inaction, it may be crucial for humanity's survival and well-being that we band together to address climate change. The Paris accord, although far from perfect, is an important step in that process. But what can science tell us about how we got in this predicament and—perhaps more important—what we can do about it?

5.1 The Greenhouse Effect

Global warming is the progressive increase of Earth's average temperature that has been occurring over the past century. There is little debate among scientists and government-appointed panels about the nature and cause of global warming. Scientists who publish in peer-reviewed journals and respected scientific panels and societies, such as the Intergovernmental Panel on Climate Control (IPCC), the National Academy of Sciences, and the American Association for the Advancement of Science (AAAS), all agree that the average temperature on Earth is increasing and that most of the warming observed in the past century is attributable to human activities.

Global warming is contributing to **global climate change,** the local changes in average temperature, precipitation, and sea level relative to historical conditions that are occurring in locations all over the planet. Although Earth's climate does fluctuate over time due to changes in Earth's orbit and solar output, human-caused (**anthropogenic**) global warming has dramatically increased the rate of change—so much so that it may be difficult for humans to adjust.

anthropo- means human.
-genic means producing.

Anthropogenic global warming is caused by recent increases in the concentrations of particular gases in the atmosphere, including water vapor, carbon dioxide (CO_2), methane (CH_4), and ozone (O_3). The accumulation of many of these **greenhouse gases** is a direct result of coal, oil, and natural gas combustion. The most abundant gas emitted by combustion of these fuels is carbon dioxide; for this reason, carbon dioxide is considered the most important greenhouse gas to control.

Earth Is a Greenhouse

The presence of carbon dioxide and the other greenhouse gases in the atmosphere leads to a phenomenon called the **greenhouse effect.** Despite this name, the phenomenon caused by these gases is not exactly like that of a greenhouse, where panes of glass allow radiation from the sun to penetrate inside and then trap the heat that radiates from warmed-up surfaces. On Earth, the greenhouse effect works like this: Warmth from the sun heats Earth's surface, which then radiates the heat energy outward. Most of this heat is radiated back into space, but some of the heat warms up the greenhouse gases in the atmosphere and then is re-radiated to Earth's surface.

In effect, greenhouse gases act like a blanket (**FIGURE 5.1**). When you sleep under a blanket at night, your body heat warms the blanket, which in turn keeps you warm. (Imagine instead sleeping under a pane of glass—it would trap some heat near your body, but it would never be able to "warm" you.) When the levels of greenhouse gases in the atmosphere increase, the effect is similar to sleeping under a thicker blanket—more heat is retained and re-radiated, and therefore the temperature underneath the atmosphere "blanket" is warmer.

The greenhouse effect is not in itself a dangerous phenomenon—in fact, it is necessary for humans and other living organisms. If Earth's atmosphere did not have some greenhouse gases, too much heat would be lost to space, and Earth would be too cold to support life. The danger imposed by anthropogenic global warming is in the speed of temperature increase—and thus the changes in weather patterns and sea levels—that exceed the ability of many living organisms to adapt. What might this mean for our own species?

Water, Heat, and Temperature

Bodies of water absorb energy and help maintain stable temperatures on Earth. You have perhaps noticed that when you heat water on a stove, the metal pot becomes hot before the water. This is because water heats more slowly than metal and has a stronger resistance to temperature change than most substances.

Heat and temperature are measures of energy. **Heat** is the total amount of energy associated with the movement of atoms and molecules in a substance. **Temperature** is a measure of the intensity of heat—for example, how fast the molecules in the substance are moving. When you are swimming in a cool lake, your body has a higher temperature than the water; however, the lake contains more heat than your body because even though its molecules are moving more slowly, the sum total of molecular movement in its large volume is much greater than the sum total of molecular movements in your much smaller body.

The formation of hydrogen bonds between neighboring molecules of water (Chapter 2) makes it more cohesive than other liquids; in other words, water molecules tend to "stick together." These same hydrogen bonds also make water resistant to temperature change, even when a large amount of heat is added. This phenomenon occurs because when water is heated, the heat energy first must disrupt the hydrogen bonds. Only after enough of the hydrogen bonds have been broken can heat cause individual water molecules to move faster, thus increasing the temperature. When water cools, hydrogen bonds re-form between adjacent molecules, releasing heat into the atmosphere. A body of water can store a large amount of heat from its surroundings while experiencing only a small increase in temperature, and vice versa (**FIGURE 5.2**).

VISUALIZE **THIS**

How would this figure change if we wanted to show more carbon dioxide added to the atmosphere?

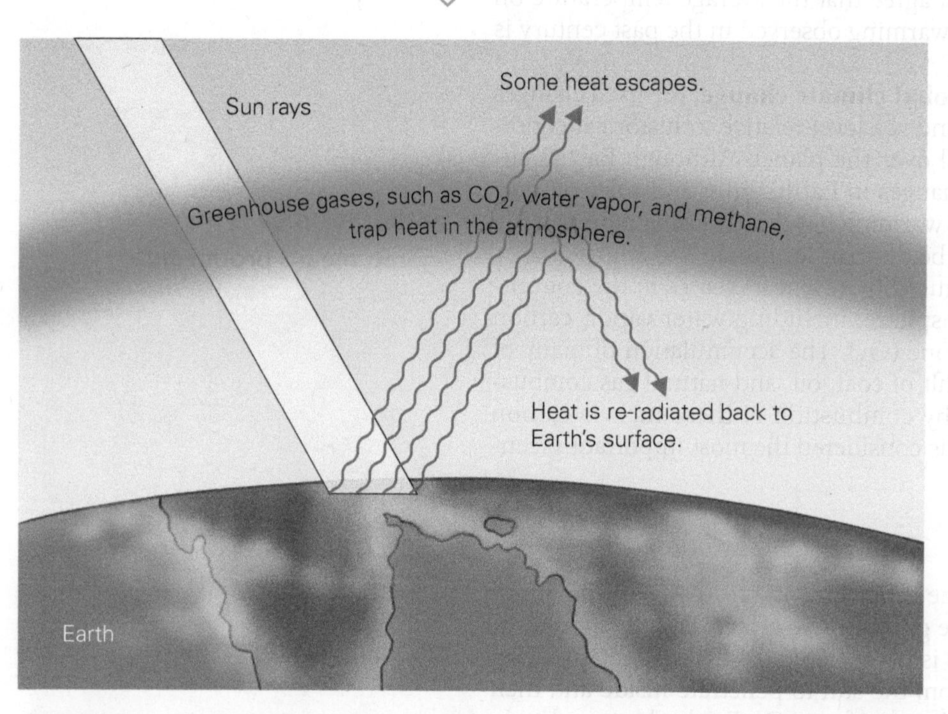
Sun rays
Some heat escapes.
Greenhouse gases, such as CO_2, water vapor, and methane, trap heat in the atmosphere.
Heat is re-radiated back to Earth's surface.
Earth

FIGURE 5.1 The greenhouse effect. Heat from the sun is absorbed in the atmosphere by water vapor, carbon dioxide, and other greenhouse gases and re-radiated back to Earth.

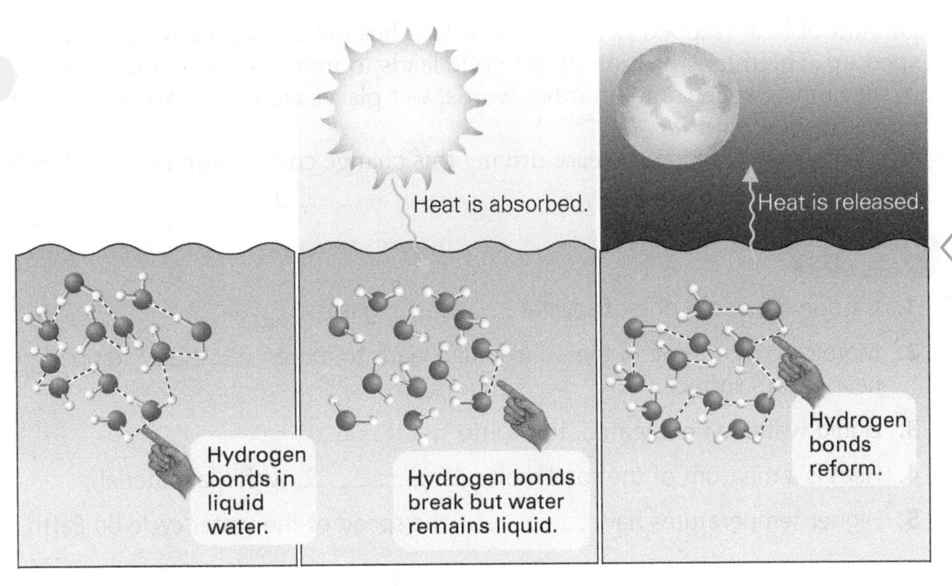

FIGURE 5.2 Hydrogen bonding in water. Hydrogen bonds break as they absorb heat and reform as water releases heat.

VISUALIZE THIS

Evaporation occurs when molecules at the surface of a liquid "escape" into a gaseous phase. Where would most of these escaped molecules appear on this figure and why?

Water's heat-absorbing capacity has important effects on Earth's climate. The vast amount of water contained in Earth's oceans and lakes keeps temperatures moderate by absorbing huge amounts of heat radiated by the sun and releasing that heat during less-sunny times, warming the air and preventing large temperature swings.

As the temperature continues to rise, individual water molecules can move fast enough to break free of all hydrogen bonds and rise into the air as water vapor. This is the basis for the water cycle that moves water from land, oceans, and lakes to clouds and then back again to Earth's surfaces (**FIGURE 5.3**). As the

VISUALIZE THIS

What would happen to the amount of water in the ocean if all of the stored ice and snow melted? How would this melting affect the shoreline?

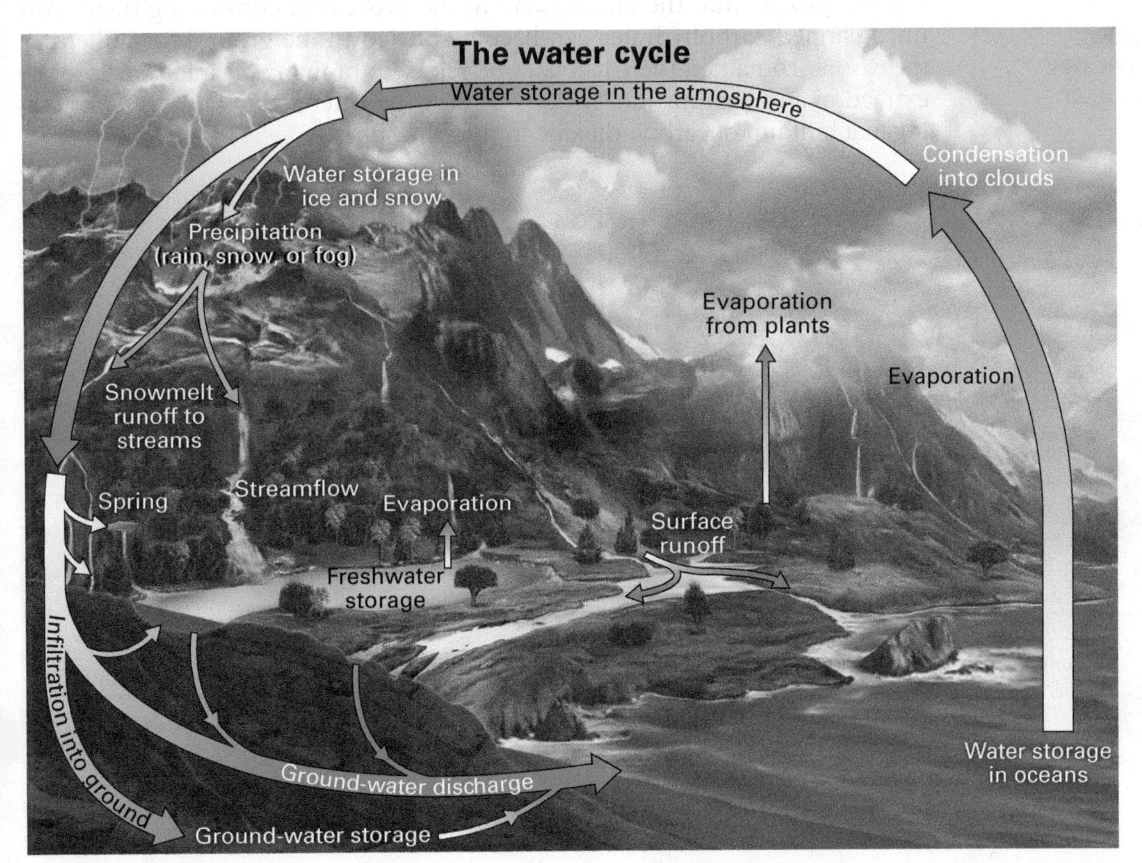

FIGURE 5.3 The water cycle. Water moves from the oceans and other surface water to the atmosphere and back, with stops in living organisms, underground pools and soil, and ice caps and glaciers on land.

amount of heat trapped near Earth's surface has increased, the water cycle has sped up. The intensification of the cycle leads to more extremes on both ends of the climate spectrum—in other words, wet places are getting wetter and dry places are getting drier.

The carbon dioxide increase driving this change comes from many different sources.

Got It?

1. Carbon dioxide is one of several ________ gases.
2. More carbon dioxide in the atmosphere leads to more ________ trapped near Earth's surface.
3. Liquid water, when heated, turns into ________.
4. Heat is a measure of the total molecular ________ within a material.
5. Higher temperatures have ________ the speed of the water cycle on Earth.

5.2 The Flow of Carbon

The atoms that make up the complex molecules of living organisms move through the environment via biogeochemical cycles. **FIGURE 5.4** illustrates how carbon, like water, cycles back and forth between living organisms, the atmosphere, bodies of water, and rock. The carbon dioxide you exhale enters the atmosphere, where it can absorb heat; these molecules can return to Earth's surface, where they can dissolve in water or be absorbed by plants or certain other organisms.

Carbon dioxide taken up by plants, algae, and some types of bacteria is converted into carbohydrates using the energy from sunlight. Most living organisms depend on these carbohydrates as a source of cellular energy and rerelease the carbon dioxide into the atmosphere in the process of consuming them. Any unconsumed carbohydrates can become buried in the ground for millennia; the carbon contained there can later be released through volcanic activity or by extraction and combustion by humans. It is the latter activity that is contributing to a buildup of carbon dioxide in the atmosphere.

VISUALIZE **THIS**

Based on the observations that carbon dioxide levels have increased in the past 100 years and that volcanic activity has remained constant, which would you predict releases more carbon dioxide into the air: volcanic or human activity?

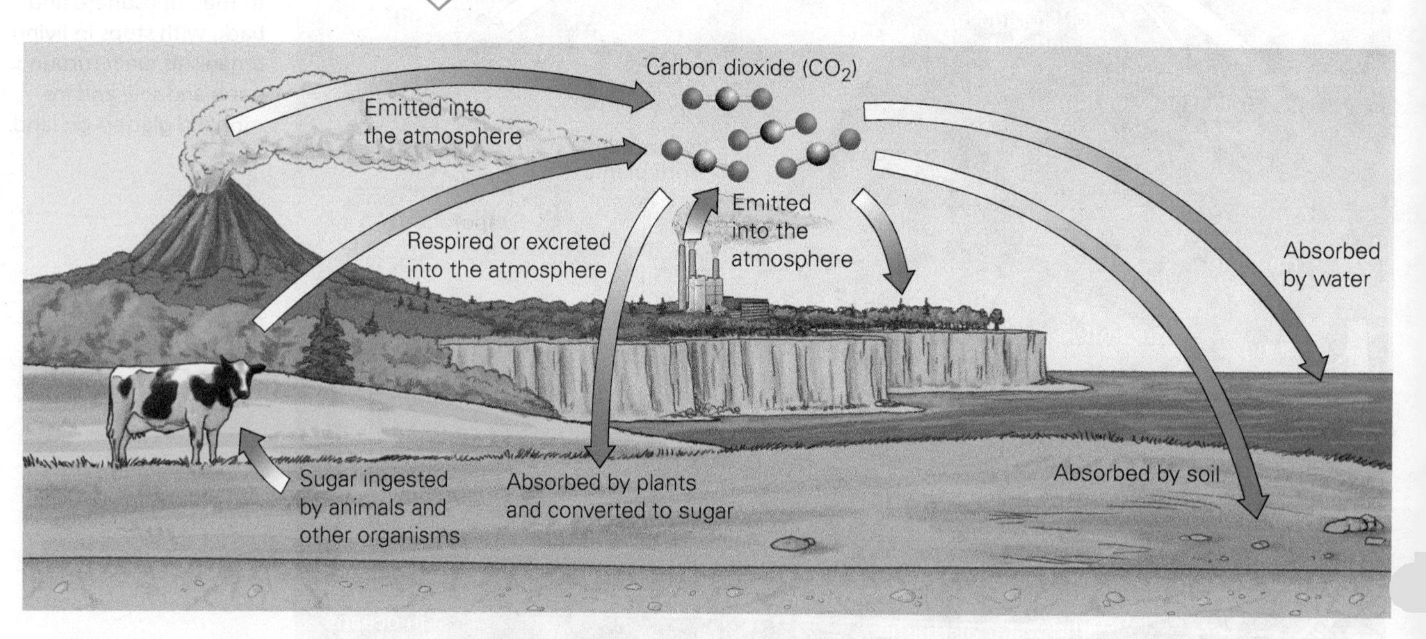

FIGURE 5.4 The flow of carbon. Living organisms, volcanoes, and fossil fuel emissions produce CO_2. Plants, oceans, and soil absorb CO_2 from the air.

FIGURE 5.5 Burning fossil fuels. The burning of fossil fuels by industrial plants and automobiles is rapidly releasing carbon dioxide that has been stored for millions of years in Earth's crust.

The stored carbohydrates discussed in the previous paragraph are known as **fossil fuels** (**FIGURE 5.5**). These fuels—petroleum, coal, and natural gas—are "fossils" because they formed from the buried remains of ancient plants and microorganisms. Over a period of millions of years, the carbohydrates in these organisms were transformed by heat and pressure deep in Earth's crust into highly concentrated energy sources. Humans now tap these energy sources to power our homes, vehicles, and businesses, but as a result of our burning of these fuels, we have released millions of years of stored carbon as carbon dioxide.

Human use of fossil fuels is having a measurable effect; increases in carbon dioxide in the atmosphere are well documented by direct measurements over the past 50 years (**FIGURE 5.6**). In addition, scientists can analyze *fossil air* trapped within ancient ice sheets to determine carbon dioxide over longer time spans. Such measurement is possible because snow falling on an ice sheet surface traps air. As snow accumulates, underlying snow is compressed into ice, and the trapped air becomes tiny ice-encased air bubbles. Thus,

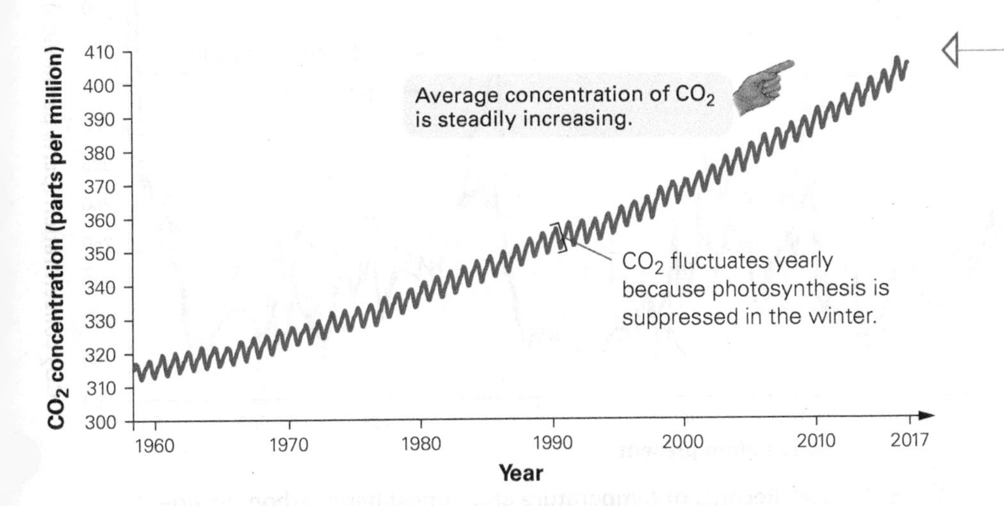

FIGURE 5.6 Increases in atmospheric carbon dioxide. Carbon dioxide levels from 1960 to present as measured by instruments at Mauna Loa observatory in Hawaii.

WORKING WITH **DATA**

What evidence in the graph demonstrates the increased rate of carbon dioxide accumulation from 2000 to 2017 compared with the 1960s?

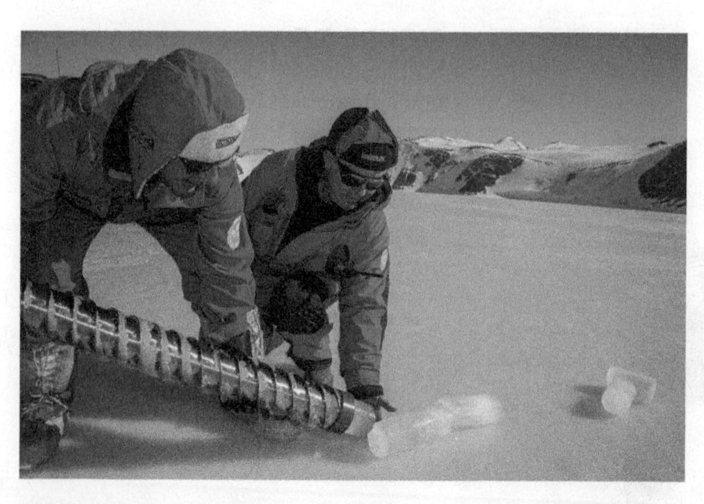

FIGURE 5.7 Ice core. By analyzing ice cores, scientists can measure past atmospheric concentrations of carbon dioxide.

these bubbles are fossils—actual samples of the gases in the atmosphere at the time they formed. Cores remove from long-lived ice sheets are analyzed to determine the concentration of carbon dioxide in the atmosphere over time (**FIGURE 5.7**). Other gases in the bubbles can provide indirect information about temperatures at the time the bubbles formed.

Ice core data from Antarctica (**FIGURE 5.8**) indicate that although Earth has gone through cycles of high carbon dioxide, the concentration of carbon dioxide in the atmosphere today is higher than at any time in human history, and in fact at any other time in the past 400,000 years. The ice core data also demonstrate that increased levels of carbon dioxide occur at the same time as increased temperatures, suggesting that carbon dioxide measurably warms Earth. Taken together, these data are quite worrisome; they tell us that Earth may soon be facing temperatures well above those we experience in the current climate, warmer than Earth has seen in millennia—and warmer than humans on Earth have ever seen.

Got It?

1. Carbon cycles among living organisms, the atmosphere, water, and __________.
2. Carbon dioxide taken up by plants and algae is converted into __________.
3. Fossil fuels include __________, __________, and __________ __________.
4. Air bubbles trapped in __________ __________ provide information about carbon dioxide levels on Earth in the distant past.
5. Compared with levels over the past 400,000 years, the amount of carbon dioxide in the atmosphere today is __________ __________.

WORKING WITH **DATA**

This graph has axes on both the left and right sides. Which line on the graph corresponds to the left axis and which to the right axis? Why are both lines included on the same graph?

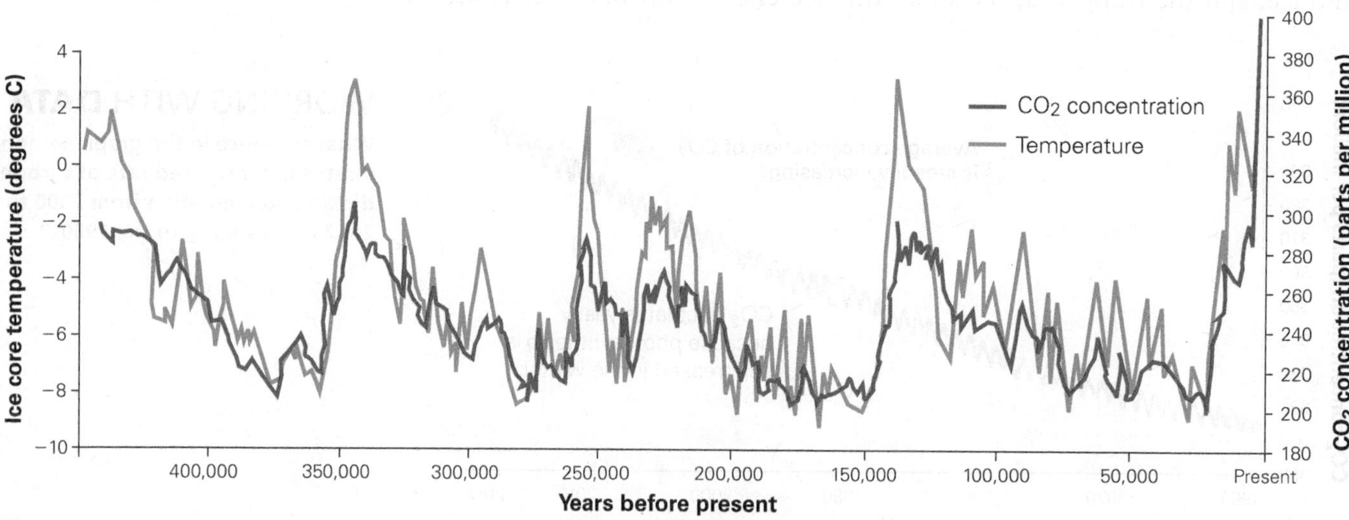

FIGURE 5.8 Records of temperature and atmospheric carbon dioxide concentration from Antarctic ice cores. These data indicate that increases in carbon dioxide levels are correlated with higher temperatures.

5.3 Can Photosynthesis Slow Down Global Climate Change?

Modern plants and certain microorganisms do as their ancient predecessors did—absorb carbon dioxide from the atmosphere and convert it to carbohydrates. Is it possible that we could depend on these modern organisms to reduce the amount of greenhouse gases released by fossil fuel burning and thus the threat posed by global climate change?

Photosynthesis is the process that traps light energy from the sun and uses it to convert carbon dioxide and water into sugar. In other words, photosynthesis transforms solar energy into the chemical energy required by all living things.

photo- means light.

Chloroplasts: The Site of Photosynthesis

Chloroplasts are the specialized organelles in plant cells where photosynthesis takes place. Chloroplasts are surrounded by two membranes (**FIGURE 5.9**). The inner and outer membranes together are called the chloroplast envelope. The chloroplast envelope encloses a compartment filled with **stroma,** the thick fluid that houses some of the enzymes of photosynthesis. Suspended in the stroma are disk-like membranous structures called **thylakoids,** which are typically stacked in piles like pancakes. The large amount of thylakoid membrane inside the chloroplast provides abundant surface area on which several of the reactions of photosynthesis can occur.

chloro- means green.

(a)

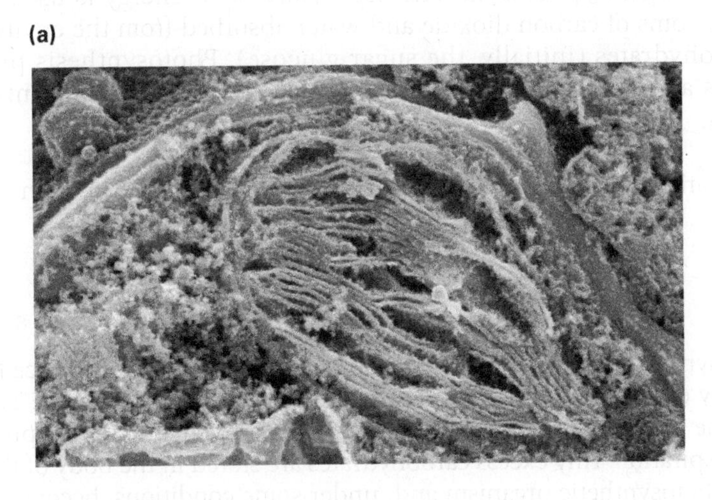

(b)

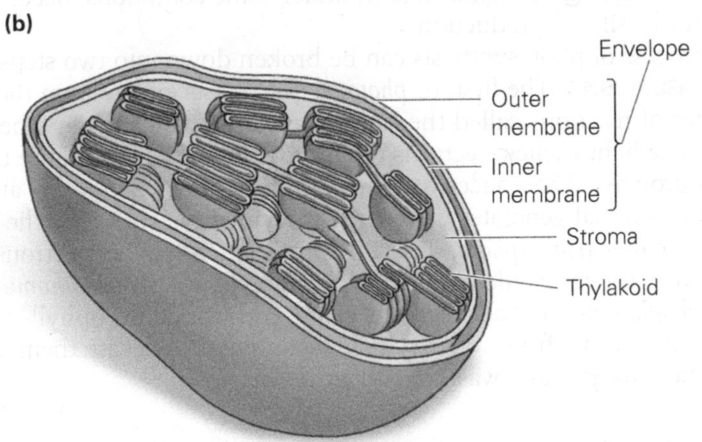

FIGURE 5.9 Chloroplasts. The cross section (a) and drawing (b) of a chloroplast show the structures involved in photosynthesis.

-phyll means leaf.

On the surface of the thylakoid membrane are millions of molecules of **chlorophyll,** a pigment that absorbs energy from the sun. It is chlorophyll that gives leaves and other plant structures their green color. Like all pigments, chlorophyll absorbs light. Light is made up of waves of varying wavelengths—to the human eye, shorter wavelengths appear violet to blue, middle wavelengths appear green, and longer wavelengths appear yellow to red (**FIGURE 5.10**). Chlorophyll looks green to human eyes because it absorbs the shorter and longer wavelengths of visible light and reflects the middle range of wavelengths.

When a pigment such as chlorophyll absorbs sunlight, electrons associated with the pigment become excited (that is, the electrons increase in energy level). In effect, light energy interacting with chlorophyll is converted to chemical energy. For most pigments that have absorbed light energy, the molecule remains in its excited state for a very brief amount of time before this chemical energy is lost as heat. (You may have noticed this in a parking lot on a sunny day. A car that is painted black, a pigment composed of all other pigments and thus absorbs all visible light wavelengths, heats quickly in the sun. A white car, painted with a pigment that does not absorb any light energy, remains relatively cooler.) Inside a chloroplast, however, the chemical energy of the excited chlorophyll molecules is not all released as heat; instead, most of that energy is captured.

WORKING WITH **DATA**

How would this graph appear for the pigments in the skin of a ripe red apple?

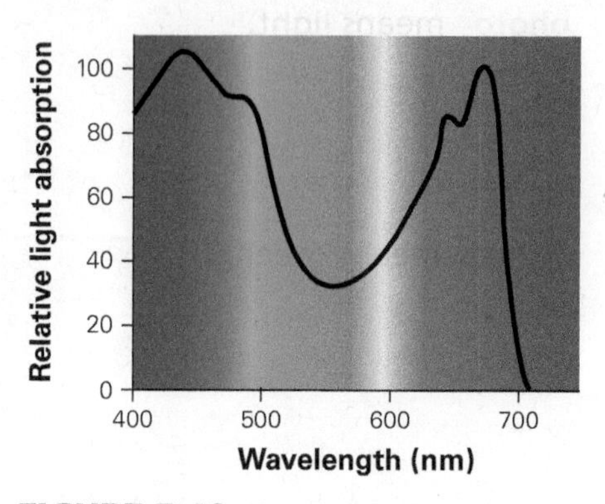

FIGURE 5.10 The absorption spectra of leaf pigments. Visible "white" light is actually made up of a series of wavelengths that appear as different colors to our eyes. The wavelengths absorbed by chlorophyll and other pigments in green plants are mainly red and blue and are thus removed from the spectrum of light that is reflected back to us from a leaf surface.

The Process of Photosynthesis

In plants and other photosynthetic organisms, solar energy is used to rearrange the atoms of carbon dioxide and water absorbed from the environment into carbohydrates (initially, the sugar glucose). Photosynthesis produces oxygen as a waste product. The equation summarizing photosynthesis is as follows:

$$\text{Carbon dioxide} + \text{Water} + \text{Light energy} \longrightarrow \text{Glucose} + \text{Oxygen;}$$

or more technically

$$6\ CO_2 + 6\ H_2O + \text{Light energy} \longrightarrow C_6H_{12}O_6 + 6\ O_2$$

Photosynthetic organisms use the carbohydrates that they produce to grow and supply energy to their cells. They, along with the organisms that eat them, liberate the energy stored in the chemical bonds of sugars via the process of cellular respiration. Any excess carbohydrates are stored in the body of the plant or other photosynthetic organism and, under some conditions, become the raw material for fossil fuel production.

The process of photosynthesis can be broken down into two steps, summarized in **FIGURE 5.11**. The first, or *photo*, step harvests energy from the sun during a series of reactions called the **light reactions,** which occur when there is sunlight. The light excites electrons of chlorophyll molecules on the thylakoids of the chloroplast. These electrons are captured and passed along an electron transport chain that generates the energy-carrying molecule ATP (the abbreviation for *adenosine triphosphate*, described in Chapter 4). The electrons are then transferred to the molecule NADPH (the abbreviation for *nicotinamide adenine dinucleotide phosphate*). The electrons removed from cholorophyll in this way are replaced by those from water, which decomposes to release them, a reaction that produces oxygen as a waste product.

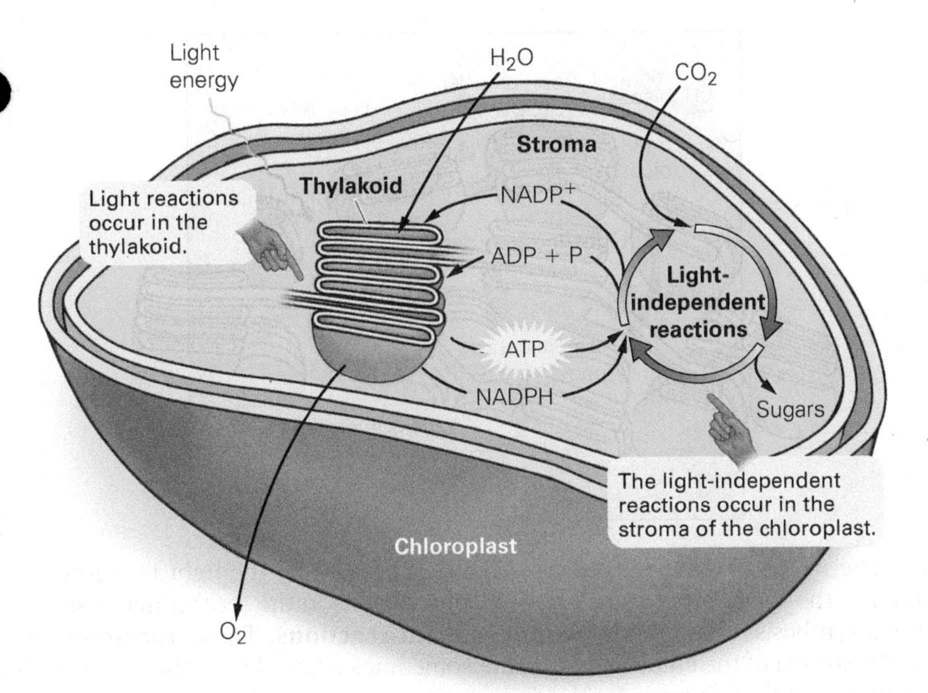

FIGURE 5.11 Photosynthesis. Sunlight drives the synthesis of glucose and oxygen from carbon dioxide and water inside the chloroplasts of plants and other photosynthetic organisms. The reaction requires two steps: the first that generates cellular energy (ATP) and harvests electrons from water (carried by NADPH) and the second that uses those products to convert carbon dioxide into sugar.

VISUALIZE THIS

How is this set of reactions similar to the electron transport chain in cellular respiration? How is it different? (To review cellular respiration, see Chapter 4.)

The details of the light reactions are illustrated in **FIGURE 5.12**. Note that the mechanism of generating ATP is similar to what occurs in the process of cellular respiration; that is, energy captured by the electron transport chain pumps protons across a membrane, in this case setting up a gradient between the inside and outside of the thylakoid. As these stockpiled protons diffuse back out of the thylakoid through an enzyme, ATP is produced.

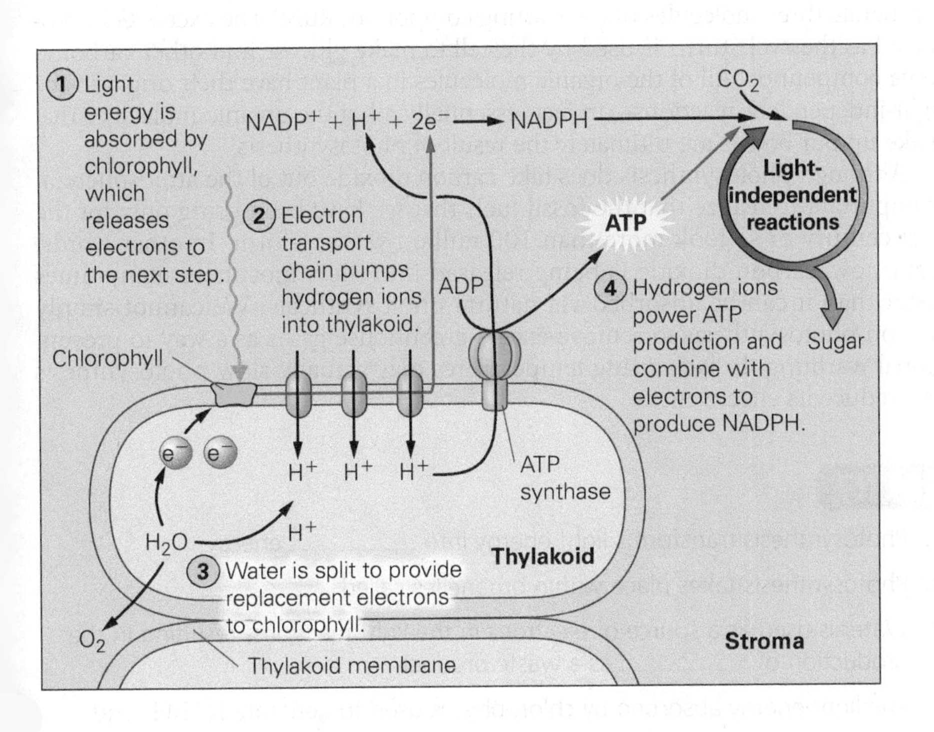

FIGURE 5.12 The light reactions of photosynthesis. (1) Sunlight strikes chlorophyll molecules located in the thylakoid membrane, exciting electrons, which then move to a higher energy level. (2) The electrons are captured by an electron transport chain, and their energy is used to pump hydrogen ions across the thylakoid membrane. (3) Water is split. Electrons removed from water are used to replace those lost from chlorophyll. Oxygen gas is released. (4) The movement of hydrogen ions out of the thylakoid power ATP production and generate NADPH. These molecules are produced in the stroma, where they will be available to the enzymes of the light-independent reactions.

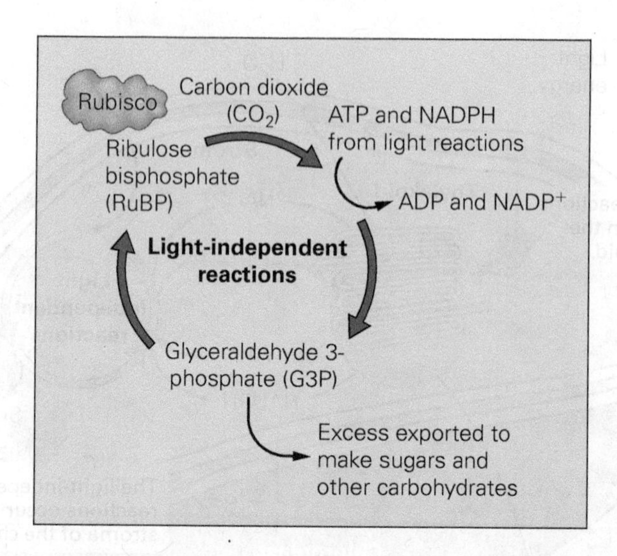

FIGURE 5.13 The light-independent reactions of photosynthesis. Carbon dioxide is incorporated into sugar in plants through a series of reactions that regenerate the initial carbon-containing starting product, the sugar ribulose bisphosphate (RuBP). The enzyme that attaches carbon dioxide to RuBP in the first step of the light-independent reactions is called rubisco. Glyceraldehyde 3-phosphate (G3P) is the simple sugar produced by these reactions. Excess G3P is exported to other pathways to produce organic molecules the plant needs.

The ATP and the electrons in NADPH generated by the light reactions are used in the synthesis of sugars from carbon dioxide in the second major step of photosynthesis, called the **light-independent reactions.** These reactions occur in the stroma of the chloroplast and are sometimes referred to as the Calvin cycle (**FIGURE 5.13**). The term "cycle" is instructive; one of the key features of this set of reactions is the regeneration of its starting molecule, a five-carbon sugar called ribulose bisphosphate, or RuBP. During the light-independent reactions, carbon dioxide is added to RuBP by the enzyme ribulose bisphosphate carboxylase oxygenase, or **rubisco,** the most abundant protein on the planet. The resulting six-carbon molecule immediately breaks down into a pair of three-carbon molecules. Input of ATP and NADPH over several subsequent steps ultimately produces the sugar glyceraldehyde three-phosphate (or G3P); because this first stable product is a three-carbon compound, this pathway is often called C_3 photosynthesis. For every six G3P molecules produced, five are rearranged to regenerate three molecules of the starting compound, RuBP. The excess G3P produced as the cycle turns is used by the cell to make glucose and other carbohydrate compounds. All of the organic molecules in a plant have their origin in the light-independent reactions—in fact, essentially all of the organic molecules that make up our bodies are ultimately the result of photosynthesis.

Although photosynthesis does take carbon dioxide out of the atmosphere, it is important to realize that the fossil fuels that we have been using only for the past century or so took more than 100 million years to form. In other words, right now, carbon dioxide is being released into the atmosphere many times faster than it can be absorbed via natural photosynthesis. We cannot simply rely on photosynthesis to remove excess greenhouse gases as a way to prevent global warming. In fact, rising temperatures may actually slow photosynthesis and reduce its effectiveness.

Got It?

1. Photosynthesis transforms light energy into ________ energy.
2. Photosynthesis takes place within organelles called ________.
3. Water is used as a source of electrons in the light reactions, resulting in the production of ________ as a waste product.
4. The light energy absorbed by chlorophyll is used to generate NADPH and ________.
5. Energy and electrons generated by the light reactions of photosynthesis are used to make ________ from ________.

5.4 How High Temperatures Might Reduce Photosynthesis

Carbon dioxide enters the leaves of land plants through tiny openings called **stomata** (**FIGURE 5.14**). Stomatal openings are surrounded by two kidney-bean-shaped cells called **guard cells.** When the guard cells are compressed against each other, the stomata are closed, thus restricting the flow of gases into or out of the plant. When the guard cells change shape to create a gap between them, the stomata are open, and carbon dioxide and oxygen gases can be exchanged.

Stomatal openings also drive water uptake from the soil and loss from the plant via a process called **transpiration** (**FIGURE 5.15**). Water evaporating via the stomata exerts a "pull" that is transmitted down through the plant and eventually draws water from the soil. On hot, dry days, the rate of evaporation exceeds the rate that the soil can supply water, so plants will close their stomata. However, closing stomatal openings also prevents carbon dioxide from entering the plant. Thus, rising temperatures are expected to limit carbon dioxide levels in leaves, and the rate of photosynthesis would be expected to decline. If this prediction is correct, then higher temperatures may have a positive feedback effect, further reducing plants' ability to pull carbon dioxide from the atmosphere.

trans- means across or to the other side.

Closing the stomatal openings may not just reduce the rate of photosynthesis; it may even counteract it. This process occurs as the result of another series of reactions known as **photorespiration.** During photorespiration, rubisco adds oxygen instead of carbon dioxide to RuBP in the first step of the light-independent reactions. Oxygen is used in this reaction when carbon dioxide levels are low inside the leaf—typically only when the stomata are closed. The compound produced by this reaction, called glycolate, cannot be used in the light-independent reactions. In fact, glycolate must be destroyed by the plant because high levels of this acid will further inhibit photosynthesis, ultimately starving the plant. The breakdown of glycolate requires energy provided by cellular respiration—thus, instead of taking up carbon dioxide, plants that are photorespiring release it.

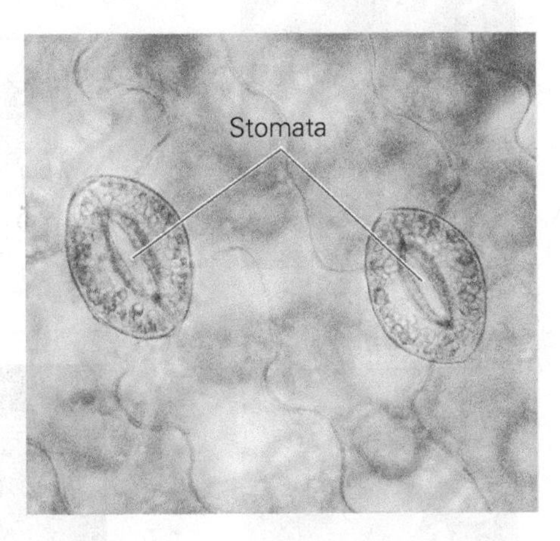

FIGURE 5.14 Stomata. Stomata are adjustable microscopic pores found on the surface of leaves that allow for gas exchange.

VISUALIZE THIS

The stomata on a plant typically have a regular pattern of opening and closing in a 24-hour period. Using this figure and your understanding of the inputs plants need to perform photosynthesis, at what periods during a 24-hour period do you think most plant stomata are open, and why?

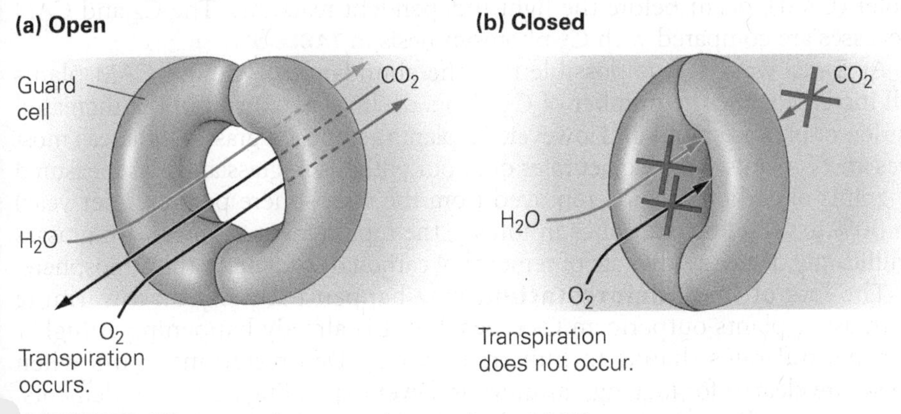

FIGURE 5.15 Gas exchange and water loss. (a) When stomata are open, oxygen and carbon dioxide can be exchanged, but water can be lost from a plant through a process called transpiration. (b) When the guard cells change shape to block the opening, gas exchange and transpiration do not occur.

TABLE 5.1 C_3, C_4, and CAM photosynthesis.

Type of Plant and Example	Stomata Status	Description
C_3 soybean	CO_2, H_2O, O_2	The light-independent reactions converts two three-carbon sugars into glucose. Photorespiration is common when temperatures are high and water levels are low.
C_4 corn	CO_2, H_2O, O_2	Enzymes scavenge CO_2 to produce four-carbon sugars, even when stomata are closed. The four-carbon sugars "pump" carbon dioxide molecules to the light-independent reactions when they break down. Photorespiration very rarely occurs.
CAM jade plant	Night: CO_2, H_2O, O_2; Day: CO_2, H_2O, O_2	Water loss is slowed by opening stomata only at night. Carbon dioxide is stored as an organic acid in the vacuole. The acid's breakdown releases this carbon dioxide to the light-independent reactions during the day. Photorespiration does not occur.

In warmer and drier environments, natural selection should favor plants that can minimize photorespiration despite having closed stomata much of the time. Two mechanisms for reducing photorespiration are known as C_4 and CAM photosynthesis, in which additional pathways that concentrate carbon dioxide in the plant (C_4) or capture it during the night when temperatures are cooler (CAM), occur before the light-independent reactions. The C_4 and CAM processes are compared with C_3 photosynthesis in TABLE 5.1.

As Earth warms, it is possible that the abundance of C_4 and CAM plants will increase while the number of C_3 plants declines as a result of the increased burden of photorespiration. However, C_4 plants are mostly grasses, whereas most trees are C_3 plants. Because net rates of photosynthesis in grasslands (as measured by grams of carbon dioxide removed from the atmosphere per acre, per year) are 30% to 60% less than rates in forests, the replacement of trees with grasses significantly decreases the rate of removal of carbon dioxide from the atmosphere.

The loss of trees, **deforestation,** may happen naturally on a warming Earth as C_4 plants outperform C_3 types, but it is already happening at higher than natural rates thanks to human activities. Deforestation occurs when forests are cleared for logging, farming, and ever-expanding human settlements. Deforestation also contributes directly to the increase in carbon dioxide within the atmosphere; current estimates are that up to 25% of the carbon dioxide introduced into the atmosphere originates from the cutting and burning of forests in the tropics alone. Clearly, one effective way to reduce global warming is to reduce deforestation and promote photosynthesis by planting trees through reforestation projects.

In the worst case, rising temperatures due to global warming will cause vegetated land to become desert, completely eliminating photosynthesis in certain areas. Conversely, warming temperatures will expose more snow- and ice-covered regions, permitting additional photosynthesis in such regions. Unfortunately, enhanced carbon dioxide uptake is likely to be offset by carbon dioxide released from the more rapid decay of carbohydrates in the soil in these formerly frozen regions. Loss of ice and snow will also lead to increased heat gain of Earth's surface, as formerly reflective surfaces become darker and more absorbent of light energy.

As we have seen, there is no guarantee that increased photosynthesis will soak up much of the excess carbon dioxide released by fossil fuel combustion. In fact, the overall rate of photosynthesis may even decline on a warmer planet.

Got It?

1. Pores called stomata on the leaf surface permit gas exchange but also the loss of ________ from the plant.
2. The size of stomatal openings can ________ due to the activities of guard cells.
3. Photorespiration is a wasteful process that results in the release of ________ ________.
4. Photorespiration is greatly reduced in plants with either ________ or ________ photosynthesis.
5. Compared with trees, grasses absorb ________ carbon dioxide from the atmosphere.

5.5 How We Can Slow Global Climate Change

Anthropogenic climate change has resulted in unprecedented droughts in some areas and torrential rains in others. Climate change will cause the extinction of animals and plants that have narrow temperature requirements. Climate change has already contributed to the spread of infectious disease and dangerous pests that were once limited to tropical environments by temperature (**FIGURE 5.16**).

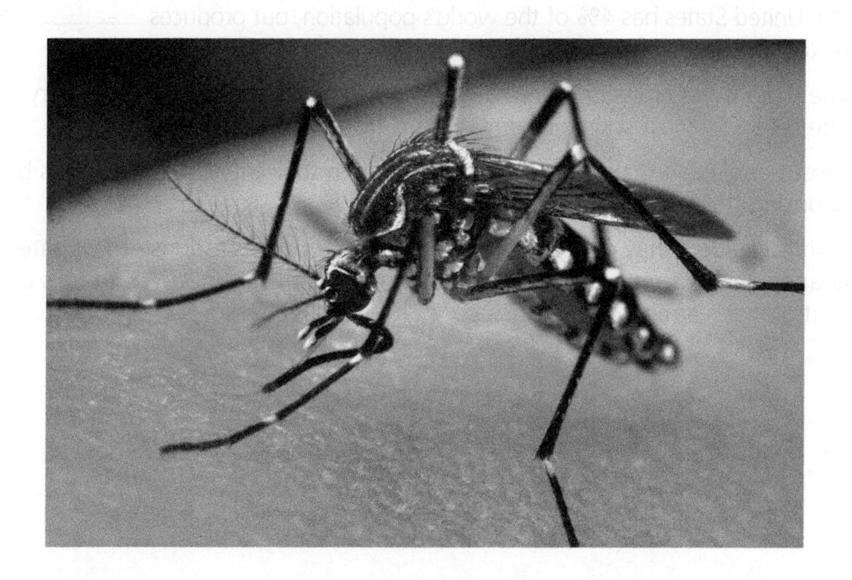

FIGURE 5.16 Emerging diseases. Certain species of mosquito are more likely to carry human diseases than other species. As Earth warms, these insects, like the *Aedes aegypti* mosquito pictured here, are beginning to move into geographical areas where they were once excluded by cold temperatures.

And in addition to the other effects of global climate change, the increasing rate of carbon dioxide absorbed by the ocean is causing it to acidify, damaging and killing coral reefs and threatening algae in the open ocean and the organisms that depend on them—and that we, in turn, depend on. Clearly, excess carbon dioxide in the atmosphere is already imposing a cost on people.

The United States has a disproportionate impact on the rate of carbon dioxide emissions into the atmosphere. Home to only 4% of the world's population, the United States produces close to 25% of the carbon dioxide emitted by fossil fuel burning. The emissions rate of carbon dioxide—what is known as a "carbon footprint"—for an average American is twice that of a Japanese or German individual, three times that of the global average, four times that of a Swede, and twenty times that of the average Indian. The average U.S. household is responsible for about 50,000 kg of CO_2 emissions yearly.

Most of the emissions for an individual country come from industry; followed by transportation; and then by commercial, residential, and agricultural emissions. All of us can work to reduce our personal contribution to global warming by decreasing residential and transportation emissions. Most residential emissions are from energy used to heat and cool homes and to power electrical appliances. Transportation emissions are affected by our choice of the vehicles we use for transport, the fuel economy of cars, and the distance traveled. TABLE 5.2 describes many ways that you can decrease your greenhouse gas emissions and indicates the number of kilograms of carbon dioxide that each action would save annually. These reductions may seem trivial in comparison with the scope of the problem, but when they are multiplied by the more than 320 million people in the United States, the savings become significant.

Having an effect on industrial, commercial, and agricultural sectors is difficult for any one individual. Changes in these areas will instead take leadership from those policy makers who are committed to reducing emissions. Making these changes requires that our leaders, and all of us, understand that even though the implications of and solutions to global warming may be open to debate, the fact that it is occurring at an unprecedented rate is not.

Got It?

1. In addition to warming the planet, excess carbon dioxide in the atmosphere is causing the oceans to become more ________.
2. The United States has 4% of the world's population, but produces ________ % of the emissions of carbon dioxide.
3. The majority of emissions from an individual country comes from industry. The second largest source of emissions is from ________.
4. Most people can make the largest reduction in their individual emissions by choosing the most efficient ________.
5. Political leaders may disagree on the most appropriate response to climate change, but it is unreasonable to disagree about the science, which tells us that global warming is ________.

TABLE 5.2 Decreasing your greenhouse gas emissions.

Here are some ideas that you can use to help slow the rate of global warming.

Action	Annual Decrease in Carbon Dioxide Production
Drive an energy-efficient vehicle. SUVs average 16 miles per gallon, whereas smaller cars average 25 miles per gallon. (An all-electric vehicle can provide the equivalent of 100 miles per gallon.)	5,900 kg (13,000 lb)
Switch to a renewable energy source (such as solar power) for your household power needs.	1,610 kg (3,550 lb)
Carpool 2 days per week.	720 kg (1,590 lb)
Recycle glass bottles, aluminum cans, plastic, newspapers, and cardboard.	385 kg (850 lb)
Walk 10 miles per week instead of driving.	270 kg (590 lb)
Use high-efficiency appliances.	180 kg (400 lb) per appliance
Buy food and other products with reusable or recyclable packaging, or reduced packaging, to save the energy required to manufacture new containers.	100 kg (230 lb)
Use a push mower instead of a power mower.	35 kg (80 lb)
Plant shade trees around your home to decrease energy consumption and to remove carbon dioxide by photosynthesis.	23 kg (50 lb)

Sounds right, but is it?

After comparing the chemical equations for aerobic respiration and photosynthesis, students are sometimes tempted to say:

Animals breathe by taking in oxygen and releasing carbon dioxide, while plants breathe by taking in carbon dioxide and releasing oxygen.

Sounds right, but it isn't.

Answer the following questions to understand why.

1. Why do plants need energy?
2. What is the molecule that provides cellular energy for these activities?
3. What is the equation for photosynthesis?
4. Does photosynthesis produce ATP that is used in the cellular activities described in your answer to question 1?
5. Where do plants get the ATP for cellular activities?
6. What is the equation for aerobic respiration?
7. Reflect on your answers to questions 1–6 and explain why the statement bolded above sounds right, but isn't.

THE BIG QUESTION

Should global warming be kept below 4°C?

As part of the 2015 Paris Agreement, participating countries agreed to reduce greenhouse gas emissions dramatically to limit global warming to 2°C. Without any action and following current trends in emissions, scientists predict that temperatures will rise at least 4°C by the middle of the 21st century.

What should I know?

What follows are some smaller questions that need to be resolved to answer the Big Question. Place a checkmark next to the questions that science can answer.

Smaller Questions	Can Science Answer?
Will a 4°C rise in temperature reduce crop production?	
Will a 4°C rise in temperature increase the number of people at risk of insect-borne disease?	
Do the actions needed to prevent a 4°C temperature increase impose an unacceptably high cost?	
Is it wrong to leave behind a world that is 4°C or more warmer for future generations?	
Do we need to eliminate all fossil fuel combustion immediately to prevent a 4°C temperature increase?	

What does the science say?

Let's examine what the data say about this smaller question:

Will a 4°C rise in temperature reduce crop production?

Scientists examined the effect of a 4°C temperature increase on corn (maize) grown in a controlled setting. They measured overall plant mass (dry weight, in grams) and grain yield (g) of plants grown per square meter (m^2) over a growing season in two conditions: normal temperature and normal temperature +4°C. Results for two runs of this experiment are shown in the table that follows.

Plant size and grain yield of corn grown in normal temperatures and in normal temperatures +4°C for the length of a growing season.

Run	Measurement	Normal Temperatures (grams per m^2)	Normal Temperatures Plus 4°C (grams per m^2)
1	Total vegetative mass	920.3	1188.0
	Grain yield	1870.0	213.9
2	Total vegetative mass	1007.0	1122.1
	Grain yield	471.2	59.9

1. Describe the results. Does it appear that increasing temperature by 4°C reduces plant size and grain yield in corn?
2. Given these data, do you think the smaller question is answered? If not, propose another study that would help answer this question.
3. Does this information help you answer the Big Question? What else do you need to consider?

Data source: J. L. Hatfield and J. H. Prueger, "Temperature Extremes: Effect on Plant Growth and Development," *Weather and Climate Extremes* 10, part A (2015): 4–10.

Chapter Review

Mastering Biology

Go to Mastering Biology to access **eText 2.0, Dynamic Study Modules**, and the **Study Area**, where you'll find practice quizzes, BioFlix™ animations, MP3 tutor sessions, current events, and more.

SUMMARY

Section 5.1

Describe the anthropogenic causes of climate change.

- Burning of coal, oil, and natural gas increase levels of greenhouse gases, such as carbon dioxide and methane, in the atmosphere. These gases in turn result in a greenhouse effect that is changing Earth's climate (p. 87).

Describe the greenhouse effect.

- Greenhouse gases, particularly carbon dioxide, increase the amount of heat retained in Earth's atmosphere, which then leads to increased surface temperatures (pp. 87–88).

Explain why water is slow to change temperature.

- Water can absorb large amounts of heat without undergoing rapid or drastic changes in temperature because heat must first be used to break hydrogen bonds between adjacent water molecules. (pp. 88–90).

SHOW YOU KNOW 5.1 The moon's average daytime temperature is 107°C (225°F), and its average nighttime temperature is –153°C (–243°F). However, the moon and Earth are the same distance from the sun. Think about the greenhouse gases on Earth to come up with a hypothesis for why temperatures on the moon fluctuate so dramatically.

Section 5.2

Compare and contrast the water cycle and the carbon cycle.

- Both water and carbon cycle among animals, plants, soil, oceans, and the atmosphere. Most of the movement of water occurs outside living organisms, whereas carbon is mostly cycled because of biological activity (pp. 89–90).

Discuss the origin of fossil fuels and their relationship to the carbon cycle.

- Fossil fuels are the buried remains of ancient plants, which took carbon dioxide out of the atmosphere and transformed it into carbohydrates. The burning of fossil fuels is returning carbon to the atmosphere (pp. 91–92).

SHOW YOU KNOW 5.2 Carbon stored by plants today or in the recent past in the form of logs or sawdust pellets can be burned for fuel. However, these types of "renewable" fuels are not considered contributors to increased global warming. What makes the carbon in these sources different from the carbon in fossil fuels?

Section 5.3

State the basic equation of photosynthesis.

- During photosynthesis, energy from sunlight is used to rearrange the atoms of carbon dioxide and water to produce sugars and oxygen. Photosynthesis is described by the equation:

$6\ CO_2 + 6\ H_2O + \text{Light energy} \rightarrow C_6H_{12}O_6 + 6\ O_2$ (pp. 93–94).

Describe the light reactions of photosynthesis.

- Photosynthesis occurs in chloroplasts. Sunlight strikes the chlorophyll molecule within chloroplasts, boosting electrons to a higher energy level. These excited electrons are used to generate ATP (pp. 94–95).
- Electrons are also passed to electron carriers (NADPH). The electrons that are lost from chlorophyll become replaced by electrons acquired during the splitting of water, and oxygen is released (p. 95).

Explain the events that occur during the light-independent reactions, and describe the relationship between the light reactions and the light-independent reactions.

- The light-independent reactions use the products of the light reactions (ATP and the electron carrier NADPH) to incorporate carbon dioxide into sugars, regenerating the starting products of a cycle and exporting excess sugars to be used as chemical building blocks for plant compounds (p. 96).

SHOW YOU KNOW 5.3 Some of the experiments that helped scientists characterize the light reactions of photosynthesis use a dye called DCPIP, which changes color when it accepts an electron—that is, when it is "reduced." What product of the light reactions does reduced DCPIP substitute for in these experiments?

Section 5.4

Explain the role of stomata in balancing photosynthesis and water loss.

- Stomata on a plant's surface allow in carbon dioxide for photosynthesis but also allow water to escape from the plant. Guard cells surrounding the stomata can change shape to close the stomata and restrict water loss (p. 97).

Define photorespiration, and explain why this process is detrimental to plants.

- Photorespiration occurs when stomata are closed, carbon dioxide declines in the plant, and oxygen is incorporated into the first step of the light-independent reactions. The resulting product is poisonous to the plant, and energy must be expended to eliminate it (p. 97).
- C_4 and CAM plants have evolved to perform photosynthesis while reducing the risk of photorespiration in dry conditions (p. 98).

SHOW YOU KNOW 5.4 Photorespiration unnecessarily uses up carbon, and thus it seems that evolution should have favored mutations in the light-independent reactions that prevent the reaction with oxygen. This hasn't happened, however. Can you think of a reason a mutation changing the enzyme binding site for carbon dioxide might actually be bad?

Section 5.5

Discuss how our own activities contribute to or reduce the risk of global climate change.

- Humans are deforesting Earth's land surface, reducing the rate of photosynthesis and thus the uptake of atmospheric carbon dioxide, so reforestation is one strategy to reduce global warming (pp. 98–99).
- Humans can reduce carbon dioxide emissions by increasing efficiency when using energy as well as reducing overall energy use (pp. 99–101).

SHOW YOU KNOW 5.5 Based on the statistic that the carbon footprint for the average American is twice that of a Japanese or German individual, three times that of the global average, four times that of a Swede, and twenty times that of the average Indian, is there a clear correlation between carbon dioxide emissions and standard of living?

ROOTS TO REMEMBER

The following roots of words come mainly from Latin and Greek and will help you to decipher terms:

anthropo- means human. Chapter term: *anthropogenic*

chloro- means green. Chapter terms: *chloroplast, chlorophyll*

-genic means producing. Chapter term: *anthropogenic*

photo- means light. Chapter terms: *photosynthesis, photorespiration*

-phyll means leaf. Chapter term: *chlorophyll*

trans- means across or to the other side. Chapter term: *transpiration*

LEARNING THE BASICS

1. Add labels to the figure that follows, which illustrates the product and reactants of photosynthesis and the relationship between its light reactions and light-independent reactions.

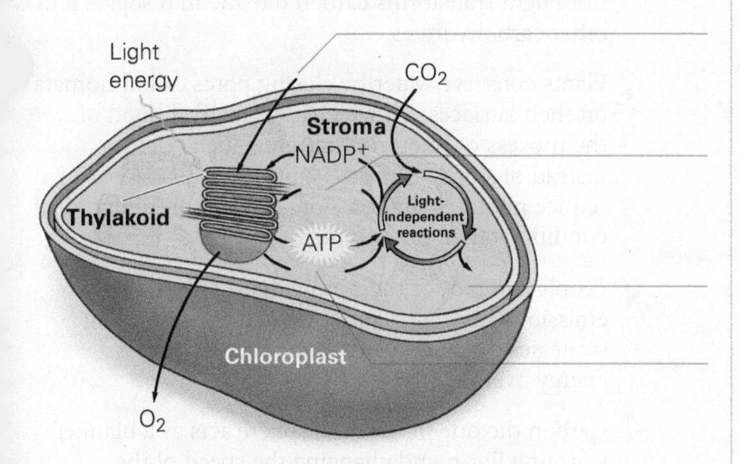

2. Carbon dioxide functions as a greenhouse gas by ________.

A. interfering with water's ability to absorb heat; B. increasing the random molecular motions of oxygen; C. allowing radiation from the sun to reach Earth and absorbing the re-radiated heat; D. splitting into carbon and oxygen and increasing the rate of cellular respiration

3. Water has a high heat-absorbing capacity because ________.

A. the sun's rays penetrate to the bottom of bodies of water, mainly heating the bottom surface; B. the strong covalent bonds that hold individual water molecules together require large inputs of heat to break; C. it has the ability to dissolve many heat-resistant solutes; D. initial energy inputs are first used to break hydrogen bonds between water molecules and only after these are broken, to raise the temperature; E. all of the above are true

4. The burning of fossil fuels ________.

A. releases carbon dioxide to the atmosphere; B. primarily occurs as a result of human activity; C. is contributing to global warming; D. is possible thanks to photosynthesis that occurred millions of years ago; E. all of the above are correct

5. Stomata on a plant's surface ________.

A. prevent oxygen from escaping; B. produce water as a result of photosynthesis; C. cannot be regulated by the plant; D. allow carbon dioxide uptake into leaves; E. are found in stacks called thylakoids

6. Which of the following does not occur during the light reactions of photosynthesis?

A. Water is released; B. Electrons from chlorophyll are moved to a higher-energy state by light; C. ATP is produced; D. NADPH is produced to carry electrons to the light-independent reactions; E. Oxygen is produced when water is split.

7. Which of the following is a false statement about photosynthesis?

A. During the light-independent reactions, electrons and ATP from the light reactions combine with atmospheric carbon dioxide to produce sugars; B. The light-independent reactions take place in the chloroplast stroma; C. Oxygen produced during the light-independent reactions is released into the atmosphere; D. Chlorophyll absorbs blue and red light and reflects green light; E. The end product of photosynthesis is a carbohydrate such as glucose.

8. Which of the following human activities generates the most carbon dioxide?

A. driving; B. cooking; C. bathing; D. using aerosol sprays

9. Photorespiration occurs ________.

A. under hot and dry conditions; B. when oxygen is incorporated in the first step of the light-independent reactions; C. when carbon dioxide levels are high inside the plant; D. A and B are correct; E. A, B, and C are correct

10. Select the true statement regarding metabolism in plant and animal cells.

A. Plant and animal cells both perform photosynthesis and aerobic respiration; B. Animal cells perform aerobic respiration only, and plant cells perform photosynthesis only; C. Plant cells perform aerobic respiration only, and animal cells perform photosynthesis only; D. Plant cells perform cellular respiration and photosynthesis, and animal cells perform aerobic respiration only.

ANALYZING AND APPLYING THE BASICS

1. During the last ice age, global temperatures were 4°C to 5°C lower than they are today. How would this temperature difference affect the water cycle?
2. Imagine an Earth without living organisms. How does this difference change the water cycle and the carbon cycle?
3. Before life evolved on Earth, there was very little oxygen in the air, primarily because oxygen is a very reactive molecule and tends to combine with other compounds. Explain why oxygen can be maintained at 16% in our atmosphere today.

GO FIND OUT

1. List five realistic actions you can take to reduce your carbon footprint. Which of these actions will have the greatest impact on your footprint? How difficult for you would it be to take this action?
2. In June 2017, President Trump announced that he was pulling the United States out of the Paris accord. He stated that the accord was placing the country at a "debilitating and tremendous disadvantage" economically and was unfair because it did not impose the same level of emissions reductions on the United States relative to other countries. Search the Internet for evidence for and against his position on the effects of the Paris Agreement. Do you think that taking the United States out of the agreement was a good choice?

MAKE THE CONNECTION

The science that you learned in this chapter has helped you better understand the real-world example used throughout this discussion. Draw a line from the statement on the left to the science that supports it on the right.

Fossil fuel emissions are changing the climate by warming temperatures and modifying rainfall patterns. ▷	◁ In the carbon cycle, carbon dioxide is removed from the atmosphere by photosynthesis, in which energy from light transforms carbon dioxide into sugars and other carbohydrates.
Fossil fuel burning increases the amount of carbon dioxide in the atmosphere. ▷	◁ Plants conserve water by closing pores called stomata on their surfaces. This increases the likelihood of the process of photorespiration, which releases, instead of stores, carbon. Plants that are adapted to reduce photorespiration store less carbon under ideal conditions than other types of plants.
Plants remove carbon dioxide from the atmosphere. ▷	◁ People can reduce their individual carbon dioxide emissions through a variety of actions, including strategies to increase efficiency and to use less energy overall.
Plants remove less carbon dioxide from the atmosphere when exposed to hot, dry conditions. ▷	◁ Carbon dioxide in the atmosphere acts as a blanket warming Earth and changing the speed of the water cycle.
Plant growth alone cannot prevent global warming; changes in human behavior are needed as well. ▷	◁ Since the evolution of photosynthesis, some carbon that has been stored in the earth is now found as coal, oil, and natural gas. Combustion of these fossil fuels releases this carbon (as carbon dioxide) back to the atmosphere.

Answers to **Got It?, Visualize This, Working with Data, Sounds Right, But Is It?, Show You Know, and Chapter Review** questions can be found in the **Answers** section at the back of the book.

6 Cancer

DNA Synthesis and Mitosis

Some cancers are caused by environmental exposures, like the skin cancers Hugh Jackman has had removed from his nose.

Cancer is a disease that will affect most people at some point in their lives. Of those who live an average life span, more than one in three will be directly affected with a diagnosis, and those who escape the disease themselves are likely to be affected by the diagnosis of a loved one.

Some types of cancer are acquired during an individual's lifetime and others are genetically inherited. Sun exposure during actor Hugh Jackman's youth led to multiple skin cancers, which he had surgically removed. Jackman credits the makeup artist on the set of an *X-men* film with spotting the initial skin cancer found on his nose.

Likewise, actress and humanitarian Angelina Jolie's battle against a genetically inherited form of breast and ovarian cancer led to her decision to have both of her breasts and ovaries surgically removed, even though she showed no signs of cancer.

Actress Angelina Jolie carries a cancer-causing gene that puts her at high risk of getting cancer.

Because Jackman's skin cancers were mostly the result of environmental exposures, his children are not at an increased risk for skin cancer, unless they, too, spend a lot of time in the sun without protecting their skin. Jolie's biological children are at increased risk of breast cancer, because she may have passed on the gene to them.

A better understanding of cancer can help us to both protect ourselves and support those we love. Lifestyle changes can lead to the prevention, delayed onset, or slowed progression of many types of cancer. When a cancer diagnosis does occur, understanding the biological mechanisms of various treatments can help us make science-based medical decisions for ourselves and help us provide support to those we love as they undergo treatment.

Lifestyle changes, like quitting smoking, can dramatically affect the risk of some cancers.

6.1 What Is Cancer?

mito- means a thread.

Cancer is a disease that occurs when a cell makes copies of itself, or replicates, when it should not. **Mitosis** is the type of cell division that occurs when one parent cell divides to form two daughter cells. This process is normally regulated so that a cell divides only when more cells are required. When this regulation fails, cancers can arise.

Tumors Can Be Cancerous

mal- means bad or evil.

meta- means change or between.

Unregulated cell division leads to a pileup of cells that form a lump or **tumor.** A tumor is a solid mass of cells that has no apparent function in the body. A cyst, conversely, is a fluid-filled lump that also has no apparent function but is not cancerous. Tumors that stay in one place and do not affect surrounding structures are said to be **benign.** Some benign tumors remain harmless; others become cancerous. Invasive tumors, or those that infiltrate surrounding tissues, are **malignant** cancers. **Metastasis** occurs when the cells of a malignant tumor break away and start new cancers at distant locations (**FIGURE 6.1**).

Cancer cells can travel virtually anywhere in the body via the lymphatic and circulatory systems. The lymphatic system collects fluid, called lymph, lost from blood vessels. The lymph is then returned to the blood vessels, a process that also allows cancer cells access to the bloodstream. Lymph nodes are structures that filter fluids released from blood vessels. When a cancer patient is undergoing surgery, the surgeon will often remove a few lymph nodes that will be analyzed for the presence of cancer cells. The presence of cancer cells in the nodes is an indication that some cells have broken away from the original tumor and might be present elsewhere in the body. Metastatic cancers are much more difficult to treat than cancers that are detected before they spread.

Risk Factors for Cancer

A **risk factor** is a condition or behavior that increases the likelihood of developing a disease. These risk factors can be affected by genetics or environmental exposures.

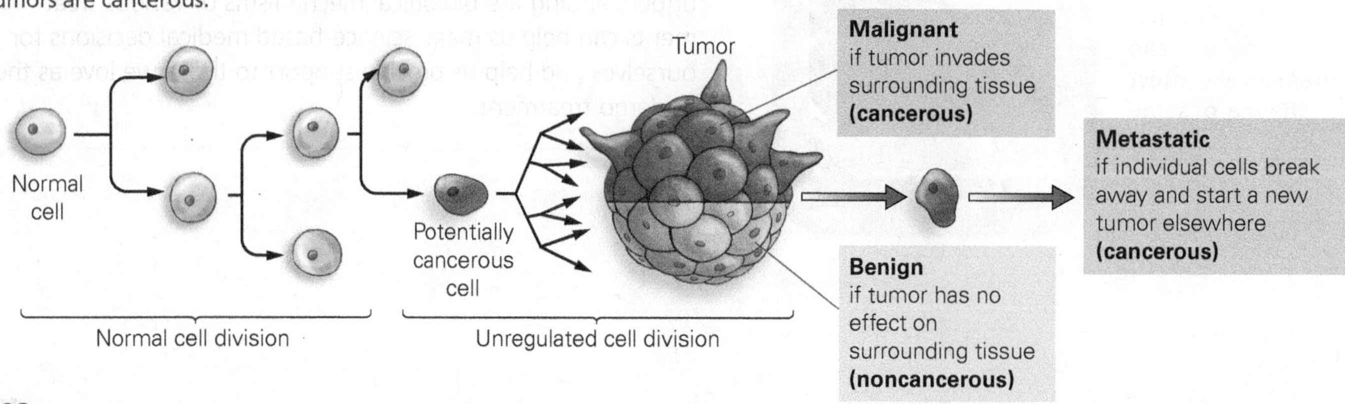

FIGURE 6.1 What is cancer?
A tumor is a clump of cells with no function. Tumors may remain benign, or they can invade surrounding tissues and become malignant. Tumor cells may move, or metastasize, to other locations in the body. Malignant and metastatic tumors are cancerous.

Inherited Cancer Risk. Angelina Jolie carries a mutated version of the *BRCA1* gene (for *br*east *ca*ncer susceptibility) that makes her far more likely to get breast cancer than other women. Jolie lost her mother and grandmother to breast cancer, and after being told that she herself had an 87% risk for breast cancer, she decided to undergo a double mastectomy—a procedure to remove both of her breasts. The mutated version of the gene that she carries also increases her risk for ovarian cancer—the disease that killed her mother. To further decrease her cancer risk Jolie recently had her ovaries and oviducts removed. How did Jolie know her risks? Because several members of Jolie's family had been diagnosed with breast and ovarian cancers, she was tested for the presence of this mutation. Because only about 1% of people carry this particular gene mutation, unless several immediate family members have been diagnosed with cancer, this testing is not usually performed.

Environmental Exposures. Exposure to particular substances, called **carcinogens,** is correlated with development of particular cancers. Everyone knows that smoking increases risk of cancer. What is less well known is that smoking combined with excessive alcohol consumption increases cancer risk at a greater level than one would expect. This is because some carcinogens enhance the activity of other carcinogens. When this occurs, the substances involved are said to be acting in a **synergistic** manner. Cigarette smoking and alcohol consumption, two practices commonly combined by college students, have a far greater effect on cancer risk than the sum of each separate risk factor combined (**FIGURE 6.2**).

Risks for skin cancer, one of the most common cancers, include exposure to the ultraviolet (UV) light from the sun and in tanning beds. Using sunscreen, avoiding getting sunburned, and never using tanning beds can help prevent this cancer. Growing up in the 1970s, when the use of sunscreen was far less common than it is today, along with spending a lot of time outside in his native sunny Australia no doubt increased Hugh Jackman's risk of obtaining skin cancer. Jackman also has light skin. Darker skin is protective against skin cancer because of the presence of more of the light-absorbing pigment melanin. An

WORKING WITH DATA

How much greater is the risk associated with the use of alcohol and tobacco versus the use of either separately? Is this increased risk additive or multiplicative? Explain your answer.

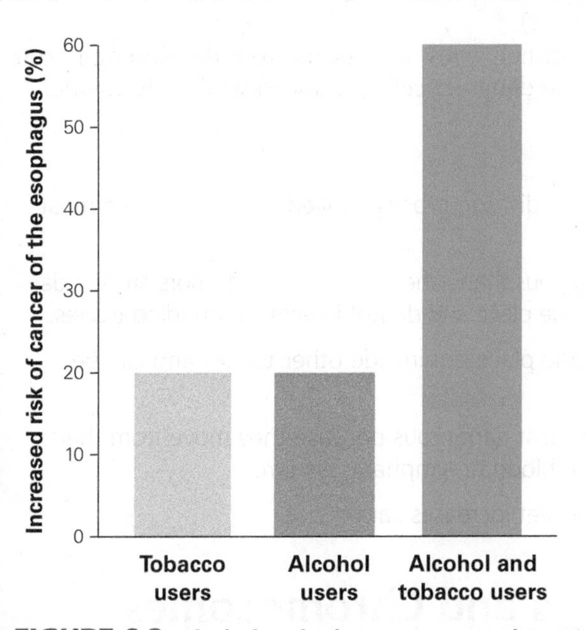

FIGURE 6.2 Alcohol and tobacco are synergists. This graph was created using data from a study published in the peer reviewed *Journal of American Gastroenterology* showing that one type of cancer of the esophagus is more likely in those that smoke cigarettes and drink alcohol than would be expected. The esophagus is a tube that connects the throat to the stomach.

TABLE 6.1 Decreasing your cancer risk.

Risk-Reducing Behavior	Specific Risk-Reduction Information	Biological Mechanism of Risk Reduction
Don't use tobacco.	The use of tobacco of any type, whether delivered via cigarettes, cigars, pipes, or chewing tobacco, increases your risk of many cancers. Electronic cigarettes, which contain nicotine only, likely carry their own risks, although studies to date have not formed a consensus.	Tobacco and cigarette smoke contain more than 20 known cancer-causing substances. Chemicals present in tobacco and cigarette smoke have been shown to increase cell division, damage DNA, inhibit a cell's ability to repair damaged DNA, and prevent cells from dying when they should.
Limit alcohol consumption.	Men who want to decrease their cancer risk should have no more than two alcoholic drinks a day, and women one or none.	When harmful chemicals dissolve in alcohol, they are able to traverse cell membranes and can damage DNA.
Eat a low-fat, high-fiber diet.	Eat at least five servings of fruits and vegetables every day as well as six servings of food from other plant sources, such as breads, cereals, grains, rice, pasta, or beans.	Plant-based foods are low in fat and high in fiber. They are also rich in antioxidants, which help prevent DNA damage.
Exercise regularly.	Engage in physical activity for at least 30 minutes, 5 days a week.	Exercise keeps the immune system functioning effectively, allowing it to recognize and destroy cancer cells.
Maintain a healthy weight.	Avoid becoming obese. If you are obese, consult a physician for a weight loss program.	Because fatty tissues can store hormones, the abundance of fatty tissue has been hypothesized to increase the risk of hormone-sensitive cancers such as breast, uterine, ovarian, and prostate cancer.

increased amount of melanin is an evolutionary adaptation to living in a geographic range with high UV light exposure. This does not mean that those of us with dark skin can get away with not using sunscreen. All of us are susceptible to skin cancer—some are just more susceptible than others.

Although UV light exposure increases risk of skin cancer, there are general risk factors that increase risk of virtually every type of cancer. These include tobacco use, excessive alcohol consumption, a high-fat and low-fiber diet, lack of exercise, and obesity (TABLE 6.1).

Limiting these exposures can help prevent cancers from developing in our cells and from being passed on to daughter cells produced when cells divide.

Got It?

1. Cells that undergo the normal division process, called __________, are most susceptible to cancer.
2. Some tumors are more dangerous than others. __________ tumors are less dangerous because they stay in one place and do not invade surrounding tissues.
3. __________ tumors stay in one place but invade other tissues and can be more dangerous tumors.
4. __________ cancers are the most dangerous because they move from their original location through the blood or lymphatic system.
5. A high-fat, __________-fiber diet increases cancer risk.

(a)

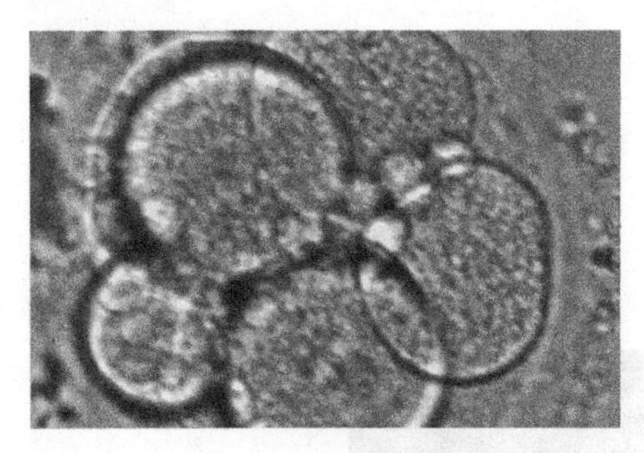

(b)

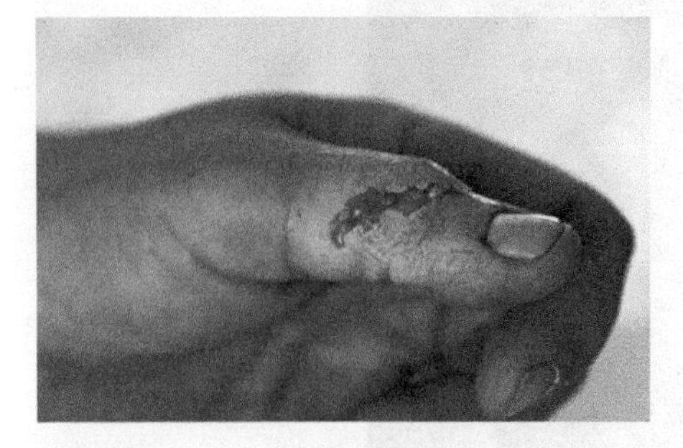

FIGURE 6.3 Why do cells divide? Cells divide to make more cells. (a) This can allow an organism to grow. Each of us begins life as a single fertilized egg cell that underwent millions of rounds of cell division to produce all the cells that comprise the tissues and organs of our bodies. (b) Cells also divide to heal wounds. As this cut heals, new cells will replace those damaged by the injury.

6.2 Passing Genes and Chromosomes to Daughter Cells

Cells have evolved to divide for a variety of reasons having nothing to do with cancer. Cell division produces new cells to allow an organism to grow, to replace damaged cells (**FIGURE 6.3**), and, in some cases, to reproduce.

Some organisms reproduce by making exact copies of themselves. Reproduction of this type, called **asexual reproduction,** results in offspring that are genetically identical to the original parent cell. Single-celled organisms, such as bacteria and amoeba, reproduce in this manner (**FIGURE 6.4a**). Some multicellular organisms can reproduce asexually also. For example, some plants can grow from clippings of stems, leaves, or roots. Reproduction from such cuttings is also a form of asexual reproduction (**FIGURE 6.4b**). Organisms whose reproduction requires genetic information from two parents undergo **sexual reproduction.** Humans reproduce sexually when sperm and egg cells each contribute genetic information at fertilization.

VISUALIZE **THIS**

(a) If one amoeba divides one time, how many amoeba will there be?
(b) If you removed one stem and leaf of the ivy plant on this tree and replanted it, would you expect it to grow into a larger stem and leaf or into two stems and leaves?

(a) Amoeba

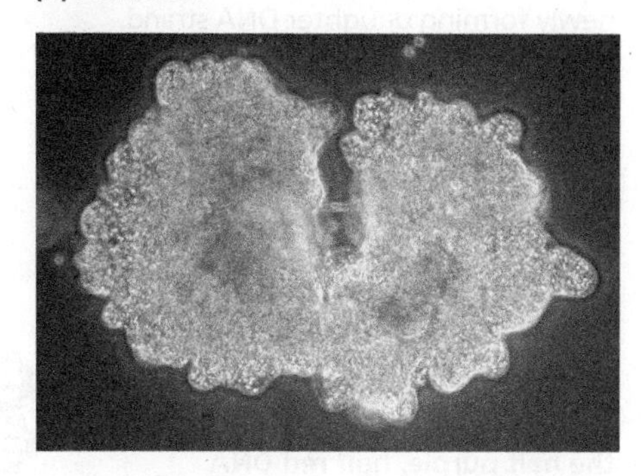

(b) English ivy

FIGURE 6.4 Asexual reproduction. (a) This single-celled amoeba divides by copying its DNA and producing offspring that are genetically identical to the original, parent amoeba. (b) Some multicellular organisms, such as this English ivy plant, can reproduce asexually from cuttings.

Genes and Chromosomes

Whether reproducing sexually or asexually, all dividing cells must first make a copy of their genetic material, the **DNA (deoxyribonucleic acid).** DNA carries the instructions, called **genes,** for building all of the proteins that a cell requires. The DNA in the nucleus is wrapped around proteins to produce structures called **chromosomes.**

Chromosomes are in an uncondensed, string-like form when a cell is not preparing to divide (**FIGURE 6.5a**). For cell division to occur, the DNA in each chromosome is compressed into a more compact linear structure that is easier to maneuver during cell division. Condensed chromosomes are less likely to become tangled or broken than are the uncondensed and string-like structures.

Each chromosome carries hundreds of genes. When a chromosome is replicated, a copy is produced that carries those same genes. The copied chromosomes, now called **sister chromatids,** are attached to each other at a region toward the middle of the replicated chromosome, called the **centromere** (**FIGURE 6.5b**). Because the centromere is not always located precisely in the center of the chromosome, it can subdivide the chromosome into one long and one short arm. Scientists have mapped the location of the *BRCA1* gene to the long arm of chromosome number 17 (**FIGURE 6.6**).

DNA Replication

During the process of **DNA replication** that precedes cell division, the double-stranded DNA molecule is copied, first by splitting the molecule in half up the middle of the helix. New nucleotides are added to each side of the original parent molecule, maintaining the A-to-T and G-to-C base pairings. This process results in two daughter DNA molecules, each composed of one strand of

(a) Unconcensed DNA

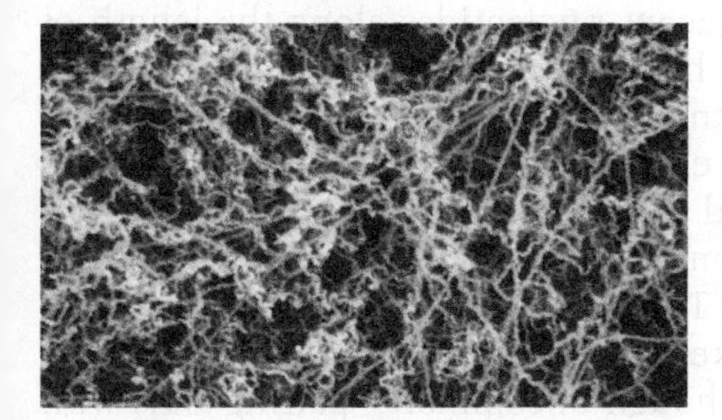

(b) DNA condensed into chromosomes

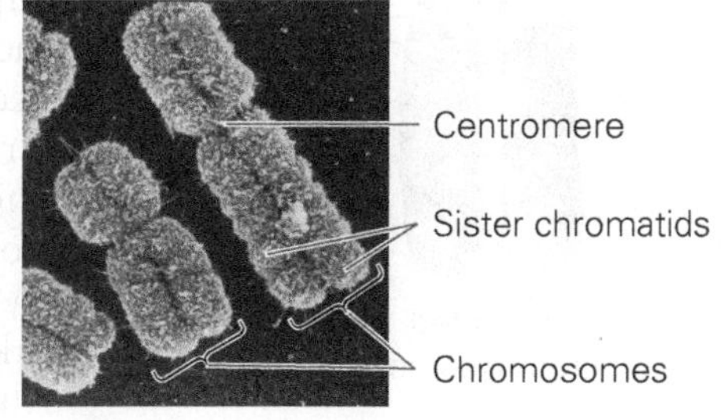

FIGURE 6.5 DNA condenses during cell division. (a) DNA in its replicated but uncondensed form prior to cell division. (b) During cell division, each copy of DNA is wrapped neatly around many small proteins, forming the condensed structure of a chromosome. After DNA replication, two identical sister chromatids are produced and joined to each other at the centromere.

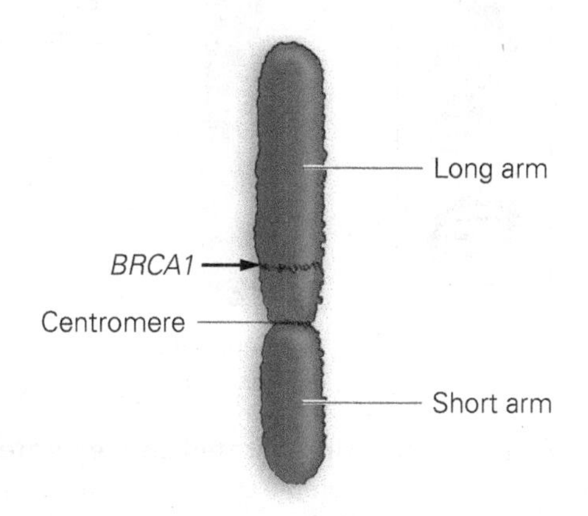

FIGURE 6.6 *BRCA1* locus. The *BRCA1* gene is located on the long arm of chromosome number 17.

FIGURE 6.7 DNA replication. (a) DNA replication results in the production of two identical daughter DNA molecules from one parent molecule. Each daughter DNA molecule contains half of the parental DNA and half of the newly synthesized DNA. (b) The DNA polymerase enzyme moves along the unwound helix, tying together adjacent nucleotides on the newly forming daughter DNA strand. Free nucleotides have three phosphate groups, two of which are cleaved to provide energy for this reaction before the nucleotide is added to the growing chain.

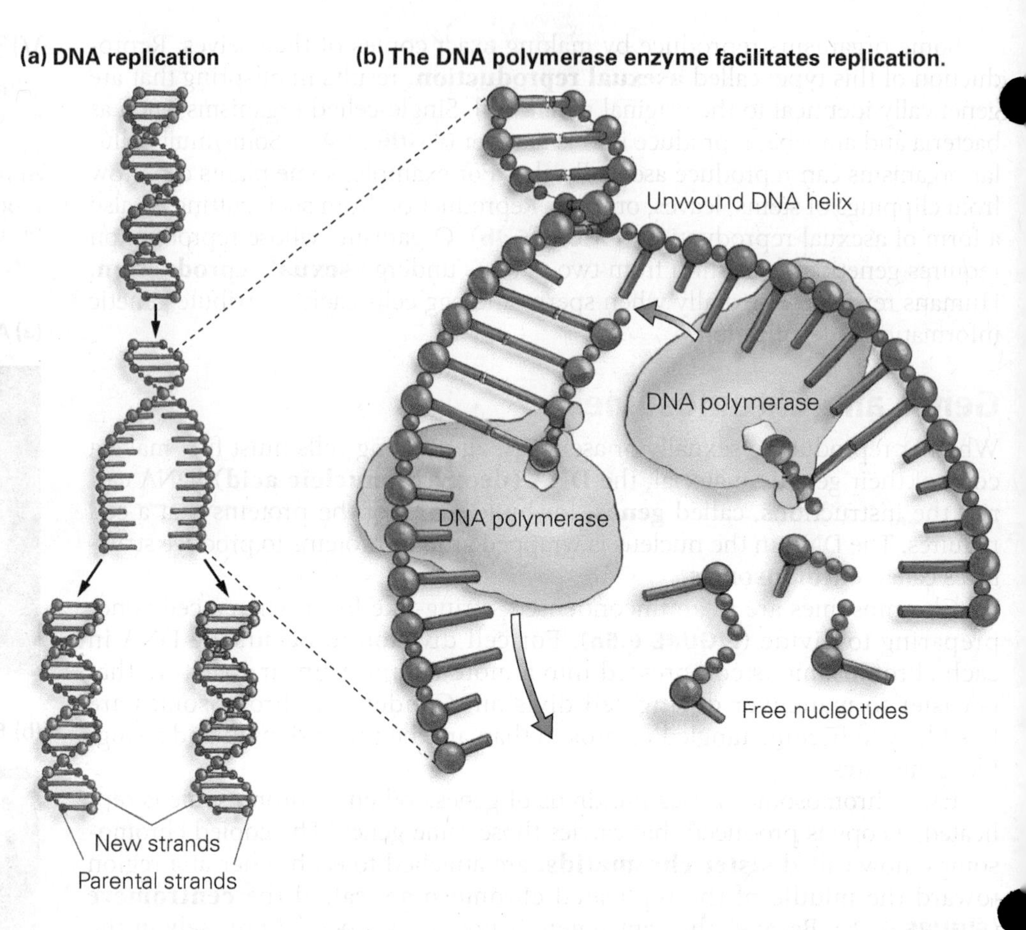

VISUALIZE THIS

Assume another round of replication were to occur to one of the half purple, half red DNA molecules shown in part (a). How many total DNA molecules would be produced? If the nucleotides being added to the newly synthesized strand are purple, what proportion of each DNA molecule would be purple?

parental nucleotides and one newly synthesized strand (**FIGURE 6.7a**). Because each newly formed DNA molecule consists of one-half conserved parental DNA and one-half new daughter DNA, this method of DNA replication is referred to as **semiconservative replication.**

Replicating the DNA requires the assistance of enzymes, in particular, a **DNA polymerase.** The DNA polymerase moves along the length of the unwound parental DNA strand to facilitate synthesis of the newly formed strand (**FIGURE 6.7b**). When free nucleotides floating in the nucleus have an affinity for each other (A for T and G for C), these complementary nucleotides bind to each other across the width of the helix and then the DNA polymerase catalyzes the formation of the covalent bond between adjacent nucleotides along the length of the helix. When an entire chromosome has been replicated, the newly synthesized sister chromatids are identical to each other and attached to each other at the centromere (**FIGURE 6.8**).

The DNA polymerase can make mistakes when facilitating base pairing. Although uncommon—pairing an A with a G, for instance—such mistakes can alter the sequence of the original gene. Changes in the DNA of a gene are called **mutations.** Normally, replication of the *BRCA1* gene produces an exact copy of the

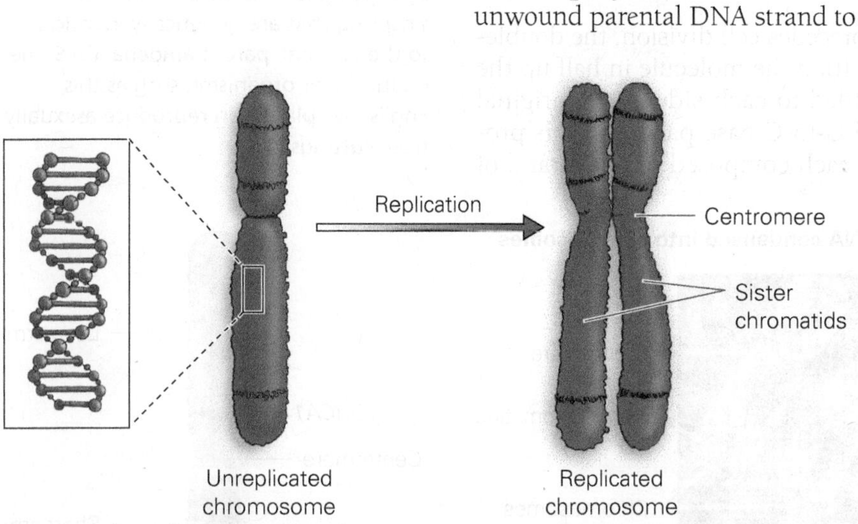

FIGURE 6.8 Unreplicated and replicated chromosomes. An unreplicated chromosome is composed of one double-stranded DNA molecule. A replicated chromosome is X-shaped and composed of two identical double-stranded DNA molecules. Each DNA molecule of the replicated chromosome is a copy of the original chromosome and is called a sister chromatid.

gene. If mistakes in the copying occur, a mutant version of the gene can be produced, which appears to be the case in the version of the gene that Jolie carries. Similarly, UV light can cause mutations by directly interfering with the replication of DNA, causing mutations.

Got It?

1. Reproduction that involves two parents producing offspring that have a unique combination of genes is called ________ reproduction.
2. Genes are located on structures called ________ and are composed of DNA.
3. A replicated chromosome is composed of two ________ chromatids.
4. The enzyme that replicates DNA is called ________ ________.
5. Sister chromatids are genetically ________ to each other.

6.3 The Cell Cycle and Mitosis

After a cell's chromosomes and DNA have been replicated, the cell is able to undergo cell division and produce daughter cells. One type of cell division, mitosis, is an asexual division that produces two daughter cells that are identical to their original parent cell and to each other. Mitosis occurs in the type of body cells called somatic cells. **Somatic cells** include any cell type that does not produce sex cells. In plants, for example, the leaves and stem are composed of somatic cells and undergo mitosis. The reproductive organs of the plant produce pollen and egg cells, which are nonsomatic cells called sex cells.

For cells that divide by mitosis, the cell cycle includes three steps: (1) **interphase,** when the DNA replicates; (2) *mitosis,* when the copied chromosomes split and move into the daughter nuclei; and (3) **cytokinesis**, when the cytoplasm of the parent cell splits (**FIGURE 6.9a**). Additionally, interphase and mitosis are further subdivided.

cyto- relates to cells.
-kinesis means motion.

Interphase

A normal cell spends most of its time in interphase (**FIGURE 6.9b**). During this phase of the cell cycle, the cell performs its typical functions and produces the proteins required for the cell to do its particular job. For example, during interphase, a muscle cell produces proteins required for muscle contraction. Different cell types spend varying amounts of time in interphase. Cells that frequently divide, like skin cells, spend less time in interphase than do those that seldom divide, such as some nerve cells. A cell that will divide also begins preparations for division during interphase. Interphase can be separated into three phases: G_1, S, and G_2.

During the G_1 (first gap or growth) phase, most of the cell's organelles duplicate. Consequently, the cell grows larger during this phase. During the S (synthesis) phase, the DNA composing the chromosomes replicates. During the G_2 (second gap) phase of the cell cycle, the cell continues to grow and prepares for the division of chromosomes that will take place during mitosis.

(a) Copying and partitioning DNA

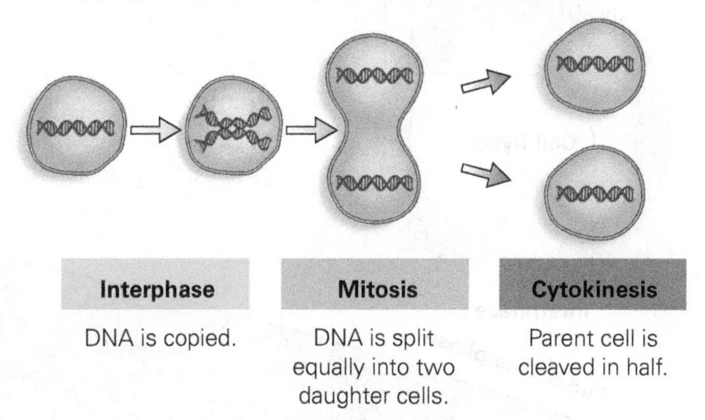

(b) Steps in the cell cycle

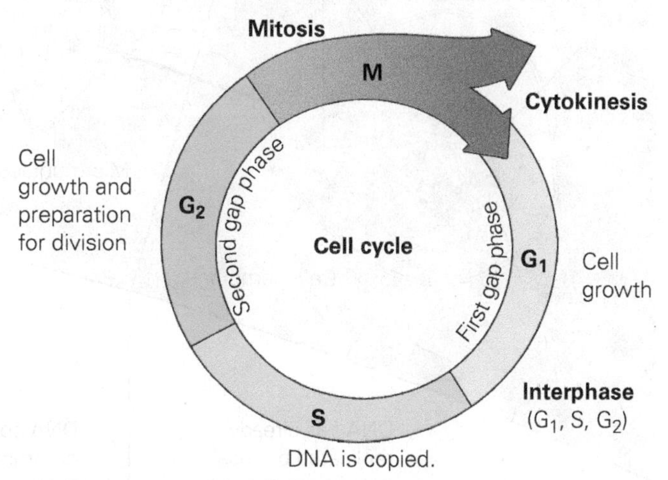

FIGURE 6.9 The cell cycle. (a) During interphase, the DNA is copied. During mitosis the copies of DNA are separated into different nuclei and cytokinesis divides the cytoplasm, creating two daughter cells. (b) During interphase, there are two stages when the cell grows in preparation for cell division, G_1 and G_2 stages, and one stage where the DNA replicates, the S stage. The chromosomes are separated and two daughter cells are formed during the M phase.

Mitosis

The movement of chromosomes from the original parent cell into two daughter cells occurs during mitosis. Whether these phases occur in an animal or a plant, the outcome of mitosis and the next phase, cytokinesis, is the same: the production of genetically identical daughter cells. To achieve this outcome, the sister chromatids of a replicated chromosome are pulled apart, and one copy of each goes into each newly forming nucleus. Mitosis is accomplished during four stages: prophase, metaphase, anaphase, and **telophase.** FIGURE 6.10 summarizes the cell cycle in animal cells. The four stages of mitosis demonstrated in animal cells are nearly identical in plant cells.

telo- means end or completion.

During **prophase,** the replicated chromosomes condense, allowing them to move around in the cell without becoming entangled. Protein structures called **microtubules** also form and grow, ultimately radiating out from opposite ends, or **poles,** of the dividing cell. The growth of microtubules helps the cell to expand. Motor proteins attached to microtubules also help pull the

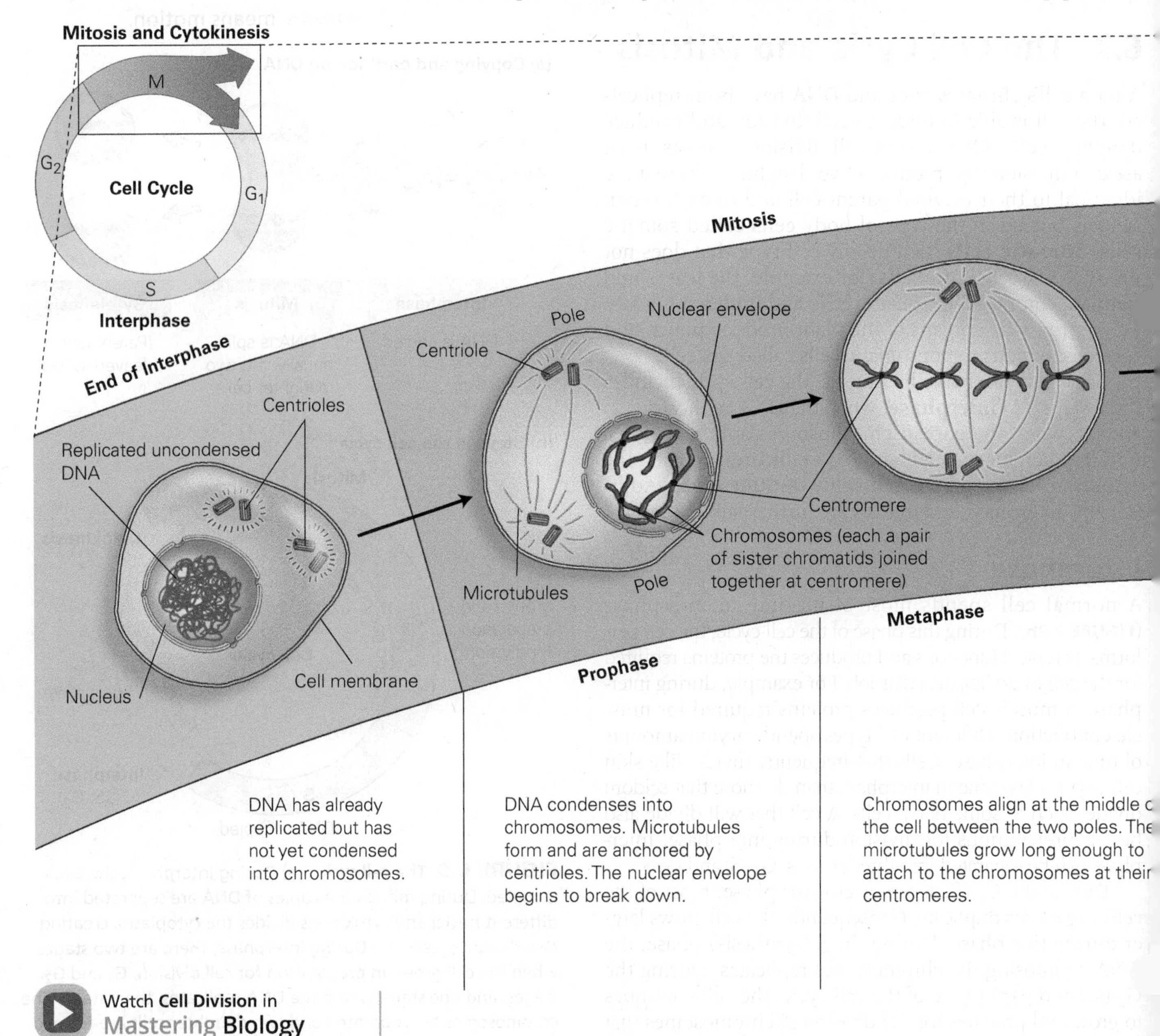

Mitosis and Cytokinesis
M
G2
Cell Cycle
G1
S
Interphase
End of Interphase
Mitosis
Centrioles
Replicated uncondensed DNA
Nucleus
Cell membrane
Centriole
Pole
Nuclear envelope
Microtubules
Pole
Centromere
Chromosomes (each a pair of sister chromatids joined together at centromere)
Prophase
Metaphase
DNA has already replicated but has not yet condensed into chromosomes.
DNA condenses into chromosomes. Microtubules form and are anchored by centrioles. The nuclear envelope begins to break down.
Chromosomes align at the middle c the cell between the two poles. The microtubules grow long enough to attach to the chromosomes at their centromeres.

Watch Cell Division in Mastering Biology

chromosomes around during cell division. The membrane that surrounds the nucleus, called the **nuclear envelope,** breaks down so that the microtubules can gain access to the replicated chromosomes. At the poles of each dividing animal cell, structures called **centrioles** physically anchor one end of each forming microtubule. Plant cells do not contain centrioles, but microtubules in these cells do remain anchored at a pole.

During **metaphase,** the replicated chromosomes are aligned across the middle, or equator, of each cell. To do this, the microtubules, which are attached to each chromosome at the centromere, lengthen and contract, pushing and pulling the chromosomes until they are aligned in single file across the middle of the cell.

During **anaphase,** the centromere splits, and the microtubules shorten to pull each sister chromatid of a chromosome to opposite poles of the cell.

In the last stage of mitosis, **telophase,** the nuclear envelopes re-form around the newly produced daughter nuclei.

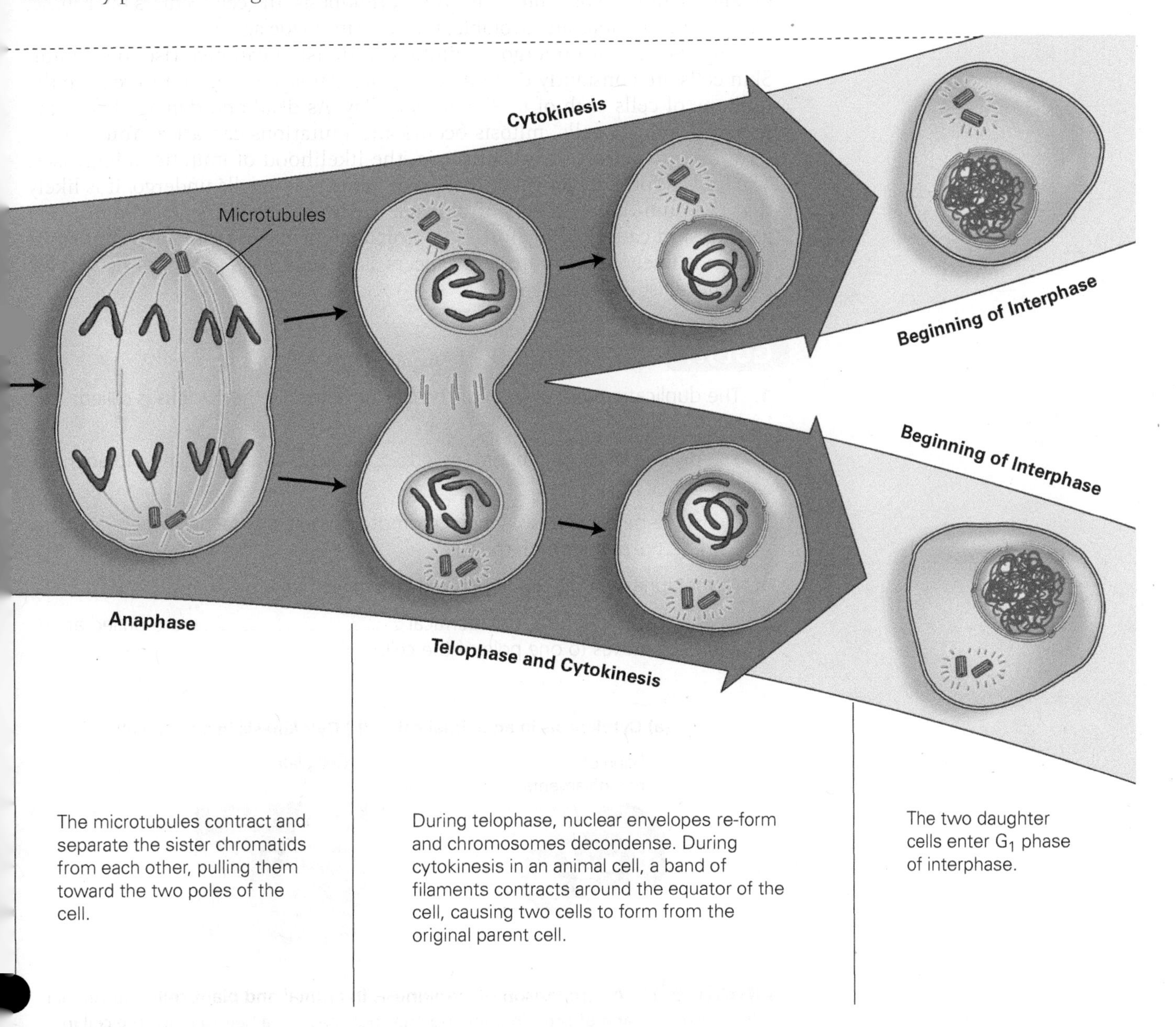

Cytokinesis
Microtubules
Beginning of Interphase
Beginning of Interphase
Anaphase
Telophase and Cytokinesis
The microtubules contract and separate the sister chromatids from each other, pulling them toward the two poles of the cell.
During telophase, nuclear envelopes re-form and chromosomes decondense. During cytokinesis in an animal cell, a band of filaments contracts around the equator of the cell, causing two cells to form from the original parent cell.
The two daughter cells enter G_1 phase of interphase.

FIGURE 6.10 Cell division in animal cells. This diagram illustrates how cell division proceeds from interphase through mitosis and cytokinesis and then back to interphase.

Cytokinesis

Cytokinesis is the division of the cytoplasm that takes place on the heels of telophase. During cytokinesis in animal cells, a band of proteins encircles the cell at the equator and divides the cytoplasm. This band of proteins contracts to pinch apart the two nuclei and the surrounding cytoplasm, creating two daughter cells from the original parent cell. In cytokinesis in plant cells, a new **cell wall,** an inflexible structure surrounding the plant cells, forms. **FIGURE 6.11** shows the difference between cytokinesis in animal and plant cells. During telophase of mitosis in a plant cell, membrane-bound vesicles deliver the materials required for building the cell wall to the center of the cell. These materials include a tough, fibrous carbohydrate called **cellulose** as well as some proteins. The membranes surrounding the vesicles gather in the center of the cell to form a structure called a **cell plate.** The cell plate and forming cell wall grow across the width of the cell and form a barrier that eventually separates the products of mitosis into two daughter cells. After cytokinesis, the cell reenters interphase, and if the conditions are favorable, the cell can divide again.

Any tissue that undergoes mitotic cell division can give rise to a tumor. Skin cells are constantly dividing to replace damaged cells and to replace the millions of cells each of us sheds every day. As dead and damaged cells are replaced by new cells, mitosis occurs and mutations can arise. You can see how, over the course of generations, the likelihood of mutations increases. The longer you live, the more rounds of mitosis your cells undergo. It is likely that the cumulative effects of damage to Hugh Jackman's DNA led to him getting skin cancer in his forties. Unfortunately, his doctors have confirmed that he is likely to get more skin cancers as he continues to age, so he sees his dermatologist every few months.

Got It?

1. The duplication and division of the chromosomes in the nucleus is called ________.
2. The division of the cytoplasm, called ________, differs in plant and animal cells.
3. The stage before mitosis, when chromosomes are replicated, is the ________ phase of interphase.
4. At metaphase of mitosis, replicated chromosomes align at the ________.
5. At anaphase of mitosis, the identical ________ ________ are pulled apart and each moves to one pole of the cell.

(a) Cytokinesis in an animal cell

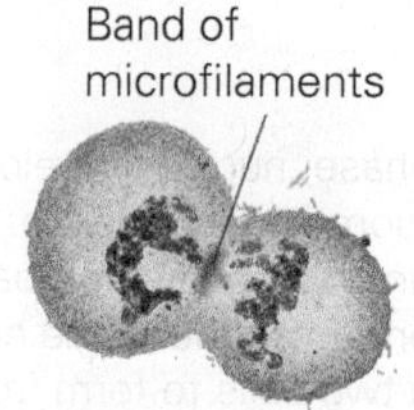

(b) Cytokinesis in a plant cell

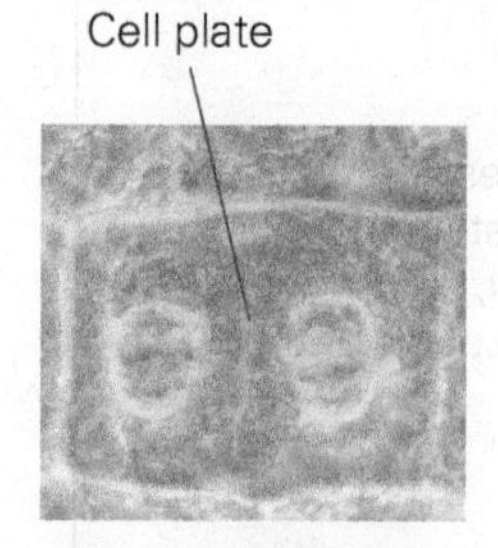

FIGURE 6.11 A comparison of cytokinesis in animal and plant cells. (a) Animal cells produce a band of protein filaments that tightens like a belt to pinch the cell in half. (b) Plant cells form a cell plate down the middle of the parent cell that gives rise to the cell wall.

6.4 Cancer Prevention, Detection, and Treatment

When cell division is working properly, it is a tightly controlled process and tumor formation is prevented. Even if one renegade cell escapes these regulatory controls and forms a tumor, mechanisms are in place to prevent the tumor from continuing to divide. Keeping these controls working properly helps prevent cancers from developing in the first place. How is such prevention accomplished?

Tumor Suppressors Help Prevent Cancer

Before allowing cell division to occur, proteins continuously survey cells to make sure that cell division is required and that the cell is not damaged in any way. If cell division is not required, or if a cell is damaged, the process will be halted at a cellular checkpoint (**FIGURE 6.12**). At the G_1 checkpoint, proteins check to determine if the cell has grown enough to be able to subdivide into two daughter cells. After the DNA has undergone replication during the S phase, the success of that replication is assessed at the G_2 checkpoint. Late in the M phase, proteins at the third checkpoint double-check that each chromosome is present in the proper duplicated configuration and attached to a microtubule.

One type of cell division regulating protein is called a **tumor suppressor.** Tumor suppressor proteins inspect newly replicated DNA. If DNA is damaged in any way, for example, if an incorrect G:T base pair exists, the cell does not continue the process of cell division (**FIGURE 6.13**). We all have many tumor suppressor and other cell cycle regulatory genes; if these genes are mutated, tumors can arise.

VISUALIZE THIS

At which checkpoint will a cell that has been treated with a chemical that prevents microtubule formation be stopped?

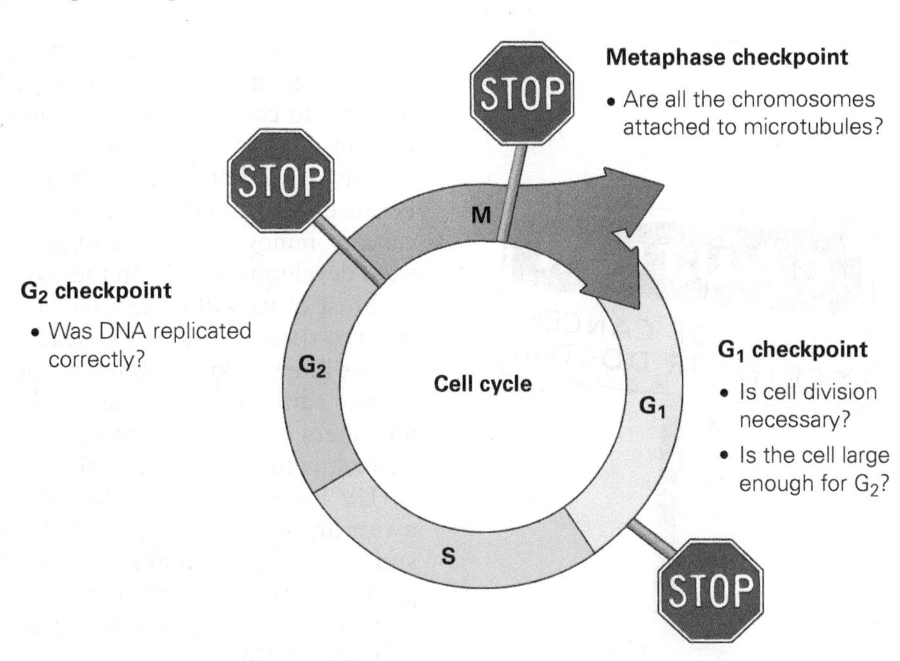
Metaphase checkpoint
• Are all the chromosomes attached to microtubules?
STOP
STOP
STOP
M
G₂ checkpoint
• Was DNA replicated correctly?
G₂
Cell cycle
G₁
S
G₁ checkpoint
• Is cell division necessary?
• Is the cell large enough for G₂?

FIGURE 6.12 Controls of the cell cycle. Checkpoints at G_1, G_2, and metaphase determine whether a cell will be allowed to divide.

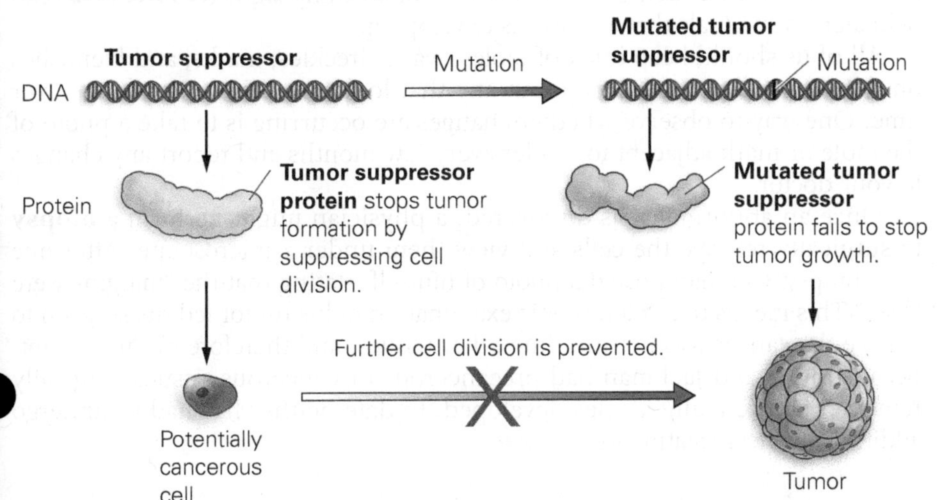
Mutations to tumor suppressor genes
Tumor suppressor
DNA
Mutation
Mutated tumor suppressor
Mutation
Protein
Tumor suppressor protein stops tumor formation by suppressing cell division.
Mutated tumor suppressor protein fails to stop tumor growth.
Further cell division is prevented.
Potentially cancerous cell
Tumor

FIGURE 6.13 Mutations to tumor suppressor genes. Mutations to tumor suppressor genes can increase the likelihood of cancer developing.

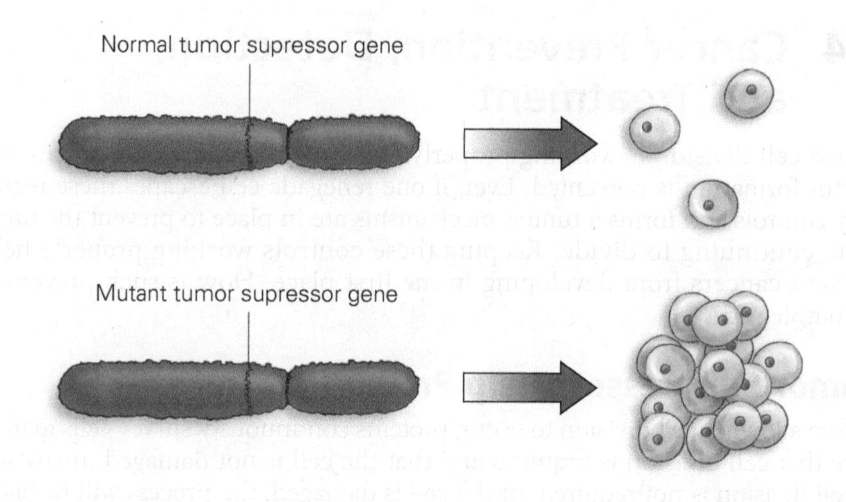

FIGURE 6.14 Normal version of a tumor suppressor gene leads to controlled cell division. The mutant version is unable to perform this job and cells divide uncontrollably. These mutations can be inherited, as was the case for Angelina Jolie. or acquired during one's lifetime, as was the case for Hugh Jackman.

The normal *BRCA1* gene encodes a protein that functions as a tumor suppressor. The mutant version of the *BRCA1* gene that Angelina Jolie inherited is not able to carry out this function (**FIGURE 6.14**). Therefore, cells where this mutant gene is expressed (i.e., cells composing breast and ovarian tissues) can divide more than they are supposed to. If this happens, tumors can form. Because Jolie inherited this mutant gene and was very likely to develop these cancers, removal of the breast and ovarian tissues was performed to prevent her from developing cancer in these organs.

Most of us will inherit few, if any, mutant cell-cycle control genes. Mutations to these genes will instead occur with exposures to environmental risk factors like smoking cigarettes, lack of exercise, and poor eating habits. A few viruses can also cause cancer. The human papilloma virus (HPV) enters cells and alters cell cycle control genes, leading to increased likelihood of cervical, vaginal, penile, anal, and oral cancers.

UV light also causes mutations to DNA. When these mutations affect a tumor suppressor gene called p53, skin cancer can result. The more sun exposure, the higher the likelihood that the DNA in the cell will undergo a mutation. Avoiding excess sun exposure, along with application of sunscreen helps prevent these mutations by preventing UV light from accessing DNA.

CAUTION
SIGNS OF CANCER
SEE YOUR DOCTOR

C hange in bowel or bladder habits

A sore that does not heal

U nusual bleeding or discharge

T hickening or lump

I ndigestion or difficulty swallowing

O bvious change in wart or mole

N agging cough or hoarseness

FIGURE 6.15 Warning signs of cancer. Self-screening for cancer can save your life. If you experience one or more of these warning signs, see your doctor.

Cancer Detection

When cancers are detected early, their progression can be halted and the odds of survival increase. Being on the lookout for warning signs (**FIGURE 6.15**) can help alert individuals that a cancer is developing.

All of us should keep track of moles, warts, freckles, marks, and blemishes on the skin and see a doctor about any that look concerning or change over time. One way to observe whether changes are occurring is to take a photo of the mole or mark adjacent to a ruler every few months and report any changes to your doctor.

Once an abnormality is discovered, a physician might perform a **biopsy** to surgically remove the cells and view them under a microscope. After one such biopsy, Jackman posted a photo of himself, stating that the "margins were clear." This means that microscopic examination of his tumor led his surgeon to believe the cancer was not invading other tissues, and therefore not malignant. Because Jolie and Jackman had precancerous or cancerous tissues surgically removed before malignancies developed, to date neither has had to undergo additional cancer treatments.

Cancer Treatment

If a cancer is difficult to remove surgically or if it may have spread, **chemotherapy** is often used to treat the remaining cancer. During chemotherapy, chemicals that selectively kill dividing cells are injected into the bloodstream. Because cancer is a disease caused by mutations, some cancer cells carry mutations that will allow them to be resistant to various chemotherapeutic agents. Therefore, treating a cancer patient with a combination of chemotherapeutic agents aimed at different cell cycle events increases the chances of destroying all the cancerous cells in a tumor. For example, some chemotherapeutic agents prevent the chromosomes from being pulled to the equator during cell division and others prevent DNA synthesis.

Unfortunately, normal cells that divide rapidly can also be affected by chemotherapy. Hair follicles, cells that produce red and white blood cells, and cells that line the intestines and stomach are often damaged or destroyed. The effects of chemotherapy therefore often include temporary hair loss, anemia (dizziness and fatigue due to decreased numbers of red blood cells), and lowered protection from infection due to decreases in the number of white blood cells. In addition, damage to the cells of the stomach and intestines can lead to nausea, vomiting, and diarrhea.

After a tumor is surgically removed, there is always a chance that some cancer cells from the tumor were not excised and remain in the body. **Radiation therapy** is the use of high-energy particles aimed at the location of the tumor in an effort to kill any cancer cells that might remain. This therapy is typically used only when cancers are located close to the surface of the body because it is difficult to focus a beam of radiation on internal organs, and tissue damage from radiation can be quite severe. It can also be used in cases where surgery is not possible.

A promising new cancer treatment called **immunotherapy** harnesses the ability of the immune system to selectively destroy cancer cells. The cell membranes of cancer cells contain high concentrations of particular unique proteins referred to as markers. Scientists can produce substances in the laboratory that seek out and destroy cancer cells by targeting these cancer cell markers. Healthy cells that do not carry many of these marker proteins are spared. Because immunotherapy destroys fewer healthy cells, it takes a less severe toll on the cancer patient's health than chemotherapy and radiation.

Immunotherapy has been used not only to treat existing cancers but, in some cases, to prevent cancer from developing. These therapies include vaccines that also harness the powers of the immune system. Cancer vaccines enable a person's immune system to quickly and forcefully respond when exposure to cancer causing viruses like HPV occurs.

Got It?

1. Proteins called __________ __________ prevent some cells from dividing when they should not.
2. When cell cycle regulating genes undergo __________, cells can divide uncontrollably.
3. To determine whether a lump is cancerous it can be removed and analyzed, a process called __________.
4. Chemicals that destroy rapidly dividing cells form the basis of a cancer treatment called __________.
5. High-energy particles are used to damage tumors during a cancer treatment called __________ therapy.

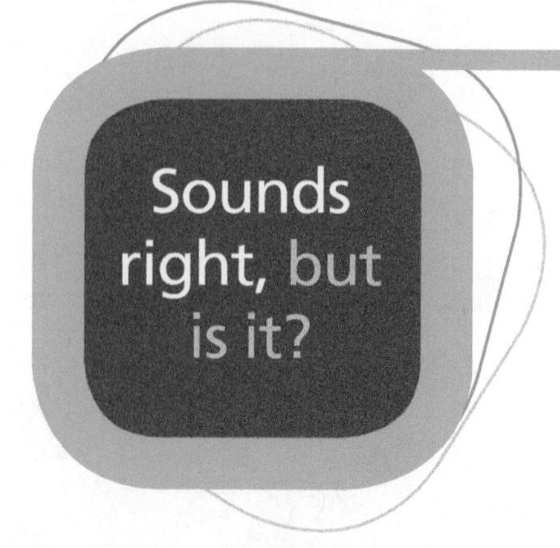
Sounds right, but is it?

A tanning salon located near campus gives a 20% discount to students. Two women on your dorm floor have been using the tanning beds. When you question the safety of this practice, your friends claim that being exposed to 20 minutes of UV light from tanning beds is actually safer than being in the sun for a few hours. When you point out that tanning beds can cause skin cancer, they say that most salons have switched to bulbs that have only one kind of UV light, making them much safer than before. They also claim that using tanning beds makes them look and feel more healthful, in part because using tanning beds helps them to get the recommended amount of vitamin D.

The use of tanning beds is not only safe, it improves health.

Sounds right, but it isn't.

Answer the following questions to understand why.

1. Even if there were evidence that tanning for 20 minutes was safer than being in the sun for a few hours, would this provide evidence that tanning beds are safe? Why or why not?
2. UV light from the sun occurs in two different wavelengths, longer UVA rays and shorter UVB rays. UVB rays cause the surface of the skin to burn while the longer UVA rays penetrate farther into the skin, making it more likely that the rays will damage DNA. Early versions of tanning beds used UVB bulbs, but most have now switched to UVA bulbs. From an economic standpoint, why might it benefit tanning bed manufacturers and salon owners to switch to UVA?
3. A tan occurs when damaged skin temporarily produces more melanin to prevent further damage. A tan is evidence that the skin has been injured. Should a tan skinned be considered evidence of good health? Why or why not?
4. Although sunlight is required for the body to synthesize vitamin D, the American Cancer Society recommends that no one be exposed to more than 15 minutes of sunlight before applying sunscreen. If you wanted to ensure that you were getting enough vitamin D without exposing yourself to UV light from any source, what could you do?
5. Consider your answers to questions 1–4 and explain why the original statement bolded above sounds right, but isn't.

THE BIG QUESTION

Can I prevent myself from getting cancer?

Most of us would be willing to consider making lifestyle changes to decrease our risk of cancer. For example, many people exercise and eat lots of fruits and vegetables, in part, to decrease their cancer risk. But how much control do we really have over this disease?

What should I know?

What follows are some smaller questions that need to be resolved to answer the Big Question. Place a checkmark next to the questions that science can answer.

Smaller Questions	Can Science Answer?
Will abstaining from smoking guarantee that I don't get cancer?	
Does the likelihood that I will get cancer increase as I age?	
Are there any cancer vaccines I should consider receiving?	
If I do get cancer, will the treatments be worse than having cancer?	
If I do get cancer, will my family be a good source of support for me?	
How many mutated cell cycle control genes did I inherit?	
What lifestyle choices can I make to decrease the likelihood of a cancer diagnosis?	

What does the science say?

Let's examine what the data say about this smaller question:

Are there any cancer vaccines I should consider receiving?

Exposure to the sexually transmitted human papilloma virus (HPV) increases risk of cervical, vaginal, penile, anal, and oral cancers. The graph below shows efficacy of the HPV vaccine against several reproductive cancers.

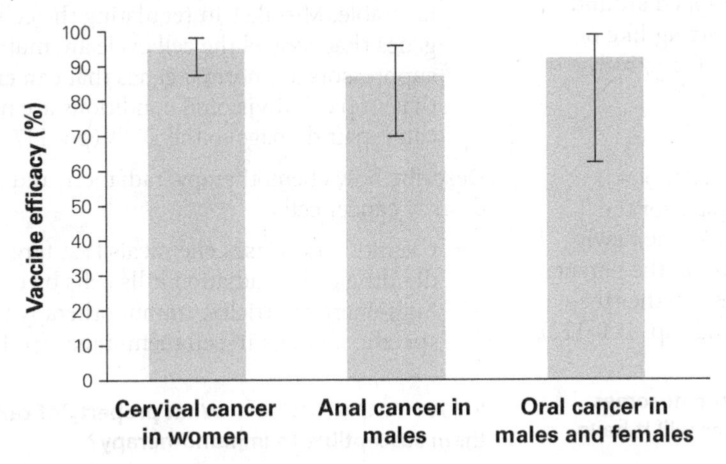

1. Describe the results. What is the approximate average rate of efficacy of these three vaccines?
2. Given these data, do you think the smaller question is answered? If not, propose another study that would help answer this question?
3. Does this information help you answer the Big Question? What else do you need to consider?

Data source: R. Herrero, P. Gonzalez, and L. Markowitz, "Present Status of Human Papilloma Vaccine Development and Implementation," *The Lancet Oncology* 16, no. 5 (2015): 206–216.

Chapter Review

Mastering Biology

Go to Mastering Biology to access **eText 2.0, Dynamic Study Modules**, and the **Study Area**, where you'll find practice quizzes, BioFlix™ animations, MP3 tutor sessions, current events, and more.

SUMMARY

Section 6.1

Describe the cellular basis of cancer.

- Unregulated cell division can lead to the formation of a tumor, which in turn can spread and become cancer (p. 108).

Compare and contrast benign and malignant tumors.

- Benign or noncancerous tumors stay in one place and do not prevent surrounding tissues and organs from functioning. Malignant tumors are those that are invasive or those that metastasize to surrounding tissues, starting new cancers (p. 108).

List several risk factors for cancer development that are under your control.

- Apart from genetics and aging, most cancer risk factors are things you can control. These include limiting UV light exposure, not smoking, eating a healthy diet, exercising, maintaining a healthy weight, and minimizing alcohol consumption (pp. 109–110).

SHOW YOU KNOW 6.1 A family member receives a cancer diagnosis. Further testing reveals that the margins of the tumor are clear. Is the latter good news or bad news? Explain your answer.

Section 6.2

List the normal functions of cell division.

- Cell division is a process cells undergo to produce new cells for growth, repair, and asexual reproduction (pp. 110–111).

Describe the structure and function of chromosomes.

- Chromosomes are composed of DNA wrapped around proteins. They can be uncondensed and string-like or condensed, depending on whether the cell is actively dividing. Chromosomes carry genes (p. 111).

Outline the process of DNA replication.

- During DNA replication, one strand of the double-stranded DNA molecule is used as a template for the synthesis of a new daughter strand of DNA. The newly synthesized DNA strand is complementary to the parent strand. The enzyme DNA polymerase ties together the nucleotides on the forming daughter strand (pp. 111–113).

SHOW YOU KNOW 6.2 If a skin cell has 46 chromosomes before it is replicated, how many chromosomes will it have after DNA replication?

Section 6.3

Describe the events that occur during interphase of the cell cycle.

- Interphase consists of two gap phases of the cell cycle (G_1 and G_2), during which the cell grows and prepares to enter mitosis or meiosis; and the S (synthesis) phase, during which time the DNA replicates. The S phase of interphase occurs between G_1 and G_2 (p. 113).

Diagram two chromosomes as they proceed through mitosis of the cell cycle.

- During prophase, the replicated chromosomes condense. At metaphase, these replicated chromosomes align across the middle of the cell. At anaphase, the sister chromatids separate from each other and align at opposite poles of the cells. At telophase, separate nuclear envelopes re-form around the linear chromosomes present at both poles of the cell (pp. 114–115).

Describe the process of cytokinesis in animal and in plant cells.

- During cytokinesis, the cytoplasm is divided into two portions, one for each daughter cell. In animal cells, this involves the pinching of one cell into two cells by a band of filaments. In plant cells, this involves the construction of a cell wall in the middle of the subdivided plant cell (p. 116).

SHOW YOU KNOW 6.3 Compare the number of DNA molecules in a replicated and unreplicated chromosome.

Section 6.4

Describe how mutations to genes that help regulate cell division can lead to tumor formation.

- When cell division is working properly, it is a tightly controlled process. Normal cells divide only when conditions are favorable. Proteins survey the cell and its environment and halt cell division if conditions are not favorable. Mistakes in regulating the cell cycle arise when genes that control the cell cycle are mutated. Tumor suppressors are normal genes that can encode proteins that stop cell division if conditions are not favorable and can repair damage to the DNA (pp. 117–118).

Describe how chemotherapy, radiation, and immunotherapy destroy cancer cells.

- Chemotherapy uses chemicals that target rapidly dividing cells. Radiation kills cells by exposing them to high-energy particles. Immunotherapy targets cancer-specific markers in cell membranes (p. 119).

SHOW YOU KNOW 6.4 What property of cancer cells makes them susceptible to immunotherapy?

ROOTS TO REMEMBER

The following roots of words come mainly from Latin and Greek and will help you to decipher terms:

cyto-	relates to cells. Chapter terms: *cytoplasm, cytokinesis*
-kinesis	means motion. Chapter term: *cytokinesis*
mal-	means bad or evil. Chapter term: *malignant*
meta-	means change or between. Chapter terms: *metastasis, metaphase*
mito-	means a thread. Chapter term: *mitosis*
telo-	means end or completion. Chapter term: *telophase*

LEARNING THE BASICS

1. Describe three ways that cancer cells differ from normal cells.
2. Add labels to the figure that follows, which illustrates duplicated chromosomes.

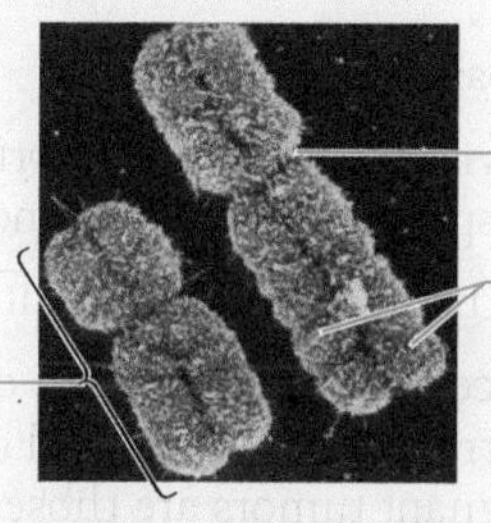

3. A cell that begins mitosis with 46 chromosomes produces daughter cells with _________ chromosomes.

A. 13; B. 23; C. 46; D. 92

4. The centromere is a region at which _________.

A. sister chromatids are attached to each other; B. metaphase chromosomes align; C. the tips of chromosomes are found; D. the nucleus is located

5. Mitosis _________.

A. occurs only in cancerous cells; B. occurs only in skin cells; C. produces daughter cells that are exact genetic copies of the parent cell; D. results in the production of three different cells

6. At metaphase of mitosis, _________.

A. the chromosomes are condensed and found at the poles; B. the chromosomes are composed of one sister chromatid; C. cytokinesis begins; D. the chromosomes are composed of two sister chromatids and are lined up along the equator of the cell

7. Sister chromatids _________.

A. are two different chromosomes attached to each other; B. are exact copies of one chromosome that are attached to each other; C. arise from the centrioles; D. are broken down by mitosis; E. are chromosomes that carry different genes

8. DNA polymerase _________.

A. attaches sister chromatids at the centromere; B. synthesizes daughter DNA molecules from fats and phospholipids; C. is the enzyme that facilitates DNA synthesis; D. causes cancer cells to stop dividing

9. If a cell at G_1 contains four picograms of DNA, how many picograms of DNA will it contain at the end of the S phase of the cell cycle?

A. 0; B. 2; C. 4; D. 8

10. In what ways is the cell cycle similar in plant and animal cells, and in what ways does it differ?

ANALYZING AND APPLYING THE BASICS

1. If your father obtained a mutation to his skin cell from UV light exposure during his youth, could he have passed that mutation on to you?

2. Why are only certain cancers treated with radiation therapy while others are treated with chemotherapy?

3. One risk of radiation therapy is an increased likelihood of new tumors emerging 5 to 15 years later. Why might this treatment increase cancer risk?

GO FIND OUT

1. Consumer genetic testing is offered by several for-profit companies. Do some research online to determine what kinds of information these tests can, and cannot, provide.

MAKE THE CONNECTION

The science that you learned in this chapter has helped you better understand the real-world example used throughout this discussion. Draw a line from the statement on the left to the science that supports it on the right.

Cancer will affect most people. ▷	◁ Smoking cigarettes increases the risk of most cancers.
Cancer can be genetically inherited, but most often it is acquired during one's lifetime. ▷	◁ 12% of women will get breast cancer in their lifetime unless they carry the *BRCA1* gene, in which case the odds of getting breast cancer increase to 87%.
Lifestyle changes can decrease cancer risk. ▷	◁ Of those who live an average lifespan, one in three will be diagnosed.
Cells that divide rapidly undergo more mutations than those that divide less rapidly. ▷	◁ Tumor suppressors help prevent tumors from forming.
All humans carry cell cycle regulating genes that, when not mutated, help prevent cancers from arising. ▷	◁ Skin cancer is the most common cancer.

Answers to **Got It?, Visualize This, Working with Data, Sounds Right, But Is It?, Show You Know,** and **Chapter Review** questions can be found in the **Answers** section at the back of the book.

7 Fertility

Infertility affects one in eight couples.

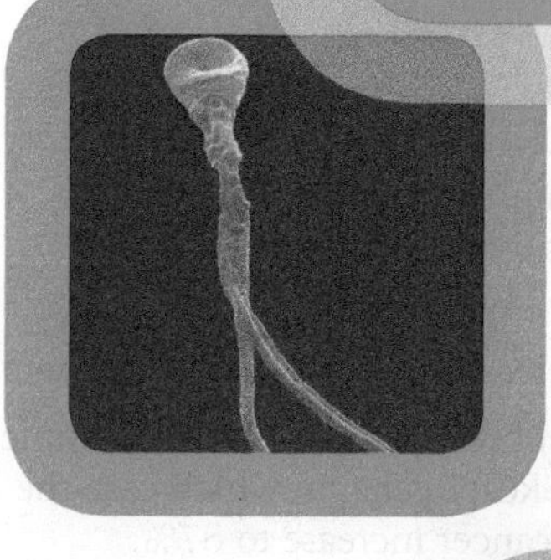

Do lifestyle choices like smoking cigarettes and marijuana alter our fertility, like this two-tailed sperm?

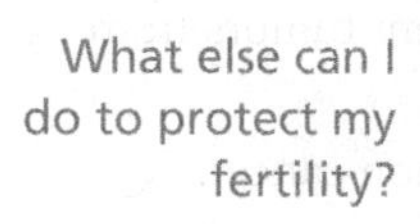

What else can I do to protect my fertility?

Meiosis and Human Reproduction

Most people who plan to have children one day assume they will be able to do so when the time seems right for them. However, this does not always work out as planned. In fact, approximately one in eight couples will have difficulty achieving a pregnancy in spite of having had unprotected intercourse several times a week for a year. Such couples are experiencing infertility and may end up needing help from medical professionals to achieve a pregnancy—and indeed, some will never be able to achieve a pregnancy.

Couples experiencing infertility often assume the underlying issue is with the woman's reproductive system. But the truth is that males and females are equally likely to be the reason for infertility. In approximately one-third of couples, infertility is due to problems in the male partner. Likewise, one-third of cases of infertility are caused by problems with the female partner. The remaining issues are thought to be caused either by problems in both partners or by phenomena medical science is not yet able to explain.

To be fertile, a couple must be able to make healthy sperm and eggs in the right number, and bring them together at the right time. Although this may seem simple in theory, our genes, age, hormones, environmental exposures, health status, and lifestyle choices can all affect our fertility. It is even possible for issues to arise before we are born. Like all organs, our reproductive organs develop while we are still in our mother's uterus. Therefore, toxins our mothers were exposed to while pregnant can alter the development of our reproductive organs. Thankfully, such developmental issues, although out of our control, are rare. The more common threats to our fertility are, at least partly, under our control. In fact, there are things we can do well in advance of attempting pregnancy that can help preserve our fertility.

7.1 Producing Sperm and Eggs: Meiosis

Sperm and egg cells are specialized male and female reproductive cells called **gametes.** The first part of the process parent cells undergo to produce gametes is called **meiosis.** After meiosis, the daughter cells undergo further modifications to become mature, functional gametes. For example, sperm cells have a tail added and undergo an increase in the number of energy-producing mitochondria. Egg cells increase in size and nutrient concentration. The entire process of meiosis and subsequent differentiation of gametes is called **gametogenesis**.

meio- means to make smaller.

-genesis means generation of or birth of.

Meiosis occurs only in the **gonads** or sex organs. In humans, the male gonads are testes, and the female gonads are ovaries. Because human **somatic,** or body, cells have 46 chromosomes and meiosis reduces that number by half, the gametes produced during meiosis contain 23 chromosomes each.

gon-, gono-, and **gonado-** means seed or generation.

soma- and **-some** means body.

Chromosomes in somatic cells occur in pairs. The 46 chromosomes in human somatic cells are actually 23 different pairs of chromosomes. In somatic cells, one member of each pair was inherited from the mother and one from the father. The members of such a **homologous pair** of chromosomes are the same size and shape and carry the same genes, although not necessarily the same versions (**FIGURE 7.1**). For example, one gene for eye color can exist in different versions, each resulting in the production of different eye colors. Different versions of a gene are called **alleles** of the gene.

homo- means the same.

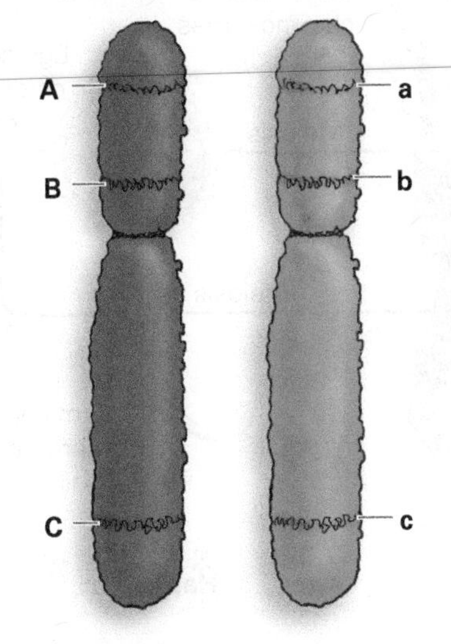

FIGURE 7.1 A homologous pair of chromosomes. Homologous pairs of chromosomes have the same genes (shown here as A, B, and C) but may have different alleles. The dominant allele is represented by an uppercase letter and the recessive allele by the same letter in lowercase. Note that the chromosomes of a homologous pair each have the same size, shape, and positioning of the centromere.

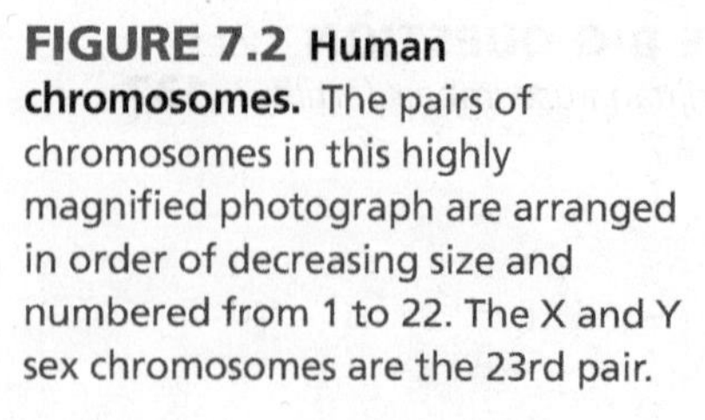

FIGURE 7.2 Human chromosomes. The pairs of chromosomes in this highly magnified photograph are arranged in order of decreasing size and numbered from 1 to 22. The X and Y sex chromosomes are the 23rd pair.

VISUALIZE **THIS**

Which sex chromosome is larger and therefore carries more genetic information?

To get a better sense of what homologous pairs of chromosomes look like, it might help to look at a highly magnified photograph showing the chromosomes from one cell of one human being, arranged in pairs (**FIGURE 7.2**). The 46 human chromosomes can be arranged into 22 pairs of nonsex chromosomes, or **autosomes,** and one pair of **sex chromosomes** (the X and Y chromosomes) to make a total of 23 pairs. Human males have an X and a Y chromosome, while females have two X chromosomes.

Once meiosis is completed, there is one copy of each chromosome (1–23) in every gamete. When only one member of each homologous pair is present in a cell, we say that the cell is **haploid (n)**—both egg cells and sperm cells are haploid. After the sperm and egg fuse, the fertilized cell, or **zygote,** will contain two sets of chromosomes and is said to be **diploid (2n)** (**FIGURE 7.3**).

-ploid means the number of sets of chromosomes—diploids have two sets.

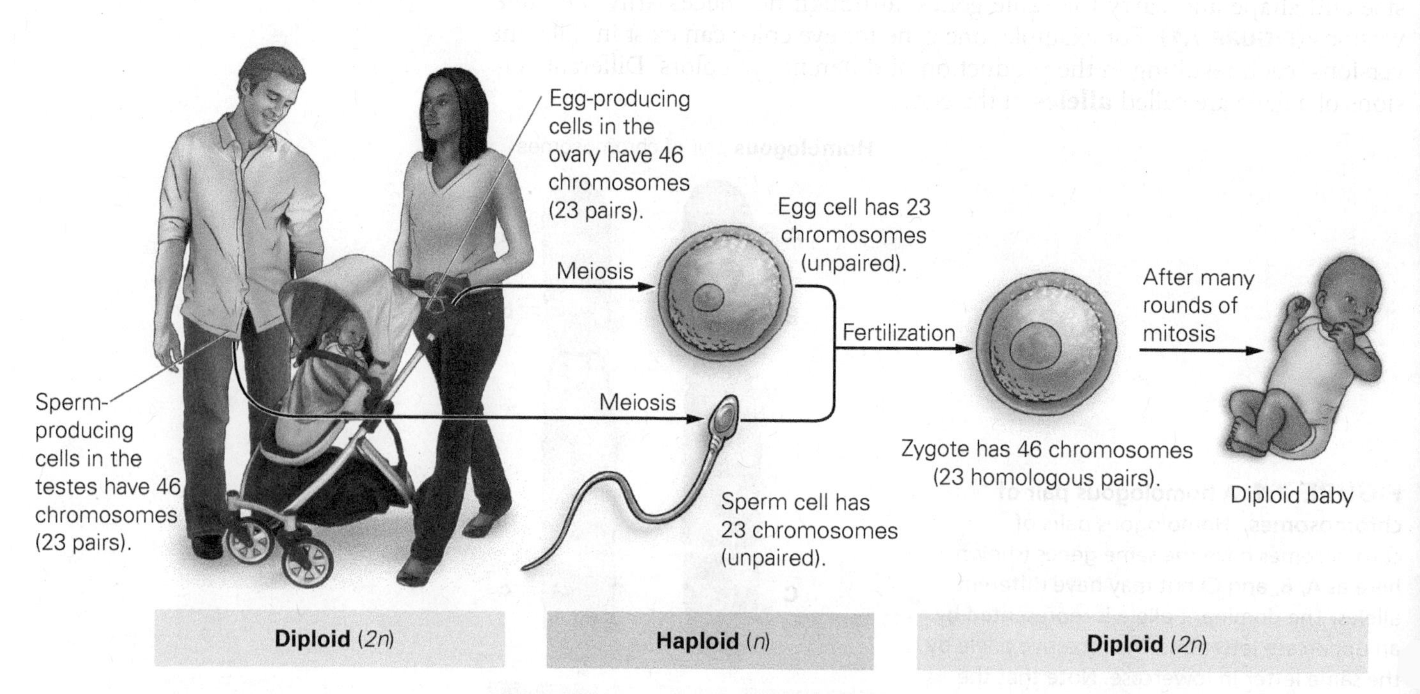

FIGURE 7.3 Gamete production. The diploid cells of the ovaries and testes undergo meiosis and produce haploid gametes. When fertilization occurs, the diploid condition is restored.

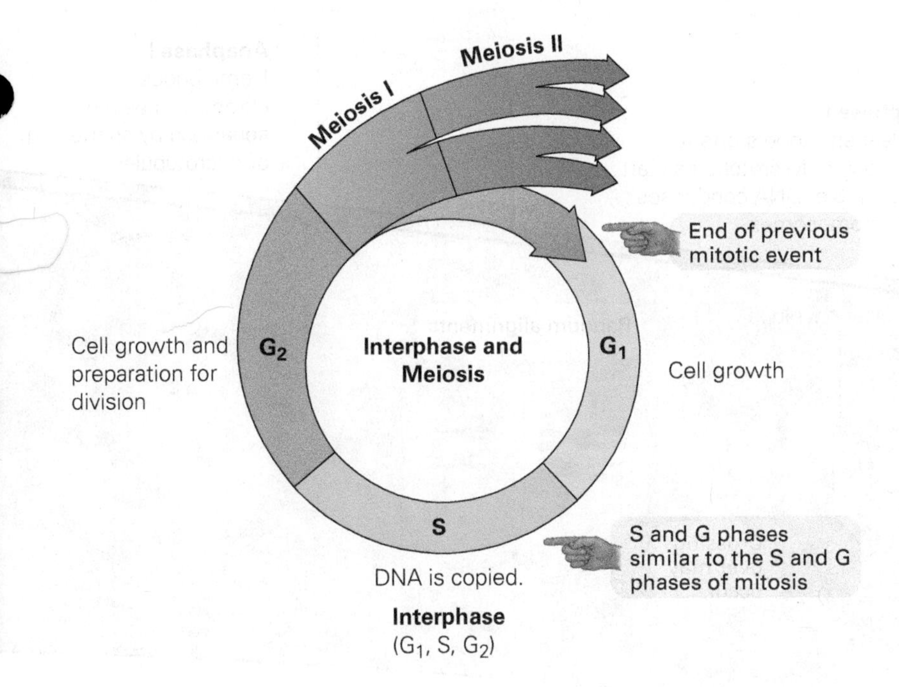

FIGURE 7.4 Interphase and meiosis. Interphase consists of G_1, S, and G_2 and is followed by two rounds of nuclear division, meiosis I and meiosis II.

Like mitosis, meiosis is preceded by an interphase stage that includes two G phases (for gap or growth) and one S phase (for synthesis of DNA) between G_1 and G_2 (**FIGURE 7.4**).

Interphase is followed by two phases of meiosis—meiosis I and meiosis II—n which divisions of the nucleus take place. Meiosis I separates the members of a homologous pair from each other. Meiosis II separates the chromatids from each other. Both meiotic divisions are followed by cytokinesis, during which the cytoplasm is divided between the resulting daughter cells. Let's look at this process in more detail.

Interphase

The interphase that precedes meiosis consists of G_1, S, and G_2. This interphase of meiosis is similar in most respects to the interphase that precedes mitosis. The centrioles from which the microtubules will originate are present. The G phases are times of cell growth and preparation for division. The S phase is when DNA replication occurs. Once the cell's DNA has been replicated, it can enter meiosis I.

Meiosis I

The first meiotic division, meiosis I, consists of prophase I, metaphase I, anaphase I, and telophase I and is shown in **FIGURE 7.5** at the top of the next page.

Prophase I. During prophase I of meiosis, the nuclear envelope starts to break down, and the microtubules begin to assemble. The previously replicated chromosomes condense so that they can be moved around the cell without becoming entangled. The condensed chromosomes can be seen under a microscope. At this time, the homologous pairs of chromosomes exchange small portions of genetic information, on a gene for gene basis, in a process called **crossing over.**

To understand the significance of crossing over, consider a gamete you could produce. For each of your paired chromosomes, a gamete will receive one copy of either the chromosome you inherited from your mom or the one you inherited from your dad. Let's say you produce a gamete with chromosome number 3

FIGURE 7.5 **Meiosis.** This diagram illustrates interphase, meiosis I, meiosis II, and cytokinesis in an animal cell.

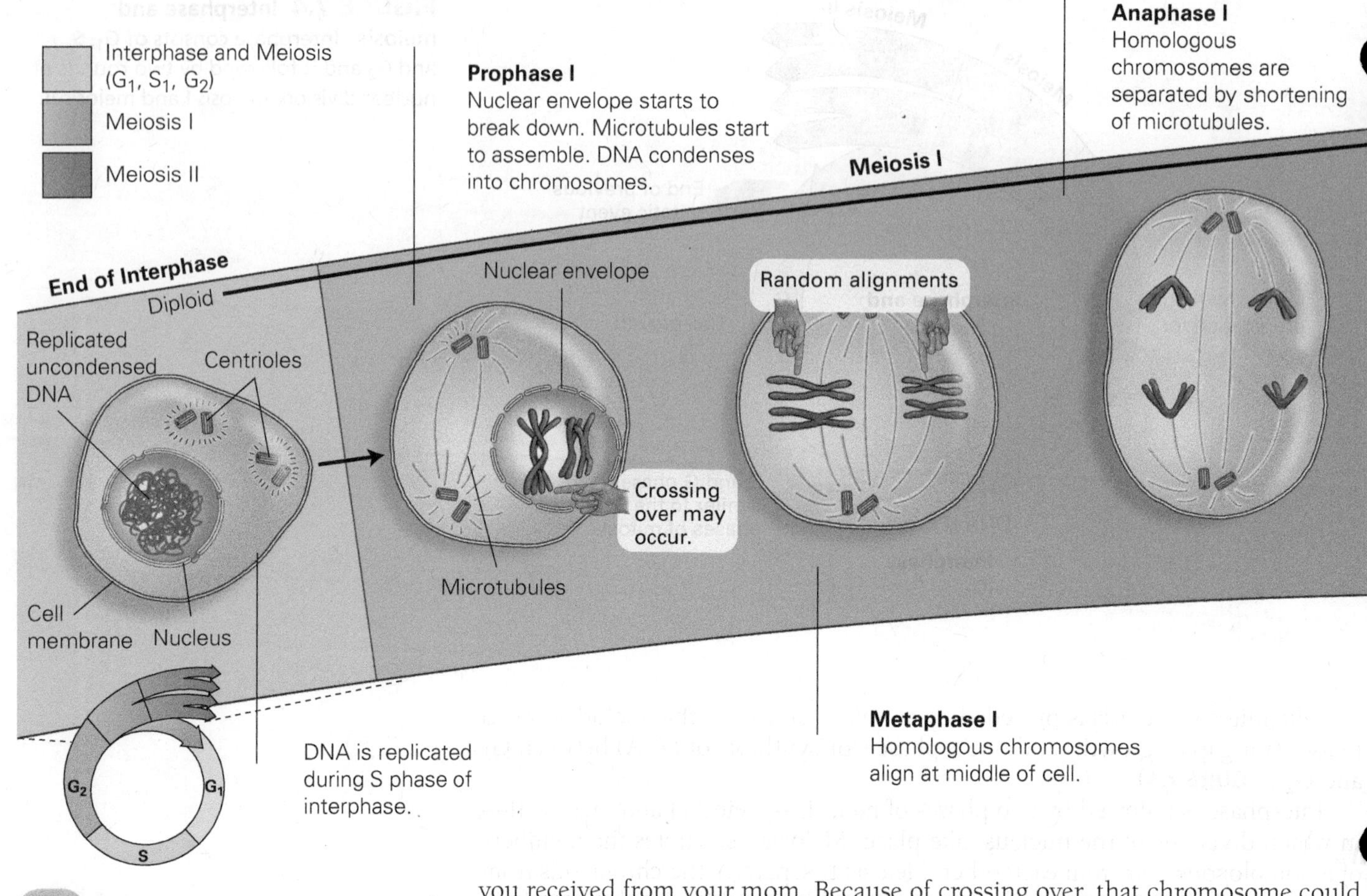
Interphase and Meiosis (G_1, S_1, G_2)
Meiosis I
Meiosis II
Prophase I
Nuclear envelope starts to break down. Microtubules start to assemble. DNA condenses into chromosomes.
Meiosis I
Anaphase I
Homologous chromosomes are separated by shortening of microtubules.
End of Interphase
Diploid
Replicated uncondensed DNA
Centrioles
Nuclear envelope
Random alignments
Crossing over may occur.
Microtubules
Cell membrane
Nucleus
Metaphase I
Homologous chromosomes align at middle of cell.
DNA is replicated during S phase of interphase.
G2
G1
S

Watch **Meiosis** in Mastering **Biology**

you received from your mom. Because of crossing over, that chromosome could also contain a small amount of genetic information from the chromosome number 3 you got from your dad (**FIGURE 7.6**).

Metaphase I. At metaphase I, the chromosomes line up at the equator of the cell, but they do so in homologous pairs. This is a key difference between meiosis and mitosis; in mitosis the chromosomes align single file at the equator. During meiosis, homologous pairs are arranged arbitrarily regarding which member faces which pole, a phenomenon called **random alignment** of the homologous pairs, which is covered in more detail in the next chapter.

For now, to better understand this, let's think of this in terms of a gamete you could produce. When the chromosome number 3 you inherited from your mom aligns with chromosome number 3 you inherited from your dad, they can align in one of two ways relative to the poles of the cell. The chromosome you inherited from your mom can face the top of the cell during one round of meiosis and the bottom the next. Because you have 23 pairs of chromosomes that can each be arranged two different ways, there are 2^{23} different alignments (more than 8 million) your chromosomes can undergo. Therefore, each of us can make more than 8 million genetically distinct gametes, which provides for a wide array of genetically diverse gametes.

Anaphase I and Telophase I. Following metaphase I of meiosis, the cell goes in to anaphase I. At anaphase I, the homologous pairs are separated from each other by the shortening of the microtubules, and at telophase I, nuclear envelopes re-form around the chromosomes. DNA is then partitioned into each of the two daughter cells by cytokinesis. Because each daughter cell contains only one copy of each member of a homologous pair, at this point the cells are haploid. Now both of these daughter cells are ready to undergo meiosis II.

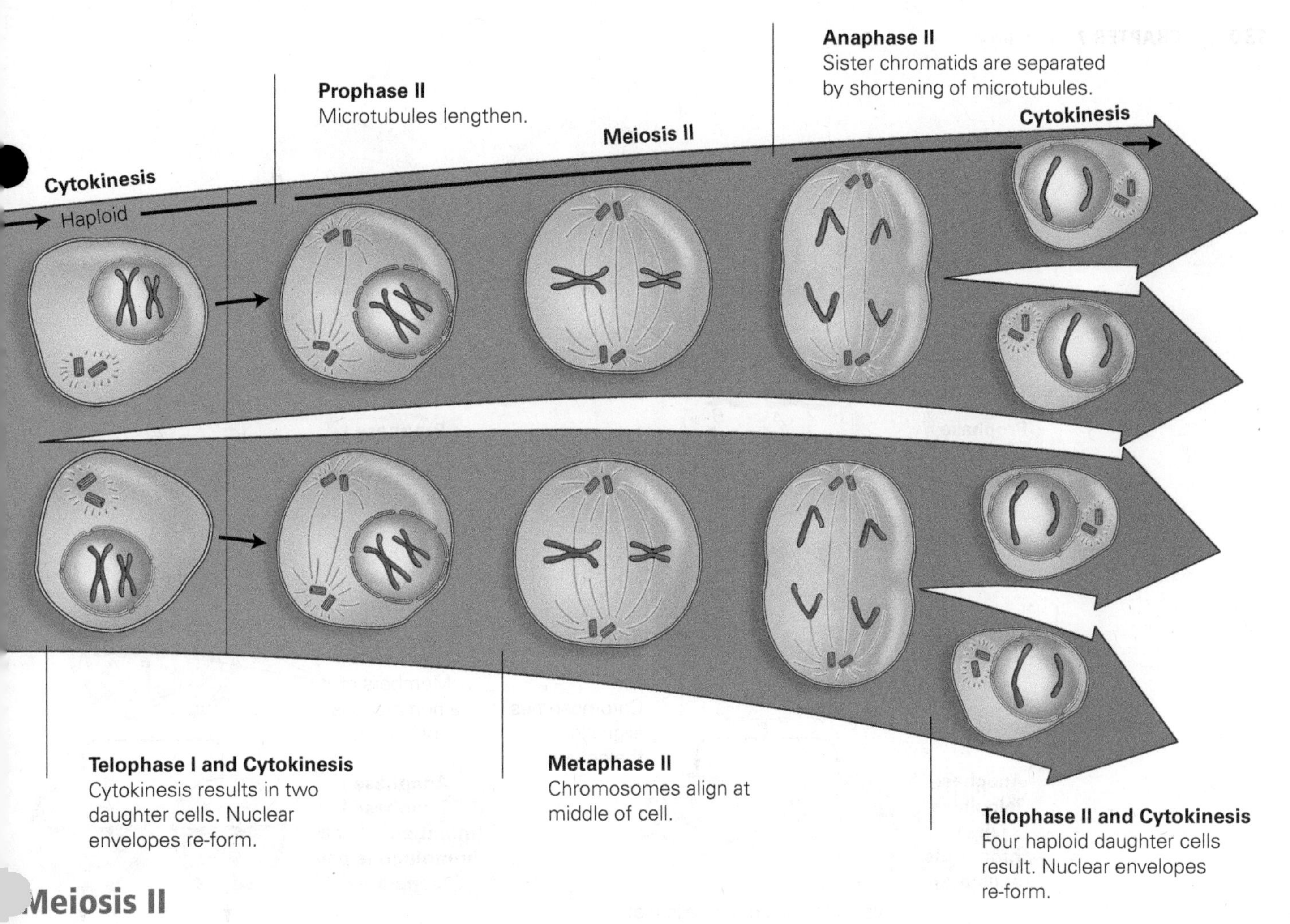
Prophase II
Microtubules lengthen.
Anaphase II
Sister chromatids are separated by shortening of microtubules.
Meiosis II
Cytokinesis
Cytokinesis
Haploid
Telophase I and Cytokinesis
Cytokinesis results in two daughter cells. Nuclear envelopes re-form.
Metaphase II
Chromosomes align at middle of cell.
Telophase II and Cytokinesis
Four haploid daughter cells result. Nuclear envelopes re-form.

Meiosis II

Meiosis II consists of prophase II, metaphase II, anaphase II, and telophase II. This second meiotic division is virtually identical to mitosis and serves to separate the sister chromatids of the replicated chromosome from each other.

At prophase II of meiosis, the cell is readying for another round of division, and the microtubules are lengthening again. At metaphase II, the chromosomes

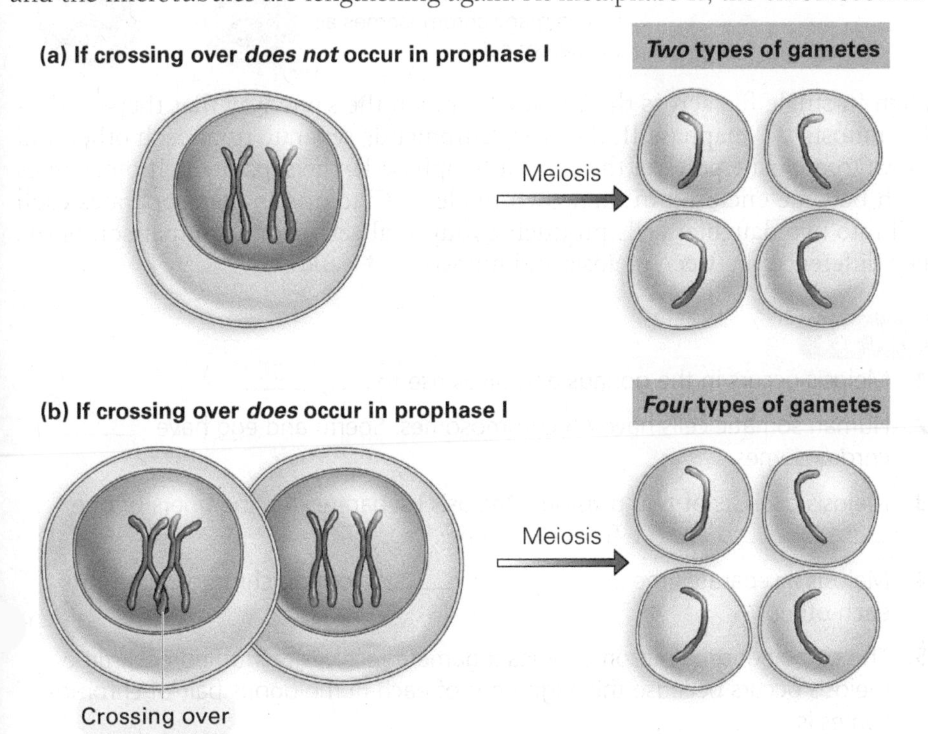
(a) If crossing over *does not* occur in prophase I
Two types of gametes
Meiosis
(b) If crossing over *does* occur in prophase I
Four types of gametes
Meiosis
Crossing over

FIGURE 7.6 Crossing over. In this example, only one homologous pair of chromosomes present in the gonads of an individual undergoing meiosis is shown. One member of this pair came from the individual's mom and one from the individual's dad. (a) If no crossing over occurs, gametes produced by this individual would contain unaltered chromosomes that are the same as were found in the individual undergoing meiosis. (b) If crossing over does occur, gametes produced can contain individual chromosomes that carry genetic information from both the individual's parents.

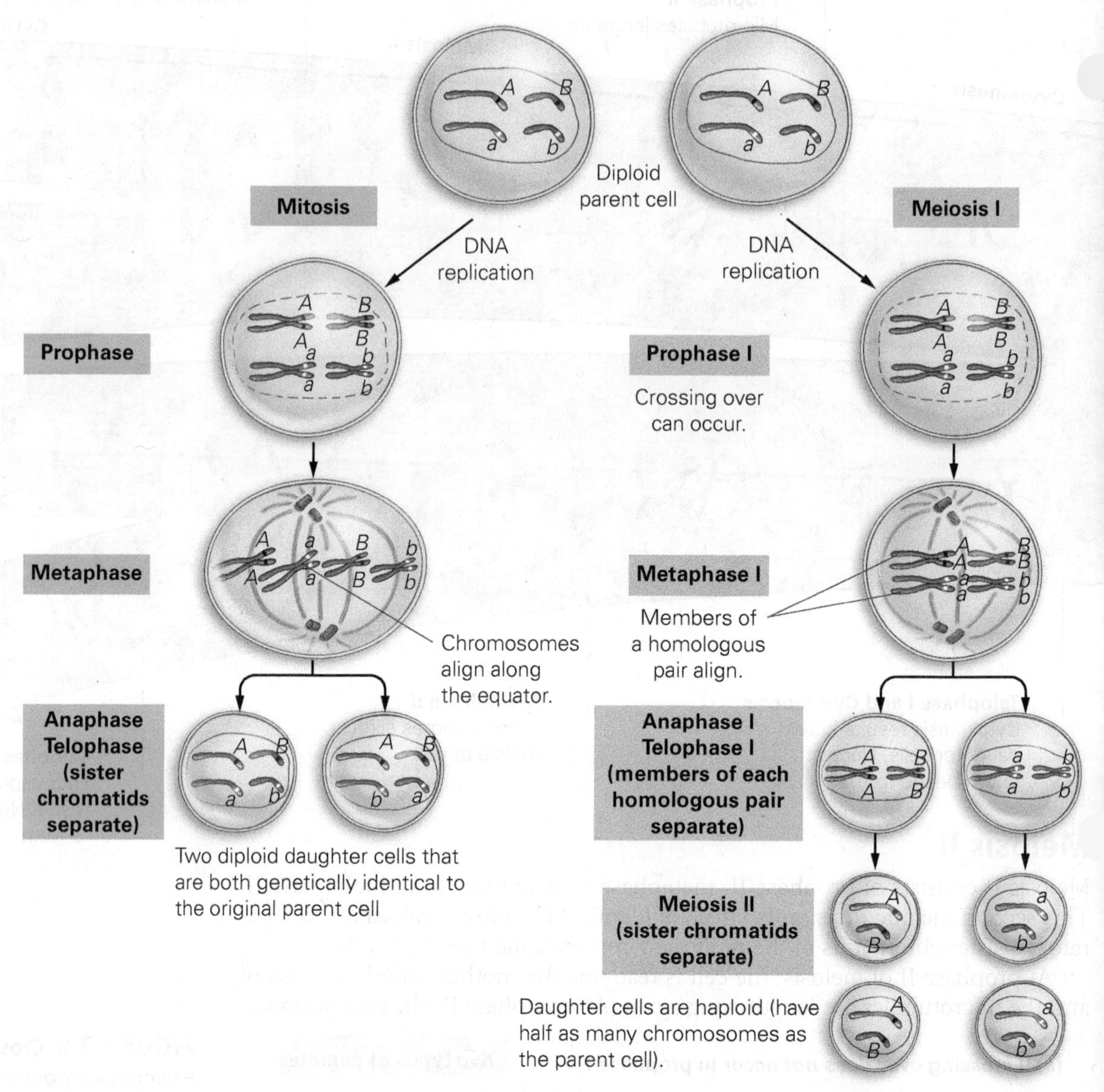

FIGURE 7.7 Comparing mitosis and meiosis. Mitosis is a type of cell division that occurs in somatic cells and gives rise to daughter cells that are exact genetic copies of the parent cell. Meiosis occurs in cells that will give rise to gametes and decreases the chromosome number by one-half. Each gamete receives one member of each homologous pair.

align in single file across the equator in much the same way that they do during mitosis. At anaphase II, the sister chromatids separate from each other and move to opposite poles of the cell. At telophase II, the separated chromosomes each become enclosed in their own nucleus. Cytokinesis then separates each cell into two daughter cells producing four total cells. For a comparison of the key differences between meiosis and mitosis see **FIGURE 7.7**.

Got It?

1. Meiosis occurs in the gonads and gives rise to ________.
2. Human somatic cells have 46 chromosomes. Sperm and egg have ________ chromosomes.
3. Meiosis consists of two divisions. Meiosis I separates the members of one ________ ________ from each other.
4. Meiosis II separates the ________ ________ of one chromosome from each other.
5. The combination of chromosomes a gamete receives is shuffled each time meiosis occurs because the alignment of each homologous pair of chromosomes is ________.

7.2 Problems with Meiosis and Lowered Fertility

Problems with meiosis can lead to lowered fertility. For example, people who have too many or too few chromosomes are less fertile than those with 46 chromosomes. For example, around 1 in 2,500 females are born with only one X chromosome, a condition called *Turner syndrome*. Likewise, around 1 in 800 males is born with a condition called *Klinefelter syndrome* in which they have two X chromosomes and one Y chromosome. This can occur when homologous pairs of chromosomes fail to separate from each other during meiosis, a phenomenon called **nondisjunction.** Nondisjunction results in sperm or egg cells with too many or too few chromosomes. If nondisjunction occurs in a male, his sperm may have both an X and a Y chromosome instead of one or the other. If that sperm fertilizes an egg cell, the individual that is born from that fertilization will be XXY and have Klienfelter syndrome. An individual with too many (or too few) chromosomes is less fertile because their mismatched chromosomes cannot pair properly at meiosis.

Not producing enough gametes is a major cause of infertility. The most common reason this occurs is aging. A man's fertility begins to decline in his mid-thirties and continues to slowly decline with age. In spite of this, many men retain the ability to father children throughout their lifetimes. One way to measure fertility rates is to study the length of time it takes to get pregnant as a function of age. **FIGURE 7.8** shows that it takes longer for older males and females to achieve a pregnancy. In addition, this figure shows that the rate at which women conceive also declines with age, but it does so at an earlier age and a faster rate than it does in men.

Sex-based differences in fertility declines can be explained, in part, by a difference that exists in when meiosis occurs in males and females. The ovaries of females actually begin meiosis while they are still developing inside their mother's uterus. Beginning at puberty, one of these "paused" cells finishes meiosis each menstrual cycle. Therefore, females are born with all the potential eggs they will ever have, already present in their ovaries. As a female enters her reproductive years, the number and quality of these potential egg cells begins to decline and continues to do so until she stops menstruating at around age 50.

WORKING WITH DATA

(a) Describe what the graph shows about male fertility before and after age 30. (b) Compare TTP in women who are in the 20–24 age range with those in the 30–34 age range. Roughly how many times longer does it take for the older women to become pregnant? (c) Why does the line for males extend into later age groups than the line for females?

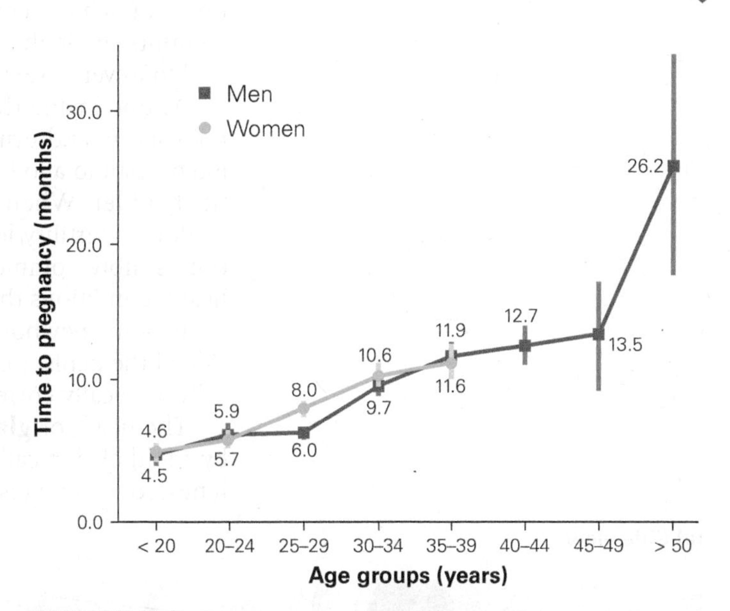

FIGURE 7.8 Time to pregnancy and age. This graph shows the average time to pregnancy (TTP) in months as a function of age.

Got It?

1. Individuals with too many or too few chromosomes are ________ fertile than those with 46 chromosomes.
2. Failure of chromosomes to separate during meiosis can occur by ________.
3. The most common cause of infertility is decreasing numbers of ________ produced.
4. Gamete production decreases with ________ in males and females.
5. Time to pregnancy ________ with age in males and females.

7.3 Bringing Sperm and Egg Together

After gametes are produced and mature, they must make their way through the male and female reproductive systems for fertilization to occur. The reproductive systems of both sexes consist of external and internal structures that produce gametes, secrete the hormones required for reproductive function, and provide a route to deliver the gametes through ducts or tubes. Improper development, disease, damage, and blockages to reproductive structures can all cause infertility.

Male Reproductive Anatomy

The external and internal components of the male reproductive system are shown in **FIGURE 7.9**. The **penis** delivers sperm to the female reproductive tract during sexual intercourse. It is composed of spongy erectile tissue. During sexual arousal, this tissue fills with blood. Pressure from the increased volume of blood in the penis seals off the veins that drain it of blood. This causes the penis to become engorged, keeping it erect. An erection is essential for insertion of the penis into the vagina, which facilitates delivery of sperm to the egg. Therefore, fertility is lower in men who have trouble maintaining an erection. This problem is more common in older men and in men who have consumed large amounts of alcohol before intercourse. Excess alcohol consumption can also lead to lowered sexual desire and an inability to reach orgasm.

A tube inside the penis, the **urethra,** provides a way out of the body for both sperm and urine. A ring of muscle at the base of the bladder tightens during orgasm to allow semen to exit the tip of the penis instead of moving toward the bladder. When the sphincter does not close, sperm are able to enter the bladder, a fertility lowering situation called retrograde ejaculation. This condition is more common in men who have suffered spinal injuries or have other health conditions than it is in healthy men.

In some newborn males, the opening of the urethra is found on the underside of the penis, instead of at the tip. This condition, called *hypospadias,* is typically surgically corrected in infancy, but if left uncorrected can lead to infertility.

The head, or **glans penis,** consists of highly sensitive skin that is covered by a fold of skin called the foreskin. In circumcised males, the foreskin has been removed. Circumcision has not been shown to affect fertility.

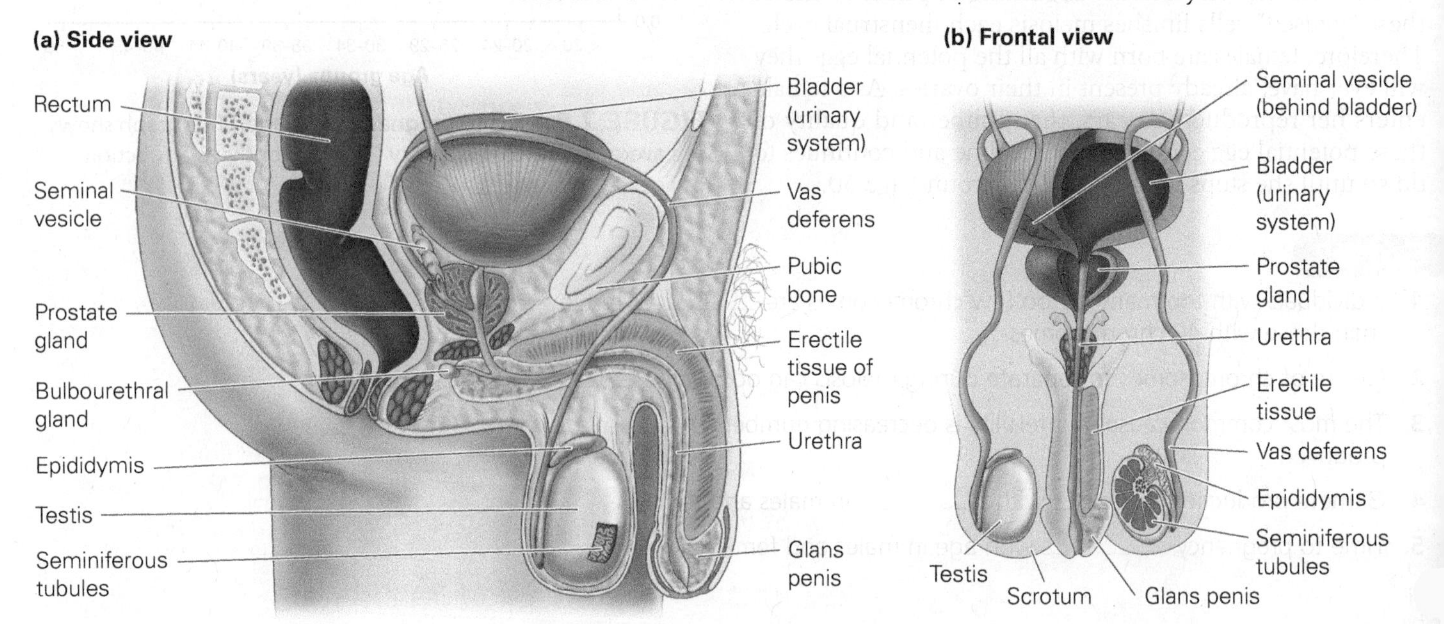

FIGURE 7.9 Male reproductive anatomy. (a) Side view and (b) front view of the male reproductive system.

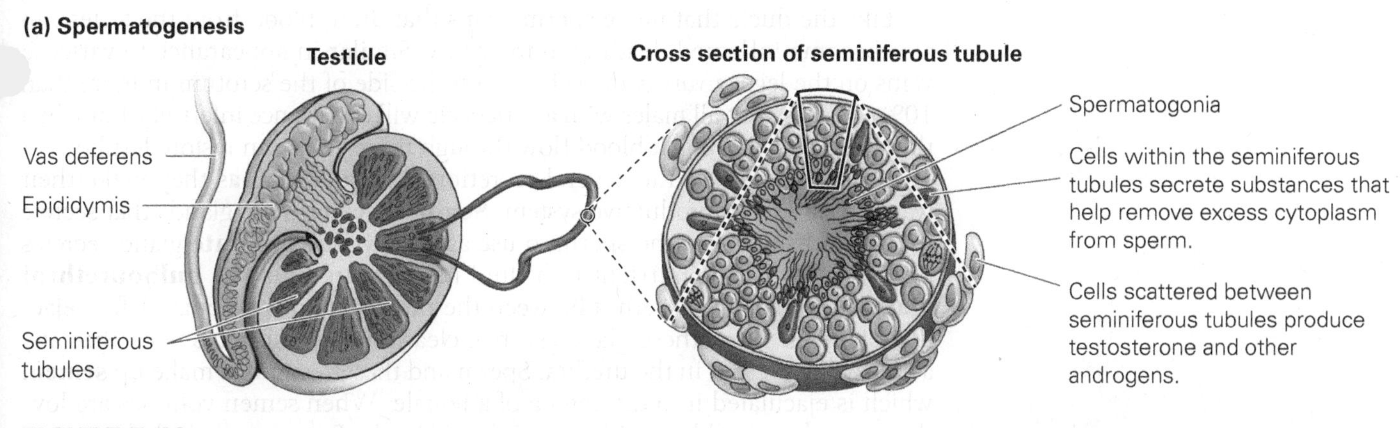
(a) Spermatogenesis
Testicle
Cross section of seminiferous tubule
Vas deferens
Epididymis
Seminiferous tubules
Spermatogonia
Cells within the seminiferous tubules secrete substances that help remove excess cytoplasm from sperm.
Cells scattered between seminiferous tubules produce testosterone and other androgens.

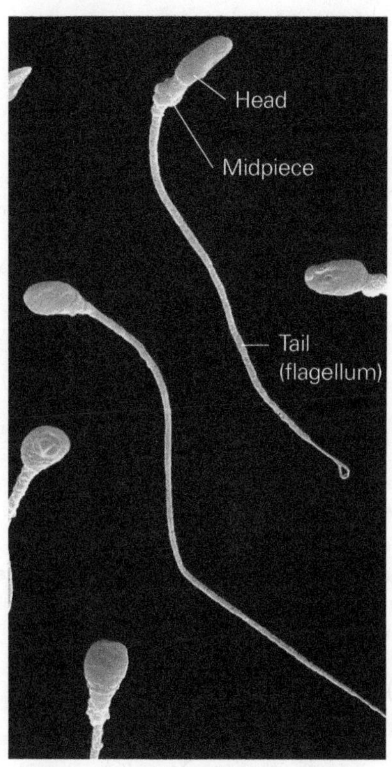
(b) Sperm
Head
Midpiece
Tail (flagellum)

FIGURE 7.10 Spermatogenesis. (a) Sperm are produced in the seminiferous tubules of each testicle. (b) The head of the sperm contains the DNA. The midpiece contains mitochondria to provide energy for the flagellum or tail.

The **scrotum** is a pouch below the penis that contains the gamete-producing and hormone-secreting testes or **testicles.** The skin of the scrotum is thin and devoid of fatty tissue. It is folded or wrinkled with little hair. Below the skin is a layer of involuntary smooth muscle that regulates the position of the testicles relative to the body to keep them at a temperature that maximizes sperm production. The scrotum contracts in the cold to keep the testes closer to the body. Because temperature is important for healthy sperm production, men who are actively trying to father children are told to limit their exposure to hot tubs and saunas.

Testicles also produce male hormones called **androgens.** When androgen levels are low, fertility is affected because fewer sperm are produced. This can be caused by medical conditions or by the use of anabolic steroids to increase muscle mass. Artificial steroid use has been shown to shrink testes, leading to lower androgen production. Androgen production in males also decreases with obesity, which can lead to low sperm counts. This may occur because fat tissues are able to convert male hormones like testosterone into estrogens or because excess body weight leads to elevated temperatures within the scrotum.

Inside each testis are many highly coiled tubes, called **seminiferous tubules,** where sperm begin their development from the time a boy reaches puberty throughout his life, until he is a very old man. The production of sperm, called **spermatogenesis** (FIGURE 7.10a), begins when cells lining the tubules first duplicate by mitosis, then one of the two daughter cells produced undergoes meiosis. Additional cells that aid the developing sperm are also located in the seminiferous tubules. These cells secrete the substances that help sperm cells develop and gain motility, or the ability to move.

A mature sperm cell is composed of a small oval head containing DNA, a midpiece that has mitochondria to provide energy for the journey to the oviduct, and a tail (or flagellum) to propel the sperm (FIGURE 7.10b). Sperm that are misshapen or not motile are less likely to be involved in fertilizing an egg cell. The percentage of properly formed, motile sperm decreases with age, as well as with marijuana, cocaine, amphetamine, tobacco, and alcohol use.

From the seminiferous tubules, sperm pass through the coiled **epididymis,** an approximately 6-meter tube that rests atop each testis. During ejaculation, sperm are propelled from the epididymis through ducts called the **vas deferens.**

Blockages to any of the tubes that carry sperm can lower the number of sperm ejaculated. When infections of male reproductive structures lead to inflammation that damages tissues, permanent blockages can occur. Some sexually transmitted infections like gonorrhea and HIV can cause permanent damage to ductal structures.

Like the ducts that move sperm, veins that drain blood from the testes can also become inflamed, leading to infertility. Similar in appearance to varicose veins on the legs, a *varicocele*, is found on the side of the scrotum in more than 10% of males. Not all males with a varicocele will experience infertility, but when they do, surgery to block blood flow through these veins can restore fertility.

Sperm are exposed to several secretions that aid them as they make their way through the reproductive system. **Seminal vesicles** are glands that secrete mucus and sugars for the sperm to use as energy. The **prostate** gland secretes a thin, milky white, nutrient-rich fluid into the urethra. The **bulbourethral glands** lie below the urethra between the prostate and the penis. Before ejaculation takes place, these glands secrete clear mucus that helps neutralize any acidic urine present in the urethra. Sperm and these secretions make up **semen,** which is ejaculated into the vagina of a female. When semen volumes are low, there may be a problem with one of these glands. Because semen helps sperm survive inside the female reproductive tract, a lower-than-normal volume of semen can cause infertility. When semen is healthy, sperm can survive for up to 5 days in the female reproductive system.

Female Reproductive Anatomy

FIGURE 7.11 shows female reproductive anatomy. The most obvious feature of the external genitalia is the **vulva.** The vulva consists of two sets of lips, or labia: the outer **labia majora,** which are fatty and have a hairy external surface, and the inner **labia minora,** which contain neither fat nor hair. At the front of the vulva, the labia minora divide around the **clitoris,** an important organ for female sexual arousal and orgasm.

Between the folds of the labia, there is an opening for the urethra, which serves as a passageway for urine from the bladder. The female urethra is about three times shorter than the male urethra. Partially due to this difference, women are more susceptible to bladder infections because bacteria have a shorter distance to travel from outside the body. Just below the urinary opening is the vaginal opening.

The organs that make up the internal genitalia are the ovaries, oviducts, uterus, and vagina. The **ovaries** are the size and shape of almonds in the shell. They secrete the female hormones estrogen and progesterone and produce egg

VISUALIZE **THIS**

Women are more susceptible to pelvic infections than men, in part because the urethra is shorter in women. Trace the path bacteria could follow from the exterior to the opening to the pelvic cavity via the oviduct.

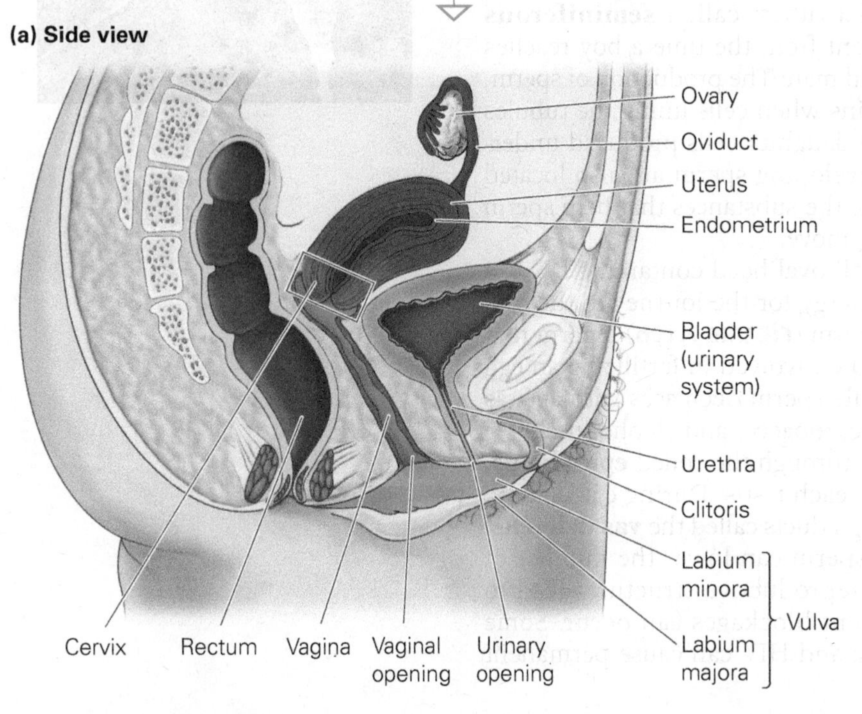

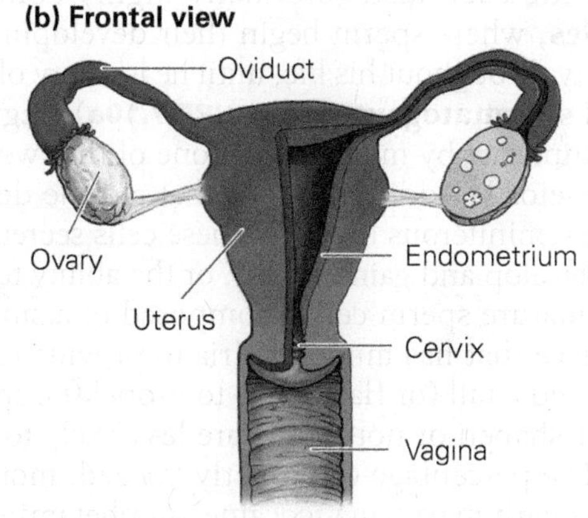

FIGURE 7.11 Female reproductive anatomy. (a) Side view and (b) front view of the female reproductive system.

(a) Follicle cycle

Uterus Oviduct Ovary

Developing egg cell

(1) The developing egg cell is surrounded by cells that produce estrogen.

Estrogen secreting cells form a protective fluid-filled sac.

Ovary

Corpus luteum

(2) The egg cell is ovulated.

(4) Unless a pregnancy has occurred, the corpus luteum degenerates.

(3) After the egg cell is ovulated, what remains of the fluid-filled sac secretes progesterone and estrogen and is now called the corpus luteum.

(b) Ovulation

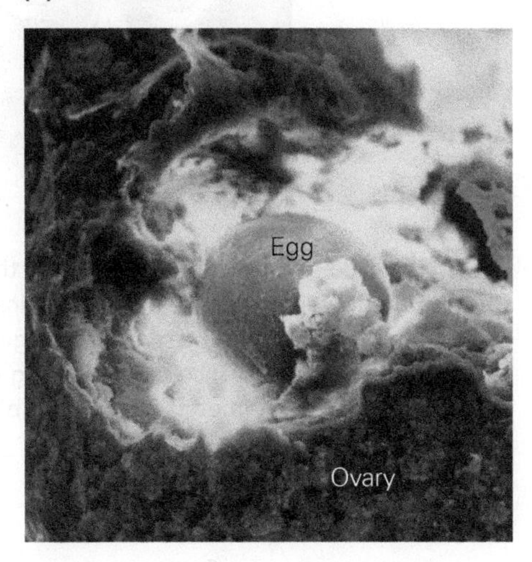

FIGURE 7.12 Oogenesis. (a) Over the course of one menstrual cycle, one follicle grows and develops in preparation for ovulation. (b) During ovulation, the egg cell bursts from the ovary.

cells. Only one functioning ovary is required for fertility. The formation and development of female gametes that occurs in the ovaries is called **oogenesis** (**FIGURE 7.12a**). Estrogen-secreting cells surround the developing egg cell, forming a sac-like structure that protects the developing egg cell. During **ovulation** this sac ruptures, much like a blister popping, to release the egg into the oviduct (**FIGURE 7.12b**). The egg cell is now about the size of a pin head. The sac that extruded the egg cell, now called the **corpus luteum,** remains in the ovary and secretes both estrogen and progesterone.

Beginning when a girl reaches puberty, she will typically ovulate one egg cell per menstrual cycle. Anything that prevents the ovaries from releasing egg cells can cause infertility. Examples include cysts in the ovaries, called *polycystic ovarian syndrome (PCOS)*. The ovaries typically produce a small amount of androgen. In PCOS, a slightly higher secretion of androgen disrupts ovulation. Being overweight affects hormone levels and can prevent ovulation. In addition, being very underweight can so damage your reproductive system that permanent infertility can result. Heavy drinking is associated with ovulation disorders, and smoking is thought to prematurely deplete the supply of eggs in the ovary.

Many young women worry that taking the birth control pill, which inhibits ovulation, will make it harder for them to have children if they decide to do so. This does not seem to be the case, even for women who stay on the pill for long periods of time.

Once a healthy egg is ovulated it moves to the oviducts. The **oviducts** are actually an extension of the top surface of the uterus. These tubes extend from the body of the uterus toward the ovaries, which are suspended within the abdominal cavity. The oviducts are not attached to the ovaries directly. Instead, they end in brushy structures that move over the surface of the ovary. These movements, along with suction inside the oviducts, direct the eggs released by the ovary into the oviducts.

oo- is from the Greek word for egg.

Scarring of the tubes involved in gamete delivery affect female fertility also. Two very common sexually transmitted infections, chlamydia and gonorrhea, can be present even in the absence of noticeable symptoms. This can lead to delays in treatment that allow bacteria time to ascend through the cervix and upper uterus to the oviducts, setting up a tissue-damaging infection that leads to permanent scarring and blockages. This condition, called *pelvic inflammatory disease*, results in permanent infertility in more than 10% of cases.

From the oviduct, the egg, which at this point in its progression could have been fertilized, moves to the uterus. The **uterus** is about the size of a fist. The wall is thick (about 1 centimeter) and is composed of some of the most powerful muscles in the human body. These muscular walls contract rhythmically during labor, childbirth, and orgasm.

The most common uterine cause of infertility is the presence of noncancerous growths called fibroids. Fibroids inside the uterus make it difficult for a fertilized egg cell to begin its development in the uterus. The incidence of fibroids increases with age and is treatable with medical intervention.

The internal surface of the uterine wall is called the **endometrium,** which changes in thickness during the course of the menstrual cycle. *Endometriosis* is a painful condition that can disrupt fertility. For unknown reasons, endometrial tissue can sometimes grow outside of the uterus and prevent the ovaries from functioning properly or block the oviducts.

The lower third of the uterus is narrower than the upper portion and is called the **cervix.** Sperm are deposited in the **vagina,** a muscular internal organ that receives the penis and serves as the birth canal during childbirth. An opening in the cervix, called the *cervical os*, dilates during childbirth, but also allows the passage of sperm from the vagina in nonpregnant women. Around the time of ovulation, the cervix produces mucus that has the consistency of raw, stringy egg white. The mucus forms parallel strands, like lane lines in a swimming pool, that help prevent sperm from becoming stuck in the folds of the cervix, thereby increasing their chance of fertilizing the egg cell in the uterus or oviduct (**FIGURE 7.13**).

Cervical mucus also helps lubricate the vagina to facilitate intercourse and helps prolong the life of sperm. Sperm die within a few hours if the stringy cervical mucus is not present, but when the mucus is present they live from 3 to 5 days.

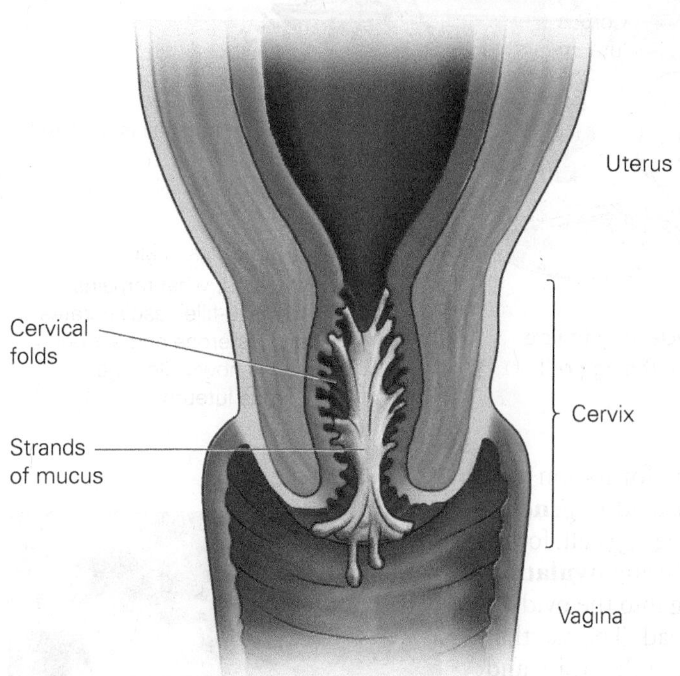

FIGURE 7.13 The cervix. The cervix is the narrow passage forming the lower end of the uterus. Cervical mucus changes before ovulation to help the sperm avoid getting stuck in the cervical folds. The cervical os is the opening that will enlarge to approximately 10 centimeters during childbirth.

Got It?

1. Cells that will produce gametes undergo meiosis in the ________.
2. Sperm mature and develop a head, midpiece, and ________.
3. Sperm and secretions from various glands together make up the ________.
4. A mature egg cell is ovulated from the ovary into the ________.
5. The ________ secretes mucus that keeps sperm alive longer, in order to fertilize an egg cell.

Sounds
right, but
is it?

Aisha and Jonah are newlyweds, both beginning graduate school programs at the same university. They are excited to start a family but would like to wait to have children until they both have jobs. Aisha has never liked hormonal birth control, and disposable methods like sponges and condoms are out of their price range. Recently Aisha came across several high-quality science resources stating that an egg cell is only capable of surviving 12–24 hours after ovulation. She also learned that cervical secretions change in response to hormones. When progesterone is high (after ovulation), the mucus thickens and is less abundant and slows the ascent of sperm. But when estrogen is high (right before ovulation), the cervical mucus is abundant, stringy, has the consistency of raw egg white, and aids sperm in traveling to the egg cell. Based on the information above, all of which is correct, she and Jonah conclude that they can avoid a pregnancy if Aisha pays close attention to her cervical secretions and if they abstain from sexual intercourse only when her cervical mucus changes to the type that facilitates sperm traveling to the oviducts.

A woman can become pregnant during only one 24-hour period every month.

Sounds right, but it isn't.

Answer the following questions to understand why.

1. Look at Figure 7.12 showing the ovary undergoing ovulation. What do the cells surrounding the developing egg cell produce?
2. As the cells surrounding a developing egg cell slowly increase in size and number, what happens to the amount of estrogen in the woman's body?
3. Although it is difficult to predict exactly when the cervical mucus would begin to change to the slippery, fertility-enhancing mucus, do you think this change would happen abruptly or begin to change slowly as estrogen increases?
4. Sperm can survive in the female reproductive tract for up to 5 days. If a couple had unprotected sex a few days before the fertility-enhancing cervical mucus was noticeably present, could the sperm already be present in the oviduct?
5. Consider your answers to questions 1–4. Explain why the statement bolded above sounds right, but isn't.

Does marijuana use impair fertility?

THE BIG
QUESTION

The use of marijuana is fairly common on college campuses. Many users believe that marijuana use is not very harmful. Legalization of marijuana for medical uses has fostered the perception that the drug can be used with few undesirable side effects. We know that marijuana use can harm a developing brain, but can it also harm one's fertility?

What should I know?

What follows are some smaller questions that need to be resolved to answer the Big Question. Place a checkmark next to the questions that science can answer.

Smaller questions	Can Science Answer?
Should marijuana be legal for medical and recreational purposes?	
Should recreational drug users not have children until they can abstain from their drug use?	
Do men who use marijuana have lower fertility than men who do not?	
Can people who use marijuana, and not other recreational drugs, be good parents?	
Do women who use marijuana have lower fertility than women who do not?	

What does the science say?

Let's examine what the data say about this smaller question:

Do men who use marijuana have lower fertility than men who do not?

Data from a study of more than 1200 men that controlled for body mass index (BMI), hours of abstinence, tobacco smoking, alcohol use, presence of sexually transmitted diseases, and use of other recreational drugs is shown below.

Frequency of marijuana use	Semen volume (mL)	Sperm concentration (millions/mL)	Total sperm count (millions)	Motile sperm (%)	Normally shaped sperm (%)
No use	3.3	48	156	58	70
Once a week or less	3.2	47	146	58	75
More than once a week	3.2	41*	107*	56*	65*

Asterisks indicate a statistically significant difference.

1. Describe the results. Which measures of sperm quality differ significantly in those that smoked marijuana more than once a week?
2. Given these data, do you think the smaller question is answered? If not, propose another study that would help answer this question?
3. Does this information help you answer the Big Question? What else do you need to consider?

Data source: T. Gundersen, N. Jorgensen, A. Bang, I. Nordkap, N. Skakkebaek, L. Prisjkorn, A. Juul, and T. Jensen, "Association Between Use of Marijuana and Male Reproductive Hormones and Semen Quality: A Study Among 1,215 Healthy Young Men," *American Journal of Epidemiology* 182, no. 6 (2015): 473–481.

Chapter Review

Mastering Biology

Go to Mastering Biology to access **eText 2.0, Dynamic Study Modules,** and the **Study Area,** where you'll find practice quizzes, BioFlix™ animations, MP3 tutor sessions, current events, and more.

SUMMARY

Section 7.1

Explain what types of cells undergo meiosis, the end result of this process, and how meiosis increases genetic diversity.

- Meiosis is a type of cell division that occurs in cells that give rise to gametes. Gametes contain half as many chromosomes as somatic cells do. The reduction of chromosome number that occurs during meiosis begins with diploid cells and ends with haploid cells (pp. 125–126).
- Meiosis is preceded by an interphase stage in which the DNA is replicated. During meiosis I, the members of a homologous pair of chromosomes are separated from each other. During meiosis II, the sister chromatids are separated from each other (pp. 127–130).

Explain the significance of crossing over and random alignment in terms of genetic diversity.

- Homologous pairs of chromosomes exchange genetic information during crossing over at prophase I of meiosis, thereby increasing the number of genetically distinct gametes that an individual can produce. The alignment of members of a homologous pair at metaphase I is random with regard to which member of a pair faces which pole. This random alignment of homologous chromosomes increases the number of different kinds of gametes an individual can produce (pp. 127–130).

SHOW YOU KNOW 7.1 Assume an organism has two homologous pairs of chromosomes. How many genetically distinct gametes can this organism produce? Disregard crossing over.

Section 7.2

Explain how altered meiosis effects fertility.

- Individuals with too many or too few chromosomes are less fertile (p. 131).
- The number of gametes decreases with age in males and females (p. 131).

SHOW YOU KNOW 7.2 If a sperm cell with no sex chromosome and an egg cell with the normal number of chromosomes join at fertilization, what would the chromosomes of the individual who is produced consist of?

Section 7.3

List the male reproductive structures and their functions.

- The penis of the male reproductive system is involved in sperm delivery; the urethra delivers both sperm and urine; the scrotum houses the androgen producing testes. Sperm are produced in the seminiferous tubules inside each testis and are stored in the epididymis before traveling through the vas deferens to the urethra. The seminal vesicles, prostate, and bulbourethral glands add secretions to sperm that help them develop and provide a source of energy. Semen is composed of the secretions from these glands combined with sperm (pp. 132–134).

List the female reproductive structures and their functions.

- The female reproductive system consists of the external vulva and clitoris, the internal vaginal passageway, the cervix at the base of the uterus that secretes sperm aiding mucus, the uterus where a fertilized egg implants, oviducts for the passage of egg cells, and ovaries that produce egg cells and hormones (pp. 134–136).

SHOW YOU KNOW 7.3 What structure must a fertilized egg cell travel through to get to the uterus?

ROOTS TO REMEMBER

The following roots of words come mainly from Latin and Greek and will help you to decipher terms:

-genesis	means generation of or birth of. Chapter terms: *oogenesis, spermatogenesis*
gon-, gono-, and gonado-	mean seed or generation. Chapter term: *gonad*
homo-	means the same. Chapter term: *homologous*
meio-	means to make smaller. Chapter term: *meiosis*
oo-	is from the Greek word for egg. Chapter terms: *oogenesis*
-ploid	means sets of chromosomes. Chapter terms: *haploid, diploid*
soma- and -some	mean body. Chapter terms: *somatic, chromosome*

LEARNING THE BASICS

1. List several ways in which meiosis differs from mitosis.
2. Add labels to the figure that follows, which illustrates female internal reproductive organs.

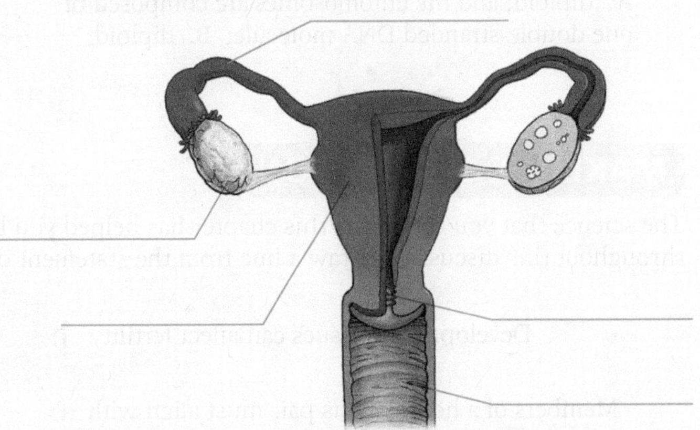

3. What happens to the egg cell and the remains of the tissue it developed in at ovulation?
4. A sperm cell follows which path?

   A. seminiferous tubules, epididymis, vas deferens, urethra; **B.** urethra, vas deferens, seminiferous tubules, epididymis; **C.** seminiferous tubules, vas deferens, epididymis, urethra; **D.** epididymis, seminiferous tubules, vas deferens, urethra; **E.** epididymis, vas deferens, seminiferous tubules, urethra

5. An egg cell that is not fertilized follows which path?

   A. ovary, oviduct, uterus, cervix; **B.** ovary, uterus, oviduct, cervix; **C.** oviduct, ovary, cervix, uterus; **D.** oviduct, ovary, uterus, cervix; **E.** ovary, oviduct, cervix, uterus

6. Which of the following is mismatched?

A. urethra: sperm passage; **B.** testes: hormone production; **C.** vas deferens: semen production; **D.** seminiferous tubules: sperm production

7. The production of gametes ________.

A. begins at puberty in males and females; **B.** requires that the testes of males produce semen; **C.** results in the production of diploid cells from haploid cells; **D.** begins at puberty in females; **E.** produces sperm and eggs that carry half the number of chromosomes as nongametes

8. If humans have 23 pairs of chromosomes, each carrying hundreds to thousands of genes, roughly how many genes are there in the human genome?

A. 23; **B.** 46; **C.** 1000; **D.** 20,000; **E.** 200,000

9. Homologous pairs of chromosomes ________.

A. are two different chromosomes attached to each other; **B.** are exact copies of one chromosome that are attached to each other; **C.** are separated from each other during meiosis I; **D.** are separated from each other during interphase; **E.** are chromosomes that carry different genes

10. After telophase I of meiosis, each daughter cell is ________.

A. diploid, and the chromosomes are composed of one double-stranded DNA molecule; **B.** diploid, and the chromosomes are composed of two sister chromatids; **C.** haploid, and the chromosomes are composed of one double-stranded DNA molecule; **D.** haploid, and the chromosomes are composed of two sister chromatids

ANALYZING AND APPLYING THE BASICS

1. The external female genitalia is often mistakenly referred to as the vagina, which is an internal passageway that abuts the cervix. What is the correct term for the female external genitalia?

2. How does untreated endometriosis result in lower fertility?

3. What hormone causes cervical secretions to become the type that aid sperm?

GO FIND OUT

1. Do an Internet search of advertisements for birth control methods. Is birth control typically marketed toward males or females? Why do you think this is the case? Propose an idea that might make the burden of birth control more fairly distributed among the sexes.

MAKE THE CONNECTION

The science that you learned in this chapter has helped you better understand the real-world example used throughout this discussion. Draw a line from the statement on the left to the science that supports it on the right.

Developmental issues can affect fertility.	Obesity affects androgen levels. Anorexia can cause permanent loss of fertility.
Members of a homologous pair must align with each other at meiosis II for healthy gametes to be produced.	Reproductive organs are formed while we are in our mother's uterus. Organs do not always form correctly.
Women are fertile from puberty until they permanently stop menstruating.	Meiosis pauses in females from their birth until puberty. Then it continues through the reproductive years.
Drug and alcohol consumption decreases fertility.	Gamete number and quality are reduced in the presence of many toxins.
A healthy body weight increases fertility.	Individuals with too many or too few chromosomes have lowered fertility.
Sexually transmitted infections can cause infertility.	Malformed sperm are not as able to fertilize an egg cell.
A couple's chances of achieving a pregnancy are reduced if either member smokes cigarettes.	Tissue damage in ductal structures can block gamete movement and prevent fertilization.

Answers to **Got It?, Visualize This, Working with Data, Sounds Right, But Is It?, Show You Know,** and **Chapter Review** questions can be found in the **Answers** section at the back of the book.

8 Does Testing Save Lives?

Every baby in the United States is screened at birth for up to 50 different conditions.

This NFL quarterback's son passed the newborn screen but was quickly diagnosed with a fatal genetic disorder.

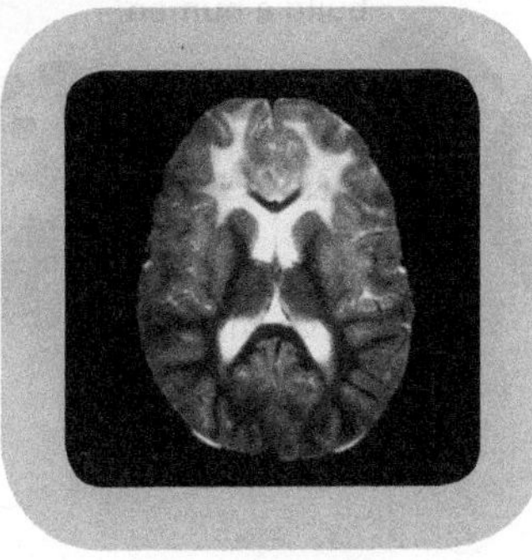

The loss of Hunter Kelly's brain tissue as a result of this condition may have been prevented if it had been caught at birth. Why isn't this disorder on the newborn screen?

Mendelian Genetics

Twenty-four hours after a baby is born in a hospital in the United States, a nurse performs a curious procedure: pricking the baby's heel with a sterile lance and then squeezing spots of his or her blood onto a form printed on filter paper. The form is labeled with the child's name and sent off to a clinical lab for analysis. This "newborn screen" is intended to catch certain disorders, the most significant effects of which may be prevented with early intervention.

The newborn screen was developed in 1963 by Dr. Robert Guthrie, a physician and microbiologist at Children's Hospital in Buffalo, New York. Dr. Guthrie's interest in newborn screening arose from a desire to prevent intellectual disabilities such as that affecting his own son, Johnny. He began by developing a test for phenylketonuria (PKU), a rare inherited disorder that can cause intellectual disability. Remarkably, the disabling effects of this condition can be avoided by careful attention to diet. Since the test's inception, the screening of millions of infants for PKU has measurably changed the life course for at least 300,000 affected individuals.

In February 1997, an infant boy born in the very same Children's Hospital was subjected to the newborn screen, which now tests for dozens of disorders in addition to PKU. The boy was Hunter Kelly, the first son of Jim Kelly, then the recently retired star quarterback for the National Football League's Buffalo Bills. Like the vast majority of babies, Hunter passed the newborn screen. But within a month, Hunter appeared noticeably irritable and at 3 months was crying nonstop. By the time he was 4 months old, Hunter was having difficulty swallowing and was experiencing regular seizures. The Kellys took him to a pediatric neurologist, where blood tests identified a terrible diagnosis: Krabbe ("crab-ay") disease, an inherited degenerative disorder of the nervous system.

There is one treatment that has the potential to halt the degeneration that marks Krabbe disease, but because so much damage had already occurred to Hunter's brain it was too late to change his prognosis. The Kellys channeled their energies

—Continued next page

into caring for their son and into building a private foundation. One mission of the Hunter's Hope Foundation is to promote early detection of Krabbe and other similar diseases. As a result of their foundation's efforts, the Kellys' home state of New York added a test for Krabbe to the newborn screen in 2006; now a handful of other states include it as well. Millions of infants have been tested for the signs of Krabbe disease in the years since.

Despite the Kellys' efforts, not every state has been convinced to add Krabbe disease to their mandated screening list, and the national review panel that evaluates diseases to add to the screen has declined to recommend it. Why would a state decline to test for a horrific disease if the test might save even one life? And how does a child like Hunter become affected by Krabbe disease?

8.1 The Inheritance of Traits

Nearly all of the 4 million babies born in the United States every year are subject to mandatory newborn screening. Each state independently determines the conditions in its screen; New York, one of the few states that screens for Krabbe disease, tests for 47 different disorders. All that is needed to screen for the majority of these conditions is a tiny blood sample, collected from a pinprick made on the child's heel. The blood is analyzed for chemical compounds that are out of range—that is, in higher or lower concentrations than would be expected in a healthy newborn.

FIGURE 8.1 Instructions inside. Both parents contribute genetic information to their offspring via their sperm and egg. The single cell that results from the fusion of one sperm with one egg contains all of the instructions necessary to produce an adult human.

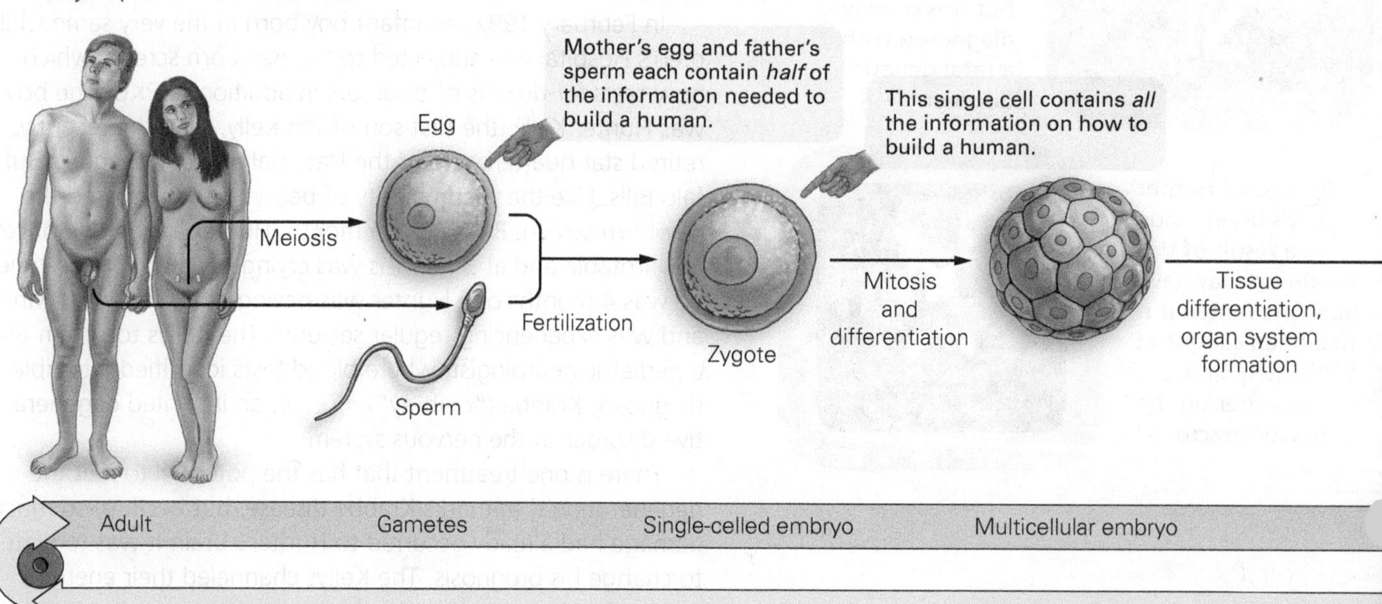
Mother's egg and father's sperm each contain *half* of the information needed to build a human.
This single cell contains *all* the information on how to build a human.
Egg
Meiosis
Fertilization
Sperm
Zygote
Mitosis and differentiation
Tissue differentiation, organ system formation
Adult
Gametes
Single-celled embryo
Multicellular embryo

As you might imagine, a positive result on a newborn screening test causes considerable anxiety among new parents. However, the majority of suspicious results for most diseases on the screen are **false positives**—in other words, they signal the possible presence of disease when one does not exist. Because false positives create anxiety and because additional tests are expensive, the federal advisory board that recommends tests to add to the screening panel is conservative about adding new conditions to its list. To be recommended, any new test must minimize the risk of false positives, must be for a condition that is well understood and effectively treatable, and/or must provide information that could affect the future reproductive decisions of the parents. To meet this final condition, the disorder tested for must have a clear heritable component.

To understand how a parents' traits are inherited by a child, you need to understand the human life cycle. A **life cycle** is a description of the growth and reproduction of an individual (**FIGURE 8.1**).

A human baby is typically produced from the fusion of a single sperm cell produced by the male parent and a single egg cell produced by the female parent. Egg and sperm (called *gametes*) fuse at **fertilization,** and the resulting cell, called a zygote, duplicates all of the genetic information it contains and undergoes mitosis to produce two identical daughter cells. Each of these daughter cells divides dozens of times in the same way. The cells in this resulting mass then differentiate into specialized cell types, which continue to divide and organize to produce the various structures of a developing human, called an embryo. Continued division of this single cell and its progeny leads to the production of a full-term infant and eventually an adult.

We are made up of trillions of individual cells, all of them the descendants of that first product of fertilization and nearly all containing exactly the same information originally found in the zygote. All of our traits, healthy or not, are influenced by the information contained in that tiny cell.

Genes and Chromosomes

Each normal sperm and egg contains information about "how to build an organism." A large portion of that information is in the form of genes, segments of DNA that generally code for proteins. Imagine genes as roughly equivalent to the words used in an instruction manual. These words are contained on chromosomes, which are roughly analogous to pages in the manual.

VISUALIZE THIS

At what stage would exposure to a mutagen that damages DNA have the greatest effect on an individual? At what stage would exposure have the least effect?

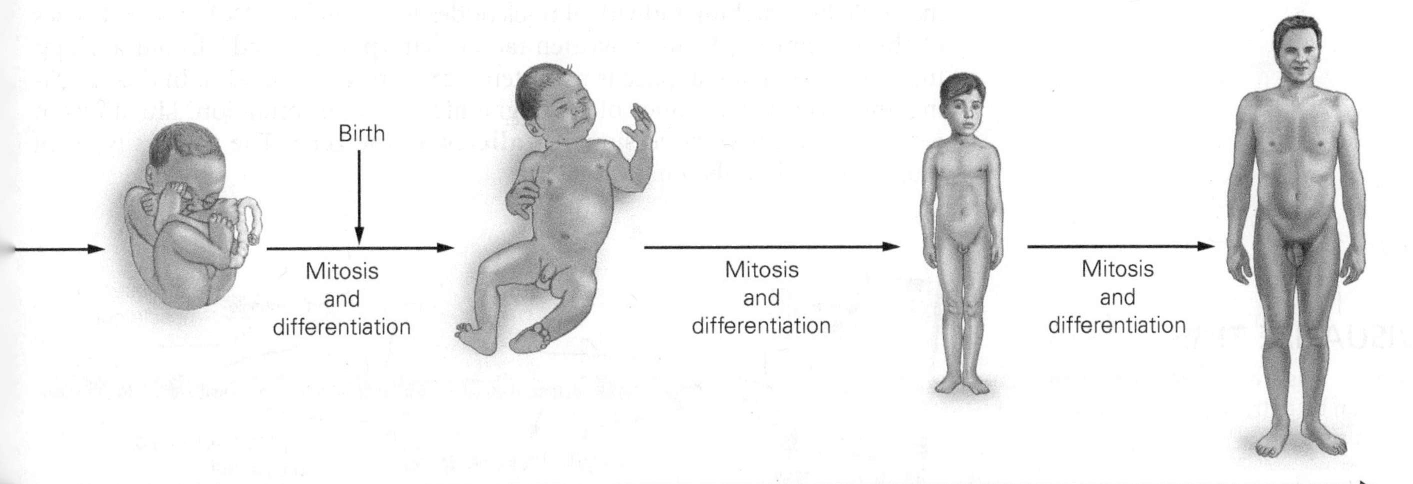

(a) Stalked adder's tongue (a fern)

(b) Single fern cell

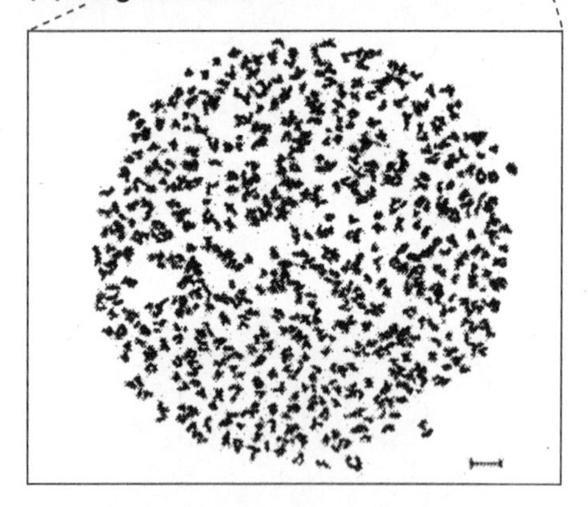

FIGURE 8.2 Variation in chromosome number. The amount of genetic information in an organism does not correlate to its complexity, as can be seen by examination of the 1260 chromosomes contained in a single cell of the stalked adder's tongue fern.

Prokaryotes such as bacteria typically contain a single, circular chromosome that floats freely inside the cell and is passed in its entirety to each offspring. In contrast, eukaryotes carry their genes on more than one linear chromosome. The number of chromosomes in eukaryotes can vary greatly, from 2 in the jumper ant (*Myrmecia pilosula*) to an incredible 1260 in the stalked adder's tongue, a species of fern (*Ophioglossum reticulatum*, **FIGURE 8.2**). Human cells contain 46 chromosomes, most of which carry thousands of genes. In our analogy, each cell has 46 pages of instructions, with each page containing thousands of words.

The instruction manual inside a cell is different from the instruction manual that comes with a kit for building a model car. You would read the manual for building the car beginning at page 1 and follow an orderly set of steps to produce the final product. A biological instruction manual is much more complicated—the pages and words to be read are different for different types of cells and may even change according to the situation. The final product of any given cell depends on the words used and the order in which the words are read from this common instruction manual.

For instance, eye cells and tongue cells in mammals both carry instructions for the protein rhodopsin, which helps detect light, but rhodopsin is produced only in eye cells, not in our tongues. Rhodopsin requires assistance from another protein, called transducin, to translate the light that strikes it into the actions of the eye cell. Transducin is also produced in cells in the tongue, but there it functions in translating the binding of certain molecules from food into the sensation of bitter flavor.

Thus, a protein may serve two or more different functions depending on its context. Because genes, like words, can be used in many combinations, the instruction manual for building a living organism is very flexible (**FIGURE 8.3**).

Producing Diversity in Offspring

Hunter Kelly's Krabbe disease resulted from the terrible coincidence of randomly occurring events. The first of these is that the genetic variant that causes the condition exists in the first place.

Gene Mutation Creates Genetic Diversity. During reproduction, genes from both parents are copied and transmitted to the next generation. The copying and transmittal of genes from one generation to the next creates **genetic variation.**

Recall the process of DNA replication, which produces a duplicate DNA molecule by matching individual nucleotides to a template DNA strand. Copies of chromosomes are thus rewritten rather than "photocopied." In our analogy, the instruction manual page is rewritten every time a cell divides. In this rewriting, there is a small chance of a typographical error, or mutation. Mutations in genes lead to different versions, or **alleles,** of the gene. The various types of mutation are described in **FIGURE 8.4**.

VISUALIZE **THIS**

Write at least two more instructions that can be extracted from this group of 14 words.

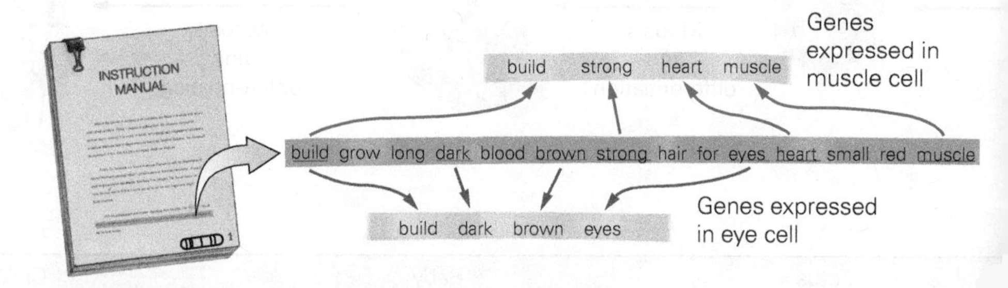

FIGURE 8.3 Genes as words in an instruction manual. Different words from the manual are used in different parts of the body, and identical words may be used in distinctive combinations in different cells.

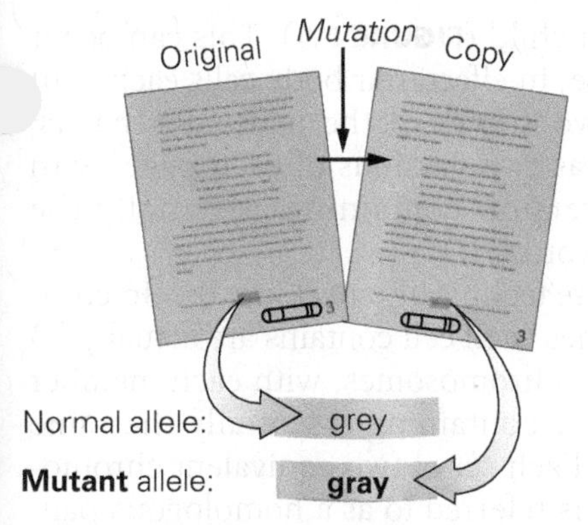

(a) The mutant allele has the same meaning (mutant allele function the same as the original allele).

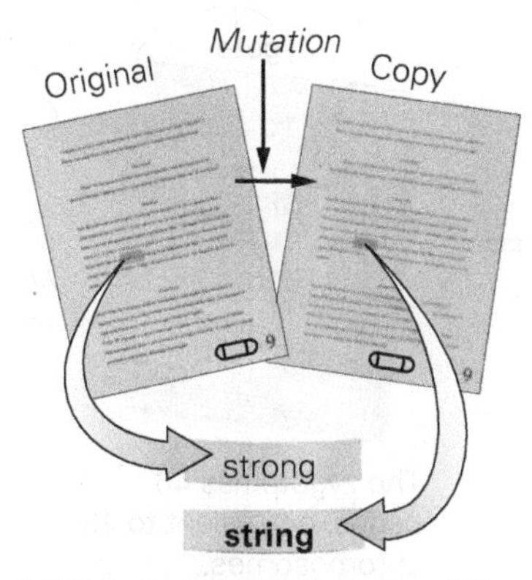

(b) The mutant allele has a different meaning (mutant allele functions differently than the original allele).

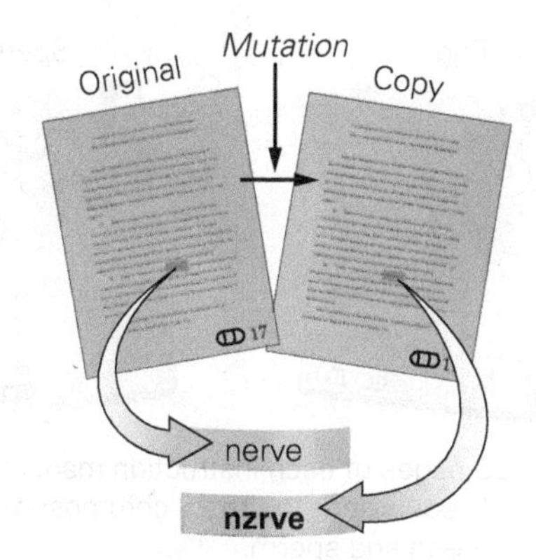

(c) The mutant allele has no meaning (mutant allele is no longer functional).

FIGURE 8.4 The formation of different alleles. Different alleles for a gene may form as a result of copying errors. In this analogy, some misspellings (mutations) do not change the meaning of the word (allele), whereas others may result in altered meanings (different allele function) or have no meaning at all (no allele function).

VISUALIZE **THIS**

Which type of mutation do you think is most likely, and why?

As you can see from this figure, many mutations result in nonsensical instructions—that is, dysfunctional alleles—and thus are often harmful to the individual who possesses them. Dysfunctional alleles tend to be lost over time because individuals with them do not function as well as those without the mutation. In short, individuals with dysfunctional mutations may not survive or may reproduce at very low rates.

However, some mutations are neutral in effect, or even beneficial in certain situations, while some harmful mutations can be hidden if the individuals who carry them also carry a functional allele. These hidden mutations tend to persist over generations. The mutations that cause Krabbe disease are of this type. Both Jim and Jill Kelly carry one copy of this mutation—but based on his athletic success, it is clear that having one mutant copy did not affect Jim Kelly's nervous system.

Because mutations occur at random and are not expected to occur in the same genes in different individuals, each of us should have a unique set of alleles reflecting the mutations passed along to us by our unique set of ancestors. By contributing to differences among families over many generations, mutation creates genetic variation in a population in the form of new alleles. When a novel characteristic increases an individual's chance of survival and reproduction, the mutation contributes to a population's adaptation to its environment (Chapter 11). Genetic misspellings are thus the engine that drives evolution itself.

In the Kelly family, only one of three children was affected by the misspelling that causes Krabbe disease. This is because, despite family similarities, each child of a set of parents is unique. The differences between siblings are a result of two factors: independent assortment and random fertilization.

Segregation and Independent Assortment Create Gamete Diversity. Both parents contribute genetic instructions to each child, but they do not contribute their entire manual. If they did, the genetic instructions carried in human cells would double every generation, making for a pretty crowded cell. Instead, the process of meiosis reduces the number of chromosomes carried in gametes by one-half (Chapter 7).

Although they are only transmitting half of their genetic information in a gamete, each parent actually gives a complete copy of the instruction manual

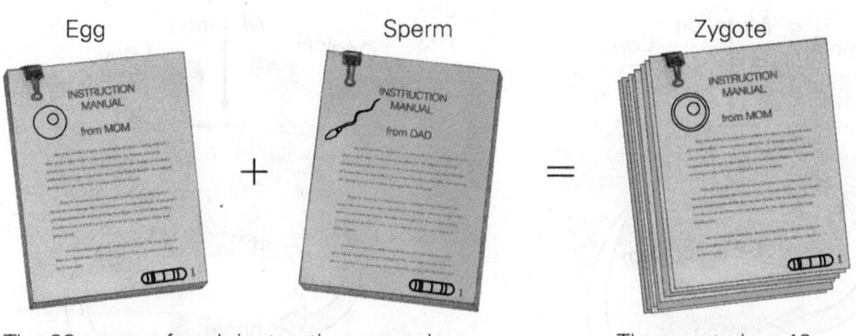
Egg
Sperm
Zygote
INSTRUCTION MANUAL
from MOM
INSTRUCTION MANUAL
from DAD
INSTRUCTION MANUAL
from MOM
+
=
The 23 pages of each instruction manual are roughly equivalent to the 23 chromosomes in each egg and sperm.
The zygote has 46 pages, equivalent to 46 chromosomes.

FIGURE 8.5 Equivalent information from parents. Each parent provides a complete set of instructions to each offspring.

to each child (**FIGURE 8.5**). This can occur because, in effect, our body cells each contain two copies of the manual—that is, each has two versions of each page, with each version containing essentially the same words.

More technically speaking, the 46 chromosomes each cell contains are actually 23 pairs of chromosomes, with each member of a pair containing essentially the same genes. Each set of two equivalent chromosomes is referred to as a homologous pair. The members of a homologous pair are equivalent, but not identical, because even though both have the same genes, each contains a unique set of alleles inherited from one or the other parent.

The process of meiosis separates homologous pairs of chromosomes and also places chromosomes independently into each gamete. These two events explain why siblings are not identical (with the exception of identical twins). When homologous pairs are separated during meiosis, the alleles carried on the members of the pair are separated as well. The separation of pairs of alleles during the production of gametes is called **segregation.** Thus, a parent with two different alleles of a gene will produce gametes with a 50% probability of containing one version of the allele and a 50% probability of containing the other version.

The segregation of chromosomes during meiosis also explains **independent assortment,** the fact that each gene is inherited (mostly) independently of other genes. Independent assortment arises from **random alignment,** the uncoordinated "lining up" of chromosome pairs before the first division of meiosis. The possible outcomes of random alignment are visible in **FIGURE 8.6**; if we consider

FIGURE 8.6 Random alignment. In this example, the organism undergoing meiosis has only four chromosomes. The organism inherited the blue chromosomes from its father and the red chromosomes from its mother. When there are two homologous pairs of chromosomes, two possible alignments, (a) and (b), can occur. These different alignments can lead to novel combinations of genes in the gametes.

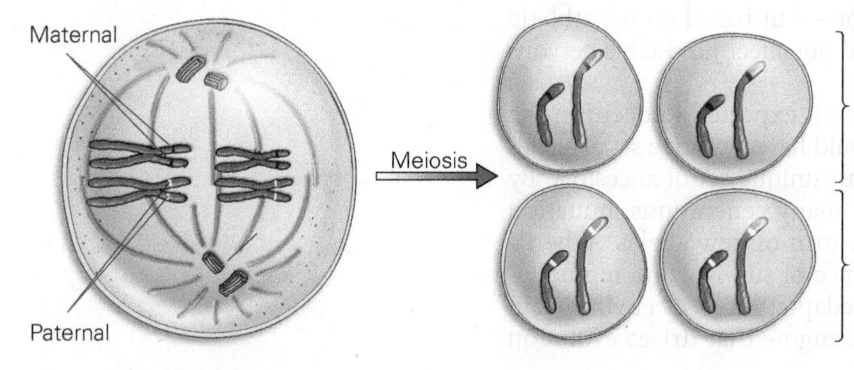
(a) One possible alignment of chromosomes in meiosis I
Maternal
Paternal
Meiosis
Two combinations of chromosomes in gametes
These gametes contain only chromosomes inherited from its mother.
These gametes contain only chromosomes inherited from its father.

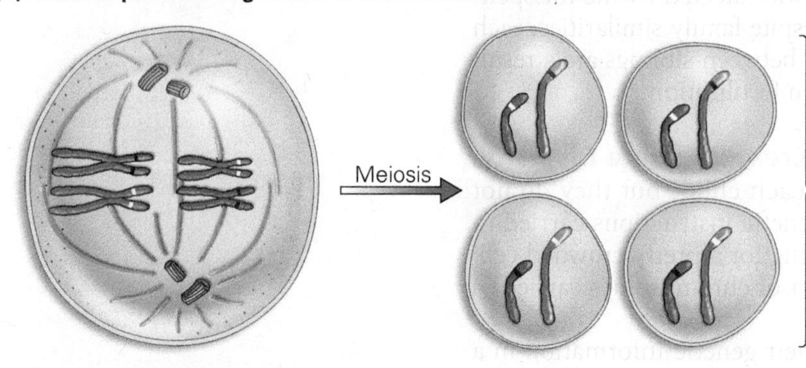
(b) Another possible alignment of chromosomes in meiosis I
Meiosis
Two additional combinations of chromosomes in gametes
These gametes contain one maternal and paternal chromosome.

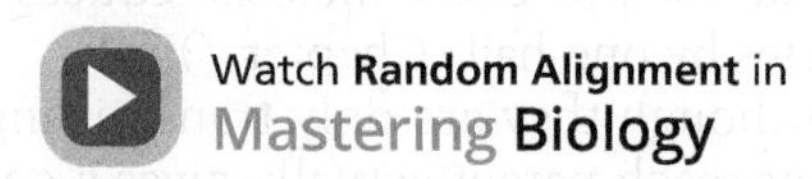
Watch Random Alignment in Mastering Biology

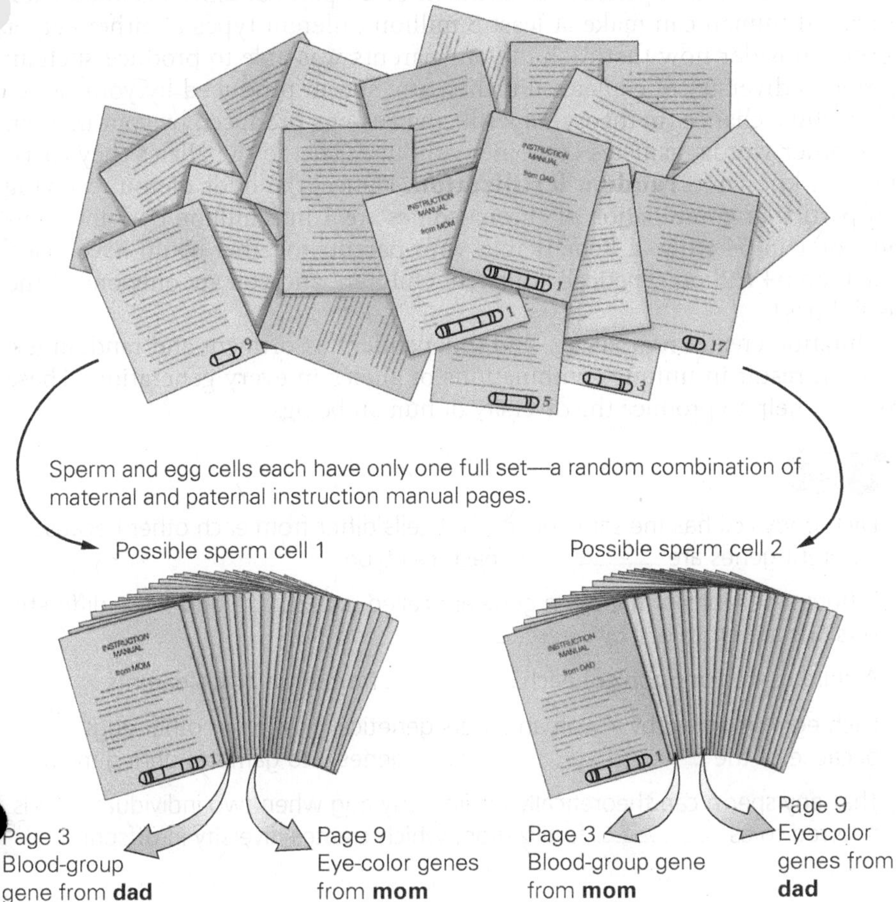
Parent cells have two copies of each chromosome—that is, two full sets of instruction manual pages, one from each parent.
Sperm and egg cells each have only one full set—a random combination of maternal and paternal instruction manual pages.
Possible sperm cell 1
Possible sperm cell 2
Page 3
Blood-group gene from dad
Page 9
Eye-color genes from mom
Page 3
Blood-group gene from mom
Page 9
Eye-color genes from dad

FIGURE 8.7 Each egg and sperm is unique. Because each sperm is produced independently, the set of chromosomes in each sperm nucleus will be a unique combination of the chromosomes that the man inherited from his biological parents.

VISUALIZE **THIS**

The man could produce four distinctly different kinds of sperm cell when considering these two genes and four alleles. List the other two possibilities not pictured here.

only two homologous pairs of chromosomes, then two different alignments are possible, and four different gametes can be produced. In this way, genes that are on different chromosomes are inherited independently of each other. As the number of chromosomes in an organism's genome increases, so does the number of possible alignments and the number of genetically distinct gametes that organism can produce.

In our analogy, the instruction manual contained in a single sperm cell is made up of a unique combination of pages from the manuals a man received from each of his parents. In fact, almost every sperm he makes will contain a unique subset of chromosomes—and thus a unique subset of his alleles. **FIGURE 8.7** illustrates this. In the figure, you can see that independent assortment causes an allele for an eye-color gene to end up in a sperm cell independently from an allele for the blood-group gene.

The independent assortment of segregated chromosomes into daughter cells is repeated every time a sperm is produced, and thus the set of alleles that a child receives from a father is different for all of his offspring. The sperm that contributed half of your genetic information might have carried an eye-color allele from your father's mom and a blood-group allele from his dad, while the sperm that produced your sister might have contained both the allele for eye color and the allele for blood group from your paternal grandmother. As a result of independent assortment, only about 50% of an individual's alleles are identical to those found in another offspring of the same parents—that is, for each gene, you have a 50% chance of being like your sister or brother.

Random Fertilization Results in a Large Variety of Potential Offspring. As a result of the independent assortment of 23 pairs of chromosomes, each individual human can make at least 8 million different types of either egg or sperm. Consider now that each of your parents was able to produce such an enormous diversity of gametes. Further, any sperm produced by your father had an equal chance (in theory) of fertilizing any egg produced by your mother.

In other words, gametes combine without regard to the alleles they carry, a process known as **random fertilization.** Hence, the odds of you receiving your particular combination of chromosomes are 1 in 8 million by 1 in 8 million—or 1 in 64 trillion. Remarkably, your parents together could have made more than 64 trillion genetically different children, and you are only one of the possibilities.

Mutation creates new alleles, and independent assortment and random fertilization result in unique combinations of alleles in every generation. These processes help to produce the diversity of human beings.

Got It?

1. Each body cell has the same genes, but cells differ from each other because different genes are ________ in each cell type.
2. Different versions of the same gene are called ________ and these different versions are produced by ________.
3. A child carries one copy of each ________ from each of his parents.
4. Each egg produced by a woman differs genetically from her other eggs because of the ________ ________ of genes into gametes during meiosis.
5. That any sperm can theoretically fertilize any egg when two individuals pair is referred to as ________ fertilization, which creates diversity in offspring.

8.2 Basic Mendelian Genetics: When the Role of Genes Is Clear

A few human genetic traits have easily identifiable patterns of inheritance. These traits are said to be "Mendelian" after Gregor Johann Mendel (**FIGURE 8.8**), who was the first person to accurately describe their inheritance. Krabbe disease, like the other genetic disorders tested for on the New York newborn panel, is a Mendelian trait.

FIGURE 8.8 Gregor Mendel. The father of the science of genetics.

Mendel was born in Austria in 1822. Because his family was poor and could not afford private schooling, he entered a monastery to obtain an education. After completing his monastic studies, Mendel attended the University of Vienna. There he studied math and botany in addition to other sciences. After leaving the university, he returned to the monastery and began his experimental studies of inheritance in garden peas.

Mendel studied close to 30,000 pea plants over a 10-year period. His careful experiments consisted of controlled matings between plants with different traits. Mendel was able to control the types of mating that occurred by hand-pollinating the peas' flowers—that is, by taking pollen, which produces sperm, from the anthers of one pea plant and applying it to the carpel (the egg-containing structure) of another pea plant. By growing the seeds that resulted from these controlled matings, he could evaluate the role of each parent in producing the traits of the offspring (**FIGURE 8.9**).

Although Mendel did not understand the physical nature of genes, he was able to determine how traits were inherited by carefully analyzing the appearance of parent pea plants and their offspring. His patient, scientifically sound

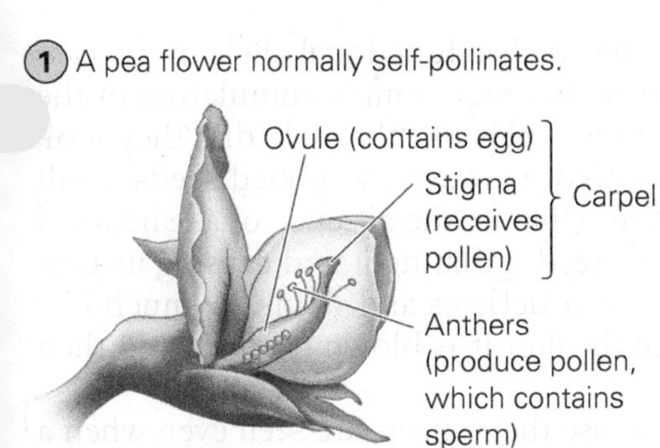
1 A pea flower normally self-pollinates.
Ovule (contains egg)
Stigma (receives pollen)
Carpel
Anthers (produce pollen, which contains sperm)

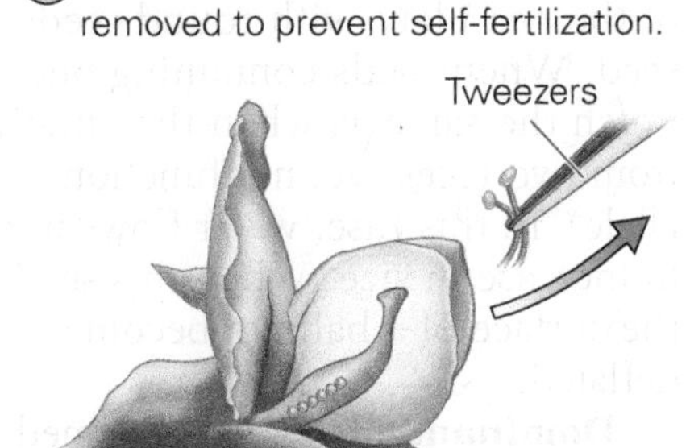
2 Pollen-containing structures can be removed to prevent self-fertilization.
Tweezers

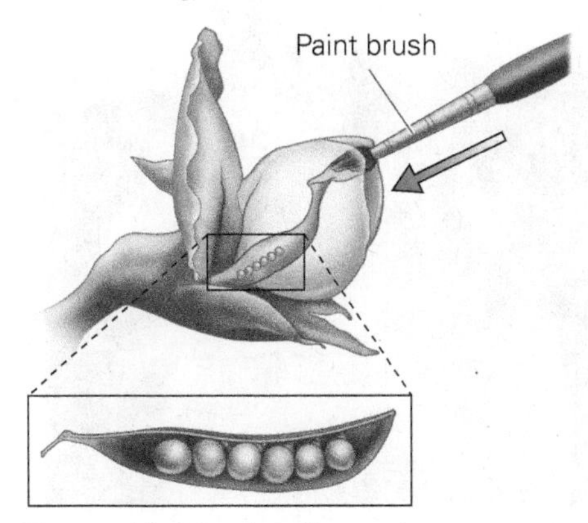
3 Pollen from another flower is dabbed on to stigma.
Paint brush

The resulting seeds will contain information on flower color, seed shape and color, and plant height from both parents.

FIGURE 8.9 Peas and genes. Pea plants were an ideal study organism for Mendel because their reproduction is easy to control, they complete their life cycle in a matter of weeks, and a single plant can produce thousands of offspring.

VISUALIZE **THIS**

Each pea develops as a result of the fertilization of a single egg. Are the peas in a single pod genetically identical?

experiments demonstrated that both parents contribute equal amounts of genetic information to their offspring.

Mendel published the results of his studies in 1865, but his scientific contemporaries did not fully appreciate the significance of his work. Mendel eventually gave up his genetic studies and focused his attention on running the monastery until his death in 1884. His work was independently rediscovered by three scientists in 1900; only then did its significance to the new science of genetics become apparent.

The pattern of inheritance Mendel described occurs primarily in traits that are the result of a single gene with a few distinct alleles. TABLE 8.1 lists some of the traits Mendel examined in peas; we will examine the principles he discovered, such as dominance and recessiveness, by looking at human genes screened on some or all state newborn panel tests.

pheno- comes from a verb meaning "to show."

hetero- means the other, another, or different.

-zygous derives from zygote, the "yoked" cell resulting from the union of an egg and sperm.

Genotype and Phenotype

We call the genetic composition of an individual his **genotype** and his physical traits his **phenotype.** The genotype is a description of the alleles for a particular gene carried on each member of a homologous pair of chromosomes (see Chapter 7, Figure 7.1). An individual who carries two different alleles for a gene has a **heterozygous** genotype. An individual who carries two copies of the same allele has a **homozygous** genotype.

The effect of an individual's genotype on his phenotype depends on the nature of the alleles he carries. Some alleles are **recessive,** meaning that their effects can be seen only if a copy of a dominant allele (described below) is not also present. For example, in pea plants the allele that codes for wrinkled seeds is recessive to the allele for round seeds. Wrinkled seeds will only appear when seeds carry only the wrinkled allele and no copies of the round allele.

Often a recessive allele is one that codes for a nonfunctional protein. Homozygotes having two copies of such an allele produce no functional protein. In contrast, heterozygotes carrying one copy of the functional allele have normal phenotypes

TABLE 8.1 Pea traits studied by Mendel.

Character Studied	Dominant Trait	Recessive Trait
Seed shape	Round	Wrinkled
Seed color	Yellow	Green
Flower color	Purple	White
Stem length	Tall	Dwarf

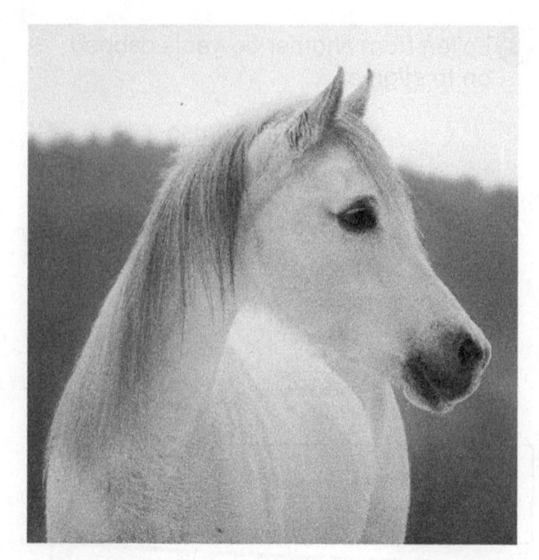

FIGURE 8.10 A dominant allele. American Albino horses occur when an individual carries an allele that prevents normal coat color development. Because the product of this allele actively interferes with a biochemical pathway, horses that have only one copy of the allele are albino.

because the normal protein is still produced. The functional allele in the case of the pea plant with round seeds prevents water from accumulating in the seed. When seeds containing one or two copies of this allele dry, they look much the same as when they first matured. However, wrinkled seeds result from two recessive, nonfunctional alleles (that is, the absence of a functional allele). In this case, water flows into the seed, inflating it and causing its coat to increase in size. When this seed dries, it deflates and wrinkles, much like the surface of a balloon becomes wrinkly after it is blown up once and then deflated.

Dominant alleles are so named because their effects are seen even when a recessive allele is present. In the wrinkled seed example, the dominant allele is the one that produces a functional protein. The dominant allele does not always code for the "normal" condition of an organism, however. Sometimes mutations can create abnormal dominant alleles that essentially mask the effects of the recessive, normal allele. For example, "American Albino" horses—known for their snow-white coats, pink skin, and dark eyes—result from an allele that stops a horse's hair-color genes from being expressed during the horse's development. Because the allele prevents normal coat color development, it has its effect even if the animal carries only one copy—in other words, albinism in horses is dominant to the normal coat color (**FIGURE 8.10**). Dominant conditions in humans include cheek dimples and the production of six fingers and toes instead of five.

Genetic Diseases in Humans

Most alleles in humans do not cause disease or dysfunction; they are simply alternative versions of genes. The diversity of alleles in the human population contributes to diversity among us in our appearance, physiology, and behaviors. However, the mutant alleles screened for in newborn panels all cause rare but serious diseases. Nearly all of these conditions are produced by recessive alleles.

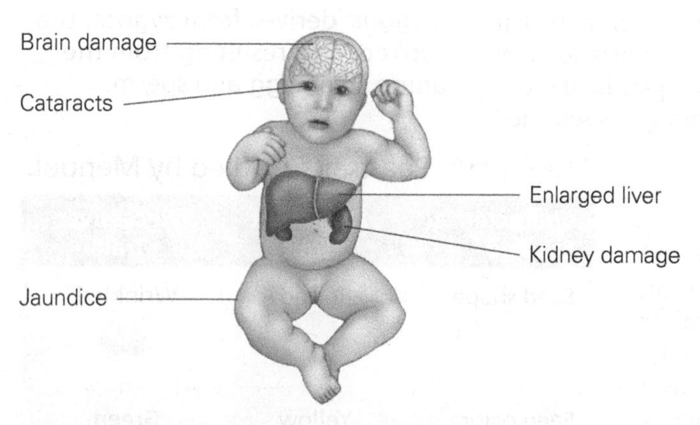

FIGURE 8.11 Pleiotropy. Galactosemia is a disorder on the New York newborn screening panel that results from an inability to digest the sugar present in milk. Babies with galactosemia exhibit a range of seemingly unrelated symptoms stemming from a buildup of sugars in various organs.

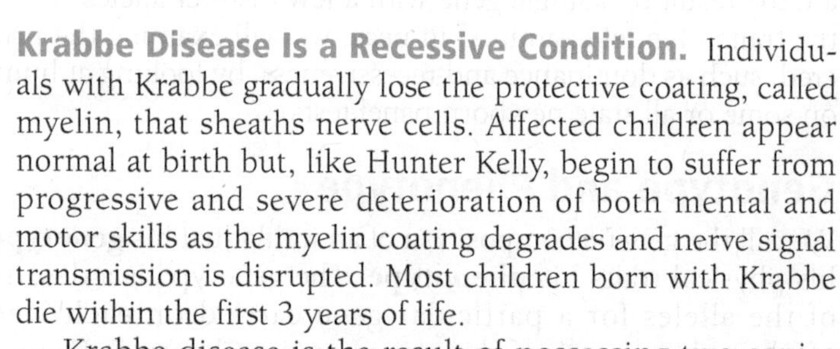

Krabbe Disease Is a Recessive Condition. Individuals with Krabbe gradually lose the protective coating, called myelin, that sheaths nerve cells. Affected children appear normal at birth but, like Hunter Kelly, begin to suffer from progressive and severe deterioration of both mental and motor skills as the myelin coating degrades and nerve signal transmission is disrupted. Most children born with Krabbe die within the first 3 years of life.

Krabbe disease is the result of possessing two copies of a mutant allele of the *GALC* gene. This gene normally codes for an enzyme that degrades certain waste products that accumulate in myelin-producing cells. These cells must function throughout life as myelin is continually refreshed and replaced. The *GALC* alleles that cause Krabbe disease are recessive because individuals who carry only one copy of the normal allele can still produce the functional enzyme. However, in individuals with no functional copies of the gene, the enzyme is not produced, waste accumulates, and the myelin-producing cells begin to die. Because the resulting loss of myelin causes multiple effects in different organs, the mutant *GALC* allele is said to exhibit **pleiotropy,** the ability of a single gene to cause multiple effects on an individual's phenotype (**FIGURE 8.11**).

pleio- comes from the Greek word for many.

Heterozygotes for a recessive disease like Krabbe are called **carriers** because even though they are unaffected, these individuals can pass the trait to the next generation. In Hunter Kelly's case, both his father and mother were carriers of Krabbe, a rare occurrence because the mutation is present in only 0.3% of the population. Because Hunter inherited mutated *GALC* genes from both, he could not produce any functional enzyme.

Many of the mutations screened for at birth are similar to the one that causes Krabbe; that is, the alleles that are responsible produce nonfunctional products. Like Krabbe, most of these conditions are rare; but a few on the screen are relatively common. For example, cystic fibrosis (CF) is among the most common genetic diseases in European populations; nearly 1 in 2500 individuals in these populations is affected with the disease, and 1 in 25 is heterozygous for the allele. Individuals with CF cannot transport chloride ions into and out of cells lining the lungs, intestines, and other organs. As a result of this dysfunction, the balance between sodium and chloride in the cell is disrupted, and the cell produces a thick, sticky mucus layer instead of the thin, slick mucus produced by cells with the normal allele. The sticky mucus accumulates in the lungs and digestive system, and as a result, most children with CF suffer from recurrent lung infections. Most children diagnosed with CF live to adulthood, but the accumulating lung damage means that their average lifespan is currently only 40 years.

Few Diseases Are Caused by a Dominant Allele. Only one condition of the 47 on the New York newborn screening panel can be caused by a dominant allele: Congenital hypothyroidism (CH) is a condition that results from inadequate production of thyroid hormones. These hormones are produced in the thyroid gland and promote energy metabolism and growth. Babies with untreated CH show low activity levels, grow slowly, and develop intellectual disabilities. CH is relatively common, affecting 1 in 4000 individuals, but the condition is easily treated by providing replacement thyroid hormones throughout an affected individual's life.

Most cases of congenital hypothyroidism are caused by recessive mutations, but a tiny fraction result from a dominant allele that was either inherited from a parent or arose spontaneously during development. (Note that the spontaneous appearance of a mutant allele is possible even when an individual has a recessive disorder—a child could inherit a nonfunctional copy from a carrier parent and unluckily experience a new mutation in the normal copy from the other parent early in their development. This is much rarer than disease caused by a spontaneous dominant mutation because it is caused by *two* independent events rather than just one.)

A few genes have been identified that can cause CH even when the affected individual carries only one mutated copy of one of them. One gene that has been well characterized in a dominant form of CH is *PAX8*. The protein produced by *PAX8* typically binds to DNA and promotes the transcription of other genes involved in the growth of the thyroid gland during embryonic development. Certain mutations in *PAX8* produce proteins that interfere with, rather than promote, the transcription of these thyroid-specific genes. Thus, even one faulty copy of this protein greatly reduces the size and development of the thyroid gland and its ability to produce thyroid hormone. A newborn screen result that indicates low thyroid hormone levels in the blood is evidence that the tested child may have CH.

A suspicious phenotype that suggests CH or any other condition on the newborn screening panel does not provide definitive information about the genotype of the tested baby. Further tests are needed to determine if a particular mutation is present. Jim and Jill Kelly had no reason to suspect that their son was at risk for Krabbe disease—in fact, they were already parents of a healthy 2-year-old daughter at the time of Hunter's birth. To understand the likelihood that Hunter was affected but his sisters were not, we can use a simple but helpful tool called a Punnett square.

Using Punnett Squares to Predict Offspring Genotypes

The inheritance of single gene traits like Krabbe disease and congenital hypothyroidism is relatively easy to understand. We can predict the likelihood of inheritance of small numbers of these traits by using a tool developed in 1905 by Reginald Punnett, a British geneticist. A **Punnett square** is a table that lists the different kinds of sperm or eggs parents can produce relative to the gene or genes in question and then predicts the possible outcomes of a **cross,** or mating, between these parents (**FIGURE 8.12**).

Using a Punnett Square with a Single Gene. The Kelly family now knows from genetic testing that both Jill and Jim carry one functional and one nonfunctional allele for the *GALC* gene. We will use a simple key to designate the Kelly's genotypes: the letters *G* and *g*, representing the dominant functional allele (*G*) and recessive nonfunctional allele (*g*). By convention, the dominant allele is always presented first in this notation; thus as carriers, Jim and Jill each have the genotype *Gg*. A genetic cross between two carriers could then be symbolized as follows:

$$Gg \times Gg$$

We know Jill could produce eggs that carried either the *G* or *g* allele because the process of meiosis will segregate the two alleles from each other. We place these two egg types across the horizontal axis of the Punnett square (as in Figure 8.12). Jim could also make two types of sperm, containing either the *G* or the *g* allele. We place these along the vertical axis. Thus, the letters on the horizontal and vertical axes represent all the possible types of eggs and sperm that Hunter's mother and father could produce by meiosis, if we consider only the gene that codes for the GALC protein.

Inside the Punnett square are all the genotypes that can be produced from a cross between these two heterozygous individuals. The content of each box is determined by combining the alleles from its corresponding egg column and sperm row.

Note that for a cross involving a single gene with two different alleles, there are three possible offspring types. The chance of this cross producing a child affected with Krabbe disease is one in four, or 25%, because the *gg* combination of alleles occurs once out of the four possible outcomes. The *GG* genotype

VISUALIZE **THIS**

In some mutations, a sperm carrying the mutant allele is less mobile than one carrying the normal allele. How would this affect the likelihood of having a child who carried two mutant alleles?

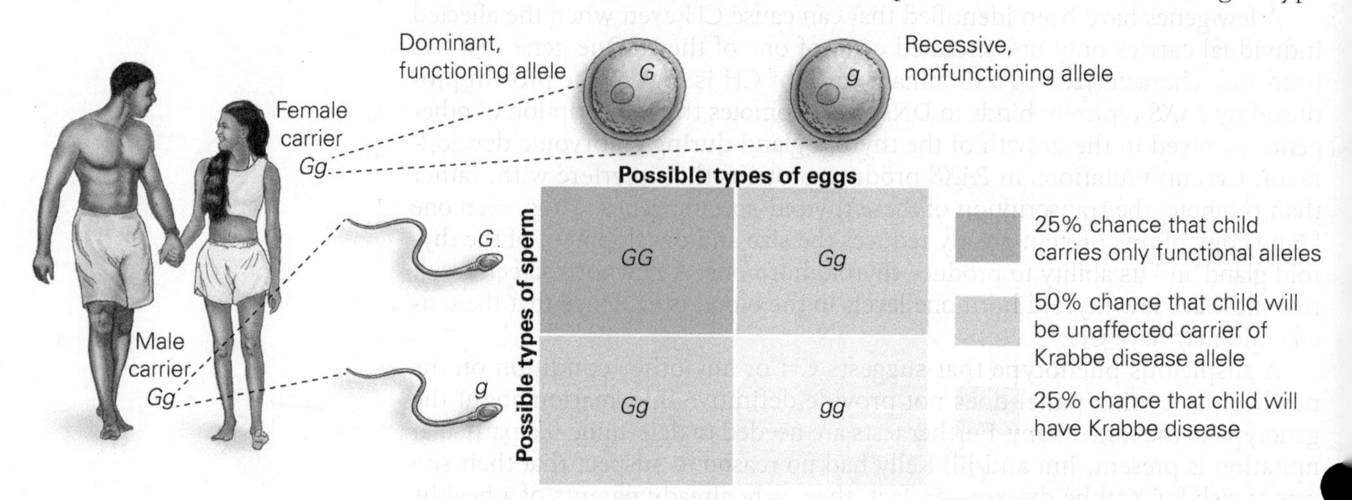

FIGURE 8.12 A cross between heterozygous individuals. This Punnett square illustrates the likelihood of having a child with Krabbe disease when both parents are carriers of a mutant *GALC* allele.

is also represented once out of four times, meaning that the probability of a homozygous unaffected child is also 25%. Genetic tests have revealed that the Kelly's oldest child is in this category. The probability of producing a child who is a carrier of Krabbe is one in two, or 50%, because two of the possible outcomes inside the Punnett square are unaffected heterozygotes—one produced by a *G* sperm and a *g* egg and the other produced by a *g* sperm and a *G* egg.

Punnett squares can be used to estimate the probability of various genotypes and phenotypes for many different genetic combinations. **FIGURE 8.13** illustrates one that analyzes the probability of a child carrying a dominant trait when one parent is affected.

You should note that the probability of a particular genotype is generated independently for each child. Each offspring of two carriers has a 25% chance of being affected. The Kelly's third child was another healthy girl, but this cannot be explained by the fact that they already had one child with Krabbe; the third child had the same risk that Hunter did of inheriting two mutated copies of the *GALC* gene. In fact, the Kelly's younger daughter does carry one copy of the mutant allele.

Punnett Squares for Crosses with More Genes. Jill Kelly often remarked that one of Hunter's most striking features was his green eyes, which were a deeper color than either hers or her husband's. Neither of the Kelly's daughters inherited this eye color, but that coincidence had nothing to do with Hunter's Krabbe disease. As with any segregated traits, the genes that determined Hunter's eye color were inherited independently of the *GALC* gene. Punnett squares can also help us understand the likelihood of Hunter's phenotype for these two independent traits.

Dihybrid crosses are genetic crosses involving two traits. Let's go back to Mendel's peas as an example. Seed color and seed shape are each determined by a single gene, and each is carried on different chromosomes. The two seed-color gene alleles Mendel studied are designated here as *Y*, which is dominant and codes for yellow color, and *y*, the recessive allele, which results in green seeds when homozygous. The two seed-shape alleles Mendel studied are designated as *R*, the dominant allele, which codes for a smooth, round shape, and *r*, which is recessive and codes for a wrinkled shape.

di- comes from the Greek word for two.

Because the genes for seed color and seed shape are on different chromosomes, they are placed in eggs and sperm independently of each other. In other words, a pea plant that is heterozygous for both genes (genotype *YyRr*) can make four different types of eggs: one carrying dominant alleles for both genes (*YR*), one carrying recessive alleles for both genes (*yr*), one carrying the dominant allele for seed color and the recessive allele for seed shape (*Yr*), and one carrying the recessive allele for color and the dominant allele for shape (*yR*).

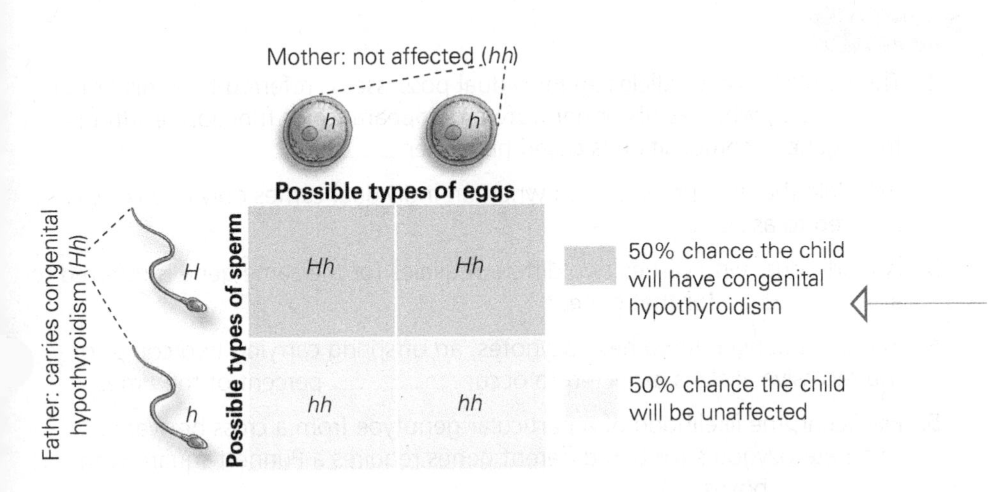

FIGURE 8.13 A cross between a heterozygous individual and a homozygous individual. This Punnett square illustrates the outcome of a cross between a man who carries a single copy of the dominant congenital hypothyroidism allele and an unaffected woman.

WORKING WITH **DATA**

If one child produced by these parents carries a dominant allele for congenital hypothyroidism, what is the likelihood that a second child produced by this couple will have the condition?

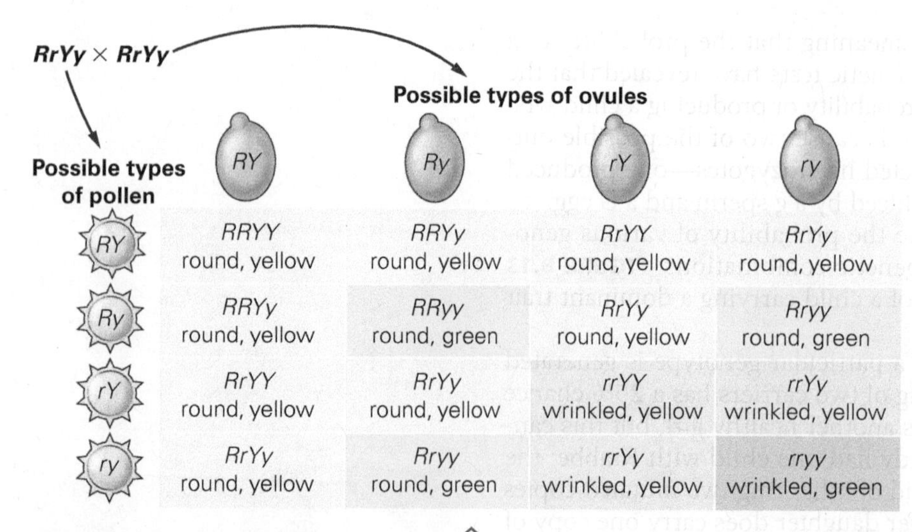

FIGURE 8.14 A dihybrid cross. Punnett squares can be used to predict the outcome of a cross involving two different genes. This cross involves two pea plants that are both heterozygous for the seed-color and seed-shape genes.

WORKING WITH **DATA**

Krabbe disease is recessive (*gg*). Eye color is actually determined by several different genes, but we'll imagine it is controlled by only one in this example—let's say brown pigmented eyes (*BB* or *Bb*) are dominant over green eyes (*bb*). What fraction of the offspring of a mother and father who are heterozygous for both of these traits would have Krabbe disease and green eyes?

As with the Punnett square discussed above, the analysis of a dihybrid cross places all possible sperm genotypes on one axis of the square and all possible egg genotypes on the other axis. Thus, a Punnett square for a cross between two individuals who are heterozygous for both seed-color and seed-shape genes would have four columns representing the four possible egg genotypes and four rows representing the four possible sperm genotypes, resulting in 16 boxes within the square describing four different possible phenotypes (**FIGURE 8.14**).

The phenotypes produced by a dihybrid cross result in a 9:3:3:1 **phenotypic ratio.** In this case, 9/16 include those genotypes produced by both dominant alleles (*Y_R_* where the dashes indicate that the second allele for each gene could be either dominant or recessive); 3/16 include those produced by dominant alleles of one gene only (*Y_rr*); 3/16 include those produced by dominant alleles of the other gene (*yyR_*); and 1/16 have the phenotype produced by possessing recessive alleles only (*yyrr*).

As you might imagine, as the number of genes in a Punnett square analysis increases, the number of boxes in the square increases, as does the number of possible genotypes. With two genes, each with two alleles, the number of unique gametes produced by a heterozygote is four, the number of boxes in the Punnett square is 16, and the number of unique genotypes of offspring that can be produced is nine. With three genes, each with two alleles, the Punnett square has 64 boxes and 22 different possible genotypes. With four genes, the square has 256 boxes, and with five genes, there are more than 1000 boxes! Predicting the outcome of a cross becomes significantly more difficult as the number of genes we are following increases.

Got It?

1. The combination of alleles an individual possesses is referred to as his or her __________, whereas his or her outward appearance or function resulting from genetic composition is called his or her __________.
2. An allele that is expressed even when an individual carries only one copy is referred to as __________.
3. An individual who carries two different alleles for the same gene is referred to as a __________ for that gene.
4. In a cross between two heterozygotes, an offspring carrying two copies of the recessive allele is expected to occur __________ percent of the time.
5. Predicting the likelihood of a particular genotype from a cross between two heterozygotes for two different genes requires a Punnett square with __________ boxes.

Flower color in snapdragons

Red = RR
Homozygote

×

White = rr
Homozygote

=

Pink = Rr
Heterozygote

FIGURE 8.15 Incomplete dominance. Snapdragons show incomplete dominance in the inheritance of flower color. The heterozygous flower has a phenotype that is in between that of the two homozygotes.

VISUALIZE THIS

If flower color were inherited such that *R* was completely dominant over *r*, what color would heterozygous flowers be?

8.3 Extensions of Mendelian Genetics

Mendel worked with traits in peas that expressed only simple dominance and recessive relationships. However, for some genes, more than one dominant allele may be produced, and for others, a dominant allele may have different effects in a heterozygote than a homozygote. These types of alleles produce inheritance patterns that are a little more complex.

When the phenotype of the heterozygote is intermediate between both homozygotes, the situation is called **incomplete dominance.** The alleles that determine flower color in snapdragons are an example: One homozygote produces red flowers, the other, presumably carrying two nonfunctional copies of a color gene, produces white flowers; the heterozygote, carrying one allele encoding red color and one allele encoding white color, produces pink flowers (**FIGURE 8.15**).

The alleles associated with sickle cell disease, a disorder on the newborn screening panel in all states, also display incomplete dominance. The mutation in this case is in a gene that codes for the protein **hemoglobin,** the oxygen-carrying protein in red blood cells. Under conditions of oxygen stress, mutant hemoglobin molecules will clump together, forming long chains that distort the cells into a sickle shape, causing them to clog small blood vessels in the body (**FIGURE 8.16**). Individuals with two copies of the mutant allele can experience these painful and debilitating sickling attacks; over the long term, repeated sickling attacks can cause significant organ damage.

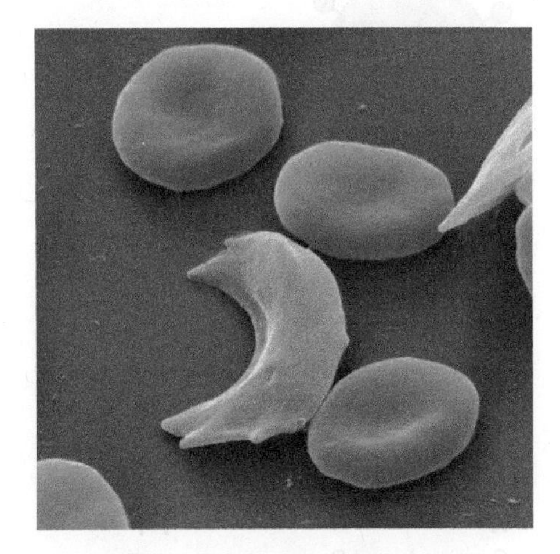

FIGURE 8.16 Sickle cell disease. The crescent-shaped red blood cell in this micrograph results from the behavior of a mutant version of the oxygen-carrying protein hemoglobin. These sickled cells are inflexible and thus become caught in the tiny blood vessels that the round cells pictured here can easily slide through.

hemo- means blood.

The sickle-cell allele is described as incompletely dominant because the mutant protein is still produced. In fact, on average, half of the hemoglobin molecules in a carrier are the mutant type. Under low oxygen levels, carriers may experience some clumping of hemoglobin molecules and symptoms of sickle-cell disease. Thus the mutant allele is not completely recessive. (The distribution of the sickle-cell allele in the human population is discussed in Chapter 13.)

In some cases, the phenotype of a heterozygote is actually a combination of both fully expressed traits, instead of a mixture. This situation, by which two different alleles of a gene are both expressed in an individual, is known as **codominance** (**FIGURE 8.17**). In cattle, for example, the allele

Coat color in cattle

Red = R^1R^1

×

White = R^2R^2

=

Roan = R^1R^2
(patchy red and white coat)

FIGURE 8.17 Codominance. Roan coat color in cattle is an example of codominance. Both alleles are equally expressed in the heterozygote, so the conventional uppercase and lowercase nomenclature for alleles no longer applies.

VISUALIZE **THIS**

How many different ABO blood system alleles are there in the human population? How many different ABO blood system alleles could one person carry?

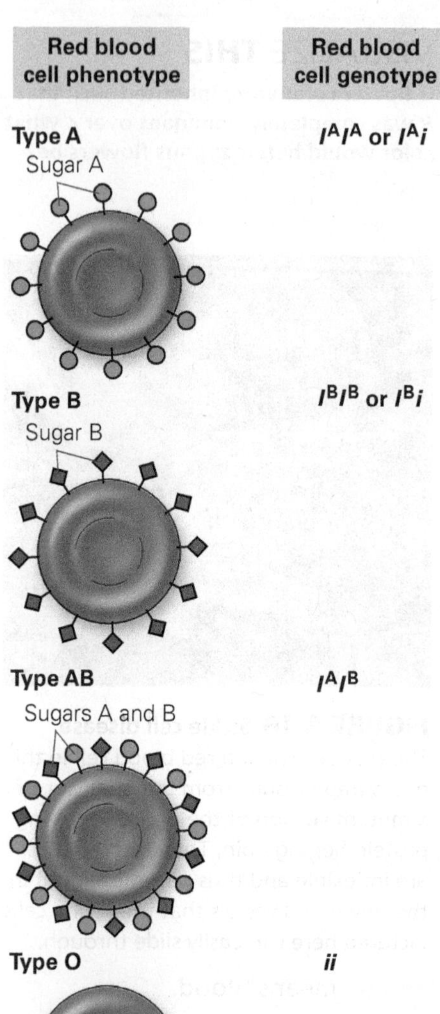

FIGURE 8.18 ABO blood system. Red blood cell phenotypes and corresponding genotypes. Alleles I^A and I^B are codominant, and both are dominant to i.

that codes for red hair color and the allele that codes for white hair color are both expressed in a heterozygote. These individuals have patchy coats that consist of an approximately equal mixture of white hairs and red hairs. None of the diseases currently on New York's newborn screening test result from codominant alleles; however, in some babies' blood samples, codominance is apparent.

Certain carbohydrates located on the surface of red blood cells determine a baby's blood type. The most well-known surface markers are part of the **ABO blood system.** The ABO blood system displays **multiple allelism,** which occurs when there are more than two alleles of a gene in the population. In fact, three distinct alleles of one blood-group gene code for the enzymes that synthesize the sugars found on the surface of red blood cells. Two of the three alleles display codominance to each other, and one allele is recessive to the other two.

FIGURE 8.18 summarizes the possible genotypes and phenotypes for the ABO blood system. The three alleles of this blood-type gene are I^A, I^B, and i. A given individual will carry only two alleles, even though three alleles are present in the entire population. In other words, one person may carry the I^A and I^B alleles, and another might carry the I^A and i alleles.

The symbols used to represent these alternate forms of the blood-type gene tell us something about their effects. The lowercase i allele is recessive to both the I^A and I^B alleles. Therefore, an infant with the genotype I^Ai has type A blood, and one with the genotype I^Bi has type B blood. A newborn with only the recessive allele, genotype ii, has type O blood. The I^A and I^B alleles are codominant in that both of these carbohydrates are found on a blood cell's surface. Thus, an infant with the genotype I^AI^B has type AB blood.

Clinicians must take ABO blood groups into account when performing blood transfusions. Persons receiving transfusions from incompatible blood groups will mount an immune response against those sugars that they do not carry on their own red blood cells. The presence of these foreign red blood cell sugars causes a severe reaction in which the donated, incompatible red blood cells form clumps, blocking blood vessels and potentially killing the recipient. **TABLE 8.2** shows the types of blood transfusions individuals of various blood types can receive.

TABLE 8.2 Blood transfusion compatibilities.

Recipient	Recipient Can Receive	Recipient Cannot Receive
Type O	Type O	Type A Type B Type AB
Type A	Type O Type A	Type B Type AB
Type B	Type O Type B	Type A Type AB
Type AB	Type O Type A Type B Type AB	None

Got It?

1. When a heterozygote has a phenotype intermediate between two homozygotes, the allele is said to be ________ dominant.
2. In sickle cell anemia, the disease-causing allele is not nonfunctional, it is present in ________ form.
3. The alleles that determine A and B blood type are best described as ________, meaning that both are expressed in the heterozygote.
4. The ABO blood system also displays another common phenomenon—that any gene may have more than ________ alleles.
5. Individuals with type O blood carry two copies of the ________ allele for blood type.

8.4 Sex and Inheritance

Rare diseases like Krabbe were first identified as resulting from genetic mutations by examining the patterns of inheritance of these conditions through families. Examining these patterns also provided insight into another phenomenon that influences the prevalence of certain conditions: the role of chromosomes in determining sex.

Sex Determination and X-Linkage

Of the 23 pairs of chromosomes present in the cells of humans, 22 pairs are **autosomes,** or nonsex chromosomes, and one pair are the **sex chromosomes.** The sex chromosomes are named for their shapes when visualized under the microscope during cell division: the X chromosome is much larger and carries many more genes than the Y chromosome. Males have 22 pairs of autosomes and one X and one Y sex chromosome. Females also have 22 pairs of autosomes, but their sex chromosomes comprise two X chromosomes.

The X and Y chromosomes are involved in establishing the sex of an individual through a process called **sex determination.** Men produce sperm cells containing one of each autosome and either an X or a Y chromosome. Females also produce gametes with 22 unpaired autosomes, but the sex chromosome in an egg cell is always one of two X chromosomes. Therefore, the sperm cell determines the genetic sex of the offspring in humans. If an X-bearing sperm unites with an egg cell, the resulting child will be female (XX). If a sperm bearing a Y chromosome unites with an egg cell, the resulting child will be male (XY) (**FIGURE 8.19**).

Among the 50 to 60 genes on the Y chromosome in humans is the *SRY* gene (for sex-determining region of the Y chromosome). The expression of this gene triggers a series of events leading to development of testes, the male sex organs, around 8 weeks after conception. In the absence of *SRY*, an embryo will develop into a female. Not all organisms have the same system of sex determination as humans do, as is illustrated in TABLE 8.3 on the next page.

Conditions that appear more common in either males or females are called **sex-linked traits.** Some of these traits have a genetic basis thanks to their location on one of the sex chromosomes. The fact that males have only one X chromosome means that they have only one copy of any **X-linked** gene; thus males are more likely to be affected by conditions caused by recessive alleles on the X chromosome. Red-green color blindness (that is, the inability to distinguish

VISUALIZE THIS

How and why would the X chromosomes in two egg cells shown below differ from each other?

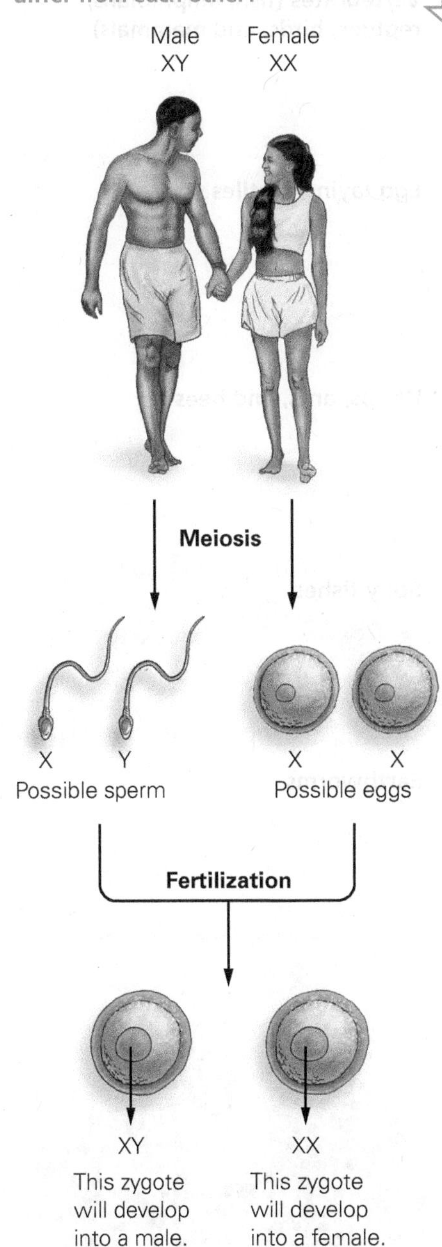

FIGURE 8.19 Sex determination in humans. Sperm and eggs each carry 22 autosomes but only one sex chromosome, resulting in a equal likelihood of male and female offspring.

TABLE 8.3 Sex determination strategies in some nonhuman organisms.

Type of Organism		Mechanism of Sex Determination
Vertebrates (fish, amphibians, reptiles, birds, and mammals)		In some vertebrates, the male has two of the same chromosomes and the female has two different chromosomes. In these cases, the female determines the sex of the offspring.
Egg-laying reptiles		In many egg-laying species, two organisms with the same suite of sex chromosomes could become different sexes. Sex depends on which genes are activated during embryonic development. For example, the sex of some reptiles is determined by the incubation temperature of the egg.
Wasps, ants, and bees		In bees, sex is determined by the presence or absence of fertilization. Males (drones) develop from unfertilized eggs. Females (workers and queens) develop from fertilized eggs.
Bony fishes		Some species of bony fishes change their sex after maturation. All individuals will become females unless they are deflected from that pathway by social signals such as displays of dominance.
Earthworms		Earthworms have both male and female reproductive organs, a condition referred to as hermaphroditism.

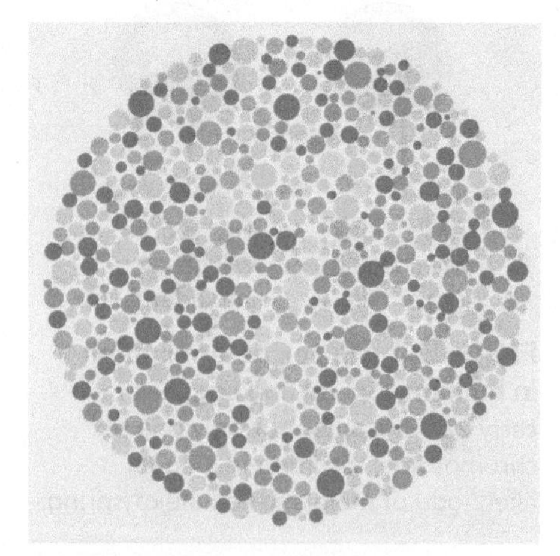

FIGURE 8.20 An X-linked trait. Individuals with red-green color blindness cannot see the number in this image. Because the gene for red-green color perception is on the X chromosome, about 8% of men have this form of color blindness, but only 0.5% of women.

between these colors), which affects about 8% of men and fewer than 1% of women, is an example of an X-linked trait (**FIGURE 8.20**).

New York's newborn screening panel tests for one exclusively X-linked disease, adrenoleukodystrophy (ALD). The cause and symptoms of ALD have similarities to Krabbe disease—a missing enzyme leads to a buildup of waste products that destroy myelin-producing cells in the brain, leading to progressive deterioration of function. Boys with ALD rarely live longer than 10 years after the onset of symptoms.

Most women carrying the ALD allele will not even realize that they are a carrier of the allele until their son is diagnosed (**FIGURE 8.21a**). ALD-affected males will always and only pass the ALD allele on to their daughters, who can then pass it on to their sons (**FIGURE 8.21b**).

Pedigrees

How do we know that ALD is associated with a gene on the X-chromosome? The sex linkage of these gene was first suspected as case reports began to accumulate in the medical literature; its location was confirmed in 1981 by tracing the lineage of ALD through families using a chart called a pedigree.

A **pedigree** is a family tree that follows the inheritance of a genetic trait for many generations of relatives. This tool is used in studying human genetics

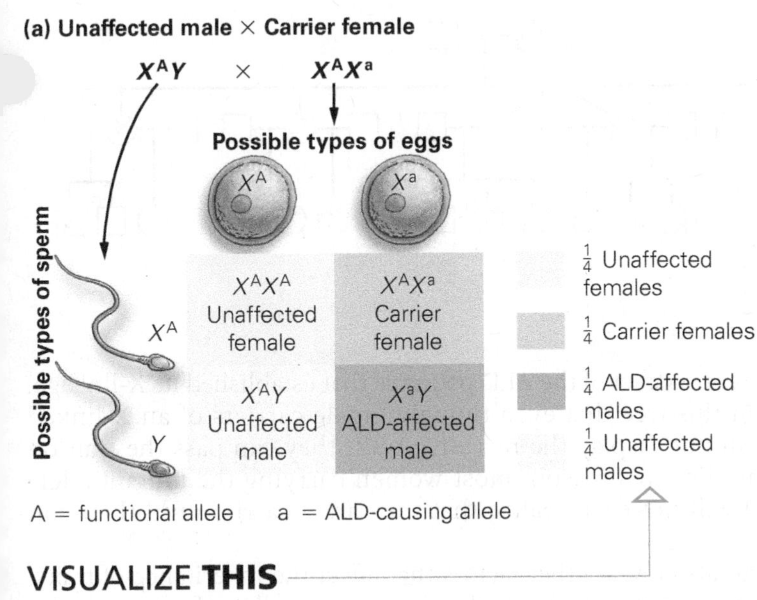

VISUALIZE THIS

What genetic cross would result in the highest frequency of affected males?

FIGURE 8.21 Genetic crosses with an X-linked trait. Cross (a) shows possible outcomes and probabilities of a mating between an unaffected male and a female carrier of adrenoleukodystrophy (ALD); (b) shows possible outcomes and probabilities of a cross between an affected male and an unaffected female.

because it is impossible to set up controlled matings between humans the way one can with fruit flies or plants; pedigrees allow scientists to study inheritance by analyzing matings that have already occurred. **FIGURE 8.22** shows how scientists can use pedigrees to determine whether a trait is inherited as autosomal dominant or recessive.

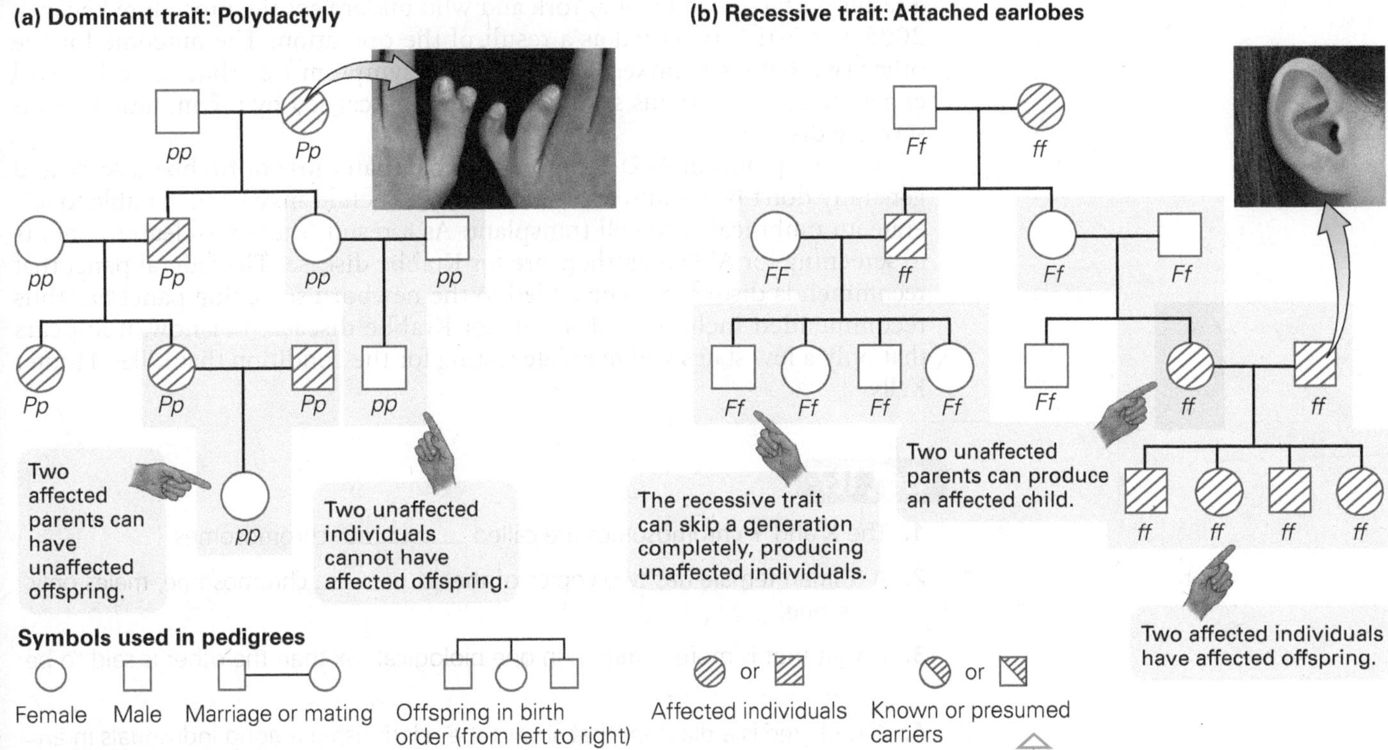

FIGURE 8.22 Pedigrees showing different modes of inheritance. (a) Polydactyly is a dominantly inherited trait. People with this condition have extra fingers or toes. (b) Having attached earlobes is a recessively inherited trait.

VISUALIZE THIS

Use the pedigree symbols to draw a pedigree of a woman who had a son with her one partner and two boys with a second partner.

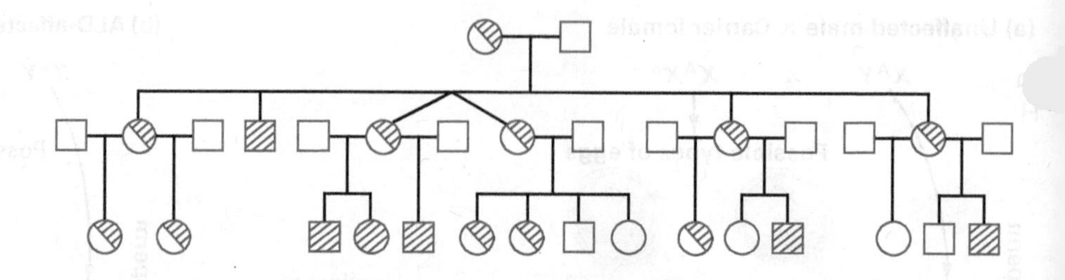

FIGURE 8.23 A pedigree of adrenoleukodystrophy. The pedigree summarized in this figure helped to firmly establish that ALD was an X-linked disorder.

FIGURE 8.23 is an extract of the ALD pedigree that established its X-linkage. You can see from this tree that even though female carriers of an X-linked recessive trait will not display the recessive trait, they can pass the trait on to their offspring. For this reason, most women carrying the mutant allele that causes ALD will not even realize that they are a carrier until their son becomes ill.

A pedigree can also reveal other factors that affect the inheritance of traits, including the fact that not all individuals who have the alleles for a particular condition actually will express that condition. (Chapter 9 examines the question of the limits on gene expression a little more closely.)

Both ALD and Krabbe disease can be potentially treated by a transplant of cells collected from the umbilical cord of an unrelated donor. The transplanted cells can "take over" normal enzyme production and thus prevent the buildup of dangerous waste products in myelin-producing cells. This treatment must occur before symptoms manifest however, because these symptoms are a sign of irreversible damage. And the transplant procedure is not without risk; of the four newborns who were found to have Krabbe disease as a result of the screen in New York and who underwent the procedure between 2006 and 2014, two died as a result of the operation. The outcome for the other two babies is mixed: one is mostly symptom free, but the other still experienced the nervous system degeneration caused by the mutation and is severely disabled.

The symptoms of ALD are more variable than those of Krabbe disease and generally don't begin until later in life, when a child may be better able to tolerate an umbilical cord cell transplant. As a result, there are clearer benefits to screening for ALD than there are for Krabbe disease. The federal panel that recommends disorders to be added to the newborn screening panel has thus recommended including ALD but not Krabbe disease. For now, it appears that only a few states will mandate testing for the condition that killed Hunter Kelly.

Got It?

1. The X and Y chromosomes are called ____________ chromosomes.
2. A human female has two copies of the ____________ chromosome; males only have one.
3. A trait that is more common in one biological sex than the other is said to be ____________ ____________.
4. A pedigree is a diagram that shows the relationship among individuals in an extended family along with their ____________.
5. If a trait is X-linked, it should show up on pedigree as appearing almost exclusively in ____________ who are children of unaffected ____________.

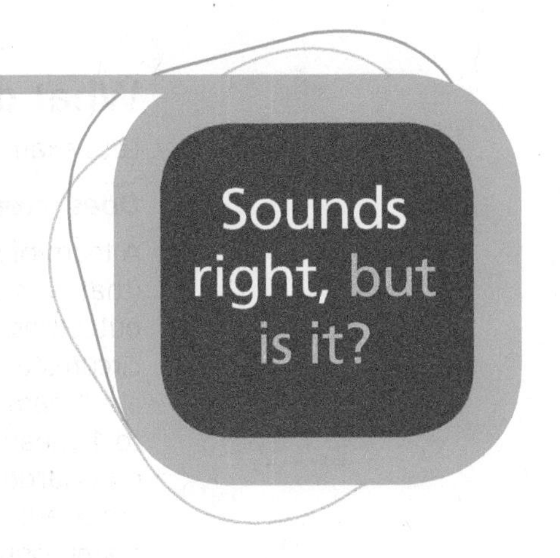
Sounds right, but is it?

In the Harry Potter books and movies, many of the characters who knew Harry's parents tell him that he resembles his mother or note his similarity to his father in his willingness to bend the rules. To many fans, these comments make sense, because a child receives half of his genetic information from his mother and half from his father. Thus, it seems fair to say that:

Harry Potter has his mother's eyes.

Sounds right, but it isn't.

Answer the following questions to understand why.

1. Do you think it is more likely that the color and shape of a person's eyes are determined by one gene or many genes?
2. Did Harry receive copies of genes that determine eye color and shape from his mother?
3. Did he receive copies of genes that determine eye color and shape from his father?
4. Think back to the Punnett squares you've viewed and drawn. Do genes for only one or both parents likely influence eye color and shape?
5. Reflect on your answers to questions 1–4. Explain why the statement bolded above sounds right, but isn't.

Should there be universal screening for killer diseases?

THE BIG QUESTION

The American Cancer Society recommends that adults undergo regular screening tests for various forms of cancer, including colon cancer in both sexes, prostate cancer in men, and cervical and breast cancer in women. The benefits of cancer screening, as with newborn screening, is that affected but symptom-free individuals may receive timely life-saving treatment; the risks include the anxiety created by false positive results and "overtreatment" of a cancer that would not have been deadly if undetected.

What should I know?

What follows are some smaller questions that need to be resolved to answer the Big Question. Place a checkmark next to the questions that science can answer.

Small Questions	Can Science Answer?
Are there health risks associated with screening for cancer?	
Does screening for cancer improve the health and survival of screened individuals?	
Does identifying affected individuals improve outcomes for other individuals in the same family?	
Is the cost of screening for cancer less than the cost of a later diagnosis?	
Do all people deserve to know whether they are possibly affected by a killer disease, regardless of cost?	

What does the science say?

Let's examine what the data say about this smaller question:

Does screening for cancer improve the health and survival of screened individuals?

A team of scientists reviewed studies on the effectiveness of breast cancer screening x-rays (that is, mammograms) on women's health. They examined the "odds ratio" for various outcomes, which is best understood as the likelihood of a particular outcome in one set of circumstances compared with another set of circumstances. In this case, the circumstances are "mammogram screenings" versus "no mammogram screenings." An odds ratio close to 1 means that there is no difference in the health outcome as a result of screening; a ratio larger than one indicates that the outcome is more common when mammograms are used; and a ratio smaller than one indicates that the outcome is less common when mammograms are used. Their analysis returned the following results:

Health Outcome	Odds Ratio, Mammogram versus None
Death due to breast cancer within 13 years	0.81
Death due to any cancer within 9 years	1.25
Death from any cause within 13 years	0.99
Cancer surgery	1.31

1. Describe the results. Does mammography screening have an effect on death rates due to cancer?
2. Given these data, do you think the smaller question is answered? If not, propose another study that would help answer this question?
3. Does this information help you answer the Big Question? What else do you need to consider?

Data source: P. C. Gøtzsche and K. J. Jørgensen, "Screening for Breast Cancer with Mammography," *Cochrane Database of Systematic Reviews* no. 6 (2013): Art. No.: CD001877.

Chapter Review

Mastering Biology

Go to Mastering Biology to access **eText 2.0, Dynamic Study Modules,** and the **Study Area,** where you'll find practice quizzes, BioFlix™ animations, MP3 tutor sessions, current events, and more.

SUMMARY

Section 8.1

Describe the relationship between genes, chromosomes, and alleles.

- Children resemble their parents in part because they inherit their parents' genes, segments of DNA that contain information about how to make proteins (pp. 143–144).
- Chromosomes contain genes. Different versions of a gene are called alleles (p. 144).
- Mutations in genes generate a variety of alleles. Each allele typically results in a slightly different protein product (pp. 144–145).

Explain why, although each cell in your body contains identical genetic information, the cells produced by your body are different from each other.

- Although nearly all cells in an individual contain the exact same genetic information, the genes that are "read" in those cells differ. Even when the same gene is expressed in two different cells, their products may have very different effects (p. 144).

Define *segregation* and *independent assortment* and explain how these processes contribute to genetic diversity.

- Segregation separates alleles during the production of eggs and sperm, meaning that a parent contributes only half of its genetic information to an offspring (pp. 145–146).

- In independent assortment, individual chromosomes are sorted into eggs and sperm independent of each other, meaning that the subset of information passed on by parents is unique for each gamete (pp. 146–147).

SHOW YOU KNOW 8.1 Not all of the DNA in a cell codes for proteins—some DNA function as binding sites for gene promoters (proteins that "turn on" genes), other segments may provide structural support for chromosomes, and other parts may be meaningless. What do you think the effect of "misspellings" is on each of these nongene segments of DNA?

Section 8.2

Distinguish between homozygous and heterozygous genotypes and describe how recessive and dominant alleles produce particular phenotypes when expressed in these genotypes.

- An individual heterozygous for a particular gene carries two different alleles for the gene, and one who is homozygous carries two identical alleles (p. 149).
- A dominant allele is expressed even when the genotype is heterozygous, and a recessive allele is only expressed when the individual carries no copies of the dominant allele—that is, when it is homozygous recessive (pp. 149–150).
- Pleiotropy occurs when a single gene leads to multiple effects (p. 150).

Demonstrate how to use a Punnett square to predict the likelihood of a particular offspring genotype and phenotype from a cross of two individuals with known genotype.

- Each column heading in a Punnett square describes one of the possible gamete genotypes one parent in a cross can produce, and each row label describes the same information from the other parent. The boxes in the square thus represent all of the possible genotypes produced by the cross, and the frequency of any genotype in the square is equal to its expected frequency in the offspring produced by the cross. (p. 152).

SHOW YOU KNOW 8.2 A man can produce four different combinations of alleles in his sperm when considering two genes on two different chromosomes. How many different allele combinations within gametes can be produced when considering three genes on three different chromosomes?

Section 8.3

Differentiate incomplete dominance from codominance.

- Incomplete dominance occurs when the phenotype of progeny is intermediate to that of both parents (p. 155).
- Codominance occurs when both alleles of a given gene are expressed (pp. 155–156).

Outline the pattern of inheritance seen in the ABO blood system.

- The ABO blood system displays both multiple allelism (alleles I^A, I^B, and *i*) and codominance because both I^A and I^B are expressed in the heterozygote (p. 156).

SHOW YOU KNOW 8.3 Imagine a new allele I^C that is codominant with I^A and I^B and dominant to *i*. List all of the possible blood-group genotypes and phenotypes in a population containing all four alleles.

Section 8.4

Describe the mechanism of sex determination in humans.

- In humans, males have an X and a Y chromosome and can produce gametes containing either sex chromosome, and females have two X chromosomes and always produce gametes containing an X chromosome (p. 157).
- When an X-bearing sperm fertilizes an egg cell, a female baby will result. When a Y-bearing sperm fertilizes an egg cell, a male baby will result (p. 157).

Explain the pattern of inheritance exhibited by sex-linked genes.

- Genes linked to the sex chromosomes show characteristic patterns of inheritance. Males need only one recessive X-linked allele to display the associated phenotype. Females can be carriers of an X-linked recessive allele and may pass an X-linked disease on to their sons (pp. 157–158).

Explain the utility of a genetic pedigree.

- Pedigrees are charts that scientists use to study the transmission of genetic traits among related individuals (pp. 158–159).
- Pedigrees can help researchers identify whether a trait is dominant, recessive, sex-linked, or has other factors influencing its inheritance (pp. 159–160).

SHOW YOU KNOW 8.4 Color blindness is an X-linked recessive trait. The normal allele codes for the production of proteins called opsins that help absorb different wavelengths of light. A lack of opsins causes insensitivity to light of red or green wavelengths. From a genetic standpoint, what must be true of the parents of a female who is color blind?

ROOTS TO REMEMBER

The following roots of words come mainly from Latin and Greek and will help you to decipher terms:

di- comes from the Greek word for two. Chapter term: *dihybrid*

hemo- means blood. Chapter term: *hemoglobin*

hetero- means the other, another, or different. Chapter terms: *heterozygous, heterozygote*

pheno- comes from a verb meaning "to show." Chapter term: *phenotype*

pleio- comes from the Greek word for many. Chapter term: *pleiotropy*

-zygous derives from *zygote*, the "yoked" cell resulting from the union of an egg and sperm. Chapter terms: *monozygotic, heterozygote*

LEARNING THE BASICS

1. What is the relationship between genotype and phenotype?
2. Add labels to the figure that follows, which illustrates a portion of the human life cycle.

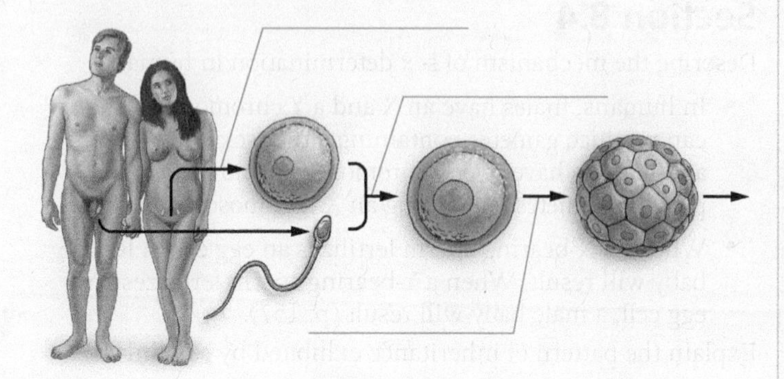

3. Which of the following statements correctly describe the relationship between genes and chromosomes?

   A. Genes are chromosomes; B. Chromosomes contain many genes; C. Genes are made up of hundreds or thousands of chromosomes; D. Genes are assorted independently during meiosis, but chromosomes are not; E. More than one of the above is correct.

4. An allele is a ______.

   A. version of a gene; B. dysfunctional gene; C. protein; D. spare copy of a gene; E. phenotype

5. Sperm and eggs in humans always ______.

   A. each have two copies of every gene; B. each have one copy of every gene; C. each contain either all recessive alleles or all dominant alleles; D. are genetically identical to all other sperm or eggs produced by that person; E. each contain all of the genetic information from their producer

6. Scientists have recently developed a process by which a skin cell from a human can be triggered to develop into a human heart muscle cell. This is possible because ______.

   A. most cells in the human body contain the genetic instructions for making all types of human cells; B. a skin cell is produced when all genes in the cell are expressed; turning off some genes in the cell results in a heart cell; C. scientists can add new genes to old cells to make them take different forms; D. a skin cell expresses only recessive alleles, so it can be triggered to produce dominant heart cell alleles; E. it is easy to mutate the genes in skin cells to produce the alleles required for other cell types

7. What is the physical basis for the independent assortment of alleles into offspring?

   A. There are chromosome divisions during gamete production; B. Homologous chromosome pairs are separated during gamete production; C. Sperm and eggs are produced by different sexes; D. Each gene codes for more than one protein; E. The instruction manual for producing a human is incomplete.

8. Among heritable diseases, which genotype can be present in an individual without causing a disease phenotype in that individual?

   A. heterozygous for a dominant disease; B. homozygous for a dominant disease; C. heterozygous for recessive disease; D. homozygous for a recessive disease; E. all of the above

9. A woman is a carrier of the X-linked recessive color blindness gene. She has children with a man with normal color vision. Which of the following is true of their offspring?

   A. All the males will be color blind; B. All the females will be carriers; C. Half the females will be color blind; D. Half the males will be color blind.

10. The pedigree in the figure below illustrates the inheritance of a sex-linked recessive trait. What is the genotype of individual II-5?

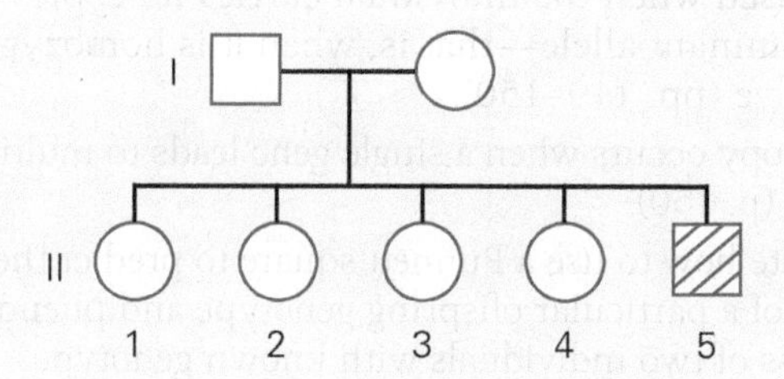

   A. X^HX^H; B. X^HX^h; C. X^hX^h; D. X^HY; E. X^hY

ANALYZING AND APPLYING THE BASICS

1. Two parents both have brown eyes, but they have two children with brown eyes and two with blue eyes. How is it possible that two people with the same eye color can have children with different eye color? If eye color in this family is determined by differences in genotype for a single gene with two alleles, what percentage of the children are expected to have blue eyes? If the ratio of brown to blue eyes in this family does not conform to expectations, why does this result not refute Mendelian genetics?
2. Children born with cystic fibrosis have higher rates of lung infection as well as difficulty absorbing nutrients from food digested in the small intestine. Explain how this disease has two very different effects.
3. Draw a pedigree of a mating between first cousins and use the pedigree to explain why matings between relatives can lead to an increased likelihood of offspring with rare recessive diseases.

GO FIND OUT

1. Use the Internet to search for a list of conditions on the newborn screening panel in your state. Choose one of these conditions to learn more about—including how common it is, what treatments exist for it, and whether it is tested for in many other states.
2. Down syndrome is not tested for in the newborn screening panel, but may be found by prenatal tests or is apparent at birth. This condition is caused by a mistake during meiosis and results in physical characteristics such as a short stature and distinct facial features as well as cognitive impairment (also known as mental retardation). Does the fact that Down syndrome is a genetic condition that results in low IQ mean that we should put fewer resources into education for people with Down syndrome? How does your answer to this question relate to questions about how we should treat individuals with other genetic conditions?

MAKE THE CONNECTION

The science that you learned in this chapter has helped you better understand the real-world example used throughout this discussion. Draw a line from the statement on the left to the science that supports it on the right.

Hunter Kelly inherited a fatal genetic disease that neither of his parents had.	Genes on different chromosomes are inherited independently of each other because of random alignment of chromosomes during meiosis.
Hunter's disease occurred because he could not eliminate a dangerous waste product from nervous system cells.	Mutations are mistakes that can occur in DNA replication and can result in the production of nonfunctional proteins.
Hunter had a 25% chance of having Krabbe disease, given the traits of his parents.	A pedigree is a diagram that illustrates family relationships and also provides information about the inheritance of a disease of interest.
Hunter's two sisters did not have Krabbe disease.	An allele can be dominant, recessive, or incompletely dominant to other alleles, which affects how the trait appears in a family. Genes on the sex chromosomes also have unique patterns of inheritance because male and female sex chromosomes differ.
That Hunter was affected by Krabbe disease was unrelated to the fact that he also inherited green eyes.	A Punnett square helps predict the likelihood of any given genotype (and phenotype) from a cross between parents of known genotype.
Not all of the genetic diseases on the newborn screening test are inherited in the same way as Krabbe disease.	A recessive condition can appear in an offspring if he or she inherits two copies of a recessive allele from unaffected parents that carry only one copy each.
Studying the families of individuals affected by a genetic disease can help us understand the nature of the genes involved.	The genotype of a child for any given gene is independent of the genotype for other children of the same parents.

Answers to Got It?, Visualize This, Working with Data, Sounds Right, But Is It?, Show You Know, and **Chapter Review** questions can be found in the **Answers** section at the back of the book.

9 Biology of Wrongful Convictions

Jennifer Thompson-Cannino misidentified Ronald Cotton as her rapist.

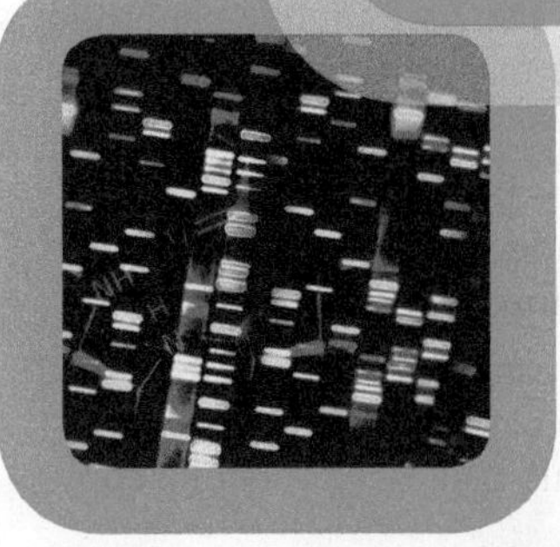

He was released after 10 years behind bars when DNA evidence proved his innocence.

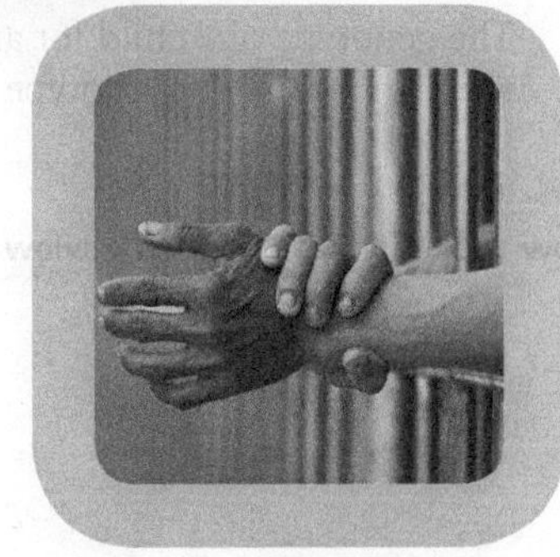

Can biology help us avoid other tragic wrongful convictions?

Complex Genetic Traits, Heritability, and DNA Profiling

In 1987, Ronald Cotton was convicted of rape and sentenced to life in prison. The case against Cotton seemed clear-cut; the victim of the assault testified that she was "100% certain" that he was her attacker. Cotton steadfastly maintained that he was not the perpetrator, and his case eventually gained the attention of The Innocence Project, an organization founded in 1992 with a mission "to free the staggering number of innocent people who remain incarcerated, and to bring reform to the system responsible for their unjust imprisonment." Thanks to the organization's help, Cotton was found innocent of the crime and was released from prison in 1995. Remarkably, he forgave his accuser and now works together with her and The Innocence Project to advocate for reforms to prevent future injustices.

Data from the National Registry of Exonerations show that 4.1% of defendants sentenced to death in the United States are later found to be innocent. Death sentence cases receive more scrutiny from the legal community than other types of criminal cases. If the number of wrongful convictions for other types of crimes is anywhere near 4% this means that tens of thousands of Americans are imprisoned for crimes they did not commit. An inmate who ends up being found not guilty—that is, who is exonerated—on average has already spent 14 years behind bars.

Wrongful convictions not only harm those convicted. Wrongful convictions allow the real perpetrator of the crime to go unpunished, possibly free to victimize others.

Why are so many individuals wrongfully convicted? The answer is in part because of the inherent flaws in eyewitness testimony. Additionally, the expectations of police, prosecutors, judges, and juries may also play a role. As the exonerated prisoners freed by the work of The Innocence Project know, the science of biology can also lead to the truth.

9.1 Eyewitness Testimony and Complex Genetic Traits

Testimony from eyewitnesses can be used in courts to secure convictions. Such testimony typically relies on physical characteristics we can see, like eye color, skin color, and height. These traits do not display the simple on-off pattern Mendel studied in his peas—discrete phenotypes determined by one gene, like round versus wrinkled seeds; instead, these traits show a large range of phenotypes in a population. This is, in part, because these complex traits are acted on by more than one gene.

Polygenic Traits

Traits that are influenced by more than one gene are called **polygenic traits.** Eye color in humans is an example of a polygenic trait because at least two genes—and perhaps as many as 15—are involved in determining this one trait. Although scientists do not completely understand all of the genetic contributions to eye color, it is clear that at least one gene is involved with the production of the brown pigment melanin and another gene helps distribute the melanin to the iris. When different alleles for these and other related genes interact, a range of eye colors, from dark brown (lots of melanin) to pale blue (very little melanin), is found in humans.

poly- means many.

As you can see from **FIGURE 9.1**, saying that someone has blue eyes does not definitively identify them for several reasons. First, because there are many people with blue eyes and second, because of the additive effects of all these genes, there are many different shades of blue eyes.

Some of the same genes that influence eye color also affect skin color, again those involving melanin production and distribution—as a result, this trait also

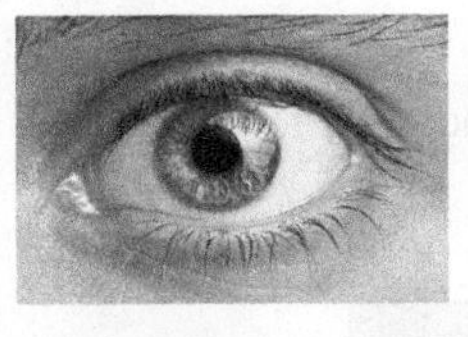
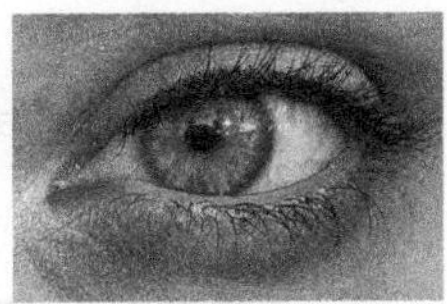
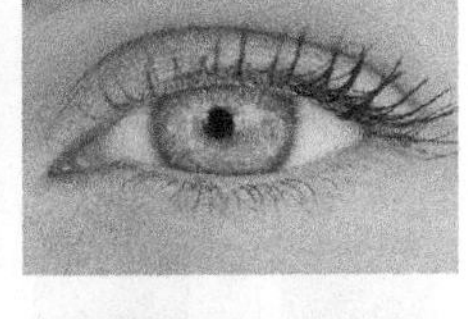
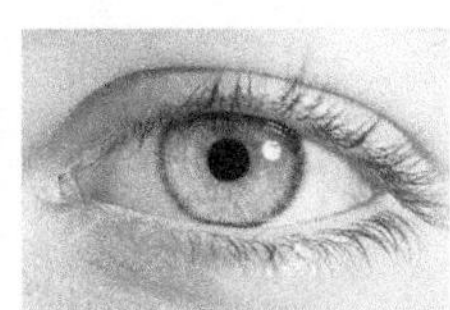
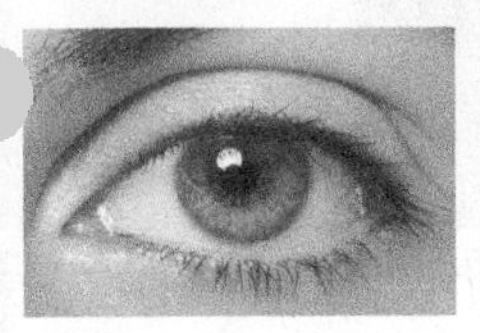
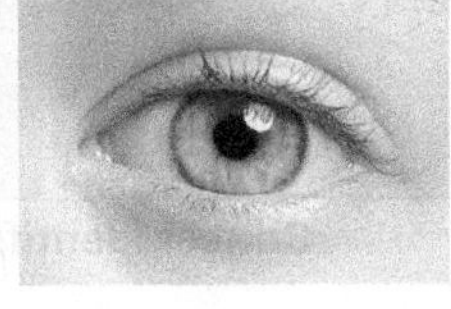
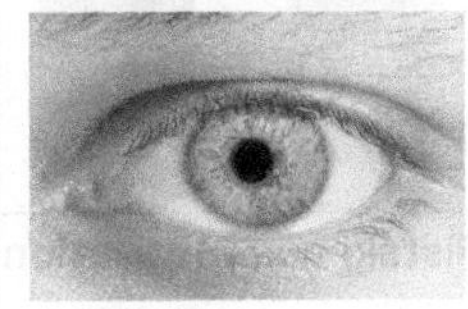
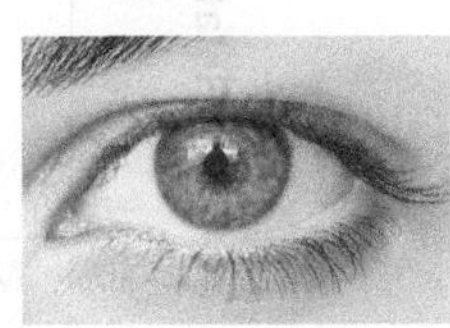
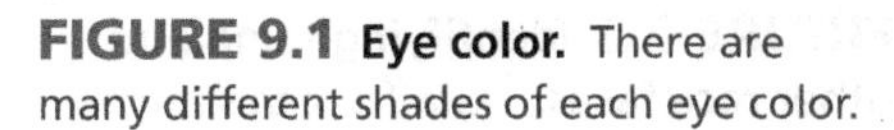

FIGURE 9.1 Eye color. There are many different shades of each eye color.

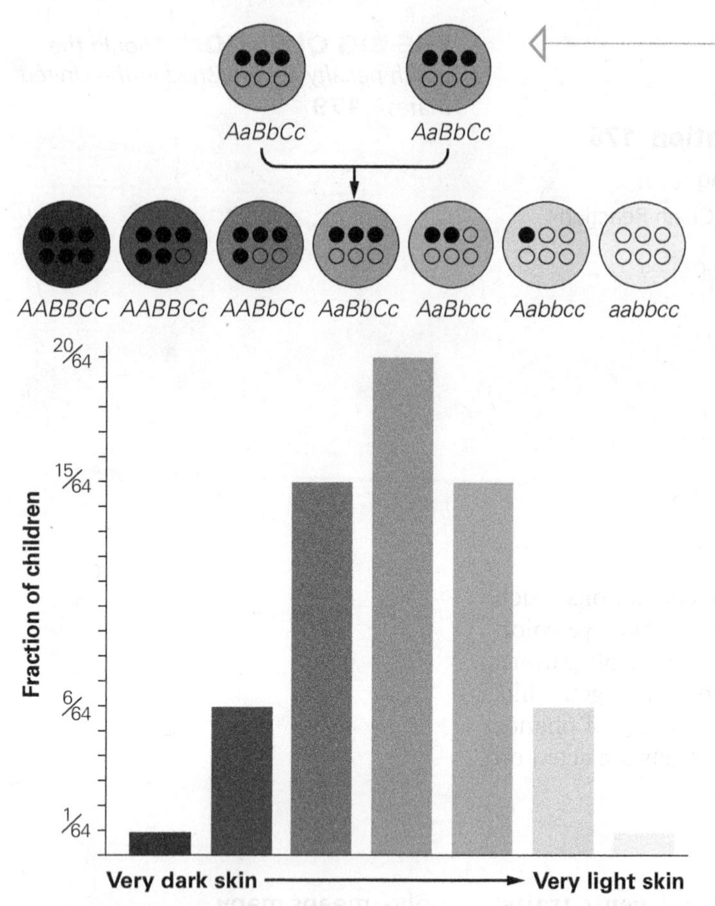

FIGURE 9.2 Skin color is influenced by at least three genes. One hypothesis for how skin color is determined involves three genes, each with two alleles. If true, we would expect a wide range of skin colors to be possible. As we saw with eye color, this explanation is not completely satisfactory. There must be more going on genetically than three genes with an additive effect because we see more than seven distinct skin colors in humans.

VISUALIZE **THIS**

If a light-skinned man has one dominant allele for skin color and his dark-skinned wife has five dominant alleles, together they can produce children with how many dominant alleles?

shows a lot of variation (**FIGURE 9.2**). Perhaps even more so than eye color, identifying an individual's skin color can be tricky. People identified as "black" in the United States possess quite a large range of skin tones. Scientists have shown that identifying someone who has a different skin color than your own can present difficulties. A recent meta-analysis involving 39 studies and more than 5000 participants showed that people are much better at correctly identifying a photo of a person of their own race than of a person who would be categorized as a different race (**FIGURE 9.3**). This phenomenon adversely affects eyewitness testimony when the victim has one skin color and the perpetrator another, as in the case of Ronald Cotton and his accuser.

Genes aren't the only factor that influences skin color, of course; skin can also be temporarily altered by the environment. Exposure to sunlight can cause the skin to darken, as melanin darkens and its production at the skin's surface increases to protect cells from the damaging effects of UV light.

Quantitative Traits

Traits that involve the actions of many genes that also interact with the environment are called **quantitative traits.** Consider height, for example: Any individual's height is a product of both the genes he or she inherited and environmental conditions such as nutrition during childhood. Therefore, there is no exact height a given set of genes will produce in a given individual. Instead there is the height the individual will become as a result of genes expressed in a particular environment. In other words, the same individual could grow to a range of different heights, based on his or her environment.

The range of different phenotypes produced in a population when a trait is controlled by many different genes is referred to as *continuous variation*, which can best be seen on a graph. These data often take the form of a curve called a *normal distribution*. **FIGURE 9.4a** illustrates the normal distribution of heights of men in a college class. Each individual is standing behind the label (at the bottom) that indicates his height. The curved line drawn across the photo summarizes these data—note the similarity of this curve to the outline of a bell, leading to its common name, a bell curve.

A bell curve contains two important pieces of information. The first is the highest point on the curve, which generally corresponds to the average, or **mean,** value for data. The mean is calculated by adding all of the values for a trait in a population and dividing by the number of individuals in that

FIGURE 9.3 Identifying same and other race faces. Data from this meta-analysis shows that when an eyewitness makes a mistaken identification of crime suspect in a series of mug shots or police lineup, it more often involves someone from a different race. In contrast, when a witness correctly identifies a face, it is more often of someone in the same race.

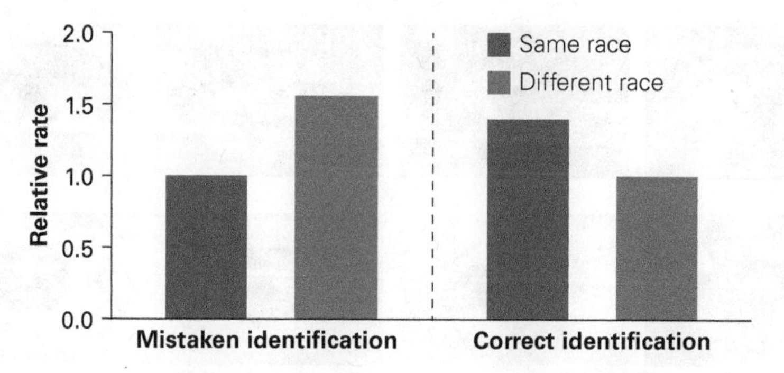

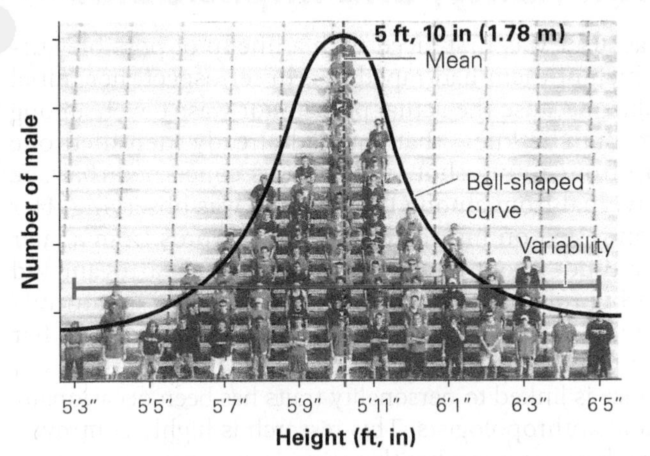

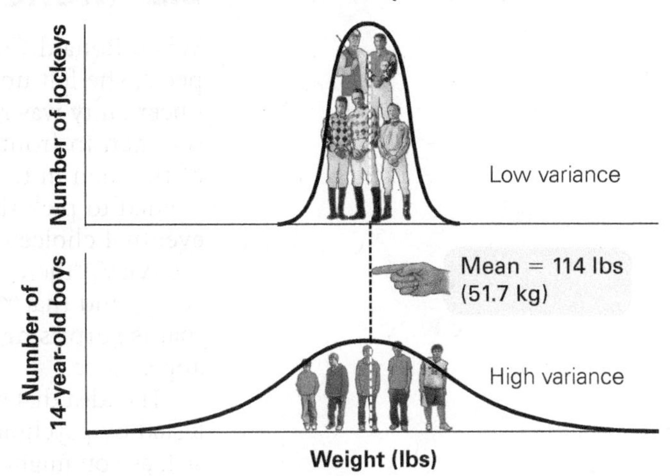

FIGURE 9.4 Height is a quantitative trait. (a) This photo of men arranged by height illustrates a normal distribution. The highest point of the bell curve is also the mean height of 5 feet, 10 inches. (b) Fourteen-year-old boys and professional jockeys have the same average weight—approximately 114 lbs. However, to be a jockey, you must be within about 4 lbs of this average. Thus, the variance among jockeys in weight is much smaller than the variance among 14-year-olds.

WORKING WITH **DATA**

(a) Does an average height of 5 feet, 10 inches, in this particular population imply that most men were this height? (b) Were most men in this population close to the mean, or was there a wide range of heights?

population. The second is in the width of the bell itself, which illustrates the variability of a population. The variability is described with a mathematical measure called **variance,** which is essentially the average distance any one individual in the population is from the mean. If a low variance for a trait indicates a small amount of variability in the population, a high variance indicates a large amount of variability (**FIGURE 9.4b**).

If a trait has low variance in a population, many individuals look relatively similar for that trait—making it less useful for eyewitness testimony. But a trait with high variance can lead to trouble, too, because focusing on a characteristic that seems uncommon may cause eyewitnesses to miss other important features. Cotton's accuser, for instance, chose him from a lineup based partially on the size and shape of his nose, which, in reality was longer and less broad than that of the actual rapist.

Because the variation among individuals in height is continuous it can be difficult for eyewitnesses to provide an accurate height estimate. As a way to combat such difficulties, many stores now have height strips on their door frames to help clerks estimate the height of a fleeing robber (**FIGURE 9.5**).

As we've seen, the continuous variation in the quantitative traits we use to identify people complicates the ability of an eyewitness to definitively identify a suspect. According to data collected by The Innocence Project, approximately 70% of exonerations were for convictions based on eyewitness misidentification.

FIGURE 9.5 An aid for eyewitnesses. A continuously variable trait like height is difficult to estimate without a reference point. Height strips in strategic locations can provide more accurate information to aid in crime solving.

Got It?

1. A trait that varies continuously, rather than having a distinct "on or off" character, is called a __________ trait.
2. The __________ point on a normal distribution typically corresponds to the mean.
3. The average distance any one individual is from the mean of the population is the __________.
4. Traits affected by more than one gene are referred to as __________.
5. Continuously varying traits may involve both many genes and the __________ interacting to produce a range of phenotypes.

9.2 Genes, Criminality, and Implicit Bias

When Ronald Cotton's accuser was first faced with a lineup of possible suspects, she felt unsure which of them had raped her. In retrospect, her initial uncertainty was reasonable, because the actual perpetrator wasn't even among the men in front of her. But she knew that she had already identified one of the men in the lineup from a mug shot, and she was concerned that she needed to pick that individual from among the men she was observing. Her eventual choice of Cotton rested on his demeanor. In her words from a later interview, "there was an attitude about him that was very, almost arrogant and smug, and that played a big role in my decision." She expressed an attitude that is surprisingly common—that you can "spot" a criminal from his or her appearance.

The idea that appearance is linked to personality traits has been occasionally tested by psychologists and anthropologists. This research is highly controversial, as you might imagine, because it implies that there is an innate component to these traits. After all, our appearance is in large part due to genetic factors, and if a particular appearance is linked to a particular behavior, that behavior must have a genetic component as well. Because a behavior like criminality is seen in a range across a population, it is another example of a quantitative trait. Few would argue that environmental influences have no effect on the development of criminality. But how can we calculate how important genes are, especially those related to appearance?

Studying Nature versus Nurture

To determine the role of genes in determining any quantitative trait, scientists calculate the **heritability** of the trait. To estimate heritability in most populations, researchers use correlations between individuals with varying degrees o genetic similarity. A correlation determines how accurately one can predict the measure of a trait in an individual when its measure in a related individual is known. For example, **FIGURE 9.6** shows a correlation between parent birds and their offspring in the strength of their response to a tetanus vaccine. An individual that responded strongly produced a large number of antitetanus proteins, called antibodies, and ones that responded weakly produced a lower number of antibodies.

WORKING WITH **DATA**

How would a graph that showed a *low correlation* between parents and offspring differ from this graph?

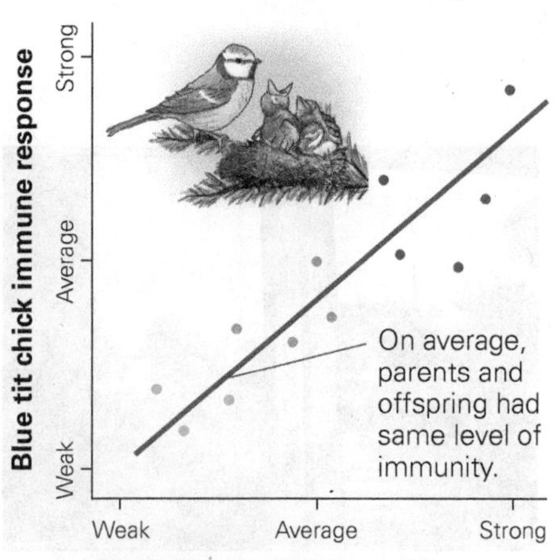

FIGURE 9.6 Using correlation to calculate heritability. The close correlation of immune system response between parents and offspring in the blue tit, a European bird, indicates that immune response is highly heritable.

As you can see from the graph, parent birds with weak responses tended to have offspring with weak responses, and parents with strong responses had offspring with strong responses. This strong correlation indicates that the ability to respond to tetanus is highly heritable—most of the difference between birds in their immune system response results from genetic differences.

Heritability in humans is measured similarly. Studies calculate how similar or different parents are to their children, or siblings are to each other, in the value of a particular trait. When examined across an entire population, the strength of a correlation can provide a measure of heritability (**FIGURE 9.7**). However, parents and the children who live with them are typically raised in a similar social and economic environment. As a result, correlations of criminality between the two groups cannot distinguish the relative importance of genes from the importance of the environment. This is the problem found in most arguments about "nature versus nurture"—do children resemble their parents because they are "born that way" or because they are "raised that way"?

To avoid the problem of overlap in environment and genes between parents and children, researchers seek situations that remove one or the other overlap. These situations are called **natural experiments** because one factor is "naturally" controlled, even without researcher intervention. Human twins are one source of a natural experiment to test hypotheses about the heritability of quantitative traits in humans.

For most traits, such as body size, people come in a wide variety of types.

How many of our differences are due to the environment, and how many are a result of different genes?

Correlations between relatives (for instance, here between fathers and sons) can provide information about the importance of genes in determining variation among individuals.

FIGURE 9.7 Determining heritability in humans. Comparisons between parents and children can help us estimate the genetic component of a quantitative trait.

Identical twins are referred to as **monozygotic** twins because they develop from one zygote—the product of a single egg and sperm. Recall that after fertilization the zygote grows and divides, producing an embryo made up of many daughter cells containing the same genetic information. Monozygotic twinning occurs when cells in an embryo separate from each other. If this happens early in development, each cell or clump of cells can develop into a complete individual, yielding twins, or in very rare cases, triplets or quadruplets who carry identical genetic information (**FIGURE 9.8a**, on next page).

mono- means one.

In contrast to identical twins, nonidentical twins (also called fraternal twins) occur when two separate eggs fuse with different sperm. These twins are called **dizygotic**, and although they develop simultaneously, they are genetically no more similar than siblings born at different times (**FIGURE 9.8b**, on next page). In humans, about 1 in every 80 pregnancies produces dizygotic twins, while only approximately 1 of every 285 pregnancies results in identical twins.

di- means two.

By comparing the prevalence of a genetic trait in monozygotic twins with the prevalence in dizygotic twins, researchers can begin to separate the effects of shared genes from the effects of shared environments. Because twins raised in the same family have similar childhood experiences, one would expect that the only real difference between monozygotic and dizygotic twins is their genetic similarity.

The estimated heritability of criminality using twin data was performed using an enormous database of the Swedish population, collected since 1961. The correlations of criminality between different types of twins yielded a heritability of 0.55. In other words, by this estimate, 55% of the variability in criminality among humans is due to differences in genotypes.

So, if criminal behavior has a significant genetic component and appearance has a genetic component, perhaps some of these genes have effects on both

FIGURE 9.8 The formation of twins. (a) Monozygotic twins form from one fertilization event and thus are genetically identical. (b) Dizygotic twins form from two independent fertilizations, resulting in two embryos who are only as genetically similar as any other siblings.

VISUALIZE **THIS**

Conjoined twins form when the two individual twins share some body parts. Which type of twin formation is more likely to result in conjoined twins?

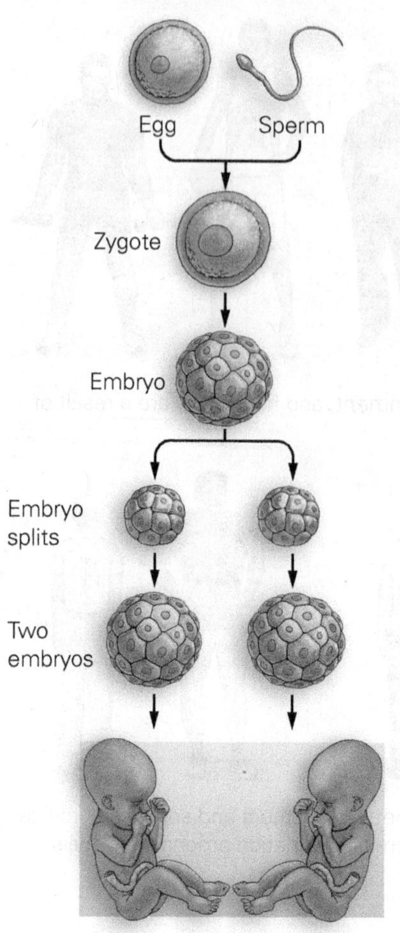

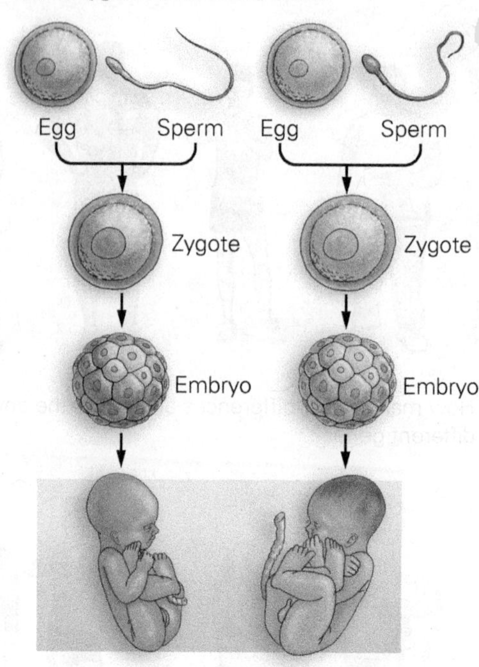

traits and thus it is possible to "see" criminality. In reality, there are very little data to support this idea. For example, a recent study measured facial characteristics in mug shots from the Michigan Department of Corrections and found no difference in face width—a trait hypothesized to be a marker of violent tendencies—between violent and nonviolent offenders. In addition, it is important to be very cautious when interpreting information about heritability, especially in traits such as behaviors that have important social consequences.

The Use and Misuse of Heritability

A calculated heritability value is unique to the population in which it was measured and to the environment of that population. Using heritability to measure the *general* importance of genes to the development of a trait requires judiciousness. The following sections illustrate why.

Differences between Groups May Be Entirely Environmental. A "thought experiment" can help illustrate this point. Body weight in laboratory mice has a strong genetic component, with a calculated heritability of about 90%. In a population of mice in which weight is variable, bigger mice have bigger offspring, and smaller mice have smaller offspring.

Imagine that we randomly divide a population of variable mice into two groups—one group is fed a rich diet, and the other group is fed a poor diet. Otherwise, the mice are treated identically. As you might predict, regardless of

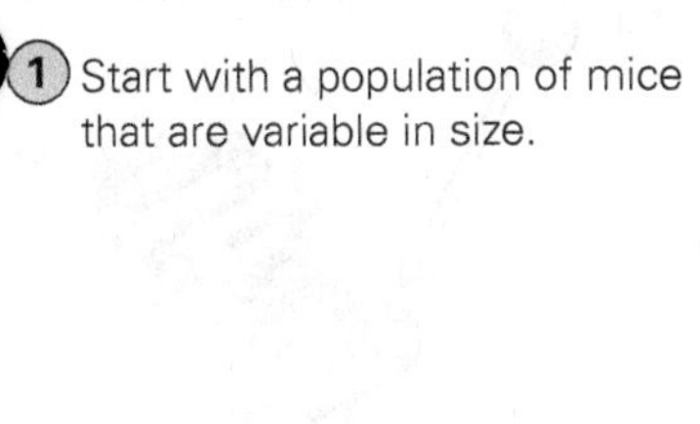

1 Start with a population of mice that are variable in size.

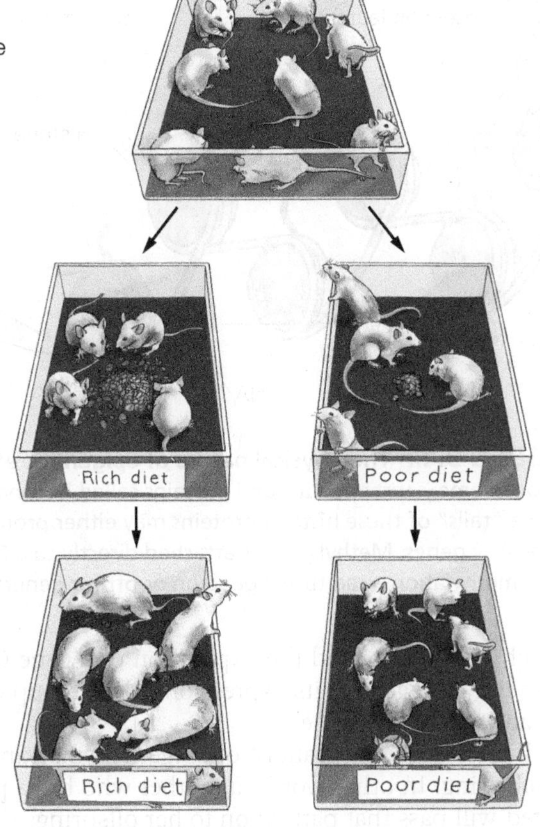

FIGURE 9.9 The environment can have powerful effects on highly heritable traits. If genetically similar populations of mice are raised in radically diverse environments, then differences between the populations are entirely due to environment.

VISUALIZE THIS

What would happen to the appearance of the mice in the next generation on both sides of this figure if all mice were switched back to the normal diet?

their genetic predispositions, the well-fed mice become fat, and the poorly fed mice become thin. Consider the outcome if we were to keep the mice in these same conditions and allowed the two groups to reproduce. Not surprisingly, the second generation of well-fed mice is likely to be much heavier than the second generation of poorly fed mice. Now imagine that another researcher came along and examined these two populations of mice without knowing their diets. Knowing that body weight is highly heritable, the researcher might logically conclude that the groups are genetically different. However, we know this is not the case—both are grandchildren of the same original source population. It is the environment of the two populations that differs (**FIGURE 9.9**).

Now extend the same thought experiment to human groups. Imagine that we have two groups of humans, and we have determined that criminality had high heritability. In this case, people in one group were affluent, and their average criminality was lower. The other group was impoverished, and their average criminality was higher. What conclusions could you draw about the genetic differences between these two populations? None. As with the laboratory mice, these differences could be entirely due to environment. The high heritability of criminality cannot tell us if two human groups in differing social environments vary in criminality because of variations in genes or because of differences in environment.

In fact, a developing understanding of the science of **epigenetics** has made it apparent that the environment plays a larger role in gene expression than scientists ever predicted. This role was convincingly described in 2003, when Dr. Randy Jirtle of Duke University demonstrated that females of genetically uniform agouti mice—a strain that was selected to be "innately" obese, and which had yellow fur—could be induced to produce offspring that were not obese and had brown fur by simply ingesting certain foods (**FIGURE 9.10**). These foods had the property of causing methyl side-groups to attach to portions of the DNA,

FIGURE 9.10 The powerful effects of environment on gene expression. These two *agouti* mice are genetically identical. This strain of mice was selected for innate obesity—the one on the (left) demonstrates the appearance of these mice when raised on a standard lab diet. The mouse on the (right) is the offspring an agouti mouse who was fed a diet supplemented with certain chemicals that changed the condition of its DNA molecules. The result is a radically different phenotype without any change in DNA sequence.

epi- means above or in addition to.

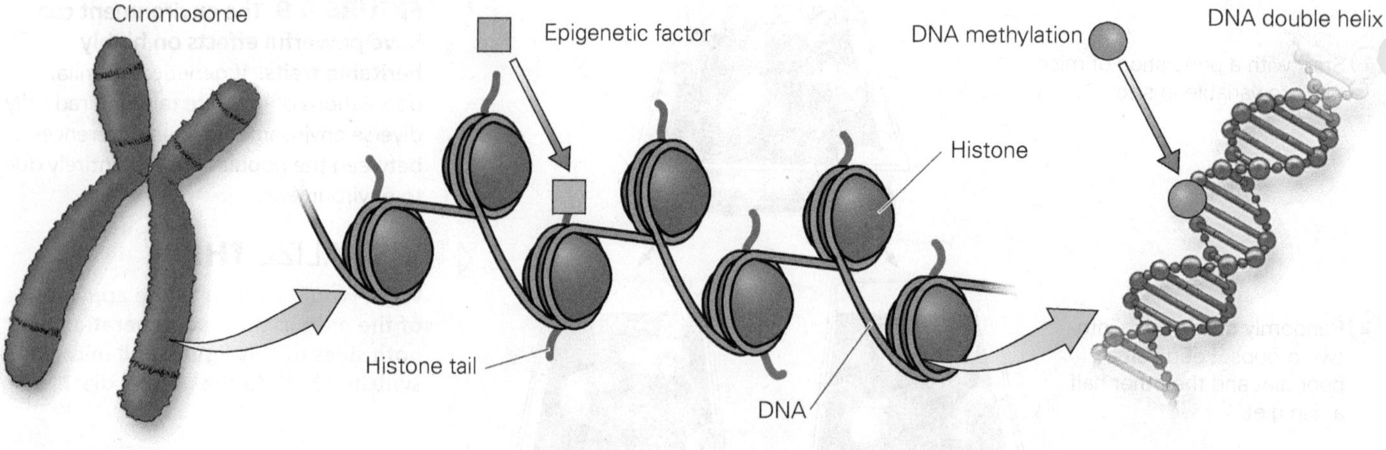

Chromosome
Epigenetic factor
DNA methylation
DNA double helix
Histone
Histone tail
DNA

FIGURE 9.11 The physical nature of epigenetic effects. The DNA in a chromosome is wrapped around proteins called histones. Epigenetic factors that attach to the "tails" of these histone proteins may either promote or inhibit the expression of nearby genes. Methyl groups attached directly to a DNA strand—a condition called DNA methylation—can turn a gene on or off for generations.

which in turn affected the expression of a gene (**FIGURE 9.11**). The gene itself does not change, but its expression is turned up or down when methyl groups are attached to the gene.

Interestingly, the pattern of epigenetic marking can be passed from parents to offspring. In other words, a mother who has a particular suite of genes methylated will pass that pattern on to her offspring.

A Highly Heritable Trait Can Still Respond to Environmental Change. A high heritability for a trait might seem to imply that this trait is not strongly influenced by environmental conditions. However, quantitative traits in other animals can be demonstrated to be both highly heritable and strongly influenced by the environment.

Rats can be bred for maze-running ability, and researchers have produced rats that are "maze bright" and rats that are "maze dull." Maze-running ability is highly heritable in the laboratory environment; that is, bright rats have bright offspring, and dull rats have dull offspring. The results of an experiment that measured the number of mistakes made by maze-bright and maze-dull rats raised in different environments are presented in **TABLE 9.1**.

In the typical lab environment, bright rats were much better at maze running than dull rats. But in both a very boring or restricted environment and a very enriched environment, the two groups of rats did about the same. In fact, no

TABLE 9.1 A highly heritable trait is not identical in all environments.

	Number of Mistakes in . . .		
Phenotype	Normal Environment	Restricted Environment	Enriched Environment
Maze-bright rats	115	170	112
Maze-dull rats	165	170	122
Explanation of Results	Maze-dull rats made more mistakes than maze-bright rats when running a maze.	Both groups made the same number of mistakes when running a maze.	Both groups made fewer mistakes when running the maze. The maze-dull rats improved the most.

FIGURE 9.12 The environment and genes. These identical twins have exactly the same genotype, but they are quite different in appearance due to environmental factors. The twin on the right was a life-long smoker and sun tanner, while the one on the left never smoked and spent less time in the sun.

rats excelled in a restricted environment, and all rats did better at maze running in enriched environments, with the duller rats improving most dramatically.

What this example demonstrates is that we cannot predict the response of a trait to a change in the environment, even when that trait is highly heritable.

Heritability Does Not Tell Us Why Two Individuals Differ. High heritability of a trait is often presumed to mean that the difference between two individuals is mostly due to differences in their genes. However, even if genes explain 90% of the population variability in a particular environment, the reason one individual differs from another may be entirely a function of environment (**FIGURE 9.12**). Currently, for example, there is no way to determine if a particular individual is antisocial because of genes, a poor environment, or some combination of both factors.

Implicit Bias

There is no reliable correlation between appearance and behavior. Nor is it likely that most criminals are "born that way." It is also important to recognize that our brains continually fool us on these points. That is, we have all developed subconscious frameworks that affect our feelings about other people based only on what we physically observe. This framework develops over our lifetimes through exposure to direct and indirect associations—it is a convenient shortcut our brains perform. This kind of shortcut makes sense—our ancestors often had to make quick decisions about, for instance, the safety of a location based on subtle physical clues. They learned the clues to look for from their own experience and the shared experience of others. This subconscious framework is known as an **implicit bias.** Implicit bias in our judgments about people takes the form of stereotypes operating below the level of our consciousness (**FIGURE 9.13**).

FIGURE 9.13 A test for implicit bias. The level of implicit bias a research participant has can be measured in a computer association test. Images or words flash on the screen that belong to defined categories. Participants tap one of two keys to classify any image or identify the position of a particular word (in the case of the test pictured, the category is "female"). If a participant has an implicit bias, he or she will take slightly longer to choose the correct key if the relationship is in opposition to their implicit bias. The test can be used to examine any stereotype. In some cases, the categories are racial groups and the attributes are "good" and "bad."

Even if we don't articulate it the way Cotton's accuser did, we all have an implicit bias that suggests to us what a criminal "looks like." This occurs whether or not we've ever met a violent criminal in our lives. Implicit bias clearly can play a role in wrongful convictions.

Got It?

1. The heritability of a trait is an estimate of the role of ________ variation in determining phenotypes for this trait.
2. Heritability of a trait can be measured by the ________ of phenotypes between relatives.
3. Twins that were produced by the fertilization of two different eggs are known as ________ twins.
4. The ________ and a high value for heritability are both important in determining phenotype.
5. Epigenetic changes are caused by environmental factors that affect gene ________.

9.3 Positive Identification

DNA evidence has been at the root of most of the several hundred exonerations obtained by The Innocence Project. DNA evidence is very powerful because it unambiguously identifies an individual—it is not subjective in the way that eyewitness accounts can be, nor does it suffer from the risk of implicit biases.

DNA Profiling

DNA profiling is a technique that uses differences in DNA sequence to positively identify an individual. When it first came into use, the technique was called DNA fingerprinting, because it helps identify people in the same way actual fingerprinting can. However, because this technology can do more than identify an individual, scientists started referring to DNA fingerprinting as DNA profiling.

Let's use another true-crime example to see the power of DNA profiling. In 1984, Earl Washington was wrongly convicted and sentenced to death for the rape and murder of a young woman in the state of Virginia. More than a decade later, his DNA was tested against DNA found in semen at the crime scene.

In Washington's case, as in all DNA profiling cases, a small subset of a person's DNA is tested. This is more efficient than analyzing all 300 billion plus base pairs of DNA found in cells. Scientists hone in on 13 sequences that all humans carry called **Short Tandem Repeats (STRs).** STRs consist of adjacent repeats of short DNA sequences (four or five bases) and are sprinkled between the gene coding sequences of all chromosomes. These 13 sites are typically used in DNA profiling because they are known to vary in number of repeats in the human population. For example, in a specific location of a particular chromosome, Washington might have had two repeats of the sequence GATC. At the same chromosomal location, the actual rapist might have five such repeats.

For a particular STR—say, GATC—scientists know the percentage of the population that carries one repeat (GATC), two repeats (GATCGATC), three repeats (GATCGATCGATC), and so on.

Polymerase Chain Reaction

To perform DNA profiling, it is often necessary to use more DNA than what is easy to remove from a suspect or what was left at a crime scene. In such a case, cells are amplified by a process involving a **polymerase chain reaction**

(PCR). It is likely that DNA from Washington's cheek cells and from semen left at the crime scene was amplified using PCR. PCR is performed in test tubes inserted in an automated machine that continuously cycles through periods of heating and cooling (**FIGURE 9.14**). Heat will **denature** DNA by breaking the hydrogen bonds between the double-stranded DNA molecule, resulting in two individual single strands of DNA.

Once the DNA is denatured, a special heat-tolerant enzyme called *Taq* polymerase uses nucleotides included in the test tube to build new DNA using each of the separated strands as a template. This enzyme was given the first part of its name (*Taq*) because it was first isolated from *Thermus aquaticus*, a bacterium discovered in Yellowstone National Park. Geysers and hot springs in this park can reach temperatures as high as 200°F, which this enzyme can withstand.

The second part of the enzyme's name (polymerase) describes its synthesizing activity—it acts as a DNA polymerase. The polymerase requires a short sequence of DNA known as a primer to be present at the beginning of the region to be copied. Scientists know the sequence of the STR to be amplified, so finding the correct primer is not difficult. This cycle of heating and cooling the tube is repeated many times, with each round of PCR doubling the amount of double-stranded DNA present in the tube, and producing millions of copies of each STR region for scientists to analyze in just a few hours (**FIGURE 9.15**).

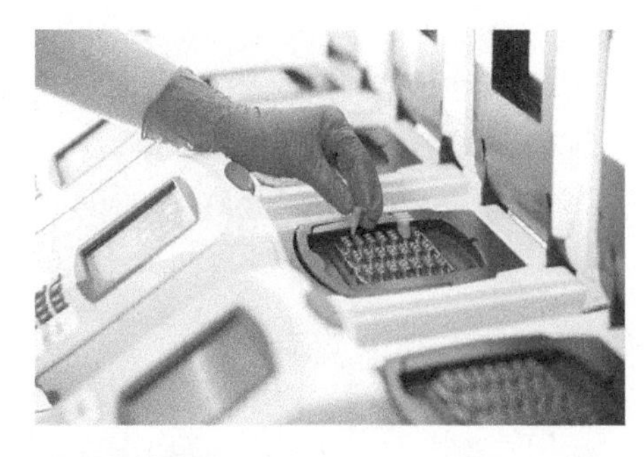

FIGURE 9.14 A thermal cycler. This machine undergoes temperature changes to allow DNA to unwind and serve as a template for the synthesis of new DNA.

① The required components are placed in a test tube and inserted into the thermal cycler.

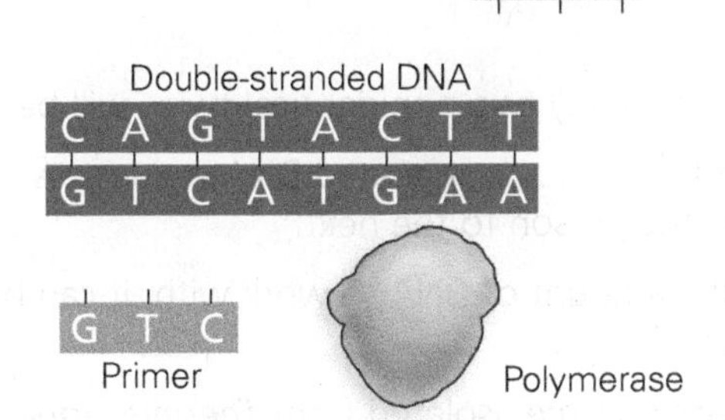

② Heating the DNA separates the double-stranded DNA into two separate single strands, allowing primers to bind to complementary sequences.

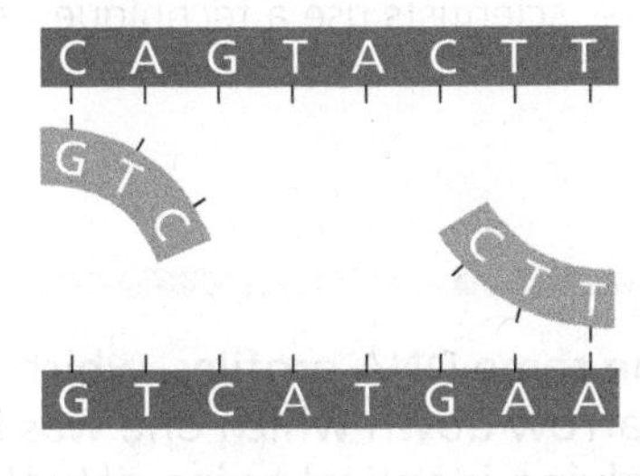

③ *Taq* polymerase uses the free nucleotides to initiate DNA synthesis, using the primer as a start site.

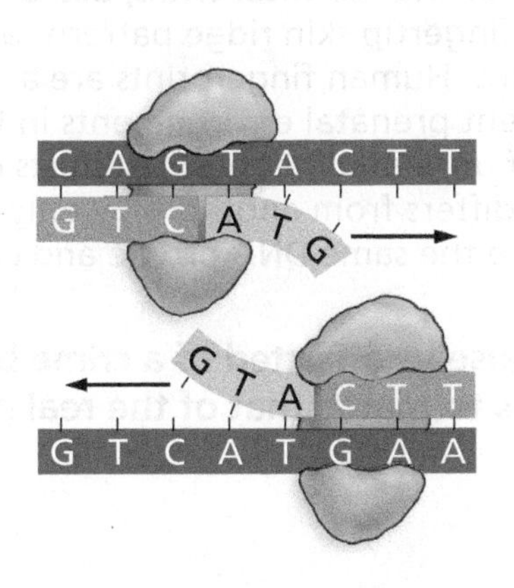

④ Two double-stranded molecules of DNA are produced from the one original DNA molecule.

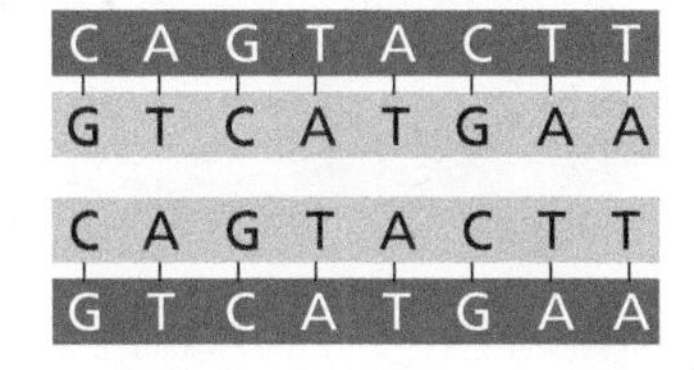

⑤ This process is repeated many times, with each cycle doubling the amount of DNA.

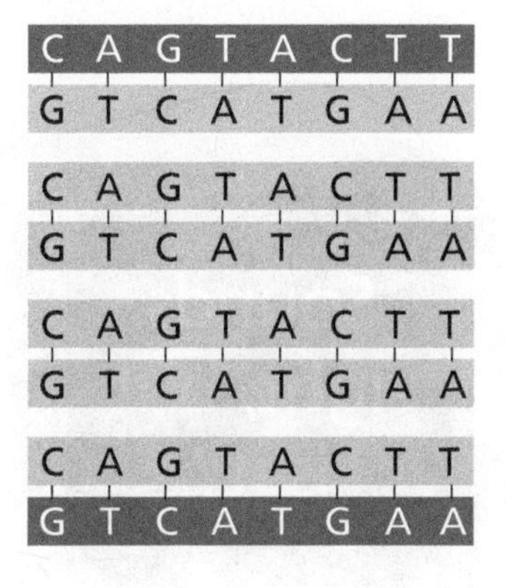

FIGURE 9.15 The polymerase chain reaction (PCR). PCR is used to make copies of DNA. Each round of PCR doubles the number of DNA molecules, yielding millions of copies of DNA in a short period of time.

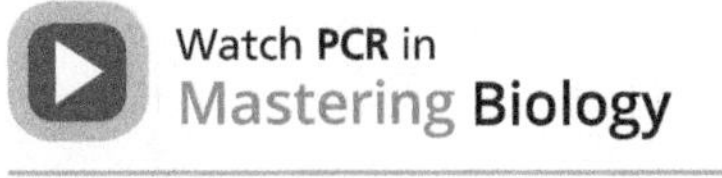

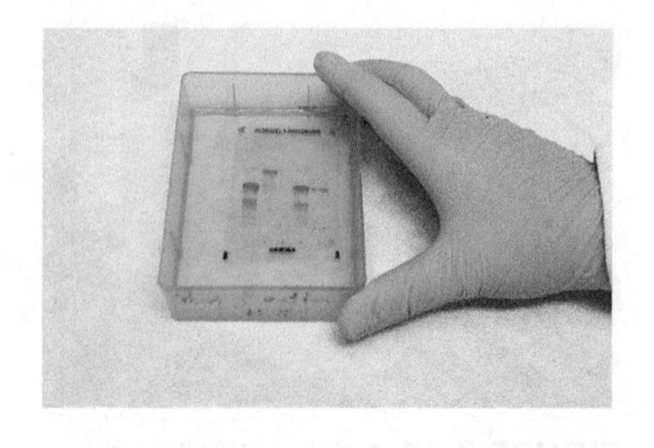

FIGURE 9.16 Agarose gel. Fragments of DNA, such as those produced by PCR, can be separated by placing them into an agarose gel and applying an electric current. Smaller DNA samples will migrate through the gel more quickly than larger ones. The DNA in this gel has been stained to show its location, which is a function of its length.

The mixture of an individual's copied STRs can then be separated by allowing the mixture to migrate through a solid support called **agarose gel,** which is similar in consistency to gelatin. When an electric current is applied, the gel impedes the progress of the larger DNA fragments more than it does the smaller ones (**FIGURE 9.16**). This type of size-based separation of DNA fragments using electric current is called **gel electrophoresis.**

In 1993, DNA profiling showed that Washington's DNA did not match the DNA found in the semen at the crime scene (**FIGURE 9.17**). After Washington's release from prison in 2000, a DNA profiling match was made to another man, who later pled guilty to the crime.

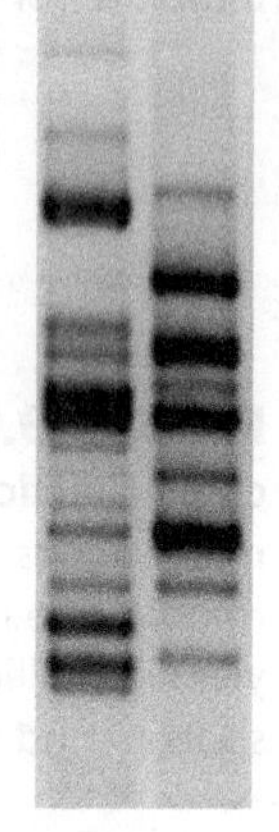

FIGURE 9.17 DNA profile. This DNA profile, while not the actual one from Earl Washington's case, shows results that could be obtained when DNA profiles are performed on two different people.

DNA profiling work by The Innocence Project has resulted in the release of 350 innocent prisoners and the identification of nearly 150 true perpetrators to date. Our understanding of biology—particularly of genes and DNA—has measurably advanced the cause of justice. By definitively identifying criminals and exonerating suspects, our growing knowledge has helped to correct for the errors of eyewitness misidentification—errors that we now understand better, thanks to science.

Got It?

1. DNA profiles from any two nonidentical twins will be ________.
2. STRs are short ________ repeated DNA sequences that differ in number of repeats from one person to the next.
3. To increase the amount of DNA to work with, it can be copied using a technique called ________ ________ ________.
4. A heat-tolerant enzyme, isolated from *Thermus aquaticus,* that synthesizes DNA is called *Taq* ________.
5. To separate DNA by size, scientists use a technique called ________ ________.

Sounds right, but is it?

Identical twins share the same DNA profiles, which makes it difficult to use DNA evidence to narrow down which one was involved in a crime. In some legal cases involving identical twins, actual fingerprint evidence can be used, because fingertip skin ridge patterns are unique even between identical twins. Human fingerprints are a quantitative trait and differ based on different prenatal environments in the womb. But what about DNA profiles of unrelated people? Scientists estimate that one human's DNA profile differs from another's by only 0.01%. Is it possible for two people to have the same DNA profile and could that result in a false conviction?

A person could be falsely convicted of a crime because his or her DNA profile happens to match that of the real perpetrator.

Sounds right, but it isn't.

Answer the following questions to understand why.

1. What nongene locations do scientists test for similar size when performing DNA profiling?
2. How many of these locations do scientists typically test?
3. For ease of calculation, let's use one average instead of 13 different STR frequencies. Assume that each of the STRs used in profiling occurs 10% of the time. What is the rough probability that two unrelated individuals would have the same 13 bands on a DNA profile?
4. There are approximately 7.4 billion people on Earth. How does this number compare with the probability of a match you calculated before?
5. Reflect on your answers to questions 1–4. Explain why the statement bolded above sounds right, but isn't.

Should the death penalty be abolished in the United States?

Capital punishment is a legal penalty in the United States and is used by the federal government and by 32 of the 50 states as punishment for the most heinous crimes. Since 1977, after the U.S. Supreme Court overturned a decade-long ban on the practice, more than 1400 people have been legally executed in the United States. However, the United States is one of only a few countries (there are 22 out of 195) in the world that permit capital punishment.

What should I know?

What follows are some smaller questions that need to be resolved to answer the Big Question. Place a checkmark next to the questions that science can answer.

Smaller Questions	Can Science Answer?
Does the death penalty reduce the incidence of murder?	
Does the death penalty cost less than alternative punishments, such as life in prison?	
Is the death penalty as currently applied "cruel and unusual punishment," which is prohibited by the U.S. Constitution?	
Is the death penalty sometimes wrongly applied to innocent people?	
Is lethal injection, the method by which the death penalty is applied, painful?	
Is it wrong for the state to impose death on any citizen, regardless of the crime they committed?	

What does the science say?

Let's examine what the data say about this smaller question:

Is lethal injection, the method by which the death penalty is applied, painful?

The lethal injection protocol—although it varies from state to state—includes an anesthetic to mask pain, a paralytic drug to prevent movement, and a chemical that stops the heart. Without anesthesia, a condemned individual during the execution process would experience burning pain, massive muscle cramping, and would be aware of gradual suffocation before the heart finally stopped. Researchers examined levels of anesthetic as measured in toxicology reports of inmates after execution. The bar graph below indicates the number of inmates at each blood level of anesthesia after execution; the text next to the bars lists what is known about the typical state of individuals at the associated anesthesia level.

—Continued next page

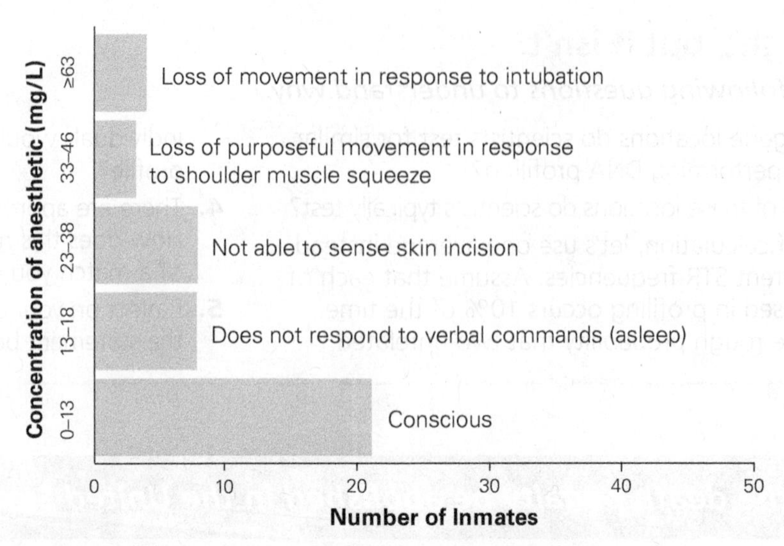

1. Describe the results. Does it appear that the levels of anesthetic used in the lethal injection protocol in these four states prevent the pain associated with the effect of the other drugs in all or most instances?
2. Given these data, do you think the smaller question is answered? If not, propose another study that would help answer this question.
3. Does this information help you answer the Big Question? What else do you need to consider?

Data Source: L. G. Koniaris, T. A. Zimmers, D. A. Lubarsky, and J. P. Sheldon, "Inadequate Anaesthesia in Lethal Injection for Execution," *The Lancet* 365, no. 9468 (2005):1412–1414.

Chapter Review

Mastering Biology

Go to Mastering Biology to access **eText 2.0, Dynamic Study Modules,** and the **Study Area,** where you'll find practice quizzes, BioFlix™ animations, MP3 tutor sessions, current events, and more.

SUMMARY

Section 9.1

Describe the nature of polygenic and quantitative traits.

- Polygenic traits are produced by the interactions of more than one gene. This interaction produces many different phenotypes (pp. 167–168).
- Quantitative traits involve interactions between many genes and the environment. The wide range of phenotypes produced show continuous variation in a distribution that approximates a bell curve (pp. 168–169).

SHOW YOU KNOW 9.1 The average height of American men has stayed the same over the past 50 years, while average heights in other countries (for example, Denmark) have increased. Height is a classic quantitative trait influenced by multiple genes, but also environmental factors such as nutrition during childhood. Provide both a genetic and an environmental hypothesis for why the height of the American population is not increasing along with other countries.

Section 9.2

Describe how heritability is calculated and what it tells us about the genetic component of quantitative traits.

- The role of genes in explaining the differences in the population seen in a quantitative trait is estimated by determining the heritability of the trait (pp. 170–171).
- Heritability is calculated by examining the correlation between parents and offspring or by comparing monozygotic twins with dizygotic twins (pp. 171–172).

Explain why a high heritability still does not always mean that a given trait is determined mostly by the genes an individual carries.

- Calculated heritability values are unique to a population in a particular environment. The environment may cause large differences among individuals, even if a trait has a high heritability (pp. 172–173).

- Environmental factors can modify DNA, causing epigenetic effects that may even be transmitted to the next generation. (pp. 173–175).

SHOW YOU KNOW 9.2 Some commentators have argued that given IQ's high heritability, policies that increase financial resources to failing schools will ultimately fail to increase achievement because such a predominantly genetic trait will not respond well to environmental change. Use your understanding of the proper application of heritability to refute this argument.

Section 9.3

Explain the significance of DNA profiling and how the process works.

- DNA profiling can be used to unambiguously identify a person. (p. 176).
- Short tandem repeats of DNA are found at specific chromosomal locations in all humans. The number of times a short fragment of DNA is repeated, however, is highly variable (p. 176).
- Many copies of the repeat sequences can be produced using the polymerase chain reaction (pp. 176–177).
- When the amplified repeated sequences are subject to gel electrophoresis, they will be separated by size and a banding pattern specific to the individual being profiled will emerge (p. 178).

SHOW YOU KNOW 9.3 Consider DNA profiles involving the analysis of eight separate STRs. If the DNA profiles of two nontwin siblings are compared, how many bands on a DNA profile would you expect them to have in common?

ROOTS TO REMEMBER

The following roots of words come mainly from Latin and Greek and will help you to decipher terms:

di- means two. Chapter term: *dizygotic*

epi- means above or in addition to. Chapter term: *epigenetics*

mono- means one. Chapter term: *monozygotic*

poly- means many. Chapter term: *polygenic*

LEARNING THE BASICS

1. Is a round yellow pea seed (genotype *RrYy*) an example of polygenic inheritance? Why or why not?

2. What factors cause quantitative variation in a trait within a population?

3. The DNA profile below is from a mother, a father, and their child. Compare the bands found in the child's profile. What is true of every band shown in the child's profile?

Mother Child Father

4. A quantitative trait ________.

A. may be one that is strongly influenced by the environment; **B.** varies continuously in a population; **C.** may be influenced by many genes; **D.** is not either off or on; **E.** all of the above

5. When graphing the phenotypes of a trait controlled by many genes and the environment, the line showing the frequency of each phenotype resembles ________.

A. a horizontal, straight line; **B.** a vertical, straight line; **C.** the letter T; **D.** a bell shape; **E.** a circle

6. When a trait is highly heritable, ________.

A. it is influenced by genes; **B.** it is not influenced by the environment; **C.** the variance of the trait in a population can be explained primarily by variance in genotypes; **D.** A and C are correct; **E.** A, B, and C are correct

7. Epigenetic changes involving methylation can directly affect the phenotype of an individual or her offspring by ________.

A. causing correlations between parents and children; **B.** causing changes to the DNA that affect gene expression but not DNA sequence; **C.** generating low heritability; **D.** increasing the likelihood of monozygotic twinning; **E.** sterilizing the DNA

8. Which of the following is *not* part of the procedure used to make a DNA profile?

A. Short tandem repeat sequences are amplified by PCR; **B.** DNA is placed in a gel and subjected to an electric current; **C.** The genes that encode DNA sequence are cloned into bacteria; **D.** DNA from blood, semen, vaginal fluids, or hair root cells can be used for analysis.

9. Add labels to the figure that follows, which illustrates the components in the PCR reaction.

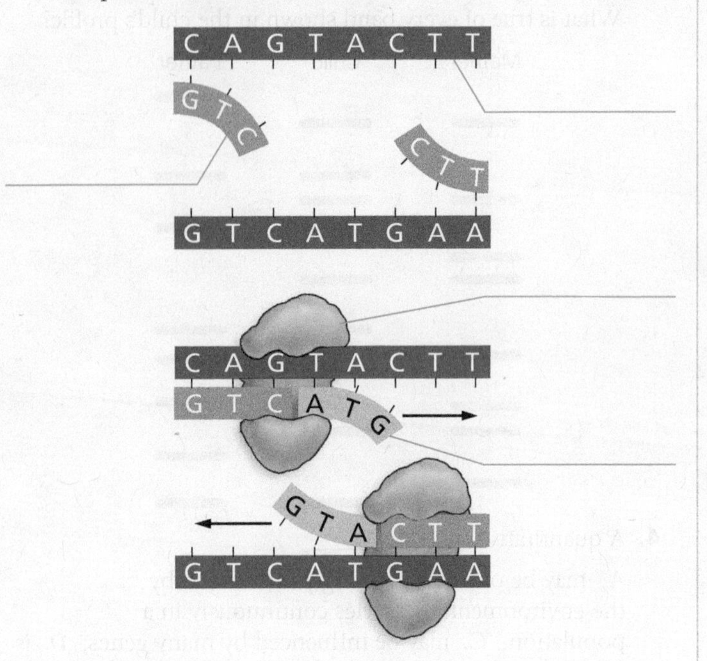

10. Which of the following statements is consistent with the DNA profile shown below?

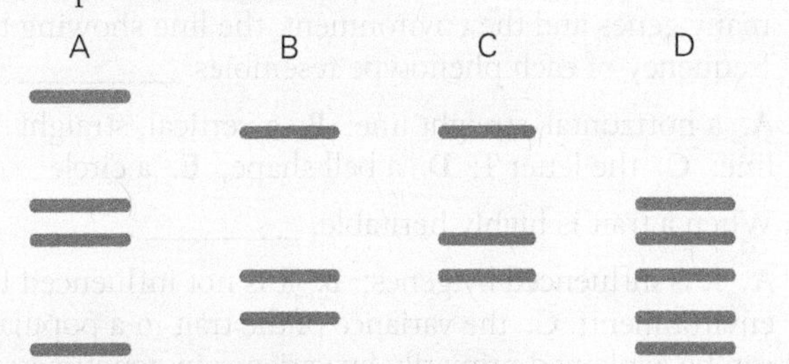

A. B is the child of A and C; **B.** C is the child of A and B; **C.** D is the child of B and C; **D.** A is the child of B and C; **E.** A is the child of C and D.

ANALYZING AND APPLYING THE BASICS

1. Why is DNA profiling analysis a much more powerful way to identify an individual than ABO blood type analysis?

2. The heritability of IQ has been estimated at about 72%. If John's IQ is 120 and Jerry's IQ is 90, does John have stronger "intelligence" genes than Jerry does? Explain your answer.

3. Draw a DNA profile showing a pattern that could be produced if samples from two identical twin sisters and their parents were analyzed.

GO FIND OUT

1. Do some Internet research on a behavioral trait that interests you to determine what is known about the heritability of that trait. Examples of behavioral traits include alcoholism, musical ability, homosexuality, shyness, and the like. Summarize what you learned in a paragraph or two.

2. You may serve on a jury where DNA evidence is presented. Research any questions you have about the veracity of DNA evidence and summarize what you learned in a paragraph or two.

MAKE THE CONNECTION

The science that you learned in this chapter has helped you better understand the real-world example used throughout this discussion. Draw a line from the statement on the left to the science that supports it on the right.

Eyewitness identification of polygenic or quantitative traits is difficult.	More false identifications and fewer positive identifications are made when the person doing the identifying is a different racial group than the one being identified.
People are better at identifying people who look like them than those who don't.	Heritability is a measure of the genetic component of a quantitative trait and is measured by correlations in phenotypes among relatives.
There is a common assumption that "bad guys" are identifiable by traits that have some innate component, such as facial structure.	The 13 STR locations differ in numbers of repeats from one individual to the next, creating unique DNA profiles.
A belief in a powerful genetic component to criminality can cause one's judgment to be distorted by bias.	Quantitative traits that have a genetic component are also highly influenced by the environment, so stereotypes have a high risk of being incorrect when applied to individuals.
Hundreds of people who had been falsely convicted have been released on the basis of DNA evidence.	There are a wide range of eye colors, skin shades, and heights that vary continuously in the human population.

Answers to **Got It?, Visualize This, Working with Data, Sounds Right, But Is It?, Show You Know,** and **Chapter Review** questions can be found in the **Answers** section at the back of the book.

10 Genetically Modified Organisms

Gene Expression, Mutation, Stem Cells, and Cloning

Many people are concerned about consuming genetically modified foods.

Some only buy foods that are labeled as not genetically modified.

Is it wrong to prevent genetically modified foods, like the golden rice shown to the right, from reaching people who could benefit from consuming it?

In the United States and Western Europe, many people choose to buy groceries at food co-ops that don't carry genetically modified products, or GMOs, spending extra money to avoid modified foods. These consumers avoid milk from cows treated with growth hormone and fruits and vegetables produced using genetic technologies.

In addition to exercising their rights at the grocery store, many protest against modified foods and work toward legislation that will prevent these foods from entering the marketplace, or at least make sure genetically modified foods are clearly labeled. But with a paucity of science to back up their claims, their actions and voices may be hurting more than their own bank account.

Are objections to genetic technologies affecting impoverished people, who have no voice in the debate but much more to lose? Take for example Golden Rice, a genetically modified rice that could help prevent deficiencies in vitamin A that cause serious health issues for the poorest people on our planet. This rice appears golden because it is rich in beta-carotene, which is necessary for the synthesis of vitamin A.

According to the World Health Organization and UNICEF, 40% of children under the age of five in the developing world are deficient in vitamin A. This deficiency not only compromises the immune system, leading to high death rates, it is also the leading cause of childhood blindness. Half a million children go blind every year, about half of whom will die within a year of losing their eyesight.

Although scientific and regulatory agencies around the world have repeatedly and consistently found modified foods to be safe, anti-GMO organizations like Greenpeace have worked to slow the development and distribution of foods, including Golden Rice.

—Continued next page

Should people in developed countries with access to a stable food supply be able to prevent technologies that could help millions of impoverished people when no science backs their claims? To better understand the debate about this and many other genetic technologies, we will look at how and why these technologies are being used as well as other potential ethical issues they raise.

10.1 Protein Synthesis and Gene Expression

One way that scientists can modify organisms is to change the amount of protein a particular gene produces. Regulating the amount of protein produced by a cell is also referred to as *regulating gene expression*. All living organisms have the ability to regulate gene expression in response to changes in the environment. This allows a particular gene to increase or decrease the amount of protein produced to meet the changing demands of the cell without wasting energy on making a protein that is not needed.

Scientists have figured out how to use this phenomenon to control gene expression in the lab. One of the first examples of scientists controlling gene expression occurred in the early 1980s when genetic engineers began to produce **recombinant bovine growth hormone (rBGH)** in their laboratories. Recombinant (r) bovine growth hormone is a protein that is made by genetically engineered bacteria. These bacterial cells have had their DNA manipulated so that it carries the instructions for, or encodes, a cow growth hormone that can be produced in the laboratory. Growth hormones act on many different organs to increase the overall size of the body. Bovine growth hormone that is produced in a laboratory can be injected into dairy cows to increase their milk production.

Production of growth hormone protein, or any protein, in the lab or in a cell, requires the use of the genetic information coded in the DNA.

From Gene to Protein

Protein synthesis involves using the instructions carried by a gene to build a particular protein. Genes do not build proteins directly; instead, they carry the instructions that dictate how a protein should be built. Understanding protein synthesis requires that we review a few basics about DNA, genes, and RNA. First, DNA is a polymer of **nucleotides** that make chemical bonds with each other based on their complementarity. Recall that adenine (A) and thymine (T) are complementary base pairs that bind to each other. Likewise, cytosine (C) and guanine (G) are complementary and bind to each other. Second, a gene is a sequence of DNA that encodes a protein. Proteins are large molecules composed

nucleo- refers to a nucleus.

of amino acids. Each protein has a unique function that is dictated by its particular structure. The structure of a protein is the result of the order of amino acids that constitute it because the chemical properties of amino acids cause a protein to fold in a particular manner. Before a protein can be built, the instructions carried by a gene are first copied. When the gene is copied, the copy is made up not of DNA (deoxyribonucleic acid) but of **RNA (ribonucleic acid).**

-ic is a common ending of acids.
-ose is a common ending for sugars.

RNA, like DNA, is a polymer of nucleotides. A nucleotide is composed of a sugar, a phosphate group, and a nitrogen-containing base. Whereas the sugar in DNA is deoxyribose, the sugar in RNA is **ribose**. RNA has the nitrogenous base uracil (U) in place of thymine; like thymine, uracil always pairs with adenine. RNA is usually single stranded, not double stranded like DNA (**FIGURE 10.1**).

When a cell requires a particular protein, a strand of RNA is produced using DNA as a guide or template. RNA nucleotides are able to make base pairs with DNA nucleotides. C pairs with G, and A pairs with U. The RNA copy then serves as a blueprint that tells the cell which amino acids to join together to produce a protein. Thus, the flow of genetic information in a eukaryotic cell is from DNA to RNA to protein (**FIGURE 10.2**).

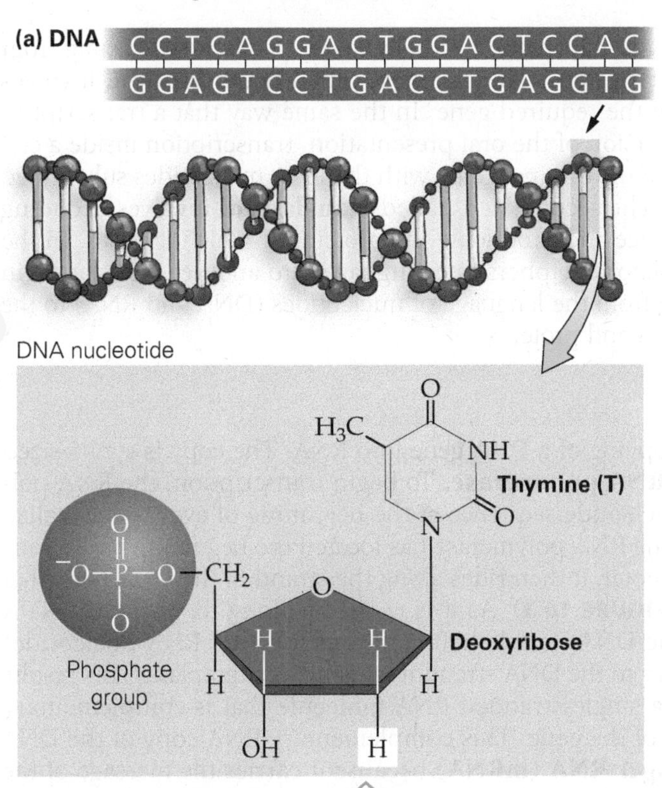

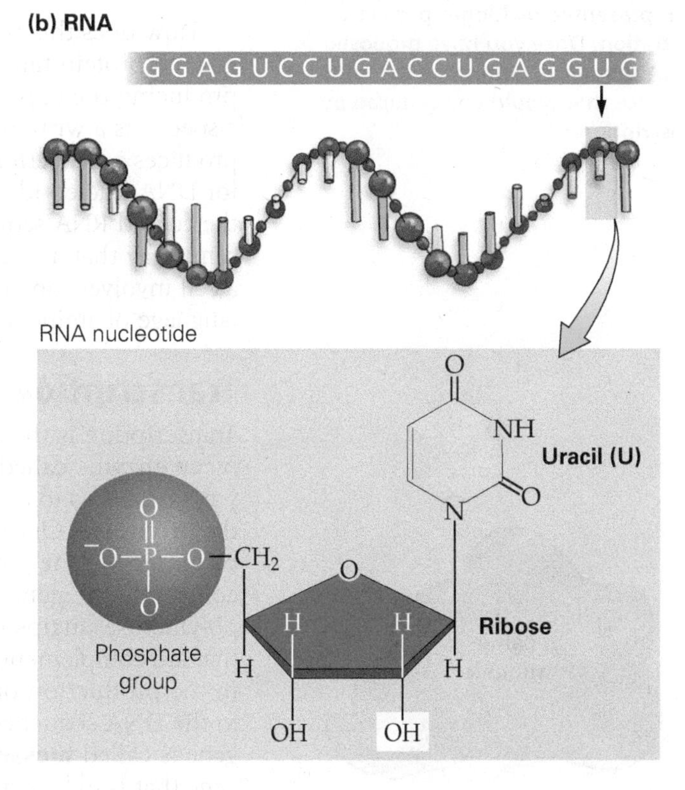

FIGURE 10.1 DNA and RNA. (a) DNA is double stranded. Each DNA nucleotide is composed of the sugar deoxyribose, a phosphate group, and a nitrogen-containing base (A, G, C, or T). (b) RNA is single stranded. RNA nucleotides are composed of the sugar ribose, a phosphate group, and a nitrogen-containing base (A, G, C, or U).

VISUALIZE **THIS**

Point out the chemical difference between the sugar in DNA and the one in RNA and the difference between the nitrogenous bases thymine and uracil.

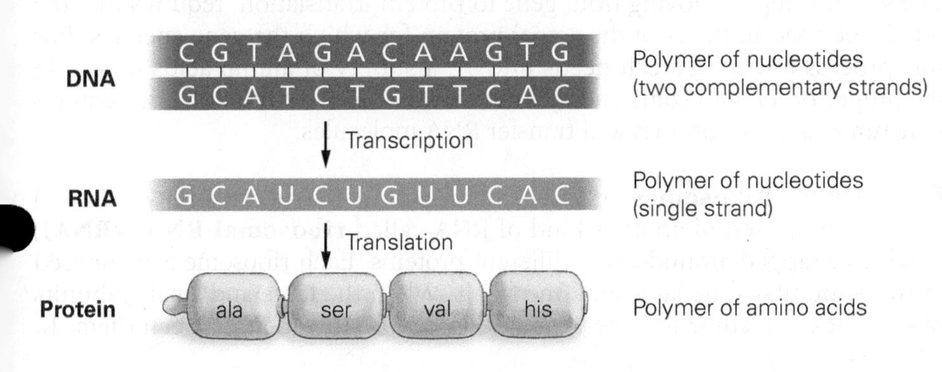

VISUALIZE **THIS**

Divide the number of nucleotides by the number of amino acids to determine how many nucleotides are required to encode each amino acid.

FIGURE 10.2 The flow of genetic information. Genetic information flows from DNA to an RNA copy of the DNA gene, to the amino acids that are joined together to produce the protein coded for by the gene.

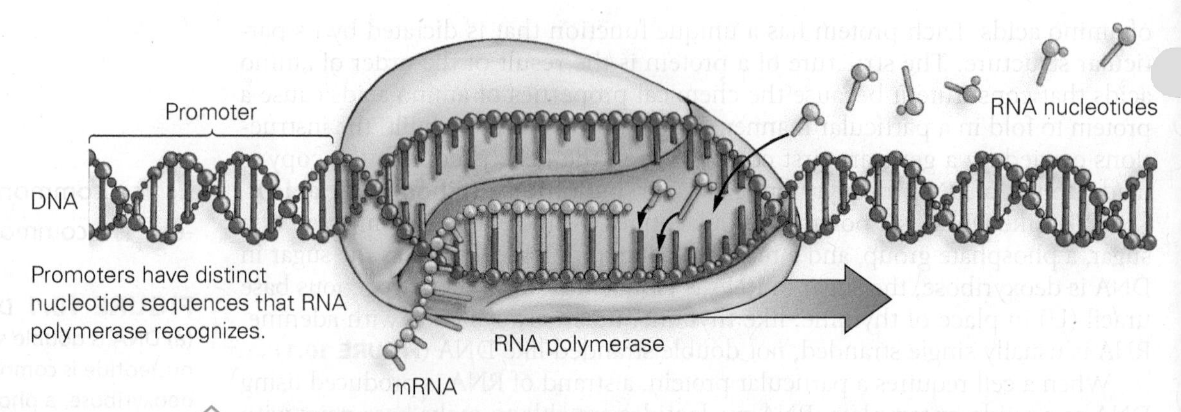

FIGURE 10.3 Transcription. RNA polymerase ties together nucleotides within the growing RNA strand as they form hydrogen bonds with their complementary base on the DNA. When the RNA polymerase reaches the end of the gene, the mRNA transcript is released.

VISUALIZE THIS

Propose a sequence for the DNA strand that is being used to produce the mRNA. Keep in mind that purines (A and G) are composed of two rings and thus are represented by longer pegs in this illustration. Once you have proposed a DNA sequence, determine the mRNA sequence that would be produced by transcription.

How does this flow of information actually take place in a cell? Going from gene to protein takes two steps. The first step, called **transcription,** involves producing the copy of the required gene. In the same way that a transcript of a speech is a written version of the oral presentation, transcription inside a cell produces a transcript of the original gene, with the RNA nucleotides substituted for DNA nucleotides. The second step, called **translation,** involves decoding the copied RNA sequence and producing the protein for which it codes. In the same way that a translator deciphers one language into another, translation in a cell involves moving from the language of nucleotides (DNA and RNA) to the language of amino acids and proteins.

Transcription

Transcription is the copying of a DNA gene into RNA. The copy is synthesized by an enzyme called **RNA polymerase.** To begin transcription, the RNA polymerase binds to a nucleotide sequence at the beginning of every gene, called the **promoter.** Once the RNA polymerase has located the beginning of the gene by binding to the promoter, it then rides along the strand of the DNA helix that comprises the gene (**FIGURE 10.3**). As it is traveling along the gene, the RNA polymerase unzips the DNA double helix and ties together RNA nucleotides that are complementary to the DNA strand it is using as a template. This results in the production of a single-stranded RNA molecule that is complementary to the DNA sequence of the gene. This complementary RNA copy of the DNA gene is called **messenger RNA (mRNA)** because it carries the message of the gene that is to be expressed.

Translation

The second step in moving from gene to protein, translation, requires that the mRNA be used to produce the actual protein for which the gene encodes. For this process to occur, a cell needs mRNA, a supply of amino acids to join in the proper order, and some energy in the form of ATP. Translation also requires structures called ribosomes and transfer RNA molecules.

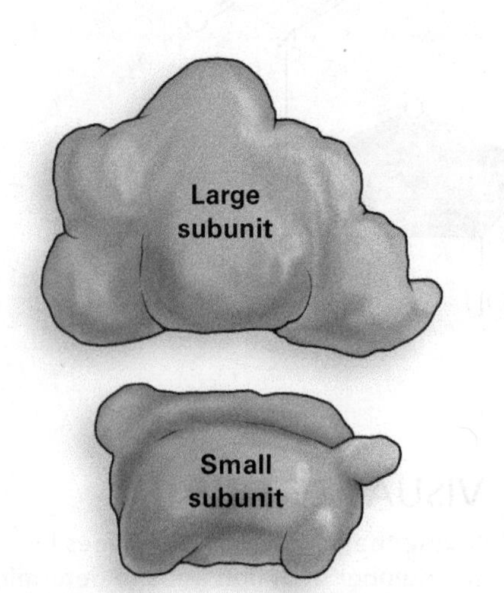

FIGURE 10.4 Ribosome. Ribosomes are composed of two subunits. Each subunit in turn is composed of rRNA and protein.

-some and **-somal** relates to a microscopically visible body in a cell.

Ribosomes. **Ribosomes** are subcellular, globular structures (**FIGURE 10.4**) that are composed of another kind of RNA called **ribosomal RNA (rRNA)**, which is wrapped around many different proteins. Each ribosome is composed of two subunits—one large and one small. When the large and small subunits of the ribosome come together, the mRNA can be threaded between them. In addition, the ribosome can bind to structures called **transfer RNA (tRNA)** that carry amino acids.

Transfer RNA (tRNA). Transfer RNA (**FIGURE 10.5**) is yet another type of RNA found in cells. An individual transfer RNA molecule carries one specific amino acid and interacts with mRNA to place the amino acid in the correct location of the growing polypeptide.

As mRNA moves through the ribosome, small sequences of nucleotides are sequentially exposed. These sequences of mRNA, called **codons,** are three nucleotides long and encode a particular amino acid. Transfer RNAs also have a set of three nucleotides, which will bind to the codon if the right sequence is present. These three nucleotides at the base of the tRNA are called the **anticodon** because they complement a codon on mRNA. The anticodon on a particular tRNA binds to the complementary mRNA codon. In this way, the codon calls for the incorporation of a specific amino acid. The ribosome moves along the mRNA sequentially, exposing codons for tRNA binding.

When a tRNA anticodon binds to the mRNA codon, a peptide bond is formed. The ribosome adds the amino acid that the tRNA is carrying to the growing chain of amino acids that will eventually constitute the finished protein. The transfer RNA functions as a sort of cellular translator, fluent in both the language of nucleotides (its own language) and the language of amino acids (the target language).

To help you understand protein synthesis, let us consider its similarity to an everyday activity such as baking a cake (**FIGURE 10.6**). To bake a cake, you would consult a cookbook (genome) for the specific recipe (gene) to make your cake (protein). You may copy the recipe (mRNA) out of the book so that the original recipe (gene) does not become stained or damaged. The original recipe (gene) is left in the book (genome) on a shelf (nucleus) so that you can make another copy when you need it. The original recipe (gene) can be copied again and again. The copy of the recipe (mRNA) is placed on the kitchen counter (ribosome) while you assemble the ingredients (amino acids). The ingredients (amino acids) for your cake (protein) include flour, sugar, butter, milk, and eggs. The ingredients are measured in measuring spoons and cups (tRNAs). (Although in baking you might use the same cups and spoons for several ingredients, in protein synthesis we use tRNAs that are dedicated to one specific ingredient.) The measuring spoons and cups bring the ingredients to the kitchen counter. Like the ingredients in a cake that can be used in many

VISUALIZE THIS

The structure of a tRNA molecule involves regions where the RNA strand forms complementary bonds with itself, causing the RNA to fold up. What nitrogenous bases might be involved in bonding in such regions of internal complementarity?

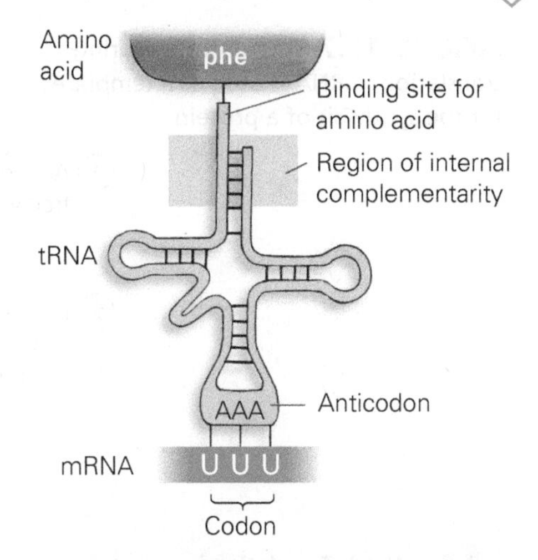

FIGURE 10.5 Transfer RNA (tRNA). Transfer RNAs translate the language of nucleotides into the language of amino acids. The tRNA that binds to UUU carries only one amino acid (phe, the three-letter abbreviation for phenylalanine).

FIGURE 10.6 Protein synthesis and cake baking. Making a protein in a cell is analogous to making a cake in your kitchen.

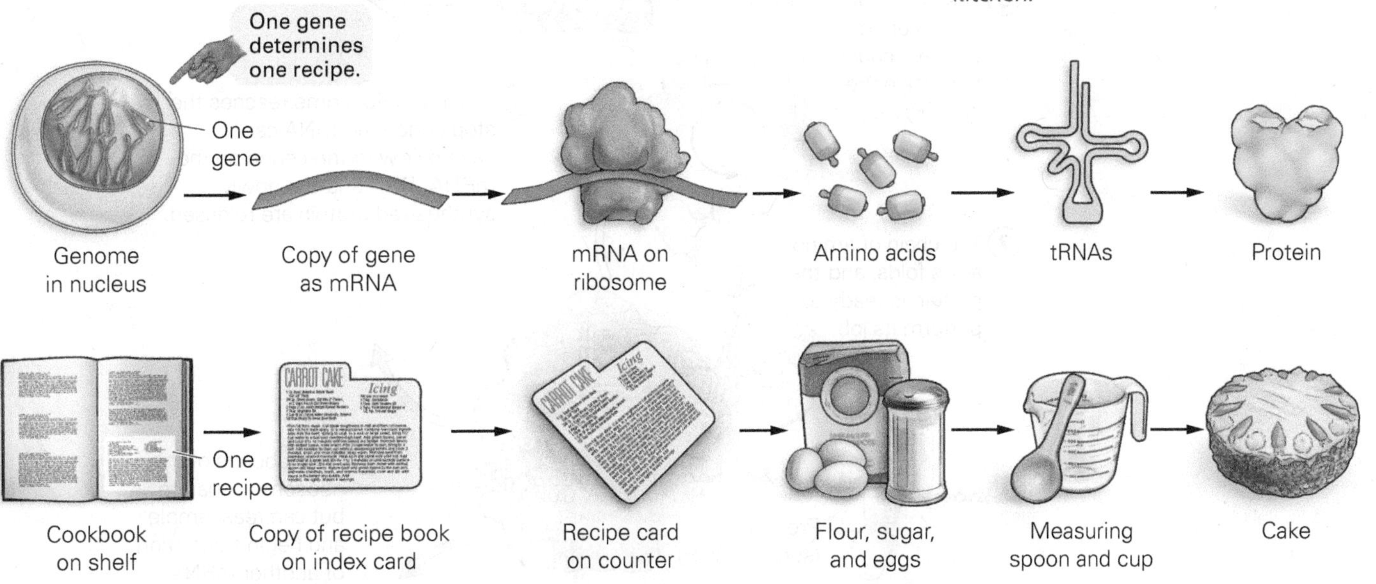

ways to produce a variety of foods, amino acids can be combined in different orders to produce different proteins. The ingredients (amino acids) are alway added according to the instructions specified by the original recipe (gene). Within cells, the sequence of bases in the DNA dictates the sequence of bases in the RNA, which in turn dictates the order of amino acids that will be joined together to produce a protein. Protein synthesis ends when a codon that does not code for an amino acid, called a **stop codon,** moves through the ribosome. When a stop codon is present in the ribosome, no new amino acid can be added, and the growing protein is released. Once released, the protein folds up on itself and moves to where it is required in the cell. A summary of the process of translation is shown in **FIGURE 10.7**.

FIGURE 10.7 Translation. During translation, mRNA is used as a template for the synthesis of a protein.

Watch Translation in Mastering Biology

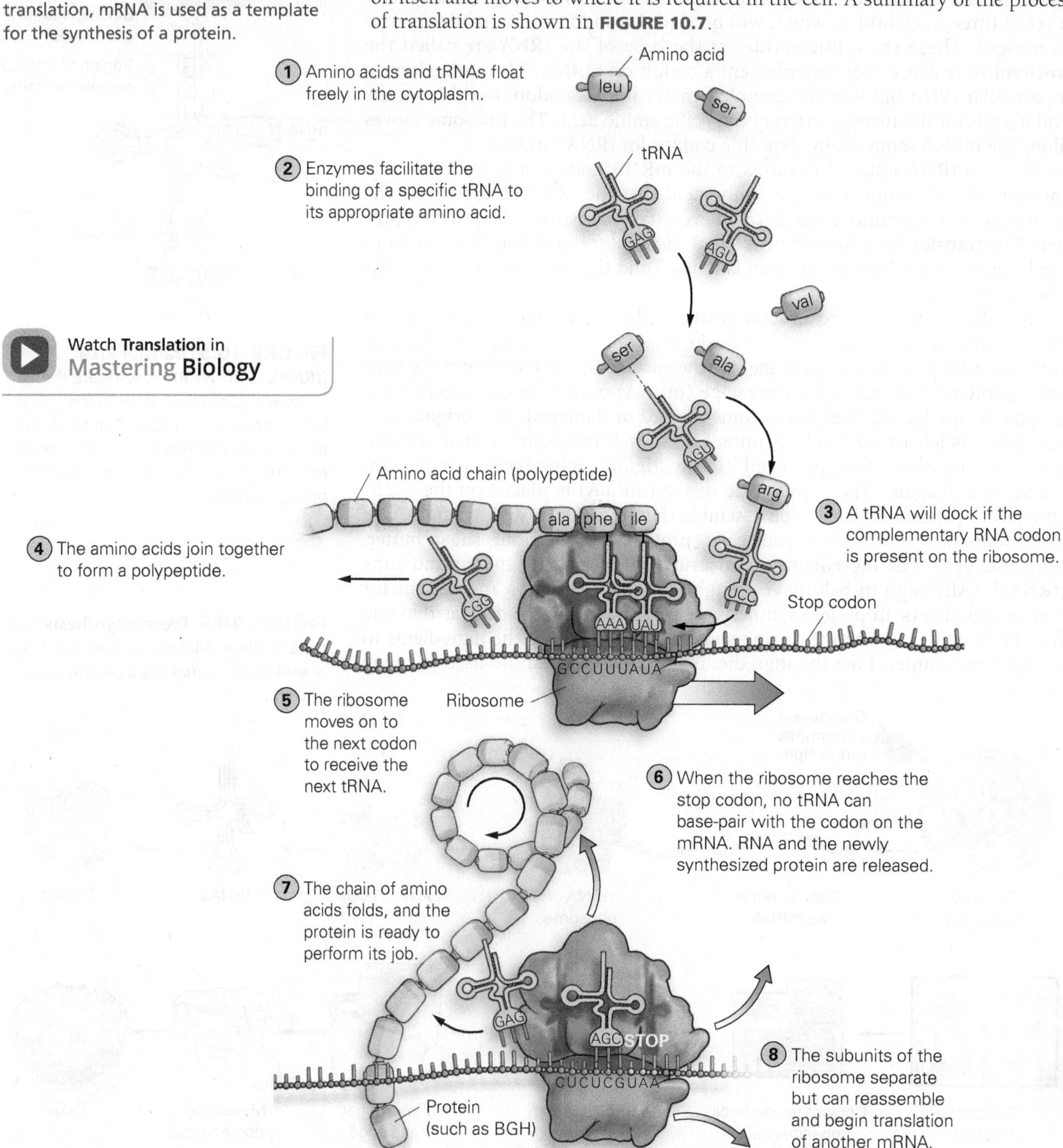

1 Amino acids and tRNAs float freely in the cytoplasm.
Amino acid
leu
ser
2 Enzymes facilitate the binding of a specific tRNA to its appropriate amino acid.
tRNA
GAG
AGU
val
ser
ala
AGU
arg
Amino acid chain (polypeptide)
ala
phe
ile
3 A tRNA will dock if the complementary RNA codon is present on the ribosome.
4 The amino acids join together to form a polypeptide.
CGG
AAA
UAU
UCC
Stop codon
GCCUUUAUA
5 The ribosome moves on to the next codon to receive the next tRNA.
Ribosome
6 When the ribosome reaches the stop codon, no tRNA can base-pair with the codon on the mRNA. RNA and the newly synthesized protein are released.
7 The chain of amino acids folds, and the protein is ready to perform its job.
GAG
AGC
STOP
CUCUCGUAA
8 The subunits of the ribosome separate but can reassemble and begin translation of another mRNA.
Protein (such as BGH)

The process of translation allows cells to join amino acids in the sequence coded by the gene. Scientists can determine the sequence of amino acids that a gene calls for by looking at the **genetic code.**

Genetic Code. The genetic code shows which mRNA codons code for which amino acids. As TABLE 10.1 shows, there are 64 codons, 61 of which code for amino acids. Three of the codons are stop codons that occur near the end of an mRNA. In the table, you can see that the codon AUG functions both as a start codon (and thus is found near the beginning of each mRNA) and as a codon dictating that the amino acid methionine (met) be incorporated into the protein being synthesized.

The genetic code has some additional properties. The code is redundant without being ambiguous, and it is also universal. The redundancy of the code can be seen in examples where the same amino acid is coded for by more than one codon. For example, the amino acid threonine (thr) is incorporated into a protein in response to the codons ACU, ACC, ACA, and ACG. There is, however, no situation where a given codon can call for more than one amino acid. For example, AGU codes for serine (ser) and nothing else. Therefore, there is no ambiguity in the genetic code regarding which amino acid any codon will call for. The genetic code is also universal in the sense that organisms typically decode the same gene to produce the same protein. This is why genes can be moved from one organism to another.

TABLE 10.1 The Genetic Code.

To determine which amino acid is coded for by each mRNA codon, first look at the left-hand side of the chart for the first-base nucleotide in the codon; there are four rows, one for each possible RNA nucleotide—A, C, G, or U. Then look at the intersection of the second-base columns at the top of the chart and the first-base rows to narrow your search. Finally, the third-base nucleotide in the codon on the right-hand side of the chart determines the amino acid that a given mRNA codon codes for. Note the three codons UAA, UAG, and UGA that do not code for an amino acid; these are stop codons. The codon AUG is a start codon, found at the beginning of most protein-coding sequences.

		Second base			
First base	**U**	**C**	**A**	**G**	**Third base**
U	UUU, UUC: Phenylalanine (phe)	UCU, UCC, UCA, UCG: Serine (ser)	UAU, UAC: Tyrosine (tyr)	UGU, UGC: Cysteine (cys)	U, C
U	UUA, UUG: Leucine (leu)		UAA: **Stop codon** UAG: **Stop codon**	UGA: **Stop codon** UGG: Tryptophan (trp)	A, G
C	CUU, CUC, CUA, CUG: Leucine (leu)	CCU, CCC, CCA, CCG: Proline (pro)	CAU, CAC: Histidine (his)	CGU, CGC, CGA, CGG: Arginine (arg)	U, C
C			CAA, CAG: Glutamine (gln)		A, G
A	AUU, AUC, AUA: Isoleucine (ile)	ACU, ACC, ACA, ACG: Threonine (thr)	AAU, AAC: Asparagine (asn)	AGU, AGC: Serine (ser)	U, C
A	AUG: Methionine (met) **Start codon**		AAA, AAG: Lysine (lys)	AGA, AGG: Arginine (arg)	A, G
G	GUU, GUC, GUA, GUG: Valine (val)	GCU, GCC, GCA, GCG: Alanine (ala)	GAU, GAC: Aspartic acid (asp)	GGU, GGC, GGA, GGG: Glycine (gly)	U, C
G			GAA, GAG: Glutamic acid (glu)		A, G

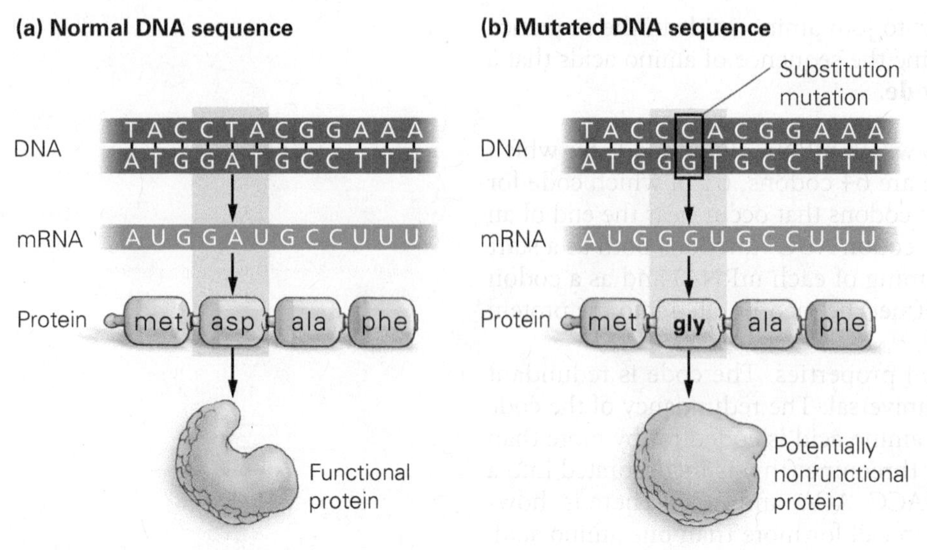

FIGURE 10.8 Substitution mutation. A single nucleotide change from the normal DNA sequence (a) to the mutated sequence (b) can result in the incorporation of a different amino acid. If the substituted amino acid has chemical properties different from those of the original amino acid, then the protein may assume a different shape and thus lose its ability to perform its job.

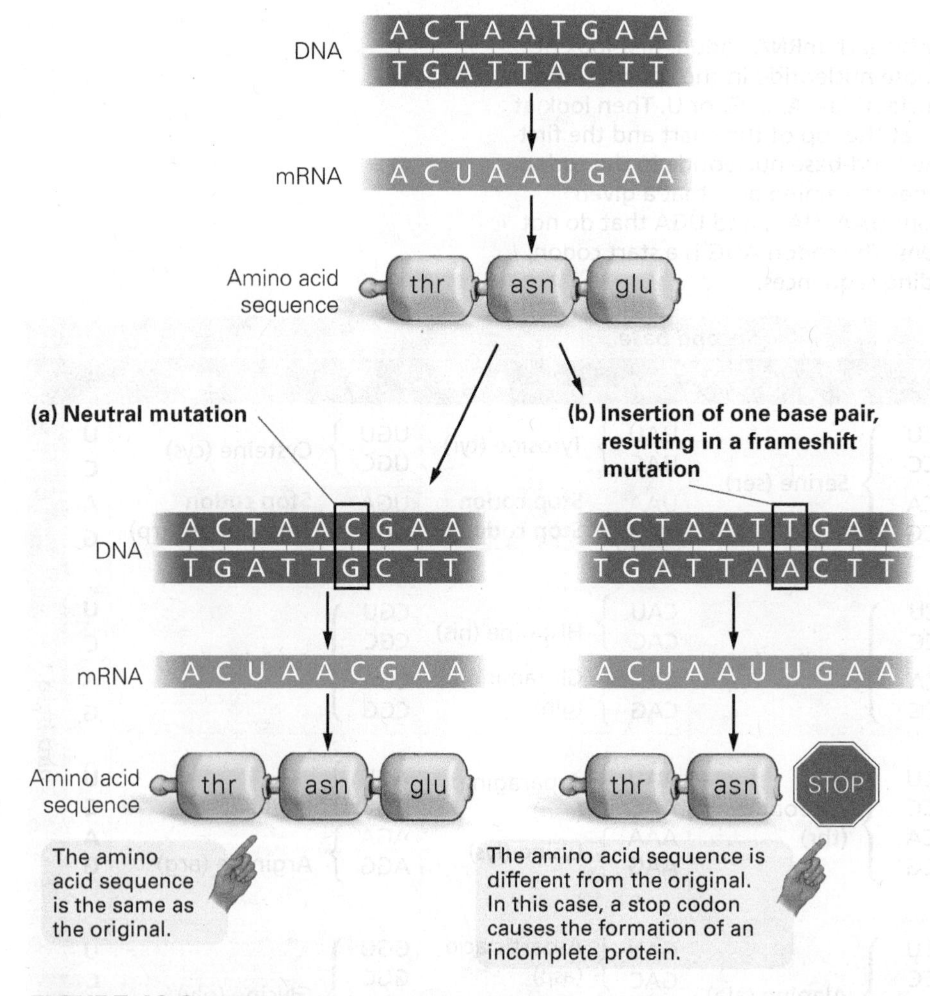

FIGURE 10.9 Neutral and frameshift mutations. (a) Neutral mutations result in the incorporation of the same amino acid as was originally called for. (b) The insertion (or deletion) of a nucleotide can result in a frameshift mutation.

Mutations

Changes to the DNA sequence, called **mutations,** can affect the order or types of amino acids incorporated into a protein during translation. Mutations to a gene can result in the production of different alleles of a gene. Different alleles result from changes in the DNA that alter the amino acid order of the encoded protein. Mutations can result in the production of either a nonfunctional protein or a protein different from the one previously required. If this protein does not have the same amino acid composition, it may not be able to perform the same job (**FIGURE 10.8**). For instance, a substitution of a single nucleotide results in the incorporation of a new amino acid in the hemoglobin protein and compromises the ability of cells to carry oxygen, producing sickle-cell disease.

There are also cases when a mutation has no effect on a protein. These cases may occur when changes to the DNA result in the production of an mRNA codon that codes for the same amino acid as was originally required. Due to the redundancy of the genetic code, a mutation that changes the mRNA codon from, say, ACU to ACC will have no impact because both of these codons code for the amino acid threonine. This is called a **neutral mutation** (**FIGURE 10.9a**). In addition, mutations can result in the substitution of one amino acid for another with similar chemical properties, which may have little or no effect on the protein.

Inserting or deleting a single nucleotide can have a severe impact because the addition (or deletion) of a nucleotide can change the groupings of nucleotides in every codon that follows (**FIGURE 10.9b**). Changing the triplet groupings is called altering the **reading frame.** All nucleotides located after an insertion or deletion will be regrouped into different codons, producing a **frameshift mutation.** For example, inserting an extra letter *H* after the fourth letter of the sentence, "The dog ate the cat," could change the reading frame to the nonsensical statement, "The dHo gat eth eca t." Inside cells, this often results in the incorporation of a stop codon and the production of a shortened, nonfunctional protein.

(a) Eukaryotic protein synthesis

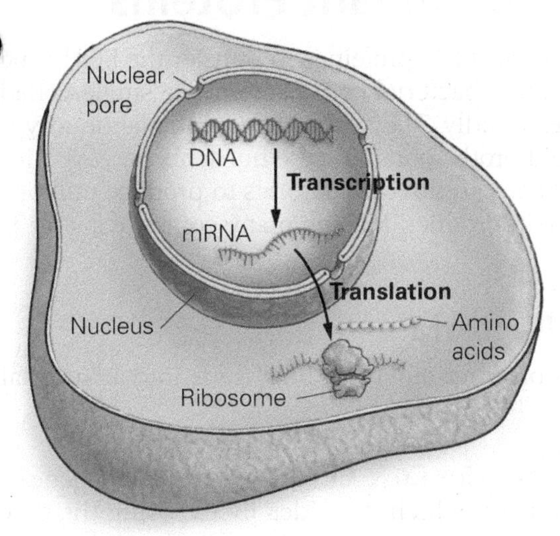

(b) Prokaryotic protein synthesis

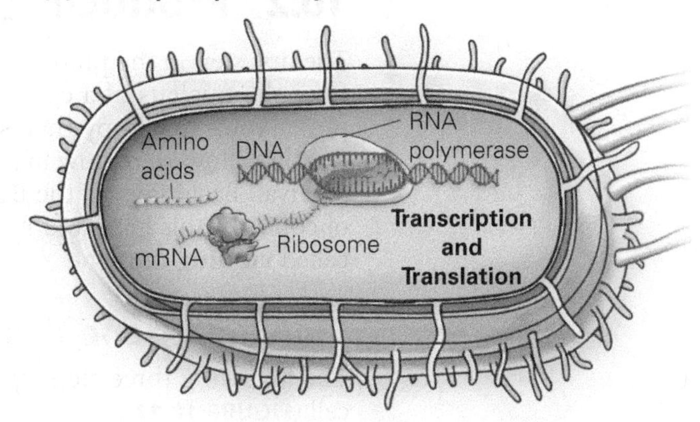

FIGURE 10.10 Protein synthesis in eukaryotic and prokaryotic cells. (a) In eukaryotes, transcription occurs in the nucleus and translation in the cytoplasm. (b) In prokaryotic cells, which lack nuclei, transcription and translation occur simultaneously.

Cells in all organisms undergo this process of protein synthesis, with different cell types selecting different genes from which to produce proteins (**FIGURE 10.10**). In eukaryotic cells, transcription and translation are spatially separate, with transcription occurring in the nucleus and translation occurring in the cytoplasm. Cells lacking a membrane-bound nucleus and organelles—in other words, prokaryotic cells like bacteria—also undergo protein synthesis. In these cells transcription and translation occur at the same time and in the same location instead of occurring in separate places. As an mRNA is being transcribed, ribosomes attach and begin translating.

Gene Expression

Each cell in your body, except sperm or egg cells, has the same complement of genes you inherited from your parents but expresses only a small percentage of those genes. For example, because your muscles and nerves each perform a specialized suite of jobs, muscle cells turn on or express one suite of genes and nerve cells another (**FIGURE 10.11**). The expression of a given gene is turned on or off, or modulated more subtly, so that the gene can respond to the cell's needs.

Genetic engineering permits precise control of gene expression in many different circumstances. In the case of rBGH, for example, farmers can simply decide how much of the protein to inject into the bloodstream of a cow. However, they must first synthesize the protein, as we will explore next.

(a) Muscle cells

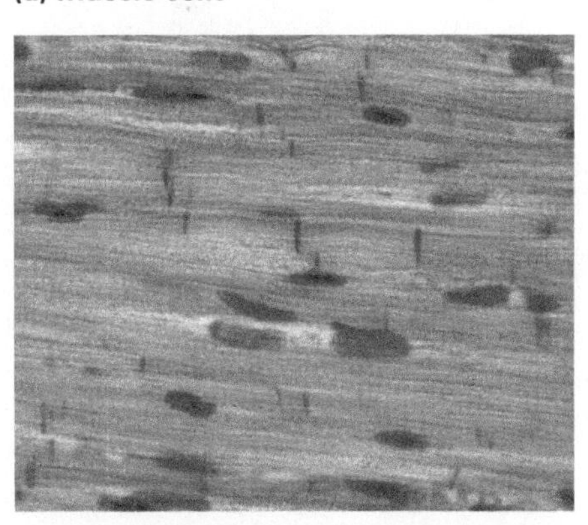

(b) Nerve cells

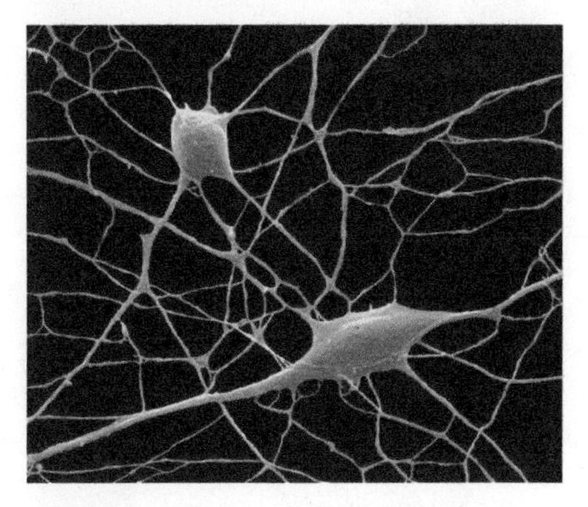

FIGURE 10.11 Gene expression differs from cell to cell. (a) A muscle cell performs different functions and expresses different genes than (b) a nerve cell. Both of these cells have the same suite of genes in their nucleus but express different subsets of genes.

Got It?

1. For a protein to be produced, first the enzyme __________ __________ binds to the promoter and undergoes transcription.
2. Transcription produces a complementary copy of the gene called __________ __________.
3. Once the complementary copy of the gene is produced it undergoes __________.
4. The decoding of the mRNA occurs on structures called __________.
5. Translation allows a cell to determine which __________ __________ to join together to make a protein.

10.2 Producing Recombinant Proteins

The first step in the production of the rBGH protein is to transfer the BGH gene from the nucleus of a cow cell into a bacterial cell. Bacteria are single-celled prokaryotes that copy themselves rapidly. They can thrive in the laboratory if they are allowed to grow in a liquid broth containing the nutrients necessary for survival. Bacteria with the BGH gene can serve as factories to produce millions of copies of this gene and its protein product. Making many copies of a gene is called **cloning** the gene.

Cloning a Gene Using Bacteria

The following three steps are involved in moving a BGH gene into a bacterial cell (**FIGURE 10.12**).

***Step 1.* Remove the Gene from the Cow Chromosome.** The gene is sliced out of the cow chromosome on which it resides by exposing the cow DNA to enzymes that cut DNA. These enzymes, called **restriction enzymes,** act like highly specific molecular scissors. Most restriction enzymes cut DNA only at specific sequences, called *palindromes*, such as

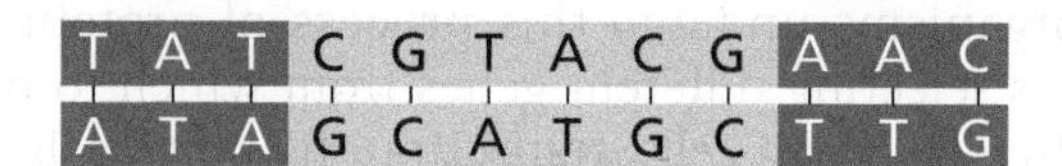

Note that the bottom middle sequence is the reverse of the top middle sequence. Many restriction enzymes cut the DNA in a staggered pattern, leaving "sticky ends," such as

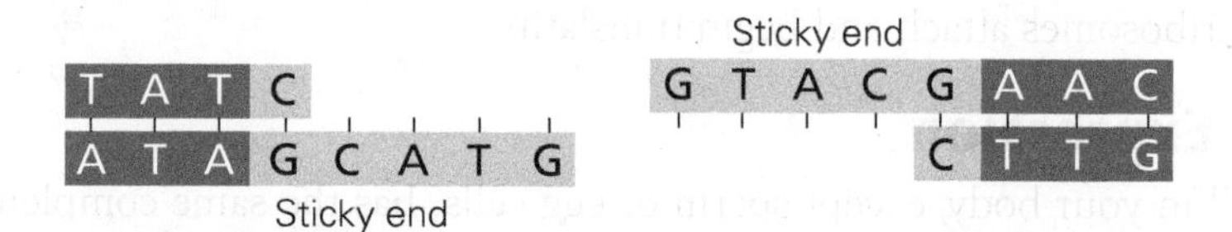

The unpaired bases form bonds with any complementary bases with which they come in contact. The enzyme selected by the scientist cuts on both ends of the BGH gene but not inside the gene.

A particular restriction enzyme cuts DNA at a specific sequence. Therefore, scientists need information about the organism's genome to determine which restriction enzyme cutting sites surround the gene of interest. Cutting the DNA generates many different fragments, only one of which will carry the gene of interest.

***Step 2.* Insert the BGH Gene into the Bacterial Plasmid.** Once the gene is removed from the cow genome, it is inserted into a bacterial structure called a plasmid. A plasmid is a circular piece of DNA that normally exists separate from the bacterial chromosome and can replicate independently of the bacterial chromosome. Think of the plasmid as a molecular ferryboat, able to carry a gene into a bacterial cell, where it can be replicated. To incorporate the BGH gene into the plasmid, the plasmid is also cut with the same restriction enzyme used to cut the gene. Cutting both the plasmid and gene with the

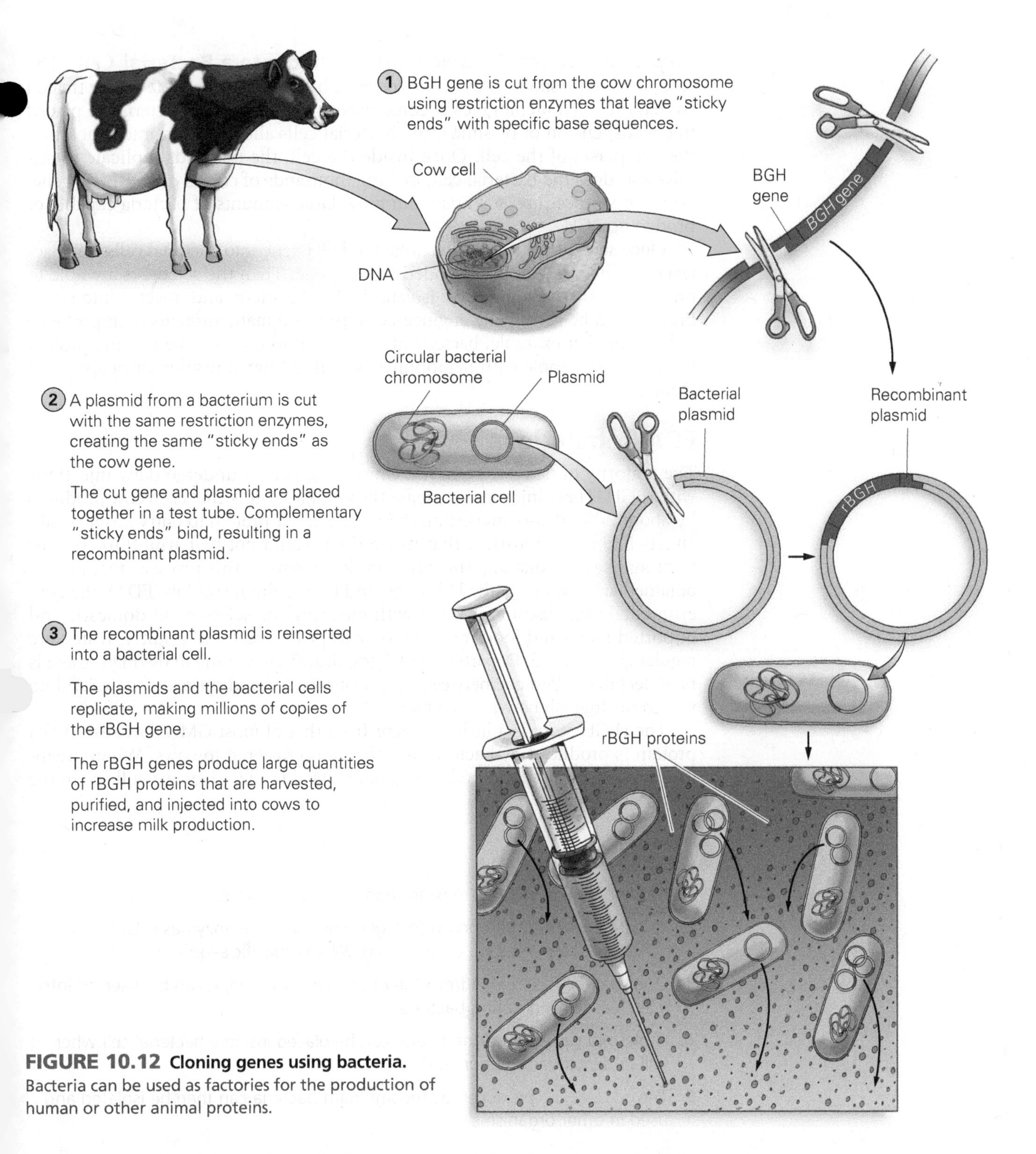

1 BGH gene is cut from the cow chromosome using restriction enzymes that leave "sticky ends" with specific base sequences.
Cow cell
DNA
BGH gene
BGH gene
Circular bacterial chromosome
Plasmid
Bacterial cell
Bacterial plasmid
Recombinant plasmid
rBGH
2 A plasmid from a bacterium is cut with the same restriction enzymes, creating the same "sticky ends" as the cow gene.
The cut gene and plasmid are placed together in a test tube. Complementary "sticky ends" bind, resulting in a recombinant plasmid.
3 The recombinant plasmid is reinserted into a bacterial cell.
The plasmids and the bacterial cells replicate, making millions of copies of the rBGH gene.
The rBGH genes produce large quantities of rBGH proteins that are harvested, purified, and injected into cows to increase milk production.
rBGH proteins

FIGURE 10.12 Cloning genes using bacteria. Bacteria can be used as factories for the production of human or other animal proteins.

same enzyme allows the sticky ends that are generated to base-pair with each other (A to T and G to C). When the cut plasmid and the cut gene are placed together in a test tube, they re-form into a circular plasmid with the extra gene incorporated.

The bacterial plasmid has now been genetically engineered to carry a cow gene. At this juncture, the BGH gene is referred to as the rBGH gene, with the *r* indicating that this product is genetically engineered, or recombinant, because it has been removed from its original location in the cow genome and recombined with the plasmid DNA.

***Step 3.* Insert the Recombinant Plasmid into a Bacterial Cell.** The recombinant plasmid is now inserted into a bacterial cell. Bacteria can be treated so that their cell membranes become porous. When treated bacteria are placed into a suspension of plasmids, the bacterial cells allow the plasmids back into the cytoplasm of the cell. Once inside the cell, the plasmids replicate themselves, as does the bacterial cell, making thousands of copies of the rBGH gene. Using this procedure, scientists can grow large amounts of bacteria capable of producing BGH.

Once scientists successfully clone the BGH gene into bacterial cells, the bacteria produce the protein encoded by the gene. Then the scientists are able to break open the bacterial cells, isolate the BGH protein, and inject it into cows. Bacteria can be genetically engineered to produce many proteins of importance to humans. For example, bacteria are now used to produce the clotting protein missing from people with hemophilia as well as human insulin for people with diabetes.

FDA Regulations

The majority of dairy cows in the United States now undergo daily injections with rBGH. These injections increase the volume of milk that each cow produces by about 20%. Before marketing the recombinant protein to dairy farmers, scientists had to demonstrate that its product would not be harmful to cows or to humans who consume the cows' milk. As part of this process, researchers obtained approval from the U.S. Food and Drug Administration (FDA), the governmental organization charged with ensuring the safety of all domestic and imported foods and food ingredients (except for meat and poultry, which are regulated by the U.S. Department of Agriculture). According to the FDA, there is no detectable difference between milk from treated and untreated cows and no way to distinguish between the two.

The rBGH story is a little different from that of most GMOs because rBGH protein is produced by bacteria and then administered to cows. When organisms are genetically modified, the genome itself is altered, as we will see in the next section.

Got It?

1. Cloning a gene involves making many __________ of it.
2. To cut genes out of an organism's genome, bacterial enzymes called __________ enzymes are used to snip DNA at specific sequences.
3. Once a gene is removed from the organism's genome, it can be inserted into a circular self-replicating bacterial __________.
4. A plasmid carrying a cloned gene can be placed inside a bacterial cell where it will undergo transcription and __________.
5. The __________ produced by recombinant bacteria can then be isolated and used in other organisms.

10.3 Genetically Modified Plants and Animals

Unless you are making a concerted effort not to, you have been eating genetically modified foods every day for the past 20 years. The majority of the corn and soy grown in the United States has been genetically modified

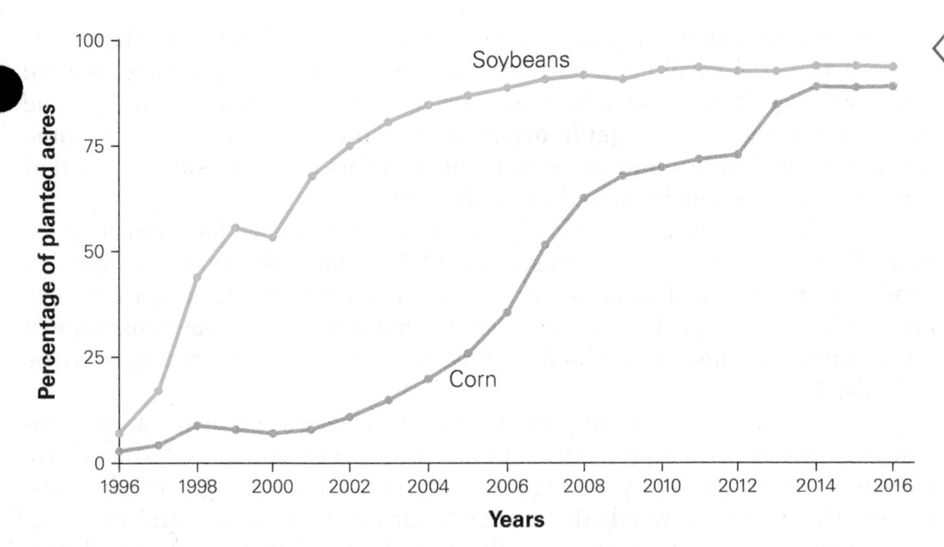

FIGURE 10.13 Genetically modified soybeans and corn. This graph, using data supplied by the U.S. Department of Agriculture (USDA), shows the increase in the percentage of acres planted with soybeans and corn that are genetically modified to be herbicide resistant.

WORKING WITH DATA

In what years did the majority (more than 50%) of (a) soybeans and (b) corn planted first shift from nongenetically modified to modified?

(**FIGURE 10.13**) and the rate of modification of other crops is increasing as well. Most processed foods contain corn syrup, soy, or vegetable oils from genetically modified plants. Some consumers find this troubling, and others find that their concerns are tempered by the fact that plants and animals have been subject to human modification by artificial selection for thousands of years. Every time a farmer chooses which plants or animals to use in producing the next generation, he or she is modifying the frequency of alleles in that population of organisms by artificial selection (**FIGURE 10.14**). In fact, Golden Rice was first produced by using artificial selection. Scientists switched to using the available genetic technologies when artificial selection techniques were only able to establish strains with a slightly higher than normal beta-carotene level over many generations. Golden Rice can be engineered to contain more than 20 times the amount of beta-carotene in only one generation.

Although crossing plants to increase nutritive value or breeding those cattle that produce more milk does not involve moving a gene from one organism to another, it does select for the propagation of particular genes.

In addition to increasing the rate at which the desirable trait can be produced, the suite of genes available for use is also increased through the use of genetic technologies. For example, two corn plants that are bred together can only draw from traits coded by genes contained in their genomes. With genetic engineering, genes from other organisms can be introduced to the genome. It is this "splicing" of genes from unrelated organisms that is the source of most of the concerns about modifying foods. To understand whether these concerns are justified, we must first consider how crops are modified.

FIGURE 10.14 Artificial selection in corn. Selective-breeding techniques resulted in the production of modern corn (right) from ancient teosinite corn.

Modifying Crop Plants

To move a novel gene into a plant, the gene must be able to gain access to the plant cell, which means it must be able to move through the plant's rigid outer cell wall. Moving genes into many agricultural crops such as corn, barley, and rice can be accomplished by using a device called a **gene gun.** A gene

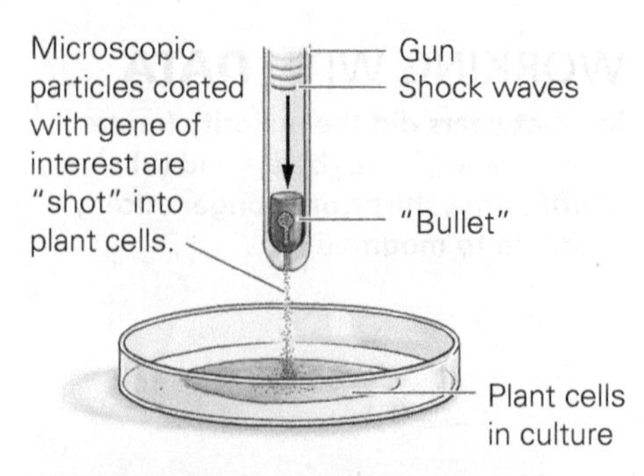
Microscopic particles coated with gene of interest are "shot" into plant cells.
Gun
Shock waves
"Bullet"
Plant cells in culture

FIGURE 10.15 Genetically modifying plants using a gene gun. A gene gun shoots a plastic bullet loaded with tiny DNA-coated pellets into a plant cell. The bullet shells are prevented from leaving the gun, but the DNA-covered pellets enter the nucleus of some cells, producing a genetically modified cell, which can give rise to a modified plant that can pass the modification to its offspring.

trans- means across.

gun shoots metal-coated pellets covered with foreign DNA into plant cells (**FIGURE 10.15**). A small percentage of these genes may be incorporated into the plant's genome. When a gene from one species is incorporated into the genome of another species, a **transgenic organism** is produced. When the embryonic plant so treated grows into an adult plant, all of its cells contain the inserted gene and the gene will be passed to its offspring.

Crop plants are genetically modified to increase their shelf life. Tomatoes, for example, have been engineered to soften and ripen more slowly. The longer ripening time means that tomatoes stay on the vine longer, thus making them taste better. The slower ripening also increases the amount of time that tomatoes will be appealing to consumers who do not want to purchase fruit that is overripe or bruised.

Genetic engineering techniques increase crop yield when plants are manipulated to be resistant to pesticides and herbicides. For centuries, farmers have tried to increase yields by killing the pests that damage crops and by controlling the growth of weeds that compete for nutrients, water, and sunlight. In the United States, farmers typically spray high volumes of chemical pesticides and herbicides directly onto their fields. This practice concerns people worried about the health effects of eating foods that have been treated by these often toxic or cancer-causing chemicals. In addition, both pesticides and herbicides may leach through the soil and contaminate drinking water. There is hope, and even some evidence (see the Big Question) that plants modified to be resistant to herbicides and pesticides would allow farmers to use less of these chemicals.

Another concern is that genetically modified crop plants may transfer engineered genes from modified crop plants to their wild or weedy relatives. Wind, rain, birds, and bees carry genetically modified pollen to related plants in nearby fields containing nongenetically modified crops. Many cultivated crops have retained the ability to interbreed with their wild relatives; in these cases, genes from farm crops can mix with genes from the wild crops—and with crops grown in nearby farms, including those that are meant to be organic (and thus free of GMO products).

Modifying Animals

Animals as well as plants undergo genetic modifications. Alaskan salmon have been engineered to carry genes from other fish that make them grow larger and faster than normal. Eggs from wild Alaskan salmon are injected with the gene that codes for a growth hormone from another faster-growing fish. Because this gene is injected into a fertilized egg cell, all the cells of the grown salmon would express this enhanced version of the growth factor.

Transgenic salmon eat less food but grow twice as fast. They can also be grown in isolation near cities, which reduces transportation costs and prevents these salmon from breeding with wild fish. If engineered salmon were released into streams they could reproduce with wild fish. If such interbreeding were to occur, the offspring of such matings might outcompete nonmodified fish and disrupt aquatic food webs. It remains to be seen whether consumers will notice any difference in the taste of modified salmon compared with the unmodified version they are used to eating.

In addition to being genetically modified for human dietary consumption, animals can be modified to help humans in other ways. Biological **pharming,** a blend of the words *pharmaceutical* and *farming*, is the production of GMOs that produce drugs used to treat or prevent disease. Tobacco plants have been engineered to produce proteins normally found in human blood. Goats have been engineered to produce a human blood-clotting protein in their milk

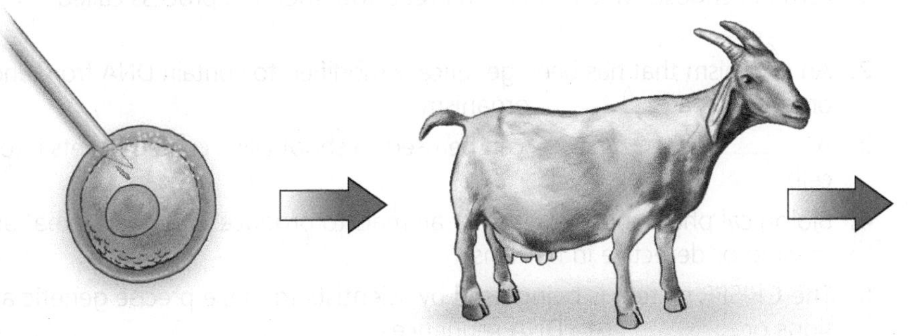

FIGURE 10.16 Goat's milk. Herds of goats that produce human blood-clotting proteins are engineered by (a) microinjecting a fertilized goat egg with a human clotting gene; (b) implanting eggs that incorporate the gene into a pregnant female goat; and (c) milking the offspring goat to obtain the human protein. Each goat can produce more clotting protein than can be isolated from tens of thousands of human blood donations.

(**FIGURE 10.16**). Concerns about pharming include the ethics of using animals in this manner and the unregulated spread of modified plant or animal genes via reproduction.

Gene Editing Using CRISPR in Plants and Animals

Instead of combining genes from species that would never mate in nature to produce transgenic organisms, scientists have recently discovered a way to directly alter, delete, or even replace a DNA sequence in an organism. The powerful new tool scientists are using for gene editing is a process known as clustered regularly interspaced short palindromic repeats, abbreviated CRISPR and pronounced "crisper." Widespread application of this technology—which is faster, more effective, and less cumbersome than its predecessor techniques—will allow precise modification of crop plants and animals to delete mutations and to add desired gene sequences.

CRISPR is a naturally occurring part of a bacterium's immune system. Its evolution allowed bacteria to thwart a viral infection by removing genetic information present in an invading virus. Scientists have co-opted this bacterial system for use in other organisms. There are two components in the CRISPR system: The first component is a guide that finds the DNA sequence it has been sent to edit, and the second is an enzyme that functions like molecular scissors to snip out unwanted DNA and replace it with the corrected DNA sequence.

Scientists are experimenting with the use of this technique to delete genes that attract pests in crop plants, decreasing the use of pesticides. In addition, crops may someday be modified to stay fresh longer or to contain more nutrients or fiber. Alterations to animals are on the horizon as well. Livestock breeders may be able to use CRISPR to produce animals with more muscle mass or leaner meat. Scientists are even attempting to edit genes in pig embryos to make their organs suitable for transplantation to humans.

Perhaps the most exciting use of CRISPR is the ability to edit genes involved in reproduction in the mosquito that causes malaria, leading to the eradication of this species of mosquito. This would prevent the deaths of close to 1 million African people a year, 70% of whom are children (**FIGURE 10.17**).

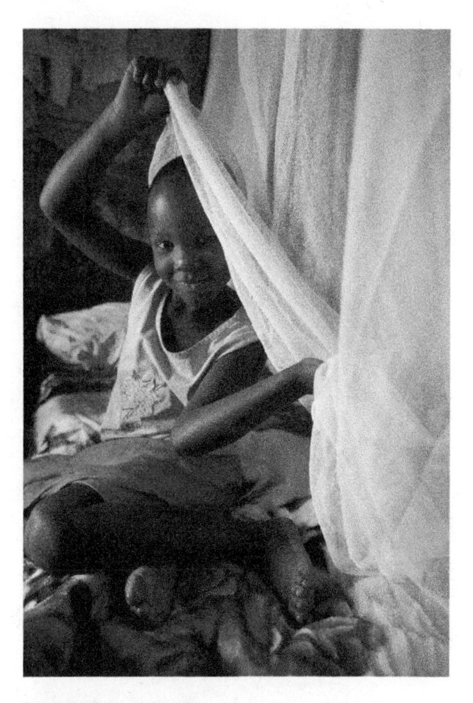

FIGURE 10.17 Malaria. Malaria is a life-threatening disease spread to humans through the bite of an infected mosquito. The use of mosquito netting has helped lower the death toll but more than half a million, mostly African children, still die from the disease annually.

Got It?

1. Farmers choose which strains to breed together in a process called __________ __________.
2. An organism that has been genetically modified to contain DNA from another organism is a __________ organism.
3. A __________ __________ can be used to shoot DNA-coated pellets into plant cells.
4. Biological pharming is the use of animals to produce __________ that are missing or defective in humans.
5. The CRISPR system is being used by scientists to make precise genetic alterations or __________ to DNA sequences.

10.4 Genetically Modified Humans

Not surprisingly, genetic modification of humans has been even more controversial than modification of other living things. From the use of stem cells to the cloning of humans, the potential risks and benefits are still emerging.

Stem Cells

Stem cells are unspecialized or **undifferentiated** precursor cells that have not yet been programmed to perform a specific function. They can become any other cell type. Imagine that you are remodeling an old home, and you have a type of material that you can mold into anything you might need—brick, tile, pipe, plaster, and so forth. Scientists believe that stem cells, when nudged in a particular direction, can become any type of cell and therefore may serve as this type of all-purpose repair material in the body.

As an embryo develops, its cells become less and less able to produce other cell types. As a human embryo grows, the early cells start dividing and forming different specialized cells such as heart cells, bone cells, and muscle cells. Once formed, specialized non–stem cells can divide only to produce replicas of themselves. They cannot backtrack and become a different type of cell.

Stem cells can be isolated from early embryos that are left over after fertility treatments. *In vitro* (Latin, meaning "in glass") fertilization procedures often result in the production of excess embryos because many egg cells are harvested from a woman who wishes to become pregnant. These egg cells are then mixed with her partner's sperm in a petri dish, often resulting in the production of many fertilized eggs that grow into embryos. A few of the embryos are then implanted into the woman's uterus. The remaining embryos are stored so that more attempts can be made if pregnancy does not result or if the couple desires more children. When the couple achieves the desired number of pregnancies, they are faced with a choice of what to do with the remaining embryos. They may choose to discard them, donate them to other couples experiencing fertility problems, or allow them to be used in stem cell research.

Stem cells can also be found in nonembryonic tissues, including the umbilical cord of a newborn and in the primary teeth of children. Adult stem cells have been found in a variety of organs and tissues, including bone marrow, some blood vessels and muscles, the brain, and the liver. The role of stem cells in the adult body is to help maintain tissues and replace damaged or diseased cells. Adult stem cell research has, in some ways, progressed more rapidly than expected due to the wide variety of tissues in which these cells can be found but has been hampered because these cells also occur in small numbers in a particular tissue and will divide only a limited number of times. These properties make establishing useful cultures of stem cells somewhat more difficult when using stem cells derived from adult tissues versus those derived from embryos.

Whether derived from embryos or the tissues of children and adults, stem cells may someday be used to replace organs damaged in accidents or organs that are gradually failing due to **degenerative diseases.** Degenerative diseases, like liver and lung diseases, heart disease, multiple sclerosis, Alzheimer's, and Parkinson's start with the slow breakdown of an organ and progress to organ failure.

Stem cells could provide healthy tissue to replace those tissues damaged by spinal cord injury or burns, for example. Using stem cells to produce healthy tissues as replacements for damaged tissues is a type of **therapeutic cloning.** New heart muscle could be produced to replace muscle damaged during a heart attack. A diabetic could be supplied with a new pancreas, and people suffering from some types of arthritis could have replacement cartilage to cushion their joints. Thousands of people waiting for organ transplants might be saved if new organs were grown in the lab.

Gene Therapy

Scientists who try to replace defective human genes with functional genes are performing **gene therapy.** One type of gene therapy, called **somatic cell gene therapy,** can be performed on body cells to fix or replace the defective protein in only the affected cells. Using this method, scientists introduce a functional version of a defective gene into an affected individual cell in the laboratory, allow the cell to reproduce, and then place the copies of the cell bearing the corrected gene into the diseased person.

Gene therapy to date has focused on diseases caused by single genes for which defective cells can be removed from the body, treated, and reintroduced to the body. For example, studies are underway to determine whether individuals with the genetically inherited lung disease cystic fibrosis may be able to breathe in harmless viruses carrying a functioning copy of the cystic fibrosis gene. If that gene gains access to lung cells and replaces the nonfunctional version with the functional one, lung function could improve (**FIGURE 10.18**).

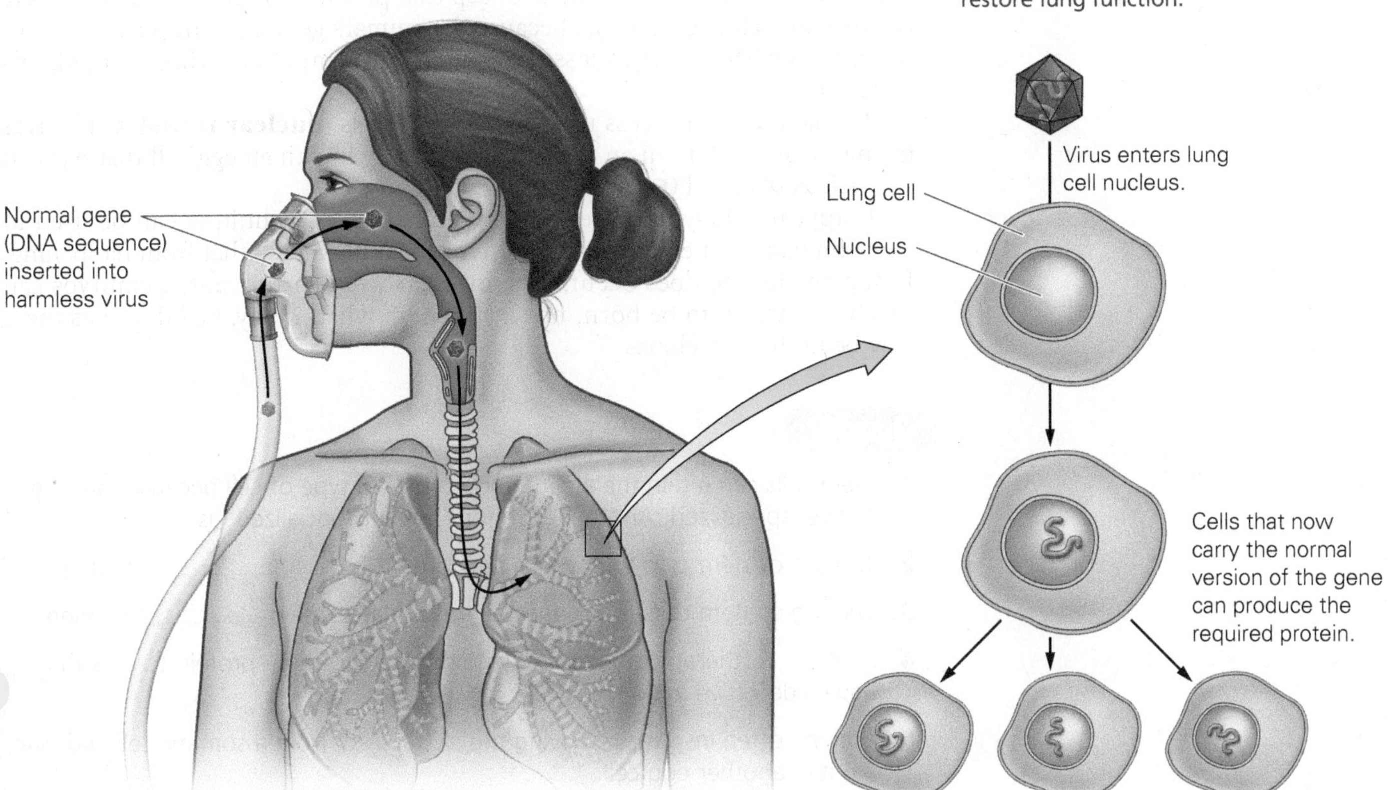
Normal gene (DNA sequence) inserted into harmless virus
Lung cell
Nucleus
Virus enters lung cell nucleus.
Cells that now carry the normal version of the gene can produce the required protein.

FIGURE 10.18 Gene therapy. A person with cystic fibrosis has defective genes that prevent normal lung function. Inhaling copies of the normal gene, inserted into a harmless virus, can help restore lung function.

One additional type of genetic engineering involves making an exact copy of an entire organism by a process called **reproductive cloning.** This process has proven very controversial, especially when the cloning involves humans.

Gene Editing in Humans

Gene editing could one day help cure or prevent human disease. CRISPR is being used to engineer tumor cells that are tested in the laboratory to determine the efficacy of cancer treatments and as a way to remove HIV from human cells. Someday, it may even be possible to fix mutant genes by editing their DNA sequence. For example, in the blood-clotting disorder hemophilia, it may be possible to remove blood cells from a hemophiliac, edit the cells outside the body, and then replace the nonmutant version back in the blood stream. Scientists have already had some luck, working with nonhuman animals, in editing the genes that cause conditions such as muscular dystrophy and cystic fibrosis. It may someday be possible to edit out a genetic disease in a sperm or egg cell during laboratory mediated *in vitro* fertilization attempts.

Cloning Humans

Although we are not used to thinking of identical twins as human clones, they are actually clones of each other. Identical twins arise when an embryo subdivides itself into two separate embryos early in development. When animals are cloned in the laboratory a different process is used, but the result is the same in the sense that the cloned organisms are twins; they are just born many years apart.

Many different animals have been cloned in the laboratory, including sheep, cattle, goats, pigs, horses, mice, cats, rabbits, camels, and monkeys. Initial attempts at cloning focused on livestock that had genetic traits that made them beneficial to farmers. Sheep that produced high-quality wool were the very first cloned animals. Because the animal's genome is duplicated, cloning prize livestock leaves less to chance than simply allowing two animals to breed.

The laboratory process used to clone animals, **nuclear transfer,** involves taking the nucleus from an adult cell and fusing it with an egg cell that has had its nucleus removed (**FIGURE 10.19**).

It appears likely that the same nuclear transfer technique can be used to clone humans, but ethical concerns have so far prevented that from happening. If human cloning does occur, it is not clear whether implanted embryos can survive gestation to be born, nor is it known what, if any, health issues there may be for human clones.

Got It?

1. Stem cells can retain the ability to become any type of cell because they are not yet specialized. Another term for "not yet specialized" is ________.
2. The use of stem cells to produce new organs is called ________ cloning.
3. The use of stem cells to produce an individual is called ________ cloning.
4. ________ therapy attempts to provide a replacement protein to cells that have a defective gene.
5. Cloning often involves removing the ________ from a somatic cell and placing it in another egg cell.

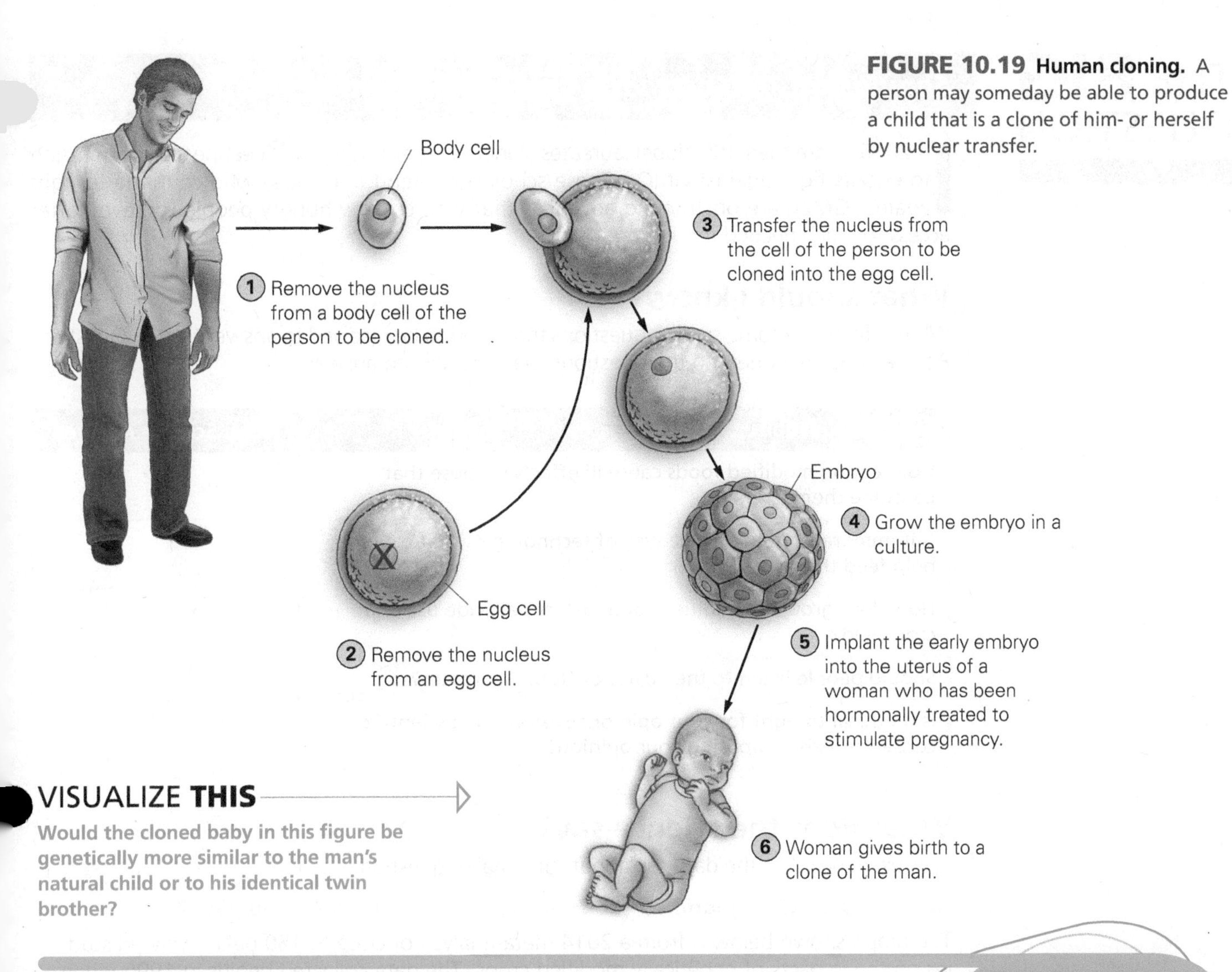

Body cell
1 Remove the nucleus from a body cell of the person to be cloned.
3 Transfer the nucleus from the cell of the person to be cloned into the egg cell.
Egg cell
2 Remove the nucleus from an egg cell.
Embryo
4 Grow the embryo in a culture.
5 Implant the early embryo into the uterus of a woman who has been hormonally treated to stimulate pregnancy.
6 Woman gives birth to a clone of the man.

FIGURE 10.19 Human cloning. A person may someday be able to produce a child that is a clone of him- or herself by nuclear transfer.

VISUALIZE THIS

Would the cloned baby in this figure be genetically more similar to the man's natural child or to his identical twin brother?

Sounds right, but is it?

Human cloning is often presented as though clones grow and develop inside a test tube until the adult human clones produced are indistinguishable from the original human they were cloned from and indistinguishable from each other as well. If you want to produce five clones of NBA player Stephen Curry, place five of his treated cells in five separate test tubes and incubate until the fully formed clones are ready to take the court. Each of these clones would have the athletic ability, height, drive, and temperament to succeed in the NBA.

Cloned humans are identical in every way.

Sounds right, but it isn't.

Answer the following questions to understand why.

1. What has to happen to an embryo produced by human cloning for continued early development to occur?
2. Would human reproductive cloning using cells from an adult immediately produce an adult clone?
3. If Stephen Curry was cloned five times, and each of the babies grew up in a household with different eating habits, ranging from very unhealthy to very healthy, do you think all five clones would grow to be exactly 6 feet, 3 inches tall?
4. Identical twins are clones of each other. List a few nongenetic factors that account for the fact that clones like identical twins do not share the same personalities, ambitions, and interests.
5. Consider your answers to questions 1–4 and explain why the original statement bolded above sounds right, but isn't.

Should anti-GMO activists give up the fight?

In 2016, more than 100 Nobel laureates signed a letter urging the Greenpeace organization to end its fight against GMOs. These scholars contend that those who continue to fight against GMOs are obstructing advances that would allow hungry people to have better access to much needed foods.

What should I know?

What follows are some smaller questions that need to be resolved to answer the Big Question. Place a checkmark next to the questions that science can answer.

Smaller Question	Can Science Answer?
Does eating modified foods cause ill effects in those that consume them?	
Is it immoral to stand in the way of technologies that could help feed the hungry?	
How does growing modified foods affect pesticide use and crop yield?	
Should people listen to the advice of Nobel laureates?	
Is it ethical to fight for your opinion, even when a scientific consensus exists opposing your opinion?	

What does the science say?

Let's examine what the data say about this smaller question:

How does growing modified foods affect pesticide use and crop yield?

The graph shown below is from a 2014 meta-analysis of close to 150 peer-reviewed studies on the impacts of genetically modified crops. The data collected begins in 1996 when a large number of farmers began switching to genetically modified crops. The asterisk indicates a significant difference from 1996 to 2014.

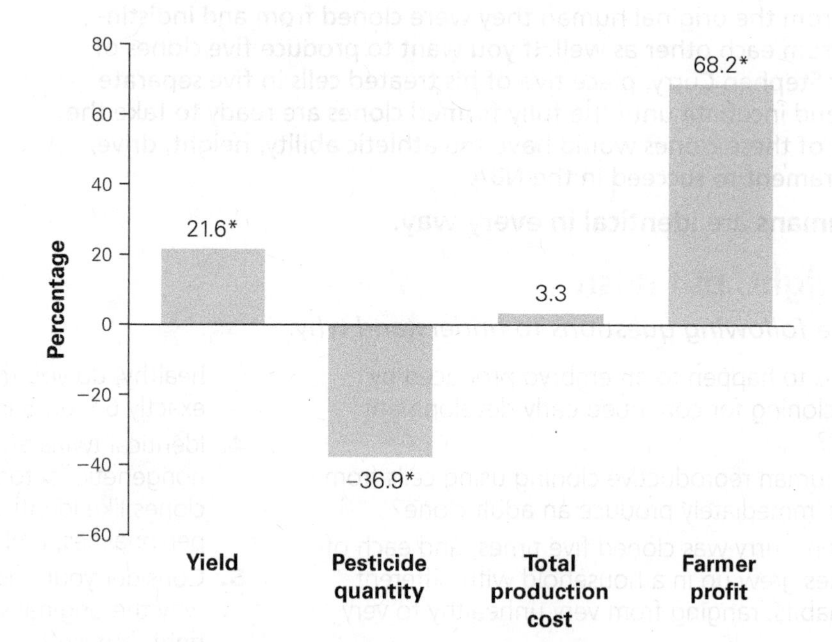

1. Describe the results. Does it appear growing modified foods affects pesticide use and crop yield?
2. Given these data, do you think the smaller question is answered? If not, propose another study that would help answer this question.
3. Does this information help you answer the Big Question? What else do you need to consider?

Data source: W. Klumper and M. Qaim, "A Meta-Analysis of the Impacts of Genetically Modified Crops," *PLoS One* 9, no. 11 (2014).

Chapter Review

Mastering Biology

Go to Mastering Biology to access **eText 2.0, Dynamic Study Modules,** and the **Study Area,** where you'll find practice quizzes, BioFlix™ animations, MP3 tutor sessions, current events, and more.

SUMMARY

Section 10.1

Define the term *gene expression.*

- Gene expression is the regulation of protein-producing genes. Such regulation can include lowering or increasing the amount of transcription and translation of the gene (p. 184).

Describe how messenger RNA is synthesized during the process of transcription.

- Transcription occurs in the nucleus of eukaryotic cells when an RNA polymerase enzyme binds to the promoter, located at the start site of a gene, and makes an mRNA that is complementary to the DNA gene (pp. 184–186).

Describe how proteins are synthesized during the process of translation.

- Translation occurs in the cytoplasm of eukaryotic cells and involves mRNA, ribosomes, and transfer RNA. Messenger RNA carries the code from the DNA, and ribosomes are the site where amino acids are assembled to synthesize proteins. Transfer RNA carries amino acids, which bind to triplet nucleotide sequences on the mRNA called codons. A particular tRNA carries a specific amino acid. Each tRNA has its unique anticodon that binds to the codon and carries instructions for its particular amino acid (pp. 186–189).

Define the term *mutation,* and explain how mutations can affect protein structure and function.

- Mutations are changes to DNA sequences that can affect protein structure and function (p. 190).
- Neutral mutations are changes to the DNA that do not result in a different amino acid being incorporated. Insertions or deletions of nucleotides can result in frameshift mutations that change the protein more drastically (p. 190).

SHOW YOU KNOW 10.1 Use the genetic code in Table 10.1 to determine the three DNA sequences that would result in the end of translation.

Section 10.2

Outline the process of cloning a gene using bacteria.

- Bacteria can be used to clone genes by placing the gene of interest into a plasmid, which then makes millions of copies of the gene as the plasmid replicates itself inside its bacterial host. Bacteria can then express the gene by transcribing an mRNA copy and translating the mRNA into a protein (pp. 192–194).

Describe how FDA approval for rBGH milk is obtained.

- Scientists showed that milk from treated cows is not harmful to cows that produce the milk or humans that drink the milk (p. 194).

SHOW YOU KNOW 10.2 Name the protein, produced by genetic engineering, that is missing or defective in people with diabetes.

Section 10.3

Describe how plants and animals are genetically modified.

- A gene gun can be used to insert a particular gene into embryonic plant cells (pp. 195–196).
- Animal cells can be modified by injection of genes into an early embryo (pp. 196–197).

List some of the health and environmental concerns surrounding genetic engineering.

- Although there have been no documented incidents of negative health effects from genetically modified food consumption, some consumers remain skeptical. Environmental concerns about genetically modified crops include their impacts on surrounding organisms, the evolution of resistances, and transfer of modified genes to wild or weedy relatives (pp. 196–197).

Describe the mechanism by which CRSIPR technology can be used to modify an organism's genome.

- Genes can be edited in any manner scientists like by first removing the DNA sequence then replacing it with a desired sequence (p. 197).

SHOW YOU KNOW 10.3 Would an organism modified using CRISPR be considered a transgenic organism? Why or why not?

Section 10.4

Describe the science behind, and significance of, stem cells and gene therapy and gene editing.

- Stem cells are undifferentiated cells that can be reprogrammed to act as a variety of cell types. Stem cells may someday allow scientists to treat degenerative diseases (pp. 198–199).
- Gene therapy involves replacing defective genes with normal genes (pp. 199–200).
- Gene editing using CRISPR may someday be used to fix defective genes in humans, including human embryos (p. 200).

Discuss the process of human cloning.

- Cloning occurs via nuclear transfer. Typically, cloning is undertaken to propagate animals with desirable agricultural traits. (p. 200).

SHOW YOU KNOW 10.4 When in the human life cycle does a defective gene need to be fixed so that it will be fixed in all cells of the adult body?

ROOTS TO REMEMBER

The following roots of words come mainly from Latin and Greek and will help you to decipher terms:

-ic	is a common ending of acids. Chapter terms: *deoxyribonucleic acid, ribonucleic acid*
nucleo-	refers to a nucleus. Chapter term: *nucleotide*
-ose	is a common ending for sugars. Chapter terms: *ribose, deoxyribose*
-some and -somal	relates to a microscopic body visible in a cell. Chapter terms: *chromosome, ribosome*
trans-	means across. Chapter terms: *transgenic*

LEARNING THE BASICS

1. List the order of nucleotides on the mRNA that would be transcribed from the following DNA sequence: CGATTACTTA.
2. Using the genetic code (Table 10.1), list the order of amino acids encoded by the following mRNA nucleotides: CAACGCAUUUUG.
3. Transcription ________.

   A. synthesizes new daughter DNA molecules from an existing DNA molecule; B. results in the synthesis of an RNA copy of a gene; C. pairs thymines (T) with adenines (A); D. occurs on ribosomes
4. Transfer RNA (tRNA) ________.

   A. carries monosaccharides to the ribosome for synthesis; B. is made of messenger RNA; C. has an anticodon region that is complementary to the mRNA codon; D. is the site of protein synthesis
5. During the process of transcription, ________.

   A. DNA serves as a template for the synthesis of more DNA; B. DNA serves as a template for the synthesis of RNA; C. DNA serves as a template for the synthesis of proteins; D. RNA serves as a template for the synthesis of proteins
6. Translation results in the production of ________.

   A. RNA; B. DNA; C. protein; D. individual amino acids; E. transfer RNA molecules
7. The RNA polymerase enzyme binds to ________, initiating transcription.

   A. amino acids; B. tRNA; C. the promoter sequence; D. the ribosome

8. A particular triplet of bases in the coding sequence of DNA is TGA. The anticodon on the tRNA that binds to the mRNA codon is ________.

A. TGA; **B.** UGA; **C.** UCU; **D.** ACU

9. RNA and DNA are similar because ________.

A. both are double-stranded helices; **B.** uracil is found in both of them; **C.** both contain the sugar deoxyribose; **D.** both are made up of nucleotides consisting of a sugar, a phosphate, and a base

10. Add labels to the figure that follows, which illustrates a transfer RNA molecule.

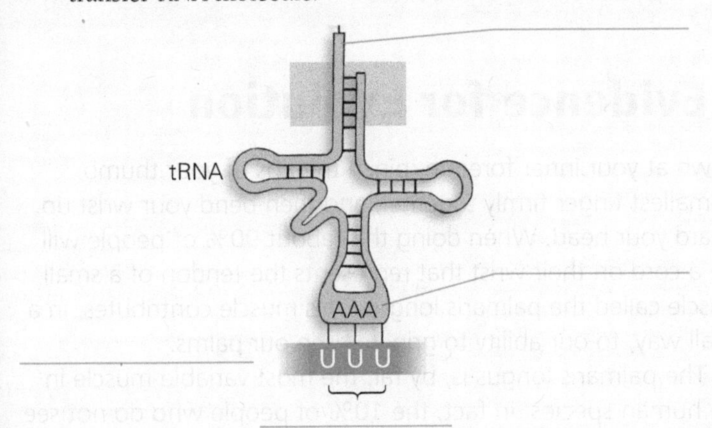

ANALYZING AND APPLYING THE BASICS

1. Take another look at the genetic code (Table 10.1). Do you see any similarities between codons that code for the same amino acid? Based on this difference, why might a mutation that affects the nucleotide in the third position of the codon be less likely to affect the structure of the protein than a mutation that affects the codon in the first position?

2. Genes encode RNA polymerase molecules. What would happen to a cell that has undergone a mutation to its RNA polymerase gene?

3. Draw a box around the six-base-pair site at which a restriction enzyme would most likely be cut.

ATGAATTCCGTCCG
TACTTAAGGCAGGC

GO FIND OUT

4. Do some Internet research to determine whether manufacturers of genetically modified products are required to label their products as modified. Do you think modified products should be labeled even if no difference has been found between a modified and nonmodified version of the product? Why or why not?

MAKE THE CONNECTION

The science that you learned in this chapter has helped you better understand the real-world example used throughout this discussion. Draw a line from the statement on the left to the science that supports it on the right.

Milk from cows treated with rBGH is safe to drink.	Bacteria can produce proteins missing in those with diabetes.
GMOs can help treat human diseases.	FDA testing shows no difference in the product of treated or untreated organisms.
Humans have been eating genetically modified foods for 20 years, with no known adverse health effects.	Artificial selection allows for the selective breeding of animals and crop plants with desired characteristics.
From ancient humans to modern ones, a low technology version of genetic modification has been a common farming practice.	Cloning organisms involves the use of embryos.
The long-term effects of GMO crops on the environment are unknown.	Most processed foods include use of genetically modified syrups and oils.
There are ethical concerns about genetically modifying humans.	Pollen from genetically modified crops can fertilize related plants.

Answers to **Got It?, Visualize This, Working with Data, Sounds Right, But Is It?, Show You Know,** and **Chapter Review** questions can be found in the **Answers** section at the back of the book.

11 Where Did We Come From?

Our ancestors used their hands to swing from branches, aided by a strong muscle in the wrist and palm.

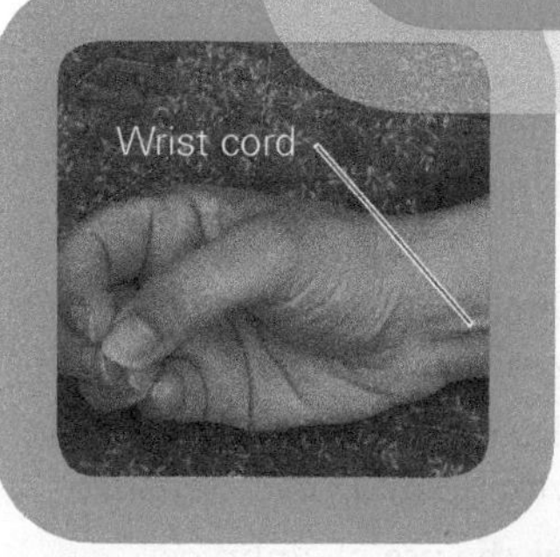

One idea about the origin of humans: special creation.

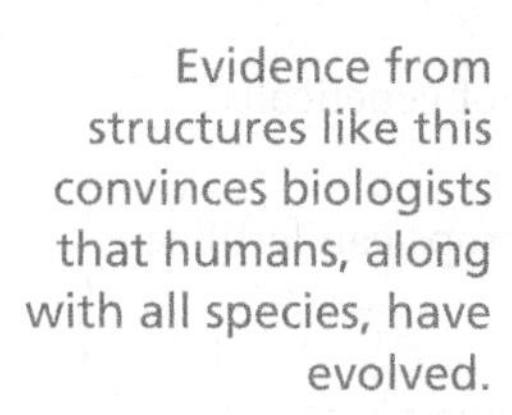

Evidence from structures like this convinces biologists that humans, along with all species, have evolved.

The Evidence for Evolution

Look down at your inner forearm, pinch the tips of your thumb and smallest finger firmly together, and then bend your wrist up toward your head. When doing this, about 90% of people will see a cord on their wrist that represents the tendon of a small muscle called the palmaris longus. This muscle contributes, in a small way, to our ability to grip items in our palms.

The palmaris longus is, by far, the most variable muscle in the human species. In fact, the 10% of people who do not see that wrist cord are likely missing this muscle entirely. Remarkably, there is no obvious difference in hand performance or wrist strength between people who have the muscle and those who do not. How can we explain why some people are missing an entire arm muscle without any obvious effect?

The palmaris longus is an important muscle for animals that move by brachiating—that is, by using their forelimbs to swing through the trees. The animals that move this way include monkeys, gibbons, and lemurs. Humans do not move in this way, but many of us have an anatomical structure that was designed to help us do so. Biologists explain the presence of a barely functional palmaris longus in humans as a product of evolution—evidence of our relationship to monkeys and lemurs and also evidence that without any advantage to possessing a well-developed muscle for brachiating, that muscle is disappearing.

Accepting this explanation for missing palmaris longus muscles means accepting the principle that human beings "evolved from less advanced life forms over millions of years." This is the wording of a question in a regularly performed poll that, in the United States, only 35% of people surveyed agree with. In the most recent version of the same poll, 47% of respondents agreed with the statement, "God created human beings in their present form in the last 10,000 years"—in other words, nearly half the people polled believe in creationism.

To biologists, these poll results are puzzling. The theory of evolution, including the concept that humans evolved from nonhuman ancestors, forms the bedrock of biological science.

The theory effectively explains not only the difference among people in palmaris longus muscles, but also our ability to move genes from one organism to another. The vast majority of scientists of all types agree with the statement that humans have evolved via natural processes. In this chapter, we examine this debate by exploring the theory of evolution and the origin of humans as a matter of science.

11.1 What Is Evolution?

Evolution really has two different meanings to biologists. The term can refer to either a process or to an organizing principle—in other words, a theory (which, as we will see, has a different definition in scientific terminology than in everyday speech).

The Process of Evolution

Generally speaking, the word *evolution* is used to mean "change," and the process of evolution reflects this definition as it applies to populations of organisms. A **biological population** is a group of individuals of the same species that is somewhat independent of other groups, often isolated from them by geography. **Biological evolution,** then, is a change in the characteristics of a biological population that occurs over the course of generations. The changes in populations that are considered evolutionary are those that are passed from parent to offspring via genes.

evol- means to unroll.

Changes that may take place in populations due only to short-term changes in their environment are not evolutionary. For example, the average dress size for women in the United States has increased from 8 to 14 over the past 50 years. This change is not genetic but occurred because of an increase in average calorie intake in the population. Thus, it is not an evolutionary change.

As an example of the process of biological evolution, consider the species of organism commonly known as head lice. Some populations of head lice in the United States have become resistant to the pesticide permethrin, found in over-the-counter delousing shampoos. In previous decades, lice infections were readily controlled through treatment with these products; however, over time, populations of lice evolved to become less susceptible to the effects of these chemicals. Unlike the example of the change in women's dress sizes resulting from increased food availability, this change in the susceptibility of the lice population to permethrin is a result of a change in their genes. The evolution of pesticide resistance can occur rapidly—a study in Israel demonstrated that populations of head lice were four times less susceptible to permethrin only

VISUALIZE **THIS**

How would this figure be different if none of the lice in this child's hair were resistant to pesticide? Could evolution occur?

Initial lice infestation consists of both susceptible (white) and resistant (orange) lice.

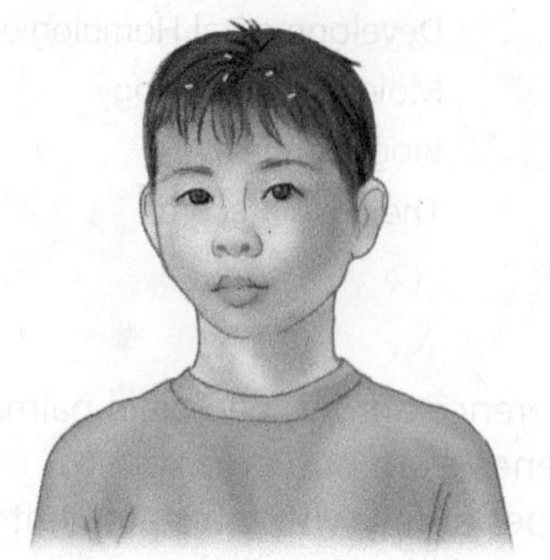

After permethrin treatment, most lice are dead, but a few that are resistant to the pesticide survive.

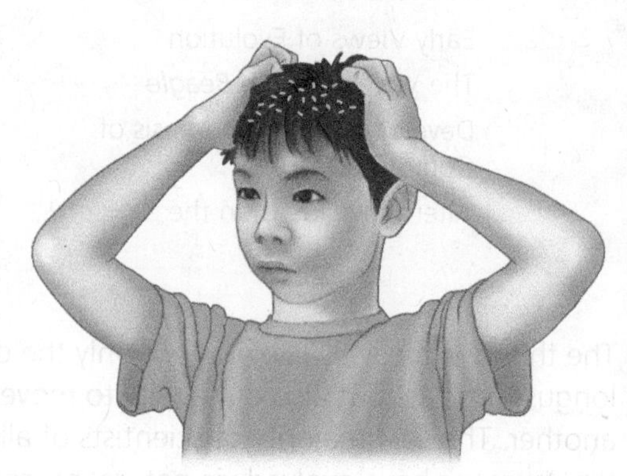

Reinfestation with the offspring of the resistant lice. The population of lice is now more resistant to permethrin.

FIGURE 11.1 The process of evolution. The evolution of pesticide resistance in lice occurs as a result of natural selection, one of the mechanisms by which traits in a population can change over time.

30 months (or 40 lice generations) after the pesticide was introduced in that country. Note that in this example, individual head lice did not "evolve" or change; instead, the population as a whole changed from one in which most lice were susceptible to the pesticide to one in which most lice were resistant to it.

The differences in resistance to permethrin among individuals in the population resulted from genetic variation. In particular, some lice carried gene variants (that is, alleles) that conferred resistance, and others carried nonresistant alleles. Because individuals with the resistant alleles survived permethrin treatment, they passed these resistant alleles on to their offspring. As a result, a population made up primarily of individual lice that carried the susceptible alleles changed into one in which most of the lice carried the resistant alleles. This change in the characteristics of the population took multiple generations (**FIGURE 11.1**).

According to the definition of evolution as a process, the population of lice has evolved. In this case, the process of **natural selection,** the differential survival and reproduction of individuals in a population, brought about the evolutionary change. Natural selection is the process by which populations adapt to their changing environment (Chapter 12). Other forces, including chance, can cause evolutionary changes in the genetic makeup of populations as well (Chapter 13).

Most people accept that traits in populations can evolve. Evolutionary change in biological populations, such as the development of pesticide resistance in insects and antibiotic resistance in bacteria (Chapter 12), has been observed multiple times. Changes that occur within a biological population are referred to as **microevolution.** The accumulation of microevolutionary changes that results in the origin of new species is known as **macroevolution.**

micro- means extremely small.
macro- means large scale.

The Theory of Evolution

Although some Americans do not accept that microevolutionary changes occur within a species, theirs is a minority view. However, as the surveys described at the beginning of the chapter illustrate, many question whether the theory of evolution, which includes both the process of microevolution and its macroevolutionary results, is "scientific truth."

Part of the issue may be a result of confusion over the use of the word *theory*. When people use *theory* in everyday conversation, they often are referring to a tentative explanation with little systematic support—in other words, a hunch. For example, a sports fan might have a theory about why her team is losing, or a gardener might have a theory about why his roses fail to bloom. But a **scientific theory** is much more substantial—it is a statement that provides the current best explanation of how the universe works (Chapter 1). Scientific theories are supported by numerous lines of evidence and have withstood repeated experimental tests. For instance, the theory of gravity explains the motion of the planets, and germ theory explains the relationship between microorganisms and human disease.

The theory of evolution is a principle for understanding how species originate and why they have the characteristics that they exhibit. The **theory of evolution** can thus be stated:

> All species present on Earth today are descendants of a single common ancestor, and all species represent the product of millions of years of accumulated microevolutionary changes.

In other words, modern animals, plants, fungi, bacteria, and other living things are related to each other and have been diverging from their common ancestor, by various processes, since the origin of life on this planet. Although the processes that cause evolution—such as natural selection—are not the subject of controversy among the general public, the part of the theory of evolution that states that all life shares a common ancestor, called the theory of **common descent,** is (**FIGURE 11.2**). It is this controversy that underlies the opposition to what is often termed *Darwinism*.

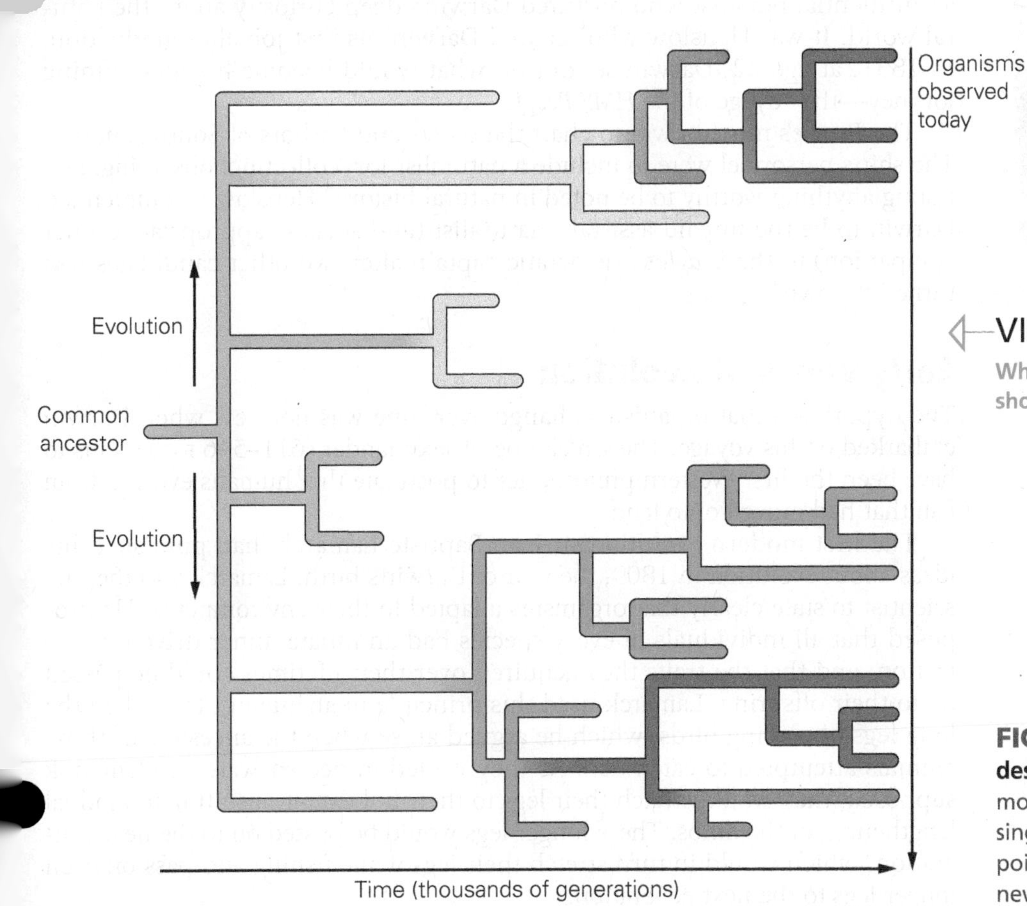

VISUALIZE THIS

Why do some branches on the tree stop short of the right side of the tree?

FIGURE 11.2 The theory of common descent. This theory states that all modern organisms are descended from a single common ancestor. Each branching point on the tree represents the origin of new species from an ancestral form.

Got It?

1. Biological evolution is the change that occurs in a population over the course of ________.
2. ________ is a process that causes evolution and occurs when some individuals in a population have greater or lower survival or reproduction than others.
3. The evolution of pesticide resistance in lice is an example of ________.
4. A scientific ________ is an explanation of how things work that is supported by multiple lines of evidence.
5. The theory of common descent states that all life shares a single ________

11.2 Charles Darwin and the Theory of Evolution

FIGURE 11.3 Charles Darwin. Beginning when he was a young naturalist, Darwin conceived of and developed a theory of evolution that remains one of the most well-supported ideas in biology.

The theory of evolution is sometimes called "Darwinism" because Charles Darwin is credited with bringing it into the mainstream of modern science (**FIGURE 11.3**).

The youngest son of a wealthy physician, Darwin spent much of his early life as a lackluster student. After dropping out of medical school, Darwin entered Cambridge University to study for the ministry at the urging of his father. Darwin hated most of his classes but did strike up friendships with several scientists at the college. One of his closest companions was Professor John Henslow an influential botanist who nurtured Darwin's deep curiosity about the natural world. It was Henslow who secured Darwin his first job after graduation. In 1831, at age 22, Darwin set out on what would become his life-defining journey—the voyage of the *HMS Beagle*.

The *Beagle*'s mission was to chart the coasts and harbors of South America. The ship's personnel were to include a naturalist for "collecting, observing, and noting anything worthy to be noted in natural history." Henslow recommended Darwin to be the unpaid assistant naturalist (and socially appropriate dinner companion) to the *Beagle*'s aristocratic captain after two other candidates had turned it down.

Early Views of Evolution

The hypothesis that organisms change over time was not new when Darwin embarked on his voyage. The Greek poet Anaximander (611–546 B.C.) seems to have been the first Western philosopher to postulate that humans evolved from fish that had moved onto land.

The first modern evolutionist, Jean Baptiste Lamarck, had published his ideas about evolution in 1809, the year of Darwin's birth. Lamarck was the first scientist to state clearly that organisms adapted to their environments. He proposed that all individuals of every species had an innate, inner drive for perfection, and that the traits they acquired over their lifetimes could be passed on to their offspring. Lamarck used this principle in an attempt to explain the long legs of wading birds, which he argued arose when the ancestors of those animals attempted to catch fish. As they waded in deeper water, so Lamarck supposed, they would stretch their legs to their full extent, resulting in gradual lengthening of the limbs. These longer legs would be passed on to the next generation, which would in turn stretch their legs while fishing and pass on even longer legs to the next generation.

FIGURE 11.4 Offspring do not inherit their parents' acquired traits. Arnold Schwarzenegger was a champion body builder before he became an actor and politician. Lamarck's theory of the inheritance of acquired characteristics suggests that Arnold's son Patrick would also have very developed muscles. Patrick's more average physique demonstrates that Lamarck's ideas were incorrect.

Lamarck's contemporaries were unconvinced by his proposed mechanism for species change—for instance, it was easily seen that the children of muscular blacksmiths had similar-sized biceps to those of bankers' children and had not inherited their fathers' highly developed muscles (**FIGURE 11.4**). Lamarck's critics were also unwilling to challenge the more socially acceptable alternative hypothesis that Earth and its organisms were created in their current forms by God and did not change over time. It was this more acceptable hypothesis that Darwin found himself questioning as a result of his around-the-world voyage.

The Voyage of the *Beagle*

In the 5 years that he spent on the expedition of the *HMS Beagle,* Darwin spent most of his time on land—luckily for him as it turns out, because he was nearly constantly seasick on board the ship. The trip was a dramatic awakening for the young man, who was awed at the sight of the Brazilian rain forest, amazed by the scantily clothed natives in the chilly climate of Tierra del Fuego, and intrigued by the diversity of animals and plants he collected (**FIGURE 11.5**).

On the ship, Darwin had ample time to read, including Charles Lyell's book *Principles of Geology,* which put forth the hypothesis that the geological processes working today are no different from those working in the past, and that large geological features result from the accumulated effect of these geological forces. In supporting this idea, Lyell argued that deep canyons resulted from the gradual erosion of rock by rivers and streams over enormous timescales. Lyell's hypothesis called into question the belief that Earth was less than 10,000 years old, an age that was deduced from a literal reading of the Bible.

Darwin was also strongly influenced by a stop in the Galápagos, a small archipelago of volcanic islands off the coast of Ecuador. Although at first look the islands seemed nearly lifeless, during the month that the *Beagle* spent sailing them, Darwin collected an astonishing variety of organisms. Many of the birds and reptiles

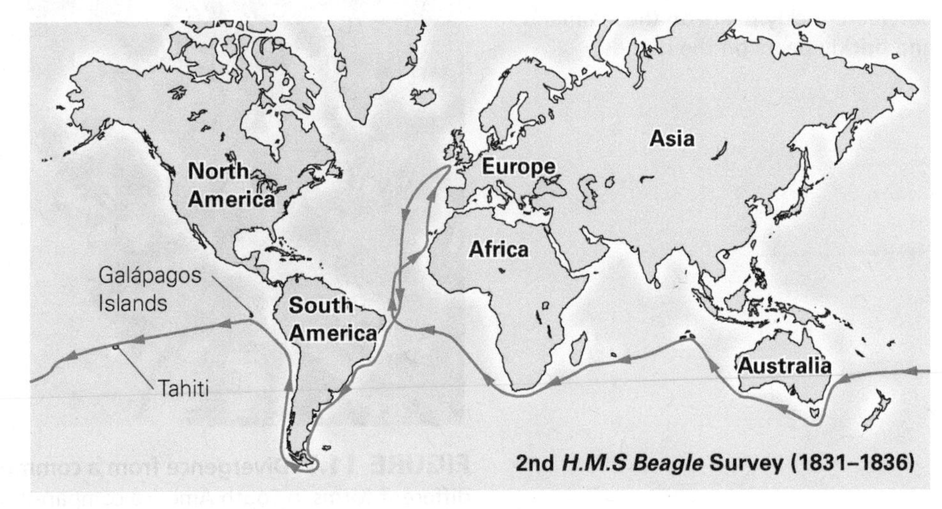

FIGURE 11.5 The voyage of HMS Beagle. Darwin's expedition took him to tropical locales from South America to Tahiti.

FIGURE 11.6 Giant tortoises of the Galápagos. The subspecies of giant tortoises on the Galápagos Islands from different environments look distinct from one another.

(a) Dome-shelled tortoise from Santa Cruz Island, an island with abundant ground-level vegetation

(b) Flat-shelled tortoise from Española Island, an island with tall vegetation

he observed appeared to be unique to each island. For instance, although all islands had populations of tortoises, the type of tortoise found on one island was different from the types found on other islands (**FIGURE 11.6**).

Darwin wondered why God would place different, unique subtypes of tortoise on islands in the same small archipelago. On his return to England, Darwin reflected on his observations and concluded that the populations of tortoises on the different islands must have descended from a single ancestral tortoise population. He noted a similar pattern of divergence between other groups of species on the closest mainland and the Galápagos Islands—for instance, prickly pear cacti in mainland Ecuador have the ground-hugging shape familiar to many of us, whereas on the Galápagos, these plants are tree size, but both the mainland and island cacti are clearly related (**FIGURE 11.7**).

Developing the Hypothesis of Common Descent

Darwin's observations and portions of his fossil collection, sent periodically back to Henslow via other ships, made Darwin a scientific celebrity even before the *Beagle* returned to England. A journal of his travels was a bestseller,

VISUALIZE **THIS**

Consider these photos along with Figure 11.6. What environmental factors might have caused the difference in evolution between prickly pears on the mainland and prickly pears on the islands?

(a) Prickly pear cactus in mainland South America

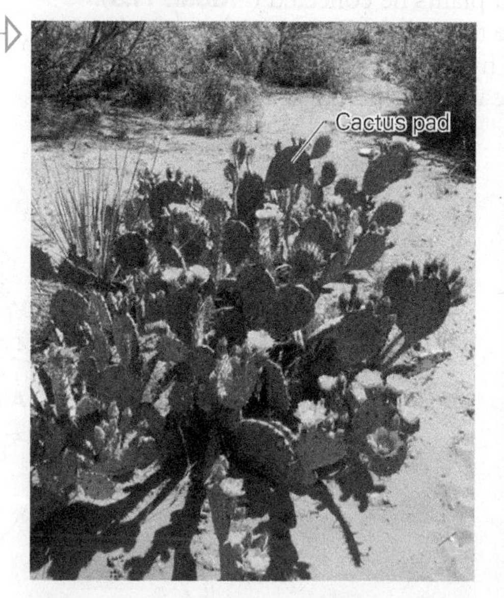

(b) Prickly pear cactus in Galápagos

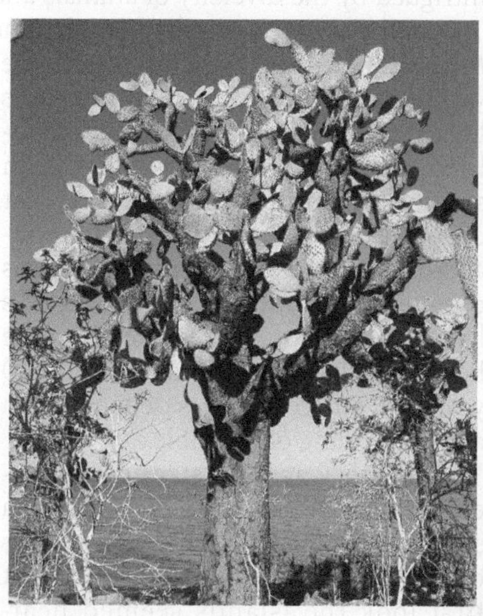

FIGURE 11.7 Divergence from a common ancestor. Prickly pear cacti have very different forms in South America compared with the Galápagos, but they clearly share ancestry, as evidenced by similar pad and flower structures.

and on returning home to England, he settled into a comfortable life with his wife Emma, heiress to a large fortune. After the voyage and the publication of his journal, Darwin reflected on his journal entries and recognized that the observations he had made on living and fossilized plants and animals supported the hypothesis of common descent. However, knowing that this hypothesis was still considered radical, Darwin shared his ideas with only a few close friends.

Instead of publishing his ideas about evolution, Darwin spent the next two decades carefully collecting evidence and further developing his theory. He was finally spurred into publishing his ideas in 1858, after receiving a letter from fellow scientist Alfred Russel Wallace. With Wallace's letter was a manuscript detailing a mechanism for evolutionary change—nearly identical to Darwin's theory of natural selection.

Concerned that his years of scholarship would be forgotten if Wallace published his ideas first, Darwin had excerpts of both his and Wallace's work presented in July 1858 at a scientific meeting in London, and the next year he published the book, *On the Origin of Species by Means of Natural Selection, or the Preservation of Favoured Races in the Struggle for Life*. The bulk of the text put forward the hypothesis of natural selection. But Darwin devoted the last several chapters of the *Origin of Species* to describing the evidence he had accumulated that supported the hypothesis of common descent.

The evidence put forth in the *Origin of Species* was so complete, and from so many different areas of biology, that the hypothesis of common descent no longer appeared to be a tentative explanation. In response, scientists began to refer to this idea and its supporting evidence as the *theory* of common descent. Indeed, most biologists today would agree that common descent is a scientific fact. Darwin's careful catalogue of evidence in his book had revolutionized the science of biology.

Alternative Ideas on the Origins and Relationships among Organisms

Let us explore the statement "common descent (or evolution) is a fact" more closely. When the *Origin of Species* was published, most Europeans believed that special creation explained how organisms came into being. According to this belief, God created organisms during the 6 days of creation described in Genesis, the first book of the Bible. This belief also states that organisms, including humans, have not changed significantly since creation. According to some biblical scholars, the Genesis story indicates that creation also occurred fairly recently, within the past 10,000 years.

Remember for a hypothesis to be testable by science, we must be able to evaluate it through observations or measurements made within the material universe (Chapter 1). Because a supernatural creator is not observable or measurable, there is no way to determine the existence or predict the actions of such an entity through the scientific method. Therefore, as it is stated, special creation is not a scientific hypothesis. In fact, any statement that supposes a supernatural cause—including intelligent design, which argues that although evolution is possible, some specific features of organisms must have been designed by a creator—cannot be considered science.

The idea of special creation does provide *some* scientifically testable hypotheses, however. For instance, the assertion that organisms came into being within the past 10,000 years and that they have not changed substantially since their creation is testable through observations of the natural world. We can call this hypothesis about the origin and relationships among living organisms the *static model hypothesis*, indicating that organisms are unchanging, in addition to being recently derived (**FIGURE 11.8a**). There are also several intermediate hypotheses between the static model and common descent. One intermediate hypothesis is that all living organisms were created, perhaps even millions of years ago, and that changes have occurred by microevolution in these species, but that brand-new species have not arisen. We will call this the *transformation hypothesis* (**FIGURE 11.8b**). Another intermediate hypothesis is that different *types* of organisms (for example, plants, animals with backbones, or insects) arose separately and since their origin have diversified into numerous species. We will call this the *separate types hypothesis* (**FIGURE 11.8c**). Are these three alternatives to the theory of common descent (**FIGURE 11.8d**) equally likely and reasonable explanations for the origin of biological diversity?

FIGURE 11.8 Four hypotheses about the origin of modern organisms. A graphical representation of the four hypotheses.

(a) Static model
Species arise separately and do not change over time.

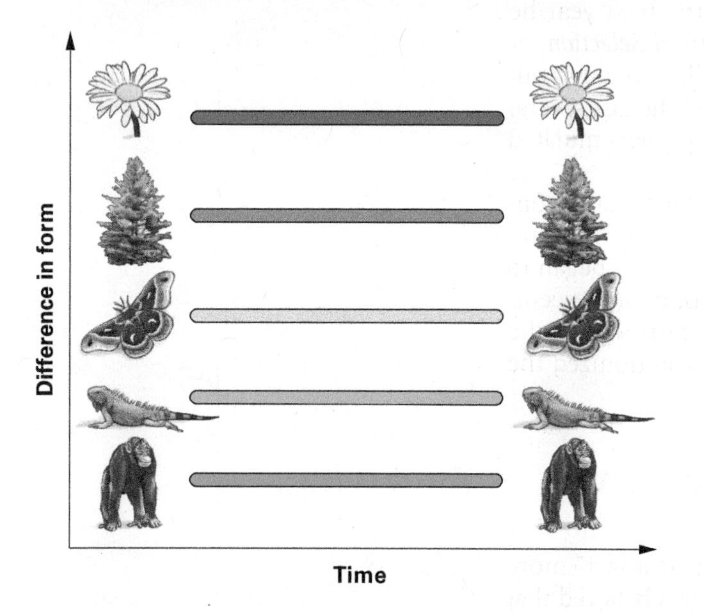

(b) Transformation
Species arise separately and change over time in order to adapt to the changing environment.

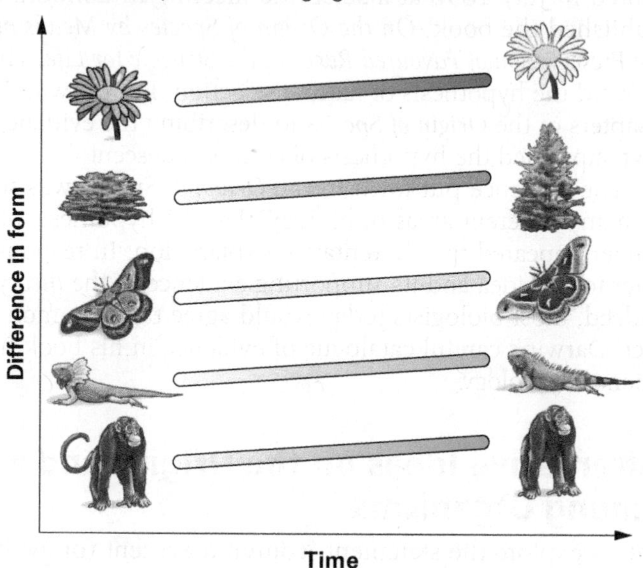

(c) Separate types
Species change over time, and new species can arise, but not from a common ancestor. Each group of species derives from a separate ancestor that arose independently.

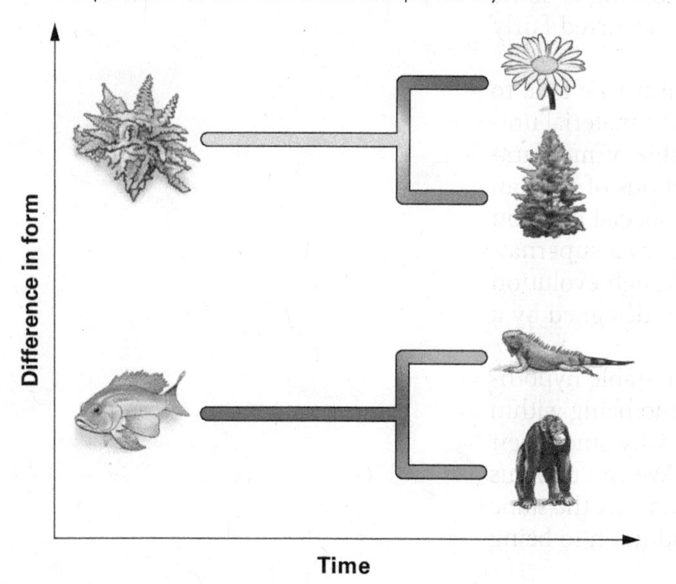

(d) Common descent
Species do change over time, and new species can arise. All species derive from a common ancestor.

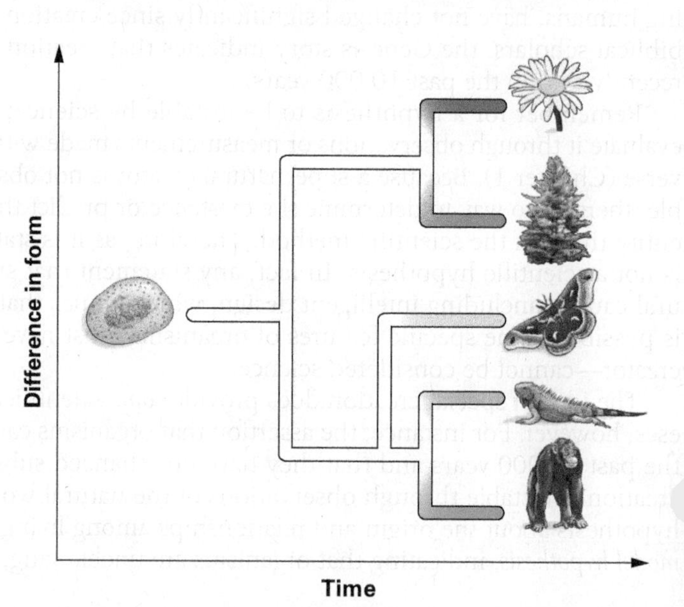

It appears from opinion polls that many Americans think there are equally likely explanations for the origins of humans. But what about all those scientists who insist that the theory of common descent is fact? Why would so many maintain this position? As you will soon see, the three alternative hypotheses are not equivalent to the theory of common descent. To understand why, we must examine the observations that help us test these hypotheses.

Got It?

1. Lamarck's theory of evolution included the supposition that traits developed over an individual's __________ are passed on to his or her offspring.
2. The last chapters of *On the Origin of Species* summarize the theory of __________ descent.
3. Darwin's essential contribution to the science of evolutionary biology was that he collected numerous lines of __________ in support of the theory.
4. A __________ creator or effect cannot be objectively observed, so hypotheses including these effects are not testable by the scientific method.
5. Competing ideas about "how the world works" are referred to as alternative __________.

11.3 Examining the Evidence for Common Descent

The evidence that all organisms share a common ancestor comes from several areas of biology and geology. However, to many people it seems unlikely that humans, with our highly developed brains, could be "just another branch" in the animal family tree. Because this notion is so common, we will address it head on—by examining the hypothesis that humans share a common ancestor with apes, including the chimpanzee and gorilla.

Any zookeeper will tell you that the primate house is among their most popular exhibits. People love apes and monkeys. It is easy to see why—primates are curious, playful, and agile. But something else drives our fascination with primates: We see ourselves reflected in them. The forward placement of their eyes and their reduced noses appear human-like. They have hands with fingernails instead of paws with claws. Some can stand and walk on two legs for short periods. They can finely manipulate objects with their fingers and opposable thumbs. They exhibit extensive parental care, and even their social relations are similar to ours—they tickle, caress, kiss, pout, and grin (**FIGURE 11.9**).

Why are primates, particularly the great apes (gorillas, orangutans, chimpanzees, and bonobos) so similar to humans? Scientists contend that it is because humans and apes are relatively recent descendants of a common biological ancestor.

FIGURE 11.9 Are humans related to apes? Biologists contend that apes and humans are similar in appearance and behavior because we share a relatively recent common ancestor.

Linnaean Classification

As the modern scientific community was developing in the sixteenth and seventeenth centuries, various methods for organizing biological diversity were proposed. Many of these classification systems grouped organisms by similarities in habitat, diet, or behavior; some of these classifications placed humans with the great apes, and others did not.

Into the classification debate stepped Carl von Linné, a Swedish physician and botanist. Von Linné gave all species of organisms a two-part, or binomial, name in Latin, which was the common language of science at the time. In fact, he adopted a Latinized name for himself—Carolus Linnaeus. The Latin names that Linnaeus assigned to other organisms typically contained information about the species' traits—for instance, *Acer saccarhum* is Latin for "maple tree that produces sugar," the sugar maple, whereas *Acer rubrum* is Latin for "red maple." The scientific name of a species contains information about its classification as well. For example, all species with the generic name *Acer* are all maples, and all species with the generic name *Ursus* are bears.

Linnaeus also created a logical and useful system for organizing diversity—grouping organisms in a hierarchy. From broadest to narrowest groupings, the classification levels he designated are

KINGDOM
PHYLUM
CLASS
ORDER
FAMILY
GENUS
SPECIES

Since Linnaeus's time, biologists have added a broader level—the Domain (Chapter 13)—as the largest grouping of organisms.

Although Linnaeus himself preceded the theory of evolution and believed in special creation, Darwin noted that Linnaeus's hierarchal classification system implied evolutionary relationships among organisms (**FIGURE 11.10**). Linnaeus's

FIGURE 11.10 The Linnaean classification of humans. All organisms within a category share basic characteristics, and as the groups become smaller subdivisions toward the bottom of the figure, the organisms have more similarities.

VISUALIZE THIS

List one or more traits that all members of the order Primates share.

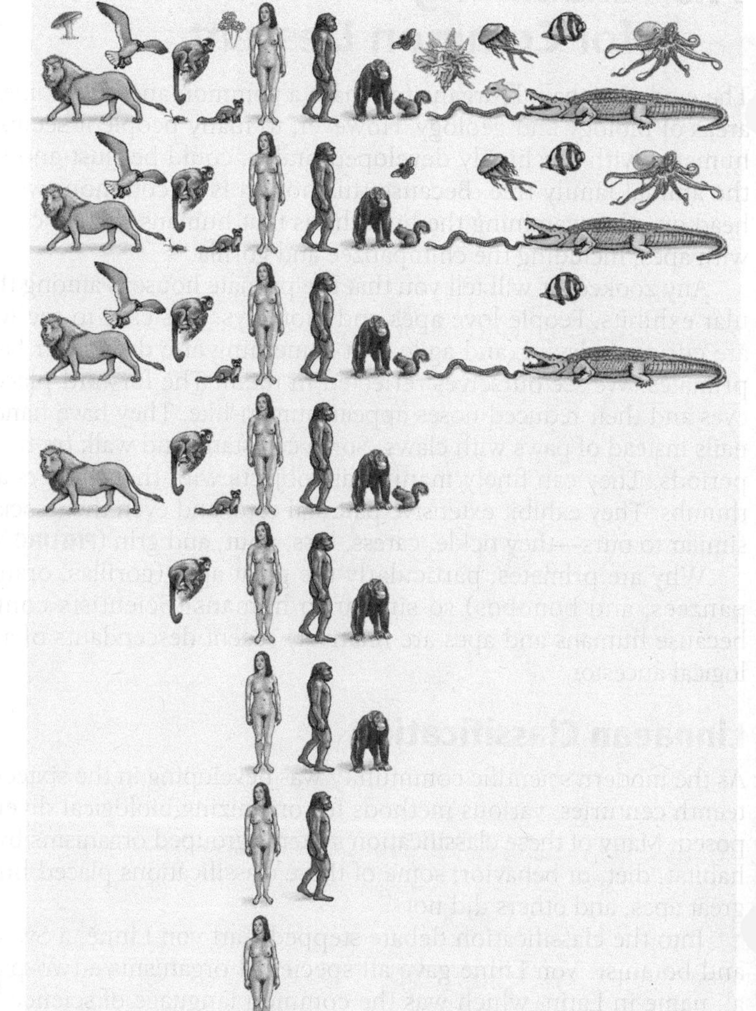

classification system was so effective at capturing these relationships that, even after the widespread acceptance of evolutionary theory, it has remained the standard for biological classification. The system has only been modified slightly, with the addition of the domain level and with "sub" and "super" levels between his categories—such as superfamily between family and order—to better represent the relationships among groups of organisms (Chapter 12).

Although Linnaeus's hierarchical categories are still in use, many of the actual groupings he proposed in the eighteenth century have been overturned or radically altered by the accumulation of additional data. However, when it comes to humans, his original classification remains largely supported by data. Linnaeus placed humans, monkeys, and apes in the same order, which he called Primates. Among the primates, humans are most similar to apes. Humans and apes share a number of characteristics, including relatively large brains, erect posture, lack of a tail, and increased flexibility of the thumb. Scientists now place humans and apes in the same family, Hominidae.

Humans and the African great apes share even more characteristics, including elongated skulls, short canine teeth, and reduced hairiness; they are placed together in the same *sub*family, Homininae. When the classification of humans, great apes, and other primates is shown in the form of a tree diagram (**FIGURE 11.11**), it is easy to see why Darwin formulated the idea that humans and modern apes evolved from the same ancestor.

VISUALIZE THIS

Add "squirrel" to this tree. How can you indicate the shared ancestor of squirrels and other members of this group?

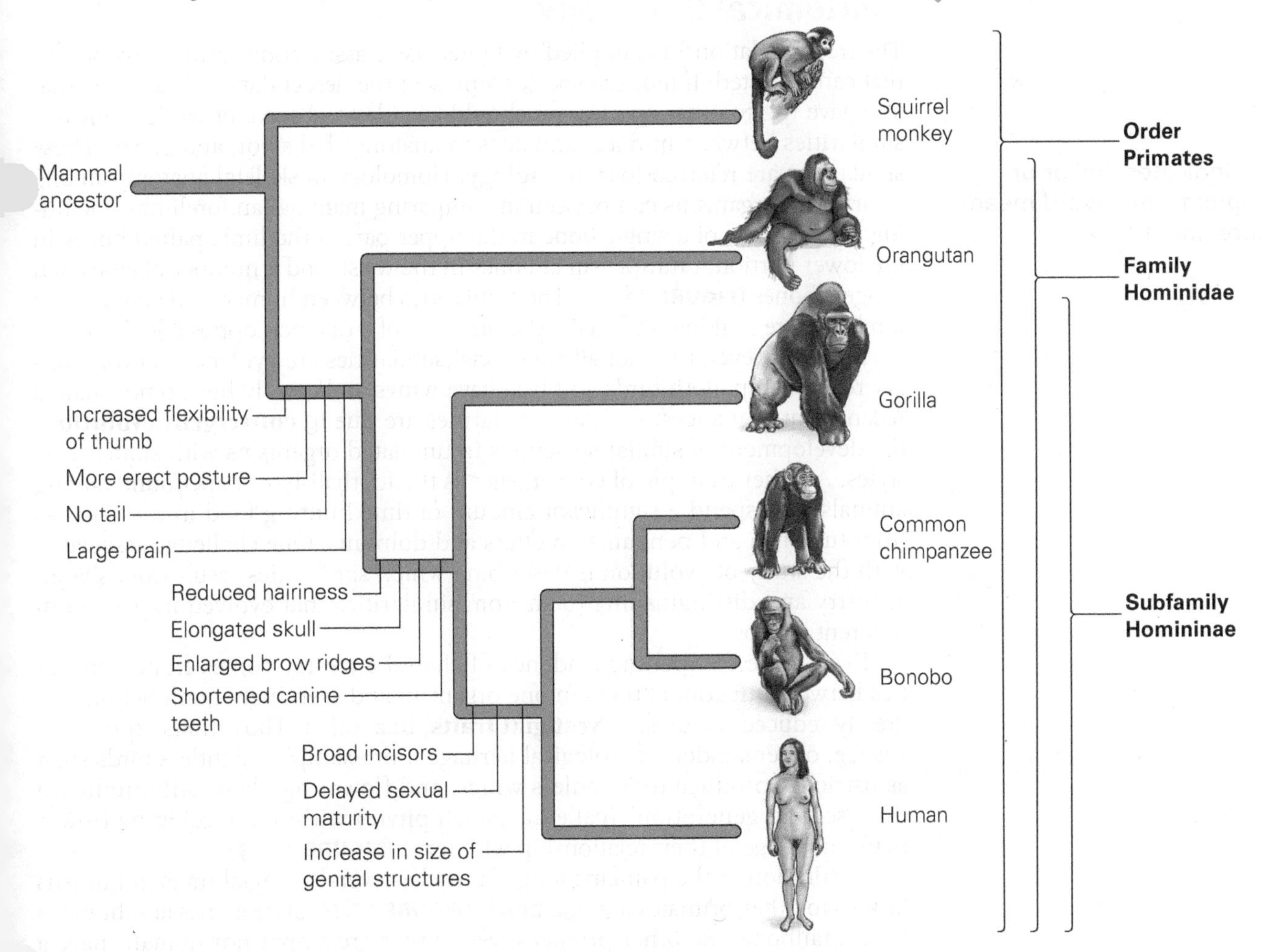

FIGURE 11.11 Shared characteristics among humans and apes imply shared ancestry. This tree diagram represents the current classification of humans and apes. Characteristics noted on the side of the evolutionary tree are shared by all of the species on that branch and those to the right.

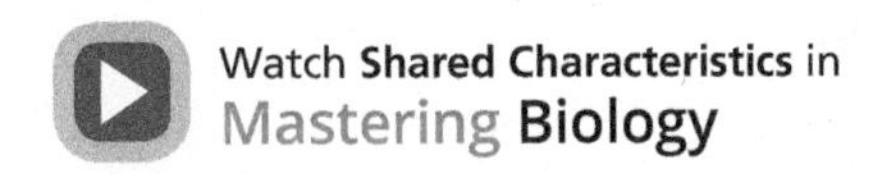

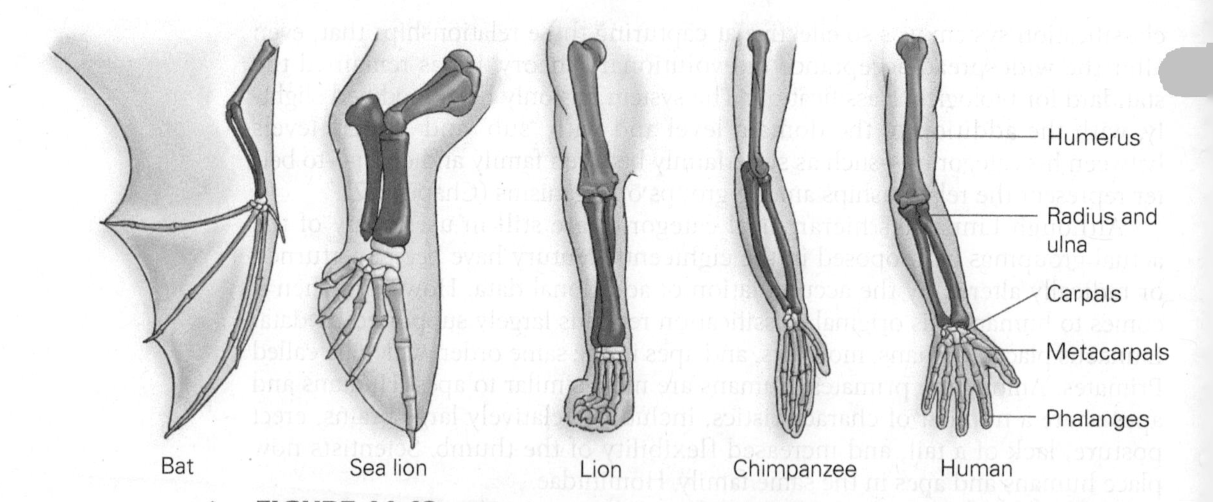

FIGURE 11.12 Homology of mammal forelimbs. The bones in the forelimbs of these mammals are very similar; equivalent bones in each organism are shaded the same color. The similarity in underlying structure despite great differences in function is evidence of shared ancestry.

VISUALIZE **THIS**

How is the structure of the bones drawn in green different among these species? How is the function of these bones different?

Anatomical Homology

The tree of relationships implied by Linnaeus's classification forms a hypothesis that can be tested. If modern species represent the descendants of ancestors that also gave rise to other species, we should be able to observe other, less obvious similarities between humans and apes in anatomy, behavior, and genes. These similarities are referred to as **homology**. Homology in skeletal anatomy among a variety of organisms can be seen in comparing mammalian forelimbs, including the presence of a single bone in the upper part of the limb, paired bones in the lower portion, multiple small bones in the wrist, and a number of elongated "finger" bones (**FIGURE 11.12**). The similarities between human and chimpanzee forelimbs are striking, especially the presence of a distinct, opposable thumb.

homolog- indicates similar or shared origin; from a word meaning "in agreement."

Note, however, that not all superficial similarities are evidence of evolutionary relationship. Both birds and bats have wings and can fly but do not share a recent common ancestor. Their similarities are due to **convergent evolution,** the development of similar structures in unrelated organisms with similar lifestyles. Another example of convergence is the football-like shape found among animals who spend a significant amount of time hunting food in the water—from tuna fish and penguins, to otters and dolphins. One challenge associated with the study of evolution is describing which similarities result from shared ancestry and distinguishing them from similarities that evolved in parallel in different groups.

Even more compelling evidence of shared ancestry comes from similarities between functional traits in one organism and seemingly nonfunctional or greatly reduced features, or **vestigial traits,** in another. These traits represent a vestige, or remainder, of biological heritage. For example, flightless birds such as ostriches produce functionless wings, and flowering plants still produce a tiny "second generation" (called a gametophyte) within a developing flower ovule, a vestige of their relationship with ferns (**FIGURE 11.13**).

In addition to the palmaris longus muscle, several vestigial traits in humans link us to other primates and mammals (**FIGURE 11.14**). Great apes and humans have a tailbone like other primates, yet neither great apes nor humans have a tail. In addition, all mammals possess tiny muscles called arrector pili at the base of each hair. When the arrector pili contract under conditions of emotional stress or cold temperatures, the hair is elevated. In furry mammals, the arrector pili help to increase the perceived size of the animal, and they increase the

(a) Fern sporophyte

(b) Fern gametophyte (an independent plant)

(c) Flowering plant sporophyte containing microscopic gametophyte

Gametophyte
generation
is found here.

FIGURE 11.13 Vestigial structures in plants. The ancestors of flowering plants were similar to modern ferns. Ferns have two independent stages in their life cycle: (a) the familiar form, which is known as the sporophyte stage, and (b) a tiny, independent, gametophyte stage. (c) Flowering plants no longer have two independent stages but do produce a tiny gametophyte within the flower itself.

insulating value of the hair coat. In humans, the same emotional or physical conditions produce only goose bumps, which provide neither benefit.

Darwin maintained that the hypothesis of evolution provided a better explanation for vestigial structures than did the hypothesis of special creation represented by the static model. A useless trait such as goose bumps is better explained as the result of inheritance from our biological ancestors rather than as a feature that appeared—or was created—independently in our species.

Developmental Homologies

Multicellular organisms demonstrate numerous similarities in the process of development from fertilized egg to adult organism. Genes that control development are similar among animals that appear as different as humans and

"Useful" trait in primate relative | Vestigial trait in human

(a) Tail bone

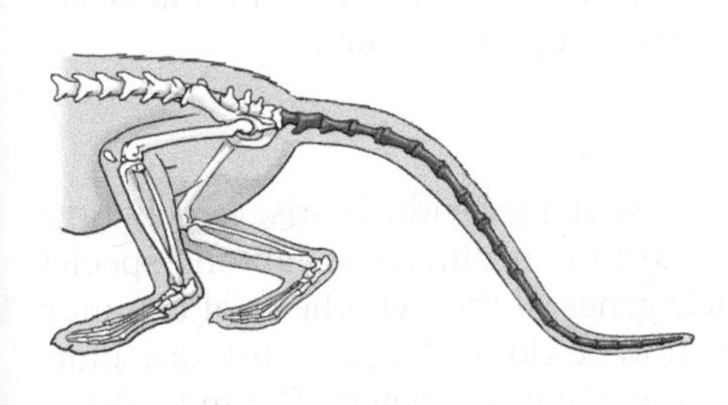

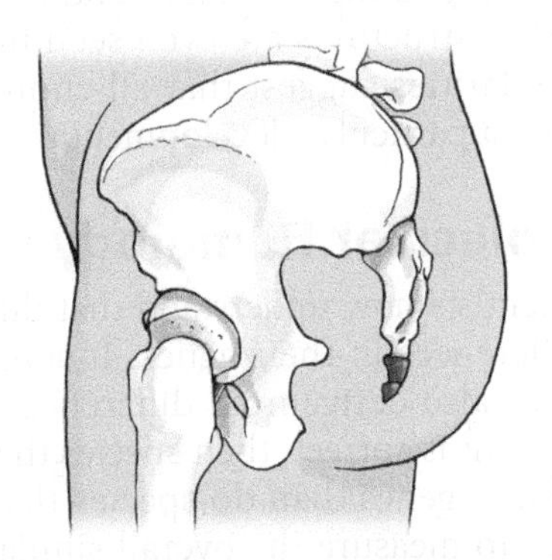

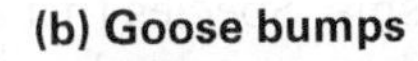

(b) Goose bumps

FIGURE 11.14 Vestigial traits reflect our evolutionary heritage. (a) Humans and other great apes do not have tails, but they do have a vestigial tailbone. (b) Goose bumps are reminders of our relatives' hairier bodies.

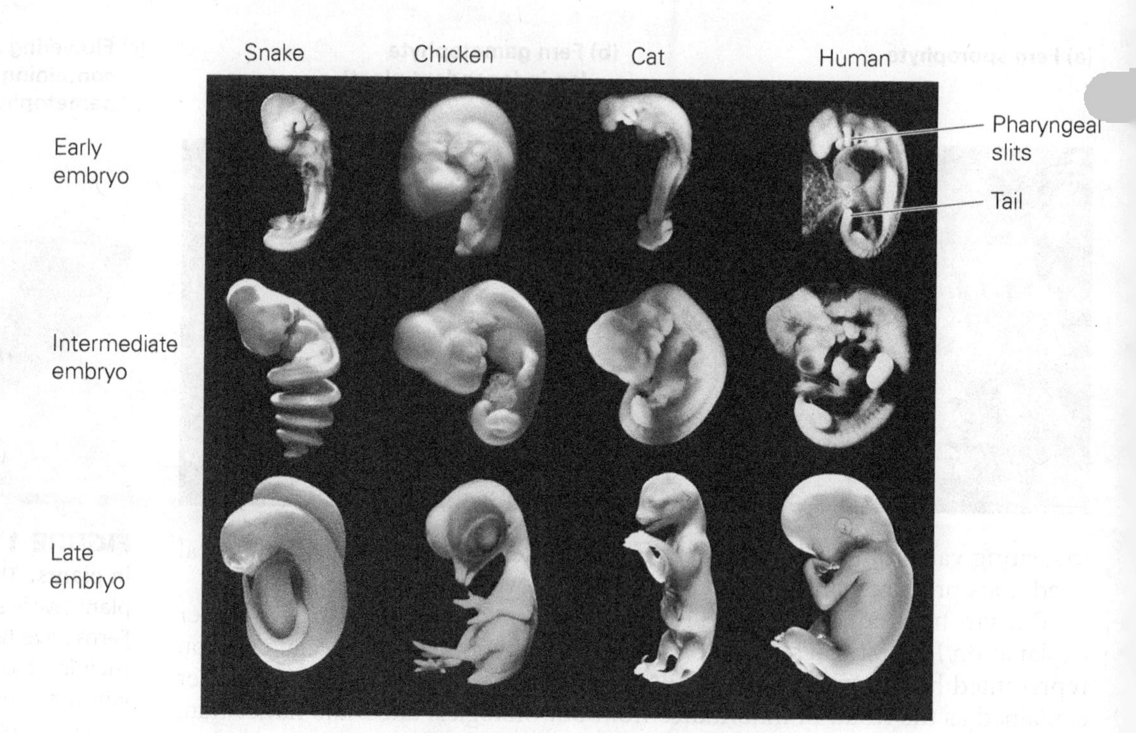

FIGURE 11.15 Similarity among chordate embryos. These diverse organisms appear very similar in the first stages of development (shown in the top row), evidence that they share a common ancestor that developed along the same pathway.

fruit flies. As a consequence of these shared developmental pathways, early embryos of very different species often look very similar. For example, all chordates—animals that have a backbone or closely related structure—produce structures called pharyngeal slits (pharyngeal means "pertaining to the throat," referring to their location) and most have tails as early embryos (**FIGURE 11.15**). These structures are even seen in human embryos early in development. The similarities suggest that all chordates derive from a single common ancestor with a particular developmental pathway that they all inherited.

Molecular Homology

Scientists now understand that differences among individuals arise largely from differences in their genes. It stands to reason that differences among species must also derive from differences in their genes. If the hypothesis of common descent is correct, then species that appear to be closely related must have more similar genes than do species that are more distantly related. The most direct way to measure the overall similarity of two species' genes is to evaluate similarities in the DNA sequences of genes found in both organisms. Species that share a more common ancestor should have more similar DNA sequences than species that share a more distant ancestor (**FIGURE 11.16**).

Many genes are found in nearly all living organisms. For instance, genes that code for histones, proteins that help store DNA neatly inside cells, are found in algae, fungi, fruit flies, humans, and all other organisms that contain linear chromosomes. Among organisms that share many aspects of structure and function, such as humans and chimpanzees, many genes are shared. However, because amino acids can be coded for by more than one DNA codon, the sequences of genes, even those that produce proteins with the same amino acid sequence, are typically not identical from one species to the next.

A comparison of the sequences of dozens of genes that are found in humans and other primates demonstrates the relationship between classification and

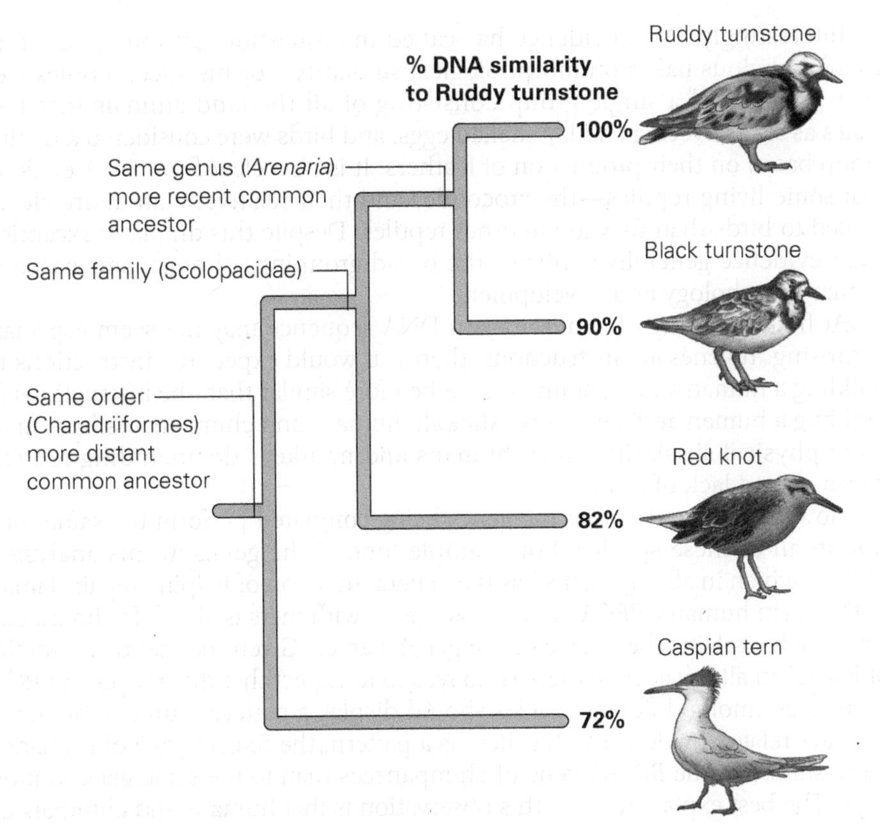

FIGURE 11.16 DNA evidence of relationship. Scientists used the physical characteristics of these four bird species to put them into different taxonomic categories, shown here in tree form. DNA studies later supported this hypothesis, because species that share a more recent common ancestor have more similar DNA sequences than species that share a more distant ancestor.

WORKING WITH DATA

Which bird is more closely related to the red knot: the Caspian tern or the black turnstone? Explain your answer.

gene sequence similarity (**FIGURE 11.17**). The DNA sequences of these genes in humans and chimpanzees are 99.01% similar, whereas the DNA sequences of humans and gorillas are identical over 98.9% of their length. More distantly related primates are less similar to humans in DNA sequence. This pattern of similarity in DNA sequence exactly matches the biological relationships implied by physical similarity in this group. This result supports the hypothesis of common descent among the primates.

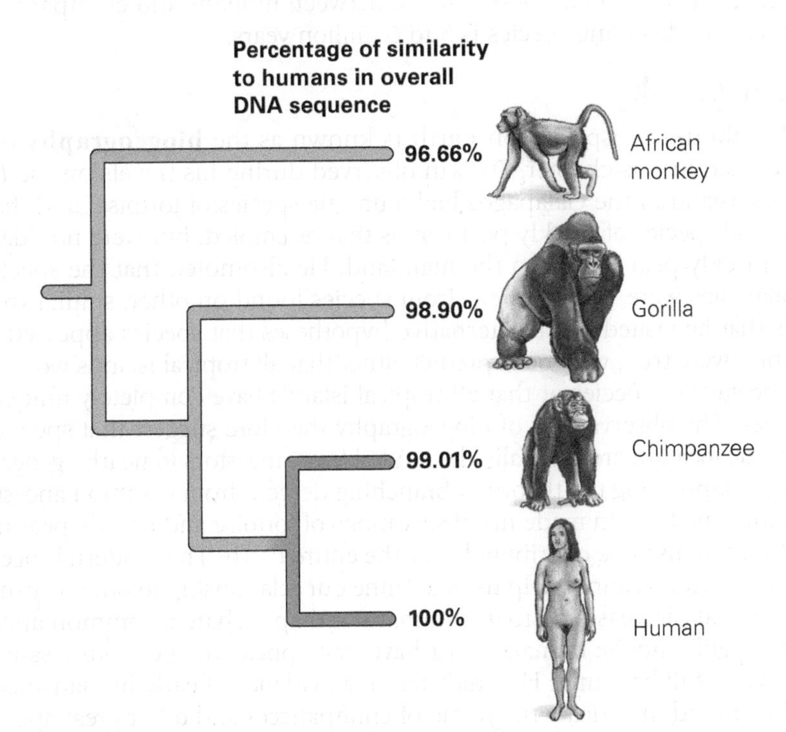

FIGURE 11.17 DNA evidence for human ancestry. Similarities in DNA sequences among the primates parallels the hypothesis of relationship implied by similar morphology and development.

WORKING WITH DATA

Looking at the percentages on this graph might lead a reader to suspect that chimpanzees and gorillas are more closely related to each other than are chimpanzees and humans. Explain why this is misreading the data.

Interestingly, DNA evidence has called into question groupings that once seemed obvious based on morphological similarity. For instance, reptiles were once considered a single group consisting of all the land animals that have scales as skin covering and lay shelled eggs, and birds were considered a distinct group based on their production of feathers. It is now clear from DNA evidence that some living reptiles—the crocodiles and their relatives—are more closely related to birds than they are to other reptiles. Despite this dramatic exception, DNA evidence generally confirms the broad groupings of organisms based on similar morphology and development.

At first, a finding of similarities in DNA sequence may not seem especially surprising. If genes are instructions, then you would expect the instructions for building a human and a chimpanzee to be more similar than the instructions for building a human and a monkey. After all, humans and chimpanzees have many more physical similarities than humans and monkeys do, including reduced hairiness and lack of a tail.

However, remember that the genes being compared perform the same function in all of these species. For example, one of the genes in this analysis is *BRCA1*, which in all organisms has the general function of helping repair damage to DNA. (In humans, *BRCA1* is also associated with increased risk for breast cancer, which explains the source of the gene's name.) Given the identical function of *BRCA1* in all organisms, there is no reason to expect that differences in *BRCA1* sequences among different species should display a pattern—unless the organisms are related by descent. But there is a pattern; the *BRCA1* gene of humans is more similar to the *BRCA1* gene of chimpanzees than to the same gene in monkeys. The best explanation for this observation is that humans and chimpanzees share a more recent common ancestor than either species shares with monkeys.

Differences in DNA sequence between humans and chimpanzees can also allow us to estimate when these two species diverged from their common ancestor. The estimate is based on a **molecular clock.** The principle behind a molecular clock is that the rate of change in certain DNA sequences, due to the accumulation of mutations that affect the DNA sequence but not the protein sequence, seems to be relatively constant within a species. According to one application of a molecular clock, the amount of time it takes for a 1% difference in DNA sequence (about the difference between humans and chimpanzees) to accumulate in diverging species is 5 to 6 million years.

Biogeography

The distribution of species on Earth is known as the **biogeography** of life. As discussed in this chapter, Darwin observed during his travels on the *Beagle* that each island in the Galápagos had a unique species of tortoise, and that the islands had species of prickly pear cactus that resembled, but were not identical to, the prickly pear found on the mainland. He also noted that the species on the Galápagos were very different from species found on other, similar tropical islands that he visited. If the alternative hypotheses that species appeared independently were true, we would predict either that all tropical islands would have the same suite of species or that all tropical islands have completely unique sets of species. The observations of biogeography therefore suggest that species in a geographic location are generally descended from ancestors in nearby geographic locations, supporting the theory of branching descent from common ancestors.

By the time Darwin made his observations of tortoise and prickly pear biogeography, humans were distributed over the entire Earth. The modern biogeography of our species cannot help us determine our relationship to other organisms. However, Darwin reasoned that if humans and apes share a common ancestor, then we highly mobile humans must have first appeared where our less-mobile relatives can still be found. He predicted that evidence of early human ancestors would be found in Africa—the home of chimpanzees and other great apes.

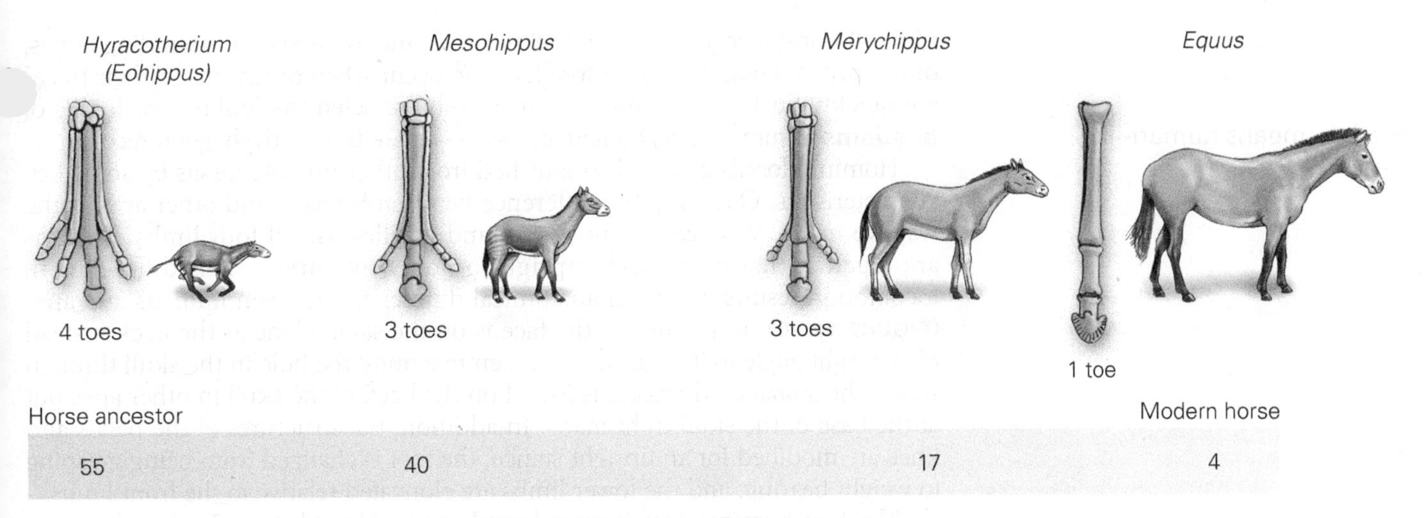

FIGURE 11.18 The fossil record of horses. Horse fossils provide a fairly complete sequence of evolutionary change from small, catlike animals with four toes to the modern horse with one massive toe.

The Fossil Record

One form of evidence of human ancestors comes from **fossils,** the remains of living organisms left in soil or rock. Human fossils and those of other species form a record of ancient life and provide direct evidence of change in organisms over time. There are many examples of fossil series that show a progression from more ancient forms to more modern forms, as in the transition between ancient and modern horses (**FIGURE 11.18**).

The chronological appearance of other groups of fossils also supports the theory of evolution. For example, evidence from anatomy and developmental biology indicates that modern mammals and other four-legged land animals evolved from fish ancestors. Correspondingly, we find the first fossil fish in ancient rocks and the first fossil mammals in younger layers of rock.

Fossils typically form when the organic material decomposes and minerals fill the space left behind (**FIGURE 11.19**). However, fossilized impressions can also

VISUALIZE **THIS**

Bones that decompose quickly do not form high-quality fossils. Based on the information from this figure, why do you think that is the case?

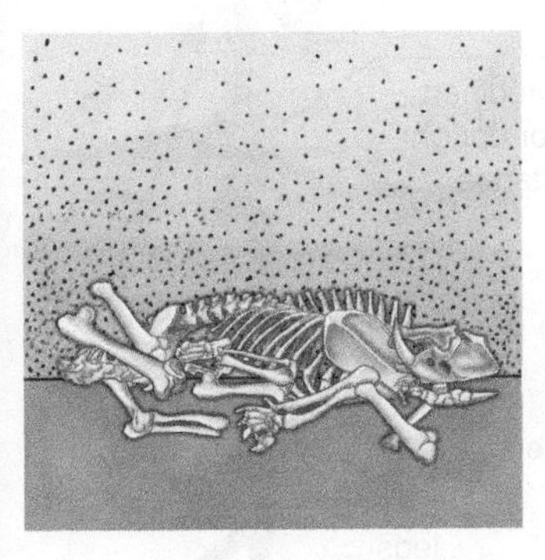

① An organism is rapidly buried in water, mud, sand, or volcanic ash. The tissues begin to decompose very slowly.

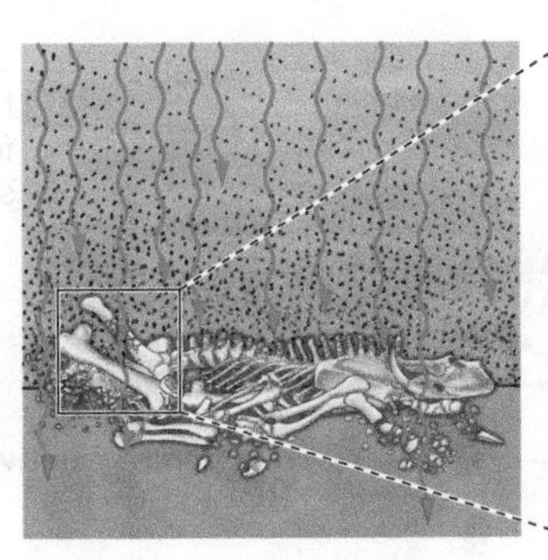

② Water seeping through the sediment picks up minerals from the soil and deposits them in the spaces left by the decaying tissue.

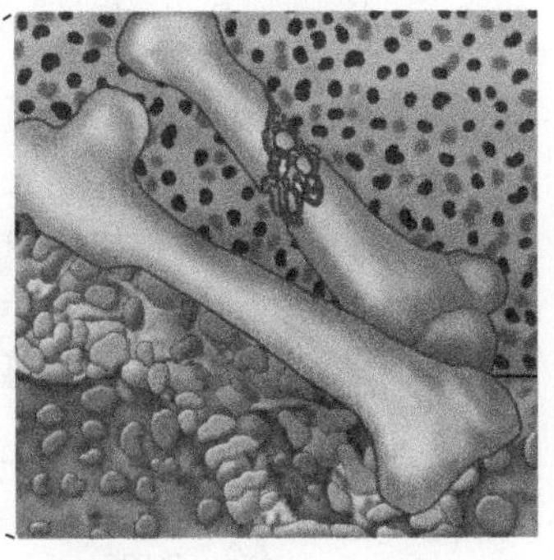

③ After thousands of years, most or all of the original tissue is replaced by very hard minerals, resulting in a rock model of the original bone.

④ When erosion or human disturbance removes the overlying sediment, the fossil is exposed.

FIGURE 11.19 Fossilization. A fossil is commonly a rock "model" of an organic structure.

form—for instance, of shells, animal burrows, the soft tissues surrounding bones, or footprints. Fossilization is more likely to occur when organisms or their traces are quickly buried by sediment. Fortunately for scientists looking for fossils of **hominins**—humans and human ancestors—there is a relatively good record.

homini- means human-like.

Hominin fossils can be distinguished from other primate fossils by some key characteristics. One essential difference between humans and other apes is the way we move. Whereas chimpanzees and gorillas use all four limbs, humans are bipedal; that is, we walk upright on only two limbs. This difference in locomotion results in several anatomical differences between humans and apes (**FIGURE 11.20**). In hominins, the face is on the same plane as the back instead of at a right angle to it; thus, the foramen magnum, the hole in the skull through which the spinal cord passes, is found on the back of the skull in other apes but at the base of the skull in humans. In addition, the structures of the pelvis and knee are modified for an upright stance; the foot is changed from being grasping to weight bearing; and the lower limbs are elongated relative to the front limbs.

The first hominin fossils were found not in Africa but in Europe. Remains of *Homo neanderthalensis* (Neanderthal man) were discovered in 1856 in a small cave within the Neander Valley of Germany. In 1891, fossils of older, human-like creatures now called *Homo erectus* (standing man) were discovered in Java (Indonesia) in 1891. It was not until 1924 that the first *African* hominin fossil, the Taung child, was discovered in South Africa. This fossil was later determined to be a much older species than Neanderthals and *Homo erectus* and was placed in a new genus, *Australopithecus*. Paleontologists continue to discover new hominin fossils in southern and eastern Africa, including the famous "Lucy," a remarkably complete skeleton of the species *Australopithecus afarensis*, discovered in 1974 in Ethiopia. Lucy's fossil skeleton included a large section of her pelvis, which clearly indicated that she walked upright.

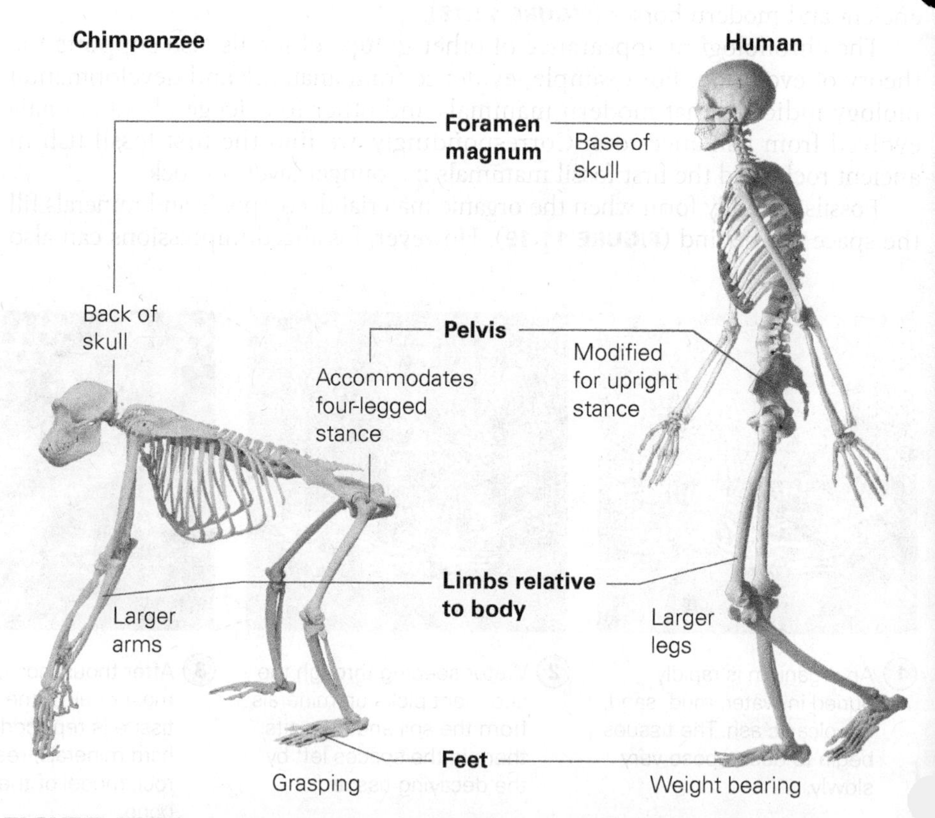

FIGURE 11.20 Anatomical differences between humans and chimpanzees. Humans are bipedal animals, while chimpanzees typically travel on all fours. If any bipedal features are present in a fossil primate, the fossil is classified as a hominin.

By determining the age of these fossils and many other hominin species, scientists have confirmed Darwin's predictions—the earliest human ancestors arose in Africa.

Scientists can determine the date when an ancient fossil organism lived by estimating the age of the rock that surrounds the fossil. **Radiometric dating** relies on radioactive decay, which occurs as radioactive elements in rock spontaneously break down into different, unique elements known as daughter products.

radio- is the combining form of radiation.

-metric means to measure.

When rock is newly formed from the liquid underlying Earth's crust, it contains a fixed amount of any radioactive element. When the rock hardens, some of these radioactive elements become trapped. Each radioactive element decays at its own unique rate, described by the element's **half-life**—the amount of time required for one-half of the amount of the element originally present to decay into the daughter product. As a trapped element decays over time, the amount of radioactive material in the rock declines, and correspondingly, the amount of daughter product increases. By determining the ratio of radioactive element to daughter product in a rock sample and knowing the half-life of the radioactive element, scientists can estimate the number of years that have passed since the rock formed (**FIGURE 11.21**). Radiometric dating has led to an estimate of the age of Earth—4.6 billion years—based on the age of the oldest rocks on the planet.

Using this technique, scientists have determined that the most ancient hominin fossil, the species *Ardepithecus ramidus*, is 5.2 to 5.8 million years old, just as predicted by the molecular clock regarding human-chimpanzee divergence. (Two even older fossil species, *Orrorin tugenensis* and *Sahelanthropus tchadensis*,

WORKING WITH DATA

How old is a rock that contains 12.5% of the original parent element?

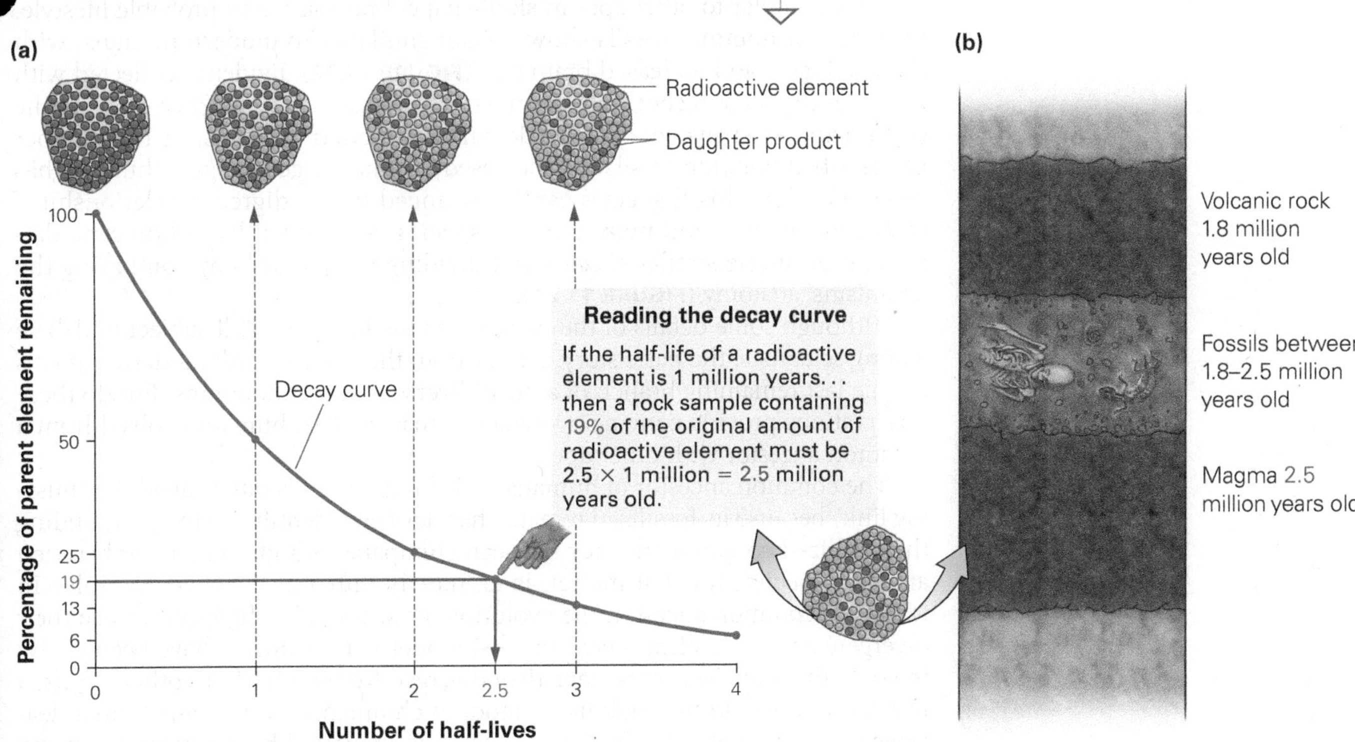

FIGURE 11.21 Radiometric dating. (a) The age of a fossil can be estimated when it is found between two layers of volcanic or magma-formed rock. (b) The age of rocks can be estimated by measuring the amount of radioactive material (designated by dark purple circles) with a known half-life and the amount of daughter material (designated by light blue circles) in a sample of rock.

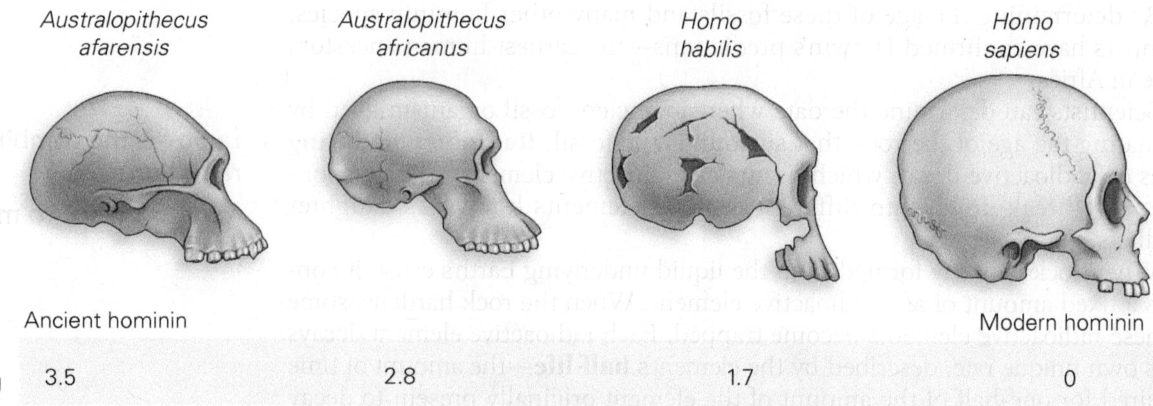

FIGURE 11.22 From ancient to modern. Fossils of ancient hominins display numerous ape-like characteristics, including a large jaw, small braincase, and receding forehead. More recent hominin fossils have a reduced jaw, larger braincase, and smaller brow ridge, much like modern humans.

have been described as 6- and 7-million-year-old human ancestors, respectively. However, most scientists are reserving judgment about whether these animals were bipedal until more examples are found.)

Like the fossil record of horses or the transition from some fish to land animals, the hominin fossil record shows a clear progression—in this case, from more "apelike" to more "human-like" ancestors over time. Besides being bipedal, humans differ from other apes in having a relatively large brain, a flatter face, and a more extensive culture. The oldest hominins are bipedal but are otherwise similar to other apes in skull shape, brain size, and probable lifestyle. More recent hominin fossils show greater similarity to modern humans, with flattened faces and increased brain size (**FIGURE 11.22**). Evidence collected with fossils of the most recent human ancestors indicates the existence of symbolic culture and extensive tool use, trademarks of modern humans. As the number of described hominin fossils has increased, a tentative genealogy of humans has emerged. These fossil species can be arranged in a pedigree of relationships, indicating ancient and more modern species determined by radiometric dating and grouping similar species and dividing unique ones by comparing the organisms' anatomy (**FIGURE 11.23**).

Although some details of the pedigree in the figure are still subject to debate among scientists, the basic story is clear from the fossil record; modern humans are the last remaining branch of a once-diverse group of hominins. But do these observations provide convincing evidence that modern humans evolved from a common ancestor with other apes?

The common ancestor of humans and chimpanzees is often called the "missing link" because its fossilized remains has not been identified. However, finding the fossilized common ancestor between chimpanzees and humans, or between any two species, for that matter, is extremely difficult, if not impossible. To identify a common ancestor, the evolutionary history of both species since their divergence must be clear. Like humans, modern chimpanzees have been evolving over the 5 million years since they diverged from humans. In other words, a missing link would not look like a modern chimpanzee with some human features or a cross between the two species—as suggested by nineteenth-century cartoons depicting Darwin as an "ape man" (**FIGURE 11.24**). The fact that scientists cannot conclusively identify a fossil species that was the common ancestor of modern humans and modern chimpanzees is not evidence that these two species are not related. The vast majority of the evidence supports the hypothesis that chimpanzees are our closest living relatives.

WORKING WITH DATA

Approximately how many years ago did the common ancestor of modern humans and *Homo habilis* exist?

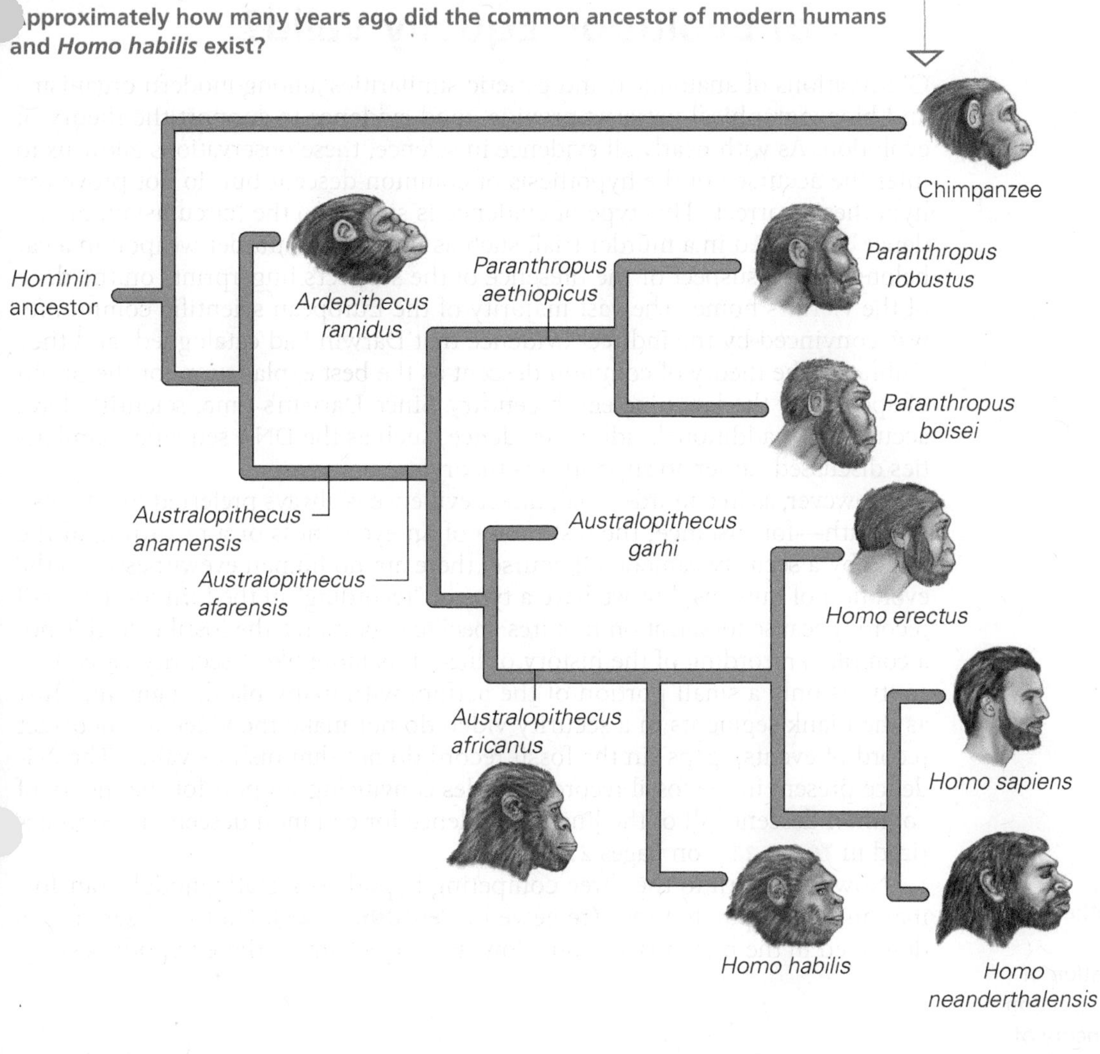

FIGURE 11.23 The evolutionary relationships among hominin species. This tree represents the current consensus among scientists who are attempting to uncover human evolutionary history.

FIGURE 11.24 The common ancestor of humans and chimpanzees? A "missing link" between humans and chimpanzees would not look half human and half ape, as this cartoon suggests.

Got It?

1. The hierarchical organization of Linneaus's classification system implies __________ relationships among organisms.
2. A __________ trait is a seemingly useless trait that appears similar to a useful trait in a related organism.
3. The evidence for evolution includes homologies in __________, __________, and __________.
4. The study of biogeography indicates that species that appear closely related tend to be geographically __________ to each other.
5. Older hominins appear more __________-like than modern hominins.

11.4 Are Alternatives to the Theory of Evolution Equally Valid?

Observations of anatomical and genetic similarities among modern organisms and biogeographical patterns provide good evidence to support the theory of evolution. As with nearly all evidence in science, these observations allow us to infer the accuracy of the hypothesis of common descent but do not prove the hypothesis correct. This type of evidence is similar to the "circumstantial evidence" presented in a murder trial, such as finding the murder weapon in a car belonging to a suspect or the presence of the suspect's fingerprints on the door of the victim's home. The vast majority of the European scientific community was convinced by the indirect evidence that Darwin had catalogued, and they embraced the theory of common descent as the best explanation for the origin of species by the late nineteenth century. Since Darwin's time, scientists have accumulated additional indirect evidence, such as the DNA sequence similarities discussed earlier, to support this theory.

However, as in a murder trial, direct evidence is always preferred to establish the truth—for instance, the testimony of an eyewitness or a recording of the crime by a security camera. Of course, there are no human eyewitnesses to the evolution of humans, but we have a type of "recording" in the form of the fossil record. Because fossilization requires specific conditions, the fossil record is not a complete recording of the history of life—it is more like a security video that captures only a small portion of the action, with many blank segments. Just as the blank segments in a security video do not make the video an incorrect record of events, "gaps" in the fossil record do not diminish its value. The evidence present in the fossil record provides convincing support for the theory of common descent. All of the lines of evidence for common descent are summarized in TABLE 11.1 on pages 229–230.

Now we return to the three competing hypotheses: static model, transformation, and separate types (reviewed in **FIGURE 11.25**). Do the observations described in the previous section allow us to reject any of these hypotheses?

FIGURE 11.25 Four hypotheses about the origin of modern organisms. A scientific evaluation of these hypotheses leads to the rejection of all of them except for the theory of common descent.

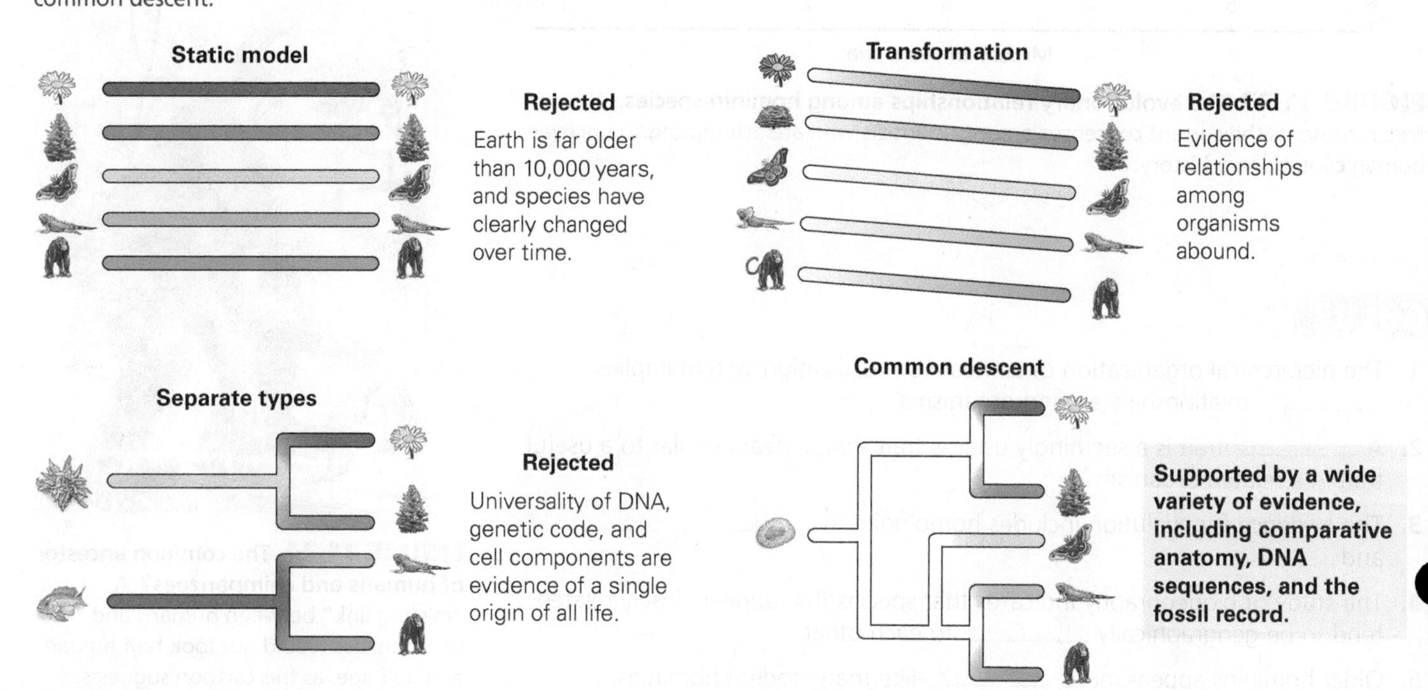

Static model
Rejected
Earth is far older than 10,000 years, and species have clearly changed over time.
Transformation
Rejected
Evidence of relationships among organisms abound.
Separate types
Rejected
Universality of DNA, genetic code, and cell components are evidence of a single origin of all life.
Common descent
Supported by a wide variety of evidence, including comparative anatomy, DNA sequences, and the fossil record.

TABLE 11.1 The evidence for evolution.

Observation	Example	Why It Suggests Common Descent
Biological classification. The most logical and useful system groups organisms hierarchically.	Humans appear most similar to the great apes. 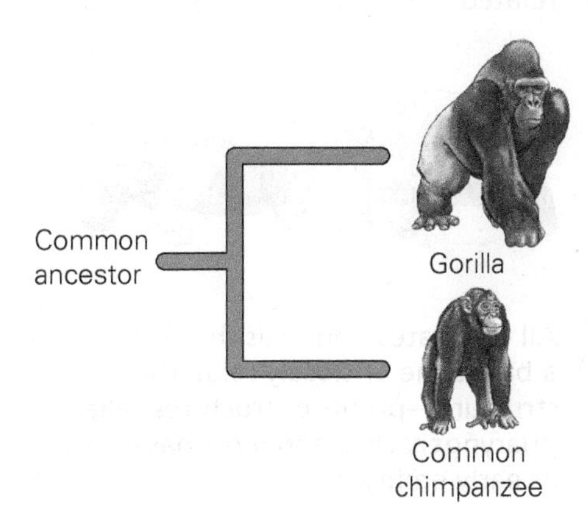	All species in the same family share a relatively recent common ancestor, and all families in the same class share a more distant common ancestor.
Anatomical homology. Organisms that look quite different have surprisingly similar structures.	Mammalian forelimbs share a common set of bones organized in the same way, despite their very different functions. 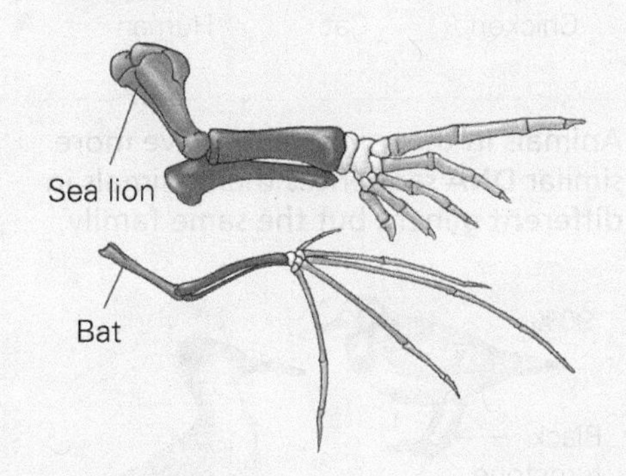	The simplest explanation is that each species inherited the basic structure from the same common ancestor, and evolution led to their modification in each group.
Vestigial traits. Some species display traits that are nonfunctional but have a functional equivalent in other species.	Flightless birds such as ostriches produce functionless wings. 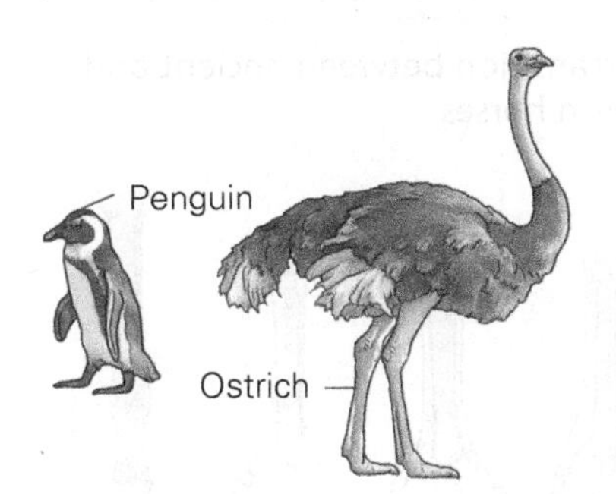	The simplest explanation is that the trait was functional in an ancestral species but lost its function over time in one branch of the evolutionary tree.

—Continued

TABLE 11.1 The evidence for evolution. *Continued—*

Observation	Example	Why It Suggests Common Descent
Biogeography. The distribution of organisms on Earth corresponds in part to the relationship implied by biological classification.	The different species of tortoises on the Galápagos Islands are clearly related. 	Similarities among species in a geographic location imply divergence from ancestors in that geographic location.
Homology in development. Early embryos of different species often look similar.	All chordates—animals that have a backbone or closely related structure—produce structures called pharyngeal slits, and most have tails as early embryos. 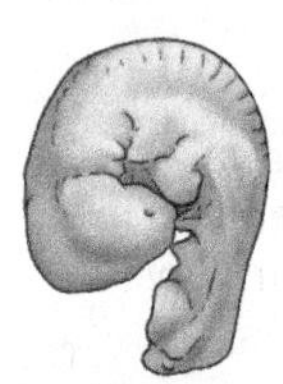 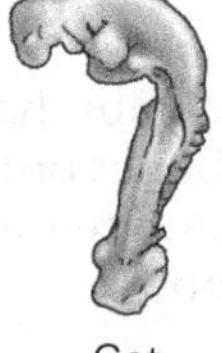 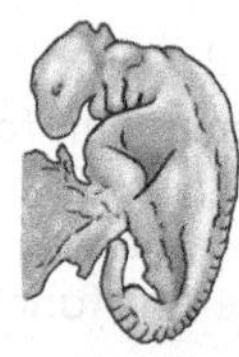 Chicken Cat Human	Similarities in early development suggest that these organisms derived from a single common ancestor that developed along a similar pathway.
Homology of DNA. The DNA sequences of species that are closely related in a taxonomic grouping are more similar than those from more distantly related groups.	Animals in the same genus have more similar DNA sequences than animals in different genera but the same family.  % DNA sequence similarity to Ruddy turnstone	Similar DNA sequences in different species imply that the species evolved from a common ancestor with a particular sequence.
The fossil record. The remains of extinct organisms show progression from more ancient forms to more modern forms.	The transition between ancient and modern horses 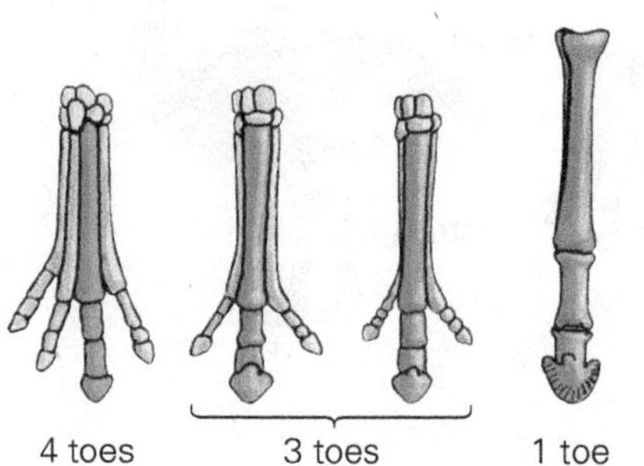	Fossils provide direct evidence of change in organisms over time and suggest relationships among modern species.

Weighing the Alternatives

The physical evidence we have discussed thus far allows us to clearly reject only one of the hypotheses—the static model. The fossil record provides unambiguous evidence that the species that have inhabited this planet have changed over time, and radiometric dating indicates that Earth is far older than 10,000 years.

Of the remaining three hypotheses, transformation is the poorest explanation of the observations. If organisms arose separately and each changed on its own path, there is no reason to expect that different species would share structures—especially if these structures are vestigial in some of the organisms. There is also no reason to expect similarities among species in DNA sequence. The hypothesis of transformation predicts that we will find little evidence of biological relationships among living organisms. As our observations have indicated, evidence of relationships abounds.

Both the hypothesis of common descent and the hypothesis of separate types contain a process by which we can explain observations of relationships. That is, both hypothesize that modern species are descendants of common ancestors. The difference between the two theories is that common descent hypothesizes a single common ancestor for all living things, whereas separate types hypothesizes that ancestors of different groups arose separately and then gave rise to different types of organisms. Separate types seems more reasonable than common descent to many people. It seems impossible that organisms as different as pine trees, mold, ladybugs, and humans share a common ancestor. However, several observations indicate that all these disparate organisms are related.

The most compelling evidence for the single origin of all life is the universality of both DNA and of the relationship between DNA and proteins. For example, genes from bacteria can be transferred to plants, and the plants will make a functional bacterial protein (Chapter 9). This is possible only because both bacteria and plants translate genetic material into functional proteins in a nearly identical manner. If bacteria and plants arose separately, we could not expect them to translate genetic information similarly.

The fact that organisms as different as pine trees, mold, ladybugs, and humans contain cells with nearly all of the same components and biochemistry is also evidence of shared ancestry (**FIGURE 11.26**). A mitochondrion could have many different possible structural forms and still perform the same function; the fact that the mitochondria in a plant cell and an animal cell are essentially identical implies that both groups of organisms inherited these mitochondria from a common ancestor.

Pine trees, mold, ladybugs, and humans *are* very different. Proponents of the hypothesis of separate types argue that the differences among these organisms could not have evolved in the time since they shared a common ancestor. But the length of time during which these organisms have been diverging is immense—nearly 2 billion years. The remaining basic similarities among all living organisms serve as evidence of their ancient relationship.

The Best Scientific Explanation for the Diversity of Life

Scientists favor the theory of common descent because it is the best explanation for how modern organisms came about. The theory of evolution—including the theory of common descent—is robust, meaning that it is a good explanation for a variety of observations and is well supported by a wide variety of evidence

FIGURE 11.26 The unity and diversity of life. The theory of evolution, including the theory of common descent, provides the best explanation for how organisms as distinct as pine trees, mold, ladybugs, and humans can look very different while sharing a genetic code and many aspects of cell structure and cell division.

from anatomy, geology, molecular biology, and genetics. Evidence for the theory of common descent demonstrates **consilience,** meaning that there is agreement among observations derived from different sources. Consilience is a feature of all strongly supported scientific theories.

The theory of common descent is no more tentative than is atomic theory; few scientists disagree with the models that describe the basic structures of atoms, and few disagree that the evidence for the theory of common descent is overwhelming. Most scientists would say that both of these theories are so well supported that we can call them fact.

Evolutionary theory helps us understand the functions of human genes, comprehend the interactions among species, and predict the consequences of a changing global environment for modern species. Describing evolution as "just a theory" vastly understates the importance of evolutionary theory as a foundation of modern biology. People who do not have a grasp of this fundamental biological principle may lack an appreciation of the basic unity and diversity of life and fail to understand the effects of evolutionary history and change on the natural world and on ourselves.

Got It?

1. The hypothesis that all living organisms appeared independently of each other is rejected because evidence of ________ among them abound.
2. The hypothesis that Earth appeared around 10,000 years ago is rejected because of the evidence from ________ dating.
3. Support for the common ancestry of all living things includes the universality of the hereditary molecule ________.
4. Another piece of evidence that all living organisms share a common ancestor is the relationship between ________ and protein.
5. Consilience of evidence for common ancestry means that we refer to this understanding as the ________ of evolution.

Macroevolution can be hard to grasp because it is so difficult to observe. Some who are skeptical of evolutionary theory might make this statement:

If the theory that humans evolved from apes is correct, then we should see chimpanzees evolving into humans today.

Sounds right, but it isn't.

Answer the following questions to understand why.

1. A woman and her sister share a common ancestor, their mother. Can both women exist at the same time?
2. Are chimpanzees successful (in other words, do they survive and reproduce) in their current environment?
3. Consider the time before humans evolved. Are the environmental conditions that existed when humans evolved the same as the environmental conditions today? Explain.
4. Is a "somewhat more human-like" chimpanzee (for example, more bipedal) likely to be more successful than a typical chimpanzee in their current environment?
5. Reflect on your answers to questions 1–4 above. Explain why the original statement bolded above, sounds right, but isn't.

THE **BIG** QUESTION

Should high school biology teachers be required to "teach the controversy"?

Some state legislatures in the United States have rejected learning standards that include teaching evolution, and others have passed laws that encourage public high school science teachers to teach the "controversy" associated with certain topics. For example, a law passed in Tennessee in 2012 encourages teachers to present the "scientific strengths and scientific weaknesses" of topics that arouse "debate and disputation" such as biological evolution (as well as the chemical origins of life, global warming, and human cloning).

What should I know?

What follows are some smaller questions that need to be resolved to answer the Big Question. Place a checkmark next to the questions that science can answer.

Smaller Questions	Can Science Answer?
Does downplaying the evidence for evolutionary theory in schools reduce the critical thinking skills of students?	
Does a lack of understanding evolutionary theory increase the likelihood of environmental damage?	
Does a lack of understanding about evolutionary theory reduce the effectiveness of health care?	
Does learning evolutionary theory negatively influence the moral behavior of children?	
Should children be exposed to all areas of human knowledge, including evolution?	
Is evolutionary theory too sophisticated to be presented to children?	

What does the science say?

Let's examine what the data say about this smaller question:

Does a lack of understanding about evolutionary theory reduce the effectiveness of health care?

The vast majority of cancer deaths occur because the cancer cells present in the body become resistant to anticancer treatment and thus "overwhelm" normal cells and body functions. The development of resistance to treatment is an evolutionary process, that is, it represents change in the cancer cell population over time. An evolutionary approach has improved treatment for antibiotic resistant diseases and HIV infection, and some argue that the same may be true for cancer treatment. Scientists examined the abstracts of more than 6000 primary research articles on treating resistant cancers to determine if an evolutionary understanding is commonly applied. The results are presented in the following graph, which shows the percent of research reports by year on therapeutic resistance or relapse that use at least one evolution term in the paper's abstract.

—Continued next page

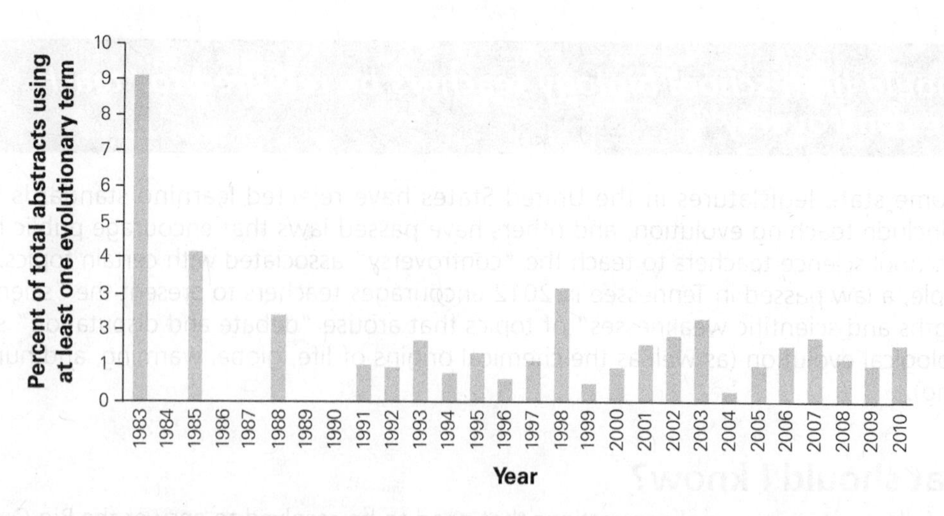

1. Describe the results. In what percent of papers published in 2010 about resistance of cancer to therapy was evolution discussed?
2. Given these data, do you think the smaller question is answered? If not, propose another study that would help answer this question.
3. Does this information help you answer the Big Question? What else do you need to consider?

Data source: C. A. Aktipis, V. S. Y. Kwan, K. A. Johnson, S. L. Neuberg, and C. C. Maley, "Overlooking Evolution: A Systematic Analysis of Cancer Relapse and Therapeutic Resistance Research," *PLoS ONE* 6, no. 11 (2011): e26100.

Chapter Review

Mastering Biology

Go to Mastering Biology to access eText 2.0, Dynamic Study Modules, and the Study Area, where you'll find practice quizzes, BioFlix™ animations, MP3 tutor sessions, current events, and more.

SUMMARY

Section 11.1

Define *biological evolution*, and distinguish it from other forms of nonevolutionary change in organisms.

- The process of evolution is the change that occurs in the characteristics of organisms in a population over time (p. 207).
- The theory of evolution, as described by Darwin, is that all modern organisms are related to one another and arose from a single common ancestor (pp. 208–209).

Illustrate the theory of common descent using a tree diagram.

- Modern organisms can be arranged on a "tree" of relationships based on similarities in morphology, development, and genes (p. 209).

SHOW YOU KNOW 11.1 From 1982 to 2008, the number of teenagers in the United States with orthodontic braces increased 99%. Was this likely the result of an evolutionary change in children's teeth? Explain.

Section 11.2

Summarize how Darwin's experiences led him to develop the outline of the theory of evolution.

- Scientists before Darwin had hypothesized that species could change over time (pp. 210–211).
- Darwin's voyage on the *HMS Beagle* led him to suspect that this hypothesis was correct. Over the course of 20 years, Darwin was able to gather enough evidence to support this hypothesis and his hypothesis of common descent so that most scientists accepted these theories as the best explanation for the diversity of life on Earth (pp. 211–213).

SHOW YOU KNOW 11.2 Darwin understood that in addition to providing evidence that evolution had occurred in *On the Origin of Species*, he needed to provide a reasonable mechanism to explain how it could occur. What is the mechanism Darwin hypothesized?

Section 11.3

Detail the modern biological classification system, and explain how it supports the theory of evolution.

- Linnaeus classified organisms based on physical similarities between them, for instance, placing humans and monkeys in the same order of animals. Darwin argued that the pattern of biological relationships illustrated by Linnaeus's classification provided strong support for the theory of common descent (pp. 215–217).

Explain how homologies in anatomy and genetics, even in useless traits, support the theory of evolution.

- Similarities in the underlying structures of a variety of organisms and the existence of vestigial structures, such as the appendix or goose bumps in humans, are difficult to explain except through the theory of common descent (pp. 218–220).
- Modern data on similarities of DNA sequences among organisms match the hypothesized evolutionary relationships suggested by anatomical similarities and provide an independent line of evidence supporting the hypothesis of common descent (pp. 220–222).

Describe how details of embryonic development support the theory of evolution.

- Similarities in embryonic development and embryonic structures among diverse organisms—for instance, the presence of pharyngeal slits in the embryos of all chordates—are best explained as a result of their common ancestry (pp. 219–220).

Define *biogeography*, and explain how it supports the theory of evolution.

- Biogeography is the study of the geographical distribution of organisms (p. 222).
- Biogeographical patterns support the hypothesis of common descent because species that appear related physically are also often close to each other geographically: for example, the tortoises and cacti on the Galápagos Islands (p. 222).

Explain how the fossil record provides direct evidence of evolutionary change in species over time.

- The fossilized remains of extinct species in many groups of organisms demonstrate a progression of forms from more ancient to more modern types. For example, early hominins appear much more ape-like than more recent hominin fossils (pp. 223–227).

SHOW YOU KNOW 11.3 Some vestigial structures persist because they may still have a function or did in the very recent past. Wisdom teeth, which erupt in early adulthood and often have to be removed because they do not "fit" in a modern person's jaw, are sometimes considered vestigial. However, they may have had a function in the recent past. Why might wisdom teeth have persisted over the course of human evolution?

Section 11.4

Articulate why the theory of evolution is considered the best explanation for the origin of humans and other organisms.

- Evidence strongly supports the hypothesis that organisms have changed over time and are related to each other (pp. 228–231).
- Shared characteristics of all life, especially the universality of DNA and the relationship between DNA and proteins, provide evidence that all organisms on Earth descended from a single common ancestor rather than from multiple ancestors (pp. 231–232).

SHOW YOU KNOW 11.4 Can you list other cellular structures or processes discussed in this text that are common to all eukaryotic organisms?

ROOTS TO REMEMBER

The following roots of words come mainly from Latin and Greek and will help you to decipher terms:

evol-	means to unroll. Chapter term: *evolution*
homini-	means human-like. Chapter terms: *hominid, hominin*
homolog-	indicates similar or shared origin; from a word meaning "in agreement." Chapter terms: *homologous, homology*
macro-	means large scale. Chapter term: *macroevolution*
-metric	means to measure. Chapter term: *radiometric*
micro-	means extremely small. Chapter term: *microevolution*
radio-	is the combining form of radiation. Chapter term: *radiometric*

LEARNING THE BASICS

1. What observations did Darwin make on the Galápagos Islands that helped convince him that evolution occurs?

A. the existence of animals that did not fit into Linnaeus's classification system; **B.** the similarities and differences among cacti and tortoises on the different islands; **C.** the presence of species he had seen on other tropical islands far from the Galápagos; **D.** the radioactive age of the rocks of the islands; **E.** fossils of human ancestors

2. Fill in the blanks in the following graph, which illustrates the decay curve of a radioactive element used to date fossil structures.

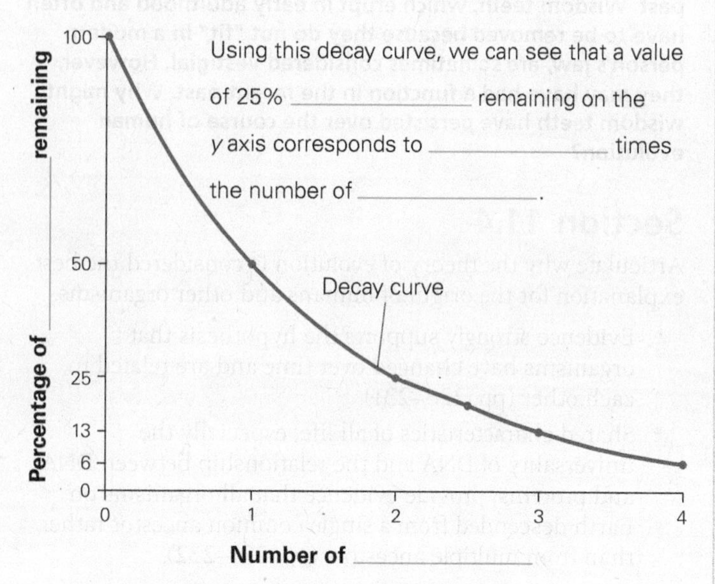

3. The process of biological evolution ______.
A. is not supported by scientific evidence; B. results in a change in the features of individuals in a population; C. takes place over the course of generations; D. B and C are correct; E. A, B, and C are correct

4. In science, a theory is a(n) ______.
A. educated guess; B. inference based on a lack of scientific evidence; C. idea with little experimental support; D. body of scientifically acceptable general principles; E. statement of fact

5. The theory of common descent states that all modern organisms ______.
A. can change in response to environmental change; B. descended from a single common ancestor; C. descended from one of many ancestors that originally arose on Earth; D. have not evolved; E. can be arranged in a hierarchy from "least evolved" to "most evolved"

6. The DNA sequence for the same gene found in several species of mammals ______.
A. is identical among all species; B. is equally different between all pairs of mammal species; C. is more similar between closely related species than between distantly related species; D. provides evidence for the hypothesis of common descent; E. more than one of the above is correct

7. Marsupial mammals give birth to young that complete their development in a pouch on the mother's abdomen. All the native mammals of Australia are marsupials, while these types of mammals are absent or uncommon on other continents. This observation is an example of ______.
A. developmental evidence for evolution; B. biogeographic evidence for evolution; C. genetic evidence for evolution; D. fossil evidence for evolution; E. not useful evidence for evolution

8. Even though marsupial mammals give birth to live young, an eggshell forms briefly early in their development. This is evidence that ______.
A. marsupials share a common ancestor with some egg-laying species; B. marsupials are not really mammals; C. all animals arose from a common ancestor; D. marsupial mammals were separately created by God; E. the fossil record of marsupial mammals is incorrect

9. A species of crayfish that lives in caves produces eyestalks like its above-ground relatives, but has no eyes. Eyestalks in cave-dwelling crayfish are thus ______.
A. an evolutionary error; B. a dominant mutation; C. biogeographical evidence of evolution; D. a vestigial trait; E. evidence that evolutionary theory may be incorrect

10. Which of the following taxonomic levels contains organisms that share the most recent common ancestor?
A. family; B. order; C. phylum; D. genus; E. class

ANALYZING AND APPLYING THE BASICS

1. The classification system devised by Linnaeus can be "rewritten" in the form of an evolutionary tree. Draw a tree that illustrates the relationship among the flowering species listed, given their classification (note that *subclass* is a grouping between class and order):
   - Pasture rose (*Rosa carolina*, Family Rosaceae, Order Rosales, Subclass Rosidae)
   - Live forever (*Sedum purpureum*, Family Crassulaceae, Order Rosales, Subclass Rosidae)
   - Spring avens (*Geum vernum*, Family Rosaceae, Order Rosales, Subclass Rosidae)
   - Spring vetch (*Vicia lathyroides*, Family Fabaceae, Order Fabales, Subclass Rosidae)
   - Multiflora rose (*Rosa multiflora*, Family Rosaceae, Order Rosales, Subclass Rosidae)

2. DNA is not the only molecule that is used to test for evolutionary relationships among organisms. Proteins can also be used, and the sequences of their building blocks (called amino acids) can be compared in much the same way that DNA sequences are compared. *Cytochrome c* is a protein found in nearly all living organisms; it functions in

the transformation of energy within cells. The percentage difference in amino acid sequence between humans and other organisms can be summarized as follows:

Cytochrome c	*Percentage difference from sequence of human sequence*
Chimpanzee	0.0
Mouse	8.7
Donkey	10.6
Carp	21.4
Yeast	32.7
Corn	33.3
Green algae	43.4

Draw the evolutionary tree implied by these data that illustrates the relationship between humans and the other organisms listed.

3. Look at the tree you generated for question 2. It does not imply that yeast is more closely related to corn than it is to green algae. Why not?

GO FIND OUT

1. Search for the biology learning standards published by your state's department of education. How do these standards address the evidence for evolution? Survey your classmates or others: What percentage of your fellow students say that evolution is still not settled science?

2. Humans and chimpanzees are more similar to each other genetically than many very similar looking species of fruit fly are to each other. What does this similarity imply regarding the usefulness of chimpanzees as stand-ins for humans during scientific research? What do you think it implies regarding our moral obligations to these animals?

MAKE THE CONNECTION

The science that you learned in this chapter has helped you better understand the real-world example used throughout this discussion. Draw a line from the statement on the left to the science that supports it on the right.

Most people accept that the characteristics within a population—say pesticide resistance in lice—can change over time. ▷	◁ Because special creation requires a supernatural event that does not obey the laws of nature, it is not a scientific hypothesis.
Some refer to the theory of evolution as "just a theory" with other competing and equally likely alternatives. ▷	◁ Homology in anatomy, development, and genetic materials provides evidence that living organisms are related with more closely related organisms having more similarities than more distantly related organisms.
People are often upset that scientists do not consider the possibility of one alternative to the theory of evolution—that is, the idea that God specially created all living organisms. ▷	◁ The fact that all known organisms have DNA, the same genetic code, and other biochemical similarities is evidence for one common ancestor.
A hypothesis that all organisms appeared separately has been rejected by science. ▷	◁ The evidence for evolution demonstrates consilience in that many lines of evidence provide support for the theory.
A hypothesis that different groups of organisms show relationship but that there is no common ancestor of life has been rejected by science. ▷	◁ A scientific theory is an idea of how the world works that is supported by evidence from many different areas of inquiry.
Fossils are the direct evidence we have to test the theory of evolution. ▷	◁ The age of a fossil can be determined by radiometric dating, and examination of the fossil remains provides information about the anatomy of organisms from the distant past that can be compared with modern organisms.
The theory of evolution is accepted by scientists as the correct explanation for the origin of humans. ▷	◁ The process of evolution can occur by natural selection—the differential survival and reproduction of individuals with different traits.

Answers to **Got It?, Visualize This, Working with Data, Sounds Right, But Is It?, Show You Know,** and **Chapter Review** questions can be found in the **Answers** section at the back of the book.

12 An Evolving Enemy

Natural Selection

A visitor from India who landed in Chicago was carrying an almost untreatable strain of an infectious disease.

In April 2015, a traveler from India landed at Chicago's O'Hare International Airport. Outwardly, she would have seemed no different from the nearly 1 million other visitors from the subcontinent that visit the United States every year. But she carried with her some dangerous baggage. After traveling throughout the Midwest—from Illinois to Tennessee to Missouri—this baggage caused the traveler to check into a hospital after returning to Chicago.

Although the visitor's identity is protected by patient confidentiality rules, we can presume that she sought medical help because she was experiencing a persistent—perhaps bloody—cough, fever and chills, and fatigue. These symptoms are likely because the United States Centers for Disease Control (CDC) announced in June that this anonymous female traveler had been diagnosed with an infection by a strain of bacteria known as XDR-TB. The deadly disease caused by this infection can be cured in only 30 percent of patients.

The potential for spread of difficult to treat diseases is a risk that can rise with the mass movement of refugees and immigrants.

In her journey, our traveler had the potential to infect thousands of people with XDR-TB—both on the airplane flight from India and in the many stops she made in the United States. There is almost no way of knowing exactly who she contacted along her route.

Thankfully, XDR-TB is not easily spread. But the story of the Indian traveler is a dramatic example of a dilemma faced worldwide. This bacterium is not normally found in the United States, only appearing when brought in by foreign travelers. Most countries' borders are not barriers to infectious disease. One response to decrease the likelihood of dangerous diseases crossing borders is to restrict visitation and immigration from countries where they are found. But the consequence of this isolationism can hurt commerce, refugees seeking a better life, and even a nation's international standing.

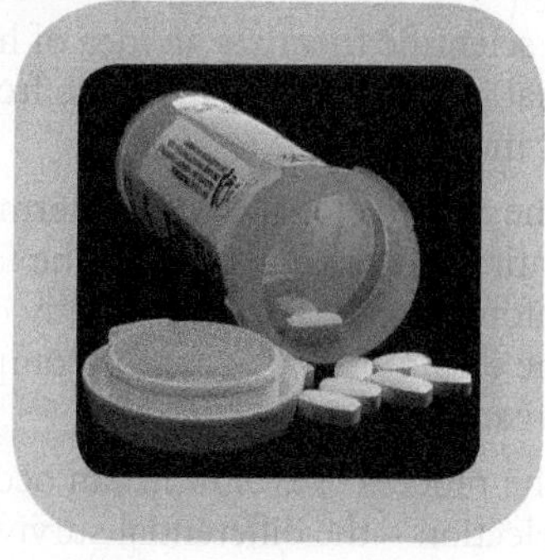

But it is the unconsciously reckless behavior of many of us that is the ultimate cause of much of the problem.

There is another approach to stopping the spread of dangerous diseases like XDR-TB: We can address our own role in preventing the evolution of untreatable diseases. Because, as it happens, the carelessness of millions of people around the world puts us at risk of XDR-TB and many other deadly microbes.

12.1 Return of a Killer

Tuberculosis (TB) has been plaguing humans for thousands of years. Tubercular decay has been found in the spines of Egyptian mummies from as long ago as 3000 B.C. In 460 B.C., the Greek physician Hippocrates identified a condition that appears to be tuberculosis as the most widespread disease of the times. As recently as 1906 in the United States, 2 of every 1000 deaths were due to TB infection.

Thanks to relatively recent advances in science and medicine, TB now accounts for only 1.5 of every 100,000 deaths in the United States. In the 1980s, the dramatic drops in infection and death due to TB around the world had many public health professionals convinced that this ancient scourge could be completely eliminated from the human population. Today, those hopes have largely faded as a result of both our own missteps and the powerful force of natural selection.

tubercul- means a small swelling.
-osis indicates a condition.

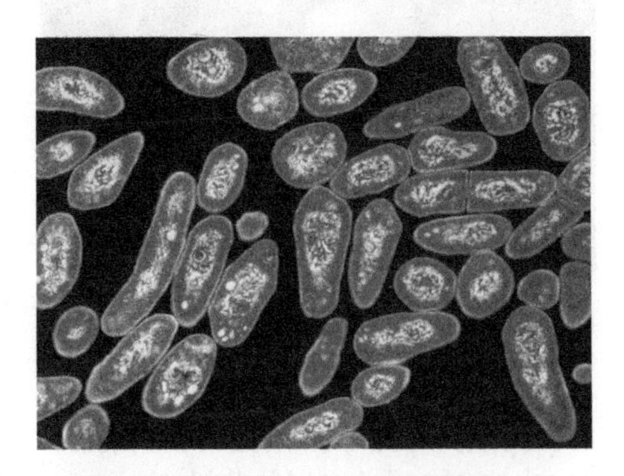

FIGURE 12.1 The organism that causes tuberculosis. The bacterial species *Mycobacterium tuberculosis*.

What Is Tuberculosis?

The bacterium *Mycobacterium tuberculosis* causes the disease tuberculosis (**FIGURE 12.1**). Tuberculosis is a public health problem not because it is especially deadly but instead because it affects so many. Two billion people, more than one-quarter of the world's population, carry *M. tuberculosis,* and new infections are estimated to occur at a rate of one per second. Ninety percent of *M. tuberculosis* infections are symptomless, and most of these resolve as the bacteria are destroyed by the infected individual's immune system. However, the remaining 10% of infected individuals develop active disease, and more than one-half of these individuals will die without treatment. Tuberculosis now causes approximately 2 million deaths worldwide every year.

The symptoms of tuberculosis include a cough that produces blood, fever, fatigue, and a long, relentless wasting in which the patient gradually becomes weaker and thinner. These symptoms explain the antiquated name for this disease, *consumption,* because the infection seemed to consume people from within.

We now understand that the consumptive symptoms of TB arise from destruction of lung tissue caused by the body's reaction to active *M. tuberculosis* infection. Colonies of the bacteria in the lung are walled off by immune system cells, creating structures called tubercles (**FIGURE 12.2**); while this response does slow the spread of the disease within the lungs, it permanently damages the tissue. The reduction in the capacity of the lungs to provide oxygen to the

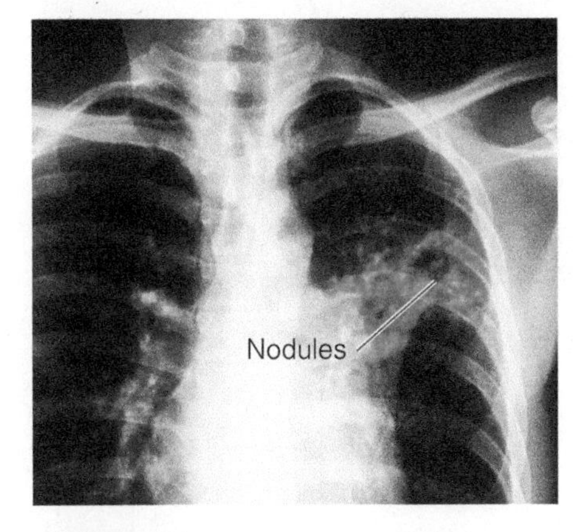

FIGURE 12.2 Lung tubercles. The red spots on this lung X-ray are nodules, or tubercles, within the lung tissue.

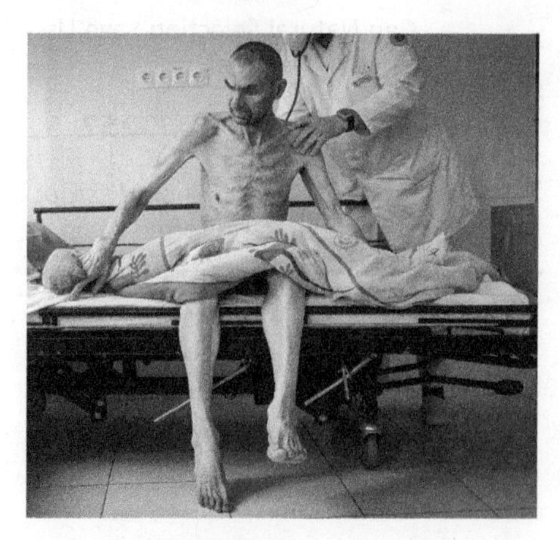

FIGURE 12.3 The effects of tuberculosis infection. As infected individuals lose more and more lung tissue to tuberculosis infection, they have difficulty obtaining enough oxygen. As a result, their tissues begin to waste away.

FIGURE 12.4 Transmission through the air. Like cold and flu viruses, the tuberculosis bacteria can be transmitted in droplets that are emitted in the sneeze or cough of an infected individual.

anti- means in opposition to.

-biotic indicates pertaining to life.

body causes the physical wasting characteristic of tuberculosis (**FIGURE 12.3**). The infection is difficult to cure because *M. tuberculosis* can lie dormant for months inside these tubercles. When the tubercles later degrade, they release bacteria back into the lung, possibly causing additional infections within the lungs and spreading to other individuals.

Although you can be infected without symptoms, transmission of tuberculosis is almost entirely from people with active disease. When individuals with an active infection cough, sneeze, speak, or spit, they expel infectious droplets. A single sneeze can release about 40,000 of these droplets (**FIGURE 12.4**). People with prolonged, frequent, or intense contact with infected individuals are at highest risk of becoming infected themselves.

Most individuals can fight off infection with *M. tuberculosis,* but some cannot. Those at highest risk of active TB are young children; the elderly; individuals in poor overall health due to poor nutrition, other illnesses, or drug abuse; and people with AIDS. Until about 60 years ago, these individuals had little chance of surviving tuberculosis. Today, their prognosis—especially in Western countries where high-quality health care is readily available—is much better. But that may be changing.

Treatment—and Treatment Failure

The treatments for tuberculosis in the nineteenth and early twentieth centuries, at least among the wealthy, consisted primarily of long stays in rural facilities where the air was fresh and unpolluted. These tuberculosis sanatoriums (**FIGURE 12.5**) were useful for two reasons. By moving patients from areas where the air was thick with lung-damaging particles to clean-air sanatoriums, doctors were able to preserve patients' lung function for longer periods of time than would otherwise be possible. And because sanatoriums isolated patients, they reduced the spread of TB to the rest of the community. In poorer communities, individuals with active TB were often forcibly isolated in much grimmer conditions.

The discovery of **antibiotics,** drugs that kill microbes, including bacteria, revolutionized tuberculosis treatment in the 1940s. Since then, infected patients with active TB are typically kept in isolation for only 2 weeks until antibiotics

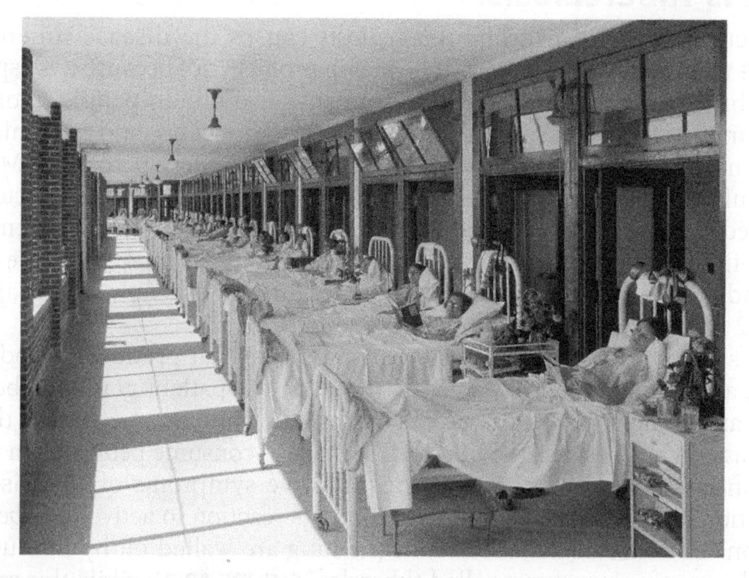

FIGURE 12.5 A tuberculosis sanatorium. Typically, patients at sanatoriums spent many hours every day exposed to fresh air, either outside or, like these patients, in large open porches.

kill off most of the *M. tuberculosis* in the lungs. At this point, the patient is no longer contagious and can return to the community. However, because *M. tuberculosis* can hide inside the body for long periods, antibiotic treatment must be maintained for 6 to 12 months to completely eliminate the organism.

Since the 1980s, however, scientists have chronicled a disturbing rise in the number of **antibiotic-resistant** tuberculosis infections—ones that cannot be cured by the standard drug treatment. According to the Centers for Disease Control and Prevention (CDC), approximately 1% of TB cases reported annually since 1993 (that is, about 2000 cases total) did not respond to standard treatments. As noted earlier, such cases are called MDR-TB, or multidrug-resistant TB. Only 76 of the 2000 cases were resistant to treatment even with second-line drugs (or XDR-TB). Even in the United States, with abundant access to resources and drugs, only an estimated 30–50% of individuals diagnosed with active XDR-TB have been cured of the disease.

In countries with fewer resources, the toll of XDR-TB could be much greater. In a recent XDR-TB outbreak in South Africa, 52 of 53 individuals diagnosed with the strain died within a month of showing signs of active disease. The resurgence of tuberculosis has now been declared a global health emergency by the World Health Organization.

Why are we losing the battle against tuberculosis? And what can be done to stop it? Answering these questions requires an understanding of an important force for evolutionary change: natural selection.

Got It?

1. The disease tuberculosis is caused by a ________.
2. The symptoms of tuberculosis are directly caused by damage to an infected individual's ________.
3. Individuals with active tuberculosis may spread the infectious organism when they ________ or ________.
4. Drugs called ________ are used against the organisms that cause diseases like tuberculosis.
5. New strains of tuberculosis have appeared that currently are essentially untreatable because they are ________.

12.2 Natural Selection Causes Evolution

In the *Origin of Species*, Charles Darwin put forth two major ideas: the theory of common descent (Chapter 11) and the theory of **natural selection.** Darwin's presentation of the theory of common descent—that all species living today appear to have descended from a single ancestor—was thorough and convincing. Within 20 years of the publication of his book, the theory of common descent had been accepted by most scientists. However, it was another 60 years before the scientific community accepted Darwin's theory of natural selection, which explains in large part *how* organisms evolved from a common ancestor to become the great variety we see today. Darwin proposed that through the process of natural selection, the physical or behavioral traits of organisms that lead to increased survival or reproduction become common within their population, while less favorable traits are lost. The changes accumulating within populations via natural selection can lead to the development of new species.

Darwin reasoned that the process of natural selection is an inevitable consequence of the competition for survival among variable individuals in a population. Today, natural selection is considered one of the most important causes

(a) Variation in coat color

(b) Variation in blooming time

FIGURE 12.6 Observation 1: Individuals within populations vary. (a) Gray wolves vary in coat color, even within a single litter of animals. (b) Flowers may vary in blooming time, with some individual plants blooming much earlier than others of the same species.

VISUALIZE **THIS** 

Under what conditions might it be an advantage for an individual plant to bloom earlier than other nearby flowers?

of evolution (although others, such as the processes of genetic drift and sexual selection as described in Chapter 13, also cause populations to change over time).

Darwin's Observations

The theory of natural selection is elegantly simple. It is an inference based on four general observations:

1. Individuals within Populations Vary. Observations of groups of humans support this statement—people do come in an enormous variety of shapes, sizes, colors, and facial features. It may be less obvious that there is variation in nonhuman populations as well. For example, in a litter of gray wolves born to a single female, individuals may vary in coat color, while in a field of flowers, one plant may bloom earlier than others (**FIGURE 12.6**). We can add all kinds of less obvious differences to this visible variation; for example, the amount of caffeine produced in the seeds of a coffee plant varies among individuals in a wild population. Each different type of individual in a population is called a **variant.**

FIGURE 12.7 Observation 2: Some of the variation among individuals can be passed on to their offspring. Darwin noted that breeders could create flocks of pigeons with fantastic traits by using as parents of the next generation only those individuals that displayed these traits.

2. Some of the Variation among Individuals Can Be Passed on to Their Offspring. Although Darwin did not understand how it occurred, he observed many examples of the general resemblance between parents and offspring. He also noticed that people took advantage of the inheritance of variation in other species. Pigeon fanciers in Darwin's time clearly recognized the inheritance of variation; they could see, for instance, that pigeons with neck ruffs were more likely to produce offspring with neck ruffs than were pigeons without ruffs. Thus, when enthusiasts wanted to produce a ruffed variety of pigeon, they encouraged breeding among the birds with this trait (**FIGURE 12.7**). Darwin hypothesized that offspring tend to have the same characteristics as their parents in natural populations as well.

For several decades after the *Origin of Species* was published, the observation that some variations were inherited was the most controversial part of the theory of natural selection. Because scientists could not adequately explain the

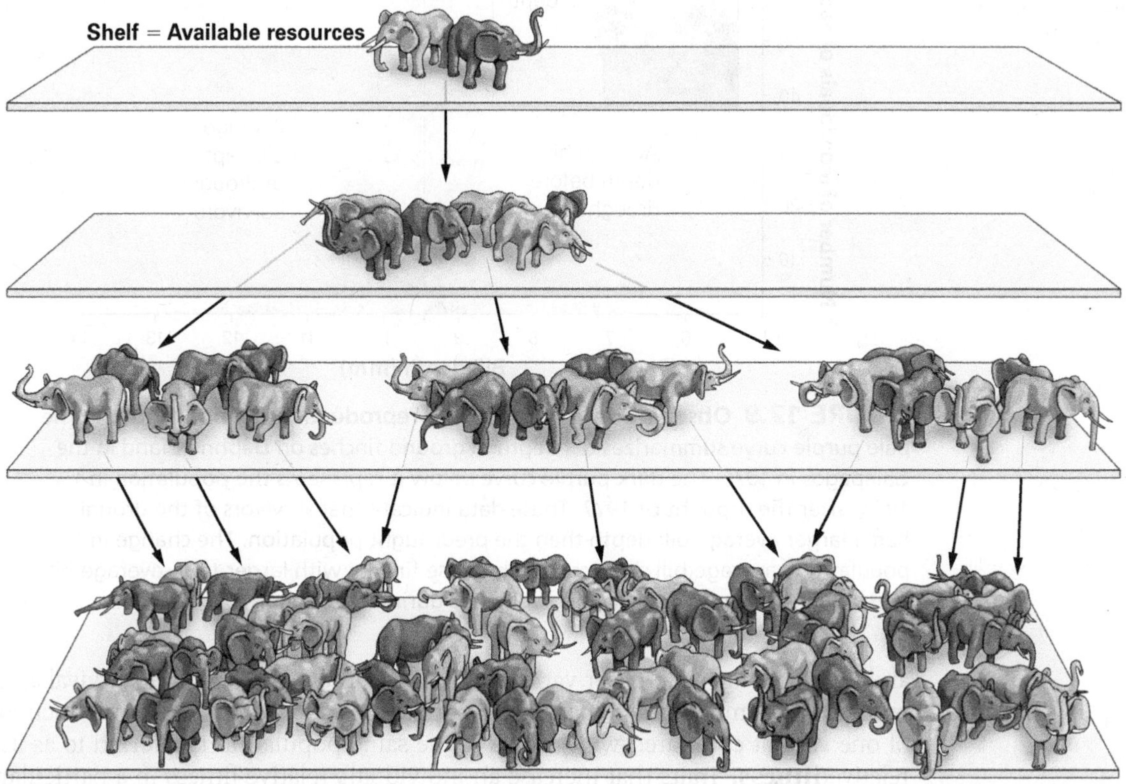

If a female elephant (colored pink) lives a full fertile lifetime, she will bear about six calves in about 90 years. On average, half of her calves will be female.

Shelf = Available resources

Generation 0 = 2 elephants

Generation 1 = 6 elephants

Generation 2 = 18 elephants

Generation 3 = 54 elephants

FIGURE 12.8 Observation 3: Populations of organisms produce more offspring than will survive. Even slow-breeding animals like elephants are capable of producing huge populations relatively quickly.

VISUALIZE **THIS**

Predict what will happen to the elephant population as a result of limited resource availability when Generation 4 is produced.

origin and inheritance of variation, many were unwilling to accept that natural selection could be a mechanism for evolutionary change. When Gregor Mendel's work on inheritance in pea plants (Chapter 8) was rediscovered in the 1900s, the mechanism for this observation became clear—natural selection operates on genetic variation that is passed from one generation to the next.

3. Populations of Organisms Produce More Offspring than Will Survive. This observation is clear to most of us—the trees in the local park make literally millions of seeds every summer, but only a small fraction of these survive to germinate, and only a few of the seedlings live for more than a year or two.

In the *Origin of Species,* Darwin gave a graphic illustration of the difference between offspring production and survival. In his example, he used elephants, animals that live long lives and are very slow breeders. A female elephant does not begin breeding until age 30, and she produces about 1 calf every 10 years until around age 90. Darwin calculated that even at this very low rate of reproduction, if all the descendants of a single pair of African elephants survived and lived full, fertile lives, after about 500 years their family would have more than 15 million members (**FIGURE 12.8**)—many more than can be supported by all the available food resources on the African continent!

4. Survival and Reproduction Are Not Random. In other words, the subset of individuals that survives long enough to reproduce is not an arbitrary group. Some variants in a population have a higher likelihood of survival

WORKING WITH DATA

Use the graph to determine how the total population size of ground finches changed between 1976 and 1978.

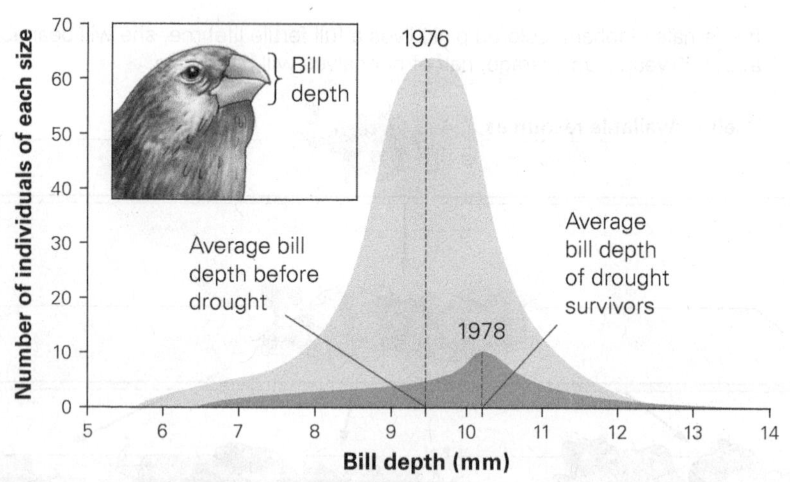

FIGURE 12.9 Observation 4: Survival and reproduction are not random. The pale purple curve summarizes bill depth in ground finches on Daphne Island in the Galápagos in 1976. The dark purple curve below it represents the population in 1978, after the drought of 1977. These data indicate that survivors of the drought had a larger average bill depth than the predrought population. The change in the population's average bill size occurred because finches with larger-than-average bills had higher fitness than did small-billed birds during the drought.

and reproduction than other variants do; that is, there is differential survival and reproduction among individuals in the population. The survival and reproduction of one variant compared with others in the same population is referred to as its relative **fitness.** Traits that increase an individual's relative fitness in a particular environment are called **adaptations.** Individuals with adaptations to a particular environment are more likely to survive and reproduce than are individuals lacking such adaptations; in other words, these individuals have higher relative fitness.

adapt- means to fit in.

Darwin referred to the results of differential survival and reproduction as natural selection. Adaptations are "naturally selected" in the sense that individuals possessing them survive and contribute offspring to the next generation. Although Darwin used the word *selection,* which implies some active choice, natural selection is a passive process that is simply determined by differences among individuals and their success in their particular environment. For example, among the birds called medium ground finches living on an island in the Galápagos archipelago, scientists have observed that when rainfall is scarce, a large bill is an adaptation that can be observed. The large bill can be explained because birds with this attribute are able to crack open large, tough seeds—the only food available during severe droughts. As shown in **FIGURE 12.9**, the 90 survivors of a 1977 drought had an average bill depth that was 6% greater than the average bill depth of the original population of 751 birds. In these environmental conditions, a large bill increases survival.

Adaptations are not only traits that increase survival. Any trait possessed by an individual that increases the number of offspring it produces relative to other individuals in a population is also an adaptation. For example, flowers in a meadow may have a relatively limited number of potential insect pollinators. More pollinator visits generally result in more seeds being produced by a single flower, so any trait that increases a flower's attractiveness to pollinators, such as a brighter color or greater nectar production, should be favored by natural selection (**FIGURE 12.10**).

FIGURE 12.10 Adaptations are not about survival only. Variations that increase a flower's attractiveness to a pollinator can increase its reproductive success by increasing the number of seeds it produces.

Darwin's Inference: Natural Selection Causes Evolution

Based on his observations, Darwin reasoned that the result of natural selection is that inherited variations that are favorable within a given environment tend to increase in frequency in a population over time, while variations that are unfavorable within a given environment tend to be lost within the population. In other words, adaptations become more common in a population as the individuals who possess them contribute larger numbers of their offspring to the succeeding generation. Natural selection results in a change in the traits of individuals in a population over the course of generations—that is, evolution. Although there are other factors, such as genetic drift and migration of individuals, that can cause populations to evolve over time, natural selection is the only force that can lead to the adaption of a population to its environment.

It is a testament to the power of the theory of natural selection that today it seems self-evident to us. But the theory of natural selection only became so powerful after it was tested and shown to work—in nature—in the manner Darwin described. Natural selection proved such a powerful idea that it has influenced how we think about many phenomena, from the success of particular brands of soft drinks to the relationships among nations. Natural selection also explains the emergence of XDR-TB.

Testing Natural Selection

Darwin proposed a scientific explanation of how evolution occurs, and like all good hypotheses, it needed to be tested. All of the tests described next illustrate that natural selection is an effective mechanism for evolutionary change.

Artificial Selection. Selection imposed by human choice is called *artificial selection*. It is artificial in the sense that humans deliberately control the survival and reproduction of individual plants and animals to change the characteristics of the population. Individuals with preferred traits are permitted to breed, whereas those that lack preferred traits are not allowed to breed.

The fancy pigeons that Darwin studied arose by artificial selection, and the great variety of domestic dogs we see today also resulted from this process. In each case, different breeds evolved through selection by breeders for various traits (**FIGURE 12.11**). These examples demonstrate that differential survival and reproduction change the characteristics of populations. However, because of the direct intervention of humans on the survival and reproduction of these organisms, artificial selection is not exactly equivalent to natural selection. Can change in populations occur without direct human intervention?

Natural Selection in the Lab. Another test of the effectiveness of natural selection is to examine whether populations living in artificially manipulated laboratory environments change over time. An example of this kind of experiment is one performed on fruit flies placed in environments containing different concentrations of alcohol.

VISUALIZE THIS

What would this sequence of changes look like if the selection was for dogs with a particular behavioral trait, for instance, pointing at a prey animal?

Artificial selection for dogs with short legs
Only these two dogs are allowed to breed.
Generation 1

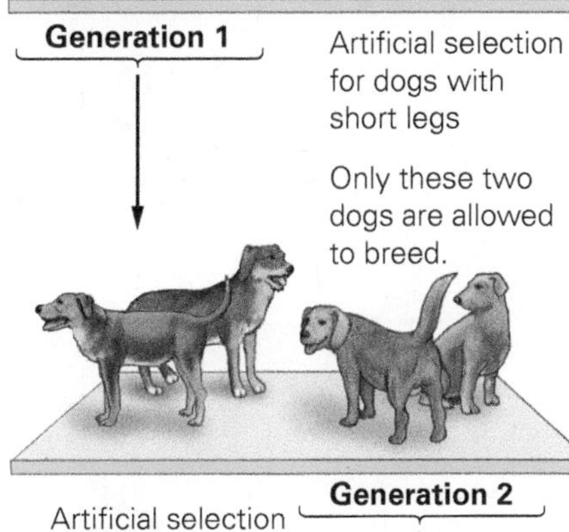
Artificial selection for dogs with short legs
Only these two dogs are allowed to breed.
Generation 2

Artificial selection for dogs with short legs
Only these two dogs are allowed to breed.
Generation 3

Dachshunds

FIGURE 12.11 Artificial selection can cause evolution. When breeders select dogs with certain traits to produce the next generation of animals, they increase the frequency of that trait in the population. Over generations, the trait can become quite exaggerated. Dachshunds are descendants of dogs that were selected for the production of very short legs.

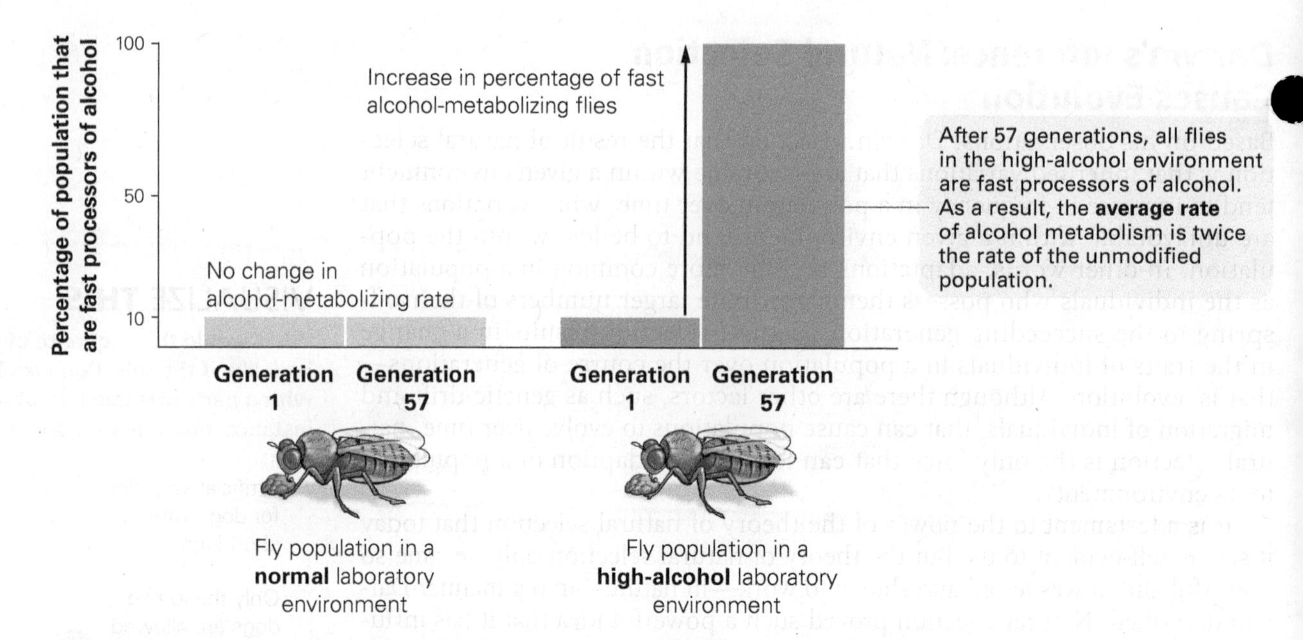

WORKING WITH DATA

This data can be represented as a line graph as well as a bar graph. Sketch the line graph.

FIGURE 12.12 Natural selection in laboratory conditions. When fruit flies are placed in a high-alcohol environment, the percentage of flies that can rapidly metabolize alcohol increases over many generations because of natural selection. In the normal laboratory environment, there is no selection for faster alcohol processing, so the average rate of alcohol metabolism does not change.

High concentrations of alcohol cause cell death in fruit flies. Many organisms, including fruit flies and humans, produce enzymes that metabolize alcohol—that is, they break it down, extract energy from it, and modify it into less-toxic chemicals. There is variation among fruit flies in the rate at which they metabolize alcohol. In a typical laboratory environment, most flies process alcohol relatively slowly, but about 10% of the population possesses an enzyme variant that allows those flies to metabolize alcohol twice as rapidly as the more common variant.

In an experiment (**FIGURE 12.12**), scientists divided a population of fruit flies into two randomized groups. Initially, these two groups had the same percentage of fast and slow alcohol metabolizers. One group of flies was placed in an environment containing typical food sources; the other group was placed in an environment containing the same food spiked with alcohol. After 57 generations, or about two years in the laboratory, the percentage of fast alcohol-metabolizing flies in the environment with only typical food sources was the same as at the beginning of the experiment—10%. But after the same number of generations, the percentage of fast alcohol-metabolizing flies in the alcohol-spiked environment was 100%. Because all of the flies in this environment were now of the fast alcohol-metabolizing variety, the *average* rate of alcohol metabolism in the population in this environment was much higher in generation 57 than in generation 1. The population had evolved.

The evolution of the fruit flies in this experiment was a result of natural selection. In an environment where alcohol concentrations were high, individuals that were able to metabolize alcohol relatively rapidly had higher fitness. Because they lived longer and were less affected by alcohol, the fast alcohol-metabolizing flies left more offspring than the slow alcohol-metabolizing flies did. Thus, each generation had a higher frequency of fast alcohol-metabolizing individuals than the previous generation did. Afte many generations, flies that could rapidly metabolize alcohol predominated in the population.

(a) Asian shore crab

(b) Blue mussels

FIGURE 12.13 Natural selection in the wild. (a) Asian shore crabs are recent invaders to the east coast; because they were unfamiliar predators to the native mussels, they initially devastated mussel beds. (b) Some populations of blue mussels have evolved the ability to recognize the presence of the crabs and respond the way they do to native predators, by adding layers to their shells.

Selection can change populations in highly regulated laboratory environments. But does it have an effect in natural wild populations?

Natural Selection in Wild Populations. The evolution of *M. tuberculosis* from being susceptible to antibiotics to being resistant is one example of natural selection in a wild population; clearly, a change in the environment (that is, the introduction of antibiotics) caused a change in the bacteria population. Dozens of other **pathogens,** organisms that cause disease, have become resistant to drugs and pesticides in the past 50 years as well. But even these changes may seem less convincing to some readers because the adaptation is to a human-imposed environmental change. Although studying adaptation to natural environmental changes in the field is a significant challenge, the effects of natural selection have also been observed in dozens of wild populations.

patho- means disease.

A classic example of natural selection in a natural setting is the evolution of bill size in Galápagos finches in response to drought (review Figure 12.9). The survivors of the drought tended to be those with the largest bills, which could more easily handle the tough seeds that were available in the dry environment. The survival of this nonrandom subset of birds resulted in a dramatic change in the next generation. The population of birds that hatched from eggs in 1978—the descendants of the drought survivors—had an average bill depth 4 to 5% larger than that of the predrought population.

A more recent example of natural selection causing evolution has occurred in the past few decades on the eastern coast of the United States, where an invasive Asian crab species is wreaking havoc on native mussels. But one species, the blue mussel, has quickly evolved the ability to thicken its shell when it grows in the presence of the Asian crab, thwarting their attacks (**FIGURE 12.13**). Scientists at the University of New Hampshire were able to demonstrate that this was an evolutionary change by comparing blue mussel populations in regions invaded by the Asian crab with those in more northerly waters, where the Asian crab cannot survive. These researchers demonstrated that while both populations of mussels thicken their shells in response to the presence of native crabs, only the mussels that had been living with the Asian crabs responded to the presence of this species. Clearly, natural selection for individual mussels that could recognize this new species of crab as a predator had caused a change in the mussel population.

Got It?

1. Two of Darwin's observations that led to the theory of natural selection are that organisms in a population ________ from each other and that traits can be ________ on their offspring.
2. The fitness of an individual is defined as its success in ________ and ________ compared with other individuals in the same population.
3. An adaptation is a trait that increases an individual's ________ relative to others in the population who do not have the trait.
4. When humans manipulate the environment and cause the evolution of traits in a population of domestic animals or plants it is called ________ ________.
5. The evolution of rapid alcohol metabolism in a population of fruit flies in a lab occurred in the presence of high alcohol levels could occur because individuals ________ in their rate of alcohol metabolism.

12.3 Natural Selection Since Darwin

Although tests of natural selection seemed to support the theory, it took 60 years for the rest of biological science to catch up to Darwin's insight and adequately explain it. Scientists working in the new field of genetics in the 1920s began to recognize that genetic principles could explain how natural selection causes the evolution of populations.

The Modern Synthesis

The union between genetics and evolution was termed "the modern synthesis" during its development in the 1930s and 1940s. The modern synthesis outlines the model of evolutionary change accepted by the vast majority of biologists.

The modern synthesis is predicated on a number of the genetic principles (discussed in Chapters 6 through 10)—namely:

- Genes are segments of genetic material (typically DNA) that contain information about the structure of molecules called proteins.
- The actions of proteins within an organism help determine its physical traits.
- Different versions of the same gene are called alleles, and variation in physical traits among individuals in a population is often due to variation in the alleles they carry.
- Different alleles for the same gene arise through mutation—changes in the DNA sequence.
- Half of the alleles carried by a parent are passed to their offspring via their eggs or sperm.

We can apply these genetic principles to the fruit flies exposed to a high-alcohol environment. In this population there are two variants, or alleles, of the gene that controls alcohol processing. One allele produces a fast alcohol-metabolizing enzyme, and the other produces a slow alcohol-metabolizing enzyme. In the high-alcohol environment, flies that made mostly fast alcohol-metabolizing enzyme had more offspring than did flies that made primarily the slow alcohol-metabolizing enzyme—because they lived longer to do so. Therefore, in the next generation, produced primarily by fast-alcohol metabolizers, a higher percentage of flies in the population inherited and thus carried at least one allele for the fast alcohol-metabolizing enzyme. This illustrates why we now describe the evolution of a population as an increase or decrease in the frequency of an allele for a particular gene.

Watch **Mutation** in Mastering **Biology**

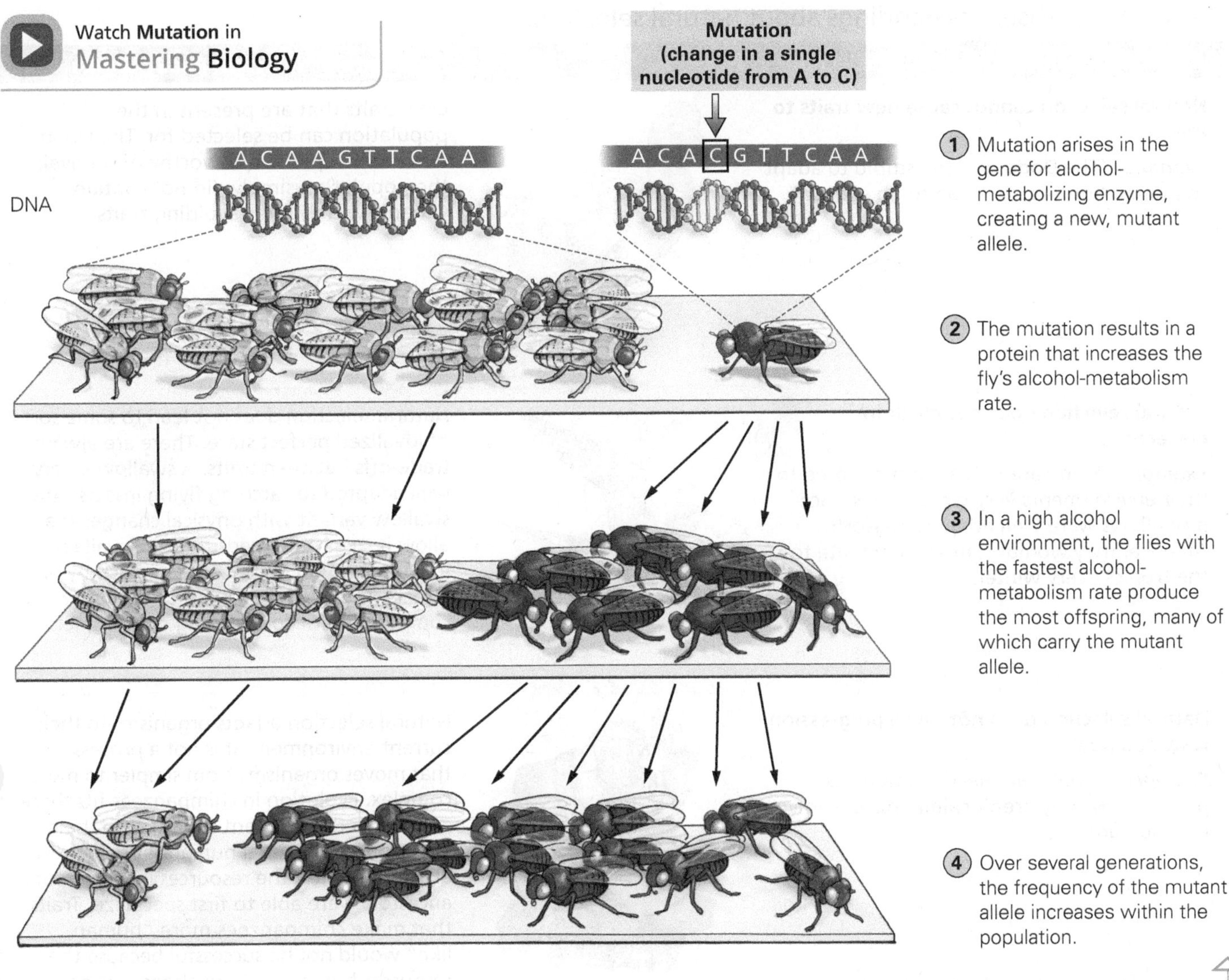

Mutation
(change in a single nucleotide from A to C)
A C A A G T T C A A
A C A C G T T C A A
DNA
1 Mutation arises in the gene for alcohol-metabolizing enzyme, creating a new, mutant allele.
2 The mutation results in a protein that increases the fly's alcohol-metabolism rate.
3 In a high alcohol environment, the flies with the fastest alcohol-metabolism rate produce the most offspring, many of which carry the mutant allele.
4 Over several generations, the frequency of the mutant allele increases within the population.

FIGURE 12.14 Mutation and natural selection. When a gene has mutated, its product may have a slightly different activity. If this new activity leads to increased fitness in individuals carrying the mutated gene, it will become more common in the population through the process of natural selection.

VISUALIZE THIS

How would the ratio of fast to slow metabolizers differ on each "shelf" of this figure if the flies were not in a high-alcohol environment?

The existence of two different alleles for alcohol metabolism in fruit flies suggests that one of these alleles is a mutated version of the other. In the normal laboratory environment, neither of these alleles appears to have a strong effect on fitness. Because the slow alcohol metabolizers are more numerous than the fast alcohol metabolizers, it appears that there might be a slight disadvantage to carrying the fast alcohol-metabolizing enzyme in a low-alcohol environment. However, in the high-alcohol environment, the mutation resulting in the fast alcohol-metabolizing allele gives a strong advantage, and its presence in the population allows for the population's evolution (**FIGURE 12.14**). Scientists now understand that the random process of gene mutation generates the raw material—variations—for evolution, and that natural selection acts as a filter that selects for or against new alleles produced by mutation.

The Subtleties of Natural Selection

Because the idea of natural selection has been applied to the realm of human society—such as to the success or failure of a particular company or technology—misunderstandings of how it works in nature are frequent. Common misunderstandings of natural selection fall into three categories: the relationship between the individual and the population, the limitations on the traits that

TABLE 12.1 Misunderstandings about natural selection.

A Misunderstanding of Natural Selection		How Natural Selection Really Works
Natural selection cannot cause new traits to arise. *Example:* "The Dodo was too stupid to adapt to human hunters, so it had to go extinct."		Only traits that are present in the population can be selected for. The Dodo was not "stupid" or unworthy of survival; the population simply did not contain variants with hunter-avoiding traits.
Natural selection does not result in perfection. *Example:* "Some animals are not adapted to their environments very well. For instance, if swallows were well adapted to North America, they wouldn't have to migrate to the tropics every winter."		Natural selection does not lead to some sort of idealized perfect state. There are always trade-offs between traits. A swallow is very well adapted to catching flying insects. Any swallow variant with physical changes that allow it to eat the seeds that are available in winter would be less able to catch flying insects. These variants would thus lose in the competition for food during the summer.
Natural selection does not cause progression toward a goal. *Example:* "If natural selection improves populations, why aren't chimpanzees evolving into humans?"		Natural selection adapts organisms to their current environment; it is not a process that moves organisms from simpler to more complex. Evolution in chimpanzees fits these animals to their current situation. In their environment, bipedal humans already exist and fully exploit the resources in which our ancestors were able to first specialize. Traits that make chimpanzees more "human-like" would not be successful because the resources humans use are already taken.

can be selected, and the ultimate result of selection. **TABLE 12.1**. provides examples of statements that you might hear in reference to natural selection and a brief explanation of how each demonstrates a misunderstanding of this process. In short, the subtleties of natural selection can be described in three statements: (1) Natural selection only acts on traits present in the population, not on individuals. (2) An adaptation may provide a benefit in most conditions faced by an organism, but not in all. (3) Natural selection results in the fit of an organism to the environment it is currently experiencing, not some future environment.

Patterns of Selection

As Darwin noted, natural selection is a force that causes the traits in a population to change over time. A more modern understanding of the process has helped scientists recognize that different environmental conditions may lead to no change in the population or even cause it to split into two species.

The type of natural selection experienced by the flies in the alcohol-lader. environment and by bacteria responding to the use of antibiotics is called **directional selection** because it causes the population traits to move in a particular direction (**FIGURE 12.15a**). Directional selection is typically the type

WORKING WITH DATA

How would the graph in part (a) differ if the red flower was avoided by the pollinator and the white flower preferred?

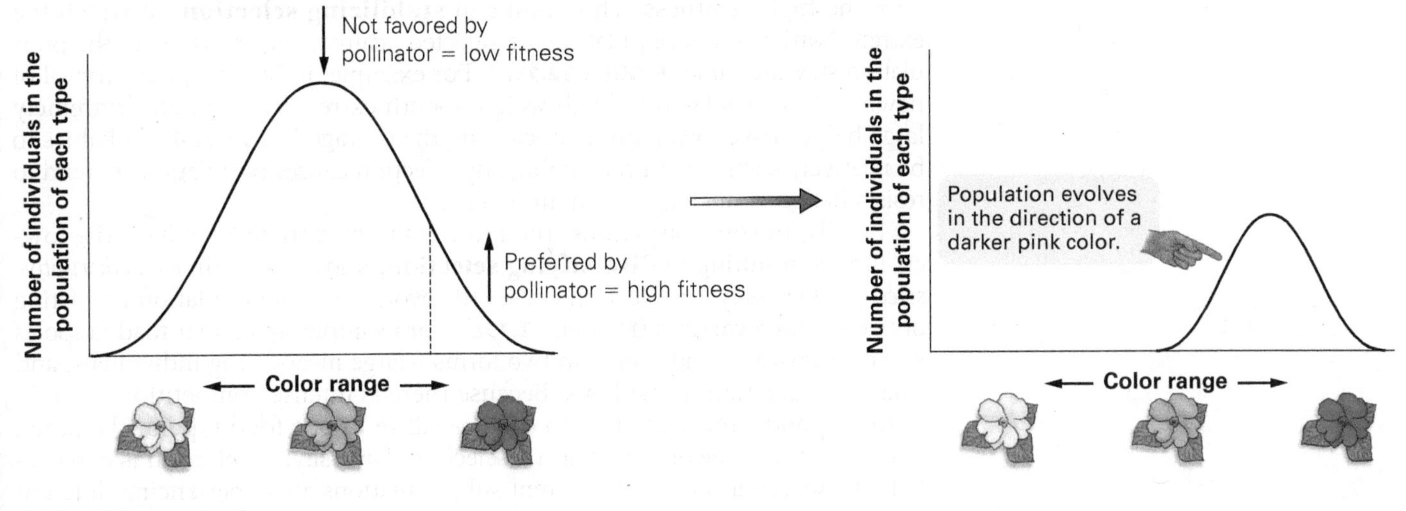

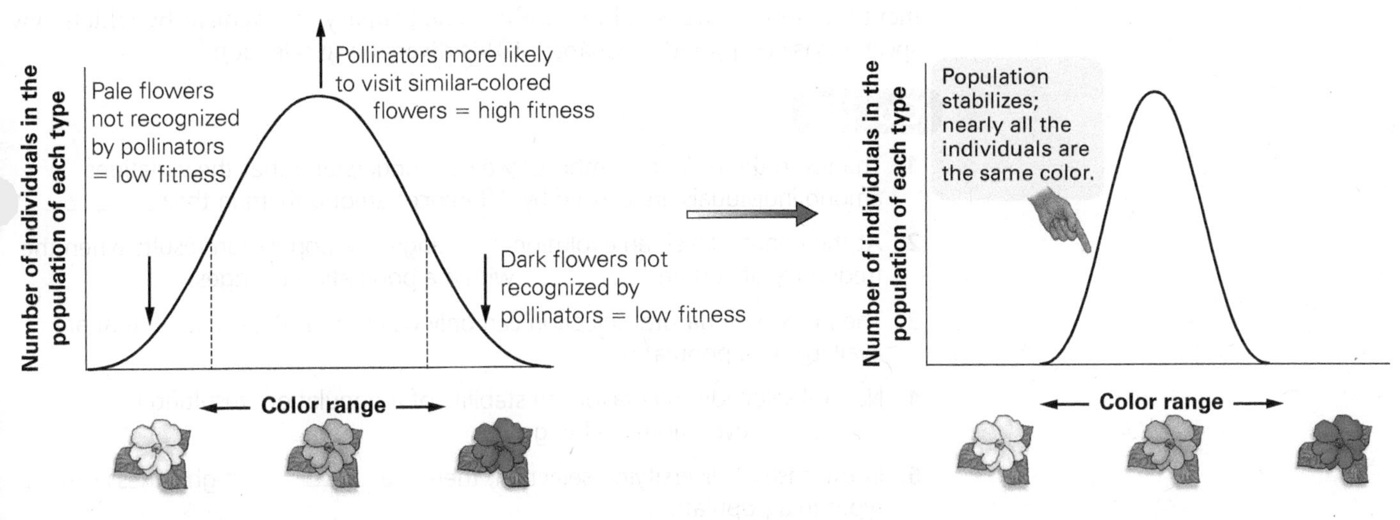

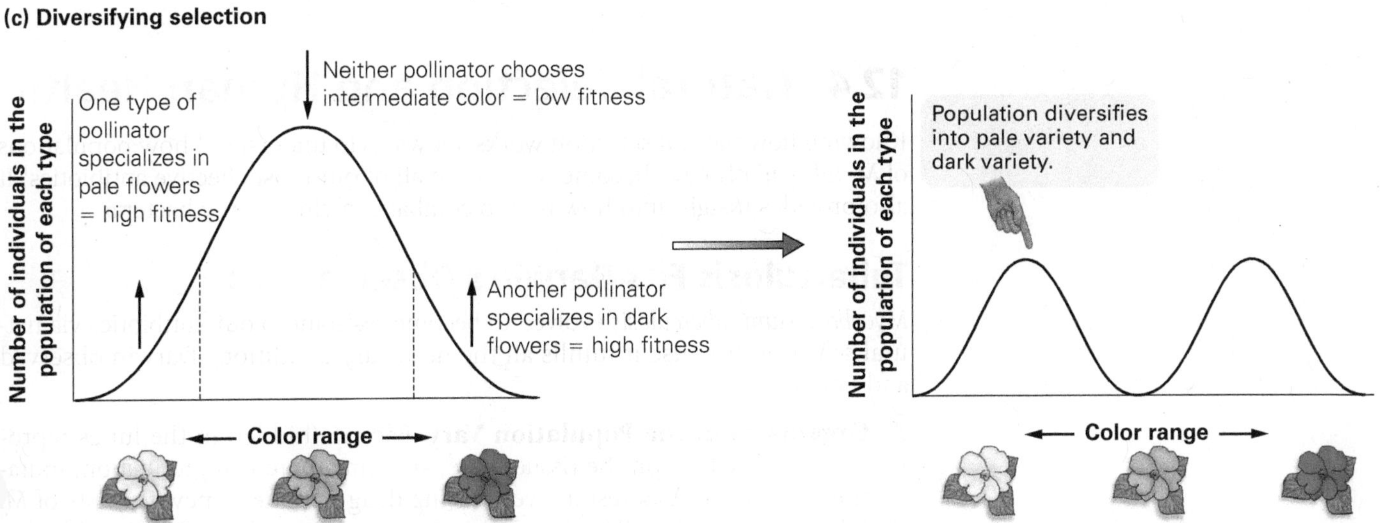

FIGURE 12.15 Directional, stabilizing, and diversifying selection. In a variable population, different environmental conditions cause different types of selection.

of selection that leads to change in a population over time—in the case of the flies, to a more alcohol-tolerant population, and in the case of bacteria, to more antibiotic-resistant strain.

In certain environments, however, the average variant in the population may have the highest fitness. This results in **stabilizing selection,** in which the extreme variants in a population are selected against, and the traits of the population stay the same (**FIGURE 12.15b**). For example, in humans, the survival of newborns is correlated to birth weight—both extremely small and extremely large babies have lower survival, causing the average birth weight of babies to be relatively stable over time. Stabilizing selection causes populations to tend to resist change in unchanging environments.

Finally, in some situations, the most common variant may have the lowest fitness, resulting in **diversifying selection,** known sometimes as *disruptive selection*. Diversifying selection causes the evolution of a population consisting of two or more variants (**FIGURE 12.15c**). For example, spadefoot toad tadpoles in New Mexican ponds come in two forms—large meat-eating individuals, and smaller, vegetarian individuals. Because there is intense competition for food in these ponds, individuals who can specialize in one food type or the other, but not both, are favored by natural selection. Diversifying selection is especially likely within a species if different subpopulations are experiencing different environmental conditions, such that the traits that bring success in one environment are not so successful in another. The primary mechanism by which new species arise (explored in Chapter 13) is diversifying selection.

Got It?

1. Thanks to the modern synthesis, we now understand that the variations among individuals are caused by differences among them in their ________.
2. At the genetic level, an evolutionary change in a population results when the frequency of certain ________ within a population changes.
3. The process of natural selection can only work on the ________ that are available in a population.
4. Natural selection may favor the stability of a population, resulting in ________ evolutionary change.
5. In the case of diversifying selection, there are ________ high-fitness variant types in a population.

12.4 Natural Selection and Human Health

Knowing how natural selection works allows us to understand how populations of *M. tuberculosis* have become resistant to all of our most effective antibiotics. It also provides insight into how best to combat evolving enemy bacteria.

Tuberculosis Fits Darwin's Observations

Mycobacterium tuberculosis evolved to become resistant to our antibiotics via natural selection because it fulfills all the necessary conditions Darwin observed and noted.

1. **Organisms in the Population Vary.** Bacteria hidden in the lungs reproduce as they feed on the tissue there; any time there is reproduction, mutation can occur. As a result, even during drug treatment, new variants of *M. tuberculosis* continually arise. Some of these variants have proteins that disable or counteract certain antibiotics, making the bacteria more resistant to these drugs.

2. **The Variation among Organisms Can Be Passed on to Offspring.** The traits that cause drug resistance are coded for in a bacterium's DNA. When a cell divides to reproduce, it copies this DNA and passes it—and the trait—on to its daughter cells. Bacteria can also pass on variation by direct transfer of fragments of DNA from one bacterial cell to another. This type of "inheritance" does not occur in multicellular organisms like ourselves, but it does not change the basic principle that variation is heritable.
3. **More Organisms Are Produced than Survive.** The antibiotic treatment eliminates most of the bacteria in the infected individual's body.
4. **An Organism's Survival Is Not Random.** Bacterial cells with traits that make them more resistant to the antibiotic are more likely to survive than those that are less resistant.

Because of the increased fitness of antibiotic-resistant variants, subsequent bacterial generations consist of a greater percentage of these variants. In other words, the population evolves to become resistant to the drug treatment.

Immediately after scientists began using antibiotics against tuberculosis, they noticed that some individuals would become ill again after seemingly successful treatment. Even more puzzling was that these recurrent infections were much more difficult to treat than the initial one. It was not until scientists incorporated their understanding of natural selection into tuberculosis treatments that effective, long-term therapies were developed.

Two characteristics of the early treatment strategies for tuberculosis actually sped up the development of drug resistance. The first was using the drugs for too short a time period. The second was using only one type of antibiotic at a time in patients with active tuberculosis disease.

Selecting for Drug Resistance

Antibiotics are effective because they buy a patient time to control a bacterial infection with his or her own immune system. By keeping the bacterial population low, antibiotics allow the body to devote energy to developing an immune response instead of losing energy to the effects of the infection.

Because most of the *M. tuberculosis* cells in an infected individual are susceptible to antibiotic, a few days of treatment will eliminate the majority of these organisms. Once most of the bacteria are dead, the patient feels much better as the most debilitating aspects of the disease (for example, fever, severe cough) are reduced or eliminated. However, a small number of bacteria that are more resistant to the antibiotic take longer to kill. If drug treatment stops as soon as the patient feels better—the typical pattern in early treatment protocols—any of the more resistant bacteria that remain can multiply and restart the infection. Thus, a drug-resistant strain is born. And because this new population is more resistant, the resurgent infection is much more difficult to control (**FIGURE 12.16**). As the number of resistant bacteria increase in the patient, the chances increase that one will appear with a mutation that makes it strongly resistant to the antibiotic. If the patient returns to the community at this point, a disease that is highly resistant to the most commonly effective drugs could begin to spread.

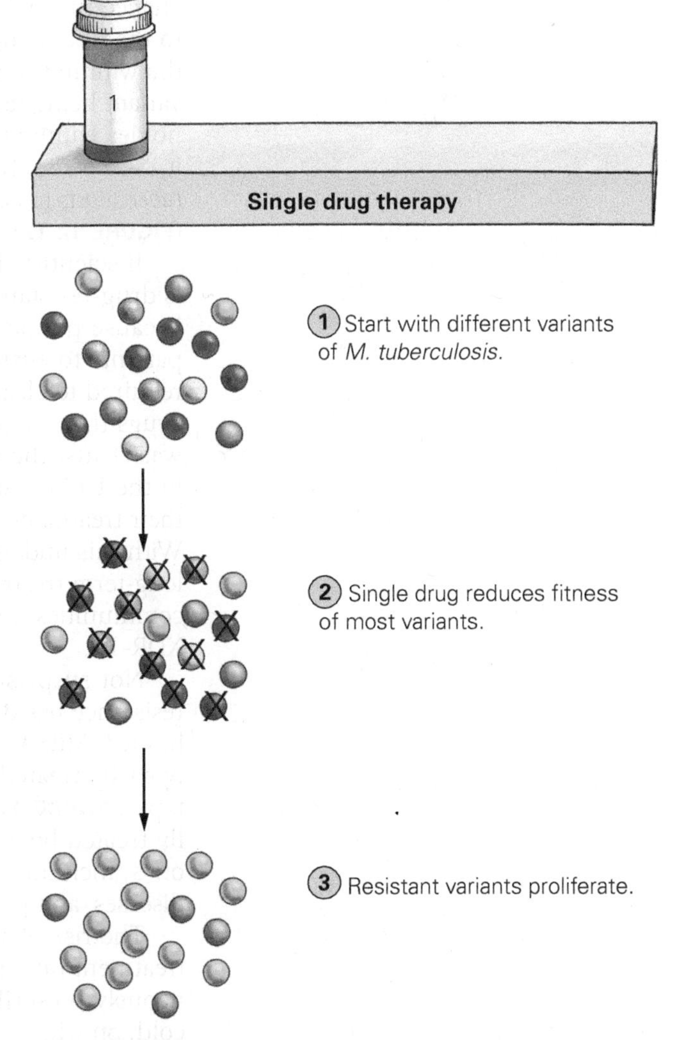
1
Single drug therapy
1 Start with different variants of M. tuberculosis.
2 Single drug reduces fitness of most variants.
3 Resistant variants proliferate.

FIGURE 12.16 Directional selection in tuberculosis. In a variable population of *M. tuberculosis* bacteria, some cells may be more resistant to a particular antibiotic. When this antibiotic is used, these resistant variants survive and continue to reproduce. The result is a population that is resistant to the antibiotic.

Stopping Drug Resistance

Combating the development of drug resistance in *M. tuberculosis* required removing the factors that promoted its development in the first place. One strategy is to maintain drug therapy for months, until all signs of the bacterial infection are cleared from the body. The other is to use multiple drugs on active infections to avoid selecting for strongly drug-resistant variants that already exist in the population.

Combination drug therapy, also called *drug cocktail therapy,* is commonly used on diseases for which resistance to a single drug can develop rapidly. HIV, the virus that causes AIDS, is one example. The effectiveness of combination therapy is based on the following fact: The greater the number of drugs used, the greater the number of changes that are required in the bacterial genome for resistance to develop.

The likelihood of a bacterial variant arising that is resistant to a single drug is relatively small but still very possible in a patient with 1 billion different bacterial variants. However, the likelihood of a bacterial variant arising with resistance to two or three drugs in a cocktail is extremely small. Put another way, the chance that any bacterium exists that is resistant to a single drug is analogous to the likelihood that in 1 billion lottery ticket holders, one person will hold the winning combination—in other words, relatively likely. The likelihood of a variant being resistant to several different drugs is analogous to that same ticket holder winning the lottery several times in a row—incredibly unlikely. Just as it is exceedingly rare to win the lottery twice in a row, it is very difficult for *M. tuberculosis* to adapt to an environment where it faces two "killer drugs" at once (**FIGURE 12.17**).

If scientists learned these lessons about drug resistance in the 1940s, why is drug-resistant *M. tuberculosis* making a comeback only now? Primarily, it is because public health researchers let down their guard and didn't follow TB patients to ensure that they continued taking their drugs for the 12 months required to clear out all of the invading bacteria. This is not surprising, as the drugs that are prescribed have difficult side effects. This reality means that it wasn't just the Indian visitor's travels that raised the risk of XDR-TB spread in the United States. It was thousands of TB patients who failed to complete their treatment regimen that led to the development of this dangerous variant. With this understanding, public health strategies that focus on maintaining the long-term treatment of individuals, and bringing that treatment to underserved communities, can go a long way in stemming the development of more cases of XDR-TB.

Not surprisingly, tuberculosis is not the only bacterial disease in which resistance has developed, largely as a result of inadequate treatment with antibiotics. MRSA, methicillin-resistant *Staphylococcus aureus,* is another formerly easily treated bacteria that has evolved into a more dangerous and deadly type (**FIGURE 12.18**). Dozens of other bacterial pathogens that were once easily treated by antibiotics now contain strains resistant to one or more antibiotics, including organisms responsible for ear infections, sexually transmitted diseases, and pneumonia.

The rise of antibiotic resistance stems not only from failure to follow drug treatment but also from the overuse of these drugs. Antibiotics are often erroneously prescribed to individuals with viral infections, such as the common cold, on which they have no effect and only serve to increase the likelihood that formerly easily controlled bacteria may develop resistance. Antibiotics are also commonly and liberally used in animal agriculture: they are given to poultry, cows, and pigs to promote growth—the mechanism by which they function is unclear, but antibiotic supplementation increases the rate of weight gain by as

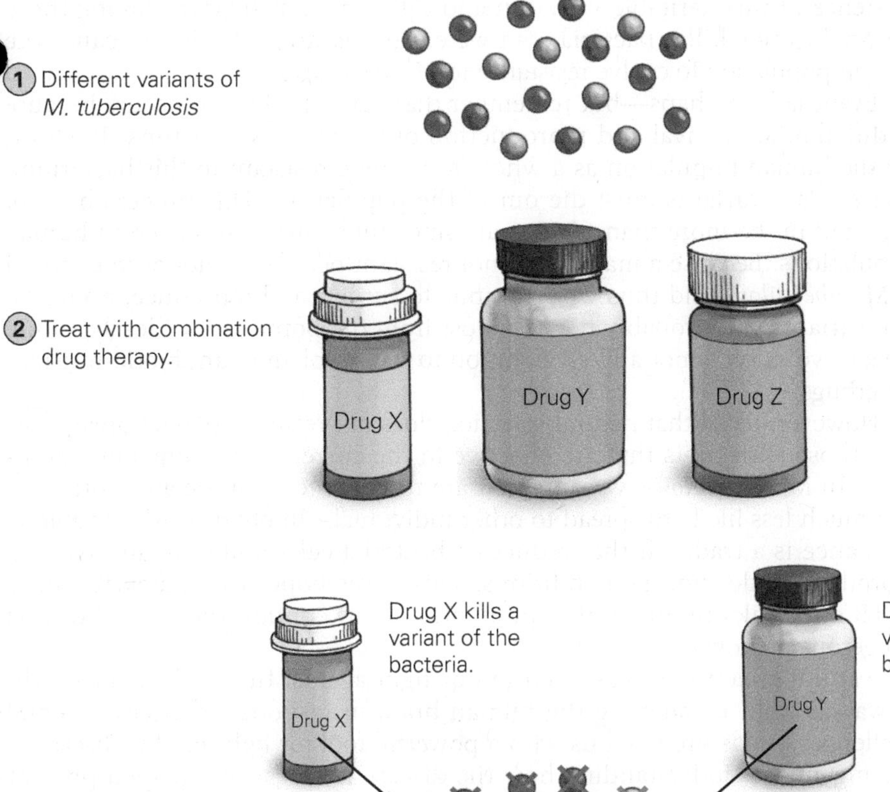
1 Different variants of M. tuberculosis
2 Treat with combination drug therapy.
Drug X
Drug Y
Drug Z
Drug X kills a variant of the bacteria.
Drug Y kills a variant of the bacteria.
Drug Z kills a variant of the bacteria.
3 New mutants are uncommon because few bacteria survive the multidrug regimen.

FIGURE 12.17 Combination drug therapy prevents antibiotic resistance. Using multiple antibiotics makes the environment much harsher for *M. tuberculosis* and decreases the likelihood that a variant with multiple resistances will evolve.

much as 10%. Drug resistance is also not confined to bacteria; the same evolutionary process has occurred in viruses, like HIV, and other pathogens, such as the protozoan that causes malaria.

To stop the development of these antibiotic-resistant "superbugs," patients and physicians must use antibiotics judiciously and wisely. But perhaps evolution has another trick up its sleeve that gives us some hope as well.

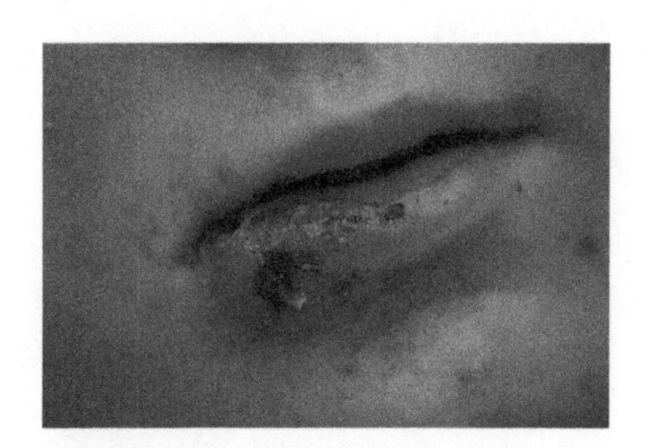

FIGURE 12.18 MRSA. Staph infections of the skin are relatively common, causing boils like these. However, if untreatable by antibiotics, staph can become a systemic disease that may be deadly.

Can Natural Selection Save Us from Superbugs?

Bacteria can rapidly evolve resistance to our antibiotics. Why can't we evolve resistance to the bacteria? Can't natural selection save us from these pathogens?

There are clearly differences among healthy individuals in their susceptibility to long-term tuberculosis infection and even to active disease. Given the existence of this heritable variation and differences in survival among those exposed to this killer bacteria, can we expect natural selection to cause the human population to evolve resistance to *M. tuberculosis*?

Eventually, perhaps—but remember that natural selection occurs because of differential survival and reproduction of individuals over time. In short, for the human population as a whole to become resistant to this bacterium, nonresistant variants must die out of the population. This process has not occurred in the more than 6000 years since tuberculosis has been in human populations. Because a majority of nonresistant people are not even exposed to *M. tuberculosis* and thus will continue to survive and reproduce, nonresistant variants will probably never be lost from the population. Clearly, future *human* evolution is not a likely solution to the problem of antibiotic-resistant superbugs.

However, recall that natural selection does not result in perfect organisms, only those with traits that are effective in the current environmental conditions. In *M. tuberculosis,* variants that are resistant to multiple antibiotics are also much less likely to spread to other individuals. In other words, antibiotic resistance is a trade-off that reduces a bacterial cell's ability to survive and reproduce under normal conditions. This raises hopes that quick response to TB outbreaks in vulnerable populations can contain the spread of this dangerous pathogen.

Natural selection has assisted in our fight against tuberculosis in another way as well. By shaping the human brain in response to environmental challenges, it has provided us with a powerful tool for fighting this disease—our ingenuity. Understanding how the disease is spread has helped prevent thousands of new infections, and the development of antibiotics has fought off millions more. A vaccine that can stimulate the immune system to prevent initial infection is in development; it may eventually eliminate the disease entirely.

The story of the XDR-TB–infected traveler who crossed the Midwest faded from the news, because it appears that she did not transmit the infection to anyone else; but the global threat of this infection remains high. But tuberculosis does not need to return to the unbeatable enemies list. And with attention to the process of natural selection guiding our use and application of antibiotics, it shouldn't become one.

Got It?

1. Antibiotic resistance evolves via natural selection because individual bacteria __________ in their resistance to antibiotics.
2. In an environment where antibiotics are applied, those bacteria that are resistant have higher __________ than those that are not, thus the population evolves.
3. Antibiotic resistance is more common when people __________ treatment before the bacteria is eliminated.
4. Antibiotic resistance is less likely to evolve in hard-to-treat pathogens when infected individuals complete their entire course of treatment and are treated with __________ drugs.
5. The ability to resist antibiotics may reduce the rate of reproduction in individual bacteria; in other words, these two traits represent a __________.

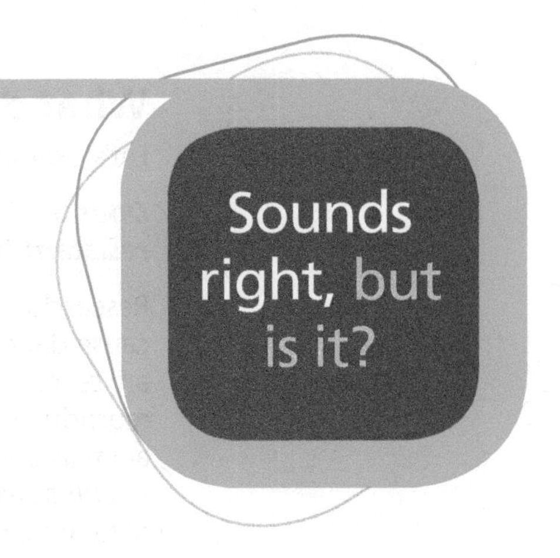

It can be difficult to imagine how complicated organisms have come about without some designer being involved. Even Darwin understood that it is difficult to see how an organ as complex as the human eye, with a pupil that regulates light, a lens that can flex to focus at different distances, and the sophisticated nervous tissue of the retina, could have evolved through the process of natural selection. But he did conceive of how it was possible. You might hear critics of evolutionary theory make a statement like this:

The human eye is too complex to have evolved by chance from nothing.

Sounds right, but it isn't.

Answer the following questions to understand why.

1. Where does the variation that natural selection acts on arise from?
2. Scientists estimate that each individual has three "new" mutations that could affect their anatomy and physiology. In a population of only 10,000 people and using our low estimate of mutations that could be subject to selection, about how many "new" variants of genes appear every generation?
3. Question 2 gives you a sense of how much genetic variety appears by chance. Is natural selection "chance"?
4. Consider the evolution of pigeon neck ruffs discussed in the chapter. What did these neck ruffs evolve from?
5. Some simple single-celled organisms can sense light and will avoid it (or move toward it). Do these organisms have eyes?
6. Simple animals, such as earthworms, can change the shape of parts of their bodies, like their mouths. Is it possible that the same mechanisms that change mouth shape could change lens or pupil shape?
7. Reflect on your answers to questions 1–6. Explain why the statement bolded above sounds right, but isn't.

Should I stop purchasing meats that are raised using antibiotics?

As of 2012, more than 80% of antibiotics used in the United States were used in animal agriculture, mainly as growth promoters. Given the rise of resistant bacterial pathogens and warnings from the Food and Drug Administration, there is increasing interest from the public in buying meat that is free from these drugs. The label "grown without antibiotics" now is used as a selling point on meats in supermarket coolers.

What should I know?

What follows are some smaller questions that need to be resolved to answer the Big Question. Place a checkmark next to the questions that science can answer.

Smaller Questions	Can Science Answer?
Is meat from antibiotic-treated animals safer for human consumption?	
Is meat from antibiotic-treated animals less expensive?	
Does antibiotic treatment of animals cause the development of antibiotic resistant human pathogens?	
Is antibiotic treatment of animals to promote faster growth a form of animal cruelty?	
Do meats from antibiotic-treated animals have a lower environmental cost (in terms of clean water and/or fossil fuel use) per pound?	

—Continued next page

What does the science say?

Let's examine what the data say about this smaller question:

Does antibiotic treatment of animals cause the development of antibiotic resistant human pathogens?

Researchers reviewed studies about antibiotic-resistant strains of *E. coli* that are known to cause dangerous infections of the urinary tract, blood, and other nonintestinal sites. There are several "reservoirs" for these strains—that is, sites where the bacteria normally live and reproduce without causing disease—including the human digestive tract, but also food animals and retail meat products. The literature review described eight antibiotic-resistant *E. coli* strains that cause human disease and identified their nonhuman reservoirs. The data are displayed in the figure below.

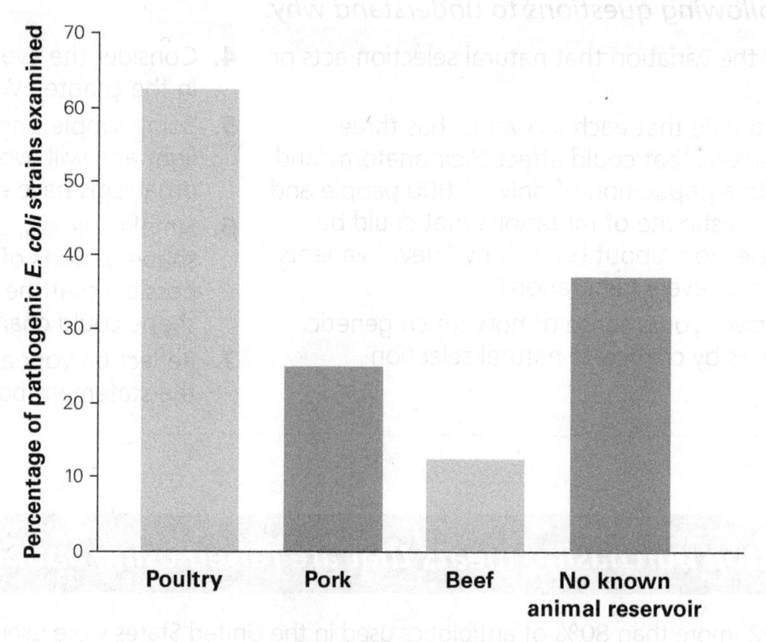

1. Describe the results. Does it appear that antibiotic resistant strains of human-pathogenic *E. coli* can be found in food animals?
2. Given these data, do you think the smaller question is answered? If not, propose another study that would help answer this question.
3. Does this information help you answer the Big Question? What else do you need to consider?

Data source: A. R. Manges and J. R. Johnson, "Food-Borne Origins of *Escherichia coli* Causing Extraintestinal Infections," *Clinical Infectious Diseases* 55, no. 5 (2012): 712–719.

Chapter Review

Mastering Biology

Go to Mastering Biology to access **eText 2.0, Dynamic Study Modules,** and the **Study Area,** where you'll find practice quizzes, BioFlix™ animations, MP3 tutor sessions, current events, and more.

SUMMARY

Section 12.1

Describe the history of tuberculosis in human populations, and explain why current treatments are ineffective against some strains of the disease.

- Tuberculosis is caused by bacterial infection and can result in lung disease and death (pp. 239–240).
- Antibiotics turned tuberculosis from a possible death sentence to a readily curable condition. However, variants of the tuberculosis bacteria have appeared that are resistant to most antibiotics (pp. 240–241).

SHOW YOU KNOW 12.1 Why is it a problem if someone who has an infectious disease seeks treatment but receives an ineffective treatment?

Section 12.2

List the four observations that led to the inference of natural selection.

- Individuals in a population vary (p. 242).
- Some of this variation can be passed on to offspring (p. 242).
- Not all individuals born in a population survive to adulthood (p. 243).
- Survival and reproduction are not random. Advantageous traits, called adaptations, increase an individual's fitness, which is his or her chance of survival or reproduction (pp. 243–244).

Explain how natural selection causes evolutionary change.

- The increased fitness of individuals with particular adaptations causes the adaptation to become more prevalent in a population over generations (p. 245).

Provide examples of evidence that supports the hypothesis that natural selection leads to the evolution of populations.

- Artificial selection, when humans deliberately control an organism's fitness, causes the evolution of different breeds of animals and varieties of plants (p. 245).
- Populations exposed to environmental changes in the lab and in nature have been shown to evolve traits that make them better fitted to the environment (pp. 245–247).

SHOW YOU KNOW 12.2 Some biologists argue that human evolution as a result of natural selection on our physical traits has nearly stopped because of our ability to use technology to compensate for poor environmental conditions or to overcome physical handicaps. What examples can you give that support this view? In what ways might humans still be subject to natural selection?

Section 12.3

Describe how natural selection works on allele frequencies in a population.

- The modern definition of evolution is a genetic change in a population of organisms (p. 248).
- Alleles that code for adaptations become more common in a population over generations as a result of natural selection (pp. 248–249).

Discuss why natural selection does not result in "perfectly adapted" organisms or drive organisms toward some ideal state.

- Natural selection can act only on the variants currently available in the population. Natural selection results in a population that is better adapted to its environment but usually not perfectly adapted as a result of trade-offs. Natural selection does not push a population in the direction of a predetermined "goal" (pp. 249–250).

List the three patterns of selection, provide examples of each, and explain how they lead to different outcomes.

- Selection can cause the traits in a population to change in a particular direction, an outcome called directional selection. However, in some environments it may cause certain traits to resist change, a process called stabilizing selection, and in other environments it may cause multiple variants to evolve, a process called diversifying selection (pp. 250–252).

SHOW YOU KNOW 12.3 The allele responsible for cystic fibrosis (discussed in Chapter 8) appears to protect carriers (individuals with one copy of the allele) from *M. tuberculosis* infection. In addition, it appears that the cystic fibrosis allele is more common in humans with ancestors from crowded, urban environments in northern Europe. Use the modern understanding of evolution via natural selection to develop a hypothesis about why the cystic fibrosis allele is more common in these populations.

Section 12.4

Using your understanding of natural selection, explain why combination drug therapy is an effective tool to combat drug resistance.

- Disease-causing organisms can evolve resistance to antibiotics because their populations consist of multiple variants that have differential survival when exposed to various drugs. Thus, a population of disease-causing organisms can evolve drug resistance via natural selection (p. 253).
- A mutant organism that is resistant to several different antibiotics is relatively unlikely, so combination drug therapy can reduce the risk of antibiotic resistance evolving by killing off the entire population (p. 254).
- As a result of trade-offs, varieties of disease-causing organisms that are multiple-drug-resistant are less likely to survive and reproduce in normal conditions than are non–drug-resistant varieties. (p. 256).

SHOW YOU KNOW 12.4 Why would some individual bacteria carry a mutation that makes them more resistant to an antibiotic even though they have never been exposed to this antibiotic?

ROOTS TO REMEMBER

The following roots of words come mainly from Latin and Greek and will help you to decipher terms:

adapt-	means to fit it. Chapter term: *adaptation*
anti-	means in opposition to. Chapter term: *antibiotic*
-biotic	indicates pertaining to life. Chapter term: *antibiotic*
-osis	indicates a condition. Chapter term: *tuberculosis*
patho-	means disease. Chapter term: *pathogen*
tubercul-	means a small swelling. Chapter term: *tuberculosis*

LEARNING THE BASICS

1. What types of drugs have helped reduce the death rate due to tuberculosis infection, and why have they become less effective more recently?
2. Define *artificial selection*, and compare and contrast it with natural selection.
3. Which of the following observations is not part of the theory of natural selection?

   A. Populations of organisms have more offspring than will survive; B. There is variation among individuals in a population; C. Modern organisms are unrelated; D. Traits can be passed on from parent to offspring; E. Some variants in a population have a higher probability of survival and reproduction than other variants do.

4. Add labels to the figure that follows, which illustrates how *Mycobacterium tuberculosis* evolves when it is exposed to an antibiotic.

Single drug therapy

The initial *M. tuberculosis* population includes both antibiotic-susceptible and ____________ variants.

Treatment with one antibiotic reduces ____________ susceptible variants.

Remaining population is made up of ____________ variants, which then proliferate.

5. The best definition of *evolutionary fitness* is ________.

   A. physical health; B. the ability to attract members of the opposite sex; C. the ability to adapt to the environment; D. survival and reproduction relative to other members of the population; E. overall strength

6. An adaptation is a trait of an organism that increases ________.

   A. its fitness; B. its ability to survive and replicate; C. in frequency in a population over many generations; D. A and B are correct; E. A, B, and C are correct

7. The heritable differences among organisms are a result of ________.

   A. differences in their DNA; B. mutation; C. differences in alleles; D. A and B are correct; E. A, B, and C are correct

8. Since the modern synthesis, the technical definition of evolution is a change in ________ in a ________ over the course of generations.

   A. traits, species; B. allele frequency, population; C. natural selection, natural environment; D. adaptations, single organism; E. fitness, population

9. Ivory from elephant tusks is a valuable commodity on the world market. As a result, male African elephants with large tusks have been heavily hunted for the past few centuries. Today, male elephants have significantly shorter tusks at full adulthood than male elephants in the early 1900s. This is an example of ________.

   A. diversifying selection; B. stabilizing selection; C. directional selection; D. chance; E. more than one of the above is correct

10. Antibiotic resistance is becoming common among organisms that cause a variety of human diseases. All of the following strategies help reduce the risk of antibiotic resistance evolving in a susceptible bacterial population except _________.

A. using antibiotics only when appropriate, for bacterial infections that are not clearing up naturally; **B.** using the drugs as directed, taking all the antibiotic over the course of days prescribed; **C.** using more than one antibiotic at a time for difficult-to-treat organisms; **D.** preventing natural selection by reducing the amount of evolution the organisms can perform; **E.** reducing the use of antibiotics in non–health-care settings, such as agriculture

ANALYZING AND APPLYING THE BASICS

1. Most domestic fruits and vegetables are a result of artificial selection from wild ancestors. Use your understanding of artificial selection to describe how domesticated strawberries must have evolved from their smaller and less sweet wild relatives. What trade-offs do domesticated strawberries exhibit relative to their wild ancestors?
2. The striped pattern on zebras' coats is considered to be an adaptation that helps reduce the likelihood of a lion or other predator identifying and preying on an individual animal. The ancestors of zebras were probably not striped. Using your understanding of the processes of mutation and natural selection, describe how a population of striped zebras might have evolved from a population of zebras without stripes.
3. Are all features of living organisms adaptations? How could you determine if a trait in an organism is a product of evolution by natural selection?

GO FIND OUT

1. In the United States, the first line of defense when attempting to stop the spread of an antibiotic-resistant pathogen is county health departments. These departments work with their state health departments and the U.S. CDC to understand and manage outbreaks. Find out what your county health department publishes publicly on the web about antibiotic resistance in your community. Can you find information on numbers of infections from your county, state, or the CDC?
2. The theory of natural selection has been applied to human culture in many different realms. For instance, there is a general belief in the United States that "survival of the fittest" determines which people are rich and which are poor. How are the forces that produce differences in wealth among individuals like natural selection? How are they different?

MAKE THE CONNECTION

The science that you learned in this chapter has helped you better understand the real-world example used throughout this discussion. Draw a line from the statement on the left to the science that supports it on the right.

The bacterium that caused tuberculosis was in decline, but now strains exist that are nearly untreatable. ▷	◁ Natural selection for resistant strains in the presence of antibiotics led to a change in tuberculosis bacteria populations.
Resistant bacteria have proteins that disable or counteract antibiotics. ▷	◁ The likelihood that an individual bacterium carries the genetic variations that allows it to disable two different antibiotics is very low.
The new strains appeared when the bacteria population evolved during incomplete antibiotic treatment. ▷	◁ Antibiotics help to keep the bacterial population low for a long enough period that an individual's immune system can finish off even drug-resistant individuals.
Combination drug therapy can help prevent the evolution of antibiotic resistance. ▷	◁ Antibiotic resistance is a genetic trait that can be passed on to offspring.
As individuals, we can prevent antibiotic resistance from developing if we are careful and thorough when using antibiotics. ▷	◁ Antibiotic resistance developed in the tuberculosis bacteria not long after the introduction of antibiotics in the 1940s.

Answers to **Got It?, Visualize This, Working with Data, Sounds Right, But Is It?, Show You Know,** and **Chapter Review** questions can be found in the **Answers** section at the back of the book.

13 Understanding Race

One milestone at the 2016 Summer Olympics was the first gold medal earned by an African American female swimmer.

Why have African Americans been underrepresented in competitive swimming?

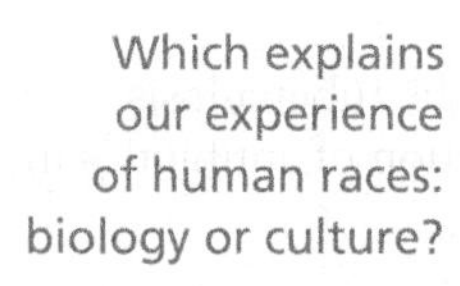

Which explains our experience of human races: biology or culture?

Speciation and Macroevolution

It was one of the most memorable images of the 2016 Summer Olympics. Simone Manuel, a swimmer for the United States who was considered a long shot to earn a medal in the 100-meter freestyle, turns to look at the scoreboard and realizes with shock and jubilation that she had won a gold medal. To many, Manuel's triumph was more than an individual victory; history will remember her as the first African American woman to win an individual gold medal in swimming in the 120-year history of the modern Olympic Games.

For decades, African Americans have achieved at the highest levels in the events of the Summer Olympics, alongside athletes of every other race. But until very recently African Americans have been conspicuously absent in swimming events. Even now, when African Americans make up approximately 12% of the U.S. population, of the more than 178,000 year-round competitive athletes who identified their race on the USA Swimming membership form in 2015, fewer than 4000 listed black or African American. That's only approximately 2% of the total. Why do so few black men and women compete in this sport?

Manuel's success undermines any misguided argument that black women (and men) are at a "natural" disadvantage in swimming. Manuel acknowledged old misconceptions at the press conference after her win, saying, "The title 'black swimmer' makes it seem like I'm not supposed to be able to win a gold medal or I'm not supposed to be able to break records." She credited a list of earlier African American Olympian swimmers, including Cullen Jones, winner of gold medals in men's swimming in the 2004 and 2008 Olympics, for inspiring her to stay with a sport in which she is an outlier.

It is clear from her statement that the misguided argument that African Americans were at a "natural disadvantage" at swimming had an effect on Manuel's confidence, despite the argument having no grounding in fact. The argument had that effect because it fits a pattern of misunderstanding that is relatively common; that is, the belief that there exists a number of innate differences among human races. Over the past several decades, biological research has examined the evidence for these supposed innate differences and found it weak or nonexistent. What this research has established is that beneath a veneer of external differences, humans are basically the same.

13.1 What Is a Species?

All humans belong to the same species. Before we can understand the concept of race, we need to understand what it means to be a "species."

In the mid-1700s, the Swedish scientist Carolus Linnaeus began the task of cataloging all of nature. Linnaeus developed a classification scheme that grouped organisms according to shared traits (Chapter 11). The primary category in his system was the **species.** Linnaeus assigned a two-part name to each species—the first part of the name indicates the **genus,** or broader group; the second part is specific to a subgroup within that genus. Lions, *Panthera leo,* are classified in the same genus with other species of roaring cats, such as the leopard, *Panthera pardus* (**FIGURE 13.1**).

Linnaeus coined the binomial name *Homo sapiens* (*Homo* meaning "man" and *sapiens* meaning "knowing or wise") to describe the human species. Modern biologists have kept the basic Linnaean classification, although they have added a **subspecies** name, *Homo sapiens sapiens,* to distinguish modern humans from earlier humans that appeared approximately 250,000 years ago. Other subspecies of human include the Neanderthals, known as *Homo sapiens neanderthalensis*.

(a) Lion

(b) Leopard

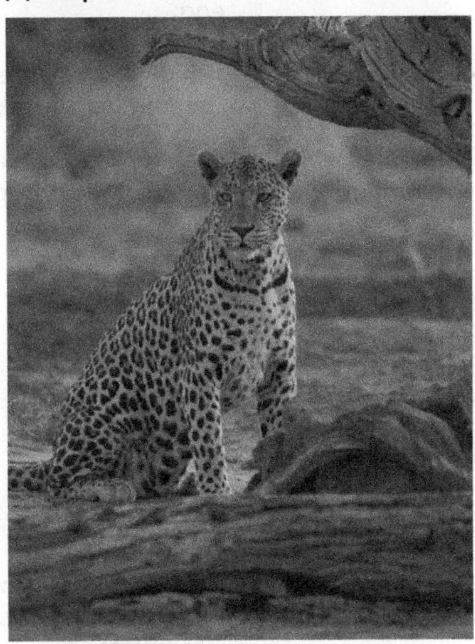

FIGURE 13.1 Same genus, different species. (a) *Panthera leo*, the lion, and (b) *Panthera pardus*, the leopard.

The Biological Species Concept

Although most people intuitively grasp the differences between most species—both lions and leopards are definitely cats but not the same species—biologists have had difficulty finding a single definition for species that can be consistently applied. The most commonly used is called the biological species concept.

Biological Species Are Reproductively Isolated. According to the **biological species concept,** a species is defined as a group of individuals that can interbreed and produce fertile offspring but cannot reproduce with members of other species. This definition can be difficult to apply. For example, species that reproduce asexually (such as most bacteria) and species known only via fossils do not easily fit into this species concept. However, the biological species concept does help us understand why species are distinct from one another.

Recall that differences in traits among individuals arise partly from differences in their genes, and that the different forms of a gene that exist are known as *alleles* of the gene. By the process of evolution, a particular allele can become more common in a species. If interbreeding does not occur, then this allele cannot spread from one species to the other. In this way, two species can evolve differences from each other. For example, various lines of evidence suggest that the common ancestor of lions and leopards had a spotted coat. The allele tha eliminated the spots from lions arose and spread within lions, but this allele is not found in leopards. The spotless allele has not been transferred to the leopard population because lions and leopards cannot interbreed.

Scientists refer to the sum total of the alleles found in all the individuals of a species as the species' **gene pool.** A change in the frequency of an allele in a gene pool can take place only *within* a biological species.

The Nature of Reproductive Isolation. The spread of an allele throughout a species' gene pool is called **gene flow.** Gene flow cannot occur between different biological species because pairing between them fails to produce fertile offspring. This reproductive isolation can take two general forms: prefertilization barriers or postfertilization barriers, which are summarized in TABLE 13.1.

TABLE 13.1 Mechanisms of reproductive isolation.

Type	Effect	Example
Prefertilization barriers prevent fertilization from occurring.		
Spatial isolation	Individuals from different species do not come in contact with each other.	Polar bear (Arctic) and spectacled bear (South America) never encounter each other in natural settings.
Behavioral	Ritual behaviors that prepare partners for mating are different in different species.	Many birds with premating songs or "dances" will not mate with individuals who do not know the ritual.
Mechanical	Sex organs are incompatible between different species, so sperm cannot reach egg.	Many insects with "lock-and-key" type genitals physically prevent sperm from contacting eggs of a different species.
Temporal	Timing of readiness to reproduce is different in different species.	Plants with different flowering periods cannot fertilize each other.
Gamete incompatibility	Proteins on egg that allow sperm binding do not bind with sperm from another species.	Animals with external reproduction, such as sponges, have specific proteins on their eggs that will bind only to sperm from the same species.
Postfertilization barriers: Fertilization occurs, but hybrid cannot reproduce.		
Hybrid inviability	Zygote cannot complete development because genetic instructions are incomplete.	A sheep crossed with a goat can produce an embryo, but the embryo dies in the early developmental stages.
Hybrid sterility	Hybrid organism cannot produce offspring because chromosome number is odd.	Mules (see Figure 13.2)

As you can see in the table, most mechanisms of reproductive isolation are invisible, occurring before mating or fertilization or soon after. In the instances when a mating between two different species results in live offspring—a **hybrid** descendant—the offspring are often sterile. A well-known example of an interspecies hybrid is the mule, resulting from a cross between a horse and a donkey. Mules are excellent farm animals, but they cannot produce their own offspring and thus don't represent a true separate species.

Hybrid sterility often occurs because hybrid individuals cannot produce viable sperm or egg cells. During the production of eggs and sperm, homologous chromosomes pair up and separate during the first cell division of meiosis (Chapter 7). Because a hybrid forms from the chromosome sets of two different species, its chromosomes are not homologous and thus cannot pair up correctly during this process.

In the case of mules, the horse parent has 64 chromosomes and therefore produces eggs or sperm with 32; the donkey parent has 62 chromosomes, producing eggs or sperm with 31. The mule will therefore have an odd number of chromosomes—63—and no way to sort these into pairs during the first division of meiosis (**FIGURE 13.2**). Although a tiny number of female mules, surprisingly, have produced offspring, this event is so rare that the gene pools of donkeys and horses have remained separate.

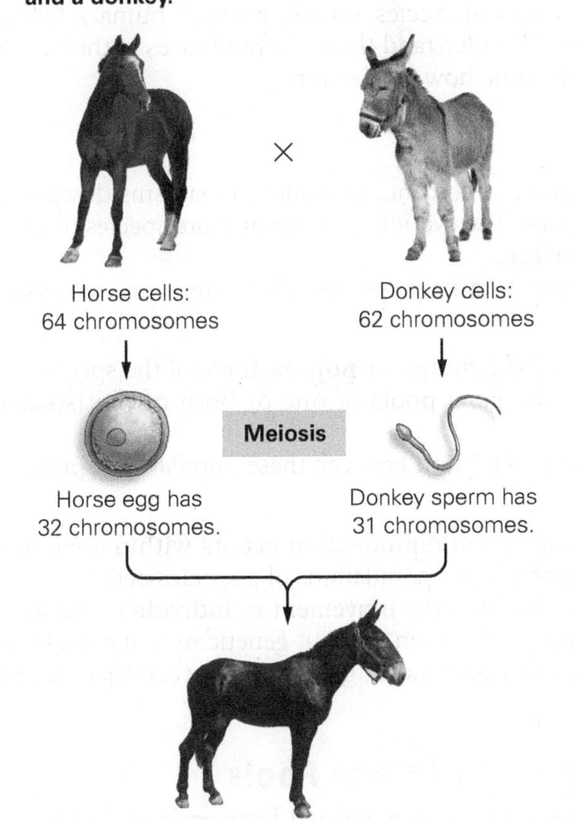

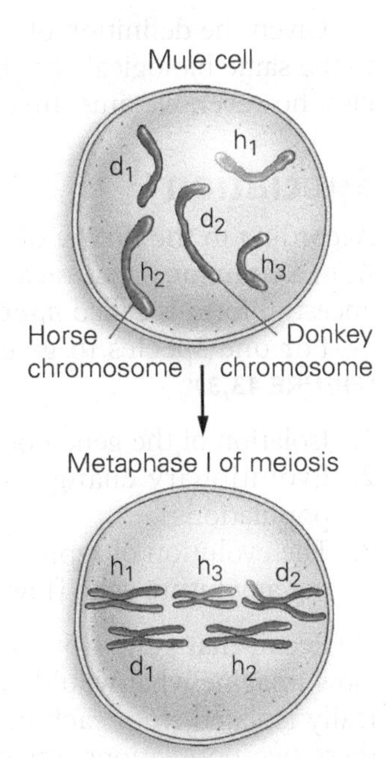

VISUALIZE THIS

If donkeys had 60 chromosomes so that their sperm contained 30, a mule would have 62 chromosomes. Would having an even number of chromosomes permit mules to be fertile?

FIGURE 13.2 Reproductive isolation between horses and donkeys. (a) A cross between a female horse and a male donkey produces a mule with 63 chromosomes. (b) Mules produce only very few eggs or sperm because their chromosomes cannot pair properly during meiosis. Only a small number of chromosomes are illustrated to simplify the drawing.

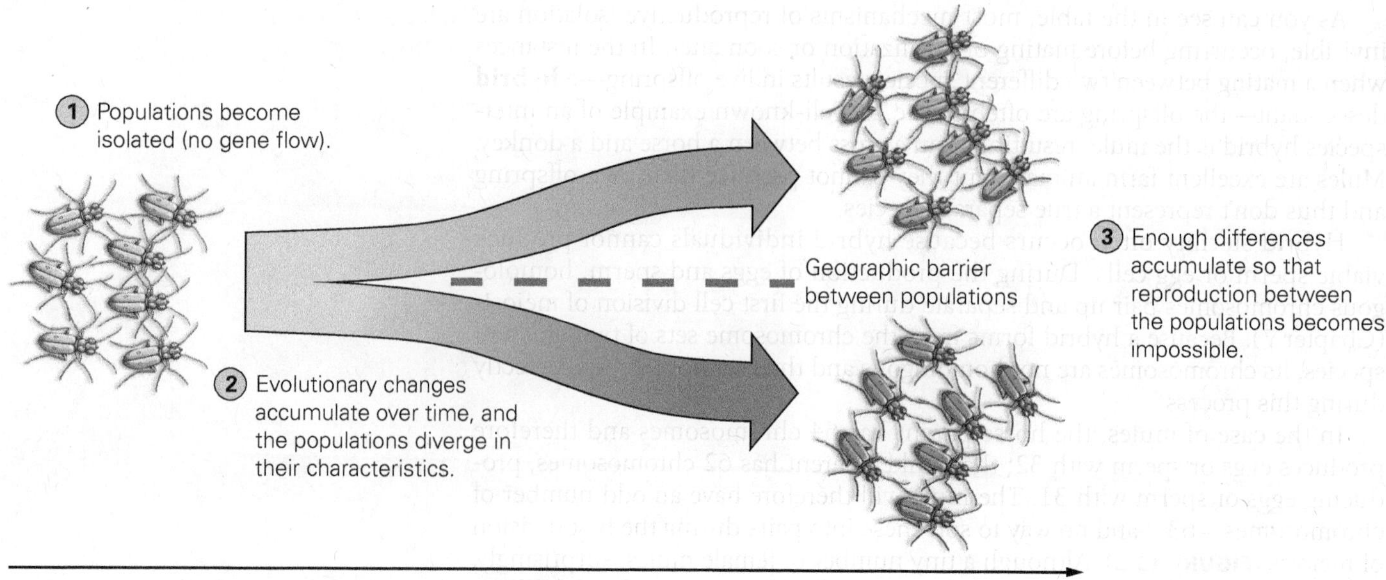

1 Populations become isolated (no gene flow).
2 Evolutionary changes accumulate over time, and the populations diverge in their characteristics.
Geographic barrier between populations
3 Enough differences accumulate so that reproduction between the populations becomes impossible.
Time

VISUALIZE THIS

Is the difference between two diverging populations always visible, as it is in this example?

FIGURE 13.3 Speciation. Isolated populations diverge in traits. Divergence can lead to reproductive isolation and thus the formation of new species.

Given the definition of a biological species, it is clear that all humans belong to the same biological species. To understand the concept of races within a species, however, we must first examine how species form.

Speciation

According to the theory of common descent, all modern organisms descended from a common ancestral species. The evolution of one or more species from an ancestral form is called **speciation.**

For one species to give rise to a new species, three steps are necessary (**FIGURE 13.3**):

1. Isolation of the gene pools of subgroups, or **populations,** of the species
2. Evolutionary changes in the gene pools of one or both of the isolated populations
3. The evolution of reproductive isolation between these populations, preventing any future gene flow

Recall that gene flow occurs when reproduction occurs within a species. Now imagine what would happen if two populations of a species became physically isolated from each other, so that the movement of individuals between these two populations was impossible. Even without genetic or behavioral barriers to mating between these two populations, gene flow between them would cease.

Isolation and Divergence of Gene Pools

The gene pools of populations may become isolated from each other for several reasons. Often, a small population becomes isolated when it migrates to a location far from the main population. This is the case on many oceanic islands. Bird, reptile, plant, and insect species on these islands appear to be the descendants of species from the nearest mainland (**FIGURE 13.4**). The original

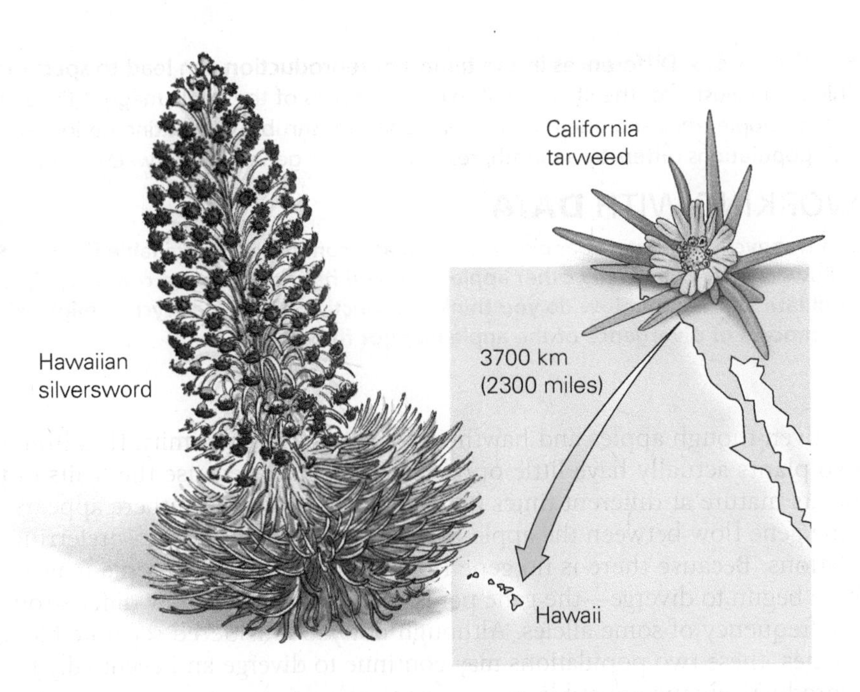

FIGURE 13.4 Migration leads to speciation. California tarweed seeds were blown or carried by birds to the Hawaiian Islands, creating an isolated population. With no gene flow between California and Hawaii, a very different group of species, Hawaiian silverswords, evolved from the tarweed colonists.

ancestral migrants arrived on the islands by chance: some by being carried by storm winds, as likely was the case for the Asian finch ancestors of Hawaiian honeycreepers; some by floating in on rafts of vegetation, as in the case of the marine iguanas of the Galápagos Islands.

The establishment of a new population far from an original population may lead to the evolution of several new species—a process known as **adaptive radiation.** According to this hypothesis, the diversity of unique species on oceanic islands, as well as in isolated bogs, caves, and lakes, resulted from colonization of these once "empty" environments by one founding species that rapidly diversified into many species.

Populations may also become isolated from each other by the intrusion of a geologic barrier. This could be an event as slow as the rise of a mountain range or as rapid as a sudden change in the course of a river. The emergence of the Isthmus of Panama separating the Pacific Ocean from the Caribbean Sea between 3 and 6 million years ago represents one such intrusion event. Scientists have described dozens of pairs of aquatic species, each made up of members living on either side of the isthmus. Over the 6 million years since their ancestral species was divided by the isthmus, genetic changes accumulated independently on each side—so much so that the members of each pair are now separate species. Populations that are isolated from each other by distance or a barrier are known as **allopatric**.

However, separation between the gene pools of two populations may occur even if the populations are living near each other, that is, if they are **sympatric.** This appears to be the case in populations of the apple maggot fly. Apple maggot flies are notorious pests of apples grown in northeastern North America. However, apple trees are not native; they were first introduced to this continent less than 300 years ago. Apple maggot flies also infest the fruit of hawthorn shrubs, a group of species that are native to North America.

allo- means different.

-patric means country or place of origin.

sym- means same or united.

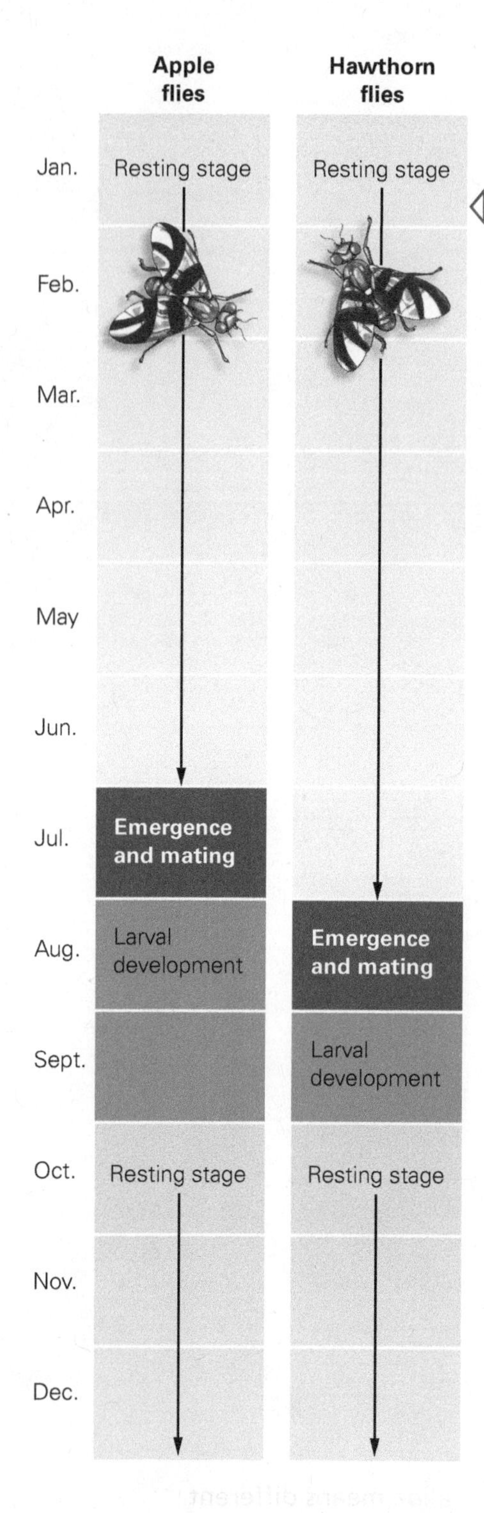

FIGURE 13.5 Differences in the timing of reproduction can lead to speciation. This graph illustrates the life cycle of two populations of the apple maggot fly: one tha lives on apple trees and one that lives on hawthorn shrubs. The mating period for these two populations differs by a month, resulting in little gene flow between them.

WORKING WITH **DATA**

The Honeycrisp variety of apple was first made commercially available in the 1990s—it flowers later than many other apple trees and its fruit does not complete ripening until late September. How do you think introduction of the Honeycrisp might affect the process of divergence of the apple maggot fly?

Even though apples and hawthorns live in close proximity, flies from these two plants actually have little opportunity to mate because the fruits of these plants mature at different times (**FIGURE 13.5**). As a result, there appears to be little gene flow between the apple-preferring and the hawthorn-preferring populations. Because there is no gene flow, the two groups of apple maggot flies have begun to diverge—the gene pools of the two groups now differ strongly in the frequency of some alleles. Although not yet considered separate biological species, these two populations may continue to diverge and eventually become reproductively incompatible.

In plants, isolation of gene pools can occur instantaneously without any barriers between populations. A simple hybrid between two plant species is typically infertile because it cannot make gametes (the same problem that occurs in mules; review Figure 13.2). However, some hybrid plants can become fertile—if a mistake during mitosis produces a cell containing duplicated chromosomes. The process of chromosome duplication is called **polyploidy**, and it results in a cell that contains two copies of each chromosome from each parent species. I polyploidy occurs inside a plant bud, all the cells of the branch that arises from that bud will be polyploid, containing two identical copies of every chromosome (**FIGURE 13.6**).

Because polyploid cells now contain pairs of identical chromosomes, meiosis can proceed, and thus flowers produced on the branch can produce eggs and sperm. A polyploid flower can then self-fertilize and give rise to hundreds of offspring, representing a brand new species that is isolated from its parent plants. Recent research suggests that this process of "instantaneous speciation" may have been a key factor in the evolution of as many as 50% of flowering plant species. Polyploidy occurs in some animal groups, such as insects and frogs, as well.

poly- means many.

-ploid- means the number of different types of chromosomes.

The Evolution of Reproductive Isolation

To become truly distinct biological species, diverging populations must become reproductively isolated either by their behavior or by genetic incompatibility. In the case of canola, genetic incompatibility occurs immediately—a cross between canola and kale does not result in offspring. In most animals, the process of divergence may be more gradual, occurring when the amount of divergence has caused numerous genetic differences between two populations.

There is no clear rule about how much divergence is required; sometimes, a difference in a single gene can lead to incompatibility, whereas at other times, populations demonstrating great physical differences can produce healthy and fertile offspring (**FIGURE 13.7**). Exactly how reproductive isolation evolves on a genetic level is still unknown and is an actively researched question in biology.

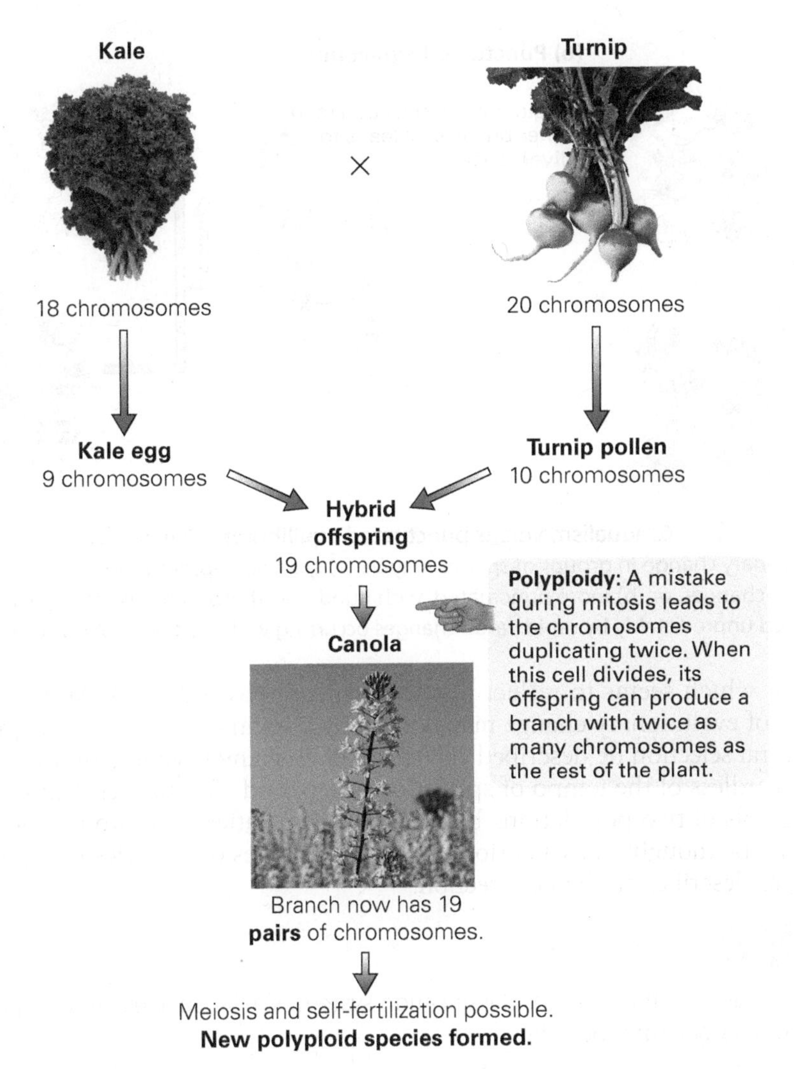
Kale
Turnip
18 chromosomes
20 chromosomes
Kale egg
9 chromosomes
Turnip pollen
10 chromosomes
Hybrid
offspring
19 chromosomes
Polyploidy: A mistake during mitosis leads to the chromosomes duplicating twice. When this cell divides, its offspring can produce a branch with twice as many chromosomes as the rest of the plant.
Canola
Branch now has 19
pairs of chromosomes.
Meiosis and self-fertilization possible.
New polyploid species formed.

FIGURE 13.6 Instantaneous speciation. Canola evolved from a hybrid of kale and turnip. Although the hybrid initially was sterile because its chromosomes could not line up during meiosis, a mistake during mitosis led to chromosome duplication in one of the plant's cells. When branches produced by this cell produced flowers, new seeds with double the number of chromosomes could form, growing into whole plants with this new chromosome number. Because it has a different number of chromosomes from either of its parents, canola pollen cannot fertilize kale or turnip plant eggs and vice versa, so this plant was instantly reproductively isolated from its parents.

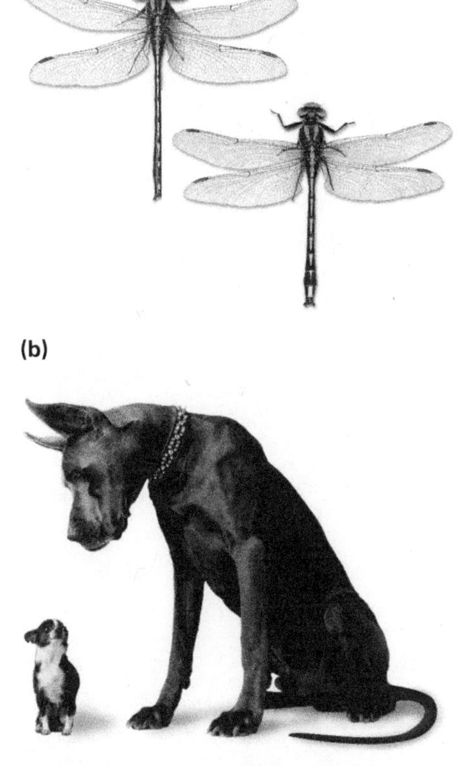
(a)
(b)

FIGURE 13.7 How different are two species? (a) These two species of dragonfly look alike but cannot interbreed. (b) Dog breeds provide a dramatic example of how the evolution of large physical differences does not always result in reproductive incompatibility.

What we do know, however, is that once reproductively isolated, species that derived from a common ancestor can accumulate many differences, even completely new genes.

Is Speciation Gradual or Sudden? Darwin assumed that speciation occurred over millions of years as tiny changes gradually accumulated. This hypothesis is known as **gradualism.** Other biologists have argued that most speciation events are sudden, result in dramatic changes in form within the course of a few thousand years, and are followed by many thousands or millions of years of little change—a hypothesis known as **punctuated equilibrium.** The hypothesis of punctuated equilibrium is supported by observations of the fossil

(a) Gradualism
Gradual natural selection (and other processes) leads to divergence.
Time
Selection
(b) Punctuated equilibrium
Rapid natural selection (and other processes) leads to divergence.
Time

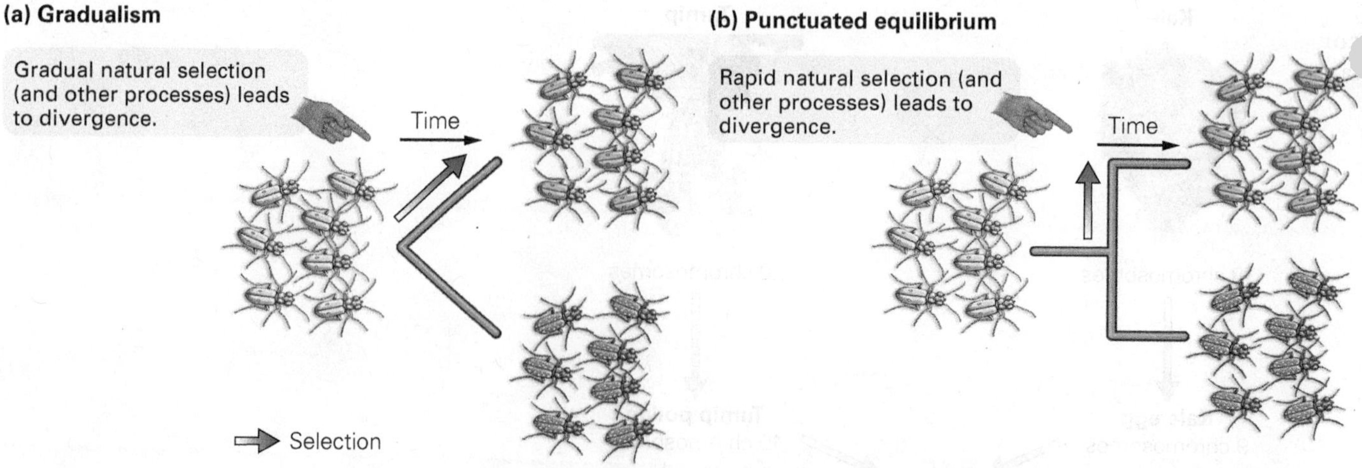

FIGURE 13.8 Gradualism versus punctuated equilibrium. The pattern of evolutionary change in groups of species may be (a) gradual, representing a constant level of small changes, or (b) more punctuated, with hundreds of thousands of years of stasis followed unpredictably by rapid, large changes occurring within a few thousand years.

record, which seems to reflect just this pattern (**FIGURE 13.8**). Although the tempo of evolutionary change may not match Darwin's predictions, the process of natural selection he described can still explain many instances of divergence.

Regardless of the tempo of speciation, the period after the separation of the gene pools of two populations but before the evolution of reproductive isolation can be thought of as a period during which races of a species may possibly form, as described in the next section.

Got It?

1. According to the Linnean classification system, similar species are grouped into a larger category, the ________.
2. Individuals within a biological species can ________ but cannot produce fertile offspring with individuals of another ________.
3. All of the alleles found within individuals in a population make up the population's gene ________.
4. The first step of speciation occurs when the gene pools of populations within a species become physically ________ from each other.
5. Speciation is complete when two populations formerly in the same species become ________ isolated from each other.

13.2 Are Human Races Biological?

Biologists do not agree on a standard definition of *biological race*. In fact, not all biologists feel that *race* is a useful term; many prefer to use the term *subspecies* to describe subgroups within a species.

However, the story that opened the chapter leads to a specific definition. That is, to say that Simone Manuel's race influenced her likelihood of being a competitive swimmer presumes that when an individual is identified as a member of a particular race, that means she is more closely related and thus biologically *more similar* to individuals of the *same* race than she is to individuals of *other* races. This definition of **biological race** describes races as populations of a single species that have diverged from each other as a result of isolation of their gene pools.

Some biologists may refer to this definition as the **genealogical species concept** because it reflects closer shared ancestry—a genealogy—among certain individuals within a biological species. With little gene flow among these isolated groups, evolutionary changes that occur in one may not occur in another. Biologists use the genealogical species concept when arguing that each isolated population of an endangered species should be preserved as a way to preserve the unique genetic characteristics found within a species.

To understand if the racial categories people in the United States identify with have a biological basis—that is, that they encompass gene pools that have long been isolated from each other—we must first understand the origin of our racial classifications.

The History of Human Races

Until the height of the European colonial period in the seventeenth and eighteenth centuries, few cultures distinguished between groups of humans according to shared physical characteristics. Historians of Ancient Greece and Rome tell us that people of those eras primarily identified themselves and others as belonging to particular cultural or social groups with different customs, diets, and languages, regardless of physical appearance.

As northern Europeans began to contact people from other parts of the world, setting groups of people apart made colonization and slavery less morally troublesome for the oppressors. This belief that human races were in fact biologically different from each other affected even scientists. Using the biological species concept, Linnaeus correctly classified all humans as a species. However, he also distinguished what he termed the "varieties" (what we would now call races) of humans, describing both physical characteristics and particular behaviors and aptitudes for each. Linnaeus did not have any biological evidence to support this notion of distinguishing human types, but the prevailing European attitudes about the "natural superiority" of their group over others influenced him (and other scientists of the day) to describe the European variety as a "naturally superior" form.

Linnaeus's classification of human races is an example of how scientists' social context has influenced their hypotheses. In this case, as in many others, this influence feeds back to society by seeming to justify and excuse injustice and brutality.

Scientists since Linnaeus continued to propose hypotheses about the number and characteristics of races of the human species. Some scientists have described as many as 26 different races of the human species, and some personal genotyping services—companies that analyze the genes of individuals to determine their ancestry—claim to be able to distinguish more than 50. But the most common number of hypothesized races is five, as is still reflected in the race categories listed on the 2010 U.S. census form: white, black, Pacific Islander, Asian, and Native American. However, do these groups truly represent biological races?

To answer this question, we can try to determine if the physical characteristics used to delineate the five supposed human races—skin color, eye shape, and hair texture, for instance—developed because these groups evolved in isolation from each other. The data needed to answer this question comes from the fossil record and from the gene pools of modern populations.

The Morphological Species Concept

The ancestors of humans are known only through the fossil record. We cannot delineate fossil species using either the biological or genealogical species concepts. Instead, **paleontologists,** scientists who study fossils, use a more practical definition: A species is defined as a group of individuals with some

paleo- means ancient.

TABLE 13.2 Comparison of three species concepts.

Species Concept	Definition	Benefits of Using This Concept	Disadvantages of Using This Concept
Biological	Species consist of organisms that can interbreed and produce fertile offspring and are reproductively isolated from other species.	Useful in identifying boundaries between populations of similar organisms. Relatively easy to evaluate for sexually reproducing species.	Cannot be applied to organisms that reproduce asexually or to fossil organisms. May not be meaningful when two populations of the same species are separated by large geographical distances.
Genealogical	Species consist of organisms that can interbreed, are all descendants of a common ancestor, and represent independent evolutionary lineages.	Most evolutionarily meaningful because each species has its own unique evolutionary history. Can be used with asexually reproducing species.	Difficult to apply in practice. Requires detailed knowledge of gene pools of populations within a biological species. Cannot be applied to fossil organisms.
Morphological	Species consist of organisms that share a set of unique physical characteristics that is not found in other groups of organisms.	Easy to use in practice on both living and fossil organisms. Only a few key features are needed for identification.	Does not necessarily reflect evolutionary independence from other groups.

morpho- means shape or form.

reliable physical characteristics distinguishing them from all other species. In other words, they look alike in some key features. This is known as the **morphological species concept.** The morphological differences among species are assumed to correlate with isolation of gene pools. TABLE 13.2 compares and contrasts the three species concepts.

Using the morphological species concept, scientists have identified the fossils of our direct human ancestors. This has allowed them to reconstruct the movement of humans since our species' first appearance.

Modern Humans: A History

The immediate predecessor of *Homo sapiens* was *Homo erectus,* a species that first appeared in east Africa about 1.8 million years ago and spread to Asia and Europe over the next 1.65 million years. Fossils identified as early *H. sapiens* appear in Africa in rocks that are approximately 250,000 years old. The fossil record shows that these early humans rapidly replaced *H. erectus* populations in Africa, Europe, and Asia.

Most data support the hypothesis that all modern human populations descended from these African *H. sapiens* ancestors within the last few hundred thousand years. One line of evidence supporting this is that humans have much less genetic diversity (measured by the number of different alleles that have been identified for any gene) than any other great ape, indicating that there has been little time to accumulate many different gene variants. Using the same reasoning, we know that African populations must be the oldest human populations because they are more genetically diverse than others around the world. Thus, African populations are likely the source of all other human populations.

Given the evidence of recent African ancestry, the physical differences we see among human populations must have arisen in the last 150,000 to 200,000 years, or in about 10,000 human generations. In evolutionary terms, this is not much time. All humans shared a common ancestor very recently; thus, the defined human races cannot be very different from each other.

Genetic Evidence of Divergence

Even with little genetic difference among populations, our question about the meaning of race is still relevant. After all, even if two races differ from each other only slightly, if the difference is consistent, then it would hold true that people are biologically more similar to members of their own race than to people of a different race. We'll explore the evidence that refutes this notion in the discussions that follow.

To determine if one race was once truly isolated from other races, researchers can examine the gene pool of populations described as a single race. Remember that when populations are isolated from each other, little gene flow occurs between them. If an allele appears in one population, it cannot spread to another. As a result, isolated populations should contain some unique alleles.

In addition to finding unique alleles, researchers should be able to observe differences among isolated populations in the percentage of the population that carries particular alleles. When a trait becomes more common in a population due to evolution, it is because the allele for that trait has become more common (Chapter 12). In other words, evolution results in a change in **allele frequency** in a population.

The study of the effect of evolutionary change on allele frequencies is known as **population genetics.** FIGURE 13.9 illustrates the relationship between individual genotypes and a population's gene frequency in a simple case—a single gene with two alleles. Here, 70% of the alleles in the population are dominant (*A*), and 30% are recessive (*a*). If populations are isolated from each other, evolutionary changes that occur in one population will not necessarily occur in another. These changes would show up as differences in allele frequency between populations—so a population isolated from the one shown in Figure 13.9 might have the same two alleles, but a different ratio; say 50% that are *A* and 50% that are *a*.

With this understanding, we can now make two predictions to test a hypothesis of whether biological races exist within a species. If a race has been isolated from other populations of the species for many generations, it should have these two traits:

1. Some unique alleles
2. Differences in allele frequency, relative to other races, for some genes

We'll next turn to the evidence and discover that the hypothesis that human races represent independent evolutionary groups is not supported.

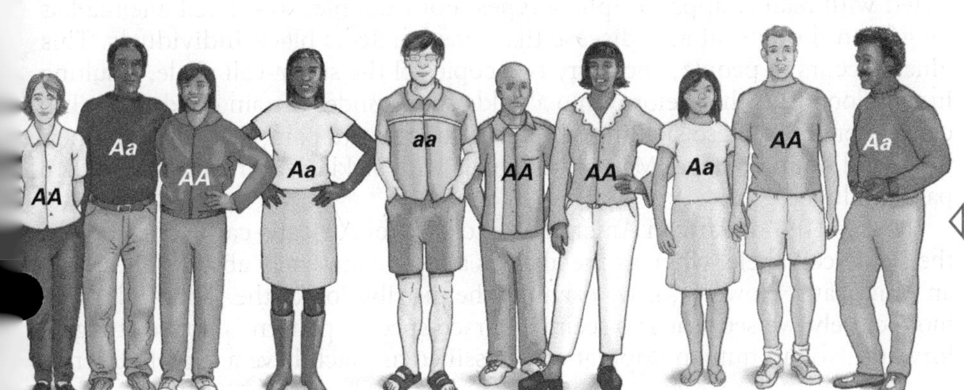

FIGURE 13.9 How to calculate allele frequency. The frequency of any allele in a population of adults can be calculated if individuals' genotypes are known.

WORKING WITH **DATA**

What is the frequency of the *a* allele in this population?

Human Races Are Not Isolated Biological Groups

Recall the five major human races described in the census: white, black, Pacific Islander, Asian, and Native American. Do these groups show the predicted pattern of race-specific alleles and unique patterns of allele frequency? Not in a way that provides convincing evidence of consistent difference among these groups. Instead, most evidence indicates that the genetic differences between unrelated individuals *within* a race are much greater than the average genetic differences *among* races.

No Alleles Are Found in All Members of a Race. The Human Genome Project has allowed researchers to scan the genome of thousands of individuals looking for evidence of alleles that are found in a single human group and not in other groups. The types of alleles that are most commonly used in this analysis are called **single nucleotide polymorphisms,** abbreviated **SNPs** ("snips"). A SNP is a single base pair in the DNA sequence of humans that can differ from one individual to another. Humans are remarkably similar, having identical gene sequences for 99% of our genome. The 1% of the genome where there is variability is primarily made up of SNPs. We can think of the different DNA bases found at a particular SNP site as different alleles.

The primary reason scientists are interested in identifying SNPs in the human genome is to understand human diversity in disease susceptibility and other traits that affect health and well-being. However, because 99% of SNPs appear to be in parts of the genome that do not produce proteins, they may have little or no effect on evolutionary fitness and thus can be readily passed on to future generations. You can imagine that when an SNP allele arises via mutation that has no negative effects on fitness—that is, which is neutral—there may be no barrier to its spread throughout a human population. For this reason, neutral SNPs are useful for understanding ancestry, but because they are neutral, they have no effect on the physical characteristics of individuals.

Researchers have identified a number of SNP alleles that are unique to particular human populations. Most genetic ancestry testing companies identify three major groups: African, European, and Asian (including Native American). Within those larger groups, certain populations display unique SNP alleles that may help identify an individual's ancestry more specifically. However, it is important to remember that the overwhelming majority of SNPs have no known effect on phenotype. And it is important to note that no SNP allele is found in every individual in any population. In fact, among groups classified in the same major race, some populations may have *no* individuals who have a particular SNP allele that is considered unique to that race.

What is true of SNP alleles may be more clearly illustrated with alleles associated with readily apparent phenotypes. For example, sickle-cell anemia has long been thought of as a disease that mostly affects black individuals. This illness occurs in people who carry two copies of the sickle-cell allele, resulting in red blood cells that deform into a sickle shape under certain conditions. The consequences of these sickling attacks include severe pain, and heart, kidney, lung, and brain damage. Many individuals with sickle-cell anemia do not live past childhood.

Nearly 10% of African Americans and 20% of Africans carry one copy of the sickle-cell allele, whereas the allele is almost completely absent in European Americans. However, if we examine the distribution of the sickle-cell allele more closely, we see that this seemingly race-specific pattern is not so straightforward. Not all human populations classified as black have a high frequency

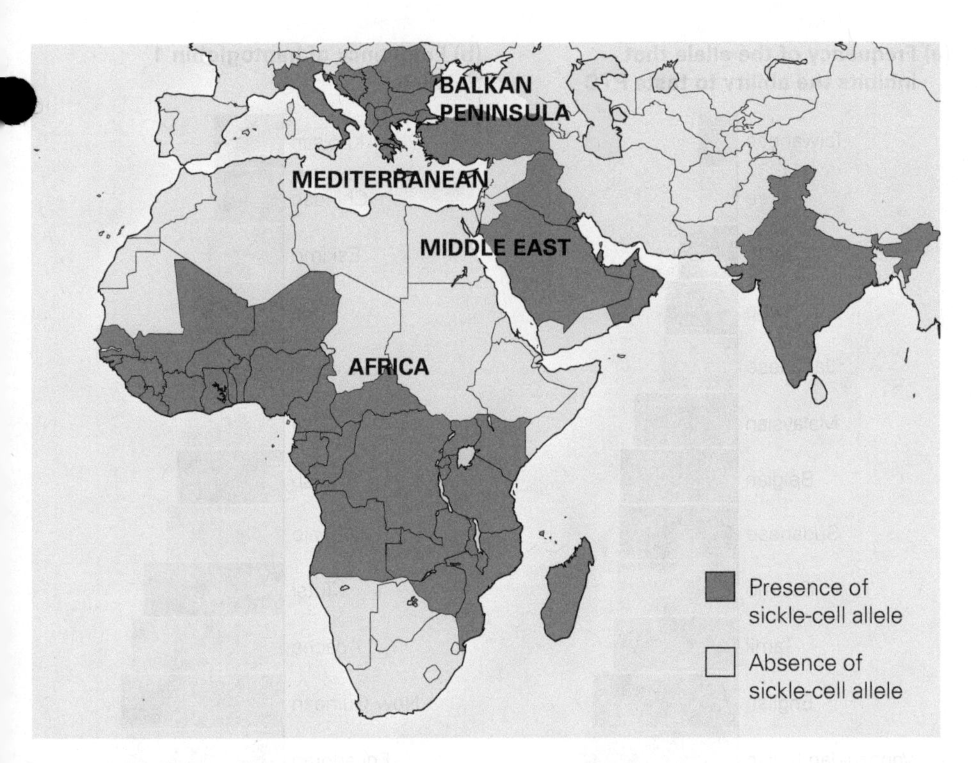

FIGURE 13.10 The sickle-cell allele. The map illustrates where the sickle-cell allele is found in human populations. Note that it is not found in all African populations but is found in some European and Asian populations.

of the sickle-cell allele. In fact, in populations from southern and north-central Africa, which are traditionally classified as black, this allele is rare or absent. Among populations who are classified as white or Asian, there are some in which the sickle-cell allele is relatively common, such as white populations in the Middle East and Asian populations in northeast India (**FIGURE 13.10**). Thus, the sickle-cell allele is not a characteristic of all black populations or unique to a supposed "black race."

In fact, scientists have not identified a single allele that is found in all (or even most) populations of a commonly described race but not found in other races. Only a tiny number of SNPs have been identified as unique to a particular human racial group, and these are never found in all populations of the race or within every individual in a population.

The hypothesis that human races represent independent evolutionary groups is not supported by these observations.

Populations within a Race Are Often as Different as Populations Compared across Races. Certain SNP alleles are useful in that they link individuals to particular population groups. This is not surprising because an ancestral population tends to be associated with a particular geography, and people living close to each other are likely to share more ancestors (and thus be more genetically similar) than people living far apart. However, if race is to be biologically meaningful, then the allele frequency for *many different* SNPs and genes should be more similar among populations within a race than among populations of different races.

Again, the pattern of SNP allele frequency among populations can be illustrated by more obvious alleles. The human populations in each part of

FIGURE 13.11 Do human races show genetic evidence of isolation? The bars on these histograms illustrate the frequency of the described allele in several different human populations. The figure illustrates that populations within these "races" are not necessarily more similar to each other than they are to populations in different races.

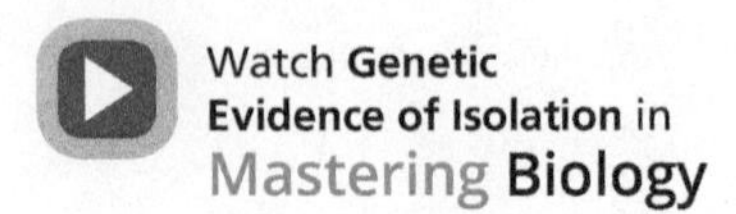

Watch **Genetic Evidence of Isolation** in Mastering **Biology**

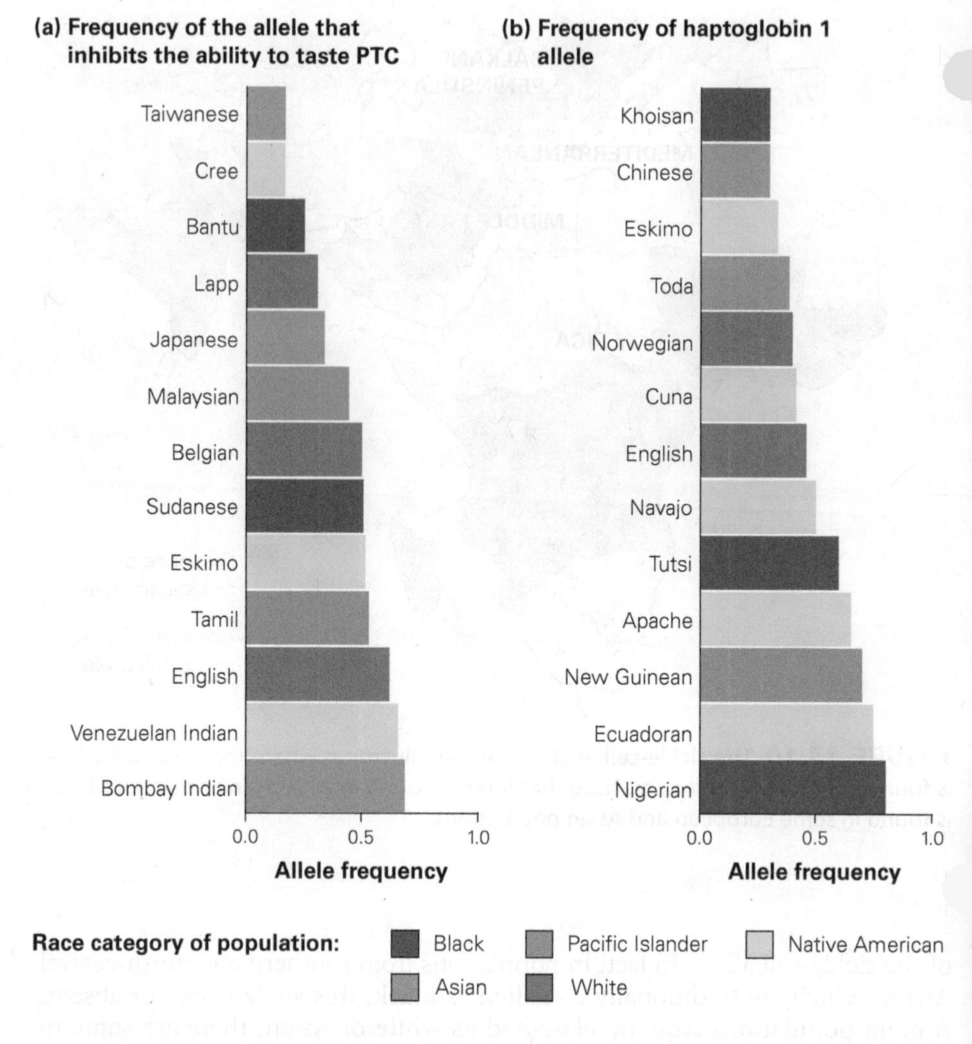

FIGURE 13.11 are listed by increasing frequency of a particular allele in the population. The color coding of each population group in each of the graphs corresponds to the racial category in which the population is typically placed. For example, at the top of Figure 13.11a, the Taiwanese population is categorized as Asian, and the Cree Indian population of eastern Canada is categorized as Native American. If the hypothesis that human racial groups have a biological basis was correct, then populations from the same racial group would be clustered together on each bar graph.

Figure 13.11a shows the frequency of the allele that interferes with an individual's ability to taste the chemical phenylthiocarbamide (PTC) in several populations. People who carry two copies of this recessive allele cannot detect PTC, which tastes bitter to people who carry one or no copies of the allele. Note that populations within a single "race," such as Asian, vary widely in the frequency of this allele.

Figure 13.11b lists the frequency of one allele for the gene *haptoglobin 1* in a number of different human populations. Haptoglobin 1 is a protein that helps scavenge the blood protein hemoglobin from old, dying red blood cells. Again, we see a wide distribution of allele frequency within the race categories.

What we see in the graphs in Figure 13.11 is that allele frequencies for these genes are *not* more similar within racial groups than among racial groups. In fact, in both cases, the populations with the highest and lowest allele frequencies belong to the same race—for these genes, there is more variability *within* a race than there are average differences *among* races. Even though certain SNPs can help us track an individual to a particular ancestral population, the pattern of diversity in other genes tells us that these "ancestral" patterns are not evidence of deep biological similarities among populations within a racial group.

In fact, the categories for human race fail to meet the criteria for identifying populations as consistently isolated from each other. Both the fossil evidence and genetic evidence indicate that the five commonly listed human racial groups do *not* represent biological races.

Human Races Have Never Been Truly Isolated

The results of genetic ancestry tests on individuals in the United States typically reflect the truth that populations have not been truly isolated for the past 500 years of European and Asian immigration and the history of the African slave trade. Recent research indicates that most African Americans have at least 20% European SNPs, and that 30% of white U.S. college students have less than 90% European SNPs. But the "melting pot" of the United States is not unique in human history. In fact, the evidence that human populations have been "mixing" since modern humans first evolved is contained within the gene pool of human populations. For instance, the frequency of the B blood group decreases from east to west across Europe (**FIGURE 13.12**). The allele that codes for this blood type apparently evolved in Asia, and the pattern of blood group distribution seen in Figure 13.12 corresponds to the movement of Asians into Europe beginning about 2000 years ago. As the Asian immigrants mixed with the European residents, their alleles became a part of the European gene pool. Populations closest to Asia experienced a large change in their gene pools, whereas populations who were more distant encountered a more varied immigrant gene pool made up of the offspring between the Asian immigrants and their European neighbors.

Other genetic analyses have led to similar maps. For example, one indicates that populations from the Middle East migrated throughout Europe and Asia about 10,000 years ago. The data from these kinds of mapping projects indicate that there are no clear boundaries within the human gene pool. Reproduction among human populations over hundreds of generations explains why the isolation required for the formation of distinct biological races never occurred.

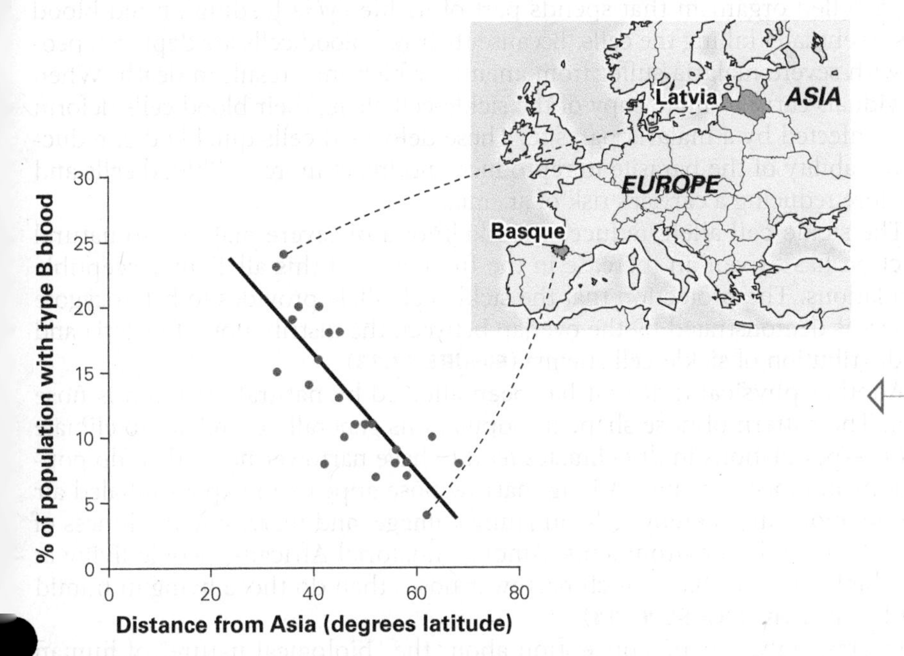

FIGURE 13.12 Genetic mixing in humans. The decline in frequency of type B blood from west to east in Europe reflects the movement of alleles from Asian populations into European populations over the past 2000 years.

WORKING WITH DATA

Type O is the most common blood type in Europe. Given the pattern of B frequency here, what do you predict the pattern of frequency of type O blood would appear on the same graph?

Got It?

1. The genealogical species concept identifies species as groups that share a recent ________ ancestry.
2. Identifying species from the fossil record requires using the ________ species concept.
3. A population that has been isolated from another population for a long time should contain some ________ alleles.
4. Populations that have been separated from each other for a long time should have different allele ________ for several genes.
5. Human populations do not show genetic evidence that they have been ________ from each other; in fact, the genetic evidence indicates significant ________.

13.3 Why Human Groups Differ

As we have learned, human races are not true "biological races." However, human populations do differ from each other in many traits. In this section, we explore what is known about why populations share certain superficial traits and differ in others.

Natural Selection

Recall the distribution of the sickle-cell allele in human populations shown in Figure 13.10. This allele is found in some populations of at least three of the typically described races. Why is it higher in certain populations?

The sickle-cell allele is higher in certain populations because in particular environments natural selection favors individuals who carry one copy of it. The sickle-cell anemia allele is an *adaptation,* a feature that increases fitness, within populations in malaria-prone areas. Malaria is a disease caused by a parasitic, single-celled organism that spends part of its life cycle feeding on red blood cells, eventually killing the cells. Because their red blood cells are depleted, people with severe malaria suffer from anemia, which may result in death. When individuals carry a single copy of the sickle-cell allele, their blood cells deform when infected by a malaria parasite. These deformed cells quickly die, reducing the ability of the parasite to reproduce and infect more red blood cells and therefore reducing a carrier's risk of anemia.

The sickle-cell allele reduces the likelihood of severe malaria, so natural selection has driven an increase in the frequency of this allele in susceptible populations. The protection that the sickle-cell allele provides to heterozygote carriers is demonstrated by the overlap between the distribution of malaria and the distribution of sickle-cell anemia (**FIGURE 13.13**).

Another physical trait that has been affected by natural selection is nose form. The pattern of nose shape in populations generally correlates to climate factors—populations in dry climates tend to have narrower noses than do populations in moist climates. A long, narrow nose appears to expose inhaled air to more moisture, thereby reducing lung damage, and increasing the fitness of individuals in dry environments. Among equatorial Africans, people living at drier high altitudes have much narrower noses than do those living in humid rain-forest areas (**FIGURE 13.14**).

Interestingly, our preconception about the "biological nature" of human races puts two populations of Africans (represented by the images in Figure 13.14) in the same race and explains differences in their nose shape as a result

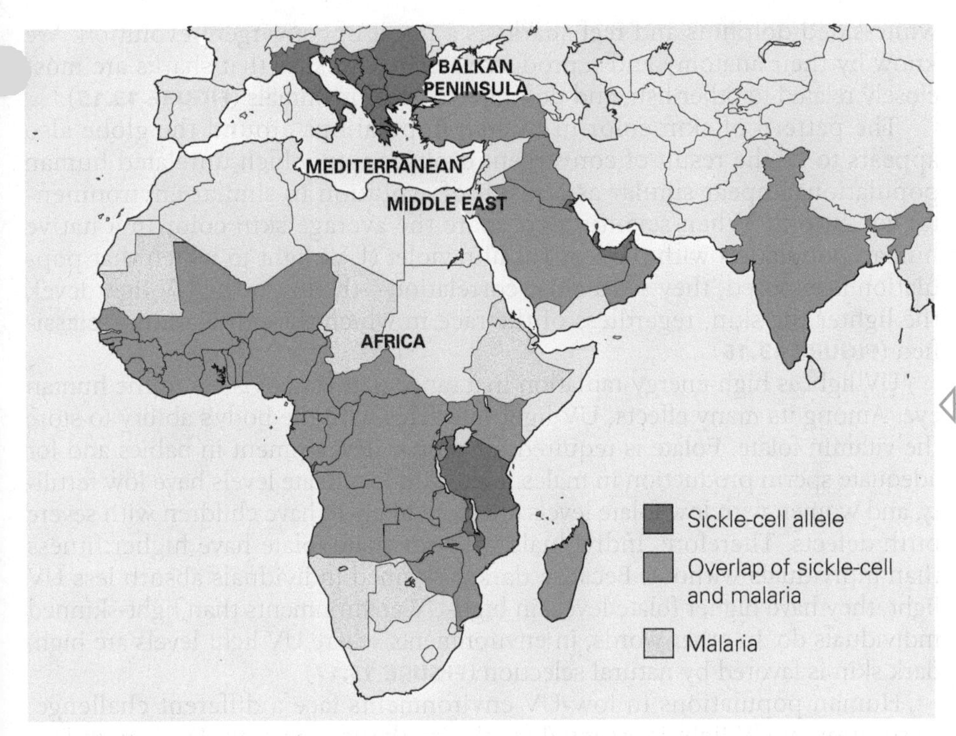

FIGURE 13.13 The sickle-cell allele is common in malarial environments. This map shows the distributions of the sickle-cell allele and malaria in human populations.

VISUALIZE THIS

Given the pattern pictured here, in which population groups in the United States would you expect to find individuals who carry the sickle-cell allele?

of natural selection, but places white and black populations into different races and explains their skin color differences as evidence of long isolation from each other. However, like nose shape, skin color is a trait that is strongly influenced by natural selection.

Convergent Evolution

Traits that are shared by unrelated populations because they share similar environmental conditions are termed *convergent*. **Convergent evolution** occurs when natural selection for similar environmental factors causes unrelated organisms to resemble each other. For example, the similarity in shape between

(a) Ethiopian with a narrow nose

(b) Bantu with a broad nose

FIGURE 13.14 Nose shape is affected by natural selection. Long, narrow noses are more common among populations in cold, dry environments apparently because they provide more humid air to the lungs than broad, flattened noses do.

FIGURE 13.15 Convergence. The similarity in shape between dolphins and sharks results from similar adaptations to life as an oceanic predator of fish, not shared ancestry.

VISUALIZE THIS

Penguins are also oceanic predators of fish. How do these animals demonstrate convergence with dolphins and sharks?

white-sided dolphins and reef sharks is a result of convergent evolution. We know by their anatomy and reproductive characteristics that sharks are most closely related to other fish, and dolphins to other mammals (**FIGURE 13.15**).

The pattern of skin color in human populations around the globe also appears to be the result of convergent evolution, in which unrelated human populations appear similar as a result of evolution in similar environmental conditions. When scientists compare the average skin color in a native human population with the level of ultraviolet (UV) light to which that population is exposed, they see a close correlation—the lower the UV light level, the lighter the skin, regardless of the race in which the population is classified (**FIGURE 13.16**).

UV light is high-energy radiation in a range that is not visible to the human eye. Among its many effects, UV light interferes with the body's ability to store the vitamin folate. Folate is required for proper development in babies and for adequate sperm production in males. Men with low folate levels have low fertility, and women with low folate levels are more likely to have children with severe birth defects. Therefore, individuals with adequate folate have higher fitness than individuals without. Because darker-skinned individuals absorb less UV light, they have higher folate levels in high-UV environments than light-skinned individuals do. In other words, in environments where UV light levels are high, dark skin is favored by natural selection (**FIGURE 13.17**).

Human populations in low-UV environments face a different challenge. Absorption of UV light is essential for the synthesis of vitamin D. Vitamin D is crucial for the proper development of bones. Women are especially harmed by low vitamin D levels—inadequate development of the pelvic bones can make giving birth deadly. There is no risk of not making enough vitamin D when UV light levels are high, regardless of skin color. However, in areas where levels of UV light are low, individuals with lighter skin absorb a larger fraction of

WORKING WITH DATA

Find two populations classified as the same race that have very different skin color.
Find two populations classified in different races that have the same skin color.

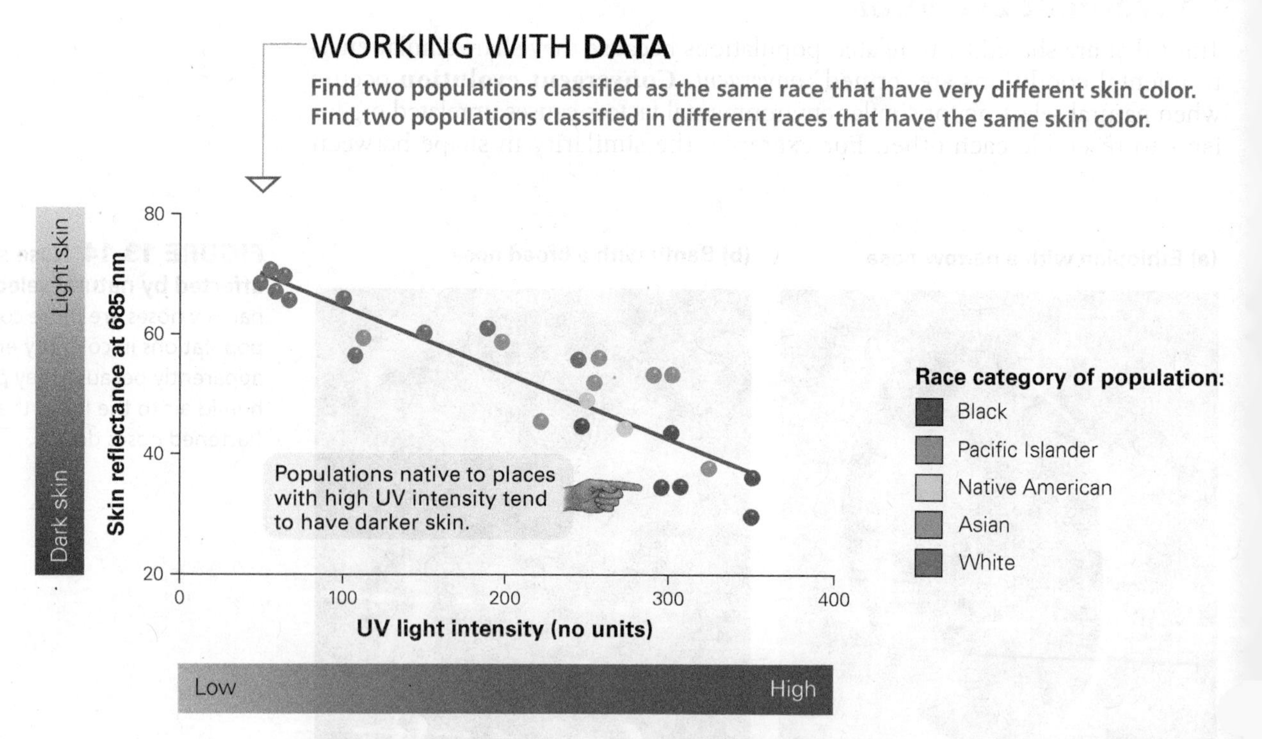

FIGURE 13.16 There is a strong correlation between skin color and exposure to UV light. Reflectance is an indication of color—higher reflectance indicates lighter skin. The color of the dots on the graph specifies the racial category of each population.

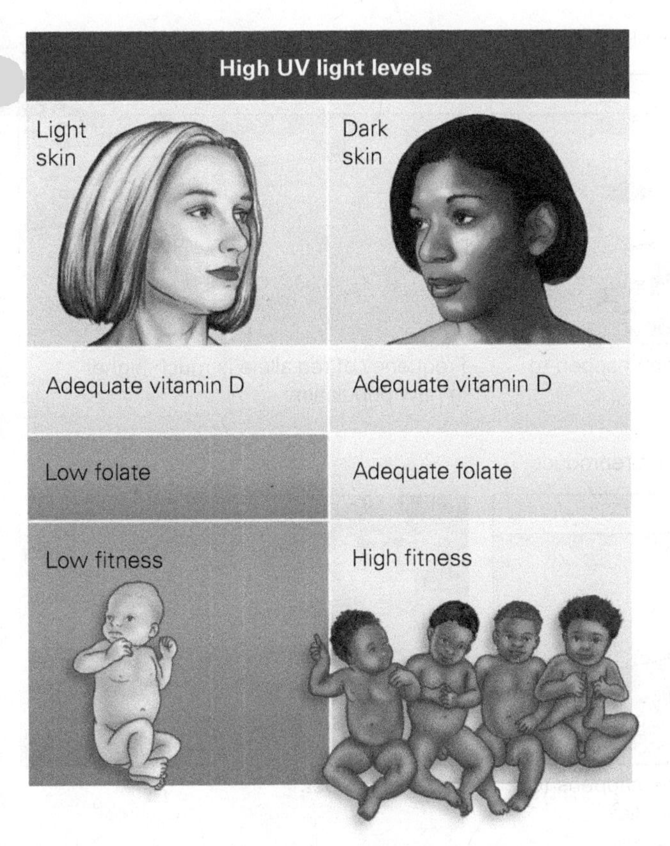

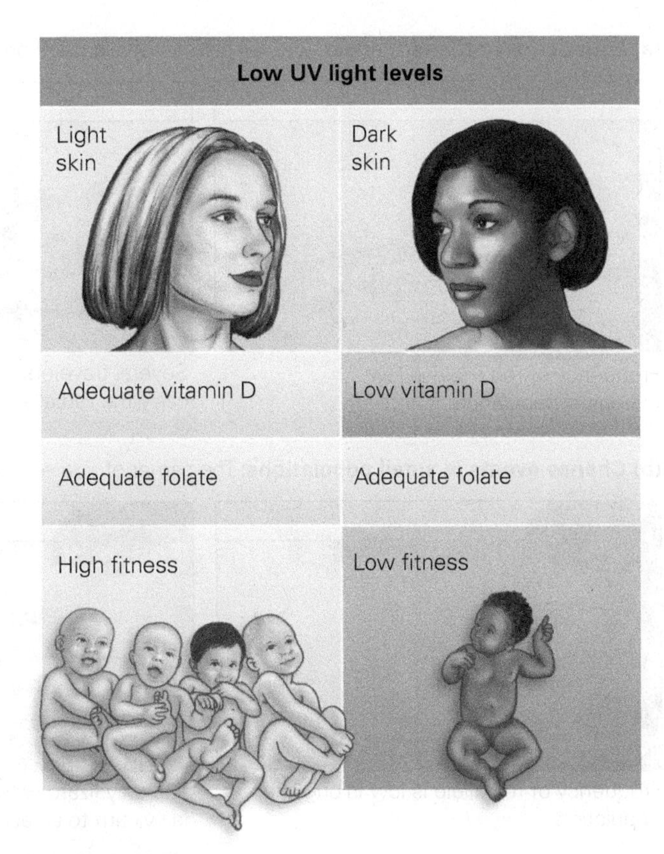

FIGURE 13.17 The relationship between UV light levels, folate, vitamin D, and skin color. Darker, UV-resistant skin is an advantage within populations in regions where UV light levels are high because individuals with darker skin have higher fitness. Lighter, UV-transparent skin is favored by natural selection where UV levels are low.

VISUALIZE THIS

What could an individual with dark skin in a low-UV environment do to compensate? What about an individual with light skin in a high-UV environment?

UV light and thus have higher levels of vitamin D than do individuals with darker skin. Thus, in less sunny environments, light skin has been favored by natural selection.

Because UV light had important effects on human physiology, it drove the evolution of skin color in human populations. Where UV light levels are high, dark skin was an adaptation, and populations became dark-skinned. Where UV light levels are low, light skin was an adaptation, and populations evolved to become light-skinned. The exception to these patterns actually support the hypothesis; the relatively dark-skinned Inuits of the northern polar regions did not experience strong natural selection for lighter skin color thanks to a vitamin D-rich fish diet. As can be observed in Figure 13.16, the pattern of skin color in human populations is a result of the convergent evolution of different populations in similar environments—generally darker skin near the equator on each continent and lighter skin near the poles—not necessarily evidence of separate races of humans.

Natural selection has caused differences among human populations, but it has also resulted in some populations superficially appearing more similar to some other human populations. In contrast, populations that appear on the surface to be similar may be quite different, simply by chance.

Genetic Drift

A change in allele frequency that occurs due to chance is called **genetic drift.** Human populations tend to travel and colonize new areas, so we seem to be

(a) Founder or bottleneck effect: A small sample of a large population establishes a new population or survives a disaster.

Frequency of red allele is low in original population.

Several travelers (or survivors) happen to carry the red allele.

Frequency of red allele is much higher in new population.

(b) Chance events in small populations: The carrier of a rare allele does not reproduce.

Frequency of red allele is low in original population.

The only lizard with red allele happens to fall victim to an accident and dies.

Red allele is lost.

FIGURE 13.18 The effects of genetic drift. A population may contain a different set of alleles because (a) its founders (or the remaining survivors of a catastrophe) were not representative of the original population; or (b) the population is so small that low-frequency alleles are lost by chance.

FIGURE 13.19 An example of the founder effect in plants. Cocklebur is a hitchhiker plant, dispersing by hooking the spikes on its fruits to the fur (or sock) of a passing animal. When these burrs are removed at a distant location, the plant can found a new population whose seeds may be different in size, shape, or color from the source population.

especially prone to evolution via genetic drift. Genetic drift occurs in two different types of situations, described below (**FIGURE 13.18**).

Founder or Bottleneck Effects. Genetic differences can occur when a small sample of a larger population establishes a new population. The gene pool of the immigrants is rarely an exact reflection of the gene pool of the source population. This difference leads to the **founder effect.**

Rare genetic diseases that are more common in certain populations may be a result of the founder effect. For example, the Amish of Pennsylvania are descended from a population of 200 German founders who immigrated to the United States more than 200 years ago. Ellis–van Creveld syndrome, a recessive disease that causes dwarfism (among other effects), is 5000 times more common in the Pennsylvania Amish population than in other German American populations. This difference is a result of a single founder in that original population who carried the allele. Because the Pennsylvania Amish usually marry others within their small religious community, the allele has stayed at a high level—1 in 8 Pennsylvania Amish are carriers of the Ellis–van Creveld allele, compared with fewer than 1 in 100 non–Amish Americans of German ancestry.

Plants with animal-dispersed seeds appear to be especially prone to the founder effect. For example, cocklebur, a widespread weed that produces hitchhiker fruit designed to grab onto the fur of a passing mammal (**FIGURE 13.19**), consists of populations that are quite variable in size and shape. The variation among populations appears to have been caused by differences in the hitchhikers that happened to be carried to new locations, where they founded new populations. A variant of the founder effect is the **bottleneck effect,** in which a disaster wipes out most of the population, leaving only a small subset of survivors.

Genetic Drift in Small Populations. Even without a population bottleneck, allele frequencies may change in a population due to chance events. When an allele is in low frequency within a small population, only a few individuals carry a copy of it. If one of these individuals fails to reproduce, or if it passes on only the more common allele to surviving offspring, the frequency of the rare allele may drop in the next generation. If the population is small enough, even relatively high-frequency alleles may be lost after a few generations by genetic drift.

A human population that illustrates the effects of genetic drift in small populations is the Hutterites, a religious sect with communities in South Dakota and Canada. Modern Hutterite populations trace their ancestry back to 442 people who migrated from Russia to North America between 1874 and 1877. Hutterites tend to marry other members of their sect, so the gene pool of this population is small and isolated from other populations. Genetic drift in this population over the past century has resulted in a near absence of type B blood among the Hutterites, as compared with a frequency of 15 to 30% in other European immigrants in North America.

Genetic drift in populations that remain small for many generations can lead to a rapid loss of many different alleles. Although this problem is uncommon in humans, the effects of genetic drift on small populations of endangered species can lead to extinction (Chapter 16).

Humans are a highly mobile species, and we have been founding new populations for millennia. Most early human populations were also probably quite small. These factors make human populations especially susceptible to the founder effect, genetic bottlenecks, and genetic drift and have contributed to the differences among modern human groups.

In addition to natural selection and random genetic change, humans' highly social nature may have enhanced certain superficial differences in appearance among different populations.

Sexual Selection

Men and women within a population may have preferences for particular physical features in their mates. These preferences, expressed over generations, can cause populations to differ in appearance. When a trait influences the likelihood of mating, that trait is under the influence of a form of natural selection called **sexual selection.**

Darwin proposed the hypothesis of sexual selection in 1871 as an explanation for differences between males and females within a species. For instance, the enormous tail on a male peacock results from female peahens that choose mates with showier tails. Because large tails require so much energy to display and are more conspicuous to their predators, peacocks with the largest tails must be both physically strong and smart to survive. In fact, tail length does appear to be a good measure of overall fitness in peacocks; the offspring of well-endowed males are more likely to survive to adulthood than are the offspring of males with scanty tails. When a peahen chooses a male with a large tail, she is ensuring that her offspring will receive high-quality genes. Sexual selection explains the differences between males and females in many species (**FIGURE 13.20**).

In humans, there is some evidence that the difference in overall body size between men and women is a result of sexual selection—namely, a widespread female preference for larger males—perhaps again because size may be an indication of overall fitness. However, some human traits may reflect simply a "social preference." For example, some scientists have suggested that lack of thick facial and body hair in some human populations resulted from sexual selection by

(a) Peacock

(b) Lion

(c) Blue morpho butterfly

FIGURE 13.20 The effects of sexual selection. Sexual selection is responsible for many unique and fantastic characteristics of organisms from (a) the peacock's tail to (b) the male lion's mane to (c) the bright colors of butterflies.

both men and women for less hairy mates. Although intriguing, there is as yet little definitive evidence to support the hypothesis that sexual selection shaped any particular human trait, and no simple way to test it.

Assortative Mating

Differences between populations may be reinforced by the ways in which individuals choose their mates. For example, in Eastern Bluebirds, individuals prefer mating with members of the opposite sex who are as colorful as themselves, a process called positive **assortative mating.**

In humans, there is a tendency for people to mate assortatively by height—that is, tall women tend to marry tall men—and by social factors such as religious preference or educational level.

When two human populations differ in obvious physical and cultural characteristics, the number of matings between them may be small if the traits of one population are considered unattractive to members of the other population. Thus, positive assortative mating tends to maintain and even exaggerate physical differences between populations. In highly social humans, assortative mating may be an important amplifier of superficial physical differences among groups.

Although human populations may display differences due to natural selection in certain environments, genetic drift, sexual selection, and assortative mating, the genetic evidence indicates that many of these differences are literally no more than skin deep.

If human races are not biologically "real," is Simone Manuel wrong to describe herself as a "black swimmer"? No, because race is a social category, not a biological one. This distinction is illustrated by another athlete, the man who is often credited with being the first African American gold medal swimmer of either sex: Anthony Ervin, who won his first Olympic medal in 2000 (**FIGURE 13.21**). Despite his designation as "first," the light-skinned, green-eyed Ervin—child of an African American father and a European Jewish mother—does not identify himself as a black individual. He told *Rolling Stone* magazine,

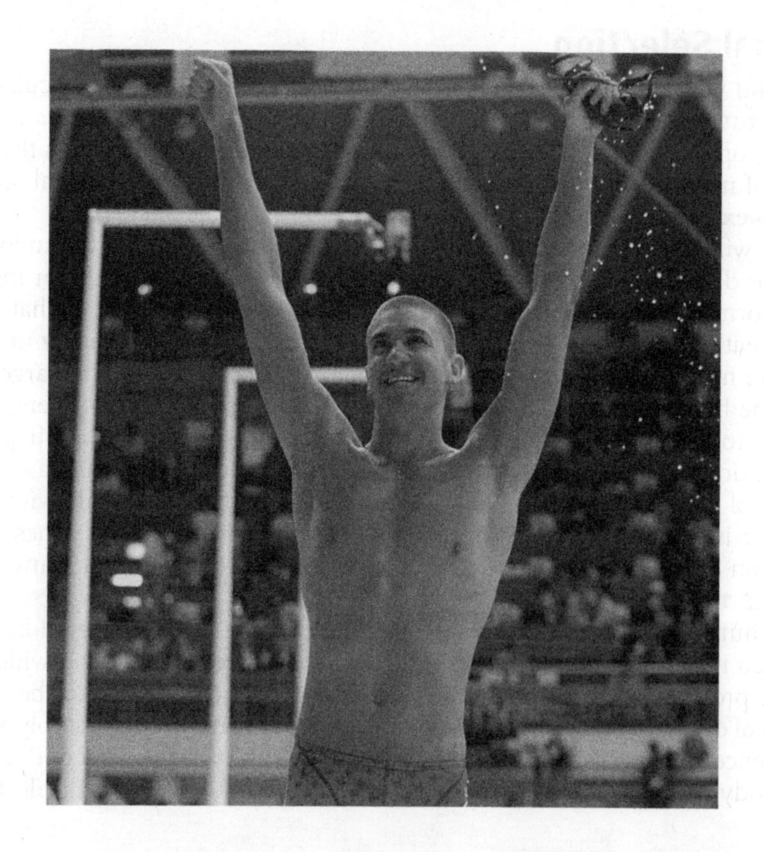

FIGURE 13.21 Race is a social construction. The man described as the first African American Olympic gold medalist swimmer does not consider himself "black," although his father was African American. His freedom to choose his racial identity is additional evidence that human racial categories are social and not biological.

"I didn't know a thing about what it was like to be part of the black experience. But now I do. It's like winning gold and having . . . people ask you what it's like to be black. That is my black experience."

Because Ervin did not appear to others as black, he did not experience the same concerns about his capabilities as a swimmer that Manuel faced. And he also did not come from a family where swimming was an unusual activity. In reality, black swimmers are not rare because African Americans are biologically less able to succeed at swimming; they are rare because pools and beaches in the United States were off-limits to black swimmers for generations (**FIGURE 13.22**). Now that those social barriers have lifted and more women like Manuel are demonstrating their excellence, it may not be long before the title "black swimmer"—just like "black baseball player" or "black President"—will be archaic, making a distinction that is not based in either biology or in social expectations.

FIGURE 13.22 Why are African Americans underrepresented in swimming? Black swimmers are attacked at a "whites-only" beach in St. Augustine, Florida, in 1964. Discrimination at public pools and beaches inhibited blacks from learning the sport and from passing swimming skills down within their families.

Got It?

1. Two unrelated populations or species responding to similar environmental pressures by evolving similar traits is known as ________ evolution.
2. Genetic drift refers to the effect of ________ on changing allele frequencies in a population.
3. Genetic drift is more likely to cause evolution in ________ populations.
4. Sexual selection refers to the effect of the choice of ________ on the evolution of traits in a population.
5. Positive assortative mating occurs when individuals choose mates that are ________ to themselves.

Sounds right, but is it?

The Hawaiian Islands are home to a group of birds found nowhere else—the Hawaiian honeycreepers. This group of about 50 species (many of which are now extinct) appears to have descended from a single species of Asian rosefinch that made it to the islands between 4 and 5 million years ago. The ancestral rosefinch likely had a short, fat bill and fed on seeds—however, many honeycreeper species have long, curved bills and feed on flower nectar; the bill shape of particular species fits perfectly with the flower shape of the plants they feed on. This fact might lead you to say:

Some species of honeycreepers evolved with long, curved bills because they needed to feed on the nectar of particular flowers.

Sounds right, but it isn't.

Answer the following questions to understand why.

1. Is it likely that variation in bill sizes and shapes existed within the population of rosefinches that inhabited the Hawaiian Islands 4–5 million years ago?
2. Is it likely that variation in foraging (that is, how an individual hunts for food) existed within this population of rosefinches?
3. Would all birds in the original population demonstrate these traits?
4. In an environment where few animals feed on a particular flower's nectar, would an individual with a slightly longer and thinner bill and the behavior of flower foraging have high evolutionary fitness?
5. How do differences among individuals in evolutionary fitness affect the traits of a population?
6. Reflect on your answers to questions 1–5, and explain why the statement bolded above sounds right, but isn't.

THE BIG QUESTION

Are affirmative action policies that favor black students applying for college admission good public policy?

"Affirmative action" is the descriptor applied to policies or measures meant to correct for past or present discrimination. As applied to college admissions, this means policies that promote the acceptance of students from underrepresented racial or ethnic groups. Some white applicants object to these policies, arguing that they lose opportunities that are unfairly awarded to individuals in other groups.

What should I know?

What follows are some smaller questions that need to be resolved to answer the Big Question. Place a checkmark next to the questions that science can answer.

Smaller Questions	Can Science Answer?
Are African Americans underrepresented in college classes when compared with the percentage they make up of the U.S. population?	
Does the injustice of past discrimination against African Americans in the United States require reparations?	
Are African Americans as capable of success in college classes as white students with equivalent high school records?	
Do African Americans perform better in college classes when they make up a larger percentage of the class population?	
Is it wrong to discriminate, either positively or negatively, on the basis of race?	
Is there a relationship between skin tone and educational attainment?	
Does a college degree provide recipients with more wealth over a lifetime than a high school degree?	

What does the science say?

Let's examine what the data say about this smaller question:

Do African Americans perform better in college classes when they make up a larger percentage of the class population?

A 2004 study at Harvard University investigated a phenomenon called "stereotype threat," which is postulated to occur whenever an individual is in a situation where negative stereotypes about his or her group might apply. This threat can lead to underperformance on particular tasks, because—paradoxically—some of the individual's unconscious attention is consumed with worry about confirming the negative stereotype. Stereotype threat may explain some of the persistent differences in black versus white performance on standardized tests. In the study, the researchers hypothesized that black students would perform more poorly on a verbal test administered by a white examiner—a situation that would tend to increase stereotype threat—compared with performance when the test was given by a black examiner. The results are displayed below:

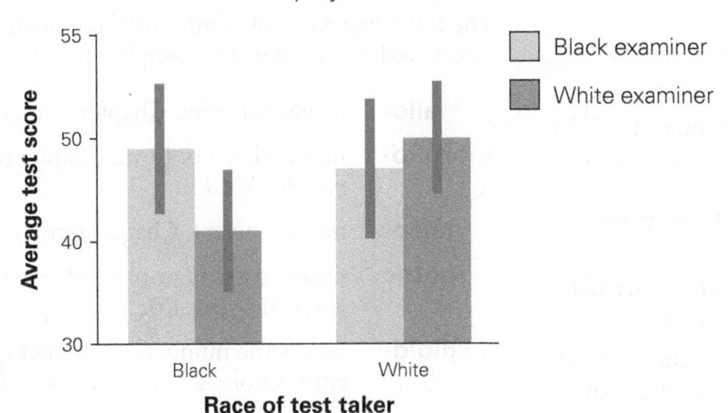

1. Describe the results. Does it appear that stereotype threat has a negative effect on the performance of black students in this study?
2. Given these data, do you think the smaller question is answered? If not, propose another study that would help answer this question.
3. Does this information help you answer the Big Question? What else do you need to consider?

Data source: D. M. Marx and P. A. Goff, "Clearing the Air: The Effect of Experimenter Race on Target's Test Performance and Subjective Experience," *British Journal Social Psychology*, 44 (2005): 645–657.

Chapter Review

Mastering Biology

Go to Mastering Biology to access **eText 2.0, Dynamic Study Modules,** and the **Study Area,** where you'll find practice quizzes, BioFlix™ animations, MP3 tutor sessions, current events, and more.

SUMMARY

Section 13.1

Define *biological species*, and list the mechanisms by which reproductive isolation is maintained by biological species.

- All humans belong to the same biological species, *Homo sapiens sapiens.* A biological species is a group of individuals that can interbreed and produce fertile offspring. Biological species are reproductively isolated from each other, thus separating their gene pools (pp. 263–264).
- Reproductive isolation is maintained by prefertilization factors, such as differences in mating behavior or timing, or postfertilization factors, such as hybrid inviability or sterility (pp. 264–265).

Describe the three steps in the process of speciation.

- Speciation occurs when populations of a species become isolated from each other. These populations diverge from each other, and reproductive isolation between the populations evolves (pp. 266–270).

SHOW YOU KNOW 13.1 The corn varieties planted by most farmers are called "hybrids" because each is produced when two different varieties of the species *Zea mays* (for example, a short variety and a sweet variety) are crossed. Do you expect that the corn produced by these crosses is sterile? Why or why not?

Section 13.2

Explain how a "race" within a biological species can be defined using the genealogical species concept.

- Genealogical races are populations of a single species that have diverged from each other but have not become reproductively isolated (pp. 270–271).

List the evidence that modern humans are a young species that arose in Africa.

- The fossil record provides evidence that the modern human species is approximately 200,000 years old (p. 272).

- Genetic evidence indicates that modern humans have limited genetic diversity and that the oldest human populations are found in Africa (p. 272).

List the genetic evidence expected when a species can be divided into unique races.

- The genetic evidence for biological race include (1) alleles that are unique to a particular race, (2) similar allele frequencies for a number of genes among populations within races, and (3) differences in allele frequencies among populations in different races (p. 273).

Summarize the evidence that indicates human races are not deep biological divisions within the human species.

- Patterns of population genetics indicate that the currently delineated races do not contain evidence that they have been isolated (pp. 274–277).
- Additional evidence indicates that human groups have been mixing for thousands of years (p. 277).

SHOW YOU KNOW 13.2 Bird watchers often distinguish different "races" of birds that vary somewhat in appearance and are geographically distant from each other. What information would help scientists determine if these different varieties are true biological races?

Section 13.3

Provide examples of traits that have become common in certain human populations due to the natural selection these populations have undergone.

- The sickle-cell allele is selected for in populations in which malaria incidence is high, and light skin is selected for in areas where the UV light level is low (pp. 278–281).

Define *genetic drift* and provide examples of how it results in the evolution of a population.

- Genetic drift is defined as changes in allele frequency due to chance events such as founder effects or population bottlenecks (pp. 281–283).

Describe how human and animal behavior can cause evolution via sexual selection and assortative mating.

- Sexual selection occurs when individuals choose mates that display some "attractive" quality (p. 283).
- Positive assortative mating, in which individuals choose mates who are like themselves, can reinforce differences between human populations (p. 284).

SHOW YOU KNOW 13.3 The word for *mother* in many different languages is a variation of the sound "ma." This appears to be an example of convergent evolution in language. Why might many different cultures share this characteristic sound for *mother*?

ROOTS TO REMEMBER

The following roots of words come mainly from Latin and Greek and will help you to decipher terms:

allo-	means different. Chapter term: *allopatric*
morpho-	means shape or form. Chapter term: *morphological*
paleo-	means ancient. Chapter term: *paleontologists*
-patric	means country or place of origin. Chapter term: *allopatric, sympatric*
-ploid-	means the number of different types of chromosomes. Chapter term: *polyploidy*
poly-	means many. Chapter term: *polyploidy*
sym-	means same or united. Chapter term: *sympatric*

LEARNING THE BASICS

1. Define "biological species."
2. Add labels to the figure that follows, which illustrates the three steps required for speciation to occur.

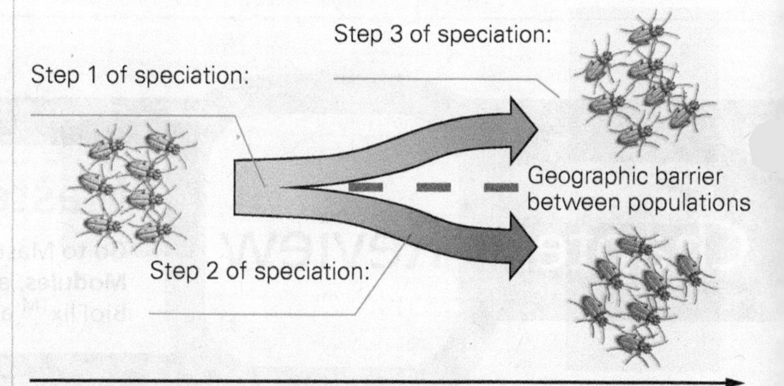

3. Describe three ways that evolution can occur via genetic drift.
4. Which of the following is an example of a prefertilization barrier to reproduction?

   A. A female mammal is unable to carry a hybrid offspring to term; **B.** Hybrid plants produce only sterile pollen; **C.** A hybrid between two bird species cannot perform a mating display; **D.** A male fly of one species performs a "wing-waving" display that does not convince a female of another species to mate with him; **E.** A hybrid embryo is not able to complete development
5. According to the most accepted scientific hypothesis about the origin of two new species from a single common ancestor, most new species arise when __________.

   A. many mutations occur; **B.** populations of the ancestral species are isolated from one another; **C.** there is no natural selection; **D.** a supernatural creator decides that two new species would be preferable to the old one; **E.** the ancestral species decides to evolve

6. For two populations of organisms to be considered separate biological species, they must be __________.

   A. reproductively isolated from each other; B. unable to produce living offspring; C. physically very different from each other; D. A and C are correct; E. A, B, and C are correct

7. The biological definition of "race" corresponds to all of the following *except*:

   A. the genealogical species concept; B. the idea that subgroups within the same species can be distinguished from each other by ancestry; C. there is a natural hierarchy of groups within a species from "lowest" to "highest" forms; D. it should be possible to identify races on the basis of shared allele frequencies among populations; E. races within a species are not reproductively isolated from each other.

8. All of the following statements support the hypothesis that humans cannot be classified into biological races *except*:

   A. There is more genetic diversity within a racial group than average differences between racial groups; B. Alleles that are common in one population in a racial group may be uncommon in other populations of the same race; C. Geneticists can use particular SNP alleles to identify the ancestral group(s) of any individual human; D. There are no alleles found in all members of a given racial group; E. There is genetic evidence of mixing among human populations occurring thousands of years ago until the present.

9. Similarity in skin color among different human populations appears to be primarily the result of __________.

   A. natural selection; B. convergent evolution; C. which biological race they belong to; D. A and B are correct; E. A, B, and C are correct

10. The tendency of individuals to choose mates who are like themselves is called __________.

    A. natural selection; B. sexual selection; C. assortative mating; D. the founder effect; E. random mating

ANALYZING AND APPLYING THE BASICS

1. Wolf populations in Alaska are separated by thousands of miles from wolf populations in the northern Great Lakes of the lower 48 states. Wolves in both populations look similar and have similar behaviors. However, the U.S. government has treated these two populations quite differently, listing the Great Lakes populations as endangered until recently but allowing hunting of wolves in Alaska. Some opponents of wolf protection have argued that the "wolf" should not be considered endangered at all in the United States because of the large population in Alaska, while supporters of wolf protection state that the Great Lakes population represents a unique population that deserves special status. Should these two populations be considered different races or species? What information would you need to test your answer?

2. Phenylketonuria (PKU) is the inability to metabolize the amino acid phenylalanine. The frequency of PKU in Irish populations is 1 in every 7000 births, while the frequency in urban British populations is 1 in 18,000 and only 1 in 36,000 in Scandinavian populations. PKU is found only in individuals who are homozygous recessive for the disease allele. Give two reasons that this allele, which can result in severe mental retardation in homozygous individuals, may be found in different frequencies in these populations.

3. A species of aster normally blooms between mid-August and late September. The range of the aster is located in a climate that usually produces freezing temperatures in late October. The pollinators (bees, ants, and other insects) of the aster are active between late April and early October. Mutations can occur in the gene that controls flowering time for some individuals of this aster, causing them to flower earlier or longer than normal. The table shows possible flowering times.

Aster Type	***Flowering Time***
Normal	August 15–September 30
Early mutation	July 1–August 15
Expanded mutation	July 15–September 30

Describe how this population of asters might split into two species.

GO FIND OUT

1. "Segregation" means "setting apart or separating." You may recall that the term is used in biology when describing the separation of paired alleles during meiosis, when sperm or eggs are produced. Segregation also occurs in human societies, when people become separated from each other based on their "group" membership. Racial segregation in schools, neighborhoods, public places, and private businesses was once legal in the United States (for example, see Figure 13.22), but is now prohibited by federal law (in particular, the Civil Rights Act of 1964). The end of legal segregation did not end the "reality" of segregation. Investigate the segregation of your community, using resources available on the web. In the United States, these include data from the Census Bureau on the racial composition of census blocks; information from local school districts about the racial makeup of the student body; and information about the racial composition of the students and faculty at your own college.

2. The only information that was collected from every household in the United States in the 2010 census was the names of residents of the household, their relationship to each other, sex, age, and race and ethnicity. Do you think it is important to collect race and ethnicity data from all citizens? Why or why not? Is there some other piece of information that you think is more useful to the government?

MAKE THE CONNECTION

The science that you learned in this chapter has helped you better understand the real-world example used throughout this discussion. Draw a line from the statement on the left to the science that supports it on the right.

Simone Manuel, an African American swimmer, competed against swimmers of other "human races" and won a gold medal.	The fossil record indicates that modern humans arose in Africa and replaced ancient human ancestors around the globe around 200,000 years ago.
The differences among human populations have arisen in the past 10,000 generations.	Natural selection related to the effects of UV light have led to a pattern of darker skin colors in populations near the equator and lighter near the pole.
Human "races" do not show the characteristics of true biological races.	The effects of genetic drift on small populations, sexual selection, and positive assortative mating may have led to differences among human populations in outward appearance that are larger than those that exist "beneath the skin."
Differences between "black" and "white" skin color are due to differences in environmental conditions.	All human beings belong to the same biological species.
The mobility of human populations and the way people choose spouses may exaggerate superficial differences among our populations.	Human races do not contain unique alleles that are not found in other human races, and populations within a race are not necessarily more genetically similar to each other than they are to populations in other races.

Answers to **Got It?, Visualize This, Working with Data, Sounds Right, But Is It?, Show You Know,** and **Chapter Review** questions can be found in the **Answers** section at the back of the book.

14 The Greatest Species on Earth?

Giant sequoia are among the most impressive structures on Earth.

But surely the species that built this pyramid is greater than any other?

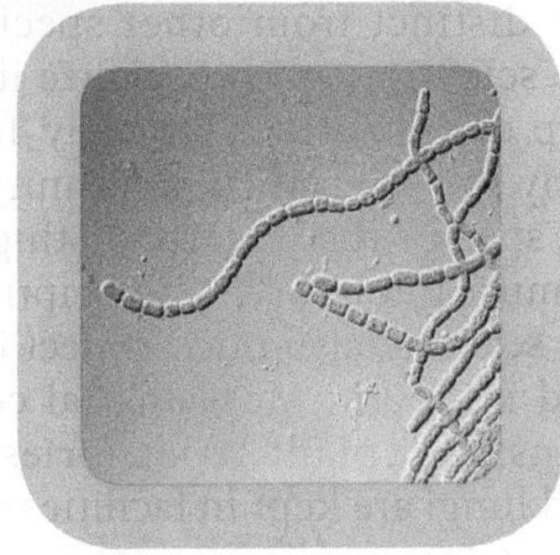

Or maybe the definition of greatest should be that which is the most self-sufficient. These cyanobacteria live on water, sunlight, and carbon dioxide.

Biodiversity and Classification

It is a tried-and-true conversation starter: "Who was the best of all time?" Ask it and hear sports fans debating the greatest athlete of all time, literature fans defending their favorite authors, students of history ranking distinguished politicians, and music lovers arguing about the world's greatest rock bands. These are topics with plenty of room for friendly disagreement and with no obvious answer. But if you interrupted any of these conversations to ask, "What is the greatest species of all time?" you'd likely get a nearly universal answer. Humans, of course. What other species built the pyramids, wrote *The Odyssey,* sent humans to the moon, or composed "Let It Be"?

It seems self-evident to many people that humans are the greatest species. We humans are highly intelligent; we use our communication skills, cooperation, and technology to solve difficult problems. We have extraordinary sensory capabilities and use them to learn about how the world works and then use that knowledge to take advantage of natural resources. We are ingenious in finding ways to survive in different environments. With these skills, we have adapted nearly every part of Earth's surface for human habitation and otherwise modified the environment to meet our needs. We dominate other species, even using the evolutionary process to shape domesticated animals and plants. What species can compete with this?

One of the few places you'd find disagreement on this question is at a gathering of biologists. Scientists know that humans are puny compared to the largest individuals ever—giant sequoia trees—many of which are thousands of years old. And are modern humans, a species only 200,000 years old, that much more impressive than horseshoe crabs, which have survived 445 million years of environmental change? Or more inventive than cyanobacteria, organisms that can make all the components they need to survive from just water, sunlight, and air? Or better adapted than Argentine ants,

—*Continued next page*

a single colony of which can spread across the globe and consist of trillions of individuals? Maybe not. In this chapter, we'll survey the range of living organisms on the planet to test the assumption that humans are the greatest species on Earth.

14.1 Biological Classification

Before we begin evaluating "greatness," we must know something about the multiple forms of living organisms. Biologists are still discovering living species and describing the fossil remnants of extinct species. But even without a complete accounting of all the different forms of life on Earth, we do have a good understanding of the range of biological diversity, both present and past.

How Many Species Exist?

A characteristic of life on Earth is that it is full of variety. Scientists refer to the variety within and among living species as **biodiversity.** Studies of biodiversity don't just provide interesting stories and fodder for arguments about the "best" species. Much more important to many biologists, these studies help us understand the evolutionary origins of different groups of organisms and their role in healthy biological systems.

We have previously (Chapter 13) detailed the challenges inherent in identifying which groups of organisms should be considered discrete **species**—in general, a species is a group of individuals that regularly breed together and are generally distinct from other species in appearance or behavior. **Systematists** are scientists who specialize in describing and categorizing a particular group of organisms. Typically, for an organism to be considered a new species by the scientific community, a systematist must create a description of the species that clearly distinguishes it from similar species, and he or she must publish this description in a professional, peer-reviewed journal. The scientist also must collect individual specimens for storage in a specialized museum. Most animal collections are found in natural history museums, and plant repositories are called herbaria (**FIGURE 14.1**); microbes and fungi are kept in facilities called type-collection centers.

Thanks to advances in computing power and communication, systematists' descriptions of species are now being collected in centralized databases; the

(a) Natural history museum

(b) Herbarium

FIGURE 14.1 Biological collections. Most collections contain several examples of each species to show the range of variation within the species. (a) A collection of mollusks in a natural history museum. (b) Plant specimens are stored in herbaria.

most recent count of these databases lists 1.3 million identified species. Using strategies that extrapolate from the well-described groups of organisms to those less well-known (**FIGURE 14.2**) leads to an estimate that the total number of species on Earth is about 8.7 million, meaning that less than 20% of the total are currently known to science. Scientists who specialize in particular groups of species provide even larger estimates: The simple fact is that we still don't know how many species exist.

WORKING WITH **DATA**

According to these graphs, about how many genera of animals have been described? About how many genera are predicted to exist?

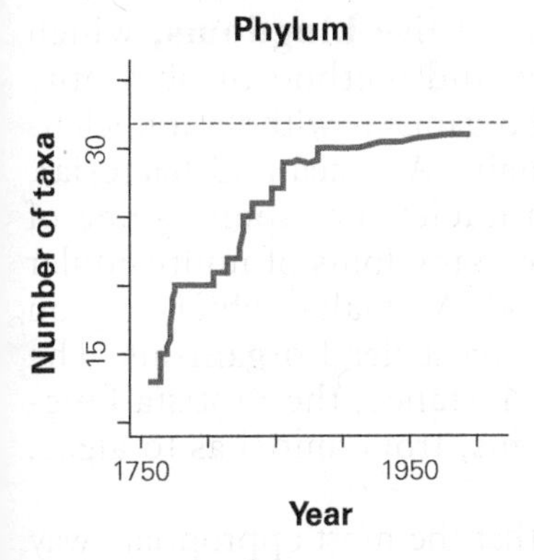

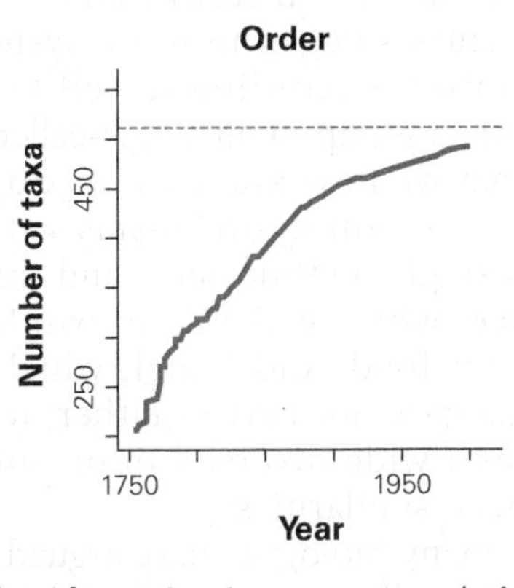

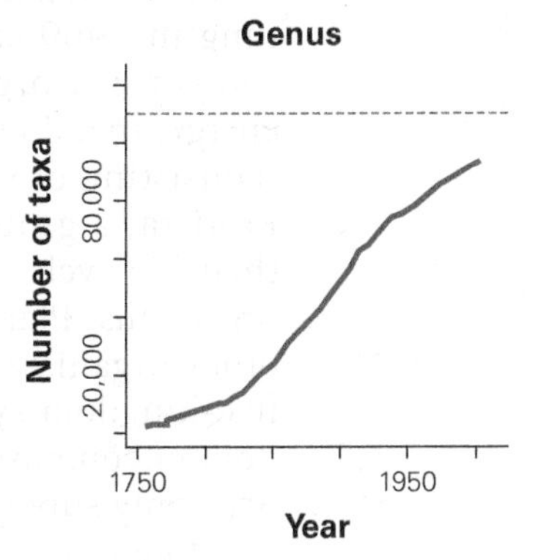

FIGURE 14.2 Estimating biodiversity. Scientists use extrapolation to determine how many species are yet to be discovered. These graphs indicate the number of new groups of animals within each taxonomic level (taxa) over time. Leveling off of the graph indicates that there are probably few new groups left to discover. The dotted lines on each graph provide an estimate of where the graph will level off, based on the historical pattern.

Scientists do know that the diversity of life on Earth today is very different from the diversity in the past. Paleontologists have been able to piece together the history of life by examining fossils and other ancient evidence. Early in this reconstruction, they recognized distinct "dynasties" of groups of organisms that appeared during different periods. The record of the rise and fall of these dynasties has allowed scientists to subdivide life's history into **geologic periods.** Each period is defined by a particular set of fossils. TABLE 14.1 gives the names of major geologic periods, their age and length, and the major biological events that occurred during each period.

Despite the long history of diversity, it is very rare for scientists to describe either a modern or fossil organism that appears unrelated to all other species. In fact, species can be grouped into a few broad categories based on shared characteristics. The most general categories scientists use today are domains.

Domains of Life

Systematists work in the field of **biological classification,** in which they attempt to organize biodiversity into discrete and logical categories. The task of classifying life is somewhat like categorizing books in a library—books can be divided into fiction or nonfiction, and within each of these divisions, more precise categories can be made (for example, nonfiction can be divided into biography, history, science, and so forth). The book-cataloging system used in most public libraries, the Dewey decimal system, is only one way of shelving books. For instance, academic and research libraries use a different system, developed by the U.S. Library of Congress. Librarians use the cataloging system that is appropriate to the collection of books they work with and the needs of the library's users. Just as there are alternative methods of organizing books, there is more than one way to organize biodiversity to meet differing needs.

Biologists have traditionally subdivided living organisms into large groups that share some basic characteristic. Sixty years ago, most biologists divided life into two categories: plants, for organisms that were immobile and apparently made their own food; and animals, for organisms that could move about and relied on other organisms for food.

Many organisms did not fit easily into this neat division of life, so beginning in 1969 scientists began to use a system of five **kingdoms,** which categorized organisms according to cell type and method of obtaining energy: the Monera, a group of all single-celled organisms without a nucleus (consisting of what we now know as the domains Archaea and Bacteria); Protista, a group containing organisms with nuclei that spend some of their life cycle as single, mobile cells; and three kingdoms of multicellular organisms: Plantae, which make their own food; Animalia, which rely on other organisms for food; and Fungi, which digest dead organisms. The five-kingdom system is not perfect either; for instance, the Protista kingdom encompasses a wide diversity of organisms, from amoebas to algae, with only superficial similarities.

More recently, many biologists have argued that the most appropriate way to classify life is according to evolutionary relationships among organisms. Recall that the theory of evolution states that all modern organisms represent the descendants of a single common ancestor. The process of divergence from early ancestors into the diversity of modern species has resulted in the

TABLE 14.1 Geological periods.

The history of life is divided into four major eras, with all but the first era divided into several periods. Periods are marked by major changes in the dominant organisms present on Earth.

Era	Period	Millions of Years Ago	Features of Life on Earth
Precambrian		4500	Formation of Earth. The first organisms, primitive bacteria, appear within 1000 million years.
		543	Life is dominated by single-celled organisms in the ocean. Ediacaran fauna appear at the end of the era.
Paleozoic	Cambrian	495	All modern animal groups appear in the oceans. Algae are abundant.
	Ordovician	439	Life is diverse in the oceans. Cephalopods like squid appear, and trilobites are common.
	Silurian	408	Life begins to invade land. The first colonists are small seedless plants, primitive insects, and soft-bodied animals.
	Devonian	354	Known as the age of fishes. Sharks and bony fish appear. Large trilobites are abundant in the oceans.
	Carboniferous	290	Land is dominated by dense forests of seedless plants. Insects become abundant. Large amphibians appear.
	Permian	251	Early reptiles appear on land. Seedless plants abundant. Coral and trilobites abundant in oceans. Permian ends with extinction of 95% of living organisms.
Mesozoic	Triassic	206	Early dinosaurs, mammals, and cycads appear on land. Life "restarts" in the oceans.
	Jurassic	144	Huge plant-eating dinosaurs evolve. Forests are dominated by cycads and tree ferns.
	Cretaceous	65	Massive carnivorous and flying dinosaurs are abundant. Large cone-bearing plants dominate forests. Flowering plants appear.
Cenozoic	Tertiary	1.8	After the extinction of the dinosaurs, mammals, birds, and flowering plants diversify.
	Quaternary	0	Most modern organisms present.

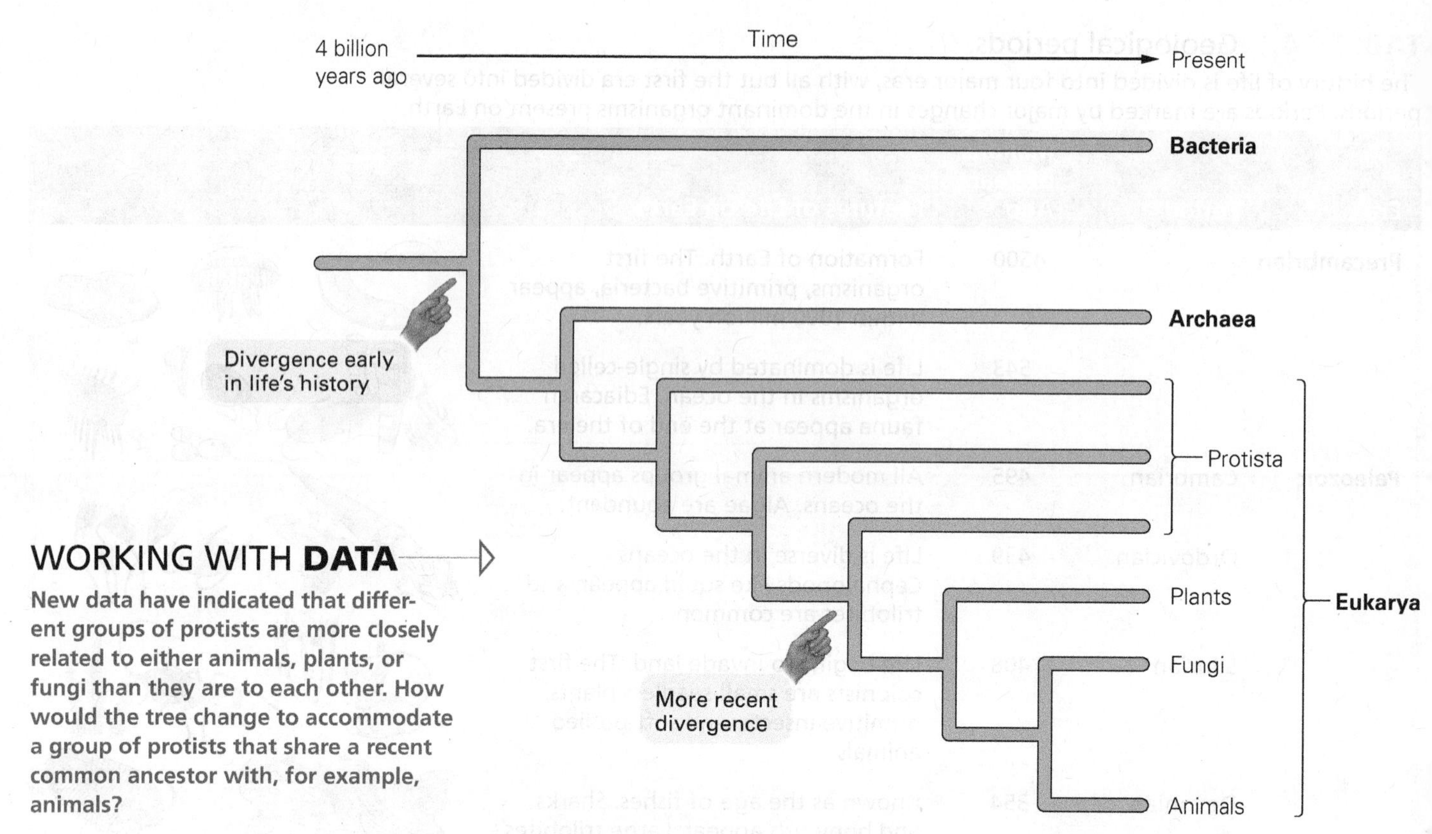

WORKING WITH DATA

New data have indicated that different groups of protists are more closely related to either animals, plants, or fungi than they are to each other. How would the tree change to accommodate a group of protists that share a recent common ancestor with, for example, animals?

FIGURE 14.3 The tree of life. This tree is a simplification of the data generated by rRNA analysis describing the possible evolutionary relationships among living organisms.

modern "tree of life" (**FIGURE 14.3**). When life is classified according to the relationships among organisms, major groupings correspond to divergences that occurred very early in life's history, and minor groupings correspond to more recent divergences. In this text, we keep the basics of the five kingdom system to facilitate our discussion, but biologists no longer consider it evolutionarily accurate.

Determining the evolutionary relationship among *all* living organisms requires comparisons of their DNA. Because each species is unique, the sequence of nucleotides within the DNA of each species is unique. However, because all species share a common ancestor, all organisms also have basic similarities in their DNA sequences. As evolutionary lineages diverged from one another, mutations in DNA sequences occurred independently in each lineage and are now a record of evolutionary relationship among organisms. In short, the DNA sequences of closely related organisms should be more similar than the DNA sequences of more distantly related organisms (**FIGURE 14.4**).

When comparing all modern species, scientists must examine genes that perform a similar function among organisms as diverse as humans, Argentine ants, and cyanobacteria. The DNA sequence that best fits this requirement contains instructions for making ribosomal RNA (rRNA), a structural part of ribosomes. Recall that ribosomes are the factory-like structures found in all cells: they are the organelles where genes are translated into proteins. Each ribosome contains several rRNA molecules in both large and small subunits. A comparison of the DNA that codes for small-subunit rRNAs from myriad organisms yielded a tree diagram similar to that shown in Figure 14.3.

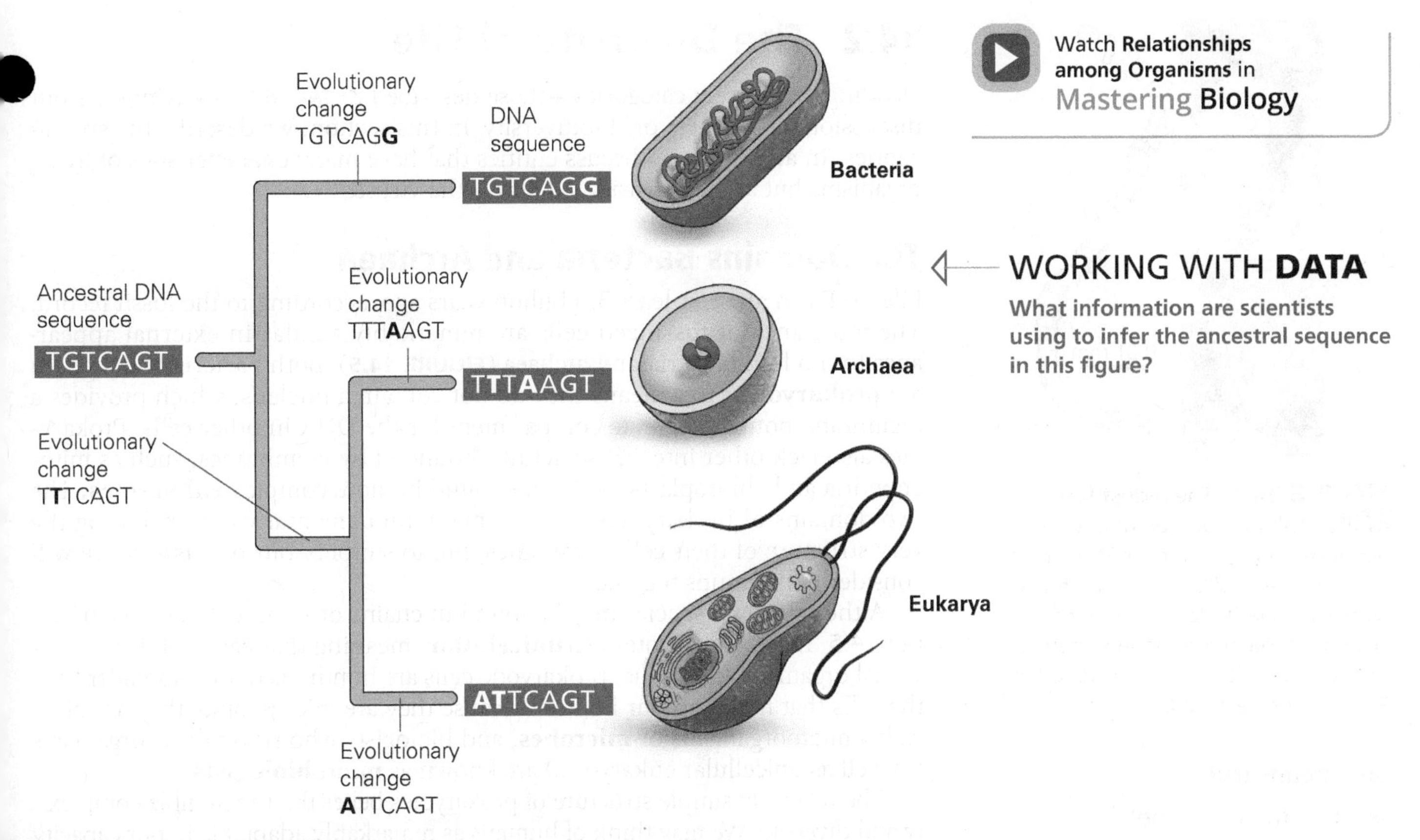

Watch **Relationships among Organisms** in Mastering **Biology**

WORKING WITH DATA
What information are scientists using to infer the ancestral sequence in this figure?

FIGURE 14.4 Determining the relationships among organisms. Comparisons of the DNA sequences that code for ribosomal RNA in various organisms have helped scientists construct the tree of life. The actual comparison consisted of thousands of base pairs; only a few are shown here for simplicity.

Note from Figure 14.3 that three of the kingdoms (Fungi, Animalia, and Plantae) represent relatively recently diverged groups of organisms. Single-celled organisms that do not contain a nucleus for storing DNA were once all placed in the same kingdom but are actually found in two different, quite distinct groups, the Archaea and Bacteria. And the kingdom Protista is a hodgepodge of many very different organisms. To better reflect such biological relationships, starting in about 1990 biologists began to categorize life into three **domains: Bacteria, Archaea,** and **Eukarya.** These domains represent the most ancient divergence of living organisms.

Got It?

1. The history of life can be subdivided into ________ ________ that are defined by a particular set of fossils.
2. Scientists estimate that about ________ percent of species currently present on Earth are still unknown.
3. According to most biologists, the most appropriate way to classify living organisms is by using ________ relationships among organisms.
4. All living things can be sorted into one of three ________, the Bacteria, Archaea, or Eukarya.
5. The next broadest category of biological classification is ________.

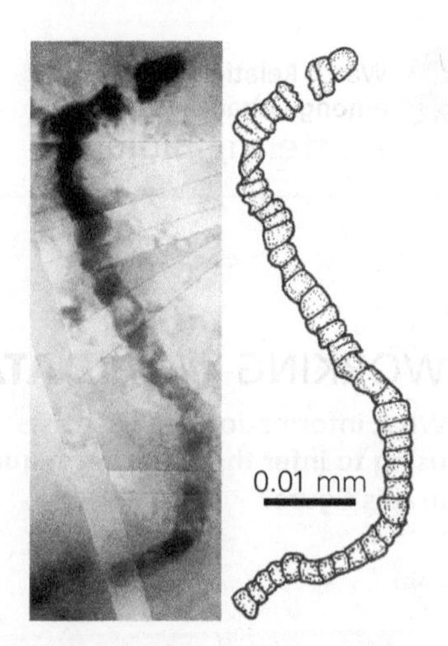

FIGURE 14.5 The oldest form of life. This photograph of a fossil is accompanied by an interpretive drawing showing the fossil's likely living form. Note its similarity to the photo of cyanobacteria in the chapter opener. The fossil was found in rocks dated at 3.465 billion years old.

14.2 The Diversity of Life

Dividing life into six categories—those described in TABLE 14.2—simplifies our discussion of the "best of" biodiversity. In this section, we describe the six categories. In addition, we discuss entities that have many characteristics of living organisms but are not considered *alive*—the viruses.

The Domains Bacteria and Archaea

Life on Earth arose at least 3.6 billion years ago, according to the fossil record. The most ancient fossilized cells are remarkably similar in external appearance to modern bacteria and archaea (**FIGURE 14.5**). Both bacteria and archaea are **prokaryotes;** this means they do not contain a nucleus, which provides a membrane-bound, separate compartment for the DNA in other cells. Prokaryotes also lack other internal structures bounded by membranes, such as mitochondria and chloroplasts, which are found in more complex **eukaryotes.** The two domains of prokaryotes differ in many fundamental ways, including the very structure of their cell membranes, but to simplify our discussion, we will consider both groups together.

eu- means true.

-karyo- means kernel.

Although some species may be found in chains or small colonies, as in Figure 14.5, most prokaryotes are **unicellular,** meaning that each cell is an individual organism. Individual prokaryotic cells are hundreds of times smaller than the cells that make up our bodies. Because they are microscopic, they are often called microorganisms or **microbes,** and biologists who study these organisms (as well as unicellular eukaryotes) are known as **microbiologists.**

The relatively simple structure of prokaryotes belies their incredible complexity and diversity. We may think of humans as remarkably adaptable in our capacity to settle anywhere from the driest deserts to the coldest tundra, but as a group prokaryotes put us to shame (**FIGURE 14.6**). Representatives of these organisms are found anywhere humans live, but some prokaryotes can survive in rocks thousands of meters beneath Earth's surface, others feed on hydrogen sulfide emitted by hydrothermal vents on the bottom of the oceans, and some live in lakes trapped underneath the Antarctic ice sheet. Many of the known Archaea specialize in surviving extreme environments, including high-salt and high-temperature habitats. Perhaps even more remarkably, many prokaryotes in both domains live on or in other organisms, evading their hosts' infection-fighting systems.

Although there are some amazing stories of human survival in the direst conditions—for example, the explorer Ernest Shackleton and his crew, who were trapped for more than two years in the Antarctic, an Uruguayan rugby team who survived a plane crash in the high Andes and hiked to safety, a young

FIGURE 14.6 A diversity of prokaryotes. (a) *Escherichia coli*, an important model organism for basic genetic studies, lives on the partially digested food in our intestines. (b) This artificially colorized electron micrograph shows *Desulforudis audaxviator,* whose name means "bold traveler." It was discovered 2 miles below Earth's surface living on the energy released by radioactive decay. (c) *Halobacterium*, a salt-loving archaean, is found in high populations in these salty ponds. The red pigment in this bacterium's cells is used in photosynthesis.

(a) *Escherichia coli*

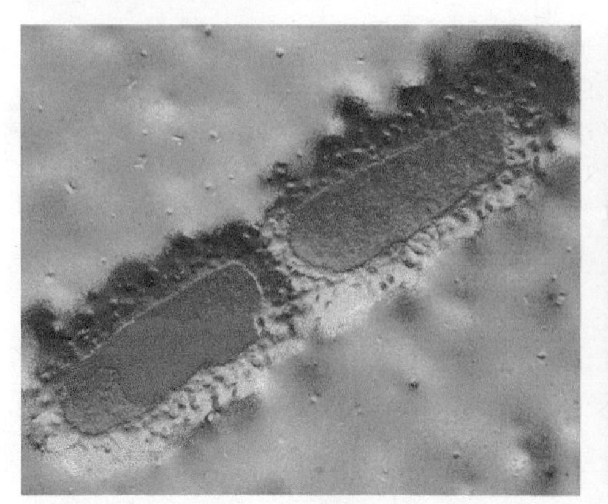

(b) *Desulforudis audaxviator*

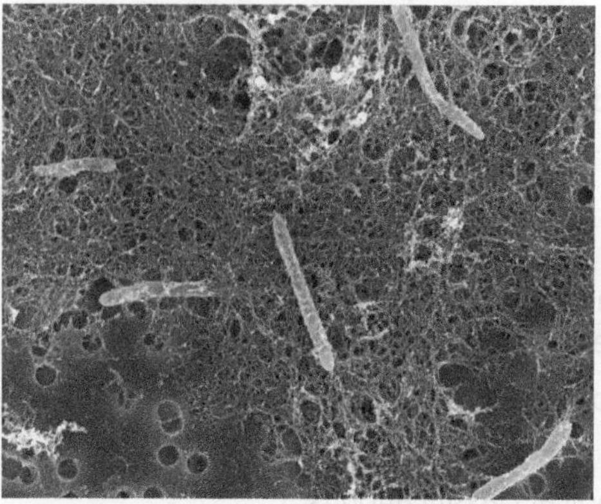

(c) *Halobacterium*

TABLE 14.2 The classification of life.

Until recently, most biologists used the five-kingdom system to organize life's diversity. Now many use a six-category system, which better reflects evolutionary relationships by acknowledging the existence of three major domains as well as four of the kingdoms.

Kingdom Name	Kingdom Characteristics	Examples	Approximate Number of Known Species	Domain Name and Characteristics
Plantae	Multicellular, make own food, largely stationary	Pines, wheat, moss, ferns	300,000	**Eukarya** All organisms contain eukaryotic cells.
Animalia	Multicellular, rely on other organisms for food, mobile for at least part of life cycle	Mammals, birds, fish, insects, spiders, sponges	1,000,000	
Fungi	Multicellular, rely on other organisms for food, reproduce by spores, body made up of thin filaments called hyphae	Mildew, mushrooms, yeast, *Penicillium*, rusts	100,000	
Protista	Mostly single-celled forms, wide diversity of lifestyles, including plant-like, fungus-like, and animal-like types	Green algae, *Amoeba*, *Paramecium*, diatoms, chytrids	15,000	
Bacteria	Prokaryotic, mostly single-celled forms, although some form colonies or filaments	*Escherichia coli*, *Salmonella*, *Bacillus anthracis*, *Anabena*, sulfur bacteria	4000	**Bacteria** Prokaryotes with cell wall containing peptidoglycan. Wide diversity of lifestyles, including many that can make their own food.
Archaea	Prokaryotic, mostly single-celled forms, although some form colonies or filaments	*Thermus aquaticus*, *Halobacteria halobium*, methanogens	1000	**Archaea** Prokaryotes without peptidoglycan and with similarities to Eukarya in genome organization and control. Many known species live in extreme environments.

girl left on a remote island who survived alone for 18 years—prokaryotes also best humans in this regard. Some bacteria have the ability to form **endospores**, resistant structures containing DNA, ribosomes, and a little bit of cytoplasm. Endospores are resistant to extreme temperatures, drying, radiation, and even the vacuum of outer space; but these structures can generate new living cells eons after they form, as demonstrated by spores revived—incredibly—after being recovered from 250-million-year-old salt deposits.

Cyanobacteria (sometimes called blue-green algae) have the unique characteristic of needing no other living organism to survive—they can manufacture all of the components of life simply from water, sunlight, and air through the processes of photosynthesis and nitrogen fixation. In fact, it is thanks to oxygen-generating activities of cyanobacteria starting around 3.6 billion years ago that Earth can now support the diverse living forms that are here today. Humans aren't the only species to have changed the very atmosphere of Earth!

Fine, so prokaryotes are more adaptable and may be more influential than humans. But humans still dominate all other species, so that must be our advantage over these microscopic competitors, right? Scientists who study *Yersinia pestis*, the bacterium that causes bubonic plague and nearly wiped out the human population of Europe in the fourteenth century might disagree, as would those who investigate the bacteria that cause tuberculosis, salmonella, typhoid, and syphilis. Granted, although humans have discovered **antibiotics** that can protect us from these often fatal diseases by killing or disabling the dangerous bacteria that cause them, we didn't create these compounds. In fact, more than half of antibiotics are created by bacteria that produce them as weapons against other, competing bacteria.

Bacterial defense against the viruses that can attack them has resulted in another class of valuable molecules called **restriction enzymes,** proteins that can chop up DNA at specific sequence sites and thus interfere with, or restrict, the growth of these viruses. Restriction enzymes are a key factor that has made much of our advances in genetic technology possible—and evolved within these seemingly simple organisms.

Finally, although Earth now teems with more than 7 billion human beings, prokaryotes are much more numerous. In fact, there are likely more prokaryotes living in your mouth right now than the total number of humans who have ever lived. And scientists still have a relatively weak understanding of exactly how many species of prokaryotes even exist: Some microbiologists estimate that the number of undescribed species could range up to 100 million.

The Origin of the Domain Eukarya

The third domain of life contains all of the organisms that keep their genetic material within a nucleus inside their cells—that is, the eukaryotes. The most ancient fossils of eukaryotic cells are approximately 2 billion years old, nearly 1.5 billion years younger than the oldest prokaryotic fossils.

The earliest eukaryotic cells likely developed from prokaryotes that produced excess cell membrane that folded into the cell itself. In some cells, these internal membranes may have segregated the genetic material into a primitive nucleus and formed channels for translating, rearranging, and packaging proteins, much like modern eukaryotes' endoplasmic reticula and Golgi bodies. According to the **endosymbiotic theory,** the mitochondria and chloroplasts found in eukaryotic cells appear to have descended from bacteria that took up residence inside larger primitive eukaryotes. When organisms live together, the relationship is known as a **symbiosis.** In this case, the symbiosis was mutually beneficial, and over time the cells became inextricably tied together (**FIGURE 14.7**).

When biologist Lynn Margulis first introduced the endosymbiotic hypothesis in the United States in 1981, many of her colleagues were skeptical. But an examination of the membranes, reproduction, and ribosomes of mitochondria and chloroplasts shows clear similarities to the same features in certain

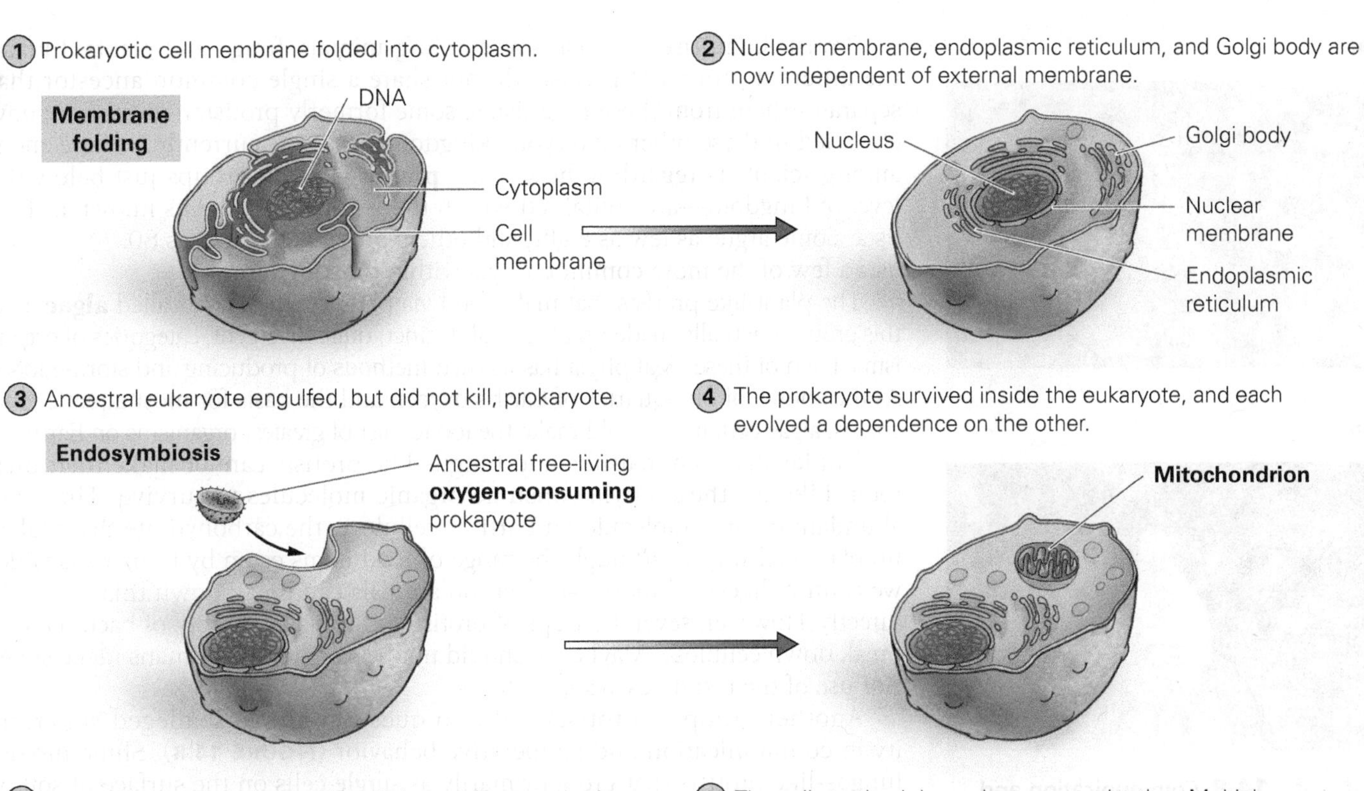

1 Prokaryotic cell membrane folded into cytoplasm.
Membrane folding
DNA
Cytoplasm
Cell membrane
2 Nuclear membrane, endoplasmic reticulum, and Golgi body are now independent of external membrane.
Nucleus
Golgi body
Nuclear membrane
Endoplasmic reticulum
3 Ancestral eukaryote engulfed, but did not kill, prokaryote.
Endosymbiosis
Ancestral free-living
oxygen-consuming
prokaryote
4 The prokaryote survived inside the eukaryote, and each evolved a dependence on the other.
Mitochondrion

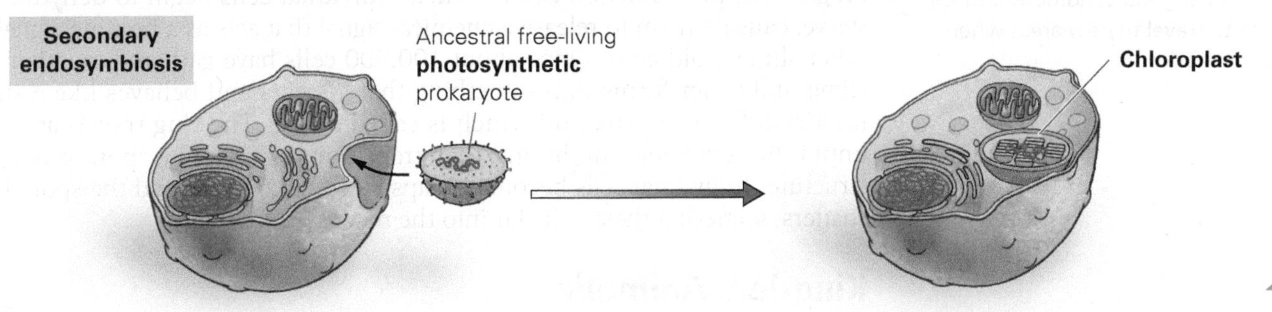

5 In the ancestors of algae and land plants, photosynthetic prokaryotes were engulfed, but not killed.
Secondary endosymbiosis
Ancestral free-living
photosynthetic
prokaryote
6 The cells evolved dependence on each other. Multiple, independent symbioses led to different algal groups.
Chloroplast

FIGURE 14.7 The evolution of eukaryotes. Mitochondria and chloroplasts appear to be descendants of once free-living bacteria that took up residence within an ancient nucleated cell.

VISUALIZE **THIS**

The bacteria ancestors of chloroplasts and mitochondria have a single membrane, but chloroplasts and mitochondria have two membranes. Use the diagram here to explain how the second membrane came about.

bacteria. Even more convincingly, some of the DNA sequence within mitochondria (mtDNA) is similar to the DNA sequence found in certain bacterial species. Today, the endosymbiotic theory is widely accepted as the best explanation for the origin of eukaryotes.

After the evolution of mitochondria, independent endosymbioses between eukaryotic cells and various species of photosynthetic bacteria appeared. These relationships led to the evolution of chloroplasts—one symbiosis led to green algae and land plants, and it appears that other relationships led to the unique chloroplasts in other modern groups, including those found in red and brown algae.

The Protista

What in the five-kingdom system is named **Protista** is made up of the simplest known eukaryotes. Most protists are single-celled creatures, although several have enormous **multicellular** (many-celled) forms. As with Bacteria and Archaea, most Protista remain unknown.

(a)

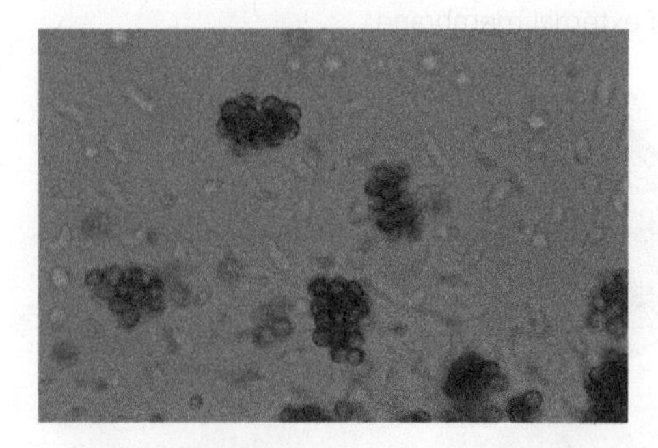

(b)

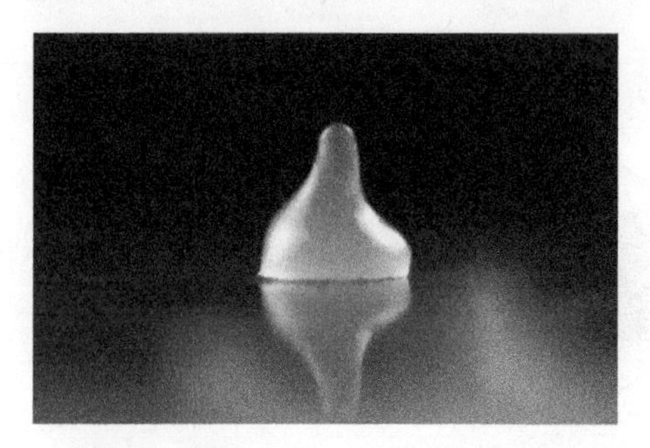

FIGURE 14.8 Communication and cooperation. Cellular slime molds (a) live as single cells in good conditions, but (b) cooperate to travel to new areas when conditions deteriorate.

Protista comprise organisms resembling animals, fungi, and plants. In fact, the different groups of protists do not share a single common ancestor that separates them from these organisms; some formerly protistan groups are now classified in these other eukaryotic kingdoms. There is currently no agreement among scientists regarding how many **phyla**—that is, groups just below the level of kingdom—are contained within the group that remains known as Protista. Some argue as few as eight, and others propose as many as 80. TABLE 14.3 lists a few of the more common phyla within the kingdom.

The plant-like protists that make food via photosynthesis are called **algae,** and this group is actually made up of several distinct, quite divergent, categories of organisms. Each of these algal phyla has its own methods of producing and storing food. As the source of one-quarter of Earth's oxygen and the basis for most aquatic food chains, algae certainly should make the top ten list of greatest organisms on Earth.

Unlike algae, animal-like and fungus-like protists cannot make their own food. Like us, these phyla consume organic molecules to survive. The most abundant organic molecule on Earth is cellulose, the carbohydrate that makes up plant cell walls. Although the range of food items eaten by humans is wide, we cannot digest cellulose—in fact, no animals can break down this molecule directly. However, several groups of protists (as well as a variety of bacteria) can break down cellulose. Maybe we should not be so sure that humans make superior use of the resources available to us.

Another group of protists call into question humans' alleged superiority in communication and cooperative behavior (**FIGURE 14.8**). Slime molds, fungus-like protists that grow primarily as single cells on the surface of soil or dead plant material, can call for reinforcements and congregate when times are tough. This phenomenon occurs when individual cells begin to dehydrate or starve, causing them to release a chemical signal that acts as a homing signal for other slime mold cells. Once about 100,000 cells have gathered together, the slime mold transforms into something that appears and behaves like a single multicellular organism, and which is called a slug. The slug travels as a unit until it finds suitable conditions, then transforms again into a spore-producing structure. Individual cells become encapsulated in cellulose and the spore head shatters, scattering these cells far into the new environment.

Kingdom Animalia

From the origin of the first prokaryote until approximately 1.2 billion years ago, life on Earth consisted only of single-celled creatures. Then multicellular organisms first began to appear in the fossil record. The ancient, many-celled creatures of 600 million years ago, called the *Ediacaran fauna*, were organisms unlike any modern species and included giant fronds and ornamented disks (**FIGURE 14.9**).

FIGURE 14.9 Ediacaran fauna. This reconstruction of multicellular organisms that lived before the Cambrian explosion is based on 580-million-year-old fossil remains.

TABLE 14.3 The diversity of Protista.

Protista contains animal-like, fungus-like, and plant-like organisms. A sampling of protistan phyla is described here.

The Protista: Common Names and Characteristics of Select Phyla		Example
Animal-like protists	**Ciliates** Free-living, single-celled organisms that use hair-like structures to move.	*Paramecium*
	Flagellates Use one or more long whip-like tail for locomotion. Most are free living, but some cause disease by infecting human organs.	*Giardia*
	Amoebas Flexible cells that can take any shape and move by extending pseudopodia ("false feet").	*Amoeba*
Fungus-like protists	**Slime molds** Feed on dead and decaying material by growing net-like bodies over a surface or by moving about as single amoeba-like cells.	*Physarum*
Plant-like protists	**Diatoms** Single cells encased in silica (glass).	Diatom
	Brown algae Large multicellular seaweeds.	Kelp
	Green algae Closest relatives to land plants. Single-celled to multicellular forms.	*Volvox*

anima- means breath or soul.

Biologists are unsure which of these ancient species is the common ancestor of modern **animals**—defined as multicellular organisms that make their living by ingesting other organisms and that are motile (have the ability to move) during at least one stage of their life cycle. What is clear from the fossil record is that by about 530 million years ago, *all* modern animal groups had emerged.

The remarkably sudden appearance of the modern forms of animals—a period comprising little more than 1% of the history of life on Earth—is referred to as the **Cambrian explosion,** named for the geologic period during which this proliferation occurred. Some scientists hypothesize that the evolution of the animal lifestyle itself—that is, as predators of other organisms—led to the Cambrian explosion.

It can be difficult to conceive of an animal as complex as a human evolving from a simple eukaryotic ancestor. However, humans are not very different from other eukaryotes. When the first cell containing a nucleus appeared, all of the complicated processes that take place in modern cells, such as cell division and cellular respiration, must have evolved. When the first multicellular animals appeared, many of the processes required to maintain these larger organisms, such as communication systems among cells and the formation of organs and organ systems, arose. Although a human and a sea star appear to be very different, the way we develop and the structures and functions of our cells and common organs are nearly identical.

In fact, there appears to be surprisingly little genetic difference between humans and starfish; most of that difference occurs in a group of genes that control **development,** the process of transforming from a fertilized egg into an adult creature. In addition, the amount of time since the divergence of the major evolutionary lineages of animals—530 million years—is quite a long time for dramatic differences among species in different phyla to evolve. Put in more familiar terms, if the time since the Cambrian explosion were one 24-hour day all of human history would fit in only the last 2 seconds.

The immense amount of time that has passed since the Cambrian explosion saw the rise and fall of the dinosaurs; meanwhile, mammals and starfish continued to diverge. Zoologists, scientists who study the kingdom **Animalia,** now describe more than 25 modern phyla in this kingdom. Some of the more well-known groups are illustrated in TABLE 14.4.

Most people typically picture mammals, birds, and reptiles when they think of animals, but species with backbones (including these animals as well as fish and amphibians), known as **vertebrates,** represent only 4% of the total species in the kingdom. Among these other vertebrates, some provide competition for the "greatest species." Many of the traits we think of as uniquely human are found in other vertebrates—and sometimes are even better developed. Beavers have the ability to dramatically modify the environment to fit their needs. Numerous vertebrates, including crows and chimpanzees, use technology in the form of tool making to better exploit the resources that surround them. Some vertebrates even appear to have a creative impulse—as evidenced by certain whale songs and the nest of the bowerbird. Dolphins have more sophisticated methods of communication, using not only auditory and visual signals like in humans but also their senses of touch and smell. And certain fish have the incredible ability to switch sex to maximize opportunities for reproduction (**FIGURE 14.10**).

(a)

(b)

(c)

(d)

FIGURE 14.10 Vertebrate skills. (a) Beavers can radically transform an environment by removing large numbers of trees and damming streams to create ponds. (b) This crow is demonstrating advanced tool use by shaping a stick to use as a fishing pole for grubs. (c) This bowerbird nest is built to attract a female, but the décor shows an artistic flair that rivals human creativity. (d) Clownfish can switch from male to female depending on the number of individuals of each sex in a group.

(a) Ant (*Linepithema humile*)

(b) Jellyfish (*Turritopsis nutricula*)

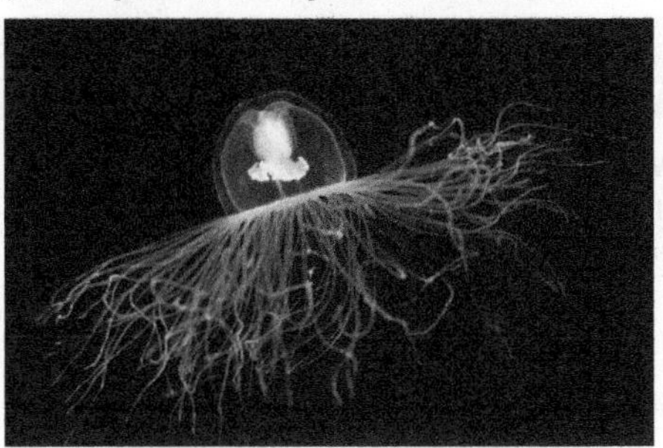

(c) Cuttlefish (*Sepia officinalis*)

FIGURE 14.11 Invertebrates. Most animals are invertebrates, including these examples.

The remaining 96% of known organisms in Animalia are the **invertebrates** (animals without backbones, as shown in **FIGURE 14.11**). In fact, the vast majority of *all* multicellular organisms on Earth are invertebrates. As a result of this diversity, there are innumerable examples of invertebrate capabilities that can outshine those of humans. Argentine ants rival humans in terms of their distribution across Earth's surface, being found on six continents and numerous oceanic islands, but outnumber humans by orders of magnitude. The jellyfish *Turritopsis nutricula* is the only animal known to be able to revert to an earlier developmental stage after becoming sexually mature, making it theoretically immortal. And octopi and their relatives have the ability to change their shape and skin to camouflage themselves in a variety of environments. Some **zoologists** estimate that there may be as many as 30 million unknown invertebrate species, especially in the oceans.

in- means without.

zoo- means pertaining to animals.

TABLE 14.4 Phyla in the kingdom Animalia.

A sampling of the diversity of animals. The rows are arranged generally in order of appearance in evolutionary time—from the more ancient sponges to the more recent chordates.

Kingdom Animalia: Major Phyla	Description	Example
Porifera	Fixed to underwater surface and filter bacteria from water that is drawn into their loosely organized body cavity.	Sponge
Cnidaria	Radially symmetric (like a wheel) with tentacles. Some are fixed to a surface as adults (e.g., corals), and others are free-floating in marine environments (e.g., jellyfish).	Anenome

—Continued next page

TABLE 14.4 Phyla in the kingdom Animalia. *Continued—*

Kingdom Animalia: Major Phyla	Description	Example
Platyhelminthes	Flatworms with a ribbon-like form. Live in a variety of environments on land and sea or as parasites of other animals.	Tapeworm
Mollusca	Soft-bodied animals often protected by a hard shell. Body plan consists of a single muscular foot and body cavity enclosed in a fleshy mantle. Phylum includes snails, clams, and squid.	Octopus
Annelida	Segmented worms. Body divided into a set of repeated segments.	Earthworm
Nematoda	Roundworms with a cylindrical body shape. Very diverse and widespread in many environments.	Roundworm
Arthropoda	Segmented animals in which the segments have become specialized into different roles (such as legs, mouthparts, and antennae). Body completely enclosed in an external skeleton that molts as the animal grows. Phylum includes insects and spiders as well as crabs and lobsters.	Shrimp

—Continued next page

TABLE 14.4 Phyla in the kingdom Animalia. *Continued—*

Kingdom Animalia: Major Phyla	Description	Example
Echinodermata	Slow-moving or immobile animals without segmentation and with radial symmetry. Internal skeleton with projections gives the animal a spiny or armored surface.	Sea urchin
Chordata	Animals with a spinal cord (or spinal cord-like structure). Includes all large land animals, as well as fish, aquatic mammals, and salamanders.	Duck-billed platypus

Kingdom Fungi

Although our ignorance about the diversity of animals is great, another kingdom of multicellular eukaryotes is even less well known—the fungi. Like plants, fungi are immobile, and many produce fruit-like organs that disperse **spores,** cells that are analogous to plant seeds in that they germinate into new individuals. In fact, because of this similarity, for generations biologists placed fungi in the plant kingdom. However, whereas plants make their own food via photosynthesis, fungi feed on other organisms by secreting digestive chemicals into their environment; these chemicals break down complex organic food sources into small molecules, which the fungi then absorb. Because they rely on other organisms for their food sources, fungi are more like animals than plants. In fact, DNA sequence analysis by **mycologists,** biologists who study fungi, indicates that Fungi and Animalia are more closely related to each other than either kingdom is to the plants.

myco- means fungal.

The mushroom you probably think of when you imagine a fungus is a misleading image of the kingdom. Most of the functional part of fungi is made up of very thin, stringy material called **hyphae,** which grows over and within its food source—a mushroom is simply the reproductive organ that appears on the surface of the food (**FIGURE 14.12**). The string-like form of hyphae maximizes

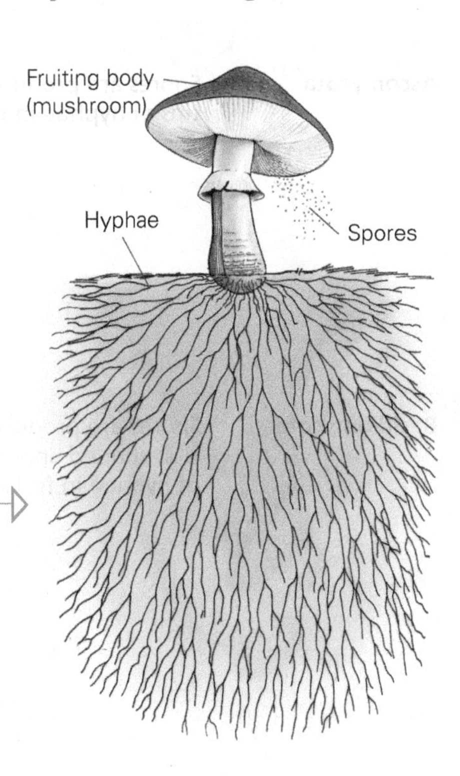

VISUALIZE **THIS**

Why is a mushroom produced outside the food resource for the hyphae?

FIGURE 14.12 Fungi. Hyphae can extend over a large area. The familiar mushroom, as well as the fruiting structures of less-familiar fungi, primarily functions only as a method of dispersing spores.

the surface over which feeding takes place. Some fungi feed on living tissue, while others decompose dead organisms. This latter role of fungi is key to recycling nutrients in biological systems.

The phyla of fungi are generally distinguished from each other by their method of spore formation (TABLE 14.5). However, convergent evolution, in which unrelated species take similar forms because of similar environments,

TABLE 14.5 Fungal diversity.

Fungi are classified into phyla based on their mode of spore production. The most common phyla are listed here.

Kingdom Fungi: Major Phyla	Description	Example
Zygomycota	Sexual reproduction occurs in a small resistant structure called a zygospore. Most reproduction is asexual—directly via mitosis.	*Rhizopus stolonifera*—bread mold
Glomeromycota	Probably do not reproduce sexually, spores formed in various manners. Unique in that all members of this phylum are mycorrhizal.	Mycorrhizal fungi
Ascomycota	Spores are produced in sacs on the tips of hyphae in fruiting structures.	Morel
Basidiomycota	Spores are produced in specialized club-shaped appendages on the tips of hyphae in fruiting structures.	*Amanita muscaria*—poisonous fly agaric

has led to the appearance of similar body shapes and lifestyles—which we call "fungal forms"—among these different phyla. One of the most commercially important fungal forms is **yeast,** a single-celled type of fungi, which lives in liquids and is found in at least two different fungal phyla. Unlike most eukaryotes, including ourselves, yeasts have a metabolic process, called alcoholic **fermentation,** that can extract significant energy from carbohydrates even in the absence of oxygen. The activity of yeasts in oxygen-poor but sugar-rich environments results in the formation of ethyl alcohol. The metabolism of yeasts within flour batter also leads to the production of carbon dioxide, which is trapped by wheat protein fibers and allows dough to "rise" during bread making.

Another form known as **mold** is found in all fungal phyla and is also commercially important. This quickly reproducing, fast-growing form can spoil fruits and other foods, although some produce certain types of flavorful cheese, including blue cheese and Camembert, as they are "spoiling" milk. The antibiotic penicillin was derived from a species of mold. In fact, about one-third of the bacteria-killing antibiotics in widespread use today are derived from fungi, which produce them because they compete with bacteria for food sources.

Although we may think humans have the monopoly in forcing other species to do our will, certain fungi have this ability as well (**FIGURE 14.13**). *Ophiocordyceps unilateralis* is a fungal parasite of ants that causes these animals to change their behavior completely. The ants the fungus infect normally live in trees—but when infected with the fungus, the ant turns into a virtual zombie. It drops from the tree, climbs the stem of a plant, and firmly attaches itself to a leaf with its jaws. There, the ant is slowly consumed from within by the fungus. When the fungus eventually produces a mushroom, it is now in a prime spot for maximal distribution of its spores. That ant is transformed into a beast of burden in the most dramatic way.

Fungi may also be the original farmers. A **mycorrhiza** is a symbiotic relationship between fungal hyphae and plant roots that benefits both partners (**FIGURE 14.14**). Perhaps 90% of all plants have mycorrhizae. A relationship

FIGURE 14.13 Slave and master. The fungus growing out of the head of this ant controlled the ant's behavior before the ant died—inducing it to crawl up a stem to maximize the fungus's dispersal.

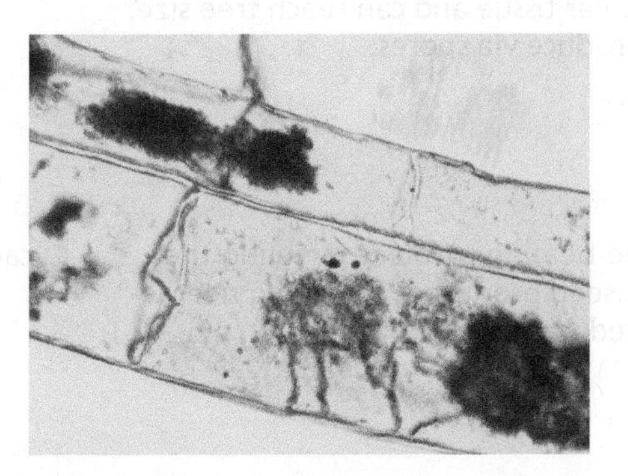

FIGURE 14.14 Fungi as farmers. Mycorrhizal fungi (seen here as dark bodies within the cells of a plant's roots) benefit by obtaining carbohydrates produced by the plant's photosynthesis. Because plants are better able to absorb certain nutrients thanks to these fungi, they can survive on poorer soils. Like human farmers, the fungus's activities benefit the growth of their own food.

between plants and fungi is evident even from the oldest fossils of land plants, and the symbiosis may have evolved when some species of fungi that were parasitizing plants did not kill these plants outright. Instead, by allowing their hosts to survive on less-fertile soils, these less-deadly fungi did millions of years ago what human farmers learned to do only in the last 20,000 years: expand and maintain habitat for a species that is their food source.

Kingdom Plantae

The kingdom **Plantae** consists of multicellular eukaryotic organisms that make their own food via photosynthesis. Plants have been present on land for more than 400 million years, and their evolution is marked by increasingly effective adaptations to the terrestrial environment (TABLE 14.6).

The first plants to colonize land were small and close to the ground. Their small size was necessary, because they had no way to transport water much distance from the surface of the soil. The evolution of **vascular tissue,** made up of specialized cells that can transport water and other substances, allowed plants to reach tree-sized proportions and to colonize much drier areas. The evolution of **seeds,** structures that protect and provide a food source for young plants, represented another adaptation to dry conditions on land.

TABLE 14.6 Plant diversity.

The four major phyla of plants are listed here in order of their appearance in evolutionary history.

Kingdom Plantae: Major Phyla	Description	Example
Bryophyta	Mosses. Lacking vascular tissue, these plants are very short and typically confined to moist areas. Reproduce via spores.	Moss
Pteridophyta	Ferns and similar plants. Contain vascular tissue and can reach tree size. Reproduce via spores.	Fern
Coniferophyta	Cone-bearing plants resembling the first seed producers, including needle-leafed trees.	Cycad
Anthophyta	Flowering plants. Seeds produced within fruits, which develop from flowers. Advances in vascular tissues and chemical defenses contribute to their current dominance on Earth.	Orchid

Several species of nonflowering seed plants would definitely rival humans for the title of Earth's greatest species. Bristlecone pine trees are among the longest-living individual organisms on Earth; the oldest known bristlecone sprouted more than 5000 years ago, at about the same time as the beginning of the Egyptian dynasty. The largest organisms on Earth are Giant Sequoias at as much as 94 m (300 ft) in height and 9 m (29 ft) in diameter—16 times bigger than the largest blue whale, the largest animal ever known to exist.

Most modern plants belong to a group that appeared about 140 million years ago, the **flowering plants.** Like their ancestors, flowering plants possess vascular tissue and produce seeds; in addition, these plants evolved a specialized reproductive organ, the flower. More than 90% of the known plant species are flowering plants (**FIGURE 14.15**).

From about 100 million to 80 million years ago, the number of distinct groups, or families, of flowering plants increased from around 20 to more than 150. During this time, flowering plants became the most abundant plant type in nearly every habitat. Among the eukaryotic kingdoms, plants are also the most well-described; many botanists believe that the number of unknown plant species is relatively small—probably a few thousand.

The rapid expansion of flowering plants is called **adaptive radiation**—the diversification of one or a few species into a large and varied group of descendant species. Adaptive radiation typically occurs either after the appearance of an evolutionary breakthrough in a group of organisms or after the extinction of a competing group. The radiation of animals during the Cambrian explosion may have been a result of the evolution of predation, and the radiation of mammals beginning about 65 million years ago occurred after the extinction of dinosaurs. The radiation of flowering plants must have been due to an evolutionary breakthrough—some advantage they had allowed them to assume roles that were already occupied.

Plant biologists, or **botanists,** still debate which traits of flowering plants give them an advantage over nonflowering plants. Some believe that the reproductive characteristics of flowering plants led to their radiation—including the ability to use other species, something we might think of as unique to humans. Flowering plants rely on the assistance of animals in transferring gametes for the process of **pollination.** For this strategy to work, plants have evolved forms

botan- means pertaining to plants.

(a) Hammer orchids

(b) Viburnum

(c) Curare vine

FIGURE 14.15 Diversity of flowering plants. These plants may rival humans for greatest species. (a) Hammer orchids manipulate their pollinators and provide them no reward. (b) Like all flowering plants, this viburnum acts like a mammal in that it only commits major resources to an egg that is fertilized. (c) The curare vine produces a deadly toxin that rivals humans' most dangerous chemical weapons.

to attract reliable pollination partners, including producing vivid and fragrant **petals** that draw insects to the plant's sugar-rich nectar. As an insect or animal becomes a reliable visitor to a flower type, the population of plants typically evolves to become more effective at "loading" the visitor with the male gamete, pollen. In turn, the plant population's modifications drive the evolution of the pollinator population, so that the pollinators become exclusive in their relationship to the flowers—by causing selection in the pollinators' traits, the flowering plants ensure that their pollen is always delivered to other flowers of the same species. The pattern of both species in a relationship evolving adaptations to each other is known as **coevolution,** and has resulted in the remarkably tight fit between certain plants and their pollinators. This has been occurring for at least 140 million years, and may have been a key factor in increasing the diversity of flowering plants.

Some flowering plant species have taken advantage of this coevolutionary process to create pollination "slaves" of male insects that will deliver pollen between flowers that look like female insects (see Figure 14.15a). The males receive nothing from this interaction, and in fact, waste a lot of time and energy in the process. Humans may have service animals, but we provide them with food, water, and shelter. It is unclear that we have ever been as effective as these flowers in tricking another animal to do our will with no reward at all!

Another trait that may have led to success among flowering plants is **double fertilization,** wherein sperm from a single pollen grain fertilize both the egg and a specialized food-producing tissue (**FIGURE 14.16**). This is similar to how mammals like humans reproduce, in that energy is only committed to offspring in large amounts after successful fertilization. This strategy protects individuals from wasting energy in the case of failed fertilization. Flowering plants evolved this conservative reproductive strategy millions of years before mammals even appeared. Humans now benefit from flowering plants' efficient seed- and fruit-making strategy—the success of the ancestors of corn, wheat, and soybeans provided the basis for human agriculture and thus the very civilizations of humans that can produce towering buildings and monuments.

VISUALIZE **THIS**

The embryo within a seed is sometimes described as equivalent to an astronaut. How is a seed plant embryo like a space traveler? How do you think it might be different?

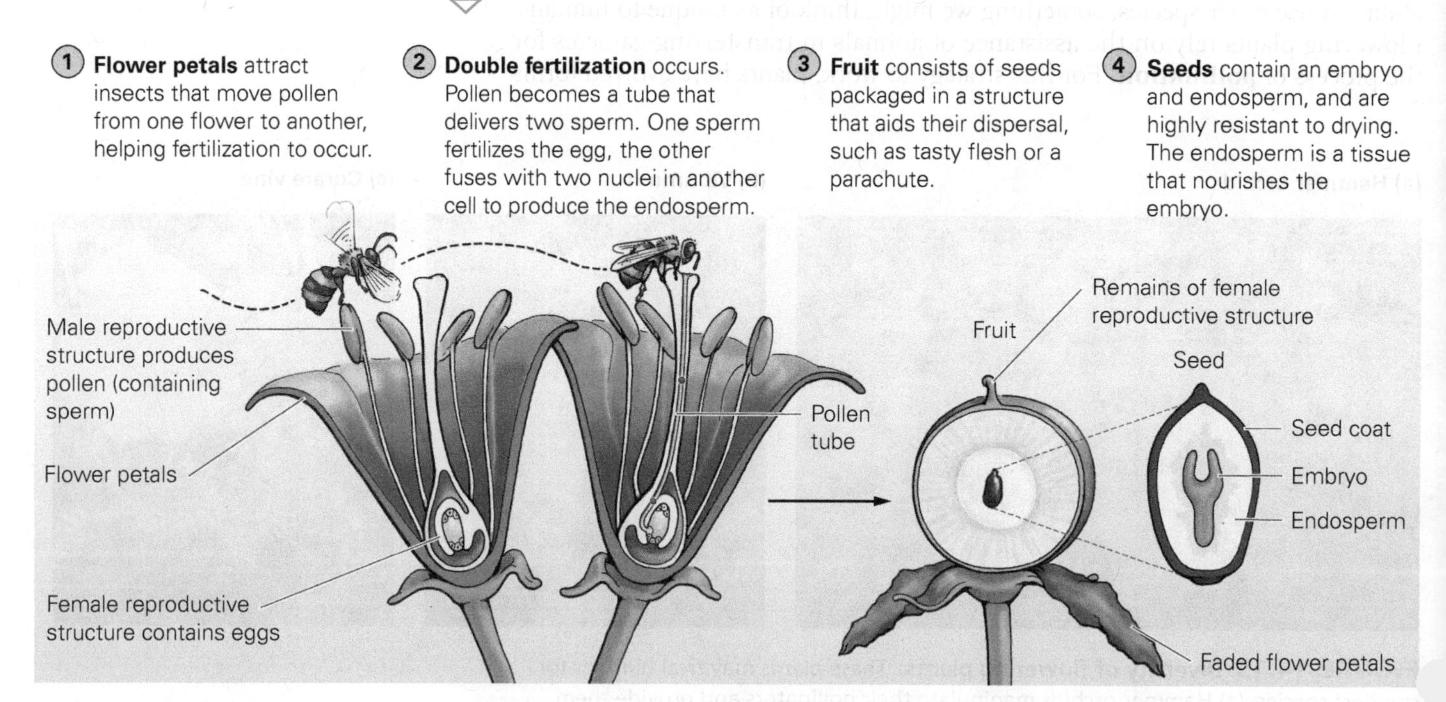

① **Flower petals** attract insects that move pollen from one flower to another, helping fertilization to occur.
② **Double fertilization** occurs. Pollen becomes a tube that delivers two sperm. One sperm fertilizes the egg, the other fuses with two nuclei in another cell to produce the endosperm.
③ **Fruit** consists of seeds packaged in a structure that aids their dispersal, such as tasty flesh or a parachute.
④ **Seeds** contain an embryo and endosperm, and are highly resistant to drying. The endosperm is a tissue that nourishes the embryo.
Male reproductive structure produces pollen (containing sperm)
Flower petals
Female reproductive structure contains eggs
Pollen tube
Fruit
Remains of female reproductive structure
Seed
Seed coat
Embryo
Endosperm
Faded flower petals

FIGURE 14.16 Sexual reproduction in flowering plants. The reproductive differences between flowering plants and other plants, including the production of fruit for dispersal, may have led to their adaptive radiation.

Flowering plants may also beat humans in their ability to defer and confound natural enemies. Because plants cannot physically escape from their predators, natural selection has favored the production of predator-deterring toxins. Most of these toxins are made via secondary reactions to primary biochemical pathways—a technological innovation, in a way. The chemicals produced are known as **secondary compounds.** For instance, curare vines produce toxins that block the connection between nerves and muscles. Animals that get this toxin, called *curarine,* in their bodies become paralyzed and can do little damage to the vine. Curarine is produced via a secondary reaction of the process of amino acid synthesis. In this case, natural selection must have favored genetic variations leading to production of not only normal amino acids but also this toxic secondary compound.

Other secondary compounds you may be familiar with are the drugs nicotine and cocaine, the poisons ricin and cyanide, and the drugs aspirin and digitalis. Although humans have learned how to extract these toxins for our own uses—sometimes not so civilized ones—flowering plants have been engaged in this form of technological chemical warfare for millions of years.

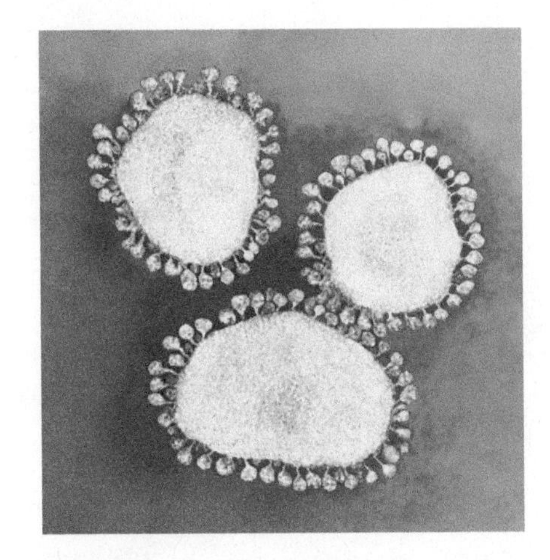

FIGURE 14.17 Viruses. The severe acute respiratory virus (the cause of severe acute respiratory syndrome, SARS), viewed under very high magnification. Visible here are protein capsids, studded with receptors that bind to host cells.

Not Quite Living: Viruses

Any accounting of biodiversity must also consider viruses, organic entities that interact with living organisms but are not quite alive themselves. Viruses are considered nonliving because they are unable to maintain homeostasis (Chapter 2) and are incapable of growth or reproduction without the assistance of another organism. Because they are not quite living, many biologists would leave viruses off of a greatest species discussion—but viruses are remarkable entities that have the power to take down human civilizations.

A virus typically consists simply of a strand of DNA or RNA surrounded by a protein shell, called a **capsid.** Some viruses are also enveloped in a membranous coat. Viruses are basically rogue pieces of genetic material that must use the cells of other organisms to reproduce (**FIGURE 14.17**). When a virus enters a host cell, it uses the transcription machinery of the cell to make copies of its DNA or RNA. The replicated viral genomes then use the ribosome, transfer RNA, and amino acids of the host to make new protein capsules and other polypeptides. If the virus is enveloped, it even uses the host cell's membrane to make envelopes for the daughter viruses.

Once a cell has been infected by a virus, it cannot perform its own necessary functions and will die. But before the infected cell dies, dozens of duplicated viruses can be released and move on to infect other cells. Some of the most devastating pathogens of humans are viruses, including polio, smallpox, HIV, and influenza. HIV is especially troublesome because it destroys immune system cells, which are essential for fighting off infection. An HIV-infected individual gradually loses the ability to combat other diseases, including the HIV infection itself.

Got It?

1. The two prokaryotic domains are __________ and __________.
2. The idea that eukaryotes evolved from mutually beneficial relationships among cells is known as the __________ theory.
3. Protists include organisms that are both unicellular and __________ and that have plant-like, __________, or fungus-like lifestyles.
4. The largest group of plants on Earth today produces reproductive structures called __________.
5. A virus consists primarily of a segment of genetic material that cannot __________ without infecting a living cell.

14.3 Learning about Species

The living world is amazingly diverse, and our knowledge of it is only fragmentary. Many biologists are attracted to the study of biological diversity for just this reason: because the variety of life is remarkable, fascinating, and largely an unexplored frontier in human knowledge. Once a new species is identified, biologists seek to place it in context with other living organisms. This may be the one place where humans have a distinct advantage over nearly all other species—in the ability to use the scientific method to learn about the past and predict the future.

Reconstructing Evolutionary History

Although we can imagine many different classification schemes, most biologists argue that one reflecting evolutionary relationships, called an **evolutionary classification,** is more useful than one based on more superficial similarities.

The relationship between certain reptiles and birds can help illustrate this point. Both Komodo dragons and alligators are large reptiles, and both are ambush predators that lay in wait for their prey. Both animals walk on four legs, are covered with scales, and are cold-blooded. Birds, on the other hand, are warm-blooded, covered with feathers, and have wings (**FIGURE 14.18**). Not surprisingly, a nonevolutionary classification would place alligators with Komodo dragons and other cold-blooded, scaly reptiles. However, overwhelming evidence from morphology, physiology, and genetics indicates that alligators are more closely related to birds than they are to Komodo dragons. If scientists wish to know more about the anatomy, physiology, behavior, and ecology of Komodo dragons, they should not use what is known about alligators—a much more distant relative—as a stepping-off point. Instead, they should focus on more closely related lizards; and it turns out scientists can learn a lot about birds by studying alligators.

Developing Evolutionary Classifications. Evolutionary classifications are based on the principle that the descendants of a common ancestor are likely to share any biological trait that first appeared in that ancestor. For example, this principle has been used to uncover the evolutionary relationship among different species of sparrows.

All species have a two-part name—the first part of the name indicates the **genus,** or broader group of relatives to which the species belongs; the second part is specific to a particular species within that genus. In effect, a species name is like a full name in which the "family group" is designated by the first name and the individual in that particular family is identified by the second part of

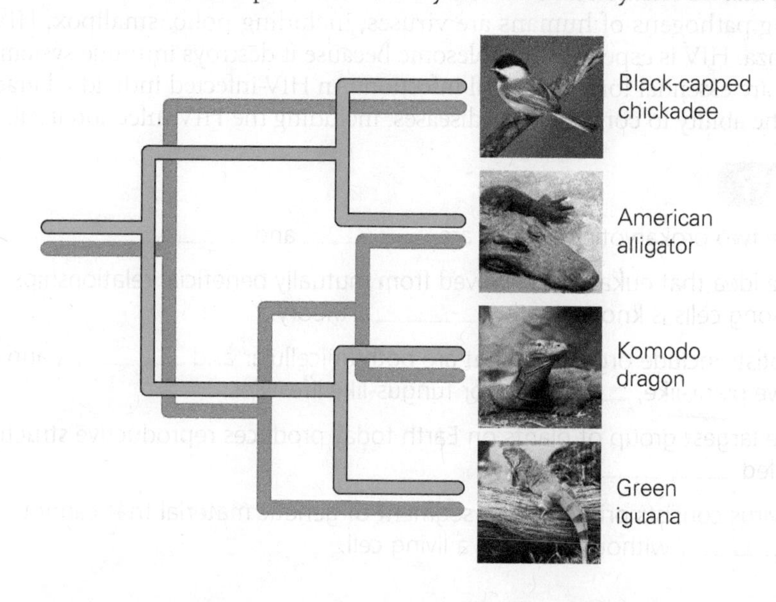

FIGURE 14.18 The challenge of biological classification. Evolutionary relationships are sometimes surprising. The phylogenetic tree in pink illustrates a relationship that you might expect based on the superficial similarities and differences among birds and various reptiles. However, the tree in green indicates the relationship that is best supported by the data.

the name. Both parts of the name are always used when identifying a particular species, although if we are discussing a group of species in the same genus, the genus name may be abbreviated to just its first letter.

FIGURE 14.19 illustrates a hypothesized **phylogeny**—the evolutionary relationship—of sparrow species in the genus *Zonotrichia*. Scientists have used a technique called **cladistic analysis,** an examination of the variation in traits relative to a closely related species, to determine this phylogeny. For example, if we examine just the heads of the four sparrow species compared with their relative, the dark-eyed junco, we see that three of the related species—all but the Harris's sparrow—have dark and light alternating stripes on the crown of their heads. This observation seems to indicate that crown striping evolved early in the radiation of these sparrows. Among the three species with crown stripes, two have seven stripes and the golden-crowned sparrow has only three stripes. An increase in the number of stripes appears to have evolved after the original striped crown pattern. Finally, of the two species with seven crown stripes, only one has evolved a distinct white patch on its throat—the aptly named white-throated sparrow. In the case of these four sparrows, it appears that every step in their radiation involved a visible change in their appearance.

Unfortunately, reconstructing evolutionary relationships is not as simple as the sparrow example suggests. Descendant species may lose a trait that evolved in their ancestor, or unrelated species may acquire identical traits via convergent evolution. You can even see convergent evolution in the phylogeny in Figure 14.19; golden-crowned sparrows, like the white-throated species, have a patch of golden feathers on their heads that appears to have evolved independently. The existence of convergent traits complicates the development of evolutionary classifications.

Testing Evolutionary Classifications. An evolutionary classification is a hypothesis of the relationship among organisms. It is difficult to test this hypothesis directly—scientists have no way of observing the actual evolutionary events that gave rise to distinct organisms. However, scientists *can* test their hypotheses by using information from both fossils and living organisms.

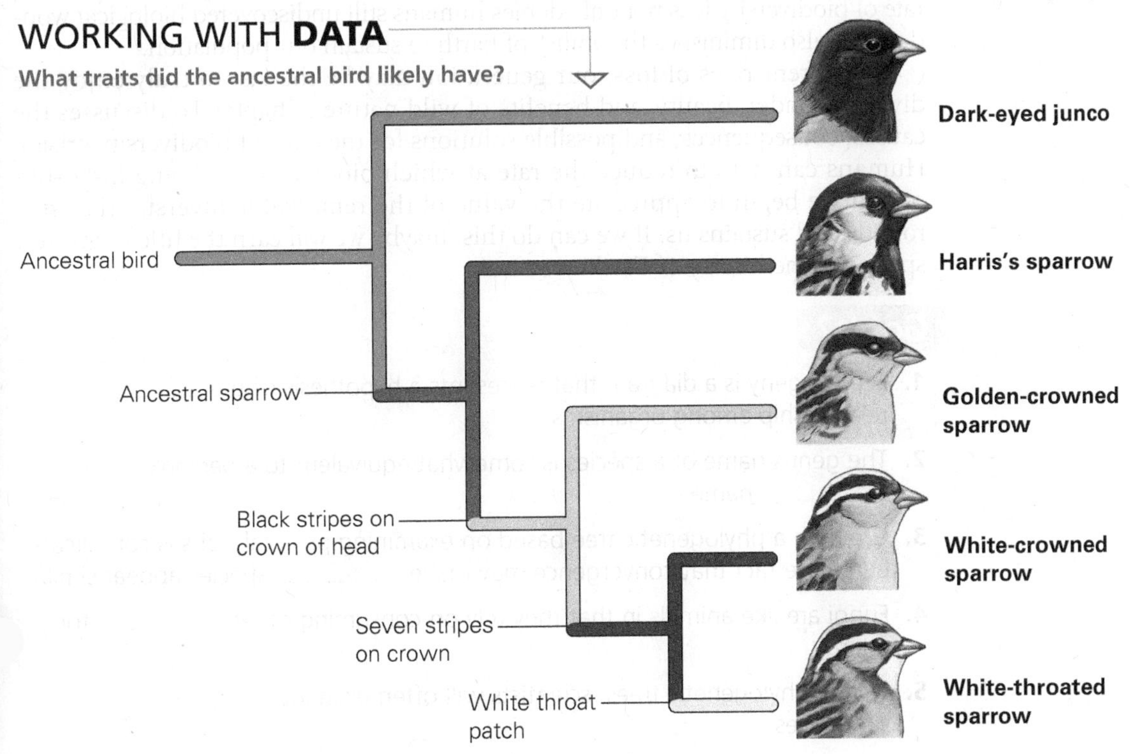

FIGURE 14.19 Reconstruction of an evolutionary history. The relationships among four species of sparrow can be illustrated by their shared physical traits compared to a more distant relative, the dark-eyed junco.

By examining the fossils of extinct organisms, scientists can gather clues about the genealogy—the record of descent from ancestors—of various organisms. For example, fossils of birds clearly indicate that these animals evolved from crocodile-like ancestors.

Information from living organisms can provide an even finer level of detail about evolutionary relationships. As illustrated in Figure 14.4, closely related species should have similar DNA. If the pattern of DNA similarity matches a hypothesized evolutionary relationship among species, the phylogeny is strongly supported. This is the case with the relationship between birds and crocodiles; DNA sequence comparisons indicate that the DNA of crocodiles is more similar to the DNA of birds than it is to the DNA of Komodo dragons.

In contrast, DNA sequence comparisons do not support the sparrow phylogeny presented in Figure 14.19. Here, the data suggest that white-crowned birds and golden-crowned birds are closely related, and white-throated sparrows are a more distant relative. In this case, more observations are needed to discern the true evolutionary relationships among the *Zonotrichia*. (In Chapter 11, we described multiple supportive tests of another phylogeny—in that case, the evolutionary relationship among humans and apes.)

The Greatest Species on Earth

Our survey of diversity provides only a small sampling of the remarkable life forms that have evolved on Earth. But even with a more detailed accounting, any argument about the greatest species (or group of species) of all time doesn't have a definitive answer. Humans are pretty great, but we can be laid low by something as simple as a not-quite-living virus, and we depend on the attributes of many other species to survive.

Where humans clearly excel compared with all other species is in our ability to use the scientific method to understand the world and to use that knowledge to change the environment. It is clear no other species on Earth has this trait. Our understanding of the world, limited as it is, also leads us to other understandings about biodiversity. We know that thousands of other living organisms are lost every year through our destruction of natural habitats. We know that the dramatic rate of biodiversity loss not only denies humans still undiscovered biological wonders but also diminishes the ability of Earth to sustain our population.

At current rates of loss, our generation may be the last to truly enjoy the diverse wonder, beauty, and benefits of wild nature (Chapter 16 discusses the causes, consequences, and possible solutions for the current biodiversity crisis). Humans can help to reduce the rate at which biodiversity is being lost—but only if we begin to appreciate the value of the remarkable diversity that surrounds and sustains us. If we can do this, maybe we *will* earn the title of greatest species in the history of Earth.

Got It?

1. A phylogeny is a diagram that represents a hypothesis of the __________ relationship among organisms.
2. The genus name of a species is somewhat equivalent to a person's __________ name.
3. Creating a phylogenetic tree based on examining physical traits is complicated by the fact that convergence may make __________ species appear similar.
4. Fungi are like animals in that they rely on consuming other __________ for energy.
5. To test phylogenetic trees, scientists will often examine __________ sequences.

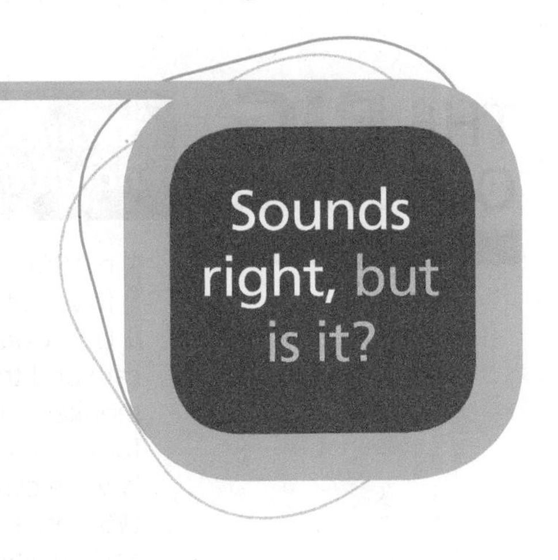

English and many other languages read from left to right and top to bottom. As a result, readers often expect summaries or conclusions on the right-hand side or bottom of a figure. This can lead to some mistaken conclusions. Consider the following phylogenetic tree describing the relationship between a few animal phyla.

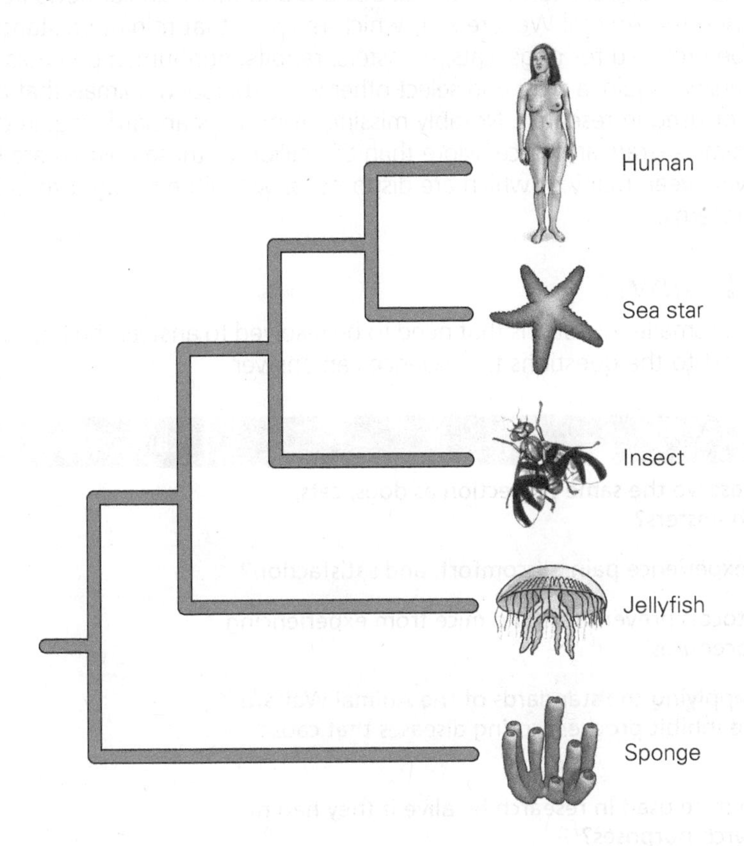

Because humans are pictured at the top of this tree, they are the most complex animal.

Sounds right, but it isn't.

Answer the following questions to understand why.

1. All of these animals share a common ancestor, the first animal. Have all of these animals been evolving for the same amount of time?
2. What traits or abilities do some or all insects have that humans do not?
3. Which of the organisms on the tree is most closely related to humans?
4. Does switching the placement of starfish and humans change the information on the tree? Explain your answer.
5. Based on your answers to questions 1–4, explain why the statement in bold sounds right, but isn't.

Should lab mice and rats have the same rights as other nonhuman animals?

In 1966, *Life* magazine published a dramatic article that described neglectful conditions found at a Maryland dog dealer's farm. As a result of this and other similar news items, the U.S. Congress passed the Animal Welfare Act, which requires that minimum standards of care and treatment be provided for dogs, cats, hamsters, rabbits, nonhuman primates such as monkeys and chimpanzees, guinea pigs, and select other warm-blooded animals that are bred for commercial sale or used in research. Notably missing from the standards required by the law are classic lab animals—rats and mice. More than 80 million of these animals are killed in U.S. research labs every year, many of which are disposed of with little thought to pain relief and quality-of-life concerns.

What should I know?

What follows are some smaller questions that need to be resolved to answer the Big Question. Place a checkmark next to the questions that science can answer.

Smaller Questions	Can Science Answer?
Do rats and mice deserve the same protection as dogs, cats, chimpanzees, and hamsters?	
Can rats and mice experience pain, discomfort, and satisfaction?	
Do current lab protocols prevent rats and mice from experiencing pain during lab procedures?	
Would the cost of applying the standards of the Animal Welfare Act to rats and mice inhibit progress curing diseases that cause human suffering?	
Would the rats and mice used in research be alive if they had not been bred for research purposes?	

What does the science say?

Let's examine what the data say about this smaller question:

Do current lab protocols prevent rats and mice from experiencing pain during lab procedures?

Mice physically respond to pain by grimacing—narrowing their eyes, puffing their cheeks, and pulling back their ears. Scientists have developed a scoring system rating the degree of pain in mice in a range from 0–2, with 2 corresponding to grimaces made under the most painful stimuli, to evaluate the pain experienced by these animals under different conditions. Scientists in this study used this pain rating technique to score photos of the faces of mice that were recovering from a surgical procedure. Mice in the experiment were exposed to varying doses of pain relieving drugs at levels above and below the dose currently suggested (but not required) for pain relief. The figure below shows the results on treatment with various levels of the pain killer, carprofen; the height of the bars indicate the difference in facial grimacing scores before surgery and after the pain killer was administered.

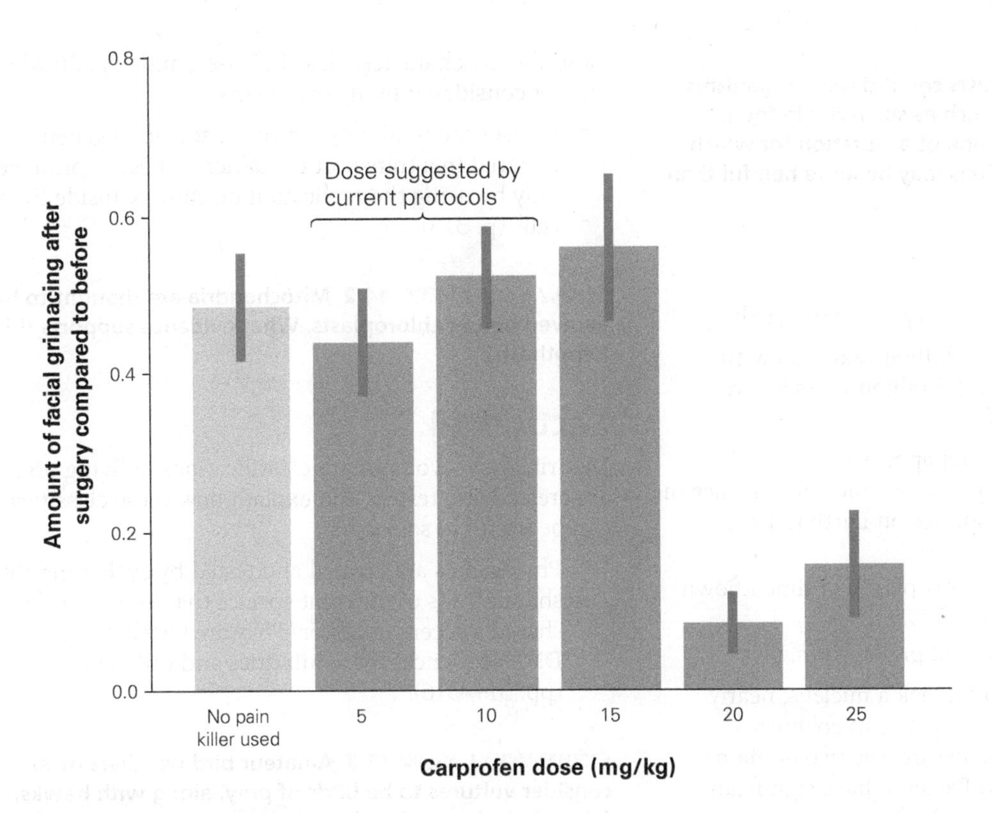

1. Describe the results. Does it appear that the administration of pain killers at currently prescribed levels controls pain in mice recovering from surgery?
2. Given these data, do you think the smaller question is answered? If not, propose another study that would help answer this question.
3. Does this information help you answer the Big Question? What else do you need to consider?

Data Source: L. C. Matsumiya, R. E. Sorge, S. G. Sotocinal, J. M. Tabaka, J. S. Wieskopf, A. Zaloum, ... J. S. Mogil, "Using the Mouse Grimace Scale to Reevaluate the Efficacy of Postoperative Analgesics in Laboratory Mice," *Journal of the American Association for Laboratory Animal Science* 51, no. 1 (2012): 42–49.

Chapter Review

Mastering Biology

Go to Mastering Biology to access **eText 2.0, Dynamic Study Modules,** and the **Study Area,** where you'll find practice quizzes, BioFlix™ animations, MP3 tutor sessions, current events, and more.

SUMMARY

Section 14.1

Cite the current estimate of the number of species described by science, and explain why this number is only an estimate.

- The number of known living species is estimated to be 1.3 million, but the total number of unknown species is likely to be several times greater (pp. 292–294).

List the three major domains of life, and describe basic characteristics of each domain.

- Bacteria share a distant common ancestor with Archaea and Eukarya, and the last two groups are more closely related. However, both Bacteria and Archaea are prokaryotic (simple single-celled organisms without a nucleus or other membrane-bound organelles), whereas organisms in the domain Eukarya are eukaryotes (pp. 294–297).
- The history of life on Earth has consisted of successive dynasties of organisms (p. 295).

SHOW YOU KNOW 14.1 Biologists could classify organisms by other shared characteristics, such as similarity in foods consumed or habitat. Can you think of a situation for which one of these types of classifications may be more helpful than an evolutionary classification?

Section 14.2

Provide a broad overview of the history of life on Earth.

- Life on Earth began about 3.6 billion years ago with prokaryotes, but it would be 1.5 billion years before eukaryotes evolved (p. 298).
- Multicellular organisms did not appear until approximately 600 million years ago, and this advance in form led to the diversity of species on Earth today (p. 300).
- Animal groups evolved in a short period of time known as the Cambrian explosion (p. 303).

Briefly describe the characteristics of prokaryotes.

- Prokaryotes are organisms that lack a nucleus; nearly all are unicellular, although some live in colonies. Despite their similarity in structure, the two domains of prokaryotes, Archaea and Bacteria, have significant differences from each other (p. 298).
- Certain prokaryotes can live in environments that appear inhospitable to life; other prokaryotes can live as spores for millions of years, waiting for ideal conditions to reanimate (p. 300).

Summarize the endosymbiotic theory for the origin of eukaryotic cells.

- Eukaryotes, cells with nuclei and other membrane-bound organelles, probably evolved from symbioses among ancestral eukaryotes and prokaryotes (pp. 300–301).

List the major characteristics of the four major kingdoms of Eukarya: Protista, Animalia, Fungi, and Plantae.

- The Protista is a hodgepodge of organisms that are typically unicellular eukaryotes. Organisms in this group may be plant-like, animal-like, or fungus-like in lifestyle (pp. 301–302).
- Animals are motile, multicellular eukaryotes that rely on other organisms for food (pp. 302–307).
- Fungi are immobile, multicellular eukaryotes that rely on other organisms for food and are made up of thin, threadlike hyphae (pp. 307–310).
- Plants are multicellular, photosynthetic eukaryotes that make their own food (p. 310).

Define *adaptive radiation*, and provide examples of where this has occurred in the history of life.

- Adaptive radiation occurs when several new species appear quickly after the evolution of a new structure or feature or the death of a competing group (p. 311).
- The diversity of flowering plants may be due to adaptive radiation resulting from their production of defensive chemicals (p. 313).

Describe the characteristics of viruses, and explain why they are not considered living organisms.

- Viruses are nonliving entities made up of genetic material in a transport container and can reproduce only by using the replication machinery inside living cells (p. 313).

SHOW YOU KNOW 14.2 Mitochondria are thought to have evolved before chloroplasts. What evidence supports this hypothesis?

Section 14.3

Describe how evolutionary classifications of living organisms are created and tested, and explain how these classifications can be useful to scientists.

- Phylogenies are created and tested by evaluating the shared traits of different species that indicate they shared a recent ancestor. They are tested by examining DNA sequences for similarities and differences (pp. 314–316).

SHOW YOU KNOW 14.3 Amateur bird watchers often consider vultures to be birds of prey, along with hawks, falcons, eagles, and owls. Vultures are scavengers rather than predators, however. What traits do vultures share with these other birds that would cause bird watchers to classify them as birds of prey? What information would help us determine the proper placement of vultures among bird groups?

ROOTS TO REMEMBER

The following roots of words come mainly from Latin and Greek and will help you to decipher terms:

anima- means breath or soul. Chapter term: *animal*

botan- means pertaining to plants. Chapter term: *botanist*

eu- means true. Chapter terms: *eukaryote, Eukarya*

in- means without. Chapter term: *invertebrate*

-karyo- means kernel. Chapter terms: *prokaryote, eukaryote*

myco- means fungal. Chapter terms: *mycologist, mycorrhizae*

zoo- means pertaining to animals. Chapter term: *zoologist*

LEARNING THE BASICS

1. How many different species have been identified by science? How many are estimated to exist?

2. Add labels to the figure that follows, which illustrates the endosymbiotic hypothesis.

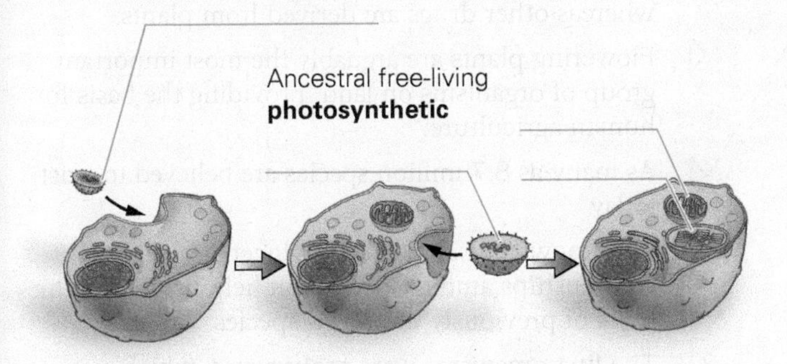

3. How are hypotheses about the evolutionary relationships among living organisms tested?

4. Which of the following kingdoms or domains is a hodgepodge of different evolutionary lineages?

   A. Bacteria; B. Protista; C. Archaea; D. Plantae; E. Animalia

5. Comparisons of ribosomal RNA among many different modern species indicate that ________.

   A. there are two very divergent groups of prokaryotes; B. the Protista represents a conglomeration of very unrelated forms; C. fungi are more closely related to animals than to plants; D. A and B are correct; E. A, B, and C are correct

6. On examining cells under a microscope, you notice that they occur singly and have no evidence of a nucleus. These cells must belong to ________.

   A. domain Eukarya; B. domain Bacteria; C. domain Archaea; D. the Protista; E. more than one of the above could be correct

7. The mitochondria in a eukaryotic cell ________.

   A. serve as the cell's power plants; B. probably evolved from a prokaryotic ancestor; C. can live independently of the eukaryotic cell; D. A and B are correct; E. A, B, and C are correct

8. Fungi feed by ________.

   A. producing their own food with the help of sunlight; B. chasing and capturing other living organisms; C. growing on their food source and secreting chemicals to break it down; D. filtering bacteria out of their surroundings; E. producing spores

9. Which of the following is/are always true?

   A. Viruses cannot reproduce outside a host cell; B. Viruses are not surrounded by a membrane; C. Viruses are not made up of cells; D. A and C are correct; E. A, B, and C are correct

10. Phylogenies are created based on the principle that all species descending from a recent common ancestor ________.

    A. should be identical; B. should share characteristics that evolved in that ancestor; C. should be found as fossils; D. should have identical DNA sequences; E. should be no more similar than species that are less closely related

ANALYZING AND APPLYING THE BASICS

1. Unless handled properly by living systems, oxygen can be quite damaging to cells. Imagine an ancient nucleated cell that ingests an oxygen-using bacterium. In an environment where oxygen levels are increasing, why might natural selection favor a eukaryotic cell that did not digest the bacterium but instead provided a "safe haven" for it?

2. Imagine you have found an organism that has never been described by science. The organism, made up of several hundred cells, feeds by anchoring itself to a submerged rock and straining single-celled algae out of pond water. What kingdom would this organism probably belong to, and why do you think so?

3. Support for the endosymbiotic theory for the evolution of eukaryotic groups includes that chloroplasts and mitochondria are surrounded by two membranes, one that derives presumably from the bacterial symbiont and one that derives from the host cell's vacuole that originally surrounded it. Interestingly, some algal phyla, for instance the Euglenophyta, contain chloroplasts composed of three membranes. Given your understanding of endosymbiosis, how do you think this extra membrane may have evolved?

GO FIND OUT

1. Research a species of plant, animal, fungi, or microbe that is not discussed in this chapter but that does something astonishing, and make a sales pitch that this organism is "greater" than humans. Include enough detail to convince your classmates.

2. Scientists initially rejected the endosymbiotic theory, which is the hypothesis that eukaryotic cells evolved from a set of cooperating independent cells. Most biologists still believe that competition for resources among organisms is the primary force for evolution. Do you think our modern culture, where competition is often valued over cooperation, might make scientists less likely to search for and see the role of cooperation in evolution, or do you think cooperation leading to evolution must be truly rare?

MAKE THE CONNECTION

The science that you learned in this chapter has helped you better understand the real-world example used throughout this discussion. Draw a line from the statement on the left to the science that supports it on the right.

Humans are not the most common species on Earth. ▷	◁ All living organisms share a common ancestor, as evidenced by the commonality of DNA and ribosomes.
Humans are not a unique form of life. ▷	◁ Antibiotics are prevalent in bacteria and fungi, whereas other drugs are derived from plants.
Humans are a relatively young species, evolving late in the history of life compared with bacteria. ▷	◁ Flowering plants are arguably the most important group of organisms on land, providing the basis for human agriculture.
Humans did not invent most drugs used to treat disease. ▷	◁ As many as 8.7 million species are believed to exist today.
The group of organisms humans belong to is relatively small. ▷	◁ Our knowledge about the phylogenetic relationships among species can help us predict the traits of previously unknown species.
Human civilization is not possible without the presence of certain other groups of organisms. ▷	◁ The first organisms were prokaryotes, which evolved at least 3.6 billion years ago.
Our ability to learn about the natural world does provide us an advantage over many species, especially if we use this advantage wisely. ▷	◁ Animals with backbones make up only 4% of all animal species.

Answers to Got It?, Visualize This, Working with Data, Sounds Right, But Is It?, Show You Know, and Chapter Review questions can be found in the Answers section at the back of the book.

15 Is the Human Population Too Large?

From space, Earth doesn't look too crowded.

But Earth's human population is 7.3 billion—and rising.

Can Earth support everyone at the same level as the average North American family?

Population Ecology

In its most recent estimate in 2015, the United Nations (UN) reported that the world's population is approximately 7.3 billion—4 billion more than the population just 50 years prior. The UN also predicted that the population would continue to grow for several more decades before stabilizing between 10 and 12.5 billion by 2100. Many observers greeted the report as another piece of bad news. The UN's estimate was not only higher than previous estimates, but it prompted many scientists and environmentalists to wonder if our planet can support 7.3 billion people for much longer, let alone a possible 5.2 billion more in the next century.

Other commentators are skeptical of doom-and-gloom statements about the perils of population growth. Environmentalists' predictions have been wrong before; in the 1960s, some predicted global food and water shortages by the year 2000. However, most measures of human health have become more upbeat since 1970, including global declines in infant mortality rates, increases in life expectancy, and a 20% rise in per capita income. By most measures, the average person is better off today than she was 50 years ago, and the UN report predicts continued increases in life expectancy even considering increasing population size.

But other UN projections give reason for concern that our large human population is rapidly reaching a real limit to growth. In 2016, an estimated 795 million people were categorized as "food insecure," meaning that they did not get enough food regularly. More than 3 million deaths among children under age 5 each year were associated with inadequate nutrition. And a rapidly changing climate affecting agricultural production around the globe poses a serious threat to future human health and survival.

So what is the truth? Is the human population larger than Earth can support for much longer? Are we headed into a global food crisis and massive famine? Or are we gradually moving toward an era when all people on Earth will be as well fed, long-lived, and affluent as the average North American?

15.1 Population Growth

popula- means multitude.

Ecology is the field of biology that focuses on the interactions among organisms as well as those between organisms and environments. The relationship between organisms and their environments can be studied at many levels—from the individual, to populations of the same species, to communities of interacting species, and finally to the effects of biological activities on the nonbiological environment, such as the atmosphere.

In ecology, a **population** is defined as all of the individuals of a species within a given area. Populations exhibit a structure, which includes the spacing of individuals (that is, their distribution) and their density (abundance). Ecologists seek to explain the distribution and abundance of the individuals within populations and to understand the factors that lead to the success or failure of a population. The interactions among species make up one set of influences on distribution and abundance (Chapter 16), but another set is the internal dynamics of the population, including the relative numbers of individuals of different sexes and ages and the numbers that are born or die in a given time period.

VISUALIZE **THIS**

Animals may become either "trap-happy"—that is, attracted to traps—or "trap-shy" after one capture. How might each of these behaviors affect a population estimate?

Population Structure

The first task of a population ecologist is to estimate the size of the population of interest. Certain populations can be counted directly, as in a census of people in the United States or a survey identifying all individuals of a particular tree species in a forest tract.

The size of more mobile or inconspicuous species can be estimated by the **mark-recapture method** (**FIGURE 15.1**). In this technique, researchers capture many individuals within a defined area, mark them in some way (for instance,

FIGURE 15.1 The mark-recapture method. 1. In the first step, a researcher captures, marks, and releases animals in a wild population. 2. Some time later, the researcher returns and determines what percentage of animals trapped again were previously marked. 3. In the second capture, the ratio of marked to total animals captured is approximately the ratio of the original number trapped to the whole population.

① Researcher captures 100 beetles in a trap, and marks each with a dab of paint.

② After one week, a trap is set again, resulting in a captured group of marked and unmarked individuals.

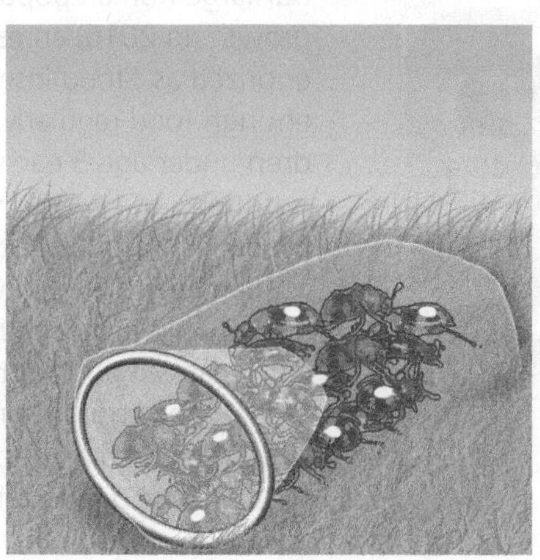

③ Total population is estimated as equivalent to the percentage of marked individuals in the second trap.

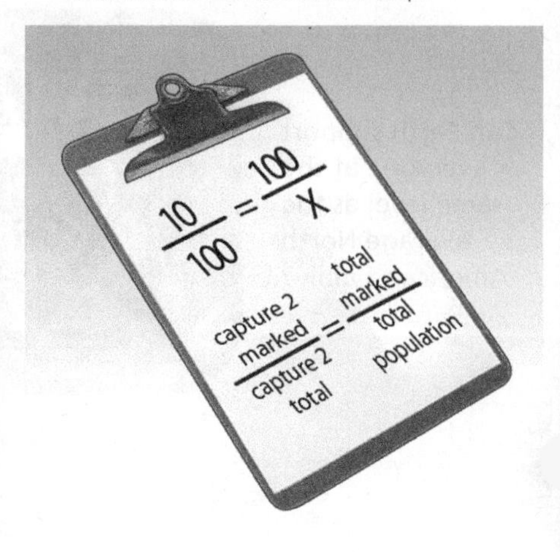

with a dab of paint), and release them back into the environment. Later, the researchers capture another group of individuals in the same area and calculate the proportion of previously marked individuals in this group. This proportion can be used to estimate the size of the total population.

For an example of the mark-recapture method, imagine that a researcher captured, marked, and released 100 beetles in a small area of forest. One week later, the researcher returns to the same place and captures another group of beetles. If she finds that 10% of the beetles caught on the second round are marked, the researcher can assume that the 100 beetles captured and marked initially represented 10% of the entire beetle population. If she assumes that the marked beetles mix evenly with the unmarked individuals after release, the total population can be estimated as near 1000 beetles.

Another basic aspect of population structure is dispersion—that is, how organisms are distributed in space. Many species show a **clumped distribution,** with high densities of individuals in certain resource-rich areas and low densities elsewhere. Plants that require certain soil conditions and the animals that depend on these plants tend to be clumped (**FIGURE 15.2a**). On a global scale, humans show a clumped distribution, with high densities found around transportation resources, such as rivers and coastlines.

The clumped distribution of humans masks a more **uniform distribution** on a local scale; for instance, the spacing between houses in a subdivision or strangers in a classroom tends to equalize the distances among individual property owners or people. Species that show a uniform distribution are often territorial —they defend their personal space from intruders. We can observe these same strong reactions among certain species of birds at a breeding site (**FIGURE 15.2b**).

Nonsocial species with the ability to tolerate a wide range of conditions typically show a **random distribution,** in which no compelling factor is actively bringing individuals together or pushing them apart. The distribution of seedlings of trees with windblown seeds is often random (**FIGURE 15.2c**).

The distribution and abundance of a population provide a partial snapshot of its current situation. The dispersion of the human population—and recent changes in that pattern—profoundly affects the natural environment (Chapter 17). However, to better understand how a population is responding to its environment, we need to determine how it is changing through time.

Exponential Population Growth

Historians have been able to use archaeological evidence and written records to determine the size of the human population in the past. For most of our history, the human population remained at very low levels. At the beginning of

(a) Clumped

(b) Uniform

(c) Random

FIGURE 15.2 Patterns of population dispersion. Individuals in a population may be (a) clumped, like these cattails growing only in soil with the correct water content; (b) uniformly distributed, like these territorial nesting penguins; or (c) randomly dispersed, like these seedlings grown from windblown seeds in a forest.

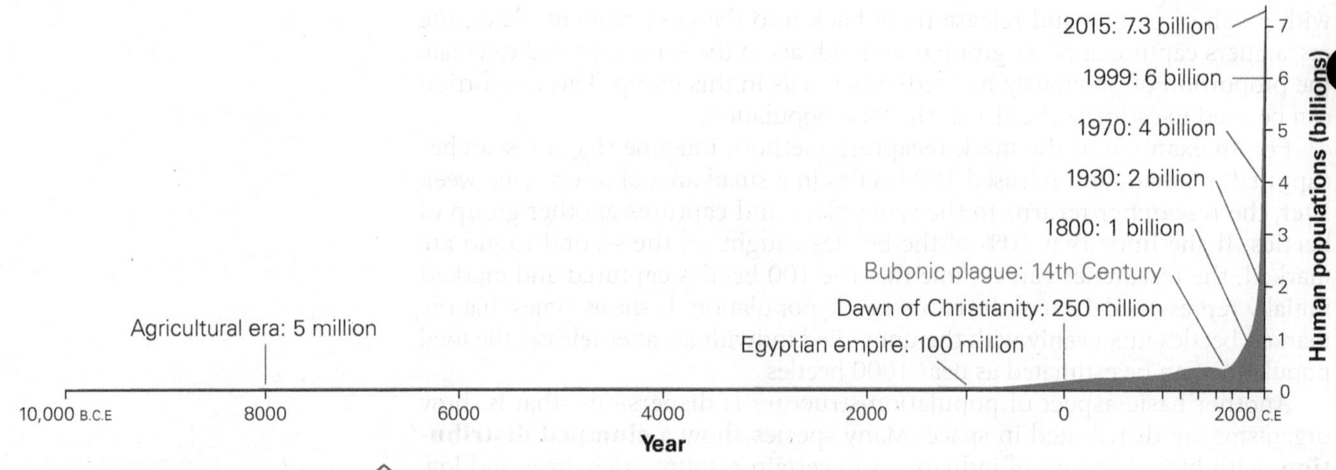

FIGURE 15.3 Exponential growth. The number of people on Earth grew relatively slowly until the eighteenth century. The rapid growth since then has occurred in proportion to the total, causing a J-shaped curve.

WORKING WITH **DATA**

The line looks relatively flat from 8000 B.C.E. to 1500 B.C.E., although the population doubled four times in that period. Was the population growing exponentially at this time?

expo- is from the word meaning to put forth.

the agricultural era, about 10,000 years ago, there were approximately 5 million humans. There were 100 million people during the Egyptian Empire (7000 years later) and about 250 million at the dawn of the Christian religion in 1 C.E. The population was growing, but at a very slow rate—approximately 0.1% per year.

Beginning around 1750, the rate at which the human population was growing jumped to about 2% per year. The human population reached 1 billion in 1800, had doubled to 2 billion by 1930, and then doubled again to 4 billion by 1970. Although the current growth rate is slower, about 1.1% per year, the rapid increase in population looks quite dramatic on a graph of human population over time (**FIGURE 15.3**).

The J-shaped line on the graph of human population growth is a striking illustration of **exponential growth**—growth that occurs in proportion to the current total. The larger a population is, the more rapidly it grows because an increase in numbers depends on individuals reproducing in the population. So, although a growth rate of 1.1% per year may seem rather small, the number of individuals added to the 7.3-billion-strong human population every year at this rate of growth is 83 million—more than the combined populations of California, Texas, and New York. Put another way, three people are added to the world population per second, and about a quarter of a million people are added every day.

What has fueled this enormous increase in human population? The annual **growth rate** of a population is the percentage change in population size over a single year. Growth rate, which is mathematically represented as *r*, is a function of the birth rate of the population (the number of births as a percentage of the population) minus the death rate (the number of deaths as a percentage of the population). For example, in the entire human population 21 babies are born per year for every 1000 people—that is, the birth rate for the population is 2.1%:

$$\frac{21}{1000} = 0.021 = 2.1\%$$

In addition, each year 10 individuals die out of every 1000 people, resulting in a death rate of 1%:

$$\frac{10}{1000} = 0.01 = 1\%$$

This results in the current growth rate of 1.1%:

$$\text{Growth rate} = \text{Birth rate} - \text{Death rate}$$

$$1.1\% = 2.1\% - 1\%$$

Today's relatively high growth rate, compared with the historical average of 0.1%, is the result of a large difference between birth and death rates.

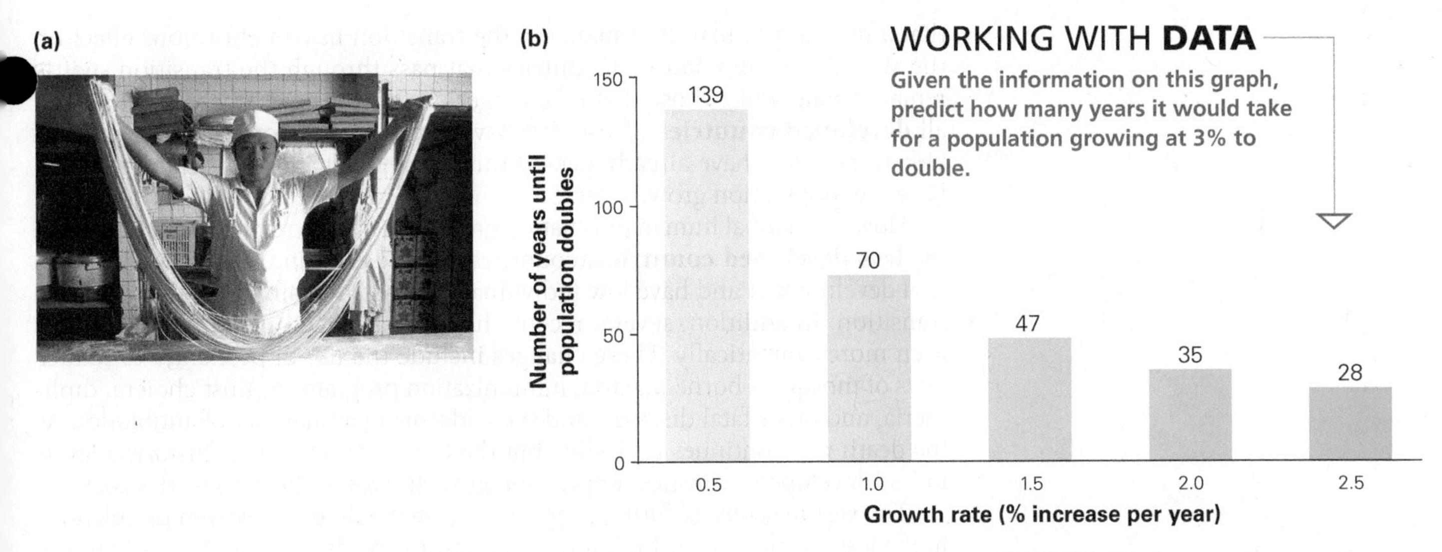

WORKING WITH **DATA**

Given the information on this graph, predict how many years it would take for a population growing at 3% to double.

FIGURE 15.4 Growth rate and doubling time. (a) Hand-pulled noodles are made by folding and pulling dough repeatedly, doubling the number of strands with every fold. A single thick noodle folded only 12 times will produce 4096 fine noodles (seen here), illustrating how rapidly growth occurs when it is exponential. (b) The time it takes a population to double in size depends on its growth rate. A population growing at 2% per year doubles in 35 years, while one growing at 1% takes twice as long to double.

It can be easier to think of exponential growth in terms of the amount of time it takes for a population to double in size. At a growth rate of 0.1, it takes 693 years for the population to double. At a rate of 1.1%, it takes only about 63 years (**FIGURE 15.4**).

The Demographic Transition

In human populations, the tendency has been for decreases in death rate to be followed by decreases in birth rate. The speed of this adjustment helps to determine population growth in the future. Before the Industrial Revolution in the late eighteenth and into the early nineteenth centuries, both birth and death rates were high in most human populations. Women gave birth to many children, but relatively few children lived to reach adulthood. Rapid population growth was triggered in the eighteenth century by a dramatic decline in infant mortality (the death rate of infants and children) in industrializing countries. In particular, advances in treating and preventing infectious disease since this period have reduced the number of children who die from these illnesses.

Not long after death rates declined in industrialized countries, birth rates followed suit, lowering growth rates again. Scientists who study human population growth refer to the period when birth rates are dropping toward lowered death rates as the **demographic transition** (**FIGURE 15.5**). The length of time

demo- is from the word meaning people.

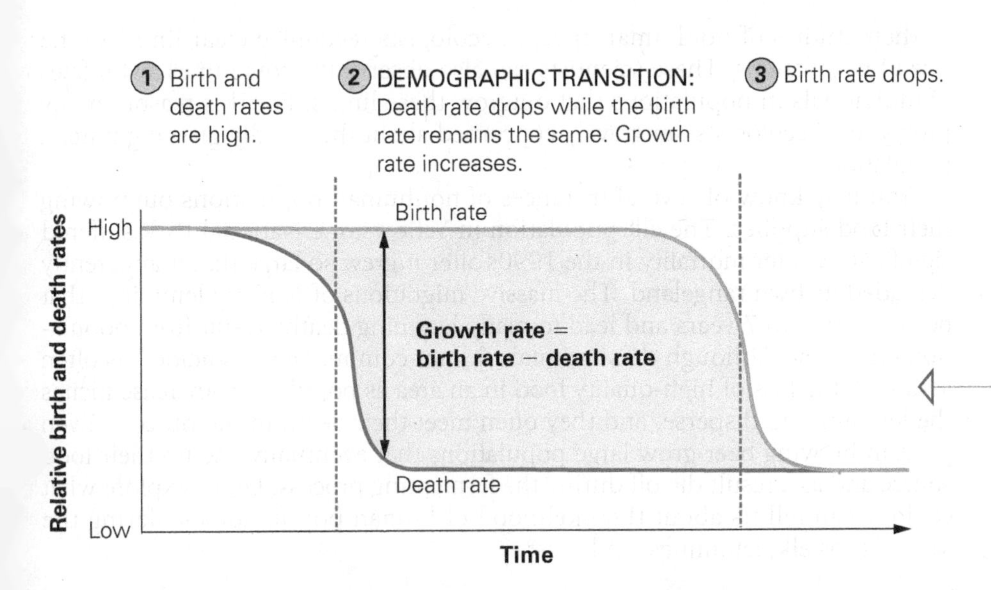

FIGURE 15.5 The demographic transition. Improvements in sanitation and medical care in human populations cause a decrease in infant mortality, dropping death rates. Because birth rates remain high, growth rates soar. Eventually, people in these populations respond by decreasing the number of children they have.

WORKING WITH **DATA**

Where on this graph is the growth rate of the population close to zero?

that a human population remains in the transition has an enormous effect on the size of that population. Countries that pass through the transition swiftly remain small, while those that take longer can become extremely large. Nearly all **developed countries** (those that have industrial economies and high individual incomes) have already passed through the demographic transition and have low population growth rates.

However, global human population growth rates have remained high because the **less developed countries** (countries that are early in the process of industrial development and have low individual incomes) remain in the demographic transition. In addition, several recent changes have decreased infant mortality even more dramatically. These changes include the use of pesticides to reduce rates of mosquito-borne malaria; immunization programs against cholera, diphtheria, and other fatal diseases; and the widespread availability of antibiotics. As the death rate continues to decline but the birth rate remains at historical levels in less developed countries, population growth rates in these countries soar.

The vast majority of future population growth will occur within populations in the less developed world, but these countries are also where the vast majority of food crises are occurring. Are the populations in these countries already too large to support themselves? Answering that question requires an understanding of the factors that limit population growth.

Got It?

1. In a mark-recapture study, the size of a population of interest is estimated by the ________ of marked individuals that are recaptured after one round of trapping has occurred.
2. More than half of the human population is found in densely populated urban areas, demonstrating a ________ distribution pattern.
3. Exponential growth is growth in proportion to the ________ population.
4. The birth rate in a population is determined by counting the number of ________ that occur in a population in a year, divided by the total ________.
5. During the demographic transition of a human population, growth rates climb as ________ rates greatly exceed ________ rates.

15.2 Limits to Population Growth

In their studies of nonhuman species, ecologists recognize clear limits to the size of populations. These scientists can also observe the sometimes awful fates of individuals in populations that outgrow these limits. For this reason, many professional ecologists are gravely concerned about the rapidly growing human population.

You may know of several instances of nonhuman populations outgrowing their food supplies. The elk population in Yellowstone National Park suffered significant winter mortality in the 1990s after it grew so large that it apparently degraded its own rangeland. The massive migrations of Norway lemmings that occur every 5 to 7 years and lead to many lemming deaths result from population crowding. Although these animals do not commit "mass suicide," as often assumed, the loss of high-quality food in an area as populations increase incites the lemmings to disperse, and they often meet their death in the process. Even yeast in brewing beer grow large populations that eventually use up their food source and as a result die off during the fermenting process. Let us explore what ecology can tell us about the likelihood of human populations suffering the same fate as elk, lemmings, and yeast.

Carrying Capacity and Logistic Growth

The examples of elk in Yellowstone and Norway lemmings illustrate a basic biological principle. Although populations have a capacity to grow exponentially, in reality, their growth is limited by the resources—food, water, shelter, and space—that individuals need to survive and reproduce. The maximum population that can be supported indefinitely in a given environment is known as the environment's **carrying capacity.**

A simplified graph of population size over time in resource-limited populations is S-shaped (**FIGURE 15.6**). This model shows the growth rate of a population declining to zero as it approaches the carrying capacity (mathematically represented as *K*). In other words, birth rate and death rate become equal, and the population stabilizes at its maximum size. Not long after ecologists first predicted this pattern of growth, called **logistic growth,** populations of organisms as diverse as flour beetles, water fleas, and single-celled protists were shown in laboratory studies to conform to this projected growth curve.

The declining growth rate near a population's carrying capacity is caused by **density-dependent factors,** which are population-limiting factors that increase in intensity as a population increases in size. Density-dependent factors include limited food supplies, increased risk of infectious disease, and an increase in toxic waste levels. Density-dependent factors cause either declines in birth rate or increases in death rate. In organisms such as fruit flies growing in laboratory culture bottles, high populations lead to increased mortality of the flies as food supplies dwindle and wastes accumulate. *Daphnia* (often known as "water fleas") living in crowded aquariums do not have enough food to support egg production, so birth rates drop. Females of white-tailed deer populations living in crowded natural habitats are less likely to have the energy reserves necessary to carry a pregnancy to term than deer in less crowded environments (**FIGURE 15.7**). Some populations of humans have experienced density-dependent mortality as well, including the epidemic of infectious bubonic plague that spread in crowded European cities in the fourteenth century, leading to the last major decline in human population (review Figure 15.3).

WORKING WITH **DATA**

Where on the graph is the number of individuals added to the population *per unit of time* highest?

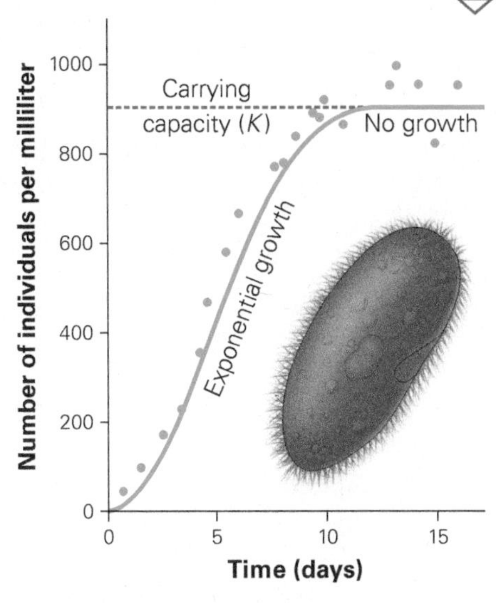

FIGURE 15.6 The logistic growth curve. The graph illustrates the change in the population of Paramecium (a single-celled protist) over time in a laboratory culture. The S-shape of the growth curve is due to a gradual slowing of the population growth rate as it approaches the carrying capacity of the environment.

log- means proportion or ratio.

(a) Fruit flies

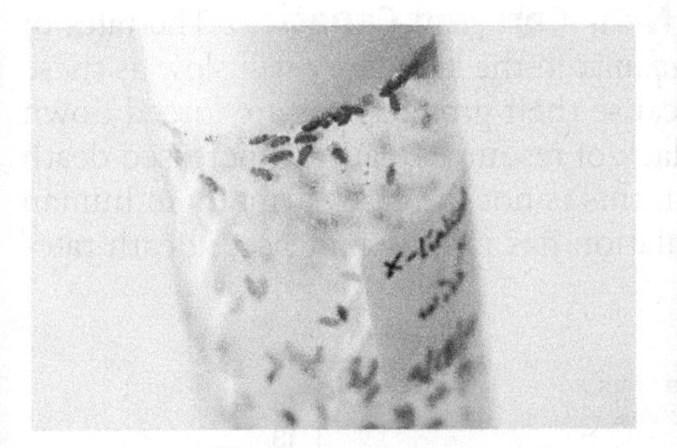

(b) *Daphnia*

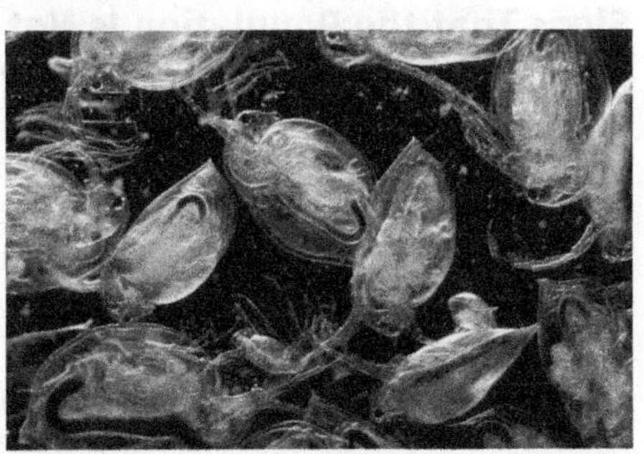

(c) White-tailed deer

(d) Humans

FIGURE 15.7 Limits to growth. Populations of (a) fruit flies in a laboratory culture, (b) *Daphnia* in an aquarium, and (c) white-tailed deer in the northeastern United States all experience high death rates or low birth rates as their populations approach the carrying capacity of the environment. (d) Do human populations face these same limits?

Density-dependent factors can be contrasted with **density-independent factors** that influence population growth rates—for instance, severe droughts that increase the death rate in plant populations regardless of their density or increased temperatures that increase the birth rate in insects. One impact of climate change may be to increase the occurrence of density-independent factors as Earth warms and dries and storms become more severe (Chapter 5). However, density-independent factors do not occur in a vacuum; they can have more or less severe effects depending on the size of a population. For example, a density-independent factor such as an unusually cold winter can be deadly to individuals in a white-footed mouse population, but the likelihood of survival is also a function of how much food each individual has stored for the winter. How much food is stored by each animal depends on the density of mice competing for that food during the autumn.

Are density-dependent factors beginning to reduce growth rates in the human population? That is, are humans nearing the carrying capacity of Earth for our population? If we are, will death rates increase as food resources dwindle and more people starve? Or will birth rates decline because fewer women will have enough food to support themselves and a developing baby?

Earth's Carrying Capacity for Humans

One way to determine if the human population is reaching Earth's carrying capacity is to examine whether, and how rapidly, the growth rate is declining. As we saw in Figure 15.6, the S-shaped curve of population size over time results from a gradually declining growth rate as the population approaches carrying capacity.

Human population growth rates were at their highest in the early 1960s, at 2.1% per year, but they have since declined to the current rate of 1.1%. This steady decline is an indication that the population, though still currently growing, is nearing a stable number. Uncertainty about the future rate of growth has led the UN to produce differing estimates of this number and how soon population stability will be reached (**FIGURE 15.8**). However, the unique characteristics of humanity make it difficult to determine exactly which population size represents Earth's carrying capacity for humans.

Signs That the Population Is Not Near Carrying Capacity. The rates of population increase of fruit flies and *Daphnia* in the laboratory will slow as these populations near carrying capacity because their growth rates are forced down by density-dependent factors, such as lack of resources, causing increased death rates or decreased birth rates. However, this is not the case currently in human populations. Even as the human population has rapidly increased, death rates

FIGURE 15.8 Projected human population growth. The United Nations report is based on a number of uncertainties and projects a range of possible population sizes by the year 2100.

WORKING WITH **DATA**

Where on this graph is the population growth rate declining toward zero?

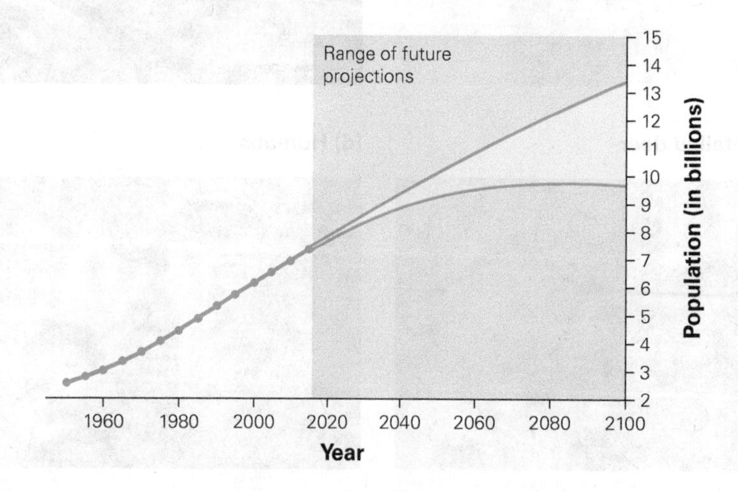

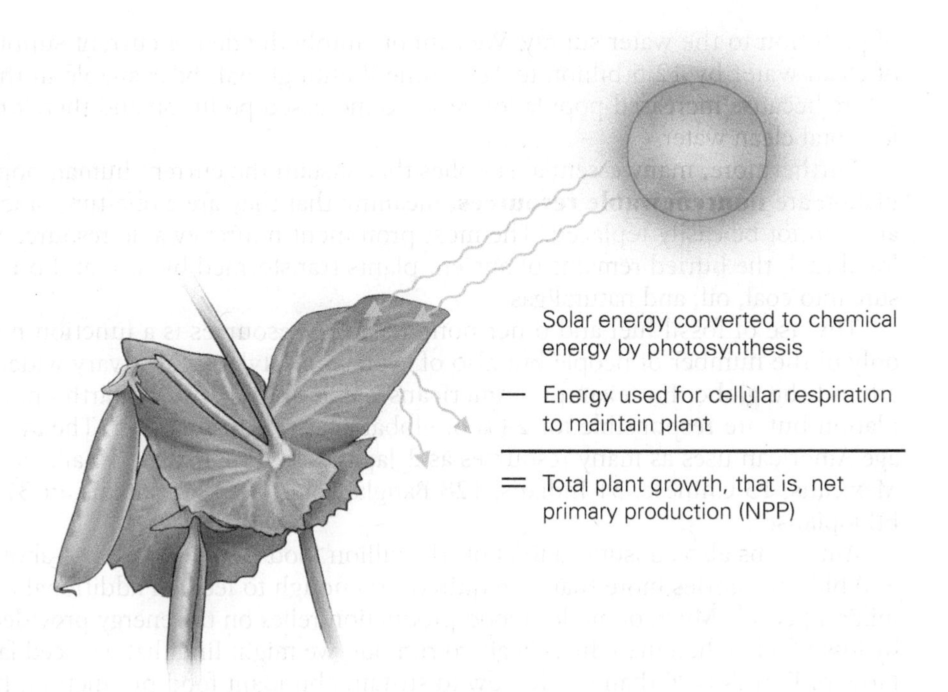
Solar energy converted to chemical energy by photosynthesis
− Energy used for cellular respiration to maintain plant
= Total plant growth, that is, net primary production (NPP)

FIGURE 15.9 Net primary production. NPP is a measure of the total calories available on Earth's surface to support consumers.

continue to decline—an indication that people are not running out of resources. Instead, growth rates are declining because birth rates are falling even faster than death rates. Unlike the water fleas and white-tailed deer, whose females are unable to have offspring when populations are near carrying capacity, birth rates in human populations are falling because women, especially those with adequate resources and opportunities to obtain education, are *choosing* to have fewer children.

Another way to determine if humans are near Earth's carrying capacity is to estimate the proportion of Earth's resources used by humans now. The amount of food energy available on the planet is referred to as the **net primary productivity (NPP).** NPP is a measure of plant growth, typically over the course of a single year (**FIGURE 15.9**).

Several different analyses of the global extent of agriculture, forestry, and animal grazing estimate that humans use between one-quarter and one-third of the total land NPP. If we accept these rough estimates, we can approximate that the carrying capacity of Earth is three to four times the present population, at least 21 billion people. This theoretical maximum is the total number of humans that could be supported by all of the photosynthetic production of the planet—leaving no resources for millions of other species. Given the dependence of humans on natural systems (explored in Chapter 16), it is unlikely that our species could survive on a planet where no natural systems remained. However, even the largest population projection by the UN, 12.5 billion, falls well short of this theoretical maximum.

Signs That the Population Is Near Carrying Capacity. Ecologists caution that the resources required to sustain a population include more than simply food, so the carrying capacity deduced from NPP estimates may be much too high. Humans also need a supply of clean water, clean air, and energy for essential tasks such as heating, food production, and food preservation.

The relationship between population size and the supply of these resources is not as straightforward as the relationship between population and food. For instance, every new person added to the population requires an equivalent amount of clean water, but every new person also introduces a certain amount of pollution to the water supply. We cannot simply divide the current supply of clean water by 12.5 billion to determine if enough will be available in the future because increased population leads to increased pollution and therefore less total clean water.

Furthermore, many essential supplies that sustain the current human population are **nonrenewable resources,** meaning that they are a one-time stock and cannot be easily replaced. The most prominent nonrenewable resource is fossil fuel, the buried remains of ancient plants transformed by heat and pressure into coal, oil, and natural gas.

The use of fossil fuel and other nonrenewable resources is a function not only of the number of people but also of average lifestyles, which vary widely around the globe. For example, Americans make up only 5% of Earth's population but are responsible for 24% of global energy consumption. The average American uses as many resources as 2 Japanese or Spaniards, 3 Italians, 6 Mexicans, 13 Chinese, 31 Indians, 128 Bangladeshis, 307 Tanzanians, or 370 Ethiopians.

Americans also consume a total of 815 billion food calories per day—about 200 billion calories more than is required, or enough to feed an additional 80 million people. Much of modern food production relies on the energy provided by fossil fuel. When these fuels begin to run out, we might find that we need far more of Earth's NPP than we do now to sustain abundant food production. In other words, the actual carrying capacity of our planet may be much lower than our approximations.

The question posed at the beginning of this section remains unanswered; there is no agreement among scientists concerning the carrying capacity of Earth for the human population. Given that uncertainty, what can ecologists tell us about the risks facing the human population that may result from massive rapid population growth?

Got It?

1. A population's carrying capacity is the ________ population size that can be supported indefinitely.
2. A decrease in the amount of food available per individual as a population gets larger is an example of a density-________ factor that affects population growth.
3. The graph of human population over time shows an S-shaped curve, indicating that the population may be nearing ________ ________.
4. Humans are an unusual species in that when resources levels are high, individuals often choose to have ________ offspring.
5. Earth's entire production of calories that could be consumed by humans is known as ________ production.

15.3 The Future of the Human Population

Unlike nearly all other species, human populations are not simply at the mercy of environmental conditions. With our ability to transform the natural world, humans have helped populations circumvent seemingly fixed natural limits. On the other hand, human ingenuity can lull people into believing that nature has an almost infinite capacity to support their ever-growing needs. Managing the growth of human populations before even the most secure of us face environmental and economic disaster requires an understanding of the risks of continued rapid growth and the strategies that help reduce it.

WORKING WITH **DATA**

Imagine that in both populations, more resources became available, raising the carrying capacity. How would this change the graphs?

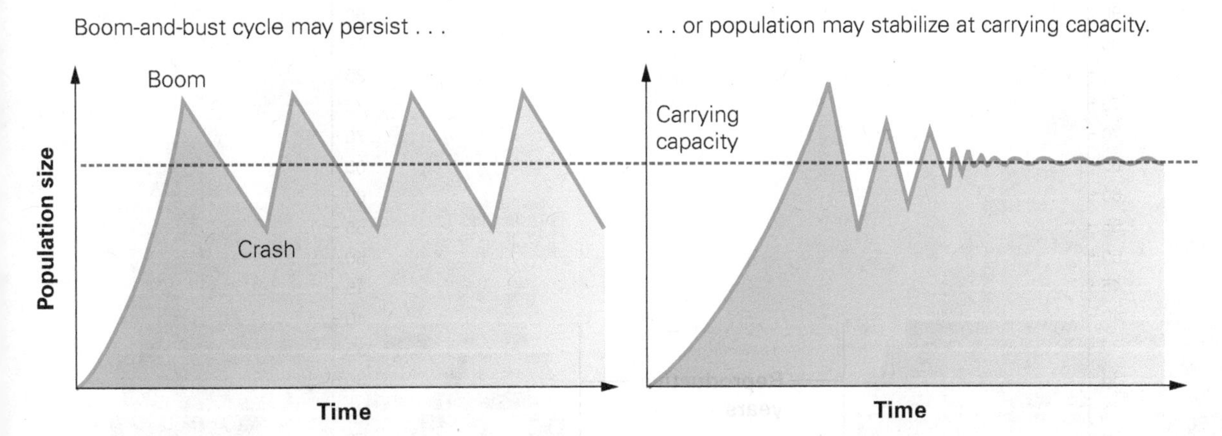

FIGURE 15.10 Overshooting and crashing. These graphs illustrate rapid population growth followed by a population crash. Over time, the population may stay in a "boom-and-bust" cycle, or it may stabilize at its carrying capacity.

A Possible Population Crash?

The use of nonrenewable resources creates a risk that the human population will overshoot a still-unknown carrying capacity. Ecologists have long known that when populations have high growth rates, they may continue to add new members even as resources dwindle. This causes the population to grow larger than the carrying capacity of the environment. The members of this large population are then competing for far too few resources, and the death rate soars while the birth rate plummets. This results in a **population crash,** a steep decline in number (**FIGURE 15.10**).

For instance, in some species of *Daphnia*, healthy offspring continue to be born for several days after the food supply becomes inadequate. This occurs because females can use their fat stores to produce additional young. The size of the population continues to rise even when there is no food left for grazing; however, when the young water fleas run out of stored fat transferred from their mothers, most individuals die. For many species with high birth rates, rapid growth followed by dramatic crashes produce a **population cycle** of repeated "booms" and "busts" in number.

Biological populations are more likely to overshoot carrying capacity when there is a time lag between when the population approaches carrying capacity and when it actually responds to that environmental limit. Scientists who study human populations note a lag between the time when humans reduce birth rates and when population growth actually begins to slow. They call this lag **demographic momentum.** The momentum occurs because while parents may be reducing their family size, their children will begin having children before the parents die, causing the population to continue growing. Even when families have an average of two children, just enough to replace the parents, demographic momentum causes the human population to grow for another 60 to 70 years before reaching a stable level.

The demographic momentum of a human population can be estimated by looking at its **population pyramid,** a summary of the numbers and proportions

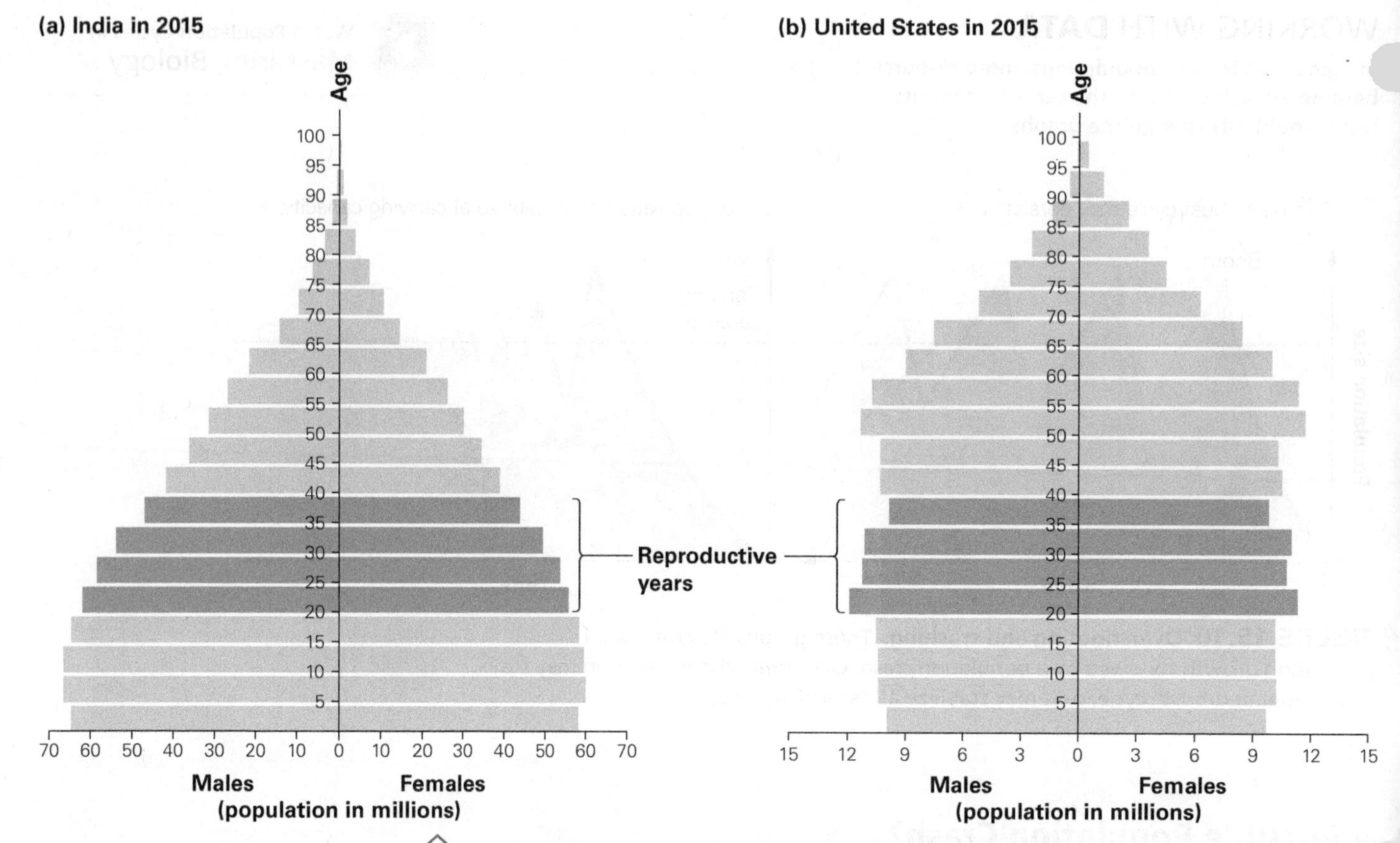

FIGURE 15.11 Demographic momentum. In a rapidly growing human population like that of (a) India in 2015, most of the population is young, and the population will continue to grow as these children reach child-bearing age. In a slower-growing or stable human population, the ages are more evenly distributed, as in (b) the United States in 2015.

WORKING WITH **DATA**

According to the graphs, what are the relative sizes of the populations of the United States and India?

of individuals of each sex and each age group. As **FIGURE 15.11** illustrates, the potential for high levels of demographic momentum occurs when the age structure most closely resembles a true pyramid, with a large proportion of young people. As this large group of young people ages, they become potential parents, with the capacity to cause the population to swell. In more stable populations, the proportion who are young is not significantly larger than the proportion who are middle aged, and the pyramid looks more like a column. The number of "parents-in-waiting" is the same as the number of current parents, keeping the population stable.

Whether our reliance on stored resources and the potential demographic momentum in human populations will result in an overshoot of Earth's carrying capacity—followed by a severe crash—remains to be seen. But human ecologists already know what factors help to slow population growth. If we act on this knowledge, a crash may become less likely.

Avoiding Disaster

As discussed previously, when death rates drop in a human population, birth rates eventually follow. Unlike any other species known to science, humans will voluntarily limit the number of babies they produce. As noted above, when more opportunities become available outside of child rearing, most women delay motherhood and have fewer children. In fact, birth rates are lowest in countries where income is high and women have access to education (**FIGURE 15.12**).

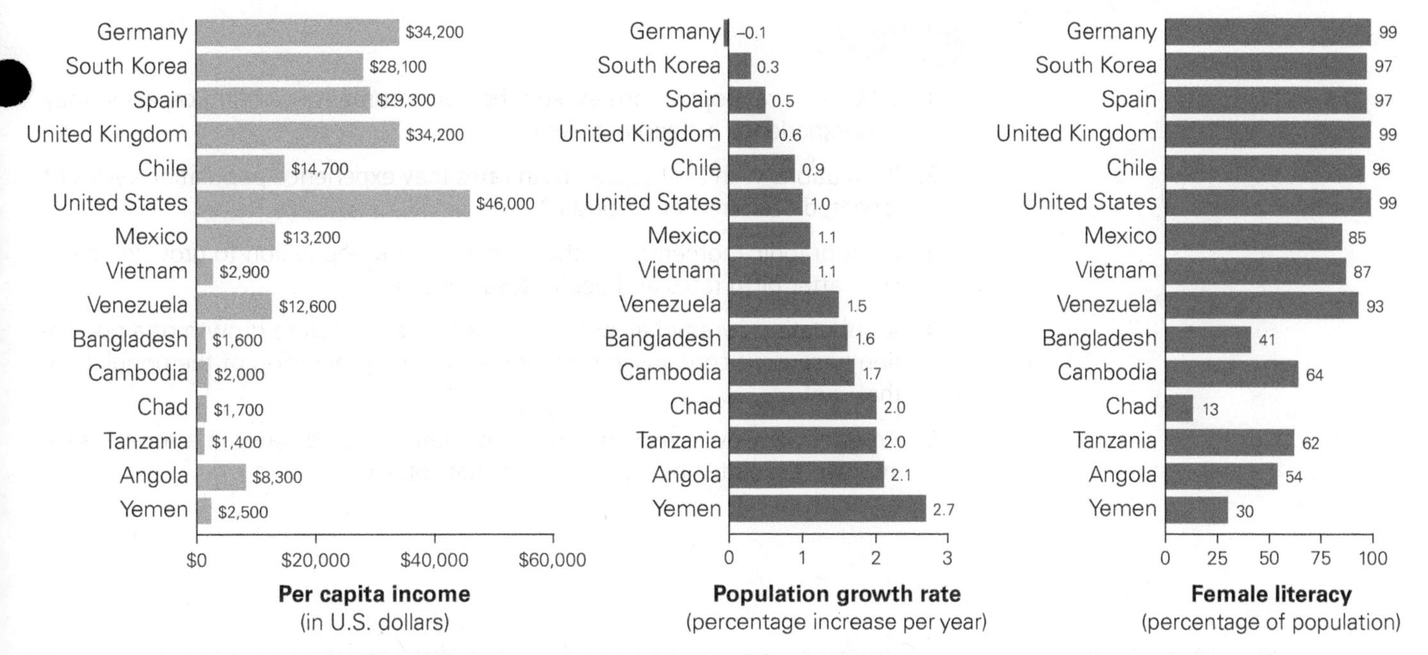

FIGURE 15.12 Income, growth rate, and women's literacy. These three graphs illustrate the relationships among income, population growth, and female literacy. Note that higher income and literacy are correlated with decreased birth rates and thus decreased population growth in most countries.

VISUALIZE **THIS**

Income and female literacy do not explain all differences among nations in growth rates. Why do you suppose the United States has a relatively high population growth rate compared with other wealthy countries?

This information provides a clear direction for public policies attempting to decrease population growth rates: improve conditions for women, including increasing access to education, health care, and the job market, and provide women with the information and tools that allow them to regulate their fertility (**FIGURE 15.13**).

Slowing growth rates before the human population reaches some environmentally imposed limit has additional benefits. Determining Earth's carrying capacity for humans as simply a function of whether food and water will be available also ignores quality-of-life issues, or what some scientists call cultural carrying capacity. An Earth that was wholly given over to the production of food for the human population would lack wild, undisturbed places and the presence of species that nurture our sense of wonder and discovery. With human populations at the limits of growth, much of our creative energy would be used for survival, taking away our ability to make and enjoy music, art, and literature. (Chapter 25 addresses issues of sustainable food production that face the human population now and that will become more pressing as the population grows.)

Limiting human population growth also leaves room for nonhuman species. Human activity is posing a direct threat to the survival of a significant percentage of Earth's biodiversity (Chapter 16)—a threat that increases in direct proportion to the size and affluence of the planet's human population.

What we have learned is that scientists cannot tell us exactly how many people Earth can support, partly because humans make unpredictable choices and partly because humans have the capacity to innovate and adjust seemingly fixed biological limits. Ultimately, the question of how many people Earth should support—and at what quality of life or including support for nonhuman species—is a question not only of science but also of our values.

FIGURE 15.13 Reducing population growth through opportunity. The Grameen Bank in Bangladesh provides small loans to help raise people from poverty. Women make up 97% of the borrowers, and successful women borrowers are less likely to have large families.

Got It?

1. When natural populations exceed their environment's carrying capacity, they may experience a steep population ________.
2. Populations with ________ birth rates may experience population cycles of repeated "booms" and "busts."
3. Demographic momentum is the tendency for a population to grow in size even after birth rates and death rates become ________.
4. A population pyramid allows one to calculate the future growth of a population because it contains information about the proportion of the population that is of ________ ________.
5. An effective way to reduce human population growth rates in a society is to improve the status of ________ in that society.

The exponential growth of populations can be challenging to comprehend, in part because exponential increases are not as familiar as simple additive increases. In fact, people often grossly underestimate the effect of exponential growth. Consider the following job offer: A corporate executive needs a temporary assistant for a 30-day assignment. The executive proposes to pay you either (a) $1000/day for 30 days, or (b) to start at 1 cent per day and then double the salary on each subsequent day (such that you earn 1 cent on the first day, 2 cents on the second day, 4 cents on the third day, 8 cents on the fourth day, and so on). You take the first option, reckoning that

$1000 per day for 30 days is a much better salary than one that starts at 0.001% of that rate and doubles every day for 30 days.

Sounds right, but it isn't.

Answer the following questions to understand why.

1. What is the definition of exponential growth?
2. Which salary proposal is an example of exponential growth?
3. What would your earnings be for the month if you choose proposal (a)?
4. What would your earnings be for the 25th day of the month if you choose proposal (b)? (You may need to make a table to calculate this, but you can also use the following equation: daily wage = 2^n, where n = the day.)
5. Reflect on your answers to questions 1–4 above and explain why the statement bolded above sounds right, but isn't.

THE **BIG** QUESTION

Should the international community continue to provide food aid to populations experiencing food crises?

Even though the world's total food production exceeds the current population's calorie needs, periodic localized food crises still occur. These crises are more likely when armed conflict interferes with the production and delivery of food, or when weather conditions cause local crop failures. The response of the international community to food crises is typically to provide aid in the form of shipments of staple foods such as rice or corn meal. However, some economists have argued that providing food aid to hungry populations is counterproductive because it undermines local farmers. For this reason, they argue that food aid should be provided in only the most extreme and limited circumstances, such as during a famine.

What should I know?

What follows are some smaller questions that would need to be resolved to answer the Big Question. Place a checkmark next to the questions that science can answer.

Smaller Questions	Can Science Answer?
Does charitable food aid provided to a population reduce the number of individuals willing to work to produce crops locally?	
Is it wrong to deny hungry impoverished people free food when we could easily provide it to them?	
Do food shortages that are not as severe as famines have negative effects on children's growth and development?	
Does providing food that is not locally produced nor part of the local diet have a negative effect on the health of a population?	
Do wealthy countries materially benefit from providing food aid to populations in food crisis?	

What does the science say?

Let's examine what the data say about this smaller question:

Do food shortages that are not as severe as famines have negative effects on children's growth and development?

Scientists surveyed the research literature that examined the relationship between "height for age" in children—a measure of nutritional status—and various aspects of motor and cognitive development. Three of the studies performed in low- and moderate-income countries examined the likelihood of walking at age 2 as a function of height by age. Pooling the results from these three studies provided the data that are plotted on the following graph:

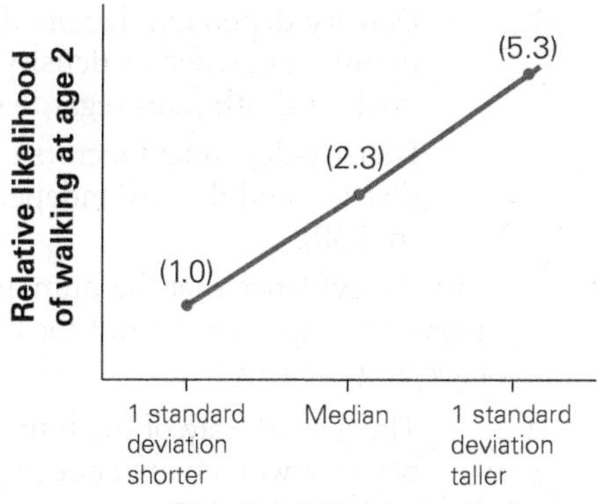

—Continued next page

1. Describe the results. Does it appear that child malnutrition negatively affects motor development?
2. Given these data, do you think the smaller question is answered? If not, propose another study that would help answer this question.
3. Does this information help you answer the Big Question? What else do you need to consider?

Data source: C. R. Sudfeld, D. C. McCoy, G. Danaei, G. Fink, M. Ezzati, K. G. Andrews, and W. W. Fawzi, "Linear Growth and Child Development in Low- and Middle-Income Countries: A Meta-Analysis," *Pediatrics* 135, no. 5 (2015): e1266–e1275.

Chapter Review

Mastering Biology

Go to Mastering Biology to access **eText 2.0, Dynamic Study Modules,** and the **Study Area,** where you'll find practice quizzes, BioFlix™ animations, MP3 tutor sessions, current events, and more.

SUMMARY

Section 15.1

Define *population,* and describe the aspects of populations that are typically measured by ecologists.

- A population is defined as a group of individuals of the same species living in a fixed area (p. 324).
- Ecologists measure a population by the number of individuals and their dispersion (p. 324).

Describe how a mark-recapture estimate of population size is performed, and estimate the size of a population from mark-recapture data.

- Population size can be estimated by capturing a sample of the population, marking the individuals, freeing them, and then capturing another sample. The proportion of marked to unmarked individuals allows calculation of total population size (pp. 324–325).

Explain how the size of a population that is experiencing exponential growth changes over time.

- Exponential growth is an increase in numbers as a function of the current population size. On a graph, it takes the form of a J-shaped curve (pp. 325–326).

Describe the demographic transition in human populations and how it contributes to population growth.

- In most populations, a decrease in death rates is followed by a decrease in birth rates. The gap between when each drop is called the demographic transition (pp. 327–328).

SHOW YOU KNOW 15.1 Life expectancy is the average age a newborn is expected to reach. Decline in infant mortality dramatically increases life expectancy in a country, even if adults are not living longer. Why is this the case?

Section 15.2

Define *carrying capacity,* and explain its relationship to logistic growth.

- The carrying capacity of the environment is the maximum population that can be supported indefinitely given current conditions. This limit leads to logistic growth, which appears as an S-shaped curve on a graph (p. 329).

Compare and contrast the effect of density-dependent and density-independent factors on population growth.

- Density-dependent factors depend on the size of the population, whereas density-independent factors affect birth or death rates regardless of population size (p. 329).
- Density-dependent factors include starvation and disease, and density-independent factors include weather (p. 330).

List the evidence that the human population may not be near carrying capacity and the evidence that it may be near carrying capacity.

- The growth rate of the human population is declining, because women are choosing to have fewer children (p. 331).

- The energy received from the sun each year could potentially support a population of 21 billion people (p. 331).
- Humans' reliance on nonrenewable resources may inflate the actual carrying capacity of Earth (pp. 331–332).

SHOW YOU KNOW 15.2 Why are infectious diseases considered a density-dependent factor rather than density-independent?

Section 15.3

Describe the conditions (in both human and other populations) under which a population crash can occur.

- Fast-growing populations that overshoot their environment's carrying capacity may experience a crash or go through periodic booms and busts (p. 333).
- It is possible that the human population will overshoot Earth's carrying capacity because of our reliance on nonrenewable resources and due to demographic momentum. (pp. 333–334).

SHOW YOU KNOW 15.3 How does demographic momentum explain why species with high birth rates are more likely to experience population cycles compared with populations with low birth rates?

ROOTS TO REMEMBER

The following roots of words come mainly from Latin and Greek and will help you to decipher terms:

demo-	is from the word meaning people. Chapter term: *demographic*
expo-	is from the word meaning to put forth. Chapter term: *exponential*
log-	means proportion or ratio. Chapter term: *logistic*
popula-	means multitude. Chapter term: *population.*

LEARNING THE BASICS

1. Add labels to the figure that follows, which illustrates the predicted pattern of changes in human population growth rates when death rates fall.

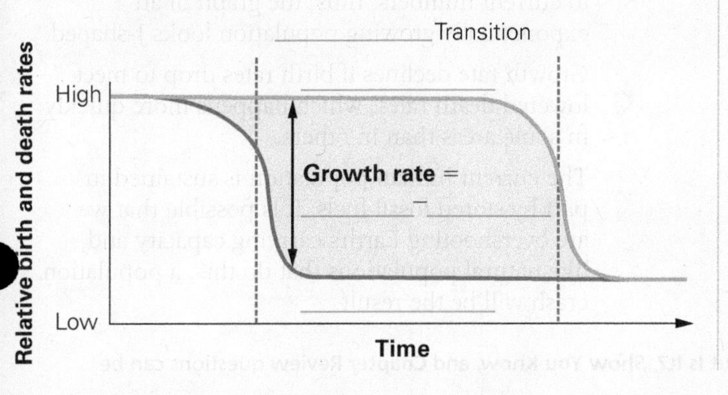

2. Explain why a decrease in population growth rate is expected as a nonhuman population approaches carrying capacity.
3. When individuals in a population are evenly spaced throughout their habitat, their dispersion is termed as ________.
   A. clumped; B. uniform; C. random; D. excessive; E. exponential
4. A population growing exponentially ________.
   A. is stable in size; B. adds a fixed number of individuals every generation; C. adds a larger number of individuals in each successive generation; D. will likely expand forever; E. will not crash
5. According to the graph shown here, the carrying capacity for fruit flies in the environment of the culture bottle is ________.

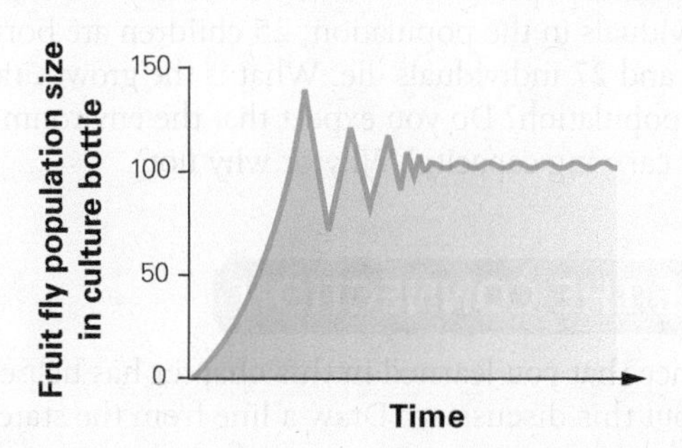

   A. 0 flies; B. 100 flies; C. 150 flies; D. between 100 and 150 flies; E. impossible to determine
6. All of the following are density-dependent factors that can influence population size *except* ________.
   A. weather; B. food supply; C. waste concentration in the environment; D. infectious disease; E. supply of suitable habitat for survival
7. In contrast to nonhuman populations, human population growth rates have begun to decline due to ________.
   A. voluntarily increasing death rates; B. voluntarily decreasing birth rates; C. involuntary increases in death rates; D. involuntary decreases in birth rates; E. density-dependent factors
8. Populations that rely on stored resources are likely to overshoot the carrying capacity of the environment and consequently experience a(n) ________.
   A. demographic momentum; B. cultural carrying capacity; C. decrease in death rates; D. population crash; E. exponential growth
9. The current carrying capacity of Earth for the human population may have been inflated by ________.
   A. demographic momentum; B. the tendency for women to want to control family size; C. an artificially low number of density-independent factors; D. our use of fossil fuels; E. recent population crashes

10. Demographic momentum refers to the tendency for ________.
   A. low population growth rates to continue to decline; B. high population growth rates to continue to increase; C. populations to continue to grow in number even when growth rates reach zero; D. populations to continue to grow in number even when women are reducing the number of children they bear; E. women to continue to have children even though they no longer wish to

ANALYZING AND APPLYING THE BASICS

1. A researcher captures 50 penguins, marks them with a spot of paint on their bills, and releases them. One month later, the researcher returns, captures another 50 penguins, and notes that only 1 has a previous mark. What is the likely size of the total penguin population in the researcher's study area?
2. Consider a population in which, for every thousand individuals in the population, 25 children are born each year and 27 individuals die. What is the growth rate of this population? Do you expect that the environment is near carrying capacity? Why or why not?
3. Review the graph shown in question 5 of Learning the Basics. How would you expect the carrying capacity of the population to change if the flies are supplied with more food? What other factors might influence the carrying capacity in this environment?

GO FIND OUT

1. Search for data from the Census about the population trends in your local community (this could be metropolitan area, state, or some other geographic boundary). How has the population in this location changed over the past 100 years? Using the resources available to you, try to determine the factors that have been most important in driving this population change (e.g., immigration rates, birth rates, etc.)
2. Africa is the only continent where increases in food production have not outpaced human population growth. Many of the most severe food crises are in African countries. Should those of us in the more developed world assist African populations? How? What factors influence your thoughts on this question?

MAKE THE CONNECTION

The science that you learned in this chapter has helped you better understand the real-world example used throughout this discussion. Draw a line from the statement on the left to the science that supports it on the right.

The number of humans on Earth has increased dramatically since 1800, from 1 billion to the current 7.3 billion. ▷	◁ A population's growth rate is equal to its birth rate minus its death rate.
The growth rate of the human population has increased thanks primarily to advancements in human health. ▷	◁ Carrying capacity is the largest population an environment can support over the long term.
The human populations of some countries are growing much faster than the human population of others. ▷	◁ Density-dependent factors are those that affect growth rates in a population as it approaches carrying capacity.
Earth's human population faces an unknown limit to growth. ▷	◁ The growth rate of the human population is slowing, but human health and wellness appears to be increasing.
Increases in famine or disease may be a sign that the human population is reaching its maximum size. ▷	◁ Unlike wild species, humans, especially women, voluntarily regulate birth rates.
Other signs indicate that humans may be able to avoid the widespread disasters associated with populations near the maximum size. ▷	◁ Exponential growth is growth in proportion to current numbers; thus, the graph of an exponentially growing population looks J-shaped.
One unknown is whether human population growth rates can slow fast enough to avoid a population crash. ▷	◁ Growth rate declines if birth rates drop to meet lowered death rates, which happens more quickly in some areas than in others.
An effective mechanism for reducing the growth rate of the human population is to increase the education and status of women. ▷	◁ The current human population is sustained in part by stored fossil fuels. It is possible that we are overshooting Earth's carrying capacity and, like natural populations that do this, a population crash will be the result.

Answers to **Got It?, Visualize This, Working with Data, Sounds Right, But Is It?, Show You Know,** and **Chapter Review** questions can be found in the **Answers** section at the back of the book.

16 Conserving Biodiversity

The rufus red knot travels 18,000 miles during its yearly migration. One tagged red knot, known as Moonbird, covered more than 400,000 miles in its lifetime.

The moonbird survived a population crash caused when an important food source was overharvested in the 1990s.

Restrictions on crab harvests seem to have stopped the species population decline but have harmed those who make a living fishing.

Community and Ecosystem Ecology

He might be the closest any single wild bird came to being a celebrity. Moonbird was a male rufus red knot (*Calidris canutus rufa*) tagged with leg bands by researchers in Argentina in early 1995. By the time Moonbird was last spotted in Delaware, in May 2014, he had traveled more than 400,000 miles over the 19 years that he had annually migrated between the Arctic and Argentina. As his name suggests, this little bird had flown the distance to the moon and back! Moonbird's story of survival inspired many and has spawned a book, countless news stories, and even a commemorative statue in Mispillion Harbor.

In addition to his stamina, Moonbird was remarkable for his survival. When he was banded, Moonbird was one of an estimated 150,000 birds of his subspecies. By 2013, the rufus red knot population was only about 30,000. In 2007, scientists were predicting the total extinction of the birds by 2013. That hasn't happened, but the population today is a fraction of its former size.

The dramatic decline in rufus red knots led to its addition to the list of threatened species in 2013. Already the decline in the rufus red knot population has led to restrictions on human activity. In particular, the Atlantic States Marine Fisheries Commission has limited the harvest of horseshoe crabs, the eggs of which are a primary food source for red knots during their migration. New Jersey went even further, banning horseshoe crab harvesting altogether. These restrictions seem to have made a difference—the population of horseshoe crabs in Delaware Bay has begun to rebound dramatically in the past several years, and rufus red knot populations, although still low, appear to have stabilized.

The ban on horseshoe crab harvesting is not without cost. The crabs are a valuable commercial species. People who harvested these crabs have lost a major source of income, thanks to concerns about a single shorebird.

—*Continued next page*

Should the survival of one or a few species come before the needs of humans? Biologists say that the concept of "birds versus people" is the wrong way to phrase the argument. Instead, they say, protecting endangered species is not about pitting birds against people; it is about protecting birds—and the crabs they depend on—to ensure the survival of all. In this chapter, we explore the causes of the loss of biological diversity and the consequences to all of us.

16.1 The Sixth Extinction

The U.S. **Endangered Species Act (ESA)** is a law passed in 1973 to protect and encourage the population growth of threatened and endangered species. Endangered organisms are at high risk of **extinction,** defined as the complete loss of a species or subspecies, and *threatened* organisms are at high risk of becoming endangered. Rapidly declining species such as the rufus red knot are exactly the type of organisms that legislators had in mind when they enacted the ESA.

The ESA was passed because of the public's concern about the continuing erosion of **biodiversity**, the entire variety of living organisms. Whooping cranes, sea otters (**FIGURE 16.1**), passenger pigeons, black-footed ferrets, and spotted skunks—once abundant species—are extinct or highly threatened in the United States as a result of human activity. The ESA was a response to the unprecedented and rapid rate of species loss at human hands. Once a species is officially listed as endangered or threatened, land necessary for the survival and recovery of the species—called "critical habitat"—is identified. It is then illegal to possess or transport these species and also to destroy critical habitat. According to a recent review, since 1972 the protection provided by the ESA has been essential in the recovery of nearly 100 species from the brink of extinction.

FIGURE 16.1 A critically endangered species. Sea otters were nearly hunted to extinction by fur traders in the 1700s. Although their populations have recovered thanks in part to the protections of the Endangered Species Act, they are still endangered by human activity.

Critics of the ESA argue that the goal of saving all species from extinction is unrealistic. Such critics argue that, after all, extinction is a natural process—the approximately 10 million species living today constitute fewer than 1% of the species that have ever existed. In the next section, we explore the scientific questions posed by ESA critics: How does the rate of extinction today compare with the rates in the past? Is the ESA just attempting to postpone the inevitable, natural process of extinction?

Measuring Extinction Rates

If the ESA critics who say that the current rate of species extinctions is "natural" and constant and not a result of human activity are correct, then the extinction rate today should be roughly equal to the rate that existed before humans evolved.

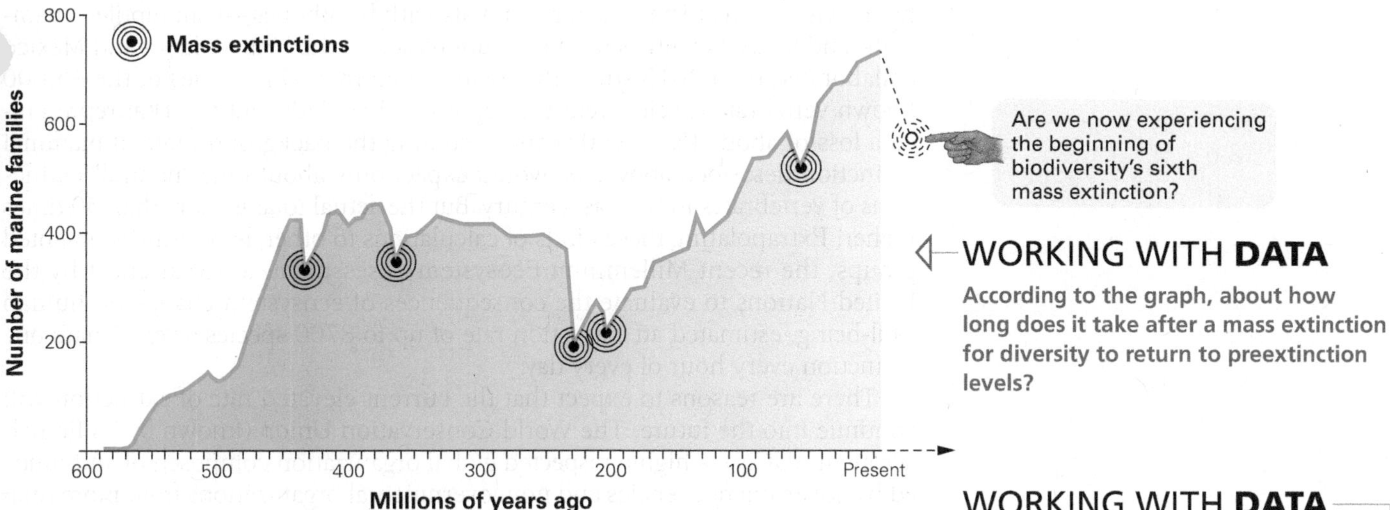
Mass extinctions
Number of marine families
800
600
400
200
600
500
400
300
200
100
Present
Millions of years ago
Are we now experiencing the beginning of biodiversity's sixth mass extinction?
WORKING WITH DATA
According to the graph, about how long does it take after a mass extinction for diversity to return to preextinction levels?

FIGURE 16.2 Mass extinction. The history of life on Earth is reflected by this graph plotting the change in the number of families of fossil marine organisms over time. The number of families has increased over the past 600 million years. However, this rise has been punctuated by five mass extinctions (marked here with black circles), each resulting in a global decline in biodiversity.

FIGURE 16.2 illustrates the diversity of marine organisms—which have the longest continuous fossil record—over time. This is a typical example of what the fossil record tells scientists about the history of biodiversity; since the rapid evolution of a wide variety of animal groups approximately 580 million years ago, the number of families of organisms has generally increased. However, this increase in biodiversity has not been smooth or steady.

The history of life has been punctuated by five **mass extinctions**—species losses that are global in scale, affect large numbers of species, and are dramatic in impact. Past mass extinctions were probably caused by massive global changes—for instance, climate fluctuations that changed sea levels, continental drift that changed ocean and land forms, and an asteroid impact that caused widespread destruction and climate change. Many scientists argue that we are now seeing biodiversity's sixth mass extinction, this one caused by human activity.

Determining whether the current rate of extinction is unusually high requires knowledge of the **background extinction rate,** the rate at which species are lost through the normal evolutionary process. Normal extinctions occur when a species lacks the variability to adapt to environmental change or when a new species arises due to the evolution of its ancestor species. The fossil record can provide clues about the background extinction rate that results from this continual process of species turnover.

The span of rock ages in which fossils of a species are found represents the life span of that species (**FIGURE 16.3**). Using these data, biologists have determined that the average life span of a marine species is around 1 million years. To maintain the current level of diversity, we would therefore expect that the background rate of extinction is about one species per million (0.0001%) per year. A recent analysis of extinction rates in the mammal fossil record indicates a similar rate of 0.0002%.

The current rate of extinction is calculated from known species disappearances. The most complete records of documented extinction occur in groups of

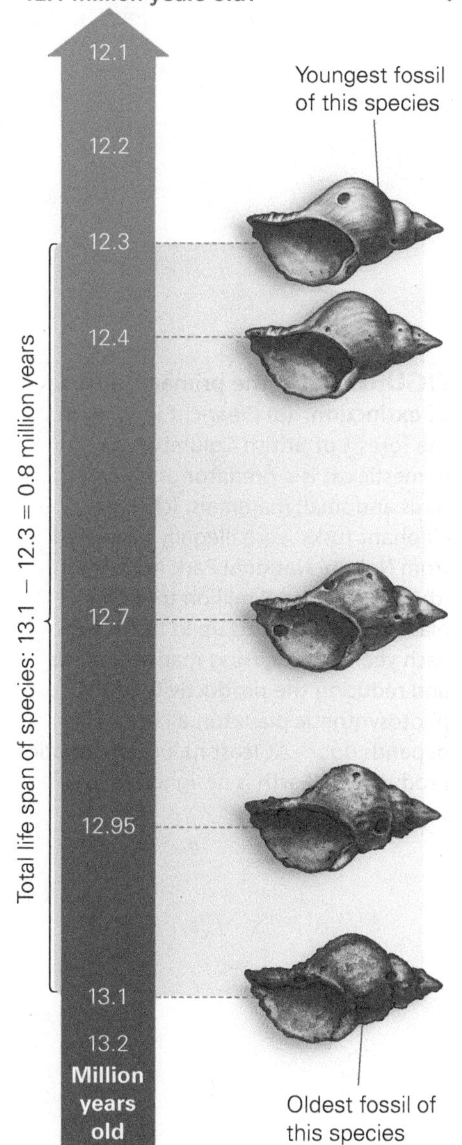
WORKING WITH DATA
What would be the new estimated life span of this species if another fossil were found in rocks dated only 12.1 million years old?
12.1
12.2
12.3
12.4
12.7
12.95
13.1
13.2
Million years old
Youngest fossil of this species
Oldest fossil of this species
Total life span of species: 13.1 − 12.3 = 0.8 million years

FIGURE 16.3 Estimating the life span of a species. Fossils of the same species are arranged on a timeline from oldest to youngest. The difference in age between the oldest and youngest fossil of a species is an estimate of the species' life span.

highly visible organisms, such as animals with backbones—fish, reptiles, mammals, and birds. Scientists from institutions across the United States and Mexico collaborated on a 2015 study that determined that 477 species of the 45,000 known vertebrate species went extinct in the twentieth century. That represented a loss of about 1% in 100 years. Assuming the background rate of mammal extinction described above, we would expect only about nine "natural" extinctions of vertebrates in the past century. But the actual total is more than 50 times higher! Extrapolating these kinds of calculations to other, less-well-documented groups, the recent Millennium Ecosystem Assessment, a 5-year effort by the United Nations to evaluate the consequences of ecosystem change on human well-being, estimated an extinction rate of up to 8700 species a year. That's one extinction every hour of every day.

There are reasons to expect that the current elevated rate of extinction will continue into the future. The World Conservation Union (known by its French acronym, IUCN), a highly respected global organization composed of and funded by government agencies and nongovernmental organizations from more than 140 countries, collects and coordinates data on threats to biodiversity. According to the IUCN's most recent analysis of species groups, 42% of amphibians, 26% of mammals, 13% of birds, and 40% of conifers are in danger of extinction (TABLE 16.1). Other taxonomic groups are less well studied, so total estimates for all groups are not available. But the overall picture is bleak; the IUCN identifies more than 23,000 species that are known to be in danger of extinction.

Causes of Extinction

The IUCN attempts to identify all the threats that put a particular species at risk of extinction. For most species at risk, some type of human activity is to blame. The most severe threats belong to one of four general categories: loss or degradation of habitat, introduction of nonnative species, overexploitation, and effects of pollution (**FIGURE 16.4**). Of these categories, the first poses the most

(a) Habitat degradation

(b) Introduced species

(c) Overexploitation of species

(d) Pollution

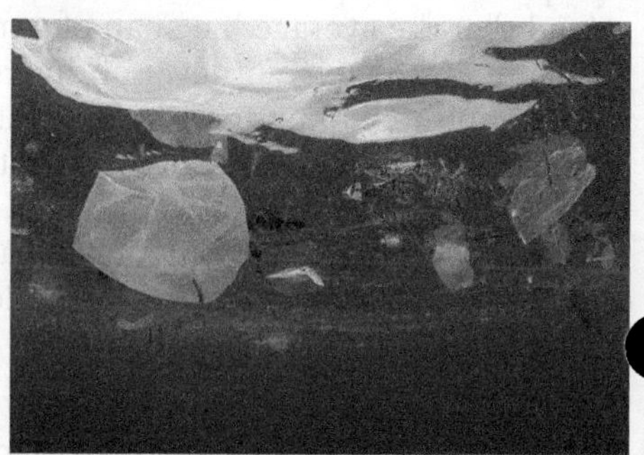

FIGURE 16.4 The primary causes of extinction. (a) Clear-cut logging in the forests of British Columbia. (b) The domestic cat is a predator of native birds and small mammals. (c) These elephant tusks were illegally harvested from Nairobi National Park in Kenya. (d) An estimated 8 million tons of plastic pollution ends up in the oceans each year, endangering marine animals and reducing the productivity of the photosynthetic plankton all ocean life depends upon. At least half of the plastic produced on Earth is designed for a single use.

serious threat; the IUCN estimates that 83% of endangered mammals, 89% of endangered birds, and 91% of endangered plants are directly threatened by damage to or destruction of the species' **habitat,** the place where they obtain their food, water, shelter, and space.

TABLE 16.1 Known threatened and endangered species.

	Estimated Number of Described Species	Number of Threatened Species in IUCN 2016 Red List	Best Estimate of % of Species Threatened in Group
Vertebrates			
Mammals	5536	1208	26%
Birds	10,424	1375	13.45%
Reptiles	10,450	989	
Amphibians	7538	2063	42%
Fishes	33,000	2343	
Invertebrates			
Insects	1,000,000	1156	
Molluscs	85,000	1967	
Crustaceans	47,000	729	
Corals	2175	237	
Arachnids	102,248	166	
Horseshoe crabs	4	1	25%
Plants			
Mosses	16,236	76	
Ferns and related	12,000	217	
Gymnosperms	1052	400	40%
Flowering plants	268,000	10,875	
Green algae	6050	0	
Red algae	7104	9	
Fungi and Protists			
Lichens	17,000	7	
Mushrooms	31,496	22	
Brown algae	3784	6	
Total	1,735,220	23,928	

Entries in red are known well enough that a reasonable estimate of the percentage of endangered in the group can be made.

Data source: International Union for Conservation of Nature and Natural Resources. 2016. The IUCN Red List of Threatened Species. Version 2016-2. Cambridge, UK: IUCN. Accessed September 4, 2016. http://www.iucnredlist.org.

FIGURE 16.5 Lost habitat = lost species. The Karner blue butterfly is native to pine barrens in eastern North America. Only about 5% of the original habitat of this type is still intact.

Habitat Destruction and Fragmentation. Habitat destruction is a major risk to species around the globe (**FIGURE 16.5**). Rates of habitat destruction caused by agricultural, industrial, and residential development accelerated everywhere throughout the twentieth century as the human population swelled from less than 2 billion in 1900 to more than 7 billion today. As the amount of natural landscape declines, the number of species supported by the habitats in these landscapes naturally decreases.

The relationship between the size of a natural area and the number of species that it can support follows a general pattern called a **species-area curve.** A species-area curve for reptiles and amphibians on a West Indies archipelago is illustrated in **FIGURE 16.6a**. Similar graphs have been generated in studies of different groups of organisms in a variety of habitats. The general pattern in all these graphs is that the number of species in an area increases rapidly as the size of the area increases, but the rate of increase slows as the area becomes very large. This rule of thumb is shown in **FIGURE 16.6b**. We can use this curve to estimate rates of extinction in regions that are rapidly being modified by human activity but difficult to survey extensively. For example, from the graph in Figure 16.6b, we can estimate that a 90% decrease in landscape area will cut the number of species living in the remaining area by half.

Using images from satellites, scientists have estimated that approximately 20% of the Brazilian Amazon rain forest has been lost since 1970: according to the species-area curve, that translates to a loss of about 5% of known species. This equals approximately 10 species of mammal, 20 species of amphibian, 65 species of bird, and 2000 plant species lost from this ecosystem just in the past half century.

Of course, habitat destruction is not limited to tropical rain forests. Freshwater lakes and streams, grasslands, and temperate forests are also experiencing high levels of modification. Many of the beaches that serve as stopover sites for the rufus red knot are affected by human development and use, contributing to the threats this species faces. According to the IUCN, if habitat destruction

WORKING WITH DATA

Using part (a) of this figure, calculate how many species of reptiles and amphibians you would expect to find on an island that is 15,000 square kilometers in area. Imagine that humans colonize this island and dramatically degrade 7500 square kilometers of the natural habitat. Using part (b), calculate the percentage of the species originally found on the island you would expect to become extinct.

(a) Species diversity increases with area.

Number of species: 0, 20, 40, 60, 80, 100

Island area (in thousands of km^2): 0, 5, 10, 15, 20, 25, 30, 35

(b) Habitat reduction is predicted to result in loss of species.

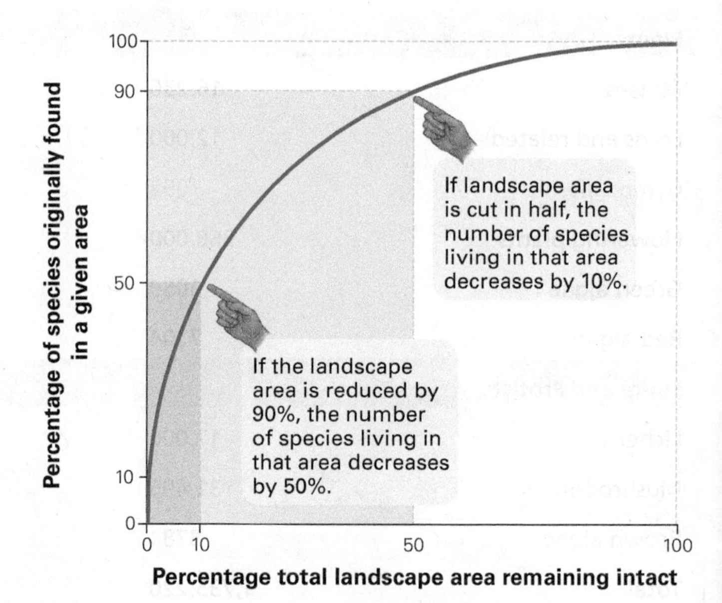

FIGURE 16.6 Predicting extinction caused by habitat destruction. (a) This curve demonstrates the relationship between the size of an island in the West Indies and the number of reptile and amphibian species living there. (b) We use a generalized species-area curve to roughly predict the number of extinctions in an area experiencing habitat loss.

around the world continues at its present rate, nearly one-fourth of *all* living species will be lost within the next 50 years.

Some critics have argued that these estimates of future extinction are too high because not all groups of species are as sensitive to habitat area as the curve in Figure 16.6b suggests. Many species may still survive and even thrive in human-modified landscapes. Other biologists contend that there are additional threats to species, including **habitat fragmentation,** and therefore the rate of species loss is likely to be even higher than these estimates.

Habitat destruction rarely results in the complete loss of a habitat type. Often what results from human activity is fragmentation, in which large areas of intact natural habitat are subdivided. Habitat fragmentation is especially threatening to large predators, such as grizzly bears and tigers, because of their need for large hunting areas.

Large predators require large, intact hunting areas due to a basic rule of biological systems: Energy flows in one direction within an ecological system along a **food chain,** which typically runs from the sun to **producers** (photosynthetic organisms), to the **primary consumers** that feed on them, to **secondary consumers** (predators that feed on the primary consumers), and so on. Along the way, most of the calories taken in at one **trophic level** (that is, a level of the food chain) are respired simply to support the activities of the individuals at that level. In other words, a substantial amount of the energy individuals take in is dissipated as heat. Each level of the food chain thus has significantly less energy available to it compared with the next lower level. You can see this in your own life; an average adult needs to consume between 1600 and 2400 kilocalories per day simply to maintain his or her current weight.

trophic means food, nourishment.

The flow of energy along a food chain leads to the principle of the **trophic pyramid,** the bottom-heavy relationship between the **biomass** (total weight) of populations at each level of the chain (**FIGURE 16.7**). Habitat destruction and fragmentation cause the lower levels of the pyramid to shrink, depriving the top predators of adequate calories for survival.

Habitat fragmentation also exposes mobile species to additional dangers. For example, the survival of the tiny population of Florida panthers—approximately 150 individuals—is threatened by high levels of direct mortality. Of the 37 recorded panther deaths in Florida in 2015, the majority (26) were hit by cars while the animals were moving from one fragment of habitat to another.

Even the species that do survive in small fragments of habitat are made more susceptible to extinction by isolation. Habitat fragmentation often makes it

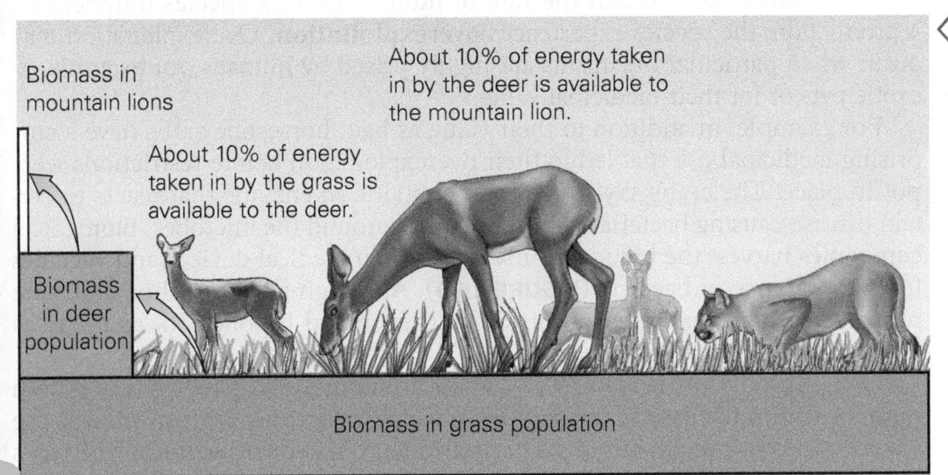

Biomass in mountain lions
About 10% of energy taken in by the deer is available to the mountain lion.
About 10% of energy taken in by the grass is available to the deer.
Biomass in deer population
Biomass in grass population

FIGURE 16.7 A trophic pyramid. Because most of the energy consumed by a trophic level is used within that level for maintenance, biomass decreases as position in the food chain increases.

WORKING WITH DATA

Imagine that the population of deer is 10,000 in this ecosystem. How many mountain lions would you expect if the relationship illustrated holds true?

impossible for individuals to follow changes in food sources or available nesting or growing sites. Isolated populations are also subject to a lack of genetic diversity resulting from inbreeding, as we discuss in detail later in the chapter.

Although habitat destruction and fragmentation are the gravest threats to endangered species, the other threats are not insignificant. According to the IUCN, activity unrelated to habitat modification plays a role in about 40% of all cases of endangerment.

Introduced Species. Another major player in species endangerment is nonnative species that are introduced to an environment. Introduced species include organisms brought by human activity, either accidentally or purposefully, to new environments. Introduced species are often dangerous to native species because they have not coevolved with them.

FIGURE 16.8 An introduced species. Pale swallow-wort (*Cynanchum rossicum)* was introduced as an ornamental plant in the late nineteenth century. It forms extensive patches that crowd out native plant species. Swallow-wort is also a "trap" for monarch butterflies, because adults mistake the plant for their normal host milkweed—eggs laid on swallow-wort do not develop.

Coevolution occurs when pairs or groups of species adapt to each other via natural selection. For instance, many birds on oceanic islands such as Hawaii and New Zealand evolved in the absence of ground predators and have not evolved behaviors or other strategies to escape or combat these predators—therefore, introduced carnivores rapidly deplete them via hunting. In Hawaii, the Pacific black rat, accidentally introduced from the holds of visiting ships, became very adept at raiding eggs from nests and contributed to the extinction of dozens of species of honeycreepers, birds found nowhere else on Earth. Even domestic cats, deliberately introduced by people who sought to control rodents around their houses, can take an enormous toll on wildlife. A recent study by the Smithsonian Conservation Biology Institute estimated that free-roaming domestic cats kill between 1.4 and 3.7 billion birds in the continental United States every year.

Introduced species may also compete with native species for resources, causing populations of the native species to decline. For example, zebra mussels, accidentally introduced to the Great Lakes through the ballast water in European trading ships, crowd out native mussel species as well as other organisms that filter algae from water, and the introduced vine pale swallow-wort, brought to the United States from Europe as an ornamental plant, takes over open landscapes in the Northeast, to the detriment of numerous native species (**FIGURE 16.8**).

Humans continue to move species around the planet. As the global trade in agriculture and other goods continues to expand, the number of species introductions is likely to increase over the next century.

Overexploitation. When the rate of human use of a species outpaces its reproduction, the species experiences **overexploitation.** Overexploitation may occur when particular organisms are highly prized by humans, for example, as exotic pets or for their medicinal value.

FIGURE 16.9 Beneficial blood. The blue tint of horseshoe crab blood derives from copper, which performs the same function as iron in our own blood. The blood from these animals is valuable as a screening tool to spot bacterial contamination in vaccines and medical devices.

For example, in addition to their value as bait, horseshoe crabs have a surprising medicinal use that led to their overexploitation before restrictions were put in place. The crab's sky-blue blood contains an enzyme that reacts to certain disease-causing bacteria by forming a clot around the microbes. Biomedical companies harvest the blood and use it to screen medical devices and vaccines for the presence of bacteria (**FIGURE 16.9**). Although the horseshoe crabs are returned to the ocean after a portion of their blood is collected, as many as 50% do not survive the procedure.

Overexploitation is also likely when an animal competes directly with humans—as in the case of the gray wolf, which was nearly exterminated in the United States by human hunters determined to protect their livestock. The rufus red knot and other species that cross international boundaries during migration are also susceptible to overexploitation because no single government regulates the total harvest; the same is true for species, like the horseshoe crab, that live in the oceans. The near extinction of several whale species in the nineteenth century was the result of the unregulated harvest of these animals by many nations. Many whales are still threatened by lack of agreement among countries on harvest limits, and stocks of cod, sea bass, and tuna are similarly at risk.

Pollution. The release of poisons, excess nutrients, and other wastes into the environment—a practice otherwise known as **pollution**—poses an additional threat to biodiversity. For example, the herbicide atrazine poisons frogs and salamanders in agricultural areas of the United States, and nitrogen pollution caused by overuse of fertilizer and car and smokestack exhaust has led to drastic declines of certain plant species within native grasslands in Europe.

Excess nutrients flowing into water threaten sea life in surprising ways. The nutrients cause increased production of algae, some of which produce toxins that can accumulate in small animals and poison the wildlife and people that consume them. Thousands of rufus red knots were killed along their migration route after feeding on mussels containing high levels of one such toxin.

Algae can be harmful even if they do not produce toxins. When a large algae population dies, bacteria feeding on dead cells consume the majority of oxygen in the water. This process of oxygen-depleting **eutrophication** results in large fish kills. Eutrophication threatens animals in hundreds of waterways all over the United States. In the Gulf of Mexico, for example, fertilizer from farm fields in the midwestern United States creates a 15,000-square-kilometer low-oxygen "dead zone" at the mouth of the Mississippi River every summer.

eu- means true or good.

Perhaps the most abundant pollutant released by humans is carbon dioxide, also a principal cause of global climate change (see Chapter 5). Computer models that link predicted changes in climate to known ranges and requirements of more than 1000 species of plants indicate that 15 to 37% of these plants face extinction in the next century as the climate changes. The rufus red knot is not immune from the threat of increasing carbon dioxide—from the impact of rising sea levels on the beaches it relies on to the dramatic changes in the Arctic tundra, where it makes its nest, climate change threatens these birds in several ways.

By nearly any measure, ESA critics who describe modern extinction rates as "natural" are incorrect. Over the past 400 years, humans have caused the extinction of species at a rate that far exceeds past rates. Human activities continue to threaten thousands of additional species around the world. Earth appears to be on the brink of a sixth mass extinction of biodiversity—and the massive global change responsible is human activity.

Many people feel a moral responsibility to minimize the human impact on other species and therefore support conservation. However, there is also a practical, human-centered reason to prevent the sixth extinction from occurring: the loss of biodiversity can hurt us as well, in ways we cannot even predict.

Got It?

1. At the current rate of species loss, Earth may be experiencing its sixth ________ ________ since the origin of life.
2. The species-area curve predicts the relationship between the size of a ________ ________ and the number of species it can support.
3. The trophic pyramid explains why the most endangered organisms in a food chain in an area experiencing habitat destruction are secondary ________ rather than those that are lower on the food chain.
4. An introduced species can be a threat to native species because it has not ________ with these natives.
5. Eutrophication leads to fish kills because the large populations of bacteria that result use up the ________ in the water.

16.2 The Consequences of Extinction

Concern over the loss of biodiversity is not simply a matter of an ethical interest in nonhuman life. Humans have evolved with and among the variety of species that exist on our planet, and the loss of these species often results in negative consequences for us.

Loss of Resources

The horseshoe crab has a clear direct benefit to humans, as already detailed. Among the other biological resources that are harvested directly from natural areas are species of wood for fuel and lumber, shellfish for protein, algae for gelatins, and herbs for medicines. The loss of any of these species affects human populations economically. One estimate places the value of wild species in the United States at $87 billion a year or about 4% of the gross domestic product.

Wild species also provide resources for humans in the form of unique biological chemicals. In fact, a number of valuable drugs, food additives, and industrial products derive from wild species. One dramatic example is the rosy periwinkle (*Catharanthus roseus*), which evolved on the island of Madagascar, one of the regions on Earth where biodiversity is most endangered (**FIGURE 16.10a**). Two drugs extracted from this plant, vincristine and vinblastine, have dramatically reduced the death rate from leukemia and Hodgkin's disease, two forms of cancer. If wild species go extinct before they are well studied, we will never know which ones might have provided compounds that would improve human lives.

Wild relatives of domesticated plants and animals, such as agricultural crops and cattle, are also important resources for humans. Genes and alleles that have been "bred out" of domesticated species are often still found in their wild relatives. These genetic resources are a reservoir of traits that can be reintroduced into agricultural species through breeding or genetic engineering. Agricultural scientists attempting to produce better strains of wheat, rice, and corn look to the wild relatives of these crops for genes conferring pest resistance and improved yields. For example, the Mexican teosinte species *Zea diploperennis* (**FIGURE 16.10b**) is an ancestor of modern corn. This species of teosinte is resistant to several viruses that plague cultivated corn; the genes that provide this resistance have been transferred to our domestic plants via hybridization.

By preserving wild relatives of domesticated crops in their natural habitats, scientists can also find resources that reduce pest damage and disease on the domestic crop. For example, the wasp *Catolaccus grandis* consumes boll weevils and has been used to control infestations of these pests in cotton fields (**FIGURE 16.10c**). *C. grandis* was discovered in the tropical forest of southern Mexico, where it parasitizes a similar pest in wild cotton populations.

Of course, introducing an insect such as *C. grandis* into a new environment carries risk, even if the introduction is meant to reduce environmental damage. We have already noted many examples of environmental disasters caused by introduced species. Often, a less risky approach to reducing pest damage to crops is to preserve nearby habitats and the ecological interactions that persist there.

(a) Rosy periwinkle

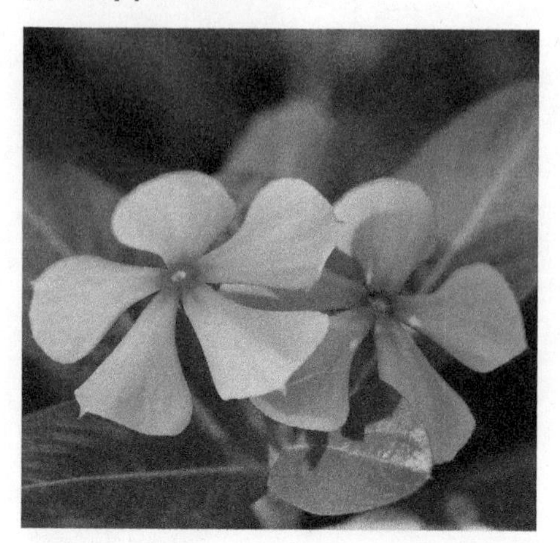

(b) Teosinte

(c) Boll weevil wasp

FIGURE 16.10 Resources from nature. (a) Anticancer drugs vincristine and vinblastine were first isolated from rosy periwinkle (*Catharanthus roseus*), a species of flower native to Madagascar. (b) Teosinte is the ancestor of modern corn, and is found in wild populations throughout Central America. This plant contains genes that might confer resistance to disease and drought in corn. (c) *Catolaccus grandis* was discovered preying on boll weevils on wild cotton in southern Mexico. This wasp is now released for the biological control of boll weevils on cotton crops throughout the world.

Predation, Mutualism, and Competition

Although humans receive direct benefits from thousands of species, most threatened and endangered species are probably of little or no use to people. Although birders and biologists would mourn its extinction, the loss of the rufus red knot would not likely cause direct harm to anyone.

In reality, most species are beneficial to humans because they are part of a biological **community,** consisting of all the organisms living together in a particular habitat area. Within a community, each species occupies a particular **ecological niche,** which can be thought of as the role or "job" of the species. The complex linkage among organisms inhabiting different niches in a community is often referred to as a **food web** (**FIGURE 16.11**). As with a spider's web, any disruption in one strand of the web of life is felt by other portions of the web. Some tugs on the web cause only minor changes to the community, and others can cause the entire web to collapse. Most commonly, losses of strands in the web are felt by a small number of associated species. The story of the

VISUALIZE THIS

Predict what would happen to the other species in this web if baleen whales went extinct.

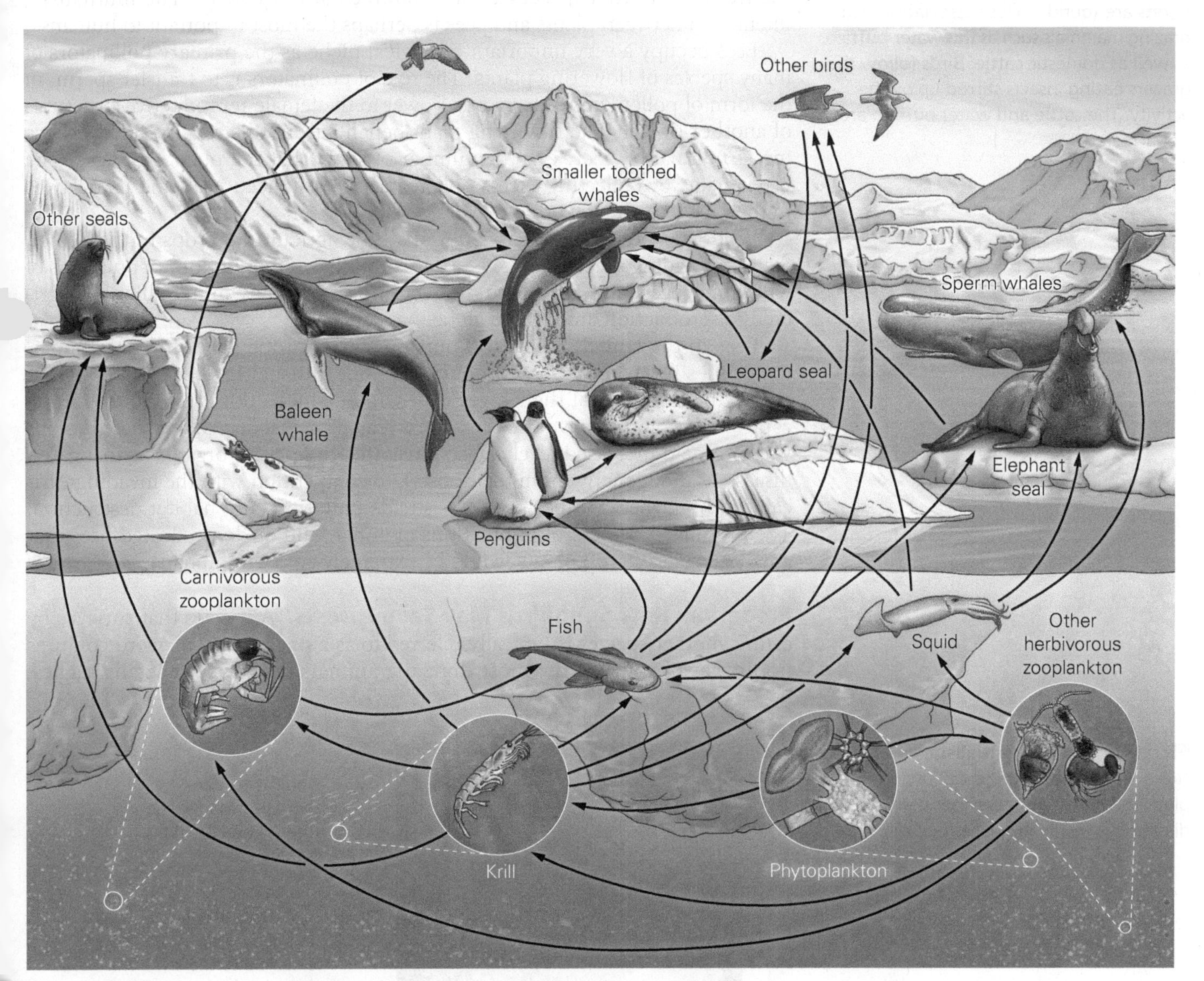

FIGURE 16.11 The web of life. Species are connected to other species in food chains—a network of food chains forms a food web. This drawing shows the feeding connections among species in the Antarctic Ocean. Black arrows represent feeding relationships; for example, penguins eat fish and in turn are eaten by leopard seals.

FIGURE 16.12 Commensalism. Cattle egrets are found in close association with grazing mammals such as this water buffalo, as well as domestic cattle. Birds follow the grazers eating insects stirred up by the activity. The cattle and water buffalo are not affected.

horseshoe crab and rufus red knot already shows us that. But some disruptions caused by the loss of seemingly insignificant species have the potential to be felt even by humans.

Mutualism: How Bees Feed the World. An interaction between two species that benefit each other is called **mutualism.** Mutualism can be contrasted with **commensalism,** a relationship in which one species benefits and the other is unaffected—for instance, the relationship between cattle egrets (a species of bird) and domestic cattle. The egrets follow the cattle as they graze, feeding on insects stirred up from the ground by these animals (**FIGURE 16.12**). The cattle do not appear to benefit or be harmed by the egrets' presence.

We find examples of mutualism in many environments. Cleaner fish that remove and consume parasites from the bodies of larger fish, fungal mycorrhizae that increase the mineral absorption of plant roots while consuming the plant's sugars, and ants that find homes in the thorns of acacia trees and defend the trees from other insects are all examples of mutualism. The mutualistic interaction between plants and bees is perhaps the most important to humans.

Bees occupy a very important ecological niche as the primary pollinators of many species of flowering plants. The role of pollinators is to transfer sperm, in the form of pollen grains, from one flower to the female reproductive structures of another flower. The flowering plant benefits from this relationship because insect pollination increases the number of seeds that the plant produces. The bee benefits by collecting excess pollen and nectar to feed itself and its relatives in the hive (**FIGURE 16.13**).

Wild bees pollinate at least 80% of all the agricultural crops in the United States, providing a net benefit of $10 to $15 billion. In addition, populations of wild honeybees have a major and direct impact on many more billions in agricultural production around the globe.

Bees in the United States and northwestern Europe have suffered dramatic declines in recent years. Although it is not unusual for beekeepers to lose approximately 20% of their hives over winter, since 2006 from 30 to 45% of captive bee colonies have been lost each year. The exact causes of these die-offs vary: from an increased level of bee **parasites** (infectious organisms that cause disease or drain energy from their hosts), to competition with the invading Africanized honeybees ("killer bees"), pesticide pollution, and habitat destruction. The prolonged decline of populations of either wild or domesticated bees that are mutualists of crop plants would be extremely costly to humans.

Predation: How Songbirds May Save Forests. A species that survives by eating another species is typically referred to as a **predator.** The word conjures up images of some of the most dramatic animals on Earth: cheetahs, eagles,

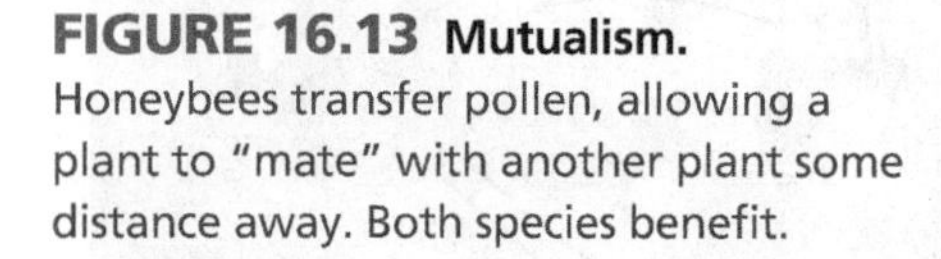

FIGURE 16.13 Mutualism. Honeybees transfer pollen, allowing a plant to "mate" with another plant some distance away. Both species benefit.

Benefit to bee:
It obtains food in the form of nectar and excess pollen.

Benefit to flower:
Its sperm (within the pollen) is carried to the female reproductive structures of another flower, enabling cross-pollination.

and killer whales. You might not picture wood warblers, a family of North American bird species characterized by their small size and colorful summer plumage, as predators; however, these beautiful songsters are voracious consumers of insects (**FIGURE 16.14a**). The hundreds of millions of individual warblers in the forests of North America collectively remove thousands of kilograms of insects from forest trees and shrubs every summer.

Most of the insects that warblers eat prey on plants. By reducing the number of insects in forests, warblers reduce the damage that insects inflict on forest plants. Reducing the amount of damage likely increases the growth rate of the trees. Harvesting trees for paper and lumber production fuels an industry worth more than $230 billion in the United States alone. At least some of this wood was produced because warblers were controlling insects in forests (**FIGURE 16.14b**).

Many species of forest warblers are experiencing declines in abundance. The loss of warbler species has several causes, including habitat destruction in their summer habitats in North America and their winter habitats in Central and South America. Warblers also face increased predation by animals whose populations benefit from human settlements, such as raccoons and housecats. Although other, less vulnerable birds may increase in number when warblers decline, these "replacement" birds are typically less insect-dependent. If smaller warbler populations correspond to lower forest growth rates and higher levels of forest disease, then these tiny, beautiful birds definitely have an important effect on the human economy.

(a) Black-throated blue warbler, predator of insects

(b) Forests suffer when insects are unchecked by predators.

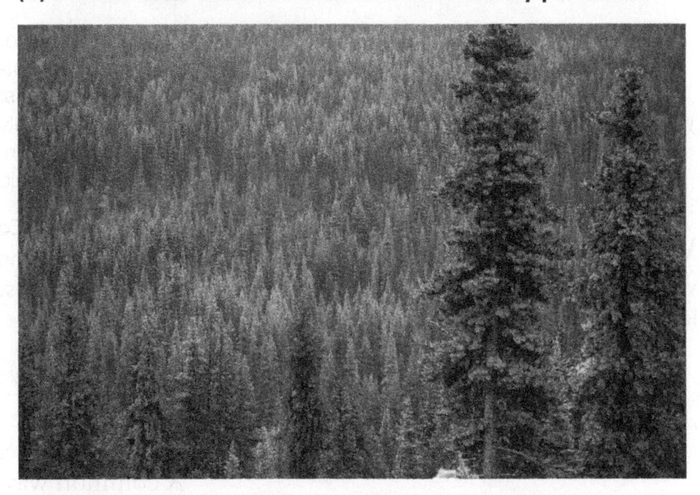

FIGURE 16.14 Predation. (a) The black-throated blue warbler is one of many warbler species native to North American forests. These birds are active predators of plant-eating insects. (b) Insects can kill trees, as seen in this photo of a spruce budworm infestation. Warblers and other insect-eating birds likely reduce the number and severity of such insect outbreaks.

Competition: How a Deliberately Infected Chicken Could Save a Life. When two species of organisms both require the same resources for life, they will be in **competition** for the resources within a habitat. In general, competition limits the size of competing populations. From a scientific perspective, to determine whether two species that seem to be using the same resource are competing, we remove one from an environment. If the population of the other species increases, then the two species are competitors.

We may imagine lions and hyenas fighting over a freshly killed antelope or weeds growing in our vegetable gardens as typical examples of competition, but most competitive interactions are invisible. The least visible competition occurs among microorganisms. However, microbial competition is often essential to the health of both people and ecological communities.

Salmonella enteritidis is a leading cause of food-borne illness in the United States. Between 2 million and 4 million people in this country are infected by *S. enteritidis* every year, experiencing fever, intestinal cramps, and diarrhea as a result. In about 10% of cases, the infection results in severe illness requiring hospitalization. Four to six hundred Americans die as a result of *S. enteritidis* infection every year.

Most *S. enteritidis* infections result from consuming undercooked poultry products, especially eggs. The U.S. Centers for Disease Control and Prevention estimate that as many as 1 in 50 consumers is exposed to eggs contaminated with *S. enteritidis* every year. Surprisingly, most of these eggs look perfectly

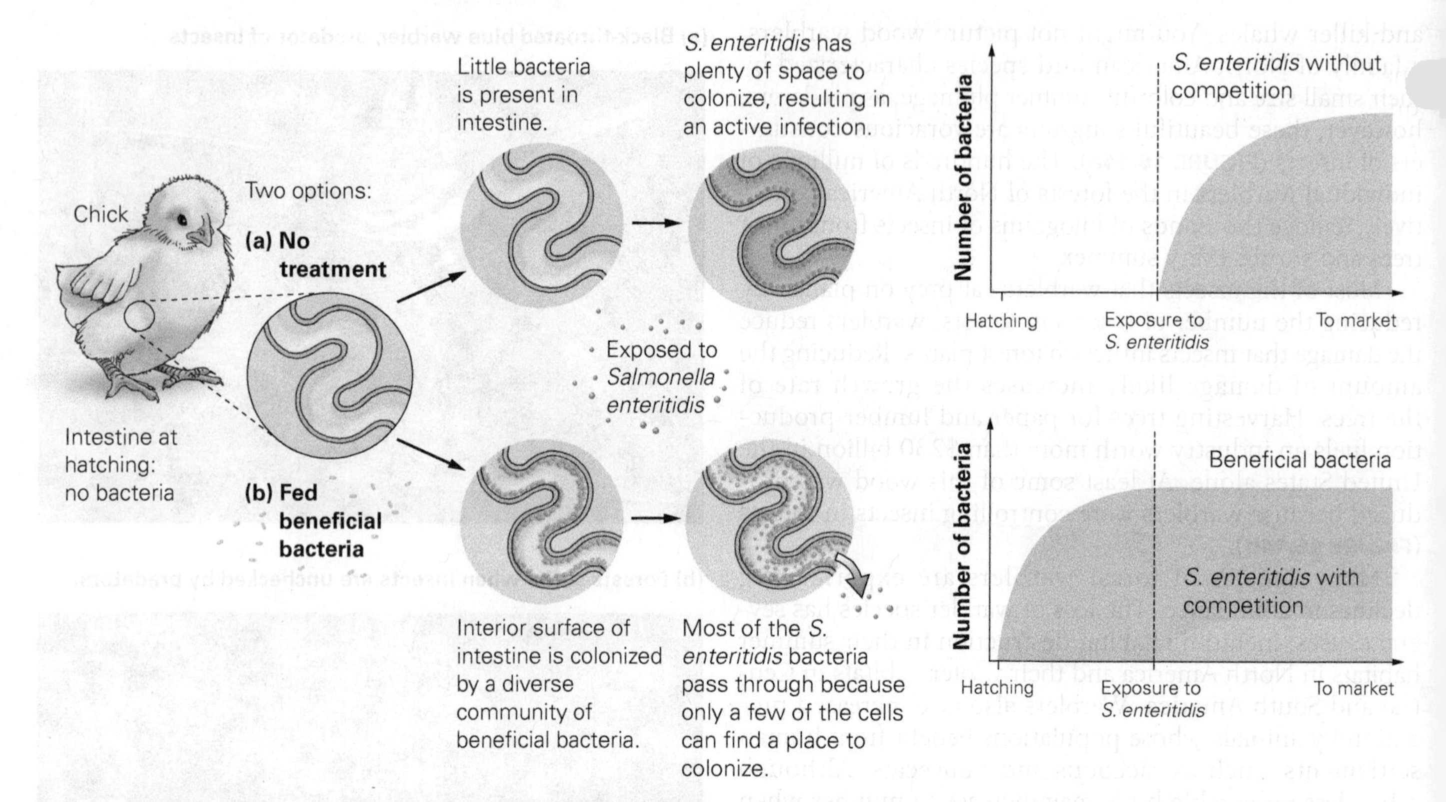

VISUALIZE THIS

Why does the total number of bacteria level off over time in both graphs?

FIGURE 16.15 Competition. If poultry producers feed very young chicks non-disease-causing (beneficial) bacteria, the beneficial bacteria take up the space and nutrients in the intestine that would be used by *S. enteritidis.*

normal and intact. These pathogens contaminate the egg when it forms inside the hen. Thus, the only way to prevent *S. enteritidis* from contaminating eggs is to keep it out of hens.

A common way to control *S. enteritidis* is to feed hens antibiotics—chemicals that kill bacteria. However, like most microbes, *S. enteritidis* strains can evolve drug resistance that makes them more difficult to kill off. But there is another way to reduce *S. enteritidis* infection in poultry: make sure another species is occupying its niche.

Most *S. enteritidis* infections originate in an animal's gut. If another bacterial species is already monopolizing the food and available space in a hen's digestive system, then *S. enteritidis* will have trouble colonizing there. Following this principle, some poultry producers now intentionally infect hens' digestive systems with harmless bacteria, a practice called **competitive exclusion,** to reduce *S. enteritidis* levels in their flocks. This technique involves feeding cultures of benign bacteria to 1-day-old birds. When the harmless bacteria become established in the niche of the chicks' intestines, the chicks will be less likely to host large *S. enteritidis* populations (**FIGURE 16.15**). There is evidence that this practice is working; *S. enteritidis* infections in chickens have dropped by nearly 50% in the United Kingdom, where competitive exclusion is common practice.

The competitive exclusion of *S. enteritidis* in hens mirrors the role of some human-associated bacteria, such as those that normally live within our intestines and genital tracts. For instance, many women who take antibiotics for a bacterial infection will then develop vaginal yeast infections because the antibiotic kills noninfectious bacteria as well, including species that normally compete with yeast. Maintaining competitive interactions between larger species can be important for humans as well. For instance, in temporary ponds, the main competitors for the algae food source are mosquitoes, tadpoles, and snails. In the absence of tadpoles and snails, mosquito populations can become quite large—potentially with severe consequences because these insects may carry deadly diseases such as malaria, West Nile and Zika viruses, and yellow fever. With frogs, toads, and their tadpoles increasingly endangered, this risk is a real one.

Keystone Species: How Wolves Feed Beavers. TABLE 16.2 summarizes the major types of ecological interactions among organisms. However, this table emphasizes the effects of each interaction on the species directly involved; it does not illustrate that many of these interactions may have multiple indirect effects.

TABLE 16.2 Types of species interactions and their direct effects.

Interaction	Example	Effect on Species 1	Effect on Species 2
Commensalism: Association increases the growth or population size of one species and does not affect the other.	1. Remora 2. Shark	+ As the shark feeds somewhat sloppily, the remora can collect the scraps.	0 The shark seems to suffer no negative effects from its hitchhiker.
Mutualism: Association increases the growth or population size of both species.	1. Ants 2. Acacia tree	+ The swollen thorns of the acacia provide shelter for the ants. The acacia leaves provide "protein bodies" that the ants harvest for food.	+ Ants kill herbivorous insects and destroy competing vegetation, benefiting the acacia.
Predation and Parasitism: Consumption of one organism by another.	1. Brown bear 2. Salmon	+ The brown bear catches the salmon and eats it, obtaining nourishment.	− The salmon does not survive.
Competition: Association causes a decrease or limitation in population size of both species.	1. Dandelion 2. Tomato plant	− The dandelion does not grow as well in the presence of the tomato plant. Dandelion produces fewer seeds and fewer offspring.	− The tomato plant does not grow optimally in the presence of the weed. Tomato plant produces fewer flowers and fruit.

(a)

(b)

FIGURE 16.16 Keystone species. (a) The keystone in an archway helps to stabilize and maintain the arch. (b) A keystone species, such as wolves in Yellowstone National Park, helps to stabilize and maintain other species in an ecosystem.

Look again at the food web pictured in Figure 16.11. None of the species in the Antarctic Ocean's biological community is connected to only one other species—they all eat something, and most of them are eaten by something else. You can imagine that penguins, by preying on squid, have a negative effect on elephant seals, which they compete with for these squid, and a more indirect positive effect on other seabirds, which compete with squid for krill. The existence of these **indirect effects** of varying importance has led ecologists to hypothesize that, in at least some communities, the activities of a single species can play a dramatic role in determining the composition of the system's food web. These organisms are called **keystone species** because their role in a community is analogous to the role of a keystone in an archway (**FIGURE 16.16a**). Remove the keystone, and an archway collapses; remove the keystone species, and the web of life collapses. It is very difficult to predict which species in an intact ecosystem may be a keystone species, but biologists can point to several examples that became apparent after a species disappeared. One example is the population of gray wolves in Yellowstone National Park (**FIGURE 16.16b**).

Gray wolves were exterminated within Yellowstone National Park by the mid-1920s because of a systematic campaign to rid the American West of this occasional predator of livestock. However, by the 1980s, thanks to insights gained from the science of ecology and a new interest in environmental health, attitudes about the wolf had changed. A new appreciation of the role of wolves in natural systems led to renewed interest in returning wolves to their historical homeland. From 1995 to 1997, 41 wolves originally trapped in Canada were released into Yellowstone National Park and surrounding areas. Thanks to protection from hunting and the wolves' own adaptability, by the end of 2015 the number had grown to at least 500 wolves in the greater Yellowstone area, and the animal was no longer considered endangered there.

During the time that wolves were extinct in Yellowstone Park, biologists noticed dramatic declines in populations of aspen, cottonwood, and willow trees. They attributed this decline to an increase in predation by elk, especially during winter when grasses become unavailable. However, just a few years after wolf reintroduction, aspen, cottonwood, and willow tree growth has rebounded in some areas of the park, even though the wolf population was still too low to make a major dent in elk populations. Besides the regions near active wolf dens, the areas of the park that saw the greatest recovery include places on the landscape where elk have limited ability to see approaching wolves or to escape. Thus, the elk will stay away from these areas to avoid wolf predation. Wolves, primarily by changing elk behavior, appear to be important to maintaining large populations of hardwood trees in Yellowstone Park.

The rebound of aspen, cottonwood, and willow populations in Yellowstone has effects on other species as well. Beaver rely on these trees for food, and their populations appear to be growing in the park after decades of decline. Warblers, insects, and even fish that depend on shelter, food, and shade from these trees are increasing in abundance as well. Wolves in Yellowstone appear to fit the profile of a classic keystone species, one whose removal had numerous and surprising effects on biodiversity.

Energy and Chemical Flows

As the examples in the previous section illustrate, the extinction of a single species may have unpredictable effects on other species in a habitat. What may be even less apparent is how the loss of seemingly insignificant species can change the environmental conditions on which the entire community depends.

Ecologists define an **ecosystem** as all of the organisms in a given area, along with their nonbiological environment. The function of an ecosystem is described in terms of the rate at which energy flows through it and the rate at which nutrients are recycled within it. The loss of some species can dramatically affect both of these ecosystem properties.

eco- means home or habitation.

Energy Flow. In nearly all ecosystems, the primary energy source is the sun. Producers convert sun energy into biomass during the process of photosynthesis. The chemical energy thus captured is passed through trophic levels in the ecosystem. Biomass is partitioned among the trophic levels, with most of it residing at the bottom of the pyramid (review Figure 16.7). As we move up the pyramid, only a portion of the energy available at one level can be converted into the biomass of the next level. The amount of biomass at the producer level effectively determines how large the population of organisms at the highest level can be.

The amount of sunlight reaching the surface of Earth and the availability of water at any given location are the major determiners of both trophic pyramid structure and energy flow through it. (Chapter 17 provides a summary of how variance in sunlight and water availability leads to differences in Earth's ecosystem types.) However, the biodiversity found in an ecosystem can also have strong effects on energy flow within it.

Studies in grasslands throughout the world have provided convincing evidence that loss of species can affect energy flow. By comparing experimental prairie gardens planted with the same total number of individual plants but with different numbers of species, scientists at the University of Minnesota and elsewhere have discovered that the overall plant biomass tends to be greater in more diverse gardens. This research indicates that a decline in diversity, even without a decline in habitat, may lead to less energy being made available to organisms higher on the food chain, including people who depend on wild-caught food.

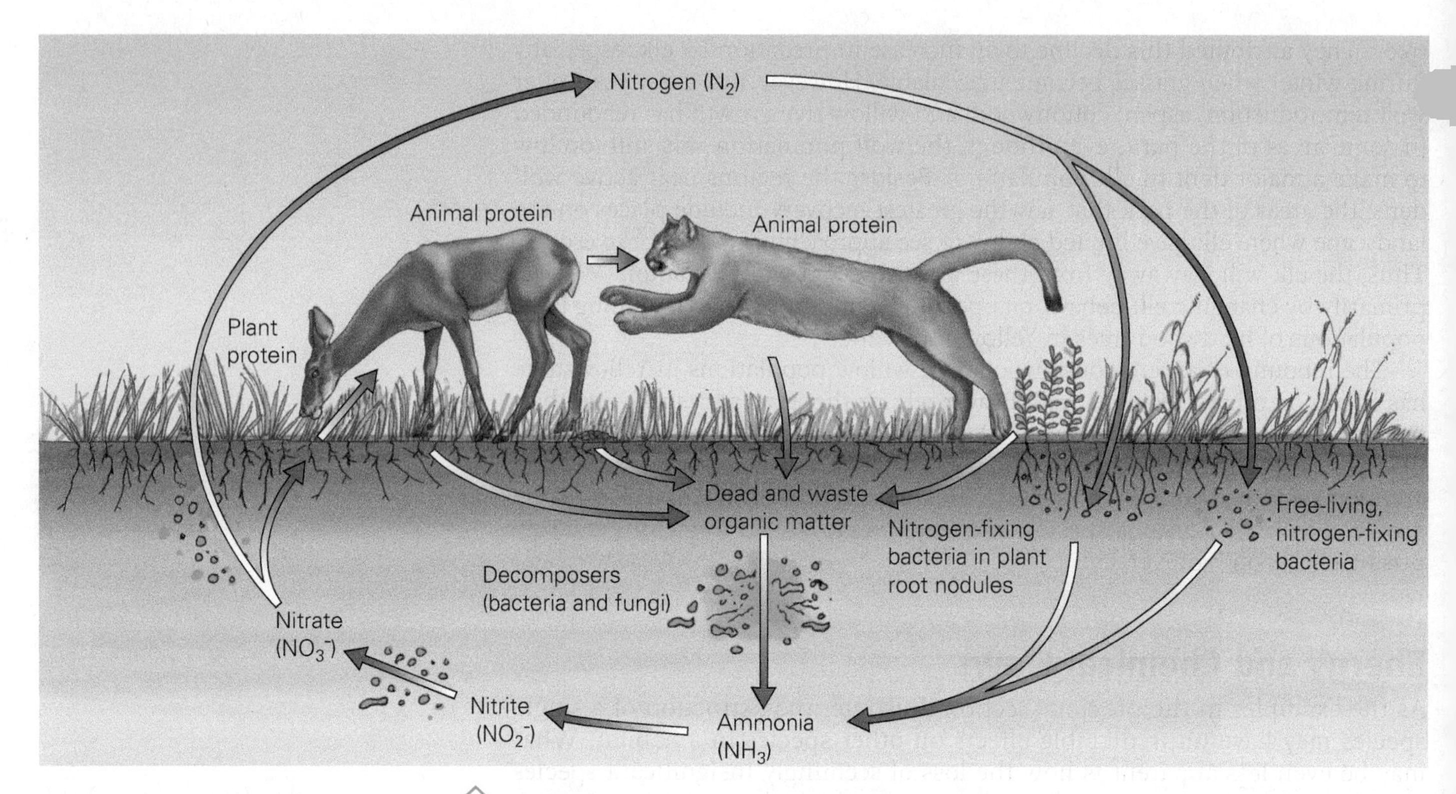

VISUALIZE THIS

Consider how people in cities obtain nutrition and what happens to their waste after digestion and remains after death. How do these features of modern human societies change nutrient flows from the natural cycle pictured here?

FIGURE 16.17 Nutrient cycling. Nutrients such as nitrogen, shown here, are recycled in an ecosystem, flowing from soil to producers to consumers and then back into the soil, where complex nutrients are decomposed into simpler forms.

Nutrient Cycling. When essential mineral nutrients for plant growth pass through a food web, they are generally not lost from the environment—hence the term **nutrient cycling.** FIGURE 16.17 illustrates the nitrogen nutrient cycle in a natural prairie. Here, the element moves from inorganic forms in the soil, such as ammonia, into plants, where it is converted to more complex organic forms. From there, the organic nitrogen typically moves from one living organism to the next as food. Nitrogen finally returns to its inorganic form thanks to the activities of **decomposers,** typically bacteria and fungi, breaking down the waste produced by these consumers.

Nitrogen is a major component of protein, and abundant protein is essential for the proper growth and functioning of all living organisms. Nitrogen is, therefore, often the nutrient that places an upper limit on production in most ecosystems—more nitrogen generally leads to greater production, whereas areas with less available nitrogen can support fewer plants (and therefore animals).

Changes in the soil community can greatly affect nutrient cycling and thus the survival of certain species in ecosystems. Scientists investigating the effects of introduced earthworms, which have invaded forests throughout the northeastern United States, have observed dramatic reductions in the diversity and abundance of plants on the forest floor (**FIGURE 16.18**). The introduced earthworms have apparently caused changes to the community of native soil organisms. As a result of these changes, the nutrient cycle is disrupted and the native plant community has suffered.

The loss of biodiversity clearly can have profound effects on the health o communities and ecosystems on which humans depend. However, controversy exists over whether the current extinction may negatively affect our psychological well-being.

Psychological Effects

Some scientists argue that the diversity of living organisms sustains humans by satisfying a deep psychological need. One of the most prominent scientists to promote this idea is Edward O. Wilson, who calls this instinctive desire to commune with nature **biophilia**.

-philia means affection or love.

Wilson contends that people seek natural landscapes because our distant ancestors evolved in similar landscapes (FIGURE 16.19). According to this hypothesis, ancient humans who had a genetic predisposition driving them to find diverse natural landscapes were more successful than those without this predisposition because diverse areas provide a wider variety of food, shelter, and tool resources. Wilson claims that we have inherited this genetic imprint of our preagricultural past.

Although there is no evidence of a genetic basis for biophilia, there is evidence that our experience with nature has psychological effects. Multiple controlled studies have measured benefits of interacting with nature, ranging from reduced stress levels to increased longevity. Individual experiences with pets and houseplants indicate that many people derive great pleasure from the presence of nonhuman organisms. Although not conclusive, these studies and experiences are intriguing because they suggest that a continued loss of biodiversity could make life in human society less pleasant overall.

The consequences of the loss of biodiversity are not confined to our generation. The fossil record illustrated in Figure 16.2 reveals that it takes 5 to 10 million years to recover the biological diversity lost during a mass extinction. The species that replaced those lost in previous mass extinctions were very different. For instance, after the mass extinction of the reptilian dinosaurs, mammals replaced them as the largest animals on Earth. We cannot predict what biodiversity will look like after another mass extinction. The mass extinction we may be witnessing today will have consequences felt by people in thousands of generations to come.

(a) Invasive earthworms absent

(b) Invasive earthworms present

FIGURE 16.18 Changes in ecosystem function. Notice how barren of living vegetation the worm-infested forest floor appears. One reason for this dramatic change may be a disruption in the native nutrient cycle by this introduced species.

FIGURE 16.19 Is our appreciation of nature innate? Humans evolved in a landscape much like this one in East Africa. Some scientists argue that we have an instinctive need to immerse ourselves in the natural world.

Got It?

1. An ecological niche is often described as the __________ of a species in a community.
2. The stinging tentacles of sea anemones provide protection for the clownfish that live within them, and the clownfish chase away predators that would eat the anemone's tentacles. This relationship is an example of __________.
3. Competition occurs when two species require the same __________ for survival.
4. A keystone species affects the populations of __________ other species in the same community.
5. The flow of energy is one way, which explains why there is more __________ at lower levels of the trophic pyramid.

16.3 Saving Species

So far in this chapter, we have established the possibility of a modern mass extinction occurring, and we have described the potentially serious costs of this loss of biodiversity to human populations. Because the sixth extinction is largely a result of human activity, reversing the trend of species loss requires political and economic, rather than scientific, decisions. But what can science tell us about how to stop the rapid erosion of biodiversity?

Protecting Habitat

On a global scale, it is difficult to know exactly which species are closest to extinction. As a result, the most effective way to prevent loss of species is to preserve as many different kinds of habitats as possible. The same species-area curve that is used to estimate the future rate of extinction also gives us hope for reducing this number. According to the curve in Figure 16.16b, species diversity declines rather slowly as habitat area declines. Thus, in theory we can lose 50% of a habitat but still retain 90% of its species. This estimate is optimistic because habitat destruction is not the only threat to biodiversity, but the species-area curve tells us that if the rate of habitat destruction is slowed or stopped, extinction rates will slow as well.

Protecting the Greatest Number of Species. Given the growing human population, it is difficult to imagine a complete halt to habitat destruction. However, biologist Norman Myers and his collaborators have concluded that 25 **biodiversity hot spots,** natural areas making up less than 2% of Earth's surface, contain up to 50% of all mammal, bird, reptile, amphibian, and plant species (**FIGURE 16.20**). Hot spots occur in areas of the globe where favorable climate conditions lead to high levels of plant production, such as rain forests, and where geological factors have resulted in the isolation of species groups, allowing them to diversify.

Stopping habitat destruction in biodiversity hot spots could greatly reduce the global extinction rate. By focusing conservation efforts on hot spot areas at the greatest risk, humans can very quickly prevent the loss of a large number of species. Of course, even with habitat protection, many species in these hot spots will likely become extinct anyway for other human-mediated reasons.

In the long term, we must find ways to preserve biodiversity while including human activity in the landscape. One option is **ecotourism,** which encourages travel to natural areas in ways that conserve the environment and improve the well-being of local people. Some hot spot countries, such as Costa Rica and

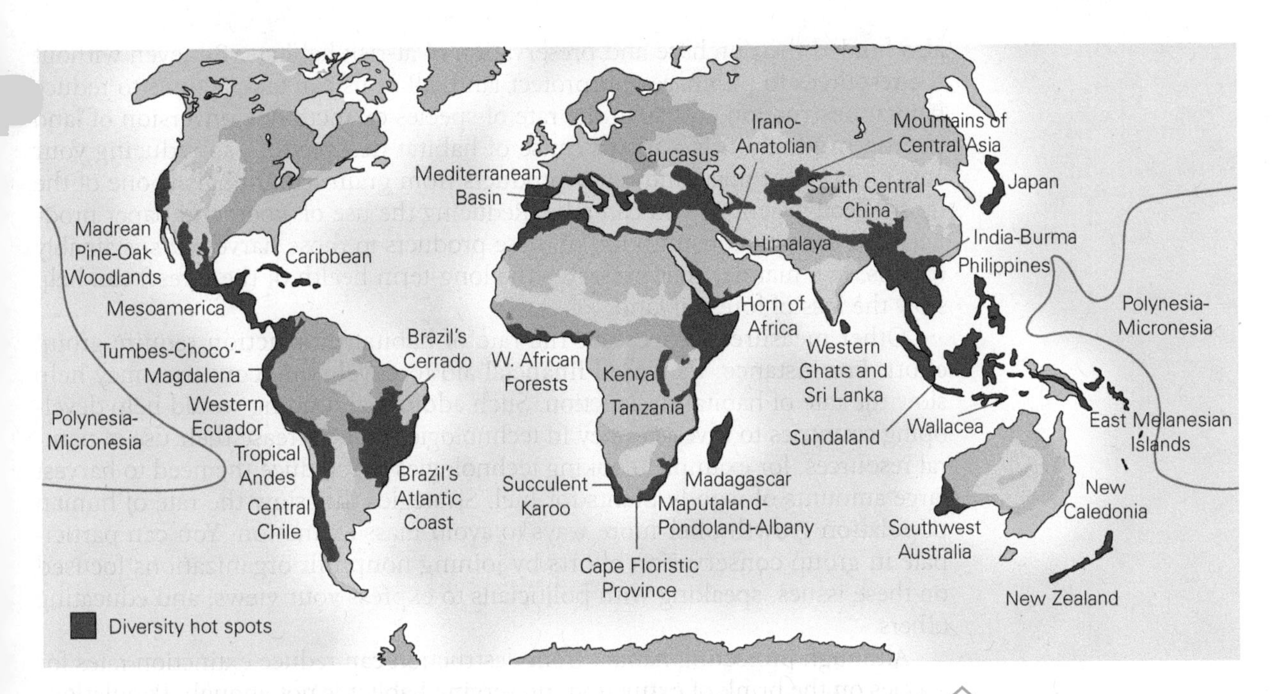

FIGURE 16.20 Diversity "hot spots." This map shows the locations of 25 regions around the world that have many unique species and that contain biodiversity hot spots. The hot spots themselves are the undeveloped areas within these regions. Notice how unevenly these regions of high biodiversity are distributed.

VISUALIZE THIS

Does it appear that there are more hot spots closer to the equator or closer to the poles?

Kenya, have used ecotourism to preserve natural areas and provide much-needed jobs to local citizens; however, this approach is not as effective for more difficult or dangerous regions.

Although preserving hot spots may greatly reduce the total number of extinctions, this approach has its critics, who say that by promoting a strategy that focuses intensely on small areas, we risk losing large amounts of biodiversity elsewhere. These critics promote an alternative approach—identifying and protecting a wide range of ecosystem *types*—designed to preserve the greatest range of biodiversity rather than just the largest number of species.

Protecting Habitat for Critically Endangered Species. Although preserving a variety of habitats ensures less extinctions, already endangered species require a more individualized approach. The ESA requires the U.S. Department of the Interior to designate critical habitats for endangered species within the United States—that is, areas in need of protection for the survival of the species. The amount of critical habitat that becomes designated depends on political as well as biological factors.

The biological part of a critical habitat designation includes conducting a study of habitat requirements for the endangered species and setting a population goal for it. The U.S. Department of the Interior's critical habitat designation has to include enough area to support the recovery population. However, federal designation of a critical habitat results in the restriction of human activities that can take place there. The U.S. Department of Interior has the ability to exclude some habitats from protection if there are "sufficient economic benefits" for doing so—a decision that is political in nature.

Decreasing the Rate of Habitat Destruction. Preserving habitat is not simply the job of national governments that designate protected areas. Private charity given to conservation organizations such as the Nature Conservancy has also funded the purchase and preservation of at-risk habitats. But even without the resources to purchase and protect land, all of us can take actions to reduce habitat destruction and stem the rate of species extinction. Conversion of land to crop production is a major cause of habitat destruction, so reducing your consumption of meat and dairy products from grain-fed animals is one of the most effective actions you can take. Reducing the use of wood and paper products and limiting consumption of these products to those harvested sustainably (that is, in a manner that preserves the long-term health of the forest) can help slow the loss of forested land.

Other measures to decrease the rate of habitat destruction require group effort. For instance, increased financial aid to developing countries may help slow the rate of habitat destruction. Such additional funding would help developing countries to invest money in technologies that decrease their use of natural resources, for example, cooking technologies that reduce the need to harvest large amounts of woody plants for fuel. Strategies that slow the rate of human population growth offer more ways to avoid mass extinction. You can participate in group conservation efforts by joining nonprofit organizations focused on these issues, speaking with politicians to express your views, and educating others.

Although protecting habitat from destruction can reduce extinction rates for species on the brink of extinction, preserving habitat is not enough. Populations can become so small that they can disappear, even with adequate living space. Recovery plans for a critically endangered species may set a short-term goal of a stable population of 500 individuals or more. To understand why at least this many individuals are required to save these species from extinction, we need to examine the special problems of small populations.

Small Populations Are Vulnerable

The growth rate of an endangered species influences how rapidly that species can attain a target population size (**FIGURE 16.21**). Horseshoe crabs have relatively high growth rates and will meet their population goals quickly if the environment is ideal. For slower-growing species, such as the California condor, populations may take decades to recover.

The rate of recovery is important because the longer a population remains small, the more it is at risk of experiencing a catastrophe that could eliminate it entirely. The story of the heath hen in the United States is a case study of just this point.

The heath hen was a small wild chicken that once ranged from Maine to Virginia, and at one time there were hundreds of thousands of individuals. European settlement of the eastern seaboard resulted in habitat loss, causing a rapid and dramatic decline of the birds' population. By the end of the nineteenth century, the only remaining heath hens lived on Martha's Vineyard, a 100-square-mile island off the coast of Cape Cod, Massachusetts. Farming on the island further reduced the habitat for heath hen breeding such that by 1907, only 50 birds survived.

In response to the decline of the heath hen, Massachusetts established a 2.5-square-mile reserve for the remaining birds on Martha's Vineyard in 1908. This solution initially seemed effective; by 1915, the population had recovered to nearly 2000 individuals. However, beginning in 1916, a series of disasters struck. First, fire destroyed much of the remaining habitat. The following winter was long and cold, and an invasion of starving predatory goshawks further reduced the heath hen population. Finally, a poultry disease introduced by imported domestic turkeys wiped out much of the remaining population. By 1927, only 14 heath hens remained—almost all males. The last surviving member of the species was spotted for the last time on March 11, 1932.

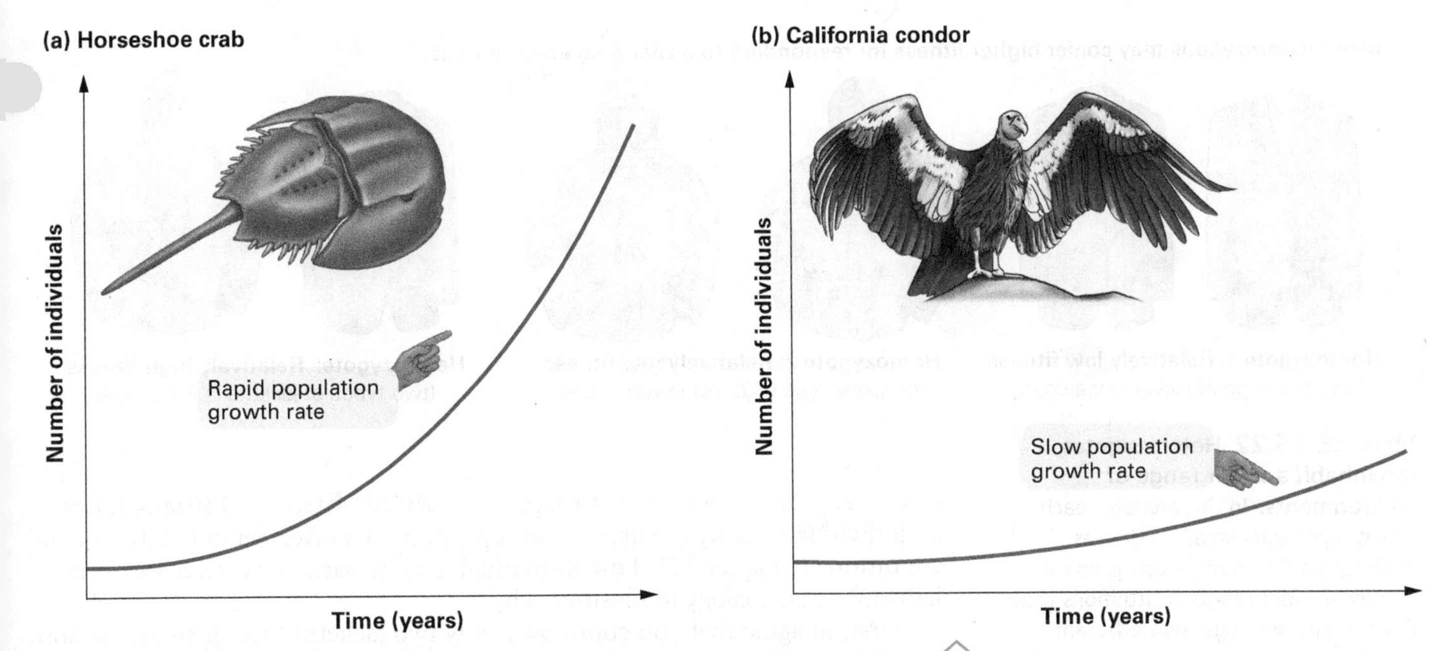

FIGURE 16.21 The effect of growth rate on species recovery. (a) This graph illustrates the rapid growth of a hypothetical population of quickly reproducing horseshoe crabs. (b) The slow growth rate of the California condor has made the recovery of this species a long process. Today, nearly 30 years after recovery efforts began, the population of wild condors numbers in the hundreds.

VISUALIZE **THIS**

Why is the growth rate of condors so much slower than the growth rate of horseshoe crabs?

The final causes of heath hen extinction were natural events—fire, harsh weather, predation, and disease. But it was the population's small size that doomed it in the face of these relatively common challenges. A population of 100,000 individuals can weather a disaster that kills 90% of its members but leaves 10,000 survivors, but a population of 1000 individuals will be nearly eliminated by the same circumstances. Even when human-caused losses to the heath hen population were halted, the species' survival was still precarious.

Small populations of endangered species can be protected from the fate that befell the heath hen. Having additional populations of the species at sites other than Martha's Vineyard would have nearly eliminated the risk that *all* members of the population would be exposed to the same series of environmental disasters. This is the rationale behind placing captive populations of endangered species at several different sites. For instance, captive whooping cranes are located at the U.S. National Biological Service's Patuxent Wildlife Research Center in Maryland, the International Crane Foundation in Wisconsin, the Audubon Center for Endangered Species Research in New Orleans, and four zoos throughout the United States and Canada.

Even with multiple habitats, if populations of endangered species remain small in number, they are subject to a more subtle but potentially equally devastating situation—the loss of genetic variability.

Conservation Genetics

A species' **genetic variability** is the sum of all of the alleles and their distribution within the species. For example, the gene that determines your ABO blood type comes in three different forms. A population containing all three of these alleles contains more genetic variability than does a population with only two alleles. The loss of genetic variability can reduce both individual and population survival.

Being heterozygous may confer higher fitness for responding to a changing environment.

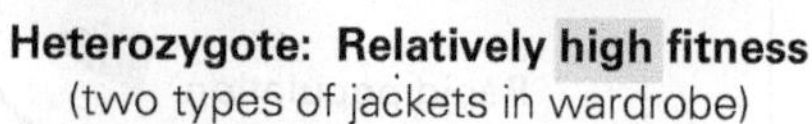

Homozygote 1: Relatively low fitness (only one type of jacket in wardrobe)

Homozygote 2: Relatively low fitness (only one type of jacket in wardrobe)

Heterozygote: Relatively high fitness (two types of jackets in wardrobe)

FIGURE 16.22 Heterozygotes can inhabit a wider range of environments. In this analogy, each jacket represents an allele. Just as having two different jackets prepares you for a wider range of situations than having only one type, two different alleles for the same gene may allow for optimal function over a wider range of conditions.

Low genetic variability reduces individual fitness. **Fitness** refers to an individual's ability to survive and reproduce in a given set of environmental conditions (Chapter 12). Low individual genetic variability decreases fitness; let's look at an analogy to illustrate why.

First, imagine that you could own only two jackets (**FIGURE 16.22**). If both are blazers, then you would be well prepared to meet a potential employer. However, if you had to walk across campus to your job interview in a snowstorm, you would be pretty uncomfortable. However, if you own one warm jacket and one blazer, you are ready for freezing weather as well as a job interview. In a way, individuals experience the same advantages when they carry two different functional alleles for a gene. If each allele codes for a functional protein, a heterozygous individual produces two slightly different proteins that perform essentially the same function. This phenomenon is known as **hetero zygote advantage.**

The second reason high individual genetic variability increases fitness is that, in many cases, one allele for a gene is **deleterious**—that is, it produces a protein that is not very functional. In our jacket analogy, a nonfunctional, deleterious allele is equivalent to a badly torn jacket. If you have a damaged jacket and an intact one, at least you have one warm covering (**FIGURE 16.23**). In this case, heterozygosity is valuable because a heterozygote still carries one functional allele. Often, deleterious alleles are recessive, meaning that the activity of the functional allele in a heterozygote masks the fact that a deleterious allele is present (see Chapter 8). An individual who is homozygous (carries two identical

VISUALIZE THIS

Homozygotes for the normal allele have high fitness, and homozygotes for the nonfunctional recessive allele have low fitness. Explain the difference in fitness in these two types of homozygotes.

FIGURE 16.23 Heterozygotes avoid the deleterious effects of recessive mutations. Again, each jacket represents an allele. If one type of jacket you can receive is nonfunctional, it is better to receive no more than one of them. Heterozygotes are likewise protected against the likelihood of having only nonfunctional alleles, as is present in homozygous recessive individuals.

Being heterozygous may confer higher fitness by masking deleterious recessive alleles.

Homozygote 1: Relatively high fitness (two functional jackets in wardrobe)

Homozygote 2: Relatively low fitness (two nonfunctional jackets in wardrobe)

Heterozygote: Relatively high fitness (one functional jacket in wardrobe)

copies of a gene) for the deleterious allele will have low fitness—in our analogy, two torn jackets and nothing else. For both of these reasons, when individuals are heterozygous for many genes (or, in our analogy, have two choices for all clothing items), the cumulative effect is greater fitness relative to individuals who are homozygous for many genes.

Heterozygosity declines in small populations over time. When related individuals mate—known as **inbreeding**—the chance that their offspring will be homozygous for any allele is relatively high. In cheetahs, high levels of inbreeding have decreased fitness by causing poor sperm quality and low cub survival, both likely due to increased expression of deleterious alleles. The costs of inbreeding are seen in humans as well; the children of first cousins have higher rates of homozygosity and higher mortality rates (thus lower fitness) than children of unrelated parents. In a small population of an endangered species, inbreeding often causes low rates of survival and reproduction and can seriously hamper a species' recovery.

Small populations also lose genetic variability because of **genetic drift,** a change in the frequency of an allele that occurs simply by chance within a population. Imagine two human populations in which the frequency of blood-type allele *A* is 1%—that is, only 1 of every 100 blood-type genes in the population is the *A* form (we use the symbol I^A for this allele). In the first population of 20,000 individuals, there are 40,000 total blood group genes (that is, two copies per individual). At 1% frequency, this population contains 400 I^A alleles.

In the second, much smaller population of only 200 individuals, only four copies of the allele are present. If the chance of passing on any given allele in both populations is equivalent to a coin flip, the chance that the I^A allele is not passed on in the population of 20,000 is equivalent to flipping 400 "heads" in succession; in the small population, it is equivalent to only four heads in a row. Genetic drift is thus much more likely to result in the complete loss of alleles in a small population (**FIGURE 16.24**).

In most populations, alleles that are lost through genetic drift have relatively small effects on fitness at the time. However, many alleles that appear to be

FIGURE 16.24 Genetic drift affects small populations more than large populations. In this graph, each line represents the average of 25 computer simulations of genetic drift for a given population size. After 100 generations, a population of 500 individuals still contains 90% of its genetic variability. In contrast, a population of 20 individuals has less than 5% of its original genetic variability.

WORKING WITH DATA
Use the graph to determine how much diversity is lost due to genetic drift in a population of 60 individuals over 100 generations.

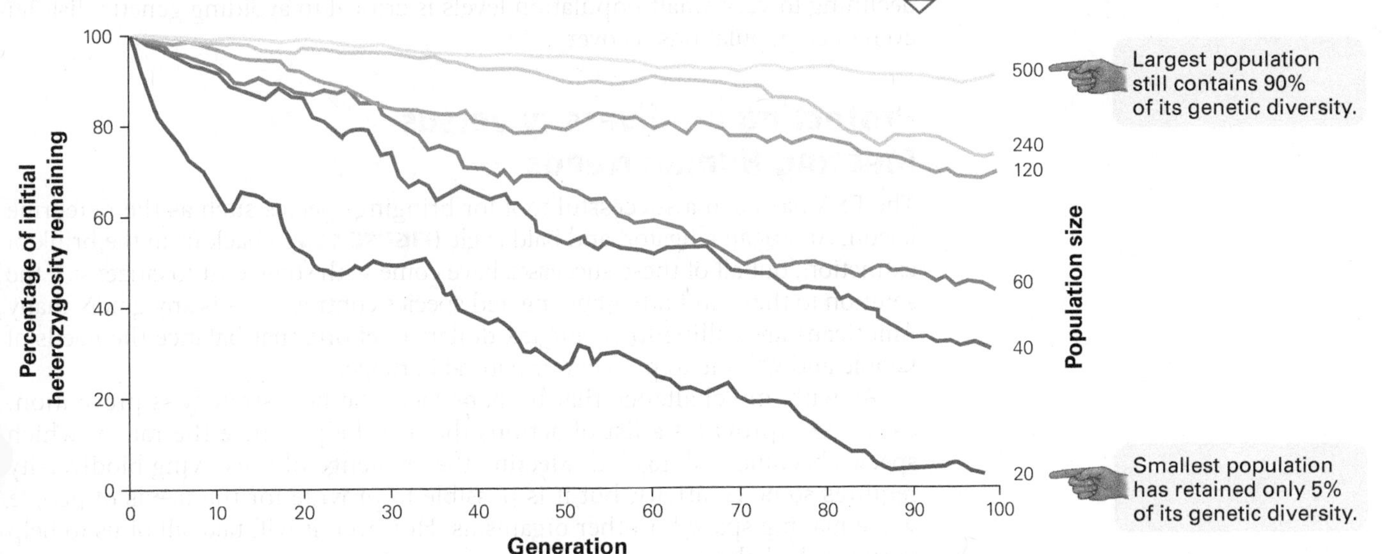

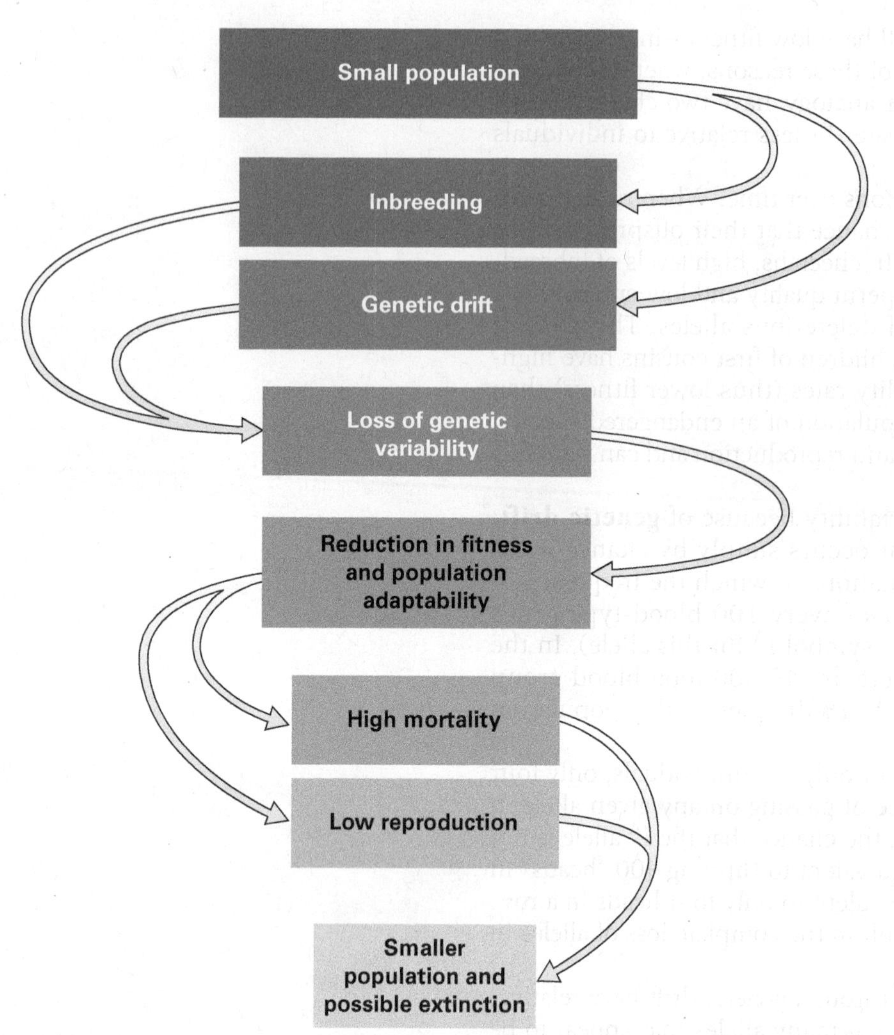

FIGURE 16.25 The extinction vortex. A small population can become trapped in a positive-feedback process that causes it to continue to shrink in size, eventually leading to extinction.

nearly neutral with respect to fitness in one environment may have positive fitness in another environment. When this is the case, the loss of these alleles may spell disaster for the entire species.

Low genetic variation puts entire populations at risk. Populations with low levels of genetic variability have an insecure future for two reasons. First, when alleles are lost, the level of inbreeding in a population increases. This means lower reproduction and higher death rates, leading to declining populations that are susceptible to all the other problems of small populations. This process is often referred to as the **extinction vortex** (**FIGURE 16.25**). The heath hen discussed in this chapter is an example of the extinction vortex; once the remnant population was reduced to small numbers by fire, disease, and predation, inbreeding depression prevented it from rebounding, thus dooming it to extinction.

Second, populations with low genetic variability may be at risk of extinction because they cannot evolve in response to changes in the environment. When few alleles are available for any given gene, it is possible that no individuals in a population possess an adaptation allowing them to survive an environmental challenge. For example, there is some evidence that people with type A blood are more resistant to cholera and bubonic plague than are people with type O or B blood. Loss of the I^A allele would make human populations more susceptible to serious declines in the face of these diseases. As a result of this possibility, preventing endangered species from declining to very small population levels is critical to avoiding genetic disaster even once populations recover.

Protecting Biodiversity versus Meeting Human Needs

The ESA has been a successful tool for bringing species such as the peregrine falcon, American alligator, and bald eagle (**FIGURE 16.26**) back from the brink of extinction, but all of these successes have come with some cost to citizens. If the solution to these and other endangered species controversies is any guide, many Americans are willing to devote tax dollars to efforts that balance the needs of people and wildlife to protect our natural heritage.

As with any challenge that humans face, the best strategy is prevention. **TABLE 16.3** provides a list of actions that can help reduce the rate at which species become endangered. Meeting the challenge of preserving biodiversity requires some creativity, but it is possible to provide for the needs of people while making space for other organisms. However, it will take all of us to help preserve the balance.

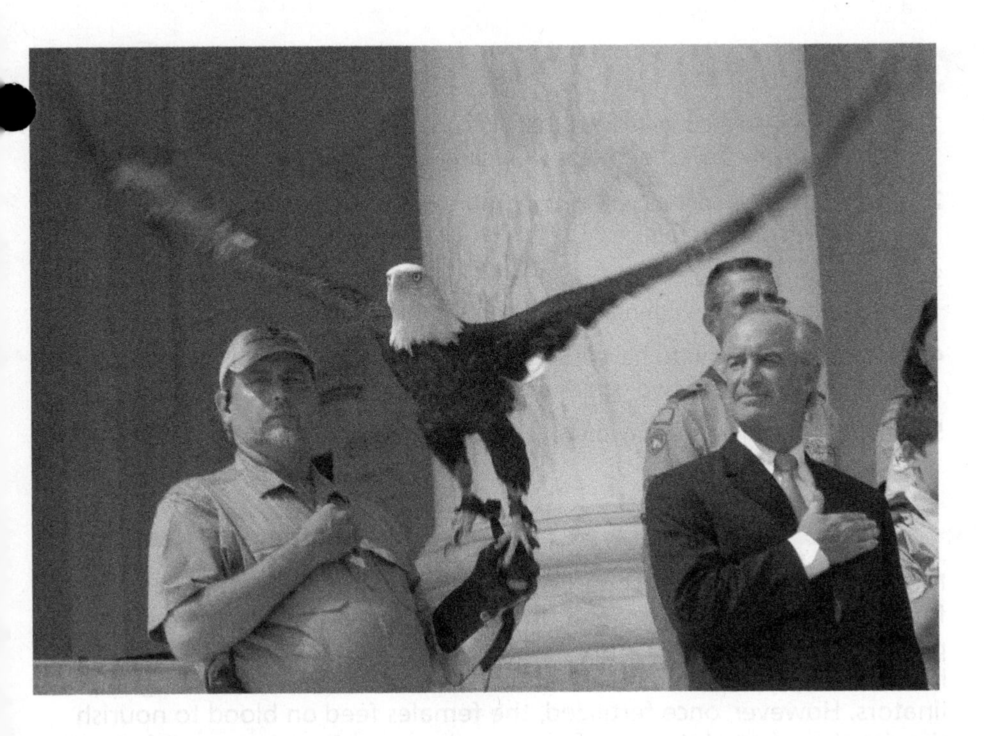

FIGURE 16.26 An endangered species success story. The bald eagle was "delisted" from endangered status in 2007, at an event celebrated at this ceremony in Washington, DC. Its dramatic recovery from near extinction was thanks in part to government protections. However, delisting is still a relatively rare event.

TABLE 16.3 Taking action to preserve biodiversity.

Objective	Why Do It?	Actions
Reduce fossil fuel use 	• Mining, drilling, and transporting fossil fuels modifies habitat and leads to pollution. • Burning fossil fuels contributes to global climate change, further degrading natural habitats.	• Buy energy-efficient vehicles and appliances. • Walk, bike, carpool, or ride the bus whenever possible. • Choose a home near school, work, or easily accessible public transportation. • Buy "clean energy" from your electric provider, if offered.
Reduce the impact of meat consumption 	• The primary cause of habitat destruction and modification is agriculture. • Modern beef, pork, and chicken production rely on grains produced on farms. One pound of beef requires 4.8 pounds of grains or about 25 square meters of agricultural land.	• Eat one or more meat-free meal per week. • Make meat as a "side dish" instead of the main course. • Purchase grass-fed or free-range meat.
Reduce pollution 	Pollution kills organisms directly or can reduce their ability to survive and reproduce in an environment.	• Do not use pesticides. • Buy products produced without the use of pesticides. • Replace toxic cleaners with biodegradable, less-harmful chemicals. • Consider the materials that make up the goods you purchase and choose the least-polluting option. • Reuse or recycle materials instead of throwing them out.
Educate yourself and others 	Change happens most rapidly when many individuals are working for it.	• Ask manufacturers or store owners about the environmental costs of their goods. • Talk to family and friends about the choices you make. • Write to decision makers to urge action on effective measures to reduce human population growth and curb habitat destruction and species extinction.

Got It?

1. The species-area curve tells us that slowing or halting ____________ ____________ will potentially slow the rate of extinction.
2. An endangered species will recover more quickly if it has a ____________ population growth rate.
3. The genetic variability of a population is measured by the number of ____________ found within a population.
4. Heterozygotes tend to have higher fitness than homozygotes because they are unlikely to have two copies of a ____________ allele.
5. Low genetic diversity may prevent a population from ____________ to environmental change.

Mosquitoes spend the first part of their lives in pools of stagnant water, feeding on algae, bacteria, and other microbes. In the water, they are preyed on by dragonflies, fish, and small crustaceans. When mosquitoes hatch, both males and females feed on flower nectar—thus acting as pollinators. However, once fertilized, the females feed on blood to nourish the development of their eggs; as a result, mosquitoes are possibly Earth's most annoying animals—at least to humans. Not only do they cause itchy welts when they bite and whine in your ear when you are trying to sleep, but they spread some of the most troublesome diseases—malaria, yellow fever, and dengue fever, among them. Malaria alone is responsible for hundreds of thousands of deaths every year. That mosquitoes are a plague on humankind might lead you to say:

Earth would definitely be better if mosquitoes were extinct.

Sounds right, but it isn't.

Answer the following questions to understand why.

1. What are the ecological relationships between (a) mosquitoes and microbes in the water; (b) mosquitoes and dragonflies; (c) mosquitoes and the plants they visit; and (d) mosquitoes and humans?
2. Sketch a food web containing mosquitoes. What organisms would likely be affected by the loss of mosquitoes, and in what way?
3. Could any of the changes listed affect organisms that are not in the web you drew? Explain.
4. What is a keystone species?
5. Are mosquitoes a keystone species? Explain.
6. Reflect on your answers to questions 1–5 above and explain why the statement bolded above sounds right, but isn't.

THE BIG QUESTION

Is wind power good or bad for birds?

Wind power seems to hold great promise as a source of nature-friendly electricity that, unlike fossil fuel sources, results in minimal habitat destruction and no ongoing pollution. But in many communities where "farms" of windmills are proposed, they face significant citizen opposition. One reason often given by opponents of these projects is concern about the deadly effects of spinning turbines on flying animals, especially birds.

What should I know?

What follows are some smaller questions that need to be resolved to answer the Big Question. Place a checkmark next to the questions that science can answer.

Smaller Questions	Can Science Answer?
Can bird deaths at wind turbines be prevented?	
If the electricity produced by wind turbines came from a fossil fuel source, would more or fewer birds be at risk?	
How do bird deaths due to collisions with wind turbines compare with bird deaths due to other causes?	
Is investing in renewable energy a bigger priority than protecting birds?	
If we take action to protect birds elsewhere, can we offset the bird mortality that occurs at wind turbines?	
Is it acceptable to build a wind farm even if we cannot know what its exact impact will be on birds?	

What does the science say?

Let's examine what the data say about this smaller question:

How do bird deaths due to collisions with wind turbines compare with bird deaths due to other causes?

A recent review examined dozens of studies that estimated bird mortality from various sources. The estimates for human causes of bird mortality in both the United States and Canada are summarized in the table below. Note that the Canadian results surveyed a wider range of causes, but because the human population of Canada is one-tenth that of the United States, total human-caused mortality in each category is lower.

	Estimated Bird Deaths per Year	
Cause	Canada	United States
Predation by domestic cats	204–348 million	1.4–4 billion
Collision with buildings	16–42 million	365–988 million
Collision with automobiles	9–19 million	200–340 million
Collision with cell phone, radio, and television towers	220,000	5–57 million
Electrocutions at power lines	160,000–802,000	Not determined
Collision with wind turbines	13,000–22,000	Not determined
Poisoning with agriculture chemicals	1–4.4 million	Not determined
Killed by fishing activities (e.g., trapped in gill nets)	2,700–45,6000	Not determined
Killed by marine oil and gas activities	2,000–4,100	Not determined

1. Describe the results. What is the relative impact of wind turbines on overall human-caused bird mortality?
2. Given these data, do you think the smaller question is answered? If not, propose another study that would help answer this question.
3. Does this information help you answer the Big Question? What else do you need to consider?

Data source: S. Loss, "Avian Interactions with Energy Infrastructure in the Context of Other Anthropogenic Threats," *The Condor* 118, no. 2 (2016): 424–432.

Chapter Review

Mastering Biology

Go to Mastering Biology to access **eText 2.0, Dynamic Study Modules,** and the **Study Area,** where you'll find practice quizzes, BioFlix™ animations, MP3 tutor sessions, current events, and more.

SUMMARY

Section 16.1

Describe how scientists can estimate current and historical rates of extinction, and provide the evidence that Earth is currently experiencing a mass extinction.

- Historical rates of extinction can be estimated by looking at the fossil record, and the rate of modern extinctions can be measured by counting the number of species known to be lost or endangered (pp. 342–344).
- The loss of biodiversity through species extinction is 50 to 100 times greater than historical rates (p. 344).

List the major causes of extinction, and describe how each is related to human activity.

- The loss of natural habitat caused by human activities is the primary cause of the extinctions occurring in the modern era (pp. 344–347).
- Additional threats of habitat fragmentation, the introduction of species to regions where they are not native, overexploitation through uncontrolled harvesting, and pollution also contribute to extinction (pp. 347–348).

Explain how the species-area curve is used to estimate species number in an area.

- Species-area curves illustrate the relationship between the number of species and the size of a geographic area (p. 346).

Explain the principle of the trophic pyramid.

- The trophic pyramid is the relationship among biomass at different levels of a food chain. In nearly all communities, there is significantly more biomass in the producer level than at higher consumer levels (p. 347).
- Species at the top of the food chain are more susceptible to extinction because less energy is available for survival at higher trophic levels (p. 347).

SHOW YOU KNOW 16.1 About one-third of the species of amphibians (frogs, toads, and salamanders) around the world are threatened or endangered, which is much higher than the rate for birds and mammals. List human activities that might be affecting these amphibious animals and describe how these human activities could cause endangerment.

Section 16.2

Define *predation*, *mutualism*, and *competition*, and provide examples of each ecological interaction.

- Species are members of communities; they can interact with one another via mutualism, where both species are benefited by the relationship; via predation, where one benefits while the other is consumed in the relationship; or via competition, where both are struggling to obtain the same resources (pp. 351–355).

Define *food web* and *ecosystem*, and explain the role of keystone and nonkeystone species in both.

- A food web is the connection among species in a community, and an ecosystem includes not only the food web but all of the nonbiological aspects of the environment occupied by a community (p. 351).
- A keystone species in a food web is one whose absence has a large effect on the entire community (pp. 355–357).
- Species can affect energy flow and nutrient cycling (pp. 357–358).

Provide an example of a nutrient cycle.

- Nutrients—for example, nitrogen—cycle through ecosystems from living organisms to inorganic forms and back again (p. 358).

SHOW YOU KNOW 16.2 All of the following species pairs are likely competitors. In each case, describe what you think the competition involves. (a) coyotes and foxes; (b) hummingbirds and bees; (c) flying squirrels and owls. How could you test your hypothesis that these animals are in competition with each other in any given environment?

Section 16.3

Explain why small population size and a loss of genetic diversity in a population are risky for both the population and individuals within the population.

- Small populations are at higher risk for extinction due to environmental catastrophes (pp. 362–363).
- Small populations are at risk when individuals have low fitness due to inbreeding and thus are less able to increase population size (pp. 363–365).
- Genetic variability is lost in small populations because of genetic drift—the loss of alleles from a population due to chance events—and thus small populations may be less able to evolve in response to environmental change (pp. 365–366).

List strategies that will help reduce the number of extinctions going forward.

- The political process enables people to develop plans for helping endangered species while minimizing the negative effects of these actions on people (p. 366).

- Eating lower on the food chain and using sustainable methods of transportation and energy generation can help conserve habitat (p. 367).

SHOW YOU KNOW 16.3 Unlike the heath hen, the passenger pigeon went extinct despite still having some sizable colonies. What factors could cause a large population to suffer the same genetic problems experienced by a small population?

ROOTS TO REMEMBER

The following roots of words come mainly from Latin and Greek and will help you to decipher terms:

eco- means home or habitation. Chapter term: *ecosystem*

eu- means true or good. Chapter term: *eutrophication*

-philia means affection or love. Chapter term: *biophilia*

trophic- means food, nourishment. Chapter terms: *eutrophication, trophic level, trophic pyramid*

LEARNING THE BASICS

1. Add labels to the figure that follows, which illustrates the interacting factors in a declining population that contribute to the extinction vortex.

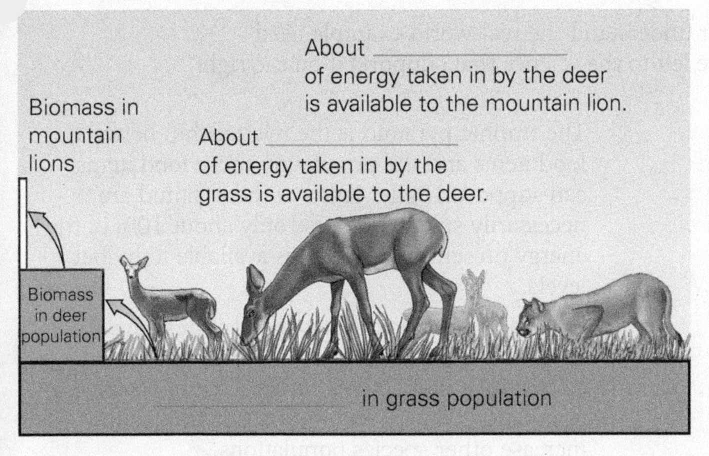

About ________ of energy taken in by the deer is available to the mountain lion.
Biomass in mountain lions
About ________ of energy taken in by the grass is available to the deer.
Biomass in deer population
________ in grass population

2. Compare and contrast the species interactions of mutualism, predation, and competition.

3. A mass extinction ________.

   A. is global in scale; B. affects many different groups of organisms; C. is caused only by human activity; D. A and B are correct; E. A, B, and C are correct

4. Current rates of species extinction appear to be approximately ________ historical rates of extinction.

   A. equal to; B. 10 times lower than; C. 10 times higher than; D. 50 to 100 times higher than; E. 1000 to 10,000 times higher than

5. According to the generalized species-area curve, when habitat is reduced to 50% of its original size, approximately ________ of the species once present there will be lost.

   A. 10%; B. 25%; C. 50%; D. 90%; E. it is impossible to estimate the percentage

6. Which cause of extinction results from humans' direct use of a species?

   A. overexploitation; B. habitat fragmentation; C. pollution; D. introduction of competitors or predators; E. global warming

7. The web of life refers to the ________.

   A. evolutionary relationships among living organisms; B. connections between species in an ecosystem; C. complicated nature of genetic variability; D. flow of information from parent to child; E. predatory effect of humans on the rest of the natural world

8. Which of the following is an example of a mutualistic relationship?

   A. moles catching and eating earthworms from the moles' underground tunnels; B. cattails and reed canary grass growing together in wetland soils; C. cleaner fish removing and eating parasites from the teeth of sharks; D. Colorado potato beetles consuming potato plant leaves; E. more than one of the above

9. The risks faced by small populations include ________.

   A. erosion of genetic variability through genetic drift; B. decreased fitness of individuals as a result of inbreeding; C. increased risk of experiencing natural disasters; D. A and B are correct; E. A, B, and C are correct

10. One advantage of preserving more than one population of an endangered species at more than one location is ________.

    A. a lower risk of extinction of the entire species if a catastrophe strikes one location; B. higher levels of inbreeding in each population; C. higher rates of genetic drift in each population; D. lower numbers of heterozygotes in each population; E. higher rates of habitat fragmentation in the different locations

11. There are fewer lions in Africa's Serengeti than there are zebras. This is principally because ________.

    A. zebras tend to drive off lions; B. lions compete directly with cheetahs, whereas zebras do not have any competitors; C. zebras have mutualists that increase their population, whereas lions do not; D. there is less energy available in zebras to support the lion population than there is in grass to support the zebras; E. zebras are a keystone species, whereas lions are not

12. Most of the nutrients available for plant growth in an ecosystem are ________.

    A. deposited in rain; B. made available through the recycling of decomposers; C. maintained within that ecosystem over time; D. B and C are correct; E. A, B, and C are correct

ANALYZING AND APPLYING THE BASICS

1. Off the coast of the Pacific Northwest, areas dominated by large algae, called kelp, are common. These "kelp forests" provide homes for small plant-eating fishes, clams, and abalone. These animals in turn provide food for crabs and larger fishes. A major predator of kelp in these areas is sea urchins, which are preyed on by sea otters. When sea otters were hunted nearly to extinction in the early twentieth century, the kelp forest collapsed. The kelp were only found in low levels, while sea urchins proliferated on the seafloor. Use this information to construct a simple food web of the kelp forest. Using the food web, explain why sea otters are a keystone species in this system.
2. In which of the following situations is genetic drift more likely and why?

   A. A population of 500 in which males compete heavily for harems of females. About 5% of the males father all of the offspring in a given generation. Females produce one offspring per season; **B.** A population of 250 in which males and females form bonded pairs that last throughout the mating season. Females produce three to four offspring per season.
3. The piping plover is a small shorebird that nests on beaches in North America. The plover population in the Great Lakes is endangered and consists of only about 60 breeding pairs. Imagine that you are developing a recovery plan for the piping plover in the Great Lakes. What sort of information about the bird and the risks to its survival would help you to determine the population goal for this species as well as how to reach this goal?

GO FIND OUT

1. What species are endangered in the region where you live or go to school? Choose one of these species and determine what factors, such as loss of habitat or overhunting, have contributed to its population decline. What actions are being taken to protect this species from extinction? What, if anything, is being done to help the population recover?
2. From your perspective, which of the following reasons for preserving biodiversity is most convincing? (a) Nonhuman species have roles in ecosystems and should be preserved to protect the ecosystems that support humans; or (b) nonhuman species have a fundamental right to existence. Explain your choice.

MAKE THE CONNECTION

The science that you learned in this chapter has helped you better understand the real-world example used throughout this discussion. Draw a line from the statement on the left to the science that supports it on the right.

The rufus red knot is one of many species that is currently threatened with extinction. ▷	◁ The trophic pyramid is the relationship between food items and the population those food items can support. Higher levels of the pyramid are necessarily smaller because only about 10% of the energy present in one level is available to higher levels.
Human modification of natural habitat has endangered many species. ▷	◁ Ecological interactions between species have an effect on a whole food web. Predatory or competitive interactions can keep certain species populations down; and mutualistic interactions can increase other species populations.
Because the rufus red knot feeds on another animal, it is especially prone to becoming endangered. ▷	◁ Small populations lose genetic diversity, which is essential for long term survival by protecting against deleterious alleles and preserving the opportunity to adapt to a changing environment.
The loss of certain species may cause an increase in other species that are harmful to humans or a decrease in populations of species that are beneficial to us. ▷	◁ Comparison of the fossil record and known modern extinctions indicates that the extinction rate is possibly 100 times higher than the "background" rate.
The longer the rufus red knot population remains small, the greater its risk of extinction. ▷	◁ The species-area curve illustrates a relationship between the size of a natural area and the number of different species that could be supported there.

Answers to **Got It?, Visualize This, Working with Data, Sounds Right, But Is It?, Show You Know,** and **Chapter Review** questions can be found in the **Answers** section at the back of the book.

17 The Human Footprint

What is humanity's footprint on Earth?

If we include the fossil fuels and stored resources used, scientists estimate that we are consuming 160% of the resources the Earth can regenerate each year.

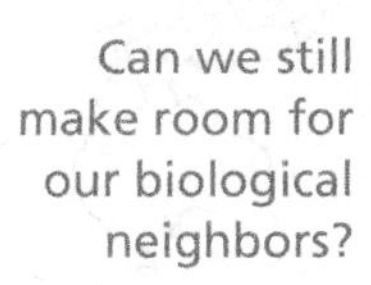

Can we still make room for our biological neighbors?

Climate and Biomes

We know the human population is growing (as described in Chapter 15). And we know how our activities can have effects on other organisms and food chains (as discussed in Chapter 16). But humans only represent a tiny fraction of this vast planet we inhabit. In fact, if every human alive today stood shoulder to shoulder, we would all fit in a plot of land that was 27 km (16.8 miles) on a side—less than the area of New York City. We might call this area the "human footprint—that is, the surface area of Earth that consists of human beings.

Of course, if we consider a footprint to be the mark that we make, the true human footprint is larger than the space our bodies take up. According to the United Nations, humans have modified 50% of Earth's land surface with our settlements and agricultural development. And this statistic accounts only for direct conversion; human activities have environmental effects far beyond their geographic boundaries, including changes to Earth's atmosphere, which receives some of our waste.

In recognition of the larger effects that human populations have on the environment, a Canadian economist in the 1990s created the concept of the **ecological footprint**—that is, the amount of land needed to support a population's activity. According to his calculations, the human population of Earth uses the equivalent of 1.6 planet's worth of energy and resources annually; in other words, it would take more than 1 year and 7 months for Earth to regenerate the resources we use in 1 year. The only way this is possible is by using up stored energy and resources—essentially by running a deficit.

Even though the 1.6 planet ecological footprint sounds bad enough, the truth is that this is an average that smooths out the differences among countries. In general, footprints for developed countries are much larger than for less developed countries—for example, the individual footprint of the average North American is nearly 13 times higher than that of the

—*Continued next page*

average Bangladeshi. If all humans lived the way the average citizen of the United States does, we would need five Earth's worth of resources!

Do opportunities exist to reduce the human footprint? It seems possible; if everyone lived the same lifestyle as the average citizen of the United Kingdom, for example, our footprint would be two-thirds the estimate for the U.S. lifestyle. In this chapter, we explore how human activities affect our biological neighborhoods and consider options for reducing our impact.

17.1 Climate Determines Habitability

Human development takes up less than half of Earth's land surface, but people can be found in nearly every environment on Earth, from the coldest polar regions to the driest, hottest deserts. Despite our widespread distribution and seeming flexibility to live in a variety of environments, we are not evenly distributed (**FIGURE 17.1**). Why are some places more crowded than others? What makes one place more suitable for humans than others? Why are some human settlements more **sustainable**—that is, able to be maintained in perpetuity—than others? The answers to these questions require, in part, an understanding of **climate,** the average conditions of a place measured over time. Climate is a broader description than **weather,** which describes the current conditions. Simply put, weather information will tell you if you have to shovel snow in the morning; climate information will tell you if you need to own a snow shovel.

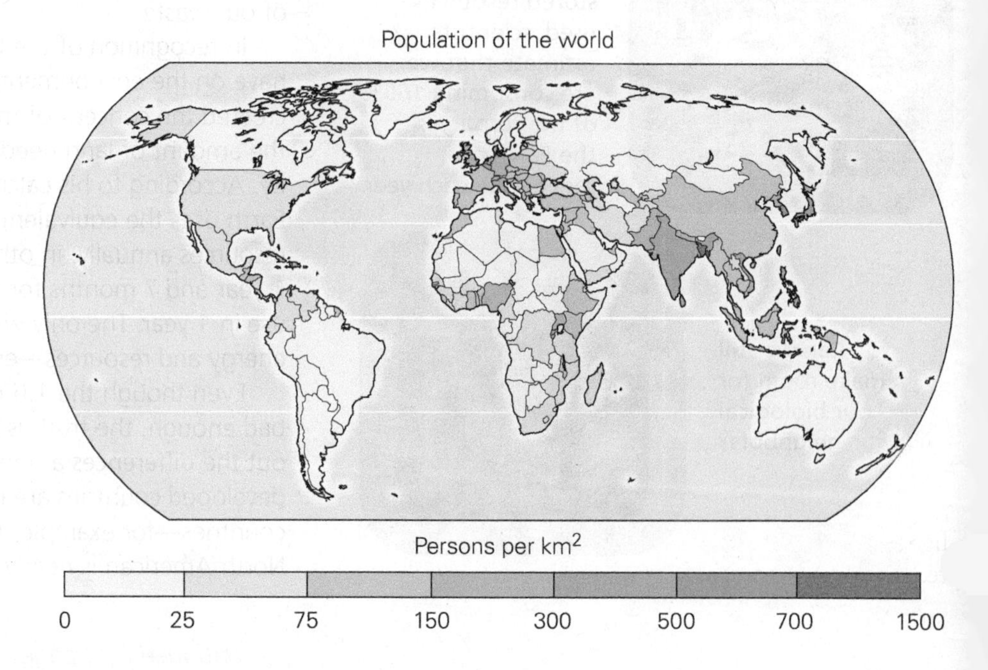

FIGURE 17.1 Distribution of people on Earth. The 7.3 billion strong human population is not evenly distributed over Earth's surface. Instead, populations are concentrated where climate conditions are most conducive to human life.

Watch **Solar Irradiance** in Mastering **Biology**

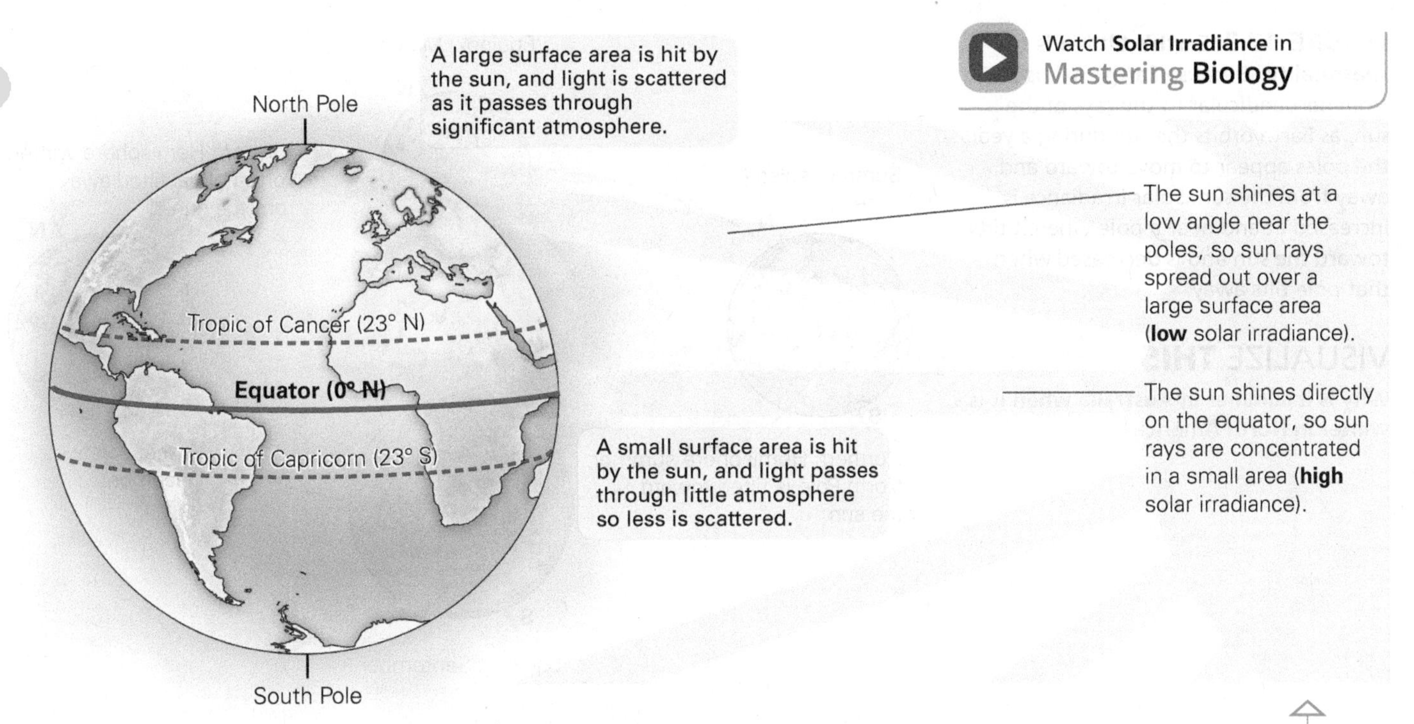

North Pole
A large surface area is hit by the sun, and light is scattered as it passes through significant atmosphere.
The sun shines at a low angle near the poles, so sun rays spread out over a large surface area (**low** solar irradiance).
Tropic of Cancer (23° N)
Equator (0° N)
The sun shines directly on the equator, so sun rays are concentrated in a small area (**high** solar irradiance).
Tropic of Capricorn (23° S)
A small surface area is hit by the sun, and light passes through little atmosphere so less is scattered.
South Pole

FIGURE 17.2 Solar irradiance on Earth's surface. The annual average temperature in a location on Earth's surface is most directly determined by its solar irradiance. Over the course of a year, areas near the equator receive the greatest amount of solar energy, whereas areas near the poles receive the least.

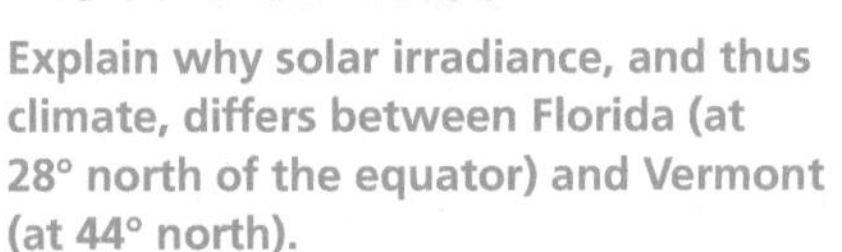

VISUALIZE **THIS**

Explain why solar irradiance, and thus climate, differs between Florida (at 28° north of the equator) and Vermont (at 44° north).

Human beings have a personal "comfort zone" that can be determined by asking people how they feel at various body temperatures. The range of comfortable core body temperatures is relatively small—from 36.5° to 37.1°C (97.7°–98.8°F), but people can maintain these core body temperatures by adapting with clothing and shelter. The more important factor limiting human settlement is food availability—and except for those populations that rely heavily on food from the oceans, that means we are concentrated in areas that are best for agricultural production. These areas are found between 20° and 60° north and south of the equator, where the temperature and **precipitation** (rainfall) are just right.

The average temperature of a region is primarily determined by the amount of **solar irradiance** it receives—the total solar energy per square meter of land or water surface. Locations that receive large amounts of solar irradiance have a higher average temperature than places receiving less.

Earth's axis is roughly perpendicular to the flow of energy from the sun. The extremes of this axis are called poles, and the circle around the planet that is equidistant to both poles is called the **equator.** The amount of solar irradiance varies between poles and equator because of the planet's shape. **FIGURE 17.2** shows two identical streams of solar energy flowing from the sun. One strikes Earth's surface directly at the equator, and the other strikes at an angle closer to the pole, where Earth's surface is "curving away" from the sun. The difference in geometry means that the surface area warmed by an amount of sunlight is much smaller at the equator than the surface area warmed at the poles. Because sunlight passing through the atmosphere can be scattered back into space, less light reaches the ground at the poles because it passes through more atmosphere at a shallow angle. In other words, the solar irradiance is greatest at the equator and lowest near the poles. This is one reason southern Florida has a warmer climate than northern Vermont.

equ- means equal.

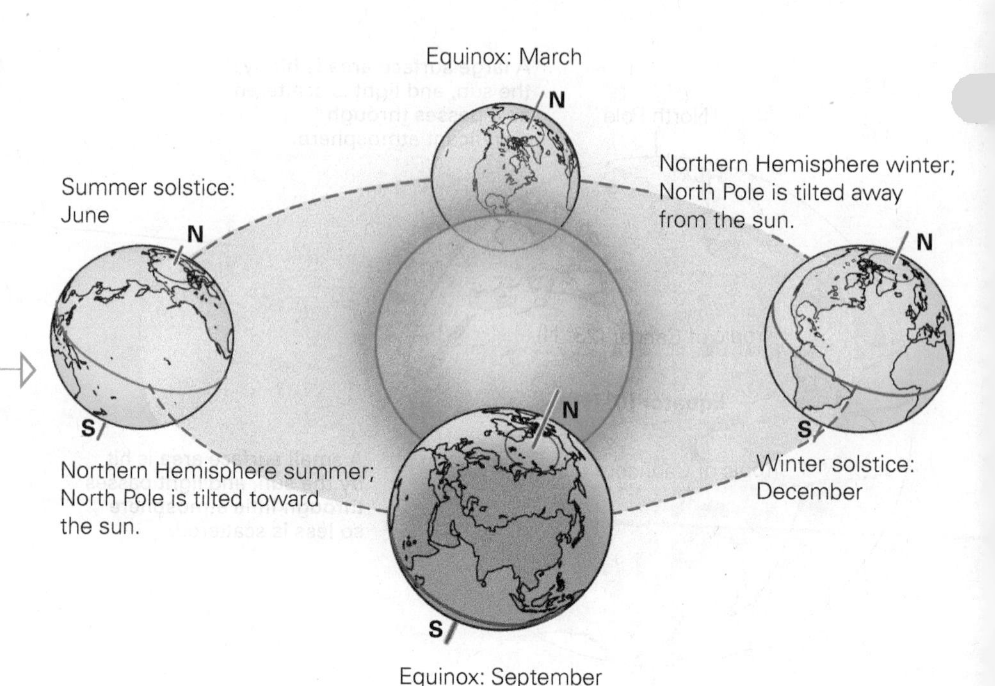

FIGURE 17.3 Earth's tilt leads to seasonality. Because Earth's axis is 23.5° from perpendicular to the rays of the sun, as Earth orbits the sun during a year, the poles appear to move toward and away from the sun. Solar irradiance is increased at and near a pole when it tilts toward the sun and is decreased when that pole tilts away.

VISUALIZE **THIS**

Why is it summer in Australia when it is winter in North America?

Solar irradiance also varies in a particular location annually. This occurs because Earth's axis actually tilts approximately 23.5° from perpendicular to the sun's rays (**FIGURE 17.3**). Because of this tilt, as Earth orbits the sun, the Northern Hemisphere (north of the equator) is tilted toward the sun during the northern summer and away from the sun during the northern winter. Solar irradiance is at its annual maximum in the Northern Hemisphere during the summer **solstice**, when the sun reaches its northern maximum and the North Pole is tilted closest to the sun. The winter solstice is the point where the sun is at its minimum. Earth's tilt also helps explain why the position of the sunrise changes over the course of a year, moving from south to north as winter turns to summer (**FIGURE 17.4**).

sol- means sun.
-stice means to stand still.

The atmosphere of Earth plays a role in determining climate as well. The naturally occurring blanket of gases surrounding Earth, including water vapor and carbon dioxide, prevents the heat absorbed from the sun during the day from escaping back into space during the night. This insulating "blanket" makes Earth habitable.

Recent human activities have changed the atmosphere of Earth. By greatly increasing carbon dioxide levels in the atmosphere via the burning of fossil fuels, the effectiveness of the blanket in retaining heat has increased. The consequences of this phenomenon are increases in global temperatures and radical changes to the climate, some of which may be unpredictable. (Chapter 5 explores human-caused global climate change in depth.)

FIGURE 17.4 The sun's travels. This image was produced by taking a picture of the sun at the same location and time approximately once each week throughout a year. When the sun is high in the sky, solar irradiance is high and temperatures are warm. When it is low, irradiance and thus temperatures are low.

Global Temperature and Precipitation Patterns

Energy from the sun is the primary driver of precipitation—that is, rain and snowfall. To understand how sunlight causes rainfall, we must first understand some of the properties of water vapor.

Condensation is the process by which molecules clump together to form liquid droplets, and evaporation occurs when molecules escape from droplets. For water molecules to remain as a vapor in the air, the rate of condensation must be less than the rate of evaporation.

The rate of evaporation depends on temperature; at high temperatures, the evaporation rate is high. The reverse is true at low temperatures. Thus, when air

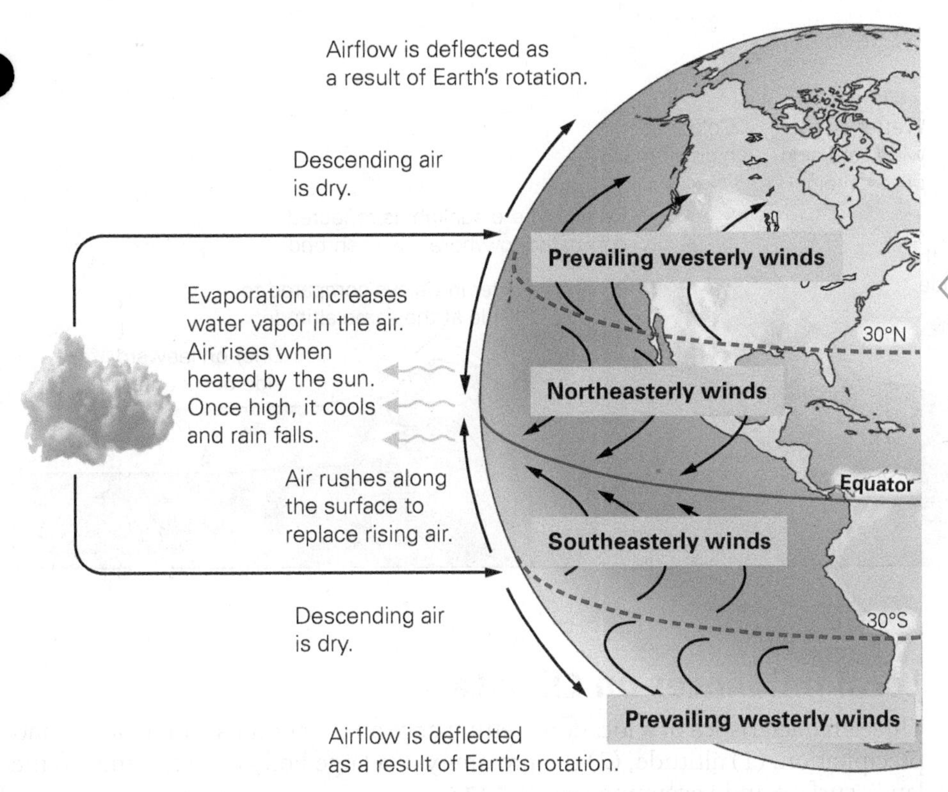
Airflow is deflected as
a result of Earth's rotation.
Descending air
is dry.
Prevailing westerly winds
Evaporation increases
water vapor in the air.
Air rises when
heated by the sun.
Once high, it cools
and rain falls.
30°N
Northeasterly winds
Air rushes along
the surface to
replace rising air.
Equator
Southeasterly winds
Descending air
is dry.
30°S
Prevailing westerly winds
Airflow is deflected
as a result of Earth's rotation.

FIGURE 17.5 Global wind patterns. High levels of solar irradiance at the solar equator lead to high levels of evaporation and rainfall. This phenomenon drives massive movements of air near the tropics into the temperate zones.

VISUALIZE THIS

The Doldrums and the Horse Latitudes are regions where there is little surface wind and thus where sailing ships are likely to stall. Based on the global wind patterns pictured here, where do you think these places are on the planet?

cools, water molecules clump into larger and larger droplets. When the droplets are large enough, concentrations of them can be seen as clouds. As clouds grow even larger, droplets can become heavy enough to fall as rain. If the temperature inside the cloud is cold enough, droplets will freeze into ice crystals, which may fall as snow. Rainfall patterns are a result of the air cycling from near Earth's surface to high in the atmosphere and back down, as illustrated in **FIGURE 17.5**.

Where solar irradiation is highest, at or near the equator, air temperatures rise quickly during the day. Because hot air is less dense than cold air, air at the equator rises. This leaves an area of low air pressure near Earth's surface that is filled by breezes blowing from the north and south.

As air rises at the equator, it cools, causing the water vapor it carries to fall as rain. This now-dry air flows in the upper atmosphere toward the poles and finally drops back to Earth's surface at about 30° north and south latitudes. This very dry falling air displaces the ground-level air at these latitudes, and that air, having picked up moisture from the surface, flows toward the poles along the surface.

The movement of air is also affected by Earth's rotation, which creates the prevailing winds in various regions of the globe. The pattern of air movement in the atmosphere that is created by the sun's heating and Earth's rotation helps explain the band of rain forests near the equator, the great deserts found near 30° north and south of the equator, and the tendency for weather patterns in North America to come from the west.

Global rainfall patterns exhibit seasonality as well. The area of maximum solar irradiance travels from 23.5° north of the equator to 23.5° south over the course of a year as a result of to the tilt in Earth's axis. Therefore, the regions of the tropics receiving the most solar energy also move. Rainy seasons occur north and south of the equator as this region of high irradiance shifts and also when this shift affects the prevailing winds. The rainy monsoon seasons in India and southern Arizona are both associated with wind shifts, as breezes move over long expanses of ocean, picking up water vapor that falls on land as rain.

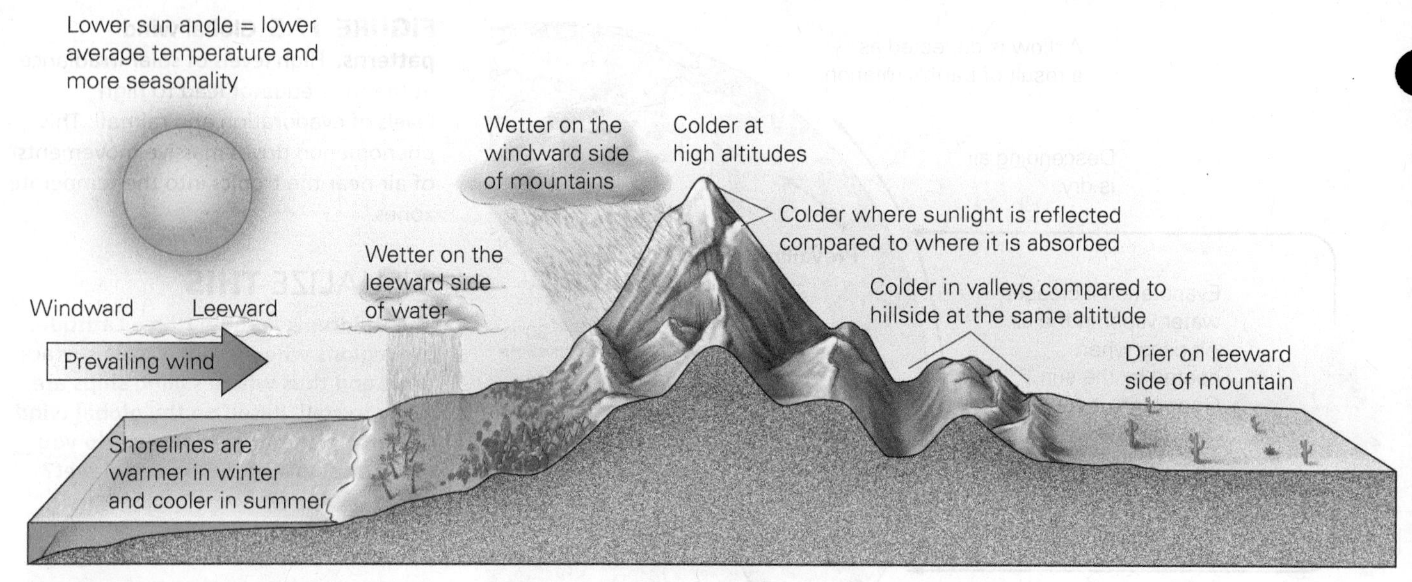
Lower sun angle = lower average temperature and more seasonality
Wetter on the windward side of mountains
Colder at high altitudes
Colder where sunlight is reflected compared to where it is absorbed
Wetter on the leeward side of water
Colder in valleys compared to hillside at the same altitude
Windward
Leeward
Prevailing wind
Drier on leeward side of mountain
Shorelines are warmer in winter and cooler in summer

FIGURE 17.6 Factors that influence local climate. Local temperature and rainfall patterns result from global climate patterns moderated by several aspects of local geography.

Local Influences on Climate

Three characteristics of a location's setting have an effect on its temperatures and precipitation: (1) altitude, (2) the proximity of a large body of water, and (3) the land's surface and vegetation (**FIGURE 17.6**).

Local Temperature Patterns. Temperature drops as altitude increases because gases expand as they rise. When this occurs over hundreds of meters and the molecules move away from each other, heat content is reduced. Temperature differences due to altitude are dramatic; for example, the summit of Mt. Everest, 8.8 km (29,035 ft) above sea level, averages –27°C (–16°F), and nearby Kathmandu, Nepal, at 1.3 km (4385 ft), averages 18°C (65°F). However, smaller differences in altitude within a region have a converse effect on air temperature. Because cold air masses are denser than warm air, pockets of cold air (for instance, air that has been in shade and not exposed to solar radiation) tend to "drain" to the lowest point on a landscape. Thus, valleys will often be colder than nearby hilltops.

Temperatures in areas near oceans, seas, and large lakes are influenced by the thermal properties of water, including its great ability to store heat (Chapter 5). Water temperature rises and falls slowly in response to solar irradiation when compared with the rate of temperature change of land surfaces. Thus, air over a large body of water is comparatively cooler in summer and warmer in winter (**FIGURE 17.7**), and nearby land areas experience more moderate temperatures than do regions further inland. Because of this phenomenon, the growing season on Long Island, New York, is as much as 30 days longer than areas in New Jersey just a few dozen miles inland from the Atlantic. Oceans also have an effect on local climates because heat is transferred through them via currents. The Gulf Stream is a current that carries water from the tropical Atlantic Ocean to the shores of northern Europe, producing a much milder climate in those regions than in other areas at the same distance from the equator. The warmth of the Gulf Stream makes Dublin, Ireland, as warm as San Francisco, even though Dublin is 1600 km (1000 mi) closer to the North Pole.

The amount of light absorbed or reflected by the surface of the land will also influence surrounding air temperature. A surface that reflects most light energy will have a lower nearby air temperature compared with a surface that absorbs most of that light energy, heats up, and radiates that heat into the air. Snow reflects more light than a forest does, so air over a snowpack remains cold. The low reflectance of asphalt pavement and most building materials contributes

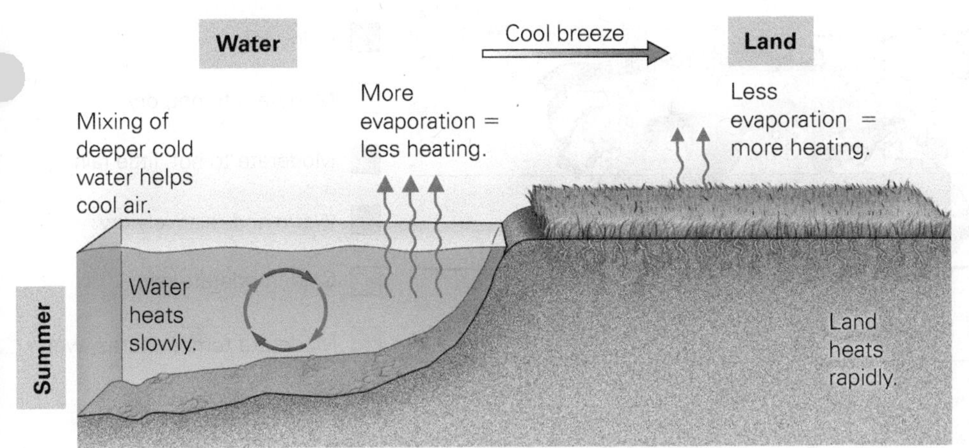
Water
Cool breeze
Land
Mixing of deeper cold water helps cool air.
More evaporation = less heating.
Less evaporation = more heating.
Water heats slowly.
Land heats rapidly.
Summer

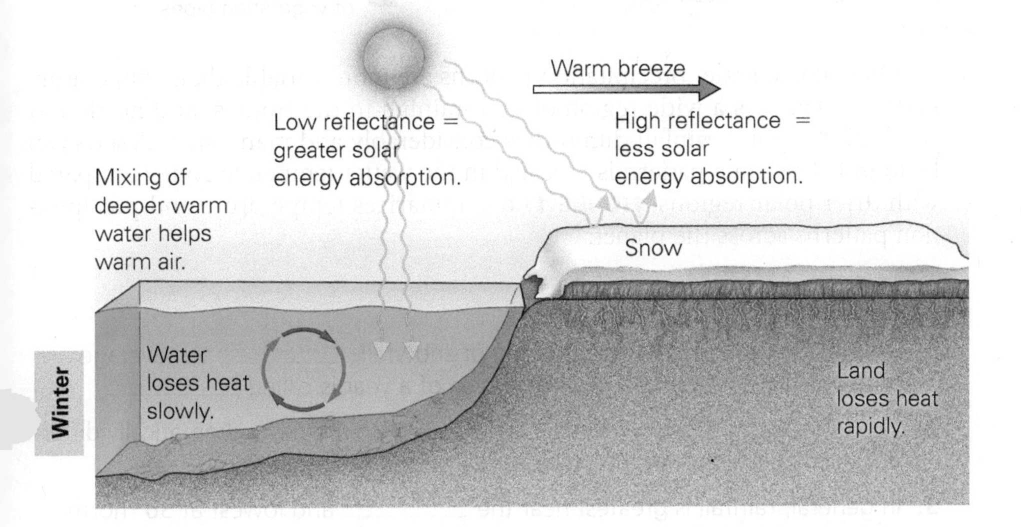
Warm breeze
Low reflectance = greater solar energy absorption.
High reflectance = less solar energy absorption.
Mixing of deeper warm water helps warm air.
Snow
Water loses heat slowly.
Land loses heat rapidly.
Winter

FIGURE 17.7 The moderating influence of water. Water heats slowly because of evaporative cooling and the mixing of warmed water with cooler deeper waters. As a result, winds blowing across water in spring and summer will cool nearby landmasses. Conversely, water absorbs heat even in winter, when the sun's rays are reflected off the snow on land, and it loses heat more slowly than land. Thus, a large body of water is a source of warmer breezes in fall and winter.

to the urban heat island effect, which is the tendency for cities to be from 0.5° to 3°C (1°–6°F) warmer than the surrounding areas. Cities are also warmer because they contain relatively little vegetation, which tends to reduce air temperature as well. Much of the solar energy absorbed by plants converts liquid water inside the plant to vapor, which also prevents the energy from being converted to heat.

Local Precipitation Patterns. The amount of precipitation that falls in a given land region is highly dependent on the context of that area—in particular, the proximity of the land to a large body of water. Wind blowing across warm water accumulates water vapor that condenses and falls when it reaches a cooler landmass. Communities surrounding the Great Lakes provide a dramatic example of this effect. For example, Toronto, on the northwest side of Lake Ontario, averages about 140 cm (55 in.) of snow per year, and Syracuse, New York, on the southeast side—receiving the prevailing wind after it crosses the lake—averages almost twice that—274 cm (108 in.).

Precipitation amounts are also affected by the presence of mountains or mountain ranges. When an air mass traveling horizontally approaches a mountain, it is forced upward. Cooling as it rises, the water vapor within it condenses to form clouds. Rain or snow then falls on the windward side of the mountain. Warming again as it drops down the other side, the dry air mass causes water to evaporate from land on the sheltered, or leeward, side of the mountain. The dry area that results is often referred to as the mountain's "rain shadow." The Great Basin of North America, encompassing nearly all of Nevada and parts of Utah, Oregon, and California, is a desert because of the rain shadow cast by the Sierra Nevada mountain range.

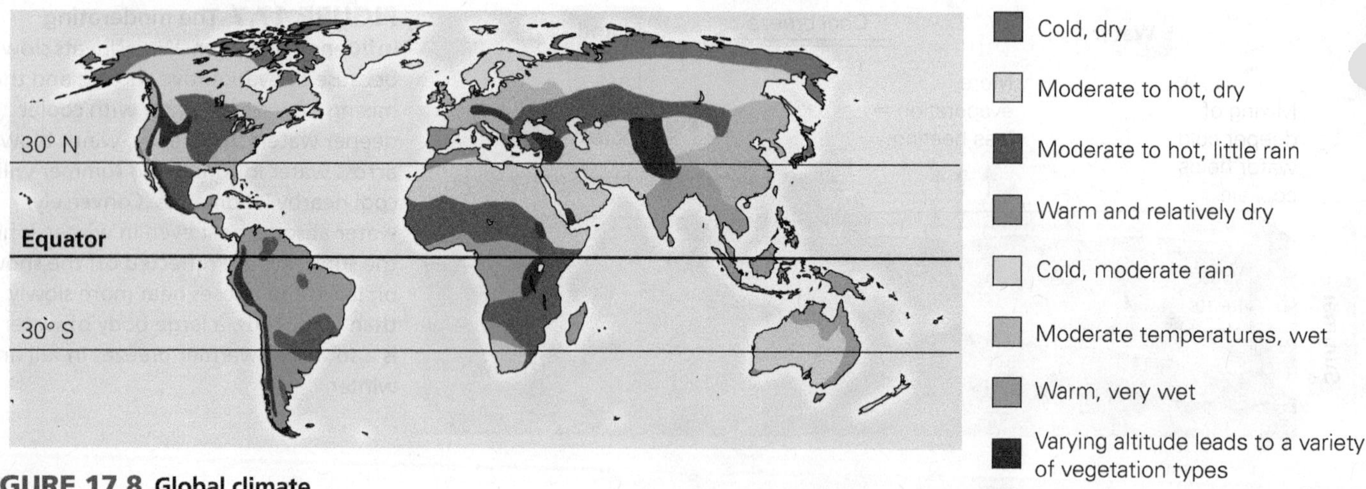

FIGURE 17.8 Global climate patterns. Average temperature in a region is somewhat predictable based on the region's distance from the equator. Precipitation is determined by some global patterns but also by local conditions, such as proximity to an ocean or mountain range.

On a global scale, precipitation patterns are more variable than temperature patterns. There is a wide region of high rainfall in the tropics, and north and south of this point rainfall diminishes considerably and many great deserts can be found. The same pattern is repeated in the wetter temperate zones compared with drier polar regions. **FIGURE 17.8** summarizes temperature and precipitation patterns across the planet.

Got It?

1. Weather is the temperature and rain or snowfall __________; the average temperature and rainfall over the course of a year is called __________.
2. The tilt of Earth produces the __________, which are more dramatically different from each other at the poles compared to the equator.
3. In general, rainfall is greatest near the __________ and lowest at 30° north and south of this line.
4. Land near large water bodies has __________ winters and __________ summers compared with land at the same latitude that is distant from water.
5. When there are mountains near a coastline, the area on the far side of the mountains is likely be exceptionally __________.

17.2 Terrestrial Biomes and the Human Footprint

Human habitation such as homes, roads, manufacturing facilities, and trash dumps are all mostly confined to Earth's **terrestrial** (land) surface. Climate plays the greatest role in determining the characteristics of a terrestrial **biome,** a geographic area defined by its primary vegetation. Plants (and animals) native to a region are adapted to the water availability and temperatures experienced there. In general, the size of the vegetation is limited by water availability—large trees require large amounts of water. Water availability is a function of total precipitation, but it is also influenced by temperature; frozen water cannot be taken up by plants.

Four basic terrestrial, or land, biomes are typically recognized: forest, grassland, desert, and tundra. Each of these basic biomes may contain variant within them; for instance, a grassland may be prairie, steppe, or savanna. The relationship between climate and biome type is illustrated in **FIGURE 17.9**, and the characteristics of the terrestrial biomes are summarized in **TABLE 17.1**.

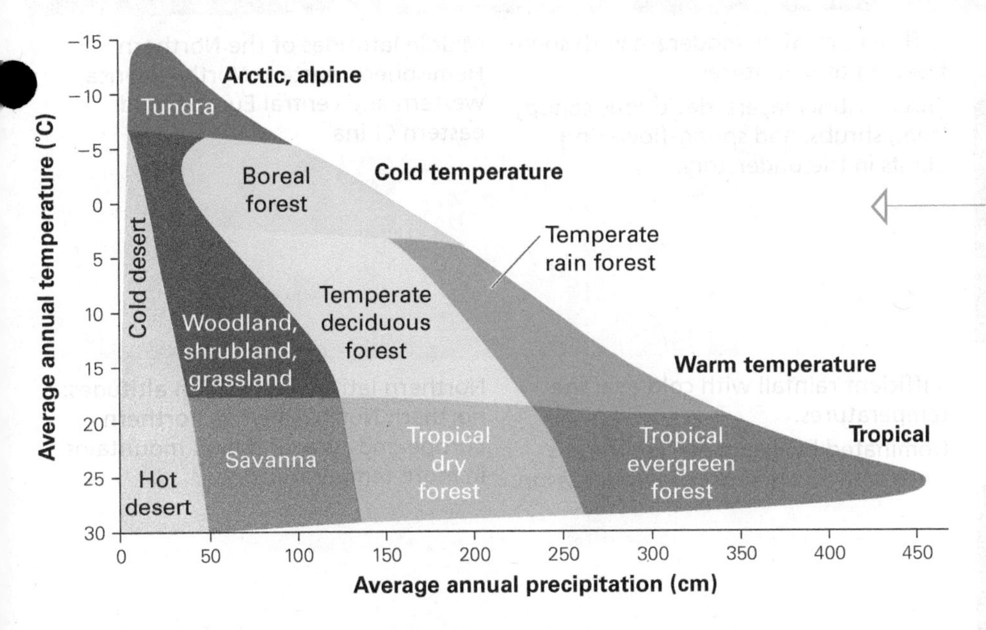

FIGURE 17.9 Biomes and climate. The primary vegetation type in a given area is determined by the region's climate.

WORKING WITH DATA

What types of biomes are possible where average annual temperature is 5°C? What factor determines exactly which biome occurs at these temperatures?

TABLE 17.1 Terrestrial biomes.

Biome	Characteristics	Location
Tundra	Very low average temperature. Very short growing season, permafrost. Plants are low to the ground and many animals migrate to warmer climates during the winter season.	Near the poles and at very high altitudes
Desert	Scant rainfall. Dry soils, very sparse vegetation. Plants often covered with spines and adapted to store water.	Near 30° north or south of the equator: northern Africa (Sahara), central Asia (Gobi), the Middle East, central Australia, and the southwestern United States
Forests and shrublands	**Dominated by woody plants**	
Tropical	High rainfall, high average temperatures. Tall trees, relatively little understory. Great diversity of species.	Around the equator: Central and South America, central Africa, India, southeast Asia, and Indonesia

—Continued

TABLE 17.1 Terrestrial biomes. *Continued—*

Biome	Characteristics	Location
Temperate (Deciduous)	Sufficient rainfall, moderate with some freezing temperatures. Three distinct layers: deciduous canopy trees, shrubs, and spring-flowering plants in the understory.	Middle latitudes of the Northern Hemisphere: eastern North America, western and central Europe, and eastern China
Boreal	Sufficient rainfall with cold average temperatures. Dominated by evergreen coniferous trees with a very short growing season.	Northern latitudes and high altitudes: northern North America, northern Europe and Asia, and high mountains in more temperate zones.
Chaparral 	Moderate temperatures, moderate seasonal rainfall, periodic fires. Dominated by spiny evergreen shrubs and maintained by frequent fires.	Areas surrounding the Mediterranean Sea and in patches in southern California, South Africa, and southwest Australia
Grasslands	**Dominated by nonwoody grasses with few or no shrubs or trees**	
Savanna 	Lower rainfall, above freezing temperatures. Scattered individual trees, maintained by periodic fires or large-animal grazing.	Tropical: about half of Africa, as well as large areas of India, South America, and Australia
Prairie and Steppe 	Lower rainfall, seasonally cold. Tall (prairie) or short (steppe) grasses, no woody plants.	Temperate areas in the middle of large landmasses: central North America, central Asia, parts of Australia, and southern South America

Tundra

The biome type where temperatures are coldest—close to Earth's poles and at high altitudes—is known as **tundra.** Here, plant growth can be sustained for only 50 to 60 days during the year, when temperatures are high enough to melt ice in the soil. As a result, agricultural production is extremely limited and humans in these regions traditionally depend on food from the ocean.

Because high temperatures are not sustained long enough to melt all of the ice stored there, places like the arctic tundra near the North Pole are underlain by **permafrost,** icy blocks of gravel and finer soil material. The permafrost layer impedes water drainage, and soils above permafrost are often boggy and saturated. The shallow saturated soil cannot support deep roots, and so the tundra cannot support tall trees.

Plants in tundra regions are adapted to windswept expanses and freezing temperatures, often growing in mutualistic multispecies "cushions" where all individuals are the same height and shelter one another. This low vegetation supports a remarkably large and diverse community of grazing mammals, such as caribou and musk oxen, and their predators, such as wolves.

Animals in tundra regions have evolved to survive long winters with structural adaptations, such as storing fat and producing extra fur or feathers. Other animals, such as ground squirrels, have adapted to such environments by evolving hibernation; they enter a sleep-like state and reduce their metabolism to maximize energy conservation. Grizzly bears and female polar bears also spend many of the coldest months in deep sleep; although this is not true hibernation, these bears are so lethargic that females give birth in this state without fully awakening. Other animals survive by migrating south to avoid the hardships of a long, frigid winter.

Although tundra is very lightly settled by humans, it experiences a significant human footprint. Large areas of tundra, especially in the arctic, are affected by our dependence on fossil fuels—oil, natural gas, and coal that formed from the remains of ancient plants. Wells and mines have caused the destruction of thousands of square kilometers of tundra with road building, oil spills, and water pollution.

Fossil fuel use in settled areas also contributes to air pollution that has impacts as far away as the tundra. The by-products of the combustion of oil, gasoline, and coal (from power plants), include nitrogen and sulfur oxides, small airborne particulates, and fuel contaminants such as mercury. When these gaseous pollutants enter the upper reaches of Earth's atmosphere, they can be carried on air currents throughout the globe. Airborne toxins such as benzene and PCBs (polychlorinated biphenyls) are taken up by plants and algae and then **bioaccumulate** as these toxins move up the food chain. For the top predators in food chains that start with trillions of individual microscopic algae cells, all carrying a little bit of toxin, bioaccumulation is a significant concern.

Air pollution in the form of greenhouse gases (see Chapter 5) also has a disproportionate effect in polar and tundra regions. Winters in Alaska have warmed by 2° to 3°C (4°–6°F), whereas elsewhere they have warmed by about 1°C (2°F). As the climate warms, permafrost melts, eventually leading to massive changes in this "unsettled" biome.

Nearly all of our footprints in the tundra could be lessened by reducing the rate of global warming. We can do this by reducing our dependence on fossil fuels and substituting renewable energy in the form of wind and solar power.

Desert

Where rainfall is less than 50 cm (20 in.) per year, the biome is called **desert**—after tundra, these areas are the most lightly populated on Earth. Although the image of a desert is often hot and barren, some deserts can be quite cold, and most have some vegetation. Plants and animals in desert regions

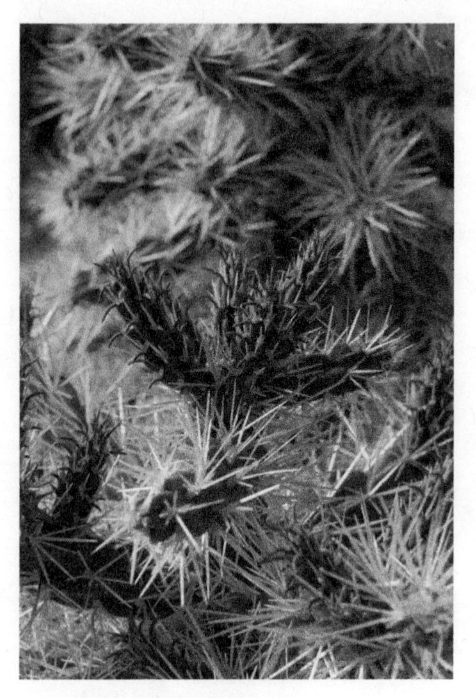

FIGURE 17.10 Desert plant adaptations to constant drought. Thick, whitish, and waxy coatings reflect light and reduce evaporation. Column-like forms reduce exposure to the high-intensity midday sun. Stems store water. The modification of leaves into spines discourages predators from accessing their water or damaging their protective surface.

have evolutionary adaptations to retain and conserve water. For example, plants are often thickly coated with waxes to reduce evaporation, contain photosynthetic adaptations that reduce water loss through leaf pores, and may be protected from predators by spines and poisonous compounds (**FIGURE 17.10**).

The dominant vegetation in many deserts are low, slow-growing woody plants with very deep roots. But deserts are also home to many flowering plants that complete their entire life cycle from seed to seed in a single season. In deserts, the wet season is quite short; many of these fast-growing plants can germinate, flower, and produce seeds in a matter of 2 or 3 weeks. The seeds they produce are hardy and adapted to survive in the hot, dry soil for many years until the correct rain conditions return.

Some animals in these dry environments have physiological adaptations that allow them to survive with little water intake. The most amazing of these animals are the various species of kangaroo rat, which apparently never consume water directly. Kangaroo rats get a small amount of water from their foods, but they also conserve water produced during the chemical reactions of metabolism and have kidneys that produce urine four times more concentrated than our own.

Although this biome does not naturally support crop production, the appeal of the sunny, warm, and dry climate of the deserts of the southwestern United States has made this region one of the fastest-growing areas in the country. The increasing population of the desert Southwest is creating a very large ecological footprint. The needs of humans and their irrigated crops is putting stress on water supplies, causing conflicts among water users, and depleting water sources for native animals. In fact, the Colorado River is so extensively used as a water source that the discharge at its outlet at the Gulf of California in Mexico averages only one-third of its historic flow. Our footprint in desert landscapes can be greatly reduced by minimizing the use of water for nonessential purposes (e.g., watering lawns) and using technology to ensure that irrigation water only goes to crop plants where and when they need it.

Forests and Shrublands

The major attribute of **forests** is the presence of trees—forests are found where rainfall is sufficient and average temperatures in the summer are above 10°C (50°F) and thus can support tree growth. Forests occupy approximately one-third of Earth's land surface and, when all forest-associated organisms are included, contain about 70% of the **biomass,** the total weight of living organisms, found on land. Forests are generally categorized into three groups based on their distance from the equator: (1) tropical forests at or near the equator, (2) temperate forests from 23.5° to 50° north and south of the equator, and (3) boreal forests close to the poles.

Tropical Forests. Extensive areas of tropical forest were once found throughout Earth's equatorial region. Tropical forests contain a large amount of biological diversity; 1 hectare (10,000 m^2) may contain as many as 750 tree species.

One hypothesis regarding why tropical forests are so rich in species is the high solar irradiance they experience. Because the energy level is high, populations of many species can be supported in a relatively small area. Think of the available energy as analogous to a pizza—the larger the pizza, the greater the number of individuals who can be fed to satisfaction.

High energy and water levels also support the growth of very large trees in tropical forests. As a result, most of the sunlight is absorbed by vegetation before it hits the ground, and most living organisms are therefore adapted to survive high in the treetops. Tropical forest animals are able to fly, glide, or move freely from branch to branch, and small plants are able to obtain nutrients and water while living on the upper branches of these huge trees (**FIGURE 17.11**).

FIGURE 17.11 Life in the canopy. This bromeliad, often found more than 60 m from the forest floor, is vase shaped as an adaptation to acquire water and nutrients from rainfall.

With warm temperatures and abundant water, the process of **decomposition,** the breakdown of waste and dead organisms, is rapid in tropical forests. Dense vegetation quickly reabsorbs the simple nutrients that are produced by decomposition, resulting in relatively little organic matter stored in the soil. Consequently, when vegetation is cleared and burned, the ash-fertilized soils can support crop growth for only 4 to 5 years. Once its soil is depleted of nutrients, a farm field carved out of tropical forest is abandoned, and a new field is cleared using the same method.

Among human populations in tropical forest areas, this slash-and-burn (or swidden) agricultural system is common. Tropical forests appear able to support swidden agriculture for many generations if the human population is small enough and abandoned plots have several decades to recover. Because they do not support intensive, permanent agriculture, tropical forests are much more lightly settled than temperate biomes. However, increasing human population levels in tropical countries have exponentially increased the human footprint. **Deforestation**—literally, the removal of forests—endangers thousands of organisms that depend on this biome for survival and threatens the sustainability of the human population (**FIGURE 17.12**).

FIGURE 17.12 Loss of tropical forest. According to the United Nations, about half of Earth's tropical forests have been cleared, like this site in the Brazilian Amazon.

Temperate Forests. Where seasonal freezing temperatures limit tropical vegetation, forests are dominated by trees that are adapted to these conditions. Temperate rain forest dominated by evergreen coniferous trees is found where rainfall is abundant, such as in the Pacific Northwest of the United States. However, in the majority of temperate forests, water is abundant enough during the growing season to support the growth of large trees, but cold temperatures during the winter limit photosynthesis and freeze water in the soil.

Trees with broad leaves can grow faster than narrow-leaved trees, so they have an advantage in the summer. However, broad leaves allow tremendous amounts of water to evaporate from a plant, leading to potentially fatal dehydration when water supplies are limited. To balance these two seasonal challenges, most broad-leaved trees in temperate forests have evolved a **deciduous** habit, meaning that they drop their leaves every autumn. In preparation for shedding its leaves, a deciduous tree seals the connection between branch and leaf, and within the leaf, chlorophyll—the green pigment essential for photosynthesis—breaks down and its components are reabsorbed by the plant. The colorful fall leaves that grace eastern North America in September and October result from secondary photosynthetic pigments and sugars that are left behind in the leaf after the chlorophyll disappears.

The short time lag between the onset of warm temperatures and the releafing of deciduous trees provides an opportunity for plants on the temperate forest floor to receive full sunlight. Spring in these forests triggers the blooming of wildflowers that flower, fruit, and produce seeds quickly before losing light and water to their towering companions (**FIGURE 17.13**). The thinner leaves of deciduous trees allow more sunlight through than do their tropical forest equivalents, permitting a shrub layer typically missing in the tropics. The forest floor and shrub layer are collectively known as the forest **understory.** In addition, animals in temperate forests are more evenly distributed throughout the forest rather than concentrated in the treetops.

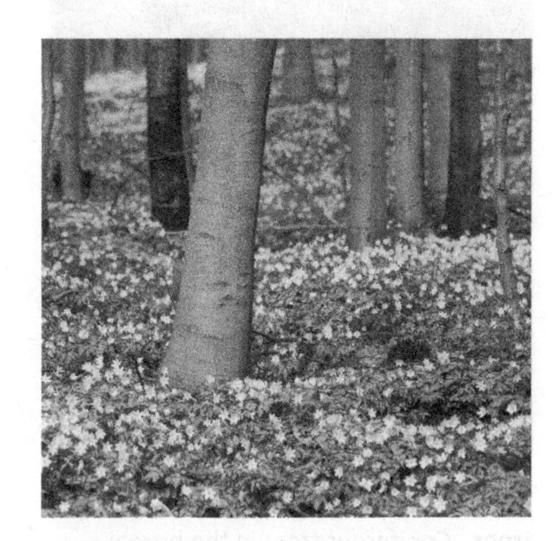

FIGURE 17.13 Springtime in a temperate deciduous forest. Forest wildflowers are adapted to take advantage of a short period between snowmelt and tree leafing by emerging from the soil, flowering, and producing fruit rapidly.

Nearly all of the deciduous forested lands in the eastern United States were converted to farmland within 100 years of the American Revolution. However, in the late nineteenth and early twentieth centuries, farms in the eastern United States were abandoned as production switched to the south and west. The regrowth of these forests provided a unique opportunity for ecologists to examine **succession,** the progressive replacement of different suites of species over time. Succession in most forests follows a predictable pattern. The first colonizers of vacant habitat are fast-growing plants that produce abundant, easily dispersed

(a)

(b)

(c)

FIGURE 17.14 Forest succession in the eastern United States. (a) Abandoned farms are quickly colonized by fast-growing herbs. (b) The second stage of succession consists of a dense landscape of shrubs and quickly growing trees. (c) The climax community in the northeastern United States is often made up of beech and maple trees.

seeds. These pioneers are replaced by plants that grow more slowly but are better competitors for light and nutrients. Eventually, a habitat patch is dominated by a set of species—called the climax community—that cannot be displaced by other species without another environmental disturbance (**FIGURE 17.14**).

The temperate forest biome is still one of the best climates for agricultural development, however, and as the human population expands, ever more of this recovering forest is converted into farms. The World Wildlife Fund estimates that worldwide, only 5% of temperate deciduous forests remain relatively untouched by humans.

The human footprint on both tropical and temperate forests could be lessened by engaging in more sustainable agricultural practices, including minimizing the consumption of meat that relies on cultivated grains to produce.

FIGURE 17.15 Coniferous trees. Coniferous trees in the boreal forest are evergreen, an adaptation that allows them to dominate this biome. These plants produce sperm and eggs on cones instead of in flowers, which may be an advantage in cold climates because they do not rely on animal pollinators but are instead wind pollinated.

Boreal Forests. Coniferous plants that produce seed cones instead of flowers and fruits dominate boreal forests (**FIGURE 17.15**). In fact, boreal forests are the only land areas where flowering plants are not the dominant vegetation type.

Climate conditions for boreal forests include very cold, long, and often snowy winters and short, moist summers. The dominant conifers in these forests are evergreens, meaning that they maintain their leaves throughout the year. Coniferous tree leaves are needle shaped and coated with a thick, waxy exterior—both adaptations reduce water loss and help shed snow during winter. Their evergreen habit likely explains conifers' dominance over flowering trees in boreal forests—the growing season is so short that the ability to begin photosynthesizing immediately after thawing gives conifers an advantage over faster-growing but slower-to-start deciduous trees.

Because they occupy regions with such cold winters and short summers, boreal forests tend to be lightly populated by humans. These areas represent some of the "wildest" landscapes on Earth—home to moose, wolves, bobcat, beaver, and a surprising diversity of summer-resident birds. However, their low human population does not protect the boreal forest from the human footprint. Trees in these landscapes are valuable for both building materials and paper products, and logging in the boreal forest is extensive. It appears that the boreal forest in North America is being cut at unsustainable rates. This impact could be lessened by reducing our consumption of paper and lumber resources—which is significantly higher in the United States than in most other countries—and more widespread and effective recycling of these products.

Both temperate and boreal forests are sites of fossil fuel extraction as well. For example, the Appalachian Mountains of the eastern United States is underlain by vast reserves of coal; mining in these regions results in the wholesale dismantling of forested mountains (**FIGURE 17.16**), and the mining of tar sands in the boreal forests of Canada has already destroyed more than 7000 square kilometers (2800 square miles), with no end in sight.

FIGURE 17.16 The cost of fossil fuel extraction. In the 1970s, mining companies developed a technique known as "mountaintop removal"—using explosives to blast off the rock above a coal deposit and dumping the waste rock into nearby valleys—to access coal. Since 1981, more than 500 square miles of West Virginia have been destroyed by this method of mining.

Chaparral. One major biome dominated by woody plants is not a forest. This landscape is known as chaparral, and its vegetation consists mostly of spiny evergreen shrubs.

Long, dry summers and frequent fires maintain the shrubby nature of chaparral. In fact, chaparral vegetation is uniquely adapted to fire. Several species have seeds that will germinate only after experiencing high temperatures. Many chaparral plants have extensive root systems that quickly resprout after aboveground parts are damaged. Chaparral will grow into temperate forest when fire is suppressed. As a result, natural selection has favored chaparral shrubs that actually encourage fire—such as rosemary, oregano, and thyme, which contain fragrant oils that are also highly flammable.

Fire can be an important contributor to the structure and function of many biomes, including chaparral. In southern California, the flammability of chaparral vegetation has come directly into conflict with rapid urbanization. In recent years, this area has experienced extensive and expensive fire seasons—for example, in 2016, fires near Carmel, California, burned nearly 140,000 acres and resulted in property damage of almost $200 million. In response to these events, support has increased for a policy of suppressing fires immediately rather than letting them burn. But in fact, such a strategy may contribute to the buildup of fuel and more intense fires that are destructive to both people and the wildlands. Fire policy in southern California, along with the long-term human modification of chaparral around the Mediterranean, makes this biome one of the most threatened on the planet.

The human footprint on chaparral landscapes could be reduced by more sensitive development for housing. Higher density settlements where a larger population lives in a smaller area would prevent loss of this habitat to **urban sprawl,** the development of suburban settlements outside the geographical limits of a city. Minimizing urban sprawl in chaparral would also make fire management much simpler.

VISUALIZE **THIS**

Think about the physical differences between elephants and cattle. Why aren't domestic cattle as effective as elephants at maintaining savanna?

Grasslands

Grasslands are regions dominated by nonwoody grasses and containing few or no shrubs or trees. These biomes occupy geographic regions where precipitation is too limited to support woody plants. These conditions are ideal for growing many of our staple grain crops, like wheat and corn—and even more so with some irrigation.

Tropical grasslands, found generally in the regions between the equatorial rain forest and the large deserts around 30°N and S latitude, are known as **savannas** and are characterized by the presence of scattered individual trees. Savannas are maintained by periodic fires or clearing. In regions where wet and dry seasons are distinct, yearly grass fires during the dry season kill off woody plants; in regions where elephants and other large grazing animals are present, the damage these animals do to trees helps to maintain the grass expanse (**FIGURE 17.17**).

FIGURE 17.17 Savanna maintenance. Elephants severely prune acacia trees when feeding on them, keeping the size of the trees relatively small; they also actively destroy thorny myrrh bushes that they do not eat, which provides additional habitat for grass to grow.

Because grazing animals eat the tops of plants, natural selection in these environments has favored plants, like grasses, that keep their growing tip at or below ground level. Grass grows from its base, so grazing (or mowing) is equivalent to a haircut—trimming back but not destroying. (Woody plants grow from the tips, so intense grazing can destroy these plants.)

Although grazing mammals are characteristic of savannas, the introduction of large numbers of cattle in these habitats has transformed the landscape from grassland to bare, sandy soil in a process known as **desertification.** For example, in the Sahel region of central Africa, overgrazing and resulting desertification has destroyed thousands of square kilometers of savanna.

Temperate grassland biomes are generally found north and south of the desert zones or in the rainshadow of temperate mountain ranges and include tallgrass **prairies** and shortgrass **steppes.** Generally, the height of the vegetation corresponds to the precipitation—greater precipitation can support taller grasses. These landscapes are generally flat or slightly rolling and contain no trees.

In the cooler temperate regions, decomposition is relatively slow, and the soil of prairies and steppes is rich with the remains of partially decayed plants. These soils provide an excellent base for agriculture. As a result, most native prairies and steppes have been plowed and replanted with crops. In North America, less than 1% of native prairie remains. Perhaps the most effective way to minimize the human footprint on grasslands is to adopt more sustainable and lower-impact agricultural practices and diets that rely less on grain-fed meat for calories.

Got It?

1. Trees are not found in areas where there is little ________.
2. Bioaccumulation occurs when pollutants become ________ in animals that are high on the food chain.
3. Tropical forests have rapid rates of decomposition, so the soil does not contain much ________ ________.
4. The deciduous growth habit, in which plants drop their leaves for part of the year, is an adaptation to low ________.
5. Grasses survive grazing (and mowing) because they grow from the ________.

17.3 Aquatic Biomes and the Human Footprint

Nearly all human beings live on Earth's land surface, but most people also live near a major body of water and are both influenced by and leave their footprints on these **aquatic** systems. Aquatic biomes are typically classified as either freshwater or saltwater, and their characteristics are summarized in TABLE 17.2.

aqua- means water.

Freshwater

Freshwater is characterized as having a low concentration of salts—typically less than 0.1% total volume. Scientists usually describe three types of freshwater biomes: lakes and ponds, rivers and streams, and wetlands.

Lakes and Ponds. Bodies of water that are inland, meaning surrounded by land surface, are known as lakes or ponds. Typically, ponds are smaller than lakes, although there is no standard for the amount of surface area required for a body of water to reach lake status. Some ponds, however, are small enough that they dry up seasonally. These vernal (springtime) ponds are often crucial to the reproductive success of frogs, salamanders, and a variety of insects that spend part of their lives in water.

TABLE 17.2 Aquatic biomes.

Biome	Characteristics
Freshwater	**Less than 0.1% total volume of salts**
Lakes and Ponds	Inland, some dry up seasonally. Yearly seasonal turnovers in larger lakes cause mixing of nutrients that would settle to the bottom.
Rivers and Streams	Flowing water moving in one direction. Different habitats along the way, from the fast-flowing, oxygen-rich regions at the headwaters to the sluggish, sediment-filled waters near the mouth.
Wetlands	Standing water with emergent plants. High species diversity due to high-nutrient levels at water–land interface.
Marine (Saltwater)	
Oceans	Contains multiple habitats, including (clockwise from top) intertidal zones, where organisms are exposed to fluctuating water levels; deep abyssal zones where no light penetrates; and open ocean.
Coral Reefs	Tropical areas in relatively shallow water. Reefs are made up of the minerals deposited by coral animals and provide habitat for a diverse array of species.
Estuaries	Where freshwater rivers flow into intertidal areas; mix of salt- and freshwater. Highly productive due to nutrients from the rivers and warmer temperatures.

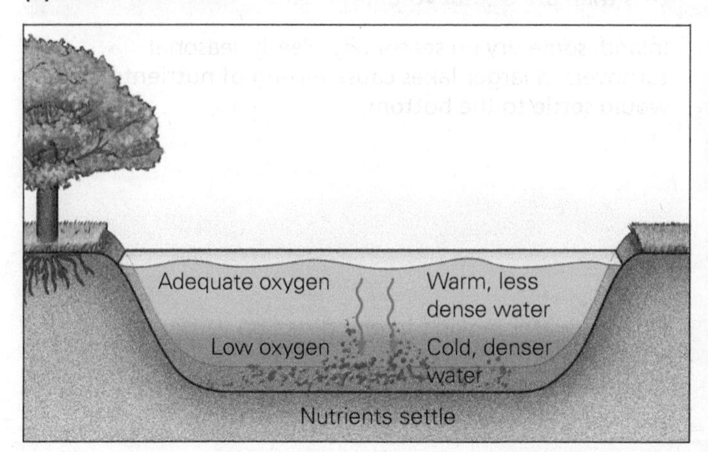

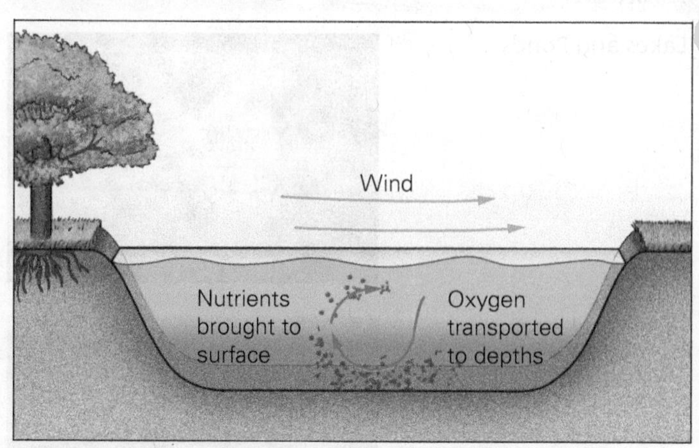

FIGURE 17.18 Nutrients in lakes. (a) A lake is stable during the summer, and oxygen levels in deeper regions diminish as nutrient levels near the surface decline. (b) Nutrients and oxygen are mixed throughout a lake during fall and spring "turnover," providing the raw materials for algae growth.

The aquatic environment of lakes and ponds can be divided into different zones: the surface and shore areas, which are typically warmer, receive more light, and are thus full of living organisms; and the deepwater areas, which are dark, low in oxygen, cold, and home to mostly decomposers. The biological productivity of lakes in temperate areas is increased by seasonal turnovers—times of the year when changes in air temperature and steady winds lead to water mixing, which redistributes nutrients from the bottom of the lake to the surface and carries fresh oxygen from the surface to deeper water (**FIGURE 17.18**).

Fertilizers applied to agricultural lands and residential lawns near lakes and ponds can increase their algae populations as these nutrients leach into the water. Humans also typically discharge their waste into nearby bodies of water, even in developed areas with sewage treatment systems (**FIGURE 17.19**), the discharge of large volumes of wastewater may occur. Ironically, having too many nutrients can lead to the "death" of these water bodies through a process known as **eutrophication.** Eutrophication occurs because large populations of algae lead to large populations of microbial decomposers, which consume a large fraction of oxygen in the water, thereby suffocating the native fish (**FIGURE 17.20**).

The life in freshwater lakes may also be threatened by **acid rain,** precipitation that has a high acid content resulting from high levels of air pollution. In lakes with little ability to buffer pH changes, acid rain lowers the pH and causes sensitive species to die off. Antipollution legislation in the United States has reduced the impact of acid rain in the northeastern states, where it was a problem in the recent past, but acid precipitation remains a significant issue in countries with less strict pollution regulations.

FIGURE 17.19 Treatment of human waste. In more developed countries, human waste is treated in expensive and technologically advanced sewage treatment plants before the wastewater is released into waterways This is a great improvement over the past, when untreated waste was dumped directly in rivers, lakes, and oceans.

Rivers and Streams. Rivers and streams are bodies of flowing water moving in one direction. These waterways can be divided into zones along their lengths: the headwaters near the source, the middle reaches, and the mouth where they flow into another body of water.

At the headwaters of a river—often by the source lake, an underground spring, or a melting snowpack—the water is cold as a result of the temperature of the source and is generally flowing quickly because its volume is small. For these reasons, water at a river's source is high in oxygen, providing an ideal habitat for cold-water fish, such as trout in the United States and Europe. Near the middle reaches of a river its width typically increases; thus, the flow rate slows and sunlight warms the water relative to the source. The diversity of fish, reptile, amphibian, and insect species that the river supports also increases in the middle reaches, because the warming water provides a better habitat for their food source of photosynthetic plants and algae. At the mouth of a river, the speed of water flow is even slower, and the amount of sediment—soil and other particulates carried in the water—is high. High levels of sediment reduce the amount of light in the water and therefore the diversity of photosynthesizers that survive at a river mouth. Oxygen levels are also typically lower there because the activities of decomposers increase relative to photosynthesis. Many of the fish found at the mouth of a river are bottom-feeders, such as carp and catfish, which eat the dead organic matter that flowing water has picked up along its way.

Rivers and streams face wholesale destruction with the development of dams and channels—dams that provide hydropower or reservoirs for cooling fossil-fuel-powered, electricity-generating plants, and channels that simplify and expedite boat traffic. These water bodies are also threatened by the same pollutants that damage lakes. Freshwater lakes and rivers represent the major source of drinking water for most human populations. The consequences of pollution can be severe—intestinal diseases such as cholera and dysentery due to contact with waste-contaminated drinking water cause the preventable deaths of more than 2 million children under 5 years old every year.

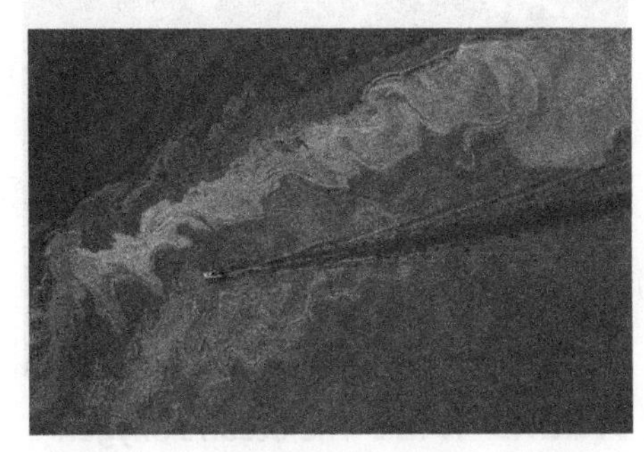

(b)

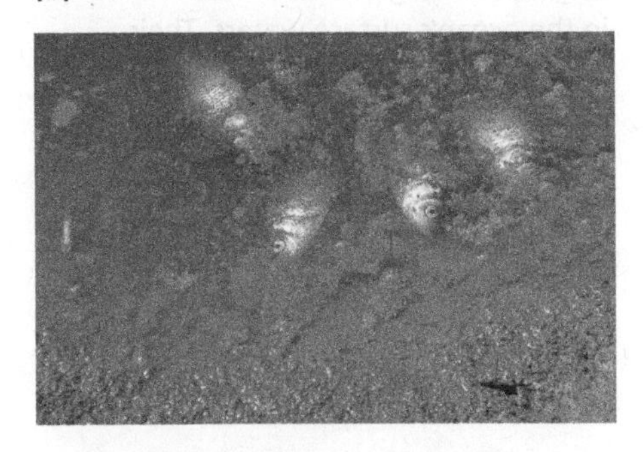

FIGURE 17.20 Eutrophication. (a) A boat travels through an algae "bloom" in Lake Erie. When these algae die, the subsequent bloom of bacteria that decompose the algae cause severe oxygen depletion, (b) leading to fish kills.

Wetlands. Areas of standing water that support emergent, or above-water, aquatic plants are called wetlands. Wetlands are comparable with tropical rain forests in the number of species supported. The high biological productivity of wetlands results from high-nutrient levels occurring when soil and organic matter washed across the landscape accumulates there (**FIGURE 17.21**).

Besides their importance as biological factories, wetlands provide health and safety benefits by slowing the flow of water. Slower water flows reduce the likelihood of flooding and allow sediments and pollutants to settle before the water flows into lakes or rivers.

Since the European settlement of the continental United States, more than 50% of wetlands have been filled, drained, or otherwise degraded. Although legislation passed in the last few decades has greatly slowed the rate of wetland loss in the United States, about 58,000 acres of this biome type are still destroyed around the country every year.

Actions that would reduce the human footprint in all of these freshwater habitats include more selective use of fertilizers, reductions in fossil fuel use, more effective management of human wastewater, and limiting urban sprawl.

FIGURE 17.21 Wetlands. Wetlands tend to be low spots on the landscape and thus accumulate nutrients running off from surrounding soils. These nutrients support the great diversity of plants and animals found in these biomes.

Saltwater

About 75% of Earth's surface is covered with saltwater, or **marine,** biomes. Marine biomes can be categorized into three types: oceans, coral reefs, and estuaries.

Oceans. The ocean covers about two-thirds of Earth's surface but is the least well-known biome of all. In truth, it can be thought of as a shifting mosaic of different environments that vary according to temperature, nutrient availability, and depth of the ocean floor. Oceanographers typically subdivide it into three regions: the open ocean, intertidal zones, and deep abyssal zones where no light penetrates.

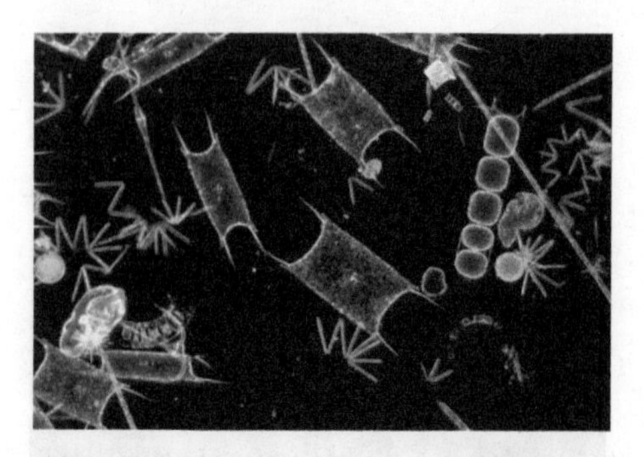

FIGURE 17.22 The ocean's oxygen factories. These phytoplankton are single-celled algae that are abundant in the ocean's surface waters. Their photosynthesis provides at least 50% of Earth's oxygen, as well as food to support an enormous diversity of life in the ocean.

More than 50% of the oxygen in Earth's atmosphere is generated by single-celled photosynthetic plankton in the open ocean (**FIGURE 17.22**). The open ocean also generates most of Earth's freshwater because water molecules evaporating from its surface condense and fall on adjacent landmasses as rain and snow, which eventually flow into lakes, ponds, rivers, and streams.

Photosynthetic plankton serve as the base of the ocean's food chain, providing energy to microscopic animals called zooplankton, which in turn feed fish, sea turtles, and even large marine mammals such as blue whales. These predators provide a source of food for yet another group of predators, including sharks and other predatory fish, as well as ocean-dwelling birds such as albatross.

Unlike lakes, oceans experience tides—regular fluctuations in water level caused by the gravitational pull of the moon. As a result of tides, ocean coasts contain unique habitats known as **intertidal zones,** which are underwater during high tide and exposed to air during low tide. Organisms in intertidal zones must be able to survive the daily fluctuations and rough wave action they experience. Adaptations such as burrow building, strong anchoring structures, and water-retaining, gelatinous outer coatings allow animals and seaweeds to take advantage of the high-nutrient environment found along the shore.

Oceans also contain a habitat known as the abyssal plain, the deep ocean floor. In these areas, sunlight never penetrates, temperatures can be quite cold, and the weight of the water above creates enormous pressure. Once thought to be lifeless because of these conditions, the abyssal plain is surprisingly rich in life, supported primarily by the nutrients that rain down from the upper layers of the ocean. In the 1970s, researchers studying the deep ocean discovered a previously unknown ecosystem supported by bacteria that use hydrogen sulfide escaping from underwater volcanic vents as an energy source. Animals in this ecosystem either use the bacteria as a food source directly or have evolved a mutualistic relationship with them, providing living space while benefiting from the metabolism of the bacteria. The abyssal plain represents the last major unexplored frontier on Earth.

Although it is vast, the ocean displays a significant human footprint; its stocks of fish are heavily exploited. Recent estimates found that species diversity in the open ocean has declined by 50% over the past 50 years as a result of fishing pressure. Seafood can be harvested much more sustainably; improvements in techniques that reduce waste and choices to focus on sustainable seafood by consumers at the market can reduce our footprint on the ocean.

Coral Reefs. Coral reefs are unusual in that the structure of the habitat is not determined by a geological feature but composed of minerals excreted by the dominant organism in the habitat: coral animals. Coral animals are very simple in structure but have a unique lifestyle—they filter food items from the water but also receive nutrients from the photosynthetic algae they host inside their bodies. Up to 90% of the nutrition of a coral is provided by the algae, which receive in return a protected site for photosynthesis and easy access to the coral's nutrients and carbon dioxide. Reef-building coral live in large clones, and each individual coral secretes a limestone skeleton that protects it from other animals and from wave action.

Coral reefs are found throughout the tropics in warm and well-lit water. Their complex structure and high biological productivity make them the most diverse aquatic habitats, rivaling terrestrial tropical rain forests in species per area.

Coral reefs are sensitive to environmental conditions and prone to "bleaching," which occurs when host coral animals lose their algae companions (**FIGURE 17.23**). Without their photosynthetic mutualists, the animals may starve. Bleaching can occur for various reasons, but recent episodes, including on Australia's Great Barrier Reef, which experienced bleaching of two-thirds of its 250-km-long northern section in 2016, seem to be associated with high

(a) Healthy coral reef

(b) Bleached coral reef

FIGURE 17.23 Threats to coral reefs. (a) A healthy coral reef teems with life and color. (b) Stressed corals lose their photosynthetic algae and become bleached. Without healthy coral, the organisms that depend on it are lost as well.

ocean temperatures. Although coral can recover from bleaching, increased global temperatures as a result of climate change may lead to more frequent bleaching and the death of especially sensitive reef systems. Reducing the effects of global warming by switching to renewable forms of power, such as solar and wind, could help reduce the human footprint on coral reefs.

Estuaries. Zones where freshwater rivers drain into salty oceans are known as **estuaries.** The mixing of fresh and saltwater combined with water-level fluctuations produced by tides creates an extremely productive habitat. Some familiar and economically important estuaries in the United States are Tampa Bay, Puget Sound, and Chesapeake Bay.

Estuaries provide a habitat for the minnows of 75% of commercial fish species and 80 to 90% of recreational fish species; they are sometimes called the "nurseries of the sea." Estuaries are also rich sources of shellfish—crabs, lobsters, and clams. Vegetation surrounding estuaries, including extensive salt marshes consisting of wetland plants that can withstand the elevated saltiness compared with freshwater, provides a buffer zone that stabilizes a shoreline and prevents erosion. Unfortunately, estuaries are threatened by human activity as well, including eutrophication from increasing fertilizer pollution and outright loss as a result of housing and resort development. Many of the same strategies that would reduce our impact on freshwater biomes would help estuaries. This includes reducing fertilizer use by lowering the production of grain-fed meat, and strategies that reduce fossil fuel consumption, such as preventing urban sprawl.

How to Leave Smaller Footprints The effect of human settlements on surrounding, and distant, natural biomes can be significant and severe, as we have discovered in this chapter. However, we have also discussed how many of these impacts can be mitigated with thoughtful planning and the use of improved technology. We have done this successfully as human populations in the past. For example, in the United States, laws such as the Clean Air Act and the Clean Water Act—passed in 1970 and 1972, respectively—have greatly reduced air and water pollution and have contributed to the recovery of once severely impaired habitats (**FIGURE 17.24**).

Motivated by these and other successes, people around the planet support actions aimed at creating sustainable communities that are both economically vital and environmentally intelligent. The environmental impact of our lifestyle is not a fixed value; nor is it inevitably tied to our economic status or happiness. Citizens of the United Kingdom are similar economically to citizens of the United States but have a per capita footprint two-thirds lower. This difference is due to two major factors: energy use, which is higher in the United States thanks to larger homes and more reliance on automobiles, and diet, which in the United States consists of 70% more beef than that consumed in the United Kingdom. The more citizens know about their own footprint, the more they may be willing to act to lessen it.

FIGURE 17.24 Fitting human needs into the bioregion. Lake Erie was once so polluted that it was considered "dead" in the 1960s. Thanks to government and citizen efforts, the lake has now recovered and provides clean swimming beaches and much healthier fish populations.

TABLE 17.3 Know your bioregion.

Environmental Factor	Your Bioregion
Climate	Describe the climate of the place where you live, including the annual average temperature, average precipitation, average number of sunny days, and names of seasons. Which global and local factors influence the climate where you live?
Vegetation	Describe the native vegetation of the place where you live. How much native vegetation remains, and what has taken its place? Which ecological factors (climate, fire, and so on) have influenced the native vegetation type?
Aquatic habitats	What are the aquatic habitats nearest to you? Can you name some of the dominant species in these habitats? What threats do these habitats face in your area?
Energy resources	What is the primary source for the electricity you use in your home? Is your bioregion rich in any less environmentally damaging energy resources?
Food resources	What agricultural products are produced in your bioregion? How easy is it to buy locally produced food?
Natural resources	What natural resources does your bioregion supply?
Wastewater and sewage	Where does your wastewater go? Does your community have any problems handling wastewater and sewage?
Trash	Where does your garbage go? What is the rate of recycling in your community, and can it be improved?
Air quality	How is the air quality in your community? What are the major air pollutants and their sources?

TABLE 17.3 poses a set of questions that will help you become more knowledgeable about your own **bioregion,** the natural components of the place where you live. Perhaps by getting to know our biological neighbors and understanding how our choices affect these organisms, we can also be inspired to reduce our own footprint.

Got It?

1. Human activities lead to eutrophication of water bodies when runoff of ______ and human ______ increases nutrient levels in the water.
2. The rate of flow and oxygen levels ______ as one moves from the source to the mouth of a river or stream.
3. Coral reefs are unique among biomes in that their structure is almost entirely dependent on the ______ that make them up.
4. Freshwater wetlands and estuaries share some similarities in that they both experience water-level ______.
5. The ecological footprint in the United States is larger than that of the United Kingdom because U.S. citizens consume more ______ and ______.

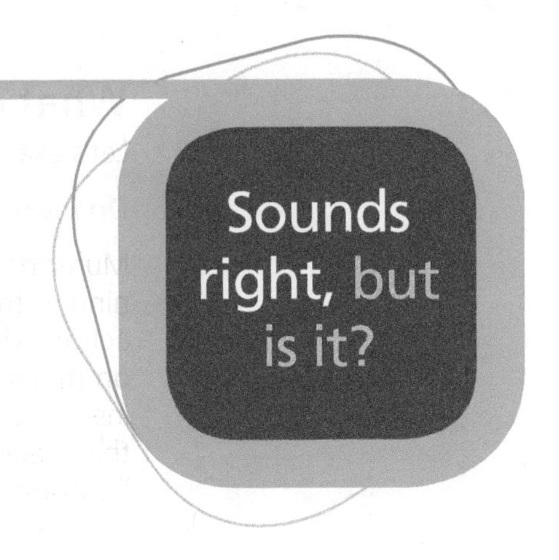
Sounds right, but is it?

In the Midwest of the United States, the sun feels hot in July but seems weak and pale in January. The quality of the sunlight is so different that it is tempting to say:

The sun is closer to Earth in summer than it is in winter.

Sounds right, but it isn't.

Answer the following questions to understand why.

1. What season (that is, winter, spring, summer, and fall) is it in January in the midwestern United States? What season is it in January in Australia?
2. What is "solar irradiance"?
3. How does the tilt of Earth's axis affect solar irradiance in January compared with July in the Midwest? How about in Australia?
4. Is the sun closer to Earth when it is summer in Australia?
5. Reflect on your answers to questions 1–4 and explain why the statement bolded sounds right, but isn't.

THE BIG QUESTION

Can my actions make a difference on global environmental issues?

When people face questions about the human impact on the planet, they can often feel overwhelmed and helpless. Each of us is only one in more than 7 billion. Our individual actions are such a tiny portion of the problem. In response, we may be tempted to give up any efforts to make a difference.

What should I know?

What follows are some smaller questions that need to be resolved to answer the Big Question. Place a checkmark next to the questions that science can answer.

Smaller Questions	Can Science Answer?
Do individual actions ever add up to a large impact on an environmental problem?	
Even if my actions have little impact on the overall situation, am I morally obligated to take actions to minimize my footprint?	
Have any countries reduced their ecological footprint through collective individual actions?	
What action could I take that has the greatest effect on reducing my individual footprint?	
Are there actions I could take that would improve my own health or security as well as reduce my ecological footprint?	

—Continued next page

What does the science say?

Let's examine what the data say about this smaller question:

Do individual actions ever add up to a large impact on an environmental problem?

Municipalities in the United States began implementing curbside recycling programs beginning in the 1980s. Before that time, recycling of household waste was entirely a voluntary action. The United States Environmental Protection Agency (EPA) has been collecting data on municipal waste generation and recycling since the 1960s—their records thus provide insight into the collective impact of individuals' decisions to recycle household trash rather than send it to a landfill or incinerator. The changes in waste generation and recycling are illustrated in in the figure below.

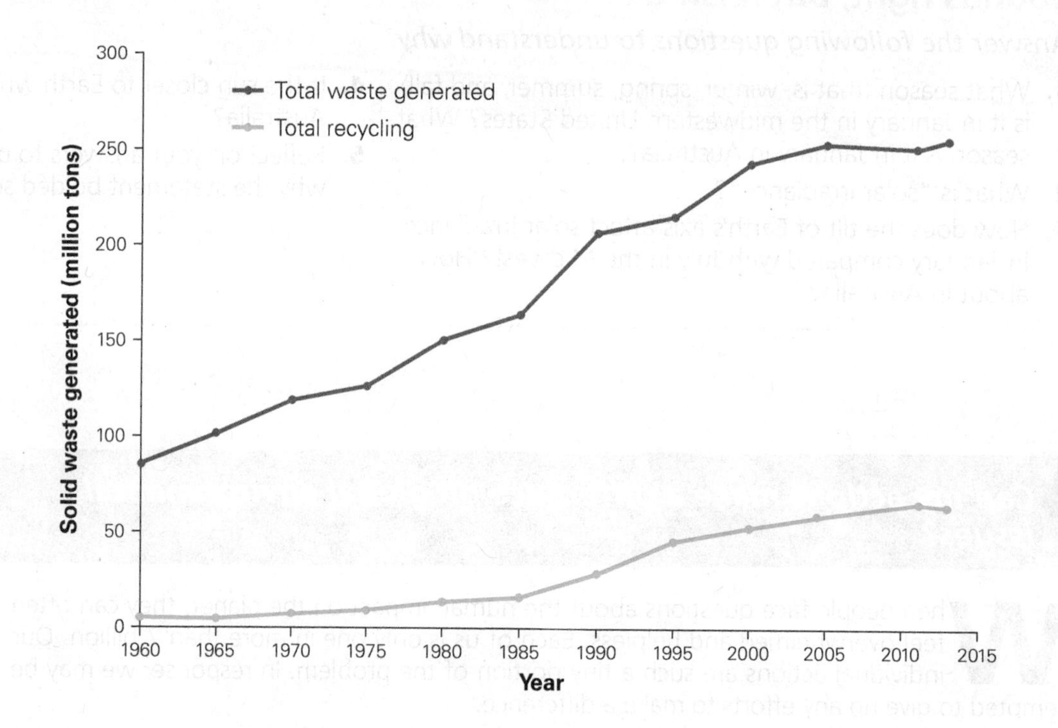

1. Describe the results. Does it appear that individual actions of recycling are having an impact on the U.S. population's footprint in terms of waste production?
2. Given these data, do you think the smaller question is answered? If not, propose another study that would help answer this question.
3. Does this information help you answer the Big Question? What else do you need to consider?

Data source: U.S. Environmental Protection Agency. 2016. Advancing Sustainable Materials Management: Facts and Figures. Washington, DC: EPA. https://www.epa.gov/smm/advancing-sustainable-materials-management-facts-and-figures.

Chapter Review

Mastering Biology

Go to Mastering Biology to access **eText 2.0, Dynamic Study Modules,** and the **Study Area,** where you'll find practice quizzes, BioFlix™ animations, MP3 tutor sessions, current events, and more.

SUMMARY

Section 17.1

Describe the factors that determine temperature and precipitation patterns on a global scale.

- Climate in an area is determined by global temperature patterns, which is driven by solar irradiance (pp. 374–375).
- Temperatures are warmer at the equator than at the poles because solar irradiance is greater at the equator (p. 375).
- Seasonal changes are caused by the tilt of Earth's axis. For example, during summer in the Northern Hemisphere, total solar irradiance during the day is greater than it is during winter (p. 376).
- Global precipitation patterns are driven by solar energy, with areas of heavy rainfall near the solar equator and dry regions at latitudes of 30° north and south (pp. 376–377).

List factors that influence climate on a local scale, and explain how they affect local temperature and precipitation patterns.

- Local temperatures are influenced by altitude, proximity to a large body of water, nearby ocean currents, the light reflectance of the land surface, and the amount of surrounding vegetation (pp. 378–379).
- Local precipitation patterns are influenced by the presence of a large body of water and the presence of a mountain range (p. 379).

SHOW YOU KNOW 17.1 There is evidence that Earth has "wobbled" on its axis throughout its history, such that the axis is more perpendicular to the sun at some times and tilted at a greater angle other times. How would a reduction in the tilt affect seasonality?

Section 17.2

Briefly describe the climate and vegetation characteristics of the terrestrial biomes, including tundra; deserts; tropical, temperate, and boreal forests; chaparral; and tropical and temperate grasslands, and give examples of animal and plant adaptations in these environments. Describe the footprint on these biomes imposed by human activity.

- Tundra occurs in areas where the growing season is 60 days or less, both near the poles and at high altitudes, and is dominated by short plants and grazing animals. Tundra is affected by climate change and oil and gas exploration (p. 383).
- Deserts are found where precipitation is 50 cm (20 in.) per year or less and contain mostly drought-resistant plants and animals adapted to the low-water environment. Deserts have a human footprint created by human settlement and water depletion (pp. 383–384).
- Forests are dominated by trees and categorized by distance from the equator into tropical, temperate, and boreal types. Tropical forests are dominated by large trees with little understory, and temperate forests have a diverse understory that flourishes when the trees drop their leaves. Boreal forest is dominated by evergreen conifers that conserve water during the long, cold winter. Forests are affected by logging and agricultural development (pp. 384–387).
- The chaparral biome is found in slightly hotter and drier areas than forests and is dominated by highly flammable woody shrubs that regenerate quickly after fire. This biome is primarily affected by human fire-suppression activities. (p. 387).
- Grasslands are found where precipitation is relatively low. These biomes are categorized by distance from the equator into tropical and temperate types. Both tropical and temperate grasslands are maintained by periodic fires and adapted to regrow quickly. Grasslands are threatened by agricultural development and overgrazing (pp. 387–388).

SHOW YOU KNOW 17.2 Explain why deciduous trees reabsorb chlorophyll, but not all pigments, from their leaves without shedding them. (Hint: Chlorophyll contains significant amounts of nitrogen.)

Section 17.3

Briefly describe the characteristics of the aquatic biomes, including freshwater lakes, rivers, and wetlands; and saltwater oceans, coral reefs, and estuaries, and provide examples of organisms found in these biomes. Describe the footprint on these biomes imposed by human activity.

- Freshwater biomes include lakes, areas of standing water dominated by aquatic plants and algae and fish; rivers, made up of flowing water with relatively few plants; and wetlands, which consist of waterlogged soil and emergent vegetation. Freshwater habitats are subject to pollution via acid rain and nutrient runoff from agricultural lands. (pp. 388–391).
- Marine biomes include oceans, which can be divided into multiple zones depending on water temperature, light, and nutrient availability. The dominant organisms in the open ocean are photosynthetic plankton, whereas the great diversity of marine life is found in coral reefs

and estuaries. Organisms in these biomes are often threatened by overharvesting (pp. 391–393).

- Increasing the awareness of citizens and communities to the nature of their bioregion and their own ecological footprint can help them devise methods for supporting human activities in a more sustainable manner (pp. 393–394).

SHOW YOU KNOW 17.3 Given the role of wetlands in slowing water flow, how might their loss contribute to eutrophication of nearby lakes and rivers?

ROOTS TO REMEMBER

The following roots of words come mainly from Latin and Greek and will help you to decipher terms:

aqua-	means water. Chapter term: *aquatic*
equ-	means equal. Chapter term: *equator*
sol-	means sun. Chapter term: *solstice*
-stice	means to stand still. Chapter term: *solstice*

LEARNING THE BASICS

1. Explain why the northern United States experiences a cold season in winter and a warm season in summer.
2. Add labels to the figure that follows, which illustrates how various factors influence local climate.

Lower sun angle = lower average ________ and more ________

________ on the windward side of mountains

________ on the leeward side of water

Windward Leeward

Prevailing wind

Shorelines are ________ in winter and ________ in summer

3. An ecological footprint ________.

A. is the position an individual holds in the ecological food chain; **B.** estimates the total land area required to support a particular person or human population; **C.** is equal to the size of a human population; **D.** helps determine the most appropriate wastewater treatment plan for a community; **E.** is often smaller than the actual land footprint of residences in a city

4. Areas of low solar irradiation are ________.

A. closer to the equator than to the poles; **B.** closer to the poles than the equator; **C.** at high altitudes; **D.** close to large bodies of water; **E.** more than one of the above is correct

5. The solar equator, the region of Earth where the sun is directly overhead, moves from 23.5°N to 23.5°S latitudes and back over the course of a year. Why?

A. Earth wobbles on its axis during the year; **B.** The position of the poles changes by this amount annually; **C.** Earth's axis is 23.5° from perpendicular to the rays of the sun; **D.** Earth moves 23.5° toward the sun in summer and 23.5° away from the sun in winter; **E.** Ocean currents carry heat from the tropical ocean north in summer and south in winter

6. Which of the following biomes is most common on Earth's land surface?

A. chaparral; **B.** desert; **C.** temperate forest; **D.** tundra; **E.** boreal forest

7. Tundra is found ________.

A. where average temperatures are low and growing seasons are short; **B.** near the poles; **C.** at high altitudes; **D.** A and B are correct; **E.** A, B, and C are correct

8. Which statement best describes the desert biome?

A. It is found wherever temperatures are high; **B.** It contains a larger amount of biomass per unit area than any other biome; **C.** Its dominant vegetation is adapted to conserve water; **D.** Most are located at the equator; **E.** It is not suitable for human habitation

9. Which of the following biomes has a structure made up primarily of the mineral deposits secreted by its dominant organisms?

A. coral reefs; **B.** freshwater lakes; **C.** rivers; **D.** estuaries; **E.** oceans

10. Which of the following actions can reduce humanity's ecological footprint?

A. reducing our dependence on fossil fuels; **B.** reducing meat consumption; **C.** living in higher density settlements; **D.** better management of human wastewater; **E.** all of the above

ANALYZING AND APPLYING THE BASICS

1. Consider the following geographic factors and predict both the climate and biome type found in the location described. Explain the reasoning that you used to determine your answer. This small city is:
   - On the coast of the Pacific Ocean
   - 20° north of the equator
   - 20 m above sea level
   - At the base of a mountain range

2. One prediction of global climate change models is that significant amounts of melting ice will change the salt content of the ocean, causing the Gulf Stream current in the Atlantic Ocean to stop altogether. How will this change likely affect Europe?
3. What can you infer about the geographical relationship among the cities in the following table and the Cascades, the primary mountain range that influences their climate?

City	*Approximate Average Annual Rainfall in Centimeters (Inches)*
Bend, OR	31 (12)
Eugene, OR	109 (43)
Portland, OR	91 (36)
Tacoma, WA	99 (39)
Walla Walla, WA	53 (21)
Yakima, WA	20 (8)

GO FIND OUT

1. Calculate your current personal ecological footprint by using the online calculator at the Global Footprint Network (www.footprintnetwork.org/en/index.php/GFN/page/calculators). Once you've calculated your current footprint, you can change your answers to determine how much of an impact different changes to your lifestyle would have on your footprint. Can you see ways to make dramatic improvements in your ecological footprint? Are you willing to make these changes?
2. How many biomes do you rely on to supply your food? Many grocery stores label the origin of their produce. The next time you go to the grocery store, try to determine the number of different countries from which your groceries come. Could you easily change your diet and shopping habits to rely on locally produced food? Why or why not?

MAKE THE CONNECTION

The science that you learned in this chapter has helped you better understand the real-world example used throughout this discussion. Draw a line from the statement on the left to the science that supports it on the right.

The human population of Earth uses about 60% more resources yearly than the Earth can generate.	Local factors that influence climate include proximity to a large body of water or to mountain ranges.
The human population is unevenly distributed on the globe; most people live in areas of the globe that have average moderate temperatures and rainfall.	Biomes that are between 20° and 40° from the equator are temperate forests and grasslands. Grassland is more common when precipitation is low.
Even in the same general area, certain regions may have cooler or wetter climate conditions than other places.	Freshwater is characterized by a salt level of lower than 0.1%.
The temperate zone is most suitable for agricultural development and thus these regions are most highly modified by human activity.	Agricultural development and fossil fuel use affect large areas of land; modifying diet and converting to renewable energy sources can reduce this impact.
However, even nontemperate terrestrial biomes exhibit significant human footprint.	Fossil fuel development and use and the harvest of building materials is common in tundra and boreal regions.
Humans depend on freshwater supplies for basic survival and on saltwater sources for many food resources.	An ecological footprint is a calculation of the amount of land providing all of the energy and resources consumed by an individual or population.
Humans can modify their activities to reduce their footprint.	Climate is the average weather of a geographical location; it is determined by position on the globe (thus exposure to solar irradiance and seasonality).

Answers to **Got It?, Visualize This, Working with Data, Sounds Right, But Is It?, Show You Know,** and **Chapter Review** questions can be found in the **Answers** section at the back of the book.

18 Organ Donation

A motorcycle accident results in severe head injuries to a young man.

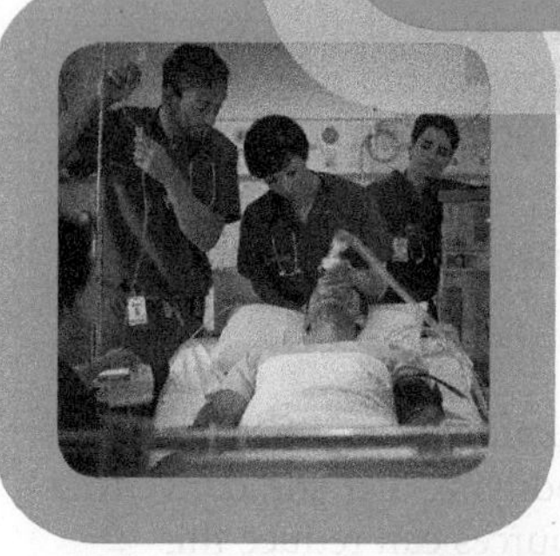

He is rushed to the hospital, where it is determined that his brain has stopped functioning.

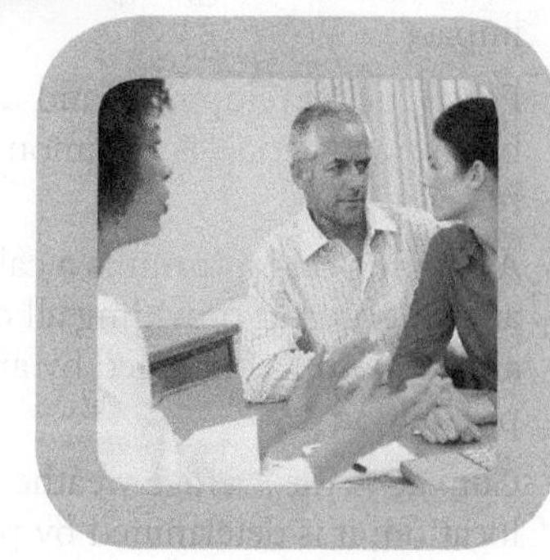

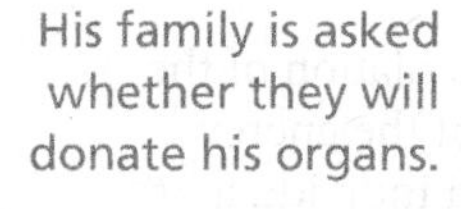

His family is asked whether they will donate his organs.

Tissues and Organs

The tragedy plays out in emergency rooms around the country thousands of times every year. A formerly healthy young man is rushed to the hospital after having been involved in a motorcycle accident. Once in the emergency room, the nonresponsive young man is placed on a ventilator because he is not able to breathe on his own, and he is given medication to keep his heart beating. Two different physicians examine him, hoping to find evidence of brain activity. One test they perform involves attempting to stimulate the young man to withdraw his hand in response to the pain produced by pressing hard on one of his fingernail beds. Another test involves investigating the pupils of the eyes to see if they constrict in response to bright light. The motorcyclist may also have electrodes placed on his head to determine whether the brain is showing any electrical activity, or he might undergo imaging studies to assess the extent of structural damage and swelling in the brain. If none of the tests performed show any sign of brain activity, he will be declared brain dead.

In the meantime, his family has assembled in the waiting room, overcome by the loss. A physician enters the waiting room and asks the family whether they would be willing to donate their loved one's tissues and organs.

The family is understandably upset, confused, and has many questions. Is there no chance that their loved one will ever be okay—even many years from now? If they agree to the transplant, will the medical staff not work as hard to keep their loved one alive? What will happen to the tissues and organs if they do decide to donate them? The family knows that their loved one would have liked the idea of helping others, but they have never really discussed organ donation. Even though his driver's license lists him as a donor, the family is not sure what to do.

What would you want your family to do? This chapter can help you decide.

18.1 Tissues

Groups of similar cell types that perform a common function make up a **tissue.** Damage to tissues can sometimes prevent them from recovering their function, which may be restored by a tissue **transplant.**

trans- means through.

We will first learn about the structure and function of different tissue types and then see some situations in which transplants can help replace damaged tissues. The human body has four main types of tissues: epithelial, connective, muscle, and nervous.

Epithelial Tissue

Epithelial tissue, or **epithelium,** is tightly packed sheets of cells that cover organs and outer surfaces and line hollow organs, vessels, and body cavities (FIGURE 18.1).

epi- means on or upon.

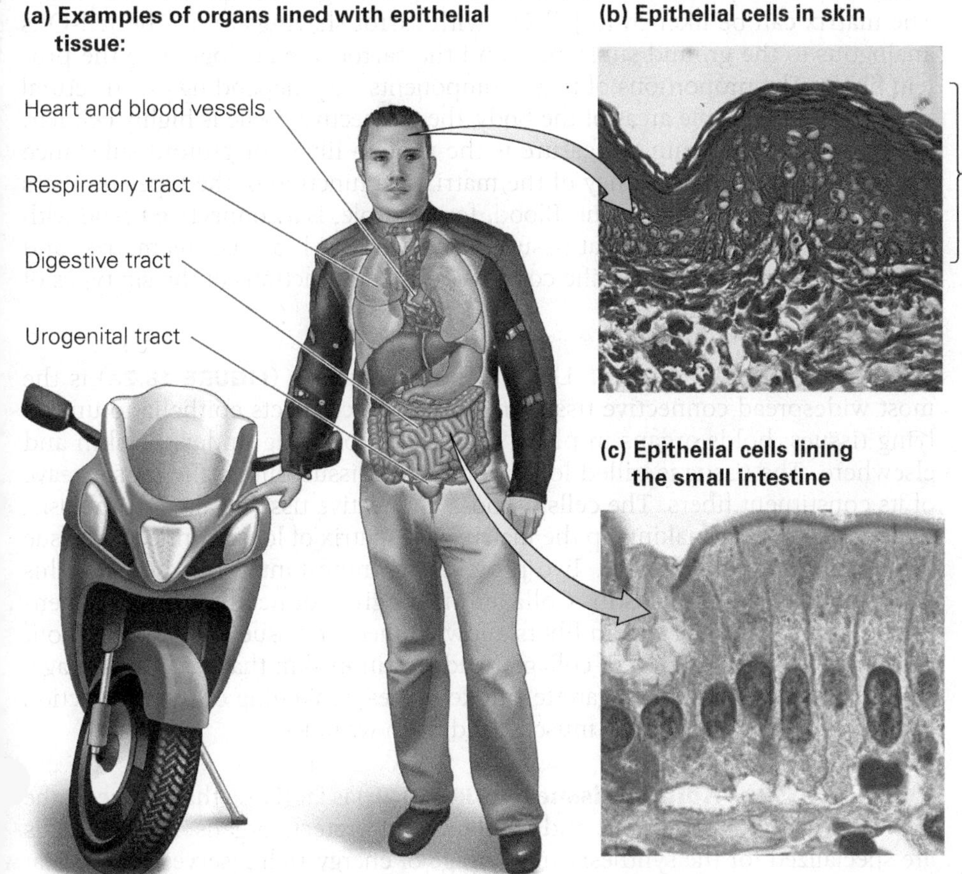

FIGURE 18.1 Epithelial tissues. (a) Epithelia line many organs. (b) Epithelial tissues are tightly packed cells that usually have one unattached surface, as seen in the epidermis of the skin. (c) Epithelia can be multilayered or a single layer, as is the epithelium that lines the small intestine.

Sheets of epithelial tissue, unlike other tissue types, are usually not attached on all surfaces. Instead, epithelial tissues are anchored on one face of the tissue but free on the other tissue face. The unattached tissue can be exposed to body fluids or the external environment. For example, the epithelial tissue that lines blood vessels is attached to the wall of the blood vessel, but its outer surface is exposed to the bloodstream. The epithelium that makes up the outer layer of skin, called the epidermis, is anchored to the underlying tissue but has a surface that is exposed. This is also true of the epithelia that line other parts of the body, including respiratory surfaces and the digestive, urinary, and genital (or urogenital) tracts. Epithelia can be a single layer or many layers thick.

Epithelial cells function in protection, secretion, and absorption. Epithelial cells in the skin help protect the body from injury and ultraviolet light. Some glands in the skin are formed by outgrowths of epithelial tissue. These glands secrete substances such as mucus, oils, and sweat. Epithelial tissues can also function to protect the body from water loss and from invading pathogens. Epithelial cells that line blood vessels and intestines function in absorbing nutrients.

Epithelial tissues constantly slough off cells, which are replaced by cell division. Skin cells are rubbed off by clothing, and cells that line the mouth and intestines are scraped off by food. These cells are replaced, resulting in a total replacement of the epidermis every 4–6 weeks.

Connective Tissue

As the name implies, connective tissues function to form connections. They usually bind organs and tissues to each other. Connective tissue is loosely organized, not tightly packed like epithelium. Our bodies contain six different types of connective tissues, all of them composed of cells embedded in a **matrix** made of protein fibers and an amorphous, gel-like ground substance. The matrix can be likened to Jell-O® with carrot shavings in it: The Jell-O is analogous to the ground substance, and the carrots are analogous to the protein fibers. The proportions of these components vary, depending on structural requirements. In some areas of the body, the connective tissue is highly cellular; in others, the predominant feature is the protein fibers or ground substance of the matrix. The consistency of the matrix is a function of the types of fibers and ground substances present. Blood, for example, is a connective tissue with a liquid matrix. Adipose or fat tissue has a thick fluid, or viscous matrix, and bone has a solid matrix. Specific compositions and functions of the six types of connective tissue follow.

Loose Connective Tissue. Loose connective tissue (**FIGURE 18.2a**) is the most widespread connective tissue in animals. It connects epithelia to underlying tissues, holds organs in place, and acts as padding under the skin and elsewhere. The tissue is called loose connective tissue due to the loose weave of its constituent fibers. The cells of loose connective tissue, called fibroblasts, secrete the proteins making up the matrix. The matrix of loose connective tissue is its chief structural feature. Two proteins are important constituents of this matrix—collagen and elastin. Collagen fibers give connective tissue tremendous strength, and the elastin fibers allow connective tissue to stretch without breaking. The degradation of collagen and elastin in skin that occurs with age, exposure to sunlight, and cigarette smoke causes weakening of the connection between skin and underlying muscle, leading to wrinkles.

adipo- means fat.

Adipose Tissue. **Adipose tissue** (**FIGURE 18.2b**) is fat tissue that connects the skin to underlying structures and insulates and protects organs. Adipose cells are specialized for the synthesis and storage of energy-rich reserves of fat. A fat

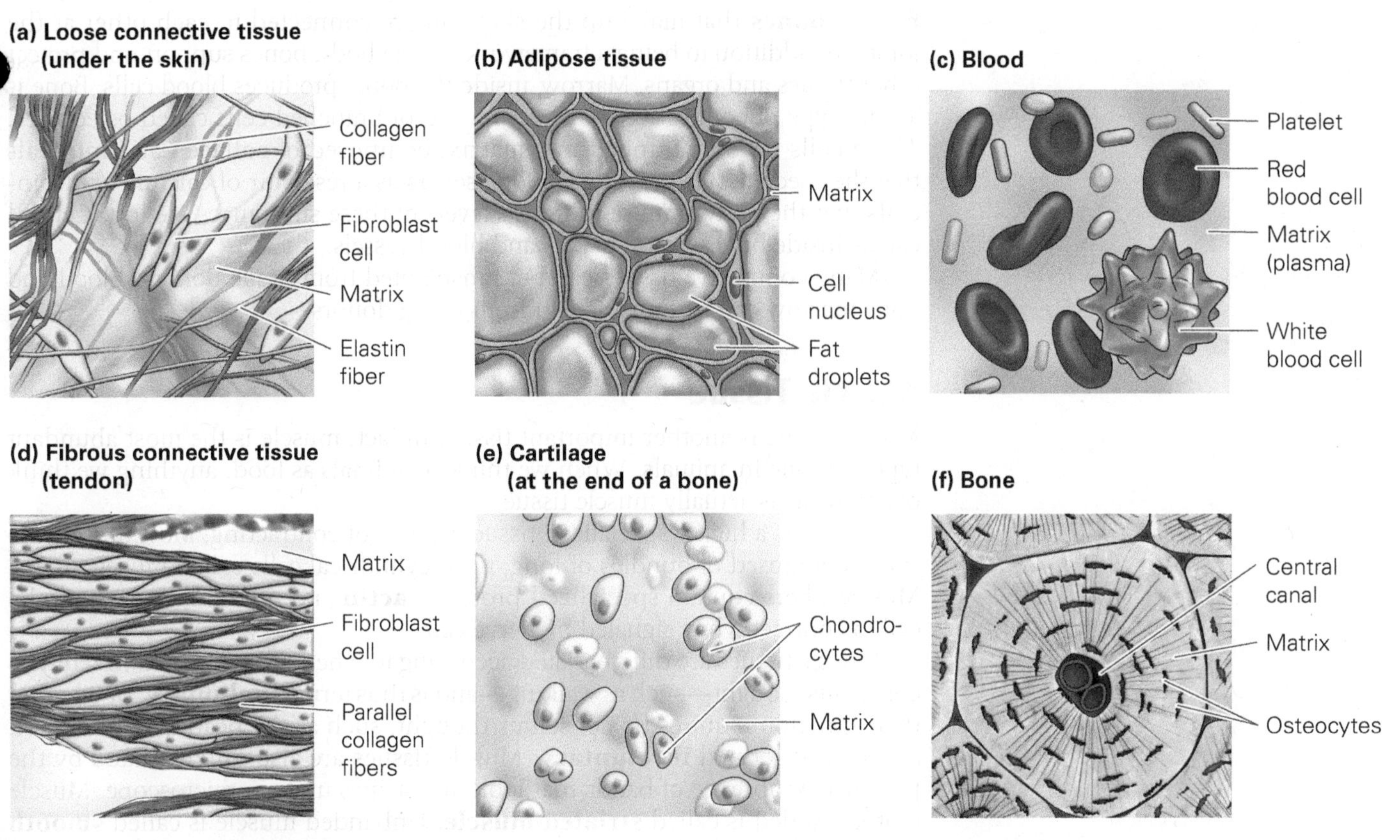

FIGURE 18.2 Connective tissues. The six different types of connective tissue are each composed of cells and a specialized matrix.

VISUALIZE **THIS**

Point out the cellular and matrix components that are the same in loose and fibrous connective tissues. Then point out what differs between the two tissues.

droplet fills the cytoplasm of these cells and shrinks when the fat is used for energy. Cells are the predominant constituent of this connective tissue; only a small amount of matrix is associated with adipose tissue.

Blood. **Blood** (**FIGURE 18.2c**) is a type of connective tissue that circulates throughout the body through arteries, veins, and capillaries and transports oxygen and nutrients to cells. The cellular component of blood tissue includes red cells, which carry oxygen; white cells, which help fight infection; and cell fragments called platelets, which function in clotting. Blood cells are suspended in a liquid- and protein-rich matrix called plasma.

Fibrous Connective Tissue. **Fibrous connective tissues** form tendons that connect muscles to bones and **ligaments** that connect bones to each other at joints. As in loose connective tissue, the cells of this tissue are called fibroblasts. The densely packed collagen fibers of the matrix, arranged in parallel, are the most conspicuous feature of fibrous connective tissue (**FIGURE 18.2d**).

lig- means to join or connect.

Cartilage. **Cartilage** is a type of connective tissue composed of cells, called chondrocytes, that secrete substances to form the dense matrix surrounding them (**FIGURE 18.2e**). The matrix is rich in collagen and other structural proteins. Cartilage connects muscles with bones, provides flexible support for the ears and nose, and allows for shock absorption. At joints, the ends of bones are covered in cartilage, which helps permit smooth gliding where joint surfaces contact each other. Cartilage has no blood vessels, so injuries heal very slowly, if at all.

Bone. **Bones** that make up the skeleton are connected to each other at the joints. In addition to being a framework for the body, bones support and protect other tissues and organs. Marrow, inside the bone, produces blood cells. Bone is a rigid type of connective tissue composed of branched cells called osteocytes. These cells become trapped in a matrix, composed of collagen and minerals that they secrete (**FIGURE 18.2f**). Bone serves as a reservoir of calcium and minerals that the body can use if dietary levels of these substances are low. Central canals inside bones house nerves and blood vessels.

Many connective tissues can be transplanted from tissue donors. Blood and bone marrow can be transplanted from living donors.

Muscle Tissue

Muscle tissue is another important tissue. In fact, muscle is the most abundant type of tissue in animals. When we think of animals as food, anything we think of as "meat" is actually muscle tissue.

Muscle is a highly specialized tissue capable of contracting. Most muscle tissue is composed of bundles of long, thin, cylindrical cells called muscle fibers. Muscle fibers contain specialized proteins, **actin** and **myosin,** that cause the cell to contract when signaled by nerve cells.

Muscle tissues are differentiated according to whether their function requires conscious thought—such as walking—and is thus termed **voluntary,** or whether their function requires no conscious thought, such as the beating of the heart, and is thus termed **involuntary.** Muscle tissues are also differentiated by the presence or absence of bands that look like stripes under a microscope. Muscle that is banded is called **striated muscle.** Unbanded muscle is called **smooth muscle.** The striated appearance is due to the banding pattern formed by actin and myosin deposits in the cells. Muscle tissue that lacks these striations also contains actin and myosin deposits, just not in a banded pattern. Humans and other vertebrates have three types of muscle tissues. These contracting tissues are skeletal, cardiac, and smooth muscle.

Skeletal Muscle. **Skeletal muscle** is usually attached to bone and produces all the movements of body parts in relation to each other. Skeletal muscle is responsible for voluntary movements such as walking. This striated muscle tissue is shown in (**FIGURE 18.3a**). Exercise causes an increase in the size of skeletal muscle cells, not an increase in the number of cells.

Cardiac Muscle. **Cardiac muscle** is found only in heart tissue. This involuntary striated tissue undergoes rhythmic contractions to produce the heartbeat. Cardiac muscle cells are highly branched and interwoven (**FIGURE 18.3b**); this structure helps propagate a contraction signal.

Smooth Muscle. **Smooth muscle,** as its name implies, is not striated (**FIGURE 18.3c**). This involuntary muscle is composed of spindle-shaped cells that make up the musculature of internal organs, blood vessels, and the digestive system. Smooth muscle contracts more slowly than skeletal muscle, but it can remain contracted for a long time. Consider the difference between the contractions of the smooth muscle lining the digestive tract to move food along as opposed to a biceps contraction. Contractions of the smooth muscles of the digestive tract are involuntary and are held for long periods of time. Contractions of the biceps are voluntary and occur more quickly. Cardiac muscle tissue can be transplanted as long as the donor heart has not been damaged. To avoid tissue breakdown in the donor heart, the transplant must take place within 4 hours after removal from the donor's body.

The last of the four types of tissues, and the least likely to be involved in a transplant, is nervous tissue.

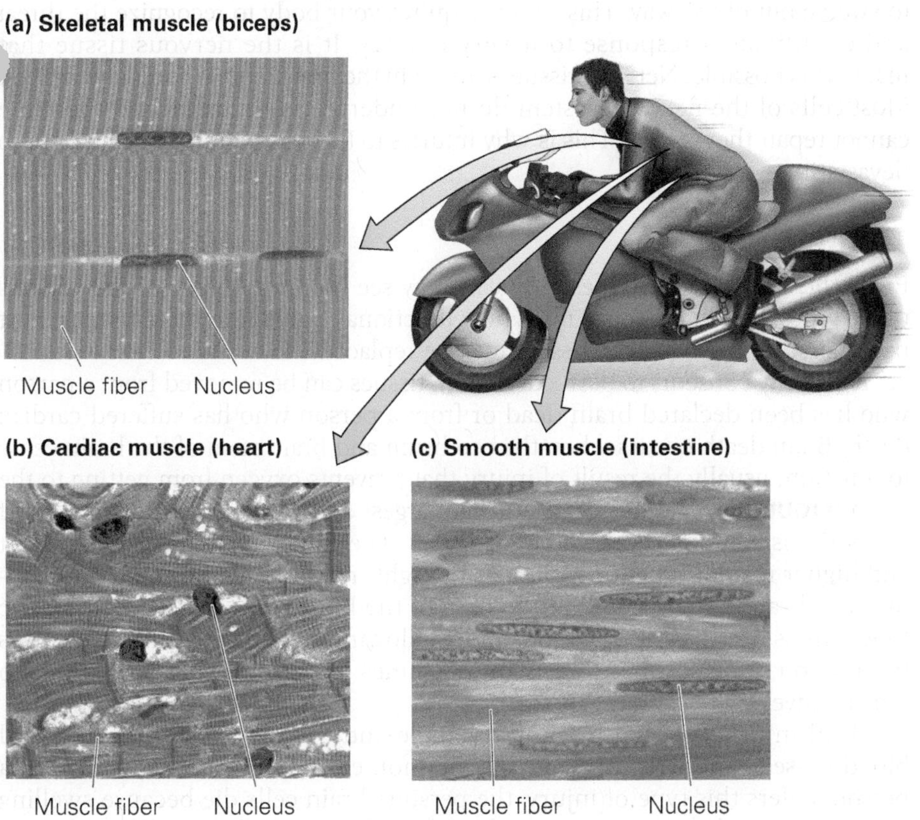

(a) Skeletal muscle (biceps)

Muscle fiber Nucleus

(b) Cardiac muscle (heart)

Muscle fiber Nucleus

(c) Smooth muscle (intestine)

Muscle fiber Nucleus

FIGURE 18.3 Muscle tissue. Skeletal (a) and cardiac (b) muscles are striated, and smooth muscle (c) is not.

VISUALIZE THIS

What causes the striped pattern seen in cardiac and skeletal muscle?

Nervous Tissue

Nervous tissue is composed mainly of cells called **neurons** (**FIGURE 18.4**) that can conduct and transmit electrical impulses. The main function of nervous tissue is to help the body sense stimuli, process the stimuli, and transmit signals from the brain out to the rest of the body and back. For example, if a car door suddenly opens in front of you while you are biking, you might react by trying

neuro- means nerve.

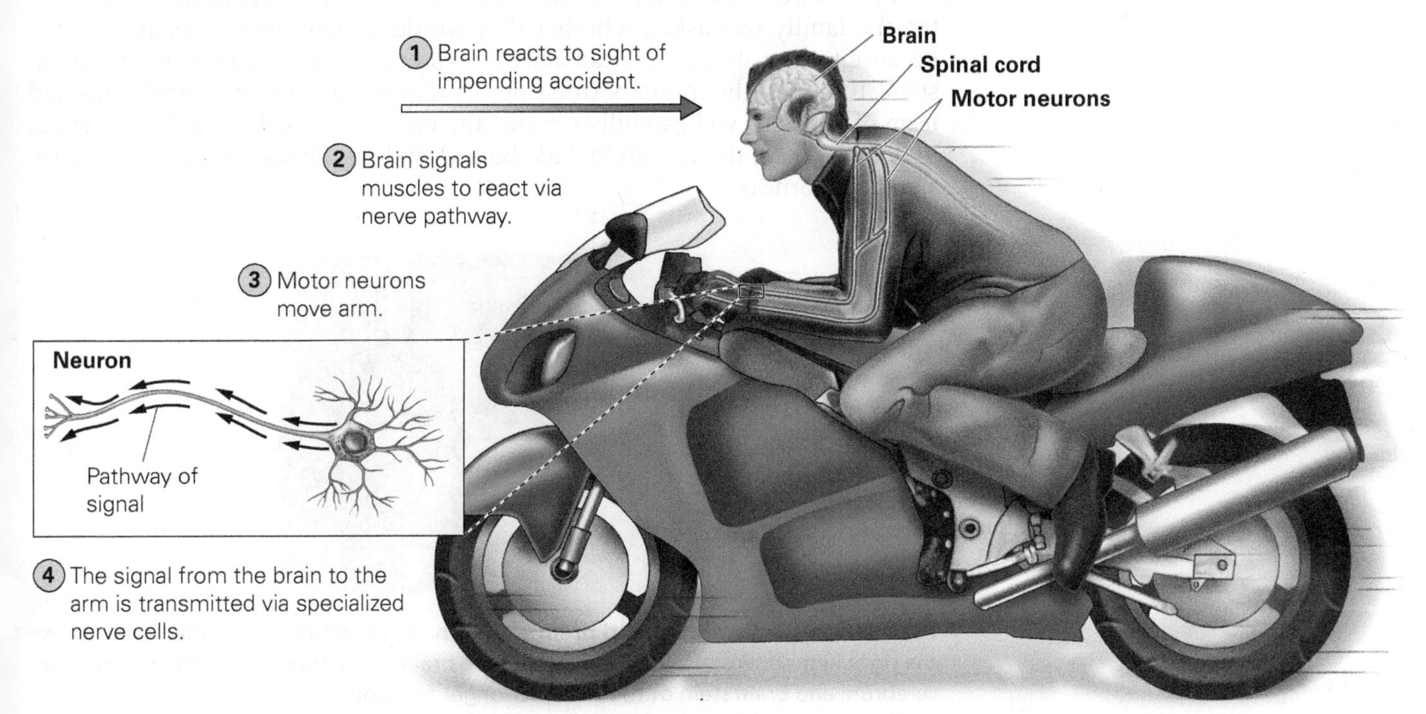

1 Brain reacts to sight of impending accident.
2 Brain signals muscles to react via nerve pathway.
3 Motor neurons move arm.
4 The signal from the brain to the arm is transmitted via specialized nerve cells.
Brain
Spinal cord
Motor neurons
Neuron
Pathway of signal

FIGURE 18.4 Nervous tissue. The brain and spinal cord are composed of nervous tissues. Nerves are composed of neurons that transmit signals to and from the brain.

to swerve out of the way. This action requires your body to recognize the threat and coordinate a response to it very quickly. It is the nervous tissue that makes this possible. Nervous tissue is found in the brain, spinal cord, and nerves. Most cells of the nervous system do not undergo cell division and therefore cannot repair themselves. This is why injuries to the brain and spinal cord are so devastating.

Tissue Donation

From the overview of tissues, you can now see the importance of the various tissue types. For an organism to be fully functional, damage to these tissues must be repaired or the damaged tissue must be replaced.

If a family consents to tissue donation, tissues can be removed from a person who has been declared brain dead or from a person who has suffered cardiac death. Brain death occurs when the cerebrum and brain stem of the brain cease to function, usually the result of injury that prevents oxygen from getting to the brain (**FIGURE 18.5**). The cerebrum, the largest and most sophisticated part of the brain, is composed of two hemispheres. The cerebrum controls movement and higher mental functions, such as thought, reasoning, emotion, and memory. The brain stem, located at the base of the brain, governs such automatic functions as heartbeat, respiration, and swallowing. When the brain stem ceases to function, a person has to rely on machines to perform these functions to remain alive.

Head injuries caused by motor vehicle and cycling accidents, ruptured blood vessels, and drowning are all common causes of brain death. When a person suffers this type of injury, the person's brain cells die because swelling and damage to the tissues prevent them from being supplied with oxygen and nutrients.

A person who is brain dead has a different prognosis from that of someone who is in a coma or vegetative state. A person in a coma or vegetative state might very well recover some functions, whereas a brain-dead person will not. This is an important distinction to make when considering organ donations.

Organ donations can come from individuals who have suffered cardiac death as well. Cardiac death occurs when the heart stops functioning.

In the case of the young motorcyclist discussed at the beginning of the chapter, the family was asked whether they would donate his tissues and organs because testing showed that all activity had ceased in his cerebrum and brain stem. If the family grants permission for tissue donation, a specially trained team of surgeons will carefully remove any usable tissues that the family agrees to donate. These tissues can include bones, tendons, ligaments, cartilage, veins, skin, and corneas.

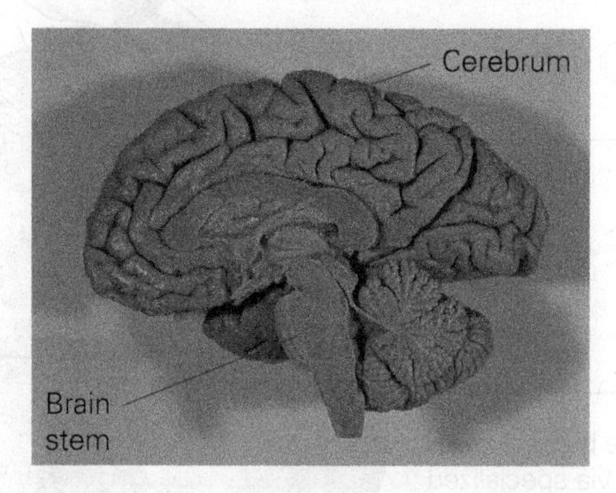

FIGURE 18.5 Cerebrum and brain stem. This cross section of a human brain shows the right hemisphere of the cerebrum and the brain stem. Brain death occurs when the cerebrum and brain stem of the brain no longer function.

The motorcyclist's bones may be donated to someone who has had bone cancer and would require an amputation without the transplant. His tendons, ligaments, and cartilage may be used to reconstruct and stabilize the damaged joints of many different people, helping them resume normal activities after accidents and injuries. Veins from the young man's legs can be used to help recipients with circulatory problems and as bypasses for people undergoing heart bypass surgery. Donated skin can be used for burn patients as a type of biological dressing, not only to help prevent infections but also to decrease blood loss and pain due to exposed nerve endings until the patients' own skin can grow back.

The transparent cornea of the eye can be transplanted to a living person who has experienced vision loss because of corneal damage. The cornea consists of a thin layer of surface epithelium overlaying a layer of fibrous connective tissue. In a healthy eye, the cornea serves as a barrier that shields the eye from dust and germs. Damage to the cornea caused by injury, infection, or chemical burn can result in loss of vision.

Once removed, the young man's tissues will be checked for the presence of bacteria and viruses that could cause disease in the recipients. Many of his tissues will then be treated with chemicals to remove any proteins that a recipient's immune system might recognize as foreign and thus attack. Because of these preventive treatments, tissue recipients do not normally reject tissue transplants and usually do not need to take drugs to suppress their immune system. Therefore, tissue recipients need not be genetically similar to the donor. After being tested and treated, the tissues will be frozen and sent to tissue and eye banks, where they will be catalogued and stored until they are needed. Most tissues can be stored for up to 5 years.

Tissue removal and treatment takes a few hours and can be performed during the time that the family begins to make plans for a funeral. One person's tissues can improve the lives of as many as 50 different people.

VISUALIZE THIS

What other organ systems might communicate with the circulatory system?

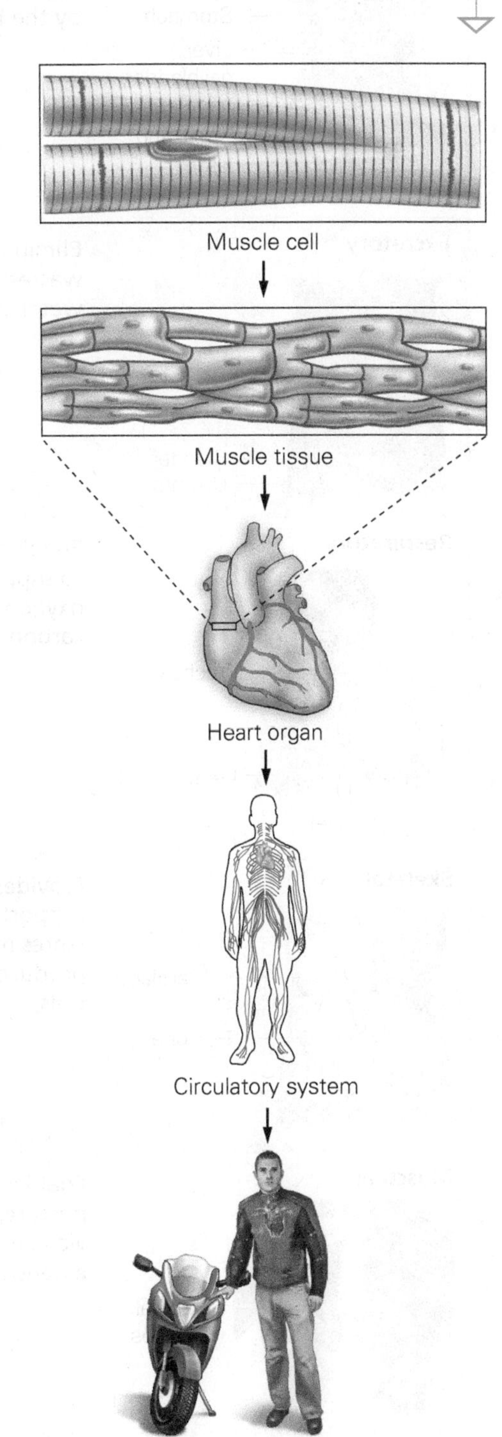

FIGURE 18.6 Levels of organization. Cells give rise to tissues. Tissues with a common function make up organs. Organs working together make up an organ system, and an entire organism often has many organ systems that communicate with each other.

Got It?

1. Tissues are composed of similar ________ with a common function.
2. Sheets of tissues that line organs and body cavities are ________ tissues.
3. ________ tissues bind organs and tissues together.
4. Cardiac, smooth, and ________ are types of muscle tissues.
5. Blood, fat, and bones are all types of ________ tissues.

18.2 Organs and Organ Systems

Living things display increasingly complex levels of organization, from cells, to tissues, to organs, to organ systems, to an entire organism (**FIGURE 18.6**). For example, the regulation of the motorcyclist's body temperature while riding through different climates requires several levels of organization. The brain (an organ) sends signals through the cells of the nervous system (an organ system) to the epidermis (tissue) that result in changes in individual units (cells) that comprise sweat glands and blood vessels, resulting in an overall change in the amount of sweat secreted and in the dilation of blood vessels.

Organs are structures composed of two or more tissues packaged together, working in concert to produce an organ's specific function. When many organs interact to perform a common function, these organs are said to be part of an **organ system.** All of the organ systems in a body work together to help an organism function. TABLE 18.1 on the next page briefly outlines the functions and component organs involved in each of the 12 organ systems.

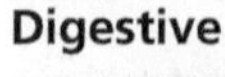

TABLE 18.1 Organ systems and functions in humans.

System	Function	System	Function
Digestive 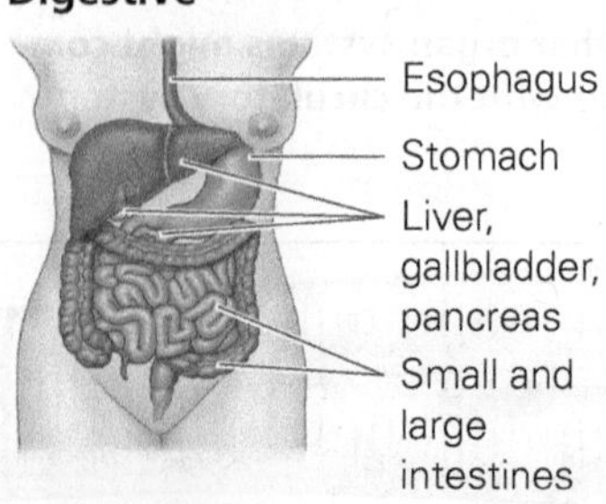	Ingests and breaks down food so that it can be absorbed by the body	**Cardiovascular** 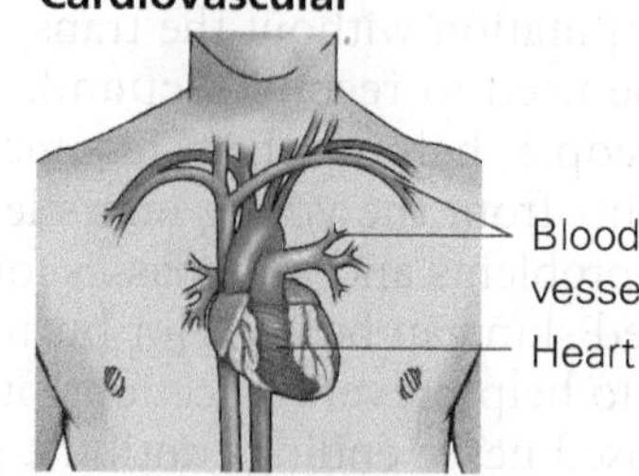	Enables the transport of nutrients, gases, hormones, and wastes to and from cells of the body
Excretory 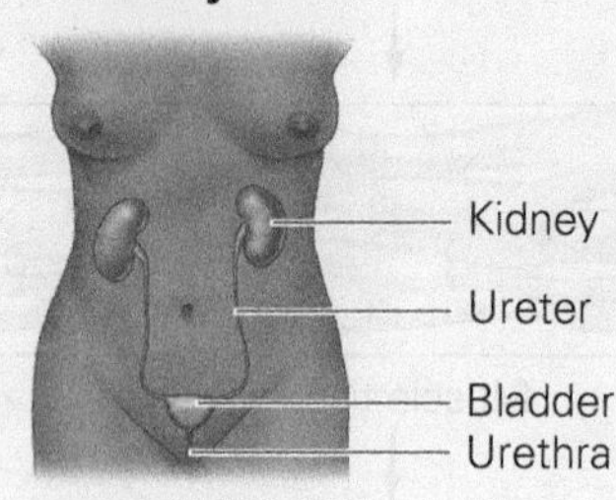	Eliminates liquid wastes; regulates water balance	**Endocrine** 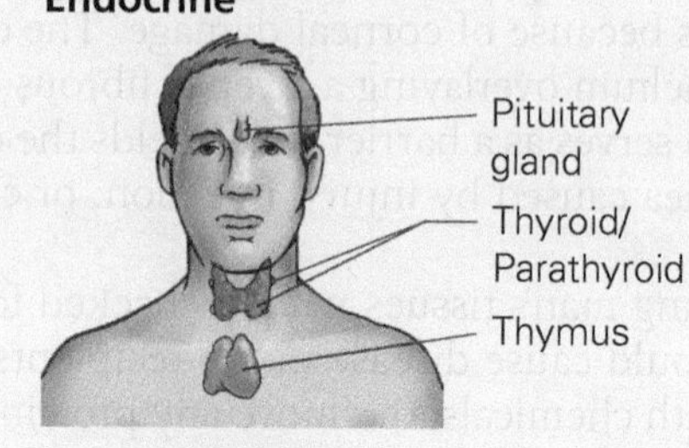	Secretes hormones into bloodstream for regulation of body activities
Respiratory 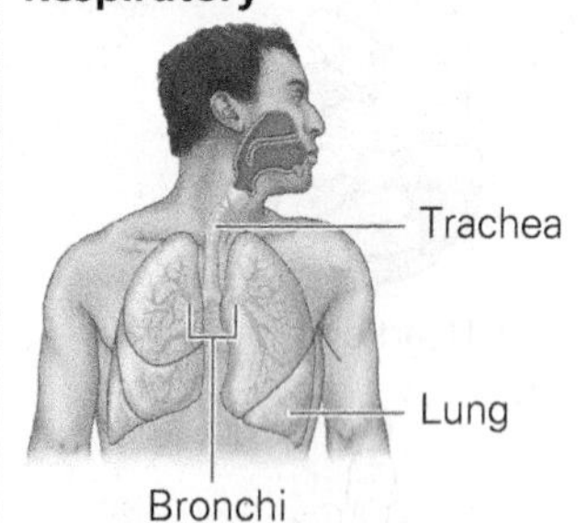	Enables gas exchange to supply blood with oxygen and remove carbon dioxide	**Nervous** 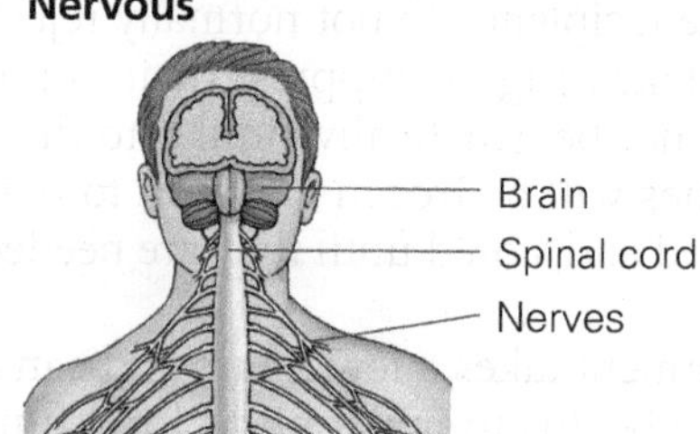	Senses environment, communicates with and activates other parts of the body
Skeletal 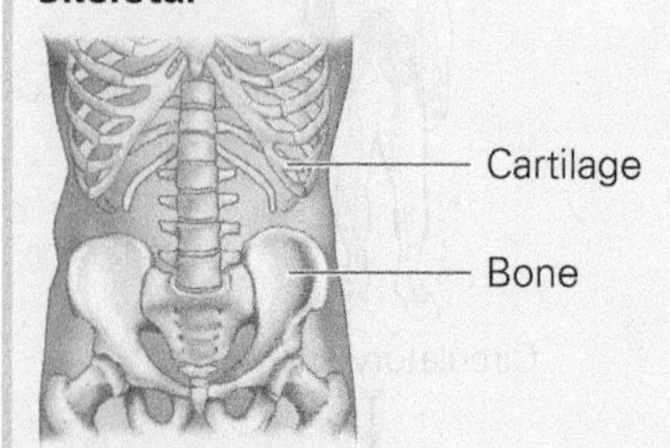	Provides mechanical support for the body; stores minerals and produces red blood cells	**Lymphatic and Immune** 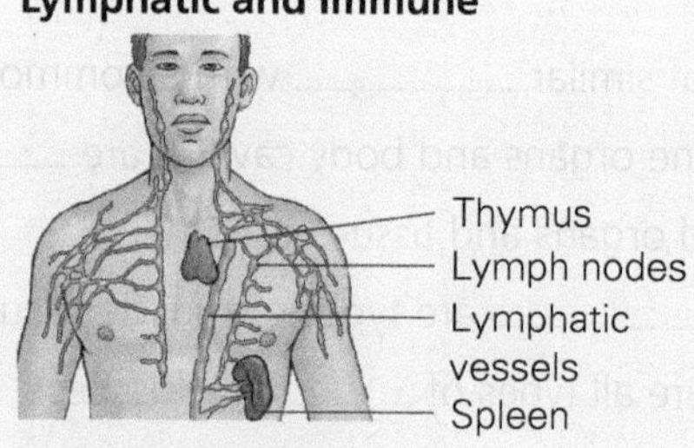	Protects against infections
Muscular 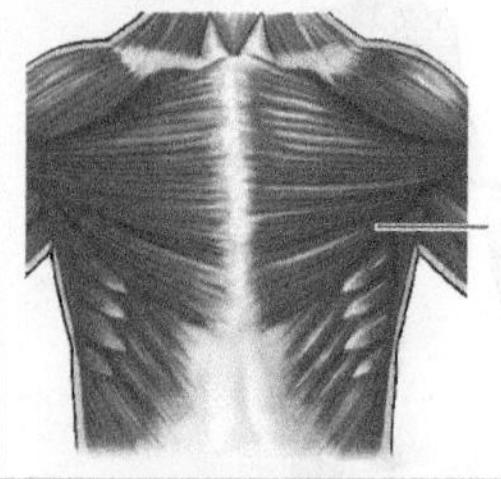	Enables movement, posture, and balance via contraction and extension of muscles	**Reproductive—female** 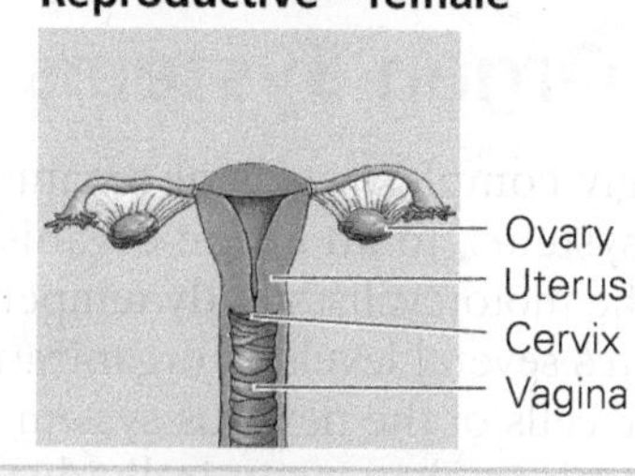	Produces eggs and supports the development of offspring
Integumentary 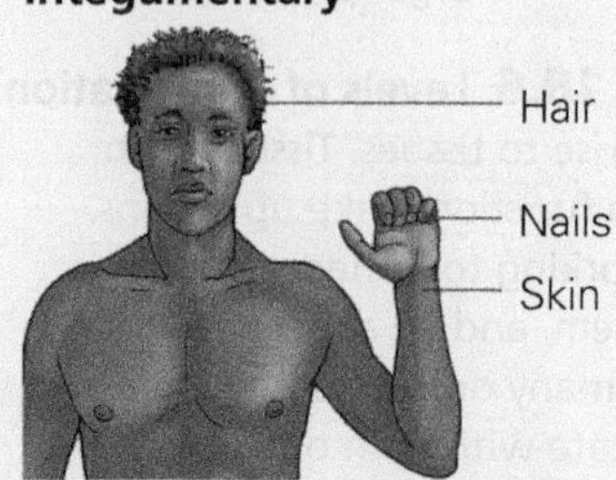	Protects body from environment, injury, and infection; stores fat	**Reproductive—male** 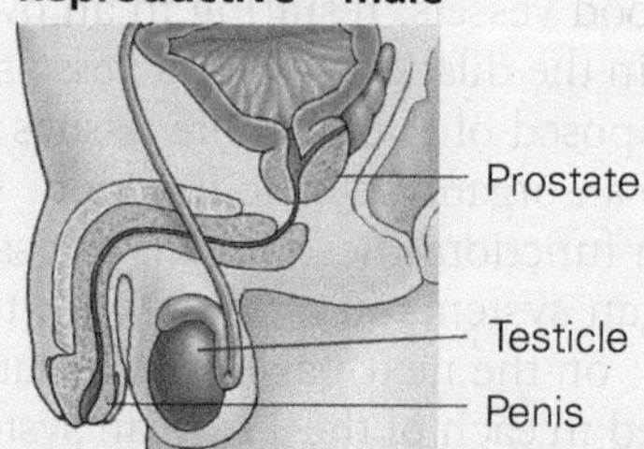	Produces and delivers sperm and associated fluids

There are many different organs in humans and other animals, such as the heart, lungs, kidney, and brain. We use the liver as a model to illustrate how an organ's ability to perform its required function is made possible by the actions of different tissue types. We also look at how an organ can function as part of an organ system.

The **liver** is a large, reddish-brown organ found on the right-hand side of the abdominal cavity below the **diaphragm.** The human liver is divided into four lobes. A greenish sac called the gallbladder is held in close contact with the liver (**FIGURE 18.7**).

Structurally, the epithelial tissue that makes up the liver is covered with connective tissue that branches and extends throughout the liver. This connective tissue acts as a sort of scaffolding to support the tissue and subdivides the tissue into hexagonal structures called **lobules.** In the middle of each lobule is a central vein that allows blood to reach all parts of the liver. Blood flows through the liver and is filtered by liver cells called **hepatocytes.** Filtering removes toxic materials, dead cells, pathogens, drugs, and alcohol from the bloodstream.

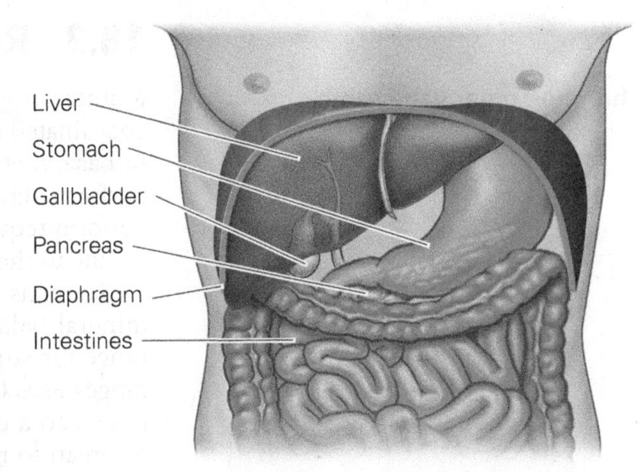

FIGURE 18.7 The liver. The human liver is located on the right side of the abdomen below the diaphragm. A duct connects the gallbladder to the liver.

dia- means through.

The liver is an important part of both the circulatory and the digestive systems. As a part of the circulatory system, the liver synthesizes blood-clotting factors, detoxifies harmful substances in the blood, regulates blood volume, and destroys old blood cells. As an accessory part of the digestive system, the liver produces bile, which helps metabolize fats.

The liver also serves as a storage site for excess glucose. Glucose monomers are joined to each other to produce a polymer of glucose called glycogen. The liver removes glucose from the blood when glucose levels are high, such as after a meal, and stores them as glycogen. When a meal has not been eaten recently, blood glucose levels fall, and the liver breaks the bonds between glucose molecules in glycogen and releases glucose monomers into the bloodstream.

Liver transplants are performed when the liver is failing. Common causes of liver failure are infection with the hepatitis C virus and chronic alcohol abuse. Both of these cause a scarring of the liver called cirrhosis.

Liver donors can be either living or brain-dead individuals. Living donors can be used for liver transplants because the liver has a remarkable characteristic that other adult organs do not share: It can regenerate itself. If a piece of the liver is removed from a living donor and placed in a compatible recipient, then both the donor and the recipient's liver will grow to full size within a year or so. Close to 7000 liver transplants occur every year in the United States.

Failure of an individual organ can affect an entire organ system or even more than one organ system. Liver failure, for instance, would have a large impact not only on the digestive system but also on the circulatory system. In addition to affecting organ systems, organ failure compromises the body's overall ability to maintain a steady state. The ability to maintain a consistent internal environment is imperative for proper health.

Got It?

1. Organs are composed of two or more __________ that work together to perform the organs function.
2. A(n) __________ __________ is two or more organs working together to perform required functions in the body.
3. The liver is part of the __________ and the __________ systems.
4. Excesses of the nutrient __________ are stored in the liver.
5. Specialized liver cells called hepatocytes function like a __________ to help remove toxins and pathogens from the blood.

18.3 Regulating the Internal Environment

homeo- means the same.

Maintaining **homeostasis,** a consistent internal environment, requires the coordinated efforts of cells, tissues, organs, and organ systems. Each cell engages in basic metabolic activities that will ensure its own survival. Cells of a tissue perform activities that contribute to the proper functioning of organs; organs perform required functions, alone or in conjunction with organ systems, to contribute to the maintenance of a stable internal environment.

Humans, for example, must keep their heart rate, blood pressure, water and mineral balance, and temperature and blood glucose levels within a narrow range for survival. Keeping physical and chemical properties within tolerable ranges uses feedback from one system to another. Feedback is information that is sent to a control center like the brain, which in turn directs a cell, tissue, or organ to respond by turning up or down a given process. Feedback can be negative or positive. In general, negative feedback negates change, and positive feedback promotes change.

Negative Feedback

Negative feedback occurs when the product of the process inhibits the process itself. A nonbiological example is the way temperature is controlled in a house. A thermostat that functions as a sort of control center turns the heater on when the room becomes too cold. When the room becomes too hot, the heater is turned off. Therefore, the negative feedback of the temperature change regulates heating systems.

Body temperature is controlled in a similar manner. When the body temperature increases, a control center in the brain sends signals to the body to dilate blood vessels near the skin and activate sweat glands, allowing heat to escape. When the body temperature cools, blood vessels near the surface of the skin constrict, muscles shiver, and body temperature increases. These negative-feedback mechanisms keep the body temperatures within the very narrow range required for survival.

Regulation of blood glucose levels by the liver provides another example of negative feedback (**FIGURE 18.8**). When food is digested, the amount of glucose in the blood increases. This stimulates the pancreas to produce **insulin,** a hormone that stimulates the liver (along with fat and muscle tissues) to remove glucose from the blood and store it as glycogen.

When most of the glucose has been removed from the blood, the pancreas secretes a hormone called glucagon, which stimulates the liver to break down glycogen and secrete glucose units into the bloodstream. Consequently, negative regulation of blood glucose levels reverses the direction of diet-induced changes in glucose levels, keeping glucose levels from swinging wildly after a meal and confining glucose levels to within a very narrow range—thus promoting homeostasis. A pancreas transplant is sometimes performed to help people with **diabetes,** a disease that occurs when insulin levels are not properly regulated by the pancreas.

Positive Feedback

Positive feedback occurs when the product of the process intensifies the process. For example, during labor and childbirth, hormones cause the muscles of the uterus to contract, pushing the baby out of its mother's body. The pressure of the baby's head against the cervix stimulates the release of more hormones, which cause the contractions to intensify in rate and duration. This in turn causes even more hormones to be released and results in even more and longer contractions. After the cervix opens completely and the baby leaves the

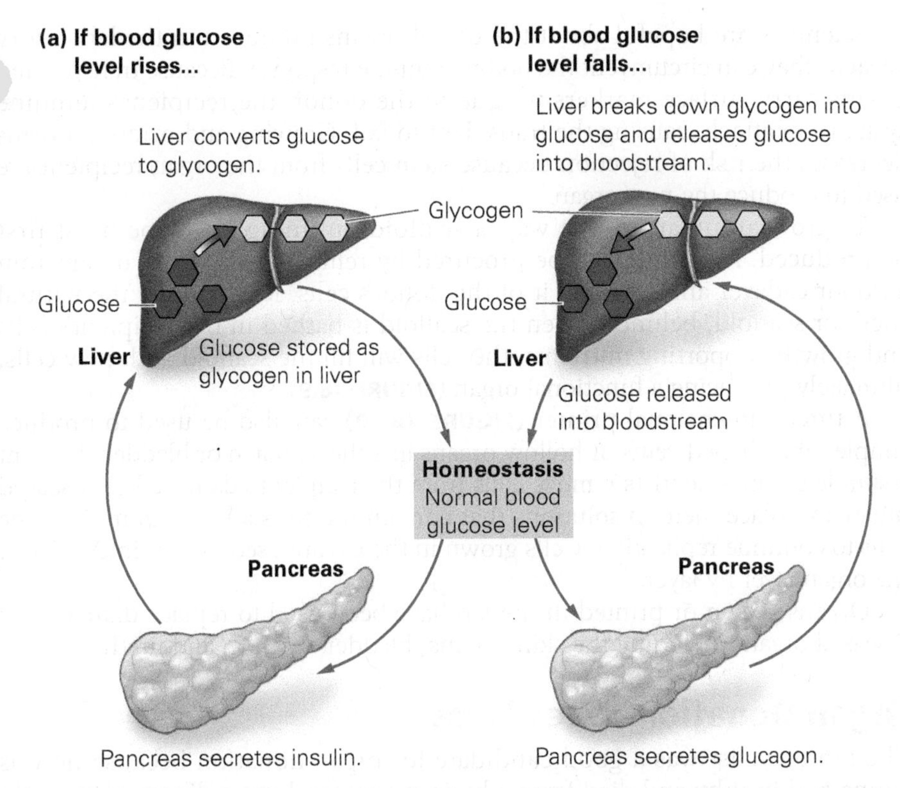
(a) If blood glucose level rises...
Liver converts glucose to glycogen.
Glucose
Liver
Glucose stored as glycogen in liver
(b) If blood glucose level falls...
Liver breaks down glycogen into glucose and releases glucose into bloodstream.
Glycogen
Glucose
Liver
Glucose released into bloodstream
Homeostasis
Normal blood glucose level
Pancreas
Pancreas
Pancreas secretes insulin.
Pancreas secretes glucagon.

FIGURE 18.8 Feedback in the regulation of blood glucose levels. (a) When intake of glucose is high, such as after a meal, the amount of glucose in the blood increases. The excess is stored as glycogen. (b) When blood glucose is low, glucose is secreted into the bloodstream.

VISUALIZE THIS

Blood glucose homeostasis can be disrupted by factors other than food you ingest. What might also cause a disruption of glucose homeostasis?

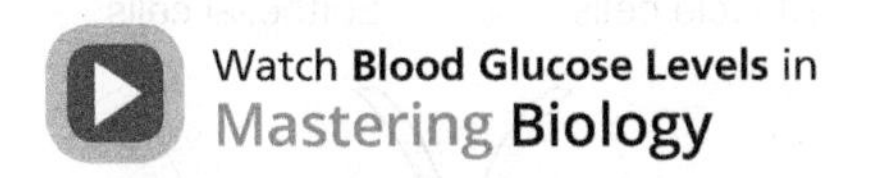
Watch **Blood Glucose Levels** in Mastering **Biology**

uterus, the contractions end. In animals, most homeostatic regulation is via negative feedback because positive feedback amplifies a response and can actually drive the process away from homeostasis. Damaged and diseased organs do not respond properly to signals that help promote homeostasis. Transplanting healthy organs to replace defective ones can restore homeostasis to a diseased individual, at least temporarily.

Got It?

1. ________ is the maintenance of a stable internal environment despite a changing external environment.
2. Negative feedback occurs when the product of the process ________ the process.
3. Insulin is produced in the ________ and helps maintain stable blood glucose levels.
4. When glucose is high, it is combined to form glycogen and stored in the ________.
5. ________ feedback promotes change.

Growing and Printing Replacement Organs

Growing and three-dimensionally printing replacement organs in a lab has only been successful in cases with organs that are relatively simple hollow- or tube-shaped organs. Solid organs like the kidney or liver are much more difficult to produce this way.

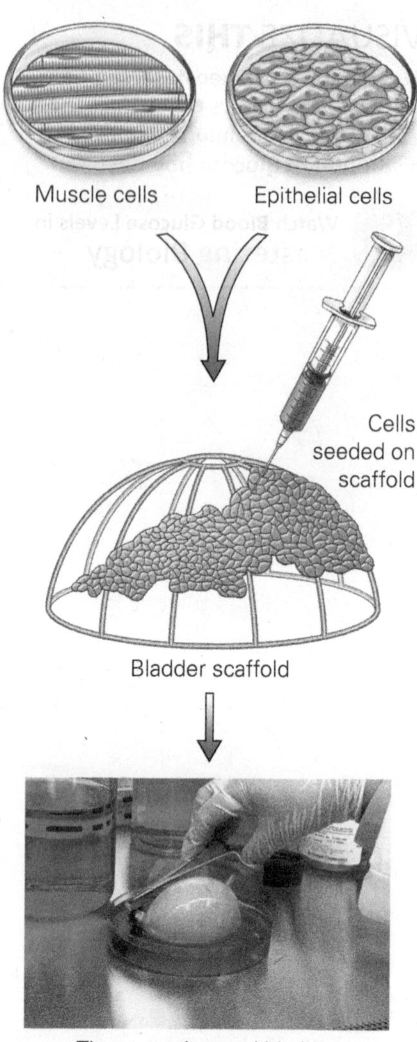

FIGURE 18.9 Growing organs. The bladder is composed of epithelial and muscle tissues. Stem cells removed from the organ recipient are grown in the lab. When layered on a scaffold shaped like a bladder, a new bladder is produced that can be implanted in the recipient.

Scientists are hopeful about the use of organs produced in the laboratory because they can circumvent the body's immune response. Because transplanted organs carry surface markers unique to the donor, the recipient's immune system can attack, causing the transplant to fail. Growing and printing organs decreases the risk of rejection because stem cells from the organ recipient are used to produce the new organ.

To grow an organ in this way, a scaffold outlining its shape must first be produced. A scaffold can be procured by removing the same organ from a donor cadaver and washing it of the donor's cells, leaving only the natural shell, or scaffold, behind. When the scaffold is bathed in the recipient's cells and growth-supporting nutrients, the cells will fill the scaffold with new cells, ultimately producing a functional organ (**FIGURE 18.9**).

A three-dimensional printer (**FIGURE 18.10**) can also be used to produce simple tube-shaped veins or hollow organs like the stomach or bladder. To print a simple organ, scientists remove cells from the recipient's damaged or diseased organ and place them in solutions that contain the substances required for the cells to continue replicating. Cells grown in the lab are used as the "ink" to print the organ layer by layer.

Organs grown or printed in the lab have been used to replace damaged or diseased organs including the skin, vagina, bladder, trachea, and urethra.

Organ Donation Saves Lives

The motorcyclist was a good candidate for organ donation because he was young and healthy and died from a brain injury (and not a disease that might have affected his organs). People who die from cardiac death can be tissue donors, but the lack of oxygen experienced during cardiac death causes organs to deteriorate, making such people less suitable organ donors. Because many more people will die of cardiac death than brain injury, tissues can be banked for future use, but there is a shortage of organs for use in transplant operations (**FIGURE 18.11**).

If the family agrees, the young man will be kept on a ventilator so that his organs continue to be nourished with oxygen and blood until the organs can be surgically removed. While the organs are being removed, medical personnel will attempt to find the best recipient. To select the recipient, a search of a computerized database is performed. Starting at the top of the list, medical staff will narrow down possible recipients, based in part on how long the recipient has been waiting. People at the top of the waiting list for a given organ tend to be very ill and will die if a transplant does not become available in a matter of days or weeks. Approximately 10% of people on waiting lists for replacement organs will die before one becomes available.

If you would like to be an organ donor, you can indicate this on your driver's license or state ID card. You can also carry a signed organ donor card or sign up using the Donate Life America app. Whether you want to be a donor or not, it is imperative that you make your feelings known to your family. Although your family cannot legally refuse to donate your organs when you have made your wishes clear, in practice relatives' desires are often honored by medical personnel. If you do not want to be a donor, let your family know this as well. That way, if you should suffer an accidental death, like the young motorcyclist, your family will not have the burden of trying to make this decision for you during a time of grief and trauma.

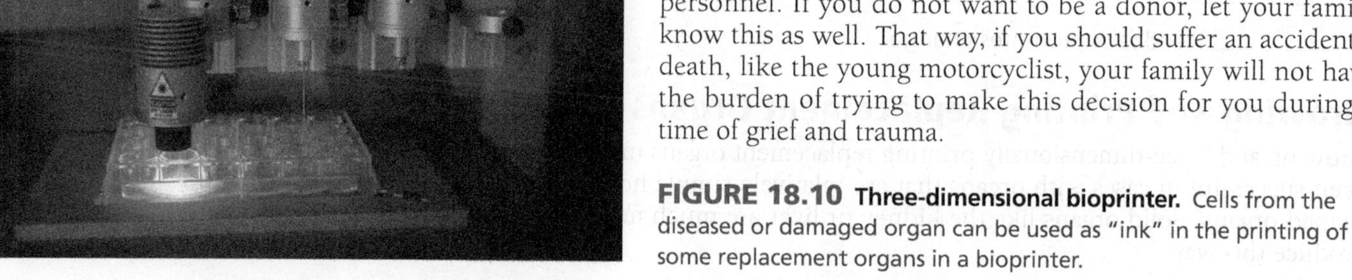

FIGURE 18.10 Three-dimensional bioprinter. Cells from the diseased or damaged organ can be used as "ink" in the printing of some replacement organs in a bioprinter.

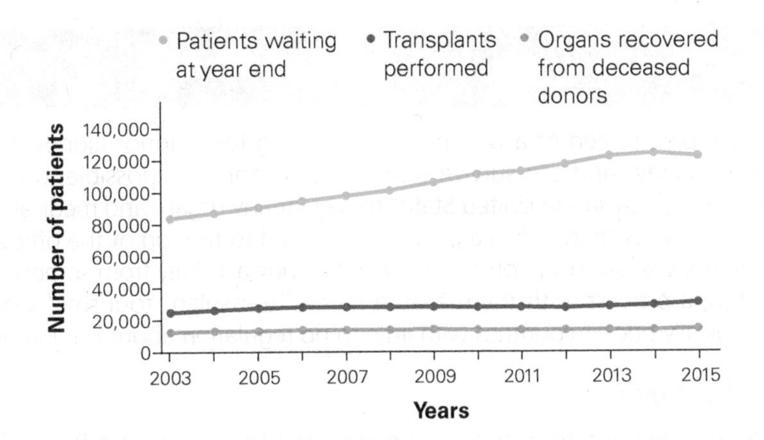

FIGURE 18.11 Organ supply and demand. There are more people awaiting donations than there are donors. Many people will die each year while awaiting a transplant.

WORKING WITH **DATA**

(a) Summarize the trend in supply versus demand between 2003 and 2015 and (b) provide an explanation for why there are more transplants performed than donor organs recovered from deceased individuals.

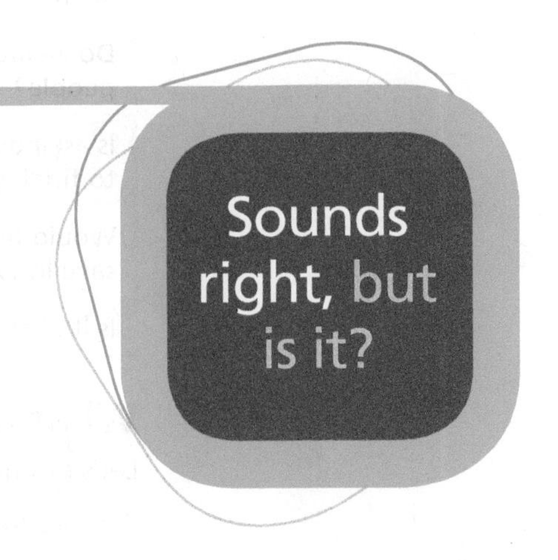

Parents facing the decision to take their brain-dead daughter off life support are understandably, overwhelmed, confused, and conflicted. They cling to the hope that their daughter may someday awaken. In fact, several years ago, a teenager they knew was in a coma for several months after a car accident and subsequently made a remarkable recovery. Besides, their loved one does not look dead; she just appears to be asleep. They have even witnessed her moving, furthering their conviction that their daughter's brain is still working. These parents refuse to deny their daughter a chance, no matter how small, at coming back to life.

There is always a chance that a brain-dead person will make a full recovery.

Sounds right, but it isn't.

Answer the following questions to understand why.

1. Brain-dead individuals show no electrical activity in the brain because brain cells are not passing electrical impulses to each other. Do comatose people show electrical activity in the brain?
2. Brain-dead people cannot breathe on their own because breathing is initiated in the brain stem, which has ceased to function. A mechanical ventilator can provide oxygen to their tissues, keeping nourished any tissues that are not dead. If the ventilator being used by a brain-dead person is removed, would that person be able to breathe on his or her own?
3. Is an unresponsive person who is breathing unaided in a coma or brain dead?
4. The regulation of body temperature, the ability to increase or decrease body temperature when the external environment changes, is a function of the brain. Would a brain-dead person be able to increase his or her body temperature in response to lowering the heat?
5. Some movements are the product of reflexes, which involve nerves between the muscles and the spine but not the brain. (When you jerk your hand away from a hot surface, the movement is done before you have time to use your brain to think about moving your hand.) Does the presence of these reflexive movements in an unresponsive person indicate that the brain is alive?
6. Consider your answers to questions 1–5 and explain why the original statement bolded above sounds right, but isn't.

THE BIG QUESTION

Should kidney donors be financially compensated?

Close to 60% of people in need of a transplant are waiting for a donor kidney. Because humans have two kidneys and can survive well with only one, it is possible to donate a kidney while alive. It is illegal in the United States to pay for any organ, and medical doctors cannot perform a transplant operation unless a person has moved to the top of the official waiting list. Some people on the wait list go to other countries and buy a kidney from someone there and have it transplanted, a process with the macabre name "transplant tourism." Countries where this occurs are typically poorer countries with little to no regulation about trading organs.

What should I know?

What follows are some questions that need to be resolved to answer the Big Question. Place a checkmark next to the questions that science can answer.

Smaller Questions	Can Science Answer?
Are there health risks associated with kidney donation?	
Do kidney donors that accept payment end up having long-term economic gains?	
Do wealthier people receive more transplants than less wealthy people?	
Is asking a friend or family member for a kidney a better approach to finding a living donor than buying one from a stranger?	
Would financially compensating kidney donors shorten wait lists and save lives?	
Is it okay for rich people to get poor people's kidneys?	

What does the science say?

Let's examine what the data say about this smaller question:

Do wealthier people receive more transplants than less wealthy people?

The data illustrated below, obtained by the World Health Organization, show the rate of kidney transplants received as a function of the wealth of the country (measured as GDP, or gross domestic product).

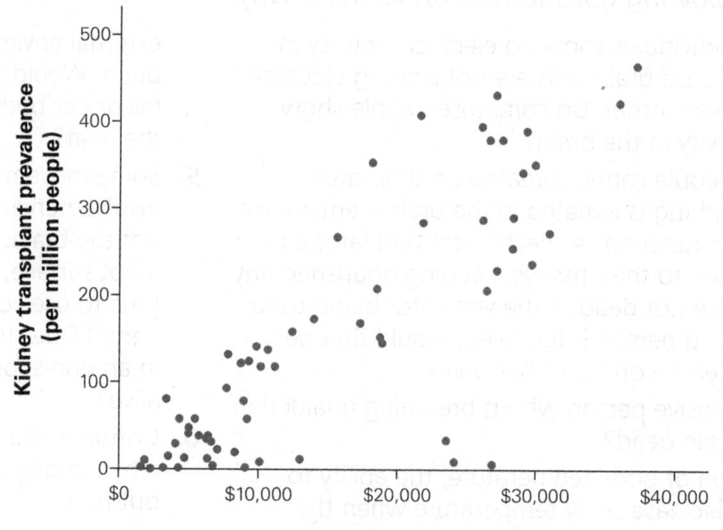

1. Describe the results. What is the overall trend in receiving a kidney transplant as it relates to GDP? If transplant tourism increases, what is likely to happen to this graph?
2. Given these data, do you think the smaller question is answered? If not, propose another study that would help answer this question.
3. Does this information help you answer the Big Question? What else do you need to consider?

Data Source: S. White, S. Chadban, S. Jan, J. Chapman, and A. Cass, "How Can We Achieve Global Equity in Provision of Renal Replacement Therapy?" *Bulletin of the World Health Organization* 86 (2008): 229–237.

Chapter Review

Mastering Biology

Go to Mastering Biology to access eText 2.0, Dynamic Study Modules, and the Study Area, where you'll find practice quizzes, BioFlix™ animations, MP3 tutor sessions, current events, and more.

SUMMARY

Section 18.1

Describe the structure and function of epithelial tissues.

- Epithelia line and cover organs, vessels, and body cavities. They are tightly packed tissues with one free surface. Outgrowths of epithelia form some glands. These tissues function in protection, secretion, and absorption (pp. 401–402).

Compare the structures and functions of the six connective tissues.

- Loosely packed connective tissues bind tissues and organs to each other. The six types of connective tissues are each composed of a characteristic cell type, embedded in a characteristic matrix (pp. 402–404).
- Loose connective tissue connects epithelia to underlying tissues and holds organs in place. The cellular components are fibroblasts, and the matrix is rich in the proteins collagen and elastin, which provide this tissue with tensile strength and elasticity (p. 402).
- Adipose tissue connects the skin to underlying structures and insulates and protects organs. Cells of this tissue synthesize and store fat and have little extracellular matrix (pp. 402–403).
- Blood is a type of connective tissue that transports oxygen and nutrients to body cells. Blood cells have a liquid matrix called plasma (p. 403).
- Fibrous connective tissue forms the tendons and ligaments. The cells of this tissue are fibroblasts, and the matrix is rich in collagen (p. 403).
- Cartilage is a flexible, shock-absorbing tissue composed of cells called chondrocytes that secrete a dense, collagenous matrix (p. 403).
- Bone tissue provides support for the body. The cells of this tissue are called osteocytes, and the matrix is rich in collagen and minerals (p. 404).

Compare the structures and functions of the three types of muscle tissues.

- Muscle tissues are composed of fibers that contract and conduct electrical impulses (pp. 404–405).
- The three types of muscle tissue can be either voluntary or involuntary and smooth or striated (p. 404).
- Skeletal muscle is attached to bones, responsible for voluntary movements, and striated (pp. 404–405).
- Cardiac muscle makes up the heart; it is involuntary and striated (pp. 404–405).
- Smooth muscle makes up the musculature of internal organs and blood vessels. Smooth muscle is involuntary and not striated (pp. 404–405).

Describe the structure and function of nervous tissues.

- Nervous tissue, found in the brain and spinal cord, is composed of neurons; it senses stimuli and transmits signals throughout the body (pp. 405–406).

SHOW YOU KNOW 18.1 What is missing from cartilage that makes it more difficult to transplant than other tissues?

Section 18.2

Explain what an organ is and how organs interact in an organ system.

- Organs are groups of tissues working in concert (p. 407).
- Organ systems are suites of organs working together to perform a function or functions (pp. 407–408).

List the structures and functions of the liver.

- The liver, like all organs, is composed of different tissues whose combined functions give rise to the organ's overall function (p. 409).
- The epithelial tissue of the liver is divided into lobes. Lobes are divided by connective tissue into lobules. Lobules contain a central vein that allows blood to reach the hepatocytes, which function as filters to remove toxins and pathogens (p. 409).
- Other functions of the liver include the production of proteins and cholesterol, secretion of bile, and storage of vitamins and glycogen (p. 409).

SHOW YOU KNOW 18.2 The integrated functions of organs and organ systems are similar to the integrated functions of car components and a car. Match the following car components—valves and pistons, fuel injection system, vehicle, and engine—with the corresponding biological component—organ system, organ, tissue, and organism.

Section 18.3

Explain how positive and negative feedback mechanisms help in the maintenance of homeostasis.

- Negative feedback is one mechanism that helps facilitate homeostasis. When negative feedback occurs, the product of the process acts to turn down the process. Negative feedback negates change (p. 410).
- Positive feedback is a less commonly used homeostatic mechanism. When positive feedback occurs, the product of the process intensifies the process. Positive feedback promotes change (p. 411).

Discuss why replacement organs that are grown or printed in the lab don't cause an immune response in those that receive them.

SHOW YOU KNOW 18.3 The pupils of your eye constrict in sunlight and dilate in the dark. Is this an example of negative or positive feedback?

ROOTS TO REMEMBER

The following roots of words come mainly from Latin and Greek and will help you to decipher terms:

adipo-	means fat. Chapter term: *adipose tissue*
dia- and **trans-**	mean through. Chapter terms: *diaphragm*, *transplant*
epi-	means on or upon. Chapter term: *epithelium*
homeo-	means the same. Chapter term: *homeostasis*
lig-	means to join or connect. Chapter term: *ligament*
neuro-	means nerve. Chapter term: *neuron*

LEARNING THE BASICS

1. Describe the four tissue types found in animals.
2. Name the six types of connective tissues.
3. Outline the differences between the three separate types of muscle tissue.
4. Epithelia ________.

   A. are loosely packed tissues that hold organs in place; **B.** can be free-floating tissues such as blood; **C.** line organs and cavities and have one surface exposed; **D.** include the tissues that hold joints together; **E.** are tissues that conduct nerve impulses
5. Muscle tissues ________.

   A. contain actin and myosin, whether or not they are striated; **B.** that facilitate movements requiring conscious thought are involuntary; **C.** increase in cell number with exercise; **D.** that make up the heart are called smooth; **E.** all of the above
6. Which of the following is not a function of the liver?

   A. storing bile; **B.** filtering toxins; **C.** producing cholesterol; **D.** storing glycogen
7. When a woman is breastfeeding, the more her infant drinks, the more milk she produces. This is an example of ________.

   A. negative feedback; **B.** positive feedback; **C.** thermoregulation; **D.** independent regulation
8. Bile ________.

   A. is stored in the pancreas; **B.** helps break down glycogen; **C.** emulsifies fats; **D.** removes water from indigestible materials in the large intestine; **E.** all of the above
9. Which of the following cell types is found in nervous tissue?

   A. osteocyte; **B.** melanocyte; **C.** leukocyte; **D.** neuron; **E.** brain and spinal cord

10. Add labels to the figure that follows, which illustrates some of the human abdominal organs.

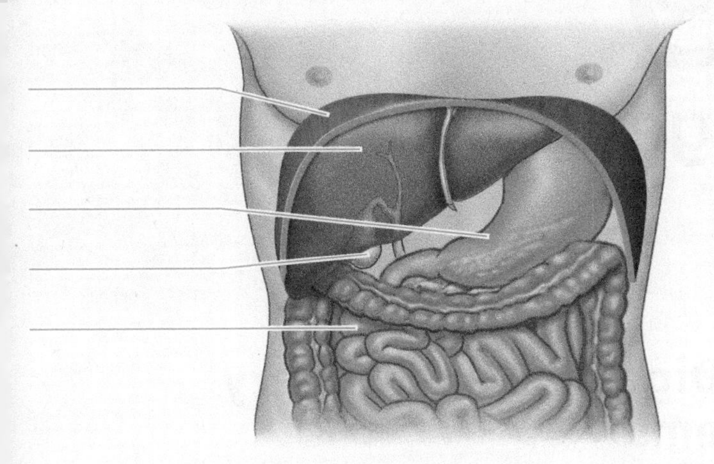

ANALYZING AND APPLYING THE BASICS

1. Propose a hypothesis for why shivering when cold may promote homeostasis.

2. How does the pancreas respond when you consume a sugary beverage?

3. Describe a nonbiological example of both negative and positive feedback.

GO FIND OUT

1. Do some research about organ donation to help determine whether you would want to be an organ donor. List the organs and tissues you would, and would not, be willing to donate.

MAKE THE CONNECTION

The science that you learned in this chapter has helped you better understand the real-world example used throughout this discussion. Draw a line from the statement on the left to the science that supports it on the right.

People in a coma are not used as organ donors. ▷	◁ The liver is an organ capable of regeneration from a slice of tissue.
Donated tissues are not likely to be rejected in recipients. ▷	◁ Donated tissues can be stored for many years before use.
More people die while waiting for organ transplants than tissue transplants. ▷	◁ The liver is involved in digestion and in helping remove toxins from the blood.
Some organs can be transplanted from living donors. ▷	◁ Hospital staff check carefully to ensure that brain death has occurred before discussing organ donation with a family.
When one organ fails, many organ systems can be affected. ▷	◁ The number of organs donated does not meet the demand. Therefore, technologies are being pursued to increase the number of available organs.
Simple hollow organs can be grown on scaffolding or via a three-dimensional printer. ▷	◁ Donated structures are chemically treated to kill viruses and bacteria.

Answers to **Got It?, Visualize This, Working with Data, Sounds Right, But Is It?, Show You Know,** and **Chapter Review** questions can be found in the **Answers** section at the back of the book.

19 Binge Drinking

A student is turning 21.

Her friend wants her to celebrate safely...

and hosts a dinner party for a small group of close friends.

The Digestive and Urinary Systems

It's Saturday night and Janelle is hosting a dinner party to celebrate the 21st birthday of her friend Lin. Lin is several years younger than Janelle. She lived with Janelle's family for two years when she was a high school exchange student. Now an international student attending college in the United States, Lin has had almost no experience with alcohol. Janelle knows that Lin is eagerly anticipating this birthday and that she will likely drink alcohol. Because she feels protective of Lin, Janelle wants to help Lin learn how to enjoy alcohol while limiting the negative consequences that can also occur.

Some of Janelle's concerns about negative consequences are based on situations she has witnessed and others on information she came across while writing a paper on alcohol abuse for a health class she took last semester.

A student who lived on Janelle's dorm floor freshman year broke his ankle when he tripped while running from campus police. Her chemistry lab partner broke his nose when he was riding with an intoxicated driver whose car hit a tree on a snow-covered road. While working on the paper for her health class, Janelle came across a government website that indicated more than 30,000 students required medical treatment for alcohol poisoning last year.

Janelle hopes to develop a plan for convincing Lin, a premed biology major, that drinking too much is bad for her body, an argument she thinks Lin may find credible. Because she has heard that eating food before drinking might help absorb some of the alcohol and that alcohol consumption causes dehydration, she plans to focus her efforts on the effects of drinking on the digestive and urinary systems.

19.1 The Digestive System

The job of the digestive system is to break down or metabolize food so that the body can use the energy stored in its chemical bonds. Food is broken down into successively smaller units as it is passed through the long tube that makes up the digestive tract. The digestive tract has openings at the mouth and anus, and along the way exposes food to the actions and secretions of various organs and glands.

Mechanical and Chemical Breakdown of Food

Digestion comprises both physical and chemical means to break down foods as they move through the digestive tract.

Mouth and Throat. The breakdown of food begins in the mouth, or oral cavity, where chewing fractures food into smaller pieces (**FIGURE 19.1**). Mechanical digestion of food occurs as teeth grind down food, which increases the surface area of the food that is exposed to enzymes within the mouth. Chemical digestion also begins in the mouth with the secretion of saliva from salivary glands. Enzymes in the saliva help break down sugars. The carbohydrate breakdown that begins in the mouth is the first step in the breakdown of foods that we consume.

Janelle knows from her research that oral cancers are more common in those who abuse alcohol than in those who do not and that alcohol consumption has been shown to increase incidence of tooth decay, gum disease, and tooth loss.

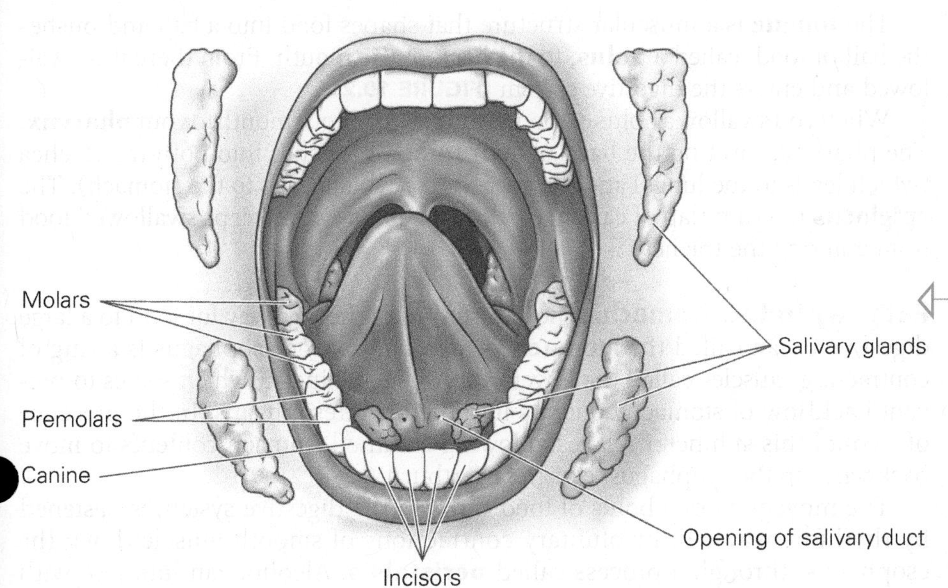

FIGURE 19.1 The human oral cavity. Food is ground into smaller pieces by the 32 adult teeth. Incisors are specialized to enable biting off pieces of food. Canine teeth are sharp enough to help rip food apart. Molars have broad surfaces that enhance grinding.

VISUALIZE THIS

The third molars are the last molars to emerge. Located at the back of the mouth, they typically appear in young adults. If a person has all of these "wisdom teeth" removed, how many adult teeth would they have?

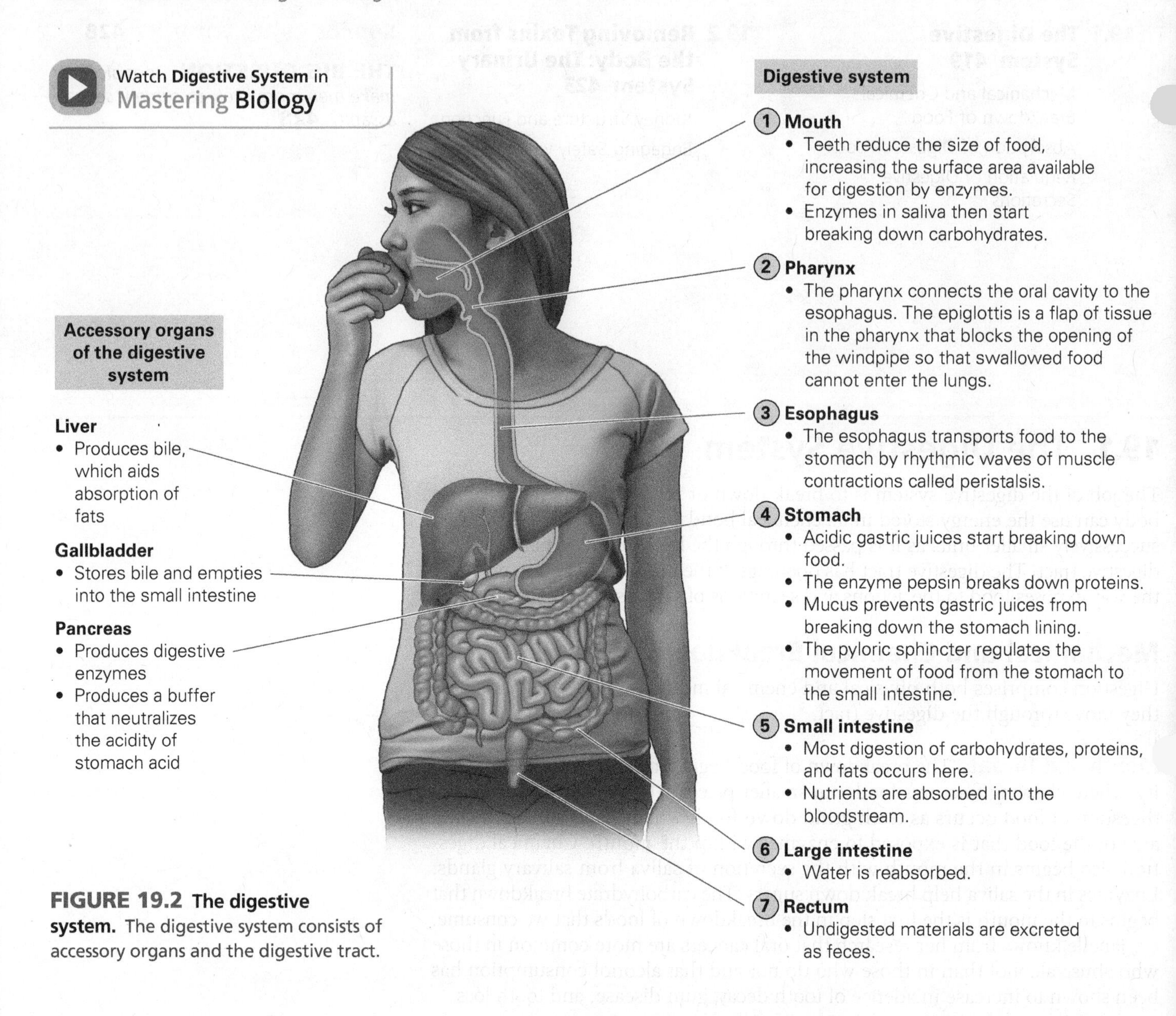

Watch Digestive System in Mastering Biology
Digestive system
1 Mouth
• Teeth reduce the size of food, increasing the surface area available for digestion by enzymes.
• Enzymes in saliva then start breaking down carbohydrates.
2 Pharynx
• The pharynx connects the oral cavity to the esophagus. The epiglottis is a flap of tissue in the pharynx that blocks the opening of the windpipe so that swallowed food cannot enter the lungs.
3 Esophagus
• The esophagus transports food to the stomach by rhythmic waves of muscle contractions called peristalsis.
4 Stomach
• Acidic gastric juices start breaking down food.
• The enzyme pepsin breaks down proteins.
• Mucus prevents gastric juices from breaking down the stomach lining.
• The pyloric sphincter regulates the movement of food from the stomach to the small intestine.
5 Small intestine
• Most digestion of carbohydrates, proteins, and fats occurs here.
• Nutrients are absorbed into the bloodstream.
6 Large intestine
• Water is reabsorbed.
7 Rectum
• Undigested materials are excreted as feces.
Accessory organs of the digestive system
Liver
• Produces bile, which aids absorption of fats
Gallbladder
• Stores bile and empties into the small intestine
Pancreas
• Produces digestive enzymes
• Produces a buffer that neutralizes the acidity of stomach acid

FIGURE 19.2 The digestive system. The digestive system consists of accessory organs and the digestive tract.

The **tongue** is a muscular structure that shapes food into a ball and pushes the ball of food, called a **bolus,** to the back of the mouth. From there it is swallowed and enters the digestive system (**FIGURE 19.2**).

When you swallow a bolus of food, it moves from your mouth to your **pharynx.** The pharynx, forming the back of your throat, branches into both the trachea (which leads to the lungs) and the esophagus (which leads to the stomach). The **epiglottis** is a thin flap of cartilage below the tongue that keeps swallowed food from entering the trachea.

epi- means upon, beside, or among.

Pathway to the Stomach. The **esophagus** is the pathway for food to a large digestive organ called the **stomach.** At the base of the esophagus is a ring of contracting muscles called the lower esophageal sphincter, which serves to prevent backflow of stomach contents into the lower esophagus. In the presence of alcohol this sphincter relaxes, allowing acidified stomach contents to move backward up the esophagus, causing heartburn.

The movement of a bolus of food through the digestive system is hastened by rhythmic waves of involuntary contractions of smooth muscle down the esophagus, through a process called **peristalsis.** Alcohol can interfere with

peri- means around.
-stalsis means to wrap or surround.

this process: When alcohol relaxes the muscles involved with peristalsis, food spends more time in the digestive tract than normal, and this increased exposure to digestive enzymes can cause diarrhea.

Once the food is inside the stomach, it is further degraded by digestive enzymes and acids secreted by specialized cells that line the stomach. The stomach undergoes peristalsis, which helps mix food into a slurry with digestive enzymes, becoming a substance called **chyme.** Note that when chyme contains a high concentration of alcohol, the stomach lining is irritated, which can trigger the vomiting reflex.

At the base of the stomach is another sphincter. The pyloric sphincter regulates the secretion of chyme into the **small intestine,** a long tube (around 20 feet long in adult humans) that serves as the major site of chemical digestion and the absorption of nutrients into the bloodstream via its epithelium. It is in the small intestine that the majority of food and alcohol is broken down and absorbed across the intestinal wall and into the bloodstream.

Metabolizing Alcohol. Only a small percentage of alcohol is metabolized in the stomach. Epithelial cells that line the stomach secrete enzymes called alcohol dehydrogenases. These enzymes help metabolize alcohol, which means there will be less alcohol in the bloodstream to cause intoxication. Women have smaller stomachs than men and thus produce less of this enzyme. What's more, women also have less of this enzyme per pound of body weight than men. This means that a man and a woman of the same size and weight will metabolize alcohol at different rates. Women metabolize alcohol more slowly, thus becoming intoxicated from less alcohol consumption. Janelle plans to share this information with Lin and encourage her not drink at the same pace as her male friends, even those who aren't any bigger than her.

There is one additional stomach-related factor Lin must consider when drinking alcohol. One of the alcohol dehydrogenase enzymes, abbreviated ALDH1, differs from the most common alcohol dehydrogenase by a single amino acid. This variant, common in people of East Asian descent, produces an enzyme that more slowly metabolizes one of the products of alcohol breakdown, called acetaldehyde. High concentrations of acetaldehyde cause flushing of the face, dizziness, nausea, and an irregular heartbeat. Because Lin comes from Chengdu, China, and thus may carry the less effective enzyme, Janelle will also warn her that she may experience this reaction when drinking alcohol.

Janelle has heard that it is good to eat a large meal before drinking. This is because the presence of food in the stomach causes the pyloric sphincter to remain closed. Because the stomach does not absorb alcohol as readily as the small intestine, preventing the alcohol from reaching the small intestine can slow the rate at which it reaches the bloodstream.

Accessory Organs. Many of the digestive enzymes used in the small intestine are produced by an organ called the **pancreas.** Secretions from the pancreas neutralize stomach acids that enter the small intestine and contain enzymes that break down carbohydrates, fats, proteins, and nucleic acids. While in the small intestine, chyme is exposed to a substance called bile, which is synthesized by the liver.

The liver, pancreas, and gallbladder are considered **accessory** organs. In a sense, they are accessories to the digestive tract because they are outside the tube but produce or secrete substances required for digestion.

The **liver** helps metabolize toxins, including some drugs and alcohol. The **gallbladder** stores and concentrates bile, which will be released into the small intestine to help dissolve fats. Concentrating bile involves removing water from it. When the gallbladder removes too much water, the bile can crystallize, causing gallstones.

The liver and pancreas are very susceptible to the effects of alcohol, both from long-term alcohol abuse and **binge drinking,** which is typically defined as the consumption of more than four drinks in a 2-hour time period. Binge drinkers quickly experience intoxicating effects of alcohol because the liver can only metabolize, on average, one drink per hour. Heavy drinking can damage the liver, in some cases causing healthy liver tissue to be replaced by scar tissue, a progressive and irreversible process called cirrhosis. Scar tissue prevents proper blood flow through the liver and can result in eventual liver failure.

-itis refers to inflammation.

Inflammation of the pancreas, called **pancreatitis**, also can be caused by excessive alcohol consumption. Pancreatitis prevents the pancreas from secreting digestive enzymes and thus disrupts digestion, which can lead to life-threatening complications.

Absorption of Digested Foods

Substances that move from the stomach to the small intestine are absorbed into the bloodstream through finger-like membranous projections of the small intestine called **villi.** An individual villus contains both a blood and a lymphatic vessel. Villi increase the surface area of the small intestine. Surface area is further increased by the presence of even smaller **microvilli** that transport nutrients into the blood vessels inside each villus (**FIGURE 19.3**). Together, the villi and microvilli in your body have as much surface area as a tennis court! Chronic alcohol abuse and binge drinking can interfere with the absorption of nutrients by damaging intestinal villi.

After traversing the small intestine, products of digestion can move across the small intestine into the bloodstream, where they are transported to individual cells. Materials that are not absorbed are passed through the **large intestine** (also known as the **colon**) through the rectum to the anus, where it will exit the body as feces. Fecal matter consists largely of indigestible plant fibers.

FIGURE 19.3 Absorption of nutrients in the small intestine. The foldings of the small intestine to produce finger-like projections called villi and microvilli increase the surface area across which nutrients can enter the bloodstream. Once in the bloodstream, nutrients can be transported to cells throughout the body.

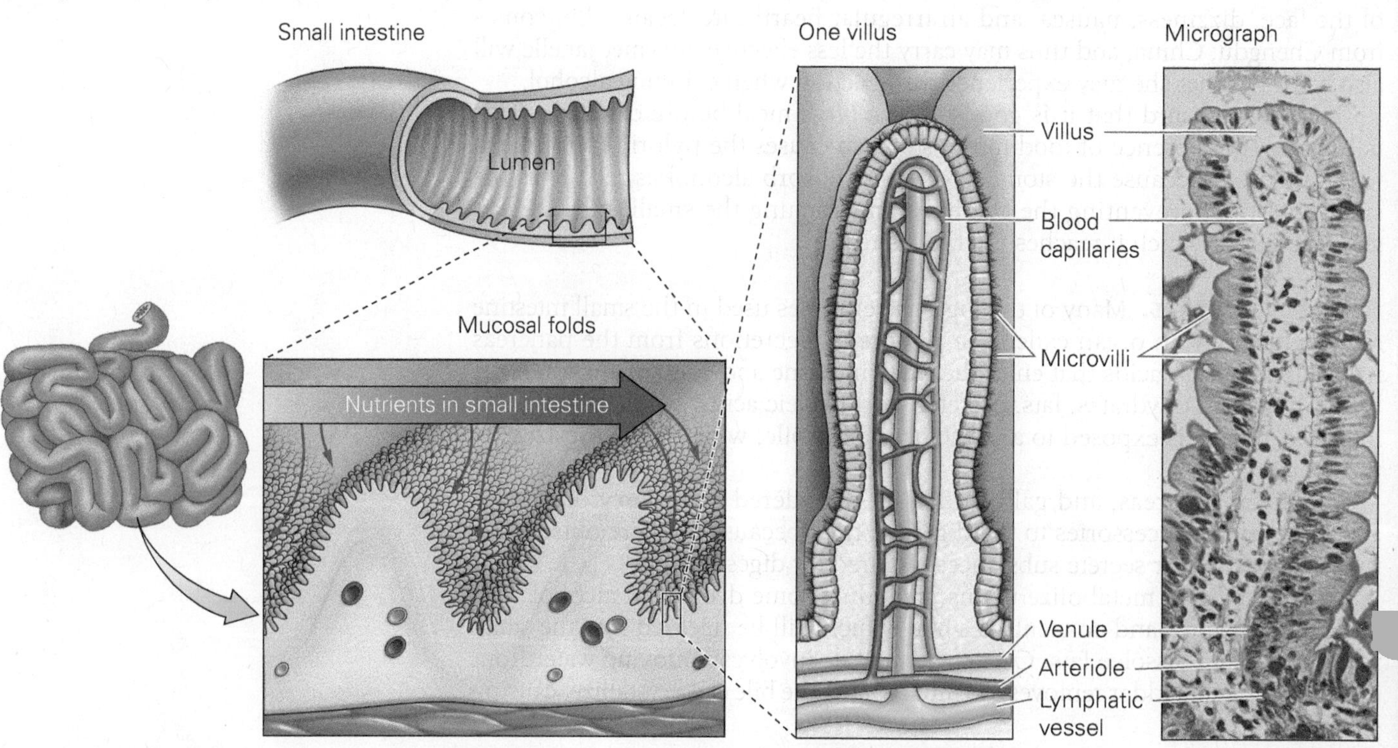

Regulation of Digestive Secretions

The secretion of digestive juices is regulated hormonally (**FIGURE 19.4**). After a meal, the stomach produces the hormone **gastrin,** which stimulates the upper part of the stomach to produce acidic gastric juices, thus facilitating digestion. The secretion of gastrin also results in release of the hormones *secretin* and *cholecystokinin* (CCK) from the small intestine. Secretin and CCK cause the pancreas and gallbladder to increase their output of digestive juices.

After exposure to the digestive secretions in the stomach, partially digested food and alcohol move into the intestines. After alcohol makes its way out of the intestine and into the bloodstream, the body begins the process of removing it from the blood. This is necessary to prevent damage such as the decrease in testosterone production that occurs in males after binge drinking. This hormonal decline can result in erectile dysfunction and lower sperm production while alcohol remains in the system.

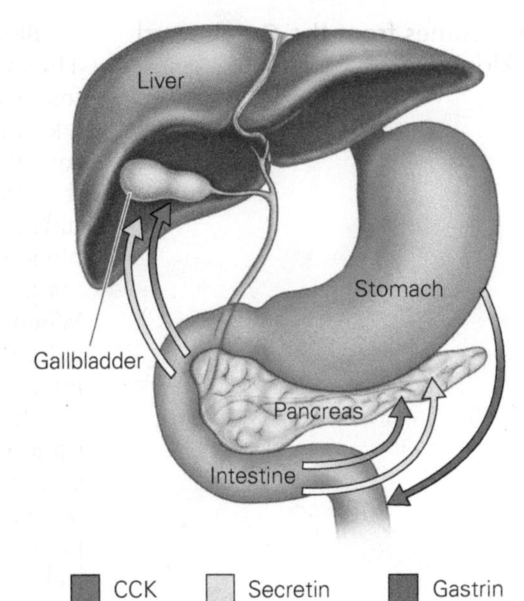

FIGURE 19.4 Hormonal control of the digestive system. The presence of food in the stomach causes the release of the hormone gastrin, which results in the intestinal release of the hormones secretin and CCK, which act on the gallbladder and pancreas to increase their output of bile and digestive enzymes.

Got It?

1. The __________ branches into both the trachea and esophagus.
2. Food moves from the mouth to the pharynx to the esophagus to the __________.
3. Rhythmic contractions of smooth muscle called __________ help move a food bolus through the digestive tract.
4. __________ is a slurry of food and enzymes in the stomach.
5. Most of the nutrients absorbed into the bloodstream move through the membranes of the __________ __________.

19.2 Removing Toxins from the Body: The Urinary System

The metabolism of any substance, including alcohol, results in waste products that circulate through the bloodstream until they are expelled from the body. The **urinary system** efficiently removes wastes while retaining valuable materials that can be reused and recycled. In humans, the major organs of the urinary system are the **kidneys,** which filter and cleanse circulating blood before sending the waste through **ureters** to the bladder. Urine is stored in the **urinary bladder** until it is expelled via the **urethra** (**FIGURE 19.5**).

VISUALIZE THIS

An untreated bladder infection can cause kidney damage. What path would the bacteria take to move from the bladder to the kidney?

Kidney Structure and Function

The kidneys are paired, approximately fist-sized organs that sit behind the liver and stomach in the upper abdominal cavity. Each kidney is densely packed with looped tubules called **nephrons.** There are close to 1,250,000 nephrons in each kidney, with a combined length in an adult human of about 145 kilometers (85 miles). Networks of capillaries surround the nephrons, allowing wastes to diffuse out of the blood and into these tubules for excretion. The entire volume of blood in the circulatory system passes through the kidneys hundreds of times per day.

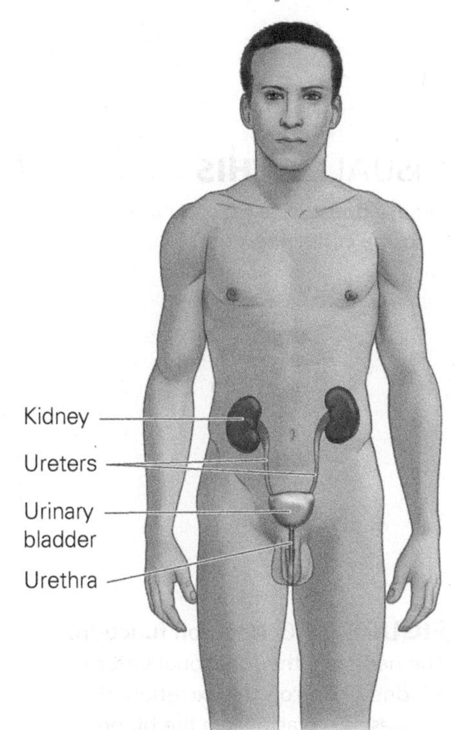

FIGURE 19.5 The human urinary system. Kidneys filter the blood and transfer waste urine to the urinary bladder, where it is held until voluntarily released through the urethra. Intoxication can result in the involuntary release of urine from the bladder.

renal comes from the Greek word for kidney.

Blood is brought to the kidneys by large vessels called the **renal arteries.** The processing of waste in the kidneys has four distinct phases, which we can follow on a diagram of a single, loop-shaped nephron embedded in the body of a kidney (**FIGURE 19.6**). The first step, **filtration,** occurs within the Bowman's capsule, a structure at the head of a nephron. The Bowman's capsule encloses the glomerulus, a compact ball of blood vessels. The capillaries in the glomerulus contain tiny pores, and blood pressure forces the plasma portion of the blood through these pores and into the upstream end of the nephron. The plasma portion of blood includes water, sugars, proteins, and other substances. When blood reaches the glomerulus, this filter allows water and small molecules through but retains large proteins in the plasma. The fluid that enters the interior of the nephron via this process is called filtrate.

One way that kidney health is assessed is by measuring the glomerular filtration rate. Preliminary studies suggest that those who drink alcohol in moderation have slightly better filtration rates than those who don't. Drinking in moderation is usually defined as the consumption of one to two alcoholic drinks per day for men and one alcoholic drink per day for women. One drink is typically defined as containing 14 g of pure alcohol. Twelve ounces of beer, 5 oz of wine, or about 1.5 oz of hard liquor like vodka, gin, and whiskey constitute one drink. If you don't drink alcohol, however, this slight potential benefit does not make it advisable to start drinking.

Because filtrate contains both wastes and valuable substances, such as sugars, amino acids, and water, the next step of waste processing is **reabsorption** of these materials across the walls of each nephron. Water moves by osmosis out of the

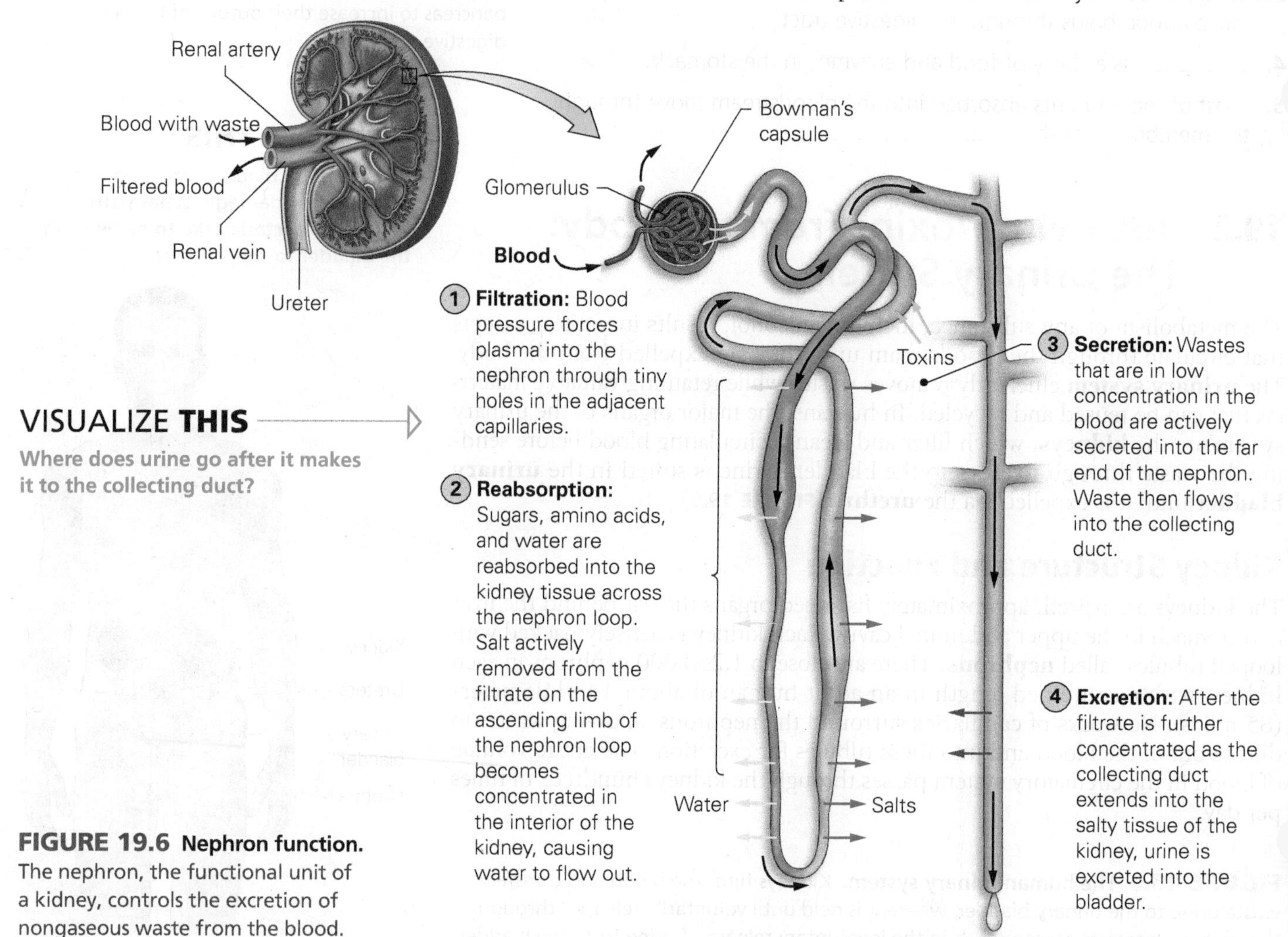

Renal artery
Blood with waste
Filtered blood
Renal vein
Ureter
Bowman's capsule
Glomerulus
Blood
Toxins
Water
Salts
1 Filtration: Blood pressure forces plasma into the nephron through tiny holes in the adjacent capillaries.
2 Reabsorption: Sugars, amino acids, and water are reabsorbed into the kidney tissue across the nephron loop. Salt actively removed from the filtrate on the ascending limb of the nephron loop becomes concentrated in the interior of the kidney, causing water to flow out.
3 Secretion: Wastes that are in low concentration in the blood are actively secreted into the far end of the nephron. Waste then flows into the collecting duct.
4 Excretion: After the filtrate is further concentrated as the collecting duct extends into the salty tissue of the kidney, urine is excreted into the bladder.

VISUALIZE **THIS**

Where does urine go after it makes it to the collecting duct?

FIGURE 19.6 Nephron function. The nephron, the functional unit of a kidney, controls the excretion of nongaseous waste from the blood.

nephron and into the kidney interior as it descends into this salty environment. On the ascending limb, the nephron walls are impermeable to water but actively transport salt into the kidney. The structure of the nephron allows for a gradient of salt concentration to be maintained, from relatively low at the top of the loop to relatively high at the base. What remains in the filtrate after the reabsorption phase is water containing a high concentration of urea, the waste product of the breakdown of amino acids.

The last segment of the nephron permits the **secretion** into the filtrate, via active transport, of certain wastes that are in low concentration in the plasma. The filtrate then moves from the nephron into a **collecting duct** that leads to the center of the kidney, called the renal pelvis. As the collecting duct crosses the salty kidney interior, more water moves by osmosis from the filtrate into the kidney tissue. Water and other materials removed from the filtrate are drawn into the capillaries, returned to the bloodstream, and eventually leave the kidney via the renal vein. The fluid that collects in the renal pelvis is called **urine** and is made up of water and organic wastes including urea and various ions. During urine **excretion,** urine leaves the kidneys and flows to the bladder.

ex- means out or out of.

Alcohol is a diuretic, which means that it promotes the formation of urine and increases the volume of urine that is released from the bladder, a process called **micturition.** Coupling the increased volume of urine produced with the deadening of awareness of the need to urinate that goes with intoxication can result in a very full bladder. Even though micturition is typically under conscious control, an intoxicated person who passes out before emptying the bladder may end up urinating on him- or herself. In this case, the body overrides the conscious control of micturition to prevent a potentially lethal bladder rupture.

In addition to being a diuretic, alcohol is also a depressant, slowing down brain function and altering perceptions, reflexes, and balance, and causing slurred speech. In an attempt to prevent the depressant effects of intoxication, some of Janelle's friends mix alcohol with energy drinks. Janelle will recommend to Lin that she does not do this because Lin should develop an awareness of when to stop drinking. This is harder to do if the depressant effects of intoxication are, in part, masked by the stimulant effects of the energy drink.

In addition to managing wastes, the urinary system also plays an important role in regulating blood volume, acidity, and salt balance. The kidneys regulate acidity and salt balance by actively excreting acids and reabsorbing salts during the secretion and reabsorption steps of urine formation. The kidneys help maintain blood pressure by regulating water excretion. When blood pressure is low, antidiuretic hormone (ADH) released by the pituitary gland increases the permeability of the collecting duct to water, allowing more water from the filtrate to return to the bloodstream. When blood pressure is high, ADH release is curtailed, and more water is excreted.

Alcohol acts on the pituitary to lessen ADH secretion. As ADH levels drop, the kidneys reabsorb less water and thus produce even more urine. Therefore, you actually urinate a higher volume of liquid than you consume when drinking alcohol. This is, in part, the cause of the dehydration experienced after an episode of heavy drinking. Dehydration can lead to nausea and headaches. For this reason, Janelle will suggest that Lin alternate drinking alcohol with drinking water.

Got It?

1. The paired ________ are the filtering organs of the urinary system.
2. Urine is stored in the urinary bladder and released through the ________.
3. The looped tubules in each kidney are ________.
4. Blood is brought to the kidneys by the ________ ________.
5. During excretion, urine leaves the kidneys and flows to the ________.

Engaging Safely with Alcohol

Janelle's research has also made it clear that moderate alcohol consumption can have some positive physiological effects. She will discuss the evidence suggesting that moderate drinkers experience lower rates of cardiovascular disease and stroke with Lin. This may be because small amounts of alcohol can relax the heart muscle, thereby lowering the rate the heart beats and lowering blood pressure. This is true whether the alcohol consumed is wine, beer, or hard liquor, but these benefits occur only with *moderate* drinking. In fact, larger doses of alcohol have the opposite effect, increasing risk of heart disease and stroke when blood pressure increases in the body's attempt to rid itself of the toxin via increased breathing, sweating, and urination. TABLE 19.1 gives other examples of how to more safely engage with alcohol.

Alcohol in the circulatory system will also make its way to the lungs, where it evaporates from the lung surface and is expelled in each breath. Because the concentration of alcohol in the air breathed out is proportional to the amount of alcohol in the blood, testing devices like the breathalyzer can be used to determine the blood alcohol concentration (BAC). You can also approximate your BAC using your weight, gender, and rate of consumption (TABLE 19.2).

The effects of alcohol in the blood are progressive. A BAC of 0.03 to 0.04 results in relaxation, mild euphoria, and decreased inhibition. Concentration is slightly impaired. By 0.05, impairment increases, affecting reasoning and motor skills, and a person should not drive. By 0.08, a person is considered legally intoxicated and it is illegal for him or her to drive. Above 0.1, emotions can become uncontrolled and behavior uncharacteristically boisterous; reflexes, reaction time, and speech are impaired. By 0.2, a person can become unconscious and experience blackouts. A BAC of 0.3 or higher results in central nervous system depression, causing impaired breathing and heart rate. Death can result. In a 140 lb female or male, ingesting three drinks in rapid succession results in a BAC of 0.1 and 0.08, respectively.

The brain is particularly sensitive to toxins during times of rapid development, most of which occurs before a person reaches their middle twenties. During these years, lifelong traits such as the ability to reason and to critically evaluate information are still developing. Binge drinking and excess alcohol consumption in general have been shown to disrupt brain activity and impair memory, decreasing the ability to ever develop these traits.

After explaining the biological effects of alcohol on the body to Lin, Janelle plans to present her with some data, because she knows biologists, including

TABLE 19.1 Safer engagement with alcohol.

Strategy	Reason
Eat before consuming alcohol.	Alcohol will pass into the intestine more slowly and thus enter the circulatory system more slowly.
Pay attention to how your body responds to alcohol.	Although most people can metabolize around one drink per hour, your metabolic rate could be slower.
Alternate drinking water with drinking alcohol.	Can lessen the severity of the dehydration caused by lowered ADH secretion and alcohol moderate intake.
Understand Medical Amnesty laws in your state.	Some students delay calling 911 when concerned about an intoxicated friend for fear of prosecution. Medical Amnesty laws prohibit prosecution of both the person who calls for help and the victim.
Never have sex with an intoxicated person.	It is a felony offense to have sex with a drunk person. An intoxicated person is not legally capable of giving consent, even if he or she expresses interest in having sex.

TABLE 19.2 Blood alcohol content in women (W) and men (M).

Subtract 0.01% for every 40 minutes of drinking.

	Body Weight in Pounds									
Drinks	90	100	120	140	160	180	200	220	240	Condition
0	.00	.00	.00	.00	.00	.00	.00	.00	.00	Only safe driving limit
1 (W)	.05	.05	.04	.03	.03	.03	.02	.02	.02	
1 (M)		.04	.03	.03	.02	.02	.02	.02	.02	Driving skills significantly affected; possible criminal penalties
2 (W)	.10	.09	.08	.07	.06	.05	.05	.04	.04	
2 (M)		.08	.06	.05	.05	.04	.04	.03	.03	
3 (W)	.15	.14	.11	.10	.09	.08	.07	.06	.06	
3 (M)		.11	.09	.08	.07	.06	.06	.05	.05	
4 (W)	.20	.18	.15	.13	.11	.10	.09	.08	.08	
4 (M)		.15	.12	.11	.09	.08	.08	.07	.06	
5 (W)	.25	.23	.19	.16	.14	.13	.11	.10	.09	
5 (M)		.19	.16	.13	.12	.11	.09	.09	.08	
6 (W)	.30	.27	.23	.19	.17	.15	.14	.12	.11	Legally intoxicated in most states; criminal penalties
6 (M)		.23	.19	.16	.14	.13	.11	.10	.09	
7 (W)	.35	.32	.27	.23	.20	.18	.16	.14	.13	
7 (M)		.26	.22	.19	.16	.15	.13	.12	.11	
8 (W)	.40	.36	.30	.26	.23	.20	.18	.17	.15	
8 (M)		.30	.25	.21	.19	.17	.15	.14	.13	
9 (W)	.45	.41	.34	.29	.26	.23	.20	.19	.17	
9 (M)		.34	.28	.24	.21	.19	.17	.15	.14	
10 (W)	.51	.45	.38	.32	.28	.25	.23	.21	.19	
10 (M)		.38	.31	.27	.23	.21	.19	.17	.16	Death possible

WORKING WITH DATA

What is the maximum number of drinks a 120 lb woman could consume over a 2-hour period to avoid becoming legally intoxicated? How many drinks can you have in a 4-hour period to avoid becoming legally intoxicated?

Lin, like to make evidence-based decisions. She will share with her study after study showing that students who do not drink, or only drink in moderation, have higher grades, better social relationships, and less involvement with law enforcement agencies; that they sleep better, suffer fewer injuries, and are less likely to engage in unplanned or unprotected sexual activity. Once Lin understands the biological basis of the damage caused by excess consumption of alcohol and the negative effects that can ensue (**FIGURE 19.7**), Janelle hopes Lin will choose to engage safely with alcohol, if she decides to drink at all.

Some unmetabolized alcohol is removed from the body during exhalation.
Alcohol progressively impairs the brain's ability to control behavior, emotions, and body functions.
Some unmetabolized alcohol is removed from the body in sweat.
Only a small percentage of alcohol is digested in the stomach. Most moves to the liver. Alcohol in the stomach causes increased gastric secretions that can irritate the stomach lining, triggering the vomiting reflex.
The liver can metabolize only ~1 drink/hour. The rest moves through the circulatory system.
Some of the alcohol that reaches the small intestine is absorbed into the circulatory system and some is metabolized by the liver.
Alcohol causes increased urine formation, leading to dehydration.

FIGURE 19.7 The destructive path of alcohol. Alcohol crosses cell membranes readily, leaving the stomach and small intestine and entering the liver and bloodstream before it can be metabolized. When a person drinks more rapidly than the liver can process the alcohol, typically one drink per hour, the excess alcohol is circulated throughout the entire body.

After a night of drinking at a party, your roommate appears to have passed out. He is seated at the end of a couch, slumped sideways on the armrest. The fingers of one hand are wrapped around the neck of a half-empty beer bottle, which also leans against the armrest. It's 2 a.m. You are tired and want to go home. You gently shake him to see if he will awaken but he does not. You are a little relieved because he was getting pretty obnoxious before he passed out, and it would likely be a struggle to get him home. You ask the people you came with—some sober, some not—for help. The consensus of the group is that because he's breathing, it is probably okay to leave him on the couch for the night. While your friends grab him under the arms and lay him on his side on the couch in case he vomits, you find a pillow and blanket for him. You alert the few remaining partygoers, all people you know fairly well, that you will return for him in the morning.

A passed-out drunk is safe as long as he is breathing and placed on his side.

Sounds right, but it isn't.

Answer the following questions to understand why.

1. If your friend had been drinking steadily all night, then had two shots of alcohol right before passing out, what is the minimum amount of time before his body will have metabolized the last two drinks he had?
2. If your friend is breathing now, is there any guarantee he will be when his blood alcohol concentration peaks?
3. If you are trying to determine whether an intoxicated person is responsive, what might you do, aside from gentle shaking, to get a response?
4. If an intoxicated person is unresponsive, what should you do?
5. Consider your answers to questions 1–4 and explain why the original statement bolded above sounds right, but isn't.

THE BIG QUESTION

Does alcohol make men more likely to commit sexual assault?

When Stanford student athlete Brock Turner was convicted of raping an unconscious woman, he pointed to alcohol as the cause. In asking for leniency from the judge in this case, he volunteered to help "change people's attitudes toward the culture surrounded by binge drinking and sexual promiscuity" on college campuses.

The argument Turner makes is not an uncommon one. Many people believe that alcohol use makes men more sexually aggressive or makes it harder for men to interpret whether consent is being given. Certainly an unconscious woman cannot give consent, but does the fact that alcohol is involved in more than 50% of sexual assaults give credibility to arguments about alcohol use and sexual promiscuity? Or would the majority of men who commit sexual assault while using alcohol commit sexual assault even when sober?

What should I know?

What follows are some smaller questions that need to be resolved to answer the Big Question. Place a checkmark next to the questions that science can answer.

Smaller Questions	Can Science Answer?
Does alcohol use make men more sexually aggressive?	
Are men who rape more violent than nonrapists?	
Does alcohol use make it easier to misunderstand whether consent has been given?	
Is it acceptable for men to ply women with alcoholic drinks to make them more likely to agree to sex?	
Are men who consider women their equals less likely to commit rape?	
Do some men drink prior to committing sexual assault to have an excuse for their behavior?	

What does the science say?

Let's examine what the data say about this smaller question:

Are men who consider women their equals less likely to commit rape?

Researchers compared responses from approximately 1500 men, about 400 of whom had been convicted of rape. All the men were asked to rate their agreement with questions about gender such as "When women work they are taking jobs away from men" or "When it really comes down to it, a lot of women are deceitful."

Mean scores from questions for rapists and nonrapists were graphed for responses to question about their attitudes toward women.

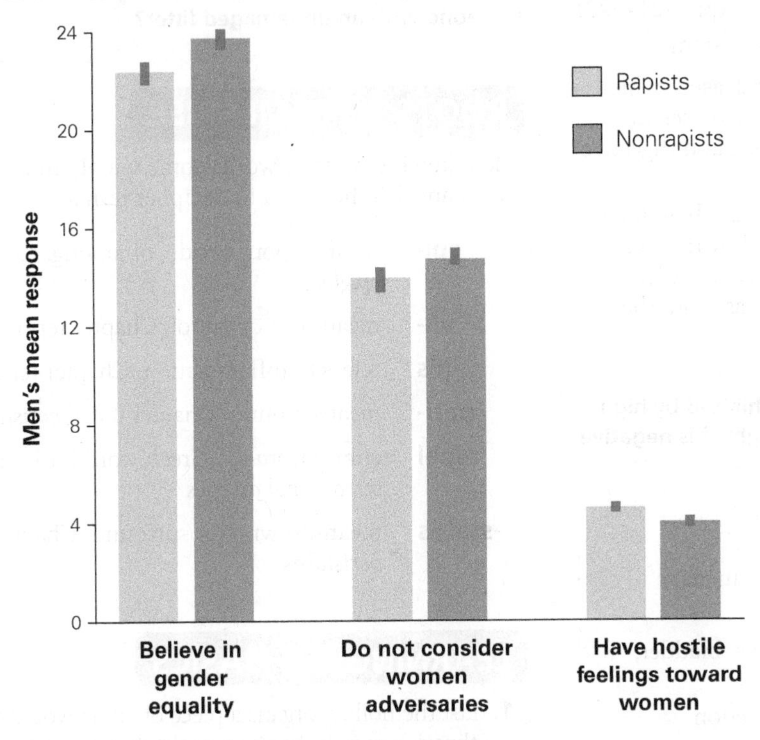

1. Describe the results. In which of the three attitude categories was there a significant difference between those who had raped and those who had not?
2. Given these data, do you think the smaller question is answered? If not, propose another study that would help answer this question.
3. Does this information help you answer the Big Question? What else do you need to consider?

Data Source: R. Jewkes, Y. Sikweyiya, R. Morrell, and K. Dunkle, "Gender Inequitable Masculinity and Sexual Entitlement in Rape Perpetration South Africa: Findings of a Cross-Sectional Study," *PLoS ONE* 6, no. 12 (2012): e29590.

Chapter Review

Mastering Biology

Go to Mastering Biology to access **eText 2.0, Dynamic Study Modules,** and the **Study Area,** where you'll find practice quizzes, BioFlix™ animations, MP3 tutor sessions, current events, and more.

SUMMARY

Section 19.1

Describe the structures and functions of the digestive system.

- The digestive system is a group of organs and glands working together to break foods into their component parts for reassembly into forms that the body can use or for use in generating energy (pp. 419–420).
- Food moves from the mouth to the pharynx and through the esophagus to the stomach to the small intestine. Digested nutrients are absorbed into the bloodstream across the small intestine and brought to cells (pp. 420–422).

List the accessory organs of the digestive system, and outline their roles in digestive processes.

- The pancreas, liver, and gallbladder are accessory organs that secrete substances that aid in digestion (pp. 421–422).

Explain the hormonal control of the digestive system.

- When food is in the stomach, gastrin is released, which signals the release of secretin and CCK. Increases in these two hormones cause increases in bile and digestive enzymes (p. 423).
- Digestive enzymes produced by the pancreas help break down most food molecules. Bile produced by the liver facilitates the breakdown of fats and the gallbladder stores and concentrates bile before its release into the small intestine (p. 423).

SHOW YOU KNOW 19.1 Gastrin release is inhibited by high concentrations of stomach acid (HCl). How might this negative feedback loop protect the stomach?

Section 19.2

List the structures composing the mammalian urinary system.

- The urinary system consists of the kidneys, bladder, ureters, and urethra (p. 423).

Describe the steps in the process of urine excretion.

- In the nephrons of the kidney, the process of filtration forces most liquid, but not cells or larger molecules, from the plasma into the kidney tubules. As the filtrate travels through the nephron, water, glucose, and other valuable molecules are reabsorbed through both active and passive mechanisms. Secretion occurs when waste materials that did not leave with the filtrate are actively brought into the nephron tubules from surrounding capillaries (pp. 423–424).
- Hormones that regulate blood pressure control the concentration of water in urine. Urine collects in the renal pelvis, then travels down the ureters to be stored in the bladder. During urination, urine is released from the bladder through the urethra and exits the body (pp. 424–425).

List a few ways to engage more safely with alcohol.

- Limit alcohol consumption (pp. 426–427).
- Eat food before drinking alcohol and alternate alcoholic beverages with water (pp. 426–427).

SHOW YOU KNOW 19.2 High blood pressure damages cells in the filtering structure of the nephron, making the pores between cells much larger. How is the urine of someone with damage to these structures likely to be different from someone with an undamaged filter?

ROOTS TO REMEMBER

The following roots of words come mainly from Latin and Greek and will help you to decipher terms:

epi-	means upon, beside, or among. Chapter term: *epiglottis*
ex-	means out or out of. Chapter term: *excretion*
-itis	refers to inflammation. Chapter term: *pancreatitis*
peri-	means around. Chapter term: *peristalsis*
renal	comes from the Greek word for kidney. Chapter term: *renal arteries*
-stalsis	means to wrap or surround. Chapter term: *peristalsis*

LEARNING THE BASICS

1. List the hollow organs a piece of apple would move through as it makes its way through the digestive system, from its ingestion in the mouth to its excretion by the anus.

2. List the accessory organs in the digestive system.

3. Describe the main function of the kidney.

4. The pharynx ________.

A. forms the connection between the small and large intestine; **B.** keeps swallowed food from entering the epiglottis; **C.** connects the esophagus to the stomach; **D.** branches to feed into the trachea and esophagus

5. Which of the following lists digestive processes in the correct order?

A. ingestion, absorption, peristalsis; B. peristalsis, absorption, ingestion; C. peristalsis, ingestion, absorption; D. ingestion, peristalsis, absorption

6. The pancreas ________.

A. secretes bile; B. produces stomach acid; C. secretes digestive enzymes; D. all of the above

7. **True/False:** Most of the digestion of nutrients occurs across the membrane that lines the interior stomach wall.

8. The villi that help absorb nutrients line the ________.

A. small intestine; B. large intestine; C. stomach; D. pharynx

9. Blood enters the kidneys from the ________.

A. duodenal vein; B. nephron; C. renal artery; D. gallbladder

10. Add labels to the figure that follows, which illustrates the urinary system.

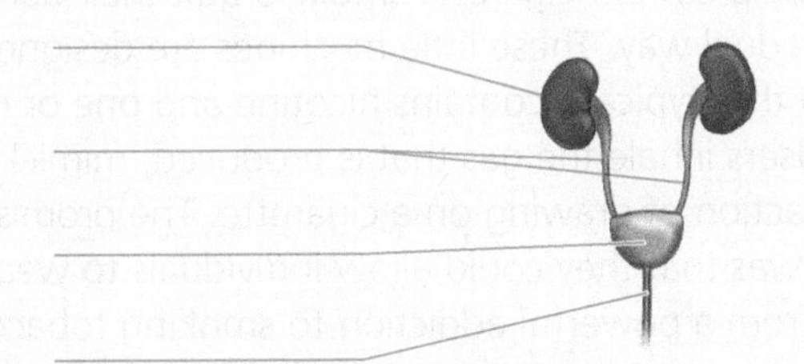

ANALYZING AND APPLYING THE BASICS

1. Studies show that aspirin inhibits the alcohol dehydrogenase family of enzymes. What advice would you give a friend who thinks he should take aspirin before drinking alcohol?

2. When food goes "down the wrong tube" after swallowing, what path does it take?

3. When food is in the digestive tract, is it actually part of the body? Why or why not?

GO FIND OUT

1. If an intoxicated person is unresponsive—in other words, not responding to gentle shaking, pinching, or yelling—you should call 911. Many students are afraid to call 911 because they worry that the drunk person or others involved will be in legal trouble. Do some research and find out what protections you and the intoxicated person have under your state's laws.

2. Rape on college campuses often involves a fellow student who intentionally gets the victim drunk and then rapes him or her. Do some research on bystander intervention and summarize what you learned. Does your college offer bystander intervention training?

MAKE THE CONNECTION

The science that you learned in this chapter has helped you better understand the real-world example used throughout this discussion. Draw a line from the statement on the left to the science that supports it on the right.

Consuming too much alcohol can cause heartburn. ▷		◁ Alcohol is a diuretic.
There are differences in the amount of alcohol people of different sexes and national origins can safely drink. ▷		◁ Alcohol weakens the esophageal sphincter.
Face flushing, dizziness, and irregular heartbeat are caused by excess alcohol consumption. ▷		◁ There are many variants of the alcohol dehydrogenase enzyme and it is present in different amounts even in same-sized individuals.
Diarrhea is caused by excess alcohol consumption. ▷		◁ Smooth muscles involved in peristalsis relax when alcohol is consumed, resulting in food spending more time exposed to stomach enzymes.
An intoxicated person produces a high volume of urine. ▷		◁ Alcohol acts on the pituitary to decrease ADH secretion.
Excess alcohol consumption leads to dehydration. ▷		◁ Acetaldehyde is produced when alcohol is broken down.

Answers to **Got It?, Visualize This, Working with Data, Sounds Right, But Is It?, Show You Know,** and **Chapter Review** questions can be found in the **Answers** section at the back of the book.

20 Clearing the Air

Respiratory and Cardiovascular Systems

E-cigarettes can look and feel like tobacco cigarettes and often give users a similar dose of nicotine.

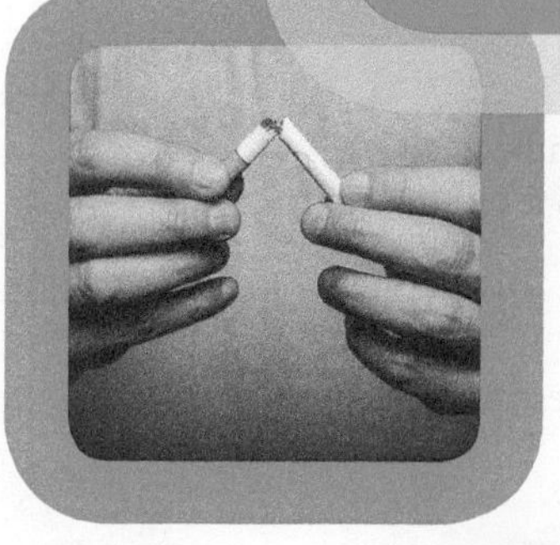

They were originally meant to help smokers quit the habit.

Many states and communities ban their use in public spaces. Is this the right approach?

When electronic cigarettes (e-cigarettes) were first made available for sale in the United States in 2006, they were hailed as a tool to help current cigarette smokers quit their dangerous habit in a gradual way. These little machines are designed to heat a liquid that typically contains nicotine and one or more flavorings. Users inhale the gas that is produced, mimicking the familiar action of drawing on a cigarette. The promise of e-cigarettes was that they could allow individuals to wean themselves from a powerful addiction to smoking tobacco. E-cigarettes also provided another benefit: the means to bypass bans on indoor cigarette use, which were proliferating in the early 2000s.

E-cigarettes enjoyed an explosive growth in popularity in the early 2010s; usage soared and stores specializing in vaporizers and liquids popped up everywhere. But in August 2016, the U.S. Food and Drug Administration (FDA) finalized a rule that regulated e-cigarettes as tobacco products. In other words, these products are now subject to the same regulations as cigarettes and chewing tobacco, including the addition of health warnings on packaging, a requirement for rigorous tests that establish product safety, and the imposition of new restrictions on sales to minors. In the years between 2006 and 2016, many cities and states had passed bans on indoor e-cigarette use (called "vaping") that paralleled the bans on indoor tobacco smoking in public areas. The FDA's new rule seemed to confirm what these local vaping bans suggested: Despite their reputation as safer alternatives to smoking, e-cigarettes pose a significant public health risk.

Critics of the e-cigarette rules and vaping bans are dismissive of the comparison with tobacco cigarettes. The vapor produced by e-cigarettes is different in many ways than the smoke produced by smoldering tobacco. These critics argue that restricting e-cigarettes removes an important tool that helps current smokers quit—or at least to reduce their consumption of much more dangerous cigarettes.

Is there a public place for vaping? In this chapter, we examine the pathway of tobacco smoke and e-cigarette vapors into the lungs and body and investigate the evidence that links exposure to these gases to preventable diseases that affect the lungs, heart, and other organs.

20.1 Effects of Smoke on the Respiratory System

Bans on indoor smoking came about as the dangers of "secondhand smoke" became clear. The smoke emitted by the lighted end of a cigarette, combined with the smoke **exhaled** by active smokers, is more accurately referred to as **environmental tobacco smoke,** or **ETS.** This smoke affects not only the **active smoker,** who is holding and **inhaling** from the cigarette, but also nonsmoking individuals in the environment. A nonsmoker in an environment high in ETS is called a **passive smoker.** ETS is similar to the smoke inhaled by an active smoker, but it has some key differences.

-hale means to breathe.

According to the National Cancer Institute, more than 4500 chemicals have been definitively identified in tobacco smoke. Passive smokers are exposed to the same mix of chemicals as active smokers, although in much lower concentrations. Surprisingly, however, certain chemical compounds are actually found in higher concentration in ETS than in actively inhaled smoke. These compounds are produced when tobacco burns incompletely on the end of a cigarette or in the bowl of a pipe in between puffs. Just like when someone blows on a smoldering fire, when a smoker takes a drag on a cigarette, oxygen is provided to the smoldering tobacco leaves, causing them to burn more completely.

Carbon monoxide is a by-product of incomplete combustion. It is also the most abundant gas in ETS. In fact, carbon monoxide is approximately five times more abundant in ETS than it is in the smoke inhaled by active smokers. You may have heard of carbon monoxide as a poisonous air pollutant that occurs inside homes with faulty furnaces or other gas-burning appliances. Although the levels of carbon monoxide in ETS alone are not deadly to passive smokers, effects of exposure to even low levels of carbon monoxide over prolonged periods can lead to serious consequences.

Also as a result of incomplete combustion, ETS contains high concentrations of airborne **particulates,** particles with a diameter less than half the width of a human hair. These particles make up the visible portion of ETS and are commonly known to cigarette smokers as tar. Because most active smokers inhale cigarette smoke through a filter tip, the amount of particles acquired by primary smoking is reduced. However, because neither active nor passive smokers breathe through a filter otherwise, both are exposed to the full concentration of particles present in ETS. Chronic exposure to airborne particulates can lead to a number of negative health effects, including emphysema and cancer.

The vapors produced by e-cigarette use are significantly different from ETS. Most important, there is no combustion in these devices, so all of the secondhand vapor is what is exhaled by the user. This vapor contains a number of different chemicals, notably nicotine, but none of the products of incomplete combustion, such as carbon monoxide. The particulates in vapor are liquids, rather than the more dangerous solids found in cigarette smoke. However, recent study of the components of secondhand e-cigarette vapors identified known cancer-causing chemicals, such as formaldehyde, and irritants such as propylene glycol among the particulates produced.

Airborne particulates enter the human body via the mouth and nose; pass through the **pharynx** (throat) and **larynx** (voice box) to the distribution network of the **trachea** (windpipe), **bronchi,** and **bronchioles;** and ultimately settle in the **lungs,** the site of gas exchange (**FIGURE 20.1**), where they cause some of their most severe consequences. However, some of the chemicals in these particles and carbon monoxide gas have the ability to cross from the lungs into the bloodstream and thus can be carried throughout the body. To understand how chemicals pass from the air into our bodies, first we must understand the structure and function of lungs.

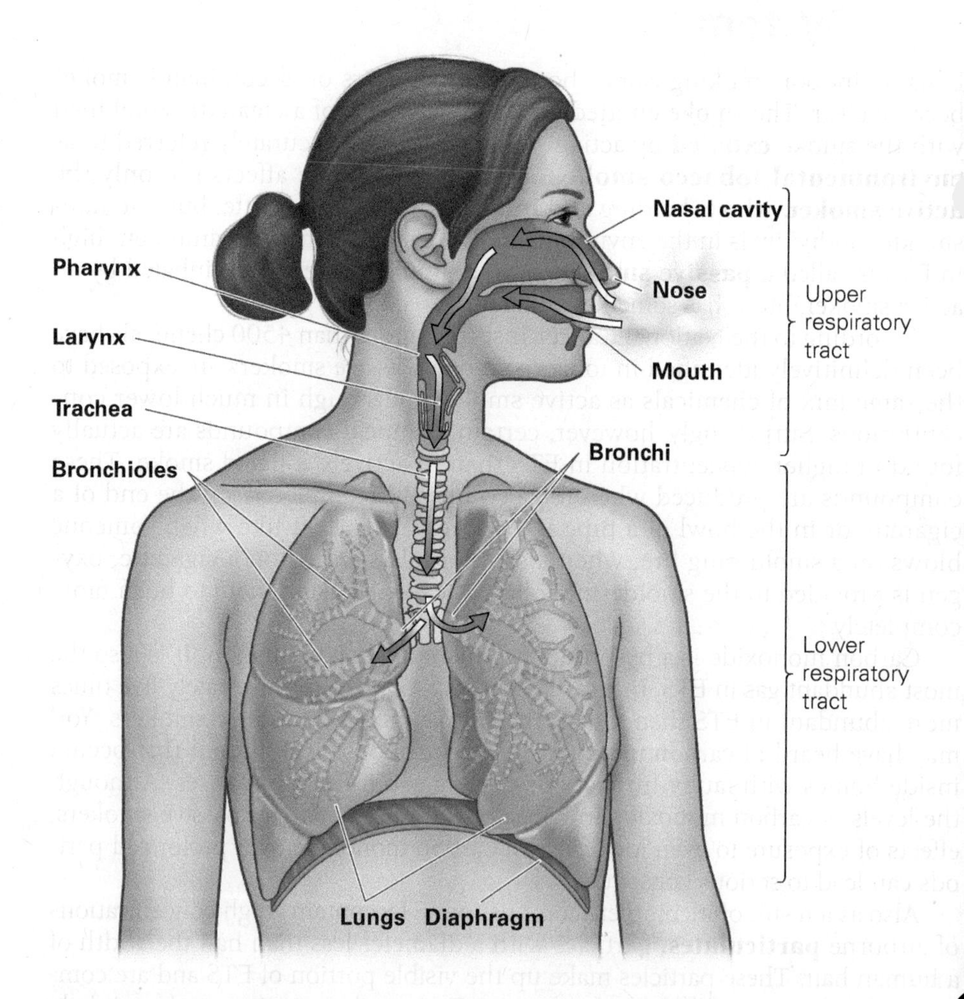

FIGURE 20.1 The human respiratory system. Smoke enters the upper respiratory system through the nasal cavity or mouth, then passes through the pharynx and larynx into the lower respiratory system. Here, it flows into a series of smaller and smaller tubes within the lungs until it comes into close contact with circulating blood.

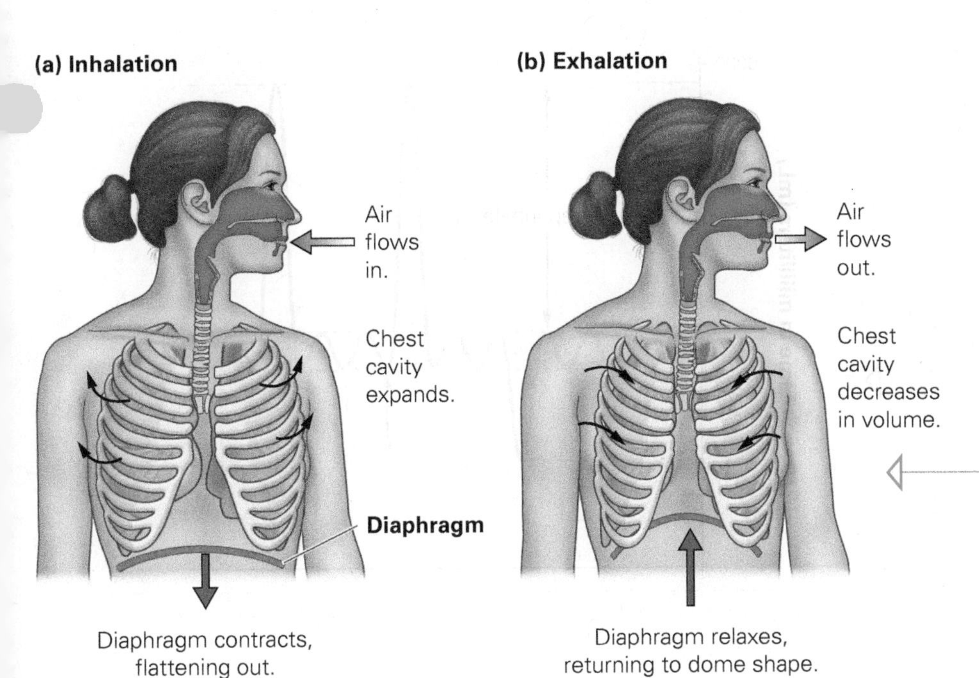
(a) Inhalation
Air flows in.
Chest cavity expands.
Diaphragm
Diaphragm contracts, flattening out.
(b) Exhalation
Air flows out.
Chest cavity decreases in volume.
Diaphragm relaxes, returning to dome shape.

FIGURE 20.2 How the diaphragm works. (a) Contraction of the dome-shaped diaphragm flattens it out, increasing the volume of the chest cavity and lowering air pressure. (b) Relaxation of the diaphragm causes the chest cavity to lose volume, increasing air pressure.

VISUALIZE **THIS**

How does a hiccup, an involuntary contraction of the diaphragm, affect normal breathing?

What Happens When You Take a Breath?

You take a breath approximately 12 times per minute, or more than 6 million times per year, without thinking much about it. In each normal breath at rest, you take in about 500 milliliters—approximately 2 cupfuls—of air and release the same amount. Over the course of a single minute, about 6 liters of air come into direct contact with the surfaces of your lungs.

Diaphragm. In contrast to many other animals, respiration in humans and other mammals is an active process. The human respiratory system is separated from the organs that make up the bulk of the digestive and reproductive systems by a strong, dome-shaped muscle called the **diaphragm.** When we inhale at rest, the diaphragm contracts and flattens out, increasing the volume of our chest cavity (**FIGURE 20.2a**). At the same time that the diaphragm is contracting, the rib cage is lifted up and outward by muscles located between the ribs, further increasing the volume of the cavity. Increasing its volume decreases the air pressure inside the chest, causing air to flow into the lungs.

The change in pressure in our chest cavity during inhalation is similar to the change in the air pressure inside a syringe when the plunger is pulled back. If the syringe is in a liquid solution, we can see the liquid flow toward the lower pressure inside the syringe, just as air flows to the lower-pressure environment inside the lungs.

When we are resting, exhalation requires no muscle contraction. The diaphragm relaxes, and the volume of the chest cavity decreases (**FIGURE 20.2b**). The air that had rushed into the lungs is squeezed into a smaller volume, increasing the pressure and causing air to flow back out of the trachea and the nasal cavity. Passive exhalation is similar to the release of air from an expanded balloon. Because the elastic walls of the balloon exert pressure on the contents, air rushes out of an open nozzle even without actively squeezing the balloon.

Exhalation becomes an active process during intense aerobic exercise, when the body must actively rid itself of carbon dioxide, one of the waste products of respiration. In this case, muscles in the abdomen contract, forcing the diaphragm upward, and muscles in the chest wall tighten, squeezing the chest cavity further. The process of active exhalation is much like the action of forcing liquid out of a syringe by depressing the plunger and reducing the volume. The

FIGURE 20.3 Lung capacity and breathing. The muscular movements of the chest and abdomen walls change the lung volume. A deeper inhale results from a strong contraction of the diaphragm and the muscles of the rib cage, leading to a larger lung volume.

WORKING WITH DATA

According to the graph, how much air is exchanged during a single inhalation/exhalation at rest?

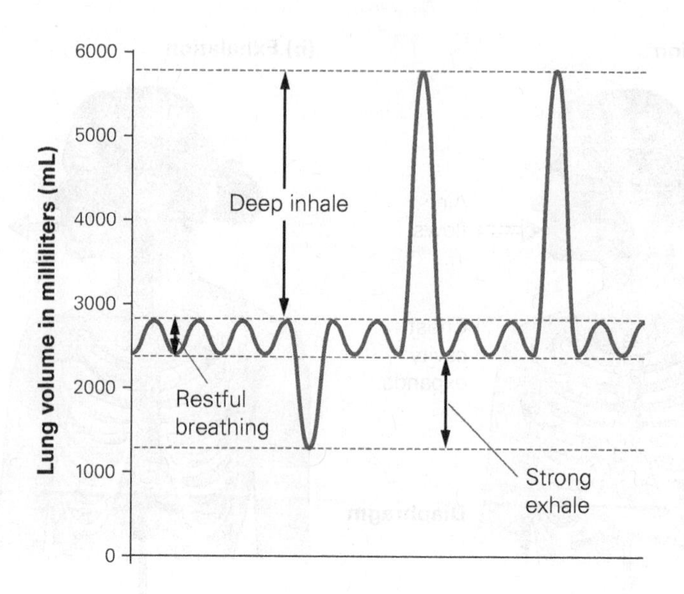

graph in **FIGURE 20.3** illustrates how the mechanics of breathing correspond to changes in lung volume, and thus the amount of air in the lungs.

Our rate of breathing is regulated by control centers in the most evolutionarily ancient part of the brain, called the brain stem. The brain stem signals the diaphragm to contract in response to carbon dioxide levels in the blood. The breathing rate increases when carbon dioxide levels in the body increase, such as during exercise. Rapid breathing moves carbon dioxide out of the body quickly while also providing oxygen for optimal ATP production.

Contraction of the diaphragm can also be controlled voluntarily. Holding your breath requires consciously overriding the signals from the brain stem—at least for a time. Taking a drag on a cigarette or e-cigarette requires stronger contraction of the diaphragm than normal breathing does to pull air through the device and into the mouth and lungs.

Speaking or making any vocal noises requires active exhalation as air forced out past the vocal cords in the larynx causes these structures to vibrate and produce sound. The amount of air forced past the vocal cords determines the volume of our speech, and muscles that control the length of the vocal cords help to determine the pitch of our speech. The shape of our mouth and lips and the shape and position of our tongue determine the actual sound that is produced. Sustained exposure to tobacco smoke can cause parts of the larynx to become covered with scar tissue, often making longtime smokers sound quite hoarse. Nonsmokers regularly exposed to environmental tobacco smoke demonstrate some changes to their vocal cords relative to other nonsmokers, although it is not clear that these changes are enough to affect the voice. Whether these changes occur in individuals inhaling e-cigarette smoke is unknown.

The diaphragm also may contract involuntarily for other reasons—for instance, when we inhale deeply before a cough. Coughing is a reflex that helps remove irritants from the trachea. Keeping air passageways into the lungs clear is crucial for maintaining adequate oxygen uptake.

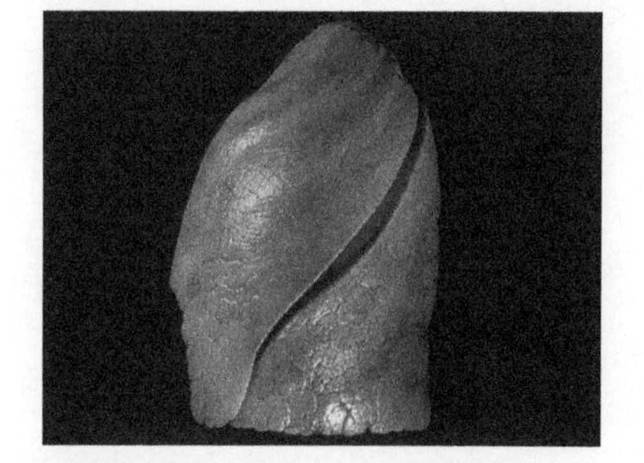

FIGURE 20.4 The respiratory surface of the human body. Lungs contain the surfaces over which gases are exchanged between the body and the environment. The total respiratory surface area contained within a pair of healthy human lungs is approximately the size of a tennis court.

Lungs. Healthy lungs are pink, rounded, and very spongy in texture (**FIGURE 20.4**). Even at rest, these organs almost completely fill the chest cavity. The external surface of the lungs is linked to the lining of the chest cavity by two thin membranes, one covering the lungs and one lining the chest wall. These membranes stick together because they are moist, much like two wet pieces of plastic wrap stick together. This ensures that when the chest cavity enlarges, the lungs expand. If the two membranes become separated—for instance, as a

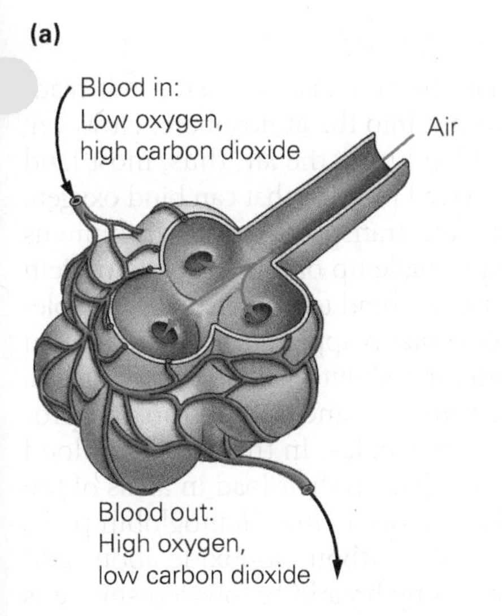

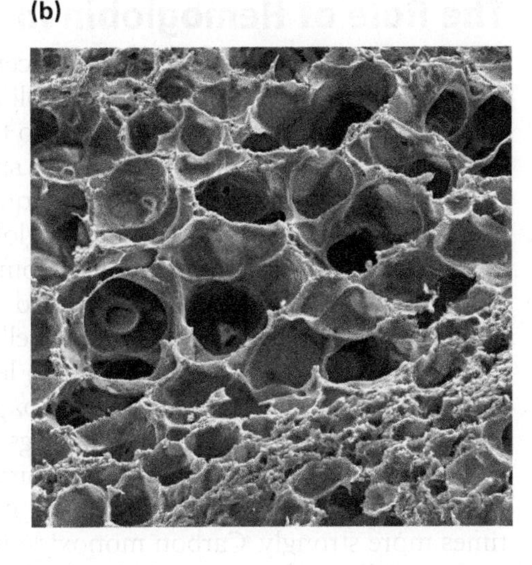

FIGURE 20.5 Alveoli. (a) Alveoli occur in clusters at the end of bronchi. The clusters are enmeshed in a dense network of thin-walled capillaries, which bring blood into close proximity with the gas in the alveoli. (b) An electron micrograph of alveoli in lungs.

result of injury to the chest wall—then air will enter the space between them, sometimes causing the underlying lung to collapse.

Once air enters the lungs, it flows through the bronchi and bronchioles, each of which dead-ends at structures called **alveoli.** The alveoli are 300 million tiny sacs within our lungs that contain the **respiratory surface** of our bodies, the tissue across which gases from inside the body are exchanged with gases in the air. The area of this respiratory surface in humans is approximately 160 square meters (1725 square feet), about the size of a tennis court. A large surface area is required to supply oxygen and remove carbon dioxide from large-bodied, active mammals. Each grape-like cluster of alveoli is surrounded by a net of tiny blood vessels, called **capillaries,** connecting the gases exchanged in these structures with the entire body (**FIGURE 20.5**).

Exposure to smoke damages and eventually destroys the respiratory surfaces within alveoli; whether this damage occurs as a result of exposure to e-cigarette vapor is still unknown. A reduction in the total amount of respiratory surface results in shortness of breath, wheezing, and an inability to participate in even moderately vigorous activity. These symptoms are all caused by the reduced capacity for gas exchange in the lungs.

VISUALIZE **THIS**

Oxygen levels in the atmosphere are low at high altitudes. How do you think this affects the rate of gas exchange in the lungs?

Gas Exchange

The primary function of the lungs is **gas exchange,** the process that allows us to acquire oxygen for cellular respiration and to expel carbon dioxide. Exchange of gases occurs between the air spaces of the alveoli and the blood contained in neighboring capillaries. The exchange works by simple diffusion across the thin membranes surrounding the blood vessels and the alveoli walls. Recall that diffusion is the passive movement of substances from where they are in high concentration to areas where they are in low concentration. Carbon dioxide in high concentration within the capillaries passes via diffusion from these structures into the alveoli, where it is maintained in low concentration by exhalation. Oxygen in high concentrations, maintained by inhalation, passes by diffusion from the alveoli to the deoxygenated blood in the capillaries (**FIGURE 20.6**).

Carbon dioxide and oxygen gas dissolve in the fluid that coats the respiratory surface prior to diffusing across the cell membrane. Because the size of the alveoli increases and decreases over the course of a single breath, this layer of fluid must be somewhat slippery to maintain constant coverage of the surface and to prevent the alveoli walls from sticking to each other. A soap-like substance called surfactant provides this slippery consistency. The composition of surfactant, and thus the flexibility of the lung tissue itself, can be affected negatively by tobacco smoke and perhaps by e-cigarette vapor.

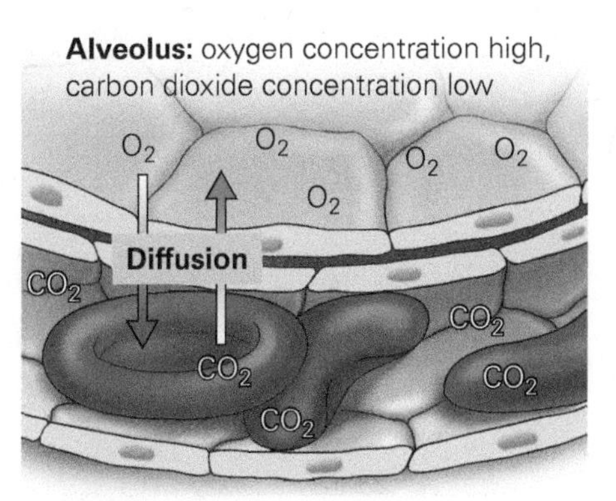

FIGURE 20.6 Gas exchange in the lungs. The movement of oxygen and carbon dioxide gas at the respiratory surface occurs via simple diffusion.

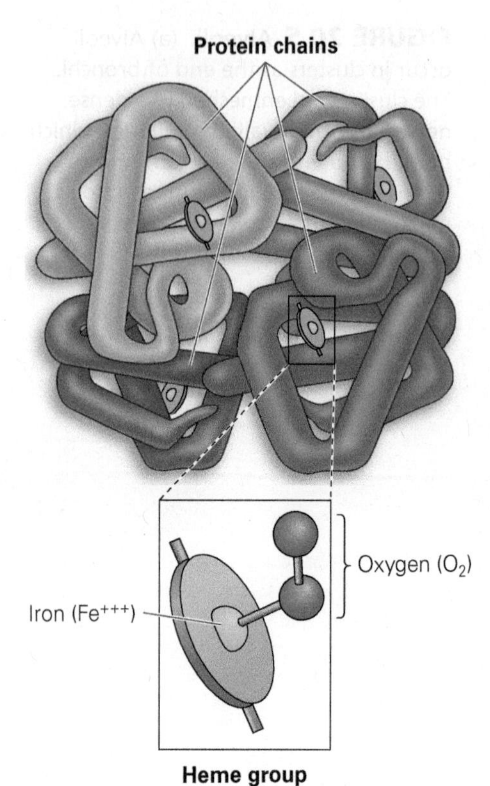

FIGURE 20.7 Hemoglobin, the blood's oxygen shuttle. The four protein chains in a single hemoglobin molecule are colored differently here, but function identically. The heme group in each chain is the oxygen-binding site.

The Role of Hemoglobin in Gas Exchange

Carbon dioxide is in relatively low concentrations in the atmosphere, so it readily moves from the blood into the alveoli and out into the atmosphere. However, oxygen does not dissolve as readily into the blood from the air; thus, most land animals (and even some plants) produce specialized proteins that can bind oxygen.

Hemoglobin is the protein that acquires and transports oxygen in humans and nearly all other vertebrates. Hemoglobin is made up of four separate protein chains, each containing a single iron atom that can bind to a single oxygen molecule (**FIGURE 20.7**). Each of our red blood cells contains approximately 250 million hemoglobin molecules; therefore, each cell can carry about 1 billion oxygen atoms.

In areas of the body where oxygen levels are low and carbon dioxide production is high, hemoglobin releases oxygen molecules. In this way, red blood cells are loaded with oxygen in the lungs and will drop their load in areas of the body where high levels of cellular respiration are occurring. Hemoglobin picks up oxygen effectively, but surprisingly, it binds to carbon monoxide about 200 times more strongly. Carbon monoxide inhaled by breathing tobacco smoke is preferentially loaded into red blood cells at the respiratory surface. Because the binding of carbon monoxide is so strong, hemoglobin is slow to release it. As a result, fewer hemoglobin molecules are available to transport oxygen. Even small amounts of carbon monoxide can tie up large amounts of hemoglobin in the body, causing severe oxygen shortages in body tissues. Although the amounts of carbon monoxide in ETS are not high enough to cause death from lack of oxygen, chronic low levels of oxygen deprivation can damage tissues and organs.

Carbon monoxide is especially damaging to developing embryos and fetuses because they must acquire the oxygen they need through exchange with their mother's blood supply. The lower-than-average birth weights of babies born to mothers who are active smokers may be due to their relative oxygen deprivation There is some evidence from animal studies that long-term exposure to the carbon monoxide levels found in ETS may contribute to diminished brain function in infants and children for the same reason—oxygen deprivation. E-cigarettes do not produce carbon monoxide. However, the biggest threat secondhand smoke poses to respiratory function does not stem from this gas. Instead, the threat comes from damage caused to lungs by microscopic particles of smoke.

Smoke Particles and Lung Function

The particulates in ETS are a significant hazard to infants and children as well as to anyone with impaired lung function, including individuals who have been chronically exposed to particulate-laden air. Although the liquid particulates from e-cigarettes appear less damaging than the solids produced by tobacco smoke, research on the effect of vapor particulates on passive "vapers" is still lacking.

Bronchitis and Asthma. Our lungs have various mechanisms for expelling and removing foreign objects. Coughing is our first response to larger particles or droplets that enter the trachea. Particles too small to trigger a cough can settle in the upper respiratory tract, trachea, or larger bronchi. These particles become trapped in mucus produced by the tissue lining these structures. The mucus and particles are swept upward toward the mouth and nose by the actions of tiny hairs called cilia on the tissue surface. When a glob of mucus reaches the tip of the larynx, it is coughed up, expelled out of the nose or mouth, or swallowed.

Airborne particles in tobacco smoke not only increase the production of mucus but also damage the cilia lining the bronchi, making it increasingly more difficult to expel these particles. This damage in turn can lead to inflammation of the bronchi, known as **bronchitis.** Active smokers may develop chronic bronchitis, which often manifests itself as a lasting cough, as abundant mucus is produced by inflammation. Particulates, even the liquid droplets in e-cigarette

-itis is used to describe inflammation (of an organ).

vapor, are also known to worsen **asthma,** an allergic response that results in the muscular constriction of bronchial walls as well as an overproduction of mucus. An acute asthma episode is often characterized by a wheezing sound during breathing, as the greatly restricted airflow creates sound oscillations. According to the U.S. Environmental Protection Agency (EPA), approximately 26,000 cases of asthma among children in the United States are caused by exposure to ETS each year. It is not clear whether a child's exposure to environmental e-cigarette vapor produced by vaping adults also increases asthma risk.

Emphysema. The physical consequences of chronic bronchitis and asthma can lead to the formation of scar tissue in the lungs, which permanently blocks bronchi. The inflammation also causes damage to alveoli walls, causing the many small alveoli to merge into fewer, larger sacs (**FIGURE 20.8**). This merging of alveoli reduces the overall area of the respiratory surface, interfering with gas exchange. The sacs of the alveoli themselves become surrounded by thick scar tissue, which interferes with the passage of oxygen into the capillaries. The buildup of scar tissue also makes the lungs less elastic, meaning that the passive process of exhalation is ineffective. As more "dead air" remains in the lungs, the lungs become overinflated, gradually increasing the size of the chest. A barrel-shaped chest can be a physical sign of the underlying disease, called **emphysema.**

Because alveoli cannot be regenerated, the lung damage that results in emphysema is permanent and irreversible. Individuals with emphysema are chronically short of breath and unable to participate in vigorous activity. The most severely affected individuals require supplemental oxygen simply to engage in the activities of daily living. The EPA estimates that adult nonsmokers exposed to ETS have a 30 to 60% higher occurrence of emphysema, asthma, and bronchitis when compared with unexposed nonsmokers.

Lung Cancer. The tiniest particulates in ETS and e-cigarette vapor can be drawn deeply into the lungs, even into the alveoli. Because alveoli lack cilia, the movement of foreign materials out of these structures is much more limited. Small particles can remain in the alveoli for long periods. We can

VISUALIZE THIS

Smaller and more numerous alveoli would increase the respiratory surface area even more. What do you think are disadvantages to a greater number of even smaller alveoli?

(a) Alveoli in a nonsmoker's lung

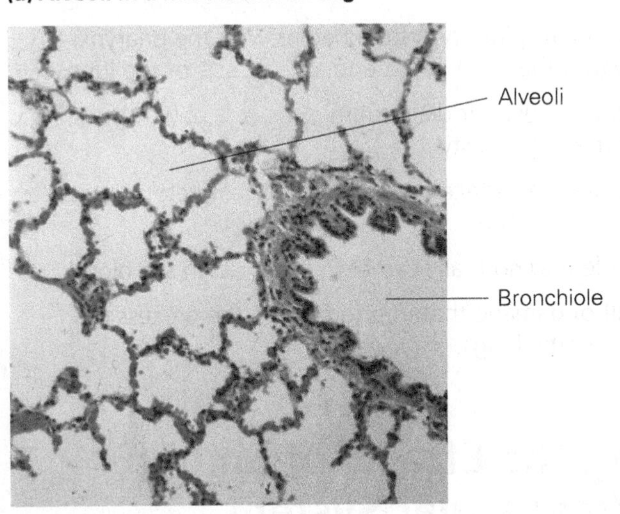

(b) Alveoli in a smoker's lung

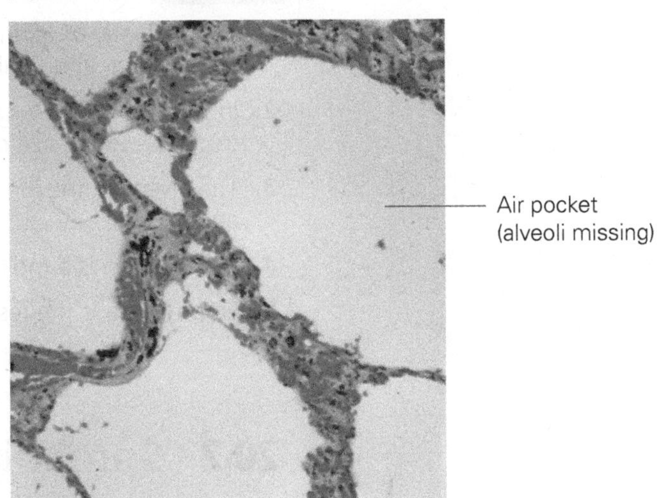

FIGURE 20.8 The effect of chronic smoke exposure on alveoli. (a) In this cross-section of a normal lung, the alveoli are tiny. The walls surrounding each air pocket are areas where gas exchange occurs. (b) In a smoker's lung, many of the alveoli walls have been destroyed, leading to larger air pockets and less surface for gas exchange. Both of these images are photographed at the same magnification.

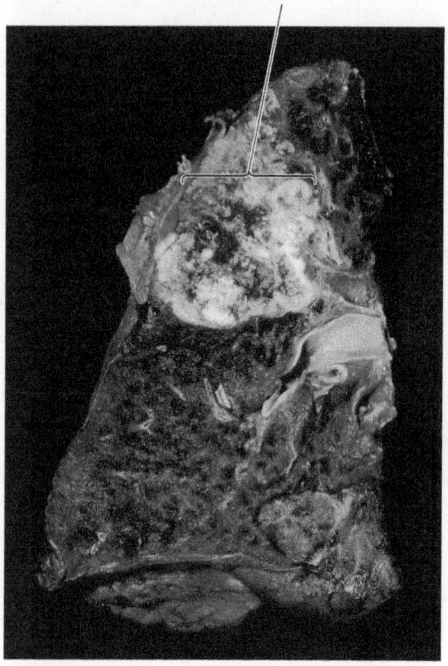

FIGURE 20.9 Tobacco tar trapped in lungs. These lungs show the dark staining caused by the accumulation of particulates. The lighter-colored mass is a cancerous tumor.

see the accumulation of particles in the lungs of a long-time smoker, which are often black with tar trapped inside the alveoli (**FIGURE 20.9**). The accumulated particles promote inflammation and fluid accumulation and generally reduce gas exchange. Even more seriously, because many components of tobacco smoke are carcinogens known to cause mutations, lung cells are susceptible to transforming into cancer cells long after smoke inhalation has stopped.

About 170,000 new cases of lung cancer are diagnosed in the United States each year, 90% in current or former smokers. However, for the 17,000 cases of lung cancer not due to active smoking, epidemiologists at the National Institutes of Health (NIH) have estimated that 3000 of these cases are directly attributable to exposure to ETS, and that passive smokers have a 20 to 30% higher risk of lung cancer than unexposed nonsmokers.

Epidemiological studies, such as the one that linked ETS exposure to lung cancer risk, have the same problem as all other correlations—it is impossible to eliminate the chance that some other difference between individuals in the study is, in reality, causing a difference between groups (see the section on hypothesis testing in Chapter 1). In the case of ETS and lung cancer, the hypothesized link is also supported by the presence of known carcinogens in secondhand smoke, by experiments using animals that showed higher rates of cancer in ETS-exposed individuals, and by a link established by hundreds of studies between active smoking and lung cancer.

Because e-cigarettes are so new, scientists have much less data about the relationship between secondhand vapor exposure and any of these lung conditions. The relationship between exposure to secondhand smoke and diseases in other organs and organ systems is also less clear from the data collected. However, given the connection between the lungs and the bloodstream, one would expect that passive smokers have a higher risk than unexposed individuals of all the diseases that are more common in active smokers. To understand how smoke and vapor breathed into the lungs can affect the entire body, we must understand first how blood and its components are moved around via the bloodstream.

Got It?

1. The pathway of air into the respiratory system begins with the pharynx, to the trachea, and finally into the ________ and ________ of the lungs.
2. Inhalation occurs when the muscular diaphragm ________, which ________ the size of the chest cavity.
3. The alveoli in the lungs are the surface across which exchange of ________ and ________ ________ occurs.
4. Hemoglobin is a molecule that normally carries ________ in the blood.
5. Emphysema is the result of damage that interferes with the process of ________ ________ in the lungs.

20.2 Spreading the Effects of Smoke: The Cardiovascular System

The risks of active smoking, in addition to lung and airway damage, include increased rates of throat, bladder, and pancreatic cancer; higher rates of heart attack, stroke, and high blood pressure; and even premature aging of skin. All of these effects occur because many of the components of tobacco smoke can cross the thin walls of alveoli into the bloodstream and move throughout the body.

(a) Plasma

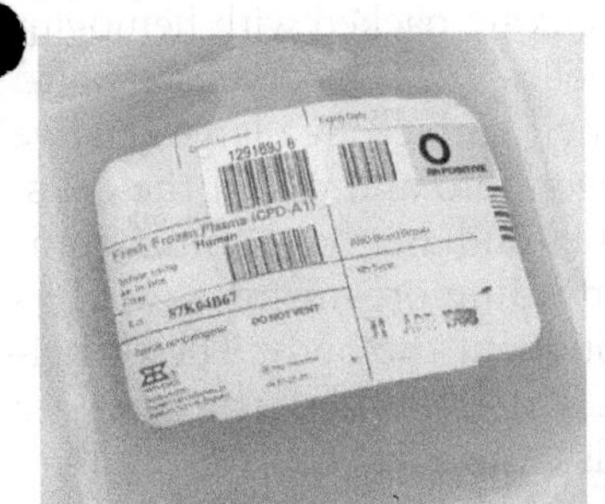

(b) Red blood cells

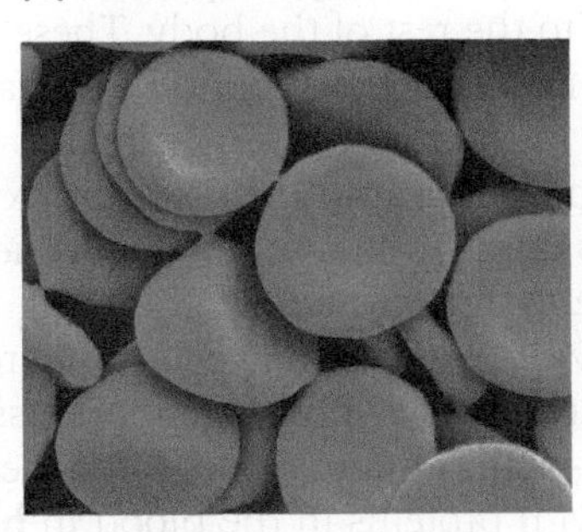

(c) White blood cells

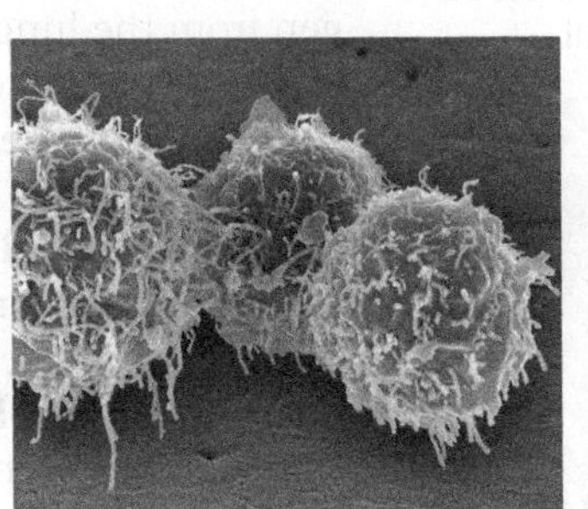

(d) Platelets

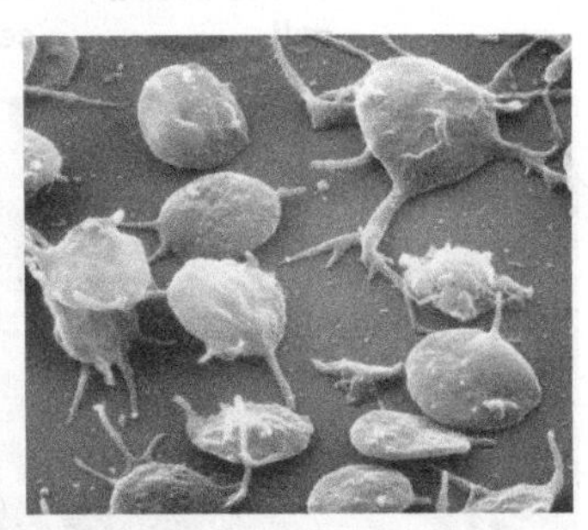

FIGURE 20.10 Components of blood. Blood consists of the fluid plasma as well as large amounts of red blood cells and lesser amounts of white blood cells and platelets.

Structure of the Cardiovascular System

Gases and other materials are distributed around the body of most animals by a **cardiovascular** system, which consists of three major components: a circulating fluid (often called blood), a pump (in most cases, this is called a heart), and a vascular system (blood vessels and capillaries).

card- and cardio- relate to the heart.

vascul- means vessel.

Blood. The 5 liters (11 pints) of blood in the vascular system of an adult human are made up of both liquid and solid (that is, cellular) portions. The liquid portion, called plasma, consists of water and dissolved proteins, salts, and gases. The cellular, or solid, portion is primarily made up of red blood cells (giving blood its color), with much smaller numbers of white blood cells and platelets (**FIGURE 20.10**). In adults, all of the cellular components of blood are produced by stem cells in the bone marrow, tissue found in the cavities of certain bones. Stem cells are unique among adult cells—instead of being completely differentiated into a single cell type, such as nerve or skin, they have the capacity to produce descendants of a variety of cell types. Blood stem cells give rise to two classes of more specialized stem cells that together can generate nine different blood cell types (**FIGURE 20.11**).

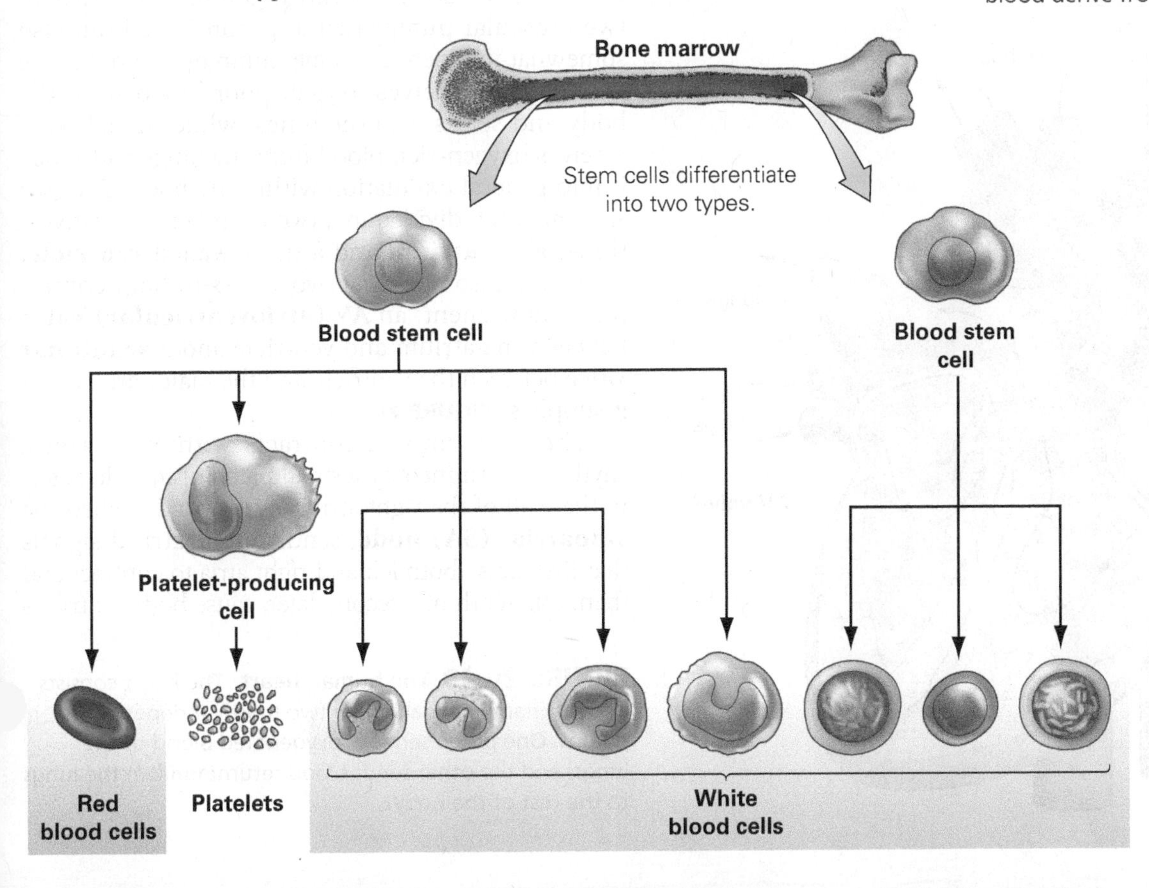

FIGURE 20.11 Blood stem cells. Blood cells are produced by stem cells found in the marrow of larger bones. Notice all cellular elements in the blood derive from these stem cells.

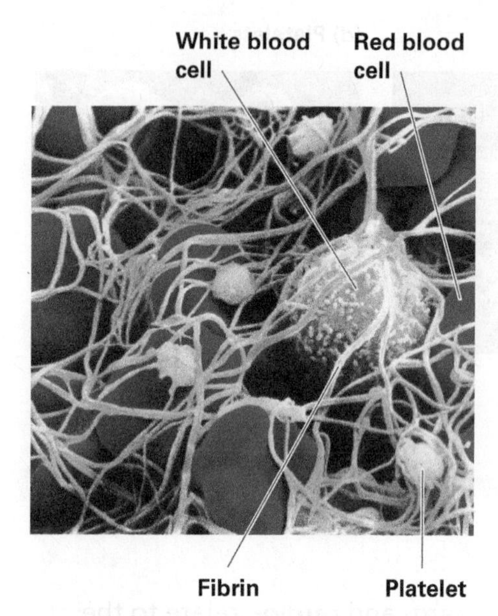

FIGURE 20.12 A blood clot. A blood clot consists of a net of protein fibers that trap blood cells and platelets, forming a temporary patch over a damaged blood vessel. As shown in the image, clots are typically red in color because the majority of cells in the patch are red blood cells.

Red blood cells are uniquely adapted to their primary task—shuttling oxygen from the lungs to the rest of the body. These cells are packed with hemoglobin molecules and lack a nucleus and other organelles. Red blood cells are also small and pinched in the middle, which makes their surface area large relative to their volume, ensuring rapid diffusion of oxygen into and out of the cells. **White blood cells** come in many different varieties and are the essential component of immune system function, attacking invading organisms as well as removing toxins, wastes, and damaged cells throughout the body. The principal function of **platelets** is to prevent blood loss. These cell-like structures are actually membrane-bound fragments of larger cells.

Platelets work with proteins in the blood in **blood clotting,** the process that stems the flow of blood out of damaged blood vessels. Immediately after injury occurs, muscles in the walls of damaged blood vessels contract to restrict the flow of blood. Within seconds, sticky platelets become attached to any breaches in the vessels, forming a temporary plug to help restrict blood flow. At the same time, chemical signals released by the damaged tissue initiate the complex clotting process. A blood clot consists of a net made up of the protein fibrin, which forms a patch over the damaged area (**FIGURE 20.12**). Cell division in the walls of the damaged vessel and surrounding tissues eventually seals the cut, and the patch dissolves.

A substance in tobacco smoke, perhaps nicotine, increases the stickiness of platelets and promotes production of fibrinogen, the precursor to fibrin. Both of these effects result in the increased risk of blood clots. When a clot forms within a blood vessel, it can block blood flow where it forms (a condition called a *thrombosis*) or break free and travel throughout the bloodstream until it becomes lodged in another blood vessel (in which case it is called an *embolism*). When a thrombus or embolus becomes lodged within the blood vessels of the heart or any other organ, it can restrict the flow of oxygen to nearby cells, killing them and causing severe damage to the organ.

VISUALIZE THIS

The left and right ventricles move the same amount of blood per heartbeat, but the left ventricle walls are thicker and more powerful. Why is this necessary?

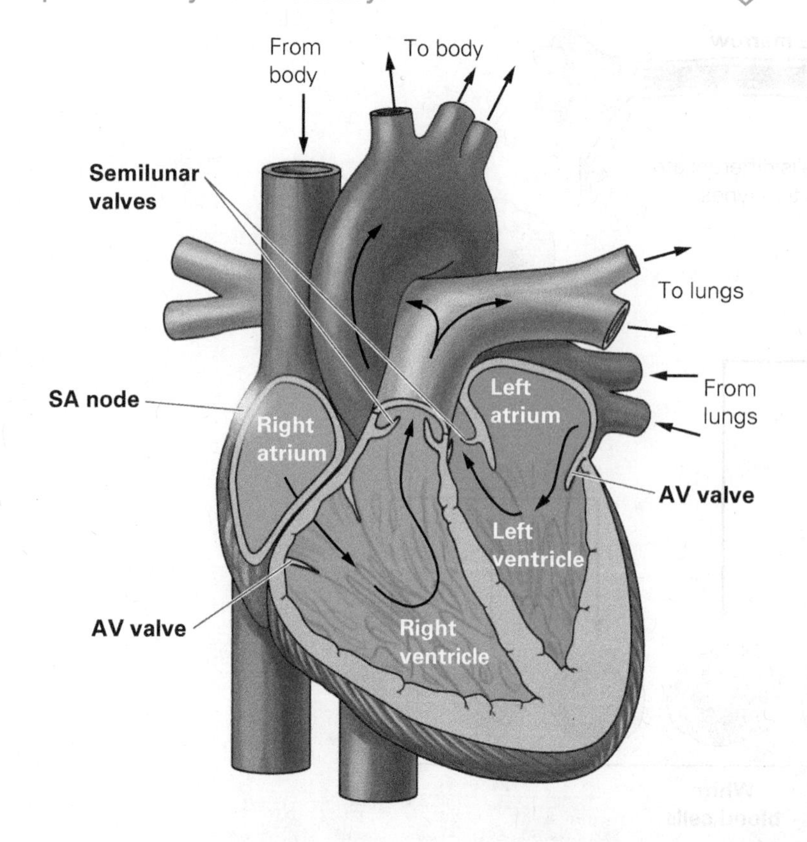

Heart. The fist-sized heart in a human consists of two muscular pumps that are coordinated but also somewhat independent. One pump on the right side of the heart receives oxygen-poor blood from the body and sends it to the lungs, while the left side receives oxygen-rich blood from the lungs and sends it into general circulation within the body. The two sides are each divided into two chambers, a relatively thin-walled **atrium** and a thick-walled **ventricle.** Each side also contains two valves to help control blood movement: an **AV (atrioventricular) valve** between the atrium and ventricle and a **semilunar valve** between the ventricle and the major artery that it supplies (**FIGURE 20.13**).

The heart muscle contracts with an intrinsic rhythm determined by a small patch of muscle tissue in the wall of the right atrium. This patch, called the **sinoatrial (SA) node,** sends out electrical signals that first cause both left and right atria to contract and then, one-tenth of a second later, cause both ventricles

FIGURE 20.13 The human heart. The heart consists of four chambers making up two mostly independent pumps. One pump sends deoxygenated blood to the lungs, and the other sends blood returning from the lungs to the rest of the body.

to contract. The complete sequence within the heart of filling with blood and then pumping is called the **cardiac cycle,** alternating between a relaxed period (called **diastole**) and a contraction phase (called **systole**).

During systole, the AV valves close as the ventricle contracts (making the first heart sound, "lubb"). At the beginning of diastole, the semilunar valves close as the ventricle relaxes ("dupp"). This cycle of contraction and relaxation leads to that distinctive heartbeat sound ("lubb-dupp, lubb-dupp").

The speed of the cardiac cycle, also known as the heart rate, is initially controlled by the SA node, which functions as an internal pacemaker (although its function can be replaced by an implanted artificial pacemaker). The speed of the pacemaker can also be affected by signals from the brain stem and spinal cord at rest and from other parts of the brain in response to experiencing danger or intense emotion. Certain drugs, such as nicotine, may also affect heart rate.

Blood Vessels. The system of tubes that carries blood to and from the heart is generally referred to as the vascular system. Components of the vascular system, broadly termed the *blood vessels,* include arteries, veins, and capillaries (**FIGURE 20.14**). **Arteries** are the branching blood vessels that carry blood from the heart, and **veins** are the converging vessels that bring it back. As in the lungs, capillaries throughout the body are tiny, thin-walled blood vessels that create a net of channels between the smallest arteries (arterioles) and the smallest veins (venules).

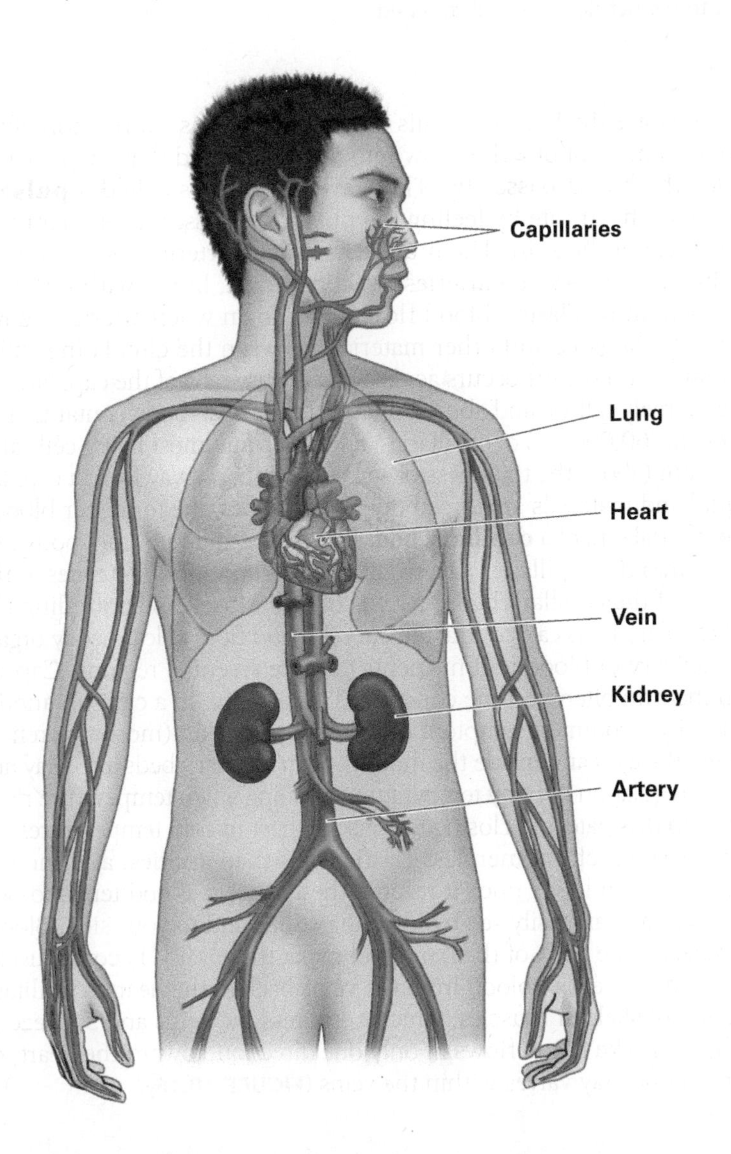

FIGURE 20.14 The vascular system. Arteries carry blood away from the heart, while veins carry blood to the heart. The two types of blood vessels are connected by nets of capillaries.

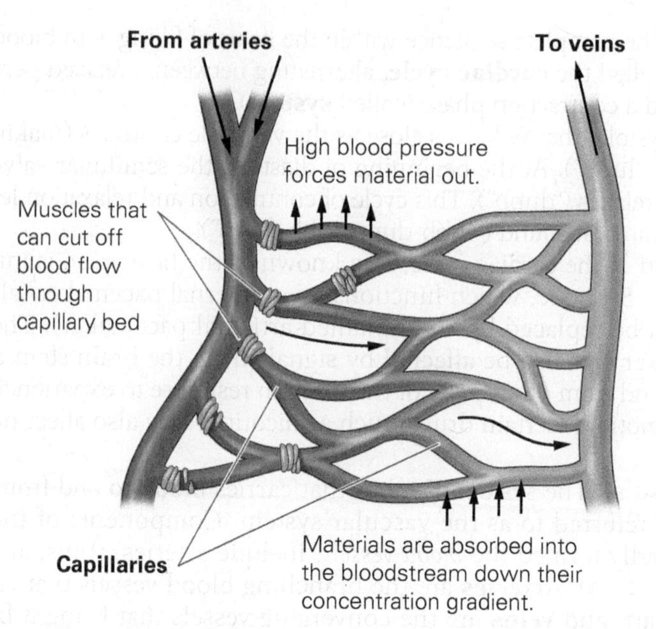
From arteries
To veins
High blood pressure
forces material out.
Muscles that
can cut off
blood flow
through
capillary bed
Capillaries
Materials are absorbed into
the bloodstream down their
concentration gradient.

FIGURE 20.15 Capillary function. Liquid forced out of the upstream end of a capillary bed carries oxygen and nutrients to the tissues. Wastes and other materials are absorbed into the blood at the downstream end. Muscles on the upstream side can contract to restrict flow through the bed.

Arteries have thick, elastic walls that balloon out as contraction of the ventricles causes a mass of blood to flow into the system and that snap back to resting size once the blood passes by. The wave of blood is called a **pulse;** you can measure your heart rate by feeling the pulse as it passes through an artery close to the surface of the skin. The muscular walls of arteries also help regulate the rate of blood flow—when arteries are constricted, blood within them is under pressure and moves faster; blood flow slows down when arteries are relaxed.

Exchange of gases and other materials between the circulating fluid and the body's tissues and organs occurs across the porous walls of the capillaries. Capillaries are extremely diffuse and abundant; an adult human body contains an estimated 100,000 km (60,000 mi) of capillaries, and as a result most living cells are no more than 0.1 mm (about the thickness of a sheet of paper) away from a capillary.

Liquid and materials are forced out of capillaries due to higher blood pressure near the arterial end of a **capillary bed.** Materials and liquid from body tissues then flow back into the capillaries as a result of concentration differences at the venous (vein) end of the capillary bed (**FIGURE 20.15**). Muscles surrounding the arterial ends of capillary beds can contract to cut off blood flow to less needy organs, allowing the delivery of blood and nutrients to more essential regions. Capillaries that nourish the skin often become constricted in response to a crisis in another organ, which is why a common symptom of illness is paler skin (most apparent in the tissues lining the eyes and inside the mouth). Skin capillary beds also play an essential role in the regulation of body temperature, opening when temperature rises to allow body heat to dissipate and closing to conserve heat in cold temperatures.

Veins have much thinner, less elastic walls than arteries, and the pressure of the blood is much lower once it reaches these vessels. Blood tends to pool within veins, a fact you can easily see by lowering your hand to your side. Blood pooling in the veins on the back of the hand will make these veins become distended and stand out. Movement of blood from the veins back to the heart is facilitated by the contraction of skeletal muscles, which compress the veins and squeeze the blood through them. The blood flows in only one direction, toward the heart, due to the presence of one-way valves within the veins (**FIGURE 20.16**).

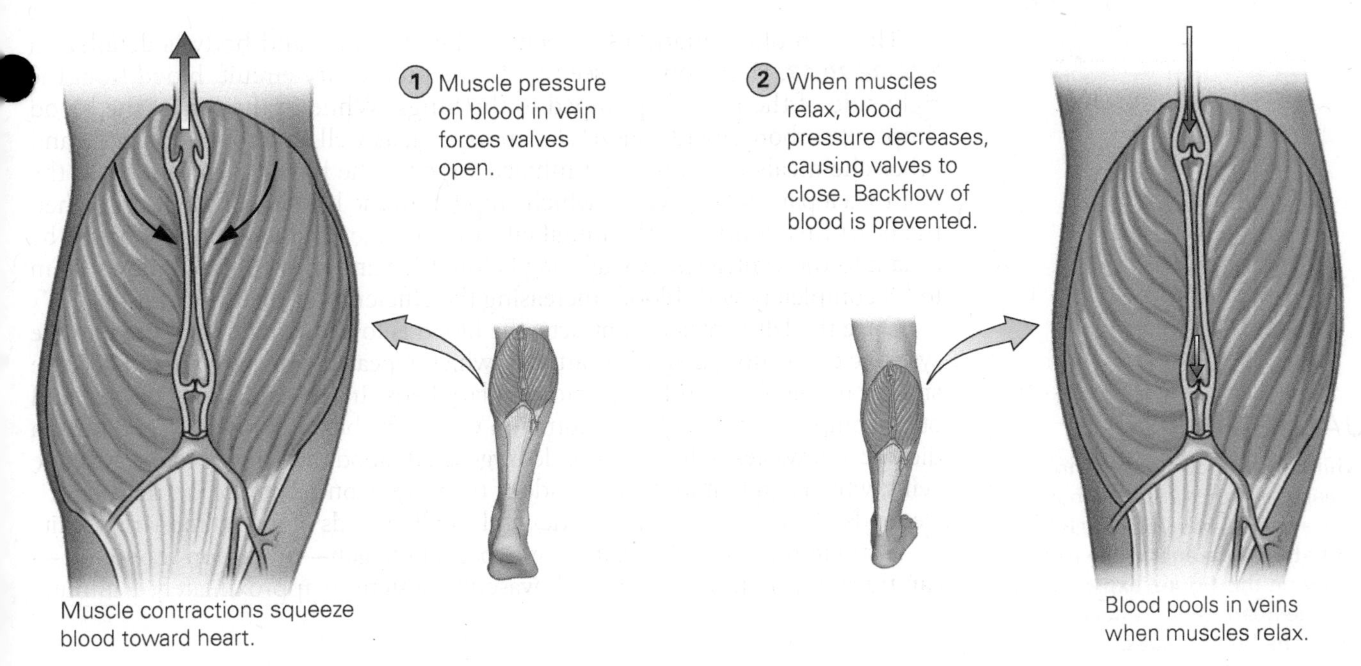

1 Muscle pressure on blood in vein forces valves open.
2 When muscles relax, blood pressure decreases, causing valves to close. Backflow of blood is prevented.
Muscle contractions squeeze blood toward heart.
Blood pools in veins when muscles relax.

FIGURE 20.16 Flow of blood in veins. Blood in the veins is under low pressure and returns to the heart due to the contraction of skeletal muscle. The blood flows in one direction, thanks to the presence of one-way valves.

VISUALIZE **THIS**

Many athletes have taken to wearing compression socks during sporting events. These socks compress the calves and shin muscles. How might compression socks affect blood circulation and oxygen delivery to the leg muscles?

The efficient flow of oxygen and materials to the body's tissues requires that the blood is maintained under enough pressure to move it throughout the vascular system. **Blood pressure,** the force of the blood against blood vessel walls, is created in part by the pulse of blood from the contracting heart and in part by the diameter of the arteries. Blood pressure rises when arteries become constricted by the action of muscles in the vessel walls. For instance, this happens when release of the stress hormones epinephrine and norepinephrine indicates that muscles need oxygen delivered quickly. In a healthy individual, blood pressure will rise in response to vigorous activity or excitement but return to normal under restful conditions.

Chronic high blood pressure, called **hypertension,** may be caused by arteries that are narrowed by constant psychological stress or by the accumulation of fatty material within the walls of the arteries. The latter condition is called **atherosclerosis** and occurs as a result of aging and a high-fat diet, as well as exposure to tobacco smoke. Hypertension can lead to damage in the walls of blood vessels, and this condition causes the heart to work much harder to push blood through the arteries.

Movement of Materials through the Cardiovascular System

The components of smoke and e-cigarette vapor can exert effects on all organ systems because materials crossing the respiratory surface enter the bloodstream, which then circulates these materials throughout the body.

In humans, the lungs and cardiovascular system are connected by a double circulation system in which blood flows in two distinct but related circuits. The **pulmonary circuit** circulates the blood into the lungs, where oxygen and other components inhaled in the air are picked up and carbon dioxide is released. The pulmonary circuit then returns this blood to the heart, where it enters a second, separate circuit. This **systemic circuit** pumps blood to the rest of the body, where the blood drops off its oxygen and picks up carbon dioxide. Blood flowing through the systemic circuit also picks up and distributes materials from the digestive system and filters nongaseous wastes via the kidneys.

The path of circulation between the heart, lungs, and body is detailed in **FIGURE 20.17**. Here you can see that in the pulmonary circuit, blood from the right side of the heart is pumped to the lungs. While in the lungs, the blood drops off carbon dioxide and picks up oxygen, as well as carbon monoxide and other chemicals in smoke and inhaled vapors. The blood then returns to the heart via a pair of large veins, which empty into the left side of the heart. When the heart first contracts, the initial effect is to force the remaining blood in the atria into the ventricles. A slight lag before the ventricles contract allows them to fill completely with blood, increasing the efficiency of the heartbeat.

Once the left ventricle contracts, the blood is forced at high pressure into the systemic circulation, first to the arteries, which repeatedly branch out and become smaller in diameter, and then into capillary beds. In the capillaries, oxygen and other components in high concentration within the blood diffuse out, and carbon dioxide and wastes diffuse in. The deoxygenated blood then travels to the systemic veins, which empty into the right side of the heart. Contraction of the right ventricle sends blood back into the arteries and capillary beds of the pulmonary circuit.

A single red blood cell carrying a load of oxygen—or carbon monoxide—can travel through the entire cardiovascular system in approximately 1 minute.

VISUALIZE **THIS**

Some children are born with a "hole in the heart," which occurs when a passageway exists between the left and right atria. Referencing the path of blood flow on this figure, explain how a hole in the heart might reduce oxygen delivery to the body tissues.

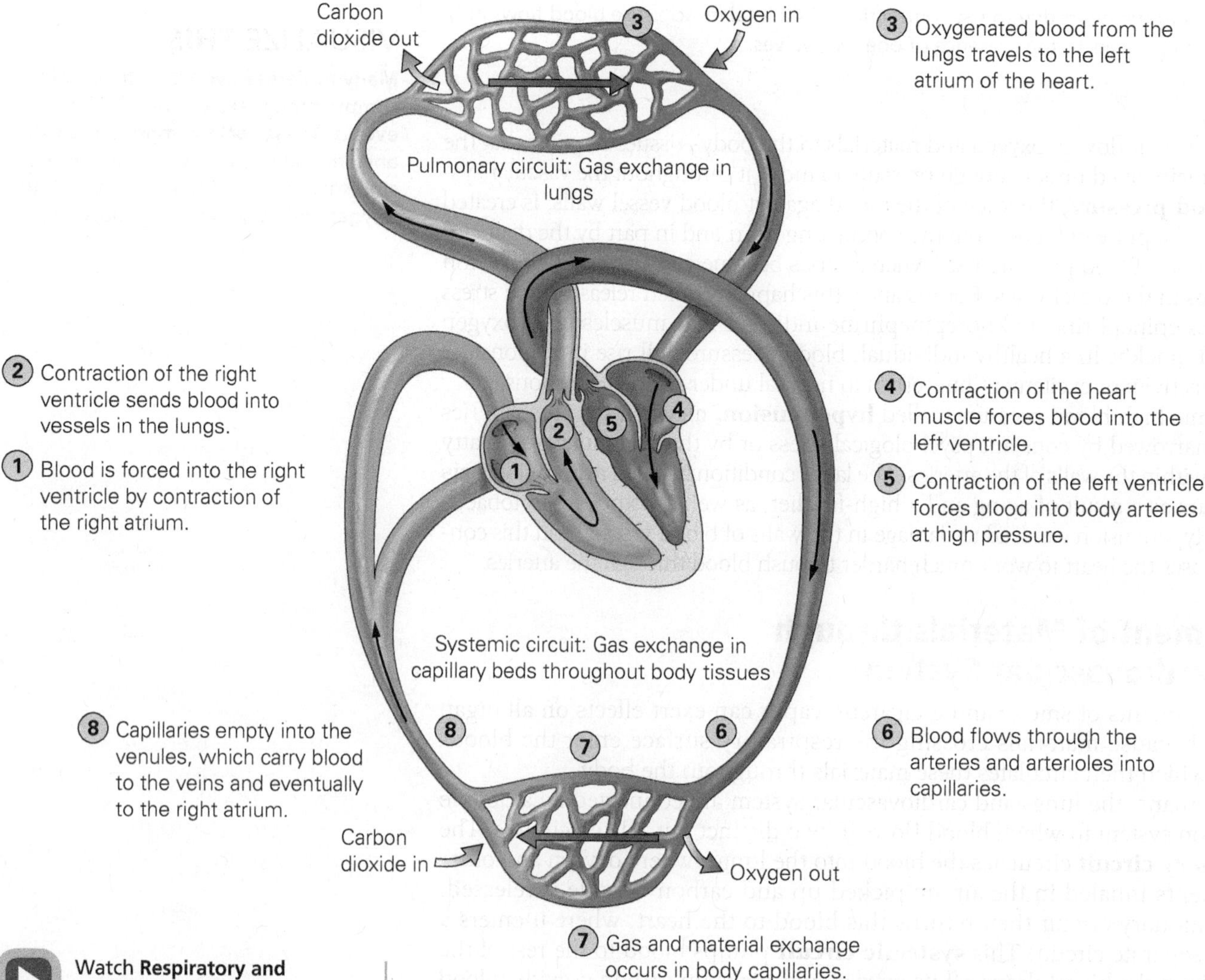

Carbon dioxide out
3
Oxygen in
3 Oxygenated blood from the lungs travels to the left atrium of the heart.
Pulmonary circuit: Gas exchange in lungs
2 Contraction of the right ventricle sends blood into vessels in the lungs.
1 Blood is forced into the right ventricle by contraction of the right atrium.
4 Contraction of the heart muscle forces blood into the left ventricle.
5 Contraction of the left ventricle forces blood into body arteries at high pressure.
1
2
5
4
Systemic circuit: Gas exchange in capillary beds throughout body tissues
8 Capillaries empty into the venules, which carry blood to the veins and eventually to the right atrium.
8
7
6
6 Blood flows through the arteries and arterioles into capillaries.
Carbon dioxide in
Oxygen out
7 Gas and material exchange occurs in body capillaries.

Watch **Respiratory and Cardiovascular System** in Mastering **Biology**

FIGURE 20.17 Connecting the respiratory and cardiovascular system. The lungs are connected to the other organ systems by the heart and blood vessels.

Thus, the chemicals we inhale have almost immediate effects on the body. These quickly occurring effects can cause more severe long-term damage as well.

Smoke and Cardiovascular Disease

Most people believe that lung cancer and other lung diseases are the primary risk associated with exposure to tobacco smoke. In reality, most of the deaths due to smoking result from heart and blood vessel damage, or **cardiovascular disease.**

Some of the cardiovascular damage that results from tobacco smoke appears to be caused by **nicotine,** the primary active ingredient in tobacco and a drug that e-cigarettes often deliver. The nicotine in tobacco smoke and e-cigarette vapor is readily absorbed across the alveoli walls and enters the bloodstream almost immediately after it is drawn into the lungs.

Nicotine is a compound produced by tobacco plants as a natural pesticide—in fact, nicotine sulfate spray is sold commercially for use by organic gardeners to combat aphids and other insects. In high doses, nicotine is toxic to humans and other mammals. In low doses, nicotine interacts with cells in the brain to stimulate the release of epinephrine, which in turn increases heart rate and blood pressure. Some of the brain cells activated by nicotine include those in the brain's "reward pathways," regions with activity that is associated with feelings of well-being and happiness. The reward pathways of the brain are an adaptation that reinforces behaviors important for our survival and reproduction; for instance, these pathways are activated in response to tasting delicious high-fat food or engaging in sexual activity.

By hijacking this pathway, nicotine and other similar drugs such as cocaine and methamphetamine become addictive, driving the user to seek out these drugs again and again to get the same response. Nicotine addiction is part of what keeps smokers smoking, despite the extensive financial and physical costs associated with tobacco use. Nicotine-containing patches, chewing gum, and nasal spray provide low doses of nicotine that can help dull the drug cravings that occur when a smoker quits. As noted earlier, e-cigarettes were first marketed as a safer alternative to smoking. One unfortunate consequence of their popularity is that they seem to have generated new nicotine users: As many as one-third of current e-cigarette users had never before tried smoking tobacco.

Nicotine has effects on other parts of the body besides the brain. It appears to increase production of LDL (low-density lipoprotein), "bad cholesterol," and reduce production of HDL (high-density lipoprotein) or "good cholesterol" (Chapter 3). As a result, individuals exposed to nicotine have higher levels of circulating lipids and thus a higher risk of atherosclerosis, the accumulation of fats and other debris on the interior walls of arteries (**FIGURE 20.18**). Recall that nicotine also may stimulate blood clot formation. A clot can cut off blood flow in arteries narrowed by both atherosclerosis and nicotine-induced hypertension. When a blockage occurs in an artery that brings blood to the brain, it results in a stroke, which causes the death of brain tissue. When a blockage stops blood flow to part of the heart muscle, it can result in the death of that muscle, that is, a **heart attack.** Stroke and heart attack are the primary cardiovascular diseases that cause the death of affected individuals.

Combining the artery-damaging and clot-inducing results of nicotine consumption with the oxygen-robbing increase in inhaled carbon monoxide, exposure to tobacco smoke represents a serious risk to heart function. According to the U.S. Centers for Disease Control and Prevention (CDC), heart disease accounts for about 147,000 deaths per year among active smokers. A smoker has a 200 to 300% greater chance of dying from heart disease than does a nonsmoker. However, the CDC also estimates that 35,000 deaths per year from heart disease among nonsmokers result from exposure to carbon monoxide in environmental tobacco smoke. There is good reason to suspect that the artery-damaging effects of smoke exposure also occur in passive smokers, but the epidemiological studies have not convincingly

(a) Normal artery

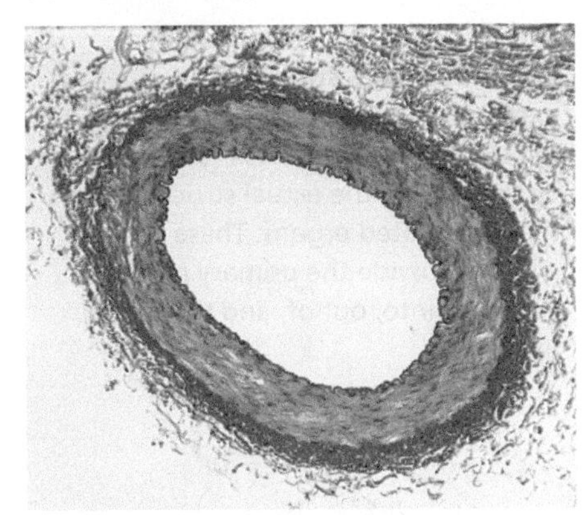

(b) Atherosclerotic artery

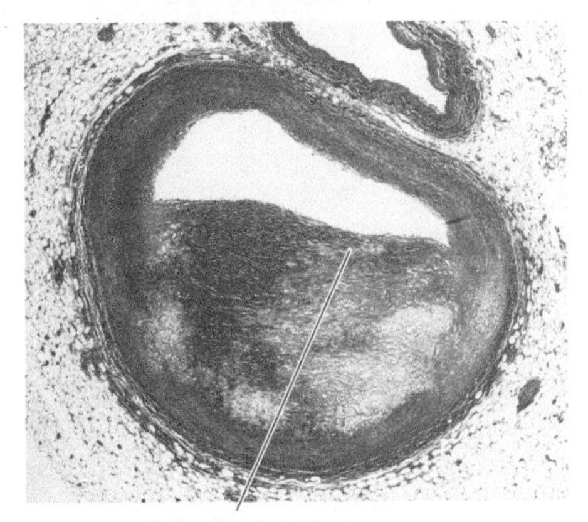

FIGURE 20.18 Atherosclerosis. Fat and cholesterol accumulate in the walls of arteries, reducing their diameter and their ability to carry blood.

demonstrated this link. And research is still minimal on the effects of nicotine delivered via e-cigarettes and the exhaled vapor nonusers are exposed to in their presence.

Unlike the permanent lung changes caused by tobacco smoke, the effects of both nicotine and carbon monoxide on cardiovascular health are more readily reversible. According to the World Health Organization (WHO), the risk of heart disease decreases by 50% in ex-smokers within 1 year of quitting, and by 15 years after quitting, the risk to ex-smokers is the same as that for people who have never smoked.

The Precautionary Principle in Public Health

The links between our respiratory and cardiovascular systems provide a mechanism by which what we inhale affects all the organs of our bodies. These systems in combination with the urinary and digestive systems (discussed in Chapter 19) make up the major pathways by which our bodies interact with the environment (**FIGURE 20.19**). Although many of the links between environmental tobacco smoke and disease have not been "proven beyond reasonable doubt," we know from our own experience that our choices about what we put into our bodies can have a direct effect on our health. The role of the respiratory system in bringing materials into the body should make environmental exposure to tobacco smoke, e-cigarette vapor, and other air pollutants concern us.

As citizens, we are often called on to make decisions in the absence of perfect knowledge. One way to approach these decisions is to use a basic precept of

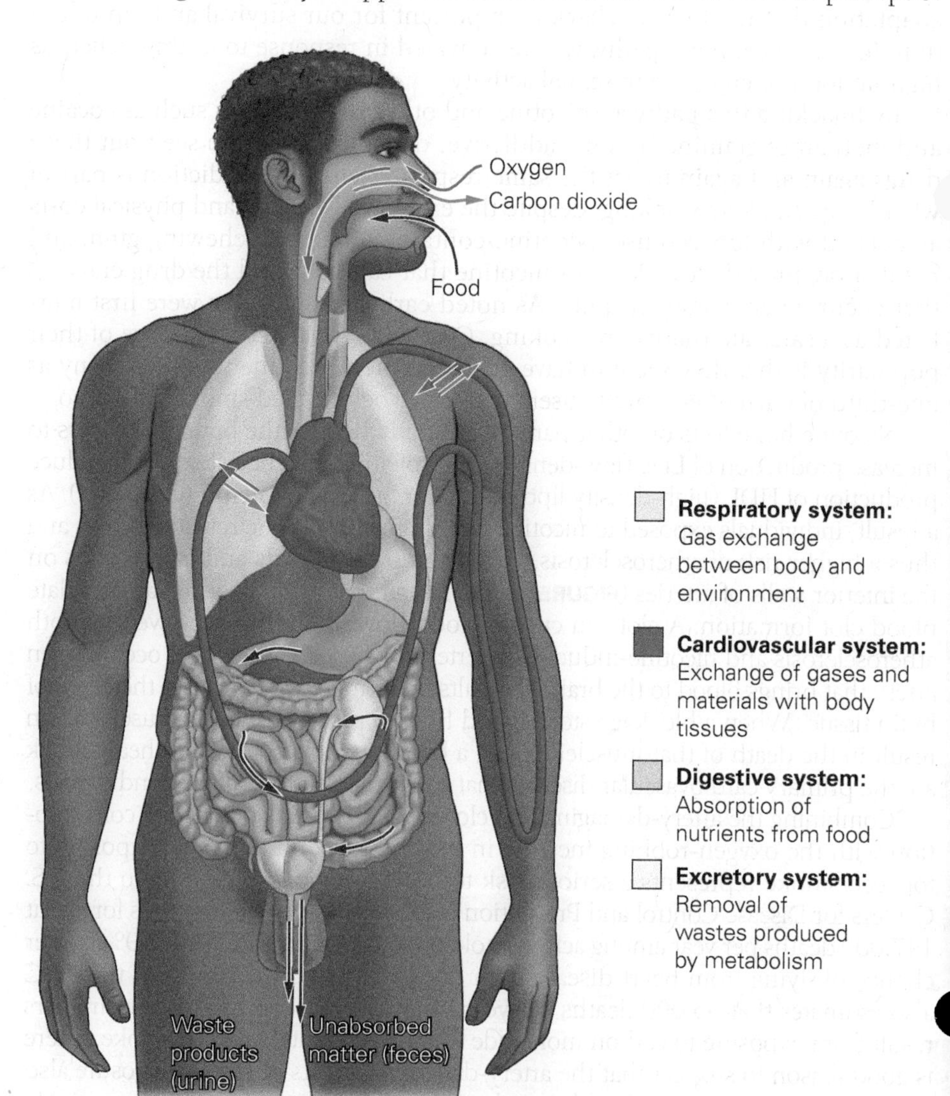

FIGURE 20.19 Respiration, circulation, digestion, and excretion. This figure illustrates the relationship among these four functions by simplifying the actual structure of their associated organs. These organ systems provide the primary routes of materials into, out of, and throughout the body.

public health called the **precautionary principle.** The precautionary principle asserts that in the absence of scientific consensus that an action does not cause harm, the wise public policy is to prohibit that action. Accumulated evidence from both epidemiology and the known risks of active smoking provide support for the hypothesis that ETS is dangerous. Thus, in most of the communities that considered imposing smoking bans, voters decided that the risk of ETS was large enough to justify the possible economic cost to bars and restaurant, so agreed to restrict smoking. Now that these bans have been in place for many years, it appears that clearing the air in formerly smoky eating establishments and entertainment venues has been mostly good for business.

Given their experience with smoking laws, many communities are now making the decision to impose e-cigarette bans. Even though the use of e-cigarettes may help some smokers quit, some ban proponents argue that permitting a public place for vaping attracts new users and thus creates new nicotine addicts. Combined with the unknown effects of environmental e-cigarette vapor on bystanders' health, the precautionary principle suggests that banning vaping in public places is a prudent course of action.

Got It?

1. Platelets in blood are a major component in the process of ____________ ____________.
2. The sound of your heart beat is due to the opening and closing of two different types of heart ____________.
3. Capillaries are the vessels where the majority of material and gas ____________ occurs between tissues and the blood stream.
4. The left side of the heart pumps blood to the ____________; the right side pumps it to the ____________.
5. Nicotine increases ____________ ____________, which increases an exposed individual's risk of heart attack and stroke.

Thanks to sustained public health campaigns, most people understand that smoking carries significant risks. As a result, rates of smoking have declined over the past few decades. However, rates of smokeless tobacco (that is, chewing tobacco and snuff) have remained the same. This phenomenon is due to a widespread belief that

Smokeless tobacco is safer than smoking.

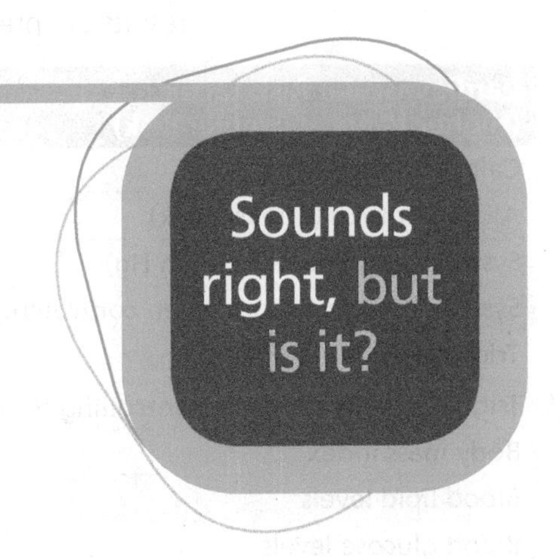

Sounds right, but it isn't.

Answer the following questions to understand why.

1. What dangerous components of tobacco are present even when it is not burned?
2. Smokeless tobacco contains nicotine. What are the effects of nicotine on the cardiovascular system?
3. The components of tobacco get into the bloodstream even if they are not inhaled. (Think back to Chapter 19.) How do materials in the organs of the digestive system, such as the mouth, stomach, and intestines, get into the blood?
4. Reflect on your answers to questions 1–3 and explain why the statement in bold above sounds right, but isn't.

Should marijuana be legalized?

Tobacco and e-cigarettes are legal substances that deliver the brain-altering and potentially addictive drug nicotine; beer, wine, and liquor are legal substances that deliver the drug alcohol. Advocates for marijuana legalization have argued that this substance, which delivers the brain-altering drug tetrahydrocannabinol (THC), should be as legally obtainable as beer and cigarettes. Half of U.S. states have made marijuana legal for medical use, and several have made it legal for "recreational" use. If your state or community has not already considered this question, it is likely to in the near future.

What should I know?

What follows are some smaller questions that need to be resolved to answer the Big Question. Place a checkmark next to the questions that science can answer.

Smaller Questions	Can Science Answer?
Is marijuana addictive?	
How much money is spent currently investigating, arresting, and jailing individuals who use or sell marijuana illegally?	
Does marijuana use lead to an increase in the use of other drugs?	
Should government prohibit people from using mind-altering substances for recreation?	
Does smoking marijuana cause an increase in cardiovascular disease?	
Does legalizing marijuana increase the percentage of teenagers who start using it?	

What does the science say?

Let's examine what the data say about this smaller question:

Does smoking marijuana cause an increase in cardiovascular disease?

Using 15 years of data from more than 3500 young adults, a team of researchers evaluated whether marijuana use was associated with a variety of cardiovascular risk factors. The results are presented in the table below.

Cardiovascular Risk Factor (Higher Value = Greater Risk)	Heavy Marijuana Users (>30% of Days)	Never Users
Calorie intake/day	3365	2746
Alcohol intake (drinks/week)	10.8	3.6
Systolic blood pressure (mm Hg)	116.5	112.7
Systolic blood pressure when controlling for differences in alcohol consumption	Not significantly different	
Triglyceride levels (mg/dL)	100	84
Triglyceride levels when controlling for differences in alcohol consumption	Not significantly different	
Body mass index	Not significantly different	
Blood lipid levels	Not significantly different	
Blood glucose levels	Not significantly different	

1. Describe the results. Does it appear that long-term marijuana use results in an increase in risk factors for cardiovascular disease? Explain your reasoning.
2. Given these data, do you think the smaller question is answered? If not, propose another study that would help answer this question.
3. Does this information help you answer the Big Question? What else do you need to consider?

Data source: N. Rodondi, M. J. Pletcher, K. Liu, S. Hulley, S. Benjamin, and S. Sidney, "Marijuana Use, Diet, Body Mass Index, and Cardiovascular Risk Factors (from the CARDIA Study)," *American Journal of Cardiology* 98, no. 4 (2006): 478–484.

Chapter Review

Mastering Biology

Go to Mastering Biology to access **eText 2.0, Dynamic Study Modules,** and the **Study Area,** where you'll find practice quizzes, BioFlix™ animations, MP3 tutor sessions, current events, and more.

SUMMARY

Section 20.1

List the structures that make up the mammalian respiratory system, and describe the path of air into the body.

- Air from the nose and mouth enters the respiratory system via the pharynx and trachea. Once in the lungs, air flows through bronchi and into alveoli, where gas exchange occurs (p. 434).

Describe the muscles involved in breathing, including the diaphragm, and explain how their movements facilitate air movement into and out of the lungs.

- The diaphragm is a dome of muscle that sits directly below the lungs (p. 435).
- Contraction of the diaphragm increases the volume of the chest cavity, decreasing air pressure and allowing air to flow in. Relaxation of the diaphragm causes the opposite to occur. Muscles surrounding the rib cage and in the abdomen can also cause changes in the chest cavity volume (pp. 435–436).

Explain the role of hemoglobin in gas exchange.

- Hemoglobin reversibly binds to oxygen molecules when oxygen is in high concentrations in the lungs. In active body tissues, the hemoglobin releases some of its oxygen to supply their work (p. 438).
- Hemoglobin binds carbon monoxide, a gas produced by smoldering materials, when it is present at a rate higher than it binds oxygen. Thus, the presence of carbon monoxide in inhaled air actually robs the body of oxygen. (p. 438).

Describe the effect of smoking on the respiratory system.

- Small particles in tobacco smoke cause cell damage in the lungs that can lead to chronic bronchitis. Chronic bronchitis then can lead to emphysema. The tiniest smoke particles are drawn into the alveoli and are difficult to expel and so expose the alveolar cells to carcinogens (pp. 438–440).

SHOW YOU KNOW 20.1 The Heimlich maneuver is a strategy for forcing an obstruction out of the trachea of a person who is choking by forcibly squeezing the abdomen directly under the rib cage. Use your understanding of inhalation and exhalation to explain why this maneuver is effective.

Section 20.2

List the organs and tissues of the cardiovascular system, and describe the function of each.

- Cardiovascular systems in animals consist of a fluid for gas and material exchange (for example, blood), "tubes" for carrying the fluid throughout the body (for example, veins and arteries), and a pump to facilitate fluid flow (the heart) (p. 443).
- Blood is made up of a liquid portion, called plasma, and a cellular portion, consisting of red blood cells, white blood cells, and platelets (pp. 441–442).
- The heart is a double pump consisting of four chambers—two atria and two ventricles. The right-side pump sends oxygen-poor blood to the lungs; the left-side pump sends oxygen-rich blood to the body (p. 442).

Describe how blood moves through the double circulation system of the heart.

- Blood from the lungs flows to the heart, where it is pumped into the systemic circulation. After dropping off a load of oxygen and picking up carbon dioxide in the body tissues, the blood returns to the heart and is pumped back to the lungs (pp. 445–447).

Explain the steps of the cardiac cycle, including the role of the heart's electrical system.

- The heart can generate its own beat via the electrical activities of the SA node. Signals from the SA node are transmitted to the atria, causing these chambers to contract and forcing blood into the ventricles. The signal is then carried by conductive fibers to the ventricles, which contract to force blood out of the heart and into circulation (pp. 442–443).

Describe the effects of the nicotine in tobacco and e-cigarettes on the cardiovascular system.

- Nicotine increases heart rate and blood pressure, putting strain on the heart muscle. Nicotine also increases the production of LDL ("bad cholesterol"), causing atherosclerosis, and increases the likelihood of blood clot formation, causing blockages in blood flow (pp. 447–448).

SHOW YOU KNOW 20.2 Varicose veins are distended veins in which blood pools. They typically form in the legs when the force of gravity on slow-moving blood causes damage to the internal anatomy of the veins. What part of the vein must be damaged to cause this condition?

ROOTS TO REMEMBER

The following roots of words come mainly from Latin and Greek and will help you to decipher terms:

card- and cardio- relate to the heart. Chapter term: *cardiovascular*

-hale means to breathe. Chapter terms: *inhale, exhale*

-itis is used to describe inflammation (of an organ). Chapter term: *bronchitis*

vascul- means vessel. Chapter term: *vascular*

LEARNING THE BASICS

1. Describe how contraction of the diaphragm allows the lungs to fill and how oxygen from inhaled air enters the lungs.
2. Add labels to the figure that follows, which illustrates breathing.

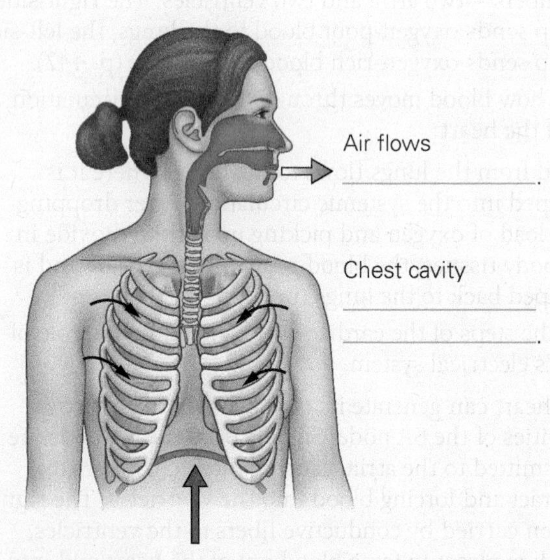

3. Alveoli are ________.

   A. small air sacs at the ends of bronchi in the lungs; B. the respiratory surface in mammals; C. surrounded by a net of capillaries; D. subject to damage because of exposure to tobacco smoke; E. all of the above are correct

4. All of the following statements about hemoglobin are true, except

   A. it is a protein that can bind oxygen; B. it is carried by red blood cells; C. it can bind carbon monoxide; D. it picks up oxygen in the body tissues and releases it in the lungs; E. it contains iron, which is responsible for its red color

5. The blood vessels that carry blood from the heart are called ________.

   A. arteries; B. veins; C. atherosclerosis; D. capillaries; E. ventricles

6. The sound of a heartbeat as heard through a stethoscope is produced by ________.

   A electrical signals from the SA node; B. rush of blood into the ventricles and out of the atria; C. the ribs expanding as the diaphragm contracts; D. the closing of valves between heart chambers and vessels; E. the pulse of blood flowing through the arteries

7. Blood flowing from body tissues to the heart ________.

   A. is next pumped via veins to the rest of the body; B. is next pumped via arteries to the lungs; C. is under relatively high pressure compared with blood leaving the heart; D. returns via millions of capillaries tied directly to the heart's two atria; E. has used up all of its hemoglobin

8. Deoxygenated blood from the body first enters the ________of the heart, and oxygenated blood from the lungs is pumped to the body by the ________.

   A. right atrium, right ventricle; B. right atrium, left atrium; C. left atrium, right ventricle; D. right atrium, left ventricle; E. right ventricle, left atrium

9. Heart attacks ________.

   A. are typically caused by a blockage in blood flow; B. result in the death of heart tissue; C. are as common in smokers who quit 15 years ago as in people who never smoked; D. A and B are correct; E. A, B, and C are correct

10. Which of the following diseases is associated with exposure to tobacco smoke?

    A. heart disease; B. lung cancer; C. stroke; D. asthma; E. all of the above

ANALYZING AND APPLYING THE BASICS

1. Some endurance athletes, such as cyclists, engage in the practice of "blood doping," in which packed red blood cells are transfused into their bloodstream immediately before an event. How might increasing the volume of red blood cells provide an advantage to an athlete? What do you think could be the risks of this practice?
2. Congestive heart failure occurs when damage to the heart results in a severely weakened heart muscle. Given your understanding of heart function, describe some likely symptoms of this condition.
3. As tobacco plants grow, radioactive minerals found in soil stick to the plants' leaves. Minerals found in phosphate fertilizer, such as radium, lead-210, and polonium-210, can also accumulate on the tobacco plant. Radioactive substances on tobacco are not removed as the tobacco is processed to make cigarettes. Therefore, each cigarette

delivers a dose of radiation along with its dose of nicotine. Over the course of a year, someone who smokes about 1.5 packs per day is exposed to a radiation dose equivalent to 300 chest X-rays. How could radioactive minerals in cigarettes produce disease throughout the human body?

GO FIND OUT

1. Exposure to tobacco smoke can cause early death, but it can also cause years or even decades of disabling, chronic health problems such as emphysema and lung cancer. Some people have argued that the best way to reduce the number of smokers is to make smoking, and the health care costs associated with it, more expensive. This strategy entails placing high taxes on cigarettes and other tobacco products and restricting current and former smokers' access to subsidized medical care. Find out what your local tax is on cigarettes and other tobacco products. Do you think the tax is high enough to discourage smoking?
2. Although tobacco causes more cases of long-term disability, alcohol is more deadly to young men and women than cigarettes because of automobile and other accidents. Do you think the success of tobacco and e-cigarette bans means that alcohol will become more restricted? Do you think this is a good policy, given the precautionary principle?

MAKE THE CONNECTION

The science that you learned in this chapter has helped you better understand the real-world example used throughout this discussion. Draw a line from the statement on the left to the science that supports it on the right.

Chemicals produced by tobacco combustion and e-cigarette machines can enter the bodies of both active and passive smokers	Air is drawn into the lungs when the chest cavity enlarges, thanks to contraction of the diaphragm. The brain stem governs this process in response to blood carbon dioxide levels.
Inhaling from a tobacco or electronic cigarette is a conscious act, but breathing in environmental smoke or vapors is generally involuntary for passive smokers.	Blood moves around the body in the cardiovascular system, powered by the pumping of the heart and the control of the arteries and capillary beds.
Exposure to tobacco smoke can damage a smoker's ability to obtain oxygen from the air.	Alveoli in the lungs are where gas exchange between the environment and the bloodstream occurs. Damage to alveoli can reduce the surface area for gas exchange, and certain chemicals in tobacco smoke can interfere with the blood's ability to carry oxygen.
Nicotine delivered via both cigarettes and e-cigarettes can affect distant parts of the body.	Blood vessels can become occluded with cholesterol-containing plaques, which reduce blood flow and increase blood pressure. Parts of a plaque can break off and clog a blood vessel, causing tissue death; the restricted blood flow also puts pressure on the heart muscle itself.
Nicotine can damage blood vessels, increasing the risk of cardiovascular disease.	Gases enter the body through the nose and mouth and travel through the respiratory system to the lungs, where materials are exchanged with the bloodstream.

Answers to **Got It?, Visualize This, Working with Data, Sounds Right, But Is It?, Show You Know,** and **Chapter Review** questions can be found in the **Answers** section at the back of the book.

21 Vaccination: Protection and Prevention or Peril?

People living in close proximity can transmit infectious diseases to each other.

Immune System, Bacteria, Viruses, and Other Pathogens

Deciding whether to be vaccinated against the flu is often one of the first health-care decisions many new college students will have to make without parental input. Many modern vaccines are made using small parts of the organism that will trigger a response from your immune system. This way, when you actually encounter the intact disease causing organism, your immune system is better prepared to fight against it.

Beginning in the late fall months, university health services personnel attempt to vaccinate as many students as possible. Many students, hoping that the few minutes and the mild discomfort spent on prevention will save them from losing several days to this illness, get vaccinated. Some students, however, are reluctant to comply. They may not want to take the time to get vaccinated, or may choose to take their chances, hoping they won't catch the flu. Other students refuse to get vaccinated because they believe the flu vaccine might make them sick or otherwise harm their health.

To help you make an informed decision about the flu vaccine, and about other recommended vaccinations, we will first learn about the causes of diseases that vaccines attempt to prevent.

Which diseases did your parents vaccinate you against?

Infectious diseases often have a devastating impact on the weakest members of society.

21.1 Infectious Agents

Tissues can be invaded and undergo damage induced by infectious agents. Once acquired, some infections can be transmitted from one individual to another, spreading disease. Such communicable or transmissible diseases differ from genetic diseases in that they are caused by organisms rather than by malfunctioning genes—although malfunctioning genes can make an organism more susceptible to infection.

Disease-causing organisms are called **pathogens.** There are many different types of pathogens, including viruses and bacteria, as well as numerous eukaryotic pathogens. When a pathogen can be spread from one organism to another, it is said to be **contagious.** Organisms that depend on other organisms to obtain nutrients and shelter required for growth and development while contributing nothing to the survival of the host organism are **parasites.** Most infectious agents are parasites. Some organisms can cause an infection in an individual, but they are not contagious if the infected individual cannot pass the infection to other individuals.

path- and patho- relate to disease.

Organisms that can be seen only when viewed through a microscope are called *microscopic organisms,* or **microbes.** The most common infectious microbes are bacteria and viruses.

Bacteria

Bacteria are a diverse group of tiny, single-celled prokaryotic organisms that are ubiquitous in the environment and in your body. In fact, the majority of cells in your body are not human cells, they are bacterial cells! Bacteria are commonly rod shaped (bacilli), spherical (cocci), or spiral (spirochetes) (**FIGURE 21.1**).

VISUALIZE THIS

Note that the bacterial cell does not contain any membrane-bound organelles such as a Golgi apparatus or an endoplasmic reticulum. Would you expect that this prokaryotic cell could have ribosomes? Why or why not?

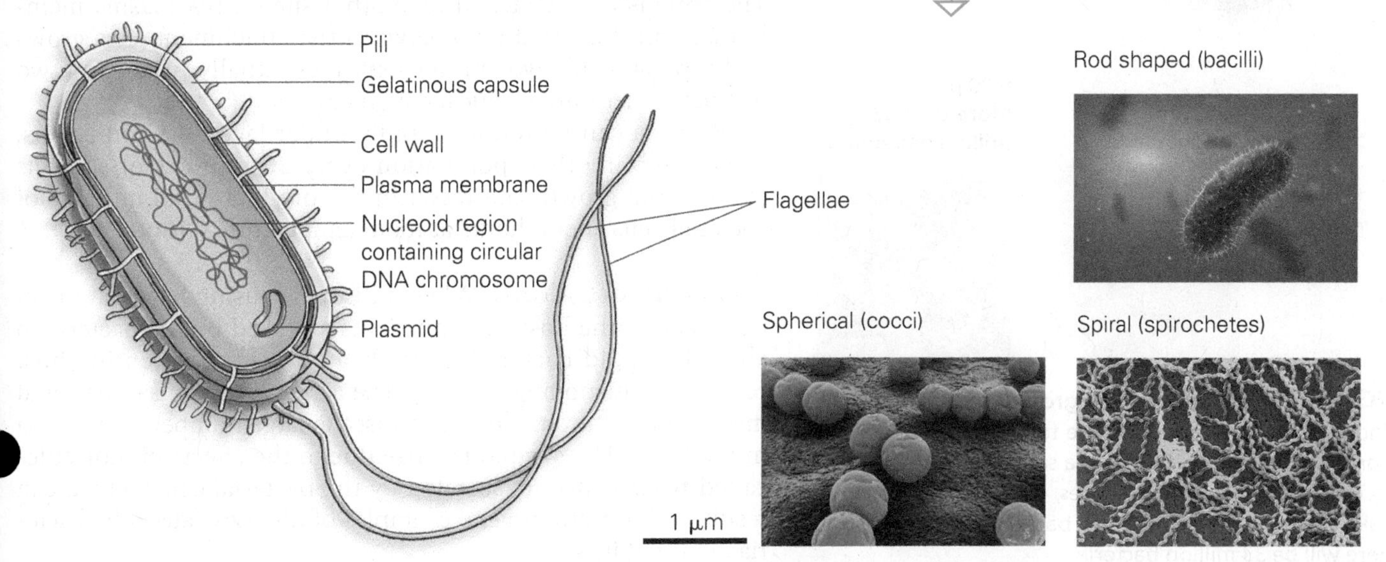

FIGURE 21.1 Bacteria. The drawing shows the structure of a typical bacterium. The photos show the three characteristic bacterial shapes.

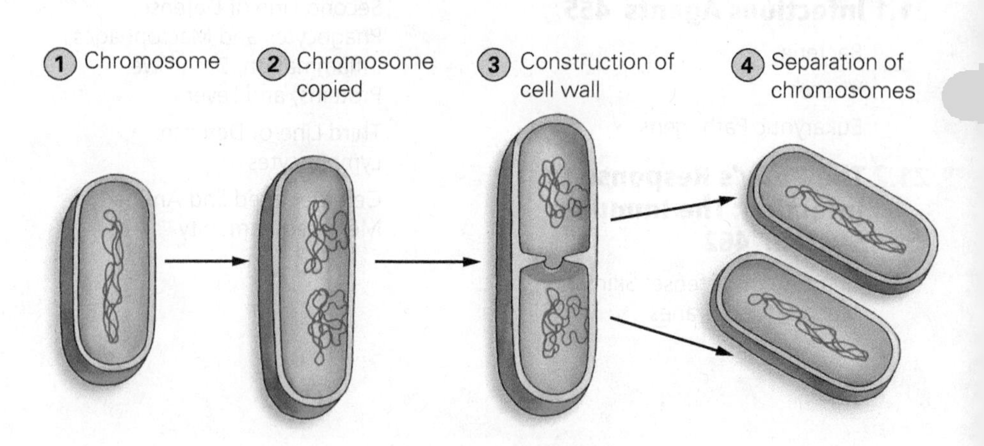

FIGURE 21.2 Binary fission. (1) The bacterial cell begins with one copy of the circular DNA chromosome that is wound around itself. (2) The chromosome is copied, and each copy is attached to the plasma membrane. (3) Continued growth separates the two chromosomes. The plasma membrane pulls inward in the middle, and a new cell wall is constructed. (4) Daughter cells separate.

Although we often think of bacteria only in their roles as disease-causing agents, many different types of bacteria are beneficial to living organisms and the environment. For example, bacteria on your skin fight invaders and boost immunity, and bacteria in the environment help break down waste and provide nutrients to other organisms.

Like other prokaryotes, bacteria lack a nucleus. Instead, they have a nucleoid region containing a double-stranded, circular DNA chromosome. In addition to the large DNA chromosome, bacteria may also contain small, circular extrachromosomal DNA (DNA that is separate from the chromosome) molecules called plasmids. Plasmids carry a few genes that are helpful to the survival of the bacteria and can be passed from one bacterium to another.

WORKING WITH DATA

How many bacteria would be present at 1:00? Sketch a line graph approximating the growth rate of bacteria over 8 hours. Add a second line to the graph approximating the growth of bacteria, when refrigerated, which slows growth, over the same time period.

Most bacterial cells are surrounded by a cell wall that provides rigidity and protection; it is composed of carbohydrate and protein molecules The cell walls of many bacteria are surrounded by a gelatinous capsule, which helps bacteria attach to cells within tissues they infect. This capsule also allows bacteria to escape destruction by cells of the immune system. Bacteria also may have one or more external flagella to aid in movement and pili, which help some bacterial cells attach to each other and pass genes.

Bacteria reproduce by a process called **binary fission** (**FIGURE 21.2**) in which a single bacterial parent cell gives rise to two genetically identical daughter cells. When a bacterial cell divides by binary fission, the single circular chromosome that is attached to the plasma membrane inside the cell wall is copied. The copy is then attached to another site on the plasma membrane, and the membrane between the attachment sites grows and separates the two chromosomes, eventually producing two distinct, genetically identical daughter cells.

Bacteria can reproduce rapidly under favorable conditions, some doubling their population every 20 minutes or so. This exponential growth can result in the production of millions of bacterial cells after 8 hours at room temperature (**FIGURE 21.3**).

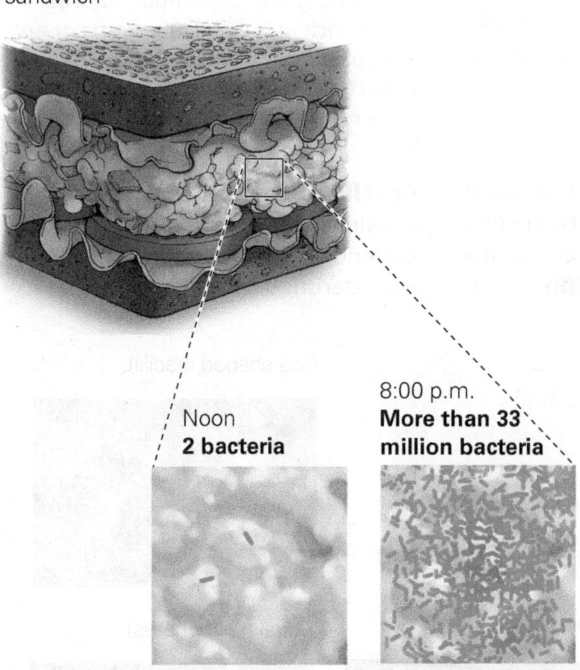

FIGURE 21.3 Exponential growth of bacteria. One kind of bacterium that can cause food poisoning reproduces every 20 minutes. If a sandwich with 2 bacteria is left out for 20 minutes, each bacterial cell will make a copy of itself, yielding 4 bacteria. Within 8 hours, there will be 33 million bacteria.

Bacterial Infections. When a disease-causing bacterial infection occurs, the host cell's nutrients are used by the bacteria to allow for rapid bacterial multiplication, thus preventing host cells from functioning properly. The actual symptoms of bacterial infection result from more than just the large numbers of bacteria in the body. The symptoms arise due to the effects of molecules called **toxins** that are secreted by the bacterial cells, which can result in cell death. Several examples of diseases caused by bacteria are listed in **TABLE 21.1**.

TABLE 21.1 Examples of diseases caused by bacteria.

Disease	Disease Information
Botulism 	*Clostridium botulinum* produces a powerful toxin that acts on the nervous system, causing paralysis in the respiratory and muscular systems. Botox—injections of the toxin produced by this bacterium—temporarily paralyzes facial muscles to prevent drooping and the appearance of wrinkles. This soil-dwelling bacterium can adhere to vegetables. Improper sterilization allows the bacteria to proliferate and produce a gas that causes telltale bulges in canned foods.
Escherichia coli infection 	*E. coli* bacteria live in the intestines of humans and other organisms. Disease can occur when human sewage or animal waste contaminates water supplies. *E. coli* infection can also occur from eating undercooked, contaminated beef or drinking contaminated, unpasteurized milk.
Staphylococcus (commonly known as staph infection) 	*Staphylococcus* bacteria that often reside on human skin and mucous membranes can invade and destroy tissue through an opening in the skin. *Staphylococcus aureus* can cause toxic shock syndrome when tampons are left in place long enough to allow bacterial growth. Methicillin-resistant *S. aureus* (*MRSA*) infection is deadly because resistances to antibiotics have evolved in these bacteria. Healthy young athletes can succumb when resistant bacteria are present on locker room floors or shared equipment enter the skin through cuts or scrapes.
Tuberculosis 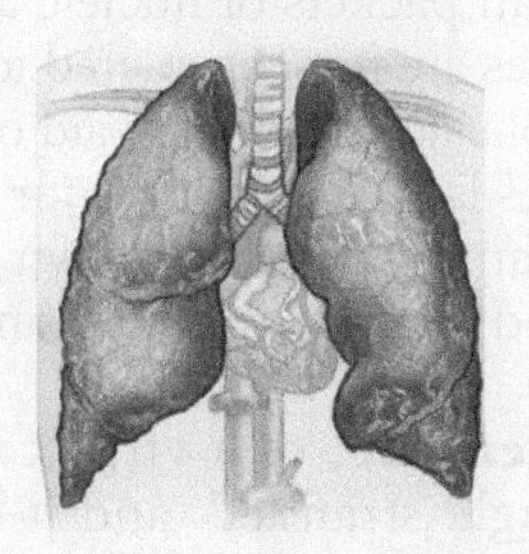	Tuberculosis is caused by the bacterium *Mycobacterium tuberculosis* (see Chapter 12). This infectious agent is spread when a person with the infection coughs or sneezes. Once inside the body, the bacteria take up residence in the lungs. The immune system responds by walling off the bacteria in a manner much like the formation of a scab. The bacteria can live within the walls for years, during which time the infection is not considered to be active, and it is not transmissible. If the bacteria escape the immune system's attempts at control, the infection becomes active.

Many bacterial infections are routinely treated with **antibiotics.** Unfortunately, the days of easily treatable bacterial infections may soon be over because some of the bacteria that cause diseases—including tuberculosis, ear infections, and gonorrhea—have become resistant to the antibiotics that used to kill them.

Antibiotic resistance develops as a result of natural selection in a bacterial population. Variation exists within any population, including within a population of bacterial cells. Some bacteria, even before exposure to an antibiotic, carry genes that enable them to resist the antibiotic. You can develop a drug-resistant infection if the resistant bacteria are selected for in your body or when you contract the resistant bacteria from someone else.

Vaccines against Bacterial Diseases. Immunizations against many bacterial diseases are recommended by the U.S. Department of Health and Human Services Centers for Disease Control and Prevention (CDC). In fact, children are supposed to have proof of vaccination to attend public school. Young children typically receive a series of injections known as the DPT vaccine, containing vaccines against three bacterial diseases. The first letter in the name of this vaccine is for diphtheria, a respiratory infection that has been eradicated in the United States but has a 10% death rate in countries that don't vaccinate. Continued vaccinations are necessary in the United States, despite the eradication of this disease, because travelers can bring the disease back into the country. The DPT vaccine also protects against pertussis, another respiratory tract infection. This potentially deadly infection results in violent coughing that makes breathing extremely difficult. The whooping sound emitted during inhalation led to the disease also being called whooping cough. The T in DPT stands for tetanus, a disease caused by a toxin produced by the *Clostridium tetani* bacterium. This bacterium can persist as a resistant spore in the soil. Puncture wounds produced by objects carrying contaminated soil allow the bacteria into tissues where it can grow, producing a toxin that has devastating, often fatal, effects on the nervous and muscular systems.

There is also a vaccine against bacterial meningitis, a disease caused by a bacterium that can cause inflammation of the membranes surrounding the brain. Around 10% of people who have meningitis will die, and many more will suffer permanent disability. Because this disease can be spread when an infected person sneezes or coughs, outbreaks of meningitis are most common in places where many people live in close quarters, like university dormitories. For this reason, the CDC recommends that the vaccine against meningitis be given to adolescents and some colleges require proof of vaccination. If you were not vaccinated against meningitis, you may want to discuss this with your doctor.

Viruses

Many of the CDC-recommended vaccines protect against viral diseases. **Viruses** (**FIGURE 21.4**) are little more than packets of nucleic acid (DNA or RNA) surrounded by a protein coat. Viruses are not considered to be living organisms for two reasons: (1) They cannot replicate without the aid of a host cell and (2) they are not composed of cells. Viruses lack the enzymes for metabolism and contain no ribosomes. Therefore, they cannot make their own proteins. They also lack cytoplasm and membrane-bound organelles. They cannot produce toxins like bacteria can.

The genetic material, or **genome,** of a virus can be DNA or RNA; it can be double-stranded or single-stranded, and it can be linear or circular. For example, the herpes virus has a double-stranded DNA genome, and the poliovirus has a single-stranded RNA genome. The genes of a virus can code for the production of all the proteins required to produce more viruses inside a host cell. RNA viruses called **retroviruses** are packaged with the enzyme reverse transcriptase, which synthesizes DNA from the RNA genome rather than the reverse as in "normal" genetic transcription, where RNA is copied from DNA.

The protein coat surrounding a virus is called its **capsid.** Many of the viruses that infect animals have an additional structure outside the capsid called the **viral envelope.** The envelope is derived from the cell membrane of the host cell and may contain additional proteins encoded by the viral genome. Viruses that are surrounded by an envelope are called **enveloped viruses.**

Viral Infection. Infection by an enveloped virus occurs when the virus gains access to the cell by fusing its envelope with the host's cell membrane (**FIGURE 21.5**). An unenveloped virus uses its capsid proteins to bind to receptor proteins in the plasma membrane of a host cell. Some capsid proteins function as enzymes that digest holes in the plasma membrane, thereby allowing the viral genome to enter cells. The cell begins to express the viral genome after the capsid is removed.

VISUALIZE **THIS**

Based only on structures shown in this figure, can you guess which parts of the virus are most likely to help it attach to a cell?

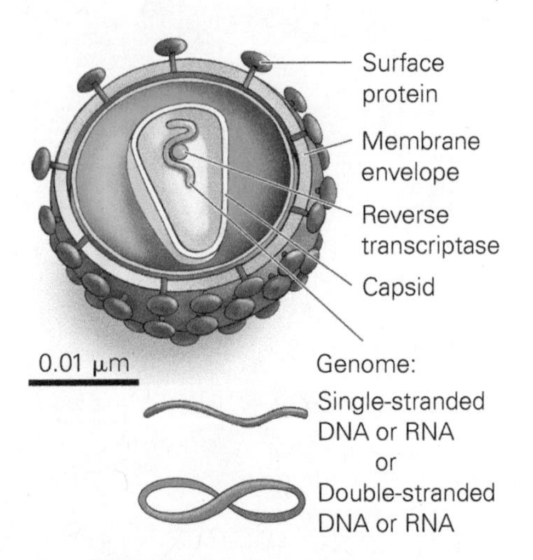

FIGURE 21.4 Viral structure. Viruses are composed of genetic material surrounded by a protein coat. Some viruses, including the one shown, are also surrounded by an envelope.

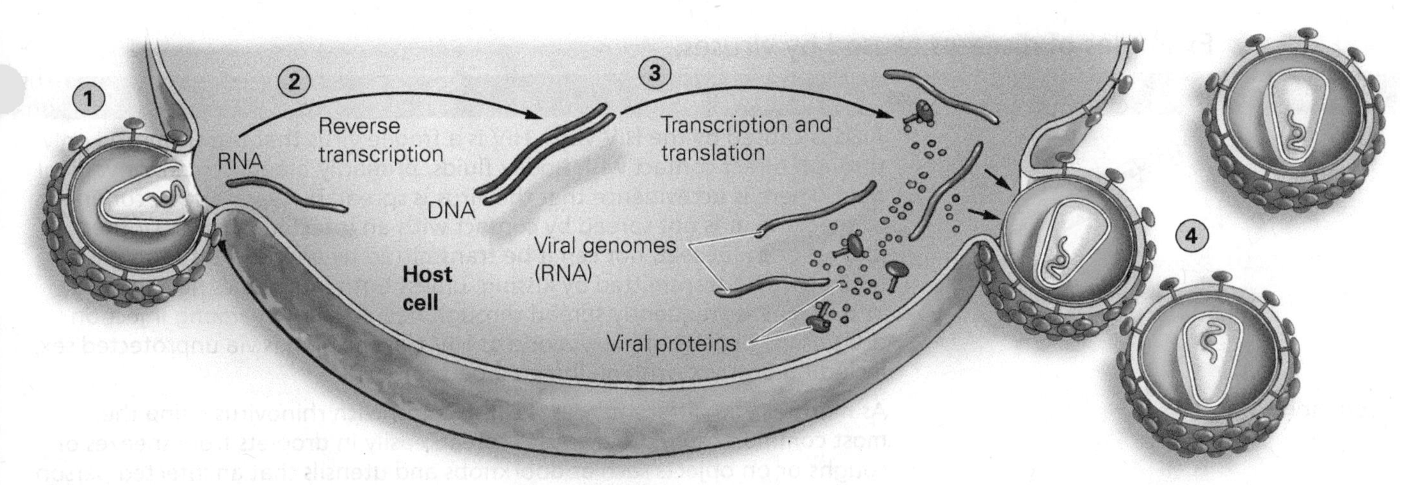

FIGURE 21.5 Viral replication. (1) Viral membrane fuses with the host cell; the capsid is removed, and the genome enters. (2) In the case of some RNA viruses, such as HIV, the virus that causes AIDS, the viral genome is used to synthesize many copies of DNA by the reverse transcriptase enzyme. (3) The DNA is transcribed and translated by the host cell. More viral proteins and copies of the viral genome are produced. (4) Once assembled, new viruses leave the host cell and infect other cells.

Regardless of how it enters, after the genome enters the host cell, the infection continues when the virus replicates itself. First, the genome is copied, then the virus uses the host-cell ribosomes and amino acids to make viral proteins for building new capsids and synthesizing some of the envelope proteins. Once assembled, the new viruses exit the cell, leaving behind some viral proteins in the host's cell membrane. The newly constructed viruses then move to other cells, spreading the infection.

Some viruses, called **latent viruses,** enter a state of dormancy in the body. During that time, the virus is not replicating. Herpes viruses, for example, which cause outbreaks of painful sores and blisters on the mouth and genitals, can undergo long periods of dormancy, during which time they are present in the body but do not cause symptoms. Herpes simplex is sexually transmitted, as are many viruses. TABLE 21.2 (page 460) lists examples of diseases caused by viruses.

Vaccines against Viral Diseases. Newborns are vaccinated against hepatitis B, a sexually transmitted virus, because this disease is easily passed from mothers to their babies during delivery. If the virus gains access to the baby, the vaccine can prevent viral replication and stop a full-blown infection from developing. When an infection is allowed to develop, the potentially fatal hepatitis virus attacks the liver and prevents it from functioning properly. Infants in the United States are also vaccinated against rotavirus, which causes severe diarrhea, fever, abdominal pain, and dehydration, and results in half a million deaths worldwide in children under the age of 5. Young children are vaccinated against polio, a disease that can cause paralysis and death. They are also given the MMR triple vaccine, a vaccine against measles, a severe and sometimes deadly rash and fever disease still common in many parts of the world; mumps, a swelling of the salivary glands that can, though rarely, cause deafness or reproductive difficulties; and rubella, a disease that causes fever and rash and can cause birth defects if a pregnant woman acquires the infection.

The MMR triple vaccine is given at around 15 months of age, which happens to coincide with the age when the first symptoms of autism appear. This correlation has led many parents of autistic children to assume that the vaccine can cause autism. However, many scientific studies have been performed to test this hypothesis, and no evidence has emerged to support it. One study published in 1998 used falsified data to show a fraudulent correlation between the vaccines and autism. This study has been retracted by the journal that published it, and the study's author has been banned from practicing medicine.

TABLE 21.2 Examples of diseases caused by viruses.

Disease	Disease Information
AIDS	AIDS is caused by the HIV virus. HIV is a fragile virus that is transmitted only through direct contact with bodily fluids, primarily blood, semen, or vaginal fluid. There is no evidence that the virus is spread by tears, sweat, coughing, or sneezing. It is not spread by contact with an infected person's clothes, phone, or toilet seat, nor can it be transmitted by an insect bite. And it is unlikely to be transmitted by kissing, unless both individuals have lesions or cuts. HIV is frequently spread through needle sharing among injection drug users, but the primary mode of HIV transmission is via unprotected sex, including oral sex, with an infected partner.
Common cold	As many as 200 different viruses cause colds, with rhinovirus being the most common cause. Cold viruses spread easily in droplets from sneezes or coughs or on objects such as doorknobs and utensils that an infected person contaminated.
Ebola 	Humans become infected when they handle infected wild animals or are exposed to the secretions of such animals. Human to human transmission is via direct contact with bodily fluids of an infected person, even if that person is recently deceased. Infection also occurs by exposure to bodily fluids present on medical supplies used to treat an infected person.
Mononucleosis 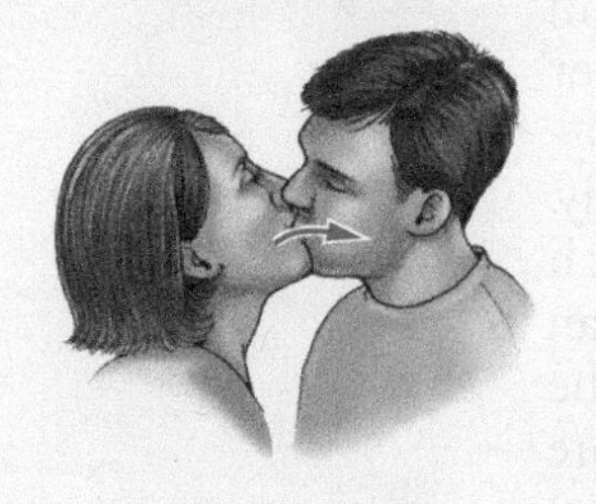	Mononucleosis is caused by the Epstein-Barr virus, a disease that commonly affects young adults. The symptoms—fatigue, weakness, fever, headache, and sore throat—can be severe and may last a month or two. The virus can be transmitted in saliva when kissing, and by coughing and sneezing or by touching an object contaminated by an infected person.
Zika 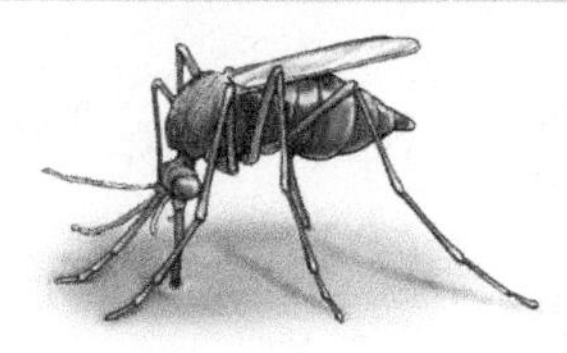	The Zika virus is transmitted when an infected mosquito bites a human. Infected humans can infect others via sexual activity. The virus can also be transmitted from a pregnant mother to her unborn child, resulting in a failure of the baby's brain to develop properly.

More recently, the CDC began recommending childhood vaccination against chicken pox, a skin rash of blister-like lesions that usually resolves itself but can cause swelling of the brain, or **encephalitis,** and pneumonia, typically when older children come down with the disease.

enceph- and **encephalo-** mean brain.

Of particular interest to college students is the vaccine against some strains of human papillomavirus (HPV), a sexually transmitted disease responsible for most cases of cancer that effects the lower third of the uterus, or cervix, in women. HPV can also cause anal and throat cancers in both men and women. Proper condom use can help lower the likelihood of transmitting this virus, but even with faithful condom use, the virus can spread.

Even with good medical care, 37% of women with cervical cancer die from the disease. In poor countries, the toll is even more dramatic—cervical cancer is the leading cause of cancer deaths in the developing world, killing more than 300,000 women every year. If a larger percentage of the population were vaccinated against HPV, most of these cervical cancer deaths could be prevented.

(a) Population where no one has been immunized

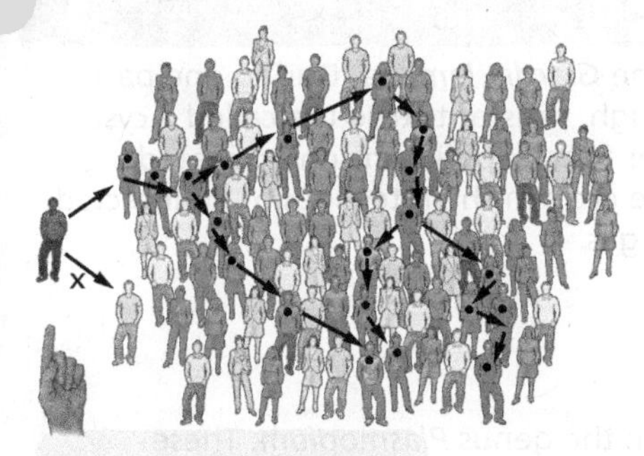

The infection can spread rapidly in the population.

(b) Population where most people have been immunized

With few susceptible people, the disease cannot spread.

Infected with disease | Susceptible to disease | Vaccinated | Genetically resistant to disease | Newly infected

FIGURE 21.6 Herd immunity. Not all members of a population need to be immunized to stop the spread of infectious disease, but the majority of the "herd" does. As long as enough individuals are protected, a person with the infection will transmit it to fewer people, and epidemics are less likely.

VISUALIZE THIS

If large numbers of people refuse vaccinations, what effect may that have on the likelihood of an epidemic occurring?

Watch **Herd Immunity** in Mastering **Biology**

The success of vaccination programs lies not with providing immunity to every susceptible individual but with reducing the total number of susceptible individuals so that the disease cannot easily spread. **Herd immunity** is an indirect way to provide protection to those individuals—like pregnant women, infants, and the immunocompromised—who cannot be vaccinated. When most of the community, or "herd," is vaccinated, the spread of the disease is disrupted and epidemics, widespread occurrences, are less likely (**FIGURE 21.6**). Individuals who choose not to get vaccinated are putting the weakest members of the "herd" at risk.

Effectively reducing the toll of cervical and other HPV caused cancers requires that as many potential carriers of HPV be immunized as possible because this will help lower the population's infection rate. The HPV vaccine also prevents infection by strains of the virus that cause genital warts. which may make it more attractive to men. If you have not had this vaccination you may want to discuss this vaccination with your physician.

The CDC also recommends annual vaccination against the flu for children and adults. The flu shot contains small parts of three or four strains of flu virus, and is designed to stimulate your body's immune system, giving it a head start against your next exposure to influenza.

Got It?

1. ________ are nonliving pathogens composed of a genome and capsid.
2. ________ are simple single-celled organisms that can be beneficial or pathogenic.
3. Some viruses are surrounded by a membrane called a(n) ________.
4. Organisms that depend on other organisms to survive but contribute nothing to the survival of the host organism are ________.
5. A virus that causes an initial infection and then later becomes dormant is called a ________ virus.

Eukaryotic Pathogens

Although many infectious diseases are caused by bacteria or viruses, some are also caused by eukaryotic organisms. Eukaryotic pathogens, such as single-celled protozoans, along with some worms and fungi, cause tremendous human suffering and death worldwide. Protozoans are often spread by water and food contaminated with animal feces. Most worms gain access to the body because of poor sanitation or by consumption of contaminated meat or fish.

TABLE 21.3 Examples of diseases caused by eukaryotic pathogens.

Disease		Disease Information
Giardiasis	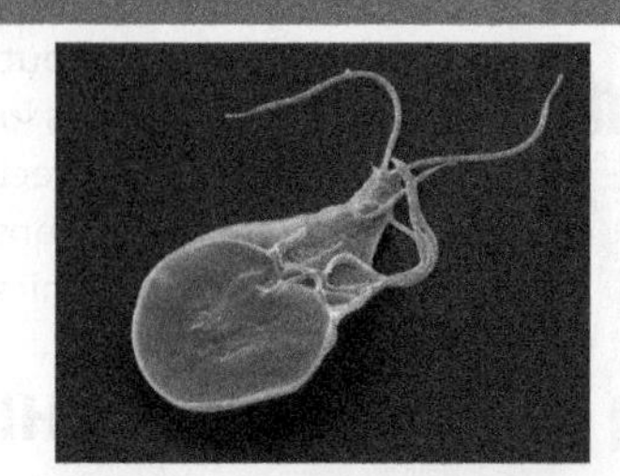	Giardiasis is caused by the waterborne *Giardia lamblia*. During one part of its life cycle, *Giardia* exists as a tough, resistant structure called a cyst. *Giardia* cysts enter bodies of water or food supplies in contaminated feces. The organism completes its life cycle in the intestines of an infected person, causing severe diarrhea and gas.
Malaria	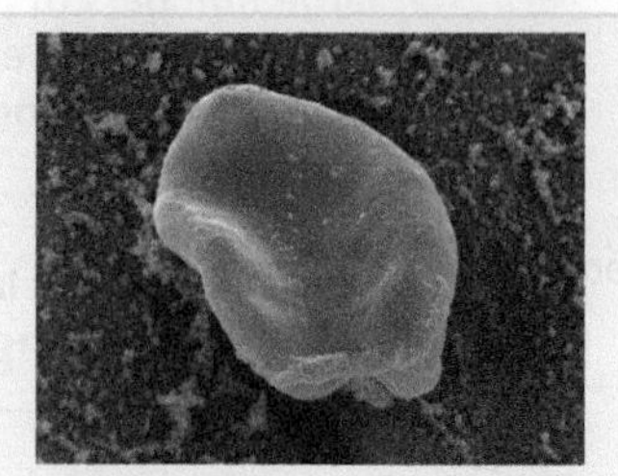	Malaria is caused by protozoans from the genus *Plasmodium*. These protozoans are transmitted when infected mosquitoes bite humans. This disease kills close to 2 million people annually.
Schistosomiasis		Freshwater snails carry this parasitic disease, caused by flatworms. Humans affected by this disease live around contaminated water, commonly found in Asia, Africa, and South America.
Tapeworm	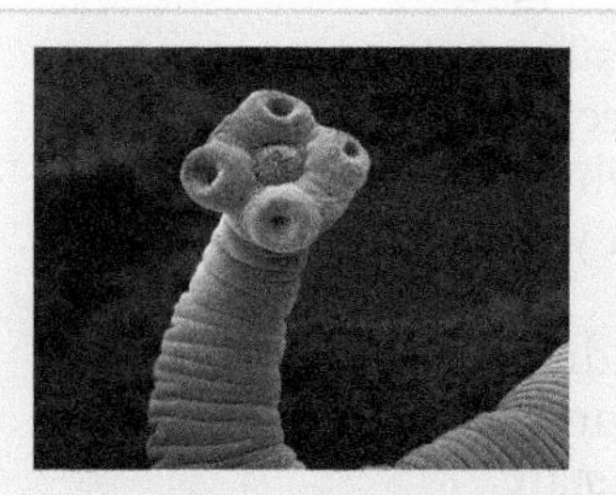	Tapeworm infections are usually caused by ingesting infected raw or undercooked pork or beef. This worm lives in the intestines and can cause abdominal pain and diarrhea.

Worms that take up residence in intestines or other organs can cause extensive damage to internal tissues. Fungi cause diseases of the skin and internal organs. They damage tissue by secreting digestive enzymes into it and absorbing the products of this digestion. Examples of diseases caused by eukaryotic pathogens are found in **TABLE 21.3**. There are no vaccines routinely given in the United States against diseases caused by these pathogens.

21.2 The Body's Response to Infection: The Immune System

In humans and most vertebrates, it is the job of the immune system to protect against infection. The immune system consists of three different lines of defense (**FIGURE 21.7**). The first line of defense helps prevent access to the body, and the remaining two lines of defense operate if the pathogen gets past the first line.

First Line of Defense: Skin and Mucous Membranes

The skin and mucous membrane secretions make up the first line of defense against pathogens. These external physical and chemical barriers are nonspecific defenses; that is, they do not distinguish one pathogen from another.

Organisms living on the body's surface are kept out by the skin. In addition to being a physical barrier, the skin sheds, taking pathogens with it, and because skin has a low pH it can help to repel microorganisms. Likewise, glands in the skin secrete chemicals that slow the growth of bacteria. For example, tears and saliva contain enzymes that break down bacterial cells, and earwax traps microorganisms.

Mucous membranes lining the respiratory, digestive, urinary, and reproductive tracts also secrete mucus that traps pathogens, which are coughed or sneezed away from the body, destroyed in the stomach, or excreted in the urine or feces. Digestive secretions, including acids, kill many microorganisms that gain access to the stomach. Vomiting can also rid the body of toxins or infectious agents. Pathogens that are able to evade the first line of defense next encounter an internal, second line of defense.

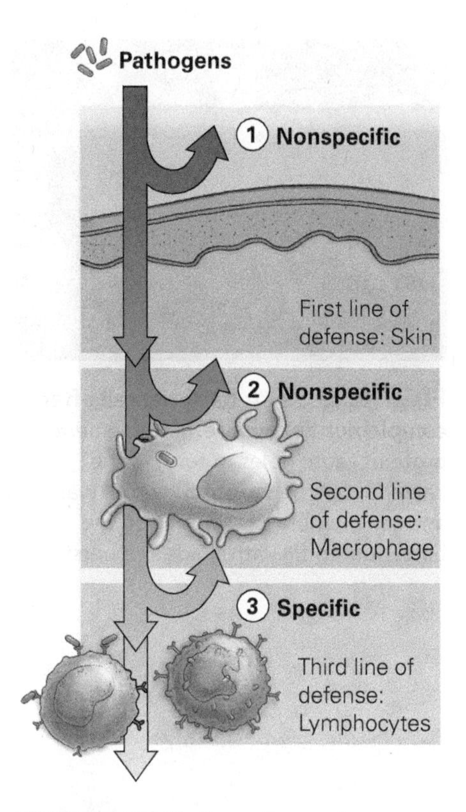

FIGURE 21.7 Three lines of defense. (1) The skin and mucous membranes serve as a nonspecific barrier to infection. (2) Macrophages attack pathogens that gain access to the body and cause an immune response. (3) Lymphocytes initiate a specific response by targeting specific pathogens and attempting to contain them.

Second Line of Defense: Phagocytes and Macrophages, Inflammation, Defensive Proteins, and Fever

If a pathogen is able to get past the first line of defense, the internal second line of defense can often stop the infection. Participants in the second line of defense are also nonspecific in that they do not target particular pathogens.

Phagocytes and Macrophages. Phagocytes are white blood cells that indiscriminately attack and ingest invaders by engulfing and digesting the invader. Macrophages are one type of phagocytic white blood cell that moves throughout the lymphatic fluid, engulfing dead and damaged cells, a process called phagocytosis (**FIGURE 21.8**). Enzymes inside the macrophage help break the invader apart. Macrophages can also clean up old blood cells, dead tissue fragments, and other cellular debris. They release chemicals that stimulate the production of more white blood cells. Much of the destruction of the offending cells occurs in the lymph nodes.

When an invader is too big to be engulfed (protozoans and worms, for example), other white blood cells cluster around the invader and secrete digestive enzymes that irritate or may even destroy the organism.

Additional white blood cells that circulate through the blood and lymph destroying invaders are called natural killer cells. These nonspecific cells attack tumor cells and virus-invaded body cells on first exposure. They release chemicals that break apart plasma membranes of their target cells, causing them to burst. Fluid, dead cells, and microorganisms accumulate at the site of the infection, producing pus. If pus cannot drain, the body may wall it off with connective tissue, producing an abscess. Another component of the second line of defense is the inflammatory response.

Inflammation. Whenever a tissue injury occurs, an inflammatory response begins. Damaged cells release chemicals that stimulate specialized cells to release **histamine.** Histamine promotes increased size of blood vessels, or vasodilation, near the injury. As more blood and cells arrive at the site of infection to speed cleanup and repair, redness, warmth, and swelling occur. The extra blood flow also brings oxygen and nutrients required for tissue healing.

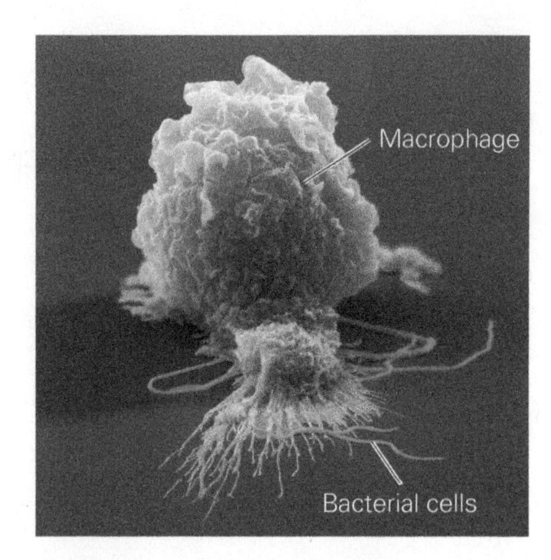

FIGURE 21.8 Phagocytosis. A macrophage attacks bacterial cells.

Defensive Proteins. In addition to white blood cells, some proteins act as nonspecific defenders. **Interferons** are proteins produced by virus-infected body cells to help uninfected cells resist infection. When infected cells die, they release interferons that bind to receptors on uninfected cells and stimulate the healthy cells to produce proteins that inhibit viral reproduction.

The **complement system** is a part of the immune system that complements or enhances the ability of the immune system to fight off invaders. There are more than 20 different defensive proteins that circulate in the blood. These proteins coat the surfaces of microbes, making them easier for macrophages to engulf. Complement proteins can also poke holes in the membranes

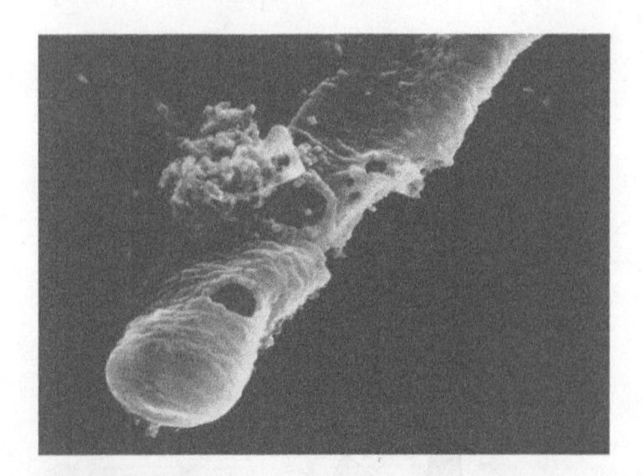

FIGURE 21.9 A bacterial cell after complement exposure. Complement proteins cause holes to form in the bacterial cell, penetrating the cell wall and cell membrane and causing the bacterium to fill with fluids and burst.

surrounding microbes, causing them to break apart (**FIGURE 21.9**) and serve to increase the inflammatory response.

Fever. A body temperature above the normal range of 97–99°F is called fever. Macrophages release chemicals that cause body temperature to increase. A slightly higher-than-normal temperature decreases bacterial growth and increases the metabolic rate of healthy body cells. This helps the body fight infection by slowing pathogen reproduction and allowing repair of tissue to occur more quickly.

Third Line of Defense: Lymphocytes

If a pathogen makes it past the first two nonspecific defense systems, the immune system presents a third line of defense. Cells of the immune system identify and attack specific microorganisms that are recognized as foreign. This **specific defense** system consists of millions of white blood cells, called **lymphocytes.** Lymphocytes travel throughout the body by moving through spaces between cells and tissues or by being transported via the blood and lymphatic systems (**FIGURE 21.10**).

The specific response generated by the immune system is triggered by proteins and carbohydrates on the surface of pathogens or on cells that have been infected by pathogens. Molecules that are foreign to the host and stimulate the immune system to react are called **antigens.** Antigens are found on invading viruses, bacteria, fungi, protozoans, and worms, as well as on the surface of foreign substances such as dust, pollen, or transplanted tissues.

Regardless of the source, when an antigen is present in the body, the production of two types of lymphocytes is enhanced: the **B lymphocytes (B cells)** and **T lymphocytes (T cells).** Like macrophages, these lymphocytes move

FIGURE 21.10 The lymphatic system. Pathogens trigger a response by the lymphatic system. The various organs of this system work to eliminate infections.

VISUALIZE THIS

If an infection is suspected, a blood sample can be analyzed. Because it is not possible to test for every pathogen, the blood is tested for cells that increase in number during an infection. What type of cell proliferates during infection?

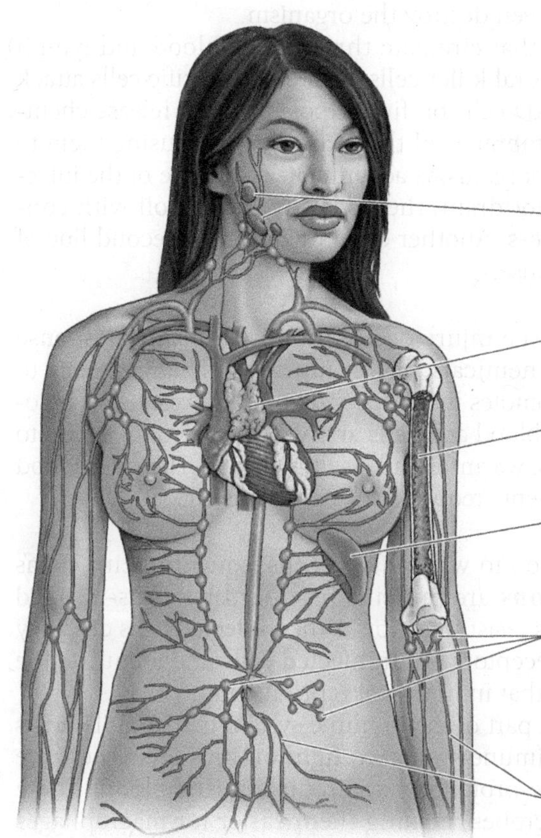

Tonsils and adenoid: Type of lymph node
Thymus: Where some lymphatic cells go to mature
Bone marrow: Produces some lymphatic cells
Spleen: Stores and purifies blood; contains high concentration of lymphocytes
Lymph nodes: Store cells and filter out bacteria and other unwanted substances to purify the lymphatic fluid; become swollen and painful when infection occurs
Lymphatic vessels: Transport fluid from tissues to lymph nodes

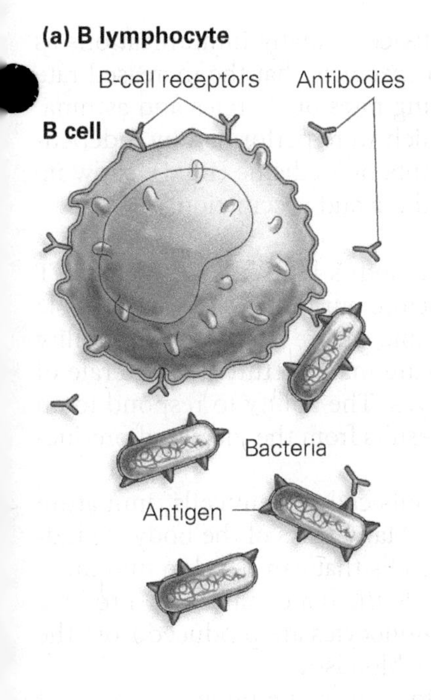

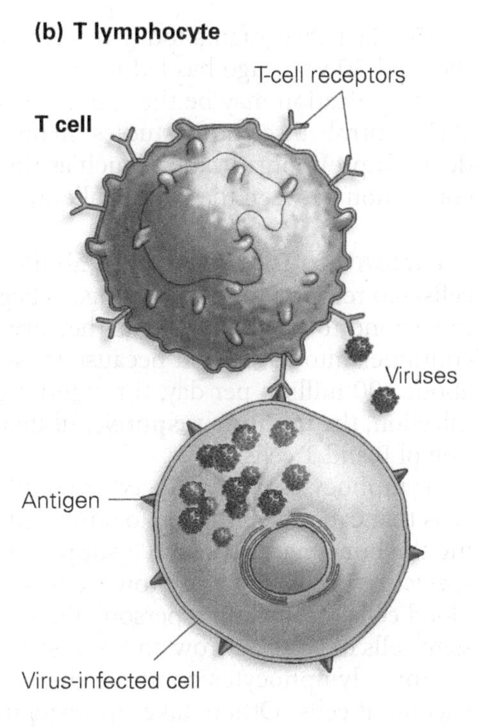

FIGURE 21.11 B cells and T cells. (a) The receptors produced by B cells function as cell-surface antigen receptors or as secreted antibodies that are present in bodily fluids. (b) T cells have surface antigen receptors only. They do not secrete antibodies. B cells and T cells differ with respect to the types of cells and organisms they respond to.

throughout the circulatory and lymphatic systems and are concentrated in the spleen and lymph nodes. Lymphocytes display **specificity** because they recognize specific antigens.

The specificity of lymphocytes is based on the presence of proteins, called **antigen receptors,** whose shape fits perfectly to a portion of the foreign molecule. The receptor is able to bind to the antigen much the way that a key fits into a lock. These receptors are either attached to the surface of the lymphocyte or secreted by the lymphocyte. B and T cells recognize different types of antigens: B cells recognize and react to small, free-living microorganisms such as bacteria and the toxins they produce (**FIGURE 21.11a**). T cells recognize and respond to body cells that have gone awry, such as cancer cells or cells that have been invaded by viruses. T cells also respond to transplanted tissues and larger organisms such as fungi and parasitic worms (**FIGURE 21.11b**).

B and T cells both recognize and help eliminate antigens, but they do so in different ways. B cells secrete **antibodies,** proteins that bind to and inactivate antigens. T cells, in contrast, do not produce antibodies but instead directly attack invaders.

Antibodies are found in the blood, lymph, intestines, and tissue fluids. They are also found in breast milk. When a baby breast-feeds, some antibodies are passed from the mother to the baby, conferring short-term **passive immunity** to the child. Because the child did not make the antibodies him- or herself, this immunity lasts only as long as the antibody remains in his or her blood. An infant's passive immunity is different from the immunity conferred by exposure to an antigen. Exposures to antigens cause the production of antibodies to combat the infection for the individual's lifetime, which is called **active immunity.**

Allergy. An **allergy** is an immune response that occurs even though no pathogen is present. The body simply reacts to a nonharmful substance as though a pathogen were present. For example, some people have allergic reactions to peanuts or ragweed pollen. Asthma may also be the result of an allergic reaction.

The fact that infants today receive about twice as many immunizations as they did 30 years ago has led many people to speculate that the increased rate of immunization may be the cause of increasing rates of allergies and asthma. Well-controlled scientific studies, many of which were performed by independent nonprofit organizations such as the Institute of Medicine (IOM), show no connection between the rates of allergy and asthma and vaccination.

Anticipating Infection. The ability of the body's lymphocytes—B and T cells—to respond to specific antigens begins before our birth. Thus, we are able to respond to infectious agents the very first time we are exposed. This ability continues into adulthood because these cells are manufactured, at the rate of about 100 million per day, throughout our lives. The ability to respond to an infection, the **immune response,** ultimately results from the increased production of B and T cells.

Lymphocytes are produced from special cells called stem cells, immature cells that can differentiate into adult cell types. Many parts of the body, including the bone marrow, retain a supply of stem cells that can develop into more specialized cells. Bone marrow stem cells enable the bone marrow to produce blood cells throughout a person's lifetime. Lymphocytes are produced from the stem cells of bone marrow and released into the bloodstream.

Some lymphocytes continue their development in the bone marrow and become B cells. Others take up residence in the **thymus** gland. The thymus gland, located behind the top of the sternum, stimulates T cells to develop. When immune cells finish their development in the bone marrow, they are called B cells; when they mature in the thymus, they are called T cells (**FIGURE 21.12**).

Another important facet of the immune system is its ability to determine whether a molecule is part of the host cell or foreign.

Self versus Nonself. The cells of a given individual have characteristic proteins on their surfaces. Developing lymphocytes are tested, in the thymus, to determine whether they will bind to these self-proteins. Any developing

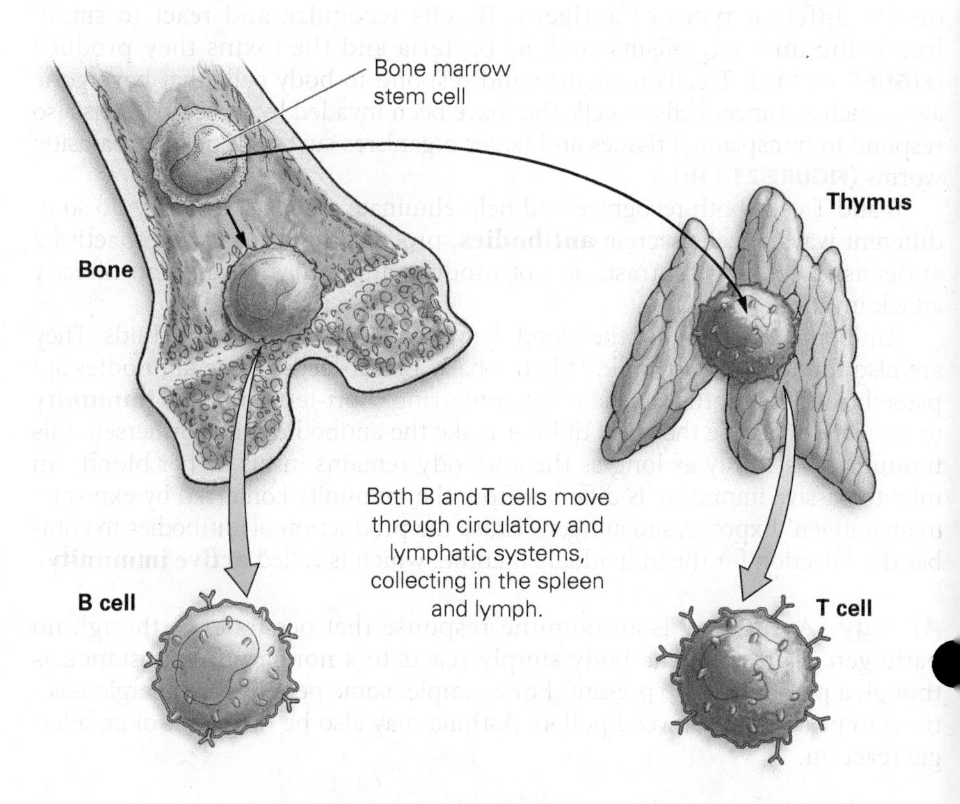

FIGURE 21.12 Lymphocyte development. Lymphocytes develop from cells in the bone marrow. Those that continue their development in the bone become B cells, while those that move to the thymus gland (located in the neck) to continue their development become T cells.

lymphocyte with antigen receptors that bind to self-proteins is eliminated, making an immune response against one's body less likely. Lymphocytes with receptors that do not bind are then allowed to develop to maturity (**FIGURE 21.13**). Thus, the body normally has no mature lymphocytes that react against self-proteins, and the immune system exhibits self-tolerance. Without this self-tolerance **autoimmune diseases,** diseases that result when a person's immune system attacks his or her own cells, can occur.

Multiple sclerosis is an autoimmune disease that occurs when T cells specific to a protein on nerve cells attack these cells in the brain. In insulin-dependent diabetes, T and B cells attack cells in the pancreas that produce the hormone insulin.

Even though we each have a single immune system, it is diversified into two subsystems so that we can combat the multitude of infectious agents encountered in our lifetimes. This diversification is a result of the differing approaches of B and T cells to ridding the body of infectious agents once they are found.

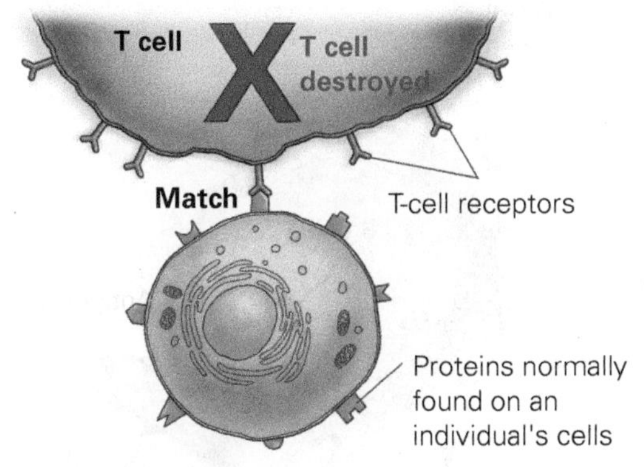

T cell
T cell destroyed
Match
T-cell receptors
Proteins normally found on an individual's cells

FIGURE 21.13 Testing lymphocytes for self-proteins. Cells have proteins and molecules on their surfaces that function as antigens. Developing lymphocytes are tested in the thymus to determine whether they are self or nonself. Lymphocytes that bind to antigens are destroyed; those that do not develop.

auto- means self.

Cell-Mediated and Antibody-Mediated Immunity

T cells provide immunity that depends on the involvement of cells rather than antibodies and is called **cell-mediated immunity.** The protection afforded by B cells is called **antibody-mediated immunity.**

Cell-Mediated Immunity. T cells respond to infection by undergoing rapid cell division to produce memory cells, which then become specialized cells (**FIGURE 21.14**) that directly attack other cells. Two of these attacking

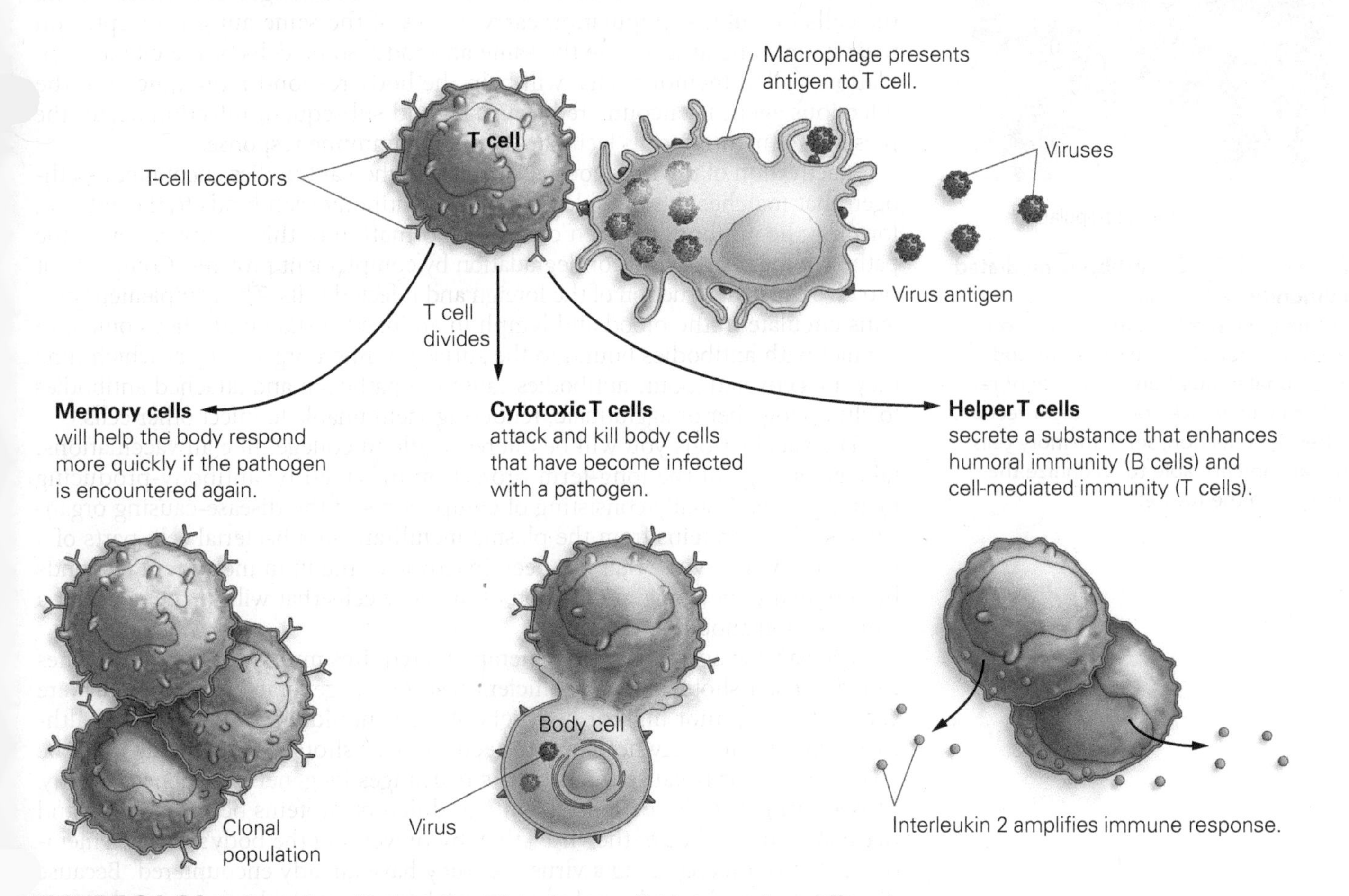

Macrophage presents antigen to T cell.
T cell
Viruses
T-cell receptors
Virus antigen
T cell divides
Memory cells will help the body respond more quickly if the pathogen is encountered again.
Cytotoxic T cells attack and kill body cells that have become infected with a pathogen.
Helper T cells secrete a substance that enhances humoral immunity (B cells) and cell-mediated immunity (T cells).
Body cell
Clonal population
Virus
Interleukin 2 amplifies immune response.

FIGURE 21.14 Cell-mediated immunity. T lymphocytes divide to produce different populations of cells: Memory cells carry the specific antigen receptor. Cytotoxic T cells attack and dismantle infected body cells. Helper T cells boost the immune response.

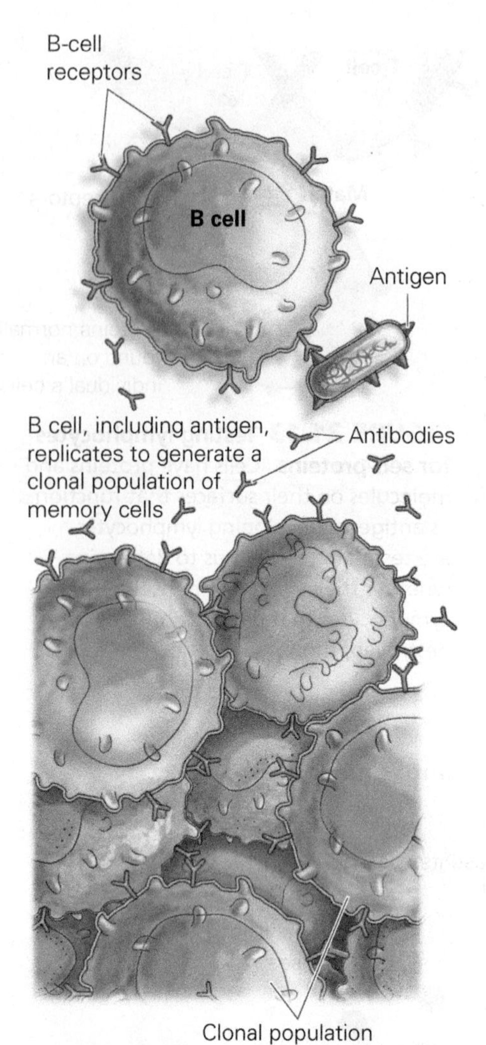

FIGURE 21.15 Antibody-mediated immunity. When a B cell binds an antigen, the B cell makes memory cells, each copy carrying the same antibody. The clonal population of such identical cells can aid in overcoming the infection. After the infection, some memory cells remain and are able to recognize the antigen in the future.

cell types are the cytotoxic T cells and helper T cells. **Cytotoxic T cells** attack and kill body cells that have become infected with a virus. When a virus infects a body cell, viral proteins are placed on the surface of the host cell. Cytotoxic T cells recognize these proteins as foreign, bind to them, and destroy the entire cell. By releasing a chemical that causes the plasma membranes of the target cell to leak, they break down the cell before the virus has had time to replicate.

Helper T cells help boost the immune response. These cells detect invaders and alert both the B and T cells that infection is occurring. Without helper T cells, there can be almost no immune response. The human immunodeficiency virus (HIV) infects and destroys helper T cells, thus crippling the body's ability to respond to any infection. Intensive efforts have been made to produce a vaccine against HIV, thus far with little success, in part because the virus mutates so rapidly. HIV causes **acquired immunodeficiency syndrome,** or AIDS.

Antibody-Mediated Immunity. Unlike T cells, B cells do not directly kill cells bearing antigens. Instead, they make and secrete antibodies that help rid the body of antigenic cells. When a B cell responds to a specific antigen, it immediately makes copies of itself, resulting in a population of identical cells, called **memory cells,** able to help fight the infection (**FIGURE 21.15**). This population of cells is called a **clonal population.** The sheer number of cells in the clonal population strengthens the immune system's ability to rid the body of the infectious agent.

The entire clonal population has the same DNA arrangement. Therefore, all the cells in a clonal population carry copies of the same antigen receptor on their membrane and secrete the same antibody. Some cells of the clonal population, called memory cells, will help the body respond more quickly if the infectious agent is encountered again. Should subsequent infection occur, the presence of memory cells facilitates a quicker immune response.

Inactivation of the infectious agent occurs when antibodies encounter a pathogen that matches the variable region. The antibody then binds to the antigen, forming the antibody-antigen complex. Formation of this complex marks the pathogen for phagocytosis or degradation by complement proteins. Complement proteins cause destruction of the foreign and infected cells. The complement proteins circulate in the blood and lymph in an inactive state until they come into contact with antibodies bound to the surface of microorganisms, at which time they are activated. Some antibodies cause the pathogen and attached antibodies to clump together or agglutinate, rendering them unable to infect other cells.

The vaccinations you will be offered while in college, like all vaccinations, take advantage of the long-term protection provided by antibody-producing memory cells. Usually consisting of components of the disease-causing organisms, such as proteins from the plasma membrane of a bacterial cell, parts of a virus, or a whole virus that has been inactivated, the immune system responds by producing the clonal population of memory cells that will be prepared for a real infection should it happen.

The vaccine against bacterial meningitis requires multiple doses, sometimes called booster shots, before a sufficient response is generated. Even if you were vaccinated in junior high or high school, you should check with your healthcare provider to see whether you need booster shots. Flu vaccines must be given every year because the flu virus rearranges its genetic sequences swiftly. This shifting genetic sequence results in different proteins being encoded and placed on the surface of the virus, thereby preventing the body's existing memory cells from recognizing a virus that they have already encountered. Because the proteins on the surface of a flu virus change so quickly, a vaccine that was prepared to protect you from last year's flu virus will not likely work against this year's flu virus.

Got It?

1. The third, specific line of defense includes white blood cells called ________.
2. Substances found on invading organisms that trigger an immune response are called ________.
3. A mother passing antibodies to her baby through breast milk is a type of ________ immunity.
4. B cells develop in the ________, and T cells reside in the ________.
5. Immune cells that help bolster an immune response are called ________ T cells.

Some good friends are having their first child in a few months. In addition to making decisions about breast-feeding, co-sleeping, and whether to let the baby use a pacifier, they are also concerned about which vaccinations to give their child. Although many of their vaccine-related questions have been answered by speaking with their doctor or looking at reliable medical Internet sources, there is one question that still bothers them. The couple came across information showing that the number of vaccines given to children today is much greater than the number given in the past. They also read on an antivaccine movement website that the developing immune system of a newborn baby or young child is not equipped to deal with this number of vaccines. They came to the following conclusion:

The number of vaccinations given to children today is too much for the average immune system to handle.

Sounds right, but it isn't.

Answer the following questions to understand why.

1. We know that the human brain undergoes rapid development during infancy and early childhood. In fact, newborns have a membrane-covered gap between their skull bones, called the fontanelle, or soft spot, that accommodates a brain growing faster than new bones can form. Does the immune system have structures that similarly change after birth or is it fully formed at birth?
2. How does the immune system get better at fighting disease, with age or with exposure to substances that cause an immune response?
3. Vaccines given a few decades ago contained whole, intact viruses that had been exposed to heat or chemicals that rendered them inactive. Do most modern vaccines contain entire viruses or small parts of a virus like a cell membrane protein?
4. Which would cause the generation of more kinds of antibodies, the whole viruses present in vaccines of old or the parts of viruses present in modern vaccines?
5. Consider your answers to questions 1–4 and explain why the original statement bolded above sounds right, but isn't.

THE BIG QUESTION

Why are females vaccinated against human papilloma virus more often than males?

HPV can cause cervical cancer, cancer of the vulva, and vaginal cancer in women and can also cause genital warts, anal cancer, and throat cancer in both males and females. In the United States, the HPV vaccination is given as a series of three injections and is recommended by the CDC for 11- to 12-year-olds (although the vaccination is available for men and women into their early to mid-20s). However, twice as many girls complete the series of vaccines required for protection against this sexually transmitted disease than do boys.

What should I know?

What follows are some smaller questions that need to be resolved to answer the Big Question. Place a checkmark next to the questions that science can answer.

Smaller Questions	Can Science Answer?
Can males protect themselves from HPV transmission by condom use alone?	
Is male to female, female to male, male to male, or female to female transmission of HPV possible?	
Is it okay for males to rely on vaccinated females to provide herd immunity?	
Can HPV be passed even when an infected person has no signs or symptoms?	
Is it immoral for males not to be vaccinated against HPV?	
Is HPV transmitted by vaginal, anal, or oral sex with someone who has the virus?	
Does the gender of the health-care provider affect the overall rate of vaccination or the ratio of males to females vaccinated?	

What does the science say?

Let's examine what the data say about this smaller question:

Does the gender of the health-care provider affect the overall rate of vaccination or the ratio of males to females vaccinated?

The data shown in the figure below is from a Cleveland Clinic study of HPV vaccination rates in 5000 patients as a function of whether their primary care provider was male or female. The asterisk indicates a significant difference between clustered bars.

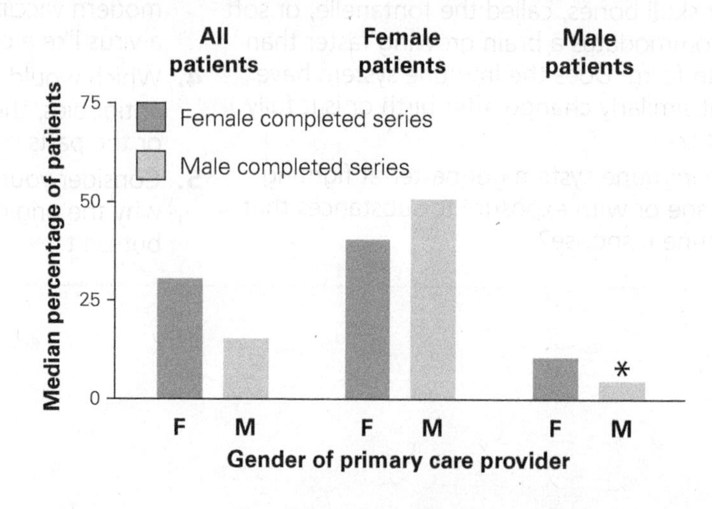

1. Describing the results irrespective of provider gender, what percentage of female and male patients were vaccinated in this study? Were male or female providers less likely to administer vaccine to male patients?
2. Given these data, do you think the smaller question is answered? If not, propose another study that would help answer this question.
3. Does this information help you answer the Big Question? What else do you need to consider?

Data Source: E. Rome, "For HPV Vaccination Rates, Gender Matters in More Ways than One." Cleveland Clinic website, available at https://consultqd.clevelandclinic.org/2015/02/for-hpv-vaccination-rates-gender-matters-in-more-ways-than-one/.

Chapter Review

Mastering Biology

Go to Mastering Biology to access **eText 2.0, Dynamic Study Modules,** and the **Study Area,** where you'll find practice quizzes, BioFlix™ animations, MP3 tutor sessions, current events, and more.

SUMMARY

Section 21.1

Compare the structure of bacteria with that of viruses.

- Bacteria are single-celled organisms with no nucleus or membrane-bound organelles. They are surrounded by a cell wall (pp. 455–456).
- Viruses are composed of nucleic acids encased in a capsid and sometimes an envelope (p. 458).

Contrast bacterial replication with viral reproduction.

- Bacteria replicate by binary fission, a process that copies the genome into two identical cells (pp. 456–458).
- Viruses replicate their genome inside host cells. The viral genome then directs the synthesis of viral components (pp. 458–460).

Explain how herd immunity protects the weakest members of society.

- The more individuals in a population that are vaccinated, the less likely an infection will spread to those that cannot be vaccinated (p. 461).

SHOW YOU KNOW 21.1 Why should people with the flu not be prescribed antibiotics?

Section 21.2

List the structures involved in the first line of defense against infection, and describe how they function in protecting the body against pathogens.

- The skin and mucous membrane secretions are nonspecific defenses that make up the first line of defense against infection (pp. 462–463).

List the participants in the second line of defense against infection, and describe their functions.

- The second line of defense consists of nonspecific internal defenses, including white blood cells such as phagocytic macrophages, which engulf and digest foreign cells, and natural killer cells, which release chemicals that disintegrate cell membranes of tumor cells and virus-infected cells (p. 463).
- Inflammation attracts phagocytes and promotes tissue healing (p. 463).
- Defensive proteins, including interferon, help protect uninfected cells from becoming infected. Complement proteins coat microbes and make them easier for macrophages to ingest (pp. 463–464).

List the two different cell types involved in the third line of defense against infection, and compare how they recognize and eliminate antigens.

- Lymphocytes are part of the third specific line of defense in response to antigens on the surface of pathogens. Exposure to antigens causes increased production of B and T lymphocytes (pp. 464–465).
- B cells secrete antibodies against pathogens, and T cells attack invaders (pp. 464–465).

Explain why allergic reactions occur.

- Allergic reactions occur when the immune system responds to a normally occurring antigen as though it were a pathogen. (pp. 465–466).

Describe the process used to determine whether a cell is foreign or native to the body and what happens when this process fails.

- The antigen receptors of B and T cells are tested for self-reactivity, and those that react against self are eliminated (pp. 466–467).

Explain the role of memory cells in helping to protect against infection.

- Memory cells carry copies of the antigen receptor and secrete antibodies against a previously encountered pathogen (pp. 467–468).

Describe the mechanism by which vaccines help confer immunity.

- Vaccines contain parts of disease-causing organisms or entire organisms that have been inactivated. The immune system produces memory cells in response to vaccination, thereby affording protection in the case of actual infection (p. 468).

SHOW YOU KNOW 21.2 After receiving a flu shot, some people report a mild flu-like illness. Use your understanding of immunology to explain why this might occur.

ROOTS TO REMEMBER

The following roots of words come mainly from Latin and Greek and will help you to decipher terms:

auto-	means self. Chapter term: *autoimmune*
enceph- and encephalo-	mean brain. Chapter term: *encephalitis*
path- and patho-	relate to disease. Chapter term: *pathogen*

LEARNING THE BASICS

1. Add labels to the figure that follows, which illustrates a virus.

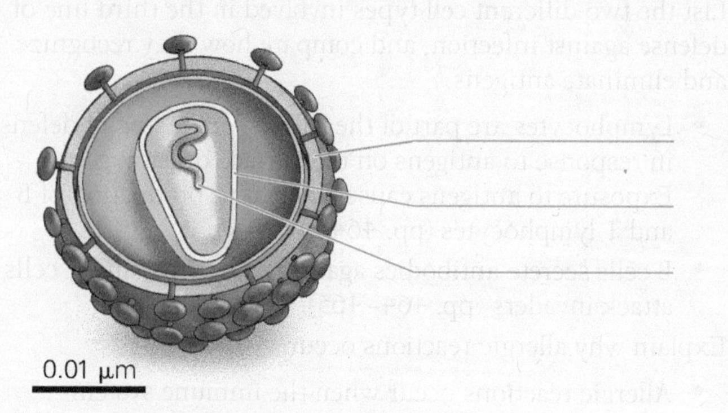

2. How do bodily secretions help protect against infection?

3. What roles do B cells and T cells play in the immune response?

4. Binary fission ________.

   A. allows bacterial cells to produce more plasmids; B. is a type of sexual cell division that viruses undergo; C. allows immune cells to replicate; D. is the method by which bacteria replicate

5. Viral replication ________.

   A. occurs only inside living cells; B. requires that copies of the viral genome be produced; C. requires that copies of viral proteins be synthesized; D. all of the above

6. Autoimmune diseases result when ________.

   A. a person's endocrine system malfunctions; B. liver enzymes malfunction; C. B cells attack T cells; D. the immune system fails to differentiate between self and nonself cells

7. Which of the following cell types divides to produce cells that make antibodies?

   A. helper T cells; B. B cells; C. cytotoxic T cells; D. all of the above

8. Helper T cells ________.

   A. help prevent leukemia; B. prevent bacteria from entering cells; C. boost immune response; D. inhibit reverse transcriptase

9. Your immune system can recognize a virus you have been exposed to once because ________.

   A. you harbor the virus for many years; B. you have genes to combat every type of virus; C. a cell that makes receptors to a virus multiplies on exposure to it and produces memory cells; D. a copy of the viral genome is inserted into a cytotoxic cell

10. Which of the following is not involved in the second line of defense?

    A. inflammation; B. complement proteins; C. macrophages; D. lymphocytes

ANALYZING AND APPLYING THE BASICS

1. Lyme disease is spread by the bite of an infected tick, but is not passed directly from one person to another. Is Lyme disease infectious, contagious, or both?

2. How are people who refuse to vaccinate protected by those who do get vaccinated?

3. Why do you need a flu shot every year but only one inoculation against some other diseases?

GO FIND OUT

1. Many states allow parents to opt out of the vaccinations required to start school. What is the protocol for being exempted from vaccination in your state? Do you think this protocol is fair to all citizens in your state? Why or why not?

2. A 21-year-old friend who was not vaccinated against HPV now wants to be vaccinated. Do some research to determine whether this friend could still be vaccinated, whether it is possible to be vaccinated on campus, and what the series of vaccines would cost.

MAKE THE CONNECTION

The science that you learned in this chapter has helped you better understand the real-world example used throughout this discussion. Draw a line from the statement on the left to the science that supports it on the right.

Antibiotic overuse is leading to concerns about disease epidemics. ▷	◁ Some infectious diseases are spread by sneezing and coughing, making those living in close quarters very susceptible.
It is recommended that students living in dorms be vaccinated against meningitis. ▷	◁ Bacteria evolve in response to medications that select for those cells that are not susceptible to the medication.
Latent viruses can be transmitted even when an infected person shows no signs of infection. ▷	◁ The HPV vaccine prevents transmission of a cancer causing virus and genital warts.
Herd immunity helps protects those that cannot be vaccinated because they are too vulnerable. ▷	◁ Individuals need to be vaccinated against the flu every year.
Males and females are recommended to get the cervical cancer vaccine. ▷	◁ The spread of infectious disease can be slowed if most of the members of a community are immunized.
Some viruses mutate rapidly. ▷	◁ Herpes is a sexually transmitted disease that affects many sexually active young adults.

Answers to **Got It?, Visualize This, Working with Data, Sounds Right, But Is It?, Show You Know,** and **Chapter Review** questions can be found in the **Answers** section at the back of the book.

22 Human Sex Differences

Endocrine, Skeletal, and Muscular Systems

Women were prohibited from Olympic ski jumping until 2014.

Are there injuries that are more common in one sex than the other?

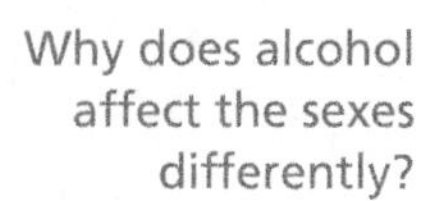

Why does alcohol affect the sexes differently?

Women's ski jumping made its Olympic debut in the 2014 Winter Olympics, making the Sochi winter games the first in which men and women competed in every sport. Men have been competing in Olympic ski jumping since 1924. Why did it take so long for women to be allowed to compete? One reason was that some of those governing and coaching the sport believed that the repeated impacts of a sport like ski jumping could damage women's reproductive organs, a contention that is entirely without medical support yet prevented women from being able to compete at the Olympic level for decades.

Although this outmoded banning of women in ski-jump competitions was baseless, there are important sex differences between females and males that are worth noting. For example, female athletes like U.S. Olympic runner Abbey D'Agostino experience a higher rate of knee injuries than their male counterparts. And beyond the realm of sports, there are known differences in the way the sexes metabolize alcohol and in how long each sex can be expected to live. Many such sex differences originate and are maintained by the actions of the endocrine system, which we will explore in this chapter, along with a look at the skeletal and muscular systems.

22.1 The Endocrine System

The **endocrine system** is an internal system of regulation and communication in the body that involves hormones, the glands that secrete them, and the particular cells that respond to the hormones. The endocrine system oversees the regulation of many different processes, including reproduction and development.

Hormones

Hormones are chemicals that travel through the circulatory system and act as signals to elicit a response in **targeted cells.** There are two general mechanisms by which hormones elicit a specific response: Hormones can either bind to receptors on the surface of the target cell and trigger a change inside the cell, or they can diffuse across the cell membrane and bind to receptors inside the cell to trigger the response.

Protein hormones (**FIGURE 22.1a**) cannot cross cell membranes. Typically, they bind to a receptor on the surface of the target cell, and that receptor stimulates a series of other proteins to help relay the message to the inside of the cell, eliciting a specific cellular response. The chain reaction that relays, or transduces, a signal from the outside of the cell to the inside is called **signal transduction.**

Steroid hormones are fat-soluble hormones that can cross cell membranes and bind to receptors inside the target cell, causing cells to turn specific genes on or off (**FIGURE 22.1b**). The steroid hormones most involved in producing anatomical sex

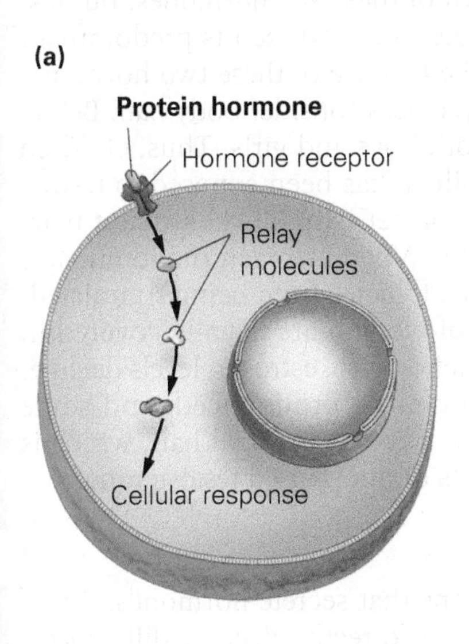

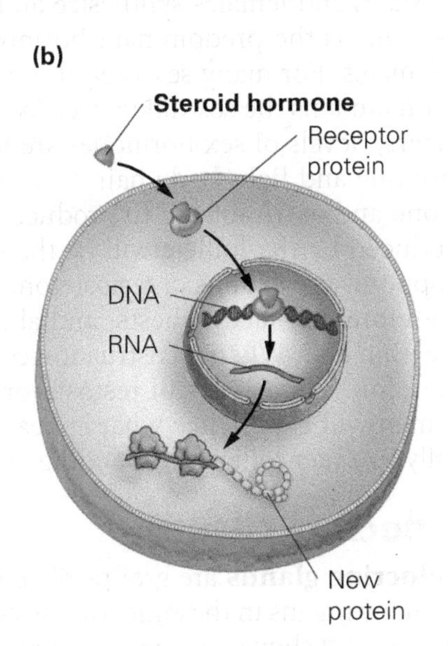

FIGURE 22.1 Hormone actions. (a) A protein hormone binds to a receptor on the cell membrane of its target cell. This binding can induce a shape change in the receptor molecule, serving as a way of transmitting a signal to the inside of the cell that some response is required. (b) Steroid hormones diffuse across the membrane and bind to receptors, which in turn regulate gene expression.

VISUALIZE **THIS**

How would the breast and larynx tissue of a female respond to the presence of testosterone?

(a) Breast tissue can respond to estrogen.

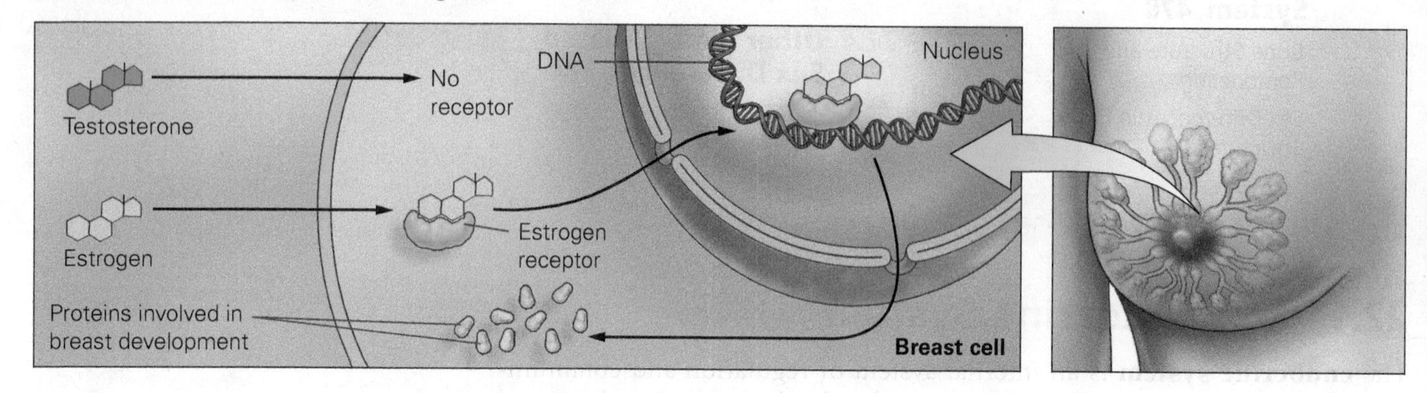

(b) The larynx responds to testosterone.

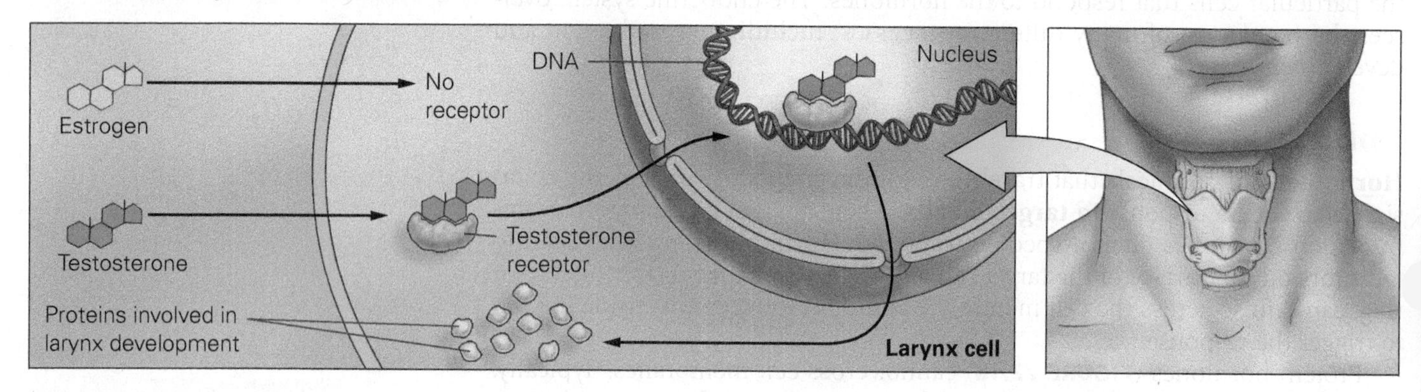

FIGURE 22.2 Sex hormones.
(a) Estrogen produced by females enters the cells that compose the breast tissue, where it binds to its receptor. The estrogen and its receptor together bind to DNA to produce proteins needed for the growth and development of the breast at puberty. (b) In males, testosterone enters the cells of the larynx and binds receptors that increase the production of proteins that change the structure of the larynx, altering the sounds it produces and resulting in voice deepening.

differences are called **sex hormones.** These include the male hormone **testosterone** and the female hormone **estrogen.** Once a sex hormone passes into the cell and binds to a receptor, the hormone and receptor bind to the DNA in the nucleus of the cell to regulate the expression of different genes (**FIGURE 22.2**).

Males and females synthesize and use both of these sex hormones, but testosterone is the predominant hormone in males and estrogen is predominant in females. For many sex-typical traits, it is the balance of these two hormones that maintains the sex difference. As an example, let's consider body hair. Before puberty, levels of sex hormones are low in both boys and girls. Thus, children have fine and light body hair. Once a hair follicle has been exposed to testosterone and is stimulated to produce thicker, coarser hair, all subsequent hairs produced by the follicle will be the same type. When estrogen predominates, it opposes the actions of testosterone and stops follicles from being stimulated. Therefore, the faces, chests, and abdomens of women are usually covered in fine, thin hair. When menstruation ceases at menopause, estrogen levels decline. Therefore the actions of testosterone become more pronounced, and some women will experience what appears to be an increase in facial hair, which is really just hair follicles responding to the effects of unopposed testosterone.

Endocrine Glands

Endocrine glands are groups of cells or organs that secrete hormones. There are many organs in the endocrine system that don't directly affect sex differences. **FIGURE 22.3** shows the locations of the most important endocrine organs.

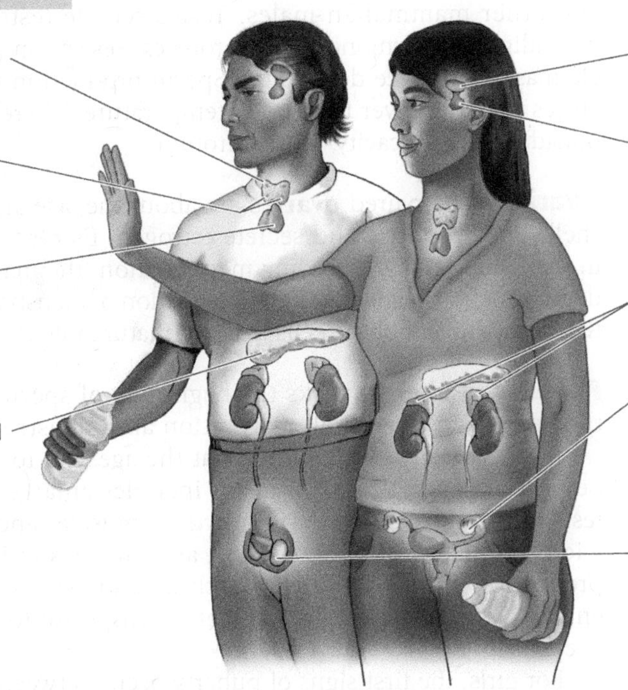
(a) Some examples of endocrine organs
Thyroid secretes calcitonin to lower blood calcium levels.
Parathyroids secrete parathyroid hormone to raise blood calcium levels.
Thymus secretes thymosin, which stimulates T cells of the immune system.
Pancreas secretes insulin and glucagon to regulate blood glucose levels.
(b) Endocrine organs involved in producing sex differences
Hypothalamus secretes gonadotropin-releasing hormone (GnRH).
Pituitary gland responds to GnRH by secreting the pituitary gonadotropins—follicle-stimulating hormone (FSH) and luteinizing hormone (LH).
Adrenal glands secrete adrenaline, testosterone (masculinizing hormone), and estrogen (feminizing hormone).
Ovaries respond to FSH and LH by secreting estrogen, which regulates menstruation, maturation of egg cells, breast development, pregnancy, and menopause.
Testes respond to FSH and LH by secreting testosterone, which aids in sperm production, increased muscle mass, and voice deepening.

FIGURE 22.3 Endocrine glands and sex differences. The locations and functions of some important endocrine glands are shown. Those involved in producing sex differences are listed on the right of the figure.

Five endocrine organs are involved in the production of biological sex differences and are investigated thoroughly in this chapter: the hypothalamus, the pituitary gland, the adrenal glands, the ovaries, and the testes.

Hypothalamus and Pituitary Gland. The **hypothalamus,** located deep inside the brain, regulates body temperature and behaviors such as hunger and thirst, in addition to reproduction. Beneath the hypothalamus is the **pituitary gland,** which secretes growth hormones that act on many organs, including bones, that help children reach adult height.

To help regulate the activities of the reproductive system, the hypothalamus secretes a hormone called **gonadotropin-releasing hormone (GnRH).** The target cells of GnRH are found in the pituitary gland that lies below the hypothalamus. GnRH causes the pituitary gland to secrete many different hormones: two in particular are involved with producing sex differences: **follicle-stimulating hormone (FSH)** and **luteinizing hormone (LH).** In males, FSH stimulates sperm production, and LH stimulates testosterone production. In females, FSH stimulates egg cell development and LH stimulates the release of an egg cell during ovulation (**FIGURE 22.4**). GnRH gets its name in part because it acts on the ovaries and testes, collectively called the gonads.

Adrenal Glands. An **adrenal gland** sits atop each kidney. These glands secrete the hormone adrenaline (epinephrine) in response to stress or excitement. The adrenal glands of both males and females also secrete a small amount of both of the sex hormones, but the majority of sex hormone synthesis and secretion occurs in the gonads.

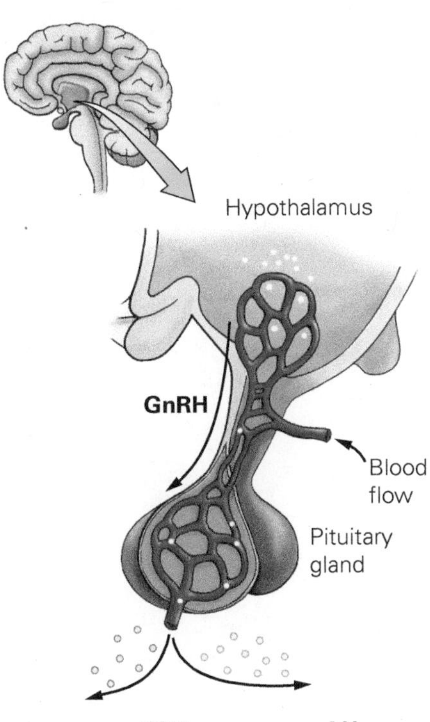
Hypothalamus
GnRH
Blood flow
Pituitary gland
FSH
LH
♂
Stimulates sperm production
Stimulates testosterone production
♀
Stimulates egg cell development
Stimulates ovulation

FIGURE 22.4 FSH and LH secretion. The hypothalamus secretes GnRH, causing the pituitary gland to secrete FSH and LH. FSH and LH are secreted by the brain of both males and females but stimulate a different response in each sex.

Testes. The paired **testes** are oval organs suspended in the scrotum of human and other mammalian males. Testes secrete testosterone, an **androgen** or masculinizing hormone. Testosterone causes sperm production, increased muscle mass, and voice deepening. Sperm production is most efficient at temperatures that are lower than body temperature. Therefore, the testes are situated outside the body cavity, in the scrotum.

Ovaries. The paired **ovaries** are about the size and shape of almonds in the shell. They produce and secrete estrogen. Estrogen regulates many functions in the female body, including menstruation, the maturation of egg cells, breast development, pregnancy, and the cessation of menstruation at menopause. Inside the ovaries are all of the cells that can mature into the egg cells for ovulation.

Puberty. **Puberty** marks the beginning of sperm production in males and the beginning of egg cell maturation and menstruation in females.

Boys typically begin puberty at the age of 9 to 14 years, with the average being 13 years. Puberty in males includes enlargement of the penis and testes; an overall growth spurt; growth of muscles and the skeleton, resulting in wide shoulders and narrow hips; and changes in hair growth, including the production of pubic, underarm, chest, and facial hair. In addition, the larynx enlarges and vocal cords lengthen in response to testosterone to produce a deeper voice.

For girls, the first signs of puberty occur between ages 8 and 13, most commonly around age 11. These signs include breast development, increased fat deposition, a growth spurt, pubic and underarm hair growth, and the commencement of menstruation.

It is at puberty that biologically induced sex differences begin to emerge, in part because the hormones produced by testes and ovaries begin to exert their effects on the skeletal system.

Got It?

1. Chemical signals that travel from glands to target organs are called ________.
2. Relaying a chemical signal from outside the cell to inside the cell can involve several steps and is called ________.
3. The ________ secretes GnRH.
4. GnRH acts on the ________ to cause the secretion of FSH and LH.
5. ________ acts on the testes and ovaries to help the production of sperm and egg cells.

22.2 The Skeletal System

The skeletal system provides support for the body, protects internal organs, aids in movement, and stores minerals. This internal framework is composed of bones and cartilage. Bones are rigid body tissues that provide support for the actions of muscles and protection for soft parts of the body—the skull, for example, which protects the brain. Bones also help produce blood cells within the bone marrow and serve as a reservoir for minerals like calcium that circulate in the body. The ends of bones are covered with cartilage, which protects against the degradation caused when two bones in a joint rub against each other.

There are 206 bones in the human skeleton, which is organized into two basic units: the axial and the appendicular skeleton (**FIGURE 22.5**). The **axial**

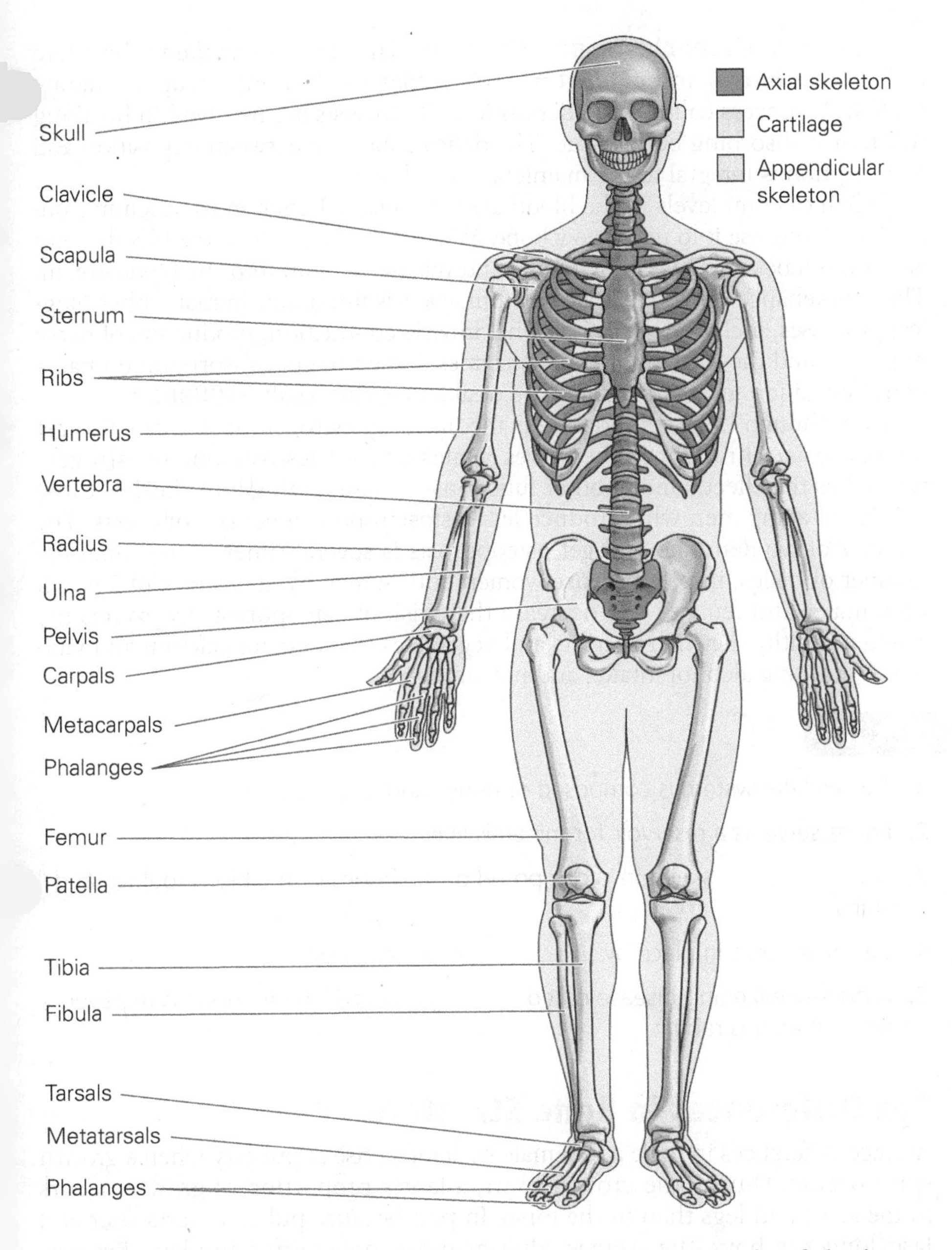

FIGURE 22.5 The human skeleton. The human skeleton can be divided into those bones that support the trunk, called the axial skeleton, and those that compose the limbs, called the appendicular skeleton. Cartilage cushions joint surfaces.

skeleton supports the trunk of the body and consists largely of the bones making up the vertebral column (or spine) and the skull. The **appendicular skeleton** is composed of the bones of the hip, shoulder, and limbs. All bones are kept in place by ligaments and moved by muscles.

Bone Structure and Remodeling

Bones are living tissues composed of cells that use the mineral calcium to continuously remodel the skeleton. The tissue composing bones can be tightly or loosely packed. **Compact bone** is the tightly packed hard outer shell of bones, and **spongy bone** is the loosely packed, porous, honeycomb-like inner bone (**FIGURE 22.6**). Because bone is a living tissue, it has blood vessels throughout. The interior of bone contains **bone marrow,** which helps to produce blood cells.

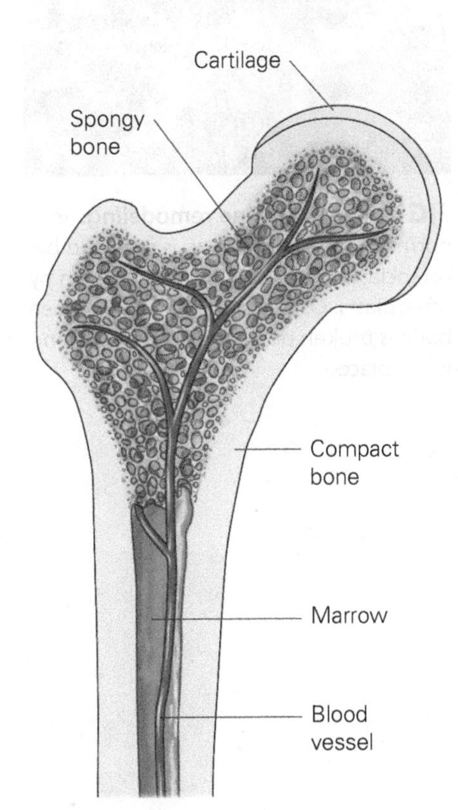

FIGURE 22.6 Bone structure. Bone is composed of compact and spongy bone. There are blood vessels running through bone, and the marrow inside bones helps produce blood cells.

osteo- means bone.
-blast refers to an early formation.
-clast means to break down.

Osteoblasts, Osteoclasts, and Calcium. The cells that make up bone are called **osteoblasts** and **osteoclasts**. Osteoblasts help bone tissue regenerate itself, by a process called bone deposition. Osteoclasts are involved in breaking down or reabsorbing bone tissue. The delicate balance between deposition and reabsorption is integral to the maintenance of bones.

When calcium levels in the blood are high, osteoblasts remove calcium from the blood and use it to make new bone. When calcium levels in the bloodstream are low, osteoclasts break down bone and release calcium into the bloodstream. The maintenance of bloodstream calcium levels is important in many physiological processes, including blood clotting, muscle contraction, production of nerve impulses, and the activity of many enzymes. When bone reabsorption outpaces bone deposition, **osteoporosis,** bone weakening, can result (**FIGURE 22.7**).

Even though bone remodeling is mediated by estrogen, males can also get osteoporosis. Throughout their lives, males convert testosterone to estrogen, which has the effect, among other functions, of maintaining bone health. Close to 20% of aging men who produce less testosterone will get osteoporosis. The number of women who will get osteoporosis is several times higher than the number of males, in part because women outlive men by an average of 5 years. Both males and females can decrease their risk of osteoporosis by exercising, eating a healthy diet rich in fruits and vegetables that contain calcium and vitamin D, limiting alcohol intake, and not smoking.

(a) Normal bone

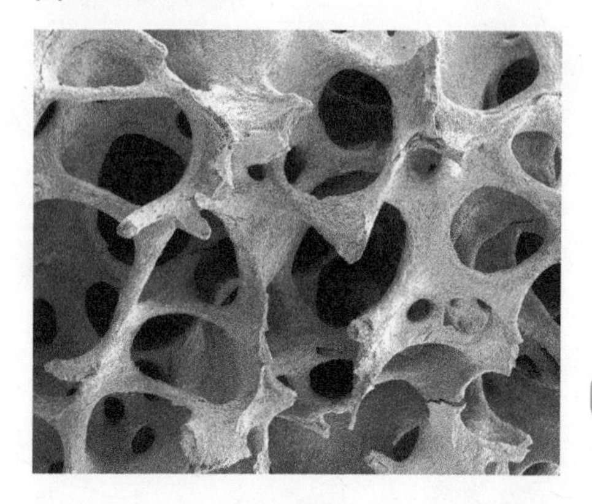

(b) Osteoporotic bone

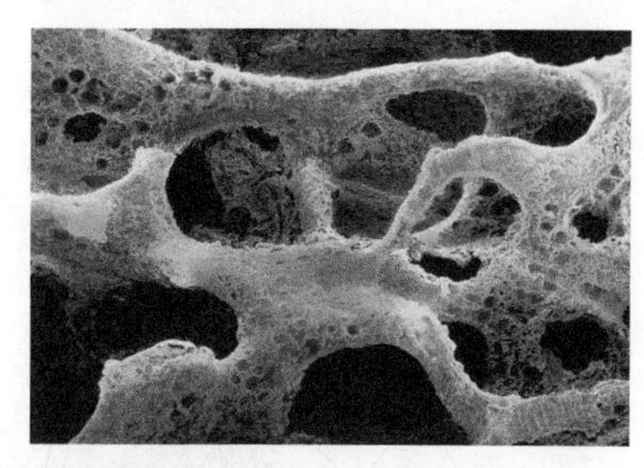

FIGURE 22.7 Bone remodeling. In normal bone (a), old bone is removed by osteoclasts and new bone is laid down by osteoblasts. Osteoporosis (b) occurs when bone is broken down more quickly than it is replaced.

Got It?

1. The skeletal system is composed of bones and ________.
2. Bones serve as a reservoir for minerals like ________.
3. The ________ skeleton is composed of the bones of the hip, shoulder, and limbs.
4. Bones are kept in place by ________ and moved by ________.
5. A bone-weakening disease called ________ occurs when bone is removed faster than it is rebuilt.

Sex Differences in Bone Structure

Average differences in male and female skeletons arise at puberty when a growth spurt occurs. During the growth spurt, a larger proportion of growth occurs in the arms and legs than in the torso. In part because puberty starts later and lasts longer in boys, the average adult man has longer arms and legs. Because men generally have shorter torsos relative to their leg length, their center of gravity, the point on the body where the weight above equals the weight below, is higher than in women (**FIGURE 22.8a**). Women stop growing after puberty while men continue growing into their early 20s, resulting in an average height difference between men and women of about 6 inches. Less height and a lower center of gravity lead to increased stability, which may help explain why women compete on the balance beam in Olympic gymnastics but men do not.

The remaining skeletal differences between males and females involve size differences in six of the 206 bones. Four of the bones that differ are found on the head. These are the **mandible,** or jawbone, which is smaller in females; the paired **temporal bones** found near the temple, which have a larger opening in males to allow for the connection of thicker muscles to support the large jaw; and the **frontal bone** of the cranium, or forehead, which is generally more rounded in females and has a less-pronounced ridge above the eyes (**FIGURE 22.8b**).

The other two bones that show general differences between men and women are the two **ossa coxae,** or hip bones, which form the bony pelvis. The round pelvic inlet that is typical of most women is produced by a bony pelvis that is

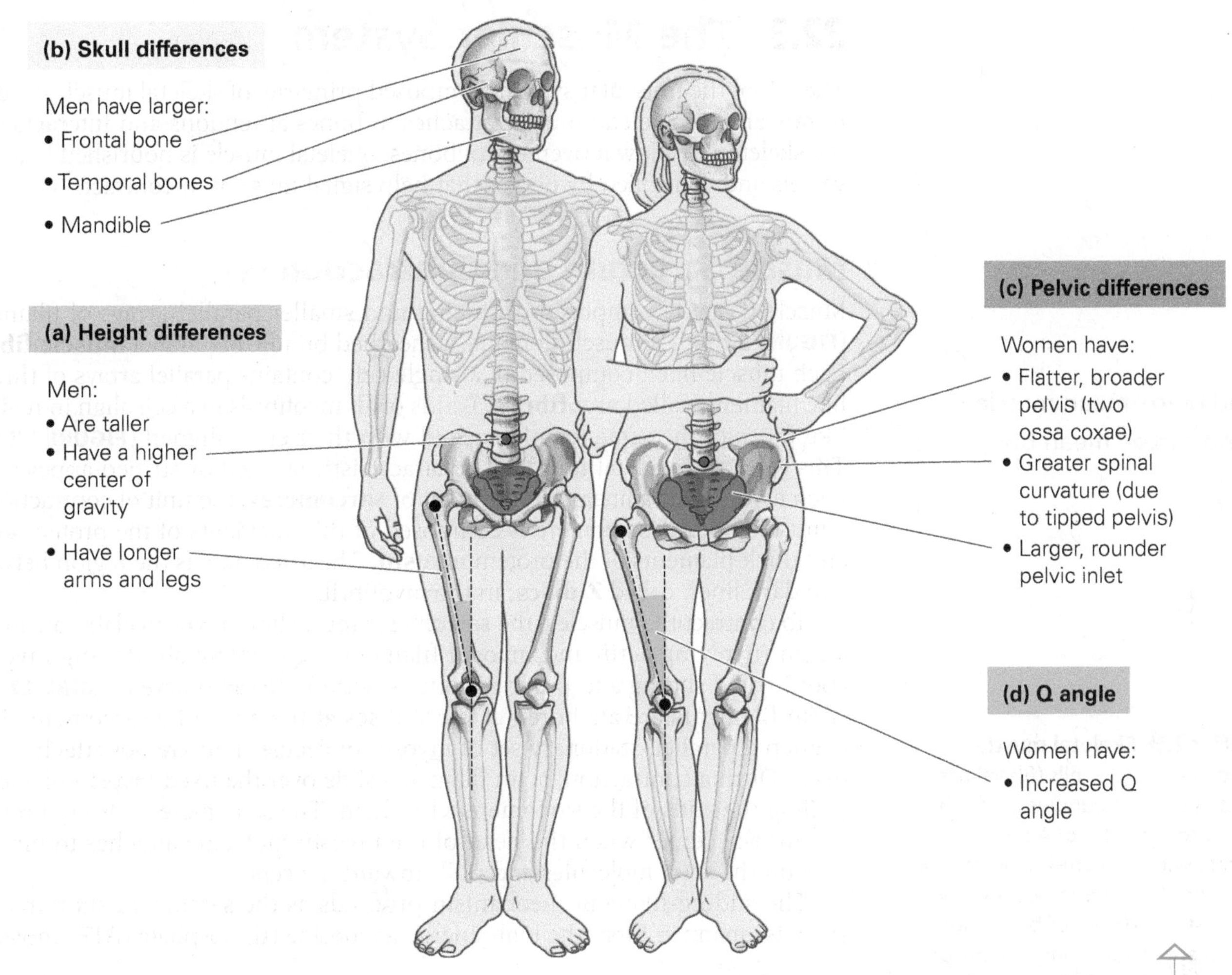
(b) Skull differences
Men have larger:
• Frontal bone
• Temporal bones
• Mandible
(a) Height differences
Men:
• Are taller
• Have a higher center of gravity
• Have longer arms and legs
(c) Pelvic differences
Women have:
• Flatter, broader pelvis (two ossa coxae)
• Greater spinal curvature (due to tipped pelvis)
• Larger, rounder pelvic inlet
(d) Q angle
Women have:
• Increased Q angle

FIGURE 22.8 Sex differences in the skeleton. Although human male and female skeletons are difficult to distinguish, consistent differences can be found in overall size (a), the size of four skull bones (b), the size and shape and tilt of the pelvic bones (c), and the Q angle (d).

VISUALIZE **THIS**

List the bones involved in determining the Q angle.

flatter and broader than a man's, which has an inlet that resembles an elongated oval. The rounder pelvic inlet in women generally permits the easier passage of a baby's head through the birth canal. A flatter pelvis also requires that the bony pelvis be tipped forward to bring the hip bones to the front of the body; this tipping forward is maintained by the curvature of the lower spine. The spinal curvature and greater pelvic tilt in individuals with broad pelvises elevate the buttocks and give a curvy appearance to the profile of women. Conversely, the male pelvis lowers the buttocks and gives a flat appearance to a man's profile (**FIGURE 22.8c**).

One last sex difference involving the skeleton has to do with the angle formed between the kneecap, femur, and the line of the tendon from the kneecap to the shinbone, called the **Q angle** (**FIGURE 22.8d**). When you stand with your feet together, your femurs, or thighbones, extend diagonally to the knees from where they attach at the hips. A broader pelvis means that the femurs are farther away from each other at the point of attachment than are femurs attached to a narrower pelvis. Women generally have greater Q angles than men do and there is some evidence that the larger Q angle in females may be partly responsible for their higher rates of knee injury.

22.3 The Muscular System

The job of the muscular system, composed primarily of skeletal muscle, is to aid in movement. Skeletal muscle attaches to bones at tendons and interacts with the skeleton to allow movement of bones. Skeletal muscle is nourished by blood vessels and controlled by nerves that help signal muscles to contract.

Muscle Structure and Contraction

my- and myo- mean muscle.
sarc- and sarco- mean body.

Muscle tissue is composed of smaller and smaller parallel arrays of filaments (**FIGURE 22.9a**). A muscle contains a sheathed bundle of parallel **muscle fibers.** Each muscle fiber, composed of a single cell, contains parallel arrays of thread-like filaments called **myofibrils.** Bands of all myofibrils in a cell align in register, like pipes of the same length stacked with their ends aligned (**FIGURE 22.9b**). This gives the skeletal muscle its characteristic striated or striped appearance. Each myofibril is a linear arrangement of **sarcomeres,** the unit of contraction of a muscle fiber. A sarcomere is composed of thin filaments of the protein **actin** and thick filaments of the protein **myosin.** The sarcomere is the region between two dark lines, called **Z discs,** in the myofibril.

To contract the muscles, the sarcomeres must shorten via an elaborate mechanism involving actin and myosin filaments. Sarcomere shortening involves coordinated sliding and pulling motions within the sarcomere (**FIGURE 22.10**). Actin filaments are anchored at the Z discs at the ends of a sarcomere. They overlap a parallel, stationary set of myosin molecules that are not attached to Z discs. During contraction, actin filaments slide over the fixed myosin filaments, pulling the ends of the sarcomere with them. The sarcomere ends are brought toward the center when the head of the myosin molecule attaches to binding sites on the actin molecules and pulls toward the center.

The sliding-filament mechanism proceeds as the sarcomere shortens in a stepwise manner. First, the high-energy adenosine triphosphate (ATP) molecule

FIGURE 22.9 Skeletal muscle structure. (a) Muscle cells (fibers) are arranged in parallel bundles inside the muscle's outer sheath. (b) A skeletal muscle cell, with its myofibrils in register, produces the characteristic banded or striated pattern. The myofibrils inside a muscle cell are linear arrangements of sarcomeres. Sarcomeres are bounded by Z discs.

(a) Skeletal muscle structure

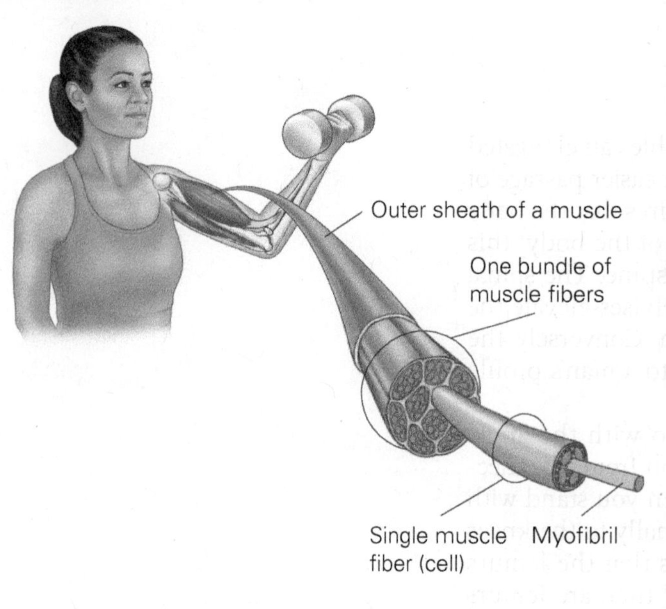

(b) Muscle fiber

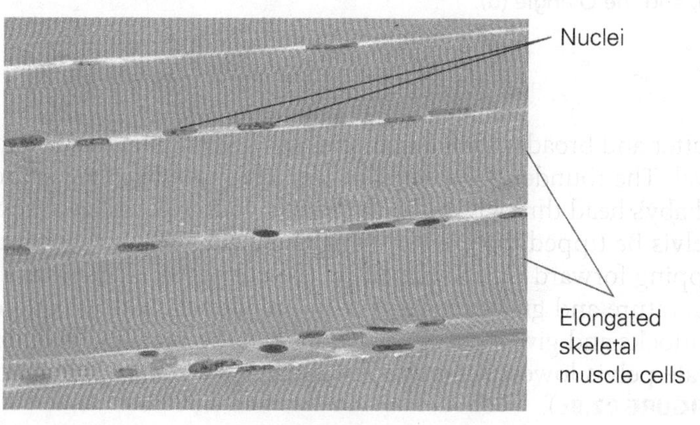

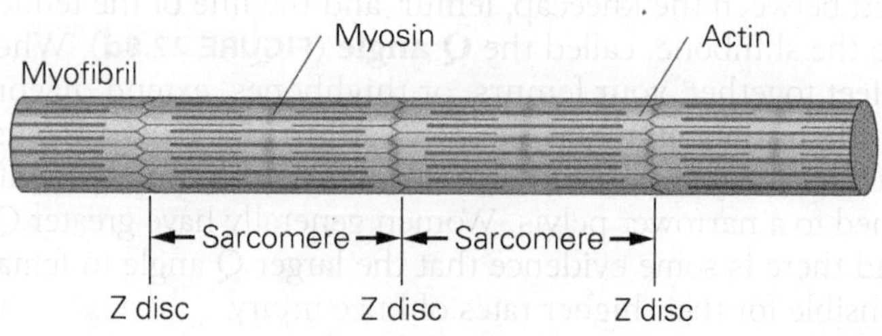

Watch **Muscle Contraction** in Mastering **Biology**

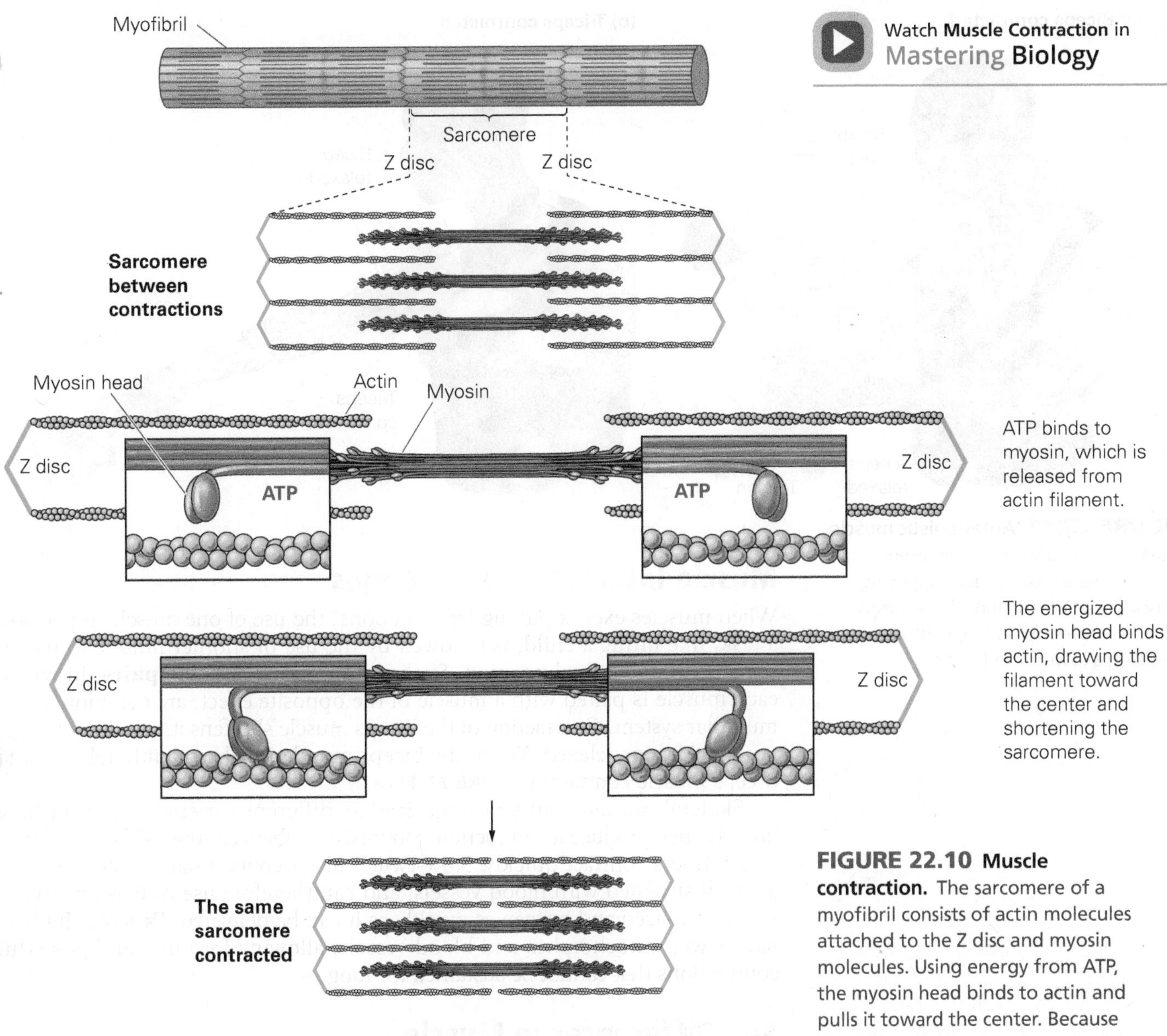

FIGURE 22.10 Muscle contraction. The sarcomere of a myofibril consists of actin molecules attached to the Z disc and myosin molecules. Using energy from ATP, the myosin head binds to actin and pulls it toward the center. Because this is happening at both ends of the sarcomere, the sarcomere shortens, allowing contraction of the muscle cell and movement of the muscle.

binds to the myosin head, causing the myosin head to detach from the binding site on the actin filament. Second, the breakdown of ATP provides energy to the myosin head, which changes shape or "cocks." Third, the myosin head binds to another actin-binding site; pulls in short, powerful strokes; releases the actin; and reattaches farther along. This process serves to pull the Z discs inward, shortening the sarcomere.

Got It?

1. A muscle ________ is composed of a single cell.
2. Parallel filaments inside the cell are called ________.
3. The unit of contraction of a muscle fiber is a ________.
4. A sarcomere is composed of the proteins ________ and ________.
5. The sarcomere is found between the ________ disks in the myofibril.

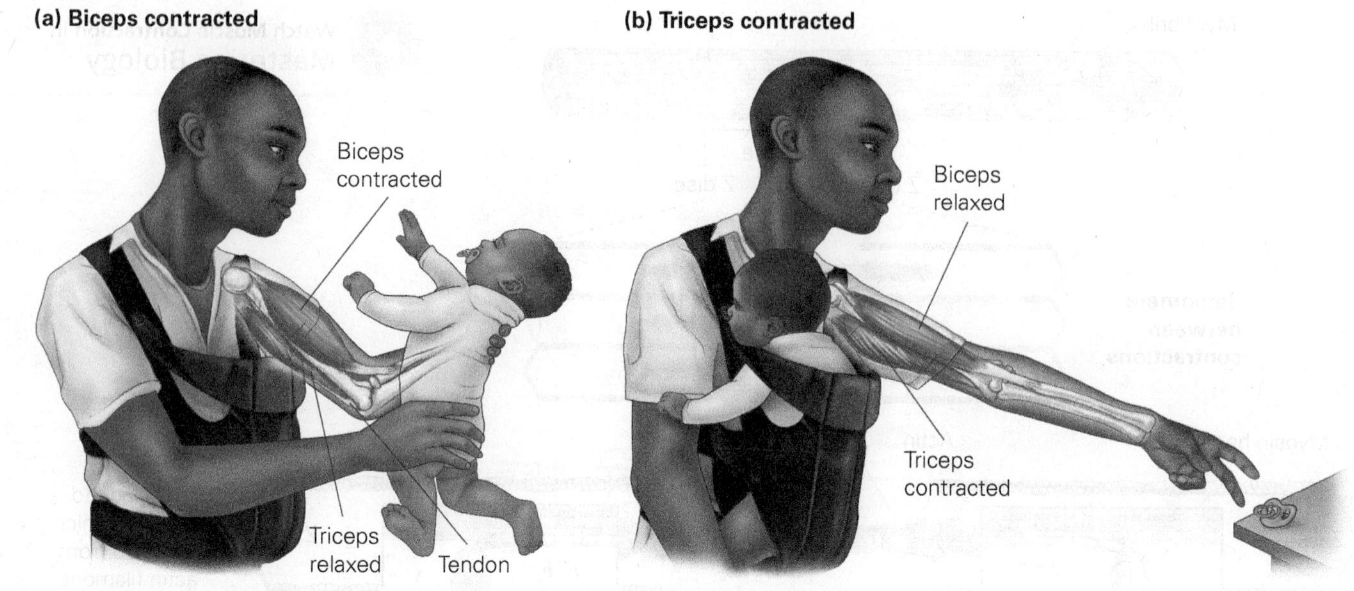

FIGURE 22.11 Antagonistic muscle pairs. Movements are often produced through the actions of two opposing muscles. (a) Contraction of the biceps raises the forearm and (b) contraction of the triceps lowers the forearm.

Muscle Interaction with Bones

When muscles exert a pulling force on bone, the use of one muscle to perform a task, like lifting a child, is followed by the use of another muscle to return the arm to its original position. Such **antagonistic muscle pairs,** in which each muscle is paired with a muscle of the opposite effect, are common in the muscular system. Contraction of the biceps muscle shortens it, while the triceps muscle remains relaxed. When the biceps muscle is subsequently relaxed, the triceps muscle contracts (**FIGURE 22.11**).

Skeletal muscles can be categorized as different varieties based on how quickly they produce a contraction. Slow-twitch fibers contract slowly, and fast-twitch fibers contract quickly. Slow-twitch fibers contain many mitochondria, are well supplied with blood vessels, and can therefore use ATP as quickly as it is synthesized and oxygen as quickly as it can be delivered. Fast-twitch fibers have fewer mitochondria and blood vessels, allowing for rapid and powerful contractions that cannot be sustained for long.

Sex Differences in Muscle

Sex differences in strength occur because testosterone released by the testes at puberty increases the size, but not the number, of muscle fibers. Muscle fibers contain testosterone receptors in their cytoplasm, and the presence of testosterone stimulates the cells to increase in mass. This is accomplished, in part, when males retain dietary nitrogen, which is used to produce more muscle proteins.

Although the ovaries and adrenal glands of both sexes release small quantities of testosterone, males have more of this hormone than do females, and the subsequent difference in muscle mass accounts for an overall difference in strength between males and females.

22.4 Other Biology-Based Sex Differences

Differences in how fat is metabolized and stored are mediated by the endocrine system. Androgens, found in higher concentrations in males, stimulate the production of proteins that help break down fats. This means that if a man and a woman of the same height and weight both eat an ice cream cone, the man is likely to burn more of the calories and the woman is likely to store more of the calories on her body as fat. In this same-sized pair, the man is likely to have more muscle mass, which requires substantial energy to maintain, than the woman, who is likely to carry more body fat. Because fat is not a very active tissue, it does not undergo contraction and flexion like muscle tissue; it costs less energy to maintain. Therefore, if this same couple were to walk while eating their ice cream cones, the man will burn calories at an even higher rate than the woman.

This more efficient breakdown of fats by males is also partly responsible for their higher rates of cardiovascular disease. Because fat is sent through the bloodstream to be metabolized by tissues instead of being stored on the body, men have higher levels of fat circulating in their bloodstreams. Fat in the bloodstreams can adhere to blood vessel walls, causing a dangerous buildup and decreasing the diameter of blood vessels. This causes blood pressure to increase and escalates risk for other cardiovascular diseases.

Fat is also stored in different locations in females and males. Women tend to have more fat cells in their hips, thighs, and buttocks. Conversely, when testosterone is high, fat is more likely to be stored in the abdomen.

Because women store more fat than men, women, on average, have an 8% thicker layer of subcutaneous fat covering muscle tissues. This means that even in a man and a woman with the same amount of muscle mass, individual muscles in the woman will appear less pronounced and more smoothed out. To maintain their fertility, women require around 10% more body fat than males (**FIGURE 22.12**). When a female does not have enough stored body fat, a hormone called leptin signals the brain that she will not be able to sustain a pregnancy and the menstrual cycle is prevented from occurring, which can cause permanent sterility, even if she increases her body fat levels.

Because fat is not very metabolically active, it does not require many blood vessels to deliver nutrients and remove wastes. Women have fewer blood vessels for their size and have less blood volume per body weight. If a man and woman of the same size both drink the same volume of the same alcoholic beverage, the alcohol will be more concentrated in her bloodstream and the effects of the alcohol will be more pronounced. Women are also less able to begin the process of alcohol metabolism in the stomach, because they have smaller stomachs and fewer digestive enzymes. These factors mean that it isn't a good idea for women to pace their drinking with that of male friends, even same-sized ones.

Another sex difference in humans is in average life expectancy. A man and woman born on the same day have different life expectancies. According to the World Bank, a woman born in the United States between 1999 and 2003 has a life expectancy of 81 years, whereas a male born in the United States during that time has a life expectancy closer to 76.

Among the reasons for the sex difference in life expectancy are that men are more likely to be murdered, commit suicide, and die in automobile accidents than are women. These reasons alone do not account for the entire difference in life expectancy, however. A hypothesis that might help explain more of the difference is based on the average age of onset of cardiovascular disease. Women most often experience heart attacks and strokes in their 70s and 80s, whereas men most often experience them a decade or so earlier. This hypothesis posits that blood loss during menstruation has a protective effect against heart disease by lowering women's overall exposure to iron. The oxygen-carrying component of blood, hemoglobin, contains iron as part of its structure. Some studies have indicated that iron produces free radicals that damage cells and their DNA, which could increase the risk of heart disease. If it turns out to be true that iron increases the risk of heart disease, males and females would be advised to limit their consumption of foods, like red meat, rich in iron as well as decrease exposure to other free radical–generating substances like cigarette smoke and alcohol.

WORKING WITH **DATA**

(a) Is there more variation in healthy body fat percentages between or within the sexes? (b) If a person has 18% body fat, can we know whether the person is male or female?

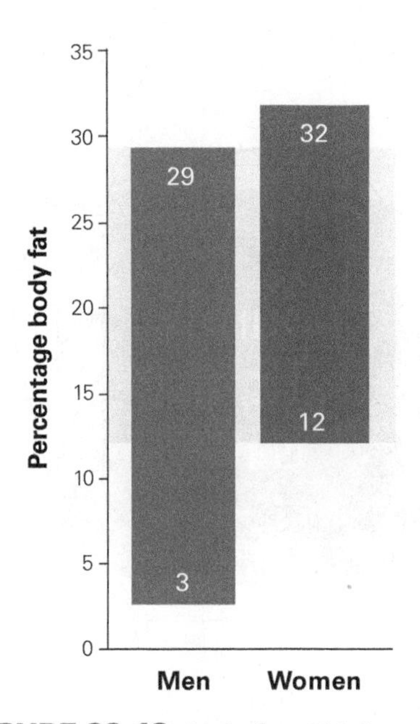

FIGURE 22.12 Body fat. This bar graph shows the range of percentages of body fat in healthy males and females.

Got It?

1. On average, males have more calorie-burning __________ tissue than females.
2. Males and females store body __________ in different locations.
3. To maintain fertility, females need more __________ than males.
4. __________ life expectancy is longer than __________.
5. Females have fewer blood vessels and less total blood __________ than males.

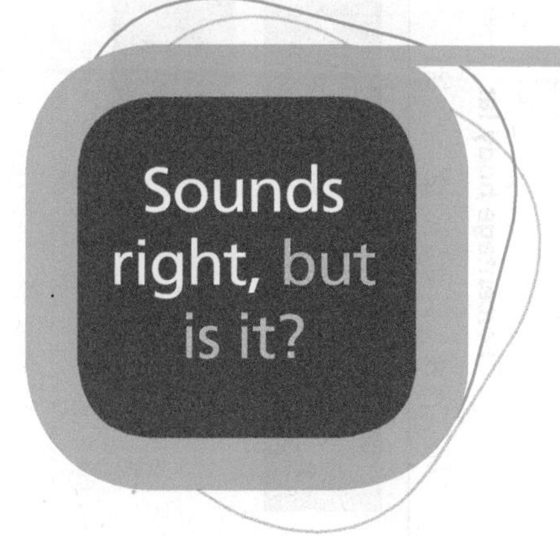

Male and female mammals have mammary glands. Mammary glands will lactate, or release milk, in response to the presence of the pituitary hormone prolactin. Prolactin can be secreted by the brains of both males and females, and acts on mammary glands. In females, prolactin is known to be released by the pituitary in response to a suckling baby.

Before the development of infant formula, wet nurses could be hired to breast-feed a baby for a mother unable to produce milk. From this we know that females can breast-feed even if they were not recently pregnant. Likewise, some adoptive mothers are able to lactate spontaneously if they manually stimulate their nipples before the adoptive child is turned over to their care.

It seems to make sense that the presence of breast tissue and prolactin would also allow human males to lactate, when a baby is allowed to suckle, which leads to the following:

Like females, human males can spontaneously lactate.

Sounds right, but it isn't.

Answer the following questions to understand why.

1. Males who are injected with prolactin or who have pituitary tumors that result in the production of excess prolactin can sometimes lactate. Would this be considered spontaneous lactation?
2. Only one nonhuman male mammal, a fruit bat from Southeast Asia, is known to lactate spontaneously. Does the fact that more recent male human mammalian ancestors do not lactate spontaneously decrease or increase the likelihood that human males would be able to lactate spontaneously?
3. During in utero development the presence of two X chromosomes sets in motion a series of biological events that help program mammary glands to be able to develop more fully at puberty. Is this likely to happen to males? Why or why not?
4. At puberty, mammary glands begin to develop more fully. Many human males do show signs of breast development during puberty. Is this breast development typically permanent or temporary in adolescent boys?
5. We know that males are able to produce prolactin. If males have less developed breasts than females, do you think they would produce the same amount of prolactin in response to suckling as females?
6. Reflect on your answers to questions 1–5 above and explain why the statement bolded above sounds right, but isn't.

THE BIG QUESTION

Is sex selection an acceptable practice?

Sex selection is the practice of choosing what sex your child will be. Sex selection techniques can include having a third child when the first two are of the same sex and you desire a child of the opposite sex. Laboratory manipulations such as fertilizing eggs with sperm bearing the desired sex chromosome or implantation of embryos of the desired sex only, aborting fetuses of the nondesired sex, and infanticide are also practices that allow for sex selection. Sex selection alters the sex ratio—the proportion of males and females in the population. The human sex ratio is normally 1.0 during the reproductive years. Is sex selection ever acceptable?

What should I know?

What follows are some smaller questions that need to be resolved to answer the Big Question. Place a checkmark next to the questions that science can answer.

Smaller Questions	Can Science Answer?
Why would parents prefer a child of a particular sex?	
Is one sex more often selected?	
How do altered sex ratios affect society?	
What are the most common methods of sex selection?	
Why does the rate of sex selection vary by geographic location?	
Does the use of ultrasound increase the rate of sex selection?	

What does the science say?

Let's examine what the data say about this smaller question:

Does the use of ultrasound increase the rate of sex selection?

Ultrasound is a medical imaging technology that relies on vibrations to help produce a picture of structures inside the body. In pregnant women, the sex of a child can be determined during an ultrasound. The bar graph below shows sex ratios in three countries before and after the widespread use of ultrasound came into use in the mid-1980s.

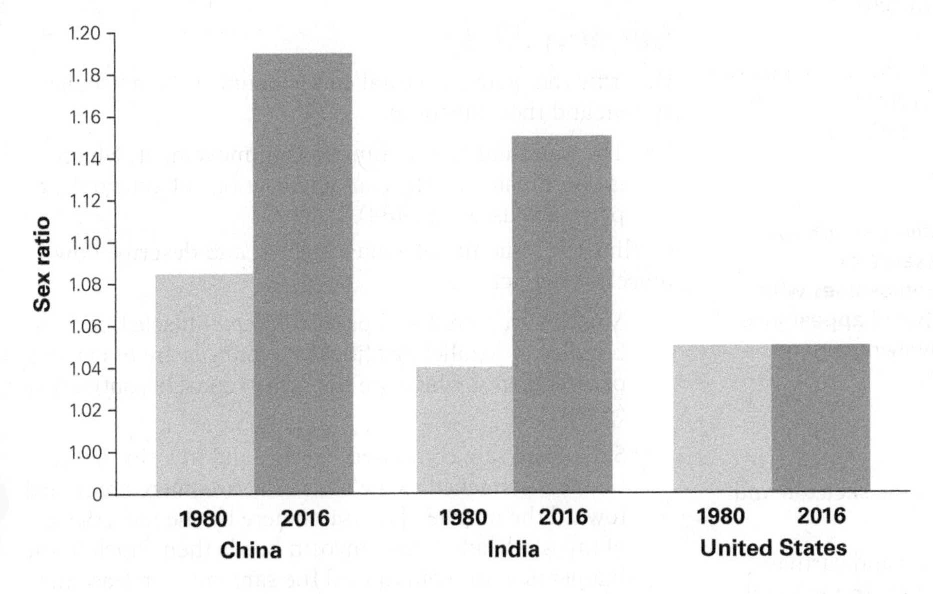

—Continued next page

1. Describe the results. What does a ratio above 1.0 mean? Which countries showed the biggest change after the more widespread use of ultrasound? Comment on what the ratio in the United States may mean.
2. Given these data, do you think the smaller question is answered? If not, propose another study that would help answer this question.
3. Does this information help you answer the Big Question? What else do you need to consider?

Data source: Population census data from India, China, and the United States.

Chapter Review

Mastering Biology

Go to Mastering Biology to access eText 2.0, Dynamic Study Modules, and the Study Area, where you'll find practice quizzes, BioFlix™ animations, MP3 tutor sessions, current events, and more.

SUMMARY

Section 22.1

Differentiate between protein and steroid hormones.

- Protein hormones trigger a response by binding to receptors on the surface of the target cell, which relay the signal to the inside of the cell (p. 475).
- Steroid hormones diffuse into the cell and act on receptors there (p. 476).

Discuss endocrine glands involved in producing sex differences.

- The hypothalamus secretes GnRH, which stimulates the pituitary gland to release FSH and LH (pp. 476–477).
- Adrenal glands secrete adrenaline, androgens, and estrogens (p. 477).
- Testes secrete testosterone, which affects the development of male characteristics. Ovaries secrete estrogen, which fosters the development of female characteristics (pp. 477–478).

SHOW YOU KNOW 22.1 Androgen insensitivity syndrome occurs when cells cannot respond to the presence of testosterone. A person with X and Y sex chromosomes with complete androgen insensitivity has the outward appearance of a typical female. Propose a hypothesis involving a gene mutation, for how this syndrome might arise.

Section 22.2

Describe the main structural components of the skeleton and their functions.

- The skeletal system is composed of bones and cartilage. It provides support, protection, movement, and mineral storage (pp. 478–479).

Discuss the process of bone remodeling.

- Bone remodeling is under the control of the endocrine system. Hormones stimulate the osteoblasts to use calcium to build bone, and the osteoclasts to break down bone and release calcium (pp. 479–480).

SHOW YOU KNOW 22.2 The declining estrogen levels associated with menopause lead to increased rates of osteoporosis in older women. Based on your understanding of bone remodeling, predict which type of bone cell is inhibited by estrogen.

Section 22.3

Describe the main structural components of the muscular system and their functions.

- The muscular system functions in movement, which is coordinated by the combined actions of antagonistic pairs of muscles (p. 484).

Outline the structure of a muscle fiber, and describe how muscles contract.

- Muscles are bundles of parallel fibers. Muscle fibers are bundles of parallel myofibrils. Myofibrils are linear arrays of sarcomeres, which are the unit of muscle contraction (p. 482).
- Sarcomeres are composed of actin and myosin. Actin filaments attached to the ends of a sarcomere are pulled toward the middle of the sarcomere by the movement of myosin heads. These myosin heads then attach to the filaments, pull them toward the sarcomere, release the

filaments, and reattach farther down the filament. This process is called the sliding-filament model of muscle contraction (pp. 482–483).

- Testosterone causes increased muscle mass (p. 484).

SHOW YOU KNOW 22.3 If a male and a female were the same height and had the same amount of total muscle mass in each bicep, would you expect the male to be able to curl a heavier dumbbell than the female? Why or why not?

Section 22.4

Describe average sex differences in body fat utilization, prevalence, and location.

- Men use fat more readily than women (p. 484).
- To maintain their fertility, women require more body fat than do men (p. 485).
- Women store fat on their hips and thighs; men store fat in their abdomen (p. 485).

Describe the sex differences in alcohol metabolism and life expectancy.

- Men metabolize alcohol more quickly than women (p. 485).
- Women live, on average, at least 5 years longer than men (p. 485).

Describe the average sex differences in the muscular and skeletal system.

SHOW YOU KNOW 22.4 As women age, they begin to store more body fat in their abdomen and less on their extremities. Propose an endocrine-based hypothesis for why this change in the location of body fat storage might occur.

ROOTS TO REMEMBER

The following roots of words come mainly from Latin and Greek and will help you to decipher terms:

-blast	refers to an early formation or to a place where something forms. Chapter term: *osteoblast*
-clast	is from the Greek verb to break or to break down. Chapter term: *osteoclast*
my- and myo-	mean muscle. Chapter terms: *myosin, myofibril*
osteo-	relates to bones (the Latin form is os- or ossi-). Chapter terms: *osteoblasts, osteoclasts, osteoporosis*
sarc- and sarco-	mean body. Chapter term: *sarcomere*

LEARNING THE BASICS

1. List the endocrine organs involved in producing sex differences and describe their functions.

2. Describe bone structure.

3. Add labels to the figure that follows, which illustrates the parts of a muscle.

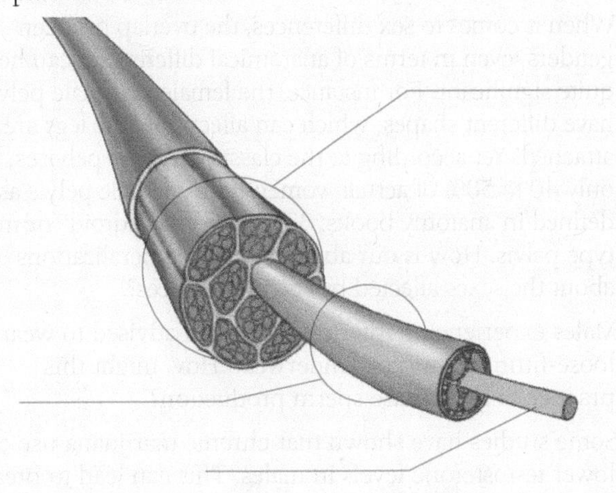

4. Steroid hormones act by ________.

   A. causing neurons to fire; B. increasing muscle contraction; C. diffusing across membranes and regulating the expression of genes; D. binding to receptors on the cell surface and transducing a response without entering the cell

5. The endocrine organ that sits atop a kidney is ________.

   A. the pituitary gland; B. the hypothalamus; C. the ovary; D. the adrenal gland; E. the testicle

6. Bones of the limbs form part of the ________.

   A. appendicular skeleton; B. axial skeleton; C. spongy bone; D. compact bone

7. Osteoblasts ________.

   A. are found in the marrow of bone; B. regulate bone deposition; C. regulate bone reabsorption; D. are more active when calcium is low

8. The antagonistic pairs of arm and leg muscles ________.

   A. allow healthy muscles to compensate for injured ones; B. allow muscles to produce opposing movements; C. allow different types of rotations of joints; D. allow myofibrils to fire in response to different stressors

9. During muscle contraction, ________.

   A. the actin heads pull the sarcomere closed; B. myosin attaches to actin, pulls it toward the center of the sarcomere, releases it, and reattaches farther along; C. myofibrils shrink due to the actions of testosterone; D. muscle fibers contract under the actions of testosterone

10. Which of the following is a true statement regarding muscle contraction?

    A. Actin filaments are stationary, and myosin heads can move; B. Z discs are pulled toward the center of the sarcomere; C. Actin is used up during this process; D. No ATP is required for movement to occur

ANALYZING AND APPLYING THE BASICS

1. When it comes to sex differences, the overlap between genders, even in terms of anatomical differences, can be quite significant. For instance, the female and male pelvises have different shapes, which can affect how the legs are attached. Yet according to the classic study on pelvises, only 40 to 50% of actual women have a female pelvis as defined in anatomy books; 33% have an "android" or male-type pelvis. How is our ability to make generalizations about the sexes affected by such differences?
2. Males experiencing infertility are often advised to wear loose-fitting pants and underwear. How might this practice help increase sperm production?
3. Some studies have shown that chronic marijuana use can lower testosterone levels in males. This can lead to breast development in males, a condition called gynomastia. Why might this occur?

GO FIND OUT

1. Do some research on the so-called son preference demonstrated in China and other regions around the world. What kinds of issues arise when sex ratios are skewed?
2. Caster Semenya is a South African Olympic runner who competes as a female. Do some research in to some of the claims made about her biological makeup. First summarize any biological differences that have been substantiated between her and other female competitors. Next comment on whether you think she has an unfair advantage when competing against women.

MAKE THE CONNECTION

The science that you learned in this chapter has helped you better understand the real-world example used throughout this discussion. Draw a line from the statement on the left to the science that supports it on the right.

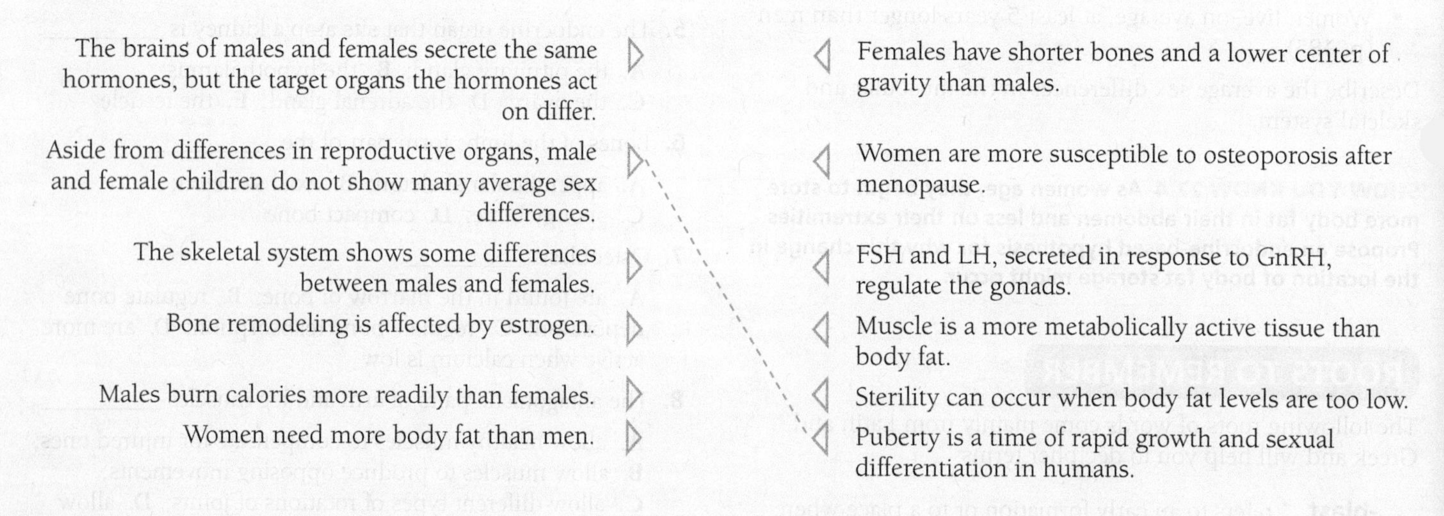

The brains of males and females secrete the same hormones, but the target organs the hormones act on differ.
Aside from differences in reproductive organs, male and female children do not show many average sex differences.
The skeletal system shows some differences between males and females.
Bone remodeling is affected by estrogen.
Males burn calories more readily than females.
Women need more body fat than men.
Females have shorter bones and a lower center of gravity than males.
Women are more susceptible to osteoporosis after menopause.
FSH and LH, secreted in response to GnRH, regulate the gonads.
Muscle is a more metabolically active tissue than body fat.
Sterility can occur when body fat levels are too low.
Puberty is a time of rapid growth and sexual differentiation in humans.

Answers to Got It?, Visualize This, Working with Data, Sounds Right, But Is It?, Show You Know, and Chapter Review questions can be found in the Answers section at the back of the book.

23
Zika in Pregnancy

A baby born with microcephaly may have been exposed to the Zika virus while in its mother's uterus.

Common species of mosquitoes carry this virus.

Prevention efforts are largely focused on killing the mosquitoes that transmit this virus to humans.

Developmental Biology, Menstruation, Birth Control, and Pregnancy

The Zika virus was first discovered in 1947, in a forest in the East African country Uganda. Until recently the virus was relatively unknown. In the past several years, however, the virus has spread at an accelerated rate, and with this outbreak there has been increasing concern over our inability to control it. The virus has been identified in more than 20 countries, including the United States.

The Zika virus is carried by the *Aedes* species of mosquito. If a mosquito that harbors the virus bites you, the virus is transferred to your bloodstream. During the time an infection is active, usually about 1 week, the virus can be transmitted to another person if the mosquito bites you, picks up the virus, and then bites someone else. Mosquito-borne infection is the most common method of transmission. The virus can also be transmitted during sex with an infected person, but it is not transmitted by coughing or sneezing.

The symptoms of Zika infection are mild enough that close to 80% of those infected do not realize that they are. When present, symptoms include a mild fever, rash, joint pain, and pink eye. If only a small percentage of those infected experience symptoms, and these symptoms tend to clear up quickly and without medical intervention, why is there so much concern about the Zika virus?

The concern about the spread of Zika stems from the ability of this virus to move from a recently infected pregnant woman to her unborn baby, with tragic results. When a woman is infected with this destructive virus during the first 3 months of her pregnancy her baby is at risk for severe disabilities. A baby born after Zika infection can have an unusually small brain and head, a condition called microcephaly. Microcephaly can also be caused by genetic disorders, other infections, or exposures to toxic substances during pregnancy. Babies with Zika-caused

—*Continued next page*

microcephaly are severely developmentally disabled, prone to seizures, have impaired vision and hearing, as well as difficulty swallowing, walking, and talking. The life span of babies with Zika-caused microcephaly is not yet known. Currently, South America is experiencing the highest rate of transmission, but transmission in North America is on the rise. Because the mosquitos that carry the virus are common worldwide, the spread of this virus is likely to be difficult to control. There is no vaccine to prevent, or medication to treat, Zika infection. Even tracking exposures is difficult because few people seek medical treatment when they experience symptoms. Current methods to prevent spread have focused on spraying insecticides to kill the mosquitos that harbor the virus in areas where active infections are known to be occurring.

Let's first look at how a Zika infection can affect a baby during the early weeks of a pregnancy and then at what you can do to protect yourself from infection.

23.1 Early Development

Women exposed to Zika in their first trimester of pregnancy are at the highest risk to have affected babies, because of the important changes that occur in the first 10 weeks after fertilization. This is a period of rapid **development,** which includes all the events that take place after fertilization and lead to the formation of a new multicellular organism.

Fertilization

When both sperm and egg are present in the oviduct, fertilization can occur (**FIGURE 23.1**). First, the sperm cells must move through the follicle cells, and then they must pass through a translucent covering on the egg cell called the *zona pellucida*. This protective layer acts as a species-specific barrier. Only sperm produced by a male of the same species as the female that produced the egg can pass through the zona pellucida. This is because getting through the zona pellucida requires binding of a specific receptor on the sperm head. This binding of the sperm head to the zona pellucida triggers the release of enzymes present in the acrosome of a sperm. The acrosomal enzymes interact with the egg cell's zona pellucida to allow the formation of a tunnel through the zona pellucida toward the egg cell's plasma membrane. Once this sperm cell makes it through the zona pellucida, its plasma membrane and the egg's plasma membrane

VISUALIZE **THIS**

At which of the steps depicted would the process be halted if the sperm and egg were from different species?

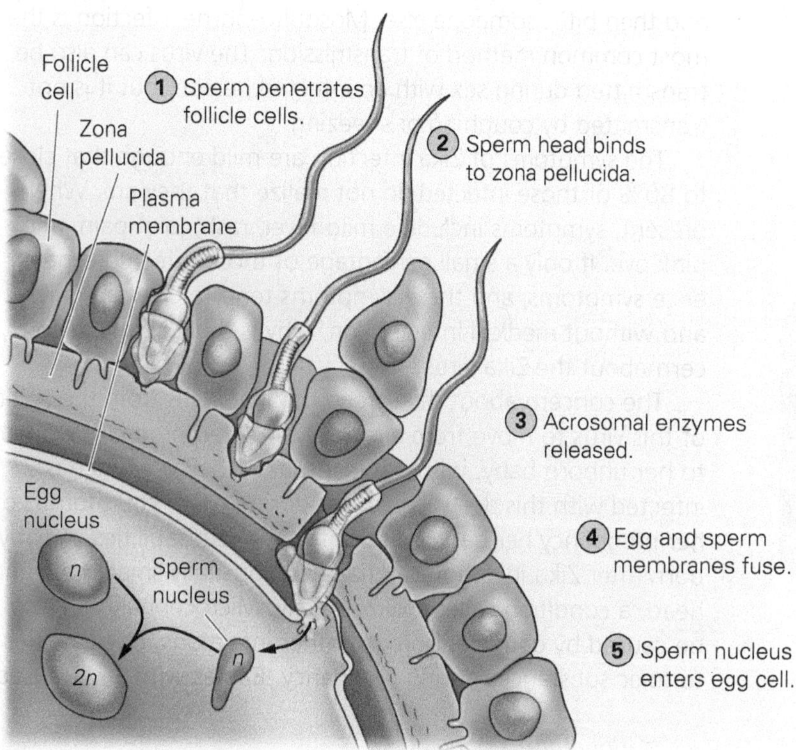

FIGURE 23.1 Fertilization. The sperm cell must traverse follicle cells and the zona pellucida before fusing with the plasma membrane and depositing its chromosomes.

fuse, and the egg cell draws the sperm cell's nucleus inward. The fertilized egg is now ready to begin the divisions of early development.

Embryonic Development

After fertilization, the fertilized egg cell, or zygote, undergoes a series of rapid cell divisions, called **cleavage,** that begin while the zygote is still in the oviduct. These divisions continue until the cluster of cells called a **morula,** which resembles a tiny raspberry, is propelled down the oviduct toward the uterus (FIGURE 23.2).

When the morula reaches the uterus, a few days after fertilization, it burrows in the uterine lining in a process called **implantation,** which gives the rapidly dividing cells access to nutrients. At this time the developing cells take the shape of a hollow ball known as the **blastocyst.** This blastocyst is divided into two clusters of cells, an inner region and outer one. The inner cluster of cells called the inner cell mass will develop into the embryo. The outer cells of the blastocyst nourish and protect the cells that will become the embryo (FIGURE 23.3). These first several weeks of development include the continued rapid cell division in the implanted blastocyst.

Weeks 3 through 9 after fertilization are considered the embryonic period of development; after that time the developing human is referred to as a **fetus.** Exposures to viruses, other pathogens, or chemicals during this critical window of development can have substantial impacts on development. This is because the embryonic period is the time the brain and other organs are developing. After the embryonic period, the fetus is basically fully formed, and for the remainder of the gestation period is mostly growing in size, with its features becoming more refined.

At the beginning of the embryonic period, when the embryo is about the size of a pencil point, the inner cell mass undergoes a dramatic reshuffling of cells to order the embryo into three layers, each of which will specialize, or **differentiate,** into the adult cells and tissues, each with their specific functions. Differentiation occurs in response to chemical signals that pass between different cell layers and adjacent cells, turning on some genes in a given cell and leaving others turned off.

The early embryo differentiates into a structure called the **gastrula,** which is composed of three layers. The process of differentiating into the three-layered structure is called **gastrulation.** The topmost outer layer, or **ectoderm,** gives rise to the brain and the rest of the nervous system, along with the skin and sense organs. The muscles, gonads, skeleton, excretory organs, and a primitive circulatory system will form in the middle layer or **mesoderm.** The innermost layer of cells, the **endoderm,** will become a tube lined with mucus membranes where the lungs, bladder, and intestines will develop.

By about 5 weeks after fertilization, the brain and face begin to develop rapidly. The embryo, which is at this time about the size of a pencil eraser, also has limb

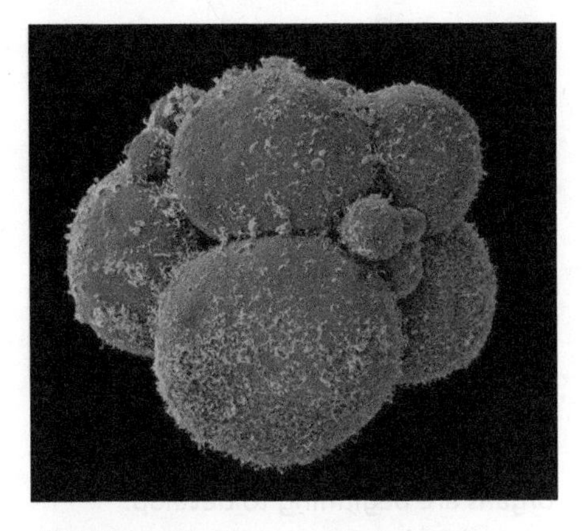

FIGURE 23.2 The morula. A fertilized egg cell begins a series of rapid cell divisions as it moves down the oviduct, producing the morula.

ecto- means outer.
meso- means intermediate.
endo- means inner.

FIGURE 23.3 Early development. The human zygote undergoes cleavage divisions to produce the blastocyst, which implants in the uterus. Continued cell divisions along with specialization result in the formation of the three-layered gastrula.

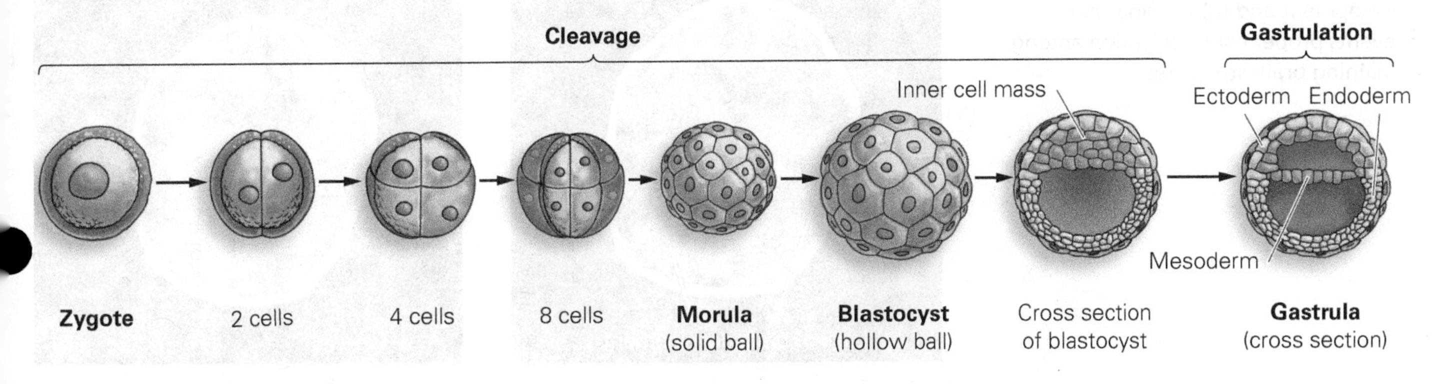

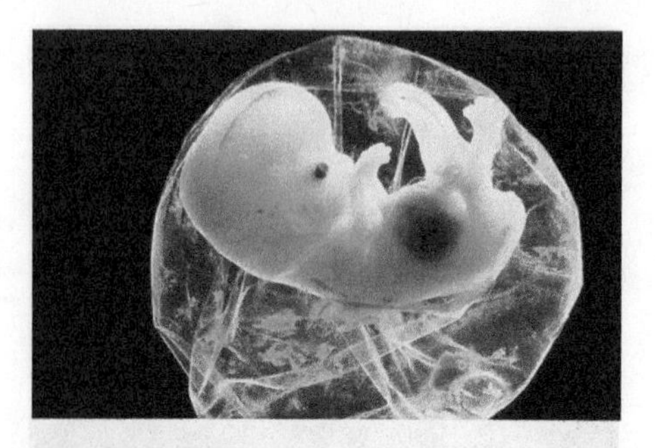

FIGURE 23.4 Human embryo 5 weeks after fertilization. The brain and other organs are beginning to develop.

micro- means small.
-cephaly or **-cephalic** means brain.

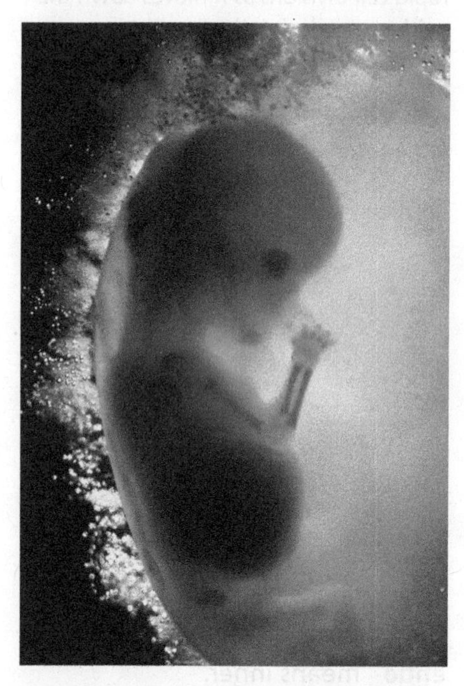

FIGURE 23.5 Human embryo 10 weeks after fertilization. The fetus after the embryonic period has mostly fully developed organs.

buds that will develop into arms and legs (**FIGURE 23.4**). During the remaining 4 weeks of the embryonic period, the ears, eyes, lips, nose, and neck (among other organs and tissues) form. The head becomes rounded and makes up about half the length of the body. At the end of the embryonic period, all of the organs and limbs are in place and the growing cells are now called a fetus (**FIGURE 23.5**).

Development of the Brain When Zika Is Present

The Zika virus preferentially infects brain cells. These can include the very early embryonic cells that will give rise to the brain as well as already differentiated brain cells. When the virus replicates inside these cells, the cells expand and eventually burst, releasing more viruses that infect nearby brain cells. The burst brain cell then dies. As the virus continues to replicate in the brain, more and more brain cells are infected and thus large parts of the brain can fail to develop. This process causes underdevelopment of the brain and it creates scars called calcifications that prevent proper communication between parts of the brain that do develop, leading to severe deformities (**FIGURE 23.6**).

The skull surrounding the brain is also affected. In a healthy pregnancy, as the brain grows in utero, pressure from growing brain tissues forces the skull to enlarge. But when this brain growth is stunted, instead of pushing the skull outward to form a normal shaped head, the skull collapses onto the brain, which results in a misshapen, smaller skull. Because the virus does not infect developing skin cells, the skin around the head continues to grow, producing the wrinkles on the face and neck characteristic of **microcephalic** babies.

Because of the tragic effects of microcephaly, some countries in which the Zika virus is most widespread are recommending couples avoid pregnancy until the virus has been controlled.

Got It?

1. The fertilized egg cell is called the __________.
2. Rapid cleavage divisions of the fertilized egg cell first produce a solid ball of cells called the __________.
3. Implantation occurs when the developing embryo attaches to the __________ lining.
4. The brain develops from the __________ layer of the gastrula.
5. __________ occurs when chemical signals turn some genes in a cell on, and others off, resulting in a cell becoming able to perform its specialized function.

FIGURE 23.6 Brain scan of microcephalic newborn. Infection of brain cells causes (a) a lack of brain development and (b) scarring that prevents proper communication among remaining brain structures.

(a)

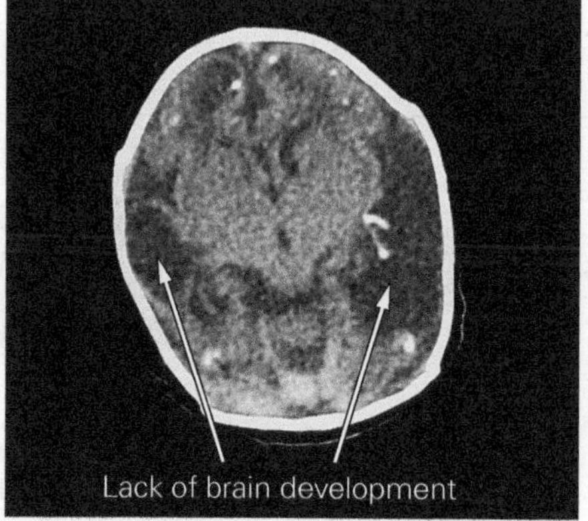

(b)

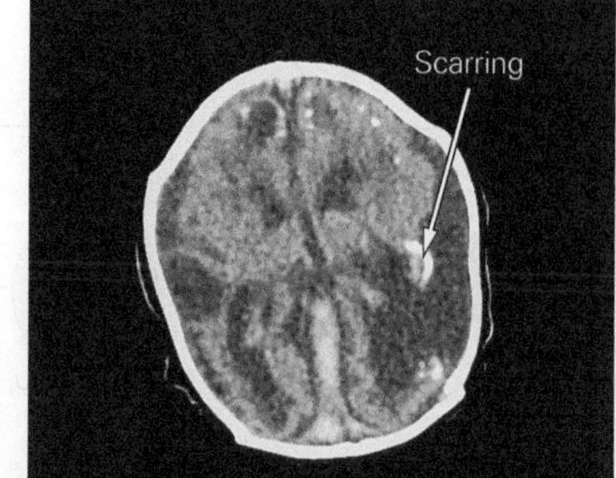

23.2 Avoiding or Delaying Pregnancy

Because the damage to a developing embryo's brain is so severe as a result of the Zika virus, many couples in areas of active transmission are choosing to delay pregnancy. Of course, this is also true for couples who have no reason to be concerned about Zika virus, but are just not ready to be parents. Avoiding or delaying pregnancy can be accomplished by abstinence or through the use of birth control. But before we discuss how different kinds of contraception work, we must first understand the menstrual cycle.

The Menstrual Cycle

The term **menstrual cycle** refers to cyclic changes that occur in fertile women. During the course of one menstrual cycle, a woman's body prepares both an egg for fertilization and the uterus for a potential pregnancy. If no pregnancy occurs, the uterine lining is excreted through the vagina, a process called **menstruation,** at which point a new menstrual cycle begins.

FIGURE 23.7 illustrates the changes in ovaries, hormone levels, and condition of the lining of the uterus, or **endometrium,** that occur throughout the typical 28-day cycle. Note that not every woman's body adheres precisely to a 28-day

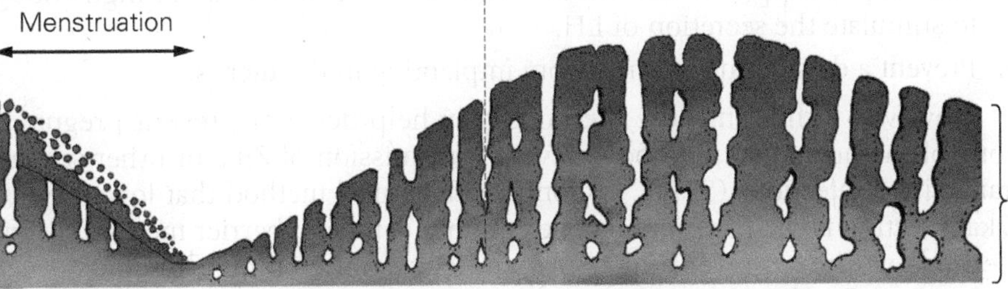

FIGURE 23.7 The menstrual cycle. Changes in hormone levels are linked to (a) the state of ovarian follicles, (b) hormone levels, and (c) uterine condition over the course of the menstrual cycle.

WORKING WITH DATA

Look at the line depicting estrogen level. Explain what causes the fluctuations in the amount of this hormone.

cycle—some women have longer cycles, and others have shorter cycles. The first day of a menstrual cycle is considered to be the first day of actual bleeding.

The hypothalamus, located in the brain, secretes gonadotropin-releasing hormone (GnRH), which stimulates another endocrine structure in the brain—the pituitary gland—to release follicle-stimulating hormone (FSH) and luteinizing hormone (LH). In males, FSH and LH are involved in the production of sperm; in females, these hormones help regulate the development of egg cells and ovulation.

The menstrual cycle operates in response to a feedback system. High levels of estrogen provide positive feedback to the hypothalamus, thereby increasing the secretion of FSH and LH. Conversely, high levels of the hormone **progesterone** provide a negative-feedback effect on the hypothalamus, thereby decreasing the secretion of GnRH, in response to which FSH and LH levels decline.

The levels of estrogen and progesterone are regulated by a woman's body:

1. Inside the ovary, the developing egg cell is surrounded by hormone secreting cells that together comprise the **follicle** (left half of Figure 23.7). The follicle begins to grow and secrete estrogen. When the follicle is large enough, it produces enough estrogen to stimulate GnRH release. This release leads to a spike in both FSH and LH levels, which lasts for about 24 hours. Ovulation occurs 10 to 12 hours after the LH peak, 14 days before menstruation in a typical 28-day cycle.
2. After ovulation, what remains of the follicle in the ovary is called the **corpus luteum** a structure that produces estrogen and progesterone (right half of Figure 23.7). Progesterone helps maintain the blood flow to the uterine lining to support early fetal development. A negative feedback loop is at work as progesterone inhibits the production of LH production. Therefore, the LH surge is inhibited by the ovarian hormone that it stimulates.

If fertilization does not occur, the corpus luteum degenerates around 10 days after ovulation. Because the corpus luteum is no longer secreting them, progesterone and estrogen levels fall, triggering the arteries that supply the uterus to spasm. This causes the lining of the uterus to be shed and often causes menstrual cramps. Decreasing levels of progesterone also release the hypothalamus from inhibitory control. Therefore, LH and FSH levels rise, and the cycle starts over.

Some birth control methods prevent the typical cycling of ovarian hormones to prevent ovulation. Others prevent the fertilized egg from implanting in the uterus and others prevent male and female gametes from encountering each other.

Birth Control Methods

The majority of U.S. couples using birth control rely on the woman to use a hormonal method like the birth control pill. (For more information on hormonal, and other, birth control methods see TABLE 23.1.) Hormonal methods typically work by altering the menstrual cycle in a few characteristic ways. These methods can do the following:

1. Change the consistency of the cervical mucus from the slick, fertility-enhancing type to the dry fertility-decreasing type, thereby impeding sperm ascent (see Chapter 7).
2. Prevent ovulation by preventing the LH surge. For example, the birth control pill supplies a continuous dose of estrogen that is not high enough to stimulate the secretion of LH.
3. Prevent a developing embryo from implanting in the uterus.

However, although these methods may help delay or prevent pregnancy, none of the hormonal methods prevent transmission of Zika or other sexually transmitted infections (STIs). The only birth control method that lowers risk of Zika infection is the male and female condom, a type of barrier method. Barrier methods work by preventing sperm and egg from coming into contact with each other. Condoms must be used from start to finish every time vaginal or

TABLE 23.1 Birth Control Methods.

The mechanism, risk, and efficacy of various birth control methods. Unintended pregnancy rates assume the method is used correctly and consistently as directed. The percentages indicate the number out of every 100 women who had an unintended pregnancy every year.

Method	Mode of Action	Risk and Efficacy
Abstinence		
	Sperm and egg never have contact.	• No risks • Unintended pregnancy rate of 0%

Method	Mode of Action	Risk and Efficacy
Hormonal		
Combination birth control pill	Synthetic estrogen and progesterone given at continuous doses prevent ovulation.	• Increased risk of heart disease and fatal blood clots for women over age 35 who smoke; slightly increased risk of breast cancer; does not protect against STIs • Unintended pregnancy rate of 6%
Minipill	Progesterone-only pill thickens cervical mucus, impedes sperm ascent, and does not allow the uterus to support a pregnancy.	• Increased risk of ovarian cysts • Does not protect against STIs • Unintended pregnancy rate of 6%
Emergency birth control pill (also known as the "morning after pill" and "Plan B")	High doses of hormones prevent sperm from reaching egg or prevent fertilized egg from attaching to wall of uterus, or prevent ovulation. This method is not the same as mifepristone, which has been dubbed the "abortion pill." Emergency birth control pills *prevent* pregnancy, and mifepristone terminates an established pregnancy.	• Does not protect against STIs • Nausea, vomiting, abdominal pain, fatigue, headache • Unintended pregnancy rate of 20%
Patch	When applied to skin, the patch delivers progesterone and estrogen to prevent ovulation and thicken cervical mucus.	• Same as combination pill • Does not protect against STIs • Unintended pregnancy rate of 9%
Vaginal ring	When inserted in vagina, the vaginal ring delivers progesterone and estrogen to prevent ovulation and thicken cervical mucus, blocking sperm ascent.	• Same as combination pill • Does not protect against STIs • Unintended pregnancy rate of 9%
Injectables	This method requires the user to have progesterone injections every 3 months, and the mode of action is the same as the minipill.	• Irregular vaginal bleeding • Does not protect against STIs • Unintended pregnancy rate of 6%
Implantables	Implantables, such as Implanon, secrete progesterone to prevent ovulation and to thicken cervical mucus.	• Irregular vaginal bleeding • Does not protect against STIs • Unintended pregnancy rate of 1%

—Continued

TABLE 23.1 Birth Control Methods. *Continued—*

Method	Mode of Action	Risk and Efficacy
Barrier		
Cervical cap	When inserted against cervix before intercourse, the cervical cap prevents sperm and egg contact. Usually used in concert with spermicidal agent.	• No known risks • Does not protect against STIs • Unintended pregnancy rate of 15%
Diaphragm	When inserted into vagina before intercourse, the diaphragm prevents sperm and egg contact.	• No known risks • Does not protect against STIs • Unintended pregnancy rate of 12%
Sponge	The spermicide-soaked sponge fits against the cervix to prevent sperm and egg contact.	Unintended pregnancy rate of 12% in women who have not had children and 24% in women who have had children
Female condom	The female condom is held against the cervix by a flexible ring to prevent sperm and egg contact.	• No known risks • Unintended pregnancy rate of 21%
Male condom	The male condom fits over the penis to prevent sperm and egg contact.	• No known risks • Unintended pregnancy rate of 18%
Other		
Spermicides	When inserted into vagina 1 hour before intercourse, sperm are killed after intercourse. Spermicide can be found in gels, suppositories, sponges and films.	• No known risks • Does not protect against STIs • Unintended pregnancy rate of 28%
Fertility awareness (also known as the rhythm method)	Fertility awareness requires abstinence for the 4 days before ovulation, the day of ovulation, and several days after ovulation. This involves charting when ovulation usually occurs based on a temperature increase that happens after ovulation and monitoring cervical mucus.	• Does not protect against STIs • Unintended pregnancy rate of 24%
Intrauterine device	This small plastic device inserted into the uterus by a physician prevents fertilization and prevents the uterus from supporting a pregnancy for many years, until its removal.	• May increase risk of pelvic inflammatory disease • If pregnancy does occur, increased risk of miscarriage or ectopic pregnancy • Unintended pregnancy rate of 1% per year
Male sterilization (vasectomy)	Each vas deferens is clamped, cut, or sealed to prevent sperm from being ejaculated in the semen.	• Does not protect against STIs • Unintended pregnancy rate of 0.15% per year

anal sex occurs. Some couples will also use dental dams (sheets of latex or other materials) during oral sex to decrease risk of transmission of sexually transmitted infections (STIs), including Zika. TABLE 23.2 lists the causes, symptoms, and treatments of the most common STIs.

TABLE 23.2 Sexually transmitted infections.

The causes, symptoms, and treatment or prevention of common STIs.

Infectious Agent	General Information	Symptoms	Treatment and Prevention
Bacterial Pathogens			
Chlamydia is caused by *Chlamydia trachomatis.*	Infects the urethra, cervix, and oviducts of women and the urethra of men.	Pelvic pain, fluid discharge	• Treat with antibiotics • Reduced transmission with condom use
Gonorrhea is caused by *Neisseria gonorrhoeae.*	Known as "the clap"; can be spread from infected mothers to babies during birth.	Thick discharge from penis or vagina	• Treat with antibiotics • Reduced transmission with condom use
Syphilis is caused by *Treponema pallidum.*	Infectious sores, or chancres, can be found on the vagina, penis, anus, rectum, lips, and mouth.	If untreated, neurological problems, paralysis, and death can occur.	Can be treated with antibiotics if caught early.

Infectious Agent	General Information	Symptoms	Treatment and Prevention
Viral Pathogens			
AIDS is caused by human immunodeficiency virus (HIV).	Transmitted via oral, anal, and vaginal sex, as well as via blood transfusion.	Over time weakens the immune system; eventually severe tissue and organ damage	• Combination-drug therapies to halt progression; no cure • Reduced transmission with condom use
Genital warts are caused by human papilloma virus (HPV).	Can be transmitted by oral, anal, and vaginal sex.	• Growths or bumps on the pubic area, penis, vulva, or vagina • Genital warts (from some types of HPV); cervical cancer (from certain types of HPV)	• Vaccination to prevent infection by many strains • Reduced transmission with condom use

—Continued

TABLE 23.2 Sexually transmitted infections. *Continued*

Infectious Agent	General Information	Symptoms	Treatment and Prevention
Viral Pathogens			
Hepatitis B is caused by hepatitis B virus (HBV).	• Transmitted via blood and other body fluids. • An infant can acquire infection from the mother during delivery.	Inflammation and scarring of the liver (cirrhosis), which may be fatal	• Vaccination for prevention • Reduced transmission with condom use
Herpes is caused by two types of herpes simplex viruses.	Transmitted via direct skin-to-skin contact or kissing, or by oral, anal, or vaginal sex.	• Cold sores or fever blisters on mouth or face • Genital herpes is sores in the genital area	• Antiviral medications to lessen symptoms; no cure • Reduced transmission with condom use

Infectious Agent	General Information	Symptoms	Treatment and Prevention
Insect, Protozoan, and Fungal Pathogens			
Pubic lice are caused by *Pediculus pubis* (insect).	Also known as "crabs"; transmitted via skin-to-skin contact or contact with an infected towel or clothing.	Itching of pubic area around 5 days after initial infection	Cured by washing the affected area with a delousing agent
Trichomoniasis is caused by *Trichomonas vaginalis* (protozoan).	Also known as "trich"; transmitted by oral, anal, and vaginal sex.	• In women, vaginal itching with a frothy yellow-green vaginal discharge • In some men, irritation in urethra after urination or ejaculation	• Treated with antibiotics • Reduced transmission with condom use
Yeast infections are caused by *Candida albicans* (fungi).	• Normal inhabitants of the female reproductive tract. • Antibiotic use may deplete normal vaginal flora allowing yeast to proliferate.	In women, thick whitish discharge from vagina and vaginal itching	Cured by antifungal medicines

Barrier methods that are worn inside the woman's body, like the cervical cap, vaginal sponge, and diaphragm, do not protect against STIs that are transmitted from the semen to the mucus membranes of the vagina. Nor do they prevent transmission of the virus present in vaginal fluids to sexual partners.

In sexually active couples living in, or traveling to, areas where Zika is active, men are advised to wait at least 6 months to try to conceive, and women are advised to wait 8 weeks. This difference in wait times is because the virus remains in the semen for longer than it does in vaginal fluids.

Even those not hoping to conceive in the near future should take precautions to help prevent infections that could harm pregnant women. For example, if you travel to an area where Zika is circulating and pick up the infection, when you come home, you can infect an unsuspecting pregnant woman if a mosquito bites you, then bites her.

To protect currently pregnant women, the CDC recommends that people who live in or travel to areas of active transmission use mosquito repellent, wear long sleeves and pants, and stay away from standing water or other mosquito breeding grounds.

Got It?

1. The dominant hormone released by the ovary early in the menstrual cycle is ________.
2. The dominant hormone released from the ovary in the second half of the menstrual cycle is ________.
3. Hormonal birth control methods can prevent ovulation by preventing the surge in ________.
4. Some hormonal birth control methods prevent the embryo from implanting in the ________.
5. ________ methods of birth control prevent sperm and egg from coming in to contact with each other.

23.3 Pregnancy and Childbirth

When sufficient levels of hCG extend the life of the corpus luteum and the corpus luteum remains intact, progesterone and estrogen levels remain high, and the endometrium is maintained, thereby allowing implantation and embryo development. After about 12 weeks of pregnancy, the **placenta,** an endocrine organ produced during pregnancy, takes over the production of hormones.

The placenta is produced by the cooperative actions of the embryo and the uterus. The outer ring of cells in the three-layered gastrula form the **trophoblast,** which secretes enzymes that make it possible for the blastocyst to implant in the wall of the uterus.

Eventually, the implanted trophoblast begins to infiltrate the uterine lining, forming fingerlike projections that are able to carry fetal blood. Maternal uterine blood vessels erode to form pools of blood around these projections (**FIGURE 23.8**). This close positioning of fetal and maternal blood supplies allows the exchange of nutrients and wastes to occur. Substances that can be freely exchanged between the fetus and the mother include oxygen, carbon dioxide, water, salts, hormones, and many drugs. Blood cells and bacteria do not

FIGURE 23.8 The placenta. The placenta is formed of maternal and fetal tissues. Projections from the embryonic trophoblast develop over time to allow the exchange of nutrients between the mother and fetus.

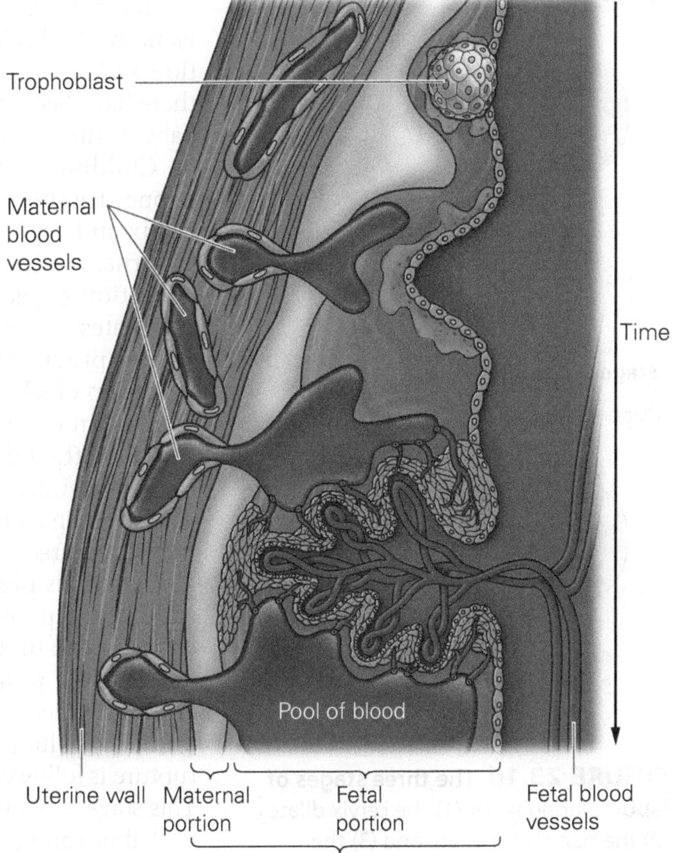

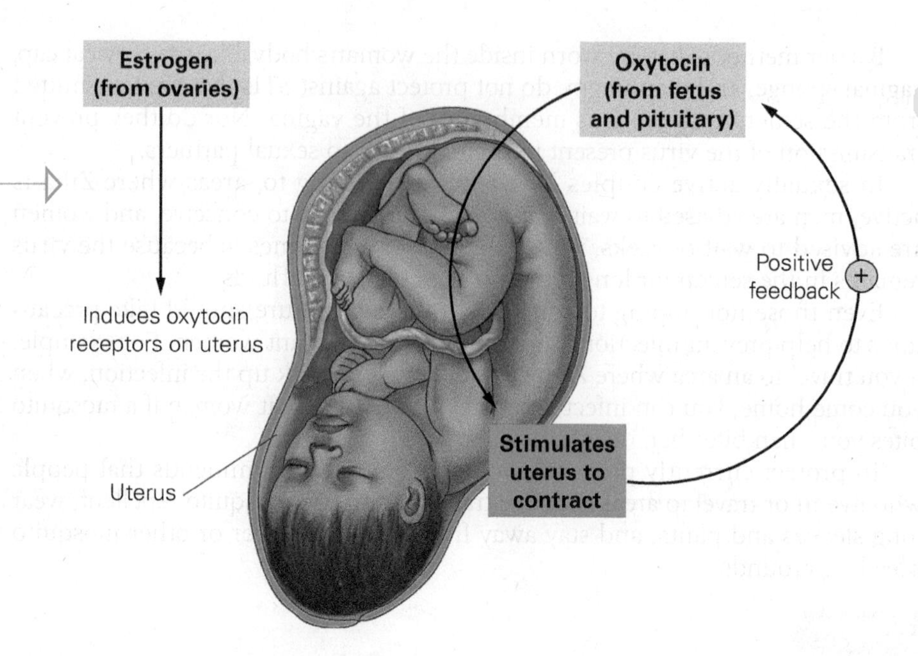

FIGURE 23.9 Induction of labor. Estrogen and oxytocin are involved in a feedback loop that stimulates labor.

VISUALIZE **THIS**

To artificially induce labor in their pregnant patients, physicians will provide an oxytocin-like drug. What effect would that have?

normally pass between fetal and maternal blood supplies, but viruses can, as can antibodies produced in response to a previous infection in the mother. Development of the placenta also seems to be affected by Zika. The nutrient and waste exchange that normally occur via the placenta are reduced in pregnancies where Zika infection has occurred. This may slow the baby's growth in the rest of the body, not just the brain.

The placenta also functions as an endocrine organ, producing estrogen and progesterone. These hormones, along with the pituitary hormone **prolactin** stimulate development of the mother's mammary glands. Shortly before delivery, the mother's estrogen level drops, and prolactin is produced by the pituitary gland of the brain. After delivery, prolactin continues to stimulate the production and secretion of breast milk, or **lactation,** during breastfeeding. To date, there have been no reported cases of a mother transmitting the Zika virus to her baby during breastfeeding.

Childbirth comprises both labor and delivery. **Labor** is characterized by strong, rhythmic contractions of the uterus that move the baby through the cervix and vagina. Induction of labor is hormonally controlled (**FIGURE 23.9**). Maternal estrogen levels increase toward the end of pregnancy, triggering the formation of receptors on the uterus for a hormone called **oxytocin,** which stimulates the smooth muscles lining the wall of the uterus to contract. Oxytocin is produced by fetal cells and by the mother's pituitary gland. During the induction of labor, oxytocin is regulated by positive feedback. During labor, the increase in oxytocin causes uterine contractions. Uterine contractions in turn stimulate the release of more oxytocin, resulting in progressively intensifying muscle contractions that ultimately expel the baby from the uterus.

A woman usually knows when labor is approaching because she begins to feel uterine contractions. Another sign that labor is approaching include the passing of the mucus plug. The mucus plug forms when cervical secretions produced early in the pregnancy thicken and block the cervix to prevent bacteria from entering the uterus. Toward the beginning of labor, when the cervix begins to dilate, the plug loosens and is passed through the vagina. This is sometimes called the *bloody show.* Another sign that labor is imminent occurs when the **amnion,** a fluid-filled sac that surrounds and protects the fetus, ruptures. This rupture is followed by the loss of fluid, occurring as either a gush or a slow trickle This stage is often referred to as a woman's *water breaking* prior to delivery.

Labor can be thought of as occurring in three stages (**FIGURE 23.10**). During the first stage of labor, the cervical opening enlarges or dilates, going from about

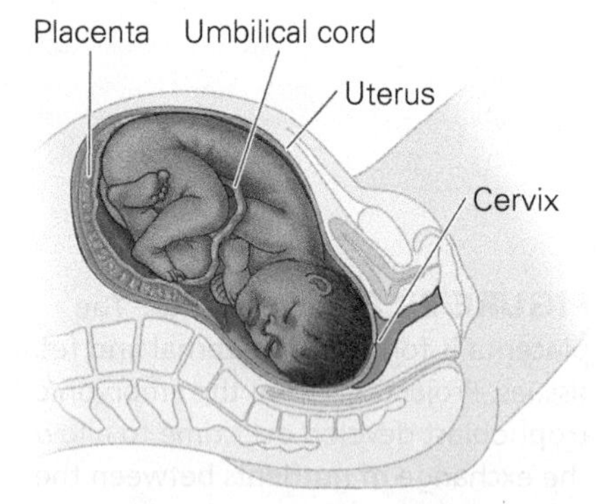

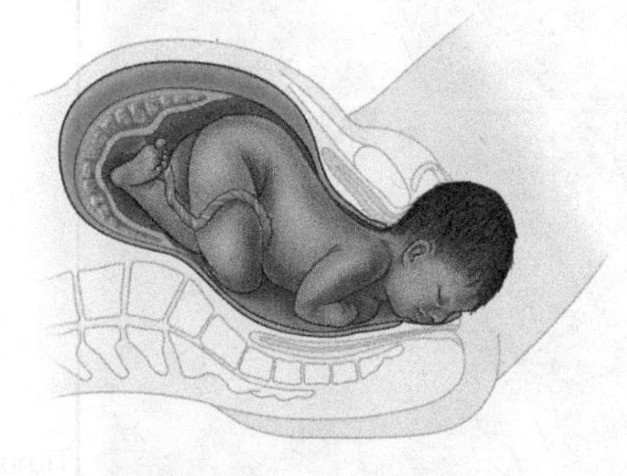

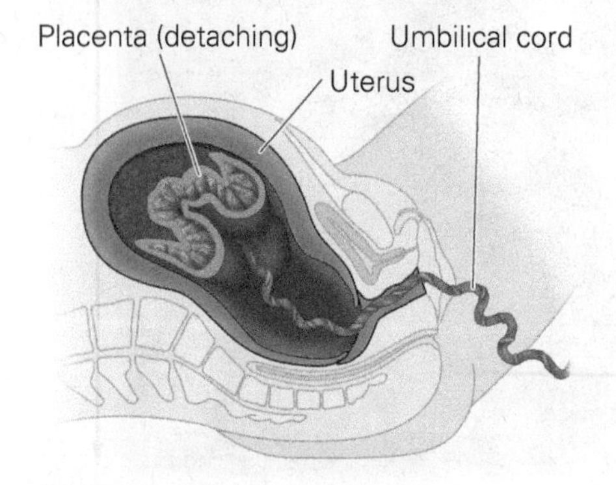

FIGURE 23.10 The three stages of labor. During labor (1) the cervix dilates, (2) the baby is expelled, and (3) the placenta is delivered.

the size of a pinhead to around 10 centimeters. Contractions of the uterine muscle tissues help to dilate the cervix because they apply pressure to the baby. The baby is forced against the cervical opening, which forces the opening to enlarge. The second stage of labor involves expulsion (delivery) of the baby through the narrow pelvic opening and birth canal, often aided by the mother's active "pushing" (contractions of the abdominal muscles); the third stage of labor is delivery of the placenta.

Zika can be transmitted from a mother to her baby during childbirth, but the outcome of a Zika infection this late in the baby's in utero development is not yet known.

Got It?

1. The __________ is the endocrine organ involved in nutrient and gas exchange that is produced by the cooperative actions of the embryo and the mother's uterus.
2. The implanted __________ infiltrates the uterine lining, forming projections that allow nutrient and waste exchange between the blood of the mother and developing fetus.
3. The placenta secretes the hormone __________, which helps stimulate lactation.
4. The hormone __________ exhibits positive feedback on the uterus during labor.
5. The membranous sac that cushions and protects the developing baby and ruptures during labor is called the __________.

Sounds right, but is it?

The normal cycling of estrogen that occurs in a fertile female involves an increase in estrogen level before the release of an egg cell from the ovary. Ovulation occurs around the 14th day of a typical 28-day menstrual cycle. After ovulation, estrogen levels decrease back to a basal level and then slowly increase again in a continuous cycle until they peak before ovulation.

When a woman takes the birth control pill, this cycling of estrogen is prevented. Instead, her body is exposed to one continuous low dose of estrogen. Low estrogen levels can lead to increased risk of osteoporosis. This risk might lead you to the following:

Taking the birth control pill leads to increased risk of osteoporosis.

Sounds right, but it isn't.

Answer the following questions to understand why.

1. Assume the ovaries of a fertile woman produce a total of 49 mg of estrogen every cycle. Assume that the basal level of estrogen produced is 1 mg/day except during the increase, plateau, and decrease that occur on days 11, 12, and 13, respectively. How many days during a 28-day cycle is a woman producing a basal level of estrogen?
2. What is the total combined mg amount of estrogen that a woman produces during all the days she is producing estrogen at the basal level?
3. How much estrogen must she produce in the 3 days before ovulation (days 11, 12, and 13)?
4. Pharmaceutical companies design birth control pills that mimic the bone-protecting effects of the normal female menstrual cycle. Even though most birth control packs have placebo pills that contain no estrogen and allow for natural menstruation, assuming that a woman takes the birth control pill every day of the month, how many mg of estrogen should be found in each pill?
5. Consider your answers to questions 1–4 and explain why the original statement bolded above sounds right, but isn't.

THE BIG QUESTION

When does human life begin?

Scientists use different terms to describe the stages of the development of a human, from zygote to embryo to fetus to baby. Some people believe human life begins at conception, others believe the early embryo is simply a mass of cells. At what point in this developmental progression does life begin?

What should I know?

What follows are some smaller questions that need to be resolved to answer the Big Question. Place a checkmark next to the questions that science can answer.

Smaller Questions	Can Science Answer?
Can any kind of cell become a viable human or only a fertilized egg cell?	
When in development does the embryo, fetus or baby display the characteristics associated with living organisms?	
Does survival in a petri dish mean cells are alive?	
Should embryos used for in vitro fertilization procedures be experimented on or thrown away?	
How old must a fetus be to survive on its own outside the womb?	
How much should society spend on keeping premature babies alive?	

What does the science say?

Let's examine what the data say about this smaller question:

How old must a fetus be to survive on its own outside the womb?

The data in the table shown is compiled from many sources and includes a range of survival rates for length of gestation. A range of rates is included because there are many factors aside from length of gestation that effect survival rates, like quality of hospital care.

Length of gestation at birth (weeks postfertilization)	Range of survival rates (percentage)
19	0
20	0–10
21	10–35
22	40–70
23	50–80
24	80–90
25	> 90
28	> 95
32	> 98

1. Describe the results. Is it possible to say with any precision how old a fetus must be to survive on its own? Why or why not?
2. Given these data, do you think the smaller question is answered? If not, propose another study that would help answer this question.
3. Does this information help you answer the Big Question? What else do you need to consider?

Data Source: T. Raju, B. Mercer, D. Burchfield, and G. Joseph Jr., "Periviable Birth: Executive Summary of a Joint Workshop by the *Eunice Kennedy Shriver* National Institute of Child Health and Human Development, Society for Maternal-Fetal Medicine, American Academy of Pediatrics, and American College of Obstetricians and Gynecologists," *American Journal of Obstetrics and Gynecology* 210 (2014): 406–417.

Chapter Review

Mastering Biology

Go to Mastering Biology to access **eText 2.0, Dynamic Study Modules,** and the **Study Area,** where you'll find practice quizzes, BioFlix™ animations, MP3 tutor sessions, current events, and more.

SUMMARY

Section 23.1

List the steps involved in fertilization.

- The zygote undergoes rapid cleavage divisions to produce the morula, which then becomes the blastocyst. The inner cell mass of the blastocyst becomes the embryo and eventually the fetus (pp. 492–493).
- The blastocyst gives rise to the three-layered gastrula. Each layer of the gastrula (ectoderm, mesoderm, and endoderm) differentiates into specific tissues (pp. 493–494).

SHOW YOU KNOW 23.1 Explain why the embryonic period is the time when the most damage can occur during development.

Section 23.2

Describe the stages of early embryonic development, from the rapid, early cell divisions through early tissue differentiation.

- During the menstrual cycle, levels of estrogen, produced by the follicle, begin to rise, eventually stimulating GnRH release and causing a spike in FSH and LH levels. LH causes ovulation to occur (pp. 495–496).
- The remnant of the follicle that stays in the ovary after ovulation is the corpus luteum, which secretes progesterone and estrogen. If fertilization does not occur, the corpus luteum degenerates, progesterone and estrogen levels fall, and menstruation occurs. Then LH and FSH levels rise, and the cycle starts over (p. 496).

Describe the hormonal regulation of the menstrual cycle.

- Some hormonal methods of birth control provide a continuous dose of estrogen, which prevents the LH surge, thereby preventing ovulation (pp. 496–498).
- Progesterone helps maintain the uterine lining and causes the cervical mucus to thicken, preventing sperm from reaching the egg in the oviduct (pp. 496–498).

SHOW YOU KNOW 23.2 Women who take the so-called combination birth control pill, which contains both estrogen and progesterone, experience a mild menstruation, called a withdrawal bleed, if they miss several doses or take placebo sugar pills for several days. Why would the withdrawal bleed occur?

Section 23.3

Describe the mechanisms of the estrogen- and progesterone-based methods of birth control.

- Pregnancy hormones are produced by the placenta, which forms in the uterus from the trophoblast of the blastocyst. These hormones help maintain the uterus and prepare the breasts for lactation (pp. 501–502).

Discuss the types of organisms that can cause sexually transmitted infections and the methods by which they are transmitted.

- Childbirth comprises labor and delivery. Labor is signaled by contractions that help expel the baby from the uterus. During labor, the cervix dilates to allow the baby's passage out of the uterus. After delivery of the baby, the mother also delivers the placenta (pp. 502–503).

Outline the hormonal controls regulating pregnancy and birthing.

Describe the stages of labor and delivery.

SHOW YOU KNOW 23.3 Pregnant women do not ovulate. By what mechanism does the progesterone produced by the placenta prevent ovulation?

ROOTS TO REMEMBER

The following roots of words come mainly from Latin and Greek and will help you to decipher terms:

-cephaly and -cephalic	means brain. Chapter term: *microcephaly*
ecto-	means external or outer. Chapter term: *ectoderm*
endo-	means internal or inner. Chapter term: *endoderm*
meso-	means intermediate. Chapter term *mesoderm*
micro-	means small. Chapter term: *microcephaly*

LEARNING THE BASICS

1. Why can two cells with the same DNA have different structures and functions?

2. What does the inner cell mass of the blastocyst give rise to?

3. Add labels to the figure that follows, which illustrates the rise and fall of hormones involved in menstruation.

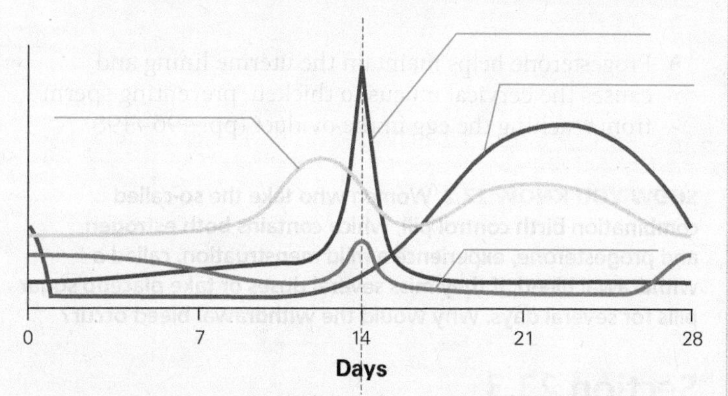

4. Ovulation occurs in response to ______.

A. a decrease in estrogen; B. an increase in FSH; C. an increase in LH; D. an increase in progesterone

5. Menstruation is regulated such that ______.

A. increasing estrogen levels have a positive feedback effect on FSH and LH; B. increasing FSH levels lead to ovulation; C. as progesterone levels increase, so do FSH and LH levels; D. ovulation occurs on the fifth day of the cycle; E. the placenta produces FSH, which stimulates ovulation

6. Which layer of the developing embryo gives rise to the brain and nervous system?

A. endoderm; B. mesoderm; C. lipoderm; D. ectoderm

7. Which hormone helps prepare the uterus for implantation?

A. estrogen; B. progesterone; C. LH; D. FSH

8. The remnant of the follicle that remains in the ovary after ovulation is called the

A. trophoblast; B. gastrula; C. corpus luteum; D. placenta

9. Projections of which structure carry fetal blood?

A. trophoblast; B. gastrula; C. placenta; D. ectoderm

10. Which hormone exhibits positive feedback on the brain during labor?

A. FSH; B. oxytocin; C. testosterone; D. estrogen

ANALYZING AND APPLYING THE BASICS

1. Assume a woman was infected with Zika when she is a teenager. In her mid-20s she becomes pregnant. Will her child be affected? Why or why not?

2. The Zika virus is present in the blood during the time of active infection, typically for about 1 week. Assume you have some of the symptoms of Zika infection and see your physician a week or so after the symptoms began. Blood is withdrawn and genetically tested for the presence of Zika. Your blood test comes back negative. Does this mean you were not infected with Zika? Why or why not?

3. Why does labor become increasingly painful as childbirth approaches?

GO FIND OUT

1. Research the effects of exposure to cigarette smoke or alcohol on embryonic development. What are the most common outcomes to a baby exposed to these chemicals during development?

2. Women who live together often report that their menstrual cycles are synchronized. In other words, they start menstruating at around the same time each month. Do some research to find out whether there is any good science to support menstrual synchrony.

MAKE THE CONNECTION

The science that you learned in this chapter has helped you better understand the real-world example used throughout this discussion. Draw a line from the statement on the left to the science that supports it on the right.

Babies infected with the Zika virus in the 10 weeks or so of development suffer the most debilitating damage.	People who have been exposed to Zika build immunity to subsequent exposures.
Some pregnant women bitten by a Zika infected mosquito will not experience an active infection, nor will their babies.	The Zika virus remains in semen longer than it does in vaginal fluids.
Men with Zika should wait longer than women with Zika to attempt to conceive.	Methods of birth control that allow exposure to semen and vaginal fluids do not protect against Zika infection.
Barrier methods and abstinence are the only birth control methods that help lower risk of sexual transmission of Zika.	If a Zika transmitting mosquito bites an infected person and transmits the virus to another person, a local outbreak can occur.
Removing all standing water from around your home can help protect against Zika infection.	Being in close proximity to mosquito breeding sites can increase risk of exposure.
Using insect repellant for 3 months after travel to areas where Zika is circulating is recommended by the CDC.	The majority of fetal brain development occurs during the first trimester of pregnancy.

Answers to **Got It?, Visualize This, Working with Data, Sounds Right, But Is It?, Show You Know,** and **Chapter Review** questions can be found in the **Answers** section at the back of the book.

24

Study Drugs: Brain Boost or Brain Drain?

Some students misuse prescription medications to help them stay awake to study.

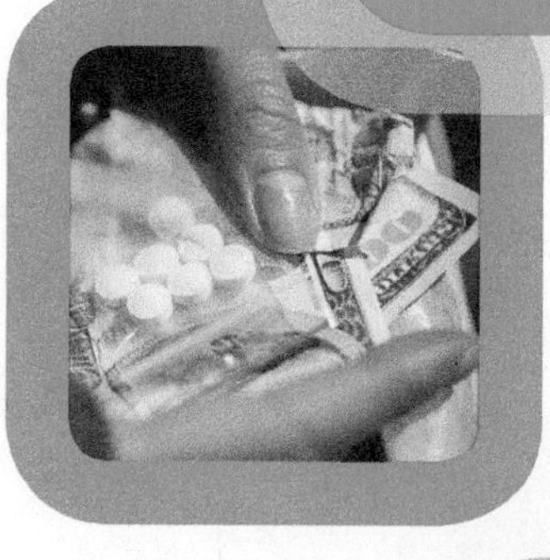

They may get these drugs from a friend with a prescription . . .

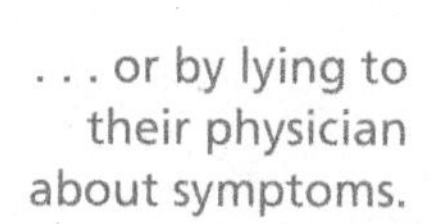

. . . or by lying to their physician about symptoms.

Brain Structure and Function

Your roommate is carrying a full load of classes and is only just learning to estimate the time she needs to complete her work. As midterms approach, she finds she does not have enough time to study for tests, finish her papers, *and* sleep. She confides in you that a mutual friend who lives down the hall has agreed to sell her some of his attention deficit disorder (ADD) medication so she can use it to pull some all-nighters. She wonders what the health risks would be, whether this is legal, and whether it would actually help her perform better on her exams and write better papers.

The use of ADD medications, like Adderall™, Vyvance, Ritalin, Dexedrine, and Concerta is legal for those who have been diagnosed with ADD or ADHD (attention deficit hyperactivity disorder). When used properly, these medications can help people function despite a propensity for forgetfulness, impulsivity, distractibility, fidgeting, impatience, and restlessness.

The misuse of stimulant medication, that is, taking the medication without a prescription, is reported by close to 30% of college students. They typically get the medications from fellow students who have been prescribed the medication to treat their ADD, or by lying to health-care providers to obtain a prescription. Distributing or buying controlled substances, like ADD medications, is an illegal activity called drug diversion. The minimum sentence for diverting ADD medications in the United States is 5 years in prison.

In addition to legal consequences, there are also health consequences to misuse of ADD medications. Misuse of ADD medications can cause insomnia, blurred vision, and increased body temperature. Misuse also increases heart rate and blood pressure, which can cause cardiovascular problems, including stroke and even death. Taking high doses of stimulants or misusing them for a long period of time can lead to paranoia, hostility, and psychosis—a mental disorder that includes the loss of contact with reality.

With the legal and medical consequences, why is the use of these so-called smart drugs so prevalent? It may be because those misusing the drugs believe that the drugs are helping them to learn more and perform better in school. But is this true? To answer this question, we must first learn about the biology of the human nervous system.

24.1 The Nervous System

Every second, millions of signals make their way to your brain and tell it what your body is doing and feeling. Your **nervous system,** which includes your brain, spinal cord, sense organs, and the nerves that interconnect those organs, receives and interprets these messages and decides how to respond. This requires the actions of specialized cells called **neurons.** Neurons, or nerve cells, carry electrical and chemical messages between your brain and other parts of your body. These electrochemical message-carrying cells can be bundled together, producing structures called **nerves.**

Central and Peripheral Nervous Systems

The brain and spinal cord together form the **central nervous system (CNS).** The central nervous system is the seat of intelligence, learning, memory, and emotion. Neurons in your eyes, ears, or skin transmit information to your CNS. There the information is processed, and the appropriate response is relayed back to a second subdivision of the nervous system, the **peripheral nervous system (PNS),** which includes the network of nerves that radiates out from the brain and spinal cord, gathering information to relay back to the CNS as well as to muscles and glands (**FIGURE 24.1**).

To carry information, the cells of the nervous system pass electrical and chemical signals to each other. Signals are

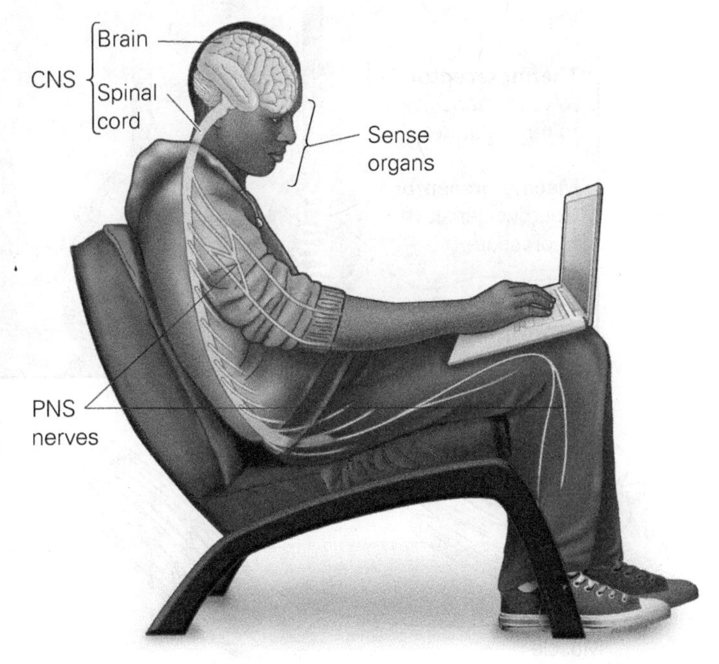

FIGURE 24.1 The central and peripheral branches of the nervous system. The brain and spinal cord of the CNS work with the PNS neurons that radiate throughout the body detecting stimuli and transmitting signals back to the CNS or to muscles and glands.

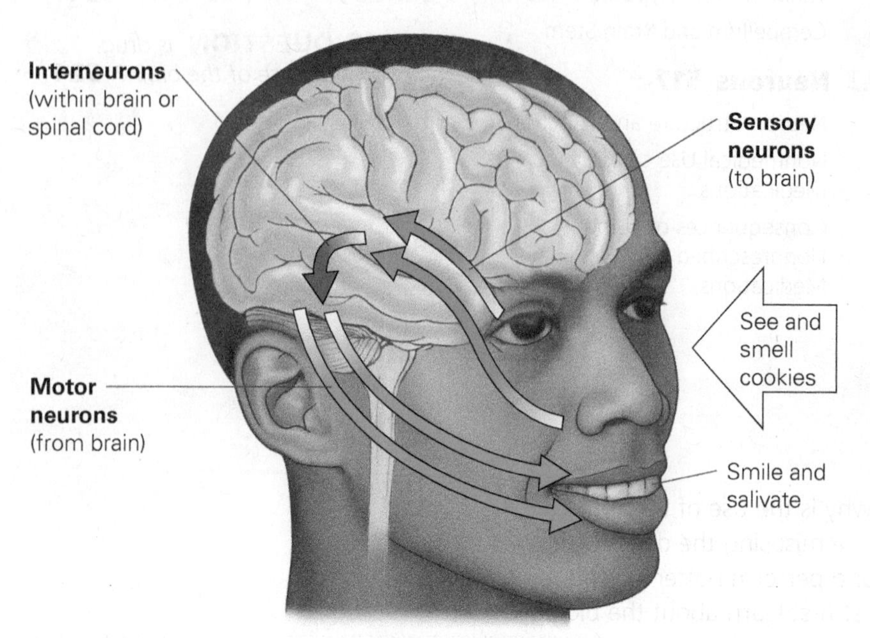

FIGURE 24.2 Neurons. The nervous system's three main functions are sensory input, integration, and motor output. Most actions, such as being tempted to eat cookies, involve input from sensory neurons, followed by integration via interneurons and motor output from motor neurons.

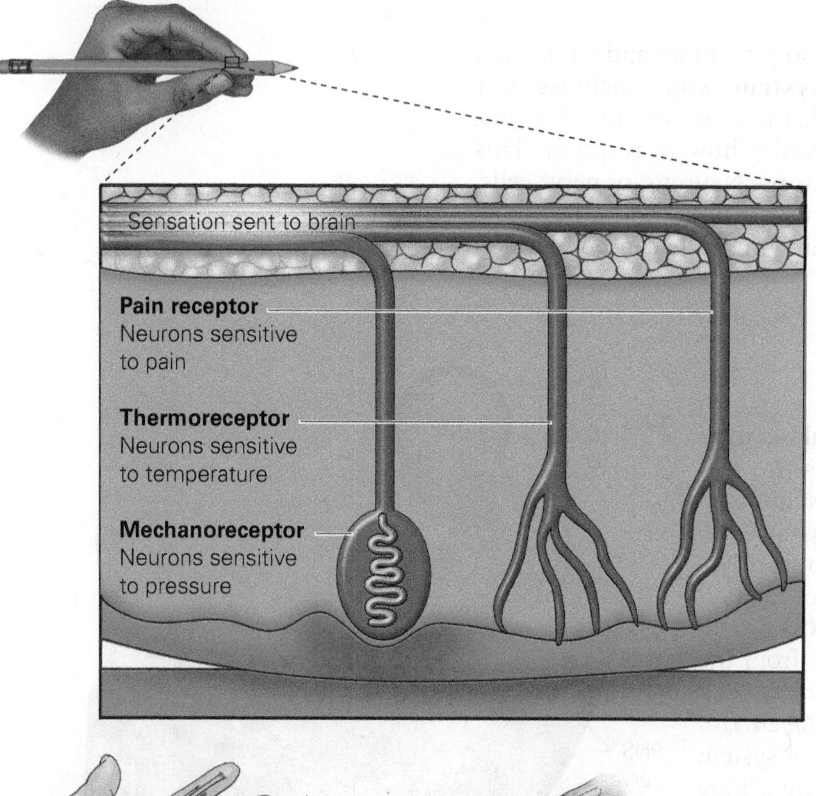

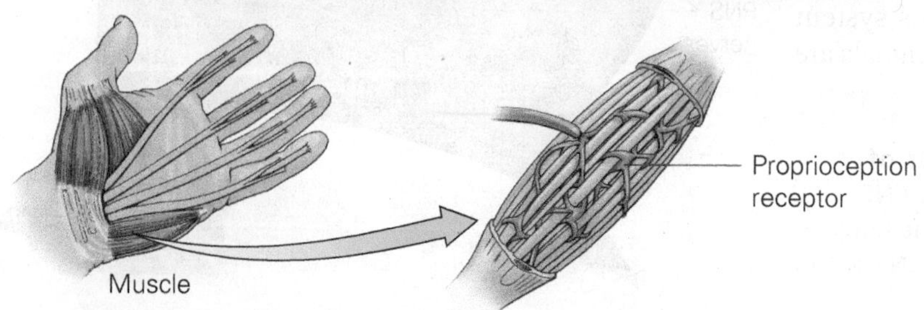

FIGURE 24.3 General senses. Holding a pencil stimulates mechanoreceptors in fingertips and proprioceptors in the hand.

transmitted from one end of a neuron to the other, between and from neighboring neurons and to the cells of tissues, organs, or glands that respond to nerve signals. These are called **effectors.** Information is carried along nerves by electrical charges called **nerve impulses** and transmitted by changes in ion concentration or by chemicals, called **neurotransmitters,** released from nerve cells.

Neurons can be grouped into three categories (**FIGURE 24.2**): (1) **sensory neurons,** which carry information toward the CNS; (2) **motor neurons,** which carry information away from the CNS toward effector tissues; and (3) **interneurons,** which are located between sensory and motor neurons within the brain or spinal cord. Most actions involve input from all sources; sensory activity is followed by integration and motor output.

The Senses

Sensory input is detected by **sensory receptors,** which are typically modified sensory neurons. Changing conditions, for example a change in the amount of light or an odor, stimulate sensory receptors to generate a signal that is carried to the brain.

The **general senses** are temperature, pain, touch, pressure, vibration, and body position (proprioception). Sensory receptors for the general senses, which are scattered throughout the body, can work together to send a message to the brain. Touch, for example, is actually many coordinated sensations generated by sensory receptors sensitive to pressure, pain, and temperature. Sensory receptors for all of these sensations are found in high density in fingertips. A message is sent to the brain when a receptor experiences a change in the shape of its membrane, like the one caused by the pressure of a pencil in your fingers. These receptors can also modulate their response based on the intensity of the stimulus—for example, they can prevent you from holding a pencil too tightly or so loosely that it drops. Holding a pencil also requires sensory receptors in the muscles of your hand. Proprioceptors in muscles send information to the brain about joint position, the amount of tension in joints, and the like (**FIGURE 24.3**).

The sensory receptors for the **special senses** are found in complex sense organs in the head. These include smell, taste, vision, and hearing (**FIGURE 24.4**). For all these senses, a change in the external world is sensed by the

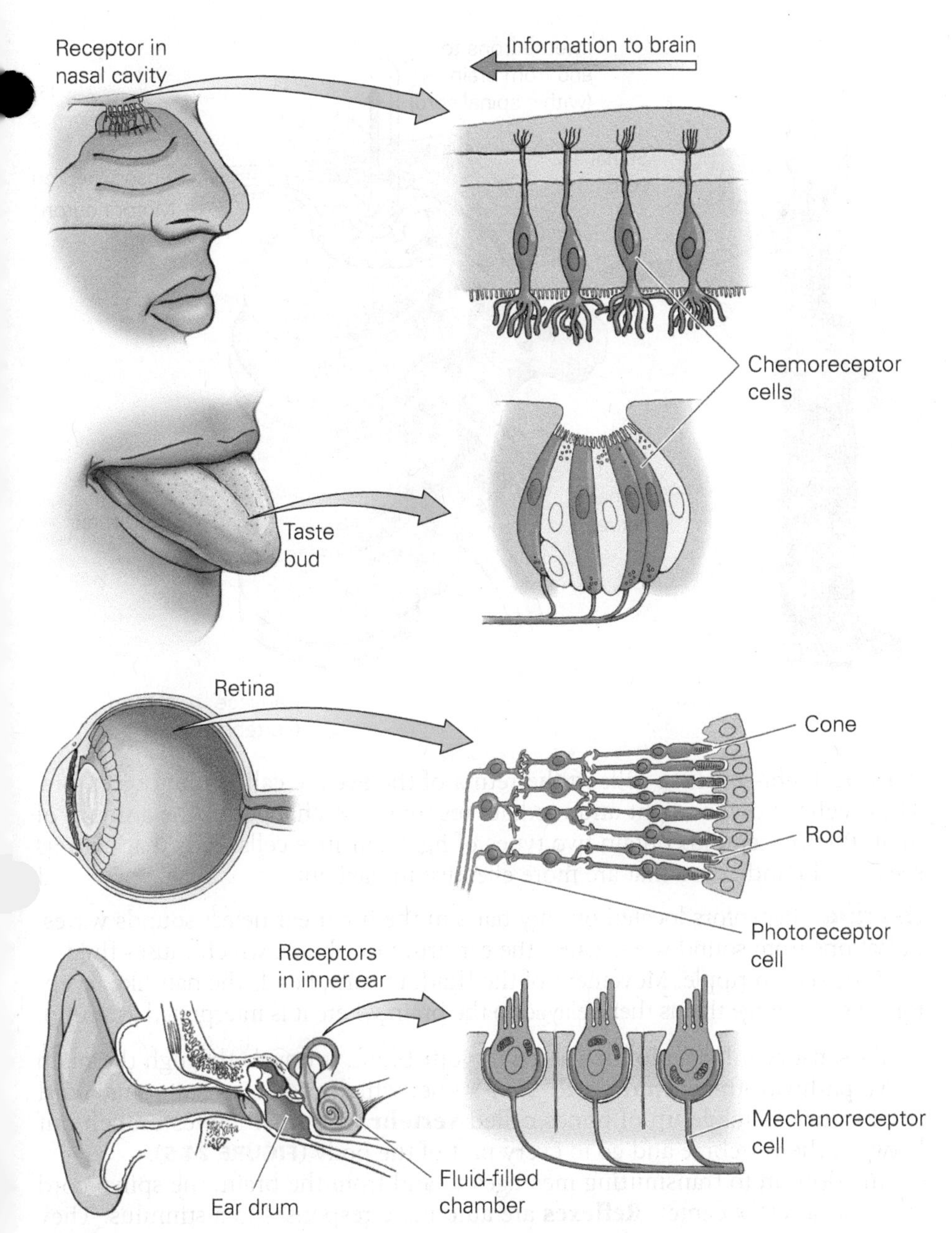

FIGURE 24.4 The special senses. Information from complex sense organs in the face are relayed to the brain by modified sensory neurons.

modified sensory neurons in the sensory receptor and passed to other neurons until it reaches the brain.

Smell. Thousands of modified sensory neurons that detect different smells are located in the nasal cavity. Called chemoreceptors, these cells contain hairlike projections that detect and bind to different chemicals in the air and then send a signal to the brain that is interpreted as a particular smell.

Taste. Taste buds on the surface of the tongue contain specialized sensory neurons, also called chemoreceptors, that respond to one of five primary taste sensations: sweet, salty, sour, bitter, or umami (savory or meaty). A wide variety of tastes occur because each food can stimulate different combinations of chemoreceptors in varying amounts.

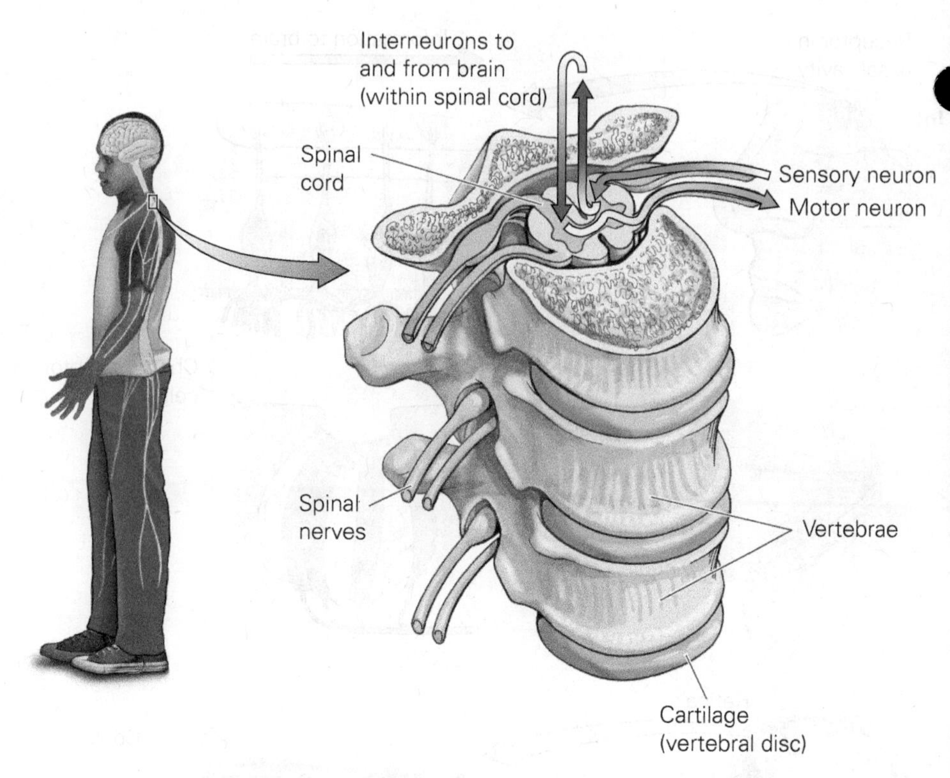

FIGURE 24.5 Spinal nerves and vertebrae. Spinal nerves branch out between the vertebrae and go to all parts of the body.

Vision. Light-sensitive cells in the retina of the eye are called photoreceptors. These cells allow for sight and mediate responses to changes in the amount of light. Photoreceptors contain two types of light sensitive cells: rods that help us see at night and cones that are more effective in daylight.

Hearing. Receptors located on tiny hairs in the inner ear detect sounds waves. Pulsations from sound waves cause the ear drum to vibrate, which causes fluid in the inner ear to ripple. Movement of the fluid, in turn, bends the hair-like receptor cells, a change that is then relayed to the brain where it is interpreted as sound.

As sensory information is passed to your brain, it travels through the main nerve pathway, the **spinal cord.** Your spine, which protects your spinal cord from injury, is made up of bones called **vertebrae.** Spinal nerves branch out between the vertebrae and go to every part of the body (**FIGURE 24.5**).

In addition to transmitting messages to and from the brain, the spinal cord serves as a reflex center. **Reflexes** are automatic responses to a stimulus. They are prewired in a circuit of neurons called a **reflex arc,** which often consists of a sensory neuron that receives information from a sensory receptor, an interneuron that passes the information along, and a motor neuron that sends a message to the muscle that needs to respond.

Reflexes allow a person to react quickly to dangerous stimuli. For example, when you touch something hot, sensory neurons from touch receptors send the message to your spinal cord (**FIGURE 24.6**). Within the spinal cord, interneurons send the message to motor neurons to withdraw your hand from the hot surface. While the spinal reflexes are removing your hand from the source of the heat, pain messages are also being sent through your spinal cord to your brain.

Amphetamines Effects

An amphetamine is a stimulant that works on the CNS by affecting nerves and chemicals in the brain. Many ADD drugs are amphetamines. Amphetamine use heightens the CNS response, causing increased alertness, mental focus, and even euphoria, but it can also lead to permanent changes in the functions the CNS controls, resulting in increased rates of depression and anxiety long after use ends.

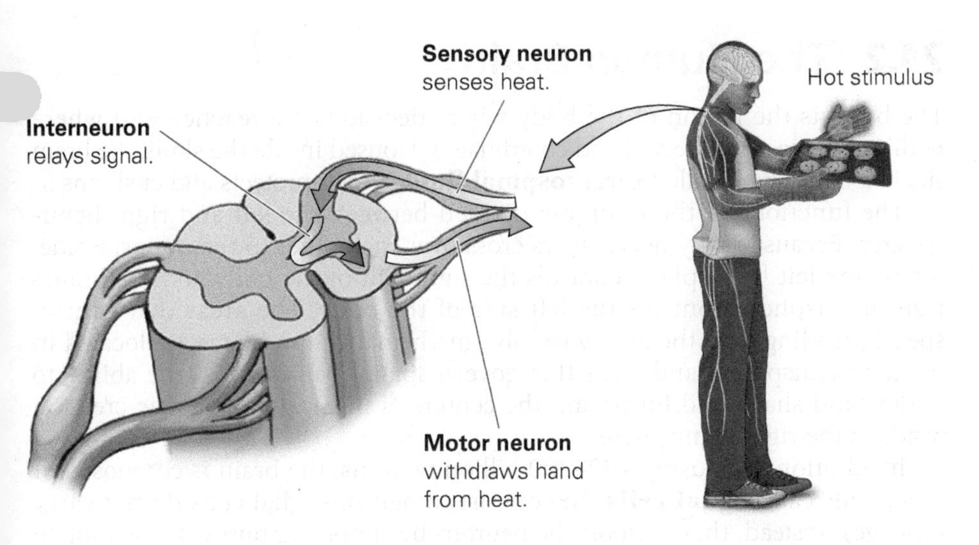
Sensory neuron
senses heat.
Hot stimulus
Interneuron
relays signal.
Motor neuron
withdraws hand
from heat.

FIGURE 24.6 A reflex arc. A reflex arc can consist of a sensory receptor, a sensory neuron, an interneuron, a motor neuron, and an effector. Touching a hot baking sheet evokes the withdrawal reflex.

Amphetamines also effect the PNS. One of the jobs of the PNS is to recognize dangerous or stressful situations and alert the CNS. Amphetamine use leading to increased alertness and mental focus can trigger the so-called fight-or-flight response, even in the absence of any threat. When in this state of hyperarousal, digestion is slowed and blood pressure and heart rate increase. These physiological changes are responsible for the anxiety, insomnia, irritability, hostility, and paranoia many users feel during misuse of amphetamines.

Some students will drink coffee or energy drinks in addition to taking ADD medications as a way to maximize the effects of nonmedical ADD drugs on focus and drive, a practice that increases the risk of increased heart attack and stroke compared with the misuse of ADD medications only. Students may also mix nonmedical ADD drugs with alcohol, hoping to be able to stay up later and party longer. Mixing amphetamines with alcohol also increase the risk of cardiovascular problems and make the user less aware of the physiological changes of both alcohol poisoning and amphetamine overdose.

It is important to note that those diagnosed with ADD are far less likely to experience the negative side effects of ADD medications than are their misusing, non-ADD peers. This avoidance of the worst side effects may be due to the manner in which the medications are prescribed. Physicians typically initially prescribe a low dose and slowly work up to larger doses, giving the physician the option of dialing back if side effects are severe. Misusers of ADD medications typically start at a higher dose, instead of slowly working up to it.

It may also be true that such medications act differently in people with ADD because their brains differ structurally from those of their non-ADD peers—and hence the need for the medication in the first place. In the next section, we take a look at the structure and function of the brain and learn what science tells us about brain differences between people with and without ADD.

Got It?

1. The specialized cells that are found in the nervous system are called __________.
2. Nerve impulses can be transmitted by chemicals released from neurons. These chemicals are called __________.
3. The __________ __________ __________ comprises the brain and spinal cord.
4. The general senses are temperature, pain, touch, pressure, and body position. Body position is also called __________.
5. An automatic response to a stimulus is called a __________.

24.2 The Human Brain

cereb- is from the Latin word for the brain.

The brain is the region of the body where decisions are reached and where bodily activities are directed and coordinated. Housed inside the skull, the brain sits in a liquid bath called **cerebrospinal fluid,** which protects and cushions it.

The functions of the brain are divided between the left and right hemispheres. Because many nerve fibers cross over each other to the opposite side, the brain's left hemisphere controls the right half of the body and the brain's right hemisphere controls the left side of the body. The areas that control speech, reading, and the ability to solve mathematical problems are located in the left hemisphere, and areas that govern spatial perceptions (the ability to understand shape and form) and the centers of musical and artistic creation reside in the right hemisphere.

In addition to housing 100–200 billion neurons, the brain is composed of other cells called **glial cells.** In contrast to neurons, glial cells do not carry messages. Instead, they support the neurons by supplying nutrients, helping to repair the brain after injury, and attacking invading bacteria. There are 10 times as many glial cells in the brain as there are neurons. Structurally, the brain is subdivided into many important anatomical regions, including the cerebrum, thalamus, hypothalamus, cerebellum, and brain stem (**FIGURE 24.7**).

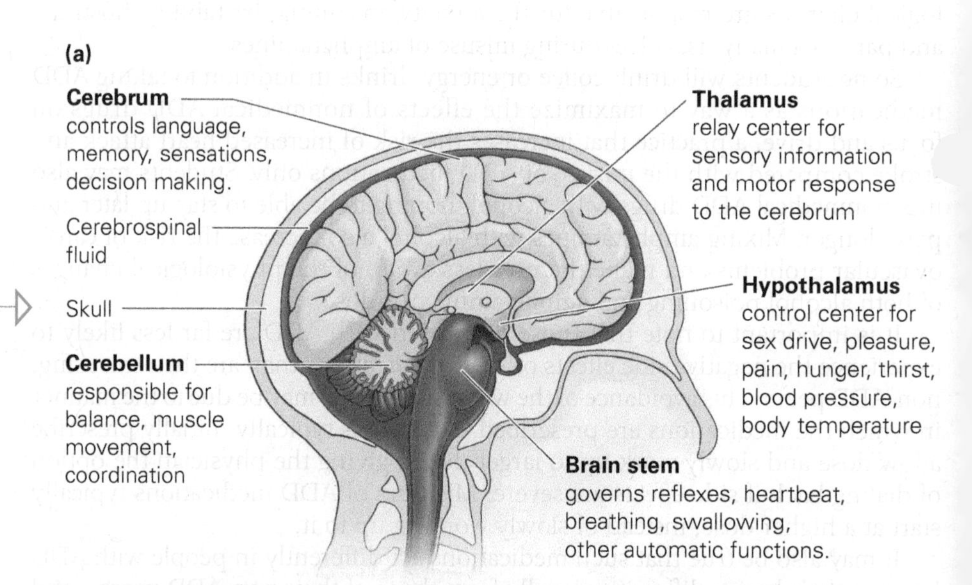
(a)
Cerebrum
controls language, memory, sensations, decision making.
Cerebrospinal fluid
Skull
Cerebellum
responsible for balance, muscle movement, coordination
Thalamus
relay center for sensory information and motor response to the cerebrum
Hypothalamus
control center for sex drive, pleasure, pain, hunger, thirst, blood pressure, body temperature
Brain stem
governs reflexes, heartbeat, breathing, swallowing, other automatic functions.

(b)

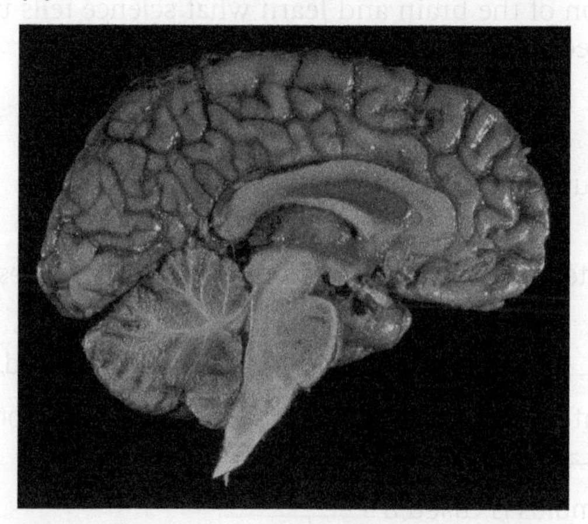

FIGURE 24.7 Anatomy of the human brain. (a) The location and function of the cerebrum, thalamus, hypothalamus, brain stem, and cerebellum are shown. (b) Photo of a cross-sectioned human brain.

VISUALIZE **THIS**

(a) Cerebellum is the diminutive of cerebrum. Which of these two structures is larger in the human brain?
(b) The root *hypo* can mean below. Is the hypothalamus above or below the thalamus?

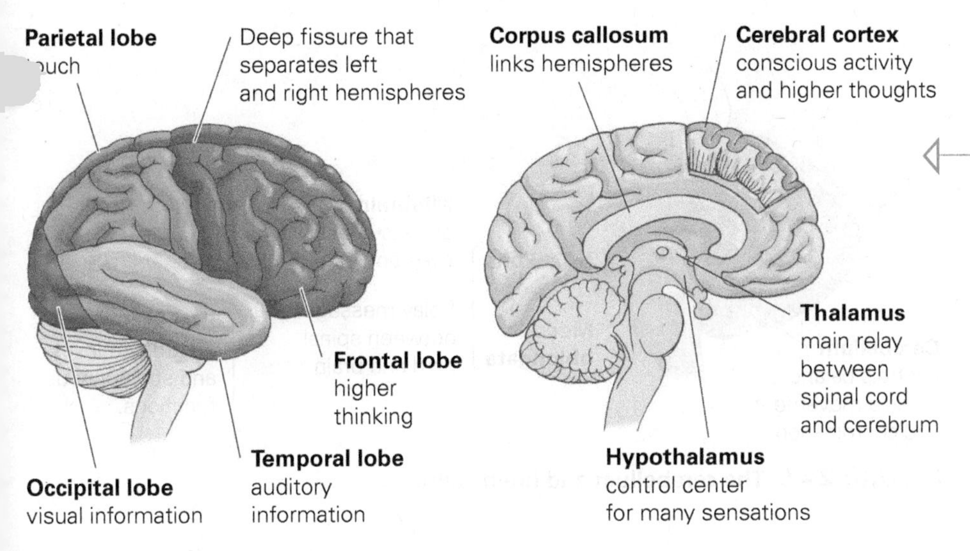

FIGURE 24.8 Structure of the cerebrum.

VISUALIZE **THIS**

Impulsivity, or acting before thinking, is a symptom associated with ADD. Based on the functions listed, an abnormality in which structure in the frontal lobe do you think might lead to impulsivity and why?

Cerebrum

The **cerebrum** fills the whole upper part of the skull. This part of the brain controls language, memory, voluntary movements, sensations, and decision making. The cerebrum has two hemispheres, each divided into four lobes (**FIGURE 24.8**):

- The **parietal lobe** processes information about touch and is involved in self-awareness.
- The **frontal lobe** processes voluntary muscle movements and is involved in planning, working memory, and impulse control.
- The **temporal lobe** is involved in processing auditory and olfactory information and is important in memory and emotion.
- The **occipital lobe** processes visual information from the eyes.

The deeply wrinkled outer surface of the cerebrum is called the **cerebral cortex**. In humans, the cerebral cortex, if unfolded, would be the size of a large (16-inch) pizza. The folding of the cortex increases the surface area and allows it to fit inside the skull. The cortex contains areas for understanding and generating speech, areas that receive input from the eyes, and areas that receive other sensory information from the body. It also contains areas that allow planning.

cortex is the Latin word for the bark of a tree.

The cerebrum and its cortex are divided from front to back into two halves—the right and left cerebral hemispheres—by a deep groove, or fissure. At the base of this fissure lies a thick bundle of nerve fibers, the **corpus callosum**, which connects the hemispheres to each other and allows communication from one side of the brain to the other.

Thalamus and Hypothalamus

Deep inside the brain, lying between the two cerebral hemispheres, are the thalamus and the hypothalamus. The **thalamus** relays information between the spinal cord and the cerebrum. The thalamus is the first region of the brain to receive messages signaling such sensations as pain, pressure, and temperature. The thalamus suppresses some signals and enhances others, which are then relayed to the cerebrum. The cerebrum processes these messages and sends signals to the spinal cord and to neurons in muscles when action is necessary. The **hypothalamus,** about the size of a kidney bean and located just under the thalamus, is the control center for sex drive, pleasure, pain, hunger, thirst, blood pressure, and body temperature. The hypothalamus also releases hormones, including those that regulate the production of sperm and egg cells as well as the menstrual cycle.

hypo- means under or below.

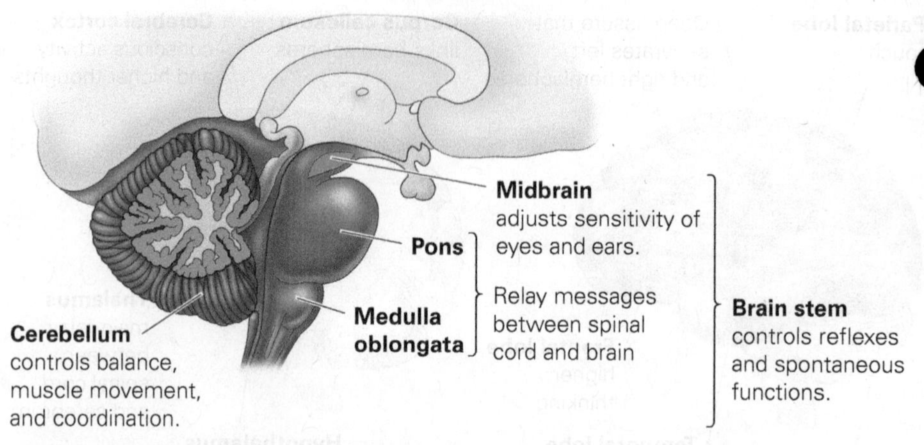

FIGURE 24.9 The cerebellum and brain stem.

Cerebellum and Brain Stem

The **cerebellum** controls balance, muscle movement, and coordination (**FIGURE 24.9**). The cerebellum looks like a smaller version of the cerebrum. It is tucked beneath the cerebral hemispheres and, like the cerebrum, has two hemispheres connected to each other by a thick band of nerves.

The **brain stem** lies below the thalamus and hypothalamus. It governs reflexes and some spontaneous functions, such as heartbeat, respiration, swallowing, and coughing.

The brain stem is composed of the midbrain, pons, and medulla oblongata. Highest on the brain stem is the **midbrain,** which adjusts the sensitivity of your eyes to light and of your ears to sound. Below the midbrain is the **pons.** The pons functions as a bridge, allowing messages to travel between the brain and the spinal cord. The **medulla oblongata** is the lower part of the brain stem. It helps control heart rate and conveys information between the spinal cord and other parts of the brain.

Some scientific studies seem to implicate differences in the frontal cortex between those with and without ADD, with the frontal cortex being smaller in those with ADD. A smaller frontal cortex may require supplementation from ADD medications, whereas a more typically sized frontal cortex does not. This may be part of the reason why ADD medications help those with ADD feel normal but give a heightened response to those using it for nonmedical reasons.

It isn't just differences in brain structure that affect functioning. There are also differences in how neurons function, which we look at next.

Got It?

1. Neurons are a major cell type in the brain, as are the __________ cells that support them.
2. The __________ controls language, memory, voluntary movements, sensations, and decision making.
3. The outer surface of the cerebrum is called the __________.
4. The __________ controls balance, muscle movement, and coordination.
5. The __________ __________ governs reflexes and other spontaneous functions.

24.3 Neurons

Neurons are highly specialized cells that usually do not divide. Therefore, damage to a neuron cannot be repaired by cell division and often results in permanent impairment. For example, damage to spinal motor neurons results in lifelong paralysis because messages can no longer be transmitted from the CNS to muscles. Likewise, injury to the brain can result in permanent brain damage if neurons in the brain are harmed.

Neuron Structure and Function

Neurons are composed of branching **dendrites** that radiate from a bulging **cell body,** which houses the nucleus and organelles, and a long, wirelike **axon,** terminating in knobby structures called the **terminal boutons** (FIGURE 24.10). A nerve impulse usually travels down the axon of one neuron and is transmitted to the dendrites of another neuron.

The axons of many neurons are coated with a protective fatty layer called the **myelin sheath.** The myelin sheath is composed of lipid-rich, neuron-supporting cells called **Schwann cells** that wrap around an axon and give tissues coated in myelin a glistening white color, called **white matter.** The myelin sheath functions like insulation on a wire by preventing sideways message transmission, thus increasing the speed at which the electrochemical impulse travels down the axon.

Unmyelinated axons, combined with dendrites and the cell bodies of other neurons, look gray in cross sections and thus are called **gray matter** (FIGURE 24.11).

Neurons transmit impulses from one part of the body to another. Many kinds of stimuli, including touch, sound, light, taste, temperature, and smell, cause a neuronal response. When you touch something, signals from touch sensors travel along sensory nerves from your skin, through your spinal cord, and into your brain. Your brain then sends out messages through your spinal cord to the motor neurons, telling your muscles how to respond. To evoke this response, nerve cells must transmit signals along their length and from one cell to the next.

Action Potential. Another name for a nerve impulse is an action potential. An action potential is a brief reversal of the electrical charge across the membrane of a nerve cell. It is propagated as a wave of electrical current down the length of the axon.

The inside of a resting neuron is negatively charged compared with the outside. Because opposite charges tend to move toward each other, the membrane serves as a source of stored energy, like a battery, by keeping these opposite charges apart. When signaled, gated protein channels along the length of the axon open to allow positively charged sodium and potassium ions to move across the membrane. The influx of positively charged sodium ions depolarizes the

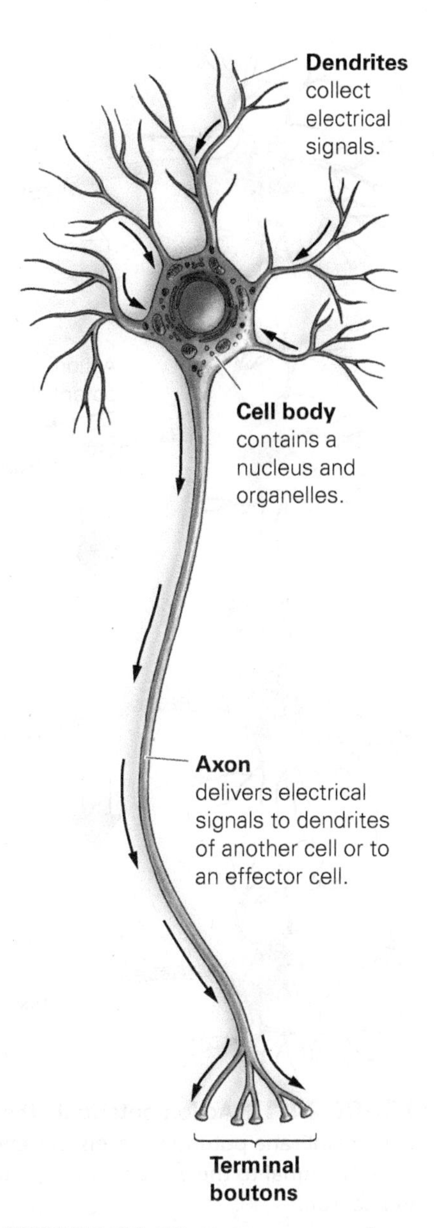
Dendrites collect electrical signals.
Cell body contains a nucleus and organelles.
Axon delivers electrical signals to dendrites of another cell or to an effector cell.
Terminal boutons

FIGURE 24.10 The structure of a generalized neuron. Nerve impulses are propagated in the direction of the arrows.

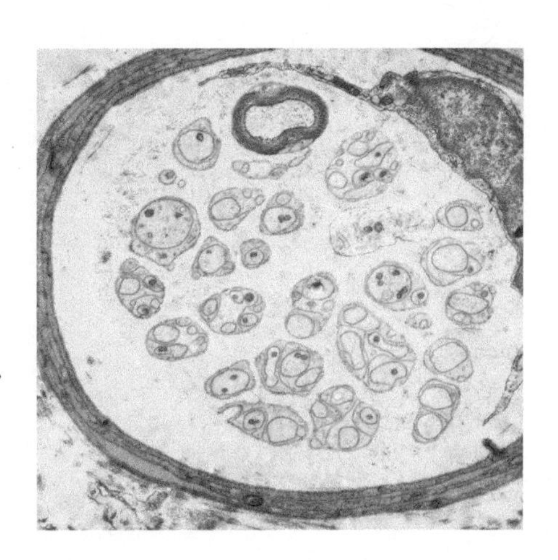

FIGURE 24.11 White and gray matter. This cross section of a nerve shows one myelinated axon (stained green) and many nonmyelinated axons (stained blue). When not stained, myelinated axons are white in appearance and unmyelinated are gray.

WORKING WITH DATA

Nerve impulses move 100 meters (m) per second along myelinated axons and 1 m per second along unmyelinated axons. If a person is 6′6″, how long would it take for a nerve impulse to travel from his head to his feet on (a) myelinated neurons (1 m is about 3.3 feet) and (b) unmyelinated neurons?

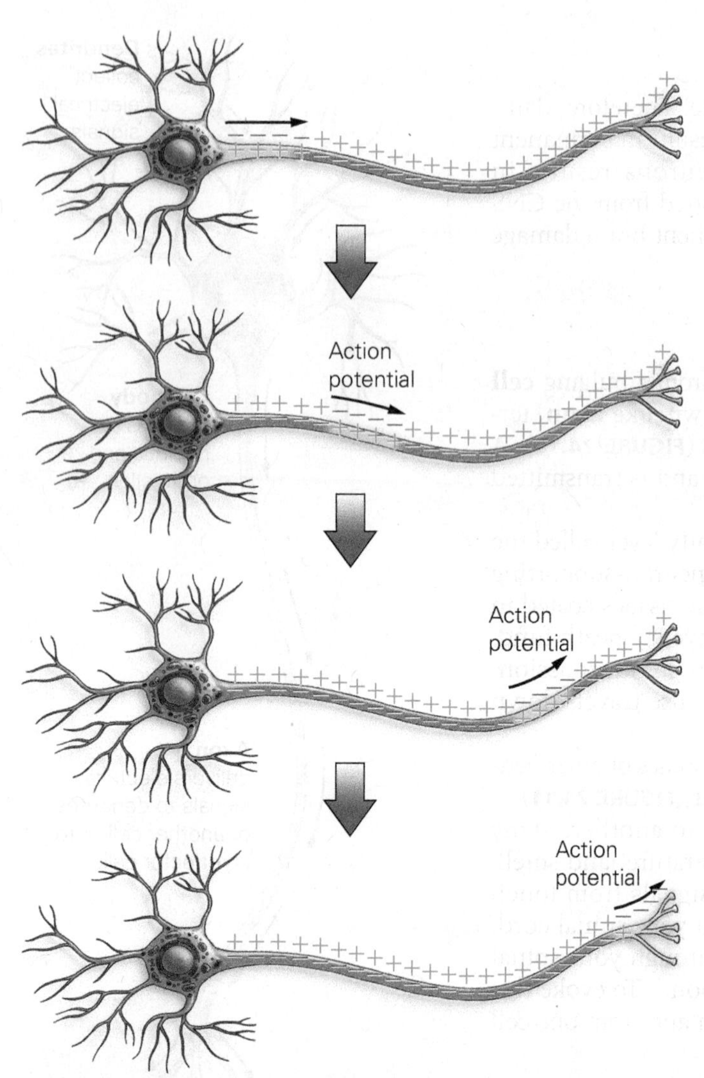

FIGURE 24.12 Action potential. The short-lived rise and fall in membrane potential travels the length of the axon, then relays the signal to the effector organ, muscle or gland, or to the next neuron.

nerve cell. A domino effect occurs as the ions spread charges toward each successive voltage-sensitive protein channel causing each to open and let in more sodium, propagating the depolarization along the length of the neuron. This impulse travels only in one direction, toward the terminal boutons (**FIGURE 24.12**). Repolarization results from diffusion of potassium ions out of the cell, causing the inside of the cell to become more negative than the outside.

Synaptic Transmission. Once a signal has traveled along the length of the axon, it must pass to the next neuron or to the effector organ, muscle, or gland. Most neurons are not directly connected to each other; the signal must be transmitted to the next neuron across a gap between the two called the **synapse** (**FIGURE 24.13a**). The synapse consists of the terminal boutons of the presynaptic neuron, the space between the two adjacent neurons, and dendrites of the postsynaptic neuron. Structures called vesicles are found in the terminal boutons of the presynaptic neuron. Each is filled with a specific chemical neurotransmitter. When an electrical impulse arrives at the terminal bouton of a nerve cell, neurotransmitters are released and they diffuse across the synapse, binding to specific receptors on the cell membrane of the postsynaptic neuron. The binding stimulates a rapid change in the uptake of sodium ions by the postsynaptic neuron. The sodium channels open, and sodium ions flow in, causing another depolarization and generating another action potential. The nerve impulse is propagated from one neuron to the next until the signal reaches a tissue.

After a neurotransmitter evokes a response, it is removed from the synapse. Some are broken down by enzymes in the synapse; others are reabsorbed by the neuron via a process called **reuptake** (**FIGURE 24.13b**). Reuptake occurs when neurotransmitter receptors on the presynaptic cell permit the neurotransmitter to re-enter. This allows the neuron to use it again. Both breakdown by enzymes and reuptake by receptors prevent continued stimulation of the postsynaptic cell. Neurotransmitters common in the human brain include serotonin, dopamine, acetylcholine, and norepinephrine. These differ from each other in chemical structure.

There is evidence linking disturbances in neurotransmission to various diseases. For example, **Alzheimer's disease,** a progressive mental deterioration that causes memory loss along with the loss of control of bodily functions that eventually results in death is thought to involve impaired function of the neurotransmitter acetylcholine in some neurons. **Depression** is a disease that involves feelings of helplessness, despair, and thoughts of suicide. Three neurotransmitters, serotonin, dopamine, and norepinephrine, seem to play a role in depression. **Parkinson's disease,** thought to be caused by low levels of dopamine, causes tremors, rigidity, and slowed movements.

The neurotransmitter dopamine also seems to play a role in drug addiction. Dopamine is typically released from the brain in response to certain pleasures, like eating a good meal or enjoying a work of art. Many drugs including some ADD medications, increase the amount of dopamine in the brain, resulting in the user experiencing intense pleasure. When the brain continues to be exposed to excessive dopamine, in an attempt at homeostasis,

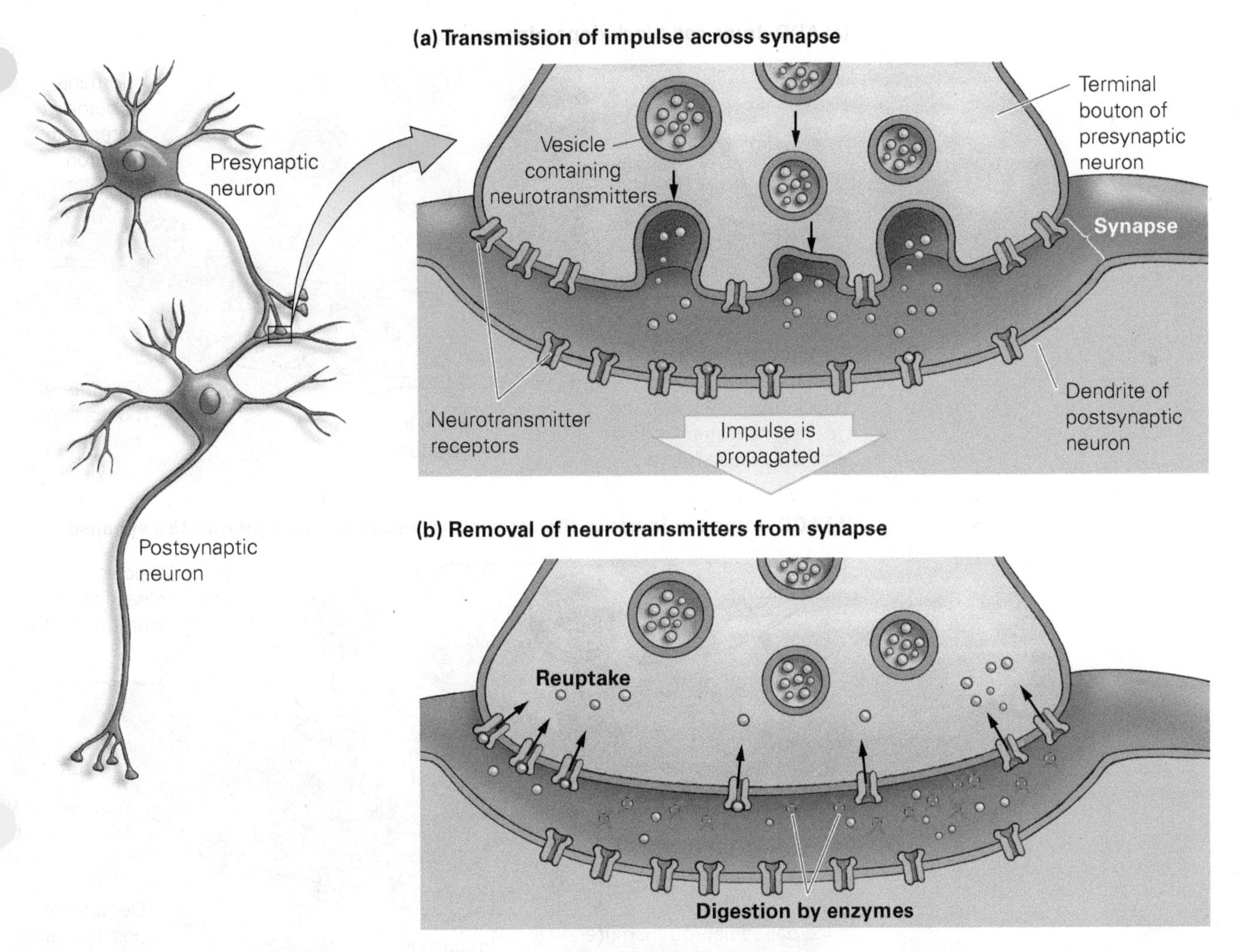

FIGURE 24.13 Propagating the nerve impulse between neurons. (a) The nerve impulse can be transmitted from the terminal bouton of one neuron to the dendrite of the next. Neurotransmitters are released from the presynaptic neuron, travel across the synapse to the postsynaptic neuron, and bind to receptors on dendrites of the postsynaptic neuron. (b) After the neurotransmitter evokes a response, it is removed from the synapse by enzymes present in the synapse that break the neurotransmitter apart or by reuptake via the presynaptic cell.

the brain responds by decreasing the number of receptors for dopamine. When the number of receptors is decreased, a drug user will require more and more drug to achieve the desired effect. Withdrawal from the drug can be extremely difficult because the brain now has a lower than normal ability to respond to the neurotransmitter. It is unclear whether the brain can ever resume its ability to respond to pleasures the way it did before the drug use and addiction occurred.

Abnormal levels of the neurotransmitters dopamine, along with norepinephrine, also seem to be involved in producing the symptoms of ADD. Dopamine controls emotions as well as complex movements, and norepinephrine helps regulate response to stress and regulates attention and focus. Therefore, someone with a low concentration of dopamine may respond impulsively to situations in which pausing to process the input would be more effective. Those with low norepinephrine may have trouble with focus and concentration.

(a) ADD drugs can block dopamine reuptake.

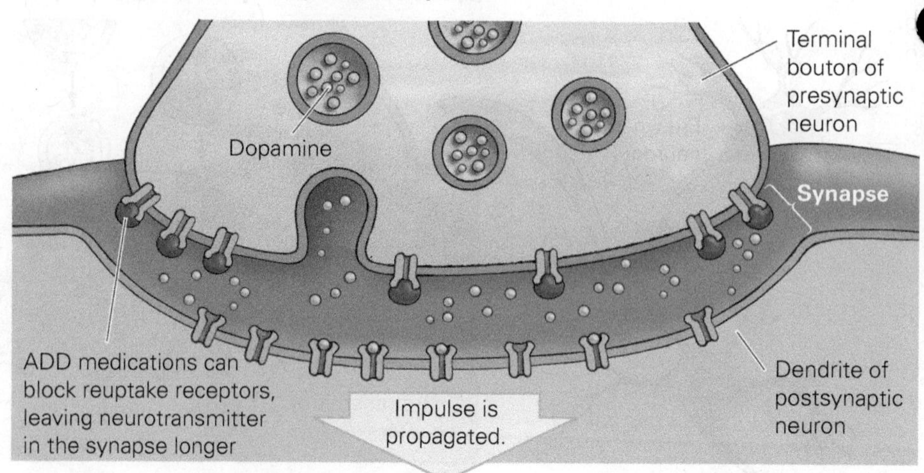

(b) ADD drugs can increase the amount of dopamine released into the synapse.

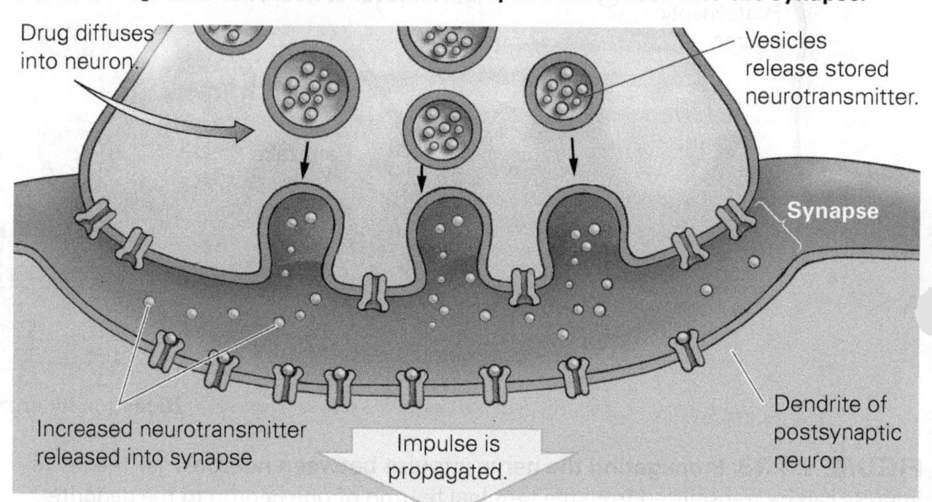

FIGURE 24.14 Mechanism of ADD prescription drug action. Some ADD drugs work by (a) preventing the reuptake of neurotransmitters, allowing longer stimulation of postsynaptic neuron or (b) increasing the release of neurotransmitters.

Although no one knows for sure what causes the lower levels of these neurotransmitters in those with ADD, drugs used to treat symptoms of ADD work to decrease the impact of these deficiencies by increasing the level of neurotransmitter in the synapse. This can be done by blocking the actions of the reuptake receptors or by causing the release of more neurotransmitter (**FIGURE 24.14**).

Nonmedical Use of ADD Medications

When a user takes ADD medications for nonmedical reasons, the brain is rapidly drained of neurotransmitter that would otherwise have been used slowly over the course of several days. The synthesis of new neurotransmitter occurs at a rate that only slowly compensates for the loss. Therefore, during the time period after the neurotransmitter is used but before the body can restock, the user will experience symptoms sometimes known as the "crash." These symptoms can include exhaustion, depression, inability to concentrate, anger, aggression, anxiety, tearfulness, and more.

For some non-ADD users, the strategy of taking ADD drugs immediately backfires when the effects of the crash are experienced during an exam. For others, one crash is enough to dissuade them from further use. The worst outcomes occur when users take too much of the drug and overdose, or when they attempt to immediately reattain or intensify the desired symptoms or alleviate crash symptoms by using more of the drug. One of the ways the body responds to the increased amount of neurotransmitter present is to decrease the number of receptors on the postsynaptic neuron, thus limiting the drug's effects. As this tolerance for the drug builds, the amount of drug required to achieve the initial effects continues to escalate, leading to a higher risk of overdose. With prolonged use, the body will respond by permanently lowering the amount of neurotransmitter produced. Now the user needs to use more just to feel normal. This is of particular concern to those under age 25, whose brains are still developing, because it may cause permanent alterations to the developing brain.

Contrary to misusers' perceptions, there is no evidence that these drugs actually boost academic performance, and they may even *decrease* performance level and grades. Although those that misuse stimulants report the perception that they are learning more, many studies suggest that this is simply the user's illusion. Studies do indicate that misuse of ADD medications may help with short-term focus and help the abuser to stay awake, but there is no evidence to suggest that these drugs help with actual learning. Misusers do not increase their exam scores or get higher grades on papers when using nonprescribed medications when compared with exam and paper grades when not using.

Consequences of Using Nonprescribed ADD Medications

Around one-third of students use ADD medications for nonmedical reasons. Therefore, the majority of students do not. This may be because of a choice not to use any nonprescription drugs or may stem from well-founded concerns about addiction and other health risks. In addition, some students consider the practice of taking ADD drugs to be cheating, in the same way that many people consider the use of steroids to improve athletic performance to be cheating. Others are concerned about the legal consequences. Giving or selling drugs for which you have a prescription is illegal, as are buying or using someone else's prescription drugs, or obtaining these drugs from a clinician under false pretenses.

Most students understand that developing time management skills and discipline now, instead of popping a pill, will be very helpful to them when they get out of college. It is hoped that the time lost to the misery of crashing, coupled with the very real health risks associated with nonmedical ADD drug use will lead students using or considering using nonprescribed medications to conclude that they are better off setting aside time to study every day instead of relying on illegal drugs to help them pull a last minute all-nighter.

Universities and professors, too, can help by understanding the scope of the problem and by helping students find strategies to manage their workload. Studies show that students are less likely to take nonmedical ADD drugs when they are given more low-stakes exams versus a few high-stakes exams, and when they submit written assignments in multiple shorter segments versus one long assignment. Whatever the mechanism, it is clear that students, faculty, and health-care professionals need to work together to decrease the consequences of drug use (TABLE 24.1) on the still-developing brains of our college students.

TABLE 24.1 Recreational drugs and the nervous system.

Drug	Mechanism of Action	Desired Mental Effect	Side Effect
Alcohol is a by-product of fermentation.	• A depressant that diffuses easily across cell membranes • Inhibits neurotransmission and interferes with the activity of neurons throughout the brain	Reduced anxiety and a sense of well-being, loss of concern for social constraints	• Impaired judgment, slurred speech, unsteady gait, slower reaction times, uncontrollable emotions • Chronic alcohol abuse damages nerve cells in frontal lobes, leading to impaired intellect, judgment, thought, and reasoning. Liver damage is also likely.
Amphetamines are used legally to treat obesity, asthma, ADD, and narcolepsy. Methamphetamines are illegal amphetamines.	• Increase dopamine and norepinephrine release • Block reuptake and inhibit breakdown of neurotransmitter	Small doses make a person feel more energetic, alert, focused, and confident.	• Effects wear off quickly, causing sudden "crashes" from depleted neurotransmitter stores and resulting in depression and fatigue. • Prolonged use results in aggressiveness, delusions, hallucinations, violent behaviors, and death.
Caffeine is a naturally occurring chemical found in plants such as coffee, tea, and cocoa.	A general stimulant of all cells, not just those of the central nervous system	Mental alertness, increased energy	Insomnia, anxiety, irritability, and increased heart rate
Cocaine is extracted from the leaves of the coca plant of South America. 	A stimulant that increases levels of dopamine and norepinephrine by decreasing reuptake	A rush of intense pleasure, increased self-confidence, and increased physical vigor	• Increased heart rate and blood pressure • Crash involves deep depression, anxiety, and fatigue.
Ecstasy is MDMA in pill form; Molly is MDMA in crystal or powder form. 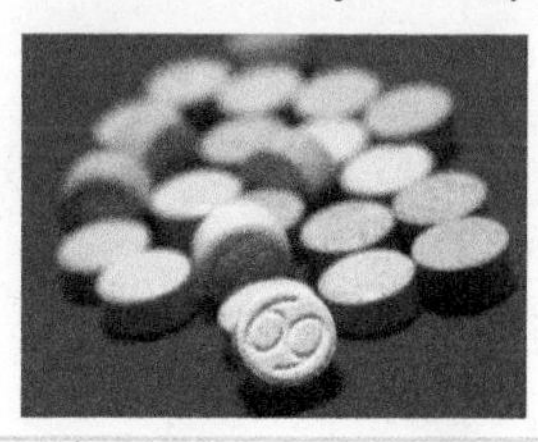	A stimulant and hallucinogenic that acts to prevent serotonin reuptake while flooding neurons with other neurotransmitters	Euphoria, enhanced emotional and mental clarity, increased energy and sexual response, heightened sensitivity to touch	• Severe dehydration, confusion, anxiety, paranoia, depression, and sleeplessness lasting several days • Permanent memory damage

—Continued

TABLE 24.1 Recreational drugs and the nervous system. *Continued—*

Drug	Mechanism of Action	Desired Mental Effect	Side Effect
Lysergic acid diethylamide (LSD) and mushrooms: LSD is a derivative of the fungus *Claviceps purpurea*, which grows on rye. 	Hallucinogen that binds to serotonin receptors in the brain, increasing the normal response to serotonin	Heightened sensory perception and bizarre changes in thought and emotion, hallucinations	• Hallucinations can lead users to dangerous actions. • Heavy use leads to permanent brain damage, including losses of memory, decreased attention span, and psychosis.
Marijuana is derived from the *Cannabis sativa* plant and contains delta-9-tetrahydrocannabinol (THC). 	• Receptors for THC are in the areas of the brain that influence mood, pleasure, memory, pain, and appetite. • THC is thought to work by increasing dopamine release.	• Altered sense of time, enhanced feeling of closeness to others, increased sensitivity to stimuli • Large doses can cause hallucinations.	• Slowed reaction time and decreased coordination • Permanently impairs short-term memory, learning, attention span, ability to store and acquire information • Decreased testosterone production and disrupted menstrual cycle
Nicotine is found in tobacco plants. 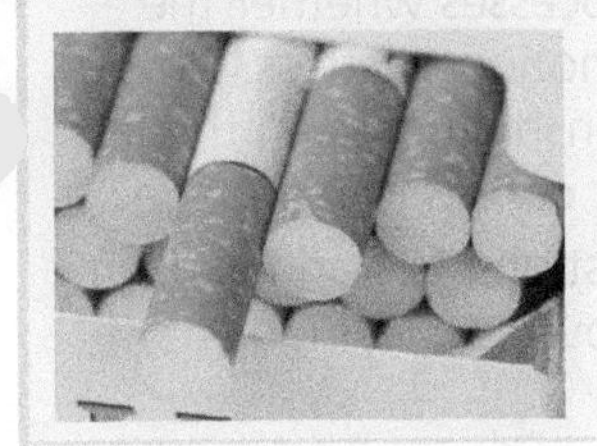	Stimulant that triggers neurons of cerebral cortex to produce acetylcholine, epinephrine, and norepinephrine	Increased alertness and awareness, appetite suppression, relaxation	• Increased odds of most cancers • Causes increased heart rate and blood pressure
Opiates—including heroin, morphine, and codeine—are derived from the opium poppy. 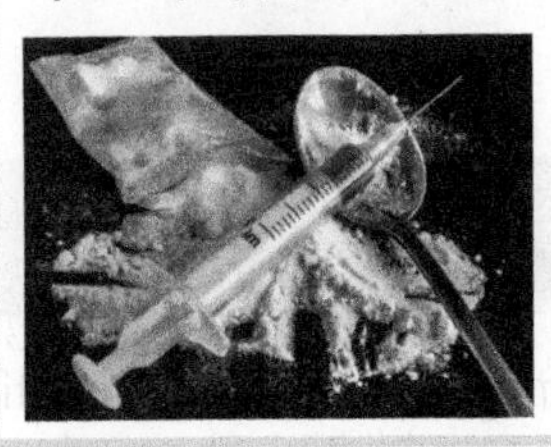	Like the endorphins produced by exercise, opiates bind receptors in neurons that control feelings of pleasure.	A quick, intense feeling of pleasure, followed by a sense of well-being, then drowsiness	• Poor motor coordination, depression, can cause coma and death

Got It?

1. Neurons are composed in part of __________ that radiate from the cell body.
2. The wire-like __________ transmits nerve impulses from the dendrite to the terminal boutons.
3. White matter is composed of neurons covered in __________ sheath.
4. An influx of positively charged ions serves to depolarize the membrane of the neuron, which typically has a more __________ charge on the inside than the outside.
5. The gap between two neurons, across which neurotransmitter flows, is called the __________.

You are flipping through a magazine and come across a quiz designed to help you determine whether you are right-brained or left-brained. The preamble to the quiz states that the human brain exhibits a phenomenon called lateralization where each side, or hemisphere, of the brain becomes dominant in particular types of thinking. According to this hypothesis, people who are right-brained are more creative, intuitive, thoughtful, and better at reading and expressing emotions, because the right side of their brain is dominant. Left-brain dominance yields people who are more analytical, logical, and better at math, reasoning, and critical thinking. With this information at hand, you might be tempted to imagine the following:

Traits like being creative or logical occur when one hemisphere completely dominates the other hemisphere.

Sounds right, but it isn't.

Answer the following questions to understand why.

1. Are there structures in one hemisphere that are not in the other hemisphere?
2. How do the two hemispheres of the brain communicate with each other?
3. Brain imaging scans taken while people are solving math problems show that both sides of the brain are active. Does this provide support for the right brain–left brain dominance hypothesis? Why or why not?
4. Evidence from neurobiological studies of people listening to spoken language suggests that the left hemisphere determines which sounds form words while the right hemisphere processes whether the words are meant to convey emotion or stress. How would you describe the relationship between the left and right hemispheres when it comes to language?
5. Consider your answers to questions 1–4 and explain why the statement bolded above sounds right, but isn't.

THE BIG QUESTION

Is drug addiction a disease of the brain?

One model of how addiction occurs, called the brain disease model, posits that genetic differences in brains can make some people more susceptible to addiction than others and can make repeated exposure to drugs more damaging in some people. These differences make it extremely difficult for addicts to be cured of their addictions. Scientists at the National Institute of Drug Abuse liken the attempts addicts make at abstaining from drug use to attempting to drive a car without brakes. You may desperately want to stop your car, and you may work very hard at stopping your car, but without brakes, you don't have much ability to stop your car. Is it true that addicts have diseased brains?

What should I know?

What follows are some smaller questions that need to be resolved to answer the Big Question. Place a checkmark next to the questions that science can answer.

Smaller Questions	Can Science Answer?
Do some individuals have a genetic predisposition to addiction?	
Does repeated exposure to drugs change the brain?	
Is drug addiction a moral failing?	
It is better to imprison drug addicts or offer them chemical dependency treatment?	
How often is drug abuse treatment successful?	
Are some people able to use drugs but not become addicted?	

What does the science say?

Let's examine what the data say about this smaller question:

Does repeated exposure to drugs change the brain?

The following graph shows data modified from a study on the volume of gray matter in cocaine addicts (on the *y* axis) with an active addiction (bottom *x* axis) and addicts during long periods of abstinence (top *x* axis).

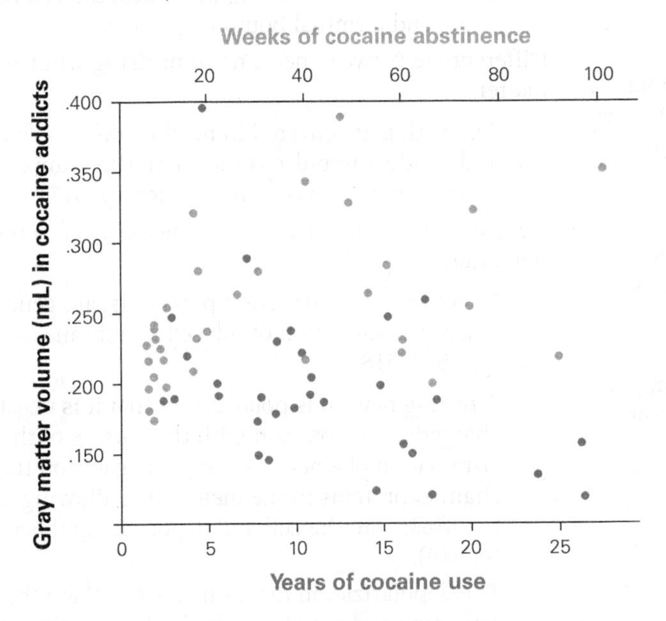

1. Describe the results. How does the overall trend in gray matter volume change (a) as length of addiction increases and (b) as length of abstinence increases?
2. Given these data, do you think the smaller question is answered? If not, propose another study that would help answer this question.
3. Does this information help you answer the Big Question? What else do you need to consider?

Data Source: C. Connolly, R. Bell, J. Foxe, and H. Garavan, "Dissociated Grey Matter Changes with Prolonged Addiction and Extended Abstinence in Cocaine Users," *PLoS ONE* 8 (2013): e59645.

Chapter Review

Mastering Biology

Go to Mastering Biology to access **eText 2.0, Dynamic Study Modules,** and the **Study Area,** where you'll find practice quizzes, BioFlix™ animations, MP3 tutor sessions, current events, and more.

SUMMARY

Section 24.1

List the components of the nervous system, and differentiate between the CNS and PNS.

- The nervous system consists of the brain and spinal cord of the central nervous system as well as the nerves of the peripheral nervous system that carry information to and from the brain (p. 509).

Describe the roles that sensory, motor, and interneurons play in the nervous system.

- Neurons are specialized cells of the nervous system that carry chemical and electrical messages (pp. 509–510).
- Sensory neurons carry information toward the brain; motor neurons carry information away from the brain; interneurons are located between sensory and motor neurons (p. 510).

Compare the general and special senses.

- Sensory receptors for general senses, found throughout the body, relay information to the brain. Receptors for the special senses are found in more complex sense organs (pp. 510–512).

Explain how reflexes work.

- Reflexes are prewired responses that carry messages from sensory receptors to interneurons to a motor neuron (p. 512).

SHOW YOU KNOW 24.1 When you touch something hot, you will often remove your hand before actually feeling the pain. Why might this be?

Section 24.2

Outline the structure and function of the cerebrum, thalamus, hypothalamus, cerebellum, and brain stem.

- The cerebrum is where most thinking occurs. The two hemispheres of the cerebrum consist of four lobes—temporal, occipital, parietal, and frontal. The outer surface of the cerebrum is the cortex (pp. 514–515).
- The thalamus and hypothalamus are located between the two cerebral hemispheres. The thalamus relays information between the spinal cord and the cerebrum, and the hypothalamus regulates many vital bodily functions (pp. 515–516).
- The cerebellum is located at the base of the brain, beneath the cerebral hemispheres. It regulates balance and coordination (p. 516).
- The brain stem is located below the thalamus and hypothalamus. It controls many unconscious functions (p. 516).

SHOW YOU KNOW 24.2 The frontal lobes of the brain are thought not to be fully developed until around age 25. Based on your understanding of the functions of the frontal lobe, propose an explanation for why people under age 25 might be more willing to use drugs than those with fully developed brains.

Section 24.3

Describe the structure of a neuron.

- Neurons are specialized cells with a structure that consists of the branching dendrites, the cell body, the axon, and terminal boutons (p. 517).

Differentiate between neurons comprising white and gray matter.

- Axons that are covered in myelin comprise white matter and conduct impulses faster than the unmyelinated axons comprising the gray matter. (p. 517).

Describe the events involved in generating an action potential.

- Nerve impulses, or action potentials, are generated when depolarization of the cell membrane occurs (pp. 517–518).
- A resting neuron is polarized in that it is negatively charged in comparison with the outside of the cell. Stimulation of a neuron opens the gates on the sodium channel proteins in the membrane, allowing sodium to diffuse into the cell and depolarizing the cell (p. 518).
- The depolarization moves in a wave down the cell, activating sodium channels in adjacent parts of the membrane and thereby moving the impulse along the length of the neuron (p. 518).

Describe how an action potential is relayed across a synapse.

- The electrical impulse is propagated to the ends of the axon, which house chemical neurotransmitters. Neurotransmitters are then released into the synapse and bind to receptors on the next neuron (pp. 518–520).

SHOW YOU KNOW 24.3 A neurotoxin carried by the poisonous pufferfish binds to sodium channels and prevents passage of ions. What effect would this neurotoxin have on the nervous system of someone who ingested it?

ROOTS TO REMEMBER

The following roots of words come mainly from Latin and Greek and will help you to decipher terms:

cereb- is from the Latin word for the brain. Chapter terms: *cerebrum*

cortex is the Latin word for the bark of a tree. Chapter term: *cerebral cortex*

hypo- means under or below. Chapter term: *hypothalamus*

LEARNING THE BASICS

1. Add labels to the figure that follows, which illustrates the parts of the brain.

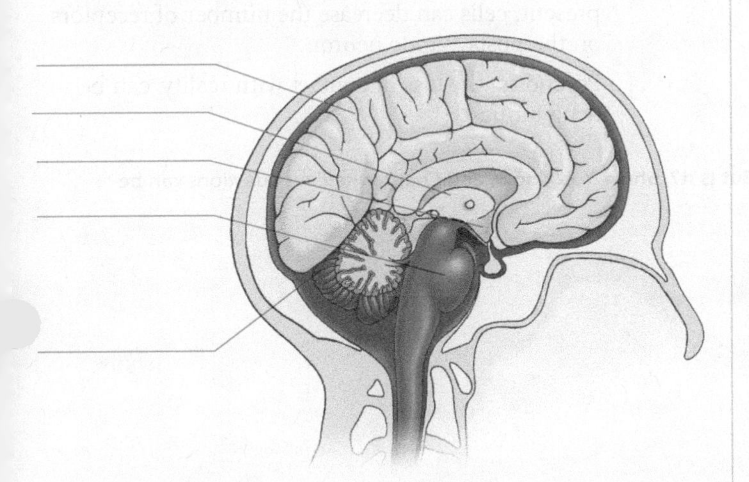

2. How do sensory, motor, and interneurons differ?
3. What are the general senses?
4. Which brain structure controls language, memory, sensations, and decision making?

   A. cerebrum; **B.** thalamus; **C.** hypothalamus; **D.** cerebellum; **E.** brain stem

5. Where in the brain is the thalamus located?

   A. in the pons; **B.** in the medulla; **C.** in the cerebral cortex; **D.** between the cerebral hemispheres; **E.** in the occipital lobe

6. Myelin ________.

   A. is a protective layer that coats some neurons; **B.** gives neurons a white appearance; **C.** prevents sideways transmission of nerve impulses, thereby increasing the rate of transmission; **D.** all of the above

7. An action potential ________.

   A. is a brief reversal of temperature in the neuronal membrane; **B.** is propagated from the terminal bouton toward the cell body; **C.** begins when sodium channels close in response to stimulation; **D.** is propagated as a wave of depolarization moves down the length of the neuron

8. Neurotransmitters ________.

   A. are electrical charges that move down myelinated axons; **B.** are released in the cell body to hasten nerve impulse transmission; **C.** are released when an electrical impulse arrives at the terminal bouton of the postsynaptic neuron; **D.** diffuse across a synapse and bind to receptors on the cell membrane of the postsynaptic neuron

9. The effects of a neurotransmitter could be increased by ________.

   A. increasing the number of receptors on the postsynaptic cell; **B.** preventing reuptake; **C.** providing more enzymes involved in synthesizing the neurotransmitter; **D.** inhibiting enzymes involved in breakdown of the neurotransmitter from the synapse; **E.** all of the above

10. **True or False:** A reflex arc is generated after multiple exposures to a negative stimulus such as heat.

ANALYZING AND APPLYING THE BASICS

1. Develop a neurobiological hypothesis for why ADD might be more severe in some people than others.
2. Nerve impulses travel more quickly at warmer temperatures and down axons that are larger in diameter. What would you predict about the axons of squid that make their home in the coldest depths of an ocean?
3. Describe how reflexes differ from other nervous system responses.

GO FIND OUT

1. Universities have started including statements about the unauthorized use of prescription medications to enhance academic performance in their policies on academic dishonesty. Does your university have such a policy?
2. People selling or abusing ADD medications sometimes request refills to replace medication they say was lost or stolen. Is it legal in your state for clinicians to ask for urine tests to determine the level of medication in the system when they suspect a patient is not taking the pills as prescribed?

MAKE THE CONNECTION

The science that you learned in this chapter has helped you better understand the real-world example used throughout this discussion. Draw a line from the statement on the left to the science that supports it on the right.

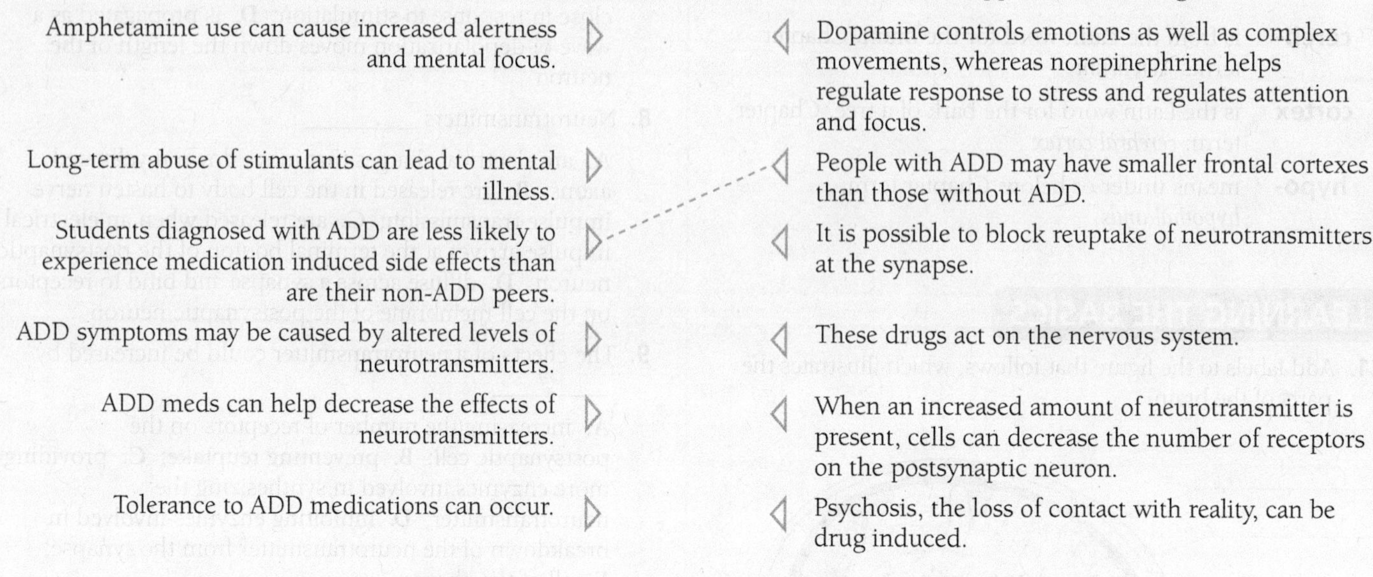

Amphetamine use can cause increased alertness and mental focus.
Long-term abuse of stimulants can lead to mental illness.
Students diagnosed with ADD are less likely to experience medication induced side effects than are their non-ADD peers.
ADD symptoms may be caused by altered levels of neurotransmitters.
ADD meds can help decrease the effects of neurotransmitters.
Tolerance to ADD medications can occur.
Dopamine controls emotions as well as complex movements, whereas norepinephrine helps regulate response to stress and regulates attention and focus.
People with ADD may have smaller frontal cortexes than those without ADD.
It is possible to block reuptake of neurotransmitters at the synapse.
These drugs act on the nervous system.
When an increased amount of neurotransmitter is present, cells can decrease the number of receptors on the postsynaptic neuron.
Psychosis, the loss of contact with reality, can be drug induced.

Answers to **Got It?, Visualize This, Working with Data, Sounds Right, But Is It?, Show You Know,** and **Chapter Review** questions can be found in the **Answers** section at the back of the book.

25 Feeding the World

Unsustainable agricultural practices in the Midwest were responsible for crop losses and the displacement of many families in the 1930s.

Similar unsustainable practices cause devastating destruction of land, even today.

Modern technology has helped reduce the impact of these disasters while feeding significantly more people. But is this production sustainable?

Plant Structure and Growth

And then the dispossessed were drawn west.... Car-loads, caravans, homeless and hungry; twenty thousand and fifty thousand and a hundred thousand and two hundred thousand. They streamed over the mountains, hungry and restless—restless as ants, scurrying to find work to do—to lift, to push, to pull, to pick, to cut—anything, any burden to bear, for food. The kids are hungry. We got no place to live.

This passage from John Steinbeck's classic novel *The Grapes of Wrath* describes the movement of people from the American Midwest to California during the era of the Dust Bowl. From 1931 to 1936, a combination of intense drought, severe storms, and damaging farming practices in the short-grass prairies of the Great Plains led to crop failures and widespread soil loss. Impoverished residents of the panhandles of Oklahoma and Texas faced a desperate choice between waiting out the bad times or pulling up stakes and leaving their homes. Nearly 25% of the population in these regions left over a period of just 4 years.

The farming practices that helped create the Dust Bowl were initially heralded as great improvements. Slow horse-drawn plows and harvesters were replaced with gasoline-powered tractors and combines, allowing more and more of the native prairie to be turned into wheat fields—but plowing prairie grasses destroyed the root systems that held the soil in place. Once these grasses were gone, soil could be picked up by fierce midwestern winds and blown thousands of miles in tremendous dust storms.

The environmental and human disaster during the Dust Bowl years was not a unique event. Today, thousands of acres of formerly productive farmland in China are turning to shifting sand. As a result, hundreds of thousands of people are moving away from this Asian Dust Bowl to cities along China's Pacific coast. Similar migrations have occurred in west-central Africa, parts of the Middle East, southern Russia, and elsewhere around the world as agriculture expands to ever-larger

—Continued next page

areas of Earth's surface. The mistakes of the Dust Bowl era are being repeated again and again.

Aggressive efforts to reverse soil erosion, combined with the return of normal rainfall levels, finally ended the Dust Bowl era in the United States. Thanks to the advances of modern agriculture, including the use of chemical fertilizers and irrigation from underground water stores, the Great Plains is now part of the most productive corn- and wheat-producing region on Earth. Can newly destroyed regions recover in the way that areas depleted in the American Dust Bowl did? Or is even the American Dust Bowl recovery an illusion maintained by huge inputs of resources that will eventually run out? To answer these questions, we must first understand the objectives and methods of agricultural practice—beginning by understanding the source of our crops and what these plants need to produce abundant food.

25.1 Plants as Food

For the first 250,000 years of human existence, our ancestors survived by hunting game and gathering fruits, seeds, roots, and other edible plant parts. Beginning about 11,000 years ago, independent human groups in what are now modern Iraq, Central America, Peru, China, New Guinea, and elsewhere began to practice **agriculture**—that is, to plant seeds, cultivate the soil, and harvest crops. What triggered the development of this radically different way of life?

The Evolution of Agriculture: Food Plant Diversity

Agriculture likely started when humans realized that nutritious and highly productive wild grasses and peas were colonizing "waste areas." As a result, they deliberately cleared additional land and scattered seed to encourage the plants' growth. Over time, the wild weeds became **domesticated** so that their growth and reproduction came under human control (**FIGURE 25.1**). The process of artificial selection, by which farmers chose which plants would produce the next generation based on their growth rates and ease of harvesting, produced crop plants that cannot survive without human intervention—and human societies that cannot survive without crop plants.

VISUALIZE **THIS**

Describe the probable actions of early farmers that would have caused teosinte to evolve via artificial selection into the familiar ear of corn.

(a) Teosinte, ancestor of modern corn

(b) Modern corn

FIGURE 25.1 The effect of domestication. (a) Teosinte, the ancestor of corn, produces a few small, loosely held kernels in a fragile "ear" that disintegrates when the kernels are ripe. (b) In contrast, corn's kernels are large, numerous, and do not detach from the cob.

Nearly all agricultural plants belong to a single phylum—the **angiosperms** (flowering plants), which produce energy-storing seeds within fruit. Around the time agriculture developed, certain angiosperms that were suitable for cultivation may have become more common due to natural climate change and increased human activity.

angio- means vessel.
-sperm means seed.

Clues about past climate conditions can be gathered from a variety of sources, including plant remains left behind in sediment, the size of annual growth rings in ancient trees and fossil wood, and air bubbles trapped in ice sheets. According to this natural record, Earth's climate changed dramatically beginning about 11,000 years ago, transitioning from a cool and wet period to a warmer, drier condition. The new climate also included more dramatic seasonality—wet seasons alternating with dry.

Annual plants, angiosperms that complete their life cycle from seed to adult plant to seed in the course of a single season, were successful in these new climate conditions. Annuals are also quite suitable for agriculture, providing a rapidly produced, abundant, and easily transported food source. Nearly all major crops are annual plants.

The annuals that provide the majority of calories to most societies evolved from weedy ancestors. These ancestors were well-suited to growing on open and cleared land, which often surrounded human settlements. The hardiest weeds in these environments were usually from the same two plant families: Poaceae, or grasses, in the class known as **monocots;** and Fabaceae, or peas and beans, from the class commonly known as **dicots.** Monocots and dicots get their names from the number of **cotyledons,** or embryonic leaves, they produce in their seeds (TABLE 25.1). The seeds of grasses (known in agriculture as cereal grains) provide a rich source of carbohydrates and 80% of the calories in a typical diet. On the other hand, the seeds of peas and beans (commonly called legumes) are rich in proteins, especially in certain amino acids that are uncommon in cereal grains. Thus, these two groups of plants together provide a reasonably balanced diet. Many cultures have their own version of this combination of plant groups. In the Middle East, the grain-legume combinations are wheat or barley with lentils or peas; in Asia, it is rice and soybeans; and in the Americas, corn and kidney beans.

TABLE 25.1 Monocot and dicot seeds.

The two major classes of angiosperms are named for the number of cotyledons contained in the seed.

Name	Number of Cotyledons (Embryonic Seed Leaves)	Example
Monocot	1 cotyledon	Corn Wheat Rice Lily
Dicot	2 cotyledons	Pea Tomato Maple Dandelion

Grains and peas are not the only domesticated plants, of course, but they are among the most important. The **staple crops,** or major sources of calories, in Europe, Asia, and Central America are wheat, rice, and corn, respectively. But in other places, people get most of their calories from other crops—for example, sweet potatoes in tropical Africa and Indonesia, white potatoes in parts of South America, and manioc (also known as cassava or tapioca) in Central and South America and western Africa. Although potatoes and manioc are also angiosperms, their food value comes not from their fruit or seeds but from other plant organs.

Plant Structure

Angiosperms have a relatively simple structure, consisting of three nonreproductive **vegetative organs** and two **reproductive organs.** The three vegetative organs are the roots, stems, and leaves; the reproductive organs are flowers and fruit (**FIGURE 25.2**).

Plant organs are made up of only four types of tissue (**FIGURE 25.3**). The primary function of **meristematic tissue** is cell division, and it is most commonly found at the tips of roots and stems. Cells arising from meristematic tissue are undifferentiated, and their adult form depends on their location in a growing plant; thus, unlike animals, a plant does not have a completely specified final form. **Epidermal tissue** serves as an outer covering of the plant body, functioning in water absorption below ground and water retention above ground. **Vascular tissue** is made up of two subtypes of liquid-conducting tissue: **xylem,** which carries water and minerals from the roots to the leaves, and **phloem,** which carries food, in the form of sap, throughout the plant. **Ground tissue** makes up the remainder of the plant body and may consist of cells that perform photosynthesis or store starch, among other roles. The relative amounts of these types of tissue in a vegetative organ often determine its nutritional value; more vascular tissue means more fiber, as in a low-calorie stalk of celery, but more ground tissue means a larger percentage of usable nutrients, as in a high-calorie potato.

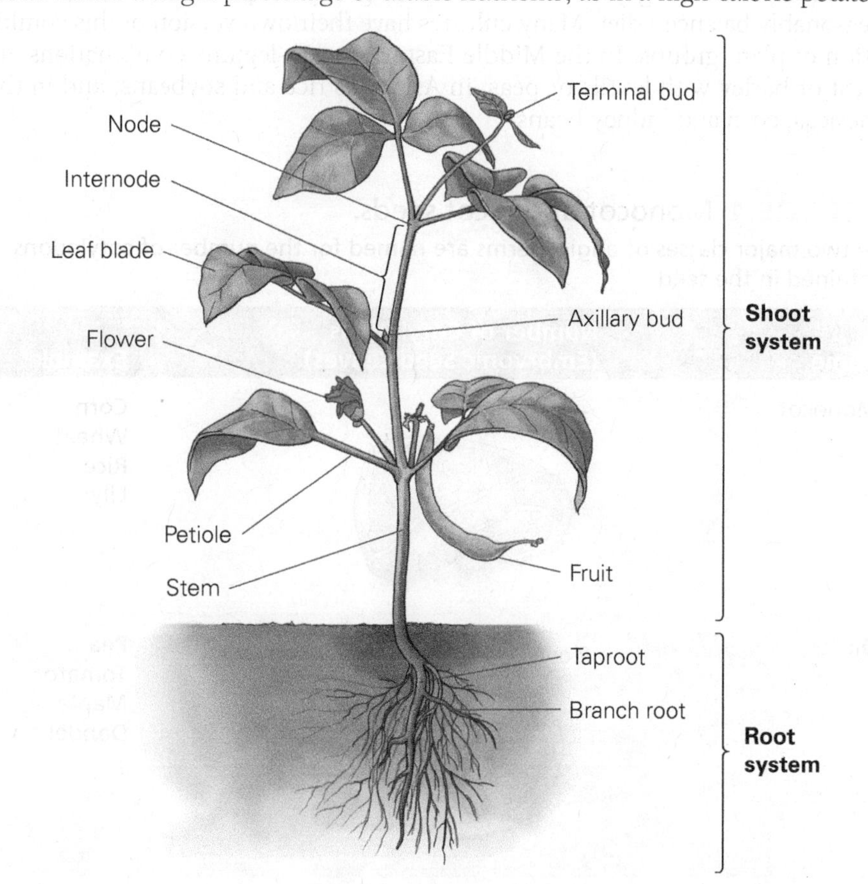

FIGURE 25.2 Plant structure. Plants consist of five organs: the vegetative structures—roots, stems, and leaves—and the reproductive structures—flowers and fruit.

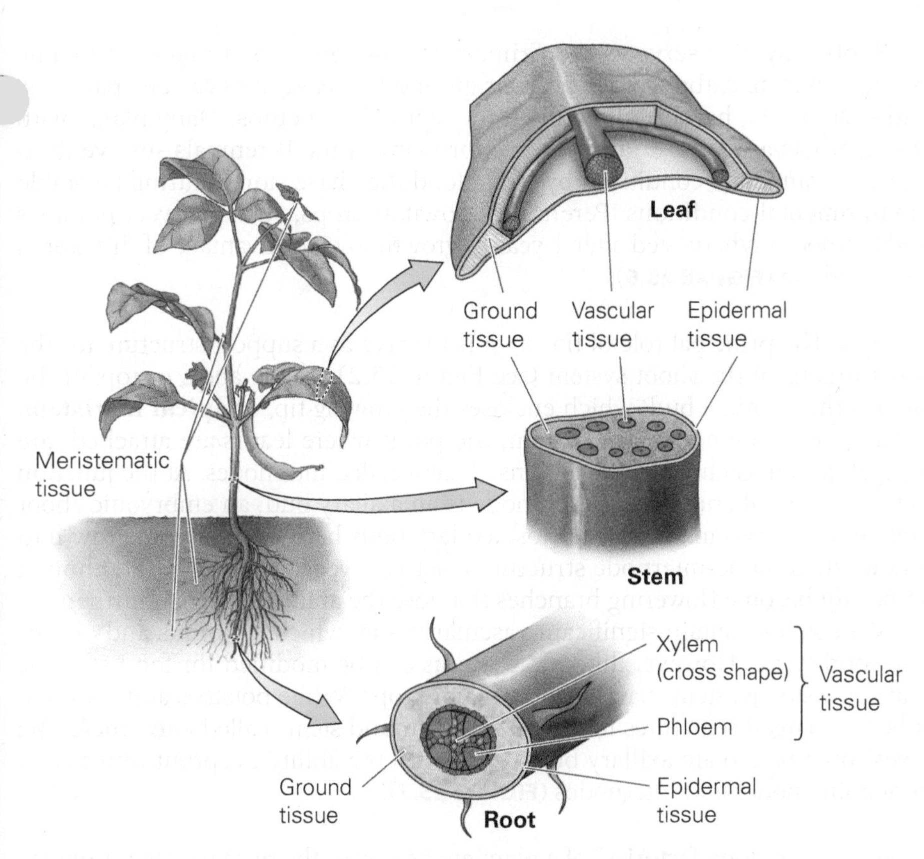

FIGURE 25.3 Plant tissue systems. Plants possess four tissue types—epidermal, ground, vascular, and meristematic.

VISUALIZE **THIS**

Based on this image, which of the four tissue types do you think is most common in the plant body?

Roots. The primary function of a plant's root system is to absorb water and minerals from the soil. The roots of most monocots are fibrous and spread out broadly, giving these plants an advantage in wetter seasons or where water drains quickly—the ability to exploit water as soon as it becomes available in the soil. However, the large central taproot of most dicots permits these plants to access minerals and water found much deeper in the soil, which is an advantage in drier places (**FIGURE 25.4**). Most of the water and minerals obtained by a plant are absorbed across the surface of root hairs, tiny projections of the epidermal cells (**FIGURE 25.5**). The meristematic tissue at the tips of roots allows the root system to extend through the soil via growth.

FIGURE 25.4 Monocot and dicot root systems. (a) Most monocots produce a fibrous root system that competes well for water but is not a strong support structure. (b) Most dicots produce a much sturdier, deeper taproot.

VISUALIZE **THIS**

How do root hairs improve a root's capacity to absorb water?

FIGURE 25.5 Root hairs. The growing ends of a plant's roots are covered with tiny, hairlike extensions of the epidermal cells.

-enni- pertains to a year.

Roots may also serve as the primary storage organ of a plant, and some plants that store carbohydrates in their ground tissue, such as carrots, parsnips, and sugar beets, have been modified into agricultural crops. Many plants with storage roots are **perennials** that live for many years. Perennials survive difficult environmental conditions by using food they have stored during favorable environmental conditions. Perennials grown as crops, such as sweet potatoes and manioc, are harvested after 1 year of growth to take advantage of this stored carbohydrate (**FIGURE 25.6**).

FIGURE 25.6 Manioc, or cassava. The storage roots of this plant are a staple in the diet of people throughout southeast Asia and elsewhere in the tropics. The roots contain large amounts of a cyanide-producing acid and must be extensively processed to make them edible.

Stems. The principal role of the stem is to serve as a support structure for the other organs of the shoot system (see Figure 25.2). At the apex, or top, of the stem is the terminal bud, which encloses the growing tip, or **apical meristem,** of the plant. The nodes of the stem, the point where leaves are attached, are separated from each other by sections of stem called internodes. At the junction between the leaf and stem at each node is an axillary bud, an embryonic shoot containing meristematic tissue. Most axillary buds have the ability to grow into a new lateral node-internode structure (that is, a vegetative branch), although some will become flowering branches that lose the ability to branch further.

Most stems contain significant vascular tissue, which is fibrous and of limited food value. However, like roots, stems can be modified for food storage, and some storage stems have become major crops. White potatoes and yams are tubers, enlarged structures found on underground stems called **rhizomes.** The "eyes" on a potato are axillary buds, each with the ability to sprout into a stem containing nodes and internodes (**FIGURE 25.7**).

rhizo- means root.

Leaves. The "food factories" of a plant are its leaves, the primary site for photosynthesis. Although no staple crops are leaves, leafy greens, including lettuce, spinach, chard, kale, and bok choy, are rich in essential vitamins and minerals and are important components of diets in many parts of the world. The leaf part that is more commonly consumed is the flat, photosynthetic blade, rather than the stiff and fibrous petiole, which consists primarily of vascular tissue. Humans have also domesticated plants that store food in modified leaves. Onions and other similar plants produce bulbs, underground storage structures that consist of short stems and fleshy, carbohydrate-filled leaves.

Humans have domesticated crops based on all imaginable plant parts (**FIGURE 25.8**). However, the vast majority of our foods are based on reproductive organs, primarily fruits and seeds.

FIGURE 25.7 A modified stem. A tuber is a modified stem, as the presence of sprouting buds on these potatoes indicates. These tubers are produced on another stem modified for underground life, a rhizome.

Plant Reproduction

We may not think of our breakfast toast and cornflakes as belonging to the same category of foods as our orange juice or bananas, but all of these foods are products of **fruit,** defined by botanists as mature flower ovaries containing seeds. Flowers and fruit are structures that are unique to angiosperms.

Flowers are the sexual organs of angiosperms (**FIGURE 25.9**). The typical flower produces both male and female structures. Male gametes are contained in pollen. The pollen is produced in structures called **anthers** on the male reproductive organ, called the **stamen.** Female gametes are contained in structures called **ovules** and produced in one or more chambers, called the **ovary,** of a flower's female reproductive organ, the **carpel.** Pollen delivered from other flowers collects on the **stigma,** a sticky pad on the top of a carpel.

The transfer of pollen from an anther to a stigma is called **pollination.** Pollination may occur via wind, as in corn and wheat, or via insects or other animals, as in soybeans. A flower's **petals** are adapted to facilitate pollination. After landing on the stigma, pollen grains germinate into tubes that extend down the length of the carpel to the ovules. A developing flower is enclosed in protective, modified leaves called **sepals** that in some plants also facilitate pollination.

FIGURE 25.8 Plants as food. Our diet includes a diversity of plant parts, including fruits, seeds, roots, and flowers.

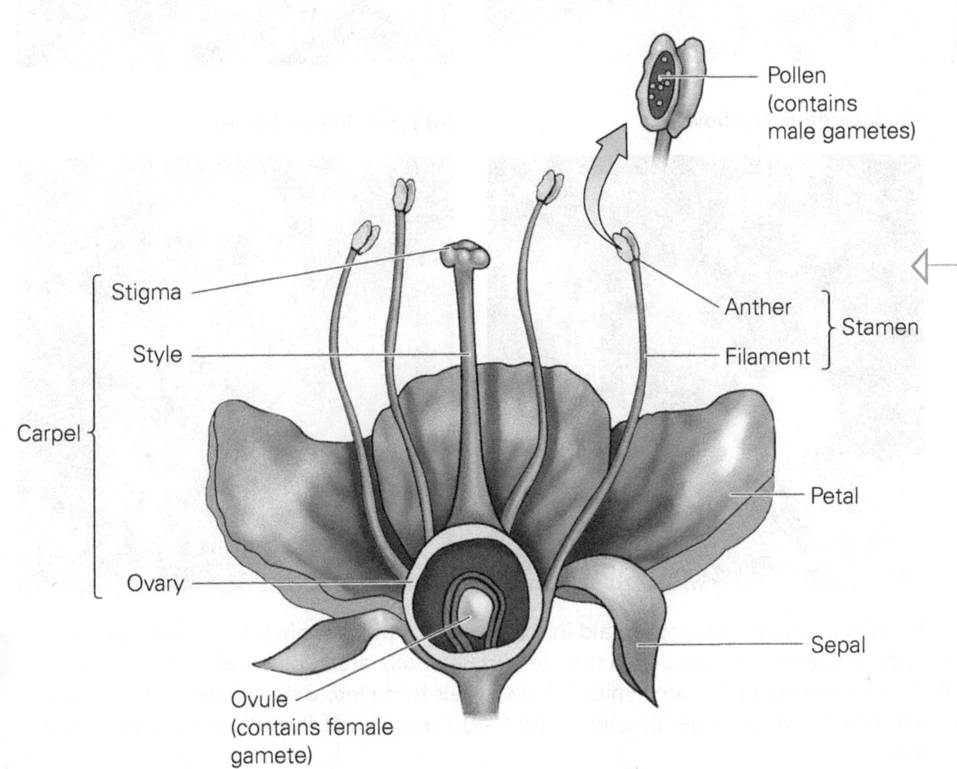

FIGURE 25.9 Flower structure. The typical flower contains both male and female reproductive organs, although many angiosperm species produce two types of flowers: male only and female only.

VISUALIZE **THIS**

What are the advantages and disadvantages of producing both male and female parts in the same flower?

In the case of wind-pollinated grass, petals are much reduced, whereas in insect-pollinated legumes, they are colorful and contain nectar to attract and reward their pollinators (**FIGURE 25.10a–b**). Many plants are pollinated only by a single species or group of species. These specific relationships evolved as pollinators selected plants that gave them the best rewards and as plants developed traits that were highly attractive to particular pollinators (**FIGURE 25.10c–e**).

Once in the ovules, the pollen tube releases two sperm. In a process known as **double fertilization,** one sperm fertilizes the egg to become the diploid embryo and the other fuses with two nuclei within the ovule to produce a triploid cell, containing three copies of each chromosome. This cell divides and develops into **endosperm,** tissue that nourishes the developing embryo. The primary component of wheat flour is the starchy endosperm of wheat seeds; whole wheat flour also contains the embryo (the germ) and accessory tissues (the bran), whereas white flour is strictly the endosperm. In some species, such as in legumes, the embryo consumes the endosperm as it develops and the seed matures. The two halves of a bean seed or peanut are enlarged cotyledons produced by the digestion of the endosperm.

VISUALIZE **THIS**

Some flowers have evolved to look like female wasps and even produce a scent that smells like the wasp's sex hormone. What animal likely pollinates these flowers?

(a) Wind-pollinated flower

(b) Bee-pollinated flower

(c) Bird-pollinated flower

(d) Bat-pollinated flower

(e) Fly-pollinated flower

FIGURE 25.10 Petals and pollination. A flower's petals are adapted to aid in the process of pollination. (a) The most successful wind-pollinated plants have reduced petals that do not interfere with pollen transfer. (b) Bee-pollinated flowers often provide a nectar reward at the base of a petal tube, and (c) bird-pollinated flowers typically have petals fused into a single deep, highly visible floral tube. (d) Flowers that are pollinated by bats often open their petals only at night. (e) The petals of fly-pollinated flowers may have the appearance—and odor—of rotting flesh.

(a) Wind-dispersed fruit

(b) Water-dispersed fruit

(c) Animal-dispersed hitchhiker fruit

(d) Animal-dispersed edible fruit

FIGURE 25.11 Seed dispersal units. Fruits are adapted to disperse seeds away from the parent plant. Many mechanisms are possible, including (a) wind, (b) water, (c) mammal fur, and (d) through an animal's digestive tract.

The embryo, endosperm, and remainder of the ovule tissues make up the **seed.** Triggered by successful fertilization, other parts of the flower—in nearly all cases, the ovary, but sometimes petals, sepals, or the flower stalk—develop into **fruit.** Most fruits are adapted to be vehicles for dispersing seeds far from the parent plant. Some fruits help seeds disperse on wind or water; other fruits help seeds disperse by using animals to carry the fruit away and drop its seeds elsewhere (**FIGURE 25.11**). Because humans have similar sensory systems and nutritional needs to those of other mammals, many of the crop plants that we rely on were selected from fruits adapted to attract mammal dispersers. Even grains are adapted to attract mammals, primarily rodents. **FIGURE 25.12** illustrates the relationship among flowers, seeds, and fruit in angiosperms.

Producing abundant fruit for human consumption requires more than simply selecting plants for optimal fruit production. The agricultural practices of people around the world are designed to maximize the growth of these plants.

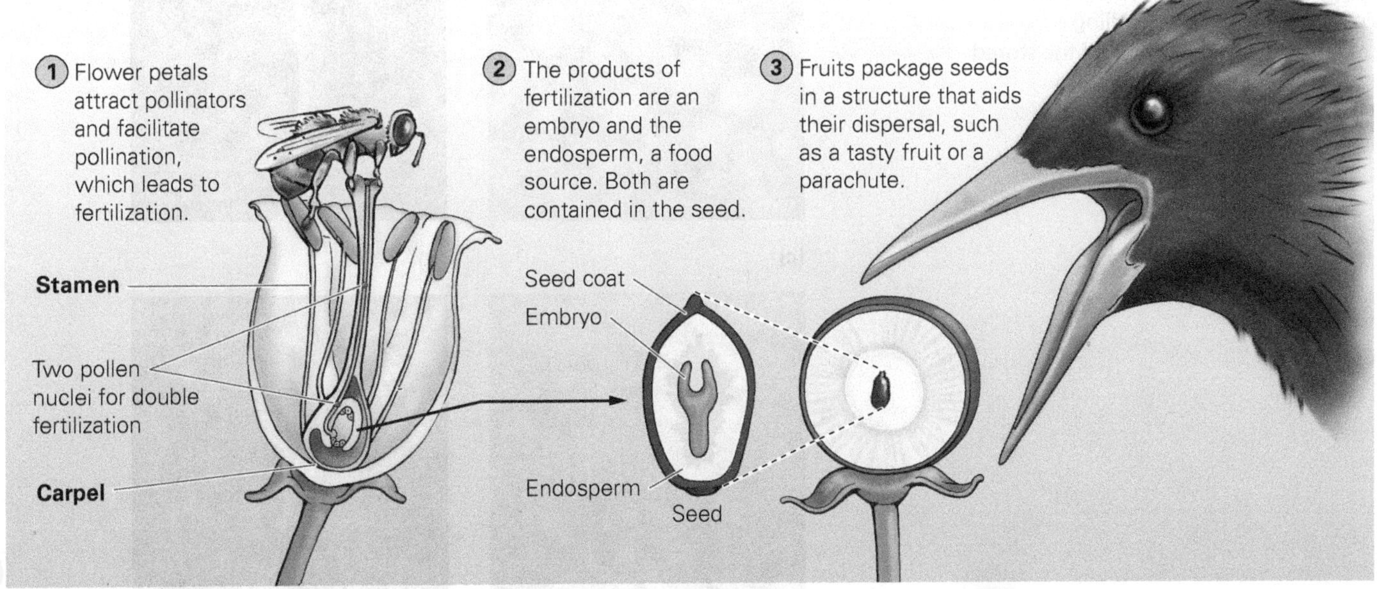

1 Flower petals attract pollinators and facilitate pollination, which leads to fertilization.
2 The products of fertilization are an embryo and the endosperm, a food source. Both are contained in the seed.
3 Fruits package seeds in a structure that aids their dispersal, such as a tasty fruit or a parachute.
Stamen
Two pollen nuclei for double fertilization
Carpel
Seed coat
Embryo
Endosperm
Seed

FIGURE 25.12 A summary of angiosperm reproduction.

Got It?

1. Annual plants complete their entire ________ ________ in one year.
2. Meristems in plants are regions of active ________ ________.
3. The primary function of a flower is to facilitate ________ reproduction.
4. An angiosperm fruit almost always develops from the ovary of the ________.
5. Double fertilization in angiosperms results in the production of a triploid endosperm, which serves as ________ for the developing embryo.

25.2 Plant Growth Requirements

Agricultural practices such as turning over soil, irrigating soil with water, and adding nutrients to soil developed independently everywhere that plants were domesticated by people. To understand why these techniques enhance agricultural production, we must understand first how plants grow.

How Plants Grow

The first stage of plant growth is the emergence of the seedling, supported by the seed's internal food stores. Later plant growth can be divided into two types—growth in length, called **primary growth,** and growth in girth or circumference, called **secondary growth.**

Germination. The emergence of an embryo from a seed is called **germination.** The process begins when the seed begins to take up water, expands, and ruptures its confining seed coat (**FIGURE 25.13a**). As long as the seed coat is

FIGURE 25.13 Germination.
(a) Seed germination begins when the seed takes up water. (b) The first organ to emerge is the root. (c) Water obtained by the root makes the cells in the embryonic shoot enlarge, allowing the shoot to emerge from the soil. (d) Germination is complete when the seedling is functioning independently of the stored food in the seed.

impermeable to water, the seed remains dormant, which means that it will not germinate. Dormancy allows seeds to act as "time travelers," passing through poor environmental conditions in a state of suspended animation. Many seeds require exposure to specific environmental conditions before they can escape dormancy.

The first structure to emerge from a germinating seed is the root, which enables the developing young plant, or **seedling,** to become anchored in the soil and absorb water (**FIGURE 25.13b**). Growth of the shoot soon follows as the seed's stored food is depleted, and photosynthesis becomes necessary (**FIGURE 25.13c**). The time between germination and when the seedling is established as an independent organism is the most vulnerable period in the life of a plant. Damage to the plant during this period is usually fatal. Once a seedling becomes established, however, damage often may be overcome by continued growth (**FIGURE 25.13d**).

Primary Growth. In plant stems, primary growth takes place through the production of additional nodes and internodes at both the shoot tips and the axillary buds. Growth in length occurs when cells at a meristem divide in a single plane, creating columns of daughter cells below them and pushing the apical meristem farther up into the air (**FIGURE 25.14**). A hormone produced by the terminal bud inhibits cell division in the axillary buds in many plants. Once the terminal apical meristem has grown away from a particular axillary bud, the bud's meristem begins to divide and produces a lateral branch, also containing axillary buds.

The node-internode system of the main stem and branches supports the **indeterminate growth** of a plant's shoot system—its ability to grow throughout its lifetime. This is in contrast to the **determinate growth** of most animals, which grow to a set size that they typically reach early in life. Not all plant growth is indeterminate—leaves and flowers tend to grow to a set size, for instance.

In roots, primary growth at the tip occurs due to cell division in the apical meristem. Cell division in the root occurs in two opposing directions, resulting in columns of cells that elongate and push the root tip forward through the soil, as well as the production of a loosely organized **root cap** that protects the tip of the root. As the root tip pushes through the soil, exterior cells of the root cap are worn off and replaced by younger cells more recently produced by the apical meristem. The primary meristematic tissue producing lateral roots is found near the center of the root, surrounding the vascular tissue. As a result, lateral roots can form anywhere along the length of a root (**FIGURE 25.15**).

Primary growth results in plants that are taller, bushier, and have more extensive root systems. In longer-lived plants, continued primary growth

VISUALIZE THIS

Identify the nodes and internode visible in this micrograph.

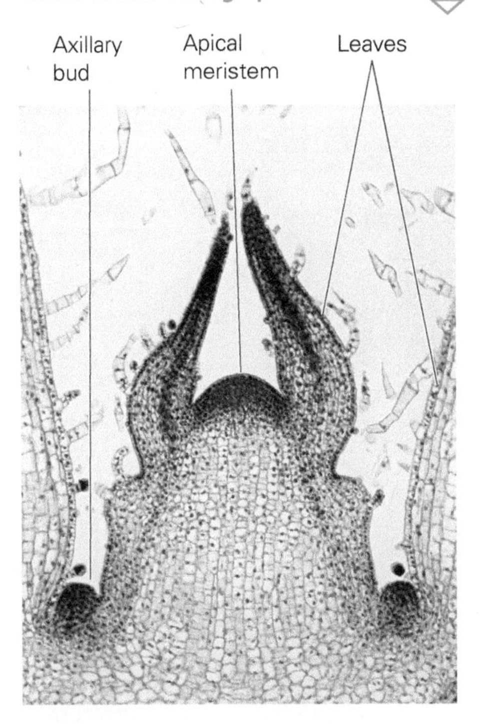

FIGURE 25.14 Primary growth in stems. This microscopic image of an apical meristem clearly shows the repeating node-internode system.

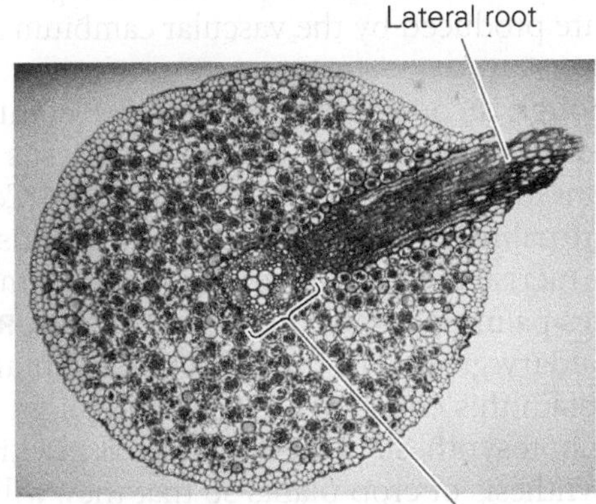

FIGURE 25.15 Lateral root formation. In this cross section, meristematic tissue in the center of the root has produced a lateral root.

VISUALIZE THIS

Why must a lateral root originate from the vascular cylinder in the center of the primary root?

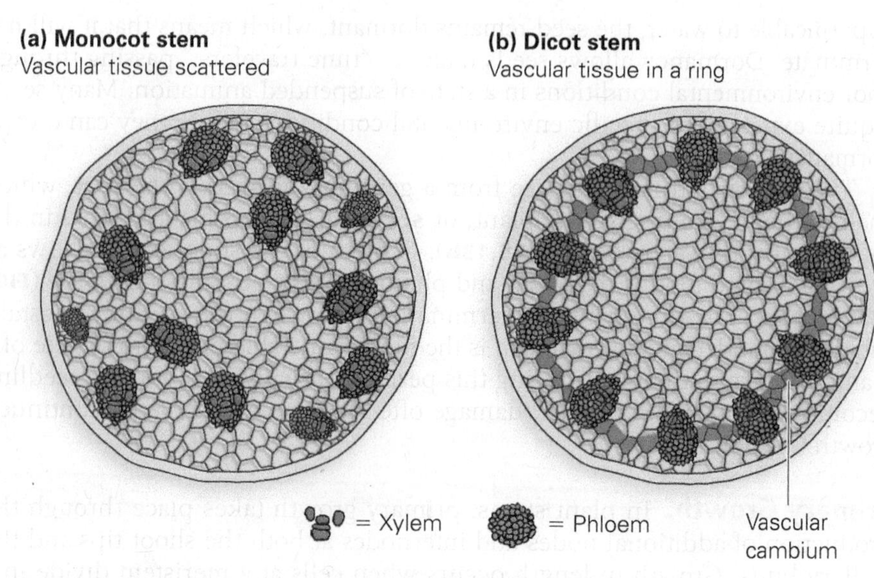

FIGURE 25.16 Monocot and dicot stems. Unlike monocot stems, which have scattered bundles of vascular tissue, dicot stems have vascular bundles arranged in a ring. This organization allows a single ring of meristematic tissue, the vascular cambium, to form between the xylem and phloem.

presents a dilemma—a tall plant cannot support itself with a skinny stem Many plants can increase their stability by making wider stems and roots via secondary growth.

Secondary Growth. All of the flowering plants that undergo true secondary growth are dicots. In these plants, a ring of cells running through the bundles of vascular tissue in the stem becomes a meristem called the vascular cambium (FIGURE 25.16). This cambium produces phloem cells to the outside of the stem and xylem cells to the inside.

The additional xylem cells produced by the cambium make up the bulk of the material in these stems, and their stiffened cell walls produce the stem's **wood** (FIGURE 25.17). Epidermis on the surface of the stem breaks apart and is shed as the stem's surface area increases. Cells immediately under the epidermis become another meristem (called the **cork cambium**). Cell division in the cork cambium results in the production of **cork,** a water-resistant outer covering. Phloem, cork, and cork cambium make up the **bark** of a woody dicot stem. All the components of bark are continually shed and replenished by secondary phloem cells, which are produced by the vascular cambium as the tree increases in diameter.

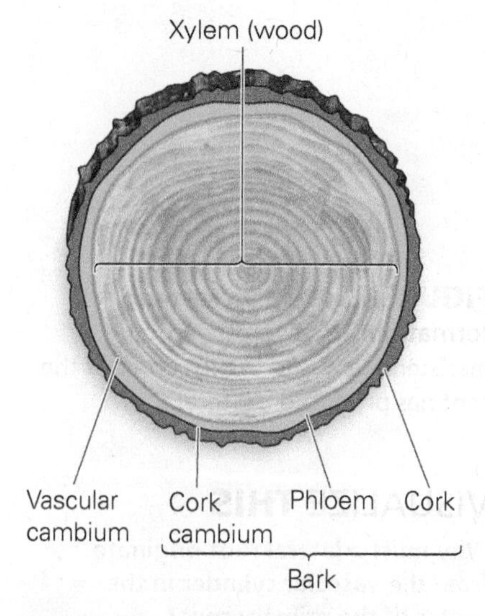

FIGURE 25.17 Dicot wood. The wood of a dicot stem is made up of xylem cells, whereas the bark consists of phloem and cork.

Interestingly, although dicot trees are much more common than monocot trees, the only tree-based staple food in human societies is the monocot coconut, which is important to communities in Indochina and Central America. The thick stem of coconut palms consists of additional strands of vascular tissue, produced by another meristem that surrounds the apical meristem. The structure of this stem makes palm trees incredibly flexible (FIGURE 25.18).

Primary and secondary growth of plants can be maintained only by abundant energy. Plants obtain this energy by transforming light into chemical energy in the process of photosynthesis. Our agricultural practices are intended to maximize the photosynthesis of crop plants so that they will produce a surplus of carbohydrates that we can harvest.

(a) Palm tree trunk cross section

(b) Palm trees bending in hurricane-force winds

FIGURE 25.18 Monocot "wood." Secondary growth in monocots, such as this palm, produces much more fibrous wood (a), which is not nearly as strong as dicot wood but is very flexible (b).

Maximizing Plant Growth: Water, Nutrients, and Pest Control

The basic equation of photosynthesis is as follows:

Carbon Dioxide + Water + Light Energy
→ Carbohydrate + Oxygen Gas

In other words, photosynthesis in plants converts energy from sunlight, carbon dioxide from the atmosphere, and water from the soil into energy-rich carbohydrate molecules. Oxygen is essentially a "waste product" of this reaction (Chapter 5).

If we look only at this chemical reaction, it is clear that plants require carbon dioxide, water, and light to produce carbohydrates. However, the process also requires a few more components. For instance, to survive, plants require much more water than that needed simply for the chemical reactions of photosynthesis. Plants also require other chemical inputs in smaller amounts. For example, photosynthesis requires the pigment **chlorophyll** to proceed, and magnesium is an essential element in chlorophyll. Inputs such as magnesium and nitrogen, an essential element in proteins, are obtained from the soil. Pest organisms that damage photosynthetic tissue and fruits and seeds also affect plant production. Therefore, to increase crop production, modern farming practices attempt to maximize the amount of light, water, carbon dioxide, and nutrients available to crop plants as well as to minimize damage caused by plant predators (**FIGURE 25.19**).

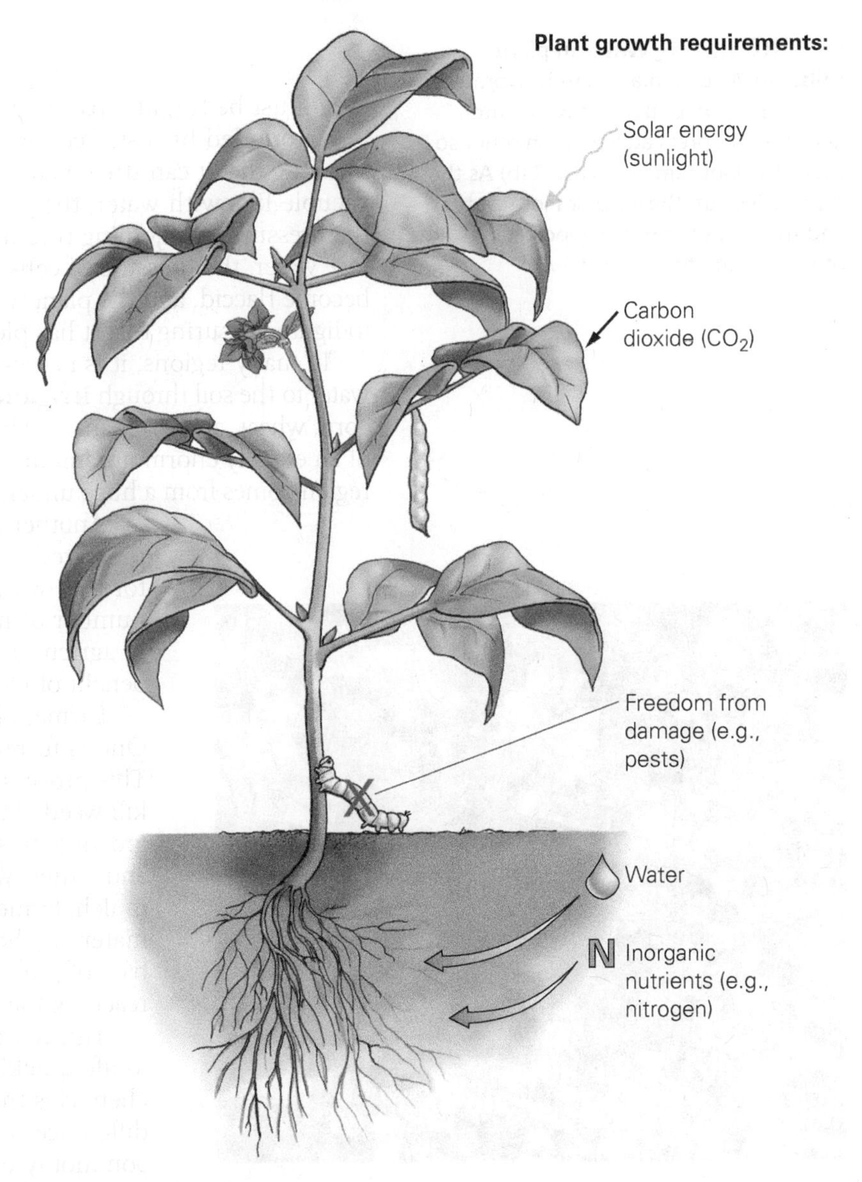

FIGURE 25.19 Five requirements for plant growth. Plants require the basic ingredients for photosynthesis—water, carbon dioxide, and light—as well as nutrients from the soil and freedom from damage to produce excess carbohydrates for human consumption.

Water, Sunlight, and Carbon Dioxide. To hold its leaves perpendicular to the sun's rays, a plant's cells must be full of water—that is,

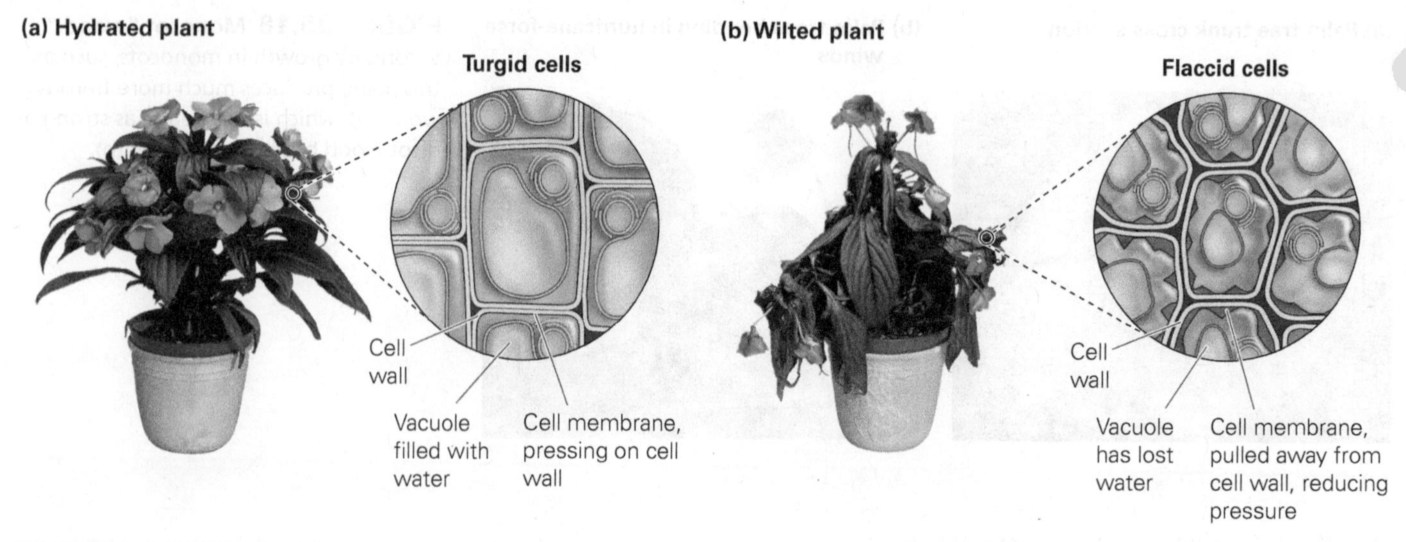

FIGURE 25.20 Water in plant cells. (a) When a plant is fully hydrated, the cell's vacuoles balloon with water, and the cells press against each other so that the plant remains "crisp." (b) As the plant dries out, the vacuoles lose water, and the cells no longer support each other, causing the plant to wilt.

they must be turgid (**FIGURE 25.20**). Recall that the membrane of a plant cell is surrounded by a stiff cell wall. The cell wall is tough, but it is also elastic, meaning that it can stretch and still hold its basic shape. As a plant cell's central vacuole fills with water, the cell wall balloons slightly, becoming turgid. It is the pressure of adjoining turgid cells that holds the leaves upright. When cells lose water, the pressure of cells pushing against each other decreases; the cells become flaccid, and the plant wilts. Thus, part of maximizing a plant's exposure to light is ensuring that it has plenty of water.

In many regions, it is necessary to prevent wilting of crop plants by adding water to the soil through **irrigation** (**FIGURE 25.21**). The enormous production of corn, wheat, and soybeans in what was the American Dust Bowl region is a result of an equally enormous amount of irrigation. The water for irrigation in this dry region comes from a huge underground reservoir called the Ogallala aquifer.

FIGURE 25.21 Irrigation. The circle pattern of lush plant growth derives from center-pivot irrigation sprayers, which spray water in an arc from the center point. The smaller circles in this image have a diameter of about 800 m.

Another way to increase the amount of water that reaches preferred plants is to reduce the amount of competition for that water. Farmers do this by trying to minimize the number of nonpreferred plants, commonly called **weeds,** in agricultural fields. Controlling weeds has the additional benefit of reducing competition for sunlight.

Farmers have several techniques for controlling weeds. One is to remove competitors before the crop is planted. This process, called tilling, involves turning over the soil to kill weeds that have sprouted from seed. After a crop begins growing, it is possible to prevent the growth of competitors and reduce water loss from the soil by mulching the crop. To mulch, farmers spread a thick layer of straw, other dead plant material, sheets of newspaper, or dark plastic around the base of preferred plants. These materials block sunlight from reaching the soil, preventing weed seeds from germinating.

Tilling does not kill all weeds, and some spring back to life quickly. In these cases, farmers may use **herbicides,** chemicals that kill plants. Herbicides that exploit chemical differences between dicot and monocot plants are the most commonly used sprays in agriculture. These herbicides are used to control dicot weeds in monocot crops such as corn and wheat. Broad-spectrum herbicides, such as glyphosate (Roundup®), kill all plants; until recently, these herbicides have been useful only for controlling weeds before planting.

Development of genetically modified organisms (GMOs) has increased the applicability of broad-spectrum herbicides to many more crop-and-weed combinations in recent years (Chapter 10).

Even when plants are fully hydrated and not wilting, their efforts to prevent water loss can interfere with growth. This is because plant strategies for conserving water can impose a cost to photosynthetic production.

The primary adaptation that plants have to reduce water loss is a waxy covering, called the cuticle, on the epidermis of the shoot system. The cuticle also restricts the diffusion of carbon dioxide into the plant. To enable plant cells to obtain adequate carbon dioxide for photosynthesis despite the cuticle, the shoot epidermis is pocked with tiny pores called **stomata** (singular: stoma). However, these pores come with a price—stomata also provide portals through which water can escape.

In most plants, each stoma is encircled by a pair of **guard cells** that regulate the size of the pore (**FIGURE 25.22**). When the guard cells have abundant water, they increase in size, and their unique shape causes the pore to open. When they are water-deprived, such as when the soil is dry, the skies are sunny, and the humidity is low, guard cells shrink and relax, causing the pore to be small. Thus, to maximize carbon dioxide uptake, plants need enough water to keep their stomata open.

Where precipitation or irrigation is sufficient to keep plants from wilting and the stomata open, the most important factor affecting plant growth is nutrient availability.

VISUALIZE THIS

Given the two competing roles of stomata, how do you expect the number of stomata to differ between plants in desert areas compared with those in tropical rain forests?

(a) When water is abundant:

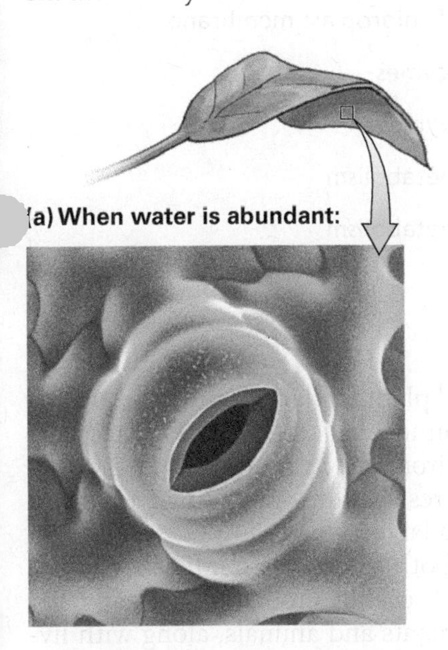

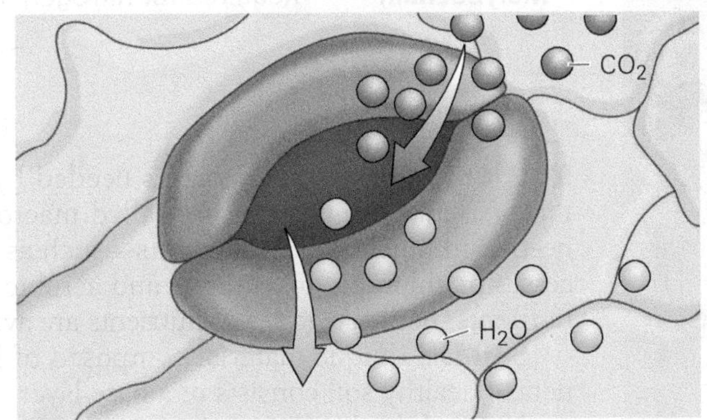

- Guard cells swell.
- Stoma (pore) is large.
- High carbon dioxide uptake
- High water loss

(b) When water is scarce:

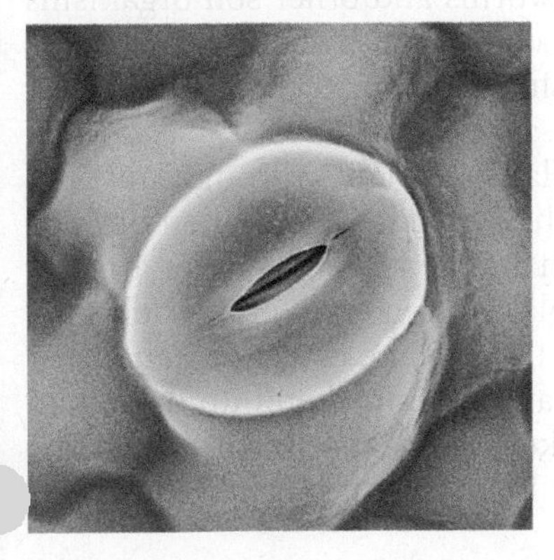

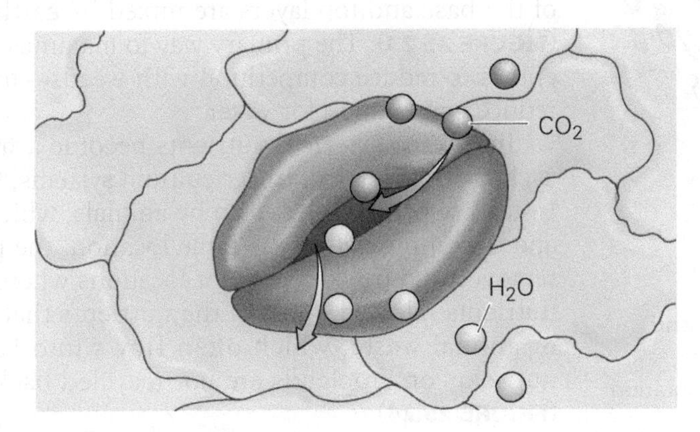

- Guard cells shrink.
- Stoma (pore) is small.
- Low carbon dioxide uptake
- Low water loss

FIGURE 25.22 Guard cells. Guard cells help to regulate the size of stomata on a plant's surface to minimize water loss or maximize carbon dioxide intake.

TABLE 25.2 Plant nutrients and their functions.

Macronutrient	
Nitrogen	Component of proteins, DNA, and RNA
Potassium	Involved in opening and closing of stomata
Calcium	Component of cell walls; involved in cell membrane permeability
Magnesium	Component of chlorophyll
Phosphorus	Component of ATP, DNA, RNA, and cell membranes
Sulfur	Component of proteins
Micronutrient	
Chlorine	Involved in moving water into and out of cells
Iron	Required for chlorophyll synthesis and aerobic respiration
Boron	Influences calcium use; required for DNA and RNA synthesis
Manganese	Required for integrity of chloroplast membrane
Zinc	Component of many enzymes
Copper	Component of some enzymes
Nickel	Required for nitrogen metabolism
Molybdenum	Required for nitrogen metabolism

FIGURE 25.23 Soil structure and development. Soil forms from the breakdown of rock and the accumulation of organic matter from dead and decaying plants, fungi, animals, and animal wastes.

Nutrients and Soil. Elements needed by plants in large amounts, such as nitrogen and magnesium, are called macronutrients (**TABLE 25.2**). Nutrients needed in much smaller amounts—such as iron, which is required in the processes of chlorophyll synthesis and aerobic respiration—are known as micronutrients. Both categories of nutrients are available in the soil.

Soil is a unique material composed of both living and nonliving components. Healthy soil consists of a base layer of recently eroded rock; a top layer made up of dead and decaying remains of plants and animals, along with living organisms that eat them, and waste; and a middle layer where components of the base and top layers are mixed by earthworms and other soil organisms (**FIGURE 25.23**). The primary way to maximize a crop plant's access to soil nutrients is to reduce competition with weeds—using the same techniques used to reduce competition for water.

In natural systems, nutrients become available to plants through nutrient cycling (Chapter 16). In agricultural systems, the nutrient cycle is often broken. Instead of plants being eaten by animals, which are then eaten by other animals and decompose all in the same location, the plants from agricultural fields are removed and trucked to other locations where they will be consumed and their nutrients released. Most of the nutrients that are consumed end up in human or animal waste, which often flows into lakes or oceans. In other words, nutrients on croplands are not recycled back into the soil but mined from it (**FIGURE 25.24**).

Because nutrient cycles are broken in agricultural fields, farmers must add nutrients to soils in the form of **fertilizer.** Fertilizer that is applied to the soil

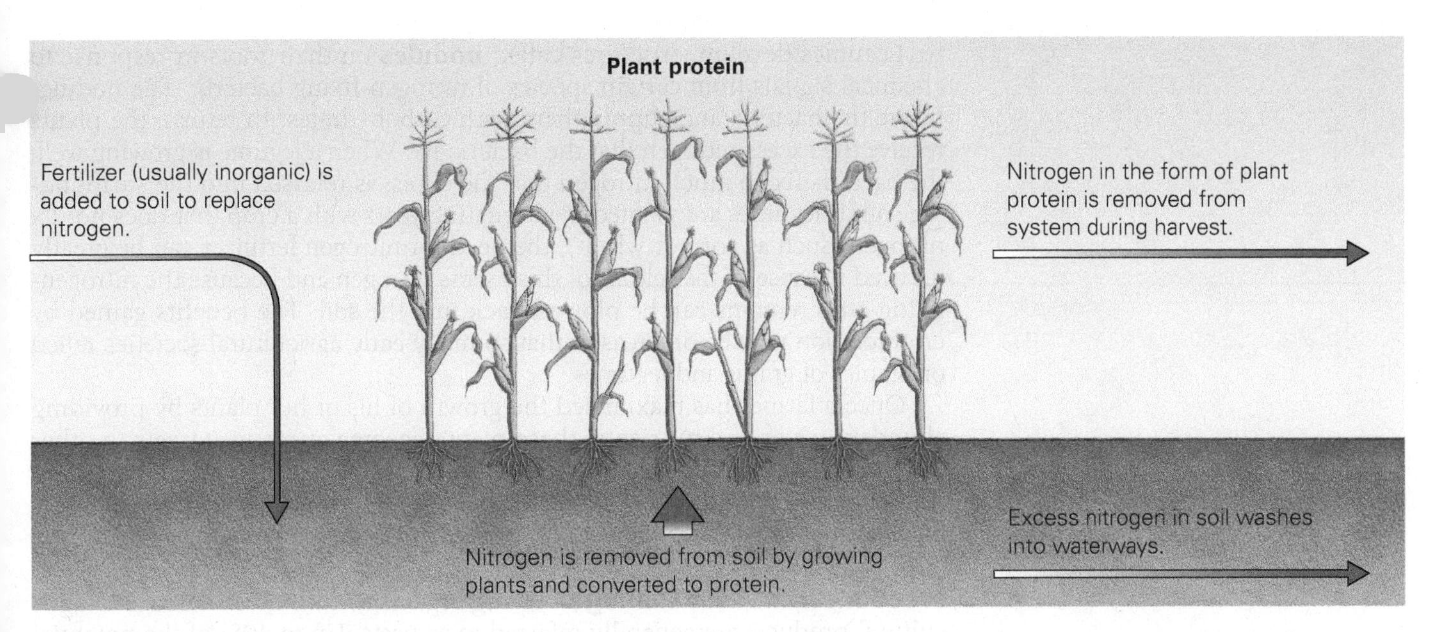

Plant protein
Fertilizer (usually inorganic) is added to soil to replace nitrogen.
Nitrogen in the form of plant protein is removed from system during harvest.
Excess nitrogen in soil washes into waterways.
Nitrogen is removed from soil by growing plants and converted to protein.

FIGURE 25.24 Nutrient mining. The movement of nitrogen in an agricultural system is illustrated here.

VISUALIZE **THIS**

How would the nitrogen flows in this image differ if the system were a natural environment?

can be **organic**—that is, made up of carbon-containing molecules from the partially decomposed waste products of plants and animals. Fertilizer can also be **inorganic**—meaning made up of simple molecules that lack carbon, such as ammonia and nitrate, and produced by an industrial process. Because plants require nutrients in inorganic form, soil organisms must first decompose organic fertilizer before its nutrients are available for plant growth.

Most farmers prefer inorganic fertilizer for two reasons: because the nutrients in it are more concentrated and much easier to transport, store, and apply than most organic fertilizers; and because plants can use these nutrients immediately. Most farmers in the United States fertilize with inorganic nitrogen, phosphorus, and potassium and occasionally will add other macronutrients such as calcium.

Farmers can also replace some nutrients in the soil through **crop rotation**—that is, by varying the crops that are planted on any one field over time. Surprisingly, some plants actually increase the levels of nutrients in the soil. These plants have a mutualistic relationship with **nitrogen-fixing bacteria** (**FIGURE 25.25**), which are microbes with the ability, unique among living organisms, to convert nitrogen gas from the atmosphere into a solid form—namely, ammonia.

(a) Alfalfa

(b) Nodules on alfalfa roots

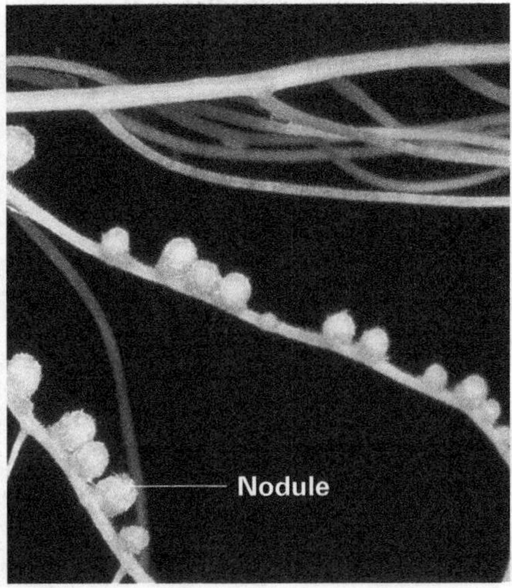

FIGURE 25.25 Nitrogen-fixing crop. (a) Alfalfa can increase the nitrogen content of soil. (b) The nodules on the roots of alfalfa contain nitrogen-fixing bacteria.

Legumes develop structures called **nodules** on their roots in response to chemical signals from certain species of nitrogen-fixing bacteria. The nodules house the bacteria and supply them with carbohydrates. In return, the plants receive the excess nitrogen that the bacteria fix. When a legume is growing well, the bacteria fix so much nitrogen that the excess is released into the surrounding soil. If legumes are planted in alternating years with a crop that does not fix nitrogen (such as corn or wheat), the need for nitrogen fertilizer can be greatly reduced because of the release of this excess nitrogen and because the nitrogen-fixing crop remains can be plowed back into the soil. The benefits gained by crop rotation may be one reason that so many early agricultural societies relied on staples of grains and legumes.

Once a farmer has maximized the growth of his or her plants by providing abundant water and nutrients, these plants become attractive targets to other consumers—including other mammals, birds, insects, fungi, and bacteria. To ensure that most of the food produced by an agricultural crop goes to feed people, a farmer must try to reduce the damage these competitors do to the crop.

Freedom from Pest Damage. Living organisms that cause damage to agricultural products are generally referred to as pests. Up to 40% of the potential food production in major crops is lost to pests every year. The primary tools farmers use to reduce pest impact are **pesticides,** chemicals designed to kill or reduce the growth of a target pest. (Pesticide is not a specific term and may refer to chemicals that kill weeds. In this text, we use the word *pesticide* only when referring to chemicals that control insect, fungal, and bacterial pests.) Pesticide use became widespread in the years following World War II. Today, millions of tons of these chemicals are applied annually to crops all over the world.

Although pesticide use has been reduced recently, thanks to integrated pest management, described in the next section, these chemicals are still in use in modern agriculture. Surprisingly, pesticide use has not considerably reduced the overall loss of crops to pests. Instead, these chemicals have dramatically influenced farming methods—primarily by allowing farmers to plant a single crop over a wide acreage. This practice is called **monoculture.** Monocultural production of crops is very efficient. Farmers have one planting, fertilizing, and pest control schedule, and they require only the planting and harvesting equipment specific to their crop (**FIGURE 25.26a**). Monoculture and the accompanying mechanization of agriculture have greatly reduced the amount of labor required to produce crops. The percentage of the population working on farms in the United States has dropped from 25% to less than 2% in the past 90 years due to this change in farming practice.

FIGURE 25.26 Monoculture and polyculture. (a) Planting a single crop over many hundreds of acres (monoculture) allows for the efficient use of massive planting and harvesting equipment, as in this Alberta wheat field. (b) Planting many different crops together (polyculture) can reduce pest problems and buffer farmers from crop failures.

(a) Monoculture

(b) Polyculture

On the other hand, monocultures are very susceptible to outbreaks of pests. If a small population of insects or disease infestation consumes or kills a host plant, another appropriate host for the pest is right next door. The easy availability of resources to pests in monocultural crops allows their populations to grow exponentially and can lead to the loss of an entire crop. Thus, although monoculture increases the efficiency of agricultural production, it results in an increased need for pesticides just to keep pest damage at historical levels.

Farmers can also use **cultural control** to minimize pest populations. Because most pests have evolved to attack a single crop, they will not cause damage to other, unrelated crops. Crop rotation, in addition to conserving soil nutrients, is a method of cultural control that moves plants away from their pests. After the growing season, a pest population spends the winter in the soil or on plant waste left in the field. If the same crop is planted the following year, the pest population has easy access to more resources and can continue to increase in population. However, if a different crop is planted, the pests must disperse in the spring in an attempt to find their required crop. Only some of these dispersers will find their host, and the population of pests in the new site will be relatively small.

Another form of cultural control is **polyculture**—planting many different crop plants over a single farm's acreage (**FIGURE 25.26b**). Unlike monocultures, polycultural planting keeps pest populations relatively small because, after the pests have consumed a patch of plants, they must disperse long distances to find another source of food. Polyculture also minimizes risk to farmers by ensuring that a pest outbreak on a single crop does not destroy all of the farm's production for that year.

Over the course of the twentieth century, the use of pesticides, fertilizer, and large farm machinery dramatically changed the way food was produced around the world. The revolution in biology—particularly the new science of genetics—has also played a role in this change.

Designing Better Plants: Hybrids and Genetic Engineering

Traditionally, farmers used artificial selection to modify the characteristics of their crop plants and animals—that is, they selected for next year's seed stock or breeding animals those that performed best under the conditions on their farm. However, this process limits farmers to only the variation that is currently available. Two developments in plant science reduced and finally eliminated this limitation.

Beginning in the 1920s, agricultural scientists began using plant breeding to produce better crops. Plant breeding generally consists of creating hybrids, the offspring of two different varieties of an agricultural crop. One of the first hybrids produced on a large scale was dwarf spring wheat. One parent variety was a nitrogen-loving, high-producing wheat that was difficult to harvest because it tended to fall over when the wheat grains were ripe. The other parent variety was a lower-producing dwarf wheat plant. The hybrid of these two varieties was high producing but short, so it responded well to fertilizer application and was easy to harvest with large machines. Because hybrids are the first-generation offspring of two distinct varieties, new hybrid seeds have to be produced every year from the parent populations. This is necessary because independent assortment occurs during meiosis (Chapter 8); thus, seeds produced by the hybrid plant will not contain the mix of chromosomes found in the first-generation plants.

In addition to the difficulty of producing hybrid seeds, plant breeding limits agricultural scientists to the genetic variation within a single species. More recently, DNA technology has permitted the creation of novel agricultural organisms containing genes from other species. These genetically modified organisms, or GMOs, include herbicide-resistant crop plants. In the United States, more than 40% of farmland planted with corn, more than 70% of the cotton crop, and nearly 80% of soybean fields are planted with genetically modified varieties. However, many questions remain about GMOs, including the likelihood that their widespread use will cause the evolution of resistance in pests, their effects on nontarget organisms, and the consequences of the "escape" of engineered genes into noncrop plants (see Chapter 10).

The concerns related to GMOs should not be downplayed. However, in several ways, these concerns pale in comparison with those created by conventional agricultural practices, including environmental degradation and loss of soil.

Got It?

1. Indeterminate growth in plants means that adult plants don't have a set final ________ and ________.
2. Abundant water is important to plant growth because it prevents ________, which interferes with the capture of sunlight.
3. Abundant water also permits stomata to stay open, facilitating ________ ________ uptake.
4. A symbiotic relationship between legumes and certain types of bacteria increases ________ levels in the surrounding soil.
5. Farmers who practice polyculture tend to have fewer serious crop losses due to pests because they are not dependent on a ________ crop, and pest populations cannot become ________ without abundant plants to feed on.

25.3 The Future of Agriculture

The use of pesticides, herbicides, fertilizer, and irrigation over the past 60 years—a phenomenon known as the "Green Revolution"—has increased food production dramatically. However, many biologists argue that current high levels of production are **unsustainable**—they compromise the ability of future generations to grow enough food to feed the human population. The future of agricultural production depends on finding ways to maximize food production while minimizing environmental damage.

Modern Agriculture Causes Environmental Damage

What evidence supports the hypothesis that current agricultural practices are unsustainable? First, the production and application of agricultural chemicals such as pesticides and fertilizers requires large amounts of nonrenewable fossil fuels; second, the water sources used to provide irrigation water are limited and under increasing demand; third, the environmental costs of fertilizer and pesticide use may be greater than the human population is able to bear.

Precious Liquids: Oil and Water. Petroleum is required not only to run farm equipment but also as a raw material for the production of fertilizer and pesticides. Although scientists and oil industry experts disagree about if and when the world's oil supply will run out, it is clear that demand for oil continues to increase, causing prices to climb as a result. Dependence on the finite and ever more costly resource of oil makes current agricultural practices unsustainable.

Some environmental scientists argue that freshwater is the "new oil." Several statistics support this view. In the United States, 65% of total water withdrawals not used in power generation are used to irrigate crops. In western states, the portion of water use devoted to irrigation exceeds 90%. It is clear to geologists studying the Ogallala aquifer (**FIGURE 25.27**) that irrigation in many of the regions above the aquifer is using water much more quickly than it is being replaced by precipitation. Analysis of the rate of water loss indicates that some areas of the aquifer will dry up within decades.

In addition to its reliance on nonrenewable or limited resources, such as oil and freshwater, many of the processes and inputs of modern agriculture can cause serious environmental damage.

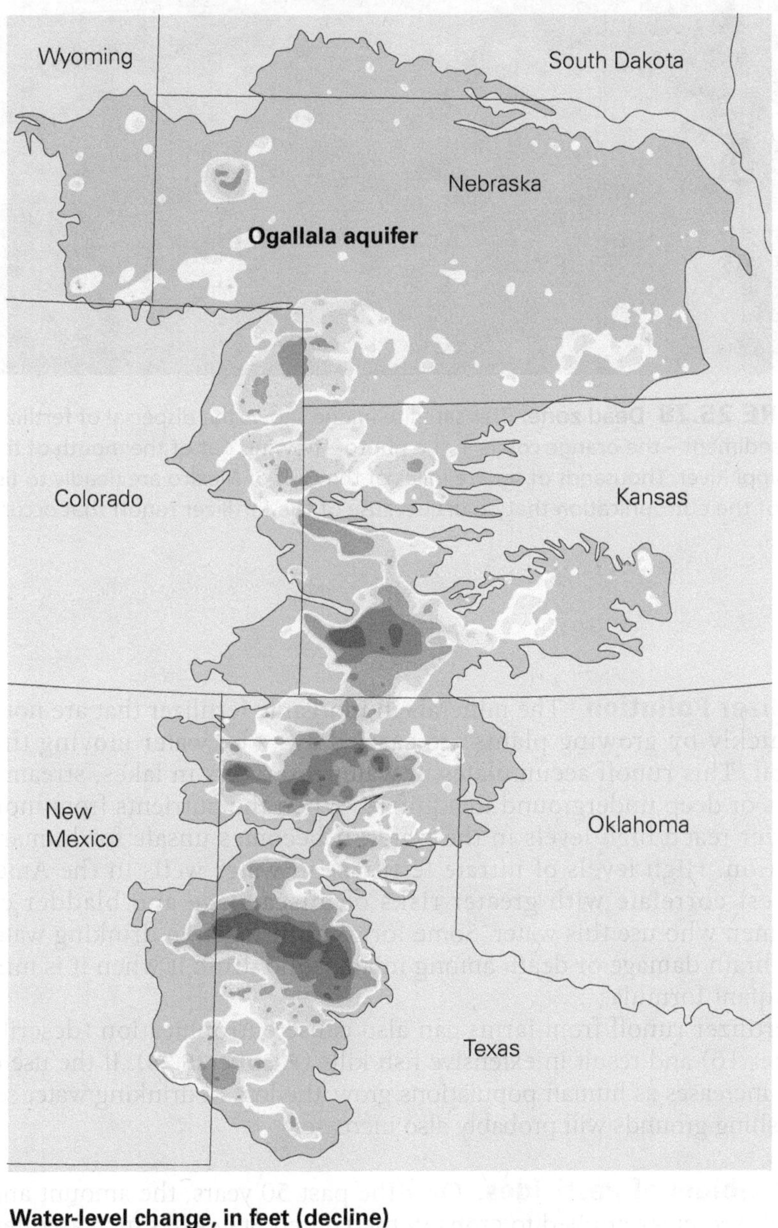

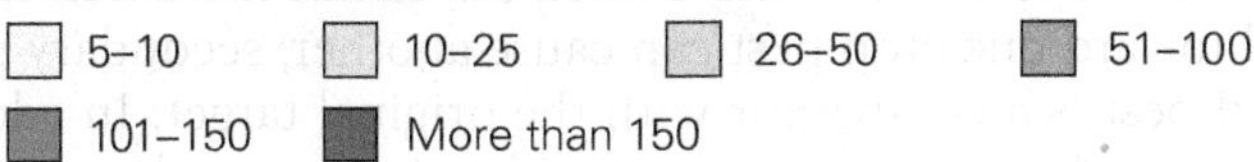

FIGURE 25.27 Unsustainable water use. The Ogallala aquifer is the largest single pool of freshwater in the world; it underlies and supplies the "bread basket" region of the United States. However, water is being removed from the aquifer much more rapidly than it refills, leading to dramatic declines in some areas. This figure illustrates the drop in water level in areas of the aquifer through 2011.

WORKING WITH **DATA**

In what regions has the aquifer lost the most volume?

FIGURE 25.28 Dead zone. This satellite image shows the dispersal of fertilizer-laden sediment—the orange colors in the photo—flowing out of the mouth of the Mississippi River. Thousands of square miles of the Gulf of Mexico are deadly to fish as a result of the eutrophication that occurs because of the fertilizer runoff that occurs each summer.

Fertilizer Pollution. The minerals in inorganic fertilizer that are not taken up quickly by growing plants are carried away by water moving through the soil. This runoff accumulates in water reservoirs in lakes, streams, and oceans or deep underground in aquifers. When the nutrients from inorganic fertilizer reach high levels in the water, it becomes unsafe for human consumption. High levels of nitrate fertilizer in water wells in the American Midwest correlate with greater risks of miscarriage and bladder cancer in women who use this water. Some forms of nitrogen in drinking water can cause brain damage or death among infants who drink it when it is mixed in with infant formula.

Fertilizer runoff from farms can also cause eutrophication (described in Chapter 16) and result in extensive fish kills (**FIGURE 25.28**). If the use of fertilizer increases as human populations grow, the loss of drinking water sources and fishing grounds will probably also increase.

The Problem of Pesticides. Over the past 50 years, the amount and toxicity of pesticides applied to crops in the United States has increased tenfold. Pesticides directed toward one crop pest can cause another, secondary pest to thrive if the second pest is a competitor with the original target. In addition, pest populations can evolve resistance to pesticides that target them. To control populations of secondary pests or now-resistant target pests, higher levels of pesticide or new, more toxic pesticides must be applied. This situation is called the pesticide treadmill because farmers must continually expend energy and resources simply to stay in place and keep pest populations under reasonable control.

One side effect of the increase in pesticide use has been an increase in the number of incidents of wildlife poisonings, especially of birds. Although trends can only be inferred from anecdotal reports, it appears that the number of wildlife poisonings has tripled since 1980. Even at low doses, many pesticides may have effects that do not cause death but may weaken individuals and make them more susceptible to predation or less able to compete—or may even affect their endocrine system (Chapter 23).

An additional problem associated with some types of pesticides is **biomagnification,** the concentration of persistent toxic chemicals at higher levels of a food web. Biomagnification is a product of the trophic pyramid, the bottom-heavy relationship between the biomass (total weight) of populations at each level of a food chain (Chapter 16). Thanks to the activism of ecologist Rachel Carson (**FIGURE 25.29**), the problem of biomagnification was recognized in the late 1960s—particularly the effects of magnified concentrations of the pesticide DDT in fish-eating birds such as bald eagles. This recognition led to a ban on DDT application. However, many persistent pesticides are still used in the United States and elsewhere, and these can accumulate in any organism that is high on a food chain—including humans.

FIGURE 25.29 Rachel Carson. *Silent Spring,* Rachel Carson's most influential book, made well known the hypothesis that pesticides like DDT accumulate in organisms by the process of biomagnification and thus can kill birds and negatively affect people.

Long-term health effects of low-level exposure to pesticides are still mostly unknown, but some evidence suggests that these effects include increased risks of cancer, birth defects, and permanent nerve damage. The pesticide treadmill demands ever more powerful and toxic chemicals, and high pesticide levels that result from biomagnification are likely to negatively affect more and more of the human population.

Wise use of fertilizers and pesticides may help minimize the environmental effects of modern agriculture; these negative consequences fade when their use is reduced. However, one consequence of modern agriculture results in environmental damage that is not easily recovered: the loss of soil via erosion.

Soil Erosion. Soil is held together by the organisms living and the roots of plants growing within it. Modern agricultural techniques destroy soil in two ways: by regularly removing the plants (through tilling or harvest) and by the application of inorganic fertilizer, which is in the form of salts and can be deadly to soil organisms. When soil loses its structure, it is easily picked up by strong winds and blown away—the Dust Bowl was a dramatic example of this type of soil erosion caused by agricultural practices.

Irrigation water running over the surface of disturbed soil can also result in erosion. In addition, irrigation can cause **salinization**—the "salting" of soil. Water from underground aquifers contains tiny amounts of mineral salts. When this water is applied to the soil, some evaporates, leaving salts behind. Eventually, so much salt accumulates that the soil becomes completely infertile. The production of soil from rock and organic material takes time, and soil lost through erosion and salinization is not easily replaced. Agricultural practices that rapidly degrade soil are unsustainable.

If we also consider the effects of rapid climate change on agriculture, it seems clear that current agricultural practices will not be sufficient to feed a growing human population indefinitely. Fortunately, some alternatives to these practices may help provide enough food more sustainably.

How to Reduce the Damage

A more **sustainable** agricultural system is one that meets the needs of the current generation of humans without compromising the ability of future generations to meet their needs. One path to sustainability focuses on requiring changes in agricultural practices. However, another path is to encourage consumers to change their eating and food-buying habits.

(a) Conventional apple

(b) Organic apple

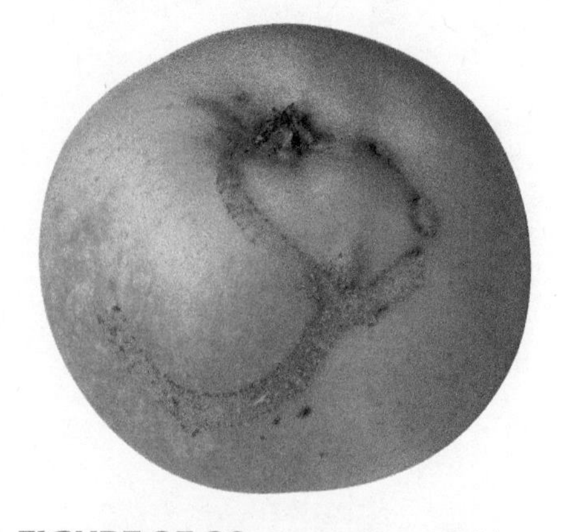

FIGURE 25.30 Cosmetic pesticide use. (a) This apple was produced on a conventional farm, with high levels of pesticides to control superficial damage. (b) This apple was produced on an organic farm, with no pesticides. The damage is only on the surface, but consumers often reject fruit that look like this.

Reducing Harmful Inputs. We have already reviewed many of the farming techniques that can reduce the environmental costs of agriculture. Minimizing monocultural plantings may be the single most effective strategy. Not only does planting polycultures reduce the need for pesticides, it also helps minimize fertilizer inputs and preserves soil by increasing crop rotation. Additionally, planting polycultures helps to insulate farmers from environmental factors that destroy a single crop. Moving away from monocultural production will not be an easy task; it must be preceded by changes in government policies and agricultural economics. For instance, government subsidies that pay farmers to produce a few commodities, such as wheat, corn, cotton, and rice, encourage monocultural production.

Even without a switch to polyculture, farmers have good reason to reduce their use of fertilizer and pesticides—after all, these inputs cost money. One technique uses technology to closely monitor the production of agricultural fields so that fertilizer and pesticide applications are targeted more effectively.

Another strategy to reduce inputs has been adoption of GMOs that produce pesticides directly. In non-GMO crops, farmers can practice **integrated pest management (IPM),** which uses releases of predator insects, introduction of pest competitors, and changes in planting techniques to reduce pesticide use. Unfortunately, neither GMO nor IPM use has been a silver bullet. Since the introduction of both strategies in conventional agriculture, pesticide and fertilizer use has still continued to rise in the United States.

The Role of the Consumer. Farmers are only one-half of the food production equation. Consumers also determine how and what food is produced. If consumers consistently select certain types of foods, farmers will try to supply them. In many cases, consumer demand encourages some of the more damaging environmental consequences of modern agriculture.

Most of the pesticides sprayed on vegetable and fruit crops are applied to reduce crop "losses" that are caused when consumers refuse to purchase fruits and vegetables with superficial signs of pest damage. The only way to produce pristine products is to apply hundreds of pounds of pesticides to completely eliminate pest insects. If consumers were willing to accept a visible but small amount of pest damage to fresh vegetables, then the amount of pesticides used on these crops could drop tremendously (**FIGURE 25.30**).

Consumer demand for meat also fuels unsustainable farming practices. Most of the beef, pork, and poultry produced in the United States comes from animals that are fed field-grown grain in enormous feed lots, called factory farms (**FIGURE 25.31**). The same production process holds for hogs, chickens, and turkeys. In fact, 66% of the cereal grain consumed in the United States, including 80% of corn and 95% of oats, is used to feed these animals. Because much of the energy consumed by an animal is used for maintenance rather than growth, the grain used as animal feed yields many fewer calories in the form of meat. Put another way, a 10-acre field of corn can support 10 people if they eat the corn directly, but only 2 people if they feed the corn to cattle and then eat the beef.

When individuals consume many of their calories as conventionally produced meat and dairy products, they are forcing the high-intensity farming that causes the environmental problems discussed in this chapter. Reducing meat consumption would also reduce the number of acres farmed—both because the acreage needed to grow grain to feed animals would be saved and because raising animals for food takes space that could be used to plant additional crops; as a result, some soils that are highly susceptible to degradation through erosion or salinization could be left undisturbed. Less land needed

FIGURE 25.31 Modern meat production. Most of the meat and dairy products available in modern supermarkets are produced in high-density "factory farms," such as the one pictured here.

for agricultural production also leaves more natural habitat for other species to survive.

Factory farming of cattle, hogs, and poultry carries environmental costs in addition to its prodigious use of crops. For example, the U.S. Department of Agriculture estimated that the meat industry produces 2 billion tons of animal waste annually; this is 100 times the amount of waste produced by the human population of the United States. Failures of manure storage ponds and inadequate disposal have led to spills, water contamination, and fish kills. In addition, the high-density populations of animals in factory farms provide prime conditions for the spread of disease. Many meat producers in the United States handle this risk by feeding large amounts of antibiotics to all of their animals, sick and healthy alike. The use of antibiotics in agricultural settings has led to the development of antibiotic resistance in several dangerous human pathogens, including *E. coli* O157:H7, a bacterium found in beef that sickens thousands of people—and kills more than 100—every year.

Organic Farming. There is little doubt that consumer demand can change agricultural practices. In recent years, many consumers with concerns about the costs of modern agriculture have changed their buying habits (**FIGURE 25.32**). As a result, some farms have switched to organic farming, a practice that shuns the use of chemical fertilizers and pesticides as well as GMOs.

The focus of organic farming is the natural maintenance of soil fertility (using crop rotation and organic fertilizers) and the development of biological diversity on the farm (using polycultures) as a method to control pests and provide income stability. Organically grown food has lower levels of pesticide residues and is produced in a way that minimizes the environmental costs of agriculture

FIGURE 25.32 A consumer movement. Sales of organic foods have grown at a rate of 20% per year over the past 15 years, thanks to changes in the buying choices of millions of individuals.

(TABLE 25.3). Organic food does often cost more for consumers; grocery bills average 60% higher according to several analyses, although the cost differences decrease when consumers purchase seasonal foods and buy directly from the grower at farm markets. And because food is relatively inexpensive in the United States, a 60% increase in food cost for a family of four totals about $3500 a year—about what an average family spends annually on meals at restaurants. Organic food does not avoid other costs seen in the conventional agricultural system, including the transport of food over long distances, although purchasing from local organic producers can minimize that environmental impact.

TABLE 25.3 Reducing the environmental impact of modern agriculture.

This table summarizes the environmental costs of modern agriculture, along with the individual and collective actions we can employ to reduce these costs.

Environmental Problem	Solution	Actions We Can Take
Biomagnification of persistent pesticides 	• Minimize monocultural plantings. • Reduce use of pesticides. • Increase use of integrated pest management.	• Accept produce that has superficial pest damage. • Purchase organic foods. • Add diversity of foods to our diets to promote polyculture. • Reduce meat consumption and thus overall crop production. • Support policies that reduce use of monocultural production (e.g., elimination of subsidies for crops). • Support integrated pest management research.
Eutrophication of surface water and nitrate pollution of groundwater 	Reduce use of inorganic fertilizer.	• Purchase organic foods. • Purchase foods grown in polyculture, where crop rotation is employed. • Reduce meat consumption.
Salinization of soil 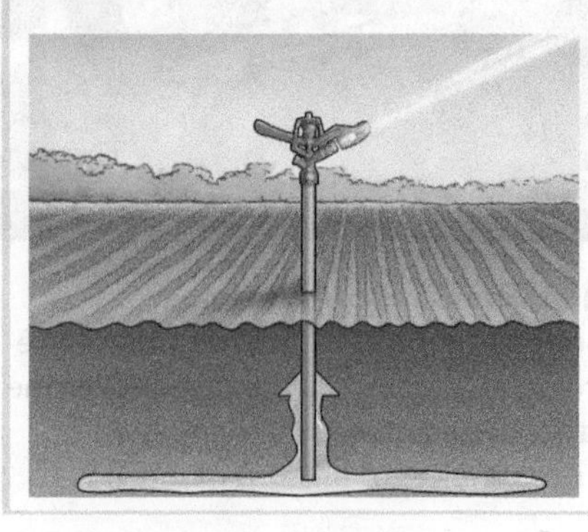	• Reduce number of acres irrigated. • Increase efficiency of irrigation.	• Reduce meat consumption. • Support research and national agricultural policies to reduce agricultural water use.

—Continued

TABLE 25.3 **Reducing the environmental impact of modern agriculture.** *Continued—*

Environmental Problem	Solution	Actions We Can Take
Soil erosion and desertification	• Reduce use of tilling on marginal lands. • Reduce amount of land used for crop production.	• Reduce meat consumption. • Support research on effective nonchemical means of weed control. • See also suggestions regarding population and aid to developing countries.
Loss of habitat for nonhuman species 	Minimize the amount of land converted to agricultural production.	• Reduce meat consumption. • Support policies that effectively lead to decreased human population growth rates. • Support policies that provide developing countries with the tools to increase crop yield sustainably.

Feeding the human population in a way that preserves environmental quality and leaves room for biodiversity will require thoughtful decision making. These decisions include those we make as individuals, such as how much meat to consume and what sort of produce we buy, and those we make collectively, such as how federal subsidies for crops are distributed. Better knowledge of plant and agricultural science and an understanding of the power and limitations of scientific knowledge can help you become an effective participant in this decision-making process.

Got It?

1. A ________ agricultural system is one that does not compromise the ability of future generations to grow food.
2. Biomagnification is the reason why animals that are at the ________ of a food chain are most negatively affected by persistent pesticides.
3. The goal of integrated pest management is to use detailed information about crop health and pest populations to ________ the use of pesticides.
4. Irrigating crops with well water can eventually make the soil ________.
5. The most effective single action you can make to reduce the environmental effect of your food choices is to eat less ________.

Nearly everyone knows that a plant that wilts needs water. But now you know that cell walls made of cellulose (a polysaccharide) also provide much of the "skeleton" of the plant. With this knowledge, you might think that watering your plant with a mixture of sugar and water (say, soda pop) will provide not only the water it needs but the building blocks for making a stronger skeleton on cell walls. In other words, **"Watering" a wilted plant with a solution of water and sugar will help the plant recover better than water alone.**

Sounds right, but it isn't.

Answer the following questions to understand why.

1. Remember what you learned earlier about transport of materials and water across cell membranes (Chapter 4). The water contained in plant cell vacuoles is nearly pure. Where would water flow in the case where you added sugar water to the soil—into or out of the cell? Explain why.
2. What does it mean that a plant cell wall is "elastic"?
3. What happens to the pressure inside plant cells that lose water? What do we call this condition?
4. How does the condition of cells relate to whether a plant is wilted or "crisp"?
5. Even if the sugar in soda pop did increase the thickness of plant cell walls, would that affect whether the plant was wilted or crisp under these conditions?
6. Given your answers to questions 1–5, explain why the statement bolded above sounds right, but isn't.

THE BIG QUESTION

Should I purchase organically grown foods?

The magazine *Consumer Reports* recently compared the costs of organic versus conventionally grown products purchased at a variety of retailers. Although there was quite a bit of variation in the prices from store to store, the authors found that, on average, organic foods were 47 percent more expensive than comparable conventionally grown foods. Is it worth the extra money?

What should I know?

What follows are some smaller questions that need to be resolved to answer the Big Question. Place a checkmark next to the questions that science can answer.

Smaller Questions	Can Science Answer?
Is organic food healthier for me than conventionally produced food?	
Is it wrong to consume food that you know is produced with unsustainable inputs?	
Is organically produced food measurably better for the environment than conventionally produced food?	
If people spend more money on food, will they be required to engage in other more unsustainable activities to survive?	
Does organic food taste better than conventionally produced food?	

What does the science say?

Let's examine what the data say about this smaller question:

Is organically produced food measurably better for the environment than conventionally produced food?

Swiss scientists performed a systematic review of published scientific studies comparing the total impact of organically and conventionally produced agricultural commodities over a product's entire life cycle. Data from the 31 peer-reviewed studies and three scientific reports in non-peer-reviewed publications they found are reproduced in the table that follows.

Relative difference in environmental impact of organic compared with conventional production, per unit of product

	Milk	Beef	Poultry	Tomatoes	Carrots	Strawberries	Soybeans	Wheat	Potatoes
Energy demand	−5%	−2%	−8%	−71%	12%	61%	−10%	−11%	−5%
Global warming potential (GWP)	−12%	−8%	−18%	−78%	−9%	39%	−12%	−9%	88%
Eutrophication potential	−13%	−1%	4%	−17%	−69%	−65%	−26%	80%	39%
Pesticide use	−100%	−99%	−100%	−53%	−100%	−96%	−100%	−100%	−100%
Water use	−69%	−76%	−73%	−28%	51%	64%	−54%	−68%	−12%
Land use	−1%	−23%	−32%	37%	−38%	−117%	−36%	−4%	1%

1. Describe the results. Does it appear that organically produced agricultural commodities have a lower life cycle cost than conventionally produced ones?
2. Given these data, do you think the smaller question is answered? If not, propose another study that would help answer this question.
3. Does this information help you answer the Big Question? What else do you need to consider?

Data Source: M. S. Meier, F. Stoessel, N. Jungbluth, R. Juraske, C. Schader, and M. Stolze, "Environmental Impacts of Organic and Conventional Agricultural Products—Are the Differences Captured by Life Cycle Assessment?" *Journal of Environmental Management* 149 (2015): 193–208.

Chapter Review

Mastering Biology

Go to Mastering Biology to access **eText 2.0, Dynamic Study Modules,** and the **Study Area,** where you'll find practice quizzes, BioFlix™ animations, MP3 tutor sessions, current events, and more.

SUMMARY

Section 25.1

List the tissues and structures found on angiosperm plants, and describe their functions.

- Nearly all major crops belong to the angiosperm phylum, which are flower- and fruit-producing plants (p. 531).
- Roots absorb water and minerals from the soil but also may be modified to store carbohydrates (pp. 533–534).
- Stems serve as support for leaves and reproductive organs but may also be modified for carbohydrate storage (p. 534).
- Leaves are the photosynthetic organs of the plant and contain some essential nutrients for humans (p. 534).
- Flowers contain both male and female reproductive organs (pp. 553–554).

Describe the process of reproduction in angiosperms, including the role of double fertilization.

- When pollen is transferred from the male to the female angiosperm, two sperm are released. One sperm fertilizes the egg and the other fuses two other nuclei inside the ovule, a process called double fertilization (p. 536).
- The resulting seed remains inside the flower structure, which becomes modified into a fruit that functions in seed dispersal (p. 537).

SHOW YOU KNOW 25.1 If a gardener removes the stem and leaf of a weedy plant without pulling out its roots, the weed will very often regrow. What does this imply about the capabilities of plant cells?

Section 25.2

Compare and contrast primary and secondary growth in dicot plants.

- Seeds may survive difficult environmental conditions by becoming dormant. The emergence of a plant embryo from a seed is called germination (pp. 538–539).
- Dicot plants undergo primary growth in the length of stems and roots, and secondary growth in the width of these organs (p. 538).
- Primary growth occurs because of cell division in the apical meristems. This meristem produces additional meristems, leading to indeterminate growth patterns in plants (pp. 539–540).
- Secondary growth in dicot stems occurs when the vascular cambium produces additional xylem, which becomes wood, and phloem, which becomes the inner part of the bark (p. 540).

Describe the role of water availability in maximizing a plant's exposure to sunlight and carbon dioxide.

- Plants require abundant water for photosynthesis, for effective interception of light, and to maximize the uptake of carbon dioxide. Farmers not only provide water through irrigation but also remove noncrop plants that compete for water (pp. 541–543).
- Plants regulate water loss but also cut off the uptake of carbon dioxide by closing the stomata on their leaf and stem surfaces (p. 543).

Provide examples of important nutrients for plant growth, and distinguish between how these nutrients occur in natural systems and how they are made available in agricultural systems.

- Soil is a matrix composed of minerals from eroded rock, dead and decaying material, and the living organisms that decompose and mix the components. Soil is the major source of nutrients for plant growth (p. 544).
- In agricultural systems, nutrients are mined from the soil. Farmers make up for the loss of nutrients by adding fertilizer and growing nitrogen-fixing crops (pp. 544–546).

SHOW YOU KNOW 25.2 Most people do not harvest their houseplants, yet many "feed" them inorganic fertilizer every few months. Why do houseplants need fertilizer?

Section 25.3

Describe the costs associated with modern agriculture, and list some possible ways these costs could be reduced.

- Widespread use of pesticides has led to monocultural production of crops. Polycultural plantings can keep pest populations down (pp. 546–547).
- Petroleum and fresh water are essential to modern agriculture and are in limited or finite supply (pp. 548–549).
- Fertilizer runs off into waterways, causing human health problems and damage to fish and other animal populations (p. 550).
- Farmers continually increase pesticide use as pests become resistant (p. 550).

- Pesticides can be persistent and accumulate in organisms at higher levels of the food web, including humans (p. 551).
- In many areas, soil is being lost through erosion and salinization faster than it is being replaced (p. 551).
- Consumers can help reduce the environmental costs of modern agriculture by accepting a small amount of pest damage on produce, eating less meat, and purchasing organically grown products (pp. 551–555).
- Organic agriculture may provide a sustainable model for future food production (pp. 553–554).

SHOW YOU KNOW 25.3 Explain how natural selection causes a pest population to become resistant to a pesticide.

ROOTS TO REMEMBER

The following roots of words come mainly from Latin and Greek and will help you to decipher terms:

angio-	means vessel. Chapter term: *angiosperm*
-enni-	pertains to a year. Chapter term: *perennial*
rhizo-	means root. Chapter term: *rhizome*
-sperm	means seed. Chapter term: *angiosperm*

LEARNING THE BASICS

1. Add labels to the figure that follows, which illustrates typical flower structure.

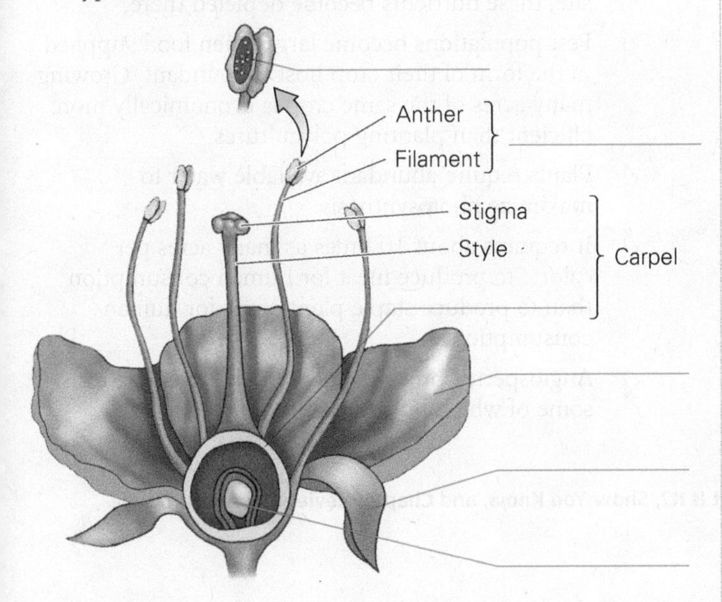

2. What does it mean to say a plant's growth is "indeterminate"?

3. All of the following plant organs have been modified in certain plants to store large amounts of carbohydrate, except ______.

A. fruit; **B.** stems; **C.** roots; **D.** leaves; **E.** apical meristems

4. One difference between roots and leaves is ______.

A. roots have vascular tissue, but leaves do not; **B.** roots can store carbohydrates, but leaves cannot; **C.** leaves have only phloem, roots only xylem; **D.** root epidermis absorbs water, leaf epidermis conserves it; **E.** leaves are only produced by flowering plants, roots by all plants

5. As a result of fertilization in flowering plants, ______.

A. pollen is picked up at the stigma by an animal for transfer to an anther; **B.** a single sperm is released by a pollen grain; **C.** a seed begins to absorb water and break free of its coat; **D.** a triploid tissue called endosperm is formed; **E.** a carpel is produced that later becomes a seed

6. In dicot woody plants, secondary growth results in ______.

A. production of new xylem and phloem; **B.** increase in the girth or width of the stem; **C.** the production of additional nodes and internodes; **D.** A and B are correct; **E.** A, B, and C are correct

7. The function of stomata is ______.

A. to prevent pests from damaging the photosynthetic surfaces of the plant; **B.** to allow carbon dioxide to enter the plant body; **C.** to sense water availability in the air; **D.** to provide stability for a growing plant; **E.** to increase nutrient availability in soil

8. The first organ to emerge from a germinating seed is the ______.

A. axillary bud; **B.** root; **C.** cotyledon; **D.** endosperm; **E.** stem

9. Soils in agricultural systems require fertilizer because ______.

A. farming mines nutrients from the soil; **B.** weeds compete with crop plants for soil nutrients; **C.** most crop plants have nitrogen-fixing bacteria in their roots and need lots of nitrogen; **D.** most nutrients run off soils into waterways, causing eutrophication; **E.** pests remove most nutrients from crop plants

10. When inorganic fertilizer ends up in waterways, ______.

A. the waterways become eutrophic; **B.** algae populations explode in surface waters; **C.** fish kills result; **D.** drinking water may become contaminated; **E.** all of the above

ANALYZING AND APPLYING THE BASICS

1. Agricultural scientists have been successful at creating only a few types of seedless fruits through traditional breeding—notably bananas, apples, grapes, pineapples, navel oranges, and a few melons. Why might it be difficult to produce seedless fruits?
2. One way to make a houseplant look fuller is to cut off its primary growing tip. Based on what you've learned about primary growth, how does this action promote a bushy form?
3. Your friend has been planting the same variety of tomatoes in the same spot in his garden for the past 8 years. In recent years, he has noticed that his plants are less healthy, and that pests damage more of the tomato fruits. Apply what you have learned in this chapter to explain why his tomato crop is declining and suggest several ways that he can improve his tomato yield.

GO FIND OUT

1. Learn about the agricultural production and issues in your local area. What are the agricultural crops most commonly grown in your area? What crops are "specialties" (these may still be relatively minor in terms of acreage they cover, but are important economically because they are sold nationwide—for example, wine grapes in northern California or the Finger Lakes of New York). What are the major pests and diseases that affect crops in the region? How extensive is organic production in your area? You may want to start your search for this information with your local or statewide cooperative extension service.
2. Have you ever considered the environmental impact of your diet? What changes would you be willing to make in your diet to reduce its environmental impact? Explain.

MAKE THE CONNECTION

The science that you learned in this chapter has helped you better understand the real-world example used throughout this discussion. Draw a line from the statement on the left to the science that supports it on the right.

Modern agriculture depends on a select group of plants. ▷	◁ Annual plants devote significantly more resources to provisioning seeds with resources for their enclosed embryo than perennial plants do.
No staple crops come from woody plants. ▷	◁ Plants need inorganic nutrients for key biochemical components, and if plants are harvested from a site, these nutrients become depleted there.
Agricultural practices include tilling soil, controlling weed plants, and irrigating crops. ▷	◁ Pest populations become large when food supplied in the form of their crop host is abundant. Growing many acres of the same crop is economically more efficient than planting polycultures.
Another common agriculture practice is fertilization. ▷	◁ Plants require abundant available water to maximize photosynthesis.
Modern agricultural practices include the use of pesticides. ▷	◁ It requires about 10 times as many acres per calorie to produce meat for human consumption than to produce staple plant crops for human consumption.
Reducing meat consumption is the single most effective strategy for reducing the total environmental cost of modern agriculture. ▷	◁ Angiosperms produce energy rich seeds and fruits, some of which humans can use for food.

Answers to **Got It?, Visualize This, Working with Data, Sounds Right, But Is It?, Show You Know,** and **Chapter Review** questions can be found in the **Answers** section at the back of the book.

26 Growing a Green Thumb

What does it take to have a garden this beautiful?

How can you make a backyard vegetable plot yield abundant produce?

It helps to have knowledge of plants, gained through experience and understanding.

Plant Physiology

Many of us know people who seem to have a "green thumb"—men and women whose homes are a jumble of thriving houseplants, who produce enough vegetables in their backyard gardens to saturate friends with excess tomatoes and zucchini, or who have flower gardens that make the neighbors, well, green with envy. How do they do it? Is it simply a gift that some are born with and some without?

A significant contributor to a gardener's success is time. Years of trial and error, weeks of preparation for a growing season, and hours of work tending to gardens and indoor plants will create an experienced grower who can get the most out of his or her plants. Part of what a gardener gains through experience is a thorough understanding of how plants work. This knowledge helps a plant lover choose the right plants for the environment, design a landscape that is always in bloom, and even shape the form of plants to fit his or her needs.

Fortunately for anyone who is an eager but inexperienced gardener, it is possible to start developing a green thumb by learning the basics of plant function. Our discussion of plant physiology fits within the context of planning a plant-filled space, such as a garden or houseplant collection. Let's get growing!

26.1 The Right Plant for the Place: Water Relations

One of the first considerations of gardening is choosing plants that can thrive in the available environment. Many factors influence plant growth in a particular space, including the soil's chemistry, light availability, and length of the **growing season,** which is the part of the year when temperature and precipitation allow plants to grow. However, one of the most important factors influencing plant production is the amount of water available to the plant. How do plants obtain and manage water, and how can we use our understanding of this process to help plants thrive?

Transpiration

Water and dissolved minerals obtained from the soil, together called **xylem sap,** travel up the xylem, the water-conducting cells inside a plant. This process occurs primarily as a result of the loss of water vapor from the leaves through evaporation. Through **transpiration,** the plant essentially pulls a column of sap through the roots from the soil, up tubes made of dead xylem cells in the stem, and into the leaves. Transpiration is effective in even the tallest trees (more than 100 meters, or 330 feet) because of the hydrogen bonds formed by water molecules (Chapter 2). Transpiration evolved in plants after they colonized land 400 million years ago, and it has allowed plants to expand their range to nearly all land on Earth's surface.

trans- means across.

The tendency for identical molecules to stick together, such as water molecules linked by hydrogen bonding, is referred to as **cohesion;** the tendency for unlike molecules to stick together, such as when water is hydrogen-bonded to other polar molecules, is called **adhesion.** Water adheres to cellulose in plant cell walls. Cohesion and adhesion together maintain the continuity of a water column during transpiration.

co- means with.

ad- means to.

The pulling force on a column of xylem sap is generated by the evaporation of water from the leaves (**FIGURE 26.1**). Stomata, the pores on leaf surfaces, control the rate of gas exchange and water loss in land plants. When water evaporates out of stomata, the forces of cohesion and adhesion on the water molecules remaining inside the leaves create **tension,** or negative water pressure. Tension is the force that allows a column of water to be pulled into a syringe when the plunger is pulled up. In plants, tension causes water molecules from the xylem to be drawn out into the leaf to replace the water that has evaporated. This in turn increases tension on the water molecules immediately below them in the xylem, causing the subsequent water molecules to move toward the leaves, and so on, all the way down to the roots and the soil.

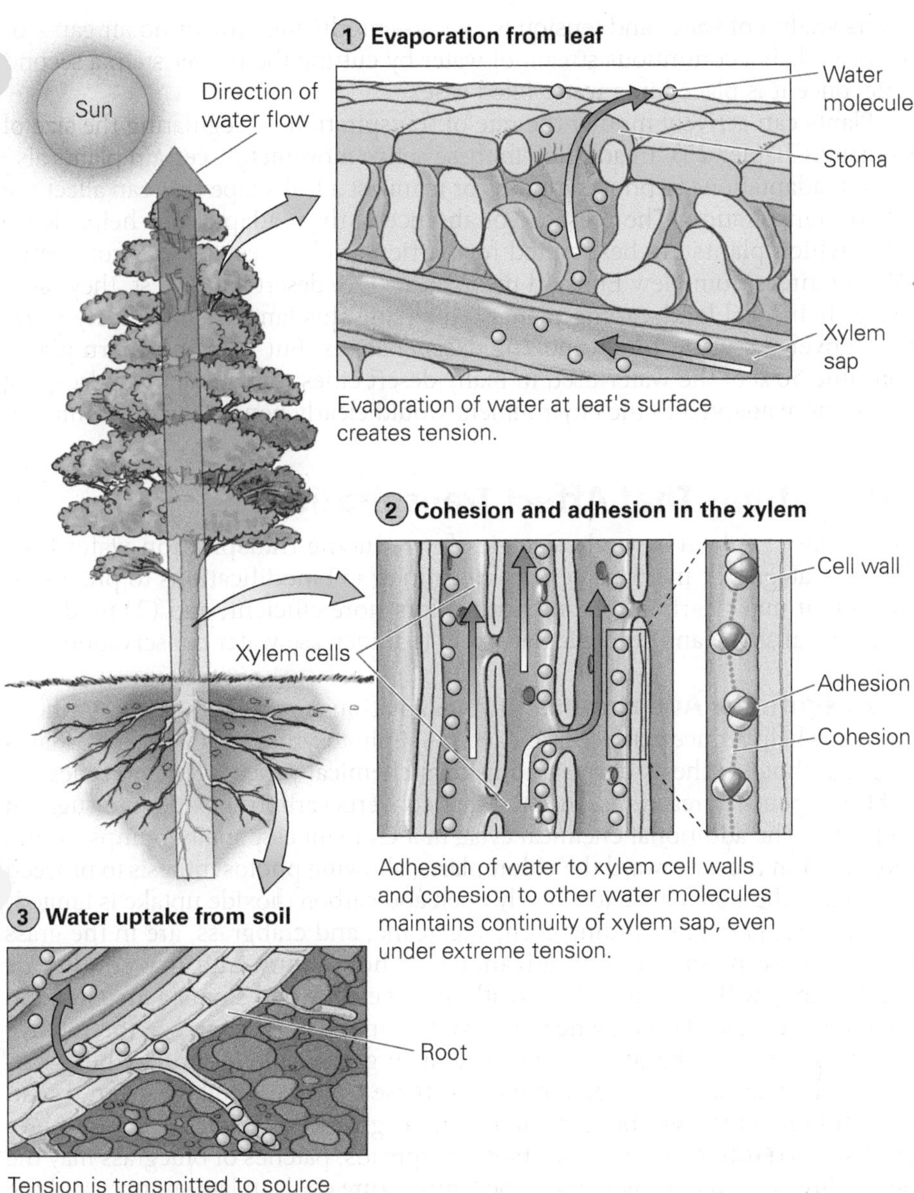

1 Evaporation from leaf
Sun
Direction of water flow
Water molecule
Stoma
Xylem sap
Evaporation of water at leaf's surface creates tension.
2 Cohesion and adhesion in the xylem
Cell wall
Xylem cells
Adhesion
Cohesion
Adhesion of water to xylem cell walls and cohesion to other water molecules maintains continuity of xylem sap, even under extreme tension.
3 Water uptake from soil
Root
Tension is transmitted to source of water, in this case the soil.

FIGURE 26.1 Transpiration. Evaporation of water from the leaves creates a tension that is transmitted to a column of water in the xylem. Water flows upward against the pull of gravity and maintains its continuity, thanks to the forces of cohesion and adhesion.

VISUALIZE THIS

How are the conditions in the tree different at night? Does transpiration occur then?

In at least 90 percent of land plants, the surface area for water (and nutrient) uptake below ground is increased by the presence of symbiotic mycorrhizae, fungal strands that in turn receive carbohydrates from the plant (**FIGURE 26.2**). The presence of mycorrhizae is greatest in soil that contains abundant organic matter and that is well aerated, which is why gardeners want you to stay out of their flower beds, so you won't compact the soil.

Another reason to refrain from walking in a garden is to avoid damaging the stems of growing plants. When a stem is injured, the continuity of the stream of water is disrupted and the tension is released. As a result, water pulls away from the cut surface. This leaves an air gap that remains even when the cut stem is placed in a water-filled vase. Because the stream of water in the xylem is no longer continuous, water from the vase cannot replace water evaporating from the xylem, and the stem will rapidly dry up and wilt. Gardeners who understand this know either to cut flowers in the early morning, when the

FIGURE 26.2 Mycorrhizae. The symbiotic relationships between the roots of many species of plants and certain kinds of fungi greatly increase the surface area for nutrient and water absorption to support plant growth.

Root
Mycorrhizae

plants are full of water and tension is lowest—producing little or no air gap—or to reestablish a continuous stream of water by cutting the flower stem a second time, once it is placed in a water-filled vase.

Plants can actively modify the rate of transpiration by regulating the size of stomata (Chapter 25). In addition to these active movements, certain plants also possess adaptations to photosynthesis or manifest a leaf shape that can affect the rate of transpiration. The presence or absence of these adaptations helps determine which plants are best suited for particular climates and environments. When retirees from New England move to the U.S. desert southwest, they often bring their local landscaping plants, preferring this familiar lush, leafy vegetation over the waxy, tough-looking desert natives. But these northern plants consume 50% of the water used in many desert cities; switching to landscaping that is more adapted to the environment would clearly benefit the environment.

FIGURE 26.3 A C_4 plant. Crabgrass has the C_4 photosynthetic pathway, which gives it an advantage over Kentucky bluegrass in hot, dry midsummer and late summer.

Adaptations That Affect Transpiration

Plants have evolved several strategies for reducing transpiration water loss. These strategies fit into two general categories: (1) modifications to photosynthesis that make carbon dioxide acquisition more efficient, and (2) modifications to leaf shape and stomata placement that increase water conservation.

Photosynthetic Adaptations. Two adaptations to a plant's photosynthetic pathways help reduce water loss: C_4 and CAM photosynthesis (Chapter 5). Plants with C_4 photosynthesis have an additional chemical process that precedes the light-independent reactions, the step that converts carbon dioxide into sugar in all plants. The additional chemical cycle in a C_4 plant essentially pumps carbon dioxide from the air toward the chloroplasts, allowing photosynthesis to proceed even when the stomata are just barely open and carbon dioxide uptake is limited.

Many C_4 plants, including corn, sugarcane, and crabgrass, are in the grass family. These plants have an advantage in hot, sunny, and dry conditions because they will continue photosynthesizing even when stomata are closed to reduce water loss. Homeowners across the United States can see the relative advantage of C_4 crabgrass over non–C_4 lawn grasses like Kentucky bluegrass during the warmest parts of summer. At these times, the lighter green, broad leaves of crabgrass will become dominant as growth of the less tolerant bluegrass slows (**FIGURE 26.3**). If the crabgrass spreads, patches of bluegrass may die off, leading to a patchy lawn when the temperature cools in late summer and the crabgrass dies.

CAM plants, including all cacti, pineapples, aloe, jade, and many orchids, accumulate carbon dioxide from the air during the night, when less water will evaporate because of the cooler temperatures. This carbon dioxide is converted to a simple carbohydrate acid and is stored in the vacuoles of leaf cells. At sunrise, the stomata close and the accumulated acid is degraded, releasing carbon dioxide for photosynthesis. The growth of a CAM plant is greatly limited by the relatively small amount of carbon dioxide it can use in a day—only what can be chemically stored. These plants are most common in extremely dry environments, although they can be successful in other environments. One of the driest environments a plant can experience is a house in winter, and many popular houseplants use the CAM process (**FIGURE 26.4**).

FIGURE 26.4 Jade, a CAM houseplant. Plants with CAM photosynthesis are adapted to very dry conditions and thus are ideal houseplants to survive dry indoor air.

So, what about the plants without either C_4 or CAM photosynthesis? Their strategies for reducing the rate of water loss through stomata are limited to the structural adaptations of their leaves.

Leaf Adaptations. The amount of water lost via transpiration is generally correlated to a plant's photosynthetic surface. That is, a plant with many large leaves tends to transpire more water than does a plant with fewer or smaller

FIGURE 26.5 Small leaves limit transpiration loss. In desert plants, small leaves help reduce the surface area for water evaporation. Some leaves may be so reduced that they are modified to spines.

leaves. Gardeners use their understanding of this relationship between leaf surface area and transpiration to help transplanted plants survive. Because removing plants from soil (no matter how carefully) inevitably results in root damage, newly transplanted plants have a reduced capacity to acquire water. To compensate for this, gardeners remove leaves when transplanting to reduce water loss.

Of course, decreased leaf area also means less photosynthesis, so plants with smaller leaves tend to grow more slowly than plants with broader leaves in the same environment. This is the trade-off experienced by needle-leaved plants, such as pines. Although their small leaves help conserve water, they cannot compete successfully in well-watered environments against broad-leaved trees. Plants that thrive in shady sites often have broad leaves because evaporation is lower and intercepting light is more crucial, whereas narrow-leaved plants tend to be more successful in bright sun. Leaf size is reduced to its extreme in cacti, which have photosynthetic stems and leaves modified into spines (**FIGURE 26.5**). Cacti grow quite slowly, but they make for another ideal houseplant that can survive most inattentive gardeners.

Some broad-leaved plants have other adaptations to reduce water loss, such as the placement of stomata on the cooler, shaded undersurface of leaves or in hair-lined pits, which trap water molecules and thus reduce water loss even when the stomata are open (**FIGURE 26.6**). Where periods of drought occur reliably on an annual basis, many broad-leaved plants have evolved to become **deciduous,** that is, to actively drop their leaves to reduce the rate of transpiration. In Section 26.2, we describe this process in greater detail.

In environments where water is abundant, plants compete for access to light. In these environments, faster-growing plants often have an advantage. Plants that have evolved strategies for maximizing transpiration, including changes in water-conducting tissues, can quickly grow to tower over their neighbors.

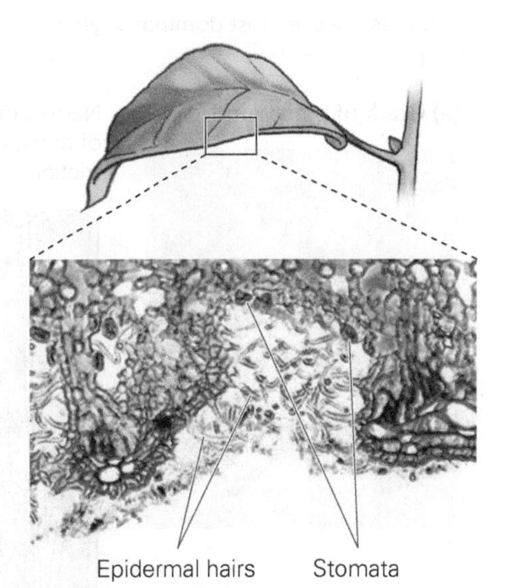

FIGURE 26.6 Shaded stomata help conserve leaf water. The stomata of this leaf are found on the shaded underside, where temperatures are lower and evaporation slower. The hair-lined pits trap evaporating water molecules and keep humidity high around the stomata.

Xylem Adaptations. When a column of xylem is under extreme tension inside a plant, the forces of cohesion and adhesion may not be strong enough to prevent the water column from breaking. Breakage of the water column results in the formation of a bubble, called an embolism, which inhibits water flow within the xylem tubes. In general, larger-diameter xylem tubes are more likely to suffer embolisms than are smaller-diameter tubes. Many drought-adapted species have only very narrow or tapering xylem tube cells, called **tracheids** (**FIGURE 26.7a**).

However, the narrow passageway produced by a column of tracheids leads to increased friction, reducing the rate of water flow—a disadvantage in moist environments. In moist conditions, plants with wider, perforated **vessel elements** are favored because they move water to their leaves up to 10 times faster, resulting in a faster growth rate (**FIGURE 26.7b**).

The diameter of the xylem tubes in any given plant thus represents a trade-off between the risk of embolism and rapid growth rate. For example, in cold climates, where liquid water may be unavailable for a large part of the year, trees are at high risk of embolism and have much narrower vessels than do trees in warmer, wetter climates where embolism is rare but competition for light is intense.

The growth of a plant in a natural landscape is strongly influenced by the balance between water availability and transpiration rate. However, in gardens and houses, water is rarely limited as long as the gardener pays attention to the plant's needs. In fact, a common problem of houseplants is not excessive dryness but too much water.

Overwatering. Excess water, by filling the air spaces in soil, prevents root cells from obtaining oxygen. As a result, the water-absorbing root hairs die, decreasing the transpiration rate and causing the plant to wilt. New gardeners may respond to an overwatered, wilting plant by adding more water, further destroying the roots and eventually killing it. As a rule of thumb, potting soil should be allowed to become completely dry down to 2 centimeters below the surface before more water is added.

Overwatering is a typical problem for houseplants, but garden and agricultural plants can also suffer from excess moisture. Soils that are high in clay, the tiniest mineral particles, are prone to becoming waterlogged because the soil's air spaces are small. Gardeners with clay soils often need to amend the soil before planting to avoid this problem. The most common soil amendment is organic matter, such as compost or peat moss. In most areas of the country, plants require about 2.5 centimeters (1 inch) of water per week, through rainfall

FIGURE 26.7 Xylem cells. Water flows through xylem tubes made up of one or two cell types. (a) Plants in dry environments have a high proportion of long, narrow tracheids. (b) In moist environments, short, wide vessel elements are the most dominant xylem cell type.

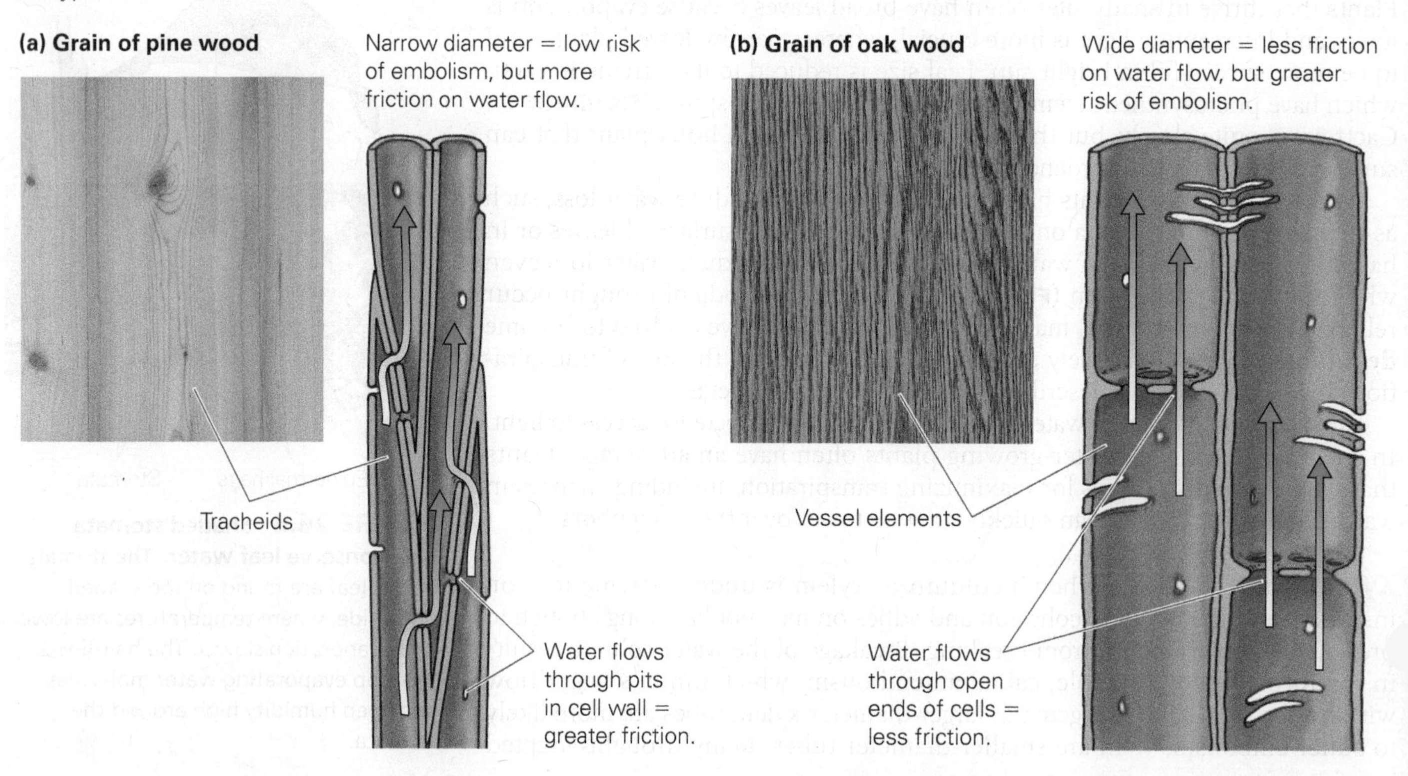
(a) Grain of pine wood
Narrow diameter = low risk of embolism, but more friction on water flow.
Tracheids
Water flows through pits in cell wall = greater friction.
(b) Grain of oak wood
Wide diameter = less friction on water flow, but greater risk of embolism.
Vessel elements
Water flows through open ends of cells = less friction.

or supplemental watering, for optimal growth. More experienced gardeners know to look for signs of water stress (limp or wilting plants, a yellow or bluish discoloration of leaves) and to check that the soil is dry a few centimeters below the surface before watering their plants.

A gardener's ability to supplement natural rainfall allows the growth of water-needy plants even in dry landscapes. However, it is far more difficult for gardeners to control the temperatures to which their landscape plants are exposed, so most garden plant choices are based on temperature requirements. Ability to tolerate freezing is one factor that determines whether a plant can survive in a particular temperature range.

Water Inside Plant Cells

Plant cells contain a central vacuole that is mostly filled with water. Water-filled vacuoles provide physical support to a plant (see Chapter 25). At low enough temperatures, this water will turn to ice, with devastating consequences. Cold tolerance is primarily a function of a plant's ability to control the freezing of water within its cells.

Ice formation in plants is damaging in three ways: (1) When water outside plant cell walls turns into ice, liquid water from inside the cells will begin flowing out by osmosis—that is, down a concentration gradient across the cell membrane (**FIGURE 26.8**). As the cell loses increasing amounts of water to the ice crystal, normal cell processes become disrupted and may cease to function. (2) When water freezes inside the central vacuole of a plant cell, it produces an ice crystal. The jagged edges of this crystal can pierce the membrane, allowing cytosol (the fluid inside the cell) to leak into the vacuole, thereby killing the cell. (3) When ice freezes within xylem tubes, air comes out of the water solution, forming embolisms and blocking future water uptake. All three of these processes can severely damage or kill plants.

A plant's tolerance of cold temperatures, known as its **hardiness,** depends on its ability to either prevent or manage freezing inside cells. Water freezes at temperatures lower than normal (0°C or 32°F) when it has solutes dissolved in it, in the same way that added antifreeze prevents water in a car's radiator from freezing. Tomato plants can survive temperatures down to −1°C (31°F), apple fruits −4°C (25°F), and cabbage as low as −9°C (16°F) because of their sugar-based cell "antifreeze." At lower temperatures, certain adaptations can improve freezing tolerance. These adaptations include controlling ice formation within cells to prevent sharp edges and internal structures that prevent xylem embolisms.

Many cold-tolerant plants require both a gradual cooling period to become maximally hardy and uninterrupted cold temperatures to remain hardy. Cold snaps that occur in late summer and early fall may damage plants that can normally thrive in even colder temperatures because they have not become **hardened,** or fully able to resist the cold. Woody plants can become damaged by cold even after they are fully hardened when bright winter sunshine warms their stems, allowing water to transpire that cannot be replaced by frozen water in the soil. This is a common occurrence in evergreens in landscaping, which can suffer from winter "burn." Freezing temperatures are most likely to cause damage during the spring, however, when intolerant leaves and flowers are beginning to emerge from formerly cold-tolerant buds.

Because of the extreme damage that freezing temperatures can cause, gardeners must know two things before planning an outdoor garden: (1) the

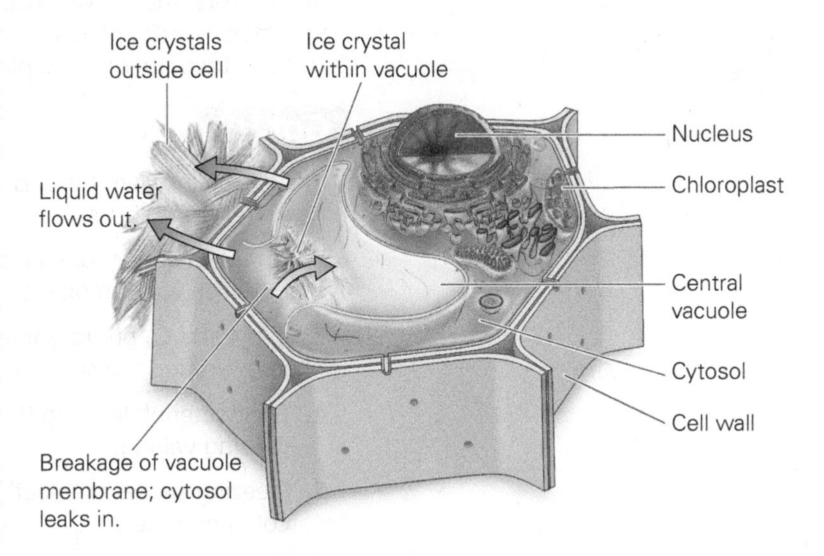

FIGURE 26.8 Freezing inside plant cells. Ice outside plant cells can cause the cell to dehydrate, interfering with normal functions. An ice crystal inside the central vacuole can pierce its membrane, allowing the cytosol to leak into the vacuole.

FIGURE 26.9 Hardiness zones. The U.S. Department of Agriculture generated this hardiness zone map based on average lowest temperatures experienced in thousands of sites across the country. The boundaries of these zones are changing rapidly as a result of climate change.

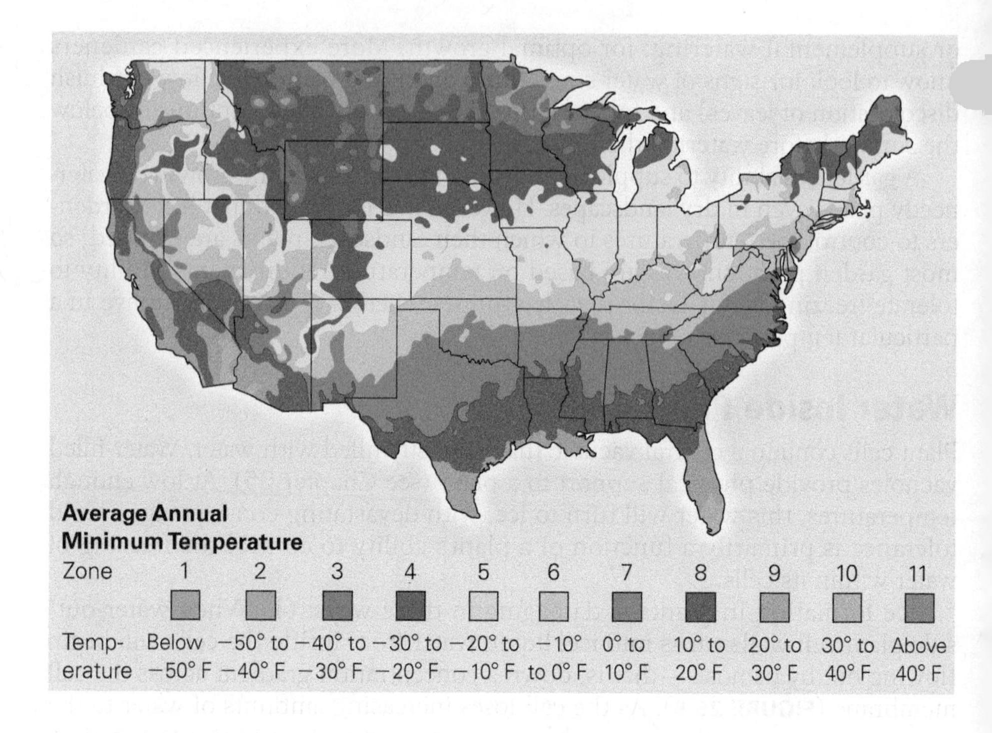

hardiness zone of their landscape—that is, how cold it gets (**FIGURE 26.9**); and (2) the rated hardiness—that is, the cold tolerance—of the plants they wish to grow. Many gardeners know, however, that with special care, plants that are less cold tolerant can survive freezing conditions. Plants that are grown outside their hardiness zone can be placed on the north sides of buildings, where they are shaded from the sun (preventing winter burn) and where cooler local temperatures slow budding and leafing until the threat of frost has decreased. Plants can also be insulated with mulch such as straw or snow, and they can be protected from unusually early or late frosts by covering with plastic tents.

Less tolerant plants are also better able to survive cold winters if they have sufficient water and sugar reserves. These reserves allow the plants to produce adequate antifreeze at the appropriate times during the hardening process. Gardeners can ensure that their sensitive plants are ready for winter by giving them plenty of water during the fall and by preventing them from using up their stored starch and sugar reserves.

To best manage a plant's sugar supply, a gardener needs to understand how plants store and move sugar. This knowledge helps gardeners not only ensure that their favorite plants survive harsh conditions but also maximizes the beauty and productivity of these plants.

Got It?

1. Tension in xylem tubes is __________ pressure caused by the __________ of water from leaves.
2. An intact water column is relatively difficult to break because of __________ between water molecules.
3. CAM and C_4 photosynthesis are both adaptations to minimize __________ loss through transpiration.
4. Overwatering leads to the death of __________ __________ in houseplants, causing wilting.
5. Freezing causes the death of individual cells but can also "break" the water column in the __________ vessels.

26.2 A Beautiful Garden: Translocation and Photoperiodism

Each gardener has personal goals for a garden—from a summer-long supply of fresh vegetables, to a continually changing palette of showy flowers, to an aesthetically pleasing but low-maintenance decor. Because plants grow throughout their lives, each plant can allocate its resources in many possible directions at any time—to greater leaf production, flower production, fruit size, and so on. By understanding how plants allocate resources, gardeners can encourage plants to transfer resources to their different plant organs at appropriate times.

Translocation of Sugars and Nutrients

Sugars and other dissolved nutrients are moved around a plant's body in phloem tissue. The solution in the phloem, or **phloem sap,** always flows from a **source,** where sugar is in high concentration, to a **sink,** where nutrients are being used or accumulated. As a result, the movement of phloem sap, called **translocation,** can be in any direction within a plant. Different organs can act as sources and sinks throughout the year; in fact, separate phloem tubes in the same branch can be translocating sugars in opposite directions at the same time. For example, some tubes in an apple tree branch may be sending sugar to the roots for storage, while others are sending sugar to the branch tips for fruit production.

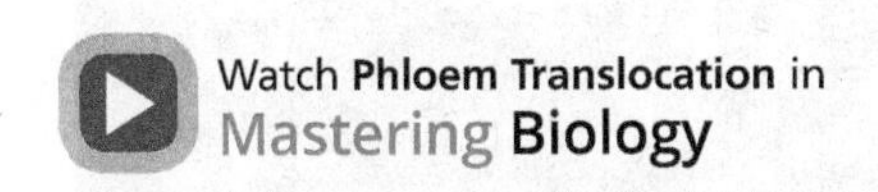

Translocation of phloem sap results from pressure differences within a phloem tube between a source and a sink. **FIGURE 26.10** illustrates the **pressure flow mechanism** that causes phloem sap movement. In the first step of pressure flow, sugar is actively loaded into a phloem tube at the source. As sugar accumulates, water flows in passively via osmosis from nearby tissues and xylem tubes. At the same time, respiring cells at the sink are acquiring sugar from the phloem tube, either through active transport or passively down a concentration gradient from the sugary phloem sap into a sugar-consuming cell. As a result of this movement into the sink cells, the sugar concentration in the nearby phloem tubes declines and water exits the phloem tube passively. Thus, water pressure in the phloem tube is high at the source and low at the sink. Just as in a garden hose, in which high water pressure at the faucet and low pressure in the open air cause water to flow out, this pressure differential causes the bulk movement of water from the high-pressure source end to the lower-pressure sink end of the phloem tubes in a plant.

VISUALIZE THIS

How would sap flow in the phloem tubes be affected by drought?

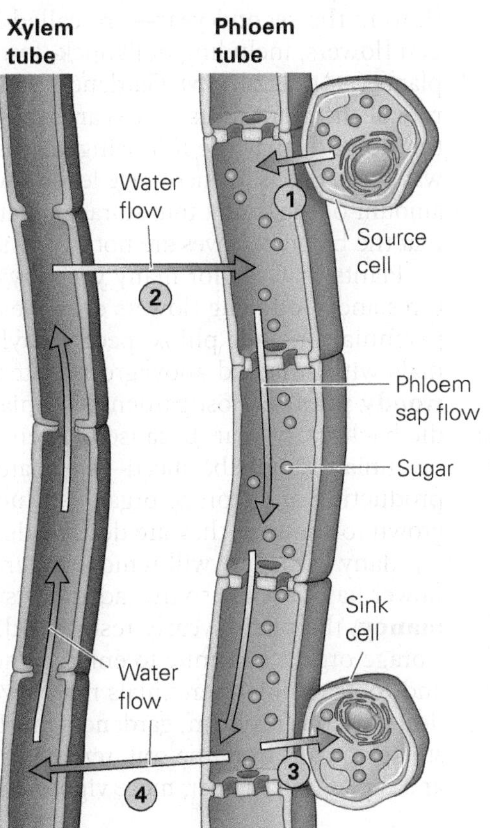

1 Sugar is actively loaded into phloem tubes at the source.

2 Water flows into phloem passively, down its concentration gradient, increasing the water pressure.

3 Sugar moves into sink cell (actively if it is a storage cell, passively if it is a growing or dividing cell).

4 Water flows out of the phloem tube passively, down its concentration gradient, decreasing the water pressure.

FIGURE 26.10 The pressure flow mechanism of phloem translocation. Phloem sap moves from source to sink because of pressure differentials in the phloem tube.

(a) Annual plants

(b) Biennial plants

(c) Perennial plants

FIGURE 26.11 Plant seasonal growth cycles. (a) Annual plants live only a single season or year, growing from a seed and producing a new seed crop at the end of their lives. Their evolutionary strategy is to put a significant amount of energy into flower production, making them ideal for garden displays. (b) Biennial plants typically make a low-profile set of leaves in the first year and send sugar into a storage root. In the second year, sugar from storage supplies a large floral display, after which the plant dies. (c) Perennials live for many years, putting energy into flower production and storage roots every year. Gardeners divide the roots of perennials after a few years of growth to promote a balance of sources and sinks.

Managing Translocation

The key feature of translocation in plants is that materials always move from source to sink. Gardeners who recognize which plant organs are functioning as sources or sinks at any given time can modify a plant's structure to promote flowering, increase the quantity and quality of produce, and prepare plants for winter.

Promoting Flowering. How a gardener manages translocation to increase flower production varies depending on the plant type.

Plants that complete their life cycle within a single year and do not survive past that year are called annuals. Many common garden flowers, including petunia, pansy, snapdragon, and cosmos, are annuals (**FIGURE 26.11a**). The evolutionary strategy of annual plants is to produce as many seeds as possible in their single year of life; to do this, annuals produce many flowers to increase their chances of pollination and fertilization. The larger the sugar source available for flower production, the greater the number of flowers (sinks) that can be produced. To manage flower production in annuals, gardeners remove, or **prune,** many of the flowers on the young transplants they purchase to encourage the plant to send sugars to different sinks—meristems that are developing more leaves.

Once an annual has reached sufficient size, the plant can be encouraged to produce flowers instead of more leaves; this can often be accomplished by restricting both water and fertilizer, which signals the plant that the growing season is coming to an end and thus flower production is a priority. If the plant is grown just for its flowers, individual blooms are removed once they fade (a process called *deadheading*) to eliminate the competitive sink of developing fruits and seeds.

Plants that live for 2 years—producing leaves the first year, surviving dormancy by storing nutrients in roots or underground stems, and producing flowers the second year—are called **biennials.** Some of the best-loved garden flowers, including hollyhock, forget-me-not, and delphinium, are biennial plants (**FIGURE 26.11b**). Gardeners who plant biennials from seed know that the role of the plant in its first year is to send sugars into the storage roots so that the floral display the following year is as large as possible. The most effective way to do this is to encourage leaf growth early in summer, thus maximizing the amount of sugar sent into storage in late summer. Typically, this means ensuring that the biennial leaves are not overly shaded by taller neighbors.

Plants that live for many years by storing nutrients in underground structures and producing flowers every year are called **perennials.** Favorite garden perennials include phlox, peony, daylilies, and tulips (**FIGURE 26.11c**). Perennials with stiffened aboveground stems, such as trees and shrubs, are called **woody plants.** Most garden perennials do not have woody stems and instead die back every year. Because of their long lives, the evolutionary strategy for perennials is more balanced—sinks are evenly divided between flower and fruit production and storage organ production. As with annuals, if perennials are grown for flowers, they are deadheaded to minimize the fruit and seed sink.

Many gardeners will remove perennial flowering stalks well before the last flower has bloomed so that adequate sugars go into the storage sink before **dormancy,** the plant's yearly rest period. In addition, because the underground storage organs continue to enlarge and become crowded over time, the stems and roots of older perennials may have difficulty accessing water and light. To deal with this problem, gardeners must regularly divide perennials by lifting the storage organs from the soil, removing old tissue that no longer produces buds, and replanting smaller, more vigorous portions.

Increasing Produce Quality and Quantity. When annuals, biennials, and perennials are grown for fruit or vegetable production, the goal is to increase the number and quality of these sinks. Many of the same techniques for flower production apply. For example, for annual crops, making sure that the plant puts energy into producing leaves and stems early ensures better production of fruit later. Other gardening strategies focus on different sinks; for instance, plants grown for leaves or stems instead of fruit (such as lettuce and asparagus) may be actively discouraged from flowering or setting seed because these activities will rob their edible parts of quality.

In many crop plants grown for fruit, gardeners must strike a balance between the number of fruits produced and their quality. If plants are allowed to continue making new fruit sinks, each fruit will be smaller and less flavorful because the sugar has to be divided among more sinks. Pruning additional tomato blossoms after a plant has a significant load of fruit and picking off a proportion of the developing apples on a tree thus will produce larger, more flavorful fruit (**FIGURE 26.12**).

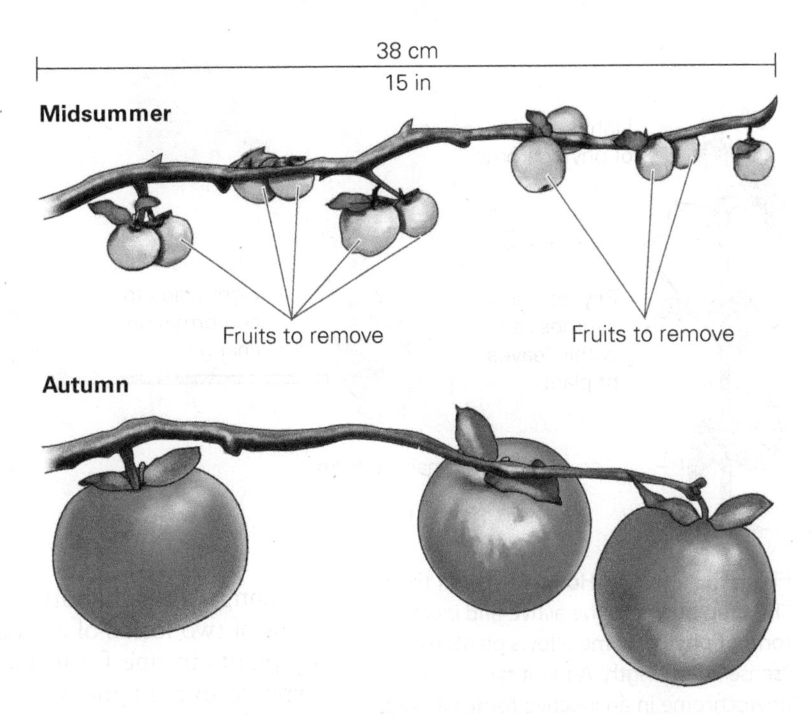

FIGURE 26.12 Maximizing fruit quality over quantity. A flavorful apple requires about 40 leaves to support it. A good rule of thumb is to thin fruit until they are spaced about the width of your hand apart.

Although most fruits do not contain vascular tissue, the same source-sink principle can be applied when harvesting fruits to maximize flavor. Once a fruit is harvested, living cells within the fruit continue to act as a sink; however, because the fruit no longer has an external source, it begins to use its own stored sugars as an energy source. As these sugars are used up, the flavor of the fruit suffers. Plant cells use sugar at higher rates when they are warm, so keeping fruit as cool as possible during harvesting helps retain maximum sweetness. To keep vegetable quality high, gardeners try to harvest during the coolest parts of the day and load their vegetables in baskets that allow for air movement.

Modifications of plant sources and sinks work for gardeners because plants develop through indeterminate growth, in which their ultimate size and shape are subject to environmental conditions rather than a genetic program. However, other responses of a plant to its environment result from innate traits. In particular, the timing of plant flowering and the beginnings of dormancy are often controlled by genetic factors.

Photoperiodism

In the early part of the twentieth century, scientists from the U.S. Department of Agriculture began studying a unique tobacco plant that had appeared in a Maryland farmer's field. This plant grew to twice the height of normal tobacco plants and continued producing leaves until the end of the growing season. Obviously, farmers were thrilled with this variety, except for one small problem—it rarely flowered in the field and thus produced little seed with which to sow an entire crop.

The scientists who worked on the strange tobacco suspected that the "Maryland mammoth" was not flowering because it was not responding to day length in the same way that other tobacco plants did. Through a process of trial and error, researchers discovered that these tobacco plants could be triggered to flower if the day length (that is, total sunlight) they experienced was shorter than a critical number of hours. Later investigations of other plants illustrated that many, like tobacco, exhibit **photoperiodism**—a biological response that is tied to a change in the proportion of light and dark within a 24-hour cycle.

photo- means light.

Tobacco, like chrysanthemums, strawberries, primroses, and poinsettias, is considered a short-day plant because it will not flower if day length surpasses a critical value. Short-day plants tend to bloom when day length is short—like the poinsettias that flower in December. Other plants will flower only if day length is longer than a critical value. These long-day plants, like irises, spinach, and lettuce, are typically summer flowers. Not all plants exhibit flowering photoperiodism. Those that do not are called day-*neutral* plants.

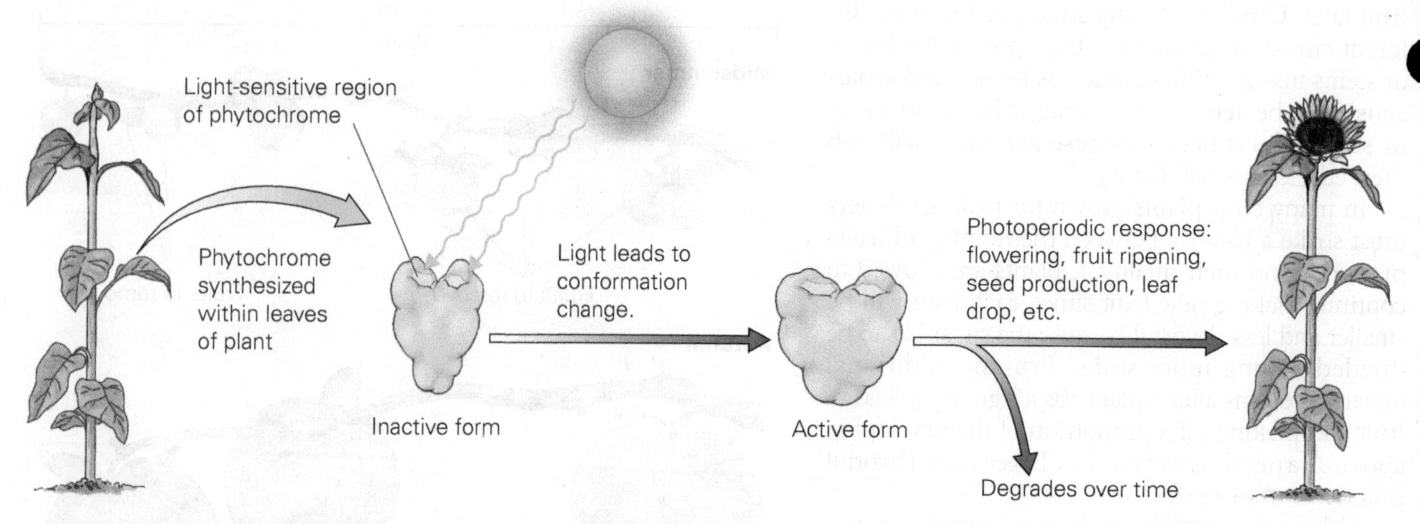

FIGURE 26.13 How plants tell time. The ratio between the active and inactive form of phytochrome allows plants to "sense" day length. A plant synthesizes phytochrome in an inactive form, whereas exposure to sunlight turns it into its active form. In the active form, phytochrome can have a number of effects, but it is slowly degraded over time.

Long-day and short-day plants are actually responding to the concentrations of two forms of a protein called **phytochrome**. Phytochrome is produced by plants in one form that changes chemical shape, or conformation, when exposed to red light, a wavelength that is abundant in sunlight (**FIGURE 26.13**). In its light-activated conformation, phytochrome acts to trigger flowering in long-day plants but delays flowering in short-day plants.

Phytochrome is produced continuously within a photoperiodic plant. The switch between the inactive and active forms of phytochrome is nearly instantaneous on exposure to light, but because proteins are regularly recycled within cells, the active form disappears after a time when no light is present. The amount of time that active phytochrome is present provides the measure of day length. Because a few seconds of light will convert phytochrome to its active form, scientists more accurately measure a photoperiodic plant's **critical night length,** the minimum amount of darkness required to inhibit the effects of phytochrome. If night is shorter than this critical length, the active form of phytochrome can exert its effect (**FIGURE 26.14**). Indoor gardeners can manipulate

phyto- means plant.
-chrome means color.

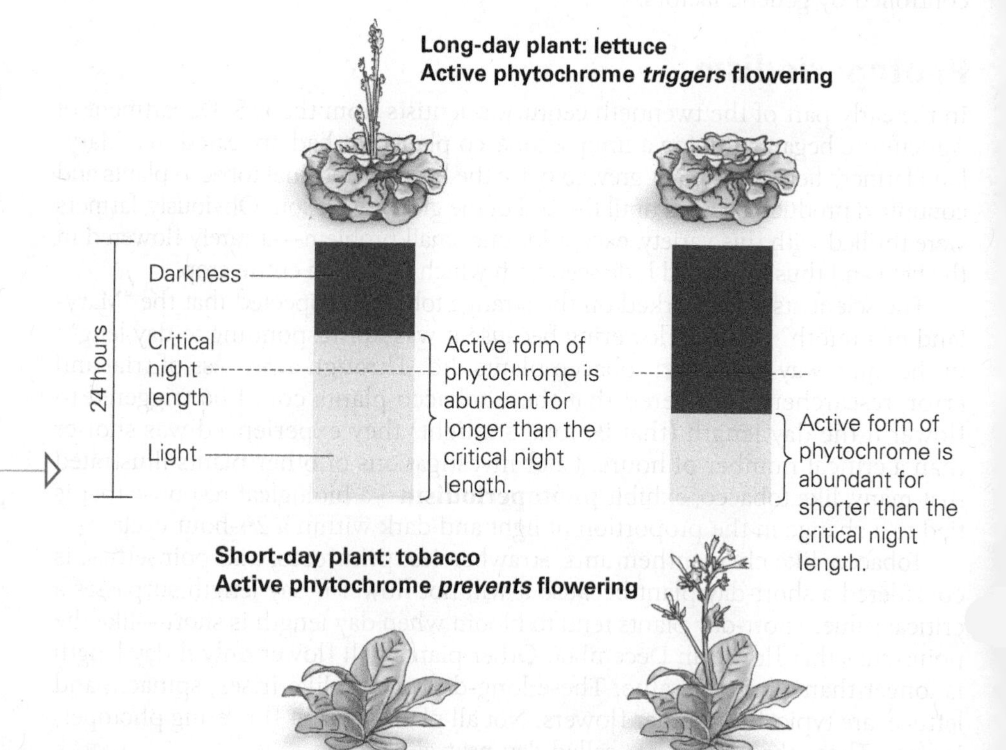

FIGURE 26.14 Phytochrome, short-day plants, and long-day plants. The presence of the active form of phytochrome has the opposite effect on short-day and long-day plants.

WORKING WITH DATA

Lettuce is typically planted so that it reaches edible size in early spring (April and May) and in fall (August to October). Given what the graph illustrates about the role of day length in determining flowering in lettuce, explain why is it less common to plant lettuce in late spring to harvest its leaves in midsummer.

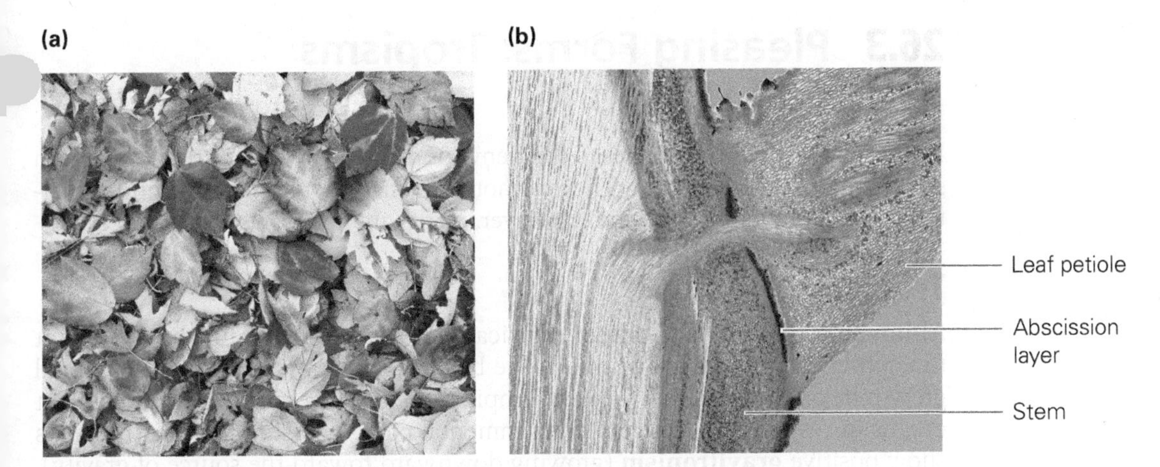

FIGURE 26.15 Abscission. (a) Autumn colors appear when deciduous plants reabsorb chlorophyll in preparation for dropping their leaves. (b) Leaf drop occurs because cells at the junction between a leaf and a stem break down, forming a weak layer that separates and leaves a layer of scar tissue.

a photoperiodic plant's exposure to artificial light during the night, or shade it during the day, to promote or prevent flowering.

Flowering is not the only activity that is triggered by light cues. The opening of leaf buds in spring and the dropping of leaves in fall are also photoperiodic in many woody plants. **Abscission,** or leaf drop, is a complex process that is first triggered by increasing night length as winter approaches. In preparation for abscission, leaves cease to produce chlorophyll for photosynthesis. As chlorophyll levels decline, other pigments in the leaf become visible. As the leaf turns, cells at the boundary between leaf and stem break down, allowing the leaf to separate from the stem and leaving a protective cap over the junction (**FIGURE 26.15**).

Experienced gardeners know that they should match the photoperiod of a plant's native environment to their own environment. The same species of plant growing at different latitudes may have different critical night lengths and will thus respond differently to light conditions. Spinach, strawberries, and onions are all especially susceptible to being planted "out of place" and thus flower, fruit, and produce bulbs too early or too late in the growing season. Southern varieties of widespread deciduous tree species can suffer from damage if planted so far north that they freeze before dropping their leaves.

In addition to finding the right plant for the site and ensuring that their plants produce when and how they wish them to, gardeners are also concerned with the aesthetic form of their gardens. For many, this means actively managing the way that plants fill and share space with each other. To modify form, we must understand how plants grow in response to environmental cues.

Got It?

1. Translocation of sap in phloem always is in the direction of __________ to __________.
2. The bulk movement of phloem sap is due to differences in __________.
3. Water flows into phloem tubes by osmosis after __________ is actively loaded into the tubes.
4. Gardeners can control how plants allocate their sugars by removing nonpreferred tissues or organs that are acting as __________.
5. Phytochrome is a protein that helps a plant measure __________ length, tuning the plant's responses to the seasons.

26.3 Pleasing Forms: Tropisms and Hormones

Plants may seem to be captives to their environment. They cannot run away when something tries to eat them. They cannot seek shade on hot days or find a sheltered spot from a freezing wind. However, plants are not as inert as they appear.

Tropisms

Although a rooted plant cannot physically move itself from one environment to another, plants can grow toward the best possible environment. Directional growth in plants is called **tropism.** Tropic responses are visible in various plant organs in response to different environmental cues. Roots in germinating seeds show positive **gravitropism** (growing downward toward the source of gravity), and stems show negative gravitropism (**FIGURE 26.16a**). This pattern ensures that emerging plant roots maximize exposure to water in the soil and that young shoots maximize exposure to light. Houseplants often demonstrate **phototropism,** directional growth in response to light, when placed on a windowsill (**FIGURE 26.16b**). Other plants grow in response to touch, exhibiting **thigmotropism.** Vining plants show thigmotropism by twining around support stakes (**FIGURE 26.16c**).

trop- means to turn.

Indoor gardeners can use their understanding of plant tropisms to maintain the beauty of their plants. Rotating houseplants regularly ensures that the phototropic growth of a plant is balanced around the whole pot. Handling plants occasionally by touching their leaves can promote a thigmotropic response of shorter, sturdier stems. It may be that the widespread belief that talking to plants is good for them comes from this response—they grow fuller not because of the kind words but because of the increased amount of touch and movement that a well-tended plant experiences.

(a) Gravitropism

(b) Phototropism

(c) Thigmotropism

FIGURE 26.16 Plant tropisms. (a) This plant stem is exhibiting negative gravitropism, orienting itself opposite the pull of gravity. The plant was allowed to reorient in the dark, indicating that the response is not to light. (b) This houseplant is growing toward a light source, exhibiting phototropism. (c) The tendril on this vine is curled around the support as a result of thigmotropism.

Plants bend toward light, or away from gravity, when cells on one side of a stem increase in length and cells on the other side do not, forcing the top of the plant to turn away from the expanding cells. This differential cell expansion is caused by different amounts of growth hormone in these cells.

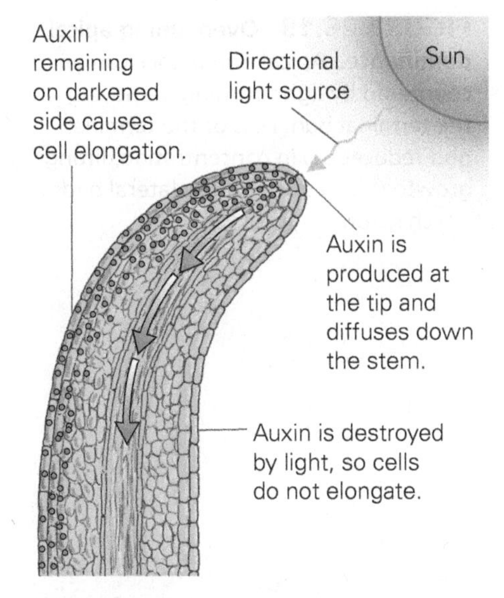

Auxin remaining on darkened side causes cell elongation.
Directional light source
Sun
Auxin is produced at the tip and diffuses down the stem.
Auxin is destroyed by light, so cells do not elongate.

FIGURE 26.17 Directional growth. Tropisms are caused when one side of a plant organ (in this case, a phototropic stem) grows more rapidly than the opposite side, causing the tip to bend.

Hormones

Plant hormones regulate a plant's internal environment and control its responses to environmental conditions. Plant hormones are similar to animal hormones in several ways: (1) Both are produced in tiny amounts but have large effects, (2) they are mobile so that they can be produced in one organ and have an effect on another, and (3) they have multiple and complex effects and interactions. TABLE 26.1 lists the five major plant hormones, their effects, and their commercial applications. We will focus on the first plant hormone identified, and the one most commonly manipulated: auxin.

Auxin is the hormone that triggers the expansion of cells and is thus responsible for causing plants to increase in size. In tropic responses, auxin is destroyed in some cells in response to an environmental signal. As the cells that retain auxin grow in length while those without it remain small, the organ itself bends away from the auxin-containing side (**FIGURE 26.17**).

Like many hormones, auxin can have opposing effects on different plant organs. In most plants, auxin causes cells to expand in a main stem but inhibits the growth of lateral buds along the sides of the stem. Thus, auxin is responsible for **apical dominance**—the tendency for the plant apex, or uppermost bud, to grow more rapidly than the lower buds. Auxin produced at the shoot tip strongly inhibits branching right below the tip. Farther from the shoot tip, auxin concentrations are lower, and its inhibitory effects are less dramatic. Auxin-mediated apical dominance is responsible, for instance, for the "Christmas tree" shape of young conifers.

Apical dominance increases the height and leaf area of a plant but suppresses both bushiness and flower and fruit production. For this reason, gardeners are often interested in reducing the amount of apical dominance that a plant exhibits. The primary method for reducing apical dominance is light pruning, also

TABLE 26.1 Plant hormones.

Hormone Type	Effects	Commercial Applications
Auxin	Promotes cell expansion, apical dominance, tropic responses, differentiation of vascular tissue; promotes roots on plant cuttings; inhibits fruit and leaf abscission; stimulates fruit development; stimulates ethylene production (in high doses)	Rooting powder for plant cuttings; prevents fruit and flower drop; promotes seedless fruit production; used as weed killer by triggering too-rapid growth
Gibberellin	Stimulates leaf and stem cell elongation and division; hastens seed germination, breaks seed dormancy; stimulates fruit development	Used to cause uniform germination of barley in brewing; prevents flower initiation; promotes seedless fruit production
Cytokinin	Promotes cell division; triggers lateral bud growth even in presence of auxin; with auxin, regulates root and shoot production; delays leaf aging	Used to trigger growth of whole plants from genetically engineered plant cells
Ethylene	Inhibits cell expansion; promotes fruit ripening; promotes abscission	Hastens ripening of tomatoes, walnuts, and grapes after harvest; used as fruit-thinning agent in commercial orchards; used to defoliate plants
Abscisic acid	Prevents seed germination; promotes dormancy	Inhibits growth and promotes flowering of potted plants for florists

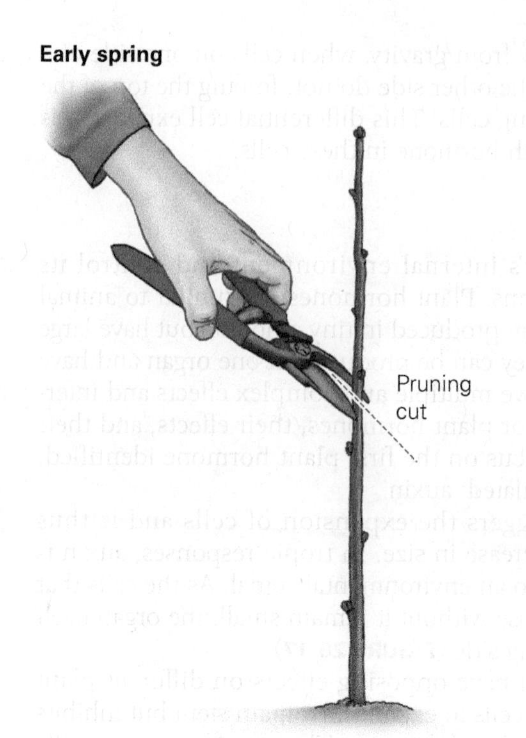

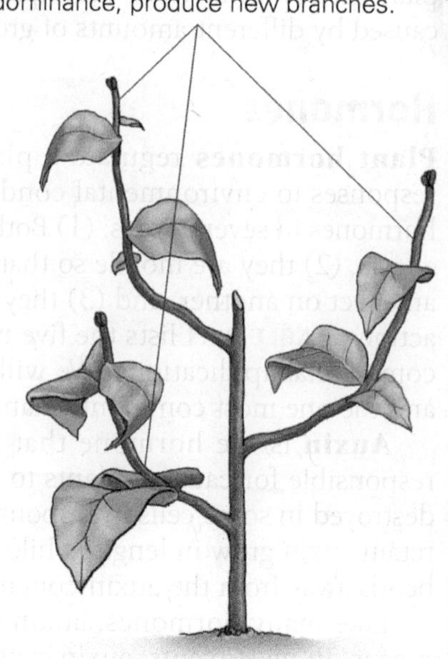

FIGURE 26.18 Overcoming apical dominance. Apical dominance can be countered by light pruning, removing the terminal bud. Loss of the terminal bud reduces auxin concentration, lifting growth suppression on the lateral buds, which sprout.

called pinching back (**FIGURE 26.18**). When the terminal bud is pruned, the plant's auxin levels are reduced, inhibition on lateral buds is lifted, and branches form. Eventually, these branches begin to exert apical dominance and must also be pinched back to maintain a plant's bushy form.

Auxin also interacts with other growth regulators. Abscission is promoted by the hormone **ethylene.** Recall that during abscission, cells at the boundary between the stem and a leaf, flower, or fruit break down so that the abscised organ can separate from the plant. Auxin reduces the sensitivity of boundary layer cells to the effects of ethylene and thus delays abscission. Levels of auxin production are positively correlated to a plant's overall health; as a result, stressed plants tend to abscise fruits and leaves earlier than do healthy plants. The antiabscission role of auxin makes it useful in commercial operations; for instance, citrus growers spray their orchards with auxin to prevent early fruit drop.

We Are All Gardeners

All of the plant hormones identified by scientists have been adopted for some commercial uses. Knowledge of the effects of these hormones, along with knowledge of transpiration, translocation, and photoperiodism, can contribute to the development of a gardener's "green thumb." But this knowledge has a deeper meaning as well. The discovery and human use of plant hormones reflects the way that evolution has created both highly complex biological processes and highly complex human brains. These brains have enabled us to modify organisms to meet human needs, and as a result, we have the opportunity to improve lives for ourselves, our descendants, and other species.

As this chapter has illustrated, a knowledgeable gardener understands that he or she gets the best results by working with natural processes, not by ignoring them. And as many of the stories in this book have demonstrated, this basic principle of working with natural processes can be a guide for all our interactions with the biological world (**FIGURE 26.19**).

FIGURE 26.19 Earth as a garden. The future of humanity depends on the wise use of our biological heritage. Learning about the natural world ideally inspires people to live within natural processes rather than trying to dominate them.

Got It?

1. Any directional growth is called ________.
2. Plants may exhibit tropisms in response to ________, ________, and ________.
3. A plant stem bends because cells on one side of the stem grow ________ than cells on the other side.
4. The hormone auxin helps trigger cell growth, except in ________ buds, where it suppresses growth.
5. Like animal hormones, plant hormones are produced in ________ amounts, can have effects on organs distant from where they are ________, and have ________ effects depending on conditions and the presence of other hormones.

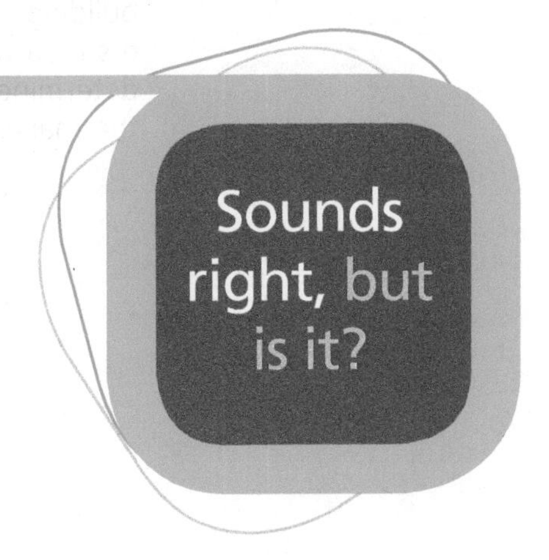

Generations of biology students have complained about having to learn about plants. "They're boring!" is a common refrain. We can summarize student reasoning about why plants are boring with the following phrase:

Plants just sit there, doing nothing.

Sounds right, but it isn't.

Answer the following questions to understand why.

1. List three characteristics of living things.
2. Choose two of the items on the list and explain how plants demonstrate each of these characteristics.
3. Using your answers to questions 1 and 2, explain why the statement in bold sounds right, but isn't.

Should I rip out my lawn?

In a typical suburban development in North America (and increasingly elsewhere), the majority of the landscape around individual houses is made up of short-mown grass lawns. Lawns are visually pleasing, provide a comfortable surface for recreation, and contribute to a sense of openness and community. However, lawns also require significant maintenance, which is time-consuming and expensive—one estimate is that Americans spend $40 billion each year on lawns. Many of the maintenance tasks carry an environmental cost, from the toxicity of pesticides needed to eliminate root-eating insects to the carbon dioxide emissions of power mowers and trimmers to the water needs of these thirsty plants.

What should I know?

What follows are some smaller questions that need to be resolved to answer the Big Question. Place a checkmark next to the questions that science can answer.

Smaller Questions	Can Science Answer?
Are lawn pesticides dangerous at typical levels of exposure?	
Is using pesticides to maintain a lawn unethical because it needlessly exposes humans and other animals to toxic chemicals?	
Does having a lawn increase fossil fuel use?	
Are lawns safer for children than other landscape choices?	
Do people have a right to maintain their property as they see fit?	
Do people generally find lawn more attractive than other landscaping choices?	
Does replacing lawn with another landscape type cause a decrease in property value?	

What does the science say?

Let's examine what the data say about this smaller question:

Does having a lawn increase fossil fuel use?

The landscaping around a building can have a significant effect on the heat gain of that building, thus affecting how much energy is needed to heat or cool the structure. Researchers measured hourly air temperature in 16 sites in Beijing, China, in the month of July to determine the effect of different land cover types (trees, lawn, and paved surfaces) on the thermal environment nearby. Their results are shown in the figure that follows.

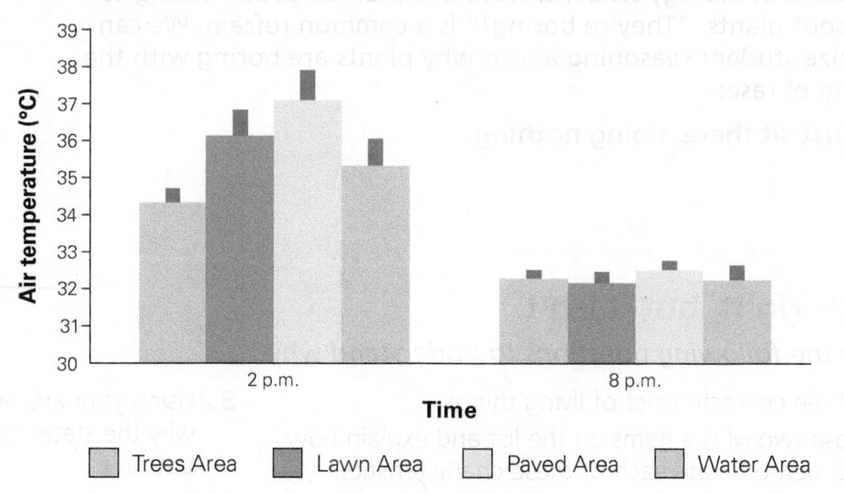

1. Describe the results. Does it appear that lawn is more or less effective than other landscape choices in reducing air temperature?
2. Given these data, do you think the smaller question is answered? If not, propose another study that would help answer this question.
3. Does this information help you answer the Big Question? What else do you need to consider?

Data Source: H. Yan and D. Li, "The Impacts of Land Cover Types on Urban Outdoor Thermal Environment: The Case of Beijing, China." *Journal of Environmental Health Science & Engineering 13* (2015).

Chapter Review

Mastering Biology

Go to Mastering Biology to access **eText 2.0, Dynamic Study Modules,** and the **Study Area,** where you'll find practice quizzes, BioFlix™ animations, MP3 tutor sessions, current events, and more.

SUMMARY

Section 26.1

Describe how water moves up a plant stem.

- Water is drawn from the soil and into a plant because of evaporation at the leaf's surface (p. 562).
- Evaporation creates tension, which pulls on the water column. The column maintains its continuity via cohesion among water molecules and adhesion between water and the xylem walls (pp. 562–563).

List the modifications of plant physiology and anatomy that reduce water loss or make plants more drought-tolerant.

- Modifications to photosynthesis can reduce water loss. These include C_4 photosynthesis, which pumps carbon dioxide to photosynthetic cells, and CAM photosynthesis, which allows carbon dioxide to be harvested from the air at night (p. 564).
- Smaller size, shaded stomata, and deciduousness are leaf adaptations to reduce transpiration (pp. 564–565).
- Narrow tracheids in xylem prevent embolisms, which destroy a plant's capacity to take up water. However, narrow vessels deliver water more slowly, a disadvantage in moist environments (pp. 565–566).
- Overwatered plants wilt because their roots die from lack of oxygen (p. 566).
- Cold-resistant plants control the rate of freezing inside cells to prevent cell damage and manage the ice-induced formation of embolisms in the xylem (p. 567).

SHOW YOU KNOW 26.1 If you have a plant that is suffering from overwatering, one solution is to remove a large number of leaves while waiting for the soil to dry out (a process that could take weeks if the soil is really soaked). Why does removing leaves help the plant survive this period?

Section 26.2

Describe the pressure flow mechanism of phloem transport and explain how source-sink relationships change in plants at different life stages and with different growth cycles.

- Sap moves in phloem from source to sink. Sugar is loaded into phloem tubes at the source, causing water to flow in. Sugar is unloaded at the sink, and water flows out. The difference in water pressure in the tube between the source the sink causes sap to move (p. 569).
- Gardeners can modify relationships between sources and sinks to ensure that energy is directed to the preferred plant organ (pp. 570–571).

Define *photoperiodism,* and explain the mechanism by which plants respond to day length.

- Many plants sense the length of daylight and respond to it. One system for sensing light relies on the phytochrome protein (pp. 571–572).
- Light-mediated responses in plants include the initiation of flower or fruit production, the emergence of leaves, and leaf abscission in deciduous trees (p. 573).

SHOW YOU KNOW 26.2 Maple syrup producers tap the phloem tubes of sugar maple trees in early spring to draw off the sugary phloem sap. Before they add the tap, what is the sugar source in the tree, and what is the sink? When a tap is inserted, how does the identity of the source or sink change?

Section 26.3

List the environmental factors that cause tropism in plants, and describe how differences in cell expansion lead to directional growth.

- Plants grow in response to environmental cues, such as the direction of a light source, gravity, and contact with other objects (p. 574).
- Directional growth is caused by differential cell expansion on opposite sides of an organ (p. 575).

Define "plant hormone," and describe ways in which various hormones affect plant growth.

- Hormones are plant growth regulators, and tune a plant to its environment. Like animal hormones, they are produced in one organ but have effects on different organs (p. 575).
- Auxin from a terminal bud causes cell expansion in stem cells but inhibits the growth of lateral buds, leading to apical dominance (p. 575).
- Auxin interacts with the hormone ethylene to control abscission (p. 576).

SHOW YOU KNOW 26.3 The herbicide known as 2,4-D is a synthetic auxin. Auxin herbicides kill plants by causing them to grow unsustainably—that is, too fast for other metabolic processes to keep up with. Leaves whose surfaces are sprayed with 2,4-D are easy to recognize within a few hours. Given the role of auxin, what do you expect they look like?

ROOTS TO REMEMBER

The following roots of words come mainly from Latin and Greek and will help you to decipher terms:

ad-	means to. Chapter term: *adhesion*
-chrome	means color. Chapter term: *phytochrome*
co-	means with. Chapter term: *cohesion*
photo-	means light. Chapter terms: *photoperiodism, phototropism*
phyto-	means plant. Chapter term: *phytochrome*
trans-	means across. Chapter terms: *transpiration, translocation*
trop-	means to turn. Chapter terms: *tropism, gravitropism, phototropism, thigmotropism*

LEARNING THE BASICS

1. Why does a column of water in a xylem tube remain intact despite being under tremendous tension?
2. How do the C4 and CAM adaptations to photosynthesis reduce water loss from plants?
3. Water moves up a plant's stem as a result of ________.

   A. the xylem pump; B. diffusion of water into roots; C. translocation of phloem sap; D. evaporation of water from the leaves; E. photosynthesis in xylem cells

4. Add labels to the figure that follows, which illustrates a tropic response.

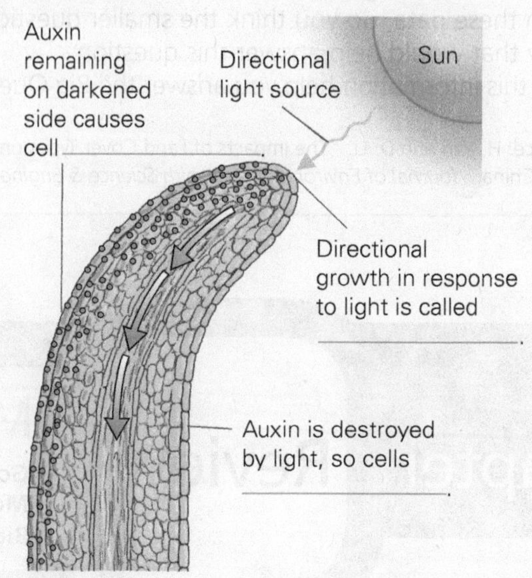

Auxin remaining on darkened side causes cell
Directional light source
Sun
Directional growth in response to light is called
Auxin is destroyed by light, so cells

5. When water is under extremely high tension in a xylem tube, ________.

   A. transpiration slows; B. water cannot be removed from the soil; C. the water column can "break," blocking further water flow; D. stomata open wide to permit greater water absorption; E. the plant is not photosynthesizing

6. Which of the following adaptations provide an advantage to plants in warm, moist environments?

   A. closing stomata in response to decreased water availability; B. large-diameter xylem vessels, permitting rapid water uptake; C. deciduousness, reducing water loss during dry periods; D. small leaf size, reducing sun exposure; E. vertically oriented photosynthetic stems, shading the stomata

7. Sap travels in phloem as a result of ________.

   A. evaporation of sugar water from the leaves; B. the phloem pump; C. a countercurrent to xylem flow; D. differences in water pressure in a phloem tube between a sugar source and a sugar sink; E. cohesion and adhesion of water to the phloem walls

8. Each of the following pairs represents a likely sugar source and its sink *except* ________.

   A. an apple seed and the fruit in which it is contained; B. mature leaves of a tomato plant and a developing tomato; C. newly emerged leaves on a bean seedling and the seedling's developing roots; D. a sugar beet root in spring and its developing flower stalk; E. the roots of an oak tree and its leaf buds in spring

9. A short-day plant can be prevented from flowering by _________.

   A. keeping it constantly in the light; B. preventing phytochrome from converting into the active form; C. ensuring that day length does not exceed the plant's critical length; D. growing it in the same environment as a long-day plant; E. more than one of the above is correct

10. Plant hormones are unlike animal hormones because _________.

    A. they are produced only in small amounts; B. they have different effects on different organs; C. they are not produced in specialized glands; D. they are mobile and can move throughout the plant body; E. interactions among different plant hormones can determine their ultimate effect on an organ

ANALYZING AND APPLYING THE BASICS

1. You have decided to divide an overgrown daylily and move a portion of it to a new spot in your yard. Thinking about water relations in plants, how should you prepare the site and the lily for transplanting?

2. Poinsettias are short-day plants with a critical night length of about 14 hours. Describe how you could induce a poinsettia to flower during the long days of summer.

3. Christmas tree farmers want to encourage apical dominance so that their plants have the "correct" form. They encourage this when plants are a few years old by pruning. What should they prune to maintain a single tip to the tree? What should they prune to ensure bushy lower branches?

GO FIND OUT

1. In even the smallest apartments, there is usually enough window space and light to grow plants. Find out what edible plants are suited for indoor growth. Could you grow some of your own food right now? Would you?

2. In the last paragraph of the chapter, we suggest taking a gardener's approach—working with natural processes and having a desire to understand—in all of our interactions with the natural world. Describe how this approach could be applied to a current issue concerning human modification of nature. Do you agree that it might be helpful to "think like a gardener," or do you see disadvantages of this attitude as well?

MAKE THE CONNECTION

The science that you learned in this chapter has helped you better understand the real-world example used throughout this discussion. Draw a line from the statement on the left to the science that supports it on the right.

Some people seem to have a "green thumb." ▷	◁ Certain plants can measure day length via the protein phytochrome and use that measurement to bloom at the proper time.
You should cut a bit off the stem of cut flowers before putting them in a vase. ▷	◁ Many desert plants have a unique type of photosynthesis that prevents water loss during the warmer parts of the day.
Some of the easiest to care for houseplants are cacti and jade plants. ▷	◁ The flow of water up the stem of a plant requires a continuous "stream" of water. Bubbles caused by releasing tension on a water column must be eliminated to restore a stream.
Most green plant parts are intolerant of freezing. ▷	◁ Phloem sap always moves from source to sink; by removing nonpreferred sinks, carbohydrates in the phloem are directed to preferred organs.
Gardeners can use pruning to ensure that their preferred plant organs receive maximal energy. ▷	◁ Knowledge of basic plant physiology can help you become a more successful gardener.
Gardeners can manipulate light exposure with some plants to promote flowering. ▷	◁ Ice in living cells disrupts the cell membranes, whereas in xylem tubes it can cause embolisms, blocking water flow.

Answers to **Got It?, Visualize This, Working with Data, Sounds Right, But Is It?, Show You Know,** and **Chapter Review** questions can be found in the **Answers** section at the back of the book.

Appendix | Metric System Conversions

To Convert Metric Units:	Multiply by:	To Get English Equivalent:
Length		
Centimeters (cm)	0.3937	Inches (in)
Meters (m)	3.2808	Feet (ft)
Meters (m)	1.0936	Yards (yd)
Kilometers (km)	0.6214	Miles (mi)
Area		
Square centimeters (cm^2)	0.155	Square inches (in^2)
Square meters (m^2)	10.7639	Square feet (ft^2)
Square meters (m^2)	1.196	Square yards (yd^2)
Square kilometers (km^2)	0.3831	Square miles (mi^2)
Hectare (ha) (10,000 m^2)	2.471	Acres (a)
Volume		
Cubic centimeters (cm^3)	0.06	Cubic inches (in^3)
Cubic meters (m^3)	35.30	Cubic feet (ft^3)
Cubic meters (m^3)	1.3079	Cubic yards (yd^3)
Cubic kilometers (km^3)	0.24	Cubic miles (mi^3)
Liters (L)	1.0567	Quarts (qt), U.S.
Liters (L)	0.26	Gallons (gal), U.S.
Mass		
Grams (g)	0.03527	Ounces (oz)
Kilograms (kg)	2.2046	Pounds (lb)
Metric ton (tonne) (t)	1.10	Ton (tn), U.S.
Speed		
Meters/second (mps)	2.24	Miles/hour (mph)
Kilometers/hour (kmph)	0.62	Miles/hour (mph)

To Convert English Units:	Multiply by:	To Get Metric Equivalent:
Length		
Inches (in)	2.54	Centimeters (cm)
Feet (ft)	0.3048	Meters (m)
Yards (yd)	0.9144	Meters (m)
Miles (mi)	1.6094	Kilometers (km)
Area		
Square inches (in^2)	6.45	Square centimeters (cm^2)
Square feet (ft^2)	0.0929	Square meters (m^2)
Square yards (yd^2)	0.8361	Square meters (m^2)
Square miles (mi^2)	2.5900	Square kilometers (km^2)
Acres (a)	0.4047	Hectare (ha) (10,000 m^2)
Volume		
Cubic inches (in^3)	16.39	Cubic centimeters (cm^3)
Cubic feet (ft^3)	0.028	Cubic meters (m^3)
Cubic yards (yd^3)	0.765	Cubic meters (m^3)
Cubic miles (mi^3)	4.17	Cubic kilometers (km^3)
Quarts (qt), U.S.	0.9463	Liters (L)
Gallons (gal), U.S.	3.8	Liters (L)
Mass		
Ounces (oz)	28.3495	Grams (g)
Pounds (lb)	0.4536	Kilograms (kg)
Ton (tn), U.S.	0.91	Metric ton (tonne) (t)
Speed		
Miles/hour (mph)	0.448	Meters/second (mps)
Miles/hour (mph)	1.6094	Kilometers/hour (kmph)

Metric Prefixes		
Prefix		**Meaning**
giga-	G	10^9 = 1,000,000,000
mega-	M	10^6 = 1,000,000
kilo-	k	10^3 = 1,000
hecto-	h	10^2 = 100
deka-	da	10^1 = 10
		10^0 = 1
deci-	d	10^{-1} = 0.1
centi-	c	10^{-2} = 0.01
milli-	m	10^{-3} = 0.001
micro-	μ	10^{-6} = 0.000001

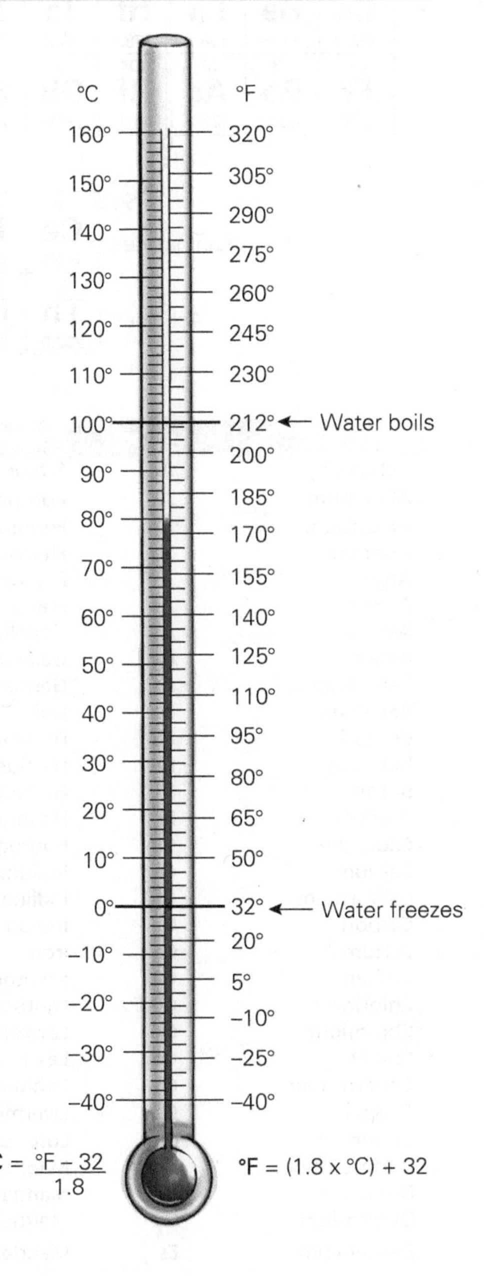

Periodic Table of the Elements

Metals   Metalloids   Nonmetals

Group number
Atomic number

Transition elements

Period	1	2											3	4	5	6	7	8
1	1 H 1.008																	2 He 4.003
2	3 Li 6.941	4 Be 9.012											5 B 10.81	6 C 12.01	7 N 14.01	8 O 16.00	9 F 19.00	10 Ne 20.18
3	11 Na 22.99	12 Mg 24.31											13 Al 26.98	14 Si 28.09	15 P 30.97	16 S 32.07	17 Cl 35.45	18 Ar 39.95
4	19 K 39.10	20 Ca 40.08	21 Sc 44.96	22 Ti 47.87	23 V 50.94	24 Cr 52.00	25 Mn 54.94	26 Fe 55.85	27 Co 58.93	28 Ni 58.69	29 Cu 63.55	30 Zn 65.41	31 Ga 69.72	32 Ge 72.64	33 As 74.92	34 Se 78.96	35 Br 79.90	36 Kr 83.80
5	37 Rb 85.47	38 Sr 87.62	39 Y 88.91	40 Zr 91.22	41 Nb 92.91	42 Mo 95.94	43 Tc (98)	44 Ru 101.1	45 Rh 102.9	46 Pd 106.4	47 Ag 107.9	48 Cd 112.4	49 In 114.8	50 Sn 118.7	51 Sb 121.8	52 Te 127.6	53 I 126.9	54 Xe 131.3
6	55 Cs 132.9	56 Ba 137.3	57 La 138.9	72 Hf 178.5	73 Ta 180.9	74 W 183.8	75 Re 186.2	76 Os 190.2	77 Ir 192.2	78 Pt 195.1	79 Au 197.0	80 Hg 200.6	81 Tl 204.4	82 Pb 207.2	83 Bi 209.0	84 Po (209)	85 At (210)	86 Rn (222)
7	87 Fr (223)	88 Ra (226)	89 Ac (227)	104 Rf (261)	105 Db (262)	106 Sg (266)	107 Bh (264)	108 Hs (269)	109 Mt (268)	110 Ds (271)	111 Rg (280)	112 Cn (285)	113 Nh (286)	114 Fl (289)	115 Mc (289)	116 Lv (293)	117 Ts (294)	118 Og (294)

Lanthanides	58 Ce 140.1	59 Pr 140.9	60 Nd 144.2	61 Pm (145)	62 Sm 150.4	63 Eu 152.0	64 Gd 157.3	65 Tb 158.9	66 Dy 162.5	67 Ho 164.9	68 Er 167.3	69 Tm 168.9	70 Yb 173.0	71 Lu 175.0
Actinides	90 Th 232.0	91 Pa 231.0	92 U 238.0	93 Np (237)	94 Pu (244)	95 Am (243)	96 Cm (247)	97 Bk (247)	98 Cf (251)	99 Es 252	100 Fm 257	101 Md 258	102 No 259	103 Lr 260

Name	Symbol	Name	Symbol	Name	Symbol	Name	Symbol
Actinium	Ac	Erbium	Er	Mercury	Hg	Rutherfordium	Rf
Aluminum	Al	Europium	Eu	Molybdenum	Mo	Samarium	Sm
Americium	Am	Fermium	Fm	Moscovium	Mc	Scandium	Sc
Antimony	Sb	Flerovium	Fl	Neodymium	Nd	Seaborgium	Sg
Argon	Ar	Fluorine	F	Neon	Ne	Selenium	Se
Arsenic	As	Francium	Fr	Neptunium	Np	Silicon	Si
Astatine	At	Gadolinium	Gd	Nickel	Ni	Silver	Ag
Barium	Ba	Gallium	Ga	Nihonium	Nh	Sodium	Na
Berkelium	Bk	Germanium	Ge	Niobium	Nb	Strontium	Sr
Beryllium	Be	Gold	Au	Nitrogen	N	Sulfur	S
Bismuth	Bi	Hafnium	Hf	Nobelium	No	Tantalum	Ta
Bohrium	Bh	Hassium	Hs	Oganesson	Og	Technetium	Tc
Boron	B	Helium	He	Osmium	Os	Tellurium	Te
Bromine	Br	Holmium	Ho	Oxygen	O	Tennessine	Ts
Cadmium	Cd	Hydrogen	H	Palladium	Pd	Terbium	Tb
Calcium	Ca	Indium	In	Phosphorus	P	Thallium	Tl
Californium	Cf	Iodiine	I	Platinum	Pt	Thorium	Th
Carbon	C	Iridium	Ir	Plutonium	Pu	Thulium	Tm
Cerium	Ce	Iron	Fe	Polonium	Po	Tin	Sn
Cesium	Cs	Krypton	Kr	Potassium	K	Titanium	Ti
Chlorine	Cl	Lanthanum	La	Praseodymium	Pr	Tungsten	W
Chromium	Cr	Lawrencium	Lr	Promethium	Pm	Uranium	U
Cobalt	Co	Lead	Pb	Protactinium	Pa	Vanadium	V
Copernicium	Cn	Lithium	Li	Radium	Ra	Xenon	Xe
Copper	Cu	Livermorium	Lv	Radon	Rn	Ytterbium	Yb
Curium	Cm	Lutetium	Lu	Rhenium	Re	Yttrium	Y
Darmstadtium	Ds	Magnesium	Mg	Rhodium	Rh	Zinc	Zn
Dubnium	Db	Manganese	Mn	Roentgenium	Rg	Zirconium	Zr
Dysprosium	Dy	Meitnerium	Mt	Rubidium	Rb		
Einsteinium	Es	Mendelevium	Md	Ruthenium	Ru		

Answers

Chapter 1

GOT IT?

p. 7: 1. hypothesis; **2.** proven false; **3.** measurements; **4.** predictions; **5.** proven

p. 15: 1. experimental treatment; **2.** hypotheses; **3.** bias; **4.** controlled; **5.** causes

p. 20: 1. statistics; **2.** sample; **3.** chance; **4.** probability; **5.** practical

p. 23: 1. primary; **2.** peer; **3.** Anecdotal; **4.** secondary sources; **5.** less

VISUALIZE THIS

Figure 1.1: Good scientists generate good hypotheses. This requires imagination, experience, and insight, as well as an understanding of science. These traits can be nurtured by taking a wide range of nonscience courses.

Figure 1.3: Even with a well-designed experiment, it is possible that an alternative factor can explain the results, and researchers need to consider if that might be the case in their experiment.

Figure 1.4: Preventing virus attachment and invasion of cells. Preventing activation of the immune system response. Other answers may be acceptable.

Figure 1.11: Exercise may reduce feelings of stress and may independently strengthen the immune system (perhaps by improving blood circulation). Thus low levels of exercise could contribute to both high stress and high cold susceptibility.

Figure 1.12: A lack of racial and ethnic diversity in an experimental sample can affect the results and interpretation of a study because it means the sample may differ from the population in systematic ways. In this case, the average hair length in the class may not reflect the entire population because different subpopulations are not represented.

Figure 1.14: Outliers (unusually high or unusually low data points) have less impact if there are more data points in the sample.

Figure 1.15: Hypothesis: true. Difference between control and experimental groups: large. Sample size: large. This result is very likely to be both statistically and practically significant.

Figure 1.16: The reviewers do not have any stake in the outcome of the experiment and can catch errors in interpretation or experimental design that the original researchers may have missed.

WORKING WITH DATA

Figure 1.7: This change in the graph makes the Echinacea tea appear much more effective.

Figure 1.10: There may not have been enough participants at a particular stress level to allow effective analysis of the correlation. If individuals at stress level 3 have the same cold susceptibility as those at stress level 4, that should cause us to question the correlation because we'd expect those at stress level 4 to have greater susceptibility.

Figure 1.13: The two data points are not "connected" to each other—that is, one does not follow the other, they are simply the same measurement in two different groups.

SOUNDS RIGHT, BUT IS IT?

1. The experiment would be a double-blind, random assignment, placebo-controlled study. In this case, patients with acne would be randomly assigned to use the cream or a placebo-alternative for enough time to see possible results in acne levels. Patients and evaluators would not know whether the cream used contained the active ingredient.

2. We don't know. Statistical significance does not equal practical significance—without information on how much difference there was between treatment and placebo samples, we don't know how large the effect of the cream was.

3. It is highly prone to bias, either by the consumer or by the clinician recording the results. Without a placebo control, these types of experiments also fail to account for other differences between nontreated and treated individuals, including the other aspects of skin care that might come along with using the cream (e.g., how many times per day the face is washed, etc.).

4. No, a single experiment could give a result that is statistically significant purely as a result of chance sampling error. In addition, even with a well-controlled experiment, there may be other explanations for a difference in acne levels between treatment and placebo groups besides the effectiveness of the active ingredient—for instance, perhaps one of the ingredients in the placebo cream actually causes additional acne. It's only after multiple tests of a hypothesis that we can come close to "proving" it.

5. Not necessarily. It means that the average amount of acne was less in the experimental group. If there was significant variability in results, it still could be that some individuals using the "real" cream had more acne or no change in acne after treatment.

6. We don't know from the statement "clinically proven" by itself if the experiment was designed to eliminate bias and/or showed a practically significant result. We also can't be confident that the hypothesis is proven without multiple tests that allow us to reject alternative hypotheses. Finally, one can't tell from the statement whether there was significant variability in the results, such that some individuals were benefitted a lot by the product and others not at all.

SHOW YOU KNOW

1.1: No, because even though Homer is a likely culprit, there are probably other individuals in your workplace who may have eaten some or all of the donuts.

1.2: Participants could have been exposed to different cold viruses; early arrivals may have been infected by a severe virus that was spreading in the hospital, while later arrivals may have been infected by a milder virus.

1.3: No, if the result is not statistically significant, we cannot conclude that hairstyles are longer in 2018.

1.4: You could use the databases available through an academic library search service. Using these databases leads to peer-reviewed research, which provides the most critically evaluated information available.

LEARNING THE BASICS

1.

Control group	Experimental group
Experiencing early cold symptoms	Experiencing early cold symptoms
Sought treatment from clinic	Sought treatment from clinic
Received placebo tea	Received **echinacea** tea

2. B; **3.** B; **4.** D; **5.** A; **6.** D; **7.** E; **8.** C; **9.** B; **10.** C; **11.** B; **12.** E

ANALYZING AND APPLYING THE BASICS

1. No, another factor that causes type 2 diabetes could lead to obesity as well, or it is possible that type 2 diabetes causes obesity, not vice versa.
2. Neither the participants nor the researchers were blind, meaning that both subject expectation and observer bias might influence the results. If participants felt that vitamin C was likely to be effective, they may underreport their cold symptoms. If the clinic workers thought one or the other treatment was less effective, they might let that bias show when they were talking to the participants, influencing the results.
3. See the answer to number 1; BDNF may increase as a result of exercise, or excess BDNF might cause an increased interest in exercise.

Chapter 2

GOT IT?

p. 32: **1.** molecules; **2.** cell; **3.** stimuli; **4.** Metabolism; **5.** homeostasis
p. 34: **1.** electrons; **2.** Positively; **3.** electronegative; **4.** partial; **5.** hydrophobic
p. 36: **1.** carbon; **2.** molecules; **3.** covalent; **4.** ionic; **5.** four
p. 41: **1.** amino acids; **2.** hydrogen; **3.** fatty acid; **4.** two; **5.** Steroids.
p. 43: **1.** prokaryotes; **2.** eukaryotic; **3.** ancestor; **4.** natural; **5.** environment

VISUALIZE THIS

Figure 2.1: 107°F

Figure 2.4: oxygen

Figure 2.9: The other locations should have similar rates of "disappearance" as the Bermuda Triangle.

Figure 2.13: sugar and phosphate

WORKING WITH DATA

Figure 2.2: a. bringing water to exams, on the *x* axis. b. standard deviation.

SOUNDS RIGHT, BUT IS IT?

1. Computers cannot grow, reproduce, or move. Neither can they metabolize or undergo homeostasis.
2. Maybe, but it seems unlikely.
3. Memorization does not allow easy extrapolation to novel questions.
4. Understanding usually beats memorization when applying knowledge to new situations.
5. Until computers can learn, they are going to require human input.

SHOW YOU KNOW

2.1: Homeostasis.

2.2: The bonds within a water molecule are covalent; the bonds between two water molecules are hydrogen bonds.

2.3: For this molecule to occur, carbon would have to be bonded to five hydrogen atoms. Carbon can bind to four other atoms, not five.

2.4: Nucleic acid, nucleotide, nitrogenous base, guanine.

2.5: A cell membrane.

LEARNING THE BASICS

1. carbohydrates, lipids, proteins, nucleic acids
2. genetic material, cell wall, cell membrane, ribosomes
3.

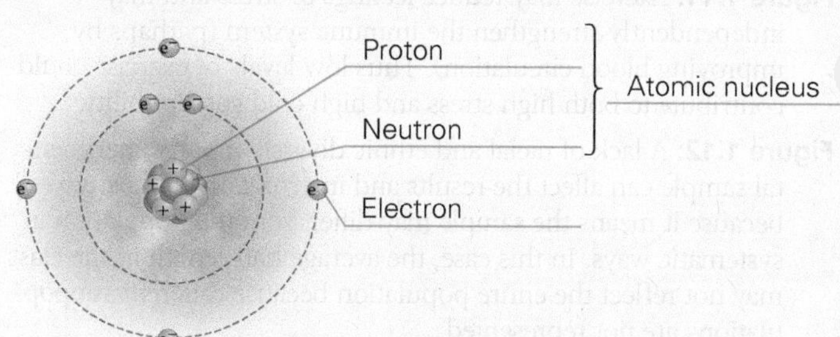

4. B; **5.** A; **6.** A; **7.** D; **8.** D; **9.** D; **10.** D; **11.** E; **12.** C

ANALYZING AND APPLYING THE BASICS

1. No. Living organisms can reproduce—viruses can't reproduce on their own.
2. Water is a good solvent and facilitates chemical reactions.
3. Only nucleic acids and proteins.

Chapter 3

GOT IT?

p. 56: **1.** energy; **2.** fats; **3.** processed; **4.** Complex; **5.** Fiber
p. 60: **1.** ribosomes; **2.** endoplasmic reticulum (or ER); **3.** cell wall; **4.** mitochondria; **5.** chloroplasts
p. 64: **1.** equilibrium; **2.** hydrophobic; **3.** charged; **4.** energy; **5.** pumps

VISUALIZE THIS

Figure 3.3: The top tail in the fat shown in part (b) is polyunsaturated.

Figure 3.4: Plant cells have a cell wall, chloroplasts, and large central vacuole and animal cells do not.

WORKING WITH DATA

Figure 3.17: a. No. b. Average amount of power used went down during training from days 1 to 6 for both groups of cyclists regardless of the type of supplement they were given. c. No, supplementing with carbohydrate and protein carbohydrate only had the same effect.

SOUNDS RIGHT, BUT IS IT?

1. High
2. Adding fertilizer would increase the concentration of solute outside the plant cell.
3. Water would move out of the plant cell.
4. The plant will wilt.
5. A wilting plant should be watered, not fertilized.

SHOW YOU KNOW

3.1: No. Water cannot be broken down to produce energy and does not play a structural role in cells.

3.2: The nuclear envelope and nuclear pores serve as added layers of protection.

3.2: No. Some larger molecules are able to move across the membrane through transport proteins.

LEARNING THE BASICS

1. Nutrients serve as energy stores and as building blocks for cellular structures.
2.

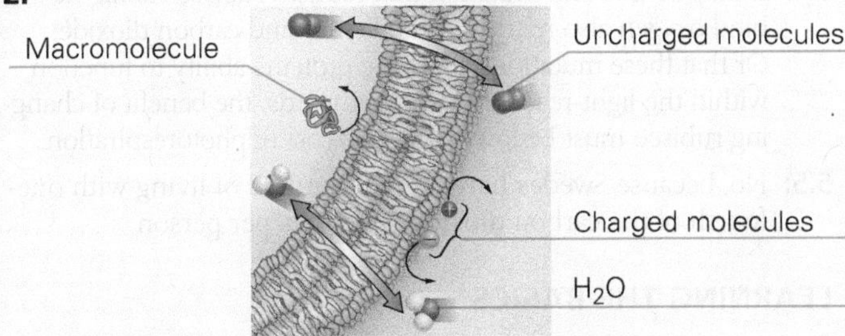

3. D; **4.** B; **5.** B; **6.** B; **7.** D; **8.** D; **9.** E; **10.** D

ANALYZING AND APPLYING THE BASICS

1. She could eat lots of protein-rich foods like beans.
2. With heart disease plotted on the *x* axis and trans fat consumption on the *y* axis, the line showing the relationship should be a diagonal line.
3. Other vitamins, minerals, antioxidants, and fiber.

Chapter 4

GOT IT?

p. 72: 1. catalyze; **2.** specificity; **3.** substrate; **4.** activation; **5.** basal
p. 74: 1. mitochondria; **2.** ATP; **3.** carbon dioxide; **4.** aerobic; **5.** genes
p. 81: 1. anorexia; **2.** bulimia; **3.** leptin; **4.** dieting; **5.** fewer

VISUALIZE THIS

Figure 4.2: No. The enzyme releases the substrate and goes back to its original shape.

Figure 4.10: Two electrons

Figure 4.12: Two turns. Each glucose molecule is converted into two three-carbon pyruvates and each of those pyruvates is decarboxylated (losing one carbon) before entering the cycle. So, each turn of the cycle releases two carbons.

Figure 4.13: Oxygen exists in cells as molecular oxygen (O_2). Because only one oxygen atom is required to make water (H_2O), the convention is to place the fraction ½ before O_2.

WORKING WITH DATA

Table 4.1: BMI = 27

Figure 4.16: Severe obesity, underweight, moderate obesity, overweight, normal weight.

Figure 4.17: a. 7.7% b. 92.3%

SOUNDS RIGHT, BUT IS IT?

1. No
2. No
3. CO_2
4. Exercise increases breathing rate. It is actually the loss of the carbon and oxygen atoms that were once part of fats that causes weight loss.
5. Weight loss occurs when fat is metabolized into carbon dioxide and released from the body. Laxatives work only on materials in the large intestine that will never become part of the body.

SHOW YOU KNOW

4.1: The enzyme would no longer be able to bind its normal substrate and convert it to product.

4.2: Aerobic exercise causes faster breathing, which brings more oxygen to the blood. Aerobic conditioning helps prevent lactic acid buildup because more oxygen is available for the conversion of lactic acid to pyruvic acid.

4.3: Body weight and calories burned during exercise are positively correlated—the more you weigh, the more calories you burn during exercise.

LEARNING THE BASICS

1. When a substrate binds to the active site of an enzyme, the enzyme clamps down on the substrate, causing increased stress on the bonds holding the substrate together.
2.

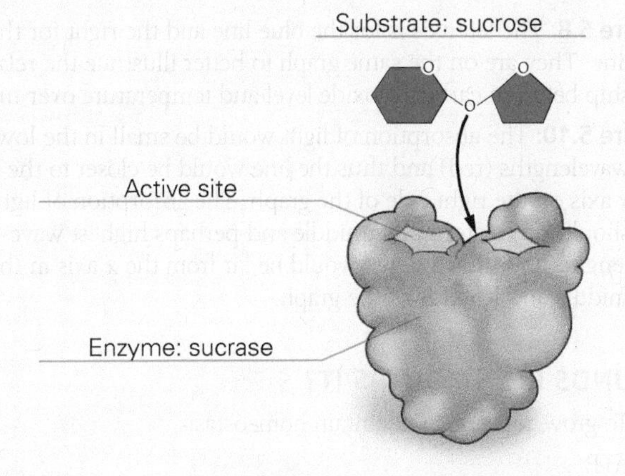

3. The reactants are glucose and oxygen. The products are carbon dioxide and water.
4. C; 5. B; 6. D; 7. D; 8. E; 9. B; 10. C

ANALYZING AND APPLYING THE BASICS

1. No. Fats and proteins also feed into cellular respiration at various points in the cycle and fats actually store more energy than carbohydrates.
2. The carbon atom would either become part of a cellular structure inside your body or be released as CO_2.

Chapter 5

GOT IT?

p. 90: 1. greenhouse; **2.** heat; **3.** vapor; **4.** movement; **5.** increased
p. 92: 1. rock; **2.** carbohydrates; **3.** oil, coal, natural gas; **4.** ice cores; **5.** much higher (or higher)
p. 96: 1. chemical; **2.** chloroplasts; **3.** oxygen; **4.** ATP; **5.** carbohydrates, CO_2
p. 99: 1. water; **2.** change; **3.** carbon dioxide; **4.** C4, CAM; **5.** less
p. 100: 1. acidic; **2.** 25; **3.** transportation; **4.** car; **5.** real

VISUALIZE THIS

Figure 5.1: The red "greenhouse gas" portion of the figure will be thicker and thus there will be an increase in the amount of heat (illustrated by more heat arrows) that is re-radiated back to Earth.

Figure 5.2: Water molecules should appear in the "air" portion of the center box. Evaporation occurs more rapidly when the water is warming because the lack of hydrogen bonds can allow the molecules to escape the liquid form.

Figure 5.3: Water levels would rise and shorelines would erode.

Figure 5.4: Human activity.

Figure 5.11: Carbon dioxide, water, and light energy are obtained from the environment, and oxygen and carbohydrates are released. NADP+ and ATP/ADP are contained within the chloroplast.

Figure 5.15: Stomata are more likely to be open during the day to allow in CO_2, because light is required for photosynthesis to proceed.

WORKING WITH DATA

Figure 5.6: The increase in concentration is steeper during the period 2000–2017 when compared with 1960–1970.

Figure 5.8: The left axis is for the blue line and the right for the red line. They are on the same graph to better illustrate the relationship between carbon dioxide level and temperature over time.

Figure 5.10: The absorption of light would be small in the lower wavelengths (red) and thus the line would be closer to the *x* axis on the right side of the graph. The absorption of light should be greater in the middle and perhaps highest wavelengths and thus the line would be far from the *x* axis in the middle and left side of the graph.

SOUNDS RIGHT, BUT IS IT?

1. To grow, reproduce, maintain homeostasis.
2. ATP.
3. $6CO_2 + 6H_2O + \text{Light energy} \rightarrow C_6H_{12}O_6 + 6O_2$
4. No, ATP produced during photosynthesis is used to produce sugars.
5. Aerobic respiration.
6. $C_6H_{12}O_6 + 6O_2 \rightarrow 6CO_2 + 6H_2O + \text{Energy (ATP)}$
7. Even though, unlike animals, plants consume carbon dioxide and release oxygen in the process of photosynthesis, because plants need to perform aerobic respiration to produce ATP for cellular activities, they also consume oxygen and release carbon dioxide, just like animals. So it is not correct to say that plants perform a sort of "opposite respiration" that the statement implies.

SHOW YOU KNOW

5.1: Because the moon has no or very little greenhouse gas, heat from the sun is not retained to be re-radiated overnight, which is why the nighttime temperatures are cold. The warm temperatures during the day occur because water is not present to absorb some of the heat energy without changing temperature.

5.2: The carbon dioxide released by the burning of logs or sawdust is essentially recycled when replacements for plants that produced these logs or sawdust are regrown. This leads to no net increase in carbon dioxide. Fossil fuel burning releases carbon that was removed from the atmosphere millions of years ago so is causing a current net increase.

5.3: NADPH.

5.4: It may be that any mutation that reduces rubisco's ability to bind oxygen also reduces its ability to bind carbon dioxide. Or that these mutations affect the protein's ability to function within the light-reactions. In other words, the benefit of changing rubisco must be lower than the cost of photorespiration.

5.5: No, because Swedes have a high standard of living with one-fourth of the carbon dioxide emissions per person.

LEARNING THE BASICS

1.

2. C; 3. D; 4. E; 5. D; 6. A; 7. C; 8. A; 9. D; 10. D

ANALYZING AND APPLYING THE BASICS

1. Because colder water has the capacity to absorb more heat before water molecules can evaporate into the atmosphere, colder temperatures mean that less water circulates in the water cycle.

2. The water cycle would remain basically unchanged. However, the carbon cycle would be quite different. The only way carbon dioxide would move from the atmosphere to Earth's surface would be by absorption into water and perhaps incorporation into rocks. It would return to the atmosphere via diffusion from water and volcanic activity.
3. High rates of photosynthesis are continually pumping oxygen into the air. So even though the molecule is being "used up" through reactions with other compounds, its continual production ensures high levels in the atmosphere.

Chapter 6

GOT IT?

p. 110: 1. mitosis; **2.** Benign; **3.** Malignant; **4.** Metastatic; **5.** low
p. 113: 1. sexual; **2.** chromosomes; **3.** sister; **4.** DNA polymerase; **5.** identical
p. 116: 1. mitosis; **2.** cytokinesis; **3.** S; **4.** equator; **5.** sister chromatids
p. 119: 1. tumor suppressors; **2.** mutations; **3.** biopsy; **4.** chemotherapy; **5.** radiation

VISUALIZE THIS

Figure 6.4: (a) There will be two amoeba after division. (b) Asexual reproduction would result in two stems with leaves, not one larger plant.

Figure 6.7: Two molecules of DNA—one 100% purple, one 50% purple.

Figure 6.12: The cell will not make it through the metaphase checkpoint without properly formed microtubules.

WORKING WITH DATA

Figure 6.2: The cancer risk to those who use both alcohol and tobacco is three times greater than it is to those who use only one of these substances. The risk is multiplicative—alcohol use increases risk by 20% as does tobacco use. An additive risk would be 40% but the actual risk is 60%.

SOUNDS RIGHT, BUT IS IT?

1. No. You would need to know the cancer risk of being in the sun for a few hours before you could decide whether that behavior was safe.
2. It will take longer to get a tan, so people will have to pay for more or longer tanning sessions.
3. No. It is evidence of skin damage.
4. Eat fruits and vegetables rich in vitamin D. Take a vitamin D supplement.
5. The use of tanning beds is not safe. It injures skin and increases cancer risk. Benefits of increased vitamin D are more safely obtained through healthy eating.

SHOW YOU KNOW

6.1: This is good news. Clear margins are an indication that the tumor is not invading surrounding tissues.

6.2: There will still be 46 chromosomes, they will be replicated and composed of two sister chromatids.

6.3: There are two DNA molecules in a replicated chromosome and one in an unreplicated chromosome.

6.4: The presence of specific marker proteins in high concentrations of the cell membrane makes cancer cells susceptible to destruction by immunotherapy.

LEARNING THE BASICS

1. Cancer cells divide when they should not, invade surrounding tissues, and move to other locations in the body.
2.

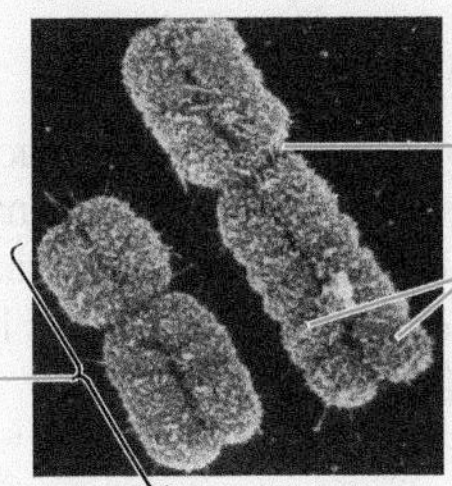

3. C; **4.** A; **5.** C; **6.** D; **7.** B; **8.** C; **9.** D
10. Plant and animal cells both have interphases followed by M phases. Cytokinesis in plant cells involves the building of a cell wall to partition one cell into two. In animal cells cytokinesis includes a pinching of the one cell into two.

ANALYZING AND APPLYING THE BASICS

1. No. Only mutations in cells that give rise to gametes are passed on. The mutation in skin cancer was acquired and will only be found in the affected skin cells.
2. Radiation can only be used on tumors in particular locations and its effects are more localized. Chemotherapy moves throughout the body attacking rapidly dividing cells.
3. Radiation causes mutations to the DNA.

Chapter 7

GOT IT?

p. 130: 1. gametes; **2.** 23; **3.** homologous pair; **4.** sister chromatids; **5.** random
p. 131: 1. less; **2.** nondisjunction; **3.** gametes; **4.** age; **5.** increases
p. 136: 1. gonads; **2.** tail (or flagellum); **3.** semen; **4.** oviduct; **5.** cervix

VISUALIZE THIS

Figure 7.2: The X chromosome.

Figure 7.11: vaginal opening → vagina → cervix → uterus → oviduct → abdomen

WORKING WITH DATA

Figure 7.8: (a) Fertility declines slowly before age 30 and then more rapidly after. (b) Almost two times as long. (c) Males remain fertile well into old age, whereas female fertility ends at menopause.

SOUNDS RIGHT, BUT IS IT?

1. Estrogen.
2. The amount of estrogen would slowly increase.
3. It begins to change slowly as estrogen increases.
4. Yes.
5. Sperm can be present in the oviduct before the change in cervical mucus.

SHOW YOU KNOW

7.1: Four.

7.2: The female produced would have 45 total chromosomes and only 1 X chromosome. This condition is called Turner syndrome.

7.3: Oviduct.

LEARNING THE BASICS

1. Meiosis is two divisions, mitosis is one; meiosis produces genetically distinct haploid daughter cells, mitosis makes exact copies of cells; meiosis can produce four daughter cells; mitosis produces one daughter cell; meiosis occurs in gonads only, mitosis occurs in most somatic cells.

2.

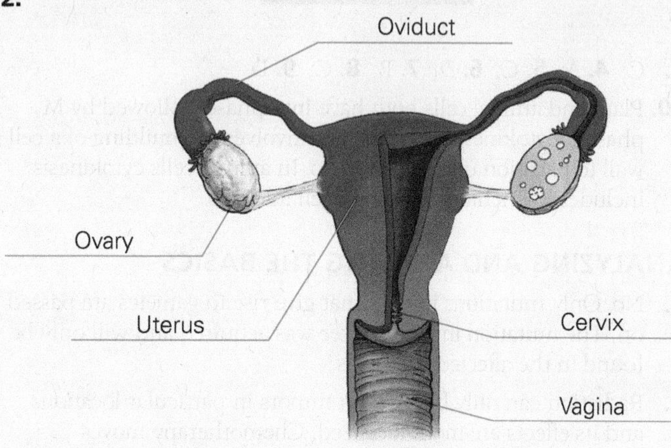

3. The egg cell leaves the ovary and enters the oviduct, leaving behind the corpus luteum.

4. A; **5.** A; **6.** C; **7.** E; **8.** D; **9.** C; **10.** D

ANALYZING AND APPLYING THE BASICS

1. Vulva.
2. Endometrial tissues block oviducts and prevent ovarian function.
3. Estrogen.

Chapter 8

GOT IT?

p. 148: 1. expressed; **2.** alleles, mutation; **3.** gene (or chromosome); **4.** independent assortment; **5.** random

p. 154: 1. genotype, phenotype; **2.** dominant; **3.** heterozygote; **4.** 25; **5.** 16

p. 157: 1. incompletely; **2.** mutated; **3.** codominant; **4.** two; **5.** recessive

p. 160: 1. sex; **2.** X; **3.** sex-linked; **4.** phenotypes; **5.** boys, mothers

VISUALIZE THIS

Figure 8.1: The most dramatic effects of a mutagen would happen if exposure occurred at the zygote stage because all cells would inherit any mutations. The least effect would occur at the adult stage, when affected cells would leave few or no offspring.

Figure 8.3: There are many possibilities.

Figure 8.4: A nonsense mutation (c) is the most likely type of mutation because most random errors would result in a misspelling that is not a "word."

Figure 8.7: Blood group gene from his dad and eye color gene from his dad; blood group gene from his mom and eye color gene from his mom.

Figure 8.9: No, they are nonidentical siblings because the genes in one egg are different from the genes in the other, and each is fertilized by a genetically unique pollen grain.

Figure 8.12: It should decrease the likelihood of having a child who carried two mutant alleles.

Figure 8.15: Red.

Figure 8.18: 3, 2.

Figure 8.19: One X chromosome would be genetically identical to the X chromosome the woman in the figure received in the egg cell her mother ovulated. The woman in the figure would have inherited the other X chromosome from her father when his sperm fertilized her mother's egg cell.

Figure 8.21: A cross involving an ALD-affected female.

Figure 8.22:

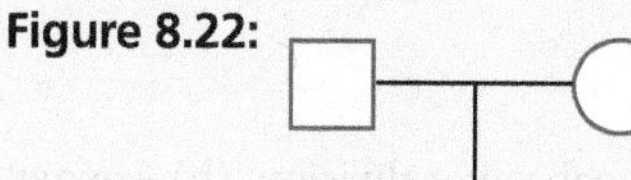

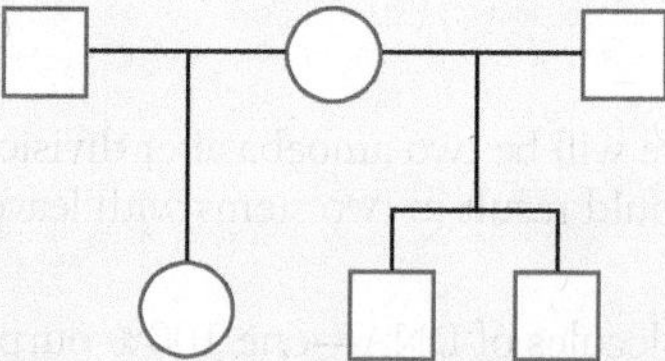

WORKING WITH DATA

Figure 8.13: 50%.

Figure 8.14: 1/16.

SOUNDS RIGHT, BUT IS IT?

1. More likely to be determined by many genes.
2. Yes.
3. Yes.
4. It is likely that genes from both parents influence eye color and shape.
5. Harry's eyes are the product of genetic information from both his mother and his father as expressed during his development. It is nearly impossible that he has the same genetic information as his mother for this trait—so although his eyes may be a similar color, they cannot be identical. (And if you are a Harry Potter fan, you know this is true—Lily Potter did not wear glasses, but Harry is dependent on his!)

SHOW YOU KNOW

8.1: A mutation in a promoter portion of the DNA may affect whether the gene is copied or translated. A mutation in a structural portion of a chromosome may weaken the chromosome or change its ability to be transcribed. A mutation in "nonsense" DNA is likely to have no effect.

8.2: Eight different gametes. The relationship is 2^n, where n is the number of chromosome pairs.

8.3: I^CI^C type C; I^AI^C type AC; I^BI^C type BC; I^CI type C; along with previously described AB types.

8.4: Her father must have the disease, and her mother is a carrier or has the disease.

LEARNING THE BASICS

1. Genotype refers to the alleles you possess for a particular gene or set of genes. Phenotype is the physical trait itself, which may be influenced by genotype and environmental factors.
2.

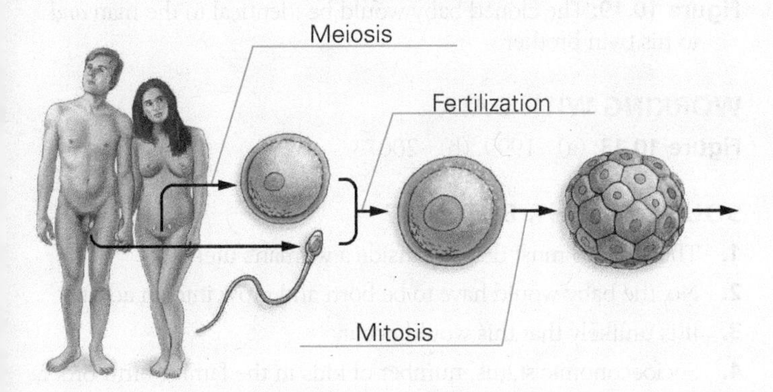

3. B; **4.** A; **5.** B; **6.** A; **7.** B; **8.** C; **9.** D; **10.** E

ANALYZING AND APPLYING THE BASICS

1. The trait could be recessive, and both parents could be heterozygotes. If the trait is coded for by a single gene with two alleles, one would expect 25% of the children (one of four) to have blue eyes. That 50% (two of four) have blue eyes does not refute Mendel because the probability of any genotype is independent for each child and is not dependent on the genotype of his or her siblings. (Consider that some families may have four boys and no girls; this does not refute the idea that the Y chromosome is carried by only 50% of a man's sperm cells.)
2. These two effects (demonstrating pleiotropy) arise because CF causes the production of sticky mucus. Mucus is important in lubricating both lungs and the digestive tract, so improper mucus production affects the function of both organ systems.
3. Pedigrees can vary, but the first cousins must share great grandparents. If one of the great grandparents carries a rare recessive allele, it can be passed down both sides of the pedigree and each of the first cousins can carry that recessive allele. If the first cousins mate, their child can have the disease encoded by the rare recessive alleles.

Chapter 9

GOT IT?

p. 169: 1. quantitative; **2.** highest; **3.** variance; **4.** polygenic; **5.** environment

p. 176: 1. genetic; **2.** correlation; **3.** dizygotic; **4.** environment; **5.** expression

p. 178: 1. different; **2.** tandem; **3.** polymerase chain reaction (or PCR); **4.** polymerase; **5.** gel electrophoresis

VISUALIZE THIS

Figure 9.2: Between two and four. The child could receive zero or one dominant allele from the dad and two or three from the mom.

Figure 9.8: Monozygotic twins are more likely to be conjoined than dizygotic twins, because the genetic similarity of their cells can promote fusion of the embryos.

Figure 9.9: They should revert to the "normal" size seen at the top of the figure.

WORKING WITH DATA

Figure 9.4: (a) No, the majority of men were not 5 feet, 10 inches. (b) There is a wide range of heights in this population.

Figure 9.6: Points would be more widely scattered, and the correlation line would be flatter.

SOUNDS RIGHT, BUT IS IT?

1. STRs.
2. 13.
3. To calculate this probability, multiply 10 times itself 13 times, i.e., 10^{13}. Therefore the probability is one in ten trillion!
4. Ten trillion is larger than 7.4 billion.
5. It is not possible for two individuals (other than identical twins) to happen to have the same DNA fingerprint. There are fewer people on Earth than there are possible DNA profiles produced using 13 STRs.

SHOW YOU KNOW

9.1: Immigration from countries where average height is shorter may be changing the genetics of height in the United States such that there are a larger number of "small" alleles today than in the past. Dietary changes may also be leading to reduced growth or early puberty, which stunts growth in height.

9.2: Even a trait that is highly heritable can be strongly influenced by the environment. If the current environment results in some individuals having very low IQ scores relative to others, it is certainly possible that a changed environment would lead to smaller differences among individuals. See the case of the "maze-dull" and "maze-bright" rats in Table 9.1.

9.3: Nonidentical twin siblings should share, on average, 50% of their genes, so they would share four STRs and four bands on a DNA profile.

LEARNING THE BASICS

1. No. *RrYy* is an example of two genes (seed shape and seed color) coding for two traits. Polygenic inheritance involves two (or more) genes coding for one trait.
2. Quantitative variation in a population can be generated by multiple genes affecting the phenotype for a trait and/or environment influences on gene expression.
3. Every band in the child's profile is also found in either the child's mother or father.

4. E; **5.** D; **6.** D; **7.** B; **8.** C

9.

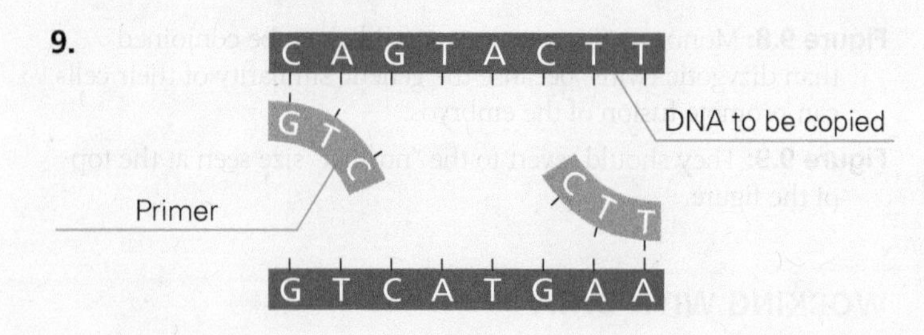

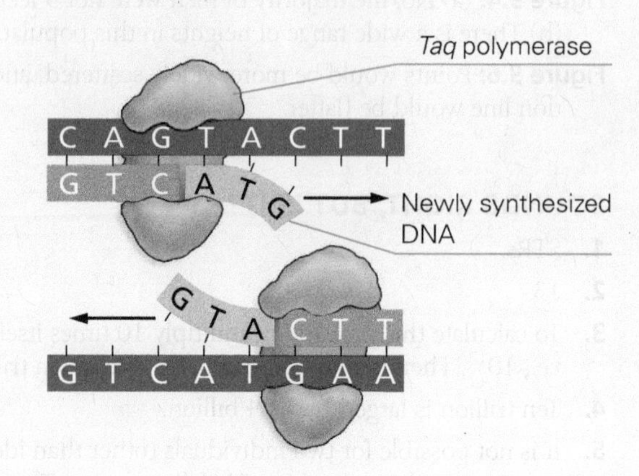

10. B

ANALYZING AND APPLYING THE BASICS

1. Blood typing analysis is not a positive identification. It can only be used to exclude someone. Assume type B blood is found at a crime scene. If the suspect has type A blood, we can exclude them. If a suspect has type B blood, it does not mean the suspect was at the crime scene, only that he or she could have been. DNA evidence positively identifies whether the suspect was at the crime scene.
2. No, heritability doesn't explain why any two individuals differ. For instance, John and Jerry could be identical twins, but Jerry might have suffered a head injury that reduced his IQ.
3. The twin sisters would have all the same bands. Each of these bands must also be present in one parent or the other.

Chapter 10

GOT IT?

p. 191: 1. RNA polymerase; **2.** messenger RNA (or mRNA); **3.** translation; **4.** ribosomes; **5.** amino acids

p. 194: 1. copies; **2.** restriction; **3.** plasmid; **4.** translation; **5.** protein

p. 198: 1. artificial selection; **2.** transgenic; **3.** gene gun; **4.** proteins; **5.** edits

p. 200: 1. undifferentiated; **2.** therapeutic; **3.** reproductive; **4.** Gene; **5.** nucleus

VISUALIZE THIS

Figure 10.1: The second carbon of the sugar, ribose, is bonded to –OH, whereas the same carbon in deoxyribose is attached to hydrogen. There is a CH3 group in thymine that is not present in uracil.

Figure 10.2: 12/4 = 3

Figure 10.3: Sample answer starting after the fork in the DNA: AGCTGGGCAGGTAC. From this particular DNA, the mRNA would be UCGACCCGUCCAUG.

Figure 10.5: Adenine paired with uracil and cytosine paired with guanine.

Figure 10.19: The cloned baby would be identical to the man *and* to his twin brother.

WORKING WITH DATA

Figure 10.13: (a) ~1999. (b) ~2007.

SOUNDS RIGHT, BUT IS IT?

1. The embryo must develop inside a woman's uterus.
2. No, the baby would have to be born and grow into an adult.
3. It is unlikely that this would occur.
4. Socioeconomic status, number of kids in the family, birth order, exposures to different activities, coaching, etc.
5. There are so many environmental exposures that would differ, even between cloned humans, that there is no guarantee that clones would have similar behavioral traits.

SHOW YOU KNOW

10.1: The stop codons are UAA, UAG, and UGA. Therefore, the DNA sequences coding for these would be ATT, ATC, and ACT.

10.2: Insulin.

10.3: No. A transgenic organism contains genetic information from an organism of a different species. CRISPR-edited organisms have had their own genome altered.

10.4: A defective gene must be fixed in the fertilized egg cell to be fixed in every cell of the resulting adult organism.

LEARNING THE BASICS

1. GCUAAUGAAU **2.** gln, arg, ile, leu **3.** B; **4.** C; **5.** B; **6.** C; **7.** C; **8.** B; **9.** D

10.

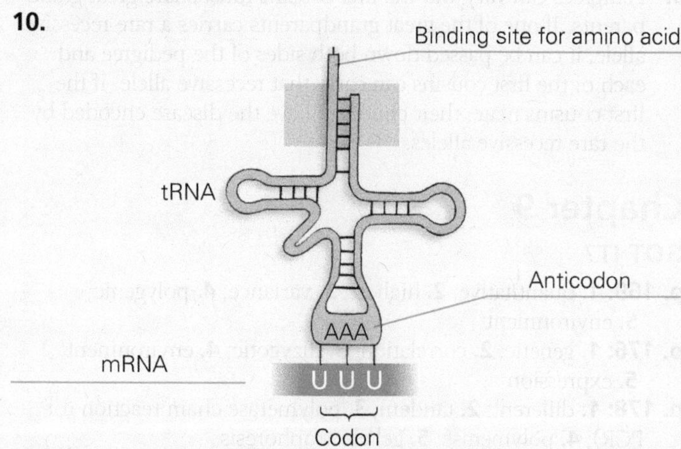

ANALYZING AND APPLYING THE BASICS

1. The same amino acid is often coded for by codons that differ in the third position. For instance, UUU and UUC both code for phenylalanine.
2. Transcription would be slowed or stopped in that cell.
3. A box should be drawn around: GAATTC; CTTAAG

Chapter 11

GOT IT?

p. 210: 1. generations; **2.** Natural selection; **3.** micro; **4.** theory; **5.** common ancestor

p. 215: 1. lifetime; **2.** common; **3.** evidence; **4.** supernatural; **5.** hypotheses

p. 227: 1. evolutionary; **2.** vestigial; **3.** anatomy, development, DNA; **4.** close; **5.** ape

p. 232: 1. relationships; **2.** radiometric **3.** DNA; **4.** DNA; **5.** theory

VISUALIZE THIS

Figure 11.1: All lice would have been eliminated by the pesticide, and the population would not have evolved.

Figure 11.2: These branches represent organisms that are extinct.

Figure 11.7: Perhaps the presence of plant-eating tortoises on the islands made those cacti with longer stems (thus out of reach of the tortoises) more likely to survive, leading to the evolution of longer stems. There also may be a lack of other trees on the islands, meaning that taller cacti were able to capture more sunlight, and thus reproduce more than smaller plants.

Figure 11.10: Hair-covered bodies, birth to live young, other characteristics of all mammals (also forward-facing eyes, and opposable thumbs).

Figure 11.11: Squirrel would be the farthest left branch, originating above the mammal ancestor. The shared ancestor is the junction point.

Figure 11.12: The green bones are longer in the bat and sea lion and function to make the limb paddle-shaped. In lions the green bones are very short and function mostly as a stable foot pad. In chimpanzees and humans, the bones are medium length and allow for fine movements of individual digits.

Figure 11.19: The minerals cannot accumulate fast enough to fill in the spaces left by the decaying tissue, so the fossil loses structure.

WORKING WITH DATA

Figure 11.16: The black turnstone shares a more recent common ancestor with the red knot than the Caspian tern does, as evidenced by greater similarity in morphology, but more convincingly, DNA sequence.

Figure 11.17: The percentages indicate the differences between humans and these two species. An analogy might help: Coffee might be 99.01% similar to cola and 89.90% similar to tea, but this does not mean cola and tea are more similar to each other than either is to coffee.

Figure 11.21: About 3 million years old.

Figure 11.23: About 2 million years ago.

SOUNDS, RIGHT, BUT IS IT?

1. Yes.

2. Yes.

3. No. Most importantly, the environment in which humans evolved did not have any humans.

4. Probably not. Humans evolved in an environment where there was an advantage to being bipedal. Given the niche of chimpanzees as forest dwellers who rely on being on the ground and in the tree canopy, it is unlikely that traits that make one individual more bipedal would provide an advantage.

5. That organisms share a common ancestor does not mean that one organism is "more evolved" than another—they are two lines of descent. In addition, the conditions under which hominins first evolved no longer exist, so it is unreasonable to assume that modern apes "should" be evolving to be more human-like.

SHOW YOU KNOW

11.1: This is very unlikely to be an evolutionary change, which results from genetic changes in a population that are inherited. The teenagers in 2008 are unlikely to have a significant genetic difference from those in 1982, instead access to (and interest in) orthodontic treatment has increased.

11.2: The theory of natural selection.

11.3: It may be that the human diet was harder on teeth in the past, so that teeth were lost due to breakage, leaving "room" for the wisdom teeth to erupt. These teeth may have been beneficial, in fact, replacing the role of missing molars.

11.4: Nucleus, ribosomes, cytoskeleton, plasma membrane, endoplasmic reticulum, Golgi apparatus, glycolysis, aerobic respiration, mitotic division (others possible).

LEARNING THE BASICS

1. B

2.

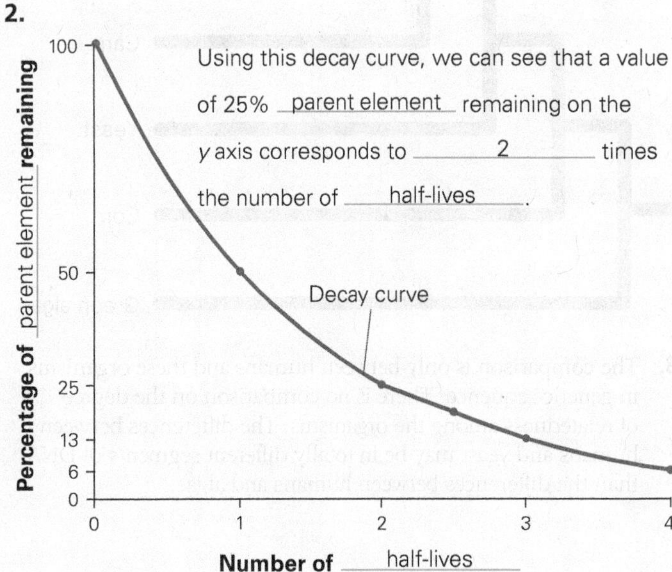

3. D; **4.** D; **5.** B; **6.** E; **7.** B; **8.** A; **9.** D; **10.** D

ANALYZING AND APPLYING THE BASICS

1.

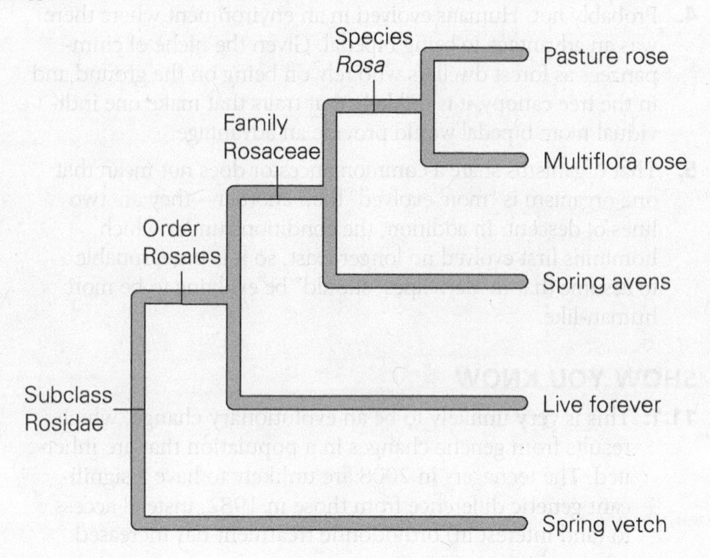

2.

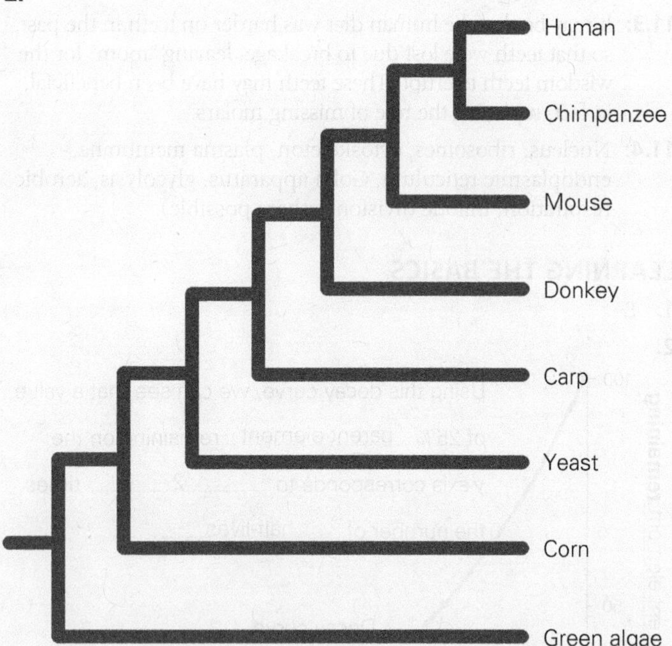

3. The comparison is only between humans and these organisms in genetic sequence. There is no comparison on the degree of relatedness among the organisms. The differences between humans and yeast may be in totally different segments of DNA than the differences between humans and algae.

Chapter 12

GOT IT?

p. 241: 1. bacterium; **2.** lungs; **3.** cough, sneeze; **4.** antibiotics; **5.** antibiotic-resistant

p. 248: 1. vary, passed; **2.** survival, reproduction; **3.** fitness; **4.** artificial selection; **5.** vary

p. 252: 1. genes (or alleles); **2.** alleles; **3.** variations (or alleles); **4.** no; **5.** multiple

p. 256: 1. vary; **2.** fitness; **3.** stop; **4.** multiple; **5.** trade-off

VISUALIZE THIS

Figure 12.6: If other flowers farther away are blooming, it may be an advantage to be able to pick up more distantly related pollen so that the plant's offspring carry novel alleles that might be advantageous.

Figure 12.8: Because the elephants are using most of their resources in Generation 3, a percentage of Generation 4 will die from starvation.

Figure 12.11: The population would change in the direction of demonstrating that behavior under the right conditions.

Figure 12.14: The individual with the fast metabolism of alcohol would not have proportionally more offspring than those without the allele, so the ratio of fast to slow metabolizers would stay the same on each "shelf."

WORKING WITH DATA

Figure 12.9: It decreased by a large amount.

Figure 12.12:

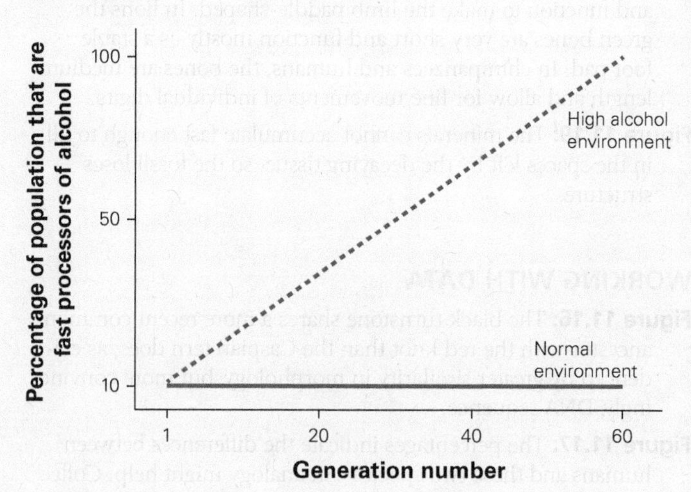

Figure 12.15: The fitness of the red-flowered plants would be lower, so the population as a whole would move toward the white direction as a result of directional selection.

SOUNDS, RIGHT, BUT IS IT?

1. Genetic mutation.
2. 10,000 × 3 = 30,000.
3. No, natural selection sorts out variations—those changes that lead to greater survival and reproduction are passed on and become more common, those that reduce survival and reproduction are lost.
4. The neck ruff evolved from normal neck feathers.
5. No.
6. Yes.
7. The number of heritable variations that can occur in a population is large. If any of these variants lead to greater survival and reproduction, they will become common in the population. Over time, structures can become more and more modified and complex as an accumulation of changes occurs, even taking on new structures and functions.

SHOW YOU KNOW

12.1: The bacteria can continue to multiply, doing damage and resulting in a larger population. More cells mean more likelihood of resistant variants, as well as a more difficult time clearing out the infection.

12.2: Humans have invented technologies and disease cures that decrease the differences in fitness among individuals with different genetics (for example, corrective lenses have increased the survival of individuals with poor eyesight). We are still subject to natural selection for traits that are less amenable to technological fixes and even in our ability to respond to technology.

12.3: The cystic fibrosis allele may provide protection from diseases that are common in more urban environments. Individuals carrying the allele in these environments were more likely to survive, thus causing the allele to become more common in those populations.

12.4: Mutations appear by chance and may not have much effect on fitness unless the environment changes. Resistance alleles regularly appear randomly, but are only selected for in the environment where they have an advantage.

LEARNING THE BASICS

1. Antibiotics have helped reduce deaths due to tuberculosis, but they have become less effective as strains of antibiotic-resistant *M. tuberculosis* have appeared in human populations.
2. Artificial selection is a process of selection of plants and animals by humans who control the survival and reproduction of members of a population to increase the frequency of human-preferred traits. Artificial selection is like natural selection in that it causes evolution; however, it differs because humans are directly choosing which organisms reproduce. In natural selection, environmental conditions cause one variant to have higher fitness than other variants.
3. C
4. **Single drug therapy**

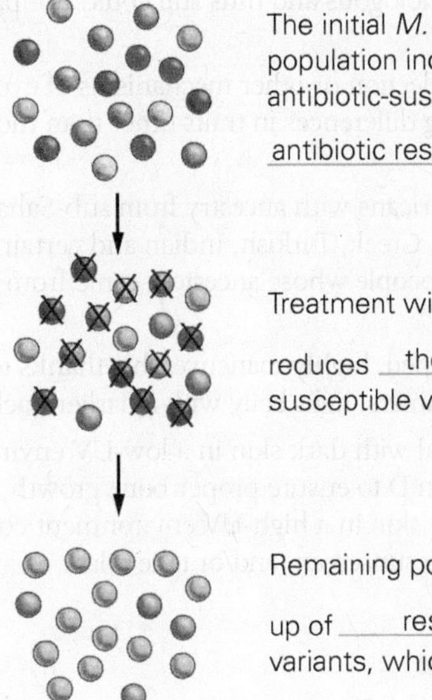

The initial *M. tuberculosis* population includes both antibiotic-susceptible and antibiotic resistant variants.
Treatment with one antibiotic reduces the number of susceptible variants.
Remaining population is made up of resistant variants, which then proliferate.

5. D; **6.** E; **7.** E; **8.** B; **9.** C; **10.** D

ANALYZING AND APPLYING THE BASICS

1. Farmers who desired larger and tastier strawberries must have saved seeds from the wild plants that produced the largest berries. Over the course of many generations of the same type of selection in domesticated strawberries, average berry size and sweetness increased dramatically. Trade-offs include the following: domestic strawberries need more energy and nutrients to devote to fruit versus other plant parts, and perhaps domesticated strawberries have fewer chemical defenses against insects and other predators.
2. Individual zebras with some sort of striping must have had greater survival than those who were not striped. Over the course of many generations of this type of selection, individuals in the population evolved dramatic stripes.
3. No, not all features are adaptations; some might arise by chance. A trait is an adaptation if it increases fitness relative to others without the trait. One can look for ways that the trait increases fitness or examine individuals who lack or have a less dramatic version of the trait to measure its fitness effects.

Chapter 13

GOT IT?

p. 270: 1. genus; **2.** interbreed, species; **3.** pool; **4.** isolated (or separated); **5.** reproductively

p. 278: 1. common; **2.** morphological; **3.** unique; **4.** frequencies; **5.** isolated, mixing (or interbreeding)

p. 285: 1. convergent; **2.** chance; **3.** small; **4.** mates; **5.** similar

VISUALIZE THIS

Figure 13.2: No, because the chromosomes from the two different parents are not homologous and thus still could not pair properly.

Figure 13.3: No, natural selection or other mechanisms of evolution could be creating differences in traits other than those related to appearance.

Figure 13.13: African Americans with ancestry from sub-Saharan Africa as well as Italian, Greek, Turkish, Indian and certain Arab American groups, and people whose ancestors came from the Balkans.

Figure 13.15: Torpedo-shaped, highly maneuverable thanks to their flippers, fast swimmers, light belly with a darker back.

Figure 13.17: An individual with dark skin in a low-UV environment could take vitamin D to ensure proper bone growth, and an individual with light skin in a high-UV environment could avoid sun exposure, use sunscreen, and/or take folate supplements.

WORKING WITH DATA

Figure 13.5: The Honeycrisp might slow divergence because newly emerged hawthorn flies may be able to mix with later-emerging apple flies on this later-flowering tree.

Figure 13.9: 6 out of 20 or 0.3.

Figure 13.12: The frequency of type O blood would likely increase with distance from Asia.

Figure 13.16: Various possibilities as visualized on the graph (for the first question, dots of the same color that are distant from each other on the *y* axis; for the second question, dots of different color that are in close proximity on the *y* axis).

SOUNDS, RIGHT, BUT IS IT?

1. Yes, as with all physical traits, it is likely variation existed for this trait.
2. Yes, as with most behavioral traits, it is likely variation existed in foraging behavior.
3. No, only those who had the alleles that coded for different bill shapes or different behaviors.
4. Yes, these traits could improve survival and reproduction of the individuals who possess them.
5. Individuals with higher evolutionary fitness leave more offspring. Thus, the traits they possess become more common in the population over time.
6. Saying the birds evolved "because they needed to" suggests a purposeful striving toward a goal. Instead, the evolution of nectar-feeding honeycreepers occurred as a result of natural selection acting on the variations present in the population. Those individuals with longer, curved bills—traits that happened to increase fitness—left more offspring, thus a portion of the population changed over time to consist of more individuals with longer, curved bills who fed on particular flowers.

SHOW YOU KNOW

13.1: No, because the varieties are the same species, so the number of chromosomes should be the same in both parents.

13.2: They should look for evidence of unique alleles that appear in one race of the birds and not the other, or look for patterns of allele frequency for several genes that are similar among subpopulations in the same proposed race, but different when comparing races with one another.

13.3: "Ma" or "mmmm" is one of the first sounds an infant can make. Because the mother is so associated with infant care, this sound became associated with her. ("Da" or "Pa" is a very early sound as well.)

LEARNING THE BASICS

1. A biological species is a group of individuals that can interbreed and produce fertile offspring. Biological species are reproductively isolated from each other, thus separating their gene pools.
2.

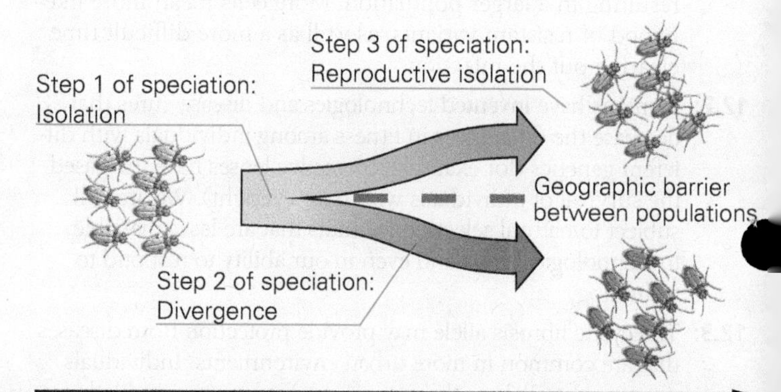

3. Genetic drift can occur as a result of founder effect (a small subpopulation moves away and contains an unrepresentative sample of the original gene pool), the bottleneck effect (the few survivors of a disaster have an unrepresentative sample of the original gene pool), or by the chance loss of alleles (death or lack of reproduction not having to do with the traits coded by the gene) from small, isolated gene pools. By changing allele frequencies, all of these events result in the evolution of a population.
4. D; **5.** B; **6.** A; **7.** C; **8.** C; **9.** D; **10.** C

ANALYZING AND APPLYING THE BASICS

1. It depends on the species definition one uses. The two populations of wolves are not different biological species. They are likely different genealogical species—the definition that corresponds to "race." To determine if they are different races, one would have to look for the standard evidence: unique alleles or unique allele frequencies in each population.

2. The allele frequency could be different as a result of genetic drift, particularly the founder effect, in these different populations. It also could be different because the allele has some yet unknown fitness advantage in the environment of Ireland relative to the environment of Britain or Scandinavia.
3. If the early mutation becomes widespread, the timing difference between the early and normal populations should cause reproductive isolation. The same is possible with the expanded mutation if certain pollinators become associated with the individuals that bloom earlier than expected.

Chapter 14

GOT IT?

p. 297: 1. geologic eras; **2.** 80; **3.** evolutionary; **4.** domains; **5.** kingdom

p. 313: 1. Archaea, Bacteria; **2.** endosymbiotic; **3.** multicellular, animal-like; **4.** flowers; **5.** reproduce

p. 316: 1. evolutionary; **2.** last (or family); **3.** unrelated; **4.** organisms; **5.** DNA

VISUALIZE THIS

Figure 14.7: The second membrane resulted from the infolding of the host cell membrane when it engulfed the bacterium.

Figure 14.12: To allow dispersal of spores away from the food source that is already in use.

Figure 14.16: The embryo is like an astronaut in that it is protected within a tough outer coating that contains all the resources it requires for at least short-term survival in a harsh environment. It's different in many ways, including that it can still exchange materials (for example, gases) with the environment, whereas an astronaut cannot get any materials required from life from space.

WORKING WITH DATA

Figure 14.2: About 84,000 genera have been described of the 110,000 or so predicted to exist.

Figure 14.3: We'd have to draw an extra branch on the tree that shows a common ancestor between animals and this protistan group. The "phylum" of Protista would also be broken up, potentially.

Figure 14.4: They use base pairs that are common to all groups as a framework and then infer that older species are more likely to possess the ancestral bases than younger species.

Figure 14.19: It probably had little head striping, like the junco, but a sparrow-colored body (brown-striped) like all of the sparrows.

SOUNDS, RIGHT, BUT IS IT?

1. Approximately, yes.
2. Flight, metamorphosis, ability to sense different chemicals or light wavelengths than we can, etc.
3. Sea star.

4. No, the tree would still indicate the same evolutionary relationship among its members—humans and starfish share a recent common ancestor, humans and insects a less recent ancestor.
5. Humans are the product of millions of years of evolution from the first animal—as are all modern animals. Although humans have many complex structures and traits, other animals have other, different complex structures and traits. An evolutionary tree can be drawn in many different ways to indicate the same relationships without implying a hierarchy—in this tree, humans do not have to be on the very top.

SHOW YOU KNOW

14.1: Grouping species by similarity in diets could help biologists understand the common needs for species of interest and perhaps simplify natural resource management. Grouping species by similarity in habitat need can help identify whether species with similar habitat needs may be at risk of extinction due to habitat loss.

14.2: Mitochondria in plant and animal cells are identical, indicating that they evolved before plants and animals diverged.

14.3: A beak that can tear meat, long wings, large size, and the ability to soar. DNA sequence analysis, detailed morphological comparisons, the fossil record.

LEARNING THE BASICS

1. Scientists have identified more than 1.3 million different species, and possibly more than 100 million different species might exist on Earth today.
2.

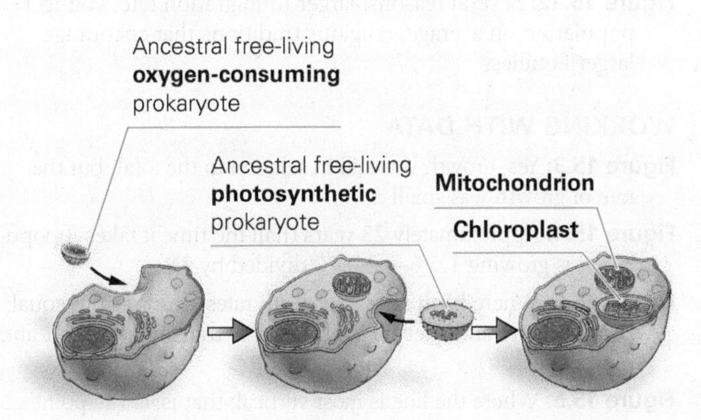

3. By comparing the DNA sequences of organisms to see if the pattern of similarity and difference matches what is predicted, and by examination of the fossil record, looking for the record of evolutionary change in the group of organisms.
4. B; **5.** E; **6.** E; **7.** D; **8.** C; **9.** D; **10.** B

ANALYZING AND APPLYING THE BASICS

1. An oxygen-sensitive cell that kept an oxygen-consuming prokaryote alive inside of it would be protected from the damaging effects of oxygen, as any nearby oxygen would quickly be "used up" by the prokaryote. This would be an advantage over cells that did not have this prokaryote.
2. Although it is not very motile, this is probably an animal because it is not photosynthetic and, unlike fungi, consumes other organisms by ingestion.
3. These chloroplasts likely evolved through a process of secondary endosymbiosis, in which the ancestral Euglena cell engulfed but did not consume a eukaryotic green algal cell. In this case the original bacterial membrane is surrounded by the secondary membrane from the engulfed cell and both of these are surrounded by the vacuole membrane from the ancestral Euglenid.

Chapter 15

GOT IT?

p. 328: **1.** percentage (or proportion); **2.** clumped; **3.** total; **4.** births, population; **5.** birth, death

p. 332: **1.** maximum; **2.** dependent; **3.** carrying capacity; **4.** fewer; **5.** net primary

p. 336: **1.** crash; **2.** high; **3.** equal; **4.** reproductive age; **5.** women

VISUALIZE THIS

Figure 15.1: Trap-happiness would cause the population estimate to be too small because a greater proportion of returnees will be marked. Trap-shyness causes the opposite effect because fewer returnees will be marked.

Figure 15.12: Several reasons: larger immigration rate, younger population on average, religious traditions that encourage larger families.

WORKING WITH DATA

Figure 15.3: Yes, growth was still in relation to the total, but the rate of growth was small.

Figure 15.4: Approximately 23 years (half the time it takes if population is growing 1.5%—also 70 divided by 3).

Figure 15.5: Where birth rates and death rates are closest to equal; that is, before and after the demographic transition (areas 1 and 3 on the graph).

Figure 15.6: Where the line is most vertical; that is, at the point where the "S" turns from rising to flattening out. Here at a population size of around 450.

Figure 15.8: In the low-growth scenario, where the graph of population size begins to level out.

Figure 15.10: The same basic pattern of population fluctuation would occur, just at the higher carrying capacity.

Figure 15.11: The United States has about one-quarter the population of India.

SOUNDS, RIGHT, BUT IS IT?

1. Growth that occurs in proportion to the current total.
2. (b), the proposal that starts at 1 cent per day.
3. $30,000.
4. $167,772.16.
5. Because the wage growth is always in relation to the previous day's wage, the number of pennies earned rises very quickly, eventually far exceeding the constant daily wage. Exponential growth causes both populations, and earnings, to rise much more rapidly than simple geometric growth. This principle can be applied to many situations, including the exponential growth of an algae bloom in a swimming pool, an invading species in a new area, or the growth of a retirement savings account.

SHOW YOU KNOW

15.1: Average life span equals the sum of the ages at death divided by the number of deaths. Deaths of infants thus greatly bring down the average life span, and reducing these deaths will raise life expectancy dramatically, even if the population health has not changed otherwise.

15.2: Infectious diseases require a population of susceptible individuals to survive and spread. If the population is small, the infectious organism cannot maintain its own population because it cannot spread from host to host quickly enough.

15.3: A large population with a high birth rate produces a large number of offspring; as these offspring have their own offspring, they can potentially overshoot the carrying capacity. A large population with a low birth rate produces many fewer young, so even when the offspring reproduce, they don't add huge numbers to the population, thus allowing a gradual approach to the carrying capacity.

LEARNING THE BASICS

1.

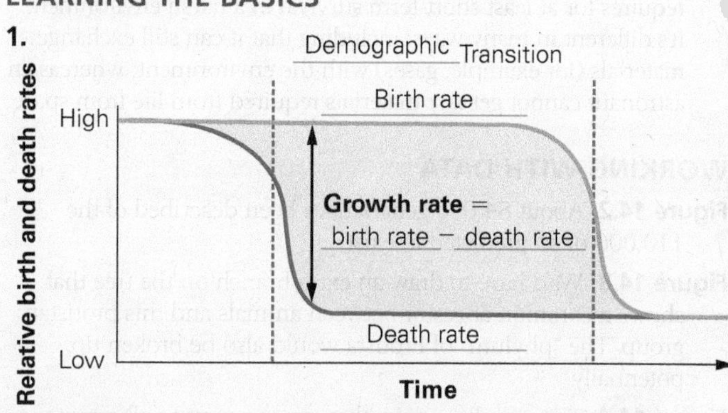

2. In most populations, growth rate declines because death rate increases (or birth rate decreases) as resources are "used up" by the population.
3. B; **4.** C; **5.** B; **6.** A; **7.** B; **8.** D; **9.** D; **10.** D

ANALYZING AND APPLYING THE BASICS

1. 2500. Solve for x: $1/50 = 50/x$. (Review Figure 15.1.)
2. −0.002 (r = Birth rate − Death rate = 25/1000 − 27/1000). If this is a nonhuman population, it is likely near carrying capacity because it is not growing, but in fact declining slightly, as would be expected if resources are limited.
3. The carrying capacity is expected to rise since more resources are available. However, because the flies are in a restricted environment (a culture bottle), the accumulation of waste may become a problem, and a large population of flies may not be able to survive this accumulation. Depending on fly behavior, there may also be an increase in aggressive interactions or a decrease in mating in more dense situations.

Chapter 16

GOT IT?

p. 349: 1. mass extinction; **2.** natural area (or habitat); **3.** consumers; **4.** coevolved; **5.** oxygen

p. 360: 1. job (or role); **2.** mutualism; **3.** resource; **4.** many (or several); **5.** biomass

p. 368: 1. habitat destruction; **2.** high; **3.** alleles; **4.** deleterious; **5.** adapting

VISUALIZE THIS

Figure 16.11: If baleen whales disappeared, krill populations would probably rise, leading to higher populations of fish and other carnivorous zooplankton that compete with whales for krill. However, it might lead to a decrease in smaller toothed whales, which prey on the baleen whales.

Figure 16.15: The total number of bacterial cells reaches the carrying capacity of the environment (that is, the intestines) in both cases—it is just the makeup of the bacterial community that changes.

Figure 16.17: In human-made systems, nutrients typically flow in one direction (from crops to humans to waste water and cemeteries), so we can no longer refer to the process as a nutrient cycle.

Figure 16.20: More are closer to the equator.

Figure 16.21: Condors have fewer offspring per pair per year and require longer development time with parental care.

Figure 16.23: Because the normal allele does not cause disease, homozygotes for the normal allele are healthy. This is not the case for individuals who are homozygous for the mutant allele, who are ill.

WORKING WITH DATA

Figure 16.2: 10 to 50 million years

Figure 16.3: 1 million years

Figure 16.6: You would expect to find approximately 80 species on an island of this size. If the island lost 50% of its habitat (7,500 of 15,000 square kilometers), you'd expect about 10% of the species to become extinct, so around 72 should remain.

Figure 16.7: About 1000 mountain lions.

Figure 16.24: About 50% is lost.

SOUNDS, RIGHT, BUT IS IT?

1. (a) Mosquitos are predators of microbes; (b) dragonflies are predators of mosquitos; (c) mosquitos and the plants are mutualists; and (d) mosquitos are predators (parasites) of humans.
2. Bacteria, algae, and other microbes in water would increase, dragonflies, some fish, and crustaceans would decrease, plants pollinated by mosquitos would decrease, and humans may increase.
3. Example answers: more microbes in water may lead to a population increase of another predator, or may lead to increased risk of disease of organisms that are hosts to these bacteria; decrease in dragonflies could negatively affect bird species that eat them, etc.
4. A keystone species is one that has a disproportionately large effect on other species in its food web. In other words, removing a keystone species can cause the rest of the system to radically change.
5. We do not know. It can be impossible to determine until a species is removed from a system how important it is to that system.
6. Mosquitos are connected to many other species and their loss would affect these species in positive or negative ways. Whether these changes would in turn negatively affect humans is nearly impossible to predict given how complicated food webs are.

SHOW YOU KNOW

16.1: Habitat destruction reducing the area where amphibians can survive, pollution interfering with growth and reproduction, introduction of exotic species that prey on or compete with native amphibian species, and even overexploitation of colorful or dramatic species for the aquarium trade.

16.2: (a) Coyotes and foxes are likely to compete over prey animals, such as rodents and rabbits; (b) hummingbirds and bees are likely to compete over flower nectar; (c) flying squirrels and owls are likely to compete for tree cavities to nest in. You could test your hypothesis by looking at population sizes when the species are found together and when they are found apart, or by introducing (or removing) one member of the pair and seeing what happens to the remaining species.

16.3: If the passenger pigeon was highly social and did not mate, hunt, or successfully produce young without the presence of a large group, the "effective population size" was likely to be much smaller than the actual population size. Additionally, the birds may have been highly choosy of mates and thus not inclined to reproduce in small groups.

LEARNING THE BASICS

1.

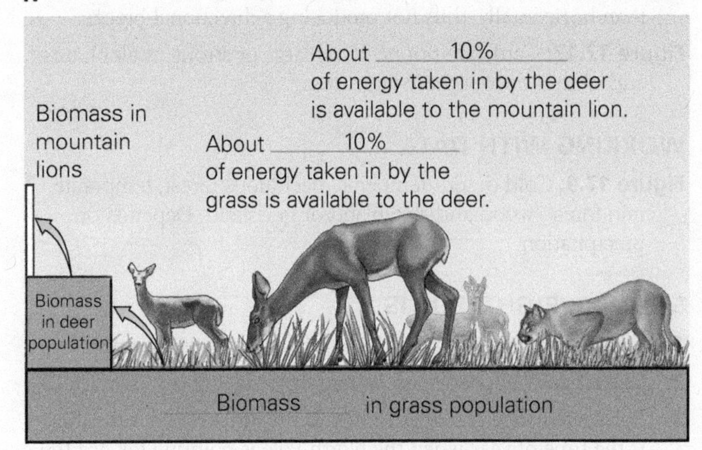
About 10% of energy taken in by the deer is available to the mountain lion.
Biomass in mountain lions
About 10% of energy taken in by the grass is available to the deer.
Biomass in deer population
Biomass in grass population

2. Mutualism is a relationship among species in which all partners benefit. Predation is a relationship among species in which one benefits and others are consumed. Competition is a relationship among species in which all partners are harmed by the presence of the others.

3. D; **4.** D; **5.** A; **6.** A; **7.** B; **8.** C; **9.** D; **10.** A; **11.** D; **12.** D

ANALYZING AND APPLYING THE BASICS

1. Kelp → sea urchins → sea otters. Kelp also provides habitat for fishes, clams, and abalones. When sea otters are removed, sea urchins devastate the kelp forest, leading to the loss of these associated species. Sea otter thus serves as a keystone.
2. Even though there are more individuals in situation A, with so few fathers contributing genes to the next generation, the likelihood of large changes in allele frequency due to drift (i.e., the founder effect) is much higher in situation A than in situation B.
3. How uncertain are the environmental conditions—that is, do density-independent factors cause the death of large numbers of individuals when they occur? What is the mating system—are a few males mating with the majority of the females, or is the species more monogamous? What is the likely carrying capacity for piping plovers in this environment?

Chapter 17

GOT IT?

p. 380: 1. today, climate; **2.** seasons; **3.** equator; **4.** warmer, cooler; **5.** dry

p. 388: 1. water; **2.** concentrated; **3.** organic matter; **4.** water; **5.** base

p. 394: 1. fertilizer, waste; **2.** decline; **3.** organisms; **4.** fluctuations; **5.** energy, beef

VISUALIZE THIS

Figure 17.2: Solar irradiance is higher in Florida compared with Vermont because it is closer to the equator. Thus Florida is warmer than Vermont.

Figure 17.3: Because when the Northern Hemisphere is tilted away from the sun, the Southern Hemisphere (where Australia is located) is tilted toward it.

Figure 17.5: Near 30° north and south and near the equator, where air is more likely to be moving vertically (up or down) rather than horizontally, thus not producing a directional breeze.

Figure 17.17: Cattle are not nearly as large or strong as elephants and cannot uproot these large trees.

WORKING WITH DATA

Figure 17.9: Cold desert, temperate deciduous forest, temperate rain forest, woodland/shrubland/or grassland. Depends on precipitation.

SOUNDS, RIGHT, BUT IS IT?

1. Midwest: January is winter. Australia: January is summer.
2. The amount of energy per square meter falling on a surface.
3. Solar irradiance is higher in July in the Midwest because this is the time of year when the North Pole is pointing toward the sun—in January, the pole points away, so solar irradiance is lower. In Australia during July, the nearest pole (the South Pole) is facing away from the sun, so solar irradiance is lower during July compared with January in the same location.
4. No.
5. Australian summer is our winter—if summer seasonality was determined by distance to the sun, both the Midwest and Australia would have summer and winter at the same time. Instead, the average temperature in an area depends on solar irradiance. The fact that Earth's axis is tilted is more important than distance to the sun in determining seasonal differences in temperatures.

SHOW YOU KNOW

17.1: A reduction in tilt would reduce seasonality because areas would be receiving the same solar irradiance throughout the year.

17.2: Chlorophyll may be "expensive" for the trees to make in some way. The most likely expense is nitrogen, which is limited in the soil and may not be easy to replace if simply dropped to the soil.

17.3: Wetlands slow water flow and thus can absorb extra nutrients before they accumulate in the open water body.

LEARNING THE BASICS

1. The tilt of Earth's axis means that as the planet revolves around the sun, the Northern Hemisphere is tilted toward the sun during part of the year and away during part of the year. The temperature at a given point on Earth's surface is determined in large part by the solar irradiance (i.e., the strength of sunlight) striking the surface. Because the Northern Hemisphere is tilted away in the winter, solar irradiance is low, as are temperatures. The opposite occurs in the summer.
2. Lower sun angle = lower average temperature and more seasonality

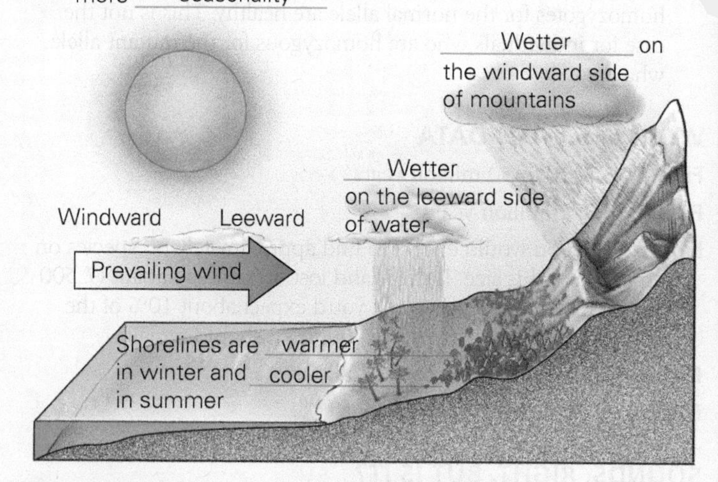
Wetter on the windward side of mountains
Wetter on the leeward side of water
Windward
Leeward
Prevailing wind
Shorelines are warmer in winter and cooler in summer

3. B; **4.** B; **5.** C; **6.** E; **7.** E; **8.** C; **9.** A; **10.** E

ANALYZING AND APPLYING THE BASICS

1. Semitropical with seasonal wetter and drier periods as a result of latitude. Relatively rainy due to location on windward side of mountain range.
2. The temperature will become less moderate because warm water will not be carried to the coasts of Europe, warming nearby landmasses. Seasonality will be more pronounced and winters colder and longer.
3. Bend, Yakima, and perhaps Walla Walla (with lower annual rainfall) are on the leeward side of the Cascades, and the other cities are on the windward side, because more rain is likely on the windward side.

Chapter 18

GOT IT?

p. 407: 1. cells; **2.** epithelial; **3.** Connective; **4.** skeletal; **5.** connective

p. 409: 1. tissues; **2.** organ system; **3.** circulatory, digestive; **4.** glucose; **5.** filter

p. 411: 1. Homeostasis; **2.** inhibits; **3.** pancreas; **4.** liver; **5.** Positive

VISUALIZE THIS

Figure 18.2: The cellular component of both is fibroblasts. The matrices of both contain collagen fibers. Fibrous connective tissue contains more collagen fibers and they are arranged in parallel. Loose connective tissue contains elastin fibers in the matrix also.

Figure 18.3: Actin and myosin.

Figure 18.6: Cardiovascular and urinary are two examples.

Figure 18.8: Damage to the liver or kidney.

WORKING WITH DATA

Figure 18.11: (a) The demand for organs continues to grow, while supply remains fairly constant. (b) Donor organs recovered from the deceased do not include donations by living donors (for organs like liver and kidney) or organs grown or printed in the laboratory.

SOUNDS RIGHT, BUT IS IT?

1. Yes.
2. No.
3. In a coma.
4. No.
5. No.
6. A brain dead person will not recover the ability to survive off life support.

SHOW YOU KNOW

18.1: The lack of blood vessels in cartilage makes it difficult for transplanted tissues to receive the nutrients they need to survive.

18.2: Organism—vehicle; organ system—fuel injection system, organ—engine; tissue—valves and pistons.

18.3: Negative.

LEARNING THE BASICS

1. Epithelia are tightly packed sheets of cells that cover and line organs, vessels, and body cavities. Connective tissues are loosely organized tissues that bind organs and tissues to each other. Muscle tissue consists of bundles of long, thin, cylindrical muscle fibers that are able to contract. Nervous tissue is composed mainly of neurons that can conduct and transmit electrical impulses.
2. Loose connective tissue, adipose tissue, blood, fibrous connective tissue, cartilage, and bone.
3. Muscle tissues differ based on the presence or absence of striations and whether the muscle is voluntary or involuntary. Cardiac muscle is striated and involuntary; skeletal is striated and voluntary; smooth muscle is not striated and involuntary.
4. C; **5.** A; **6.** A; **7.** B; **8.** C; **9.** D
10.

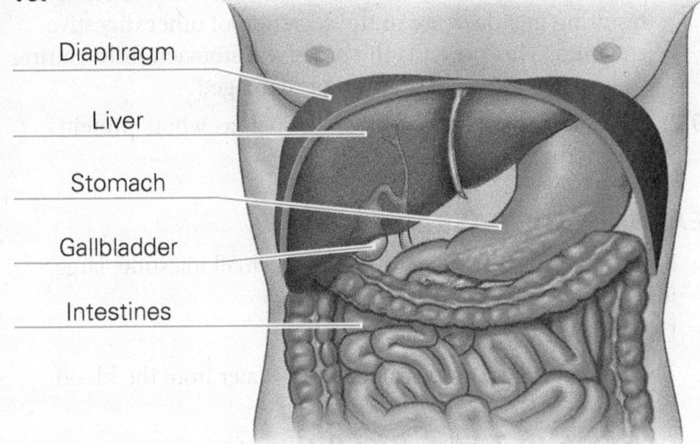

ANALYZING AND APPLYING THE BASICS

1. When muscles contract and expand during shivering, heat energy is produced.
2. It secretes insulin.
3. Answers will vary. An example of negative feedback is regulating home temperature. An example of positive feedback is getting a good grade makes a student work harder.

Chapter 19

GOT IT?

p. 423: 1. pharynx; **2.** stomach; **3.** peristalsis; **4.** Chyme; **5.** small intestine

p. 425: 1. kidneys; **2.** urethra; **3.** nephrons; **4.** renal arteries; **5.** bladder

VISUALIZE THIS

Figure 19.1: 28; third molars are found on both sides of the upper and lower jaws.

Figure 19.5: Bacteria can move from bladder through the ureters to kidney.

Figure 19.6: To the bladder.

WORKING WITH DATA

Table 19.2: 120 lb woman could only have two drinks in 2 hours because three drinks in 2 hours would give her a BAC of 0.8, which is legally drunk. Answers for second question will vary for each student.

SOUNDS RIGHT, BUT IS IT?

1. He won't metabolize those drinks for 2 hours.
2. No.
3. Slapping, pinching, and yelling.

4. Call 911.
5. Because his BAC was not at its peak when he passed out, you cannot know whether he is in need of medical attention.

SHOW YOU KNOW

19.1: An increase in HCl would cause gastrin release to slow, resulting in a decrease in the secretion of other digestive enzymes. This prevents the pH of the stomach from getting so low that the stomach lining is damaged.

19.2: It would contain larger materials such as whole proteins and perhaps some blood cells.

LEARNING THE BASICS

1. Mouth, pharynx, esophagus, stomach, small intestine, large intestine.
2. Liver, pancreas, gallbladder.
3. The kidneys remove waste and excess water from the blood, producing urine.
4. D; **5.** D; **6.** C; **7.** False; **8.** A; **9.** C
10.

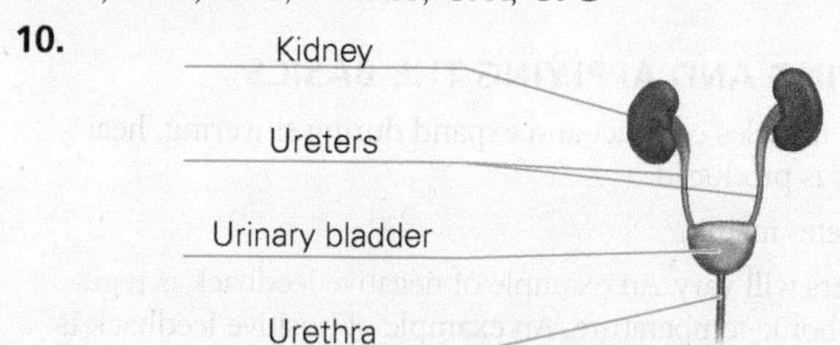

ANALYZING AND APPLYING THE BASICS

1. Taking aspirin before drinking might make the symptoms experienced after drinking worse, not better.
2. Food moved from the pharynx to the trachea instead of the esophagus.
3. No; food is just moving through a tube inside the body.

Chapter 20

GOT IT?

p. 440: 1. bronchi, bronchioles; **2.** contracts, increases; **3.** oxygen, carbon dioxide; **4.** oxygen; **5.** gas exchange

p. 449: 1. blood clotting; **2.** valves; **3.** exchange; **4.** body, lungs; **5.** blood pressure

VISUALIZE THIS

Figure 20.2: It causes a rapid intake of air.

Figure 20.6: Less oxygen crosses the respiratory membrane, meaning that individuals must breathe more rapidly to obtain adequate oxygen.

Figure 20.8: The disadvantages would be greater likelihood that the alveoli walls will stick together, the creation of more friction impeding the exchange of gases with the outside air, and by devoting more energy to build cells in the lung there is less energy for other organs.

Figure 20.13: More force is required to move blood to the body extremities than to the lungs, and too much force sending blood to the lungs can cause fluid to leak from capillaries into alveoli, interfering with breathing.

Figure 20.16: By preventing blood from pooling in the veins of the lower leg, compression socks can increase blood circulation and thus the rate of oxygen delivery.

Figure 20.17: Deoxygenated blood from the right atrium can flow into the left atrium, meaning that the blood pumped to the body has an overall lower oxygen content.

WORKING WITH DATA

Figure 20.3: About 500 mL is exchanged at rest.

SOUNDS RIGHT, BUT IS IT?

1. Numerous carcinogens, including radioactive compounds, in the plant itself.
2. It increases heart rate, blood pressure, and levels of "bad" cholesterol, all of which increase the risk of heart attack and stroke.
3. They are absorbed across the cells lining these organs.
4. Smokeless tobacco contains many of the compounds known to result in negative health outcomes as found in tobacco smoke. These compounds are still absorbed into the bloodstream and can have the same effect, regardless of the delivery system.

SHOW YOU KNOW

20.1: By forcing the diaphragm upward violently, the volume of the lungs is quickly reduced, resulting in a large exhalation of air.

20.2: The valves that prevent backflow.

LEARNING THE BASICS

1. Contraction of the diaphragm flattens out this dome-shaped muscle at the base of the chest cavity. This has the effect of increasing the size of the lung cavity, lowering air pressure, and allowing outside, oxygen-rich air to flow through the upper respiratory system and into the lungs.
2.

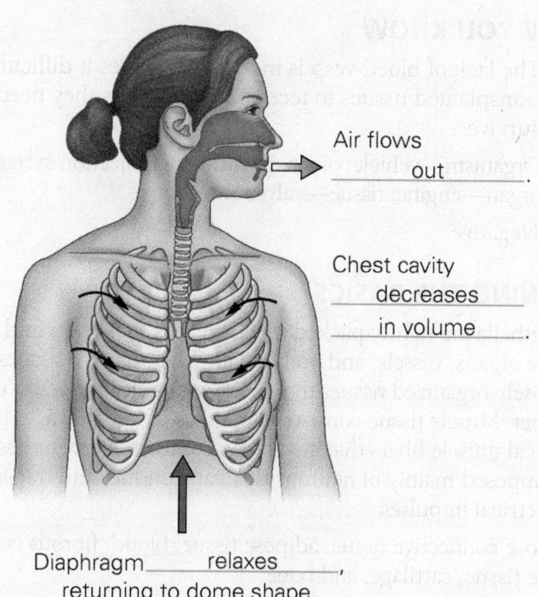

3. E; **4.** D; **5.** A; **6.** D; **7.** B; **8.** D; **9.** D; **10.** E

ANALYZING AND APPLYING THE BASICS

1. The blood after doping has greater oxygen-carrying capacity and can supply muscles with oxygen more effectively. The danger is the risk of clots forming from excess blood cells.
2. Oxygen delivery to tissues will fail to meet the body's demand, resulting in fatigue and shortness of breath (breathing rate will be triggered to increase in response to low O_2 levels).
3. By crossing into the bloodstream at the respiratory membrane and then being carried throughout the body by the cardiovascular system.

Chapter 21

GOT IT?

p. 461: 1. Viruses; **2.** Bacteria; **3.** envelope; **4.** parasites; **5.** latent
p. 469: 1. lymphocytes; **2.** antigens; **3.** passive; **4.** bone, thymus; **5.** helper

VISUALIZE THIS

Figure 21.1: Yes, because ribosomes are not bounded by membranes.

Figure 21.4: Surface proteins.

Figure 21.6: The odds of an epidemic increase as the numbers of vaccinated people decreases.

Figure 21.10: White blood cells.

WORKING WITH DATA

Figure 21.3: The 2 bacteria present at noon would double to 4 at 12:20, double again to 8 bacteria by 12:40, and double again at 1:00 giving 16 bacteria; exponential growth curve is linear at first then shows a rapid increase (concave line) as cells multiply. the line showing growth at cooler temp would have this same concave shape, but would grow much less rapidly.

SOUNDS RIGHT, BUT IS IT?

1. The immune system is fully formed at birth.
2. The immune system gets stronger with exposure to antigens.
3. Modern vaccines contain small parts of viruses.
4. Whole viruses would cause the generation of more kinds of antibodies.
5. It is not true that modern children are exposed to more antigens in vaccines because the whole virus vaccines of old would have presented more of an immune challenge than modern vaccines.

SHOW YOU KNOW

21.1: Antibiotics are not effective against viruses, they only kill bacteria.

21.2: As the immune system responds to the vaccine, some mild symptoms may occur, but they will prevent more severe symptoms from occurring when exposure to the actual virus occurs.

LEARNING THE BASICS

1.

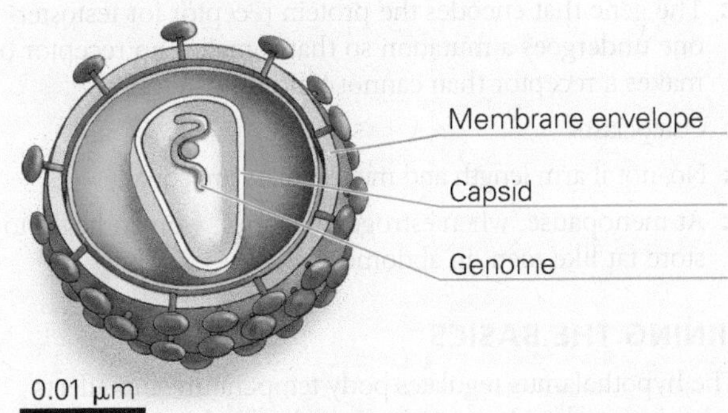

2. They can help remove infectious organisms from the body.
3. B cells produce receptors that are found on cell surfaces or secreted as antibodies and destroy invaders. T cells also produce receptors but not ones that are secreted.
4. D; **5.** D; **6.** D; **7.** B; **8.** C; **9.** C; **10.** D

ANALYZING AND APPLYING THE BASICS

1. Lyme disease is infectious but not contagious.
2. The greater the number of individuals in a population that are vaccinated, the lower the likelihood of transmission to the unvaccinated.
3. The flu virus mutates so rapidly that last year's vaccine will likely be ineffective against this year's flu strains.

Chapter 22

GOT IT?

p. 478: 1. hormones; **2.** signal transduction; **3.** hypothalamus; **4.** pituitary; **5.** FSH
p. 480: 1. cartilage; **2.** calcium; **3.** appendicular; **4.** ligaments, muscles; **5.** osteoporosis
p. 483: 1. fiber; **2.** myofibrils; **3.** sarcomere; **4.** actin, myosin; **5.** Z
p. 486: 1. muscle; **2.** fat; **3.** fat; **4.** Women's, men's; **5.** volume

VISUALIZE THIS

Figure 21.2: The breast tissue would develop less and the larynx tissue more.

Figure 21.8: The femur (thigh), patella (kneecap), and shinbone.

WORKING WITH DATA

Figure 22.12: (a) Within; (b) no.

SOUNDS RIGHT, BUT IT ISN'T

1. No.
2. Decrease.
3. No; because human males have one X and one Y chromosome.
4. Temporary.
5. Not normally.
6. Because the mammary glands of human males were not programmed in utero to develop as fully as those in females, they are unable to lactate spontaneously.

SHOW YOU KNOW

22.1: The gene that encodes the protein receptor for testosterone undergoes a mutation so that it makes no receptor or makes a receptor than cannot bind testosterone.

22.2: Osteoclasts.

22.3: No, not if arm length and muscle mass are equal.

22.4: At menopause, when estrogen declines, women begin to store fat like men, in abdomen.

LEARNING THE BASICS

1. The hypothalamus regulates body temperature and affects behaviors such as hunger, thirst, and reproduction. The pituitary gland secretes two hormones involved with producing sex differences: follicle-stimulating hormones (FSH) for sperm production and egg cell development, and luteinizing hormones (LH) for testosterone production and the release of an egg cell during ovulation. Adrenal glands secrete the hormone adrenaline (epinephrine) in response to stress or excitement and androgens, including testosterone and estrogen. Testes secrete testosterone, the hormone that aids in sperm production, hair thickness and distribution, increased muscle mass, and voice deepening. Ovaries produce and secrete estrogen, which regulates menstruation, the maturation of egg cells, breast development, pregnancy, and menopause.
2. Bone tissue can be tightly or loosely packed. Compact bone forms the hard outer shell of bones, and spongy bone is the porous honeycomb-like inner bone. Blood vessels run throughout bone. Bone marrow inside bones helps to produce blood cells.
3.

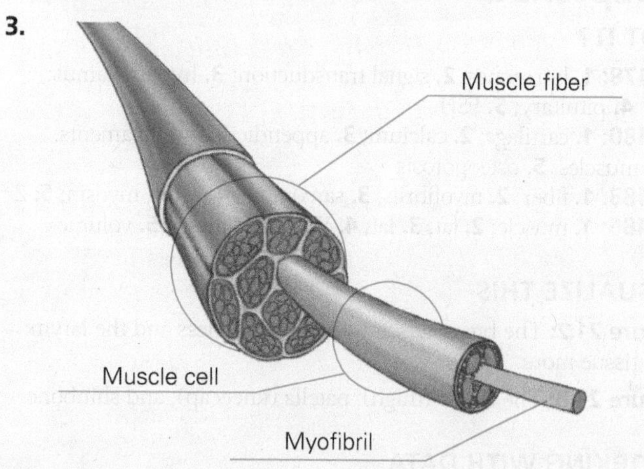

4. C; **5.** D; **6.** A; **7.** B; **8.** B; **9.** B; **10.** B

ANALYZING AND APPLYING THE BASICS

1. Differences between members of the same sex prevent generalizations from having much meaning.
2. Sperm production is highest at temperatures slightly lower than body temperature. Looser clothing keeps the testes cooler.
3. Lowering testosterone levels may allow the estrogen present in males to have a larger effect, leading to development of male breast tissue.

Chapter 23

GOT IT?

p. 494: 1. zygote; **2.** morula; **3.** uterine; **4.** ectoderm; **5.** Differentiation

p. 501: 1. estrogen; **2.** progesterone; **3.** LH; **4.** uterus; **5.** barrier

p. 503: 1. placenta; **2.** trophoblast; **3.** prolactin; **4.** oxytocin; **5.** amnion

VISUALIZE THIS

Figure 23.1: Step 2: The binding of the sperm head to the zona pellucida is the species-specific step.

Figure 23.9: The drug would cause the uterus to contract, which would stimulate the release of oxytocin via positive feedback.

WORKING WITH DATA

Figure 23.7: Estrogen increases as the follicle cells develop, which provides positive feedback on the hypothalamus to increase FSH and LH, furthering follicle development and increasing estrogen levels. When the egg cell is ovulated, the remnant of the follicle turns into the corpus luteum that secretes less estrogen.

SOUNDS RIGHT, BUT IS IT?

1. 25 days.
2. 25 mg.
3. 24 mg.
4. 1.75 mg.
5. The birth control pill can be manufactured to deliver the same monthly amount of estrogen as the normal menstrual cycle delivers.

SHOW YOU KNOW

23.1: Cells are rapidly dividing and forming more complex structures. After this period, it is mainly just growth that occurs.

23.2: Low levels of progesterone cause the lining of uterus to weaken enough that some bleeding occurs.

23.3: Progesterone elicits a negative feedback response in the hypothalamus, preventing FSH and LH secretion and development of an egg cell.

LEARNING THE BASICS

1. A different suite of genes is activated in each cell type.
2. An embryo.
3.

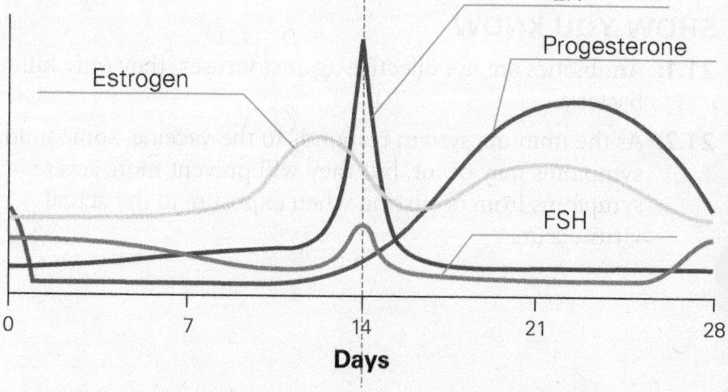

LH
Progesterone
Estrogen
FSH
0 7 14 21 28
Days

4. C; 5. B; 6. D; 7. B; 8. C; 9. A; 10. B

ANALYZING AND APPLYING THE BASICS

1. No. The mother will have built up immunity to the virus during the initial infection.
2. No. A negative test means that the virus is not there at the time of testing, not that it was never there.
3. Oxytocin causes the uterus to contract, which in turn causes the release of more oxytocin in a positive feedback loop between the uterus and brain, intensifying muscle contractions and making labor and delivery increasingly more painful.

Chapter 24

GOT IT?

p. 513: 1. neurons; **2.** neurotransmitters; **3.** central nervous system; **4.** proprioception; **5.** reflex

p. 516: 1. glial; **2.** cerebrum; **3.** cerebral cortex; **4.** cerebellum; **5.** brain stem

p. 523: 1. dendrites; **2.** axon; **3.** myelin; **4.** negative; **5.** synapse

VISUALIZE THIS

Figure 24.7: (a) The cerebrum; (b) below.

Figure 24.8: Cerebral cortex because altered functioning in this structure could lead to some of the symptoms of ADD.

WORKING WITH DATA

Figure 24.11: (a) If an impulse travels 100 m per second it would travel 2 m in 1/50 of a second; (b) 1/500 second.

SOUNDS RIGHT BUT IT ISN'T

1. No
2. Through the corpus callosum, a band of nerve fibers that connects the two hemispheres of the brain.
3. No: If math skills are located in only one side of the brain, you would expect only one side of the brain to be active when solving math problems.
4. The two hemispheres complement each other.
5. The two hemispheres of the brain contain the same structures, communicate with each other and perform functions that complement each other.

SHOW YOU KNOW

24.1: The distance to the brain is longer than the distance to the spinal cord. Therefore, by the time the pain message reaches your brain, your spinal cord has already sent a message to your hand to remove it from the hot surface.

24.2: A less well-developed frontal lobe can lead to less impulse control and less thoughtful planning.

24.3: Blocking sodium channels prevents action potentials from being generated.

LEARNING THE BASICS

1.

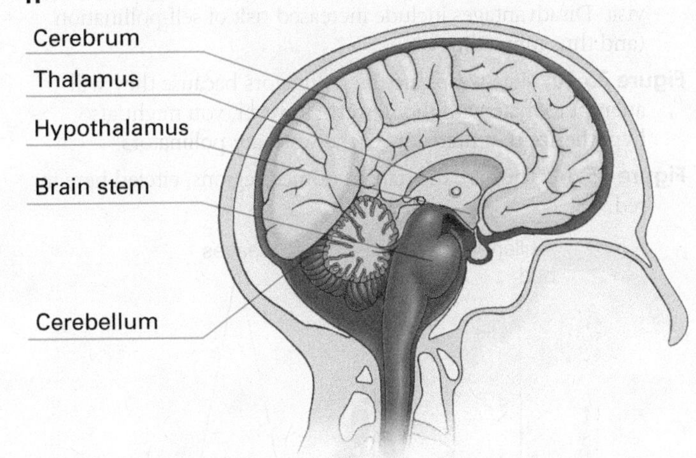

2. Sensory neurons carry information to the brain. They are connected to the motor neurons, which carry information away from the brain, by interneurons.
3. General senses do not have special organs, but are found all over the body. These include sense of touch, pain, and temperature.
4. A; **5.** D; **6.** D; **7.** D; **8.** D; **9.** E; **10.** False

ANALYZING AND APPLYING THE BASICS

1. A more strongly affected individual may have less dopamine or norepinephrine; they may have fewer dopamine or norepinephrine receptors; they may have less well-developed frontal lobe leading to less planning and impulse control.
2. They could have axons that are larger in diameter.
3. Reflexes are prewired and automatic responses, whereas other nervous system responses, require sensory activity followed by integration and motor output.

Chapter 25

GOT IT?

p. 538: 1. life cycle; **2.** cell division; **3.** sexual; **4.** flower; **5.** food

p. 548: 1. size, shape; **2.** wilting; **3.** carbon dioxide; **4.** nitrogen; **5.** single, abundant

p. 555: 1. sustainable; **2.** top; **3.** reduce; **4.** salty; **5.** meat

VISUALIZE THIS

Figure 25.1: By saving seeds from the plants that produced the largest ears and those ears that kept their seeds, farmers drove the evolution of the crop toward larger and larger, tightly bound fruits.

Figure 25.3: Ground tissue.

Figure 25.5: By increasing the surface area for absorption manyfold.

Figure 25.9: Advantages are that a pollinator can both pick up pollen for dispersal and facilitate fertilization of eggs in a single visit. Disadvantages include increased risk of self-pollination (and thus inbreeding).

Figure 25.10: Male wasps are the pollinators because they will attempt to mate with the flower! However, you might also hypothesize that predators of the wasp are pollinators.

Figure 25.14: They are the darker, domed regions, circled here in red.

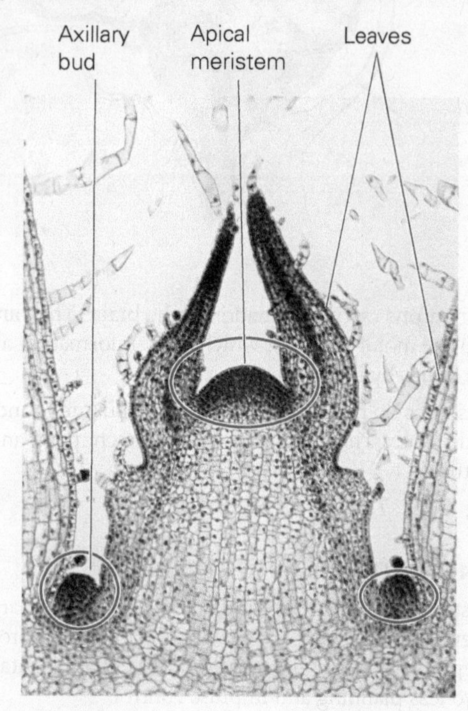

Figure 25.15: So that the vascular tissue is continuous from the branch root to the stem.

Figure 25.22: There should be fewer stomata on the leaves of plants in desert areas compared with tropical areas because plants in the desert need to conserve water.

Figure 25.24: Nutrients would return to the soil after consumption by organisms present on the site.

WORKING WITH DATA

Figure 25.27: The panhandles of Texas and Oklahoma.

SOUNDS RIGHT, BUT IS IT?

1. Water will flow out of the cell, from a region where it is high concentration to where it is in lower concentration.
2. It will stretch and hold its shape, putting pressure on the cell contents.
3. It goes down. Cells become flaccid.
4. Flaccid cells do not support each other within an organ, so the plant will wilt.
5. No, because the elasticity is the important part, not the overall thickness.
6. The thickness of the cell walls is not much of a factor in determining whether a plant will wilt—instead it is the water in the cells that matters. Adding a sugary solution to the soil will not strengthen the plant, instead it will draw water out of the plant, causing it to wilt further.

SHOW YOU KNOW

25.1: Plant cells can transform from one type (e.g., root) into others (e.g., stem).

25.2: They do get larger and flower and thus use up nutrients. They also do not get additional nutrients from a natural cycle.

25.3: If some individuals in a population are resistant to the pesticide, they survive and produce the next generation, passing the "resistance" trait on to the next generation.

LEARNING THE BASICS

1.

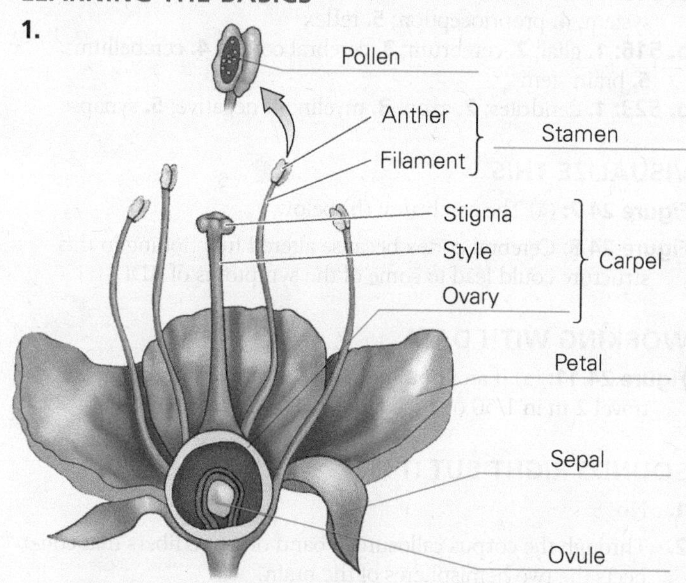

2. Plants with indeterminate growth do not have a final defined size; they will reach and always have the capability to grow under the right conditions.
3. E; **4.** D; **5.** D; **6.** D; **7.** B; **8.** B; **9.** A; **10.** E

ANALYZING AND APPLYING THE BASICS

1. Most plants will only produce fruit if the ovules are fertilized. It is difficult to trick a plant into thinking it has been fertilized when it has not.
2. New growth will have to occur at the axillary buds, causing the formation of side branches.
3. The soil may be depleted of nutrients that are especially helpful to tomatoes, and pests specific to tomatoes may have established populations in that area of his garden so they are ready to attack the plants as soon as the plants appear. The easiest solution is to move the tomatoes to a new area of the yard or garden. He could also add organic fertilizer to the soil to replenish it.

Chapter 26

GOT IT?

p. 568: 1. negative, evaporation; **2.** cohesion; **3.** water; **4.** root hairs or roots; **5.** xylem

p. 573: 1. source, sink; **2.** pressure; **3.** sugar; **4.** sinks; **5.** day (night is acceptable as well)

p. 577: 1. tropism; **2.** light, gravity, touch (order does not matter); **3.** longer; **4.** lateral; **5.** small, produced, multiple

VISUALIZE THIS

Figure 26.1: Evaporation is much reduced and the stomata are most likely closed (because the light reactions of photosynthesis cannot proceed without sunlight). Transpiration is unlikely.

Figure 26.10: Sap flow would be reduced because water would be less "available" to move into the phloem tubes.

WORKING WITH DATA

Figure 26.14: Because days are long in midsummer, the lettuce does not experience the night length required to lose enough active phytochrome. Because active phytochrome promotes flowering in lettuce, plants sown in summer will flower before they are large enough to harvest.

SOUNDS RIGHT, BUT IS IT?

1. Answers are numerous. They exchange materials with the environment, transform energy, evolve, respond to environmental conditions, etc.
2. Again, variable answers. We are looking for students to connect what you've learned about plants in Chapters 25 and 26 to the basic characteristics of life.
3. Another variable answer. Here we are asking you simply to state how plants are more than just…potted plants.

SHOW YOU KNOW

26.1: If many of the roots have died as a result of oxygen deprivation, the rate of transpiration may be too high for them to adequately replace water lost from the plant. Removing leaves reduces this rate.

26.2: Before the tap is inserted, the sugar source is roots and the sink is tree leaf buds; after the tap is inserted, the sugar source is still the roots, but some of the sink is the tap.

26.3: The cells on the surface of the leaf grow more quickly than the cells underneath, causing the leaf to curl.

LEARNING THE BASICS

1. Hydrogen bonding among water molecules (cohesion) and between water molecules and xylem cell walls (adhesion) provides a force that counters the strong negative pressure and keeps the water column together.
2. The C_4 adaptation pumps carbon dioxide toward actively photosynthetic cells, allowing stomata to remain only partially open. The CAM adaptation allows plants to accumulate carbon dioxide during the cooler night when less water will evaporate so that they may keep their stomata closed during the hottest parts of the day.
3. D
4.

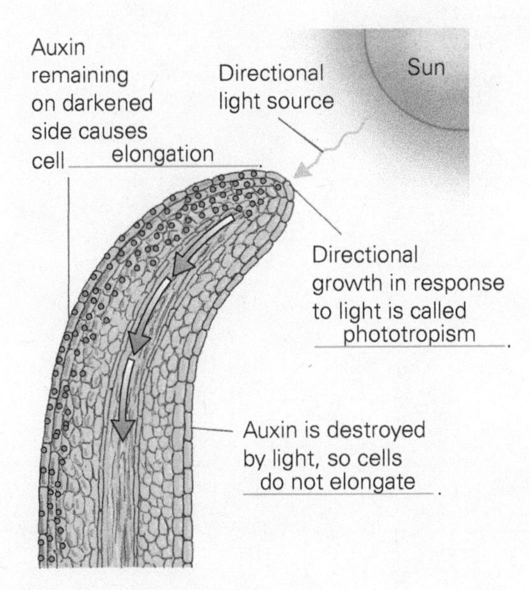
Auxin remaining on darkened side causes cell elongation
Directional light source
Sun
Directional growth in response to light is called phototropism.
Auxin is destroyed by light, so cells do not elongate.

5. C; **6.** B; **7.** D; **8.** A; **9.** A; **10.** C

ANALYZING AND APPLYING THE BASICS

1. Make sure that the soil in the site is well watered and mulched. The lily should have some of its leaves removed to reduce transpiration.
2. Keep it in a dark closet for a good portion of each day until it produces flower buds.
3. Prune competing apical meristems so that only one remains to promote a single tip. Prune apical meristems of side branches to encourage these side branches to branch more to improve bushiness.

Glossary

ABO blood system A system for categorizing human blood based on the presence or absence of carbohydrates on the surface of red blood cells.

abscisic acid A hormone in plants associated with dormancy in seeds and buds.

abscission The dropping off of leaves, flowers, fruits, or other plant organs.

accessory organs Organs including the pancreas, liver, and gallbladder, which aid the digestive system.

acetylcholine A neurotransmitter with many functions, including facilitating muscle movements, and thought to be involved in the development of Alzheimer's disease.

acid A substance that increases the concentration of hydrogen ions in a solution.

acid rain Rain (or other precipitation) that is unusually acidic; caused by air pollution in the form of sulfur and nitrogen dioxides.

acquired immune deficiency syndrome (AIDS) Syndrome characterized by severely reduced immune system function and numerous opportunistic infections. Results from infection with HIV.

acrosome An organelle at the tip of the sperm cell containing enzymes that help the sperm penetrate the egg cell.

actin A protein found in muscle tissue that, together with myosin, facilitates contraction.

action potential Wave of depolarization in a neuron propagated to the end of the axon—also called a nerve impulse.

activation energy The amount of energy that reactants in a chemical reaction must absorb before the reaction can start.

active immunity Immunity that results from the production of antibodies, as differentiated from passive immunity.

active site Substrate-binding region of an enzyme.

active smoker An individual who smokes tobacco.

active transport The ATP-requiring movement of substances across a membrane against their concentration gradient.

adaptation Trait that is favored by natural selection and increases an individual's fitness in a particular environment.

adaptive radiation Diversification of one or a few species into large and very diverse groups of descendant species.

ADD *See* attention deficit disorder.

adenine Nitrogenous base in DNA.

adenosine diphosphate (ADP) A nucleotide composed of adenine, a sugar, and two phosphate groups. Produced by the hydrolysis of the terminal phosphate bond of ATP.

adenosine triphosphate (ATP) A nucleotide composed of adenine, the sugar ribose, and three phosphate groups that can be hydrolyzed to release energy. Form of energy that cells can use.

adhesion The sticking together of unlike materials—often water and a particular surface.

adipose tissue Fat-storing connective tissue.

adrenal gland Either of two endocrine glands, one located atop each kidney, that secrete adrenaline in response to stress or excitement, help maintain water and salt balance, and secrete small amounts of sex hormones.

aerobic An organism, environment, or cellular process that requires oxygen.

aerobic respiration Cellular respiration that uses oxygen as the electron acceptor.

agarose gel A jelly-like slab used to separate molecules on the basis of molecular weight.

agriculture Cultivation of crops, raising of livestock; farming.

algae Photosynthetic protists.

allele Alternate versions of the same gene, produced by mutations.

allele frequency The percentage of the gene copies in a population that are of a particular form, or allele.

allergy An abnormally high sensitivity to allergens such as pollen or microorganisms. Can cause sneezing, itching, runny nose, and watery eyes.

allopatric Geographic separation of a population of organisms from others of the same species. Usually in reference to speciation.

alternative hypothesis Factor other than the tested hypothesis that may explain observations.

alveoli (singular: alveolus) Sacs inside lungs, making up the respiratory surface in land vertebrates and some fish.

Alzheimer's disease Progressive mental deterioration in which there is memory loss along with the loss of control of bodily functions, ultimately resulting in death.

amenorrhea Abnormal cessation of the menstrual cycle.

amino acid Monomer subunit of a protein. Contains an amino, a carboxyl, and a unique side group.

amnion The fluid-filled sac in which a developing embryo is suspended.

anaerobic An organism, environment, or cellular process that does not require oxygen.

anaerobic respiration A process of energy generation that uses molecules other than oxygen as electron acceptors.

anaphase Stage of mitosis during which microtubules contract and separate sister chromatids.

androgen A masculinizing hormone, such as testosterone, secreted by the adrenal glands.

anecdotal evidence Information based on an individual's personal experience.

angiosperm Plant in the phyla Angiospermae, which produce seeds borne within fruit.

animal An organism that obtains energy and carbon by ingesting other organisms and is typically motile for part of its life cycle.

Animalia Kingdom of Eukarya containing organisms that ingest others and are typically motile for at least part of their life cycle.

annual growth rate Proportional change in population size over a single year. Growth rate is a function of the birth rate minus the death rate of the population.

annual plant Plant that completes its life cycle in a single growing season.

anorexia Self-starvation.

antagonistic muscle pair A set of muscles whose actions oppose each other.

anther The pollen-containing structure on the stamen of a flower.

anthropogenic Human caused.

antibiotic A chemical that kills or disables bacteria.

antibiotic resistant Characteristic of certain bacteria; a physiological characteristic that permits them to survive in the presence of particular antibiotics.

antibody Protein made by the immune system in response to the presence of foreign substances or antigens. Can serve as a receptor on a B cell or be secreted by plasma cells.

antibody-mediated immunity Immunity that occurs via secreted antibodies, versus immunity mediated by T cells.

anticodon Region of tRNA that binds to an mRNA codon.

antigen Short for antibody-generating substances; an antigen is a molecule that is foreign to the host and stimulates the immune system to react.

antigen receptor Protein in B- and T-cell membrane that bind to specific antigens.

antioxidant Certain vitamins and other substances that protect the body from the damaging effects of free radicals.

antiparallel Feature of a DNA double helix in which nucleotides face *up* on one side of the helix and *down* on the other.

apical dominance The influence exerted by the terminal bud in suppressing the growth of axillary or lateral buds. Mediated by the hormone auxin.

apical meristem The actively dividing cells at the tip of stems and roots in plants.

appendicular skeleton The part of the skeleton composed of the bones of the hip, shoulder, and limbs.

aquaporin A transport protein in the membrane of a plant or animal cell that facilitates the diffusion of water across the membrane by osmosis.

aquatic Of, or relating to, water.

Archaea Domain of prokaryotic organisms.

artery Blood vessel that carries oxygenated blood from the heart to body tissues.

artificial selection Selective breeding of domesticated animals and plants to increase the frequency of desirable traits.

asexual reproduction A type of reproduction in which one parent gives rise to genetically identical offspring.

assortative mating Tendency for individuals to mate with other individuals who are like themselves.

asthma A respiratory disease characterized by spasmodic inflammation of the air passages in the lungs and overproduction of mucus. Often triggered by air contaminants.

atherosclerosis Accumulation of fatty deposits within blood vessels; hardening of the arteries.

atom The smallest unit of matter that retains the properties of an element.

ATP synthase Enzyme found in the mitochondrial membrane that helps synthesize ATP.

atrium An upper chamber of the heart that receives blood from the body or lungs and pumps it to a ventricle.

attention deficit disorder (ADD) Syndrome characterized by forgetfulness; distractibility, fidgeting; restlessness; impatience; difficulty sustaining attention in work, play, or conversation; or difficulty following instructions and completing tasks in more than one setting.

autoimmune disease Any of the diseases that result from an attack by the immune system on normal body cells.

autosome Any chromosome not involved in sex determination. Chromosomes 1–22 are autosomes. The X and Y chromosomes are not.

auxin A class of plant hormones that control cell elongation, among other effects.

AV valves Heart valves between the atria and the ventricles.

axial skeleton Part of the skeleton that supports the trunk of the body and consists largely of the bones making up the vertebral column or spine and much of the skull.

axillary bud A bud located at the junction between a plant stem and leaf.

axon Long, wire-like portion of the neuron that ends in a terminal bouton.

B lymphocyte (B cell) The type of white blood cell responsible for antibody-mediated immunity.

background extinction rate The rate of extinction resulting from the normal process of species turnover.

bacteria Domain of prokaryotic organisms.

bark The outer layer of a woody stem, consisting of cork, cork cambium, and phloem.

basal metabolic rate Resting energy use of an awake, alert person.

base A substance that reduces the concentration of hydrogen ions in a solution.

base-pairing rule A rule governing the pairing of nitrogenous bases. In DNA, adenine pairs with thymine and cytosine with guanine.

behavioral isolation Prevention of mating between individuals in two different populations based on differences in behavior.

benign The type of tumor that stays in one place and does not affect surrounding tissues.

bias Influence of research participants' opinions on experimental results.

biennial Plant that completes its life cycle over the course of two growing seasons.
bile Mixture of substances produced in the liver that aids in digestion by emulsifying fats.
binary fission An asexual form of bacterial reproduction.
binge drinking The consumption of more than four drinks in a two-hour time period.
bioaccumulation A phenomenon that results in the concentration of persistent (i.e., slow to degrade) pollutants in the bodies of animals at high levels of a food chain.
biodiversity Variety within and among living organisms.
biodiversity hot spot Region that contains both significant species diversity and species that are rare or endangered by human destruction.
biogeography The study of the geographic distribution of organisms.
biological classification Field of science attempting to organize biodiversity into discrete, logical categories.
biological diversity Entire variety of living organisms.
biological evolution *See* evolution.
biological population Individuals of the same species that live and breed in the same geographic area.
biological race Populations of a single species that have diverged from each other. Biologists do not agree on a definition of *race*. *See also* subspecies.
biological species concept Definition of a species as a group of individuals that can interbreed and produce fertile offspring but typically cannot breed with members of another species.
biology The study of living organisms.
biomagnification Concentration of toxic chemicals in the higher levels of a food web.
biomass The mass of all individuals of a species, or of all individuals on a level of a food web, within an ecosystem.
biome A broad ecological community defined by a particular vegetation type (e.g., temperate forest, prairie), which is typically determined by climate factors.
biophilia Humans' innate desire to be surrounded by natural landscapes and objects.
biopsy Surgical removal of some cells, tissue, or fluid to determine whether cells are cancerous.
bioregion A geographical region determined by shared natural features rather than human-defined boundaries.
bipedal Walking upright on two limbs.
blade The broad part of a leaf.
blastocyst An embryonic stage consisting of a hollow ball of cells.
blind experiment Test in which subjects are not aware of exactly what they are predicted to experience.
blood The combination of cells and liquid that flows through blood vessels in the cardiovascular system; made up of red blood cells, plasma, white blood cells, and platelets.
blood clotting The process by which a blood clot, a mass of the protein fibrin and dead blood cells, forms in the region of blood vessel damage.
blood pressure The force of the blood as it travels through the arteries; partially determined by artery diameter and elasticity.
blood vessel The structure that carries blood via the circulatory system throughout the body; arteries, capillaries, and veins.
body mass index (BMI) Calculation using height and weight to determine a number that correlates to an estimate of a person's amount of body fat with health risks.
bolus In digestion, a soft ball of chewed food.
bone A type of connective tissue consisting of living cells in a matrix rich in collagen and calcium.
bone marrow Network of soft connective tissue that fills bones and is involved in the production of red blood cells.
boreal forest A biome type found in regions with long, cold winters and short, cool summers. Characterized by coniferous trees.
botanist Plant biologist.
bottleneck effect Dramatic but short-lived reduction in population size followed by an increase in population.
brain stem Region of the brain that lies below the thalamus and hypothalamus; it governs reflexes and some involuntary functions such as breathing and swallowing.
bronchi The large air passageways from the trachea into the lungs.
bronchiole The branching air passageway inside the lungs.
bronchitis Inflammation of the bronchi and bronchioles in the lungs.
bulbourethral gland Either of the two glands at the base of the penis that secrete acid-neutralizing fluids into semen.
bulimia Binge eating followed by purging.

C_3 plant Plant that uses the light-independent reactions of photosynthesis to incorporate carbon dioxide into a 3-carbon compound.
C_4 plant Plant that performs reactions incorporating carbon dioxide into a 4-carbon compound that ultimately provides carbon dioxide for the light-independent reactions.
calcium Nutrient required in plant cells for the production of cell walls and in humans for bone strength and blood clotting.
calorie Amount of energy required to raise the temperature of one gram of water by 1°C.
CAM plant A plant that uses Crassulacean acid metabolism, a variant of photosynthesis during which carbon dioxide is stored in sugars at night and released during the day to prevent water loss.
Cambrian explosion Relatively rapid evolution of the modern forms of multicellular life that occurred approximately 550 million years ago.
cancer A disease that occurs when cell division escapes regulatory controls.
capillary The smallest blood vessel of the cardiovascular system, connecting arteries to veins and allowing material exchange across their thin walls.

capillary bed A branching network of capillaries supplying a particular organ or region of the body.
capsid Protein coat that surrounds a virus.
capsule Gelatinous outer covering of bacterial cells that aids in attachment to host cells during an infection.
carbohydrate Energy-rich molecule that is the major source of energy for the cell. Consists of carbon, hydrogen, and oxygen in the ratio CH_2O.
carcinogen Substance that causes cancer or increases the rate of its development.
cardiac cycle The cycle of contraction and relaxation that a normally functioning heart undergoes over the course of a single heart beat.
cardiac muscle Muscle that forms the contractile wall of the heart.
cardiovascular disease Malfunction of the cardiovascular system, including, but not limited to, heart attack, stroke, and hypertension.
cardiovascular system The organ system made up of the heart and circulatory system, including arteries, capillaries, and veins.
carpel The female structure of a flower, containing the ovary, style, and stigma.
carrier Individual who is heterozygous for a recessive allele.
carrying capacity Maximum population that the environment can support.
cartilage Connective tissue found in the skeletal system that is rich in collagen fibers.
catalyst A substance that lowers the activation energy of a chemical reaction, thereby speeding up the reaction.
catalyze To speed up the rate of a chemical reaction. Enzymes are biological catalysts.
cell Basic unit of life; an organism's fundamental building-block units.
cell body Portion of the neuron that houses the nucleus and organelles.
cell cycle An ordered sequence of events in the life cycle of a eukaryotic cell from its origin until its division to produce daughter cells. Consists of M, G_1, S, and G_2 phases.
cell division Process a cell undergoes when it makes copies of itself. Production of daughter cells from an original parent cell.
cell plate A double layer of new cell membrane that appears in the middle of a dividing plant cell and divides the cytoplasm of the dividing cell.
cell wall Tough but elastic structure surrounding plant and bacterial cell membranes.
cell-mediated immunity Type of specific immune response carried out by T cells.
cellular respiration Metabolic reactions occurring in cells that result in the oxidation of macromolecules to produce ATP.
cellulose A structural polysaccharide found in cell walls and composed of glucose molecules.
central nervous system (CNS) Includes brain and spinal cord and is responsible for integrating, processing, and coordinating information taken in by the senses. It is the seat of functions such as intelligence, learning, memory, and emotion.
central vacuole A membrane-enclosed sac in a plant cell that functions to store many different substances.
centriole A structure in animal cells that helps anchor for microtubules during cell division.
centromere Region of a chromosome where sister chromatids are attached and to which microtubules bind.
cereal grains Staples of the human diet that are from the grass family, namely, wheat, corn, oats, rye, and barley.
cerebellum Region of the brain that controls balance, muscle movement, and coordination.
cerebral cortex Deeply wrinkled outer surface of the cerebrum where conscious activity and higher thought originate.
cerebrospinal fluid Protective liquid bath that surrounds the brain within the skull.
cerebrum Portion of the brain in which language, memory, sensations, and decision making are controlled. The cerebrum has two hemispheres, each of which has four lobes.
cervix The lower narrow portion of the uterus at the top end of the vagina.
chaparral A biome characteristic of climates with hot, dry summers and mild, wet winters and a dominant vegetation of aromatic shrubs.
checkpoint Stoppage during cell division that occurs to verify that division is proceeding correctly.
chemical reaction A process by which one or more chemical substances are transformed into one or more different chemical substances.
chemotherapy The use of chemicals to try to kill rapidly dividing (cancerous) cells.
chlorophyll Green pigment found in the chloroplast of plant cells.
chloroplast An organelle found in plant cells that absorbs sunlight and uses the energy derived to produce sugars.
cholesterol A steroid found in animal cell membranes that affects membrane fluidity. Serves as the precursor to estrogen and testosterone.
chondrocyte A type of cartilage cell that produces collagen and other substances.
chromosome Subcellular structure composed of a long single molecule of DNA and associated proteins, housed inside the nucleus.
chyme Partially digested food and enzymes that are passed from the stomach to the intestine.
circulatory system The vessels that transport blood, nutrients, and waste around the body.
citric acid cycle A chemical cycle occurring in the matrix of the mitochondria that breaks the remains of sugars down to produce carbon dioxide.
cladistic analysis A technique for determining the evolutionary relationships among organisms that relies on identification and comparison of newly evolved traits.

classification system Method for organizing biological diversity.

cleavage In embryology, the period of rapid cell division that occurs during animal development.

climate The average temperature and precipitation as well as seasonality.

climax community The group of species that is stable over time in a particular set of environmental conditions.

clitoris Sensitive erectile tissue found in the external genitalia of females that functions in sexual arousal.

clonal population Population of identical cells copied from the immune cell that first encounters an antigen. The entire clonal population has the same DNA arrangement, and all cells in a clonal population carry the same receptor on their membrane.

cloning Producing copies of a gene or an organism that are genetically identical.

clumped distribution A spatial arrangement of individuals in a population where large numbers are concentrated in patches with intervening, sparsely populated areas separating them.

codominance Two different alleles of a gene that are equally expressed in the heterozygote.

codon A triplet of mRNA nucleotides. Transfer RNA molecules bind to codons during protein synthesis.

coenzyme (or cofactor) Substances such as vitamins that help enzymes catalyze chemical reactions.

coevolution Occurs when change in one biological species triggers a change in an associated species. Coevolution commonly occurs between predator and prey or parasite and host.

cohesion The tendency for molecules of the same material to stick together.

collecting duct A structure in the kidney that accepts filtrate from multiple nephrons and transmits it to the renal pelvis. Each kidney contains hundreds of collecting ducts. The amount of water retained by the kidneys is largely controlled at the collecting ducts.

colon Tube between the small intestine and anus of the digestive system.

combination drug therapy The use of more than one drug simultaneously to treat a disease. Often used for disease organisms that mutate quickly or are difficult to control to combat the problem of drug resistance.

commensalism In ecology, a relationship between two species in which one is benefitted and the other is neither harmed nor benefitted.

common descent The theory that all living organisms on Earth descended from a single common ancestor that appeared in the distant past.

community A group of interacting species in the same geographic area.

compact bone The hard, outer shell of bones.

competition Interaction that occurs when two species of organisms both require the same resources within a habitat; competition tends to limit the size of populations.

competitive exclusion Reduction or elimination of one species in an environment resulting from the presence of another species that requires the same or similar resources.

complement protein Type of blood protein with which an antibody-antigen complex can combine in order to kill bacterial cells. Enhances the immune response on many levels.

complement system A part of the immune system that enhances the abilities of phagocytes and antibodies to ward off invaders.

complementary Complementary bases pair with each other by hydrogen bonding across the DNA helix. Adenine is complementary to thymine and to guanine.

complementary base pair Nitrogenous bases that hydrogen bond to each other. In DNA, adenine is complementary to thymine, and cytosine is complementary to guanine. In RNA, adenine is complementary to uracil and guanine to cytosine.

complete protein Dietary protein that contains all the essential amino acids.

complex carbohydrate Carbohydrate consisting of two or more monosaccharides.

connective tissue Animal tissue that functions to bind and support other tissues. Composed of a small number of cells embedded in a matrix.

consilience The unity of knowledge. Used to describe a scientific theory that has multiple lines of evidence to support it.

contagious Spreading from one organism to another.

control Subject for an experiment who is similar to experimental subject except is not exposed to the experimental treatment. Used as baseline values for comparison.

convergent evolution Evolution of the same trait or set of traits in different populations as a result of shared environmental conditions rather than shared ancestry.

coral reef Highly diverse biome found in warm, shallow salt water, dominated by the limestone structures created by coral animals.

cork A tissue produced by the cork cambium that provides protection to a woody stem by forming outer bark.

cork cambium A meristematic tissue in bark that arises from phloem cells and creates cork cells before being shed.

corpus callosum Bundle of nerve fibers at the base of the cerebral fissure that provides a communication link between the cerebral hemispheres.

corpus luteum Hormone-producing tissue (the ovarian follicle after ovulation) that makes progesterone and estrogen and degenerates about 12 days after ovulation if fertilization does not occur.

correlation Describes a relationship between two factors.

cotyledon The first leaves produced by an embryonic seed plant.

covalent bond A type of strong chemical bond in which two atoms share electrons.

CRISPR A gene editing system (for "clustered regularly interspaced palindromic repeats") that evolved in bacteria but can be used to edit a gene in any species.
critical night length The night length that a plant must experience in order to trigger a photoperiodic response.
crop rotation The practice of growing different crops at different times on the same field in order to maintain soil fertility and decrease pest damage.
cross In genetics, the mating of two organisms.
crossing over Gene for gene exchange of genetic information between members of a homologous pair of chromosomes.
cultural control Control of agricultural pests through environmental techniques, such as crop rotation and maintenance of native vegetation.
cuticle The waxy layer on the outer surface of plant epidermal cells.
cyst Noncancerous, fluid-filled growth.
cytokinesis Part of the cell cycle during which two daughter cells are formed by the cytoplasm splitting.
cytoplasm The entire contents of the cell (except the nucleus) surrounded by the plasma membrane.
cytosine Nitrogenous base.
cytoskeleton A network of tubules and fibers that branch throughout the cytoplasm.
cytosol The semifluid portion of the cytoplasm.
cytotoxic T cell Immune-system cell that attacks and kills body cells that have become infected with a virus before the virus has had time to replicate.

data Information collected by scientists during hypothesis testing.
daughter cells The offspring cells that are produced by the process of cell division.
death rate Number of deaths averaged over the population as a whole.
deciduous Pertaining to woody plants that drop their leaves at the end of a growing season.
decomposer An organism, typically bacteria and fungi in the soil, whose action breaks down complex molecules into simpler ones.
decomposition The breakdown of organic material into smaller molecules.
deductive reasoning Making a prediction about the outcome of a test; *if/then* statements.
defensive protein Nonspecific protein of the immune system including interferons and complement proteins.
deforestation The removal of forest lands, often to enable the development of agriculture.
degenerative disease Disease characterized by progressive deterioration.
dehydration Loss of water.
deleterious In genetics, said of a mutation that reduces an individual's fitness.
demographic momentum Lag between the time that humans reduce birth rates and the time that population numbers respond.
demographic transition The period of time between when death rates in a human population fall (as a result of improved technology) and when birth rates fall (as a result of voluntary limitation of pregnancy).
denature (1) In proteins, the process where proteins unravel and change their native shape, thus losing their biological activity. (2) For DNA, the breaking of hydrogen bonds between the two strands of the double-stranded DNA helix, resulting in single-stranded DNA.
dendrite Short extension of the neuron that receives signals from other cells.
density-dependent factor Any of the factors related to a population's size that influence the current growth rate of a population—for example, communicable disease or starvation.
density-independent factor Any of the factors unrelated to a population's size that influence the current growth rate of a population—for example, natural disasters or poor weather conditions.
deoxyribonucleic acid (DNA) Molecule of heredity that stores the information required for making all of the proteins required by the cell.
deoxyribose The five-carbon sugar in DNA.
dependent variable The variable in a study that is expected to change in response to changes in the independent variable.
depolarization Reduction in the charge difference across the neuronal membrane.
depression Disease that involves feelings of helplessness and despair, and sometimes thoughts of suicide.
desert Biome found in areas of minimal rainfall. Characterized by sparse vegetation.
desertification The process by which formerly productive land is converted to unproductive land, typically by overgrazing of cattle or unsustainable agricultural practices, but also increasingly by the effects of climate change.
determinate growth Growth of limited duration.
developed country A term used by the United Nations to describe a country with high per-capita income and significant industrial development.
development All of the progressive changes that produce an organism's body.
diabetes Disorder of carbohydrate metabolism characterized by impaired ability to produce or respond to the hormone insulin.
diaphragm (1) Dome-shaped muscle at the base of the chest cavity. Contraction of this muscle helps draw air into the lungs. (2) A birth control device consisting of a flexible contraceptive disk that covers the cervix to prevent the entry of sperm.
diastole The stage of the cardiac cycle when the heart relaxes and fills with blood.
dicot The class of angiosperms characterized by having two cotyledons.
differentiation Structural and functional divergence of cells as they become specialized.

diffusion The spontaneous movement of substances from a region of their own high concentration to a region of their own low concentration.
dihybrid cross A genetic cross involving the alleles of two different genes. For example, AaBb × AaBb.
diploid cell A cell containing homologous pairs of chromosomes (2n).
directional selection Natural selection for individuals at one end of a range of phenotypes.
disaccharide A double sugar consisting of two monosaccharides joined together by a glycosidic linkage.
diverge *See* divergence.
divergence Occurs when gene flow is eliminated between two populations. Over time, traits found in one population begin to differ from traits found in the other population.
diversifying selection Natural selection for individuals at both ends of a range of phenotypes but against the *average* phenotype.
dizygotic twins Fraternal twins (nonidentical) that develop when two different sperm fertilize two different egg cells.
DNA *See* deoxyribonucleic acid (DNA).
DNA polymerase Enzyme that facilitates base pairing during DNA synthesis.
DNA profiling A technique used to identify individuals on the basis of DNA sequences.
DNA replication The synthesis of two daughter DNA molecules from one original parent molecule. Takes place during the S phase of interphase.
domain Most inclusive biological category. Biologists group life into three major domains.
domesticated Referring to animals or plants that are now largely dependent on humans for reproduction and survival as a result of artificial selection.
dominant Applies to an allele with an effect that is visible in a heterozygote.
dopamine Neurotransmitter active in pathways that control emotions and complex movements.
dormancy A condition of arrested growth and development in plants.
dormant In a state of dormancy, that is, alive, but not growing or developing.
double blind Experimental design protocol when both research subjects and scientists performing the measurements are unaware of either the experimental hypothesis or who is in the control or experimental group.
double fertilization The fusion of egg and sperm and the simultaneous fusion of a sperm with nuclei in the ovule of angiosperms.

ecological footprint A measure of the natural resources used by a human population or society.
ecological niche The functional role of a species within a community or ecosystem, including its resource use and interactions with other species.
ecology Field of biology that focuses on the interactions between organisms and their environment.
ecosystem All of the organisms and natural features in a given area.
ecotourism The visitation of specific geographical sites by tourists interested in natural attractions, especially animals and plants.
ectoderm The outermost of the three germ layers that arise during animal development.
effector Muscle, gland, or organ stimulated by a nerve.
egg cell Gamete produced by a female organism.
electron A negatively charged subatomic particle.
electron shell An energy level representing the distance of an electron from the nucleus of an atom.
electron transport chain A series of proteins in the mitochondrial and chloroplast membranes that move electrons during the redox reactions that release energy to produce ATP.
electronegative The tendency to attract electrons to form a chemical bond.
element A substance that cannot be broken down into any other substance.
embryo The developmental stage commencing after the first mitotic divisions of the zygote and ending when body structures begin to appear; from about the second week after fertilization to about the ninth week.
emphysema A lung disease caused by the breakdown of alveoli walls; characterized by shortness of breath and an expanded chest cavity.
encephalitis Pathology, or disease, of the brain.
Endangered Species Act (ESA) U.S. law intended to protect and encourage the population growth of threatened and endangered species, enacted in 1973.
endocrine gland Any of the glands that secrete hormones into the bloodstream.
endocrine system The internal system of chemical signals involving hormones, the organs and glands that produce and secrete them, and the target cells that respond to them.
endocytosis The uptake of substances into cells by a pinching inward of the plasma membrane.
endoderm The innermost of the three germ layers that arise during animal development.
endometrium Lining of the uterus, shed during menstruation.
endoplasmic reticulum (ER) A network of membranes in eukaryotic cells. When rough, or studded with ribosomes, it functions as a workbench for protein synthesis. When devoid of ribosomes, or smooth, it functions in phospholipid and steroid synthesis and detoxification.
endosperm The triploid structure formed by the fusion of a sperm and two nuclei in the ovule of an angiosperm plant.
endospore Resistant form of certain bacterial species. Produced as a resting stage to bring cell through poor conditions.
endosymbiotic theory Theory that organelles such as mitochondria and chloroplasts in eukaryotic cells evolved from prokaryotic cells that took up residence inside ancestral eukaryotes.
enveloped viruses A virus that is surrounded by a membrane.

environmental tobacco smoke (ETS) The tobacco smoke in the air that results from smoldering tobacco on the lit ends of cigarettes and pipes as well as the smoke exhaled by active smokers.

enzyme Protein that catalyzes and regulates the rate of metabolic reactions.

epidemic Contagious disease that spreads rapidly and extensively among many individuals.

epidemiology The branch of medical science that studies the cause, spread, and control of disease.

epidermal tissue The outermost layer of cells of the leaf and of young stems and roots.

epididymis A coiled tube located adjacent to the testes where sperm are stored.

epigenetics The study of changes in phenotypes caused by changes in gene expression rather than changes in DNA sequence.

epiglottis Flap that blocks the windpipe so food goes down the pharynx, not into the lungs.

epithelial tissue (epithelia) Tightly packed sheets of cells that line organs and body cavities.

equator The circle around Earth that is equidistant to both poles.

esophagus Tube that conducts food from the pharynx to the stomach.

essential amino acid Any of the amino acids that humans cannot synthesize and thus must be obtained from the diet.

essential fatty acid Any of the fatty acids that animals cannot synthesize and must be obtained from the diet.

estrogen Any of the feminizing hormones secreted by the ovary in females and adrenal glands in both sexes.

estuary An aquatic biome that forms at the outlet of a river into a larger body of water such as a lake or ocean.

ethylene A plant hormone involved in fruit ripening and abscission.

Eukarya Domain of life consisting of all organisms containing cells with nuclei.

eukaryote Cell that has a nucleus and membrane-bounded organelles.

eutrophication Process resulting in periods of dangerously low oxygen levels in water, sometimes caused by high levels of nitrogen and phosphorus from fertilizer runoff, that result in increased growth of algae in waterways.

evolution Changes in the features (traits) of individuals in a biological population that occur over the course of generations. *See also* theory of evolution.

evolutionary classification System of organizing biodiversity according to the evolutionary relationships among living organisms.

excretion The release of urine from the kidneys into the storage organ of the bladder.

exhale To emit breath or vapor, to breathe out.

exocytosis The secretion of molecules from a cell via fusion of membrane-bounded vesicles with the plasma membrane.

experiment Contrived situation designed to test specific hypotheses.

exponential growth Growth that occurs in proportion to the current total.

extinction Complete loss of a species.

extinction vortex A process by which an endangered population is driven toward extinction via loss of genetic diversity, increased vulnerability to the effects of density-independent factors, and inbreeding depression.

facilitated diffusion The spontaneous passage of molecules, through membrane proteins, down their concentration gradient.

false positive A test result that incorrectly indicates that a disease or condition is present.

falsifiable Able to be proved false.

fat Energy-rich, hydrophobic lipid molecule composed of a three-carbon skeleton bonded to three fatty acids.

fatty acid A long, energy rich chain of hydrocarbons. Fatty acids vary on the basis of their length and on the number and placement of double bonds.

femur Bone extending from the pelvis to the knee; also called thighbone.

fermentation A process that makes a small amount of ATP from glucose without using an electron transport chain. Ethyl alcohol and lactic acid are produced by this process.

fertilization The fusion of haploid gametes (in humans, egg and sperm) to produce a diploid zygote.

fertilizer Any of a variety of growth medium amendments that increase the level of plant nutrients.

fetus The term used to describe a developing human from the ninth week of development until birth.

fever Abnormally high body temperature.

fibrin A protein produced during the process of blood clotting that makes up a net to trap and block blood flow from a damaged blood vessel.

fibroblast A protein-secreting cell found in loose connective tissue.

fibrous connective tissue A dense tissue found in tendons and ligaments composed largely of collagen fibers.

filtration In the kidneys, the removal of plasma from the bloodstream through capillaries surrounding a nephron.

fitness Relative survival and reproduction of one variant compared with others in the same population.

flagellum (*plural*: flagella) A long cellular projection that aids in motility.

flower Reproductive structure of a flowering plant.

flowering plant Member of the kingdom Plantae, which produce flowers and fruit. *See also* angiosperm.

follicle Fluid-filled sac in the ovary that contains the developing egg cell and secretes estrogen.

follicle-stimulating hormone (FSH) Hormone secreted by the pituitary gland involved in sperm production, regulation of ovulation, and regulation of menstruation.

food chain The linear relationship between trophic levels from producers to primary consumers, and so on.
food web The feeding connections between and among organisms in an environment.
foramen magnum Hole in the skull that allows passage of the spinal cord.
forest Terrestrial community characterized by the presence of trees.
fossil Remains of plants or animals that once existed, left in soil or rock.
fossil fuel Nonrenewable resource consisting of the buried remains of ancient plants that have been transformed by heat and pressure into coal and oil.
fossil record Physical evidence left by organisms that existed in the past.
founder effect Type of sampling error that occurs when a small subset of individuals emigrates from the main population and begins a new population, leading to differences in the gene pools of both.
frameshift mutation A mutation that occurs when the number of nucleotides inserted or deleted from a DNA sequence is not a multiple of three.
free radical A substance containing an unpaired electron that is therefore unstable and highly reactive, causing damage to cells.
frontal bone Upper front portion of the cranium, or the forehead.
frontal lobe The largest and most anterior portion of each cerebral hemisphere.
fruit A mature ovary of an angiosperm, containing the seeds.
Fungi Kingdom of eukaryotes made up of members that are immobile, rely on other organisms as their food source, and are made up of hyphae that secrete digestive enzymes into the environment and that absorb the digested materials.

gall bladder Organ that stores bile and empties into the small intestine.
gamete Specialized sex cell (sperm and egg in humans) that contains half as many chromosomes as other body cells and is therefore haploid.
gamete incompatibility An isolating mechanism between species in which sperm from one cannot fertilize eggs from another.
gametogenesis The production and maturation of gametes.
gas exchange The passage of gases, as a result of diffusion, from one compartment to another.
gastrin Hormone that stimulates the secretion of stomach acid.
gastrula The two-layered, cup-shaped stage of embryonic development.
gastrulation The process, during embryonic development, that results in a three layered embryo.
gel electrophoresis The separation of biological molecules on the basis of their size and charge by measuring their rate of movement through an electric field.
gene Discrete unit of heritable information about genetic traits. Consists of a sequence of DNA that codes for a specific polypeptide—a protein or part of a protein.
gene expression Turning a gene on or off. A gene is expressed when the protein it encodes is synthesized.
gene flow Spread of an allele throughout a species' gene pool.
gene gun Device used to shoot DNA-coated pellets into plant cells.
gene pool All of the alleles found in all of the individuals of a species.
gene therapy Replacing defective genes (or their protein products) with functional ones.
genealogical species concept A scheme that identifies as separate species all populations with a unique lineage.
general sense Also called proprioception. Any of the senses including temperature, pain, touch, pressure, and body position, with sensory receptors scattered throughout the body.
genetically modified organisms (GMOs) Organisms whose genome incorporates genes from another organism; also called transgenic or genetically engineered organisms.
genetic code Table showing which mRNA codons code for which amino acids.
genetic drift Change in allele frequency that occurs as a result of chance.
genetic variability All of the forms of genes, and the distribution of these forms, found within a species.
genetic variation Differences in alleles that exist among individuals in a population.
genome Entire suite of genes present in an organism.
genotype Genetic composition of an individual.
genus Broader biological category to which several similar species may belong.
geologic period A unit of time defined according to the rocks and fossils characteristic of that period.
germination The beginning of growth by a seed.
gibberellin A plant hormone that causes the elongation of stem cells.
glans penis The head of the penis.
glial cell A type of cell within the brain that does not carry messages but rather supports neurons by supplying nutrients, repairing the brain after injury, and attacking invading bacteria.
global climate change Changes in regional patterns of temperature and precipitation that are occurring around Earth as a result of increasing greenhouse gases in the atmosphere.
global warming Increase in average global temperatures as a result of the release of increased amounts of carbon dioxide and other greenhouse gases into the atmosphere.
glycolysis The splitting of glucose into pyruvate, which helps drive the synthesis of a small amount of ATP.
Golgi apparatus An organelle in eukaryotic cells consisting of flattened membranous sacs that modify and sort proteins and other substances.

gonadotropin-releasing hormone (GnRH) Hormone produced by the hypothalamus that stimulates the pituitary gland to release FSH and LH, thereby stimulating the activities of the gonads.

gonads The male and female sex organs; testicles in human males or ovaries in human females.

gradualism The hypothesis that evolutionary change occurs in tiny increments over long periods of time.

grassland Biome characterized by the dominance of grasses, usually found in regions of lower precipitation.

gravitropism Directional plant growth in response to gravity.

gray matter Unmyelinated axons, combined with dendrites and cell bodies of other neurons that appear gray in cross section.

greenhouse effect The retention of heat in the atmosphere by carbon dioxide and other greenhouse gases.

greenhouse gas Atmospheric gas such as water vapor, carbon dioxide (CO_2), methane (CH_4), and ozone (O_3) that absorb heat and thus contribute to the greenhouse effect.

ground tissue The plant tissues other than the vascular tissue, epidermis, and meristem.

growing season The length of time from last freeze in spring to first freeze in autumn.

growth rate Annual death rate in a population subtracted from the annual birth rate.

guanine Nitrogenous base in DNA.

guard cell Either of the paired cells encircling stomata that serve to regulate the size of the stomatal pore in leaves.

habitat Place where an organism lives.

habitat destruction Modification and degradation of natural forests, grasslands, wetlands, and waterways by people; primary cause of species loss.

habitat fragmentation Threat to biodiversity caused by humans that occurs when large areas of intact natural habitat are subdivided by human activities.

half-life Amount of time required for one-half the amount of a radioactive element that is originally present to decay into the daughter product.

haploid Describes cells containing only one member of each homologous pair of chromosomes; in humans, these cells are eggs and sperm.

hardened Relating to plants that have undergone the appropriate physiological changes that allow them to survive freezing temperatures.

hardiness In plants, the ability to survive freezing temperatures.

heart The muscular organ that pumps blood via the circulatory system to the lungs and body.

heart attack An acute condition, during which blood flow is blocked to a portion of the heart muscle, causing part of the muscle to be damaged or die.

heat The total amount of energy associated with the movement of atoms and molecules in a substance.

helper T cell A type of immune-system cell that enhances cell-mediated immunity and humoral immunity by secreting a substance that increases the strength of the immune response. *See also* T4 cell.

hemoglobin An iron-containing protein that carries oxygen in red blood cells.

hepatocyte Liver cell.

herbicide A chemical that kills plants.

herd immunity Immunity indirectly provided to those who cannot be vaccinated, including infants and pregnant women, when the majority of community members are vaccinated. The spread of the disease is contained and outbreaks are less likely in highly vaccinated populations.

heritability The amount of variation for a trait in a population that can be explained by differences in genes among individuals.

heterozygote Individual carrying two different alleles of a particular gene.

heterozygote advantage The tendency in individuals that are heterozygous for a larger number of genes to have higher fitness.

heterozygous Said of a genotype containing two different alleles of a gene.

high-density lipoprotein (HDL) A cholesterol-carrying particle in the blood that is high in protein and low in cholesterol.

histamine Chemical released from mast cells during allergic reactions. Causes blood vessels to dilate and become more permeable and lowers blood pressure.

HIV *See* human immunodeficiency virus (HIV).

homeostasis The steady-state condition an organism works to maintain.

hominin Referring to humans and human ancestors.

homologous pair Set of two chromosomes of the same size and shape with centromeres in the same position. Homologous pairs of chromosomes carry the same genes in the same locations but may carry different alleles.

homology Similarity in characteristics as a result of common ancestry.

homozygous Having two copies of the same allele of a gene.

hormones A protein or steroid produced in one tissue that travels through the circulatory system to act on another tissue to produce some physiological effect.

human immunodeficiency virus (HIV) Agent identified as causing the transmission and symptoms of AIDS.

hybrid Organism with parents from two different species. Offspring of two different strains of an agricultural crop.

hydrocarbon A compound consisting of carbons and hydrogens.

hydrogen atom One negatively charged electron and one positively charged proton.

hydrogenation Adding hydrogen gas under pressure to make liquid oils more solid.

hydrogen bond A type of weak chemical bond in which a hydrogen atom of one molecule is attracted to an electronegative atom of another molecule.

hydrogen ion The positively charged ion of hydrogen (H+) formed by removal of the electron from a hydrogen atom.
hydrophilic Readily dissolving in water.
hydrophobic Not able to dissolve in water.
hypertension High blood pressure.
hyphae Thin, stringy fungal material that grows over and within a food source.
hypothalamus Gland that helps regulate body temperature; influences behaviors such as hunger, thirst, and reproduction; and secretes a hormone (GnRH) that stimulates the activities of the gonads.
hypothesis Tentative explanation for an observation that requires testing to validate.

immune response Ability of the immune system to respond to an infection, resulting from increased production of B cells and T cells.
immune system The organ system that produces cells and cell products, such as antibodies, that help remove pathogenic organisms.
immunotherapy Cancer treatment that targets cancer cells for destruction by the immune system.
implantation The process in early pregnancy whereby a fertilized egg cell attaches to the uterine lining.
implicit bias The attitudes or stereotypes that affect our understanding, actions, and decisions in an unconscious manner.
inbreeding Mating between related individuals.
inbreeding depression Negative effect of homozygosity on the fitness of members of a population.
incomplete dominance A type of inheritance where the heterozygote has a phenotype intermediate to both homozygotes.
independent assortment The separation of homologous pairs of chromosomes into gametes independently of one another during meiosis.
independent variable A factor whose value influences the value of the dependent variable, but is not influenced by it. In experiments, the variable that is manipulated.
indeterminate growth Unrestricted or unlimited growth.
indirect effect In ecology, a condition where one species affects another indirectly through intervening species.
induced fit A change in shape of the active site of an enzyme so that it binds tightly to a substrate.
inductive reasoning A logical process that argues from specific instances to a general conclusion.
infant mortality Death rate of infants and children under the age of five.
infectious Applies to a pathogen that finds a tissue inside the body that will support its growth.
infertility The inability to conceive after one year of unprotected intercourse.
inflammatory response A line of defense triggered by a pathogen penetrating the skin or mucus membranes.
inhale To breathe in.
inner cell mass A cluster of cells in the blastocyst that eventually develops into the embryo.
inorganic Said of chemical compounds that do not contain carbon.
insulin A hormone secreted by the pancreas that lowers blood glucose levels by promoting the uptake of glucose by cells and the storage of glucose as glycogen in the liver.
integrated pest management (IPM) A strategy for pest control that relies on a mix of pesticides, cultural control, and better knowledge of pest populations.
interferon A chemical messenger produced by virus-infected cells that helps other cells resist infection.
intermembrane space The space between two membranes; for example, the space between the inner and outer mitochondrial membrane.
interneuron A neuron located between sensory and motor neurons that functions to integrate sensory input and motor output.
internode The region of a plant stem between nodes.
interphase Part of the cell cycle when a cell is preparing for division and the DNA is duplicated. Consists of G_1, S, and G_2.
intertidal zone The biome that forms on ocean shorelines between the high tide elevation and the low tide elevation.
introduced species A nonnative species that was intentionally or unintentionally brought to a new environment by humans.
invertebrate Animal without backbone.
involuntary muscle Muscle tissue whose action requires no conscious thought.
ion Electrically charged atom.
ionic bond A chemical bond resulting from the attraction of oppositely charged ions.
irrigation The technique of supplying additional water to crop plants, typically via flooding or spraying.

keystone species A species that has an unusually strong effect on the structure of the community it inhabits.
kidney Major organ of the excretory system, responsible for filtering liquid waste from the blood.
kingdom In some classifications, the most inclusive group of organisms; life is typically categorized into five or six. In other classification systems, the level below domain on the hierarchy.

labia majora Paired thick folds of skin that enclose and protect the labia minor of the vulva.
labia minora Paired thin folds of the vulva that enclose the urinary and vaginal openings and clitoris.
labor Strong rhythmic contractions that force a developing baby from the uterus through the vagina during childbirth.
lactation Production of milk to nurse offspring.
lake An aquatic biome that is completely landlocked.

large intestine Colon; portion of the digestive system located between the small intestine and the anus that absorbs water and forms feces.

larynx A portion of the upper respiratory tract made up primarily of stiff cartilage. Also known as the voice box.

latent virus Dormant virus.

leaf The primary photosynthetic organ in plants.

legumes A family of plants that produces seeds in pods, like peas and beans.

leptin A hormone by fat cells that may be involved in the regulation of appetite.

less developed country A term used by the United Nations to describe a country with low per-capita income and often poor population health and economic prospects.

life cycle Description of the growth and reproduction of an individual.

ligament A band of fibrous connective tissue joining bones.

light-independent reactions A series of reactions that occur in the chloroplast during photosynthesis and that utilize NADPH and ATP to reduce carbon dioxide and produce sugars.

light reactions A series of reactions that occur in chloroplasts during photosynthesis and serve to convert energy from the sun into the energy stored in ATP. These reactions also produce oxygen gas.

lipid Hydrophobic molecule, including fats, phospholipids, and steroids.

liver Organ with many functions, including the production of bile to aid in the absorption of fats.

lobule Subdivision of the lobes of the liver.

logistic growth Pattern of growth seen in populations that are limited by resources available in the environment. A graph of logistic growth over time typically takes the form of an S-shaped curve.

loose connective tissue Connective tissue that serves to bind epithelia to underlying tissues and to hold organs in place.

low-density lipoprotein (LDL) Cholesterol-carrying substance in the blood that is high in cholesterol and low in protein.

lung The primary organ of the respiratory system; the site where gas exchange occurs.

luteinizing hormone (LH) Hormone involved in sperm production, regulation of ovulation, and regulation of menstruation.

lymph node Organ located along lymph vessels that filters lymph and help defend against bacteria and viruses.

lymphatic system A system of vessels and nodes that return fluid and protein to the blood.

lymphocyte White blood cells that make up part of the immune system.

lysosome A membrane-bounded sac of hydrolytic enzymes found in the cytoplasm of many cells.

macroevolution Large-scale evolutionary change, usually referring to the origin of new species.

macromolecule Any of the large molecules including polysaccharides, proteins, and nucleic acids, composed of subunits joined by dehydration synthesis.

macronutrient Nutrient required in large quantities.

macrophage Phagocytic white blood cell that swells and releases toxins to kill bacteria.

malignant Describes a tumor that is cancerous, whether it is invasive or metastatic.

mandible Bone of the lower jaw.

marine Of, or pertaining to, salt water.

mark-recapture method A technique for estimating population size, consisting of capturing and marking a number of individuals, releasing them, and recapturing more individuals to determine what proportion are marked.

mass extinction Loss of species that is rapid and global in scale, and affects a wide variety of organisms.

matrix (1) In a mitochondrion, the semifluid substance inside the inner mitochondrial membrane, which houses the enzymes of the citric acid cycle. (2) In connective tissue, a nonliving substance between cells, ranging from fluid blood plasma to fibrous matrix in tendons to solid bone matrix.

mean Average value of a group of measurements.

mechanical isolation A form of reproductive isolation between species that depends on the incompatibility of the genitalia of individuals of different species.

medulla oblongata Region of the brain stem that is a continuation of the spinal cord and conveys information between the spinal cord and other parts of the brain.

meiosis Process that diploid sex cells undergo to produce haploid daughter cells. Occurs during gametogenesis.

memory cell Cell that is part of a clonal population, programmed to respond to a specific antigen, that helps the body respond quickly if the infectious agent is encountered again.

menopause Cessation of menstruation.

menstrual cycle Changes that occur in the uterus and depend on intricate interrelationships among the brain, ovaries, and lining of the uterus.

menstruation The shedding of the lining of the uterus during the menstrual cycle.

meristematic tissue Undifferentiated plant tissue from which new plant cells arise.

mesoderm The middle of three germ layers that arise during animal development.

messenger RNA (mRNA) Complementary RNA copy of a DNA gene, produced during transcription. The mRNA undergoes translation to synthesize a protein.

metabolic rate Measure of an individual's energy use.

metabolism All of the physical and chemical reactions that produce and use energy.

metaphase Stage of mitosis during which duplicated chromosomes align across the middle of the cell.

metastasis When cells from a tumor break away and start new cancers at distant locations.

microbe Microscopic organism, especially Bacteria and Archaea.
microbiologists Scientists who study microscopic organisms, especially referring to those who study prokaryotes.
microevolution Changes that occur in the characteristics of a population.
micronutrient Nutrient needed in small quantities.
microorganism *See* microbe.
microtubule Protein structure that moves chromosomes around during mitosis and meiosis.
microvillus Fine fingerlike projection composed of epithelial cells that function in absorption.
micturition Release of urine from the bladder. Also known as urination.
midbrain Uppermost region of the brain stem, which adjusts the sensitivity of the eyes to light and of the ears to sound.
mineral Inorganic nutrient essential to many cell functions.
mitochondria Organelles in which products of the digestive system are converted to ATP.
mitosis The division of the nucleus that produces daughter cells that are genetically identical to the parent cell. Also, portion of the cell cycle in which DNA is apportioned into two daughter cells.
model organism Any nonhuman organism used in genetic studies to help scientists understand human genes because they share genes with humans.
model systems *See* model organisms.
mold A fungal form characterized by rapid, asexual reproduction.
molecular clock Principle that DNA mutations accumulate in the genome of a species at a constant rate, permitting estimates of when the common ancestor of two species existed.
molecule Two or more atoms held together by covalent bonds.
monocot One of the two classes of flowering plants, also called narrow-leaved plants. Monocot seeds contain one leaf (cotyledon).
monoculture Practice of planting a single crop over a wide acreage.
monomer An individual molecule that binds to other molecules to form a polymer.
monosaccharide Simple sugar.
monozygotic twins Identical twins that developed from one zygote.
morphological species concept Definition of species that relies on differences in physical characteristics among them.
morphology Appearance or outward physical characteristics.
morula A ball of cells formed by rapid cell division after fertilization.
motor neuron Neuron that carries information away from the brain or spinal cord to muscles or glands.
mulching Covering the ground around favored plants with light-blocking material to prevent weed growth.
multicellular The condition of being composed of many coordinated cells.
multiple allelism A gene for which there are more than two alleles in the population.
muscle fiber Single cell that aligns with others in parallel bundles to form muscles.
muscle tissue Specialized contractile tissue.
mutation Change to a DNA sequence that may result in the production of altered proteins.
mutualism Interaction between two species that provides benefits to both species.
mycologist Scientist who specializes in the study of fungi.
mycorrhiza A symbiotic relationship between fungal hyphae and plant roots that benefits both partners.
myelin sheath Protective layer that coats many axons, formed by supporting cells such as Schwann cells. The myelin sheath increases the speed at which the electrochemical impulse travels down the axon.
myofibril Structure found in muscle cells, composed of thin filaments of actin and thick filaments of myosin.
myosin A type of protein filament that, along with actin, causes muscle cells to contract.

natural experiment Situation where unique circumstances allow a hypothesis test without prior intervention by researchers.
natural killer cell A cell that attacks virus-infected cells or tumor cells without being activated by an immune system cell or antibody.
natural selection Process by which individuals with certain traits have greater survival and reproduction than individuals who lack these traits, resulting in an increase in the frequency of successful alleles and a decrease in the frequency of unsuccessful ones.
negative feedback A mechanism of maintaining homeostasis in which the product of the process inhibits the process.
nephron The functional structure within a kidney where waste filtration and urine concentration occurs.
nerve Bundle of neurons; nerves branch out from the brain and spinal cord to eyes, ears, internal organs, skin, and bones.
nerve impulse Electrochemical signal that controls the activities of muscles, glands, organs, and organ systems.
nervous system Brain, spinal cord, sense organs, and nerves that connect organs and link this system with other organ systems.
nervous tissue Tissue composed of neurons and associated cells.
net primary production (NPP) Amount of solar energy converted to chemical energy by plants, minus the amount of this chemical energy plants need to support themselves. A measure of plant growth, typically over the course of a single year.

neuron Specialized message-carrying cell of the nervous system.

neurotransmitter One of many chemicals released by the presynaptic neuron into the synapse, which then diffuse across the synapse and bind to receptors on the membrane of the postsynaptic neuron.

neutral mutation A genetic mutation that confers no selective advantage or disadvantage.

neutron An electrically neutral particle found in the nucleus of an atom.

nicotinamide adenine dinucleotide Intracellular electron carrier. Oxidized form is NAD+; reduced form is NADH.

nicotine The active drug in tobacco that stimulates dopamine receptors in the brain.

nitrogen-fixing bacteria Organisms that convert nitrogen gas from the atmosphere into a form that can be taken up by plant roots; some species live in the root nodules of legumes.

nitrogenous base Nitrogen-containing base found in DNA: A, C, G, and T and in RNA: U.

node Point on a plant stem where a leaf and axillary bud arise.

nodule Compartment housing nitrogen-fixing bacteria, produced on the roots of legume plants such as beans and alfalfa.

nondisjunction Failure of chromosomes to separate properly during meiosis.

nonpolar Won't dissolve in water. Hydrophobic.

nonrenewable resource Resource that is a one-time supply and cannot be easily replaced.

nonspecific defenses Defense system against infection that does not distinguish one pathogen from another. Includes the skin, secretions, and mucous membranes.

normal distribution Bell-shaped curve, as for the distribution of quantitative traits in a population.

nuclear envelope The double membrane enclosing the nucleus in eukaryotes.

nuclear transfer Transfer of a nucleus from one cell to another cell that has had its nucleus removed.

nucleic acids Polymers of nucleotides that comprise DNA and RNA.

nucleoid region The region of a prokaryotic cell where the DNA is located.

nucleolus A spherical structure found inside the nucleus where ribosomes are produced.

nucleotides Building blocks of nucleic acids that include a sugar, a phosphate, and a nitrogenous base.

nucleus Cell structure that houses DNA; found in eukaryotes.

nutrient cycling Process by which inorganic nutrients become available to plants. Nutrient cycling in a natural environment relies on a healthy community of decomposers within the soil.

nutrients (1) Substances that provide nourishment. (2) Atoms other than carbon, hydrogen, and oxygen that must be obtained from an organisms environment.

objective Without bias.

occipital lobe The posterior lobe of each cerebral hemisphere, containing the visual center of the brain.

ocean A biome consisting of open stretches of salt water.

oogenesis Formation and development of female gametes, which occurs in the ovaries and results in the production of egg cells.

organ A specialized structure composed of several different types of tissues.

organ systems Suites of organs working together to perform a function or functions.

organelle Subcellular structure found in the cytoplasm of eukaryotic cells that performs a specific job.

organic When pertaining to agriculture, refers to products grown without the use of manufactured pesticides and inorganic fertilizer.

organic chemistry The chemistry of carbon-containing substances.

osmosis The diffusion of water across a selectively permeable membrane.

ossa coxae The paired bones that form the bony pelvis.

osteoblast Bone-forming cell responsible for the deposition of collagen.

osteoclast Bone-reabsorbing cell that liberates calcium.

osteocyte Highly branched cell found in bone.

osteoporosis A condition resulting in an elevated risk of bone breakage from weakened bones.

ovary (1) In animals, the paired abdominal structures that produce egg cells and secrete female hormones. (2) In plants, a chamber of the carpel containing the ovules.

overexploitation Threat to biodiversity caused by humans that encompasses overhunting and overharvesting.

oviduct Egg-carrying duct that brings egg cells from ovaries to uterus.

ovulation Release of an egg cell from the ovary.

ovule Structure in flowering plants consisting of eggs and accessory tissues. After fertilization, will develop into a seed.

oxytocin Pituitary hormone that stimulates the contraction of smooth muscle of the uterus during labor and facilitates secretion of milk from the breast during nursing.

pacemaker A patch of heart tissue or an implanted device that produces a regular electrical signal, setting the rhythm for the cardiac cycle.

paleontologist Scientist who searches for, describes, and studies ancient organisms.

pancreas Gland that secretes digestive enzymes and insulin.

pancreatitis Inflammation of the pancreas, an organ involved in digestion and the regulation of glucose levels.

parasite An organism that benefits from an association with another organism that is harmed by the association.

parathyroid gland One of four endocrine glands located on the thyroid that secrete parathyroid hormone in order to regulate blood calcium levels.

parietal lobe Part of the brain that processes information about touch and is involved in self-awareness.

Parkinson's disease Disease that results in tremors, rigidity, and slowed movements. May be due to faulty dopamine production.

particulate Tiny airborne particle found in smoke and other pollutants.

passive immunity Immunity acquired when antibodies are passed from one individual to another, as from mother to child during breast-feeding.

passive smoker An individual who inhales environmental tobacco smoke but who does not actively consume tobacco.

passive transport The diffusion of substances across a membrane with their concentration gradient and not requiring an input of ATP.

pathogen Disease-causing organism.

pedigree Family tree that follows the inheritance of a genetic trait for many generations.

peer review The process by which reports of scientific research are examined and critiqued by other researchers before they are published in scholarly journals.

penis The copulatory structure in males.

peptide bond Covalent bond that joins the amino group and carboxyl group of adjacent amino acids.

perennial *See* perennial plant.

perennial plant Plant that lives for many years.

peripheral nervous system (PNS) Network of nerves outside the brain and spinal cord that links the CNS with sense organs.

peristalsis Rhythmic muscle contractions that move food through the digestive system.

permafrost Permanently frozen soil.

pest Any organism that competes with humans for agricultural production or other resources.

pesticide Chemical that kills or disables agricultural pests.

pesticide treadmill The tendency for pesticide toxicity and total applications to increase once a farmer begins to employ pesticides.

petal A flower part that typically functions to attract pollinators.

petiole The stalk of a leaf.

pH A logarithmic measure of the hydrogen ion concentration ranging from 0 to 14. Lower numbers indicate higher hydrogen ion concentrations.

pharming Genetic engineering to make pharmaceuticals.

pharynx Tube and muscles connecting the mouth to the esophagus; throat.

phenotype Physical and physiological traits of an individual.

phenotypic ratio Proportion of individuals produced by a genetic cross who possess each of the various phenotypes that cross can generate.

phloem The portion of a plant's vascular tissue modified for multidirectional transport of nutrients throughout the body of the plant.

phloem sap The liquid, often containing dissolved carbohydrates, carried in the phloem tubes in a plant.

phospholipid One of three types of lipids; phospholipids are components of cell membranes.

phospholipid bilayer The membrane that surrounds cells and organelles and is composed of two layers of phospholipids.

phosphorylation To introduce a phosphoryl group into an organic compound.

photoperiodism Response to the duration and timing of day and night length in plants.

photorespiration A series of reactions triggered by the closing of stomatal openings to prevent water loss.

photosynthesis Process by which plants, along with algae and some bacteria, transform light energy to chemical energy.

phototropism Growth in response to directional light.

phyla (singular: phylum) The taxonomic category below kingdom and above class.

phylogeny Evolutionary history of a group of organisms.

phytochrome The light-sensitive chemical in plants responsible for photoperiodic responses.

pituitary gland Small gland attached by a stalk to the base of the brain that secretes growth hormone, reproductive hormones, and other hormones.

placebo Sham treatments in experiments.

placenta Membrane produced by a developing fetus that releases a hormone to extend the life of the corpus luteum.

plant hormone Substance in plants that helps tune the organisms' response to the environment.

Plantae Multicellular photosynthetic eukaryotes, excluding algae.

plasma The liquid portion of blood.

plasma membrane Structure that encloses a cell, defining the cell's outer boundary.

plasmid Circular piece of bacterial DNA that normally exists separate from the bacterial chromosome and can make copies of itself.

platelet Cell in the blood that carries constituents required for the clotting response.

pleiotropy The ability of one gene to affect many different functions.

polar Describes a molecule with regions having different charges; goes into solution in water.

polarization The difference in charge between the inside and outside of the resting neuron cell.

poles Opposite ends of a sphere, such as of a cell or of a planet such as Earth.

pollen The male gametophyte of seed plants.

pollination The process by which pollen is transferred to the female structures of a plant.

pollinator An organism that transfers sperm (pollen grains) from one flower to the female reproductive structures of another flower.

pollution Human-caused threat to biodiversity involving the release of poisons, excess nutrients, and other wastes into the environment.

polyculture Practice of planting many different crop plants over a single farm's acreage.

polygenic trait A trait influenced by many genes.

polymer General term for a macromolecule composed of many chemically bonded monomers.

polymerase An enzyme that catalyzes phosphodiester bond formation between nucleotides.

polymerase chain reaction (PCR) A laboratory technique that allows the production of many identical DNA molecules.

polyploidy A chromosomal condition involving more than two sets of chromosomes.

polysaccharide A carbohydrate composed of three or more monosaccharides.

polyunsaturated Relating to fats consisting of carbon chains with many double bonds unsaturated by hydrogen atoms.

pond An aquatic biome that is completely landlocked.

pons A structure located on the brain stem, between the brain and spinal cord.

population Subgroup of a species that is somewhat independent from other groups.

population crash Steep decline in number that may occur when a population grows larger than the carrying capacity of its environment.

population cycle In some populations, the tendency to increase in number above the environment's carrying capacity, resulting in a crash, following by an overshoot of the carrying capacity and another crash, continuing indefinitely.

population genetics Study of the factors in a population that determine allele frequencies and their change over time.

population pyramid A visual representation of the number of individuals in different age categories in a population.

positive feedback A relatively uncommon homeostatic mechanism in which the product of a process intensifies the process, thereby promoting change.

postsynaptic neuron The neuron that responds to neurotransmitter released from the presynaptic neuron.

prairie A grassland biome.

precautionary principle The principle that the introduction of a new product or activity whose ultimate effects are disputed or unknown should be resisted to prevent unpredicted harmful effects.

precipitation When water vapor in the atmosphere turns to liquid or solid form and falls to Earth's surface.

predation Act of capturing and consuming an individual of another species.

predator Organism that eats other organisms.

prediction Result expected from a particular test of a hypothesis if the hypothesis were true.

pressure flow mechanism The process by which phloem sap moves from a sugar source to a sugar sink within a plant.

presynaptic The neuron that secretes neurotransmitter into a synapse, transmitting a signal.

primary consumer Organism that eats plants.

primary growth Growth occurring at the tips of a plant, originating in the apical meristems.

primary source Article reporting research results, written by researchers, and reviewed by the scientific community.

probability Likelihood that something is the case or will happen.

processed food Food that has been modified from its original form to increase shelf life, transportability, or the like.

producer Organism that produces carbohydrates from inorganic carbon, typically via photosynthesis.

product The modified chemical that results from a chemical or enzymatic reaction.

progesterone Ovarian hormone. High levels have a negative feedback effect on the hypothalamus, causing GnRH secretion to decrease.

prokaryote Type of cell that does not have a nucleus or membrane-bounded organelles.

prolactin A hormone produced by the pituitary gland that stimulates the development of mammary glands.

promoter Sequence of nucleotides to which the polymerase binds to start transcription.

prophase Stage of mitosis during which duplicated chromosomes condense.

prostate A gland that secretes an acid-neutralizing fluid into semen.

protein Cellular constituent made of amino acids coded for by genes. Proteins can have structural, transport, or enzymatic roles.

protein synthesis Joining amino acids together, in an order dictated by a gene, to produce a protein.

Protista Kingdom in the domain Eukarya containing a diversity of eukaryotic organisms, most of which are unicellular.

proton A positively charged subatomic particle.

prune To trim back the growing tips of a plant.

pseudoscience Information presented as scientific but does not hold up under scientific scrutiny.

psychosis Abnormal condition of the mind.

puberty The point in human development when male and female hormones are triggered in the body. Males produce sperm at this time, and females begin the menstrual cycle.

pulmonary circuit The path of blood through vessels from the heart, through the lungs, and back to the heart.

pulse The volume of blood that passes into the arteries as a result of the heart's contraction.

punctuated equilibrium The hypothesis that evolutionary changes occur rapidly and in short bursts, followed by long periods of little change.

Punnett square Table that lists the different kinds of sperm or eggs parents can produce relative to the gene or genes in question and predicts the possible outcomes of a cross between these parents.

pus Fluid, dead cells, and microorganisms that accumulate at the site of the infection.

pyruvic acid The three-carbon molecule produced by glycolysis.

Q angle Angle of the femur in relation to a horizontal line drawn through the kneecap.

quantitative trait Trait that has many possible values.

race *See* biological race.

radiation therapy The use of high-energy particles to destroy cancer cells.

radioactive decay Natural, spontaneous breakdown of radioactive elements into different elements, or *daughter products*.

radiometric dating Technique that relies on radioactive decay to estimate a fossil's age.

random alignment When members of a homologous pair line up randomly with respect to maternal or paternal origin during metaphase I of meiosis, thus increasing the genetic diversity of offspring.

random assignment Placing individuals into experimental and control groups randomly to eliminate systematic differences between the groups.

random distribution The dispersion of individuals in a population without pattern.

random fertilization The unpredictability of exactly which gametes will fuse during the process of sexual reproduction.

reabsorption Reuptake of water and other essential substances by a nephron from the filtrate initially squeezed into the kidney.

reactant Any starting material in a chemical reaction.

reading frame The grouping of mRNAs into three base codons for translation.

receptor (1) Protein on the surface of a cell that recognizes and binds to a specific chemical signal. (2) *See also* sensory receptor.

recessive Applies to an allele with an effect that is not visible in a heterozygote.

recombinant Produced by manipulating a DNA sequence.

recombinant bovine growth hormone (rBGH) Growth hormone produced in a laboratory and injected into cows to increase their size and ability to produce milk.

red blood cell Primary cellular component of blood, responsible for ferrying oxygen throughout the body.

reflex Automatic response to a stimulus.

reflex arc Nerve pathway followed during a reflex consisting of a sensory receptor, a sensory neuron, an interneuron, a motor neuron, and an effector.

renal artery The vessel that carries blood to the kidney for filtering.

renal vein The vessel that carries filtered blood from the kidney.

repolarization The restoration of a charge difference across a membrane.

reproductive cloning Transferring the nucleus from a donor adult cell to an egg cell without a nucleus in order to clone the adult.

reproductive isolation Prevention of gene flow between different biological species due to failure to produce fertile offspring; can include premating barriers and postmating barriers.

reproductive organ Internal and external genitalia involved in production and delivery of gametes.

respiratory surface Body surface across which gas exchange occurs.

respiratory system The organ system involved in gas exchange between an animal and its environment. In humans, the lungs and air passages.

restriction enzyme An enzyme that cleaves DNA at specific nucleotide sequences.

retrovirus RNA virus that synthesizes DNA using reverse transcriptase.

reuptake In neurons, the process by which neurotransmitters are reabsorbed by the neuron that secreted them.

reverse transcriptase Enzyme in RNA viruses that produces DNA by transcription of viral RNA.

rhizomes Underground stems in plants.

ribonucleic acid (RNA) Information-carrying molecule composed of nucleotides.

ribose The five-carbon sugar in RNA.

ribosomal RNA (rRNA) RNA molecules that are part of the structure of the ribosome, found in all organisms that translate RNA into protein.

ribosome Subcellular structure that helps translate genetic material into proteins by anchoring and exposing small sequences of mRNA.

risk factor Any exposure or behavior that increases the likelihood of disease.

Ritalin Stimulant used to treat ADD.

river Aquatic biome characterized by flowing water.

RNA *See* ribonucleic acid (RNA).

RNA polymerase Enzyme that synthesizes mRNA from a DNA template during transcription.

root cap Loosely organized cells that cover the growing tip of a root and are continually shed as the root pushes through the soil.

root hair Extension of an epidermal cell on young roots, which maximize the surface area for water and nutrient uptake.

root system The plant organ system responsible for water and nutrient uptake and anchorage.

rough endoplasmic reticulum Ribosome-studded subcellular membranes found in the cytoplasm and responsible for some protein synthesis.

rubisco Abbreviation for ribulose bisphosphate carboxylase oxygenase, the enzyme that catalyzes the first step in the light-independent reactions of photosynthesis.

runoff Water moving across a land surface, picking up contaminants, and eventually flowing into lakes, streams, or groundwater.

salinization Degradation of soil by mineral salts deposited as a result of irrigation.

salt A charged substance that ionizes in solution.

sample Small subgroup of a population used in an experimental test.

sample size Number of individuals in both the experimental and control groups.

sampling error Effect of chance on experimental results.

sarcomere Any of the repeating units in a myofibril of striated muscle, bounded by Z discs.

saturated fat Type of lipid rich in single bonds. Found in butter and other fats that are solids at room temperature. This type of fat is associated with higher blood cholesterol levels.

savanna Grassland biome containing scattered trees.

Schwann cell Cells that form the myelin sheath along the axons of nerve cells in the peripheral nervous system.

scientific method A systematic method of research consisting of putting a hypothesis to a test designed to disprove it, if it is in fact false.

scientific theory Body of scientifically accepted general principles that explain natural phenomena.

scrotum The pouch of skin that houses the testes.

secondary compounds Chemicals produced by plants and some other organisms as side reactions to normal metabolic pathways and that typically have an antipredator or antibiotic function.

secondary consumers Animals that eat primary consumers; predators.

secondary growth Growth in girth of plants, as a result of cell division in lateral meristems.

secondary sources Books, news media, and advertisements as sources of scientific information.

secretion A step in waste removal in the kidney, in which contaminants in low concentration in the blood are actively absorbed into the urine.

seed A plant embryo packaged with a food source and surrounded by a seed coat.

seedling A young plant that develops from a germinating seed.

segregation Separation of pairs of alleles during the production of gametes. Results in a 50% probability that a given gamete contains one allele rather than the other.

semen Sperm and energy-rich associated fluids.

semiconservative replication DNA replication results in the production of two daughter DNA molecules, each with one conserved and one new strand of DNA.

semilunar valve Heart valve controlling blood flow from the ventricles into blood vessels leading away from the heart.

seminal vesicle Either of two pouchlike glands located on both sides of the bladder that add a fructose-rich fluid to semen prior to ejaculation.

seminiferous tubule Highly coiled tube in the testicles where sperm are formed.

semipermeable In biological membranes, a membrane that allows some substances to pass but prohibits the passage of others.

sensory neuron A neuron that conducts impulses from a sense organ to the central nervous system.

sensory receptor Any of the cellular systems that collect information about the environment inside or outside the body and transmit that information to the brain.

sepal The outermost floral structure, usually enclosing the other flower parts in a bud.

sex chromosome Any of the sex-determining chromosomes (X and Y in humans).

sex determination Determining the biological sex of an offspring. Humans have a chromosomal mechanism of sex determination in which two X chromosomes produce a female and an X and a Y chromosome produce a male.

sex hormone Any of the steroid hormones that affect development and functions of reproductive structures and secondary sex characteristics.

sex-linked trait A trait that is linked to a gene carried on one of the sex chromosomes.

sexual reproduction Reproduction involving two parents that gives rise to offspring that have unique combinations of genes.

sexual selection Form of natural selection that occurs when a trait influences the likelihood of mating.

shoot system The organs of a plant that are modified for photosynthesis: stem and leaf.

short tandem repeat Small repeat sequences of DNA used in profiling.

signal transduction When a change in a cell or its environment is relayed through various molecules and results in a cellular response.

single nucleotide polymorphism (SNP) A DNA sequence variation that occurs when members of species differ from each other at a single nucleotide (A, T, C, or G) locus.

sink A plant organ that is using carbohydrate, either for growth and metabolism or for conversion into starch.

sinoatrial (SA) node Region of the heart muscle that generates an electrical signal that controls heart rate. *See also* pacemaker.

sister chromatid Either of the two duplicated, identical copies of a chromosome formed after DNA synthesis.

skeletal muscle Striated muscle involved with voluntary movements.

small intestine The narrow, twisting, upper part of the intestine where nutrients are absorbed into the blood.

smooth endoplasmic reticulum The subcellular, cytoplasmic membrane system responsible for lipid and steroid biosynthesis.

smooth muscle Nonstriated, spindle-shaped muscle cells that line organs and blood vessels.

soil Medium for plant growth made up of mineral particles, partially decayed organic matter, and living organisms.

soil erosion Loss of topsoil.

solar irradiance The amount of solar energy hitting Earth's surface at any given point.

solstice When the sun reaches its maximum and minimum elevation in the sky.

solute The substance that is dissolved in a solution.

solution A mixture of two or more substances.

solvent A substance, such as water, that a solute is dissolved in to make a solution.

somatic cell Any of the body cells in an organism. Any cell that is not a gamete.

somatic cell gene therapy Changes to malfunctioning genes in somatic or body cells. These changes will not be passed to offspring.

source Plant organs that are actively generating sugar.

spatial isolation A mechanism for reproductive isolation that depends on the geographic separation of populations.

special creation The hypothesis that all organisms on Earth arose as a result of the actions of a supernatural creator.

special senses Senses that have specialized organs. These include sight, hearing, equilibrium, taste, touch, and smell.

speciation Evolution of one or more species from an ancestral form; macroevolution.

species A group of individuals that regularly breed together and are generally distinct from other species in appearance or behavior.

species-area curve Graph describing the relationship between the size of a natural landscape and the relative number of species it contains.

specific defense Defense against pathogens that utilizes white blood cells of the immune system.

specificity Phenomenon of enzyme shape determining the reaction the enzyme catalyzes.

sperm Gametes produced by males.

spermatogenesis Production of sperm.

spermatozoa Mature sperm composed of a small head containing DNA, a midpiece that contains mitochondria, and a tail (flagellum).

spinal cord Thick cord of nervous tissue that extends from the base of the brain through the spinal column.

spongy bone The porous, honeycomblike material of the inner bone.

spore Reproductive cell in plants and fungi that is capable of developing into an adult without fusing with another cell.

stabilizing selection Natural selection that favors the average phenotype and selects against the extremes in the population.

stamen Male flower part, producing the pollen.

standard error A measure of the variance of a sample; essentially the average distance a single data point is from the mean value for the sample.

staple crop One of a number of agricultural crops that make up the majority of calories in human societies.

statistical test Mathematical formulation that helps scientists evaluate whether the results of a single experiment demonstrate the effect of treatment.

statistically significant Said of results for which there is a low probability that experimental groups differ simply by chance.

statistics Specialized branch of mathematics used in the evaluation of experimental data.

stem The part of vascular plants that is above ground, as well as similar structures found underground.

stem cell Cells that can divide indefinitely and can differentiate into other cell types.

steppe Biome characterized by short grasses; found in regions with relatively little annual precipitation.

steroid Naturally occurring or synthetic organic fat-soluble substance that produces physiologic effects.

stigma Sticky pad on the tip of a flower's carpel where pollen is deposited.

stomach A sac-like structure in the abdomen that is involved in digestion.

stomata Pores on the photosynthetic surfaces of plants that allow air into the internal structure of leaves and green stems. Stomata also provide portals through which water can escape.

stop codon An mRNA codon that does not code for an amino acid and causes the amino acid chain to be released into the cytoplasm.

stream Biome characterized by flowing water, sometimes seasonal. Typically smaller than rivers.

striated muscle A voluntary muscle made up of elongated, multinucleated fibers. Typically skeletal and cardiac muscle that is distinguished from smooth muscle by the in-register banding patterns of actin and myosin filaments.

stroke Acute condition caused by a blood clot that blocks blood flow to an organ or other region of the body.

stroma The semi-fluid matrix inside a chloroplast where the light-independent reactions of photosynthesis occurs.

subspecies Subdivision of a species that is not reproductively isolated but represents a population or set of populations with a unique evolutionary history. *See also* biological race.

substrate The substance upon which an enzyme reacts.

succession Replacement of ecological communities over time since a disturbance, until finally reaching a stable state.
sugar-phosphate backbone Series of alternating sugars and phosphates along the length of the DNA helix.
supernatural Not constrained by the laws of nature.
sustainable Any human activity that can be maintained without depletion of or permanent damage to the resources on which it depends.
symbiosis A relationship between two species.
sympatric In the same geographic region.
synapse Gap between neurons consisting of the terminal boutons of the presynaptic neuron, the space between the two adjacent neurons, and the membrane of the postsynaptic neuron.
synergistic The creation of a whole that is greater than the sum of the parts.
systematist Biologist who specializes in describing and categorizing a particular group of organisms.
systemic circuit Flow of blood from the heart to body capillaries and back to the heart.
systole Portion of the cardiac cycle when the heart is contracting, forcing blood into arteries.

T lymphocyte (T cell) Immune-system cell that develops in the thymus gland; it facilitates cell-mediated immunity.
target cell A cell that responds to regulatory signals such as hormones.
telophase Stage of mitosis during which the nuclear envelope forms around the newly produced daughter nucleus, and chromosomes decondense.
temperate forest Biome dominated by deciduous trees.
temperature A measure of the intensity of heat or kinetic energy.
temporal bone Either of the paired bones forming the sides and base of the cranium.
temporal isolation Reproductive isolation between populations maintained by differences in the timing of mating or emergence.
temporal lobe Part of the cerebral hemisphere that processes auditory and visual information, memory, and emotion. Located in front of the occipital lobe.
tension In plants, negative water pressure.
terminal bouton Knoblike structure at the end of an axon.
terminal bud Apical meristem on the tip of a stem or branch.
terrestrial Of or on dry land.
testable Possible to evaluate through observations of the measurable universe.
testicle (*plural:* testes) The paired male gonads, involved in gametogenesis and secretion of reproductive hormones.
testosterone Masculinizing hormone secreted by the testes.
thalamus Main relay center between the spinal cord and the cerebrum.
theory *See* scientific theory.
theory of evolution Theory that all organisms on Earth today are descendants of a single ancestor that arose in the distant past. *See also* evolution.
therapeutic cloning Using early embryos as donors of stem cells for the replacement of damaged tissues and organs in another individual.
thigmotropism Growth in response to contact with a solid object.
thylakoid Flattened membranous sac located in the chloroplast stroma. Function in photosynthesis.
thymine Nitrogenous base in DNA.
thymus An endocrine gland located in the neck region that helps establish the immune system.
thyroid gland An endocrine gland in the neck region that stimulates metabolism and regulates blood calcium levels.
tilling Turning over the soil to kill weed seedlings, with the goal of removing competitors before a crop is planted.
tissue A group of cells with a common function.
tongue A muscular structure in the mouth that has taste buds that help you taste food. It aids in breaking down food for the digestive process.
toxin poisonous substance produced by bacteria or other cell.
trachea Air passage from upper respiratory system into lower respiratory system. Also called windpipe.
tracheid Narrow xylem cell with pitted walls.
transcription Production of an RNA copy of the protein coding DNA gene sequence.
trans fat Contains unsaturated fatty acids that have been hydrogenated, which changes the fat from a liquid to a solid at room temperature.
transfer RNA (tRNA) Amino-acid-carrying RNA structure with an anti-codon that binds to an mRNA codon.
transgenic organism Organism whose genome incorporates genes from another organism; also called genetically modified organism (GMO).
translation Process by which an mRNA sequence is used to produce a protein.
translocation Movement of phloem sap around the body of a plant.
transplant Organ transplantation involves removing an organ from the body of one person and placing it in the body of another person.
transpiration Movement of water from the roots to the leaves of a plant, powered by evaporation of water at the leaves and the cohesive and adhesive properties of water.
trophic level Feeding level or position on a food chain; for example, producers and primary consumers.
trophic pyramid Relationship among the mass of populations at each level of a food web.
trophoblast The outer layer of a developing blastocyst that supplies nutrition to the embryo.
tropical forest Biome dominated by broad-leaved, evergreen trees; found in areas where temperatures never drop below the freezing point of water.
tropism In plants, directional growth.

tuberculosis (TB) Degenerative lung disease caused by infection with the bacterium *Mycobacterium tuberculosis*.
tumor Mass of tissue that has no apparent function in the body.
tumor suppressor Cellular protein that stops tumor formation by suppressing cell division. When mutated leads to increased likelihood of cancer.
tundra Biome that forms under very low temperature conditions. Characterized by low-growing plants.
turgid In plants, cells that are filled with enough water so that the cell walls are deformed.

understory The level of a forest below the canopy trees and shrub cover, typically consisting of herbaceous perennial plants and tree and shrub seedlings.
undifferentiated A cell that is not specialized.
unicellular Made up of a single cell.
uniform distribution Occurs when individuals in a population are disbursed in a uniform manner across a habitat.
unsaturated fat Type of lipid containing many carbon-to-carbon double bonds; liquid at room temperature.
unsustainable Relating to practices that may compromise the ability of future generations to support a large human population with an adequate quality of life.
uracil Nitrogenous base in RNA.
urban sprawl The tendency for the boundaries of urban areas to grow over time as people build housing and commercial districts farther and farther from an urban core.
ureter Tube that delivers urine from a kidney to the bladder.
urethra Urine-carrying duct that also carries sperm in males.
urinary bladder An organ of the excretory system that stores urine after it is excreted from the kidneys.
urinary system The organ system responsible for the filtering, collection, and excretion of liquid waste; consisting of the kidneys, ureters, bladder, and urethra.
urine Liquid expressed by the kidneys and expelled from the bladder in mammals, containing the soluble waste products of metabolism.
uterus Pear-shaped muscular organ in females that can support pregnancy and that undergoes menstruation when its lining is shed.

vaccination A preparation of a weakened or killed pathogen, or portion of a pathogen, that will stimulate the immune system of a recipient to prepare a long-term defense (memory cells) against that pathogen.
vagina Muscular canal in females leading from the cervix to the vulva.
variable A factor that varies in a population or over time.
variance Mathematical term for the amount of variation in a population.
variant An individual in a population that differs genetically from other individuals in the population.
vascular cambium Meristematic tissue that forms in vascular bundles of dicot roots and stems and permits secondary growth.
vascular system Plant tissue system responsible for the delivery of water and dissolved solutes throughout the plant body.
vascular tissue Cells that transport water and other materials within a plant.
vas deferens Either of the two ducts in males that carry sperm from the epididymis to the urethra.
vegetative organ Plant organ that is not involved in sexual reproduction.
vein Vessel that carries blood from the body tissues back to the heart.
ventricle Chamber of the heart that pumps blood from the heart to the lungs or systemic circulation.
vertebra (*plural*: vertebrae) Bone of the spinal column through which the spinal cord passes.
vertebral column Also called the spine; the series of vertebrae and cartilaginous disks extending from the brain to the pelvis.
vertebrate Animal with a backbone.
vesicle Membrane-bounded sac-like structure. In neurons, these structures are found in the terminal bouton and store neurotransmitters.
vessel element Wide xylem cell with perforated ends, found only in angiosperms.
vestigial trait Modified with no or relatively minor function compared to the function in other descendants of the same ancestor.
villus (*plural*: villi) Small fingerlike projections on the inside of the small intestine that function in nutrient absorption.
viral envelope Layer formed around some virus protein coats (capsids) that is derived from the cell membrane of the host cell and may also contain some proteins encoded by the viral genome.
virus Infectious intracellular parasite with its own genetic material that can only reproduce by forcing its host to make copies of it.
vitamin Organic nutrient needed in small amounts. Most vitamins function as coenzymes.
voluntary Muscle normally under conscious control. Mainly skeletal muscle.
vulva The outer portion of the female external genitalia including the labia majora, labia minora, clitoris, and vaginal and urethral openings.

wastewater Liquid waste produced by residential, commercial, or industrial activities.
water One molecule of water consists of one oxygen and two hydrogen atoms.
weather Current temperature and precipitation conditions.
weed Common term for nonpreferred plant.

wetland Biome characterized by standing water, shallow enough to permit plant rooting.
white blood cell General term for cell of the immune system.
white matter Nervous system tissue, especially in the brain and spinal cord, made of myelinated cells.
whole food Any food that has not undergone processing. Includes grains, beans, nuts, seeds, fruits, and vegetables.
wood Xylem cells produced by secondary growth of stems and roots.
woody plant Plant that produces stiffened stems via secondary production of xylem.

***x*-axis** The horizontal axis of a graph. Typically describes the independent variable.
X-linked gene Any of the genes located on the X chromosome.
xylem Plant vascular tissue, dead at maturity, that carries water and dissolved minerals in a one-way flow from the roots to the shoots of a plant.
xylem sap Water and dissolved minerals flowing in the xylem vessels.

***y*-axis** The vertical axis of a graph. Typically describes the dependent variable.
yeast Single-celled eukaryotic organisms found in bread dough. Often used as model organisms and in genetic engineering.
Y-linked gene Any of the genes located on the Y chromosome.

Z disc The border of a sarcomere in muscle.
zoologist Scientist who specializes in the study of animals.
zygote Single cell resulting from the fusion of gametes (egg and sperm).

Credits

Cover

Dimitrios/Shutterstock
S-F/Shutterstock
Pandora Studio/Shutterstock
10kPhotography/RooM the Agency/Alamy
Johner Images/Alamy
Science Picture Co/Alamy
Ragnar Th Sigurdsson/Arctic Images/Alamy

Text and Art Credits

Chapter 1 Figure 1-7. Source: Data from: Lindenmuth, G. Frank, Ph.D. and Elise B. Lindenmuth, Ph.D., "The efficacy of echinacea compound herbal tea preparation on the severity and duration of upper respiratory and flu symptoms: A randomized, double-blind placebo-controlled study," *The Journal of Alternative and Complementary Medicine*, Vol. 6: 4, pp. 327–334, 2000, Mary Ann Liebert, Inc.

Figure 1-10. Source: Based on Cohen, Sheldon, PhD et al., "Psychological Stress and Susceptibility to the Common Cold," *New England Journal of Medicine*, Vol. 325: 9, pp. 606–612, fig. 1, Aug. 29, 1991. Massachusetts Medical Society.

Figure 1-13. Data source: Mossad, Sherif B., MD et al., "Zinc Gluconate Lozenges for Treating the Common Cold, A Randomized, Double-Blind, Placebo-Controlled Study," *Annals of Internal Medicine*, Vol. 125: 2, July 15, 1996.

Chapter 4 Figure 4-17. Data source: *After 'The Biggest Loser,' Their Bodies Fought to Regain Weight* by Gina Kolata. Published by The New York Times Company, 2016.

Chapter 5 p. 86. Source: *Nations Approve Landmark Climate Accord in Paris* by Coral Davenport. Published by The New York Times Company, © 2015.

Figure 5-6. Data source: "Trends in Atmospheric Carbon Dioxide—Mauna Loa," NOAA/ESRL Global Monitoring Division, www.esrl.noaa.gov/gmd/ccgg/trends/.

Figure 5-8. Source: Based on Macmillan Publishers Ltd., from J.R. Petit et al., "Climate and atmospheric history of the past 420,000 years from the Vostok ice core, Antarctica," *Nature*, 399 (3):429–436, June 1999. Nature Publishing, Inc.

Chapter 9 p. 166. Source: The Innocence Project Mission Statement.

p. 170. Source: *What Jennifer Saw* by Jennifer Thompson. Published by Public Broadcasting Service (PBS).

Chapter 10 Figure 10-13. Data source: http://www.ers.usda.gov/media/185551/biotechcrops2016_d.html.

Chapter 11 p. 210. Source: Charles Darwin, *The Voyage of the Beagle* (1809–1882).

Chapter 13 p. 285. Source: Constantine Markides, Olympics 2012: *Gold Medal Swimmer Anthony Ervin Is Out to Reclaim His Title*, Rolling Stone magazine, July 27, 2012.

Chapter 15 Figure 15-8. Source: United Nations, Department of Economic and Social Affairs, Population Division (2015). World Population Prospects: The 2015 Revision. http://esa.un.org/unpd/wpp/

Chapter 18 Figure 18-11. Source: U.S. Department of Health and Human Services https://optn.transplant.hrsa.gov.

Chapter 25 p. 529. Source: John Steinbeck, (1939). *The Grapes of Wrath*, Viking Press.

Photo Credits

Chapter 1 Opener (T), Focus Pocus LTD/Fotolia; opener (C), ZenShui/Laurence Mouton/Getty Images; opener (B), Michaeljung/Shutterstock; 1.2a Eye of Science/Science Source; 1.2b Anders Wiklund/Scanpix/Reuters; 1.5 Danil Semyonov/AFP/Getty Images; 1.6 Zina Seletskaya/Shutterstock; 1.9a Heiti Paves/Alamy; 1.9b Stefan Klein/Getty Images; 1.9c Thomas Deerinck/NCMIR/Science Source; 1.17 SamuelBrownNG/iStock/Getty Images; 1.18 GVictoria/Shutterstock

Chapter 2 Opener (T), AF archive/Alamy; opener (C), Suzanne Tucker/Shutterstock; opener (B), Txking/Shutterstock; 2.1 Elizalebedewa/Fotolia; 2.14a SunnyS/Fotolia; 2.14b Valentina R./Fotolia; 2.14c Elena Larina/Shutterstock; 2.14d Lsantilli/Fotolia; 2.14e MovingMoment/Fotolia; 2.14f Joe Gough/Fotolia; 2.14g Pixelrobot/Fotolia; 2.15a Juergen Berger/Science Source; 2.16 (1) Dr. Gary Gaugler/Science Source; 2.16 (2) Eye of Science/Science Source; 2.16 (3) Jan Krejci/Shutterstock; 2.16 (4) Christopher Meder/Fotolia; 2.16 (5) Joefotofl/Fotolia; 2.16 (6) M. I. Walker/Science Source; 2.16 (7) Ivonne Wierink/Fotolia; 2.16 (8) Claude Huot/Shutterstock

Chapter 3 Opener (T), Andersen Ross/Blend Images/AGE Fotostock; opener (C), Alex Segre/Alamy; opener (B), Franckreporter/E+/Getty Images; 3.1a Akulamatiau/Fotolia; 3.1b Seralex/Fotolia; 3.7 Dr. David Furness/Keele University/Science Source; 3.8 CNRI/Science Source; 3.9 Callista Images/Cultura RM/Alamy; 3.10 Don W. Fawcett/Science Source; 3.11 Science Source; 3.12 Science Source; 3.13 SPL/Science Source; 3.14 David M. Phillips/Science Source; 3.15 Don W. Fawcett/Science Source; 3.16 Ron Boardman/Life Science Image/FLPA/Science Source

Chapter 4 Opener (T), Silvia Jansen/E+/Getty Images; opener (C), PeopleImages/Getty Images; opener (B), Ariel Skelley/DigitalVision/Getty Images; 4.9a Professors Pietro M. Motta & Tomonori Naguro/Science Source; 4.15a Lorimer Images/Shutterstock; 4.15b SCIMAT/Science Source

Chapter 5 Opener (T), Richard Green/Alamy; opener (C), Redmond Durrell/Alamy; opener (B), WIS Bernard/Paris Match/Getty Images; 5.5 KYTan/Shutterstock; 5.7 Paul Nicklen/National Geographic/Getty Images; 5.9a Dr. David Furness/Keele University/Science Source; 5.14 Winton Patnode/Science Source; Table 5.1a Flariviere/Shutterstock; Table 5.1b GJones Creative/Shutterstock; Table 5.1c SVF74/Shutterstock; 5.16 CDC

Chapter 6 Opener (T), Pixplus/Bauer-Griffin/GC Images/Getty Images; opener (C), Rex Features/AP Images; opener (B), Humannet/

Shutterstock; 6.2 Image Source/Getty Images; 6.3a Cortier/BSIP/Alamy; 6.3b Plougmannn/E+/Getty Images; 6.4a Biophoto Associates/Science Source; 6.4b Rossco/Shutterstock; 6.5a Biophoto Associates/Science Source; 6.5b Science Source; 6.11a Dr. Gopal Murti/Science Source; 6.11b Kent Wood/Science Source; p.112. Science Source

Chapter 7 Opener (T), Larry Williams/Corbis/Getty Images; opener (C), Steve Gschmeissner/Science Photo Library/Getty Images; opener (B), Priscilla Gragg/Blend Images/Getty Images; 7.2 Mediscan/Alamy; 7.10b Eye of Science/Science Source; 7.12b Profs. P.M. Motta & J. Van Blerkom/Science Source

Chapter 8 Opener (T), J. Scott Applewhite/AP Images; opener (C), Tony Tomsic/AP Images; opener (B), Living Art Enterprises, LLC/Science Source; 8.2a Gerald McCormack; 8.8 Pictorial Press Ltd/Alamy; 8.10 Skmj/Shutterstock; 8.15 (1) Krutar/Shutterstock; 8.15 (2) Jlarrumbe/Shutterstock; 8.15 (3) Maksim Shebeko/Fotolia; 8.16 Eye of Science/Science Source; 8.17 (1) Andrea Cerri Ferrari/Fotolia; 8.17 (2) Luri/Shutterstock; 8.17 (3) Sam Wirzba/Design Pics Inc/Alamy; 8.2 Siraanamwong/Fotolia; Table 8.3 (1) Johan Pienaar/Shutterstock; Table 8.3 (2) Catchlight Lens/Shutterstock; Table 8.3 (3) Dancestrokes/Shutterstock; Table 8.3 (4) Arnon Ayal/Shutterstock; Table 8.3 (5) Henrik Larsson/Shutterstock; 8.22a Biophoto Associates/Science Source; 8.22b Kwangmoo/Fotolia

Chapter 9 Opener (T), Chuck Burton/AP Images; opener (C), Isak55/Shutterstock; opener (B), Sakhorn38/Fotolia; 9.1 (1) Lea Thomas/EyeEm/Getty Images; 9.1 (2) Macroworld/E+/Getty Images; 9.1 (3) Philippe Bigard/OJO Images/Getty Images; 9.1 (4) Anthony Lee/Caiaimage/Getty Images; 9.1 (5) Andersphoto/Fotolia; 9.1 (6) Anton/Fotolia; 9.1 (7) Anton/Fotolia; 9.1 (8) Muratart/Shutterstock; 9.5 Jeff Lenard/National Association of Convenience Stores; 9.10 Randy Jirtle and Dana Dolinoy; 9.12 Darrick E. Antell, M.D., F.A.C.S; 9.14 Vit Kovalcik/Fotolia; 9.16 BSIP/Newscom; 9.17 Zmeel/E+/Getty Images

Chapter 10 Opener (T), Alex Milan Tracy/Sipa USA/Newscom; opener (C), David Grossman/Alamy; opener (B), International Rice Research Institute (IRRI); 10.11a, Biology Pics/Science Source; 10.11b, David McCarthy/Science Source; 10.14, John Doebley; 10.17, Jake Lyell/Alamy

Chapter 11 Opener (T), Ralph Lee Hopkins/Getty Images; opener (C), Scala/Art Resource, NY; opener (B), William Campbell/Sygma/Getty Images; 11.3 FineArt/Alamy; 11.4 (1) Interfoto/Alamy; 11.4 (2) Mle/Zoj Wenn Photos/Newscom; 11.6a Daniel Padavona/Shutterstock; 11.6b Van Der Meer Marica/Arterra Picture Library/Alamy; 11.7a Jeffrey M. Frank/Shutterstock; 11.7b Paul Sterry/Worldwide Picture Library/Alamy; 11.9 Fuse/Corbis/Getty Images; 11.13a Karloss/Shutterstock; 11.13b Jerome Wexler/Science Source; 11.13c Martin Shields/Alamy; 11.14b (1) Tom Brakefield/Stockbyte/Getty Images; 11.14b (2) Bele Olmez/ImageBroker/Alamy; 11.15 Richardson, M. K./National Museum of Health & Medicine; 11.19 (4) Javier Trueba/Madrid Scientific Films/Science Source; 11.24 SPL/Science Source; 11.26 (1) Thorsten Rust/Shutterstock; 11.26 (2) Dario Sabljak/Shutterstock; 11.26 (3) Denis Tabler/Fotolia; 11.26 (4) Ross Petukhov/Fotolia

Chapter 12 Opener (T), Gurpreet Singh/EyeEm/Getty Images; opener (C), Antonio Masiello/ZUMA Press/Newscom; opener (B), Michael Coddington/Shutterstock; 12.1 Kwangshin Kim/Science Source; 12.2 Puwadol Jaturawutthichai/Alamy; 12.3 Florian Bachmeier/ImageBroker/Alamy; 12.4 Nick Gregory/Alamy; 12.5 University of Louisville Special Collections; 12.6a Dennis W Donohue/Shutterstock; 12.6b Sil63/Shutterstock; 12.7 PetStockBoys/Alamy; 12.10 Blickwinkel/Alamy; 12.13a Frank Hecker/Alamy; 12.13b Stuart Pearce/Alamy; 12.18 Smileitsmccheeze/Getty Images

Chapter 13 Opener (T), ZUMA Press, Inc/Alamy; opener (C), H. Mark Weidman Photography/Alamy; opener (B), Design Pics Inc./Alamy; 13.1 Keith Levit/Shutterstock; Table 13.1 (1) Mariko Yuki/Shutterstock; Table 13.1 (2) Jeff Yonover; 13.2 (T) Eric Isselee/Shutterstock; 13.2 (B) Dorling Kindersley, Ltd.; 13.6 (1) Suzifoo/E+/Getty Images; 13.6 (2) Brand X Pictures/Stockbyte/Getty Images; 13.6 (3) Peter Grosch/Shutterstock; 13.7a Professor Forrest L. Mitchell; 13.7b David Allan Brandt/Iconica/Getty Images; 13.14a Buena Vista Images/Digital Vision/Getty Images; 13.14b Ariadne Van Zandbergen/Alamy; 13.15 (1) VisionDive/Shutterstock; 13.15 (2) Hiroshi Sato/Shutterstock; 13.19 Wasanajai/Shutterstock; 13.20a Shawn Hempel/Shutterstock; 13.20b Bobkeenan Photography/Shutterstock; 13.20c Tomas Kvidera/Shutterstock; 13.21 Nick Wilson/Getty Images; 13.22 Rolls Press/Popperfoto/Getty Images

Chapter 14 Opener (T), Mark Wilkinson/Alamy; opener (C), Merydolla/Shutterstock; opener (B), Michael Abbey/Science Source; 14.1a Imagebroker/Alamy; 14.1b Mauro Fermariello/Science Source; 14.5 University of California at Los Angeles (Asian Am Studies); 14.6a Custom Medical Stock Photo/Alamy; 14.6b NASA; 14.6c David Wall/Alamy; 14.8a Ray Simons/Science Source; 14.8b Alex Wild; 14.9 Chase Studio/Science Source; Table 14.3 (1) Jubal Harshaw/Shutterstock; Table 14.3 (2) MedicalRF/Alamy; Table 14.3 (3) Lebendkulturen.de/Shutterstock; Table 14.3 (4) Torsten Lorenz/Shutterstock; Table 14.3 (5) Dee Breger/Science Source; Table 14.3 (6) Jfybel/iStock/Getty Images; Table 14.3 (7) Lebendkulturen.de/Shutterstock; 14.10a PhotosByNancy/Shutterstock; 14.10b Jean-Paul Ferrero/Mary Evans Picture Library Ltd/AGE Fotostock; 14.10c Crisod/Fotolia; 14.10d Steffen & Alexandra/Mary Evans Picture Library Ltd/AGE Fotostock; 14.11a Thomas J. Peterson/Alamy; 14.11b Images & Stories/Alamy; 14.11c Ekkapan Poddamrong8/Shutterstock; Table 14.4 (1) Jolanta Wojcicka/Shutterstock; Table 14.4 (2) Mark Aplet/Shutterstock; Table 14.4 (3) Dickson Despommier/Science Source; Table 14.4 (4) Photonimo/Shutterstock; Table 14.4 (5) D. Kucharski K. Kucharska/Shutterstock; Table 14.4 (6) Carsten Medom Madsen/Shutterstock; Table 14.4 (7) Paytai/Shutterstock; Table 14.4 (8) Natalie Jean/Shutterstock; Table 14.4 (9) Tom McHugh/Science Source; Table 14.5 (1) Biophoto Associates/Science Source; Table 14.5 (2) Robin Treadwell/Science Source; Table 14.5 (3) Nmelnychuk/Fotolia; Table 14.5 (4) Henk Bentlage/Shutterstock; 14.13 Patrick Landmann/Science Source; 14.14 Robin Treadwell/Science Source; Table 14.6 (1) Lucie Zapletalova/Shutterstock; Table 14.6 (2) Chursina Viktoriia/Shutterstock; Table 14.6 (3) Debbie Aird Photography/Shutterstock; Table 14.6 (4) Digital Vision/Photodisc/Getty Images; 14.15a Christian Ziegler/Minden Pictures/Newscom; 14.15b Konzeptm/Fotolia; 14.15c Photo by Peter Goltra, Courtesy National Tropical Botanical Garden; 14.17 Dr. Linda M. Stannard/University of Cape Town/Science Source; 14.18 (1) Bruce MacQueen/Shutterstock; 14.18 (2) Boggy22/iStock/Getty Images; 14.18 (3) Scooper Digital/iStock/Getty Images; 14.18 (4) Iphoto5151/Fotolia

Chapter 15 Opener (T), NASAopener (C), Atlantide Phototravel/Corbis Documentary/Getty Images; opener (B), Rawpixel/Shutterstock; 15.2a Onepony/Fotolia; 15.2b Kevin Schafer/Alamy; 15.2c Matauw/iStock/Getty Images; 15.4a Pixel 8/Alamy; 15.7a Martin Shields/Alamy; 15.7b Robert Pickett/Papilio/Alamy; 15.7c Wojciech Nowak/Fotolia; 15.7d Hadyniak/E+/Getty Images; 15.13 John Van Hasselt/Sygma/Corbis/Getty Images

Chapter 16 Opener (T), lrh847/Getty Images; opener (C), Jana Shea/Shutterstock; opener (B), Elizabeth Robertson/Philadelphia Inquirer/KRT/Newscom; 16.1 Kevin Schafer/Alamy; 16.4a Chris Cheadle/Alamy; 16.4b Mejzer Annamarie/EyeEm/Getty Images; 16.4c Thomas Mukoya/Reuters; 16.4d Rich Carey/Shutterstock; 16.5 Steve Apps/Alamy; 16.8 Bill Brooks/Alamy; 16.9 Steve Helber/AP Images; 16.10a Norman Chan/Shutterstock; 16.10b John Doebley; 16.10c Juan A. Morales-Ramos; 16.12 Pjmalsbury/E+/Getty Images; 16.13 Anton Ryabtsov/Shutterstock; 16.14a G. Ronald Austing/Science Source; 16.14b Tracy Ferrero/Alamy; 16.16a H46it/Getty Images; 16.16b Jupiterimages/Getty Images; 16.18 Dave Hansen/University of Minnesota Agricultural Experiment Station; 16.19 Ostill/Shutterstock; 16.26 Petty Officer 2nd Class Molly A. Burgess, USN/U.S. Department of Defense Visual Information Center

Chapter 17 Opener (T), YAY Media AS/Alamy; opener (C), Chungking/Fotolia; opener (B), Accent Alaska/Alamy; 17.4 Frank Zullo/Getty Images; Table 17.1 (1) Stillfx/iStock/Getty Images; Table 17.1 (2) R. S. Ryan/Shutterstock; Table 17.1 (3) TT/iStock/Getty Images; Table 17.1 (4) Mguntow/123RF; Table 17.1 (5) Mac99/iStock/Getty Images; Table 17.1 (6) JosjeN/Shutterstock; Table 17.1 (7) Charles Krebs/Photodisc/Getty Images; Table 17.1 (8) Jennifer Stone/Shutterstock; 17.10 Liz Every/AGE Fotostock; 17.11 Erhard Nerger/Image Broker/Alamy; 17.12 Eco Images/Universal Images Group/Getty Images; 17.13 Gucio_55/Shutterstock; 17.14a BARNpix/Alamy; 17.14b Natali26/Shutterstock; 17.14c Mares Lucian/Shutterstock; 17.15 Jochen Tack/Alamy; 17.16 Alan Gignoux/Alamy; 17.17 AfriPics/Alamy; Table 17.2 (1) Robert Pickett/Papilio/Alamy; Table 17.2 (2) Ernest Manewal/Photolibrary/Getty Images; Table 17.2 (3) Lynda Lehmann/Shutterstock; Table 17.2 (4) Shime/Fotolia; Table 17.2 (5) Bernard Castelein/Nature Picture Library; Table 17.2 (6) OAR/National Undersea Research Program (NURP); NOAA; Table 17.2 (7) Fotolia; Table 17.2 (8) Martin H Smith/FLPA/AGE Fotostock; 17.19 Pulsar Images/Alamy; 17.20a Aurora Photos/Peter Essick/Alamy; 17.20b A Hartl/Blickwinkel/AGE Fotostock; 17.21 (1) Przemyslaw Wasilewski/Shutterstock; 17.21 (2) Roberto Tetsuo Okamura/Shutterstock; 17.22 D P Wilson/FLPA/AGE Fotostock; 17.23a Borisoff/Shutterstock; 17.23b Tubuceo/Shutterstock; 17.24 Kansas City Star/Tribune News Service/Getty Images

Chapter 18 Opener (T), Deepblue4you/Getty Images; opener (C), Martin Bararud/Caiaimage/OJO/Getty Images; opener (B), Shutterstock; 18.1b Ed Reschke/Photolibrary/Getty Images; 18.1c Nina Zanetti/Pearson Education, Inc.; 18.3 Robert Tallitsch/Pearson Education, Inc.; 18.5 Biophoto Associates/Science Source; 18.9 Brian Walker/AP Images; 18.10 BSIP/UIG/Getty Images

Chapter 19 Opener (T), Robyn Mackenzie/Shutterstock; opener (C), Frederic Cirou/PhotoAlto/Alamy; opener (B), Pressmaster/Shutterstock; 19.3 CNRI/Science Source

Chapter 20 Opener (T), Jane Khomi/Moment/Getty Images; opener (C), Marc Bruxelle/iStock/Getty Images; opener (B), Ian Dagnall/Alamy; 20.4 VideoSurgery/Science Source; 20.5b Eye of Science/Science Source; 20.8 Biophoto Associates/Science Source; 20.9 Science Source; 20.10a Science Source; 20.10b Mediscan/Alamy; 20.10c Steve Gschmeissner/SPL/Getty Images; 20.10d NIBSC/Science Source; 20.12 CNRI/Science Source; 20.18a Chuck Brown/Science Source; 20.18b OGphoto/iStock/Getty Images

Chapter 21 Opener (T), Ken Hawkin/Alamy; opener (C), Science Photo Library/AGE Fotostock; opener (B), FatCamera/Getty Images; 21.1 (1) Cliparea/Fotolia; 21.1 (2) Sebastian Kaulitzki/Shutterstock; 21.1 (3) H.S. Photos/Alamy; Table 21.3 (1) Eye of Science/Science Source; Table 21.3 (2) David Scharf/Science Source; Table 21.3 (3) The Science Picture Company/Alamy; Table 21.3 (4) Mediscan/Alamy; 21.8 Eye of Science/Science Source; 21.9 Scanpix Sweden AB

Chapter 22 Opener (T), Technotr/Getty Images; opener (C), Johannes Eisele/Getty Images; opener (B), Sigrid Olsson/PhotoAlto Agency RF Collections/Getty Images; 22.7a Steve Gschmeissner/SPL/AGE Fotostock; 22.7b Professor Pietro M. Motta/Science Source; 22.9b Nin Zanetti/Pearson Education, Inc.

Chapter 23 Opener (T), Ueslei Marcelino/Reuters; opener (C), Prof. Frank Hadley Collins, Dir., Cntr. for Global Health and Infectious Diseases, Univ. of Notre Dame/Centers for Disease Control and Prevention; opener (B), Joe Raedle/Getty Images; 23.2 Dr. Yorgos Nikas/Science Source; 23.4 Biophoto Associates/Science Source; 23.5 R. Bevilacqua/CNRI/Science Source; 23.6 Mario Tama/Getty Images; Table 23.1 (2) Don Farrall/Photodisc/Getty Images; Table 23.1 (3) Anurak Khuntapol/123RF; Table 23.1 (4) Toons17/Shutterstock; Table 23.1 (5) Garo/Phanie/AGE Fotostock; Table 23.1 (6) Image Point Fr/Shutterstock; Table 23.1 (7) Kumar Sriskandan/Alamy; Table 23.1 (8) Dr P. Marazzi/Science Source; Table 23.1 (11) Garo/Phanie/Science Source; Table 23.1 (13) Yang5i/Getty Images; Table 23.1 (15) Ever/iStock/Getty Images; Table 23.1 (16) Image Point Fr/Shutterstock; Table 23.1 (17) Nikuwka/Shutterstock; Table 23.2 (1) Vem/BSIP SA/Alamy; Table 23.2 (2) Stocktrek Images, Inc./Alamy; Table 23.2 (3) Science Picture Company/Alamy; Table 23.2 (4) Martynowi.cz/Shutterstock; Table 23.2 (5) Scott Camazine/Alamy; Table 23.2 (6) Cavallini James/BSIP SA/Alamy; Table 23.2 (7) Callista Images/Cultura RM/Alamy; Table 23.2 (8) Warren Rosenberg/Fotolia; Table 23.2 (9) Science Picture Co/Alamy; Table 23.2 (10) Cavallini James/BSIP SA/Alamy

Chapter 24 Opener (T), Dean Drobot/Shutterstock; opener (C), Syda Productions/Shutterstock; opener (B), Andrei_R/Shutterstock; 24.7b Martin M. Rotker/Science Source; 24.11 Jessica Wilson/Biophoto Associates/Science Source; Table 24.1 (1) Gilmanshin/Shutterstock; Table 24.1 (2) Toriasdream/iStock/Getty Images; Table 24.1 (3) Istarif/iStock/Getty Images; Table 24.1 (4) Photopixel/Shutterstock; Table 24.1 (5) Andrew Burns/Shutterstock; Table 24.1 (6) George Post/Science Source; Table 24.1 (7) Iurii Konoval/123RF; Table 24.1 (8) Kiselev Andrey Valerevich/Shutterstock; Table 24.1 (9) Fotomaximum/Fotolia

Chapter 25 Opener (T), Photo12/Universal Images Group/Getty Images; opener (C), Andrew McConnell/Alamy; opener (B), Digital Vision/Photodisc/Getty Images; 25.1a John Doebley; 25.1b Zonesix/Shutterstock; 25.4 John Kaprielian/Science Source; 25.5 Dr. Jeremy Burgess/Science Source; 25.6 Joachim E. Röttgers/ImageBroker/Alamy; 25.7 Kathy Burns/Shutterstock; 25.8 (1) Ranald MacKechnie/Dorling Kindersley, Ltd.; 25.8 (2) Susanna Price/Dorling Kindersley, Ltd.; 25.8 (3) Roger Phillips/Dorling Kindersley, Ltd.; 25.8 (4) C Squared Studios/Photodisc/Getty Images; 25.8 (5) Denis Pepin/Shutterstock; 25.8 (6) Siede Preis/Photodisc/Getty Images; 25.8 (7) Lew Robertson/Corbis/Getty Images; 25.8 (8) Dave King/Dorling Kindersley, Ltd.; 25.10a Jerome Wexler/Science Source; 25.10b Arvind Balaraman/Shutterstock; 25.10c Ondrej Prosicky/Shutterstock; 25.10d Dr. Merlin D. Tuttle/Science Source; 25.10e Anthony Mercieca/Science Source; 25.11a Steve Bloom Images/Alamy; 25.11b Kevin Schafer/Alamy; 25.11c Scott Camazine/Alamy; 25.11d Gary Meszaros/Avalon/Bruce Coleman Inc/Alamy; 25.13a Shane Kennedy/Shutterstock; 25.13b Dr. Jeremy Burgess/Science Source; 25.13c Kellis/Shutterstock; 25.13d Adam Hart-Davis/Science Source; 25.14

Ed Reschke/Photolibrary/Getty Images; 25.15 Patrick J. Lynch/Science Source; 25.18a Gabriel Rif/Alamy; 25.18b Tim Aylen/AP Images; 25.20 Nigel Cattlin/Science Source; 25.21 NASA; 25.22 Science Source; 25.25a Jens Ottoson/Shutterstock; 25.25b Science Source; 25.26a Terry W. Egers/CorbisDocumentary/Getty Images; 25.26b Herbert Kehrer/Getty Images; 25.28 NASA; 25.29 Erich Hartmann/Magnum Photos; 25.30b Nigel Cattlin/Alamy; 25.31 Gary Kazanjian/AP Images; 25.32 Juice Images/Alamy

Chapter 26 Opener (T), Essenin/Getty Images; opener (C), Geri Lavrov/Moment/Getty Images; opener (B), Fotosmurf03/iStock/Getty Images; 26.3 Picsunv/iStock/Getty Images; 26.4 Elena Elisseeva/Shutterstock; 26.5 Radius Images/Alamy; 26.6 Dr. Keith Wheeler/Science Source; 26.7a Ultramarinfoto/Getty Images; 26.7b Daniil Kirillov/Hemera/Getty Images; 26.11a PrairieArtProject/Getty Images; 26.11b Aga7ta/Fotolia; 26.11c Peter Anderson/Dorling Kindersley, Ltd.; 26.15a Dr. Keith Wheeler/Science Source; 26.15b Gino Santa Maria/Fotolia; 26.16a Martin Shields/Alamy; 26.16b John Kaprielian/Science Source; 26.16c Charles D. Winters/Science Source; 26.19 Samot/Shutterstock; 28 UN.28.29 Ed Reschke/Photolibrary/Getty Images

Index

Page numbers in bold indicate definitions. An *f* indicates a figure and a *t* indicates a table.

A

C

D

E

G

H

N

Q

R

S

T

U

V

W

X

Y

Z